College Physics

Fifth edition

College Physics

Fifth edition

Franklin Miller, Jr.

Kenyon College, Emeritus

HARCOURT BRACE JOVANOVICH, INC.

New York   San Diego   Chicago   San Francisco   Atlanta

London   Sydney   Toronto

Cover photo: Crystalline material used in solid-state diode laser.
Manfred Kage from Peter Arnold.

Printed in the United States of America

Library of Congress Catalog Card Number: 81-84627

ISBN: 0-15-511737-8

Picture Credits

1 Lynn McLaren/Photo Researchers, Inc.; **22** Popper-foto/Pictorial Parade; **55** © David Glaubinger/Jero-boam, Inc.; **73** Thomas B. Greenslade, Jr.; **108** Thomas B. Greenslade, Jr.; **117** NASA Photographs; **125** AMF Incorporated; **148** Dr. Harold Edgerton, MIT, Cambridge, MA; **151** Mike Grone Photo/Cedar Point Amusement Land; **158** Fermi National Accelerator Laboratory; **165** Sikorsky Aircraft; **170** Courtesy of the Archives, California Institute of Technology; **177** Jet Propulsion Laboratory; **178** E. Leybold's Nachfolger, courtesy J. Klinger; **192** Edith Reichmann; **219** © Richard McCullough/Black Star; **240** Photo from Monkmeyer Press; **249** Picker Corporation; **251** William B. Joyce; **255** Harlow W. Ades; **267** Thomas B. Greenslade, Jr.; **296** The Daily News Photo; **310** Franklin Miller, Jr.; **333** Ellen Robinson; **354** Thomas B. Greenslade, Jr.; **371** Cooper-Bessemer; **390** Kenneth Jewell; **410** Syd Greenberg/DPI; **436** Harbrace Photo; **454** Gregory Spaid; **482** Franklin Miller, Jr.; **534** International Business Machines; **535 (both)** Radio Corporation of America; **553** John R. Seavolt; **570** Bausch & Lomb; **571** Helen Faye; **602** National Bureau of Standards; **608** E. Leybold's Nachfolger, courtesy J. Klinger; **613 (left)** Franklin Miller, Jr.; **614** Bausch & Lomb; **616** Gaertner Scientific Corporation; **621** Dr. Henry Milch; **630 (top)** Courtesy E. N. Leith and J. Upatnieks, *Scientific American*; **630 (bottom)** William Vandivert, *Scientific American*; **661** Wide World Photo; **666** Fundamental Photographs; **694 (Fig. 29-9)** Franklin Miller, Jr.; **697** Dr. Albert Rose; **712** Mount Wilson and Las Companas Observatories, Carnegie Institution of Washington; **714** Bausch & Lomb; **725** Franklin Miller, Jr., & Thomas B. Greenslade, Jr.; **732** Thomas B. Greenslade, Jr.; **750 (Fig. 31-12)** Franklin Miller, Jr.; **752** Brookhaven National Laboratory; **781** Courtesy Argonne National Laboratory; **797** U.S. Atomic Energy Commission; **798** Fermi National Accelerator Laboratory; **800** Photo CERN; **806** Brookhaven National Laboratory; **811 (both)** C. F. Powell, Physical Society, London.

Preface

The author of a text in elementary physics is, properly, under some obligation to define the audience and justify his or her outlook, aims, and methods. My primary aim has been simply to prepare a text having maximum clarity and teachability.

This book is intended to be used in an elementary physics course to which the student brings no mathematics beyond simple algebra. It reflects my conviction that all students can and should become familiar with the basic ideas of physics; for this reason, I have emphasized principles rather than technological applications. This is not to say, however, that the text has no points of contact with everyday life and practical situations. On the contrary, a great many worked examples, questions, and problems are scattered throughout the book. Each of these has been chosen to illustrate and "bring down to earth" some physical concept. Three chapters deal specifically with applications but retain close correlation with basic principles: Applied Electricity, Applied Optics, and Applied Nuclear Physics.

This is a basic text aimed at a wide audience with diverse goals. A significant number of students using the book are in some program within the life sciences; for this reason, examples and problems related to the life sciences are introduced throughout the book.

The organization of the elementary physics course is under active study by many teachers. The changes that may come about within the next decade will not necessarily be along the lines one might predict today. For this reason, I have presented the topics of physics in a semitraditional sequence that preserves a logical structure found useful by many teachers over a period of many years. But I have introduced so-called modern physics early and often within this framework; in this way physics can be presented as a growing subject, and the intimate relationship of classical physics to modern physics can be made clearer.

The organization of this fifth edition has been somewhat changed from that of the fourth edition. Density and the basic structure of matter have been brought forward into Chap. 1 in order to get the course off to a fast start, with kinematics now being introduced in Chap. 2. The conservation laws of linear momentum and energy are placed in separate chapters. An elementary treatment of frames of reference and inertial forces remains early in the book (Chap. 5), but the study of relativity as such has been moved back to a separate Chap. 28, just prior to quantum physics. In the treatment of geometrical optics, image location by the method of curvatures has been retained, but ray tracing and the traditional lens equation are given as alternative methods for those who prefer them. Material on atomic and nuclear physics remains in five chapters to give a less-concentrated presentation. Chapter 31, on atomic structure, offers material often studied at the beginning of chemistry courses; here it is presented to the student who has acquired an adequate background in elementary physics through study of this book. The entire chapter can be treated as "For Further Study," at the teacher's discretion. The material on subatomic particles in Chap. 33 has been updated.

In general, additional material for this edition has been kept to a bare minimum, in accordance with my desire to present fundamentals without an excessive array of applications. Electron tubes have been further deemphasized, but a section has been added on that useful electron device, the cathode-ray tube. The material on thermodynamics has been reorganized and strengthened. Every effort has been made to keep the book of manageable length.

The book is completely metric; no problems or examples are stated in British units. Symbols and abbreviations are those of the International System of units (SI). I believe there is merit in a "soft" conversion approach, so in several places, especially in mechanics, familiar non-SI units are parenthetically introduced following the perhaps less familiar SI units. The calorie, a non-SI unit, has been retained for some use in heat flow; I do not think the time is ripe for a complete break with the calorie as a unit of energy in heat transfer problems, although the joule is given increased prominence. No matter which system of units is used, the student is encouraged to look for dimensional consistency in equations and in solutions of problems. A convention regarding significant figures has been adopted, as discussed on pages 12–13.

Throughout the book I have given emphasis to the conservation laws and the concepts of waves, fields, and particles. Although I have not adopted a straight historical approach, I have included many brief historical references and in a few areas have gone rather fully into the development of physical concepts.

Physics is a rewarding subject that requires a student's best effort, and I have refrained from spoon-feeding a watered-down or sugar-coated subject matter. Nevertheless, every possible aid to digestion has been included in the book. There are numerous worked examples in the body of the text. At the end of each chapter are a summary, a check list of key concepts, a set of questions to help in reviewing on the verbal level, some multiple-choice review questions, and a series of carefully graded problems (see the explanation on page 19) to assist the student in the mastery of principles through practice. With very few exceptions, every problem is either new or has new numerical data. An important study aid is the review of simple high school mathematics in the Appendix.

For added flexibility, sections labeled "For Further Study," with accompanying problems, follow many of the chapters; these are set apart in such a way that the continuity of the primary material is not disturbed. Note, though, that the For Further Study sections are not considered to be less important than those in the main body of the text. These sections are not mere appendices to chapters; they are placed where they are in order to allow a teacher greater flexibility in arranging the course. The better students should be encouraged to dip into these sections on their own; and, of course, a teacher is free to assign any of these sections to an entire class. The text is, however, complete and self-sufficient without this extra material.

It is becoming increasingly apparent that for maximum insight into physics at any level, a student should not be deprived of the help offered by the use of limits. The concept of limits is developed and explained as the need arises, requiring only a knowledge of high school algebra. Graphical interpretations are emphasized. Formulas for derivative and integral are not used in the main body of the text; some formal calculus is used in some of the optional For Further Study sections.

Electric current is described in terms of conventional positive current, for two reasons. If, as is customary, electric field and potential difference are defined in terms of a positive test charge, then there is a logical advantage in using conventional current directed "downhill" from a region of high potential. Also, a significant number of students using this book will encounter conventional current in their later study in fields outside physics.

I have not felt that an encyclopedic text is desirable for the type of course for which this book is intended. I have preferred to approach fundamentals in a leisurely way, allowing full opportunity for development of background and assimilation of principles. To achieve this treatment in depth, the discussion of certain time-honored subjects has been greatly reduced or even omitted. In most cases, the References section at the end of each chapter will serve the needs of students wishing to explore such areas. The References also give opportunity for further reading, both in the history of science and in contemporary developments.

The References at the end of chapters include mention of certain films. This is not the place to attempt a critical evaluation of longer films. References are included, however, to selected short "single-concept" films that are distributed in cartridges and are, in many colleges and universities, available for selection and viewing by individual students outside regular class hours. Cartridge films are listed that are judged to be helpful at the level of this text. These films are available separately and in sets from several distributors.

It is fitting to acknowledge my great indebtedness to my teachers, colleagues, and students who have influenced me throughout the years. Of particular value has been the constructive criticism of many teachers, users of the previous editions, who responded to my request for comments and suggestions. Thanks are also due to all who furnished photographs for the text, and to the persons mentioned in the preface to the fourth edition. I also thank the following, who have given assistance in the preparation of this new edition: S. I. Ben-Abraham,

P. J. Collings, T. B. Greenslade, Jr., M. Iona, D. E. McBride, P. H. McClean, H. P. Stephenson, T. G. Trippe, and F. T. Worrall. The residual errors in the book are, of course, entirely my own.

FRANKLIN MILLER, JR.
Gambier, OH 43022

Contents

13
Temperature and Expansion
296

14
Heat and Heat Transfer
310

15
Thermal Behavior of Gases
333

The Nature of Physics

1-1 Science— Endless Frontier of the Mind

The representation of science as a frontier of the mind is a useful description of one of man's finest intellectual activities. The geographical frontier—the boundary between known and unknown territory—has always existed. Pioneers establish themselves at the very edge of the wilderness, consolidate their position, and strike out anew. During the centuries, the forms of geographical exploration have been as many and varied as have been the explorers: the Phoenician sailor, whose frontier was the sea; the Roman legionary, who garrisoned the distant islands of Britain; the Indian scout, who opened up the Western territories of the United States. Today, the geographical frontier is above our heads, extending to the moon and beyond.

It is no mere play on words to liken scientific investigation to exploration. We shall, in this course, repeatedly see how a scientific beachhead is won by an intrepid and imaginative thinker; how the frontier is made secure by developments

in related fields and by the application of new basic knowledge to practical problems; and how the scientific frontier is extended a bit at a time by scientists building upon past achievements. We see how appropriate are two phrases that have often been used to describe science. James B. Conant, chemist and former president of Harvard University, considers science to be a "series of conceptual schemes resulting from observation and experiment and leading to further experiment and observation."* Percy W. Bridgman, a Nobel prizewinner in physics, puts it more forcefully: science means "doing one's damnedest with one's mind, no holds barred."† Each of these statements, modified in phraseology, would be a perfect description of the act of geographical exploration.

Not many decades ago, it seemed that the geographical world was limited and that frontiers must soon disappear. We are now well past

---

* *On Understanding Science*, Yale University Press, 1946.
† "Prospect for Intelligence," *Yale Review*, **34**, 450, 1945; reprinted in *Reflections of a Physicist*, Philosophical Library, 1950, p. 342.

Gathering data in the laboratory is an essential part of the study of physics.

the threshold of a space age, and once again we are presented with what may be an endless series of frontiers. Similarly, about 90 years ago, some scientists thought that the last frontier of physical science had been reached; they thought all that remained was to work out the details and extend the accuracy of measurement another decimal place or two. How wrong this turned out to be! Unforeseen new frontiers were reached and opened to investigation. By 1900 the free electron, x rays, and radioactivity had been discovered, and quantum theory was being developed. Major breakthroughs continue to be made; in the language of exploration, scouting parties are returning with exciting glimpses beyond the present frontier. There is every evidence that scientific frontiers, like geographical ones, form an endless sequence. How are these frontiers being penetrated? What methods are used in scientific investigation?

It is amazing and somewhat reassuring that while the content of physics has become enlarged and specialized, there has been no corresponding radical change in the underlying methodology of scientific investigation since the days of Galileo (1564–1642). However, if we look for a tried and true routine that will always yield results, we shall be disappointed. It is just the essence of scientific methods that the intellect is *freed* from routine. Any scientist, including the physicist, must combine inspiration, intuition, concentration, and hard work if he is to advance knowledge. Reliance on set formulas or routine procedure is not likely to give rise to either new discoveries or new understanding of old discoveries. Yet scientists would be foolish indeed to forgo the many helps and advantages that are their inheritance from a continuous chain of fellow scientists that reaches back several thousand years. You who are beginning the study of physics will find this point illustrated in the text. We shall discuss the basic laws of physics, which are relatively few in number, and use them continually to illuminate everyday experience. We shall approach new ideas as much as possible from the point of view of their

relationship to topics covered earlier in the course.

In this book, illustrations of scientific methods will be by implication rather than by detailed exposition. There are many ways of approaching scientific knowledge—perhaps as many ways as there are scientists. But all of them have several things in common: collecting facts by observation; constructing a *hypothesis**; testing the hypothesis. It will be seen that any method that incorporates these three procedures is especially suitable for the development of *natural* science, where facts can often be collected by controlled experiment. In the *social* sciences it often proves necessary to construct hypotheses without benefit of controlled experiment. The data of social science are those of human relationships and cannot be gathered by a series of experiments in which one variable at a time is changed. Thus social scientists almost always must rely for their data on close observation of happenings not under direct control. Natural scientists often suffer from a similar limitation, especially when a new field of investigation is opened up.

Whatever the source of a hypothesis—an act of imaginative thinking, an educated guess, or a carefully constructed working hypothesis based on hundreds of precise experiments—the next step is to test it. It is in the devising of tests that the skill and ingenuity of the experimenter lie. A hypothesis is rejected if even a single previously known fact is at variance with it,[†] but the scientist goes further. She actively seeks new facts, perhaps by performing experiments suggested by the hypothesis or by mathematically analyzing experiments performed by others. If A (our hypothesis) is true, then B should be true, so let

---

* Webster defines *hypothesis* as "a proposition tentatively assumed in order to draw out its logical or empirical consequences and so test its accord with facts that are known or may be determined."

† The picturesque expression, "the exception proves the rule," does not mean what it seems to. The word "prove" originally meant "test," as in "proving ground." So the statement really reads, "the exception *tests* the rule"—and the exception proves the rule false, not true.

us make an experiment to see whether or not B is true. If B is not true, then our hypothesis is incorrect and must be modified or abandoned. If B *is* true, then we are one step closer to verifying our hypothesis A, but we have not proved it. Indeed, no matter how many different B's are proved true, we have not proved the truth of A beyond all possible doubt.

When a hypothesis leads to the design of experiments that would earlier have been meaningless or never thought of, and when these experiments in turn lead to a new hypothesis that can similarly be tested, then we may speak of a *theory*. All scientific "principles" are really theories, for the scientist is always ready to modify his or her most cherished concepts if new experimental evidence turns up, or even if another concept will explain the evidence equally well. From hypothesis to principle of nature may be a long or a short trail, but the terms are different only in degree. In the 5th century B.C. the Greek Demokritos made a *conjecture*, without any experimental evidence, of the existence of small indivisible particles that he called atoms. In 1808 Dalton's atomic *hypothesis* made its appearance, based upon certain experimental evidence. In the late 1800s the atomic *theory* was refined and expanded. Now, in the late 1900s the *principle* of atoms is completely accepted, but it is subject to such extensive experimentation and reinterpretation that the very use of the word "atom" may be misleadingly naive.

We have deliberately kept our discussion of scientific methods brief, for two reasons. First, people mean different things when they speak of "scientific method"; you will discover this for yourself if you read some of the references listed at the end of this chapter. Second, and more important, we hope that you will derive from this book and others your own understanding of what science is. We believe that scientific attitudes should grow out of work in science rather than be laid down as arbitrary, almost meaningless, rules. So we leave this discussion with a final note: scientific methods have limitations, the

most severe being an unspoken presumption that "knowledge" means "scientific knowledge." It would be most *un*scientific to assume that religious knowledge, or faith, cannot exist simply because it cannot be shown to be true by scientific methods. One should not expect a method devised specifically to cope with experimental or observed facts to do anything other than that for which it was invented. Nevertheless, when properly applied, scientific methods have proved to be the most dependable and fruitful means of dealing with the physical universe and have made possible the degree of control over our physical environment that is so characteristic of our civilization.

1-2 *Physics as a Natural Science*

Attempts to understand and control the forces of nature are a characteristic trait of the human intellect dating back to prehistory. Yet only comparatively recently has the special discipline we call *physics* been a distinct field of study. Aristotle, Archimedes, and even Galileo and Newton called themselves natural philosophers, taking all nature for their province. As the body of knowledge grew, it became necessary to specialize, for not even a genius could hope to keep up with developments in astronomy, biology, medicine, geology, chemistry, mathematics, and a host of other specialties. Those who are concerned mainly with the applications of science to the betterment of human environment have been the engineers. Physicists and chemists generally have been concerned with the more basic aspects of energy and nonliving matter. Chemistry deals primarily with molecular changes and the rearrangements of the atoms that form molecules, but the dividing line between chemistry and physics can hardly be drawn in a dogmatic fashion. A list of some journals currently published in the United States illustrates the delicate shades of emphasis among physical scientists. Running the gamut from "pure physics" to "pure chemistry" we have the *Physical Review, Journal of Chemical Physics, Journal of Physical Chemistry,* and *Journal of the American Chemical Society.* The titles of some other research journals indicate a great deal of specialization even within the broad field of physics: *Journal of the Institute of Metals* (Great Britain); *Revue d'Optique* (France); *Kristallografiya* (U.S.S.R.); *Revista de Geofísica* (Spain); *Przegląd elektrotechniczny* (Poland). Many journals cover the whole field of physics, including *Zeitschrift für Physik* (Germany) and the *Canadian Journal of Physics.* These are just a few of hundreds of journals through which original research in physical science is presented to the scientific community. In the year 1980, no less than 109,577 technical papers were summarized in *Physics Abstracts;* this flood of reports came from 2673 journals published throughout the world. Statistics can, of course, be twisted, and yet other lines of evidence converge to the same conclusion: the all-round scholar of the 17th and 18th centuries, a person of the first rank in several disciplines, could exist only rarely now, if at all. No one scholar can hope to assimilate all the rapidly growing source material of even one science such as physics.

We are thus led to the idea of partial understanding: the acceptance of limited, but concentrated, knowledge as being better than no knowledge at all. The greatest research physicists of today have probed deeply and brilliantly into their own fields of interest, but they have acquired less profound knowledge outside their specialty.

Like other sciences, physics has developed through the combined efforts of great numbers of dedicated workers in many countries. We must realize that the greatest of these scientists were, each in his or her own day, at the then-existing frontier of science. We tend to accept Newton's laws of motion, which we use so often, as "handed down from on high," forgetting that Newton had to come to grips with nature, "no holds barred," in order to arrive at his formulation of the laws of mechanics. Newton's laws

were, in fact, "modern physics" in the early 1700s.

We see that modern physics is as old as physics itself. Yet today we give Galileo and Newton a prominent place in our study of mechanics not for historical reasons alone; Galileo and Newton happened also to be correct, and their methods are still tremendously useful. Present-day physicists—modern modern physicists if you will—rely on Newtonian mechanics in all their work with large-scale (visible) objects moving at ordinary speeds (small compared with the speed of light). The orbits of space vehicles are exactly calculated from Newton's laws, which were formulated 300 years ago; the law of conservation of energy is still, after more than a century, a fundamental statement about nature that is used daily by research physicists and engineers. We would limit our understanding of physics if we neglected the contributions of those who went before us, and we would be unable to understand the generalizations into which some of the earlier laws of physics have evolved.

1-3 The Nature of the Modern Physics Course

One objective of a physics course is to impart useful knowledge. If you are going on in medicine, dentistry, agriculture, architecture, home economics, science teaching, psychology, or geology, you will find professional use for the facts and methods you study with this book. However, many of the students for whom this book is written do not intend to be scientists and are studying physics for its "cultural" value. It is all very well to discuss scientific methodology as an interesting example of human intellectual development, but this sort of scrutiny might equally well be given in a philosophy class. In our study of physics, we frankly stress the value of factual, detailed knowledge of laws, principles, and applications. Our culture is one in which applied science plays an ever-increasing role, and it is truly of cultural value to be able to understand a little of what goes on. Naturally, you will not be able to design a nuclear reactor as a result of diligent study of this book, but you *will* be able to appreciate some of the design problems. You will not be able to repair your TV set, but you *will* understand the basic laws of photoelectricity. Most important, you will come to appreciate the openness of scientific knowledge. What one group of men and women has discovered, another group can as easily find out. No country can have a monopoly on hydrogen bombs or TV sets. Whether you intend to be a writer, politician, business executive, homemaker, or musician, the factual knowledge of physics that you gain from your college course will help you feel at home in our science-oriented culture.

You will be aided in your study of physics by your teacher's lectures, your laboratory work, and, of course, this book. A physics textbook can be organized in many ways. To a large extent, we shall find it profitable to stay with certain specific aspects of physics long enough to gain some firm knowledge and the ability to reason out solutions to problems. Although we shall follow a time-honored sequence of subjects, we shall introduce ideas of "modern" physics wherever possible in the earlier work, rather than save them all for a final few chapters. To be sure, certain aspects of radiation, radioactivity, and nuclear physics should be studied in a concentrated and systematic way. But many other aspects of recent physics will be brought in as illustrative material for the "classical" or old-fashioned physics. The dividing line cannot be drawn sharply. After all, in 1687 the law of gravitation was "modern" physics, and in 1867 the molecular theory of heat was "modern" physics. In the second half of the 20th century, we have arrived at a point at which much that has been segregated as modern physics can be assimilated into the standard areas of mechanics, heat, and the like.

We begin with the study of mechanics, since the ideas of mechanics are used over and over again in the development of the other branches of physics. Indeed, the sequence of mechanics, sound, heat, electricity, light, and atomics has, for good reasons, become almost standard. The phenomena of sound are those of the mechanical vibrations of material bodies, and so we should study mechanics before sound. Heat and temperature are related to the mechanical motions of molecules, and so we should study mechanics before heat. We can calculate the pressure exerted by the molecules of a gas, using the laws of motion developed in our study of mechanics. Electricity can also best be studied with the aid of mechanical concepts: the forces acting on moving electrons or protons cause acceleration just as the force of gravitation causes acceleration of a satellite or a planet. We study light, which exhibits the properties of an electromagnetic wave, after electricity; indeed, light waves and radio waves are identical in nature and differ only in wavelength. We use mechanics, wave theory, electricity, and optics, as well as quantum ideas, in the interpretation of the light emitted by atoms and molecules; thus interpreted, the spectra give clues to the structure of the outer atom and to the Periodic Table. Our text concludes with a close study of the inner atom (the nucleus) and a look at some applications of nuclear physics.

As you study physics, you will notice that it is a cumulative process. Much of what you learn is carried over and used in later parts of the course. This relatedness means that you should keep up to date in your assignments. Several study aids have been included in this book to help you. Use them. The summary, check list, and questions at the end of each chapter will help you review and consolidate your knowledge. Avoid excessive underlining, which can lead to reliance on memory rather than on understanding. Problems are graded in difficulty; all serve the purpose of aiding you to master *principles* through *practice*. Sections marked "for further study" appear at the close of many of the chapters; these should give you a chance to spread your wings if you so desire. Don't overlook the references at the end of each chapter; the original writings of the pioneers of physics have a flavor and a lasting value that cannot be duplicated. Their biographies are revealing and often inspiring. Through other references, you can make contact with the growing edge—the frontier—of contemporary physics.

And now, let us get on with the task at hand, which is to learn the facts and methods of physics, a typical science. During the next few months you will many times approach the frontiers of physics. You will stand alongside pioneers of the past as they penetrate the barriers of their time, and you will, yourself, see where your contemporaries are striking forward, "no holds barred." May physics come alive for you, as it has for them.

1-4 Measurements in Physics— SI Units

It is the function of science to correlate precise measurements of physical quantities. By such correlations basic laws are discovered. The basic units of measurement must be well defined, and it is helpful if measurements are made in a consistent set of units. For instance, if we wish to find the distance traveled by a car, we would not multiply its speed in kilometers per hour by the elapsed time in minutes. Common sense dictates that we either convert the speed to kilometers per minute or convert the time to hours. Such quantities as speed (kilometers/hour) and area (square meters) are *dimensional quantities*; a measured dimensional quantity such as the area of a field has different numerical values in different systems of units. The same area can be expressed as 20,000 square centimeters or 2 square meters. The conversion of such quantities from one unit to another unit of the same dimensions will be made clear in illustrative examples throughout the book.

It is found that all mechanical quantities, such as energy, force, momentum, volume, and velocity, can be expressed in terms of three basic dimensional quantities. The choice of the three quantities can be made in several ways. In scientific work it is customary to take as the basic

dimensions (written in square brackets) mass [M], length [L], and time [T]. Velocity, for example, can be measured in kilometers per hour, feet per second, or meters per second— always a length divided by a time. We write [velocity] = [L]/[T], where the square brackets indicate that the equation relates *dimensions* only and is not necessarily a complete numerical equation. A dimensionless numerical factor may or may not be present. Thus the complete equation for the volume of a cylinder of radius r and height h is $V = \pi r^2 h$. The factor π is dimensionless, and the dimensions of r and h are each [L]. Hence we have the dimensional equation [volume] = $[L^2][L]$ = $[L^3]$. If the cylinder has diameter d and height h, the volume formula is $V = (\pi/4)d^2 h$, in which the numerical factor is different but the dimensional equation is still [volume] = $[L^3]$, as it must be for *any* volume.

Dimensional checks can be of great help in testing equations for possible error. In algebra you learned not to add "dollars" to "doughnuts" if you wanted a meaningful answer. In physics, if two quantities are to be added or subtracted, they must be of the same dimensions; likewise, the quantities on both sides of an equal sign must have the same dimensions.

Example 1-1

A student derives an equation for the distance traveled by a falling body:

$$s = \tfrac{1}{2}v_0 t^2 + vt$$

where s is the distance fallen, v_0 is the original velocity of the body, t is the elapsed time, and v is the final velocity of the body. Is the equation a possibly correct one?

Making a dimensional check and ignoring the factor $\frac{1}{2}$, which has no dimensions, we have the following:

Using fractions:

$$[L] \overset{?}{=} \frac{[L]}{[T]}[T^2] + \frac{[L]}{[T]}[T]$$

$$[L] \overset{?}{=} [LT] + [L]$$

Table 1-1 Important Metric Prefixes*

Prefix	Abbreviation	Meaning	Typical Examples
tera	T	$\times 10^{12}$	1 TeV (particle-accelerator energy) = 10^{12} electron volts
giga	G	$\times 10^{9}$	1 gigahertz (radar frequency) = 10^{9} Hz = 10^{9} cycles/s
mega	M	$\times 10^{6}$	1 megaton (equivalent TNT strength of nuclear weapons) = 10^{6} tons
kilo	k	$\times 10^{3}$	1 kilogram = 1000 g
deci	d	$\times 10^{-1}$	1 decibel = 0.1 bel
centi	c	$\times 10^{-2}$	1 centimeter = 0.01 m
milli	m	$\times 10^{-3}$	1 milliampere = 0.001 A
micro	μ	$\times 10^{-6}$	1 microvolt = 10^{-6} V
nano	n	$\times 10^{-9}$	1 nanosecond = 10^{-9} s
pico	p	$\times 10^{-12}$	1 picofarad = 10^{-12} F
femto	f	$\times 10^{-15}$	1 femtometer (approximate size of a proton) = 10^{-15} m

* Other metric prefixes not in common use are peta (P) for 10^{15}, exa (E) for 10^{18}, and atto (a) for 10^{-18}. The micron, a unit of length used in older literature, is 1 micrometer (mi'crometer) = 1 μm = 10^{-6} m = 0.001 mm.

Using negative exponents:

$$[L] \stackrel{?}{=} [LT^{-1}][T^2] + [LT^{-1}][T]$$
$$[L] \stackrel{?}{=} [LT] + [L]$$

Note that the dimensions, in square brackets, are combined just like ordinary algebraic quantities. We see that the dimensions are incorrect and so the equation is *wrong*. How to make it *correct* is another matter, something into which we are not yet prepared to go.

To cover *all* fields of physics, seven dimensional quantities are used. The *International System of Units* (abbreviated SI, for Système International) uses the meter, kilogram, and second for mechanics; other SI base units are the ampere (electric current), the kelvin (temperature), the mole (amount of substance), and the candela (luminous intensity).

Scientists use SI as the preferred system of units. In SI the standard of length was formerly defined as the distance between two marks on a carefully protected bar of platinum-iridium alloy preserved at the International Bureau of Weights and Measures at Sèvres, France. This "international prototype meter" and all copies of it in the various countries of the world were sub-

ject to unknown aging effects, oxidation, or even destruction through natural or man-made catastrophe. In 1960 the *meter* (m) was defined as exactly 1,650,763.73 wavelengths of a certain orange line in the spectrum of the krypton isotope of atomic mass number 86. This standard is probably good to about 4 parts in 10^9. An even more reproducible standard will, in the foreseeable future, be based on light waves from lasers that are stabilized by vibrations of iodine or methane molecules. For convenience, multiples and submultiples of the meter are used, such as the kilometer* (1000 m), the centimeter ($\frac{1}{100}$ m), and the millimeter ($\frac{1}{1000}$ m).

The important metric prefixes, shown in Table 1-1, place emphasis on units in successive ratios of 1000[†]; note that multiples greater than 10^3 are abbreviated with capital letters. As indicated in the examples, metric prefixes can be used for multiples and submultiples of any unit, even those not in the metric system itself.

---

* The word should be pronounced with the accent on the first syllable: kil'ometer, just as for kil'ogram.

[†] If you are not completely familiar with the use of exponents and with scientific notation (in powers of 10), this would be a fine time to review Secs. A-2 and A-3 of the Mathematical Appendix.

The SI unit of mass is the *kilogram* (kg), which is the mass of a certain block of platinum-iridium alloy known as the "international prototype kilogram" preserved at Sèvres. It is well known that all material bodies tend to stay at rest or, if in motion, tend to resist any change in the magnitude or direction of the motion. This attribute of matter is called inertia, and the *mass* of a body is a quantitative measure of its inertia. The *weight* of a body, on the other hand, is the force of gravity upon it. Weight and mass have different dimensions and should not be confused with each other. We shall discuss mass in detail in Chap. 3, and in Chaps. 5 and 9 we shall see how it is possible to compare two masses by observing changes in motion. Such methods of measuring mass (inertia) are fundamental but are not capable of the highest precision. Fortunately, the weights of two bodies, at any given point on the surface of the earth, are in strict proportion to their masses (Sec. 3-5), and therefore masses can be compared by weighing. Two bodies that weigh the same on an equal-arm balance or a spring balance also have equal masses. Other metric units for mass are the gram ($\frac{1}{1000}$ kg) and the metric ton (1000 kg). The metric ton is also called a *tonne*.

The SI unit of time is the *second* (s), formerly defined in terms of the length of the day, later in terms of the length of the year. These are not quite constant. The present standard, accurate to a few parts in 10^{11}, is the duration of 9,192,631,770 periods of the radiation corresponding to a certain transition between energy levels in the cesium atom of atomic mass number 133. The cesium standard was adopted in 1964 but is designated as "temporary" because research in the next few years may indicate that a standard based on some other atom such as hydrogen or thallium would be even more reproducible.

It is interesting to note the historical basis of the metric units. When the French authorities established the metric system in 1791, they intended that the meter should be exactly one 10-millionth of the distance from the equator to either pole. It was also intended that the gram should be the mass of one cubic centimeter of pure water at 4°C (a temperature near which the density of water is relatively independent of temperature, hence allowing the most precise measurement). These goals, however, were not quite realized, and so the prototype meter and kilogram based upon these procedures did not have exactly the intended values.

The adoption of the metric system by the French revolutionaries was a manifestation of the Age of Reason; they wished to establish a "rational" basis for every aspect of life, and the decimal system fitted into such a scheme. In 1795 the centime was adopted (it should properly have been called the centifranc). The division of the right angle into 100 equal parts or "grads," instead of 90 degrees, was proposed but never gained wide acceptance until very recently, when it began to be used for some engineering calculations. The revolutionary calendar had months of 30 days, divided into three 10-day "decades" instead of 7-day weeks; this calendar was abolished by Napoleon in 1805. It is significant that the names of the months in the revolutionary calendar were based on the seasons and the fruits of the earth; how logical it seemed to turn to the earth itself for the basis of a new system of measurement!

Of the three fundamental quantities, two—length and time—can now be defined in terms of atomic behavior: the wavelength of a certain light, and the frequency of a certain atomic vibration. It is natural to ask if an atomic definition of mass can be constructed. The masses of individual atoms are presumably suitable for the purpose, but it is too difficult to make an accurate count of the large number of atoms contained in a weighable amount of matter. The best we could do is to define a kilogram as the mass of 5.018412×10^{25} atoms of the carbon isotope of atomic mass number 12. The experimental uncertainty in the count is about 1 part in 10^6. Since two masses can be compared with far greater accuracy (1 part in 10^9) by weighing, it appears that we shall have to wait for improved experimental methods before an atomic definition of the kilogram would be any improvement over the present definition in terms of the international prototype kilogram.

The *meter-kilogram-second* (mks) *system*, a subset of SI, is called an absolute system of units, since the kilogram is defined in a way that has nothing to do with the gravitational field of the earth or any other planet. It is mass (inertia), not weight, that is used.

The *centimeter-gram-second* (cgs) *system* differs from the mks system only in using the centimeter and the gram, which are submultiples of the meter and the kilogram. It is a subset of SI, and it, too, is an absolute system of units.

The *foot-pound-second* (fps) *system* is another system used in English-speaking countries. The origins of the foot, yard, and pound are lost in antiquity. Prototype yards and pounds were formerly maintained in England and the United States, but now the primary standards for all countries are metric, and the British standards are secondary standards, defined in terms of the metric standards.

The metric system is used in all countries for scientific work and is the common system of measurement in almost all major countries. In the English-speaking countries the fps system is used in ordinary daily life, and it is the basis for the British engineering system that is used extensively in engineering work. However, the growth of international trade is a strong force toward adoption and full use of the metric system, already legally permitted in the United States. Business and industry leaders are becoming aware of the monetary losses, chiefly in loss of trade, caused by the complex fps system and other irrational units such as quarts, miles, inches, and tons.

For two reasons, therefore—the already prevalent use of the metric system in science, and the strong possibility of a changeover in business and industry—we will use metric units in this book.

Table 1-2 shows that the physicist is concerned with an enormous range of values of mass, length, and time. A galaxy, containing some 10^{11} stars, has a mass of about 10^{41} kg, and the observable universe (some 10^{10} galaxies) has a mass of about 10^{51} kg. The mass of an electron is only about 10^{-30} kg, so we see that observable masses span a range of more

Table 1-2 Ranges of Mass, Length, and Time Studied by Physicists
(All values are approximate, rounded to the nearest power of 10.)

Mass (kg)		Size or Length (m)		Duration or Age (s)	
10^{51}	totality of observable universe	10^{26}	totality of observable universe	10^{18}	oldest galaxies
				10^{16}	life on earth
10^{41}	the Milky Way galaxy	10^{21}	the Milky Way galaxy	10^{13}	human race
10^{30}	sun	10^{12}	sun	10^{11}	half-life of carbon-14 isotope
10^{24}	earth	10^{7}	earth	10^{9}	human life span
10^{10}	pyramid	10^{2}	pyramid	10^{7}	academic year
10^{6}	elephant	10^{1}	elephant	10^{3}	class session
10^{2}	human	10^{0}	human	10^{0}	a heartbeat
10^{-4}	flea	10^{-3}	flea	10^{-3}	period of vibration of guitar string
10^{-9}	bacterium	10^{-5}	bacterium	10^{-6}	lifetime of a muon
10^{-12}	red blood cell	10^{-6}	wavelength of visible light	10^{-8}	period of a TV electromagnetic wave
10^{-20}	poliomyelitis virus			10^{-15}	period of visible light
10^{-25}	iron atom	10^{-8}	poliomyelitis virus	10^{-23}	lifetime of a very unstable subatomic particle
10^{-30}	electron	10^{-10}	iron atom		
		10^{-14}	iron nucleus		

than 80 *orders of magnitude* (the ratio is $10^{51}/10^{-30} = 10^{81}$, each power of 10 being an order of magnitude). It is interesting, and possibly significant, that a human being's mass, size, and heartbeat are about "average." One line of reasoning runs as follows: It is not possible for a group of, say, 100 to 10,000 atoms to have the complexity of arrangement needed for the advanced biochemical reactions of self-awareness and consciousness. On the other hand, a living entity the size of a galaxy would be ruled out because interaction between its component parts could proceed no faster than the speed of light, so life "as we know it" would be impossibly slow-paced. The human organism's position in the scale of things may well be related to considerations such as these.

In our study of physics we make the assumption that the *same* laws apply throughout the range of sizes, masses, and times of the components of the universe. In the 17th century, Sir Isaac Newton was strongly motivated by a desire to prove that the same law of gravitation that describes the fall of an apple is also manifested in the motions of the planets around the sun. We shall see in Chaps. 3 and 8 how Newton's law of *universal* gravitation fulfilled this expectation. We should not be overconfident, but so far there seems to be no reason to doubt that all the laws of physics apply throughout the entire domain of the natural universe.

1-5   *What Is Matter?*

A significant development in science took place when the earliest scientists began to consider the properties of matter rather than the behavior of particular objects. It is one thing to know the weight of an iron axhead or an iron suit of armor, but axheads are large and small, heavy and light, and it is surely of greater importance to study the nature of the substance iron than to limit oneself to specific iron objects, be they ball bearings or bridge girders.

Substances have many properties; in addition to a host of mechanical characteristics, we can list thermal conductivity, electrical conductivity, optical index of refraction, shielding effect for x rays, and neutron absorption, to name only a few. Tables of properties of matter are contained in handbooks that may run to several thousand pages. We presume that *all* properties of matter are, in principle, ultimately to be correlated with a relatively few basic *attributes* of matter, such as inertia, gravitational forces, electric forces, and nuclear forces. Let us take a preview of these fundamental attributes of matter. As in all preliminary surveys, certain terms will have to be used loosely. You have undoubtedly studied general science, and so you will have a feeling for words such as "force" and "charge" that we shall use now but cannot define precisely until later chapters.

A first and basic attribute of matter is that any piece of it, which we call a "body," tends to resist any change in its motion. This attribute of matter is called *inertia*. A forward force must be applied to a car to cause it to pick up speed. If the car is already moving, a backward force must be applied to cause it to slow down. Unless a force is applied, the car tends to remain as it is, either at rest or in uniform (steady) motion in a straight line. We call such an object a *material body*, by which we mean that all matter has inertia. In Chap. 3 we shall see how inertia can be measured quantitatively, and we shall use the word *mass* to describe the amount of inertia that a material body has.

A second attribute of matter is its weight. We all know that the *weight* of an apple is the *force of gravitation* between the earth and the apple. Weight is an *interaction* between two bodies, and gravitational force is known to fall off rapidly as the distance between the bodies increases. A given apple has no unique weight—its weight depends on its position relative to a second body (the earth). It is a property of every particle of matter to exert gravitational attraction on every other particle of matter anywhere else in the universe. We shall discuss gravitation more fully in Chap. 8.

Most matter exerts *electric force* on other matter.* Electric forces are varied and complex, but to a first approximation several distinct aspects of such forces can be studied separately. Coulomb's law, to be discussed in Sec. 18-5, describes the force between stationary charged bodies: like charges repel each other; unlike charges attract each other; and the magnitude of the force varies inversely as the square of the distance between the charges. Coulomb forces are also called *electrostatic forces*. Actually, as we shall see in Chap. 18, electrostatic forces are intrinsically far greater than gravitational forces. It is only because most ordinary matter is neutral (since atoms contain equal numbers of positive and negative charges) that the electric forces between ordinary objects cancel out sufficiently to unmask such relatively small gravitational forces as the weight of a 10-ton truck. Another aspect of electric force is presented by the force exerted between charges in motion. Such forces are called *magnetic forces*; they are evident in the behavior of magnets, electric motors, and many other engineering devices. *Elastic forces*, which may be either attractive or repulsive, involve a third aspect of electric force, which has to do with the overlapping of the charged parts of atoms. When an archer stretches her bow, she is calling into play electric forces between neutral atoms. Such forces are, of course, effective only when atoms are nearly in contact with each other, say within about 5×10^{-10} m or less. We often call these forces *short-range electric forces*, because at greater distances they are negligible compared with Coulomb forces or magnetic forces. If you have studied chemistry, you may know that many *chemical forces* (the covalent

bonds) are also short-range electric forces. We see that various aspects of electric forces are characteristic attributes of matter. It is important to realize that electric forces and gravitational forces are exerted entirely independently of each other; there is no known connection between them.[†]

Finally, two kinds of *nuclear forces* are exerted between particles of matter. These are extremely short-range forces and are negligibly small unless the particles are about 10^{-14} m or less from each other. Consequently, these forces are unfamiliar in the large-scale, everyday world. They are of vital importance, however, for without some sort of cohesive nuclear force the particles that make up the nuclei of atoms would not stick together. With modern high-energy apparatus it is possible to shoot nuclei at each other with sufficient speed to overcome the electric forces of repulsion and allow positively charged particles to approach closely the positively charged nuclei. In this way nuclear forces are being studied in laboratories all over the world. As the picture unfolds, it is found that nuclear forces are at least as complex and varied as electric forces. In Chap. 33 we shall discuss the two kinds of nuclear force, known as the *strong interaction* and the *weak interaction*.

These, then, are the attributes of what we call matter. Matter has inertia, and matter exerts gravitational, electric, and nuclear forces upon other matter. These are the underlying physical facts that, properly interpreted, must give us an amount of the many tremendously varied properties of matter encountered in nature.

1-6 Density and Specific Gravity

Perhaps the most obvious property of matter is its density. Aluminum is a "light" metal; gold is a "heavy" metal. We define *density d* as the mass

---

* Certain neutral subatomic particles such as the neutron and the neutral mesons exert no electrostatic forces, although they do exert nuclear forces and gravitational forces. All such particles that have a finite rest mass are unstable when far removed from other particles and cannot long exist by themselves (see Table 33-5).

[†] The theoretical physicist Albert Einstein (1879–1955) devoted the last decades of his life to an unsuccessful search for such a connection.

per unit volume of a substance. Thus, since each cubic centimeter of water has a mass of 1 g, its density is 1 g/cm^3. The density of this most familiar substance is expressed in the mks system as 1000 kg/m^3. In Table 1-3 are given the densities of some representative solids, liquids, and gases. Just as for water, the density of any substance in kg/m^3 is found by multiplying the cgs value by 1000.

Example 1-2

A block of metal has dimensions 5 cm × 7 cm × 20 cm, and its mass is 5 kg. Out of which of the various metals listed in Table 1-3 is the block probably made?

The density of the block must be expressed in grams per cubic centimeter in order to compare it with the table. We must first express the mass in grams. Since 1 kg = 1000 g, the mass is 5 × 1000 = 5000 g.

$$\text{Volume} = (5 \text{ cm})(7 \text{ cm})(20 \text{ cm}) = 700 \text{ cm}^3$$

We now calculate the density:

$$\text{Density} = \frac{\text{mass}}{\text{volume}} = \frac{5000 \text{ g}}{700 \text{ cm}^3} = \boxed{7.14 \text{ g/cm}^3}$$

The block is probably made of zinc.

A Note on Significant Figures. Strictly speaking, writing the mass of the block in Example 1-2 as 5 kg implies that the mass lies somewhere between 4.50 kg and 5.50 kg. (See the discussion of significant figures in Sec. A-4 of the Appendix.) However, in this text we arbitrarily assume that all data given in the examples and problems are known to three significant figures, unless otherwise stated. Thus, in this text, $m = 5$ kg may be thought of as $m = 5.00$ kg for purposes of computation. This is not good engineering practice, and it would not be allowed in a write-up of a laboratory experiment. However, our purpose in solving problems is to illustrate basic laws and methods of physics, not to illustrate laboratory practice. When we write $m = 5$ kg in this book, we mean that the mass

Table 1-3 Density

Substance	Density g/cm^3
Solids	
aluminum	2.70
copper	8.93
tin	7.29
brass	8.44
zinc	7.14
nickel	8.90
magnesium	1.75
iron	7.86
lead	11.3
gold	19.3
uranium	18.7
wood (pine)	0.35–0.6
concrete	2.7
ice	0.917
limestone	2.7
bone	1.7–2.0
diamond	3.51
cork	0.22–0.26
steel	7.8
Liquids	
water	1.00
gasoline	0.66–0.69
glycerin	1.26
mercury	13.6
alcohol (ethyl)	0.791
sea water	1.025
carbon tetrachloride	1.595
olive oil	0.918
blood	1.1
Gases*	
air	1.293×10^{-3}
hydrogen	0.0899×10^{-3}
oxygen	1.429×10^{-3}
helium	0.1785×10^{-3}
carbon dioxide	1.977×10^{-3}
tungsten hexafluoride	12.9×10^{-3}
methane	0.717×10^{-3}

*All at 0°C and 1 atm pressure.

may be considered to lie between 4.995 kg and 5.005 kg. This would be written as 5.00 kg in a laboratory report.

Example 1-3

If the mass of an iron engine block is 157 kg, how much mass would an airplane designer save if the same block were cast in magnesium?

From Table 1-3, we determine that the density of iron is 7860 kg/m³. Since density = mass/volume, we can solve for the volume of the block:

$$\text{Volume} = \frac{\text{mass}}{\text{density}} = \frac{157 \text{ kg}}{7860 \text{ kg/m}^3} = 0.020 \text{ m}^3$$

When the block is cast in magnesium, its volume is still 0.020 m³.

$$\text{Mass} = (\text{density})(\text{volume})$$
$$= (1750 \text{ kg/m}^3)(0.020 \text{ m}^3) = 35 \text{ kg}$$

The designer will save $157 - 35 = \boxed{122 \text{ kg}}$

For many purposes it is convenient to compare substances with one another. Pure water at 0°C (or at some other standard fixed temperature) is often used as a standard substance, and we define *specific gravity* as relative density, given by a ratio:

$$\text{Specific gravity} = \frac{\text{density of substance}}{\text{density of water}}$$

Any given volume of aluminum has 2.70 times as much mass as the same volume of water, and its specific gravity is 2.70. Note carefully that specific gravity is a pure number, having no units, since it is a ratio of two similar quantities.* Because of its definition in terms of water, and because water has a density of 1.00 g/cm³, the specific gravity of any substance is *numerically equal to* the density in cgs units.

---

* To avoid small numbers, specific gravity for gases is sometimes tabulated relative to a standard gas such as air at 0°C and 1 atmosphere pressure.

1-7 The Structure of Matter

We have seen the advantages of considering the *substance* iron rather than a multitude of different iron objects of various sizes and shapes. Can we simplify still further and consider what, if anything, is common to both iron and aluminum, for instance? What is the difference between a block of iron and an equally large block of aluminum that makes the iron block weigh so much more? Let us compare iron and aluminum in detail.

As a result of much experimentation by many scientists, a picture of the structure of a substance such as iron has gradually unfolded. According to this picture, iron is a collection of atoms in a symmetrical arrangement called a *crystal*. The crystal structures of various elements, alloys, and compounds differ in both the arrangement and spacing of their atoms (Fig. 1-1). Thus the density of a solid depends in part on its crystal structure. By way of illustration, the atomic weight of iron is 56 and that of aluminum is 27, meaning that each iron atom is just over twice as heavy as each aluminum atom. Since the density of iron is almost 3 times that of aluminum (Table 1-3), we see that the fact that each iron atom is twice as heavy as each aluminum atom is not enough to account for the difference in density. Therefore, the iron atoms must be packed more closely. Detailed x-ray analysis of the crystal structures of iron and aluminum shows this to be the case, and an exact numerical check is obtained for the densities.

Going one step further, we ask why the atomic weights of iron and aluminum differ. Here the planetary model[†] of an atom is a good first approximation. The heavy *nucleus* contains more

---

[†] To the scientist, the term *model* does not usually mean a physical rendering, such as a model airplane; a model is a *representation* that allows us to think more concretely about some aspect of the real world. The model is not the reality; it simply offers us a useful way of dealing with reality. Models, like hypotheses, are always subject to correction as we gather more evidence from experiment or observation.

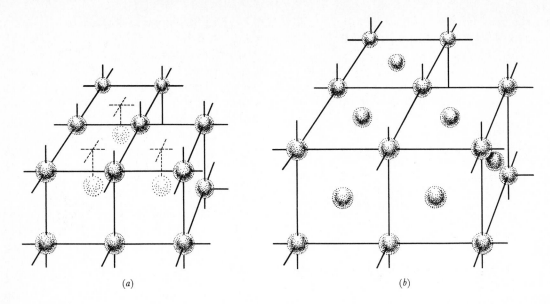

(a)   (b)

Figure 1-1   Crystals of solid bodies differ both in the distance between atoms and in the geometrical form of the crystal structure. (*a*) Iron crystal in body-centered cubic structure. (*b*) Aluminum crystal in face-centered cubic structure. Approximate scale: 1 cm $= 1.6 \times 10^{-10}$ m.

than 99.9% of the total mass (and weight) of the atom. The rest of the mass of an atom is due to a number of *electrons* that travel around the nucleus in orbits, like planets around the sun. The nucleus contains "heavy" particles known as *nucleons*, which may either be electrically neutral or have a positive charge. A neutral nucleon is called a *neutron*, and a positively charged nucleon is called a *proton*. These heavy particles, the neutron and the proton, are considered to be merely different states of the same particle. They have almost identical masses, and it is entirely possible for a proton to change into a neutron, and vice versa. The probability of such a change-over within the nucleus depends on many factors, but when the change takes place spontaneously, the nucleus is said to be *radioactive*, or *unstable*.

Each electron has a charge that is negative and numerically equal to that of the proton. A neutral atom, therefore, has the same number of electrons in the electron cloud as it has protons in the nucleus. This number is called the *atomic*

number of the element. Chemical properties are determined by the number of electrons in the cloud. The "iron-ness" of iron, for instance, is summarized by the many chemical reactions in which iron takes part, and these reactions are governed by the number of electrons in the electron cloud, that is, by the atomic number.

Figure 1-2 shows schematically several common atoms. Note that two kinds of iron atom are shown. One has a total of 54 nucleons in its nucleus, and the other has a total of 56 nucleons. Both have 26 protons—the difference is in the number of neutrons. These atoms are called *isotopes* of iron; the atomic number of each iron isotope is seen to be 26. As a matter of notation, a symbol such as $^{54}_{26}\text{Fe}$ is used, where the *mass number*, equal to the total number of nucleons, is written as a superscript at the upper left of the chemical symbol, and the atomic number is written as a subscript at the lower left.*

---

* Since the atomic number of any atom is known whenever the chemical symbol is known, the writing of the atomic number is superfluous and is often omitted.

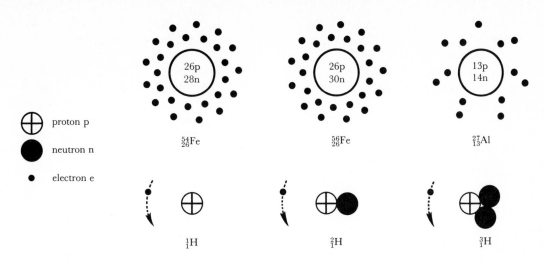

Figure 1-2 Schematic representation of several atoms (not drawn to scale).

Normally, an iron nucleus draws to itself a cloud of 26 electrons, equal to the positive charge of the 26 protons within the nucleus. If, somehow, only 25 electrons surround a $_{26}$Fe nucleus, we have a net positive charge of 1 unit, and the atom is called an *ion*, written as Fe^+ in this case. A doubly charged iron ion, written as Fe^{2+}, would have only 24 electrons in its cloud, and so on. Naturally, since the electron cloud determines chemical properties, iron ions behave quite differently from neutral iron atoms. It is also possible to have negative ions; thus a $_{17}Cl^-$ ion has 18 electrons in its cloud, one more than the 17 that would be equal to the atomic number.

The isotopes of the simplest element, hydrogen, are of special interest. The $_1^1H$ nucleus is simply a proton; most (99.985%) hydrogen atoms in nature have a nuclei of this sort. The nucleus of "heavy hydrogen"—*deuterium*, $_1^2H$— is made up of one proton and one neutron. (This combination is called a *deuteron*.) Finally, $_1^3H$, or *tritium*, has one proton and two neutrons. However, tritium is radioactive, with a 50–50 chance that one neutron will change to a proton within about 12 years; when this happens, an electron is formed and ejected from the nucleus.

Many elements have more than one stable isotope. For instance, there are ten stable iso-

topes of tin, ranging from $_{50}^{112}Sn$ to $_{50}^{126}Sn$. Chemically, the isotopes of an element—tin, for instance—are identical, since the atomic numbers are the same. Separation of isotopes is thus difficult and depends upon physical rather than chemical methods.

In this book we follow the recommendations of the International Union of Pure and Applied Chemistry. By *atomic weight* is meant the *relative atomic mass* of the element, defined as the ratio of average mass per atom of a naturally occurring composition of the element to $\frac{1}{12}$ of the mass of an atom of ^{12}C. Atomic weight, being a ratio of masses, is a pure number and has no units; thus the atomic weight of tin is 118.69. (It would be wrong to say that the atomic weight of tin is 118.69 g.) A similar situation exists for the unitless quantity specific gravity, which is relative density (the ratio of the density of a substance to that of water). The chemical atomic weight is the average value of the weights of all the stable isotopes of an element, with due regard for relative abundance.

It is impossible to draw atoms to scale. Most of the atom is empty space, with the distances between nucleus and electrons vastly larger than the sizes of the particles. It also follows that the density of the nuclear matter is enormous, for only in that way can the overall density of matter be as great as it is. If the electron cloud of the iron atom ^{56}Fe were expanded to the size of a football field, the nucleus would be represented by a pea-sized ball 4 mm in diameter, weighing 6 million tons, and the electrons would be represented by 26 mosquitoes weighing 120 tons each, flying around the pea at distances ranging up to 50 meters. This is an extreme case of "model-itis"*—the uncritical acceptance of models as true descriptions of nature. In some respects the model is valid, especially as regards the rather definite boundary for the nuclear matter. On the other hand, the behavior of the electrons is often more closely approximated by a model of a continuous cloud of charge, rather than by a small group of mosquito-like or planet-like particles.

Experiment shows that the sizes of the nuclei increase as the number of particles increases. In fact, the *volume* of the ^{238}U nucleus is just about $\frac{238}{56}$ times as great as the volume of the ^{56}Fe nucleus. This means that at long last we have come upon a property of matter that is remarkably uniform from element to element: all nuclei

---

* See the footnote on page 81 for further description of this dread disease.

have about the same density. Our model of a nucleus as a collection of incompressible neutrons and protons therefore represents this phase of nuclear behavior fairly well. The variations in density of various solids, which are shown in Table 1-3, are to be attributed to several causes: (1) The atomic weights of the elements vary from 1 to over 200; (2) the sizes of the individual atoms differ because of variations in the sizes of the electron clouds; and (3) atoms themselves are packed together in different ways to form crystals. The atoms in a crystal are held apart by electric forces, due to the repulsion between the like positive charges of neighboring nuclei. If somehow these repulsive forces could be annulled, the nuclei could be packed closely and matter of tremendous density would result. The nuclei of all the atoms in the earth would make a ball only 200 m in radius! Although such densities are unknown on earth, certain stars (known as white dwarfs) have partially collapsed and have densities ranging up to 50,000 g/cm^3 (almost a ton per cubic inch)—still far less than the universal density of nuclear matter, which is about 2×10^{14} g/cm^3. The superdense, small, rapidly rotating stars known as pulsars are fully collapsed, with densities in the range of 10^{14} g/cm^3.

Summary   Physics deals mainly with the more fundamental aspects of energy and nonliving matter. Physical knowledge comes to us through application of scientific methods: the gathering of data, formulating of hypotheses, and testing of hypotheses by means of controlled experiment. The division of physics into compartments such as mechanics and electricity is merely an aid to study, and we shall cross the boundaries many times in our further work.

Mechanical quantities can be expressed in terms of three fundamental dimensional quantities, such as mass, length, and time, or force, length, and time. Scientists are now able to define length and time in terms of atomic standards that are imperishable and universally reproducible, but such a procedure for mass is not yet practicable, although it is theoretically possible. Dimensional analysis is often useful in checking equations for possible error.

Matter is characterized by inertia and by the occurrence of gravitational, electric, and nuclear forces between material particles. Gravitational force and electric force act at long range. In addition, the short-range effects of electric force are responsible for elastic forces and for certain cohesive and repulsive forces between atoms in molecules. Nuclear forces, which act at still shorter range, are not yet fully understood.

All the many properties of matter would be, in theory at least, predictable if these basic attributes were fully understood. Mathematical difficulties are so formidable that approximate models are often used as an aid to (but not a substitute for) the description of nature. One such model is discussed in this chapter: the atomic model for correlating the densities of substances.

Density is mass per unit volume. Specific gravity is a dimensionless quantity equal to the density of a substance divided by the density of a standard substance, usually water. Differences in density of various substances are not entirely due to differences in the masses of the individual atoms. The sizes of atoms, determined mainly by the sizes of their electron clouds, and differences in crystal structure also affect the density.

The planetary model of an atom has proved valuable. The nucleus contains a close-packed cluster of nucleons, known as neutrons and protons, with total neutrality of charge for the atom as a whole maintained by an electron cloud that contains as many electrons as there are protons in the nucleus. This number is the atomic number of the element. Chemical properties of an atom are determined by the number of electrons in the cloud. Isotopes of a given element are chemically alike, have the same nuclear charge, and differ only in mass due to different numbers of neutrons in the nucleus. The nuclei of various elements differ in size in such a way that the density of nuclear matter is approximately constant for all elements.

Check List

hypothesis	inertia	neutron
theory	mass	proton
law	weight	atomic number
dimensional quantities	matter	isotope
SI	Coulomb electric forces	ion
meter	short-range electric	deuterium
kilogram	forces	tritium
second	nuclear forces	atomic weight
mks system	density	modelitis
cgs system	specific gravity	
absolute system	nucleon	

References

At the end of each chapter of this book will be found references to sources where various aspects of the material covered in the text are explored more deeply. In addition, we list here several works that are applicable throughout the course.

GENERAL REFERENCES

W. F. Magie, *A Source Book in Physics* (McGraw-Hill, New York, 1935). Contains extracts from the original writings of the great experimental physicists from 1600 through the year 1900. A valuable reference with which every student should have at least a nodding acquaintance.

F. Cajori, *A History of Physics* (Macmillan, New York, 1906; Dover, New York, 1929).

H. Butterfield, *The Origins of Modern Science: 1300–1800* (Macmillan, New York, 1957; Collier, New York, 1962 [revised ed.]; Free Press, New York, 1965).

G. Sarton, *A History of Science* (Havard University Press, Cambridge, Mass., 1952).

L. W. Taylor, *Physics: the Pioneer Science* (Houghton Mifflin, Boston, 1941; Dover, New York, 1959).

G. Holton and S. G. Brush, *Introduction to Concepts and Theories in Physical Science* (Addison-Wesley, Reading, Mass., 1973). Excellent historical and philosophical background material.

E. M. Rogers, *Physics for the Inquiring Mind* (Princeton University Press, Princeton, 1960).

J. Walker, *The Flying Circus of Physics, with Answers* (Wiley, New York, 1977).

STANDARD TEXTS

The student should also have available some standard text that is designed for those specializing in physics, chemistry, and engineering. It will be of value to refer from time to time to a book of a more technical nature than this one, for more detail, more rigor, more mathematics, or more diversified applications. Some of the standard texts in the field are the following:

R. P. Feynman, R. B. Leighton, and M. Sands, *The Feynman Lectures on Physics* (Addison-Wesley, Reading, Mass., 1963).

Berkeley Physics Course—Vol. I, *Mechanics*, C. Kittel, W. D. Knight, and M. A. Rudermann; Vol. II, *Electricity and Magnetism*, E. M. Purcell; Vol. III, *Waves*, F. S. Crawford, Jr.; Vol. IV, *Quantum Physics*, E. H. Wichmann; Vol. V, *Statistical Physics*, F. Reif (McGraw-Hill, New York, 1967).

F. W. Sears, M. W. Zemansky, and H. D. Young, *University Physics*, 5th ed. (Addison-Wesley, Reading, Mass., 1976).

R. Resnick and D. Halliday, Part I, and D. Halliday and R. Resnick, Part II, *Physics*, 3rd ed. (Wiley, New York, 1978).

P. A. Tipler, *Physics* (Worth, New York, 1976).

M. Alonso and E. J. Finn, *Physics* (Addison-Wesley, Reading, Mas., 1970).

R. T. Weidner and R. L. Sells, *Elementary Physics: Classical and Modern* (Allyn and Bacon, Boston, 1975).

This is not meant to be an all-inclusive list of good books in this classification; your instructor no doubt has favorites that he or she will be glad to recommend.

PHYSICS AND THE LIFE SCIENCES

Any of the following books will serve as a reference for students interested in biology, medicine, or dentistry.

R. W. Stacy, D. T. Williams, R. E. Worden, and R. O. McMorris, *Essentials of Biological and Medical Physics* (McGraw-Hill, New York, 1955).

P. Davidowits, *Physics in Biology and Medicine* (Prentice-Hall, Englewood Cliffs, N. J., 1975).

L. H. Greenberg, *Physics for Biology and Pre-Med Students* (Saunders, Philadelphia, 1975).

S. G. G. MacDonald and D. M. Burns, *Physics for the Life and Health Sciences* (Addison-Wesley, Reading, Mass., 1975).

H. J. Metcalf, *Topics in Classical Biophysics* (Prentice-Hall, Englewood Cliffs, N.J., 1980).

J. J. I. Shonle, *Environmental Applications of General Physics* (Addison-Wesley, Reading, Mass., 1975).

A. M. Saperstein, *Physics: Energy in the Environment* (Little, Brown, Boston, 1975).

J. M. Fowler, *Energy and the Environment* (McGraw-Hill, New York, 1975).

J. Priest, *Energy for a Technological Society*, 2nd ed. (Addison-Wesley, Reading, Mass., 1979).

L. C. Ruedisili and M. W. Firebaugh, eds., *Perspectives on Energy* (Oxford Press, New York, 1978).

Questions

Note: Each chapter in this book is concluded with a set of questions and problems. The *questions* are designed to help fix ideas on a verbal level. The *problems* are arranged into three groups: Type A problems are the easiest and often merely test knowledge of the definitions and equations given in the text. Type B problems usually require some understanding of physical laws and methods. They are "standard" problems, well within the grasp of the student who has understood the chapter. Type C problems are of several kinds: they may be more difficult or time-consuming, or they may depend on more mathematics than the others, or they may follow and deal with the additional text that is presented at the end of some chapters under the heading For Further Study (see page 52).

1-1 Make a list of the titles of the physics journals in English that are carried by your college library. Check also for *Science, Nature, Scientific American, American Scientist, Sky and Telescope, Environment, Bulletin of the Atomic Scientists*, and *Isis*.

1-2 Suppose we could communicate with little green men on Mars only by means of radio signals. Compose a text for their message that would give enough information to enable us to tell how long their unit of length, the retem, is in comparison with our meter. (Although they spell their words backward, they are supposed to be not at all backward intellectually.) Could we find out the size of their unit of mass?

1-3 What is the formula for the area of a triangle? What is the dimensional formula for the area of a triangle?

1-4 The radius of a circle inscribed in any triangle whose sides are a, b, c, is given by

$$\frac{\sqrt{s(s-a)(s-b)(s-c)}}{s}$$

where s is an abbreviation for $\frac{1}{2}(a + b + c)$. Check this formula for dimensional consistency.

1-5 A student derives the following formula for the distance traveled by an accelerated body:

$$s = vt - \tfrac{1}{2}at^2$$

where s = distance, in meters; v = final velocity of body, in meters per second; t = elapsed time, in seconds; and a = acceleration, in meters per second per second. Check this formula for dimensional consistency. What statement can you make about the correctness of this formula?

1-6 The French centime was defined to have a mass of 2 g of copper. Explain how the mass of this coin is related to the circumference of the earth and the physical properties of water.

1-7 An object has a mass of 1 metric ton. Show that the weight of this object is about equal to an ordinary ton (2000 lb), within about 10%. (See Appendix Table 4.)

1-8 Consider the copper isotope $^{63}_{29}$Cu. How many electrons are in the electron cloud if the atom is neutral? How many if the atom is a doubly charged positive ion? How many neutrons are in this nucleus? How many neutrons are in the nucleus of $^{65}_{29}$Cu? How many protons are in the nucleus of $^{65}_{29}$Cu? How many nucleons are in the nucleus of $^{65}_{29}$Cu? What is the atomic weight of $^{65}_{29}$Cu? What is the atomic number of $^{63}_{29}$Cu? of $^{65}_{29}$Cu?

1-9 Since like charges repel each other, why don't the 26 positively charged protons in an iron nucleus fly apart instead of remaining together in a small region of space?

1-10 The deuterium nucleus contains two particles; name them. What is the nature of the attractive force between the two particles?

1-11 What is the difference between the structure of the uranium isotopes $^{235}_{92}$U and $^{238}_{92}$U?

MULTIPLE CHOICE

1-12 The dimensions of density are   (a) $[ML^{-3}T]$;   (b) $[ML^{-3}]$;   (c) the same as those of specific gravity.

1-13 The density of ice, in kg/m³, is   (a) 0.917;   (b) 91.7;   (c) 917.

1-14 The mass of a tissue sample was found to be 0.0427 g. The number of significant figures in this result is   (a) 3;   (b) 4;   (c) 5.

1-15 The number of protons in the nucleus of $^{65}_{30}$Zn is   (a) 30;   (b) 35;   (c) 65.

1-16 The electron cloud of a neutral atom has as many electrons as the nucleus has   (a) nucleons;   (b) neutrons;   (c) protons.

1-17 The atomic number of hydrogen is   (a) 1;   (b) 1 or 2;   (c) 1, 2, or 3.

Problems

Note: Use the Periodic Table (Appendix Table 6), when necessary, to find atomic weights. Also see the table of equivalents and conversion factors, Appendix Table 4.

1-A1 Express the number of seconds in a century in the form 3.156×10^x.

1-A2 The circumference of the earth at the equator is 40,067 km. Express this in meters, in the form $x.xx \times 10^x$.

1-A3 Calculate the mass of air (at 0°C and 1 atm pressure) above an ice-hockey rink that is in a building 40 m long × 30 m wide × 15 m high. Express your answer in metric tons.

1-A4 What is the specific gravity of methane, relative to air? (Use Table 1-3.)

1-A5 What is the mass of the helium in a balloon if the volume is 40 m³ and the specific gravity (relative to air) is 0.14?

1-A6 What is the specific gravity relative to air of the "heaviest" gas in Table 1-3?

1-A7 What is the approximate volume of a human brain of specific gravity 1.040 and mass 0.940 kg?

1-A8 The Great Pyramid of Cheops has a volume of 2.4×10^{12} cm³ and is made of limestone. What is the mass of the pyramid   (a) in grams?   (b) in kilograms?   (c) in metric tons?

1-A9 If a cube of material 0.01 mm on an edge (barely visible to the naked eye) had the density of nuclear matter, what would be its mass? Could you lift it?

1-B1 If 20 liters of water and 8 liters of glycerin are mixed to form 28 liters of antifreeze solution, find the specific gravity of the mixture, assuming that no chemical reaction takes place.

1-B2 Of the two stable isotopes of copper, $^{63}_{29}$Cu occurs 69% of the time and $^{65}_{29}$Cu occurs 31% of the time. What is the average atomic weight of copper? Check your answer by referring to the Periodic Table (Appendix Table 6).

1-B3 Stars known as pulsars show variations in light during times much shorter than 1 s, apparently related to rapid rotation. They are therefore assumed to be very small and dense. Calculate the diameter of a completely collapsed neutron star of density equal to that of nuclear matter. Assume that the star's mass is equal to that of the sun, 2×10^{30} kg.

References

1. G. P. Thomson, "Nature of Physics and Its Relation to Other Disciplines," *Am. J. Phys.* **28**, 187 (1960).
2. G. Holton, "The Relevance of Physics," *Physics Today* **23**(10), 40 (Nov. 1970).
3. H. Butterfield, "The Scientific Revolution," *Sci. American* **203**(3), 173 (Sept. 1960).
4. M. Nicholson, "Resource Letter SL-1 on Science and Literature," *Am. J. Phys.* **33**, 175 (1965). Gives 96 references, many of them available in college libraries.
5. P. Morrison, *Sci. American* **233**(4), 132 (Oct. 1975). A book review of *The International Bureau of Weights and Measures*, edited by C. H. Page and P. Vigoreux. Useful comments on the nature and use of standards.
6. A. V. Astin, "Standards of Measurement," *Sci. American* **218**(6), 50 (June 1968).
7. L. K. Nash, *The Atomic-Molecular Theory* (Harvard University Press, Cambridge, Mass., 1950); also in *Harvard Case Histories in Experimental Science*, Vol. 1 (Harvard University Press, Cambridge, Mass., 1957), pp. 215–321.

Kinematics—
The Description of Motion

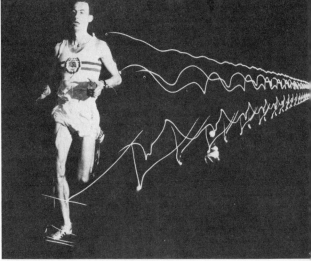

We continue our study of physics by considering motion. We first attempt to develop a way of describing the motions of physical bodies, ignoring for the time being the possible causes of motions. The study of motion as such, divorced from its origin or cause, is called *kinematics*.* Fortunately, our study of kinematics will prove useful in its own right, since some interesting problems can be solved without considering the causes of motion. Even more important, without the ability to think clearly about motion itself, we would be prevented from getting an adequate grasp of *dynamics*,† the study of forces and the changes in motion that they cause.

The Italian natural philosopher Galileo Galilei (1564–1642) became the most gifted experimental physicist of his time. Indeed, it was largely because of the special insight and example of Galileo that experimental methods be-

came widely accepted as a part of physical science. He argued eloquently for the Copernican theory of the solar system—which asserted that the earth and the planets are revolving around the sun, as opposed to the Ptolemaic system, according to which the earth was the center of the universe—and was persecuted for his teachings. Typically, although Galileo did not invent the telescope, it was he who first used it to observe the satellites of Jupiter, thus providing experimental evidence for the Copernican theory. Galileo's experimental outlook, his reliance on observation as the final authority, and his skill in explaining his ideas to others make him one of the great figures in the history of physics. His outlook pervades all of present-day science.

Galileo realized that an adequate description of motion requires a quantitative, or mathematical, outlook, but he was hampered by the primitive condition of mathematics in his day. Even with simple high-school algebra, we are better off than Galileo. But we will find that for maximum insight into the nature of motion we

---

* From the Greek *kinein*, to move, the same root from which come *cinema*, motion picture, and *kinetic*, pertaining to motion.

† From the Greek *dynamis*, power, the same root from which come dynamo, dynamite, and dynasty.

Small lights attached to the runner's body give a picturesque description of motion. Can you interpret the details of the shapes of the various curves?

must also use the concepts of functions and limits.

After we have expressed the laws of kinematics in algebraic form, we shall be able to proceed to the study of dynamics, with its numerous applications to basic science and to everyday life. The study of dynamics is a foundation for all of our later work, including the study of atomic structure and high-energy particle accelerators, as well as of older subjects, such as pendulums and collisions between molecules.

2-1 Types of Motion

Consider these motions:

(a) A coffee cup near the edge of a table slips off, falls to the floor, and breaks.

(b) A parachutist falls freely for 260 m (800 ft), then opens his chute and slows down to a steady speed of 12 m/s (27 mi/h); he continues to fall at this speed and eventually strikes the ground and comes to rest.

(c) A golf ball dropped from a window rebounds from the sidewalk a number of times, always rising to successively lesser heights, and eventually comes to rest on the sidewalk.

(d) A hockey puck glides across the ice, traveling in a straight line toward the goal without any appreciable loss of speed.

(e) The earth moves around the sun at variable speed, in an elliptical orbit, moving slightly faster when it is slightly closer to the sun (in January).

(f) A swimmer dives off a diving board, and the end of the board continues to vibrate up and down for a while.

(g) A truck driver avoids a collision by jamming on her brakes and comes to a screeching halt without swerving to either side.

(h) An electron leaves the electron gun in the neck of a television picture tube, passes immediately through the deflection coils, where it is bent sideways, and then proceeds at constant speed in a straight line until it strikes the face of the tube and makes a spot of light.

(i) An artificial satellite travels at a steady speed of 7 km/s (15,700 mi/h) around the earth, in a circular orbit of radius 8000 km (5000 mi).

(j) A boy on a swing travels along a circular arc, picking up speed as he approaches the lowest point in his motion, and losing speed as he rises again.

We have only scratched the surface; countless additional types of motion easily suggest themselves for analysis.

Which of these motions is "simplest"? The first thing to notice is that most of the descriptions involve several distinct types of motion. For instance, in (b) there are four distinct periods: (1) while the man is freely falling; (2) the short interval of time while the parachute has just opened and he is slowing down; (3) the remainder of the trip in the air; and (4) the small fraction of a second while he is striking the earth, during which his speed is reduced to zero. Considering the *whole* motion, we must say that (d) is perhaps the simplest, since neither direction of motion nor speed changes. Motion (g) is probably the next most simple (assuming from the meager description that the truck loses speed at a steady rate). Motion such as (d) is called *uniform motion in a straight line*, and motion such as (g) is called *uniformly accelerated motion*.* Motions such as (a) and (c) are combinations of the two basic motions described above, but (e), (f), (h), (i), and (j) involve changing directions and are more complicated. Motion (b) is hard to describe exactly, since just after the parachute opens the man loses speed, but not the same amount each second.

It is tempting to contemplate the causes of some of the described motions. For instance, in (b), is the effect of air resistance constant? Such questions belong in the realm of dynamics and will be studied in Chap. 3.

---

* In this case "deceleration" might seem a more descriptive word, but in physics the term "negative acceleration" is usually preferred.

2-2 Uniform Motion in a Straight Line

The quantitative description of uniform motion in a straight line is simple: by definition, velocity is the time rate of change of position, that is, displacement divided by time. In symbols, for *uniform* velocity,

$$v_0 = \frac{s}{t} \quad \text{or} \quad s = v_0 t \qquad (2\text{-}1)$$

The meaning of the symbols is important, since we shall use the same symbols later on. Here t is the elapsed time, s is the *displacement* (the distance measured from the starting point to the final point that the body reaches in time t), and v_0 is the constant velocity of the body. Since the body's speed* does not change during uniform motion, v_0 is also the initial velocity; we shall use v_0 for initial velocity in our later study of nonuniform motion. Equation 2-1 is valid only for uniform motion at constant velocity.

The dimensional equation for velocity is $[\text{velocity}] = [\text{L}]/[\text{T}] = [\text{LT}^{-1}]$. Velocities are measured in m/s, ft/s, mi/h, and so on—always a length unit divided by a time unit.

A simple example of uniform motion will repay close study.

Example 2-1

In a television tube, an electron travels in a straight line from the electron gun to the fluorescent screen, a distance of 35 cm (14 in.), in 6 nanoseconds (6 ns) (see the table of prefixes on page 7). What is its velocity in meters per second? in kilometers per hour? in miles per second?

$$\boxed{\begin{aligned} s &= 35 \text{ cm} \\ t &= 6 \text{ ns} \\ v_0 &= ? \end{aligned}}$$

$$s = v_0 t$$

$$v_0 = \frac{s}{t} = \frac{(35 \text{ cm})(1 \text{ m}/100 \text{ cm})}{(6 \text{ ns})(10^{-9} \text{ s}/1 \text{ ns})}$$

$$= \boxed{5.83 \times 10^7 \text{ m/s}}$$

To change to km/h, we make use of the fact that 3600 s is the same as 1 hour:

$$v_0 = (5.83 \times 10^7 \text{ m/s})\left(\frac{1 \text{ km}}{1000 \text{ m}}\right)\left(\frac{3600 \text{ s}}{1 \text{ h}}\right)$$

$$= \boxed{2.1 \times 10^8 \text{ km/h}}$$

To change to mi/s, we use a ratio factor to convert kilometers to miles (see equivalents in Appendix Table 4):

$$v_0 = (5.83 \times 10^7 \text{ m/s})\left(\frac{1 \text{ km}}{10^3 \text{ m}}\right)\left(\frac{1 \text{ mi}}{1.609 \text{ km}}\right)$$

$$= \boxed{3.62 \times 10^4 \text{ mi/s}}$$

"Tricks of the Trade." We have used this simple problem to illustrate some of the "tricks of the trade" in problem solving:

(*a*) First we read the problem carefully and make a list of knowns and unknowns, setting them off in a box.

(*b*) Often a simple diagram can be used to label some of the knowns and unknowns.

(*c*) We then write down an equation, preferably a basic one that actually defines a quantity that is of interest in the problem. Here such an equation is $s = v_0 t$; the linear relationship between s and t is the basis for defining uniform velocity.

(*d*) Since we are asked not for s but for v_0, which is buried in the equation, we next solve algebraically for the desired unknown, in this

---

* In Sec. 2-8 we shall distinguish carefully between *speed*, which simply describes how fast a body is moving, and *velocity*, which describes both how fast and in what direction it is moving. A similar distinction is made between *distance* and *displacement*. In the next few sections we deal only with motion along a straight line. Such motion can be in only two directions, forward and backward (or up and down, and so on), so we can take care of change in direction by the use of algebraic signs, $+$ and $-$. See Example 2-4.

case obtaining $v_0 = s/t$. We usually do this before substituting any numbers.

(*e*) We substitute into the revised equation the information from the box containing the knowns and unknowns. The factor 1 m/100 cm in the numerator does not change the value of the answer, since 1 m *is the same as* 100 cm, and hence 1 m/100 cm is just another way of writing 1. The solver thinks along these lines: "I have 35 cm in the numerator, but I want m in the numerator. Therefore I write $35 \text{ cm} \times \left(\dfrac{\text{m}}{\text{cm}} \right)$ so that I can later cancel out the centimeters and have meters left. Finally, I fill in the numbers, knowing that 1 m equals 100 cm, getting 35 cm × (1 m/100 cm). Note that for the purposes of canceling units, it is better to write $\dfrac{\text{m}}{\text{s}}, \dfrac{\text{km}}{\text{h}}$, and so on, than to use the slant line to indicate division, as in m/s (although for convenience of typesetting we shall often use the latter style in this text).

(*f*) The final answer, with its units, is clearly indicated by some device such as a box or underlining.

It is simply taking advantage of past experience and study to use a basic equation that you fully understand, such as $s = v_0 t$. It would be wrong, however, to rely on memory and blind substitution into formulas. Problem solving is intended to help you master the meaning of physics; it should not become a battle of wits between the student and a collection of formulas pulled out of a magician's hat. *The most important single step is the first one—careful study of the problem to see what information is pertinent and just what is asked for.* The expert may carry these bits of information for as long as it takes to come up with just the right equation or equations; the beginner would do well to emulate the text examples and place the given data in a list, perhaps enclosed in a box.

Uniform motion can be given a graphical interpretation if we plot velocity v or displacement s as a function of time t (see the Mathematical Appendix for a discussion of graphs). The time is plotted horizontally (axis of abscissas), since time is considered the *independent variable*. That is, we choose a certain time at will and then determine the corresponding velocity or displacement, which we call *dependent variables*. Consider first the case of uniform motion in a straight line (Fig. 2-1)—say, a train moving at 15 m/s. The velocity is constant ($v = v_0$), so the

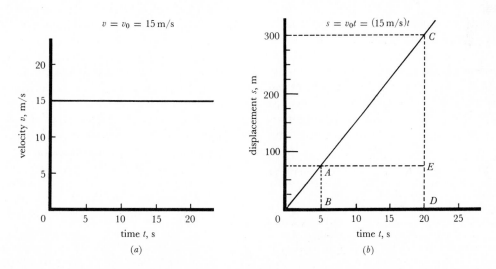

Figure 2-1 (*a*) Velocity graph for uniform motion. (*b*) Displacement graph for uniform motion.

velocity graph is a horizontal straight line (Fig. 2-1a). The displacement increases steadily as time goes by, so the displacement graph is a straight line passing through the origin (Fig. 2-1b). There is a relationship between these two graphs: the *slope** of the displacement graph is a constant, and this slope equals the velocity v_0. In Fig. 2-1a a train's velocity is a constant 15 m/s. The displacement 5 s after time $t = 0$ is 75 m, as read from the graph of Fig. 2-1b, and the displacement 20 s after time $t = 0$ is 300 m. The slope of segment AC of the displacement graph is

$$v_0 = \frac{300 \text{ m} - 75 \text{ m}}{20 \text{ s} - 5 \text{ s}} = \frac{225 \text{ m}}{15 \text{ s}} = 15 \text{ m/s}$$

The slope of a straight-line graph is everywhere the same; the slope of segment OA is also 15 m/s, as can easily be calculated. (Note that triangle OAB is similar to triangle ACE.)

2-3 Average Velocity

As an example of a motion that is more complicated than that in Example 2-1 but is not unusual, consider a car that traveled on a straight turnpike in the United States where markers indicate distances along the road, measured from the state border. If the car moved from the 215-km marker to the 275-km marker, the displacement was 275 km − 215 km = 60 km. If this portion of the journey began at 11:10 A.M. and ended at 11:58 A.M., the elapsed time was 48 min, or 0.8 h. If the car moved at a steady rate, its velocity was constant with magnitude 60 km/0.8 h = 75 km/h. If the car did not move

---

* The slope of a straight line is the increase in the dependent variable divided by the corresponding increase in the independent variable. Thus in Fig. 2-1b, the slope is 300 m/20 s = 15 m/s. Slopes usually have units—in this case, m/s. The angle that the line makes with the horizontal axis depends on the scale to which the graph is plotted. This is arbitrary and has nothing to do with the slope as here defined.

at a steady rate, the velocity we have just calculated is called the *average velocity*. In general, average velocity $\bar{v}$, or v_{av}, is given by

$$\bar{v} = v_{av} = \frac{s - s_0}{t - t_0} \qquad (2\text{-}2)$$

where s_0 is the initial displacement (at time t_0) and s is the final displacement (at time t). Usually we choose to measure displacement in such a way that s_0 and t_0 are both zero; this amounts to calling $s = 0$ when $t = 0$. For instance, at 11:10 A.M. we could have set our watch to read 0 and renumbered the markers so that the 215-km marker, which we passed at that instant, was "0." Then we would have passed the "60" marker at "0:48" by our watch.

With $s_0 = 0$ and $t_0 = 0$, Eq. 2-2 becomes

$$\bar{v} = \frac{s}{t} \quad \text{or} \quad s = \bar{v}t \qquad (2\text{-}3)$$

Note that we now have two equations for displacement. The equation $s = \bar{v}t$ is always true, for *any* motion; the similar equation $s = v_0 t$ is true for *uniform* motion, where v_0 is constant.

Let us study the motion of the car along the turnpike more closely. So far, we have found the average velocity to be 75 km/h, but we know nothing about the details of the motion. Such details are available in the graph of Fig. 2-2a, which gives the displacement of the car as a function of time. The journey is idealized in such a way that the displacement graph is a series of straight-line segments.[†] During any interval, the car's displacement increased steadily, and its velocity was constant. This constant velocity was also the average velocity during the interval. According to the definition expressed in Eq. 2-2,

$$\bar{v} = \frac{\Delta s}{\Delta t}$$

[†] Physically, this amounts to assuming that changes in velocity are abrupt, taking place during intervals of, say, a few seconds, too short to show clearly on the graph.

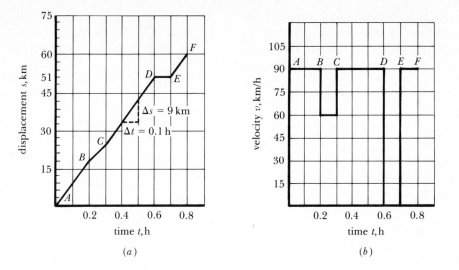

Figure 2-2 (*a*) Displacement graph for car on turnpike. (*b*) Velocity graph for car on turnpike.

where $\Delta s / \Delta t$* is the slope of a straight segment of the displacement graph in Fig. 2-2*a*. Thus, during any part of the interval *CD*, the velocity was 90 km/h, as given by the illustrated values of Δs and Δt; here $\bar{v} = \Delta s / \Delta t = 9 \text{ km}/0.1 \text{ h} = 90 \text{ km/h}$.

The constant value $\bar{v} = 90$ km/h is plotted as the horizontal segment *CD* of the velocity graph in Fig. 2-2*b*. As may easily be verified from the graphs, the car's velocity was 60 km/h in a tunnel (*BC*), 0 km/h during a 0.1 h stop for a coffee break (*DE*), and 90 km/h during segments *AB* and *EF*. The average velocity for the whole journey was 60 km/0.8 h = 75 km/h, although (as it happens) for no segment of the journey was the velocity equal to this value.

The discussion above makes it evident that a velocity graph, such as Fig. 2-2*b*, can be found by taking the slope of the displacement graph at several points. This statement has, of course, been proved so far only for motion with constant

velocity; what to do if the displacement graph is not made up of straight-line segments is discussed in the next section.

2-4 Instantaneous Velocity

If the displacement graph is curved, the ratio $\Delta s / \Delta t$ is not constant, and hence the average velocity depends on the magnitude of the time interval used. The *instantaneous velocity* is defined by a limiting process in which smaller and smaller time intervals are used, each interval containing the instant at which the velocity is desired. As smaller time intervals are used, both Δt and Δs approach zero, but their ratio $\Delta s / \Delta t$ approaches a limit. This means that Δt can be made small enough so that a further decrease of Δt will change the value of $\Delta s / \Delta t$ by an amount that is less than any arbitrary preassigned value, however small. We define instantaneous velocity v by such a limit:

$$v = \lim_{\Delta t \to 0} \left(\frac{\Delta s}{\Delta t} \right) \qquad (2\text{-}4)$$

---

* The symbol Δ (Greek delta) is used to indicate a change in any quantity. Thus Δs (read "delta *s*") means the change in *s*. We shall use this notation many times in our future work.

which is read "v equals the limit of $\Delta s/\Delta t$ as Δt approaches zero." This limit, if it exists,* is so useful that it is given a special name, the *derivative* of s with respect to t, written as ds/dt. Thus we have

definition of derivative
$$\lim_{\Delta t \to 0} \left(\frac{\Delta s}{\Delta t}\right) = \frac{ds}{dt} \qquad (2\text{-}5)$$

In another notation often used, if s is some function of t written as $s(t)$, then the derivative of s is another (related) function of t called $s'(t)$; $s'(t)$ and ds/dt mean exactly the same thing. Using our new terminology, we say that *instantaneous velocity is the time derivative of displacement*. Graphically, the derivative is the slope of the tangent to the displacement curve, drawn at the time in question.

Example 2-2

The displacement of a body varies according to the equation $s = 10t^3$, where s is in cm and t is in s. (Such a displacement function is possible, although not very common in physics.) What is the instantaneous velocity at $t = 2$ s?

First approximation: If we choose $\Delta t = 0.1$ s, we find

t	$s = 10t^3$
2.0	$10(2.0)^3 = 80.00$
2.1	$10(2.1)^3 = 92.61$

$$\bar{v} = \frac{\Delta s}{\Delta t} = \frac{92.61 - 80.00}{0.1}$$

$$= \frac{12.61}{0.1} = 126.1 \text{ cm/s}$$

This is surely somewhat too large, since Fig. 2-3 shows that the chord PQ whose slope we have obtained is steeper than the tangent PR.

---

* All functions used in elementary physics are sufficiently "well behaved" so that ratios such as we are considering *do* approach a limit.

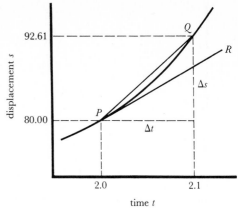

Figure 2-3

Second approximation: As a next try, we choose $\Delta t = 0.01$ s:

t	$s = 10t^3$
2.00	$10(2.00)^3 = 80.00000$
2.01	$10(2.01)^3 = 81.20601$

$$\bar{v} = \frac{\Delta s}{\Delta t} = \frac{81.20601 - 80.00000}{0.01}$$

$$= \frac{1.20601}{0.01} = 120.601 \text{ cm/s}$$

Third approximation: For a still smaller Δt, we choose $\Delta t = 0.001$ s:

t	$s = 10t^3$
2.000	$10(2.000)^3 = 80.00000000$
2.001	$10(2.001)^3 = 80.12006001$

$$\bar{v} = \frac{\Delta s}{\Delta t} = \frac{80.12006001 - 80.00000000}{0.001}$$

$$= \frac{0.12006001}{0.001} = 120.06001 \text{ cm/s}$$

The calculations seem to indicate that $\Delta s/\Delta t$ does indeed approach a limit, and we conjecture that this limit is, for $t = 2$ s, exactly 120 cm/s. Whatever the exact value of the limit, we know that it *does exist*, because we know that the tangent PR has a definite slope. We can calculate this limit as precisely as we wish by choosing a

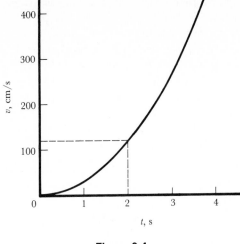

Figure 2-4

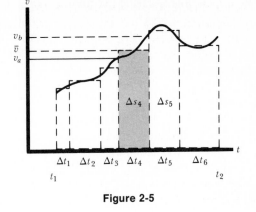

Figure 2-5

smaller and smaller Δt. Thus we have (laboriously) found *one* point on the velocity graph: when $t = 2\,\text{s}$, $v = 120\,\text{cm/s}$. We could now repeat the whole process to find $\lim_{\Delta t \to 0} (\Delta s/\Delta t)$ at any other time, and thus build up the entire velocity graph. The result, v as a function of t, is shown in Fig. 2-4.

2-5 Calculating the Displacement

We have found that there is a close relationship between displacement and instantaneous velocity; in Sec. 2-3 we saw that whenever the displacement graph is known, we can construct the velocity graph. Each ordinate of the velocity graph equals the slope of the displacement graph at the corresponding time. This construction amounts to finding the derivative of the displacement with respect to time, which can be done graphically, if necessary.*

---

* If the displacement is a known algebraic function of the time, the velocity function (derivative) can be found by using a formula. We will not need to use any formal calculus in the main sections of this book. It is important to understand the meaning of the basic concepts of calculus, but our emphasis will be on graphical interpretations rather than on formulas.

It is natural to ask if the process can be reversed: Given the velocity graph, can the displacement graph be found? The answer is yes, and we illustrate the method in Fig. 2-5 and the following discussion. Once again we use a limiting process, but this time we find the limit of a sum.

To find the total displacement during the time interval extending from t_1 to t_2, we divide the total interval into many subintervals having durations $\Delta t_1, \Delta t_2, \ldots, \Delta t_n$. From the definition of average velocity (Eq. 2-3), it follows that the displacement during any subinterval is given by $\Delta s = \bar{v}\Delta t$, where $\bar{v}$ is "in the neighborhood of" v_a (start of subinterval) and v_b (end of subinterval). (Note that $\bar{v}$ for a subinterval does not necessarily lie between v_a and v_b; for instance, during the fifth subinterval, $\bar{v}$ is greater than both v_a and v_b.) The displacement during any subinterval is thus represented by the area of a rectangle, as shown by the shaded area for the fourth subinterval. The total displacement from t_1 to t_2 is the sum of the areas of the rectangles. The crux of the problem is in choosing the proper values for the average velocities $\bar{v}$. The difficulty is lessened if we use very many segments, letting each Δt approach zero, for then v_a, v_b, and $\bar{v}$ tend to coincide. The total displacement equals the sum† of

---

† The symbol Σ (Greek sigma) indicates a summation process.

the displacements during the subintervals:

$$s = \sum_{i=1}^{n} \Delta s_i = \Delta s_1 + \Delta s_2 + \cdots + \Delta s_n$$

$$= \bar{v}_1 \Delta t_1 + \bar{v}_2 \Delta t_2 + \cdots + \bar{v}_n \Delta t_n$$

$$= \sum_{i=1}^{n} \bar{v}_i \Delta t_i \qquad (2\text{-}6)$$

For $n = 6$, this sum of rectangles is the area under the broken line in Fig. 2-5. In the limit, as $n \to \infty$ and each $\Delta t \to 0$, the broken line coincides with the curve, and the sum in Eq. 2-6 equals the area under the curve. Thus we have shown that the displacement equals the area under the velocity graph. We indicate that the displacement equals the limit of a sum by writing

$$s = \lim_{\substack{n \to \infty \\ \Delta t_i \to 0}} \sum_{i=1}^{n} v_i \Delta t_i \qquad (2\text{-}7)$$

where v_i is any value of v in the ith interval. As a matter of terminology, the limit of the sum, as written in Eq. 2-7, is called the *definite integral* of v with respect to t, between the limits t_1 and t_2, and is denoted by

definition of integral
$$\lim_{\Delta t \to 0} \sum v \, \Delta t = \int_{t_1}^{t_2} v \, dt \qquad (2\text{-}8)$$

The integral sign $\int$ is a script "s," the initial letter of the word "sum."

By way of illustration, let us compute the displacement graph of Fig. 2-2a, starting with the velocity graph of Fig. 2-2b. First let us find the value of s at the time $t = 0.6$ h. The area under the velocity graph (from $t = 0$ to $t = 0.6$ h) consists of three rectangles, and thus

$$s = v_1 \Delta t_1 + v_2 \Delta t_2 + v_3 \Delta t_3$$
$$= (90 \text{ km/h})(0.2 \text{ h}) + (60 \text{ km/h})(0.1 \text{ h})$$
$$+ (90 \text{ km/h})(0.3 \text{ h})$$
$$= 18 \text{ km} + 6 \text{ km} + 27 \text{ km} = 51 \text{ km}$$

Thus we have found point D on the displacement graph: when $t = 0.6$ h, $s = 51$ km. Every other point on the displacement graph can be found by a similar summing of the area under the velocity graph, and the entire displacement function be-

comes known, starting with the known velocity function.

Our brief study of the basic ideas of calculus has centered around two related functions: the displacement function $s(t)$, and the velocity function $v(t)$. The relationship between these two functions can be expressed in two equivalent ways:

$$v(t) = \frac{ds}{dt} \qquad (2\text{-}9a)$$

and

$$s(t) = s(0) + \int_{0}^{t} v \, dt \qquad (2\text{-}9b)$$

Here $s(0)$ is the value of s when $t = 0$; this initial displacement could also be written as s_0. As pointed out at the start of Sec. 2-3, we usually choose our reference point so that $s = 0$ when $t = 0$; for this choice $s(0) = 0$. Graphical methods can always be used to calculate the approximate numerical value of a derivative or an integral; but if one function (s or v) is given in algebraic form, tables such as those in the Mathematical Appendix can often be used to find the related function (v or s). In summary,

> The derivative of a function at any point is the slope of its graph at that point.

> The definite integral of a function between any two points is the area under that part of its graph which extends between the two points.

Graphical interpretations such as these are probably the most useful way of expressing the basic concepts of calculus.

2-6 *Uniformly Accelerated Motion*

In common language, acceleration is the picking up of speed. The technical definition involves the *rate* at which velocity changes. If the velocity changes rapidly, the acceleration is greater than if the same velocity change takes place during a longer time interval. *Acceleration* is the *time rate of change of velocity* and is therefore equal to the derivative of velocity with respect to time. In symbols,

$$a(t) = \lim_{\Delta t \to 0} \left(\frac{\Delta v}{\Delta t} \right)$$

or

$$a(t) = \frac{dv}{dt} = v'(t) \qquad (2\text{-}10)$$

Note that acceleration is, in general, a function of the time. However, in uniformly accelerated motion the acceleration is constant, and for this case $dv/dt = \Delta v/\Delta t$ regardless of the size of the interval Δt. The velocity at a time t seconds after the start of an interval is found from

$$a = \frac{dv}{dt} = \frac{\Delta v}{\Delta t} = \frac{v - v_0}{t - 0}$$

which can be rearranged to give

$$v = v_0 + at \qquad (2\text{-}11)$$

where a is the acceleration, v_0 is the initial velocity, and v is the final* velocity after a time t has elapsed. The dimensional equation for acceleration is

$$[\text{acceleration}] = \frac{[\text{velocity}]}{[\text{time}]}$$

$$= \frac{[LT^{-1}]}{[T]} = [LT^{-2}]$$

Acceleration is measured in m/s², ft/s², km/h·s, and so on, always a distance unit divided by two time units.[†]

In the following example we use some of the tricks of the trade mentioned on pages 24–25.

Example 2-3

A sports car traveling on a straight road has an initial velocity of 10 m/s; it gains velocity steadily, and 5 s later its velocity is 25 m/s. What is the acceleration?

---

* The adjective "final" refers to the end of the time interval. The motion may actually continue for an indefinite time.
† The unit m/s² is read as "meters per second per second," or sometimes as "meters per second squared." The unit km/h·s is read as "kilometers per hour per second." An area unit such as m² is usually read as "square meters."

First we read the problem carefully and make a list of knowns and unknowns, setting them off in a box. We decide that the statement of the problem indicates that the car moves with uniform acceleration (it gains velocity "steadily").

$v_0 = 10$ m/s	$v = v_0 + at$
$v = 25$ m/s	Solving for the unknown
$t = 5$ s	a, we get
$a = ?$	$a = \dfrac{v - v_0}{t}$

$$= \frac{25 \text{ m/s} - 10 \text{ m/s}}{5 \text{ s}}$$

$$= \frac{15 \text{ m/s}}{5 \text{ s}} = \boxed{3 \text{ m/s}^2}$$

After a time interval t, a body is found to be displaced from its original position. If the acceleration is positive (velocity increasing), the displacement is greater than it would have been if the velocity had remained constant at its initial value. We can calculate the displacement by finding the area under the velocity graph. For uniformly accelerated motion, the velocity graph is a straight line of slope $\Delta v/\Delta t = a$ (Fig. 2-6). To find the displacement during a time interval from $t = 0$ to $t = t$, we need to calculate

$$s = \lim_{\Delta t \to 0} \sum v \, \Delta t$$

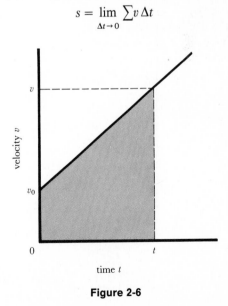

Figure 2-6

or

$$s = \int_0^t v \, dt$$

Evaluation of the sum is easy; it is the area of the shaded trapezoid in Fig. 2-6. Using plane geometry, we obtain the formula

$$s = \frac{v + v_0}{2} t = \bar{v} t \qquad (2\text{-}12)$$

where $\bar{v}$ is the average velocity.

Actually, the two equations for v and s (Eqs. 2-11 and 2-12) are sufficient to solve all problems concerning uniformly accelerated motion. However, for convenience, two additional equations are often used. These equations, which can be obtained by algebraic manipulation of the basic equations, give us two new equations that relate v, v_0, s, t, and a.

Solve Eq. 2-11 for t and substitute into Eq. 2-12, getting

$$s = \left(\frac{v + v_0}{2}\right)\left(\frac{v - v_0}{a}\right)$$

which simplifies to

$$v^2 = v_0^2 + 2as \qquad (2\text{-}13)$$

Again, substituting the right-hand side of Eq. 2-11 for v into Eq. 2-12, we find

$$s = \frac{(v_0 + at) + v_0}{2} t$$

which simplifies to

$$s = v_0 t + \tfrac{1}{2}at^2 \qquad (2\text{-}14)$$

Let us collect these four equations together:

$$v = v_0 + at \qquad (2\text{-}11)$$

$$s = \frac{v_0 + v}{2} t \qquad (2\text{-}12)$$

$$v^2 = v_0^2 + 2as \qquad (2\text{-}13)$$

$$s = v_0 t + \tfrac{1}{2}at^2 \qquad (2\text{-}14)$$

You will note a certain systematic property of these equations. We have four different variables

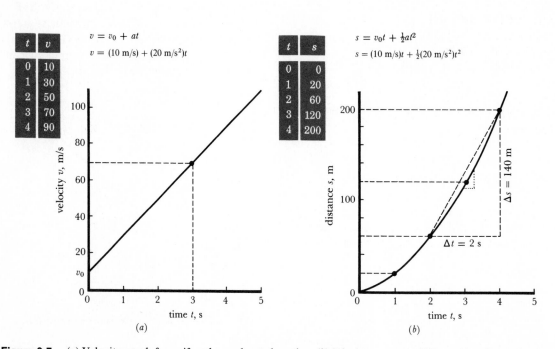

Figure 2-7 (a) Velocity graph for uniformly accelerated motion. (b) Displacement graph for uniformly accelerated motion. The instantaneous velocity at $t = 3$ s is 70 m/s. Between $t = 2$ s and $t = 4$ s, the sled travels 140 m.

that are related to each other (v_0 is not a variable; the initial velocity is a constant for any given problem). The variable s is missing from Eq. 2-11, a from Eq. 2-12, t from Eq. 2-13, and v from Eq. 2-14. For any given problem, try to select a suitable equation from the set 2-11 through 2-14, to avoid as much wasted motion as possible.

Uniformly accelerated motion includes uniform motion as a special case. In uniform motion there is no acceleration, and by setting a equal to zero in Eq. 2-14 we get $s = v_0 t$, which is the same as Eq. 2-1.

In uniformly accelerated motion, the velocity graph (plotted from Eq. 2-11) is a straight line, and the displacement graph (plotted from Eq. 2-14) is a parabola. These graphs are shown in Fig. 2-7 for the motion of a rocket sled being given an acceleration test; the sled has initial velocity 10 m/s (22 mi/h) and accelerates at 20 m/s².

Figure 2-7 illustrates a property of uniformly accelerated motion: The average velocity over any time interval (not necessarily a small interval) is equal to the instantaneous velocity at the mid-time of the interval. Thus $\bar{v} = 140$ m/2 s = 70 m/s from Fig. 2-7b, which is exactly the value of v at $t = 3$ s, the mid-time of the 2-second interval. *Caution:* This simple result applies only to uniformly accelerated motion in which a is constant.

2-7 Acceleration Due to Gravity

By far the most familiar example of uniformly accelerated motion is that of a freely falling body. If air resistance can be ignored,* a falling

---

* Air resistance *can* be ignored if the falling body is heavy and has a relatively small surface area. Air resistance can certainly not be ignored for a falling raindrop or for a skydiver-parachute combination. But in these cases, the motion is not uniformly accelerated. Our study of dynamics in later chapters will reveal the effects of friction, including air resistance, more clearly.

body picks up speed at the rate of about 9.8 m/s each second (32 ft/s each second). This constant is called the *acceleration due to gravity* and is denoted by the symbol g. The value of g varies slightly from place to place on the earth; at 45° latitude and sea level, g is 9.80600 m/s² (32.169 ft/s²). In problem work g is taken to be exactly 9.80 m/s² unless otherwise stated.

Example 2-4

On Jan. 16, 1980, a steel TV tower was under construction. At 4:15 P.M. a girder broke loose and fell 122 m (400 ft) to the ground. How long did it take to reach the ground?

First we make a table of knowns and unknowns, assuming the positive direction to be upward; we also assume that $s = 0$ at the top of the tower, 122 m above the ground, since that is where the girder is when $t = 0$. For this choice of positive direction, the acceleration, which is downward, is *negative*. The displacement, which is downward from the starting point, is also negative. We therefore use -9.80 m/s² for a and -122 m for s.

$$s = -122 \text{ m}$$
$$t = ?$$
$$a = -9.80 \text{ m/s}^2$$
$$v_0 = 0$$

We indicate these facts in diagram (Fig. 2-8a).

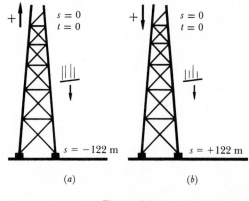

(a)   (b)

Figure 2-8

We now use Eq. 2-14 to solve the problem:

$$s = v_0 t + \tfrac{1}{2} a t^2 = 0 + \tfrac{1}{2} a t^2$$

$$t = \sqrt{\frac{2s}{a}} = \sqrt{\frac{2(-122 \text{ m})}{-9.80 \text{ m/s}^2}}$$

$$= \sqrt{24.9 \text{ s}^2} = \boxed{5.0 \text{ s}}$$

Do not worry about making the proper choice of signs; merely take care that $s = 0$ at the location of the object at the start of the problem, when $t = 0$. Either direction may be taken as positive providing you stick with it throughout the problem. For instance, the same problem can be solved in this way (Fig. 2-8b):

$$
\begin{array}{l}
s = +122 \text{ m} \\
t = ? \\
a = +9.80 \text{ m/s}^2 \\
v_0 = 0
\end{array}
$$

$$t = \sqrt{\frac{2s}{a}} = \sqrt{\frac{2(+122 \text{ m})}{+9.80 \text{ m/s}^2}}$$

$$= \sqrt{24.9 \text{ s}^2} = \boxed{5.0 \text{ s}}$$

If you are confident that the correct answer will be obtained regardless of the mechanical work of algebra and signs, you are free to concentrate on the really important part of problem solving—setting up the problem. Cultivate the ability to grasp the *physical* essentials and ignore the irrelevant details (such as the time of day or the fact that the tower is made of steel). Our next example shows the advantage of choosing the proper equation in order to make the algebraic work easiest.

_____ Example 2-5

A ball is thrown straight up with initial velocity 20 m/s. After reaching its maximum height, on the way down the ball strikes a bird 10 m above the ground. How fast is the ball moving when it strikes the bird?

The problem could be solved in three steps: First find the maximum height (it turns out to be 20.4 m). Then find the time required for the ball to fall 10.4 m down to the 10-m level starting from rest at the highest point (it turns out to be 1.46 s). Finally, compute the velocity acquired in this time interval. This method of solution hides the essential simplicity of the action that takes place. The motion is in one continuous sequence, and the acceleration is constant in both magnitude and direction, being 9.8 m/s² downward at all times. Even when the ball is (momentarily) motionless at the highest point, the acceleration is 9.8 m/s² downward, since the velocity is changing from upward to downward at this instant. We solve the problem in one step, without finding the maximum height.

We choose the positive direction to be upward, with $s = 0$ at the ground level. We seek the velocity of the ball when the displacement is $+10$ m.

$$
\begin{array}{l}
v_0 = 20 \text{ m/s} \\
v = ? \\
s = +10 \text{ m} \\
a = -9.8 \text{ m/s}^2
\end{array}
$$

$$v^2 = v_0{}^2 + 2as$$

$$= \left(20 \frac{\text{m}}{\text{s}}\right)^2 + 2\left(-9.8 \frac{\text{m}}{\text{s}^2}\right)(+10 \text{ m})$$

$$= 400 \text{ m}^2/\text{s}^2 - 196 \text{ m}^2/\text{s}^2$$

$$v = \sqrt{204 \text{ m}^2/\text{s}^2} = \boxed{\pm 14.3 \text{ m/s}}$$

We interpret the $\pm$ sign as follows: $v = +14.3$ m/s when the ball is at the $+10$-m level on the way up, and $v = -14.3$ m/s when the ball is at the $+10$-m level on the way down. In view of the wording of the problem, we select the $-$ sign for our answer: $v = -14.3$ m/s when the ball strikes the bird.

2-8 Vectors

Our study of simple kinematical problems would hardly be complete if we discussed only motion in a straight line. The motion of a baseball after it leaves the bat on its way into home-run territory; the motion of a particle of water after it leaves a nozzle on its way to a blazing building—these are examples of projectile motion, in which the body has two *simultaneous* velocities. The up-and-down part of the motion is due to gravity, and the forward motion is practically constant (if air resistance can be neglected). Other types of motion also involve two simultaneous velocities—for example, a man walking across a moving train or swimming across a river. To study the resulting motion, each part going on independently of the other, we need to use the concept of vectors and components of vectors.

A *vector* quantity has both magnitude and direction, whereas a *scalar* quantity has only magnitude. We speak of an *upward* force, an *eastward* velocity, a displacement *toward the northwest*; these are vector quantities. On the contrary, it is meaningless to think of "4 hours toward the south" or "100 cubic meters downward"; these quantities are scalars. A partial list of vector and scalar quantities is given in Table 2-1.

A speedometer on a car is correctly named since it indicates only the *magnitude* of the velocity. You cannot tell by looking at the speedometer what the *direction* of the velocity is. Likewise, when a salesman extols a car's pickup, he doesn't care whether the acceleration is acting northward or eastward; he is interested only in

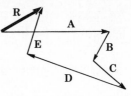

Figure 2-9 Head-to-tail method for addition of vectors. **R** is the sum of vectors **A**, **B**, **C**, **D**, and **E**.

its magnitude. On the other hand, weight (the downward gravitational force on a body) is a vector, like all forces. Vectors are usually represented by arrows. The length of the arrow represents (to some scale) the magnitude of the vector, and the direction of the arrow indicates the direction of the vector. In printing equations involving vectors, **bold face** type is used to indicate vectors,* and ordinary *italic* type is used to indicate scalars as well as the magnitudes of vectors. Thus the vector equation $\mathbf{s} = \mathbf{v}_0 t$ conveys more information about uniform motion than does the scalar equation $s = v_0 t$; since $\mathbf{s}$ and $\mathbf{v}_0$ are related by the scalar factor t, if t is positive they have the same direction and are parallel vectors.

The essential feature of vectors is that the combined effect (known as the *resultant*) of two or more vectors may be found by a geometrical rule known as *vector addition*. We place the vectors head to tail, maintaining their magnitudes and directions, and the resultant is the vector drawn from the tail of the first vector to the head of the final vector (Fig. 2-9). Applied to displacements, it works out as in the following example.

Example 2-6

A girl delivering papers covers her route by traveling 3 blocks west, 4 blocks north, then 6 blocks east. What is her final displacement? What is the total distance she travels?

Table 2-1 Vector and Scalar Quantities*

Vector Quantities		Scalar Quantities
displacement	⟷	distance
velocity	⟷	speed
acceleration	⟷	pickup
force		time
momentum		volume
torque (moment of force)		work
magnetic field strength		mass (inertia)

* The first three of these vector quantities are related to the corresponding scalar quantities in the second column; for instance, speed is the magnitude of the velocity. Such a correlation is not to be made for the remaining vectors.

---

* In handwritten work, a vector quantity is denoted by an arrow above the symbol: $\vec{s} = \vec{v}_0 t$.

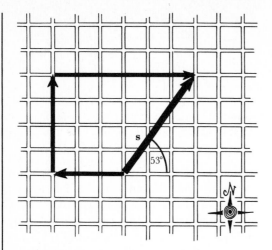

Figure 2-10 Vector addition of displacements.

The displacements are laid off on a scale diagram (Fig. 2-10), and by measurement of the length of the resultant (or by simple geometry), the displacement is found to be of magnitude 5 blocks. A protractor is used to find that the direction of the resultant is 53° north of east. (Can you see how to apply the Pythagorean theorem to this example?) The displacement is a vector and must be fully described: the girl's displacement is

5 blocks, 53° N of E

The distance traveled by the girl is a total of 13 blocks, found by adding up the various legs of her journey. This distance is a scalar quantity, and no particular direction can be associated with the figure of 13 blocks.

This example illustrates the distinction between displacement, which is a vector, and distance, which is a scalar. As another example, after a summer of vacation driving, the displacement is zero as the family car eases back into its garage, although the distance covered (and the fuel consumed!) may be considerable.

Vector addition of velocities is necessary if a body has two or more velocities simultaneously, due perhaps to separate causes.

Example 2-7

A helicopter heads due northeast and has an air speed of 120 km/h. Simultaneously, a wind of 50 km/h blows from the north. What is the displacement of the helicopter 2 h after leaving the heliport?

In 2 h the helicopter would have gone 240 km due to its own air speed and 100 km due to the wind. The two displacements take place simultaneously. Laying off the two displacements to scale (Fig. 2-11), we find (graphically) the resultant displacement to be

183 km, 22.3° N of E

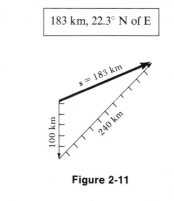

Figure 2-11

Example 2-8

Let us look at the same problem from a slightly different viewpoint. Instead of adding displacements, we can add (vectorially) the two velocities to get the magnitude and direction of the resultant velocity (Fig. 2-12). In this case the "sum" of 50 km/h and 120 km/h is only 91.7 km/h. The magnitude of the resultant displacement is found, as before, from

$$s = v_0 t$$
$$= (91.7 \text{ km/h})(2 \text{ h}) = \boxed{183 \text{ km}}$$

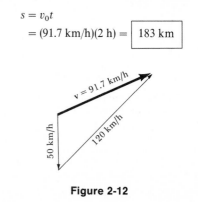

Figure 2-12

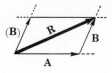

Figure 2-13 Parallelogram method for addition of two vectors.

If only two vectors are to be added, it is often convenient to originate two arrows at the same point. Then the diagram looks like Fig. 2-13. The resultant is the diagonal of a parallelogram, and this method is called the *parallelogram method*. It is not so general as the head-to-tail method, since it can be used for only two vectors at a time.

2-9 Relative Velocity

We are all familiar with the idea of relative velocity. A motorcycle officer traveling at 120 km/h (75 mi/h) overtakes a car traveling at 100 km/h (62 mi/h); we say that the velocity of the motorcycle relative to the car is 120 − 100, or 20 km/h. In this case, speeds are subtracted. Sometimes speeds are added, as in the case of a head-on collision.

For a complete treatment of relative velocity, useful for situations in which the motions are not parallel to each other, vector methods are required. A ship is going southward at 40 km/h, and a deckwalker moves northward at 3 km/h relative to the deck. It is not hard to see that the velocity of the walker relative to the ocean is 37 km/h, southward. To avoid confusion in more complicated cases, let us put this simple problem into a standard form. We call northward the positive direction:

Velocity of passenger relative to ocean

$\qquad$ = (velocity of passenger relative to ship)
$\qquad\quad$ + (velocity of ship relative to ocean)

$\qquad$ = (+3 km/h) + (−40 km/h)

$\qquad$ = −37 km/h (southward)

In general such additions are to be carried out in a vector sense. If we use the symbol $\mathbf{v}_{PO}$ for the velocity of the passenger (*P*) relative to the ocean (*O*), and so on, we can write the vector sum as

$$\mathbf{v}_{PO} = \mathbf{v}_{PS} + \mathbf{v}_{SO} \qquad (2\text{-}15)$$

Note carefully the order of the subscripts: the second subscript of $\mathbf{v}_{PS}$ is the same as the first subscript of $\mathbf{v}_{SO}$. This scheme can be extended to include any number of bodies in relative motion, as in our next example.

Example 2-9

A train *T* is moving eastward at 8 m/s; a waiter *W* is walking toward the rear of the train at 1 m/s; and a fly *F* is charging toward the north across the waiter's tray at 1.5 m/s relative to the tray. What is the velocity of the fly relative to the earth *E*?

$$\mathbf{v}_{FE} = \mathbf{v}_{FW} + \mathbf{v}_{WT} + \mathbf{v}_{TE}$$

The vector diagram (Fig. 2-14) expresses the vector equation written above. If a carefully constructed, large (full-page) diagram is used, the velocity of the fly relative to the earth measures out to be

$\boxed{\text{7.2 m/s, directed } 12° \text{ N of E}}$

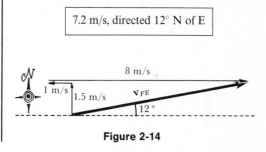

Figure 2-14

Another example involving relative velocities is thought-provoking, mainly because one of the given vectors must be reversed in direction before it is used in a vector sum.

Example 2-10

A bicyclist moves northward at 15 m/s, and a wind of 20 m/s is blowing directly from the east. What is the apparent magnitude and direction of the wind experienced by the bicyclist?

The actual wind may be considered as the velocity $\mathbf{v}_{AE}$ of the air A relative to the earth E. The apparent wind is the velocity $\mathbf{v}_{AB}$ of the air A relative to the bicyclist B. We want $\mathbf{v}_{AB}$, so we build a vector sum with $\mathbf{v}_{AB}$ on the left-hand side; there is only one order of subscripts that is suitable for the right-hand side of the equation:

$$\mathbf{v}_{AB} = \mathbf{v}_{AE} + \mathbf{v}_{EB}$$

We are given $\mathbf{v}_{BE}$, the velocity of the bicyclist relative to the earth (15 m/s northward), but the equation calls for $\mathbf{v}_{EB}$, the velocity of the earth relative to the bicyclist. However, $\mathbf{v}_{EB} = -\mathbf{v}_{BE}$, so we rewrite the vector equation

$$\mathbf{v}_{AB} = \mathbf{v}_{AE} + (-\mathbf{v}_{BE})$$

We add $\mathbf{v}_{AE}$ and $(-\mathbf{v}_{BE})$, laying them off to scale as in Fig. 2-15, with $(-\mathbf{v}_{BE})$ equal to 15 m/s southward. From measurement of the drawing we obtain the magnitude and direction of $\mathbf{v}_{AB}$:

> 25 m/s, coming from 37° N of E

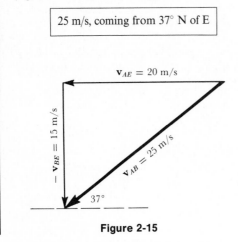

Figure 2-15

2-10 Components of a Vector

When one vector such as $\mathbf{C}$ is the resultant of two other vectors such as $\mathbf{A}$ and $\mathbf{B}$, then we call $\mathbf{A}$ and $\mathbf{B}$ the *components* of $\mathbf{C}$. Any given vector can be resolved into components in an infinite number of ways (Fig. 2-16). Usually we find it useful to look for components that are perpendicular to each other. For instance, if a ball is traveling through the air in a direction making an angle θ with the horizontal, we can show in a diagram

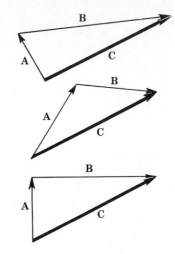

Figure 2-16 In each diagram, $\mathbf{A}$ and $\mathbf{B}$ are components of C.

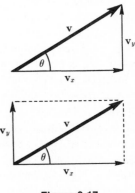

Figure 2-17

(Fig. 2-17) the horizontal component of $\mathbf{v}$ (denoted by $\mathbf{v}_x$) and the vertical component of $\mathbf{v}$ (denoted by $\mathbf{v}_y$). Since the components are at right angles to each other, the Pythagorean theorem can be used, and the magnitude of $\mathbf{v}$ is given by $v = \sqrt{v_x{}^2 + v_y{}^2}$.

At this point we recall some simple trigonometry, namely, the definitions of the *sine*, *cosine*, and *tangent* (abbreviated sin, cos, and tan). For the right triangle* having sides a, b, c (Fig. 2-18), sin θ is an abbreviation for the ratio

---

* These definitions are true only for right triangles.

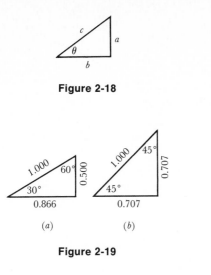

Figure 2-18

Figure 2-19

(a)   (b)

a/c (opposite side over hypotenuse). Similarly, $\cos \theta$ is an abbreviation for b/c (adjacent side over hypotenuse), and $\tan \theta$ is an abbreviation for a/b (opposite side over adjacent side). In any $30°$-$60°$-$90°$ triangle (Fig. 2-19a), the short side is half the hypotenuse, and thus $\sin 30° = 0.500$. Other sines, cosines, and tangents of common angles can be read directly from the figures; for other angles, you will need to use Appendix Table 5 or a pocket calculator.

The sine and cosine are particularly useful in finding components of vectors, as in the following examples.

Example 2-11

A home-run ball moving at 50 m/s is caught by a fan in the bleachers. Its path as it approaches the fan makes an angle of $30°$ with the horizontal (Fig. 2-20). What are the horizontal and vertical components of the ball's velocity?

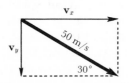

Figure 2-20   Components of velocity.

$$\frac{v_x}{50 \text{ m/s}} = \cos 30°$$

$$v_x = (50 \text{ m/s})(\cos 30°)$$

$$= (50 \text{ m/s})(0.866)$$

$$= \boxed{43.3 \text{ m/s}}$$

$$\frac{v_y}{50 \text{ m/s}} = \sin 30°$$

$$v_y = (50 \text{ m/s})(\sin 30°)$$

$$= (50 \text{ m/s})(0.500)$$

$$= \boxed{25 \text{ m/s}}$$

In this example, v_y would be called -25 m/s or $+25$ m/s, depending on which direction was chosen as the positive direction.

Example 2-12

A sled coasts down a $40°$ hill with an acceleration of 5 m/s² (Fig. 2-21). What is the vertical component of its acceleration?

$$\frac{a_y}{5 \text{ m/s}^2} = \sin 40°$$

$$a_y = (5 \text{ m/s}^2)(\sin 40°)$$

$$= (5 \text{ m/s}^2)(0.643)$$

$$= \boxed{3.22 \text{ m/s}^2}$$

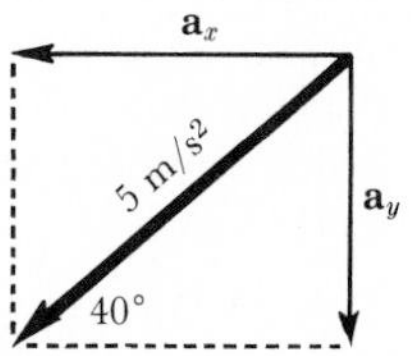

Figure 2-21   Components of acceleration.

With a little practice, you will be able to get the component directly as a product, skipping the step involving the proportion. Think of $\sin \theta$ and $\cos \theta$ as fractions that are used to calculate

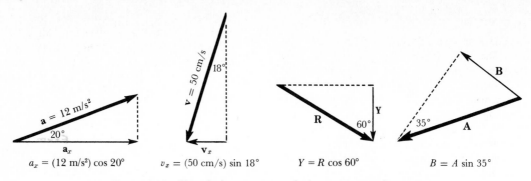

$$a_x = (12 \text{ m/s}^2) \cos 20°$$ $$v_x = (50 \text{ cm/s}) \sin 18°$$ $$Y = R \cos 60°$$ $$B = A \sin 35°$$

Figure 2-22   Use of trigonometry to find components of vectors.

the sides of a right triangle when you know the hypotenuse. The side is always less than the hypotenuse, and the sine or cosine is always less than 1. To get the side *opposite* the angle, simply multiply the hypotenuse by the *sine* of the angle. To get the side *adjacent* to the angle, multiply the hypotenuse by the *cosine* of the angle. Thus, in Fig. 2-22, the marked component is found in each case by multiplying the magnitude of the vector by a sine or a cosine.

The tangent is useful in finding the direction of a vector when its components are known, as in the following example.

Example 2-13

Find the direction of the velocity of the fly relative to the earth in Example 2-9.

Here we know the forward component of the fly's velocity, which is 7 m/s eastward, and the sideways component of the velocity, which is 1.5 m/s toward the north. Putting these together in a vector triangle (Fig. 2-23), we find

$$\tan \theta = \frac{1.5 \text{ m/s}}{7 \text{ m/s}} = 0.214$$

Figure 2-23

Looking this up in a table of trig functions, we find that the angle θ is about 12°. (We use the angle in the table whose tangent is nearest to 0.214.) Thus we get the same answer that was obtained in Example 2-9 by a scale diagram—the fly moves at an angle of

> 12° N of E

2-11 Projectiles

In projectile motion we really have two motions occurring at the same time. The only thing the two motions have in common is that they take place *during the same time interval*. The horizontal part of the motion is uniform motion in a straight line, since there is no horizontal resisting force. The vertical part of the motion is uniformly accelerated motion, since it is governed by the laws of falling bodies. We now make the well-justified assumption that a projectile obeys the Superposition Principle, to be discussed in Sec. 10-7. As applied to projectile motion, this means that the resultant displacement (velocity, acceleration) equals the vector sum of the separate displacements (velocities, accelerations) due to the separate causes. Thus it is merely necessary to resolve the velocity into its horizontal and vertical components at any instant and treat each motion separately.

Example 2-14

A stone is thrown horizontally from a cliff 30 m high. The initial velocity is 6 m/s. How far from the base of the cliff does the stone strike the ground (Fig. 2-24)?

Consider first the *vertical* motion in order to find how long the stone is in the air. The vertical component of the initial velocity is zero, since the stone is thrown horizontally; therefore $v_0 = 0$, as far as the vertical motion is concerned. We choose upward as the positive direction for this part of the problem and show the choice in the diagram.

$$v_0 = 0$$
$$a = -9.8 \text{ m/s}^2$$
$$s = -30 \text{ m}$$
$$t = ?$$

$$s = v_0 t + \tfrac{1}{2} a t^2$$
$$-30 \text{ m} = 0(t) + \tfrac{1}{2}(-9.8 \text{ m/s}^2)(t^2)$$
$$t = 2.47 \text{ s}$$

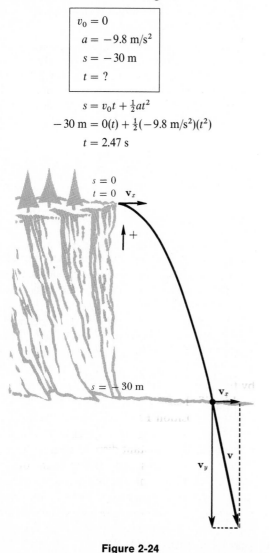

Figure 2-24

Next consider the *horizontal* motion. Since $a = 0$ (there is no horizontal acceleration since air resistance is neglected), $s = v_0 t$ for this motion, and v_0 is the full 6 m/s since it is directed horizontally.

$$s = v_0 t = (6 \text{ m/s})(2.47 \text{ s}) = \boxed{14.8 \text{ m}}$$

Example 2-15

In the above example, calculate the velocity of the stone as it hits the ground.

Since velocity is a vector, we must find both its magnitude and its direction in order to have a complete answer.

Vertical motion:

$$v_0 = 0$$
$$a = -9.8 \text{ m/s}^2$$
$$v = ?$$
$$s = -30 \text{ m}$$

$$v^2 = v_0{}^2 + 2as$$
$$v^2 = 0^2 + 2(-9.8 \text{ m/s}^2)(-30 \text{ m})$$
$$v = \pm 24.2 \text{ m/s}; \, v_y = -24.2 \text{ m/s}$$

We interpret this answer as v_y, the vertical component of the final velocity.

Horizontal motion: The horizontal component of the final velocity is the original horizontal component, which has remained unchanged, since air resistance is neglected. The magnitude of the resultant final velocity is found by combining v_x and v_y:

$$v = \sqrt{v_x{}^2 + v_y{}^2} = \sqrt{6^2 + (-24.2)^2} = 24.9 \text{ m/s}$$

The angle θ is found from a scale diagram or by trigonometry:

$$\cos \theta = \frac{6 \text{ m/s}}{24.9 \text{ m/s}} = 0.241$$

$$\theta = 76°$$

Thus our answer is

$$\mathbf{v} = \boxed{\begin{array}{l} 24.9 \text{ m/s at an angle} \\ 76° \text{ below horizontal} \end{array}}$$

In Example 2-14, we used the same symbols for different quantities. Thus in the box, $v_0 = 0$, whereas in the last line of the same problem, $v_0 = 6$ m/s. Similarly, in the box, $s = -30$ m, whereas on the last line, $s = 14.8$ m. These apparent contradictions will be cleared up if you remember that we are actually solving two separate problems when we solve a projectile problem. The only symbol that is necessarily the same for the two motions is t, the time interval, which is 2.47 s for both parts of Example 2-14.

Example 2-16

A package of medical supplies is released from a small plane flying a mercy mission over an isolated jungle settlement. The plane flies horizontally with a speed of 20 m/s at an altitude of 20 m. Where will the package strike the ground?

In this case the package carries with it the horizontal velocity of the plane. Just after it is released, the package is traveling forward at 20 m/s.

We consider the vertical and horizontal motions separately, as in all projectile problems.

Vertical motions:

$$\begin{array}{l} v_0 = 0 \\ s = -20 \text{ m} \\ a = -9.8 \text{ m/s}^2 \\ t = ? \end{array}$$

$$s = v_0 t + \tfrac{1}{2}at^2$$

Since the original velocity has no vertical component, $v_0 = 0$, and so

$$t = \sqrt{\frac{2s}{a}} = \sqrt{\frac{2(-20 \text{ m})}{-9.8 \text{ m/s}^2}} = 2.02 \text{ s}$$

Horizontal motion: Since the horizontal acceleration is zero,

$$s = v_0 t = (20 \text{ m/s})(2.02 \text{ s}) = \boxed{40 \text{ m}}$$

The package strikes 40 m ahead of the point directly below the point of release.

Since the plane's velocity is 20 m/s and the package's horizontal component of velocity is also 20 m/s, the package always remains directly beneath the plane as it travels. We can say that the package always has the same *horizontal* position as the plane does, because its *horizontal* velocity relative to the plane is always zero.

So far, our treatment of projectiles has been confined to cases in which the initial velocity is horizontal. If this is not the case, we resolve the initial velocity into its horizontal and vertical components. The vertical component is used to find the time in the air, and the horizontal component of the initial velocity is used to find the horizontal distance that the projectile travels.

Example 2-17

A football is thrown at a velocity of 24 m/s at an angle of 30° above the horizontal. Calculate (a) the maximum height of the ball; (b) the time in the air; and (c) the horizontal distance (range) to the point where the ball is caught by the receiver.

Resolving the initial velocity into its components (Fig. 2-25), we find that v_0 in the vertical direction is 24 sin 30° = 12 m/s.

(a) We choose the upward direction to be positive and make a table of knowns and unknowns for the vertical motion of the ball. At the highest point, the vertical velocity of the ball is 0.

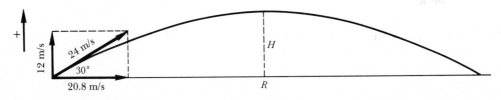

Figure 2-25

$$\boxed{\begin{array}{l} s = ? \\ v = 0 \\ v_0 = +12 \text{ m/s} \\ a = -9.8 \text{ m/s}^2 \end{array}}$$

$$v^2 = v_0{}^2 + 2as$$

$$s = \frac{v^2 - v_0{}^2}{2a}$$

$$= \frac{0 - (+12 \text{ m/s})^2}{2(-9.8 \text{ m/s}^2)} = \boxed{7.34 \text{ m}}$$

This is the maximum height, H.

(b) The ball is in the air for a time sufficient for its initial upward velocity component $+12$ m/s to become 0 (at the highest point) and then become -12 m/s as the ball is caught by the receiver.

$$\boxed{\begin{array}{l} v_0 = +12 \text{ m/s} \\ v = -12 \text{ m/s} \\ a = -9.8 \text{ m/s}^2 \\ t = ? \end{array}}$$

$$v = v_0 + at$$

$$t = \frac{v - v_0}{a}$$

$$= \frac{(-12 \text{ m/s}) - (+12 \text{ m/s})}{-9.8 \text{ m/s}^2} = \boxed{2.45 \text{ s}}$$

(c) The horizontal component of the initial velocity is $24 \cos 30° = 20.8$ m/s. Since we are neglecting air resistance, the horizontal motion is uniform, and the distance to the point where the ball is caught by the receiver is

$$s = v_0 t = (20.8 \text{ m/s})(2.45 \text{ s})$$

$$= \boxed{51.0 \text{ m}}$$

This is the range, R.

Suppose we want to make a table showing how the horizontal range of a gun depends on its angle of elevation θ. Let us derive an equation for this purpose and check the dimensions on both sides of our final equation.

Although the mathematical solution of this problem is not difficult, let us first look at the physics of the situation. There are two extreme cases: (a) If the elevation is 90° (gun pointed straight up), the range R is obviously zero. (b) If the gun is pointed nearly horizontally ($\theta = 0$), the bullet is in the air for only a short time interval t before striking the ground, since the vertical component of the initial velocity is so small. Therefore the forward velocity component acts for only a very short time, and the range will be very small in this case also.

In some respects it is easier to work with symbols than with numbers. We follow the same scheme as in Example 2-17, using symbols instead of numbers.

Vertical motion:

$$v_y = v_0 \sin \theta \tag{2-16}$$

$$t = \frac{v_0 \sin \theta - (-v_0 \sin \theta)}{g} = \frac{2v_0 \sin \theta}{g}$$

Horizontal motion:

$$v_x = v_0 \cos \theta \tag{2-17}$$

$$R = v_x t = (v_0 \cos \theta)\left(\frac{2v_0 \sin \theta}{g}\right)$$

$$= \frac{2v_0{}^2}{g} \cos \theta \sin \theta \tag{2-18}$$

This is the equation we wished to derive. We now proceed to check our result in two ways. Any physical equation must have the same dimensions on both sides of the equals sign. In this case, we make the check as follows ($\cos \theta$ and $\sin \theta$ are ratios of lengths and hence have no dimensions):

$$[\text{L}] \overset{?}{=} \frac{[\text{LT}^{-1}]^2}{[\text{LT}^{-2}]} \overset{?}{=} \frac{[\text{L}^2\text{T}^{-2}]}{[\text{LT}^{-2}]}$$

$$[\text{L}] = [\text{L}]$$

By this check, we have shown that our result is at least not obviously wrong, but we have of course only verified its dimensions, not the exact form. At any rate, we have passed the first hurdle. Our answer is "not wrong" dimensionally, and thus it is more likely to be "right."

A second check on our derivation is to see whether the answer is reasonable. Does it fulfill the expected behavior listed under (a) and (b) at the beginning of our solution? If the angle is 90°, condition (a) is fulfilled, since cos 90° = 0, and the factor cos θ therefore makes R become zero. Likewise, condition (b) is fulfilled since the factor sin θ becomes zero when θ is 0°. Thus both of the preliminary expectations (a) and (b) are fulfilled, another indication that our derivation is probably correct.

Similar checks should be made for any physical derivation.

In actual practice, the effects of air resistance are by no means negligible in projectile problems. Air resistance lengthens the time in the air and also causes the horizontal velocity to decrease somewhat from the original value. The exact study of projectile motion is called *ballistics*, and its mathematical complexities are such that large computers are often used.

We have by no means exhausted the study of kinematics, for we have concerned ourselves chiefly with constant motion and with uniformly accelerated motions, or combinations of such motions. We can immediately extend our analysis of straight-line motion to include some aspects of curvilinear motion. Consider a car moving on a winding turnpike. Even if its speed is constant, its velocity (a vector) changes since its direction changes. Acceleration is defined as the rate of change of velocity, and so we see that the curvilinear motion always is accelerated motion. However, if we consider only components of displacement, velocity, and acceleration measured along the curved turnpike, the one-dimensional equations 2-11 through 2-14 are valid, just as if the path were stretched out into a straight line. We shall return to kinematics in Chap. 7, when we investigate circular motion. For the present, we have a sufficiently precise description of motion to be able to proceed to dynamics, the study of the causes of motion.

Summary   Kinematics is the description of motion, without regard to causes. Dynamics (to be studied in Chap. 3) is the study of the causes of motion, in terms of force and inertia. For straight-line motion, average velocity is displacement divided by elapsed time; it equals the slope of a chord drawn between two points on the graph of displacement versus time. Instantaneous velocity is the limit of the average velocity as the time interval approaches zero; it equals the slope of the tangent to the displacement graph drawn at the point representing the time at which the velocity is desired. Instantaneous velocity is the time derivative of displacement, and displacement is the integral of the velocity function.

In uniformly accelerated motion, the variables are related to each other by the following equations:

$$v = v_0 + at \qquad s = \frac{v + v_0}{2}\,t = \bar{v}t \qquad v^2 = v_0{}^2 + 2as \qquad s = v_0 t + \tfrac{1}{2}at^2$$

Motion with constant velocity is a special case of accelerated motion, in which $a = 0$. In the absence of air resistance, the acceleration of a freely falling body is approximately 9.80 m/s² (or 32 ft/s²). For uniformly accelerated motion the displacement graph is a portion of a parabola, and the velocity graph is a straight line whose slope is the acceleration.

Vector quantities have both magnitude and direction; scalars have magnitude only. The resultant of any number of vectors is found by the head-to-tail rule; for two vectors at a time the parallelogram rule is also convenient. Components of vectors are found by scale drawings or by the use of the sine and cosine.

If v_{AB} represents the velocity of A relative to B, and so on, then the velocity of A relative to another body Z can be found by vector addition:

$$v_{AZ} = v_{AB} + v_{BC} + v_{CD} + \cdots + v_{YZ}$$

The motion of a projectile may be regarded as two motions that are independent and yet simultaneous. In the absence of air resistance, the vertical motion is uniformly accelerated, and the horizontal motion is one of constant velocity.

Check List

kinematics	acceleration	$v_{AC} = v_{AB} + v_{BC}$
dynamics	pickup	resultant
vector	average velocity	component
scalar	instantaneous velocity	sine
displacement	slope of a graph	cosine
distance	derivative	tangent
velocity	definite integral	
speed	relative velocity	

Questions

2-1 Which of the following equations cannot be correct, because of dimensional inconsistency? Are the others necessarily correct?
(a) $s = v_0 t + 3at^2$ (b) $v = v_0^2 - \frac{1}{2}as^2$ (c) $s + \frac{1}{2}at^2 = vt$ (d) $at^2 + v/t = 2s^2/t^3$

2-2 How would you measure "the velocity" of a jet airplane at the instant it passes by a control tower in which you are stationed?

2-3 Every automobile is equipped with an odometer on the dashboard. Does this measure a scalar or a vector quantity?

2-4 What is the description, in words, of the motion of the falling package in Example 2-16 as seen from the plane? What is the description, as seen from a speeding automobile moving at 30 m/s parallel to the plane's motion and in the same direction? The automobile was directly beneath the plane when the package was released.

2-5 A base runner sometimes takes a jump at the bag as he approaches first base. Is this more likely to result in an out than if he makes a level approach?

MULTIPLE CHOICE

2-6 The dimensions of acceleration are (a) $[LT^2]$; (b) $[LT^{-1}]$; (c) $[LT^{-2}]$.

2-7 Starting from rest, during the first second a ball falls (a) 2.45 m; (b) 4.90 m; (c) 9.80 m.

2-8 A ball is thrown upward at 20 m/s. During the first second, its average velocity is (a) 10.2 m/s; (b) 15.1 m/s; (c) 15.9 m/s.

2-9 A ball is thrown straight up and later caught. (a) The velocity at the top of the motion is zero; (b) the acceleration at the top of the motion is zero; (c) both of the above.

2-10 A football is thrown in an arc from one player to another. Relative to the initial horizontal velocity component, when the ball is at its highest point, its horizontal velocity component is (a) greater; (b) the same; (c) less.

2-11 If displacement is plotted versus time, the slope of the graph is the (a) instantaneous acceleration; (b) instantaneous velocity; (c) rate of change of velocity.

Problems

Note: Use $g = 9.80$ m/s^2 and neglect air resistance. Use Appendix Table 4 for metric equivalents. It is useful to remember that 1 mile = 1.61 km.

2-A1 A car travels the length of a turnpike that is 480 km long in 5 h, 30 min. What is its average speed   (*a*) in km/h?   (*b*) in mi/h?

2-A2 If while you are driving along at the legal limit of 88.5 km/h (55 mi/h), your attention wanders for 0.50 s, how far (in m) do you travel "blind" during that half-second?

2-A3 A snail traveling at a snail's pace (5 m/day) decides to slow down to only 1 m/day and allows himself 2 min in which to make the change.   (*a*) Express his initial velocity in km/h; in mi/h; in ft/s.   (*b*) Compute the acceleration in m/day·min; in m/s^2.

2-A4 The displacement graph for a subway train traveling from one station to another along a straight track is shown in Fig. 2-26.   (*a*) Estimate the instantaneous velocity 40 s after leaving the station.   (*b*) What is the instantaneous velocity 100 s after leaving the station?   (*c*) What is the average velocity during the first 100 s?

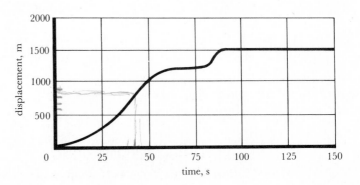

Figure 2-26   Graph of displacement vs. time for the motion of a subway train.

2-A5 Refer to Fig. 2-26.   (*a*) What is the average velocity of the train during the first 50 s? (*b*) What is the instantaneous velocity 75 s after leaving the station?

2-A6 The velocity graph for the motion of a playful dolphin is shown in Fig. 2-27.   (*a*) At what time was the dolphin's velocity greatest?   (*b*) When was the forward accelera-

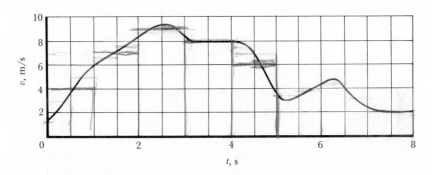

Figure 2-27   Velocity graph for the motion of a dolphin.

tion greatest? (*c*) Was the acceleration constant at any time during the motion? (*d*) What was the acceleration at $t = 1.5$ s? (*e*) Use graphical integration to estimate the total distance traveled by the dolphin during the first 5 s.

2-A7 At 3:15 P.M. a motorboat was moving at 13.2 m/s eastward; at 3:18 P.M. its velocity was 9.6 m/s eastward. What were the magnitude and direction of the average acceleration of the motorboat?

2-A8 A flower pot falls off the roof of a tall building. (*a*) What is the instantaneous velocity of the pot at the end of 2 s of its fall? (*b*) What is the velocity at the end of 3 s of fall? (*c*) What is the average velocity, averaged over the third second of the fall? (*d*) What is the distance traveled during the third second?

2-A9 A sprinter was timed at 10.0 s for the 100-m dash; another runner in the same race was timed at 10.4 s. (*a*) Calculate the average speed of each runner. (*b*) How far apart were the two runners when the winner crossed the finish line?

2-A10 A ball is thrown upward at 13.8 m/s. Calculate the magnitude and direction of the ball's velocity (*a*) 1 s after being thrown; (*b*) 2 s after being thrown.

2-A11 In Example 2-7 add the same two vectors, in the other sequence, starting with the 240-km displacement.

2-A12 Using numerical values of sine or cosine, calculate the magnitudes of the components a_x and v_x in Fig. 2-22.

2-A13 On copies of each triangle of Fig. 2-22, draw an arrowhead on the remaining leg of the triangle so that the hypotenuse is the vector sum of the two components, arranged head to tail.

2-A14 Using a careful scale diagram, add the vectors of Fig. 2-9 in the sequence

$$A + C + B + E + D$$

2-A15 A child in a city where the blocks are square starts from the corner of West Third Avenue and North Second Street, and walks south 5 blocks along West Third Avenue, crossing Main Street and reaching South Third Street. She then walks 5 blocks eastward, crossing Center Avenue to reach East Second Avenue. She then runs to the corner of North First Street and East Second Avenue, and finally she goes along North First Street to West Fourth Avenue. (*a*) What is the distance traveled? (*b*) What is the displacement for the entire trip?

2-A16 An airplane in a power dive is moving at 400 m/s downward at an angle of 60° below the horizontal. If the sun is directly overhead, how fast is the plane's shadow moving along the ground?

2-A17 A plane is traveling toward the southwest at 280 km/h. (*a*) What is the westward component of its velocity? (*b*) What is the eastward component of its velocity?

2-A18 Two children 10 m apart are playing catch on board a train that is moving at 40 m/s in a straight line. Child *A* throws the ball horizontally at 15 m/s relative to himself toward the rear of the train. (*a*) What is the velocity of the ball relative to child *B*? (*b*) What is the velocity of the ball relative to a stationary observer on the ground? (*c*) How long a time is required for the ball to go from *A* to *B*?

2-A19 All points on the equator are moving eastward at about 470 m/s due to the earth's rotation. (*a*) If a rocket is launched at the equator, aimed horizontally toward the east with a velocity of 3600 m/s relative to the earth, what will its velocity be relative to the center of the earth? (*b*) Repeat part (*a*) assuming the rocket is aimed toward the west. (*c*) Why are most U.S. space launchings made from the Florida coast?

2-B1 A plane flying directly into a head wind proceeded from city A to city B, a distance of 900 km, at an average speed of 500 km/h. The plane paused 24 min on the ground while refueling, then flew back to city A (aided by the tail wind) at an average speed of 600 km/h. (*a*) What was the average speed for the entire trip, including the layover at city B? (*b*) What was the magnitude and direction of the wind velocity?

2-B2 A bullet is fired horizontally at 800 m/s and strikes a target that is 160 m from the gun. If the marksman hears the sound of the impact on the target 0.67 s after he fires the gun, what is the speed of sound?

2-B3 A girl drops a stone from a bridge that is 20 m above a river. If the speed of sound is 340 m/s, how long after dropping the stone does she hear the splash?

2-B4 Nolan Ryan, formerly with the California Angels, pitched a fast ball that crossed home plate at a speed of 45.06 m/s. (*a*) Express this speed in mi/h. (*b*) Home plate is 18.5 m from the pitcher's mound. How long did it take for the ball to travel this distance, assuming constant speed? (*c*) If the catcher allowed his mitt to recoil backward 9.0 cm while catching the ball, what was the negative acceleration of the ball while it was being slowed down by the catcher?

2-B5 A truck traveling 108 km/h (67 mi/h) is slowed down at a uniform rate of 2.5 m/s². How far will it travel before its speed is half the original value?

2-B6 A speedboat increases its speed at the rate of 3 m/s². How much time is required for the speed to increase from 9 m/s to 21 m/s? How far does it travel during this time?

2-B7 With what vertical velocity does a kangaroo take off if it can jump vertically upward a height of 2.5 m?

2-B8 A stone is dropped from a bridge, and 2 s later another stone is dropped. How far apart are the two stones by the time the first one has reached a speed of 24.5 m/s?

2-B9 A ball is thrown vertically upward at 14.7 m/s. (*a*) How high does it rise? (*b*) How long a time is required for it to reach its maximum height? (*c*) How long is it in the air? (*d*) With what speed does it strike the ground?

2-B10 A small mailbag is released from a helicopter that is descending steadily at 2 m/s. After 2 s, (*a*) what is the velocity of the bag? (*b*) how far is it below the helicopter?

2-B11 A small mailbag is released from a helicopter that is rising steadily at 2 m/s. After 2 s, (*a*) what is the velocity of the bag? (*b*) how far is it below the helicopter?

2-B12 With what upward velocity should a package be thrown in order to be caught easily by a person on a balcony 6 m above the ground?

2-B13 Two students are on a balcony 24.5 m above the street. One student throws a ball vertically downward at 19.6 m/s; at the same instant the other student throws a ball vertically upward at 19.6 m/s. The second ball just misses the balcony on the way down. (*a*) How long after one ball strikes the street does the other ball strike? (*b*) With what velocity does each ball strike the street? (*c*) How far apart are the balls 1 s after they are thrown?

2-B14 A book is held 4.9 cm above the table top and released. How long is the book in the air?

2-B15 Reaction time can be estimated by the "dollar-bill-catch" method. A dollar bill is held by A with its long side in a vertical plane. Observer B is poised ready to catch the bill; B's fingers are 13 cm below A's. The bill is released without warning; B can keep it if it can be caught, but A never has to pay off. What is the minimum neurological reaction time for the signal to get from the eye to B's finger muscles?

2B-16 In a test of automobile bumpers, a car moving at 7 m/s collides head-on with a stone wall. From what height would the car have to fall straight down in order to make an equally hard collision?

2B-17 Highway safety engineers consider $-30g$ to be an acceptable acceleration during a crash. On this basis, what is the minimum acceptable stopping distance during a frontal barrier crash if the initial speed is 24.6 m/s (55 mi/h)?

2B-18 A ball is thrown vertically to a height of 40 m and allowed to strike the ground. If it loses one-fifth its speed while in contact with the ground, how high does it rise on the rebound?

2-B19 A certain car has a maximum positive acceleration (when starting up) of $+2$ m/s^2 and a maximum negative acceleration (when stopping) of -4 m/s^2. The speed limit is 16 m/s (35 mi/h). What is the shortest legal time in which the driver of the car can go from one stop sign to another, a distance of 400 m?

2-B20 A football is given a forward velocity of 10 m/s by a passer whose arm moves through a horizontal distance of 1 m. Compute the average forward acceleration of the ball.

2-B21 A speeding car passes a highway patrol checkpoint and then decelerates at a constant rate; 5 s later, the car is 175 m from the checkpoint, and its speed is then 24 m/s. (a) What was the car's velocity when it passed the checkpoint? (b) What was the acceleration of the car?

2-B22 A sailor drops a pocket knife from the top of a mast on a ship sailing eastward at 8 m/s. The mast is 19.6 m high. Where does the knife hit the deck?

2-B23 A plane has an air speed of 181 km/h, and the pilot notices that although she is headed due east, she is actually traveling northeast and covers 128 km in 30 min. What is the magnitude and direction of the velocity of the wind that is blowing her off course?

2-B24 A helicopter headed due north has an air speed of 80 km/h, and the wind is 60 km/h from the west. (a) What are the magnitude and direction of the plane's velocity relative to the ground? (b) How far does the helicopter travel in 12 min?

2-B25 A bicyclist finds that when he is moving at 5 m/s along a southbound bicycle trail, the wind appears (relative to him) to come from the southeast toward the northwest at 10 m/s. What are the magnitude and direction of the wind velocity, relative to the earth?

2-B26 A jogger is out on a day when the wind is blowing at 12 m/s from north to south. (a) If the jogger's velocity is 3 m/s north to south, what is the velocity of the wind relative to the jogger? (b) Still maintaining a ground speed of 3 m/s, through what angle should the jogger swerve so that the wind will seem to be blowing directly broadside, perpendicular to the jogger's motion?

2-B27 A swimmer can swim 2 m/s in still water. (a) Starting from the west bank of a river that flows southward at 1 m/s, in what direction should he head (somewhat upstream) so that he will travel directly across? (*Hint:* The velocity of the swimmer relative to the earth must be a vector directed toward the east.) (b) What will be the magnitude of the swimmer's resultant velocity? (c) How much time will be required for the trip if the river is 200 m wide? (d) If the swimmer swims 200 m directly downstream and then swims back again to his starting point, how does his time for the round trip compare with what it would have been if he had swum to a point directly across from his starting point and back again, which is also a round trip of 400 m?

2-B28 If the swimmer of Prob. 2-B27 heads straight across the river so that he lands somewhat downstream, how long will it take him to cross?

2-B29 A Ping-Pong ball rolls with a speed of 0.9 m/s toward the edge of a table that is 0.80 m above the floor. The ball rolls off the table; how long is it in the air? How far out from the edge of the table does the ball hit the floor?

2-B30 (a) With what *vertical* component of velocity must a jumping frog take off if it is in the air for 0.50 s? (b) How high does the frog jump? (c) How far ahead of its starting point (on a level surface) does the frog land if its takeoff is at 30° above the horizontal?

2-B31 Why does a hunter raise the barrel of his rifle when aiming at a distant target? If he aims directly at a target 200 m away, by how much will he miss the target if the muzzle velocity of the bullet is 500 m/s?

2-B32 A bullet strikes a target that is 100 m from a gun and level with it. During its flight, the maximum height of the bullet above the horizontal line between gun and target was 2.00 cm. Calculate the muzzle velocity of the bullet.

2-B33 A diver leaves a springboard horizontally with a velocity of 3.0 m/s. The board is 3 m above the surface of the pool. (a) How far out from a point below the end of the springboard does she strike the water? (b) What are the horizontal and vertical components of the diver's velocity when she strikes the water?

2-B34 How much time for maneuvering in air does a diver gain by taking off horizontally from a 3-m board instead of horizontally off a 1-m board?

2-B35 A stone is thrown horizontally from a bridge 40 m above the water. (a) How long is the stone in the air? (b) What must be the initial velocity of the stone if the line joining the bridge and the splash is inclined 45° below the horizontal? (c) At what angle does the stone strike the water?

2-B36 A boy can throw a ball a maxmium horizontal distance R on a level field. How high can he throw the same ball vertically upward? Assume that his muscles give the ball the same speed in each case. (Is this a valid assumption?)

2-B37 A boy throws a ball horizontally at 8 m/s straight out from a window in a building; the ball is caught by a friend on the street who is 14 m from the base of the building. How high above the street is the window?

2-B38 An archer shoots an arrow with a horizontal velocity component of 30 m/s and a vertical velocity component of 19.6 m/s. Calculate (a) the time in the air; (b) the horizontal range; (c) the maximum height of the arrow; (d) the initial speed of the arrow.

2-B39 On level ground, a ball is thrown forward and upward. The ball is in the air 2 s and strikes ground 30 m from the thrower. With what speed, and at what angle, was the ball thrown?

2-B40 What would the range become if the football passer of Example 2-17 threw the ball at an angle of 40° with the horizontal?

2-B41 A toy rocket is fired upward at an initial speed of 49 m/s. At the very top of its rise, a second motor fires briefly, giving the rocket a horizontal velocity of 49 m/s. (a) How long is the rocket in the air? (b) Where does it strike the ground? (c) Where is the rocket 7 s after it leaves the ground?

2-B42 On the moon the acceleration due to gravity is $\frac{1}{6}$ that on the earth. Prove that a broad jumper can jump 6 times as far on the moon as on the earth if he takes off with the same speed at the same angle.

2-B43 A broad jumper takes off at an angle of 20° with the horizontal and reaches a maximum height of 65 cm at mid-flight. (a) What is the forward component of his velocity? (b) How far does he jump in the forward direction?

2-C1 A late passenger, sprinting at 7.5 m/s, is 27 m away from the rear end of a commuter train when it starts out of the station with an acceleration of 1 m/s^2. Can the passenger catch the train if the platform is long enough? (*Note:* This problem requires solution of a quadratic equation. Can you explain the significance of the two answers you get for the time?)

2-C2 A mile runner traveling at constant speed finds himself with 1480 ft still to go when 3 min, 10 s have already elapsed. If for his home-stretch "kick" he accelerates at 1.00 ft/s^2 for 10.0 s and then holds the new speed until the finish line, can he still run a 4-minute mile? *Note:* One mile is 5280 ft.

2-C3 A car and a truck are each traveling at 20 m/s, and the car is 36 m behind the truck. The car driver decides to pass the truck, and he steps on the gas, producing an acceleration of 2 m/s^2. (*a*) How long will it be before the car is alongside the truck? (*b*) How fast will the car be moving relative to the truck when they are side by side? (*c*) How far will the car travel while reaching the truck? (*d*) How far will the truck travel during the same time?

2-C4 The equation $s = vt - \frac{1}{2}at^2$ was studied *dimensionally* in Ques. 1-5 (page 19) and can be shown to be *not incorrect* since the dimensions on each side of the equals sign are those of length. By algebraic manipulation of the equations of Sec. 2-6, show that the given equation is correct not only dimensionally, but also algebraically. (The symbols are as defined in Ques. 1-5.)

2-C5 A ball is thrown vertically upward at 19.6 m/s. (*a*) How long a time is required for it to reach a level 14.7 m above its starting point? (*b*) How fast is it then moving? (*c*) Explain your two answers to part (*a*).

2-C6 A flower pot topples from a penthouse window sill. A tenant on a lower floor observes that the pot passes a picture window 2.0 m tall in 0.1 s. How far above the top of the picture window is the penthouse window sill?

2-C7 A diver takes off with a speed of 8.00 m/s from a diving board 3 m high, at an angle of 25° above the horizontal. How much later does she strike the water?

2-C8 A cat jumps off a piano that is 1 m high. The initial velocity of the cat is 2.5 m/s at an angle 37° above the horizontal. How far out from the piano does the cat strike the floor?

2-C9 A pilot cuts loose his fuel tanks in an effort to gain altitude. At the time of release, he was 200 m above ground and traveling upward at an angle of 30° above the horizontal, with a speed of 98 m/s. For how long were the tanks in the air?

2-C10 On a ski jump, a skier takes off in a direction 15° above the horizontal with a speed of 24 m/s. She lands at a point 20 m below the level of the takeoff point. (*a*) Through what distance, measured horizontally, does the skier jump? (*b*) With what speed does she strike the ground?

2-C11 An Olympic basketball player shoots toward a basket that is 6.000 m horizontally from him and 3.048 m above the floor. The ball leaves his hand 2.000 m above the floor at an angle 60° above horizontal. What speed should the player give the ball? (*Hint:* Set up a pair of equations, one for horizontal motion and one for vertical motion, with v_0 and t as unknowns.)

2-C12 A ball is projected horizontally from the edge of a table that is 1 m high, and it strikes the floor at a point 1.5 m from the base of the table. (*a*) What is the initial velocity of the ball? (*b*) How high is the ball above the floor when its velocity makes an angle of 45° with the horizontal?

2-C13 A motorcycle daredevil picked up speed coming down a ramp, then took off from a point on a parking lot. His trajectory was at 20° above the horizontal at the moment when he just cleared the first of 14 school buses, each 3 m wide and 3 m tall, parked with sides touching. He just cleared the last bus and landed on the parking lot at the same height from which he became airborne. (a) What were the horizontal and vertical components of his take-off velocity? (b) How long was he in the air? (c) What total horizontal distance was covered in the jump?

2-C14 Prove that the path of the stone in Fig. 2-24 is a parabola. [*Hint:* Eliminate t from the equations for $x(t)$ and $y(t)$ to obtain y as a function of x.]

2-C15 Derive a formula for the maximum height of a projectile shot at an elevation θ with a muzzle velocity $\mathbf{v}_0$. Check your answer dimensionally.

2-C16 For what elevation angle θ above the horizontal is the range R of a projectile a maximum, for a given muzzle velocity $\mathbf{v}_0$? The target is at the same level as the gun.

2-C17 A projectile is shot at an elevation θ, and the line joining the gun to the point where the projectile has maximum height makes an angle ϕ with the horizontal. Derive a trigonometric relationship between θ and ϕ.

For Further Study

Note: In sections titled For Further Study, which are found at the ends of many chapters, we will go beyond the material required of most students. These additional sections are of different types: Some are fairly rigorous proofs of statements assumed without proof earlier in the chapter; some are interesting topics that are sidelines rather than integral parts of the course; and some are extensions of material in the text, more difficult or more mathematical in nature. Some For Further Study sections, like the following one, make use of formal calculus to go beyond the level of material in the main body of the text.

2-12 Accelerated Motion from the Viewpoint of Formal Calculus

In our study of uniformly accelerated motion (pages 30–33), we found it helpful to use the *concepts* of calculus. Thus instantaneous velocity was defined as the limit of a ratio, and displacement was given as the limit of a sum. In our earlier work we stressed the graphical interpre-

tations of derivative (as the slope of a curve) and integration (as the area under a curve).

If you have studied even a little elementary calculus, you will find that the *formulas* for differentiation and integration make possible a greater insight into some of the basic ideas of motion. (A brief review of formal calculus is in Appendix E on pages 838–840.) Let us use calculus to derive some of the equations of uniformly accelerated motion in a systematic fashion. We start with a definition:

$$\text{acceleration} = v'(t) = a = \text{constant}$$

The velocity $v(t)$ at time t is the area under the acceleration graph $v'(t)$ plus the initial value $v(0)$ at the time $t = 0$. We obtain

$$v(t) = \int_0^t v'(t)\,dt + v(0) = \int_0^t a\,dt + v(0)$$

$$= a \int_0^t t^0\,dt + v(0) = a\left(\frac{t^1}{1} - \frac{0^1}{1}\right) + v(0)$$

or

$$v = v_0 + at$$

where $v(0) = v_0 =$ initial velocity. This is Eq. 2-11. Proceeding, we find $s(t)$ by a similar integration:

$$s(t) = \int_0^t s'(t)\,dt + s(0) = \int_0^t v(t)\,dt + s(0)$$

$$= \int_0^t (v_0 + at)\,dt + s(0)$$

$$= \int_0^t v_0 t^0\,dt + \int_0^t at^1\,dt + s(0)$$

$$= \frac{v_0 t^1}{1} + \frac{at^2}{2} + s(0)$$

If we choose $s(0) = 0$, as is usual, we have

$$s(t) = v_0 t + \tfrac{1}{2}at^2$$

This is Eq. 2-14. Thus, starting with the acceleration, we have, in sequence, found the velocity and the displacement by integration.

In reverse, we can by differentiation verify that $s'(t) = v(t)$, and $v'(t) = a$. Acceleration is the derivative of the derivative of displacement; it is called the second derivative of s with respect to t and is written as $s''(t)$, or as d^2s/dt^2, or as $\dfrac{d^2}{dt^2}(s)$.

We start with the equation for displacement,

$$s = s_0 + v_0 t + \tfrac{1}{2}at^2$$

where we have added the term s_0, which is the initial displacement—the value of s when $t = 0$. Since s_0, v_0, and a are constants independent of t, differentiation with respect to t gives

$$s'(t) = \frac{ds}{dt} = v_0 + \tfrac{1}{2}a(2t)$$

or

$$v = v_0 + at$$

This is Eq. 2-11. Another differentiation gives

$$v'(t) = s''(t) = \frac{d^2s}{dt^2} = a(1t^0)$$

or

$$a = a$$

which confirms that the acceleration is constant, as expected. Thus we have found velocity and acceleration, by successive differentiation, starting with the displacement.

Problems   2-C18   For the displacement function $s(t) = 4t^2 - 5t$, evaluate $\Delta s/\Delta t$ at $t_0 = 3$ s by the method of Example 2-2, using (a) $\Delta t = 0.1$ s; (b) $\Delta t = 0.01$ s; and (c) $\Delta t = 0.002$ s. (d) Does $\Delta s/\Delta t$ approach a limit? (*Hint:* Organize your calculations in neat tabular form as you work. Check your result by differentiation.)

2-C19   A body's displacement is given by the equation $s = 200 + 32t - 5t^2$. (a) If s is in meters and t in seconds, what are the dimensions of the quantities represented by the numbers 200, 32, and -5? (b) What is the velocity at $t = 2$ s? (c) How far does the body travel during a 1-s interval, beginning at $t = 2$ s? (d) What is the acceleration at $t = 2$ s?

2-C20   A particle moves according to the equation $v(t) = t^2 - 3$ (in m/s). Estimate graphically the displacement of the particle from $t = 2.0$ s to $t = 3.0$ s. (*Hint:* Use ten subintervals Δt, each of size 0.1 s. Check your result by formal integration.)

References   1. I. Bernard Cohen, "Galileo," *Sci. American* **181**(2), 40 (Aug. 1949). A biography.
2. R. B. Lindsay, "Galileo Galilei, 1564–1642, and the Motion of Falling Bodies," *Am. J. Phys.* **10**, 285 (1942).
3. W. F. Magie, *A Source Book in Physics* (McGraw-Hill, New York, 1935; also Harvard University Press, Cambridge, Mass., 1963). Selections from Galileo's *Two New Sciences:* acceleration and laws of falling bodies, pp. 2–17; projectile motion, pp. 19–22.

4. G. Holton and D. H. D. Roller, *Foundations of Modern Physical Science* (Addison-Wesley, Reading, Mass., 1958), pp. 24–28, 34–38. Selections from Galileo's writings, with commentary.
5. S. Drake and J. MacLachlan, "Galileo's Discovery of the Parabolic Trajectory," *Sci. American* **232**(3), 102 (Mar. 1975).
6. W. S. Porter, "The Range of a Projectile," *Phys. Teach.* **15**, 358 (1977).
7. H. Van Dael and H. Bert, "Range of a Projectile," *Am. J. Phys.* **47**, 466 (1979).
8. S. Chapman, "Catching a Baseball," *Am. J. Phys.* **36**, 868 (1968).
9. F. L. Friedman, *The Velocity Vector; The Acceleration Vector; Velocity and Acceleration in Free Fall* (films).
10. J. Stull, *Constant Velocity and Uniform Acceleration* (film).

3

Dynamics

In our study of certain simple motions in Chap. 2, we made a beginning in the study of kinematics, a science for which Galileo (1564–1642) was largely responsible. Galileo studied projectile motion in detail and had a clear idea of the concept of inertia. His methods of investigation placed emphasis on experiment and on the mathematical description of physical laws, such as those concerned with the motions of falling bodies and projectiles. Galileo's methods are, of course, characteristic of science as we know it today. Building upon the work of Galileo and others and making full use of the "new" outlook in his methods, Isaac Newton (1642–1727) created a system of mechanics that ranks as one of the great intellectual achievements of all time. In this chapter we shall study the laws of motion as set forth by Newton, according to which the foundations of the science of dynamics were laid.

Born within a year of Galileo's death, Isaac Newton was the most influential scientist of modern times. According to his own opinion, Newton was "in the prime of my age for inven-

tion" during the plague years 1665 and 1666, when he left Cambridge University and studied mathematics, mechanics, and optics at his farm home in Lincolnshire. Shortly after he returned to the university, his own teacher, Isaac Barrow, resigned his professorship so that the 26-year-old Newton could have it. By this time the young genius had made profound discoveries in many fields, but he became so involved in controversy concerning his first published work (*Theory About Light and Colors*) that he came to care little about public recognition for his work. In 1684 he was persuaded by his friend, Edmund Halley, the astronomer, to organize and publish his system of mechanics, including the theory of gravitation. Thus, in 1687, Newton's *Principia*— probably the most important book in physics ever written—made its appearance. Newton lived for 40 years after the *Principia* was published and was a legend in his own time, but he devoted himself mainly to theology and to the conscientious performance of his duties as Warden (later Master) of the Mint and as president of the Royal Society. The story of Newton's life

Newton's third law is illustrated by two equal and opposite forces: the hose pushes the water toward the left, and the water pushes the hose toward the right.

is best told at length; consult the references at the end of this chapter.

3-1 Force—The Cause of Acceleration

The role of force in everyday life is a familiar one. Indeed, it seems almost superfluous to try to define such a self-evident concept as force. Pushes and pulls of all sorts—especially those exerted by muscles—are easily tagged with the word. A horse pulls a wagon; a baby throws a rattle; a motor raises an elevator cage. These are common examples of force involving motion. However, the existence or nonexistence of motion is really no indication of the existence of force. The horse can exert a force and yet not move a heavy block of stone; the baby can (and often does) push on an immovable wall and produce no motion; and there must be a force (tension) exerted by the elevator cable to support the cage even when it is stopped at the sixth floor.

If we are to get anywhere in our study of dynamics, we must first have a precise definition of force. It is obvious that forces *can* cause motion, but they do not seem to be *always* correlated with motion. Where is the motion, if any, "caused" by the pull of gravity on a parked tractor-trailer? And where is the force, if any, causing the motion of the star Regulus as it drifts majestically through space? Questions such as these puzzled the natural philosophers of many centuries. Newton clearly stated that *change* of motion is caused by forces and only by forces; unchanging, "uniform" motion requires no force. However, all bodies have inertia, by which we mean they "resist" being accelerated. Let us then define *force* as *that which causes the acceleration of a material body.*

No object is ever acted on by only a single force, although this situation is often closely approximated when one of the acting forces is much larger than all the others. The essential factor causing acceleration is the *net force*, which

is the unbalanced, or resultant, force. For instance, if two equal and opposite forces act on a body, there is no net force to cause acceleration, and the body's velocity remains constant. The condition of constant velocity is called *equilibrium*; this definition includes uniform motion in a straight line ($\mathbf{v} = $ constant $\neq 0$) as well as a state of rest ($\mathbf{v} = $ constant $= 0$).

3-2 Newton's First Law

The motion of a body on which no net force acts is described by Newton's first law, often called the law of inertia:

Newton's first law of motion:

A body remains at rest, or if in motion it remains in uniform motion with constant speed in a straight line, unless it is acted on by an unbalanced external force.

The first law is, of course, merely a statement of the qualitative definition of force that we have already given and is a recognition of inertia as an essential property of matter. Nevertheless, Newton's first law was of great value, for it refuted a common misconception that had been prevalent since the time of Aristotle. It had been supposed that the "natural" state of matter was one of no motion; indeed this seemed borne out by experiments in which an object sliding across a floor came to rest with no apparent force acting on it. Newton interpreted this same observation by saying that the force of friction caused the change of motion, or acceleration. When a horse pulls a wagon along the road with constant velocity, the older view was that the force exerted by the horse is needed in order to *cause* the (uniform) motion. Newton's interpretation was that the forward force of the horse is balanced by a backward force of friction, the net force is zero, and hence, once the wagon has been set in motion, it continues to move forward because of its inertia.

3-3 Newton's Second Law

We have seen that Newton's first law tells us what does not happen when no net force is applied to a body. To extend the laws of motion, we naturally seek a law describing what *does* happen when a net force *is* applied. Our first step is to quantify the concept of force. To do this, we assume that the magnitude of a force is measured by how much acceleration it can impart to a given body. Thus, if a net force F gives a certain body 6 times as much acceleration as does a net force F', then we *define* F to be 6 times as large as F'. In general, the ratio of two forces is defined to be equal to the ratio of the accelerations they produce. In symbols, for a given test body,

$$\frac{(\text{net } F)}{(\text{net } F)'} = \frac{a}{a'} \qquad (3\text{-}1)$$

This equation gives us a way of comparing the magnitudes of two forces, that is, by letting them act, one after the other, on the same test body. We are, in effect, defining the magnitude of force by describing an experimental procedure for measuring it. This "operational" definition is a satisfying and useful one, since experiment shows that it gives the same ratio for any given pair of forces F and F', no matter what test body is used for the acceleration experiments. Now we multiply both sides of Eq. 3-1 by (net F)/a, obtaining

$$\frac{(\text{net } F)}{a} = \frac{(\text{net } F)'}{a'}$$

$$= \text{constant for any test body}$$

To complete the definition of force, we define the direction of a force as being parallel to the acceleration that it gives an object. Thus net $\mathbf{F}$ and $\mathbf{a}$ are parallel vectors, and for any test body we can write

$$\text{net } \mathbf{F} = \text{constant} \times \mathbf{a} \qquad (3\text{-}2)$$

Our next step is to quantify the concept of inertia. The constant in Eq. 3-2 is surely some-

how related to the inertia of the test body, for if the body has a large inertia, a large net force is needed to produce a given acceleration. Let us then define the constant in Eq. 3-2 as being proportional to the mass m of the body. That is,

$$\text{net } \mathbf{F} \propto m\mathbf{a} \qquad (3\text{-}3)$$

where we use a proportion (symbolized by $\propto$) rather than an equality, since we have not yet established units for $\mathbf{F}$ and m. Mass is a quantitative measure of inertia, and by our definition the masses of two bodies can be compared by acceleration experiments. For example, suppose we wish to find the value of the mass of an unknown object in relation to the mass of some standard object, such as that of the international prototype kilogram. Our experiment consists of applying a given force (any force) to each mass in turn and observing the accelerations produced. If the force gives the standard mass an acceleration of 3 m/s^2, and later it is found that the same force gives the unknown mass an acceleration of 10 m/s^2, we use Eq. 3-3 to conclude that the unknown body has less inertia than the standard body; its mass is $\frac{3}{10}$ that of the standard body.

We have now seen how to compare two forces with each other and how to compare two masses with each other, although we have said nothing about the units in which force and mass might be measured. Placing the emphasis on acceleration, we can rewrite Eq. 3-3 as

$$\mathbf{a} \propto \frac{\text{net } \mathbf{F}}{m} \qquad (3\text{-}4)$$

In words, we have

Newton's second law of motion:

The acceleration produced by an unbalanced force acting on a body is proportional to the magnitude of the net force, in the same direction as the force, and inversely proportional to the mass of the body.

In Eq. 3-4, $\mathbf{a}$ and net $\mathbf{F}$ represent vectors that are parallel to each other, and m is a scalar quantity.

You are earnestly advised to learn why we take the trouble to write the word "net" in front of the force in Newton's second law. Failure to write (or worse yet, to think) *net* force is the most prevalent single source of error in problem solving. For example, if a railroad engine pulls with a force of 20 tons in a forward direction and there is a retarding frictional force of 3 tons acting toward the rear of the train, the net force is a vector sum, or resultant, of four forces. The downward pull of gravity is balanced by the upward push of the tracks on the train, and the remaining two forces act in the same line. Therefore, ordinary algebraic addition can be used: $20 + (-3) = 17$ tons. The use of the word "net" or the summation symbol Σ will help remind you that you must look carefully for all the forces on a body, not just the obvious one. Many simple problems apparently involving only one force have been in fact highly simplified by neglecting friction. This is permissible, but we should know what we are neglecting. Then, too, in many problems the weight of a body is balanced out by an upward push of floor, table, or earth, as in the case just discussed. We tend to forget that these forces are acting on the body, since they cancel out in so many cases.

Let us look for a moment at the relationship between Newton's first and second laws. In the proportion $\mathbf{a} \propto$ (net $\mathbf{F}$)$/m$, if we substitute 0 for the net force, we get an acceleration equal to 0. Thus Newton's second law predicts no change in motion if no net force acts. But this is Newton's first law. Hence we say that the first law is just a *special case* of the second law; it is obtained from the second law by using a special value of the net force, namely zero.

3-4 Newton's Third Law— Action and Reaction

Forces occur in pairs. Everyone who has played tug of war knows from experience that he or she can exert force only if there are people on the other end of the rope. Replace the opposing team by a post fixed in the ground and we have a one-person tug of war. If a boy exerts force on molecules at his end of a rope, the rope molecules pull back on him. Facts such as these are summarized in

Newton's third law of motion:

Whenever one body exerts a force on a second body, the second body exerts a force on the first body; these forces are equal in magnitude and oppositely directed.

Newton called these paired forces *action* and *reaction*. It is essential to realize that the two forces mentioned in Newton's third law always act on different bodies and hence cannot add up to zero. *Only if forces act on the same body can they be combined into a single net force on that body.*

Pairs of forces (action-reaction pairs) are all around us, but sometimes the reaction force is not obvious. Any action-reaction pair can be described in the form "A acts on B and B acts on A." If a book rests on a table, the book presses down *on the table* and the table pushes up *on the book*. When a strong wind blows against a wall, molecules of air push against the wall and the wall pushes back against the molecules of air.

A more complicated (and subtle) example is that of a ball hanging by a rope from the ceiling. The ball weighs 50 N [the force unit *newton* (N) is defined in the next section]. Very many molecular action-reaction pairs are involved here, but if the rope is not accelerated, or if it has negligible mass, the net effect is that the rope "transmits" the force of the ball to the ceiling and vice versa. We say that there are two complete action-reaction pairs involved here, a "tension" pair and a "weight" pair (four forces in all, each equal to 50 N). In Fig. 3-1, $\mathbf{T} = -\mathbf{T}'$ and $\mathbf{W} = -\mathbf{W}'$, in accordance with Newton's third law. (Test them by reading the description of $\mathbf{T}$ backward, and see if it properly describes $\mathbf{T}'$.) However, in spite of the fact that $\mathbf{T}$ and $\mathbf{W}$ are equal and opposite, they do not constitute an action-reaction pair. Bear in mind that action

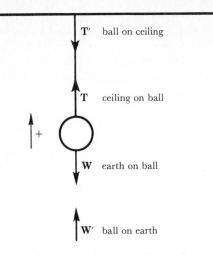

Figure 3-1 Action-reaction pairs.

and reaction always act on different bodies, whereas **T** and **W** both act on the ball. Hence they must be equal and opposite for some other reason, not the third law. This reason is Newton's second law: The ball is in equilibrium, its acceleration is zero, and hence the vector sum of the forces acting *on the ball* must be zero. Only two forces, **T** and **W**, act on the ball, and hence **T** + **W** = 0, or **T** = −**W**. This has nothing whatever to do with the third law. If now we pull down slightly on the ball, the rope stretches slightly, and **T** (hence also **T′**) increases, say to 52 N. Releasing the ball, we find that it is no longer in equilibrium, since net **F** = −50 + 52 = 2 N (upward). But even though the ball is not in equilibrium, Newton's third law still holds. **T** = 52 N, **T′** = −52 N; **W** = −50 N, **W′** = 50 N. If we cut the rope, so that the ball is freely falling, it still weighs 50 N, and it still exerts an upward force of 50 N on the earth.

It will help you identify pairs of action-reaction forces if you cast the description of each force in the form "_____ acting on _____." Thus, in the example of the ball hanging from the ceiling, the force **W** is that of *earth acting on ball*, and its reaction force **W′** is the force of *ball acting on earth*. This is in the standard form, "*A* acts on *B*, and *B* acts on *A*."

Newton's third law is an important statement about the nature of force. In general terms, a force is an *interaction* between two bodies. Whatever the nature of the interaction—whether it is gravitational, electrostatic, magnetic, nuclear, or some other type of interaction—we must recognize that it takes two bodies to make a force. For example, during a tug of war, short-range electric forces act between neighboring atoms of the rope. Such forces, called elastic forces in Chap. 9, occur as action-reaction pairs. Action and reaction are simply two aspects of the same force—two sides of the same coin, so to speak. It is impossible to have one without the other. The word "*interaction*" describes this situation nicely. Either force could equally well be called the action, and the other force would then be the reaction.

3-5 *Units for Force and Mass*

To be useful, Newton's second law must be made into an equation. We can do this by writing Eq. 3-3 as net **F** = *K*m**a**, where *K* is a constant of proportionality. You will be relieved to know that it is possible to make this constant *K* equal to 1, by suitable choices of force units described below. Therefore we will write Newton's second law as

$$\text{net } \mathbf{F} = m\mathbf{a} \tag{3-5}$$

In many applications of Newton's second law, motion is in a straight line, and the vector directions can be taken care of by use of + and − signs. In straight-line motion, therefore, we can write Newton's second law as an equation between the magnitudes of force and acceleration:

$$\text{net } F = ma \tag{3-6}$$

We shall use this equation, in which *K* = 1, to define the units of force in the two systems of metric units that we use.

In the **mks system,** based on the meter, kilogram, and second, the newly defined SI force unit is the *newton* (N). *A newton is that force which will give a mass of one kilogram an acceleration of one meter per second per second.*

To gain a feeling for this unfamiliar force unit, the newton, let us imagine a 1-kg ball falling freely, acted on only by the force of gravity (its weight). We assume negligible air friction. The ball has a mass of 1 kg. Experiment shows that the acceleration of this freely falling body is 9.8 m/s². Hence

$$\text{net } F = ma$$
$$W - 0 = (1 \text{ kg})(9.8 \text{ m/s}^2)$$
$$W = 9.8 \text{ kg} \cdot \text{m/s}^2 = \boxed{9.8 \text{ N}}$$

We see that the weight of the ball is 9.8 N.*

The dimensional equation for force is found by substituting dimensions for mass and acceleration into Newton's second law:

$$\text{net } F = ma$$
$$[F] = [M][LT^{-2}]$$

Thus $[F] = [MLT^{-2}]$ for any system of units. In particular, for the mks system, we have just seen that the collection of units "kg·m/s²" is equivalent to the single unit "newton." Since we know that a 1-kg ball weighs about 2.2 lb, our argument shows that 9.8 N is about 2.2 lb, or

---

* We do not say that "1 kg equals 9.8 N." This is incorrect, since the kilogram is a unit of mass and the newton is a unit of force. However, it *is* correct to say that a body whose mass is 1 kg weighs 9.8 N on the earth.

1 N is about 0.23 lb. Typical example: a quarter-pound stick of butter weighs about one newton.

Example 3-1 illustrates the use of mks units in solving a simple problem in dynamics with the help of Newton's second law.

Example 3-1

A car having a mass of 1000 kg comes to a stop in 40 m. If the initial speed was 20 m/s, what average stopping force (assumed to be constant) was supplied by the road acting on the car?

First we assume the positive direction to be toward the right (Fig. 3-2). Then, using the equations of kinematics (Chap. 2), we find the acceleration:

$v_0 = +20$ m/s
$v = 0$
$s = +40$ m
$a = ?$

$$v^2 = v_0^2 + 2as$$
$$a = \frac{v^2 - v_0^2}{2s}$$
$$= \frac{0 - (20 \text{ m/s})^2}{2(+40 \text{ m})}$$
$$= -5 \text{ m/s}^2$$

Now, using Newton's second law, we find the total backward force of friction f exerted by the road on the car (Fig. 3-2):

$$\text{net } F = ma$$
$$f = (1000 \text{ kg})(-5 \text{ m/s}^2)$$
$$= \boxed{-5000 \text{ N}}$$

The answer appears as a negative force, which indicates that it acts in the negative direction, or backward. If we had chosen the positive direction

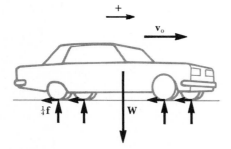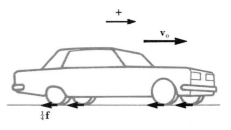

Figure 3-2

to be toward the left, the force would have come out positive, but the acceleration would also have been positive. The *physical meaning* of the answer is the same regardless of which direction is chosen as positive.

The first diagram in Fig. 3-2 shows the downward force of gravity (the weight **W** of the car) and also the upward forces exerted on the four wheels of the car by the pavement. The sum of the four upward forces exerted by the pavement is equal to the downward force **W**, and so all the vertical forces have a vector sum equal to zero. The second diagram, simplified by omitting these vertical forces, which cancel out, shows only the force of friction, $\frac{1}{4}\mathbf{f}$ on each wheel. In either diagram, the net force is **f**, and it is **f**, whose magnitude is f, that is used in applying Newton's second law. (The simplifying assumptions are made that $\frac{1}{4}$ of the weight is borne by each wheel, and the total backward force of friction is divided equally among the four wheels.)

Example 3-2

During a rescue at sea, a man of mass 70 kg dangles on the end of a rope attached to a helicopter (Fig. 3-3). If the helicopter accelerates upward at

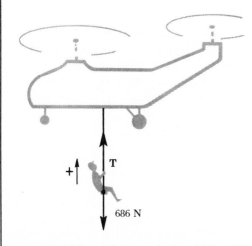

Figure 3-3 The net force on the man is $T - 686\ N$.

400 cm/s^2, what tension must the rope be able to withstand?

Before setting up an equation, we decide to use mks units. We find the weight of the man in newtons by multiplying his mass by 9.8, and we change the acceleration to meters per second per second by dividing by 100. Then, from the second law,

$$\text{net } F = ma$$
$$T - (70\text{ kg})(9.8\text{ m/s}^2) = (70\text{ kg})(+4.00\text{ m/s}^2)$$

$$T = \boxed{966\text{ N}}$$

This force is greater than the weight of the man (which is 686 N)—an expected result, since there would have to be a net upward force on the man in order to cause his upward acceleration.

In this example, note that by Newton's third law the downward force of the man on the rope is equal to the upward force of the rope on the man. Hence the man must be able to grip the rope with a force of 966 N, which is considerably greater than his own weight.

In the **cgs system,** based upon the centimeter, gram, and second, the *dyne* (dyn) is the newly defined force unit. *A dyne is that force which will give a mass of one gram an acceleration of one centimeter per second per second.*

The dyne is a very small unit of force. A large pea has a mass of about a gram. If the pea is allowed to fall freely, acted on only by gravity, the only appreciable force acting will be the weight W of the pea.

$$\text{net } F = ma$$
$$W - 0 = (1\text{ g})(980\text{ cm/s}^2)$$

$$W = 980\text{ g}\cdot\text{cm/s}^2 = \boxed{980\text{ dyn}}$$

Thus a pea weighs almost a thousand dynes! In fact, a good-sized mosquito weighs about a dyne. Our discussion also shows that the collection of unit "g·cm/s^2" is equivalent to the single unit "dyne." From its definition, we calculate the SI equivalent of the dyne:

$$1\text{ dyn} = (10^{-3}\text{ kg})(10^{-2}\text{ m/s}^2) = \boxed{10^{-5}\text{ N}}$$

Thus a dyne is 10 μN.

The cgs system of units was formerly used for almost all physical work in the metric system, but it is gradually being replaced by the mks system. However, cgs units are sometimes still used where very small forces are involved, as in our next example.

Example 3-3

In an approximate model of mammalian heart action, the speed of about 20 cm³ of blood increases from 30 cm/s to 40 cm/s during a portion of the heartbeat 0.08 s long. What force is exerted by the heart muscle?

We find in Table 1-3 that the density of blood is 1.1 g/cm³. Therefore the mass of blood is

$$m = (20 \text{ cm}^3)(1.1 \text{ g/cm}^3) = 22 \text{ g}$$

The average acceleration is

$$a = (40 \text{ cm/s} - 30 \text{ cm/s})/0.08 \text{ s} = 125 \text{ cm/s}^2$$

$$\text{net } F = ma$$

$$= (22 \text{ g})(125 \text{ cm/s}^2) = \boxed{2750 \text{ dyn}}$$

In SI this force could be expressed as 27.5 mN.

Force and mass have different dimensions and can never have the same units. Consistent sets of units for use in Newton's second law may be summarized as follows:

$$\text{net } F = ma$$
$$\text{N} = (\text{kg})(\text{m/s}^2)$$
$$\text{dyn} = (\text{g})(\text{cm/s}^2)$$

In the metric system, these units, and these only, should be used in Newton's second law or in any equation derived from the second law. However, in common language people *do* say "the mouse weighed 250 g at the start of the experiment" or "a weight of 50 kg is applied at one end of a lever." These statements are wrong, and you should guard against such looseness of expression. The mouse had a *mass* (inertia) of 250 g, and its weight, a force, was

$$250 \times 980 = 245,000 \text{ dyn}$$

A Formula for Weight. Imagine a body of mass m that is acted on only by the force of gravity (its weight, a downward force **W**). Such a body is a freely falling body, and its acceleration is **g**, a downward vector. We now apply Newton's second law to this hypothetical situation in order to derive a formula for the weight of the body. Substituting into

$$\text{net } \mathbf{F} = m\mathbf{a}$$

gives

$$\mathbf{W} = m\mathbf{g} \qquad (3\text{-}7)$$

Although derived by a "thought experiment," the result is general. In any system of units, the magnitude of the weight of a body is given by $W = mg$. For instance, the weight of a 10-kg turkey is $10 \times 9.8 = 98$ N; a 5-g Ping-Pong ball weighs $5 \times 980 = 4900$ dyn.

In any system of units, we have shown that the weight of a body is given by $W = mg$. Thus an object of mass 1 kg weighs 9.8 N (on the earth). If this equation is solved for the mass, we obtain

$$m = \frac{W}{g} \qquad (3\text{-}8)$$

Newton's second law then becomes

$$\text{net } F = \left(\frac{W}{g}\right) a \qquad (3\text{-}9)$$

In any equation where mass m occurs, you can substitute (W/g) for m. Often the numerical value of the mass is not required, and questions about the acceleration of a body can be answered by use of the equation $m = W/g$, as in the following examples.

Example 3-4

A man whose weight is 800 N is on ice skates and is pushed horizontally by the wind with a net force of 80 N. What is his horizontal acceleration?

Since the weight of the man is given, we will use W/g for the mass.

$$\text{net } F = \left(\frac{W}{g}\right)a$$

$$80 \text{ N} = \left(\frac{800 \text{ N}}{9.8 \text{ m/s}^2}\right)a$$

$$a = \left(\frac{80 \text{ N}}{800 \text{ N}}\right)(9.8 \text{ m/s}^2) = \boxed{0.98 \text{ m/s}^2}$$

This result is in agreement with intuition—note that the man is pushed horizontally by a force numerically equal to $\frac{1}{10}$ his weight, and his acceleration is numerically $\frac{1}{10}$ as much as if he were freely falling.

Example 3-5

What is the horizontal force on a bowling ball that weighs 70 N if the bowler gives it a velocity of 5.6 m/s in 0.40 s?

The acceleration is found first:

$$
\begin{aligned}
&v = 5.6 \text{ m/s} & v &= v_0 + at \\
&v_0 = 0 & & \\
&a = ? & a &= \frac{v - v_0}{t} \\
&t = 0.40 \text{ s} & & \\
& & &= \frac{5.6 \text{ m/s} - 0}{0.40 \text{ s}} \\
& & &= 14 \text{ m/s}^2
\end{aligned}
$$

We now use Eq. 3-9:

$$\text{net } F = \left(\frac{W}{g}\right)a$$

$$= \left(\frac{70 \text{ N}}{9.8 \text{ m/s}^2}\right)(14 \text{ m/s}^2)$$

$$= \boxed{100 \text{ N}}$$

In this case, the net force is *numerically equal to* $\frac{100}{70}$ the weight of the ball, because the acceleration is *numerically equal to* $\frac{100}{70} g$. The net force on the ball is, of course, horizontal, since the acceleration is horizontal.

These problems have nothing to do with falling bodies, even though g appears in the equation for mass. Our next example shows how we can calculate the acceleration of a body whose weight is zero. Bear in mind that when W/g is used for mass, it is the weight of the body *on earth* that is divided by the value of g measured *on earth*.

Example 3-6

In a region far removed from the earth's gravitational attraction, workers are assembling a space station. A section that weighed 70,000 N (about 8 tons) on earth is floating through space at a speed of 0.3 m/s toward a worker who is 1 m in front of the very massive section of the station that has already been assembled (Fig. 3-4). If the worker can exert a force of 600 N, should she try to stop the oncoming section, or should she quickly move out of the way?

First, we find the acceleration of the moving section.

$$\text{net } F = \left(\frac{W}{g}\right)a$$

whence

$$a = \frac{(\text{net } F)\,g}{W} = \frac{(-600 \text{ N})(9.8 \text{ m/s}^2)}{70{,}000 \text{ N}}$$

$$= -0.084 \text{ m/s}^2$$

While the section is being brought to a stop, it travels a distance

$$s = \frac{v^2 - v_0{}^2}{2a} = \frac{0 - (0.3 \text{ m/s})^2}{2(-0.084 \text{ m/s}^2)} = \boxed{+0.54 \text{ m}}$$

The worker has no need to worry.

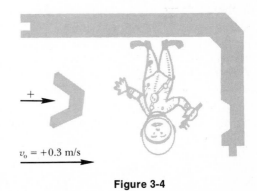

$v_0 = +0.3 \text{ m/s}$

Figure 3-4

3-6 Isolation of Bodies in Problem Solving

Newton's second law applies to any complete system or to any part of a system of bodies. When solving any specific problem, it is essential that you decide clearly and definitely just what body or system of bodies is being considered. Obviously, if all forces on a system are considered, the total mass of the system must be used; if forces on only part of the system are considered, then the mass of only that part must be used. The technique of isolating (in the mind's eye) a part of a system is called the *free-body method;* by ignoring all forces and masses outside the selected part, that part becomes like a body free of other forces not acting directly on it. Let us illustrate this point by considering a typical pulley system known as Atwood's machine.

_____ **Example 3-7**

A light cord connecting objects of mass 10 kg and 6 kg passes over a light, frictionless pulley (Fig. 3-5). (a) What is the acceleration of the system? (b) What is the tension in the cord?

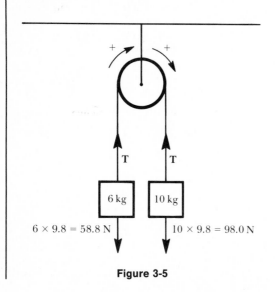

Figure 3-5

First, we consider the system as a whole; that is, we take the free body to be the two masses and the (massless) cord connecting them. The total mass that must be set into motion is 16 kg. Two external forces act on the system: $+(10 \text{ kg})(9.8 \text{ m/s}^2) = +98.0 \text{ N}$ and $-(6 \text{ kg})(9.8 \text{ m/s}^2) = -58.8 \text{ N}$, where the signs are in accordance with the assumed positive direction shown in the diagram. The net force on the system is $98.0 \text{ N} - 58.8 \text{ N} = 39.2 \text{ N}$. The tension is an internal force, acting between parts of the system, and does not enter into the calculation of the net force *on* the system.

$$\text{net } F = ma$$
$$98.0 \text{ N} - 58.8 \text{ N} = (16 \text{ kg})(a)$$

$$a = \frac{39.2 \text{ N}}{16 \text{ kg}} = \boxed{2.45 \text{ m/s}^2}$$

Next, to find the tension, we must consider one or the other object by itself. To indicate this procedure, we draw a dotted line around the 10-kg object (Fig. 3-6a) and apply Newton's second law. We disregard everything outside the dotted line. From this point of view, we do not know whether the tension is caused by a 6-kg object on the other side of the

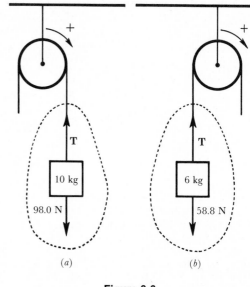

(a)   (b)

Figure 3-6

pulley, or by a spring attached to the cord, or perhaps, for all we know, by a little green man from Venus situated just outside the dotted line and pulling upward on the 10-kg object. The essential point is that we do not care *what* causes the tension; we merely wish to compute its value, whatever the cause. Applying the second law to the 10-kg object, we find

$$\text{net } F = ma$$
$$98.0 \text{ N} - T = (10 \text{ kg})(2.45 \text{ m/s}^2)$$
$$T = \boxed{73.5 \text{ N}}$$

As a check, we can compute the tension by isolating the 6-kg object (Fig. 3-6b).

$$\text{net } F = ma$$
$$T - 58.8 \text{ N} = (6 \text{ kg})(2.45 \text{ m/s}^2)$$
$$T = \boxed{73.5 \text{ N}}$$

Let us look at some of the features of the solution to Example 3-7. (1) It should scarcely be surprising that the tension in the cord on the left side of the pulley is numerically the same as that on the right side of the pulley. According to Newton's third law, the force of the 6-kg object on the 10-kg object is equal and opposite to the force of the 10-kg object on the 6-kg object. The cord merely serves to transmit this force.* (2) In Example 3-7, the acceleration is $\frac{1}{4}g$, and each object receives a *net* force of $\frac{1}{4}$ its weight. Thus the 10-kg object experiences a net downward force of $98.0 \text{ N} - 73.5 \text{ N} = 24.5 \text{ N}$, which is $\frac{1}{4}$ of its weight. Likewise, the 6-kg object experiences a net upward force of $73.5 \text{ N} - 58.8 \text{ N} = 14.7 \text{ N}$, which is numerically equal to $\frac{1}{4}$ of *its* weight, al-

---

* If the cord had mass, then the tension would not be constant, since each particle of the cord would require a net force to accelerate it. If the pulley had mass, the tensions on the two sides of the pulley would have to be different, in order to accelerate the particles of the pulley. In this text, unless stated otherwise, all connecting cords and pulleys have negligible mass, and under these circumstances the tension is uniform.

though directed upward, as it must be to cause upward acceleration. (3) An alternative solution to the problem is to use two unknowns, the acceleration a and the tension T, and set up two equations, one for each object:

$$98.0 \text{ N} - T = (10 \text{ kg})(a)$$
$$T - 58.8 \text{ N} = (6 \text{ kg})(a)$$

Solving these equations simultaneously[†] gives $a = 2.45 \text{ m/s}^2$ and $T = 73.5 \text{ N}$.

Example 3-8

A tractor of mass 1000 kg is attached by means of a horizontal, massless chain to a log whose mass is 400 kg. The tension in the chain is 2000 N, and the backward force of friction exerted by the ground on the log is 800 N. How far will the log move in 2 s, starting from rest?

Our problem is really two problems: We compute the acceleration, using Newton's second law; then we have a straightforward problem in kinematics to find the displacement.

First we make a diagram (Fig. 3-7), showing on it all the known forces and masses, as well as other forces whose values we do not know. Then we isolate one of the bodies for consideration; we choose the log for this purpose, since all the horizontal forces on it are known. We also choose a positive direction and indicate it on the diagram. Now we apply Newton's second law to the log:

$$\text{net } F = ma$$
$$2000 \text{ N} - 800 \text{ N} = (400 \text{ kg})(a)$$
$$a = 1200 \text{ N}/400 \text{ kg} = +3 \text{ m/s}^2$$

To find the displacement, we use the equations of kinematics:

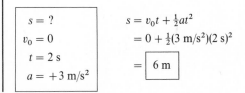

$s = ?$	$s = v_0 t + \frac{1}{2}at^2$
$v_0 = 0$	$= 0 + \frac{1}{2}(3 \text{ m/s}^2)(2 \text{ s})^2$
$t = 2 \text{ s}$	$= \boxed{6 \text{ m}}$
$a = +3 \text{ m/s}^2$	

[†] See Sec. B-2 of the Mathematical Review in the Appendix.

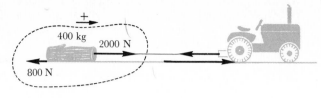

400 kg

2000 N

800 N

Figure 3-7

Note that the mass of the tractor is not used in this problem, since we are considering the *log's* motion. Similarly, the forward force of the ground on the tractor is not used, since this force does not act on the log.

Summary Newton's first law serves as a qualitative definition of the concepts of force and inertia. Mass is a quantitative measure of inertia—that property of matter which causes it to resist linear acceleration. Newton's second law, net $F = ma$, tells us how much unbalanced or net force is required to give acceleration to an object having mass. This kind of mass is called inertial mass. Newton's third law states that forces occur in pairs. Action and reaction always act on different bodies.

Problem solving requires a clear understanding of exactly which body or system of bodies is considered in Newton's second law. This body or system must be carefully isolated, and *all* forces acting *on* the body must be considered. Forces exerted *by* the body or system are to be ignored. Consistent sets of units must be used when applying Newton's second law. In the mks system, the units are the newton, kilogram, and meter per second per second. In the cgs system, the acceptable units are the dyne, gram, and centimeter per second per second. Force is not mass, and mass is not force; these quantities cannot have the same units in an equation. The equation $m = W/g$ can be used when solving problems in which the weight is given.

Check List

Newton's first law of motion
Newton's second law of motion
Newton's third law of motion

newton
dyne
free-body diagram

Questions

3-1 Is the following statement true or false? "When a book is at rest lying on a table, the downward force of gravity on the book is equal and opposite to the upward force of the table on the book. This is an example of Newton's third law." Explain your answer.

3-2 Two sleds, *A* and *B*, are connected by a light but stiff spring and then separated far enough apart, on smooth ice, so that the spring is stretched somewhat. The sleds and their passengers are then released, and the spring contracts, pulling the unequally loaded sleds together. During the motion of *A* toward *B* and *B* toward *A*, (*a*) are the forces on the sleds equal and opposite? If so, is this an example of Newton's third law? (*b*) Do the sleds have equal and opposite accelerations? (*c*) Do the sleds acquire equal speeds? (*d*) How can the mass ratio of the two sleds be found?

3-3 A horse is pulling a heavy cart, and the cart (and horse) are both being accelerated. Is the force of the cart on the horse equal and opposite to that of the horse on the cart? If your answer is "yes," explain how it is that these equal and opposite forces give rise to the "net force" that is necessary to cause acceleration. If your answer is "no," explain whether Newton's third law is true while the cart is accelerating.

3-4 In Ques. 3-3, explain how the horse becomes accelerated. What is the origin of the "net force" on him? In which direction is the net force on the horse?

3-5 Which of the following units are suitable for expressing the mass of a body? Which are suitable for weight? Kilogram, pound, dyne, gram, metric ton, newton, ounce.

3-6 If Galileo dropped two objects of unequal weight from the Leaning Tower, why is it that the heavier one did not reach ground first, since its greater weight would be expected to cause it to be accelerated more rapidly?

3-7 A freight train of 100 cars, each of mass 10 tons, is held together by couplings between the cars. Ignore friction as far as the cars are concerned. Is the tension in the coupling between the third and the fourth cars the same as the tension in the coupling between the thirty-third and the thirty-fourth cars? Answer this question for two conditions: (*a*) if the train has constant velocity; (*b*) if the train is accelerating.

3-8 In Example 3-8, where is the reaction force that, by Newton's third law, is equal and opposite to the 800 N force of friction? Express your answer in the form "_____acting on_____."

3-9 An iron ball is hanging by a thread, and a similar thread is attached to the lower side of the ball and hangs loosely. If the loose thread is given a sharp jerk, it may break, whereas if it is given a slow and steady pull, the top thread breaks. Explain.

3-10 A light rope hangs over a frictionless pulley, and a 15-kg monkey is counterbalanced by a 15-kg mirror that is just opposite the monkey (Fig. 3-8). As the monkey climbs up and down his side of the rope, he cannot escape his reflection, since the mirror is always just opposite him. Explain. [After thorough discussion of this problem, you may wish to verify your reasoning by reference to *Am. J. Phys.* **16**, 248 (1948) and **16**, 320 (1948).]

MULTIPLE CHOICE

3-11 The dimensions of force are (*a*) $[ML^{-1}T]$; (*b*) $[ML^2T^{-1}]$; (*c*) $[MLT^{-2}]$.

3-12 A net force of 1 N acts for 1 s on a mass of 1 kg that is initially at rest. The displacement of the mass is (*a*) 0.5 m; (*b*) 1.0 m; (*c*) 9.8 m.

3-13 A man and a boy, initially at rest on frictionless ice, push each other apart. After a short time, which skater is farther from the starting point? (*a*) the man; (*b*) the boy; (*c*) neither.

Figure 3-8

3-14 While a body is freely falling, Newton's third law is (*a*) exactly true; (*b*) approximately true, if air resistance can be neglected; (*c*) untrue until the falling body comes to rest on the earth's surface.

3-15 In a free-body diagram, the force on the selected body is always (*a*) 0; (*b*) *mg*; (*c*) the resultant force.

3-16 Action and reaction act (*a*) on two different bodies; (*b*) in opposite directions; (*c*) both of these.

Problems **3-A1** What is the net force acting on a spider whose weight is 9000 dynes, if he is partially supported by a strand of silk that exerts a force of 7000 dynes?

3-A2 What is your own mass, in kilograms? What is your own weight, in newtons?

3-A3 During reentry of a 1000-kg satellite, it is acted on by air resistance that supplies an upward force of 20 kN (kilonewtons). What is the net force on the satellite?

3-A4 A body of mass 4 kg is acted on by a net force of 20 N. What acceleration is produced?

3-A5 A force of 400 dyn acting on a body produces an acceleration of 50 cm/s^2. What is the mass of the body?

3-A6 What is the force, in newtons, required to give a ball of mass 3.5 kg an acceleration of 4 m/s^2?

3-A7 What net force is needed to give a mass of 110 g an acceleration of 30 cm/s^2?

3-A8 A child can exert a forward push of 30 N. What acceleration can she give to a loaded go-cart that weighs 98 N?

3-A9 A sled of mass 24 kg coasts over the ice with a backward acceleration of 0.5 m/s^2. What is the retarding force of friction?

3-B1 A falling coconut has a mass of 1.5 kg, and the upward force of air resistance is 8.7 N. Calculate the magnitude and direction of the acceleration of the coconut.

3-B2 A parachutist who weighs 800 N is partially supported by a 600-N upward force of air resistance. What is the acceleration of the parachutist?

3-B3 An elevator weighs 39.2 kN (kilonewtons), and the upward tension in the supporting cable is 34.4 kN. (a) What is the acceleration? (b) Starting from rest, how far will the elevator move (in which direction) in 2 s?

3-B4 A truck of mass 2000 kg is moving at 20 m/s and is acted on by two forces (assumed constant): the forward force of 800 N due to the engine, and a 200-N retarding force of friction. (a) At what rate is the truck gaining speed? (b) How far will it travel in 8 s?

3-B5 (a) Assuming constant negative acceleration, how far does a supertanker of mass 200,000 metric tons travel while coasting to a stop from an initial speed of 30 km/h (16 knots) in 21 min? (b) What force is supplied by fluid friction of the water while the tanker is stopping?

3-B6 A car of mass 1000 kg traveling at 10 m/s is 15 m from the start of an intersection 55 m wide when the traffic light changes from green to yellow; 6 s later the light changes to red. Can the driver avoid being in the intersection on a red light? Available forces are 3000 N (in a panic stop) and 1000 N (for forward acceleration).

3-B7 A golf ball of mass 60 g is struck by a golf club and acquires a speed of 80 m/s during the impact, which lasts for 2×10^{-4} s. What force (assumed constant) is exerted on the ball?

3-B8 After falling from a height of 30 m, a ball of mass 0.5 kg rebounded upward, leaving the ground with 75% of the speed with which it struck. If the contact between ball and ground lasted for 0.002 s, what average force was exerted on the ball?

3-B9 A rifle bullet of mass 10 g has a muzzle velocity of 720 m/s, and the length of the rifle barrel is 80 cm. What is the net force accelerating the bullet, assuming it to be constant?

3-B10 A basketball player wishes to pass the 0.62-kg ball to a teammate who is 5.0 m away, measured horizontally. She draws her arm back to a position 5.8 m from the teammate and releases the ball after pushing it forward 0.8 m. (a) What force must she apply to the ball if it is to reach the teammate 0.2 s after release? (b) How far does the ball "droop" down during its trajectory?

3-B11 A baseball of mass 160 g moving horizontally at 30 m/s toward the batter is struck by the bat and leaves in the opposite direction at 42 m/s. The contact between the ball and the bat lasted for 1 ms $(10^{-3}$ s). Calculate (a) the average force of the bat on the ball; (b) the average force of the ball on the bat. Express your answers in kilonewtons (kN).

3-B12 When an arrow of mass 0.07 kg is shot from a bow, it experiences an average force of 210 N applied through a distance of 0.6 m. (a) With what speed does it leave the bow? (b) How high would the arrow go if it were fired straight up? (Ignore air resistance.)

3-B13 During a test of seat belts, a car of mass 1.1 metric tons traveling at 12 m/s collides with a heavy truck. The car moves 4 m forward while being brought to rest. (a) What force (assumed constant) is exerted on the car by the truck? (b) What force (assumed constant) is exerted on a 90-kg driver by the seat belt?

3-B14 A person wearing seat belts and shoulder harness is in a car that is struck from behind, causing a forward acceleration of 1215 m/s^2. The person's head, of mass 3.4 kg, must, if it is to follow the body, be acted on by a large forward force exerted by neck muscles and the upper spine. Calculate the magnitude of the force involved in this "whiplash" accident. (*Note:* The same event is studied in a different frame of reference in Chap. 5, where Galilean relativity is discussed; see Fig. 5-11 on page 119.)

3-B15 A 75-kg aviator in "free fall" acquires a velocity of 50 m/s and then opens his parachute. After falling an additional 20 m, his velocity has been reduced to 8 m/s. (a) What is the average acceleration of the aviator while his fall is being checked? (b) What is the average force exerted by the parachute?

3-B16 A woman of mass 50 kg wishes to escape a burning building by sliding down an improvised rope made of nylon stockings tied together. The maximum upward force that the stockings can exert without tearing is 300 N. (a) Can the woman slide down at constant speed? (b) What is the least acceleration with which the woman can slide down the rope?

3-B17 A vertical rope is attached to a trunk of mass 25 kg. What tension in the rope is needed to cause the trunk to acquire an upward velocity of 3 m/s in 0.6 s?

3-B18 A 20-kg sled is coasting with constant velocity at 5 m/s over perfectly smooth, level ice. It enters a rough stretch of ice 20 m long in which the force of friction is 8 N. With what speed does the sled emerge from the rough stretch?

3-B19 A 5-kg iron ball is jerked upward by a rope; it rises 20 cm in 0.1 s. Assuming uniform acceleration, calculate the tension in the rope.

3-B20 A rocket of mass 50 metric tons is acted on by an upward thrust of 550 kN. If the rocket is 30 m tall, how long a time is required for it to rise off the launching pad a distance equal to its own height?

3-B21 An elevator of mass 1000 kg is supported by a cable that can sustain a force of 15,000 N. What is the greatest upward acceleration that can be given the elevator without breaking the cable?

3-B22 A tractor of mass 800 kg is attached by a horizontal massless chain to a log whose mass is 400 kg. The backward force of friction exerted by the ground on the log is 600 N. The system has a forward acceleration of 2 m/s^2. Calculate (a) the tension in the chain; (b) the forward force of the ground on the tractor.

3-B23 A horse of mass 600 kg is setting into motion a loaded stoneboat of mass 400 kg (Fig. 3-9). The force of friction is 500 N. (a) What forward force F is exerted on the horse by the ground if the system has a forward acceleration of 2 m/s^2? (*Hint:* Consider the

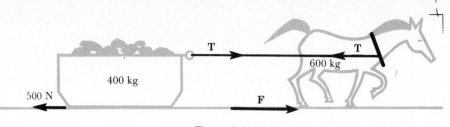

Figure 3-9

whole system in solving this part of the problem.) (*b*) What is the tension in the connecting rope? (*Hint:* Consider either the horse *or* the stoneboat for this part; as a check, do it both ways.)

3-B24 Objects of mass 3 kg and 2 kg are connected by a light cord that passes over a horizontal frictionless rod. (*a*) What is the acceleration of the system? (*b*) What is the tension in the cord on the 3-kg side? (*c*) What is the tension in the cord on the 2-kg side?

3-B25 A bag of feed of mass 32 kg is being hoisted at a steady speed of 1 m/s by a farmer who uses a 17-kg counterweight and also supplies a downward force to a rope that passes over a frictionless pulley. (*a*) What force does the farmer apply to the rope? (*b*) If the farmer releases the rope, what will be the velocity of the feed bag after 2 s?

3-B26 In Fig. 3-10 the frictional force between the 30-kg block and the table is negligible. (*a*) If the peg is removed, what will be the acceleration of the system? (*b*) How long will it take the block to hit the pulley? (*c*) What is the tension in the cord while the block is moving? (*d*) What is the tension in the cord after the block ceases to move?

3-B27 Objects of mass 3 kg and 4 kg are connected by a light cord that passes over a frictionless pulley. (*a*) How long a time will be required for the 4-kg object to fall 4 m, starting from rest? (*b*) What is the tension in the cord on the 3-kg side? (*c*) What is the tension in the cord on the 4-kg side?

3-B28 A 100-kg track star, at the start of a sprint, pushed on the ground with a measured force 1800 N at an angle of 60°, as shown in Fig. 3-11. What forward acceleration was produced?

3-C1 The driver of a 500-kg sports car, heading directly for a railroad crossing 250 m away, applies the brakes in a panic stop. The car is moving (illegally) at 40 m/s, and the brakes

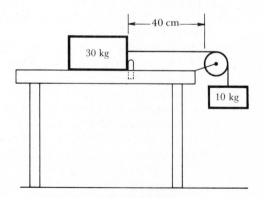

Figure 3-10

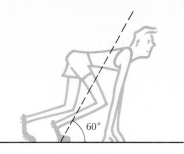

Figure 3-11

can supply a force of 1200 N. (*a*) How fast will the car be moving when it reaches the crossing? (*b*) Will the driver escape collision with a freight train that, at the instant the brakes are applied, is blocking the road and still requires 8.0 s to clear the crossing?

3-C2 Two boxes of mass 20 kg and 12 kg are in contact on a horizontal, frictionless surface. A horizontal force of 128 N is applied to the 20-kg box, pushing it against the other box. (*a*) Compute the acceleration of the system. (*b*) Compute the force with which the two boxes push against each other. (*c*) If the 128-N force is applied to the 12-kg box instead of the 20-kg box, find the acceleration and the force with which the boxes push against each other.

3-C3 Objects of mass 2 kg and 3 kg are connected by a light cord that passes over a frictionless pulley 5 m above the floor. Initially the masses are each at rest 2 m above the floor, with the cord taut. After the system is released, what maximum height above the floor does the 2-kg object reach?

3-C4 An open paint bucket of mass 12 kg is attached to another paint bucket of mass 9 kg by a light rope 21 m long. The light bucket is at rest on the ground, and the rope goes up and over a small frictionless pulley and is attached to the heavy bucket, which is on a window sill 9 m above the bucket on the ground. There is no slack in the rope. If the heavy bucket slips off the window sill, after what time will the first big splash occur when the buckets collide?

3-C5 Solve Problem 3-B26 under the assumption that the table is rough and a 38-N force of friction acts toward the left.

3-C6 Two small balls of equal mass are connected by a string 1 m long and are laid out on a smooth table top with one ball (*A*) just at the edge of the table and the other ball (*B*) 1 m from the edge. The table is 0.6 m high. Ball *A* is nudged gently over the edge of the table, and things begin to happen. (*a*) How much time is required for ball *B* to strike the floor? (*b*) How far from the base of the table does ball *B* strike?

3-C7 Three cars of a model train, each of mass 0.6 kg, are connected together and are pulled by an engine of mass 1.2 kg. The track supplies a forward force of 12 N on the engine, and there is a negligible frictional drag. (*a*) Calculate the acceleration of the system. (*b*) Calculate the tension in the coupling that connects the front car to the middle car.

3-C8 The system shown in Fig. 3-5 is held with the 6-kg block just opposite the 10-kg block, both blocks being 2 m above the floor and 3 m below the pulley. The 6-kg block is then given a downward initial velocity sufficient to cause it to just reach the floor. How long after the initial push will the blocks again be opposite each other?

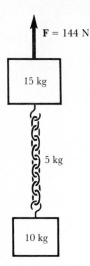

F = 144 N

15 kg

5 kg

10 kg

Figure 3-12

3-C9 The two blocks in Fig. 3-12 are connected by a heavy chain of mass 5 kg. An upward force of 144 N is applied to the top block. (a) How far and in which direction will the system move in 0.50 s, starting from rest? (b) What is the tension at the top link of the chain?

3-C10 A sack of feed weighing 300 N is hung from a rope that passes over a frictionless pulley to a man on the ground. If the breaking strength of the rope is 320 N, what is the shortest time in which the man can raise the bag 2 m, starting from rest?

3-C11 Bob and Joe, two construction workers on the roof of a building, are about to raise a keg of nails from the ground by means of a rope passing over a pulley 10 m above the ground. Bob weighs 900 N, Joe 600 N, the keg 300 N, and the nails 600 N. They slip off the roof, and the following unfortunate sequence of events takes place: Bob and Joe, hanging on the rope together, strike the ground just as the keg hits the pulley. Unnerved by his fall, Bob lets go of the rope, and the keg pulls Joe up to the roof where he cracks his head against the pulley but gamely hangs on. However, the nails spill out of the keg when it strikes ground, and the empty keg rises as Joe returns to the ground. Finally, Joe has had enough and lets go of the rope and remains on the ground, only to be hit by the empty keg again. Ignoring the possible mid-air collisions that merely add insult to injury, how long did it take this little drama to unfold?

References

1. I. B. Cohen, "Newton," *Sci. American* **193**(6), 73 (Dec. 1955). A biography.
2. J. W. N. Sullivan, *Isaac Newton, 1642–1727* (Macmillan, New York, 1938).
3. R. F. Metzdorf, "Sir Isaac Newton, 1642–1727—A Study of a Universal Mind," *Am. J. Phys.* **10**, 293 (1942).
4. N. R. Hanson, "History and Philosophy of Science in an Undergraduate Physics Course," *Physics Today* **8**(8), 4 (Aug. 1955). See especially remarks on pages 7 and 8 on the law of inertia.
5. C. G. Adler and B. L. Coulter, "Aristotle: Villain or Victim?" *Phys. Teach.* **13**, 35 (1975). A discussion of some of Aristotle's ideas on motion.
6. A. Franklin, "Principle of inertia in the Middle Ages," *Am. J. Phys.* **44**, 529 (1976).
7. A. Franklin, "Inertia in the Middle Ages," *Phys. Teach.* **16**, 201 (1978). Somewhat less detailed than Ref. 6.
8. C. G. Adler and B. L. Coulter, "Galileo and the Tower of Pisa experiment," *Am. J. Phys.* **46**, 199 (1978).
9. D. Sciama, "Inertia," *Sci. American* **196**(2), 99 (Feb. 1957). A defense of Mach's position that inertia depends on the interaction between any body and the rest of the universe rather than being a property of a body.
10. F. G. Karioris, "Inertia demonstration revisited," *Am. J. Phys.* **46**, 710 (1978). The first two sections give an elementary discussion of the ball-and-thread demonstration (see Question 3-9).
11. R. A. Serway, "Lecture demonstration of the 'bosun's chair,'" *Am. J. Phys.* **44**, 882 (1976). An application of free-body methods.
12. J. Stull, *Newton's First and Second Laws* (film).
13. J. Stull, *Newton's Third Law* (film).

4

Statics

4-1 Equilibrium

We have seen that Newton's laws of motion are adequate to determine the behavior of a body or a system of bodies, since the acceleration can be calculated from a knowledge of the net force and the mass on which it acts. In particular, we now ask what conditions are necessary to cause a body to have zero acceleration. Such a body is said to be in equilibrium, and the study of forces in equilibrium is called *statics*. We have seen that Newton's first law is a special case of Newton's second law; thus statics is a special case of dynamics, the case in which acceleration = 0. For the engineer who is concerned with the stability of structures and the safe loads that can be carried by girders, it is an important special case, and the problems of statics are of vital concern. For us, however, the important thing is to recognize that the study of statics presents nothing really new; it is merely a systematic application of one aspect of Newton's second law.

Equilibrium is the condition of a body whose velocity is constant in magnitude and direction.

This includes, of course, the case where the velocity is zero and remains so. If a body remains at rest, it is said to be in *static equilibrium*. If a body remains in steady motion in a straight line, it is said to be in *dynamic equilibrium*. These are *definitions* of equilibrium; it is one thing to define a condition, and another to state how to obtain that condition. In order to obtain equilibrium, we must see to it that no unbalanced force is applied to the body in question. It is rare indeed that *no* force acts on a body. One could approximate such a situation only in interstellar space, where the gravitational attraction of the stars would be negligibly small. Far more commonly, a number of forces are acting, but they are balanced to give a resultant force of zero.

We shall first consider the situation in which all the acting forces pass through a single point (the dimensionless "particle" of the philosophers and mathematicians). In general, Newton's second law says that net $\mathbf{F} = m\mathbf{a}$; if the acceleration is zero, then

$$\text{net } \mathbf{F} = m(0)$$

A truck that is being towed is in equilibrium when the resultant force and the resultant torque are each equal to zero.

which means that

$$\text{net } \mathbf{F} = 0$$

or

$$\sum \mathbf{F} = 0$$

This vector equation gives the condition for equilibrium of a particle. In words, a particle is in equilibrium if the vector sum of all forces acting on it is zero.

4-2 Center of Gravity

Actual physical objects are never mathematical points; for instance, a solid plastic phonograph record is composed of 10^{23} or so atoms, on each of which there is a downward force of gravity. A few of these parallel downward forces are shown in Fig. 4-1a. The *center of gravity* (c.g.) is defined as the point in the body through which a single downward force equal to the weight of the body may be considered to act. For the *uniform* disk in Fig. 4-1b, the center of gravity is seen to be at the geometrical center of the disk. In other words, if $\mathbf{w}_1, \mathbf{w}_2, \mathbf{w}_3, \ldots$ are the weights of the individual molecules and $\mathbf{W}$ is the total weight (equal to $\mathbf{w}_1 + \mathbf{w}_2 + \mathbf{w}_3 + \cdots$), then the many tiny forces $\mathbf{w}_1$ and the others are equivalent to the single force $\mathbf{W}$ acting through the center of gravity. For a uniform body of sufficient symmetry, such as a disk, sphere, cube, square, or rectangle, the c.g. is at the geometric center. (We

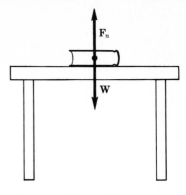

Figure 4-2

shall see in Sec. 4-6 how to compute the position of the c.g. of more complicated bodies.) This trick of replacing many parallel forces by a single force often makes it possible to apply the condition for equilibrium of a particle (net $\mathbf{F} = 0$) to an actual body whose c.g. is known.

A simple example will show how we tend to take these ideas for granted. Consider a book lying at rest on a table. We say that the downward force of gravity is balanced by the upward force of the table. Actually, the downward force of gravity is the sum of all the downward forces of gravity on the individual molecules; this "effective" or "resultant" force passes through the c.g. Likewise the upward force $\mathbf{F}_n$ of the table is the resultant of the upward forces exerted by many molecules of the table top pressing upward on the many places where the book is in contact with the table; the line of action of the resultant of all these upward forces also passes through the c.g. of the book. We can draw a picture (Fig. 4-2) showing only two forces acting on one particle, even though we really have many forces acting on many particles.

4-3 Equilibrium of a Particle

In Chap. 2 we learned how to add vectors by the head-to-tail method as well as by the com-

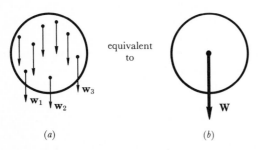

(a)   (b)

Figure 4-1 The center of gravity of a uniform disk is at its center.

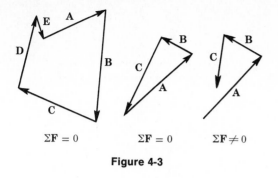

$\Sigma \mathbf{F} = 0$ $\Sigma \mathbf{F} = 0$ $\Sigma \mathbf{F} \neq 0$

Figure 4-3

ponent method. The resultant force $\Sigma \mathbf{F}$ is a vector, and for equilibrium this vector is zero. If $\Sigma \mathbf{F} = 0$, the forces must form a closed polygon (a triangle if there are only three forces) (Fig. 4-3). A vector can be zero only if each of its perpendicular components is zero*; thus the single vector equation $\Sigma \mathbf{F} = 0$ is equivalent to three component equations:

$$\sum F_x = 0 \qquad \sum F_y = 0 \qquad \sum F_z = 0$$

When a body is in equilibrium and all but one of the forces acting on it are known, the unknown force can be found by a graphical construction, trigonometry, or the component method.[†] Let us solve a typical equilibrium problem by each

---

* By the Pythagorean theorem, the magnitude of a vector is given by the square root of the sum of the squares of its perpendicular components. Since the squares of the components will always be positive, regardless of their directions, $\sqrt{F_x^2 + F_y^2 + F_z^2}$ can be zero only if F_x, F_y, and F_z are each zero.

† Our preferred unit of force in applying Newton's second law ($\Sigma \mathbf{F} = m\mathbf{a}$) is the newton in the mks system of units; we use the dyne ($= 10^{-5}$ N) in the cgs system. We have somewhat more freedom in statics, where $\mathbf{a} = 0$, for then $\Sigma \mathbf{F} = 0$ is true for *any* force units, even the pound, ton, and ounce. A conversion factor between the pound and the newton, for instance, would cancel out because of the 0 on the right-hand side of the equation.

method. It is a matter of choice (and convenience) which method to use.

Example 4-1

A boy who weighs 400 N (90 lb) sits in a swing. What force **F**, applied horizontally, will hold the swing and boy out at an angle of 30° with the vertical? What is then the total tension in the ropes?

To avoid confusion, we use two diagrams. The *space diagram* is a picture of the boy, tree limb, ropes, and so on. The forces involved are all shown acting *on* the body whose equilibrium we are considering—

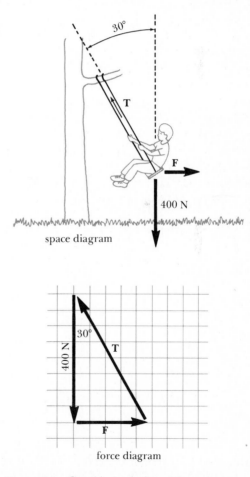

Figure 4-4   Graphical solution of Example 4-1.

in this case, the boy. There are three such forces: (1) the weight of the boy (400 N acting downward, through the c.g. of the boy); (2) the force **F** (of unknown magnitude and directed horizontally along a line passing through the c.g.); and (3) the total tension **T** exerted by the ropes (of unknown magnitude, directed along the ropes and passing through the c.g. of the boy). All these forces act at the same point, the c.g., and so we can apply the condition for equilibrium of a particle. To construct the *force diagram*, we place the three force vectors head to tail to make a closed triangle. *The fact that the forces form a closed triangle expresses the condition* $\Sigma\mathbf{F} = 0$. It is now easy to find the unknown forces from the force diagram.

First method: Graphical solution (Fig. 4-4): We lay off the force diagram to scale—letting 400 N be represented by 8 cm, for instance—and find by measurement that F is about 4.6 cm, which corresponds to

$$F \approx \boxed{230 \text{ N}}$$

and T is about 9.2 cm, which corresponds to

$$T \approx \boxed{460 \text{ N}}$$

Second method: Using trigonometry (Fig. 4-5): From the definitions of $\tan\theta$ and $\cos\theta$ (page 39) we have

$$\frac{F}{400 \text{ N}} = \tan 30°$$

$$F = (400 \text{ N})(\tan 30°)$$

$$= (400 \text{ N})(0.577) = \boxed{231 \text{ N}}$$

$$\frac{400 \text{ N}}{T} = \cos 30°$$

$$T = \frac{400 \text{ N}}{\cos 30°} = \frac{400 \text{ N}}{0.866} = \boxed{462 \text{ N}}$$

Third method: Using components (Fig. 4-6): Let us look at this same problem from the point of view of components of force. The space diagram is the same as before. For the force diagram, we choose a

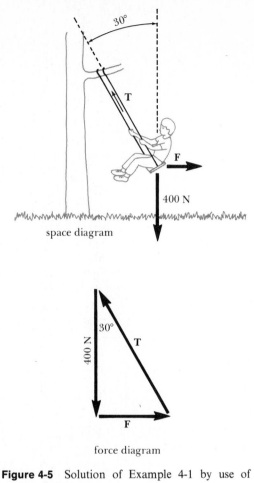

space diagram

force diagram

Figure 4-5 Solution of Example 4-1 by use of simple trigonometry.

set of axes with the x axis horizontal and the y axis vertical, and we lay out all the forces acting *on* the boy in a free-body diagram. We then replace the tension **T** by its components, T_x and T_y. To indicate that **T** itself is no longer to be considered, we draw a wavy line through the vector **T** and ignore it from now on. To solve the problem, we use the conditions for equilibrium in the component form. (There is no z component of force, so $\Sigma F_z = 0$ is automatically satisfied and is not even written down.)

From $\Sigma F_y = 0$,

$$T_y - 400 \text{ N} = 0$$

$$T_y = 400 \text{ N}$$

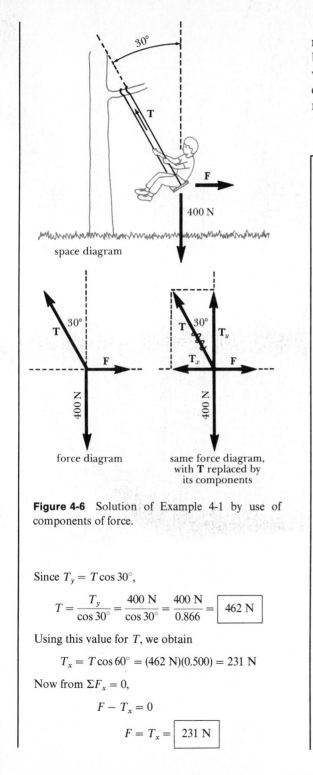

space diagram

force diagram

same force diagram,
with **T** replaced by
its components

Figure 4-6 Solution of Example 4-1 by use of components of force.

Since $T_y = T \cos 30°$,

$$T = \frac{T_y}{\cos 30°} = \frac{400 \text{ N}}{\cos 30°} = \frac{400 \text{ N}}{0.866} = \boxed{462 \text{ N}}$$

Using this value for T, we obtain

$$T_x = T \cos 60° = (462 \text{ N})(0.500) = 231 \text{ N}$$

Now from $\Sigma F_x = 0$,

$$F - T_x = 0$$

$$F = T_x = \boxed{231 \text{ N}}$$

A body need not be at rest to be in equilibrium. All that is necessary is that the acceleration be zero; however, this does not require the velocity to be zero. In our next example, the conditions for equilibrium are applied to a moving body in dynamic equilibrium.

Example 4-2

A trunk weighing 500 N is being pushed up a smooth inclined plane making an angle of 30° with the horizontal. What force, applied parallel to the plane, is necessary to allow the trunk to move up the plane with steady speed of 1 m/s?

Since the motion is uniform, the trunk is in dynamic equilibrium, and the same diagrams are to be used as if the trunk were held stationary. The

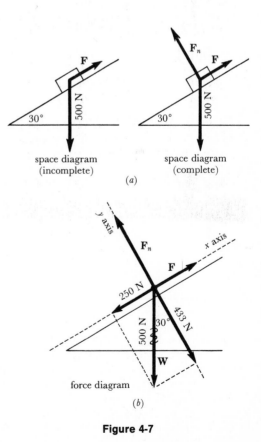

space diagram
(incomplete)

space diagram
(complete)

(a)

force diagram

(b)

Figure 4-7

magnitude of the constant speed, 1 m/s, is of no importance for the solution of the problem; nor is it of any consequence for us to know how and by what forces the trunk was accelerated to this speed.

The first space diagram (Fig. 4-7a) is incomplete, because one force was overlooked. A little reflection will show that the trunk cannot be in equilibrium if the weight **W** and the force **F** are the only forces acting. These two forces are not in the same straight line, and hence, regardless of their magnitudes, they could never cancel each other. The third force is not hard to find. The plane helps hold up the trunk. Since the plane is smooth, the force F_n of the plane on the trunk is normal (perpendicular) to the surface of the plane (that is, there is no frictional drag, which would be parallel to the surface). We may simplify the vector diagram by choosing the x axis along the plane and the y axis perpendicular to the plane (Fig. 4-7b). Then the components of the weight are (500 N)(sin 30°) = 250 N parallel to the plane, and (500 N)(cos 30°) = 433 N perpendicular to the plane. Hence,

from $\Sigma F_x = 0$ we get $F = \boxed{250 \text{ N}}$

from $\Sigma F_y = 0$ we get $F_n = \boxed{433 \text{ N}}$

In solving problems, Newton's third law is of great help. It often happens that we are asked for a certain force, but the equal and opposite reaction force is the one that acts on the body in question and that can be computed from the conditions for equilibrium. In this case we compute the force *on* the body and then use the third law to find the force exerted *by* the body.

―――――――――――― **Example 4-3**

A rope attached to one wall is passed over an equally high smooth wall 20 m away, and it is then attached to a heavy box on the ground (Fig. 4-8). If a downward force of 10 N is applied at the center of the rope as shown and the rope sags 1 m, what force is exerted by the rope on the box?

Although we are asked for the force T' on the box, we can compute the force T exerted *by* the box

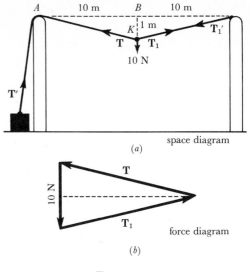

(a) space diagram

(b) force diagram

Figure 4-8

on the knot K, which is the particle whose equilibrium we are considering. Newton's third law tells us then that T and T' are numerically equal, since they are an action-reaction pair. Putting the forces head to tail, we make the force diagram a triangle, half of which is similar to the space triangle ABK. By the Pythagorean theorem, $AK = \sqrt{10^2 + 1^2} = 10.05$ m. Hence, using similar triangles, we have

$$\frac{5 \text{ N}}{T} = \frac{1 \text{ m}}{10.05 \text{ m}}$$

$$T = \boxed{50.3 \text{ N}}$$

This is the force of the box on the knot. By Newton's third law, the force of the knot on the box (transmitted by the rope) is also 50.3 N. Note also that since the force triangle is isosceles, T_1 is also 50.3 N.

4-4 Friction

When one body slides or tends to slide over another body, the force that acts to oppose the tendency to move is called the force of friction.

This frictional force is always *parallel* to the surfaces that are in contact. The experimental facts about friction may be summarized by the "laws" of friction:*

1. The maximum frictional force on a body resting on another body is proportional to the normal force pushing the two surfaces together.

For bodies at rest:

$$f_s = \mu_s F_n \qquad (4\text{-}1)$$

where f_s is the *maximum* force of static friction, F_n is the *normal force* (resultant of all forces perpendicular to the surfaces), and μ_s (the Greek letter mu) is the *coefficient of static friction*. In words,

Coefficient of static friction:

$$\mu_s = \frac{\text{maximum force of friction}}{\text{normal force}}$$

2. If one body is sliding on another body, the force of friction is constant, independent of the relative velocity of the two surfaces.

For bodies in motion:

$$f_k = \mu_k F_n \qquad (4\text{-}2)$$

where f_k is the *constant* force of kinetic friction (also called sliding friction), F_n is the normal force, and μ_k is the *coefficient of kinetic friction*.

Coefficient of kinetic friction:

$$\mu_k = \frac{\text{force of friction}}{\text{normal force}}$$

---

* These are not laws in the same sense as Newton's first law, which is exactly true under all circumstances. The "laws" of friction, like many so-called laws of physics, merely conveniently represent certain experimental facts well enough for making a start in the design of machines, and for other purposes. For an interesting discussion of the failures of the laws of friction, especially the second law, see Ref. 8 at the end of the chapter.

To a good approximation, static and kinetic frictional forces are independent of the area of the surfaces that are in contact. Kinetic friction is also almost independent of the relative velocity of the surfaces, if the velocity is not too great. A coefficient of friction, which is the ratio of two forces, is a dimensionless quantity and has no units. Because Eqs. 4-1 and 4-2 are very similar, you will have to be careful to distinguish between static and kinetic friction in any given situation and use the proper coefficient of friction. The coefficient of kinetic friction is less than the coefficient of static friction (Table 4-1). This means that the moment a body starts to slip, the force of friction decreases, usually by a considerable amount.

The force of *rolling friction* is much less than that of kinetic friction. When a ball rolls on a table, the table and ball are slightly deformed at the points of contact, with the result that the ball is perpetually climbing a small hill as it rolls along. The hill exerts a slight backward force on the ball, which accounts for the friction. Imperfect recovery of the ball or table after the point of contact moves on to another part of the ball is also responsible for some losses due to internal molecular friction. Reduction of friction by means of ball bearings often makes the difference between success and failure of modern machines.

Table 4-1
Coefficients of Static and Kinetic Friction[*]

Surfaces	Coefficient of Static Friction	Coefficient of Kinetic Friction
steel on steel, dry	0.6	0.3
steel on wood, dry	0.4	0.2
steel on ice	0.1	0.06
wood on wood, dry	0.35	0.15
metal on metal, greased	0.15	0.08

[*] Typical values; actual coefficients of friction vary greatly and depend on such factors as condition of surfaces, moisture, grain of wood, lubrication, and duration of contact.

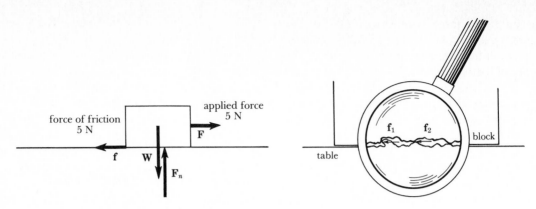

force of friction
5 N

applied force
5 N

F

force of friction
5 N

f   **W**

$\mathbf{F}_n$

f_1 f_2 block

table

Figure 4-9   The force of friction acts at the surface between two bodies.

To make the laws of friction more reasonable, let us look at the phenomenon of static friction from a microscopic point of view. Consider a metal block of weight 100 N resting on a table, and let the coefficient of static friction be 0.30. Then the maximum force of friction is, according to our definition,

$$f_s = \mu_s F_n = (0.30)(100 \text{ N}) = 30 \text{ N}$$

How can such a force arise? On account of the roughness of the surface, contact is actually made at only a relatively few points. At these points, we can imagine little hooklike protuberances that come into action as soon as any force, however small, is applied (Fig. 4-9). The hooks are deformed slightly, and they exert elastic forces* against each other. The hooks are probably actually "welded" together where they are in intimate contact. If, say, a force of 5 N is applied to the right, the block remains in static equilibrium, which means that all the little elastic frictional forces $\mathbf{f}_1$, $\mathbf{f}_2$, and so on, add up to exactly 5 N. If now the applied force is increased

to 10 N, the block does not start to move, but the small hooks "give" a little. Being more deformed than before, they exert a correspondingly larger horizontal force, amounting now to exactly 10 N. (We know that the force of friction must be *exactly* 10 N; if it were 10.01 or 9.99 N, for instance, there would be a resultant unbalanced force of 0.01 N one way or the other on the block, and it would start to move.) As the applied force is increased, there finally comes a time when the little hooks break down. This happens when the total force reaches 30 N, and so we say that 30 N is the "maximum" force of static friction.

If the block weighs 200 N instead of 100 N, it sinks into the table a little further, and (presumably) twice as many little hooks come into play. Hence the maximum force of friction is twice as much as before. In general, we expect the maximum force of static friction to be in proportion to the normal force pressing the surfaces together.

We have arrived at an interpretation of the relationship $f_s = \mu_s F_n$. Our procedure has been to make a *model* of a physical phenomenon (static friction) in terms of a more easily visualized or better-understood phenomenon (in this case, elasticity). The making of models is a favorite pastime of physicists. For instance, in Chap. 1

---

* We shall discuss elastic forces in more detail in Chap. 9; they are intimate short-range electric forces of attraction or repulsion between atoms that are close to each other.

we described the atom in terms of a familiar model, a miniature solar system. We must bear in mind that the models are not in themselves reality, but are merely suggestive. They are to be used only as long as their use helps predict new phenomena and correlate old observations. Sooner or later, we expect that even a good model may break down, for by definition a model is large-scale, drawn from everyday experience, and it is not necessarily true that small-scale phenomena must follow the large-scale laws.* Often, of course, a model is only an approximate analogy (with limited uses) and fails to describe the phenomenon fully even at the outset.

In considering problems involving frictional force, it is necessary to remember that the normal force is the resultant of all forces acting normal to the surfaces, not merely the weight or a component of the weight.

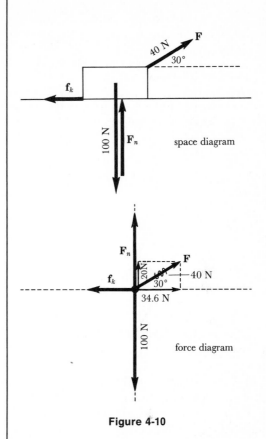

Figure 4-10

Example 4-4

A box that weighs 100 N is being steadily dragged along the floor by a rope that makes an angle of $30°$ with the horizontal (Fig. 4-10). If the tension in the rope is 40 N, what is the force of friction? What is the coefficient of friction?

Since the box is actually in motion, the force of kinetic friction is operative here. The box is in equilibrium (constant speed), and so $\Sigma F_x = 0$ and $\Sigma F_y = 0$. The condition $\Sigma F_x = 0$ tells us that the horizontal component of the force of the rope,

(40 N)(0.866), equals the force of friction f_k. Hence $f_k = 34.6$ N. Considering next the vertical forces, we find several such forces: the weight, the upward component of the tension in the rope, and the upward force of the floor on the box. The latter force F_n is the normal force causing the frictional drag. The condition $\Sigma F_y = 0$ gives us $F_n + 20 \, N = 100$ N, whence $F_n = 80$ N. We may say that the upward component of the tension in the rope somewhat "eases up" the force between box and floor. If we know both f_k and F_n, it is a simple matter to find the coefficient of kinetic friction:

$$\mu_k = \frac{f_k}{F_n} = \frac{34.6 \text{ N}}{80 \text{ N}} = \boxed{0.43}$$

---

* The model of an atom as a miniature solar system broke down in 1926. This does not mean that the atom really "is" or "is not" like a solar system. We still use the model and speak of the nucleus around which the electrons move, but now we realize that the model has its limitations and some predictions of the model are simply not experimentally true, nor need they be. The planetary model is an approximation, but a useful approximation. Uncritical acceptance of models as reality (a disease known in some circles as "modelitis") has led to numerous blind alleys in physics.

Example 4-5

An otter of weight $\mathbf{W}$ is at rest on a 30° plane (Fig. 4-11). If the otter is about to slip, what is the coefficient of static friction?

From the space diagram, it is evident that the otter is acted on by three forces: the weight $\mathbf{W}$, the normal force $\mathbf{F}_n$ of the plane on the otter, and the force of static friction $\mathbf{f}_s$. We resolve $\mathbf{W}$ into two components as shown, and then apply the conditions for equilibrium:

$$F_n = W \cos 30°$$
$$f_s = W \sin 30°$$

Hence

$$\mu_s = \frac{f_s}{F_n} = \frac{W \sin 30°}{W \cos 30°} = \frac{0.500\ W}{0.866\ W} = \boxed{0.577}$$

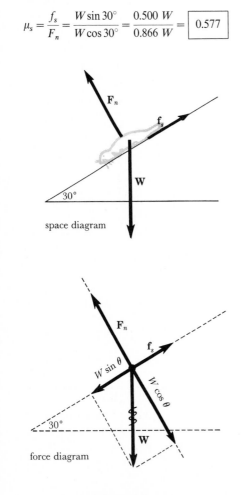

space diagram

force diagram

Figure 4-11

Note that in the solution of this problem, the weight of the otter canceled out. Only the angle of the plane was important. This illustrates a common way of measuring the coefficient of static friction: Place the object on a plane and increase the angle of the plane until the object just begins to slip. The angle so measured is called the *angle of repose*, and the coefficient of static friction equals the tangent of the angle of repose (Fig. 4-11). If the otter in Example 4-5 moves along a plane without acceleration, it is in dynamic equilibrium, and the tangent of the angle of the plane equals the coefficient of kinetic friction.

Bear in mind that the equation $f_s = \mu_s F_n$ gives the *maximum* possible force of static friction for any given pair of surfaces and for a given normal force. In general, $f_s \leqslant \mu_s F_n$; unless motion is taking place or is about to take place, the force of friction is less than $\mu_s F_n$. Until the maximum value is reached, the force of friction is self-adjusting, only as much as is required to give equilibrium. For the block and table in Fig. 4-12, the coefficient of static friction is 0.40, and the coefficient of kinetic friction is 0.36. The normal force is 100 N in each case, equal to the weight of the block. The static force of friction varies from 0 to 40 N as the applied force $\mathbf{F}$ increases. Of course, in Fig. 4-12e the block is no longer in equilibrium, since net $\mathbf{F} \neq 0$. Since it is sliding, we know that we are obtaining the force of kinetic friction, which is 0.36 × 100 N = 36 N. In this case, the unbalanced force of 14 N causes an acceleration that can be computed using Newton's second law.

Example 4-6

Compute the acceleration of the block in each case of Fig. 4-12. The block weighs 100 N, and its mass is 100/9.8 kg.

For (a), (b), (c), and (d), the net force is zero, and Newton's second law gives zero for the acceleration. The block is in equilibrium.

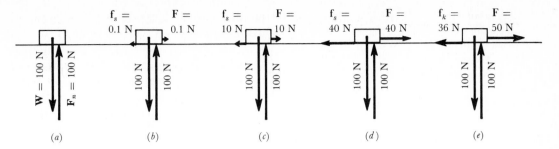

Figure 4-12 (*a*) through (*d*) Force of static friction equals the applied force; $f_s \leqslant 40$ N. (*e*) Force of kinetic friction is constant; $f_k = 0.36F_n = 36$ N. For clarity, the two vertical forces are drawn slightly separated.

For case (*e*),

$$\text{net } F = ma$$

$$50\text{ N} - 36\text{ N} = (100/9.8\text{ kg})(a)$$

$$a = \frac{14\text{ N}}{100/9.8\text{ kg}} = \boxed{1.37\text{ m/s}^2}$$

The acceleration is in the direction of the net force, that is, toward the right.

The following example illustrates how the force of friction acting on a body can cause an acceleration of that body relative to an outside observer.

Example 4-7

An iron casting of mass 2000 kg (2 metric tons) is being transported by a truck that is moving at a steady velocity (Fig. 4-13*a*). The driver sees a red light and applies his brakes, causing the truck to decelerate (have negative acceleration) at a rate of 3 m/s². Will the casting remain in place, or will it slide forward and crush the driver? The coefficient of static friction is 0.4, and the coefficient of kinetic friction is 0.3.

The question really is, *can* the casting be decelerated at 3 m/s²? (All velocities and accelerations are relative to the ground, as seen by an outside observer standing on the sidewalk.) The casting

tends to slide forward on the bed of the truck, but the frictional force acts to oppose this tendency.

Consider the forces on the casting (Fig. 4-13*b*): Only one horizontal force acts on the casting—the force of friction.

$$\Sigma F = ma$$

$$f = (2000\text{ kg})(-3\text{ m/s}^2)$$

$$= -6000\text{ N} \quad \text{(backward)}$$

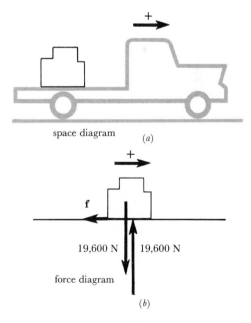

space diagram

(*a*)

force diagram

(*b*)

Figure 4-13

A frictional force of 6000 N would be needed to cause the casting to decelerate along with the truck. The weight of the casting is 2000 kg × 9.8 m/s² = 19,600 N. The maximum force of static friction is 0.4(19,600 N) = 7840 N, so there is no reason for alarm. The force of friction in this example is self-adjusting and is equal to 6000 N, which is all that is required; the maximum value of 7840 N is not needed.

4-5 Equilibrium of a Rigid Body—Torque

We have considered the equilibrium of an idealized particle and have found that a single vector equation, $\Sigma \mathbf{F} = 0$, expresses the conditions that must be fulfilled in order to obtain equilibrium. We also have seen that in many cases in which all applied forces pass through the c.g., a body may be considered to be a particle even though it has extension (that is, a size and shape). Let us consider the types of motion possible for a body that has measurable extension. Such a body may have rotation as well as translation. In *translational motion* all particles of a rigid body travel in parallel paths (not necessarily straight lines), whereas in *rotational motion* (discussed more fully in Chap. 7) the particles describe circles of various sizes about some axis. Often the two types of motion are combined, as in a football that is end-over-ending (Fig. 4-14). The only type of motion having meaning for

an ideal particle is translation. Our previously discussed condition that the resultant force be zero is a condition for translational equilibrium.

In considering rotational equilibrium (no change in rotational velocity), we find from experience that a new quantity called *torque* is the deciding factor. The effectiveness of a force in changing the rotational state of a body is governed not only by the magnitude of the force but also by its direction and its point of application. The *lever arm* is the perpendicular distance from the axis of rotation to the line of action of the force. The product, force × lever arm, is defined as the torque caused by the force; in order to have rotational equilibrium it is necessary that the torques of all forces add up to zero. The dimensional equation for torque is

$$[\text{torque}] = [F][L],$$

or

$$[\text{torque}] = [MLT^{-2}][L] = [ML^2T^{-2}].$$

Torques are expressed in units such as N·m, dyn·cm, lb·ft, and oz·in.—the product of any force unit and a distance unit. No entirely satisfactory symbol for torque has been universally agreed on; we shall use the symbol τ (the Greek letter tau) for torque to avoid confusion with symbols for tension or time.*

---

* Torque is often called *moment of force*, but this term sometimes leads to confusion. If you learned about "moments" in a high-school course, rest assured that torque and moment of force are identical in meaning and use.

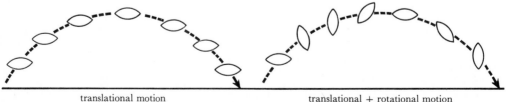

<table>
<tr><td>translational motion</td><td>translational + rotational motion</td></tr>
</table>

Figure 4-14

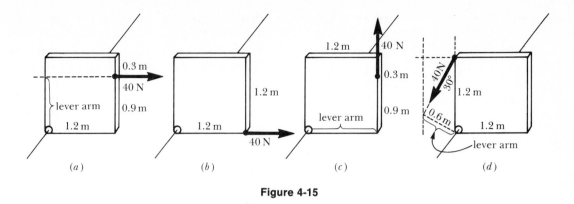

Figure 4-15

The torque caused by a given force depends greatly on the point of application and the direction of the force. In Fig. 4-15 is shown a piece of plywood 1.2 m on an edge, with an axis of rotation through one corner. This axis is perpendicular to the plane of the plywood, and the force in each case is applied parallel to the plane of the plywood. In each case the magnitude of the force is 40 N, but the lever arms are quite different, and so are the torques. In (a), the distance from the axis to the point of application of the force is 1.5 m, but it is the distance to the *line of action* (0.9 m) that counts. We shall consider torques as plus or minus, according to whether they tend to cause counterclockwise or clockwise rotation, respectively. If you have trouble in determining whether a given force produces a clockwise or counterclockwise torque, draw an imaginary circle tangent to the line of action (Fig. 4-16) and with its center coincident with the axis of rotation. The radius of this circle is the lever arm, and the direction in which a point on the circumference of the circle would be moved by the force is the direction associated with the torque. Thus in Fig. 4-16, the direction of the torque is clockwise, or negative.

Torque may be given a vector interpretation, with the direction of a torque vector being parallel to the axis of rotation. All the torques of Fig. 4-15 can be represented by vectors directed along the axis of rotation, either toward or away from the reader. To make this more definite we use the *right-hand screw rule* to define the direction of a torque vector. We imagine turning an ordinary (right-handed) screw either clockwise or counterclockwise in the direction of rotation produced by the torque. The direction of advance of the screw is defined as the direction of the torque vector. Thus in Fig. 4-16 the torque vector is in the direction shown, away from the reader, which is the direction a screw would advance if turned clockwise.

With this background, we now give a formula

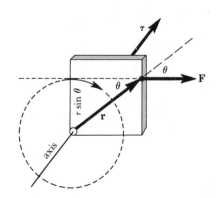

Figure 4-16 Finding the direction of a torque.

for torque. In Fig. 4-16 the vector **r** is drawn from the axis of rotation to the point of application of the force **F**; the magnitude of the lever arm is less than r and is given by $r \sin \theta$, where θ is the angle between the direction of **r** and that of **F**. The magnitude of the torque, force × lever arm, is

$$\tau = (r \sin \theta)(F)$$

and the direction of the torque is perpendicular both to **r** and **F**, as specified by the right-hand rule. To give a single compact formula for both magnitude and direction of the torque, we use the *cross product* of two vectors:

$$\tau = \mathbf{r} \times \mathbf{F}$$

where the "cross" multiplication symbol (×) is an abbreviation for the factor $\sin \theta$, and it is understood that the direction of τ is given by the right-hand rule for the advance of a screw rotated from the direction of **r** toward the direction of **F**. (The order of writing the factors is important: **F** × **r** is the negative of **r** × **F** because the screw would advance in the opposite direction if turned from **F** toward **r**.) Other cross products will be useful in our study of electromagnetism. In our present study of statics, we do not need the full power of vector methods, because in simple problems the torques are all along one axis and can be represented algebraically as + or −. (We did much the same thing in Chap. 2 where velocities and accelerations, all along the same line, were given + and − values.)

For a rigid body to be in equilibrium, two conditions must be satisfied. Net force must be zero, and net torque must be zero.* We can write these conditions in the form of two vector equations:

Translational equilibrium: $\quad \Sigma \mathbf{F} = 0$

Rotational equilibrium: $\quad \Sigma \tau = 0$

These are equivalent to six simultaneous equations. $\Sigma \mathbf{F} = 0$ means $\Sigma F_x = 0$, $\Sigma F_y = 0$, and $\Sigma F_z = 0$ (forces along each of three mutually perpendicular directions must add up to zero). Similarly, $\Sigma \tau = 0$ means $\Sigma \tau_x = 0$, $\Sigma \tau_y = 0$, and $\Sigma \tau_z = 0$ (torques around each of three mutually perpendicular axes of rotation must add up to zero). However, in most practical problems, we need consider torques about only one axis, since usually all the forces either act in one plane or can be replaced by forces acting in a single plane. If we call this plane the xy plane, all axes of rotation are parallel to the z axis, and so we have just three equations for equilibrium of a rigid body:

Horizontal forces to the right = horizontal forces to the left:

$$\Sigma F_x = 0$$

Upward forces = downward forces:

$$\Sigma F_y = 0$$

Clockwise torques = counterclockwise torques:

$$\Sigma \tau = 0$$

Since we have three equations to be solved simultaneously, each problem could have as many as three unknowns (either forces or angles). The solution of problems in statics involving forces and torques calls for ingenuity and common sense. There are no simple rules of proce-

---

* Although we have stated the torque condition as based on experience (page 84), it is not an independent law of nature. The torque condition can be derived from Newton's first and third laws, with the additional (very reasonable) assumption that the force between any pair of particles of a rigid body acts along the line joining those two particles.

dure, and the following examples will give an idea of some of the tricks of the trade. The most common source of error is failure to identify the object whose equilibrium is being considered. You must learn to consider *all* the forces acting *on* the body. Of course, Newton's third law will be of great help when you are given (or asked for) forces exerted *by* the body. Use the third law to obtain the corresponding forces on the body, and use this information in the solution of the problem.

Example 4-8

A load weighing 60 N is to be supported by a force **F** applied at the end of a weightless lever, as shown in Fig. 4-17a. What force is necessary? What is the force on the fulcrum A when the lever is in equilibrium?

The body whose equilibrium is being considered is the lever. Draw in all the forces acting *on the lever* (including the upward force exerted on the lever by the fulcrum) (Fig. 4-17b).

From $\Sigma\tau = 0$, with axis at A,

Clockwise torques = counterclockwise torques

$$(60 \text{ N})(40 \text{ cm}) = (F)(120 \text{ cm})$$

$$F = \frac{(60 \text{ N})(40 \text{ cm})}{120 \text{ cm}}$$

$$= \boxed{20 \text{ N}}$$

To find the force on the fulcrum, we must first find the force exerted *by* the fulcrum on the lever. We have just found that $F = 20$ N.
From $\Sigma F_y = 0$,

Upward forces = downward forces

$$P + 20 \text{ N} = 60 \text{ N}$$

$$P = \boxed{40 \text{ N (upward)}}$$

Now by Newton's third law, the force exerted by the lever on the fulcrum is equal and opposite to P.

$$P' = \boxed{40 \text{ N (downward)}}$$

We do not show P' in the diagram because it does not act *on* the lever whose equilibrium we are considering.

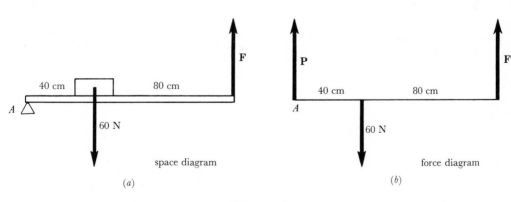

space diagram

(a)

force diagram

(b)

Figure 4-17

Example 4-9

An upraised forearm (Fig. 4-18) is pulling a motion picture screen downward with a force of 50 N. The biceps muscle is attached to the radius as shown. For simplicity, neglect the weight of the radius and ignore the effect of the ulna. Calculate the force exerted by the biceps muscle.

It is the radius that we are concerned with, and in the vector diagram we are careful to show only the forces on that bone. We ignore forces due to the radius. Thus at the elbow, **H** and **V** are the unknown horizontal and vertical components of the force exerted jointly (!) by the humerus and the ulna on the radius. The resultant force of the elbow on the radius will be downward and to the left. The screen exerts an upward force of 50 N.

Although it is actually immaterial where the axis of rotation is assumed to be, in practice a judicious choice of the axis will simplify the work considerably. By putting the axis at the joint, two of the three unknowns (**H** and **V**) have no lever arm and hence exert no torque. The torque equation then involves only one unknown force:

Counterclockwise torque = clockwise torque

$$(50 \text{ N})(21 \text{ cm}) = F(2.8 \text{ cm})$$

$$F = \boxed{375 \text{ N}}$$

Now using $\Sigma F_x = 0$ gives

$$H = F = \boxed{375 \text{ N}}$$

and $\Sigma F_y = 0$ gives

$$V = \boxed{50 \text{ N}}$$

If we wish, we can combine the two forces due to the joint into a single force (Fig. 4-19):

$$R = \sqrt{H^2 + V^2} = \sqrt{375^2 + 50^2} = 378 \text{ N}$$

$$\tan \theta = \frac{50 \text{ N}}{375 \text{ N}} = 0.133$$

$$\theta = 8° \text{ below the horizontal}$$

Either way we look at it there are three unknowns in this problem. Either we are asked for F, H, and V, or we are asked for F, R, and θ. We are assured of being able to solve the problem, since we have three general equations out of which to solve for any three of the unknowns.

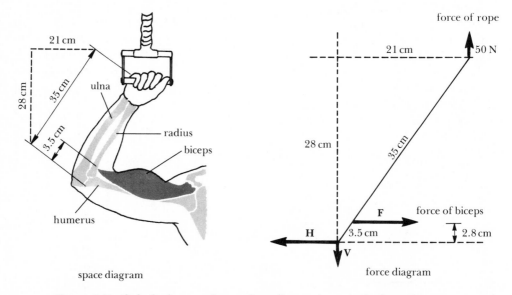

space diagram

force diagram

Figure 4-18 Only the forces acting on the radius are drawn in the force diagram.

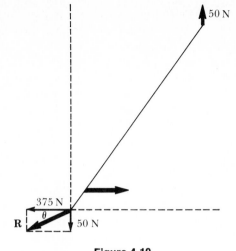

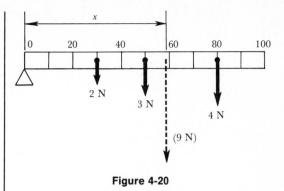

Figure 4-20

is at the 80-cm mark (Fig. 4-20). Where is the center of gravity of the system?

Imagine the meter stick supported horizontally, with an axis through the 0-cm mark. We equate the torque due to the weight of the stick and its loads to the torque that would act if all the weight were at the center of gravity, at an unknown lever arm x.

$$(2 \text{ N})(30 \text{ cm}) + (3 \text{ N})(50 \text{ cm}) + (4 \text{ N})(80 \text{ cm})$$
$$= (9 \text{ N})(x)$$

$$x = \frac{530 \text{ N} \cdot \text{cm}}{9 \text{ N}} = \boxed{58.9 \text{ cm}}$$

The c.g. of the loaded meter stick is at the 58.9-cm mark.

375 N

R θ 50 N

Figure 4-19

Let us note one interesting feature of the solution. The force of the joint on the radius is not directed along the bone. The bone makes an angle of 53° with the horizontal (the angle whose tangent is $\frac{28}{21}$), while the force of the joint makes an angle of 8° with the horizontal (Fig. 4-19).

4-6 Computation of the Center of Gravity

The definition of center of gravity can be expressed in terms of torque: the total weight of a body, if acting vertically through the c.g., would produce the same total torque about any axis as do the weights of the component parts of the body, acting at their actual positions. The c.g. of a geometrically regular body is at its center (for instance, the c.g. of a uniform meter stick is at the 50-cm mark). The c.g. of an irregular body composed of a few heavy spots at definite places can be found by calculating torques, as in the following example.

─────────────────────────────
Example 4-10

A uniform meter stick 100 cm long weighing 3 N is loaded with two small heavy objects as follows: A load of 2 N is at the 30-cm mark, and a load of 4 N

The initial choice of axis at the 0-cm mark was entirely arbitrary. If a body is in equilibrium about one axis, it must also be in equilibrium if the torques are computed about any other axis. To illustrate this, let us solve Example 4-10 using an axis through the 20-cm mark instead of the 0-cm mark (Fig. 4-21).

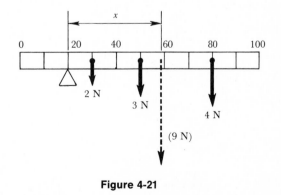

Figure 4-21

$$(2 \text{ N})(10 \text{ cm}) + (3 \text{ N})(30 \text{ cm}) + (4 \text{ N})(60 \text{ cm})$$
$$= (9 \text{ N})(x)$$

$$x = \frac{350 \text{ N·cm}}{9 \text{ N}} = \boxed{38.9 \text{ cm}}$$

This is the location of the c.g. *measured from the 20-cm mark.* Physically, it is the same point that was obtained previously, 58.9 cm from the 0-cm mark.

The c.g. of a body having a continuous distribution of weight (such as a semicircular sheet of plywood) can be computed in a similar way, except that since there are infinitely many weight particles instead of just a few as in Example 4-10, the methods of integral calculus must be used in summing up the torques due to gravity. See Sec. 4-9 for further discussion of center of gravity.

4-7 Types of Equilibrium

If a body is supported at a single point, it is acted on by two forces: the force of gravity (its weight) acting at the c.g., and the upward force of the pivot. The body will be in equilibrium if its c.g. is somewhere on the vertical line passing through the point of support, for then it is possible for the two forces to balance. There are three possible cases:

1. *Stable equilibrium.* If the c.g. is directly below the point of support, the body tends to return to its original position after being given a slight displacement. When displaced, the weight, acting at the c.g., gives rise to a restoring torque. A typical example of a body in stable equilibrium is a pendulum bob at its lowest point (Fig. 4-22).

2. *Unstable equilibrium.* If the c.g. is directly above the point of support, the body tends to move farther away from its original position when it is given a slight displacement. When displaced, the weight, acting at the c.g., gives rise to a torque that tends to increase the displacement. A typical example of a body in unstable equilibrium is a pole vaulter at the very top of his motion (Fig. 4-23).

3. *Neutral equilibrium.* If the c.g. coincides with the point of support, the body remains in whatever position it is placed. The weight, acting

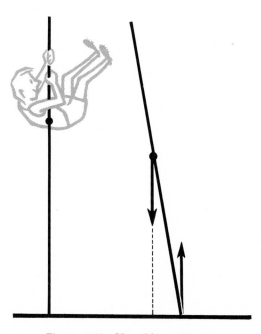

Figure 4-23 Unstable equilibrium.

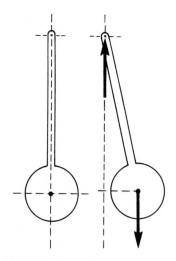

Figure 4-22 Stable equilibrium.

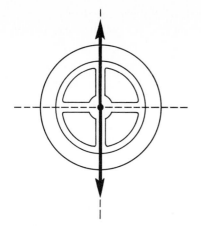

Figure 4-24 Neutral equilibrium.

at the c.g., produces no torque since the lever arm is zero. A well-balanced wheel on a horizontal axis is an example of a body in neutral equilibrium (Fig. 4-24).

You can easily balance a horizontal ruler on a spherical ball, on a cylindrical can, or on a pencil lying on its side. Although the c.g. of the ruler is above the point of support, this is different from case 1 discussed above because the body is not "supported at a single point." If the ruler is tilted slightly, the point of contact with the surface shifts, and there is a restoring torque about the new axis if the ruler is not too thick. The curved surface must be rough enough to prevent slipping.

An example of a different type of equilibrium is afforded by a chair in its normal position. The vertical line through the c.g. intersects the floor at a point that lies within the area defined by the points of support (the legs). The chair is in stable equilibrium with respect to tilting but in neutral equilibrium with respect to sliding along the floor. Unstable equilibrium can easily be obtained (but not maintained) by tilting back the chair.

If a body has uniform velocity in a straight line, it is in dynamic equilibrium. Here, too,

stable, unstable, and neutral equilibria can be discussed. As an example of stable dynamic equilibrium, consider a raindrop falling steadily. If for some reason its speed increases, air resistance also increases (the drop strikes more air molecules per second), and so the speed returns to the original value.

4-8 Solution of Problems Involving Center of Gravity

In solving problems, the weight of a body, really consisting of almost infinitely many tiny forces of gravitation, is replaced by a single force acting at the c.g. From this point of view, discussed more fully in Sec. 4-2, the weight becomes a single force, to be treated like any other force. A little reflection will show that this is a tremendously important simplification, although one that is usually taken for granted. Our next examples make use of this simplifying concept.

Example 4-11

A man who weighs 800 N (180 lb) climbs to the top of a 6-m ladder that is leaning against a smooth wall at an angle of 60° with the horizontal. The ladder weighs 400 N (90 lb), and its c.g. is 2 m from the bottom end (Fig. 4-25). What must be the coefficient of static friction at the ground if the ladder is not to slip?

In our mind's eye, let us isolate the ladder as a free body and consider all forces on the ladder. They are as follows: the weight of the man; the weight of the ladder; the push of the wall **P**; and the two components of the force of the ground on the ladder, **H** and **V**. Since the wall is smooth, there can be no frictional force parallel to the surface of the wall, so the push **P** must be horizontal, as shown. We next fill in the dimensions in the space diagram, using the length of the ladder and the known 60° angle. For instance, the vertical height AB is found from $AB/6 \text{ m} = \sin 60°$; $AB = 5.20 \text{ m}$.

By taking torques about an axis passing through C, we eliminate the torques due to **H** and **V**.

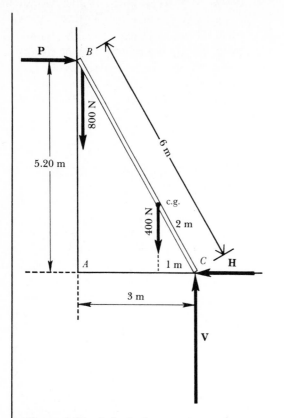

Figure 4-25 Only the forces acting on the ladder are drawn in the diagram.

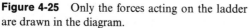

$$(800 \text{ N})(3 \text{ m}) + (400 \text{ N})(1 \text{ m}) = P(5.20 \text{ m})$$

$$P = \frac{2800 \text{ N·m}}{5.20 \text{ m}} = \boxed{538 \text{ N}}$$

Now, from $\Sigma F_x = 0$,

$$H = P = \boxed{538 \text{ N}}$$

and, from $\Sigma F_y = 0$,

$$V = 800 \text{ N} + 400 \text{ N}$$
$$= \boxed{1200 \text{ N}}$$

We interpret H, the force exerted by the ground parallel to the surface of the ground, as the frictional force. Our solution shows that we need 538 N of frictional force. Since, by the wording of the prob-

lem, the ladder is about to slip, we know that the 538 N of friction represents the maximum force of static friction. Hence,

$$\mu_s = \frac{\text{maximum force of static friction}}{\text{normal force holding surfaces together}}$$

$$= \frac{538 \text{ N}}{1200 \text{ N}} = \boxed{0.45}$$

There are two points to note in the solution to Example 4-11. When we isolated the ladder, we carefully excluded forces exerted by the ladder. There *is* a horizontal force to the left, at the top of the ladder, but this force acts on the wall, not on the ladder; hence we ignore it. Newton's third law tells us that the ladder pushes against the wall with a force of 538 N toward the left, since *P* is 538 N toward the right. Also, we were able to calculate the coefficient of static friction only because the ladder was just about to slip. If the coefficient of static friction were 0.50, the ladder would still have been in equilibrium, and the force of static friction would still have been 538 N. The maximum possible force of static friction would have been $\mu_s F_n = (0.50)(1200 \text{ N}) = 600 \text{ N}$, but we would have been getting only as much friction as was actually needed for equilibrium.

Sometimes, as in the following example, it is convenient to resolve all forces into horizontal and vertical components before computing torques.

Example 4-12

A uniform piece of plywood weighing 200 N is in a vertical plane, with a 120-cm side resting on the floor and a 90-cm side vertical (Fig. 4-26). What force **F** applied as shown will cause the plywood to just start to rotate about the pivot at the corner *P*?

By use of a little trigonometry we find the lever arm for the force **F** about an axis through *P*. The diagonal *PB* is 150 cm, and the angle *PBC* is found from sin *PBC* = 90 cm/150 cm = 0.600; $\angle PBC = 36.9°$. Hence the angle *PBD* is 66.9°, and *PD* =

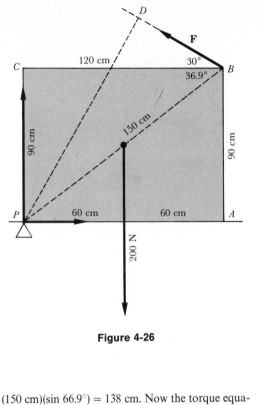

Figure 4-26

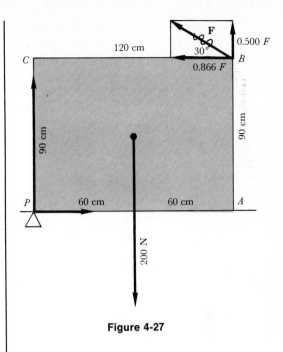

Figure 4-27

$(150 \text{ cm})(\sin 66.9°) = 138 \text{ cm}$. Now the torque equation is used:

$$(200 \text{ N})(60 \text{ cm}) = F(138 \text{ cm})$$

$$F = \frac{12,000 \text{ N} \cdot \text{cm}}{138 \text{ cm}} = \boxed{87 \text{ N}}$$

Perhaps you will find it easier to solve this problem by first resolving **F** into its components before applying the torque equation. The advantage is that now the lever arms are more easily found (Fig. 4-27).

$$(200 \text{ N})(60 \text{ cm}) = (0.866F)(90 \text{ cm})$$
$$+ (0.500F)(120 \text{ cm})$$
$$12,000 \text{ N} \cdot \text{cm} = 78F \text{ cm} + 60F \text{ cm}$$
$$= 138F \text{ cm}$$

$$F = \frac{12,000 \text{ N} \cdot \text{cm}}{138 \text{ cm}} = \boxed{87 \text{ N}}$$

Summary   Equilibrium is the condition in which the magnitude and direction of the velocity of a body remain constant. In static equilibrium, the velocity remains zero and the body is at rest. In dynamic equilibrium, the body moves with constant velocity in a straight line.

For a particle to be in equilibrium, it is necessary for the vector sum of the forces on the particle to be zero. In many cases, it is possible to treat a body that is composed of many particles as a single particle, provided all externally applied forces act, or can be considered to act, through the center of gravity of the body. This is true because the c.g. is defined as that point of a body through which the resultant of all the tiny parallel forces of gravity passes.

Frictional forces act parallel to surfaces that are in contact and are dependent upon the normal force pushing the surfaces together. For any given normal force, the force of static friction is variable, of whatever amount is needed, up to a maximum force that is equal to the coefficient of static friction × the normal force. After motion starts, the force of kinetic friction is constant, regardless of the velocity, and is equal to the coefficient of kinetic friction × the normal force. The force of kinetic friction is usually considerably less than the maximum force of static friction.

Torque, defined as force × lever arm, is a vector quantity whose direction is given by the right-hand screw rule. The cross product is used to define the magnitude and direction of a torque vector.

For a rigid body to be in equilibrium, two conditions must be fulfilled: The vector sum of the forces on the body must be zero (as for a particle), and also the vector sum of the torques on the body must be zero. In applying the second condition, any convenient axis may be chosen for computing the torques due to the applied forces.

In solving problems in statics, it is vital to be sure just which particle or body it is whose equilibrium is being considered. In applying the conditions for equilibrium, only the forces acting on the body are to be considered.

Check List

equilibrium (definition)
static equilibrium (definition)
dynamic equilibrium (definition)
equilibrium of a particle (conditions)
normal force
coefficient of static friction
coefficient of kinetic friction
angle of repose
translational motion

rotational motion
torque
lever arm
right-hand screw rule
cross product
center of gravity
equilibrium of a rigid body (conditions)
stable, unstable, and neutral equilibria
     (definitions and conditions)

Questions

4-1 A freight train moves along a straight track with a constant speed of 20 m/s. Is it in equilibrium? Explain why the condition of equilibrium of a particle can be applied to a freight train that is a kilometer in length.

4-2 A baseball is thrown straight upward and momentarily comes to rest at the highest point in its path. Is it in equilibrium at this instant?

4-3 Give several examples each of stable, unstable, and neutral equilibria.

4-4 Discuss the nature of the equilibrium of the Leaning Tower of Pisa.

4-5 Give an example of a body that is in motion, but is in equilibrium. Give an example of a body that is at rest, yet is not in equilibrium.

4-6 What is meant by the statement that statics is a special case of dynamics?

4-7 List some beneficial uses of the force of static friction.

4-8 On what does the force of friction between two surfaces depend?

4-9 On what does the coefficient of friction between two surfaces depend?

4-10 In Sec. 4-2 the c.g. of a uniform, *solid* phonograph record was stated to be at the center. Where is the c.g. of a 45-rpm record that is 17.3 cm in diameter and has a central hole 3.8 cm in diameter?

MULTIPLE CHOICE

4-11 A body that is in equilibrium could have (*a*) uniform velocity in a straight line; (*b*) motion in a circular path at uniform speed; (*c*) neither of these.

4-12 A 200-N force acts horizontally toward the east. The component of this force along an axis directed toward the northeast is (*a*) 141 N; (*b*) the same in magnitude as the component in a southeast direction; (*c*) both of these.

4-13 In Fig. 4-12*e*, the acceleration of the block is (*a*) 0.14*g*; (*b*) 0.36*g*; (*c*) 0.50*g*.

4-14 A sports car goes over the top of a hill at a steady speed of 88 km/h. The motion is one of (*a*) translation only; (*b*) rotation only; (*c*) both translation and rotation.

4-15 Which of the following are scalar quantities? (*a*) weight; (*b*) torque; (*c*) neither of these.

4-16 If a ladder leans against a wall at an angle of 60° with the horizontal, the force of the ground on the ladder makes an angle with the ground that is (*a*) greater than 60°; (*b*) 60°; (*c*) less than 60°.

Problems

Note: Force and torque are vector quantities. Hence an answer to a problem in which force or torque is calculated is, in general, incomplete unless both magnitude and direction are specified.

4-A1 A horizontal force of 80 N acts toward the northeast. (*a*) What is the component of this force along an east–west line? (*b*) What is its component along a line running toward a point 15° east of north?

4-A2 A plane in level flight has a velocity of 400 km/h at an angle of 30° S of W. (*a*) What is the component of this velocity along an east–west line? (*b*) What is the velocity component along a line that runs northeast–southwest?

4-A3 In an experiment in atomic physics, an electron is acted on by two forces: an electric force of 80 units directed horizontally, and a magnetic force of 60 units directed upward. What are the magnitude and direction of the net force acting on the electron?

4-A4 A boat is acted on by a force of 400 N directed toward the north due to the motor and by an eastward force of 300 N caused by the wind. What are the magnitude and direction of the net force on the boat?

4-A5 A vertical force of 80 N is required to raise the upper sash of a classroom window. How much force will be needed if the force is applied by a pole that makes an angle of 25° with the window?

4-A6 Whenever an electric power line follows a bend in a road, a guy wire must be used to prevent the pole from leaning over. Try to observe this in your community. Show by a vector diagram the three forces that act at the top of the pole, and explain how the horizontal component of the force of the guy wire balances the two forces exerted on the pole by the power line. What is the direction of the resultant force exerted by all three wires on the top of the pole?

4-A7 A guy wire 8 m long extends from the top of a telephone pole to a point on the ground 4 m from the base of the pole. The tension in the wire is 2000 N. What is the horizontal component of the force on the pole?

4-A8 A child who weighs 500 N sits on a sled that weighs 60 N. If the coefficient of kinetic friction between the steel runners and the ice is 0.08, with how much force must a playmate pull horizontally in order to keep the child and sled moving over the level ice at constant velocity?

4-A9 In Prob. 4-A8 it is found that a horizontal force of 56 N is needed to start the child and sled moving. What is the coefficient of static friction?

4-A10 With what force must an ice skater push a fellow skater who weighs 300 N in order to start her gliding over the ice? (Use data from Table 4-1.)

4-A11 Suppose that the plywood in Fig. 4-15 weighs something. Is the torque due to the weight of the plywood about the indicated axis clockwise or counterclockwise?

4-A12 Again referring to Fig. 4-15, compute the torque due to the 40-N force in each case, if the axis passes through the upper right-hand corner of the plywood square.

4-A13 Calculate the torque of the 400-N force about an axis through B in Fig. 4-25. Is this torque clockwise or counterclockwise? What is the direction of the vector representing this torque?

4-A14 What is the torque exerted by a jeweler using a microwrench if his fingers supply a force of 1.2 N at a distance of 7.5 cm from the axis of rotation? The force is applied in a direction making an angle of 70° with the handle of the wrench.

4-A15 When a rigid body is in equilibrium, $\Sigma\tau = 0$ when the torques are computed about *any* axis. In Example 4-11 the axis was chosen to pass through the bottom of the ladder. Using the values of H, V, and P found in the solution for Example 4-11, verify that $\Sigma\tau = 0$ when torques are computed about an axis through point A.

4-B1 Three horizontal forces act on a particle: 6 N toward the north, 4 N toward the east, and 4 N toward the southwest. Using a scale of 1 N = 2 cm or 1 in., lay off the forces head to tail and find the magnitude and direction of the resultant force on the particle. (Use graph paper and a protractor.)

4-B2 Repeat Prob. 4-B1, laying off the vectors in a different order.

4-B3 Repeat Prob. 4-B1, using the component method. Resolve each of the given forces into x and y components; find the resultant x component by addition; find the resultant y component by addition; and combine these two vectors by the Pythagorean theorem to find the resultant of all the forces. (Express the direction of the resultant with respect to east.)

4-B4 Add graphically the following vectors: 400 N toward the north, 200 N toward the west, and 300 N toward the southwest.

4-B5 Repeat Prob. 4-B4, using the component method. Resolve each of the given forces into x and y components; find the resultant x component by addition; find the resultant y component by addition; and combine these two vectors by the Pythagorean theorem to find the magnitude and direction of the resultant of all the forces.

4-B6 Solve Example 4-2 by the head-to-tail (triangle) method.

4-B7 A traffic light of weight 200 N is suspended by two wires as in Fig. 4-28.   (*a*) What is the tension in wire AB?   (*b*) What is the tension in wire AC?

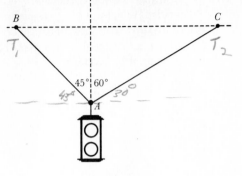

Figure 4-28

4-B8 A light horizontal wire is stretched between two posts 30 m apart. A bird weighing 6 N alights at the center of the wire, and the wire sags 8 cm. What is the tension in the wire when the bird is sitting on it?

4-B9 In Fig. 4-29, the ball's mass is 5 kg. Calculate the tension T in a horizontal string that holds the ball in position A. (The dashed lines are for use in Problem 4-C8.)

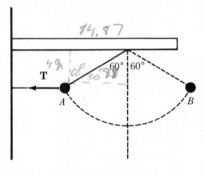

Figure 4-29

4-B10 Suppose that the trunk in Example 4-2 is moving *down* the plane at a constant speed of 1 m/s. What force F, directed upward parallel to the plane, will allow this to happen?

4-B11 Solve Example 4-2 if the applied force is horizontal instead of parallel to the plane.

4-B12 A 10-kg block on a table for which $\mu_s = 0.5$ is connected to a knot K by a horizontal string (Fig. 4-30). Another block hangs from point K, and a string at 30° goes upward

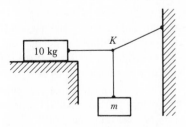

Figure 4-30

to a fixed wall. What maximum value can m be before the block on the table starts to move to the right?

4-B13 In Fig. 4-31, the coefficients of friction for both surfaces are $\mu_s = 0.6$ and $\mu_k = 0.4$. What force F is needed to keep the bottom block moving to the left with a constant acceleration of 2.5 m/s²?

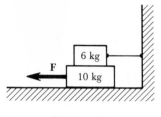

Figure 4-31

4-B14 What force, applied upward and at an angle of 30° with the vertical, is needed to move a scrub brush of weight 20 N upward with uniform velocity along a vertical wall for which $\mu_k = 0.2$ and $\mu_s = 0.3$?

4-B15 A book weighing 16 N rests on a wooden table top. The table is 80 cm on an edge, and the coefficient of static friction is 0.3. The table is then tilted so that one edge is 40 cm higher than the opposite edge, and the book begins to slide. With how much force should one press against the book, perpendicularly to the surface of the table, in order to keep it from sliding? (*Hint:* First find out how much frictional force is needed.)

4-B16 The center of gravity of an empty automobile is 1 m ahead of the rear axle, and the empty car weighs 6000 N. A 400-N driver sits 1.6 m ahead of the rear axle, and three passengers, each weighing 700 N, sit 0.6 m ahead of the rear axle. How far from the rear axle is the c.g. of the loaded car?

4-B17 A collapsible fishing pole consists of three sections, each 40 cm long and uniform. The sections weigh 6 N, 4 N, and 2 N, respectively. When assembled with the 4-N section in the middle, how far from the heavy end is the c.g. of the system?

4-B18 Tom and Dick are carrying their young friend Harry on a uniform horizontal plank 8 m long that weighs 140 N. Harry, who weighs 400 N, is sitting 5 m from Tom and 3 m from Dick. How much weight does each man support?

4-B19 A movie stunt woman of weight 700 N walks out to the end of a uniform horizontal plank that projects perpendicularly over the edge of a roof. The plank is 6 m long and weighs 500 N. How far from the roof can the plank overhang?

4-B20 A trunk weighing 500 N is on a frictionless inclined plane that rises 60 cm for every 100 cm measured along the plane. A force of 450 N is applied, directed upward along the plane. (*a*) Calculate the net force along the plane. (*b*) What is the acceleration of the trunk? (*c*) What is the normal push of the plane on the trunk?

4-B21 A sledge weighing 1400 N is being dragged along a rough road for which the coefficient of kinetic friction is 0.6. The forward tension in the rope is 900 N. Compute the resultant force on the sledge. Is the sledge in equilibrium? If not, what is its acceleration?

4-B22 A mountain climber weighing 800 N is held on a 60° slope by a rope attached to a tree at the top of the peak. The tension in the rope is 480 N. Compute (*a*) the normal force on the climber; (b) the magnitude and direction of the force of friction on the climber.

4-B23 A box of mass 100 kg is on a rough horizontal pavement for which the coefficient of kinetic friction is 0.3. The box is acted on by gravity and by a horizontal rope in which the tension is 400 N. How far does the box move in 2 s, starting from rest?

4-B24 A box of mass 80 kg is on a rough horizontal pavement for which the coefficient of kinetic friction is 0.3. The box is acted on by gravity and by a rope in which the tension is 400 N. The rope makes an angle of 30° with the horizontal. How far does the box move in 2 s, starting from rest?

4-B25 The lever of Fig. 4-32 is a flat piece of metal in a vertical plane. The lever weighs 80 N, with its c.g. at C. What horizontal force applied at B will hold the 60-N load at A in equilibrium?

4-B26 What vertical force applied at B in Fig. 4-32 (replacing the horizontal one) will hold the 60-N load at A in equilibrium? Is the required force upward or downward?

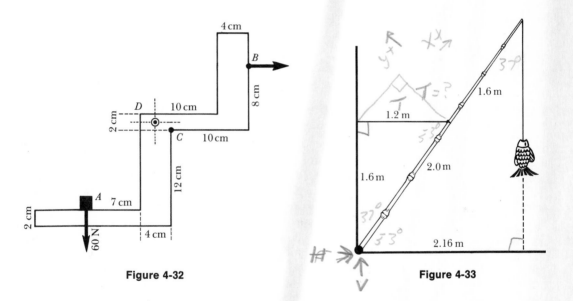

Figure 4-32   Figure 4-33

4-B27 A light bamboo fishing pole 3.6 m long is supported by a horizontal string as shown in Fig. 4-33. A 3-kg fish hangs from the end of the pole, and the pole is pivoted at the bottom. What is the tension in the supporting string, and what are the components of the force of the pivot on the pole?

4-B28 A cylindrical barrel of weight 600 N and diameter 52 cm is lying on a pavement with its curved surface snugly against a curb 16 cm high. (a) Make a careful diagram, then use the Pythagorean theorem to find the distance from the foot of the curb to the bottom of the barrel. (b) What force, applied horizontally at the top of the barrel, is needed to just ease it up off the pavement, so that the barrel is pivoted on the edge of the curb?

4-B29 A gardener pushes a hoe that is 1.8 m long into the ground by applying forces shown in Fig. 4-34. She applies with one hand a horizontal force **F** at the top of the handle, and with the other hand she applies a force of 40 N perpendicularly to the handle, 0.6 m from the top. The handle makes an angle of 45° with the ground. (*a*) What is the value of *F*? (*b*) Calculate the vertical component of the force exerted on the ground by the hoe. (Ignore the weight of the hoe.) (*c*) Calculate the horizontal component of the force of the hoe on the ground. Is this force directed toward the left or toward the right?

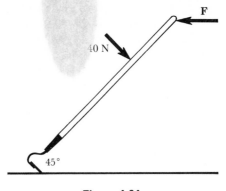

40 N

F

45°

Figure 4-34

4-B30 A cook holds a 2-kg carton of milk at arm's length (Fig. 4-35). What force **B** must be exerted by the brachialis muscle? (Ignore the weight of the forearm.)

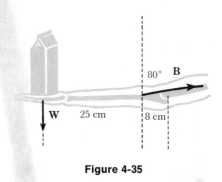

80° **B**

W 25 cm 8 cm

Figure 4-35

4-B31 A uniform pole 6 m long weighs 200 N and is attached by a pivot at one end to a vertical wall. The pole is held at an angle of 30° above the horizontal by a horizontal guy wire between the wall and the upper end of the pole. A load of 600 N hangs from the upper end of the pole. (*a*) Find the tension in the guy wire. (*b*) What is the resultant force of the wall on the pole?

4-B32 A yogini seeking to locate her center of gravity supported herself in a horizontal supine position with her head on one chair and her heels on a spring scale on another chair. The two points of contact were 165 cm apart, and a friend observed the scale to read 18 × 9.8 N. Later the yogini stood upright on the scale, which then read 50 × 9.8 N. How far from her heels was her center of gravity?

4-B33 A patient 200 cm tall weighs 700 N, and his center of gravity is 70 cm from the top of his head. He is carried on a uniform stretcher 200 cm long, of weight 200 N. What upward force must each stretcher bearer exert?

4-B34 A uniform ladder 5 m long weighing 200 N is leaning against a smooth, vertical wall with its base 3 m from the wall. The coefficient of static friction between the bottom of the ladder and the ground is 0.5. How far, measured along the ladder, can a 600-N man climb before the ladder starts to slip?

4-B35 A ladder 5 m long weighs 400 N, with its c.g. 2 m from the lower end. The ladder leans against a smooth wall and makes a 60° angle with the horizontal. What coefficient of static friction is needed to keep the ladder from slipping? (*Hint:* First compute the components of the force exerted by the ground on the ladder.)

4-B36 A uniform rectangular sign 0.8 m tall and 2 m wide weighing 600 N is held in a vertical plane, perpendicular to a wall, by a horizontal pin through the top inside corner, and by a guy wire that runs from the outer top corner of the sign to a point on the wall 1.5 m above the pin. (*a*) Calculate the tension in the wire. (*b*) Calculate the magnitude and direction of the force on the pin.

4-B37 A sign in the form of a uniform rectangular board 1 m wide and 0.6 m high weighs 200 N. The sign is hung as shown in Fig. 4-36 from a horizontal uniform pole 2 m long that weighs 240 N, and a guy wire *CD* makes a 30° angle with the horizontal. Calculate (*a*) the tension in the guy wire; (*b*) the horizontal and vertical components of the force exerted by the wall at *A*. (*Hint:* Each vertical rope at *B* and *C* supports half the weight of the sign.)

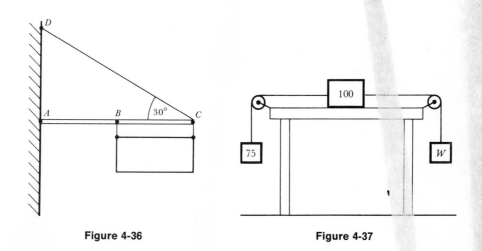

Figure 4-36   Figure 4-37

4-B38 A block that weighs 100 N rests on a rough table for which the coefficient of static friction is 0.40 and the coefficient of kinetic friction is 0.30. The block is counterbalanced by a 75-N weight and a weight *W*, as shown in Fig. 4-37. (*a*) What is the force of friction if *W* = 60 N? (*b*) For what range of values for *W* can the system remain at rest?

4-B39 In Fig. 4-37, assume the weight W to be 25 N and the coefficient of kinetic friction between the 100-N block and the table to be 0.30. What speed will the system acquire after moving 0.6 m, starting from rest?

4-B40 Show that the casting of Example 4-7 will not slip, whatever its mass (other data for the problem remaining the same).

4-C1 Using the definitions of sine and cosine given on pages 38–39, prove that for any angle, $\tan\theta = \sin\theta/\cos\theta$. Check this equation by referring to the table of sines, cosines, and tangents in Appendix Table 5, using any angle of your choice.

4-C2 Prove that the tangent of the angle of repose for a block resting on an inclined plane is equal to the coefficient of static friction (see Example 4-5).

4-C3 Tell how the coefficient of kinetic friction is measured by determining the angle of a plane down which a block will slide at constant velocity.

4-C4 A uniform ruler of length L and weight W leans without slipping against a smooth vertical wall at an angle θ with the vertical, with its base on a rough table. Show that the coefficient of static friction between the table top and the ruler must be at least equal to $\frac{1}{2}\tan\theta$. Interpret your result in the limiting cases as $\theta \to 0$, and as $\theta \to 90°$.

4-C5 Calculate the torque of the 40-N force in Fig. 4-15d if the axis passes through the lower right-hand corner of the plywood square.

4-C6 Given a table with four light legs equally spaced around the circumference of a uniform circular top weighing 100 N, find the smallest weight that, when placed on the table, will be able to upset it.

4-C7 A four-legged desk weighing 400 N is being dragged along a rough floor at uniform velocity (Fig. 4-38). The top surface of the desk is 75 cm above the floor, and the distance between the front legs and back legs is 120 cm. The coefficient of kinetic friction is 0.20. The c.g. of the desk is several centimeters below the center of the top, and the force is applied horizontally at the center of one edge. (*a*) What force is needed to drag the desk? (*b*) What is the normal force exerted by the floor on each front leg of the desk? (*c*) What is the normal force exerted by the floor on each rear leg of the desk? (*d*) What are the magnitude and direction of the force of one of the front legs on the floor?

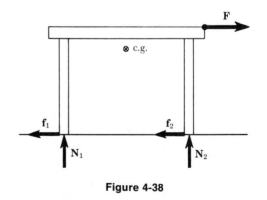

Figure 4-38

4-C8 A ball on the end of a light string (Fig. 4-29) is held at a 60° angle by means of the horizontal thread T. The thread is burned, and the ball swings over to position B, also 60°

from the vertical. What is the ratio of the tension in the string when the ball is at position *A*, before burning the thread, to the tension when the ball has swung to position *B*?

4-C9 An 800-N man stands on tiptoe on one foot so that all his weight is borne by the ground beneath the ball of the foot (Fig. 4-39*a*). If the foot and ankle are considered as a free body (Fig. 4-39*b*), the three forces that are in equilibrium are the reaction **W'** of the ground, the pull **T** of the Achilles tendon (exerted by the gastrocnemius muscle), and the compression **C** of the tibia. The force **C** is downward, at 15° from the vertical, and the force **T** is upward, at 21° from the vertical. **W'** is an upward force of 800 N. (*a*) Make a carefully drawn scale diagram, and estimate the values of *C* and *T*. (*b*) Calculate the values of *C* and *T*. (*Hint:* Use two simultaneous equations, one expressing the equilibrium of horizontal components of force, and the other expressing the equilibrium of vertical components of force. An alternative method of solution is to use the law of sines; see the Mathematical Appendix.)

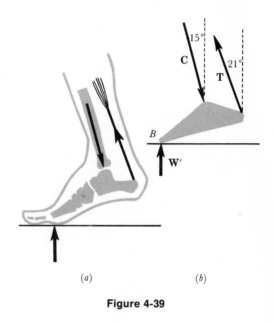

(*a*)   (*b*)

Figure 4-39

4-C10 A cake of ice starting from rest slides down a chute that is 6 m long and inclined 30° to the horizontal; the ice reaches the bottom of the chute with a speed of 7.2 m/s. Calculate the value of the appropriate coefficient of friction.

4-C11 A wood block is on a horizontal wood plank (see Table 4-1). The plank is tilted until the block starts to move; the plank is then held at this angle. How far down the plank does the block move in 1.2 s?

4-C12 A small boy is playing inside an empty, horizontal, cylindrical storage tank of radius 10 m. If the coefficient of static friction between his sneakers and the tank surface is 0.61, how high above the bottom of the tank (measured vertically) can the boy climb without slipping?

4-C13 A dog of mass m is on icy ground (coefficient of friction 0.05) and sees a squirrel in a tree 3 m away. How long will it take the dog to reach the tree if it accelerates as rapidly as possible?

4-C14 Two ropes support a load that weighs 500 N. The ropes are perpendicular to each other, and one rope has twice the tension of the other. Find the tension in each rope, and the angle that each rope makes with the vertical.

For Further Study

4-9 Center of Mass

Let us look a little more closely at our statement of Newton's second law (Sec. 3-3). We used the purposely vague term "body" when we said that "the acceleration produced by an unbalanced force acting on a body is proportional to the magnitude of the net force. . . ." In actual practice we do not often deal with particles, but rather with large collections of particles on which various forces are exerted, and which may exert various forces on each other. For instance, a book falls through the air after being dropped from some height. In Fig. 4-40 we show just 7 of the 10^{23} or so atoms in the book, and we show the externally acting forces on these atoms: forces due to gravity and to air resistance. We can combine all these small forces into one net force ($\Sigma\mathbf{F}$) that is vertical and has magnitude $W - f$. We can also combine the masses of all the molecules into a total mass (Σm). It can be shown that there exists a point called the *center of mass* (c.m.) such that if the net external force $\Sigma\mathbf{F}$ is applied at the c.m., then the c.m. moves just as a *particle* of mass Σm would move if it were acted on by a force $\Sigma\mathbf{F}$. Thus the center of mass is defined as that point at which all the *mass* of a body can be considered to be concentrated, when using Newton's second law. This definition is not to be confused with that for the center of gravity, which is the point at which all the *weight* of a body can be considered to be concentrated, when applying the torque condition or when calculating the resultant force of gravity.

It is for such reasons that we have been free and easy about treating extended bodies (trains, cars, and balls, for instance) as if they were particles. We use the concept of center of mass implicitly throughout the book. For instance, a bomb bursts in mid-air (Fig. 4-41). At B, fragments of the bomb start to move in various directions. The fragments are of unequal mass. However, the c.m. of the fragments continues to move along the original parabolic path. Since no new *external* force has been applied to the

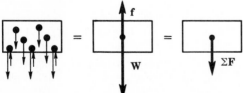

Figure 4-40 Forces acting on the molecules of a falling body, equivalent to a single force $\Sigma\mathbf{F}$ acting at the center of mass of the body.

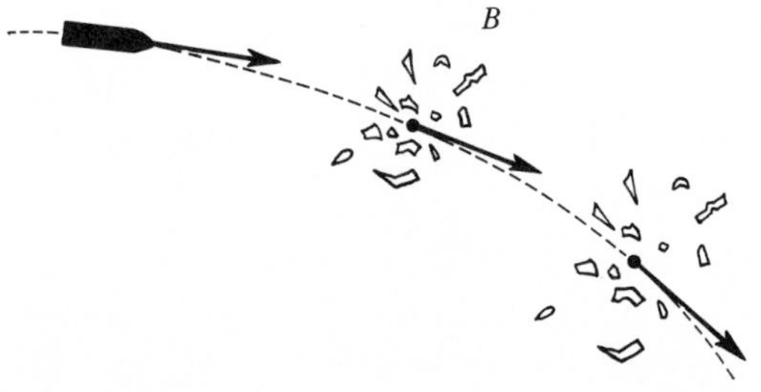

Figure 4-41 After the explosion, the center of mass of the fragments continues to move in a parabolic path.

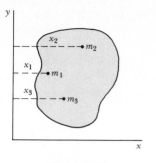

Figure 4-42

bomb, the fragments behave just as if the force of gravity were still acting on the total mass concentrated at the c.m. of the system. Newton's first law (a special case of the second law) also should be stated in terms of center of mass: "The center of mass of a body remains at rest, or moves with uniform velocity in a straight line, unless acted on by an unbalanced external force."

A formula for calculating the position of the c.m. can be derived rather easily. For simplicity we consider motion along the x axis; vector notation need not be used. For the various particles of the system (Fig. 4-42), we have $F_1 = m_1 a_1$, $F_2 = m_2 a_2$, $F_3 = m_3 a_3$, and so on. Adding these equations, which are Newton's second law applied to each particle in turn, we get

$$F_1 + F_2 + F_3 + \cdots$$
$$= m_1 a_1 + m_2 a_2 + m_3 a_3 + \cdots$$

Each acceleration is the second derivative of the corresponding displacement, so we can write

$$F_1 + F_2 + F_3 + \cdots$$
$$= m_1 \frac{d^2 x_1}{dt^2} + m_2 \frac{d^2 x_2}{dt^2} + m_3 \frac{d^2 x_3}{dt^2} + \cdots$$

$$\Sigma F = \frac{d^2}{dt^2} (m_1 x_1 + m_2 x_2 + m_3 x_3 + \cdots)$$

$$\frac{\Sigma F}{\Sigma m} = \frac{d^2}{dt^2} \left(\frac{m_1 x_1 + m_2 x_2 + m_3 x_3 + \cdots}{m_1 + m_2 + m_3 + \cdots} \right)$$

$$(4\text{-}3)$$

From this we see that if the x coordinate of the center of mass is defined as

$$x_{\text{c.m.}} = \frac{m_1 x_1 + m_2 x_2 + m_3 x_3 + \cdots}{m_1 + m_2 + m_3 + \cdots} \qquad (4\text{-}4)$$

then Eq. 4-3 becomes

$$\frac{\Sigma F}{\Sigma m} = \frac{d^2}{dt^2} (x_{\text{c.m.}})$$

or

$$\frac{\Sigma F}{\Sigma m} = a_{\text{c.m.}} \qquad (4\text{-}5)$$

Similar formulas can be written for the coordinates $y_{\text{c.m.}}$ and $z_{\text{c.m.}}$ of the center of mass. We have proved that the acceleration of the c.m. is the same as that of a single particle of mass Σm acted on by a single force ΣF.

Let us now derive a formula for the position of the center of *gravity* (c.g.). On the earth, where **g** is nearly constant in magnitude and direction, it turns out that the same formula serves both for $x_{\text{c.m.}}$ and for $x_{\text{c.g.}}$! Indeed, many authors use the terms interchangeably. As in Sec. 4-6, the c.g. is that point at which all the *weight* of the body can be considered to be concentrated when applying the torque condition or when calculating the resultant force of gravity on an extended object. In Fig. 4-43, we see that each particle contributes a torque (= force × lever arm). We equate the torque exerted by the total weight of all the particles acting at the center of gravity to

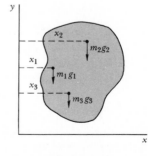

Figure 4-43

the sum of the torques exerted by the weights of the individual particles:

$$(m_1g_1 + m_2g_2 + m_3g_3 + \cdots)x_{c.g.}$$
$$= (m_1g_1)x_1 + (m_2g_2)x_2 + (m_3g_3)x_3 + \cdots$$

$$x_{c.g.} = \frac{m_1g_1x_1 + m_2g_2x_2 + m_3g_3x_3 + \cdots}{m_1g_1 + m_2g_2 + m_3g_3 + \cdots}$$

If, and only if, the body is in a uniform gravitational field, we can cancel the g's and obtain

$$x_{c.g.} = \frac{m_1x_1 + m_2x_2 + m_3x_3 + \cdots}{m_1 + m_2 + m_3 + \cdots} \quad (4\text{-}6)$$

For all practical purposes, the c.m. is located at the same point as the c.g. However, this is strictly true only if the body is in a *uniform* gravitational field. Consider a uniform cube of stone (a mountain) a kilometer on an edge. The c.m. is at the exact geometrical center of the cube, but the c.g. is slightly below the center, since the bottom half of the cube is closer to the center of the earth and consequently weighs more than the upper half. Fine points such as these help sharpen our thinking and help point out the real difference in concept between center of mass and center of gravity.

Problems   **4-C15** Body A of mass 3 kg is on a smooth, horizontal floor at the origin of coordinates, and body B of mass 5 kg is on the x axis, 8 m east of A. (a) Where is the c.m. of the system? (b) If a force of -6 N acts on A for 4 s, and a force of $+5$ N acts on B for 4 s, where is the new c.m.? (c) Through what distance $s_{c.m.}$ has the c.m. moved? (d) Calculate the acceleration of the c.m., using $s_{c.m.} = \frac{1}{2}at^2$. (e) Verify for this case that the acceleration found in (d) is also given by $\Sigma F/\Sigma m$.

4-C16 Calculate the three coordinates of the c.g. of the system of three uniform cubes, each side of length 4 cm, arranged as shown in Fig. 4-44. The density of the cube at the left is twice that of the other two cubes.

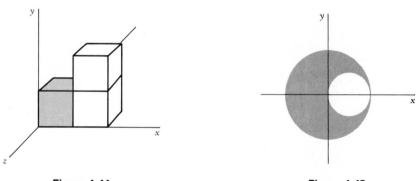

Figure 4-44   Figure 4-45

4-C17 A uniform disk of radius R has a hole of radius $\frac{1}{2}R$ cut from it as shown in Fig. 4-45. How far from the center of the disk is the c.g. of the remaining portion? (*Hint:* From symmetry, the c.g. is on the x axis. Apply Eq. 4-6 with one mass a complete disk and the other mass a smaller disk of appropriate negative mass.)

References   1. D. B. Steinman, "Bridges," *Sci. American*, **191**(5), 60 (Nov. 1954). Excellent photographs, including one of the Tacoma Narrows bridge in the act of collapsing.
2. L. A. Strait, V. T. Inman, and H. J. Ralston, "Sample Illustrations of Physical Principles Selected from Physiology and Medicine," *Am. J. Phys.* **15**, 375 (1947); also *Am. J. Phys.*

19, 173 (1951). An analysis of the action of the human body, considered as a system of levers and forces applied through muscles and tendons.

3. F. A. Smith, Jr., "Biomechanics using a crane beam," *Phys. Teach.* **16**, 220 (1978).

4. R. M. Sutton, "Two Notes on the Physics of Walking," *Am. J. Phys.* **23**, 490 (1955).

5. E. Rabinowicz, "Friction" (Resource Letter F-1), *Am. J. Phys.* **31**, 897 (1963). Gives an extensive bibliography.

6. F. Palmer, "What About Friction?" *Am. J. Phys.* **17**, 181, 327, 336 (1949).

7. C. A. Maney, "Experimental Study of Sliding Friction," *Am. J. Phys.* **20**, 203 (1952).

8. F. Palmer, "Friction," *Sci. American* **184**(2), 54 (Feb. 1956).

9. E. Rabinowicz, "Stick and Slip," *Sci. American* **194**(5), 109 (May 1956).

10. R. A. Bartels, "Braking distance versus mass for automobiles," *Am. J. Phys.* **45**, 398 (1977). The distance is essentially independent of mass over a 3-fold range of mass; slightly shorter minimum stopping distances are found for "some or all wheels locked" in a skid.

11. J. D. Edmonds, Jr., "Speed and displacement relations for tire skids," *Am. J. Phys.* **48**, 253 (1980). Realistic approach taking account of variation of μ_k with speed. Simple computer program (note an obviously misplaced parenthesis in line 20).

12. R. C. Smith, "General physics and the automobile tire," *Am. J. Phys.* **46**, 858 (1978). The actual tire-road interaction is more complex than would be expected from a simple model using coefficients of friction.

13. W. F. Magie, *A Source Book in Physics* (McGraw-Hill, New York, 1935), p. 22. An early treatment of the inclined plane, by Stevin.

5

Conservation of Momentum

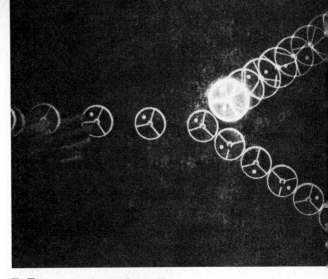

In this chapter and the next we shall study two of the three* great conservation laws of mechanics that deal with isolated systems. The component parts of such a system may interact with each other, but there is no interaction of any sort with any particles or objects outside the system. Thus we have this definition:

An isolated system is one that is not acted on by external forces.

Wherever possible we use conservation laws as statements of basic and general truths about the physical world. They are indeed of great philosophical significance for an understanding of the symmetries of space and time. A more practical reason for using conservation laws is that they apply to whole *systems* of particles, making it possible to get useful results without being overwhelmed by detailed consideration of every particle or part of the system individually.

---

* Conservation of angular momentum is the subject of Sec. 7-8.

The law of conservation of momentum was known to Newton, and it is still considered to be valid, even in areas where Newton's original formulation of the laws of mechanics has been replaced by Einstein's more inclusive theory. As a general principle, the law of conservation of momentum continues to be of utmost value in many fields of physics and engineering, including space navigation, quantum mechanics, and relativity.

Another unifying principle, the law of conservation of energy, was harder to come by. Not until about 1845 was it realized that the energy content of an isolated system remains constant. For example, when a falling ball approaches the earth, and the earth moves slightly to meet the ball, the system (earth + ball) has a constant total amount of energy; the energy of the system is simply transformed from one kind into another during the process. Similarly, if a sufficiently broad definition of energy is used, the total energy of a system is conserved during a chemical reaction between two atoms or during the collision of two particles in a high-energy research accelerator.

Momentum is conserved in this air-table experiment in which a puck moving toward the right makes a glancing collision with a stationary puck.

5-1 Definition of Momentum

The momentum of a moving body is so fundamental that early writers such as Newton called it "quantity of motion" or simply "motion." We define the linear *momentum* of a body as the product of its mass and its velocity:

Momentum = (mass)(velocity)

Since velocity is a vector quantity, momentum is also a vector quantity, having the same direction as the velocity of the body. We use the symbol **p** for linear momentum and write

$$\mathbf{p} = m\mathbf{v}$$

In many problems concerned with motion in a straight line, a separate symbol for momentum is not needed, and the vector nature of momentum can be taken care of by writing $+mv$ and $-mv$. The dimensions of momentum are

$$[\text{momentum}] = [M][LT^{-1}] = [MLT^{-1}]$$

There are no special names for units of momentum; $kg \cdot m/s$ or $g \cdot cm/s$ may be used—any mass unit multiplied by a velocity unit.

5-2 Newton's Second Law in Terms of Momentum

We may restate Newton's second law in terms of change of momentum. In fact, Newton's own statement of the law involved rate of change of "motion," which was his term for momentum.

Let us consider the special case of a constant force, which gives rise to a constant acceleration. As we saw in Chap. 2, for constant acceleration the equation

$$\mathbf{a} = \frac{\mathbf{v} - \mathbf{v}_0}{t - t_0} = \frac{\Delta \mathbf{v}}{\Delta t}$$

is valid for a time interval Δt of any size, however large. Therefore,

$$\text{net } \mathbf{F} = m\mathbf{a}$$

$$\text{net } \mathbf{F} = m\frac{\mathbf{v} - \mathbf{v}_0}{t - t_0} = m\frac{\Delta \mathbf{v}}{\Delta t}$$

$$\text{net } \mathbf{F} = \frac{m\mathbf{v} - m\mathbf{v}_0}{t - t_0} = \frac{\Delta \mathbf{p}}{\Delta t} \qquad (5\text{-}1)$$

Thus *net force equals the rate of change of momentum.** Multiplying both sides by the time interval Δt, we obtain

$$(\text{net } \mathbf{F})\Delta t = \Delta \mathbf{p} \qquad (5\text{-}2)$$

$$(\text{net } \mathbf{F})\Delta t = \mathbf{p} - \mathbf{p}_0 = m\mathbf{v} - m\mathbf{v}_0$$

This product of net force and the time during which it acts is defined as *impulse*; Eq. 5-2 tells us that the impulse equals the change of momentum of a body. Impulse is measured in units of force multiplied by time, such as $N \cdot s$ or $dyn \cdot s$. We see from Eq. 5-2 that these units are also suitable for momentum. As far as total change of momentum is concerned, a large force acting for a short time may produce the same effect as a small force acting for a long time. In combination with Newton's third law, we have here an interpretation of several common phenomena, such as the impact of a stream of water, the force on the piston of an automobile cylinder, the propulsion of a jet rocket, and even the pressure due to the impact of gas molecules on the wall of a container.

Example 5-1

A fire hose sends 20 kg of water each second onto a burning building (Fig. 5-1). If the water leaves the nozzle at 60 m/s and does not bounce back from the wall, what is the force (assumed constant) on the wall of the building?

In 1.00 s, a certain amount of momentum of the water is canceled; to do this, the wall exerts a force to the left. The force of the wall on the water is

$$\text{net } F = \frac{mv - mv_0}{\Delta t}$$

$$= \frac{(20 \text{ kg})(0) - (20 \text{ kg})(60 \text{ m/s})}{1.00 \text{ s}}$$

$$= -1200 \text{ kg} \cdot \text{m/s}^2 = 1200 \text{ N to the left}$$

---

* If the force is not constant but is instead some function of time, the conclusion is still true that net force equals the rate of change of momentum. To prove this we take the limit of $\Delta \mathbf{p}/\Delta t$ as $\Delta t \to 0$ and obtain a derivative: net $\mathbf{F} = d\mathbf{p}/dt$. Similarly, Eq. 5-2, in the limit, leads to a definite integral for impulse: $\int_{t_0}^{t} (\text{net } \mathbf{F}) \, dt = m\mathbf{v} - m\mathbf{v}_0$.

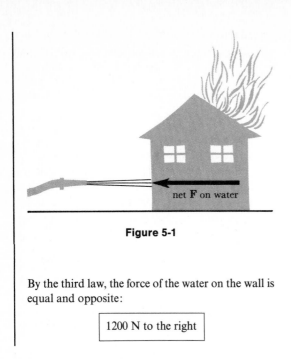

Figure 5-1

By the third law, the force of the water on the wall is equal and opposite:

$$\boxed{1200 \text{ N to the right}}$$

Example 5-2

The force of an explosion on the piston of an engine is due to the change of momentum of the gas particles. What must be the speed of the molecules of gas if 0.6 g of gas exerts a force of 1400 N on the piston in an explosion that lasts 0.001 s?

To have a consistent set of mks units, we express the mass of 0.6 g as 0.6×10^{-3} kg. Each molecule has its velocity changed from $+v$ to $-v$ by the piston, and the change of velocity is therefore $\Delta v = v - (-v) = 2v$.

$$\text{net } F = \frac{\Delta(mv)}{\Delta t} = \frac{m\,\Delta v}{\Delta t}$$

$$1400 \text{ N} = \frac{(0.6 \times 10^{-3} \text{ kg})(2v)}{10^{-3} \text{ s}}$$

$$v = \boxed{1170 \text{ m/s}}$$

Example 5-3

A rocket of mass 10^4 kg, starting from rest, is acted on by a net force of 2×10^5 N for 20 s. What is the final velocity of the rocket?

$$F\,\Delta t = mv - mv_0$$
$$(2 \times 10^5 \text{ N})(20 \text{ s}) = (10^4 \text{ kg})(v) - 0$$
$$v = \boxed{400 \text{ m/s}}$$

In practice, the mass of the rocket would decrease as fuel is used up; our solution is approximate, based on an *average* mass of 10^4 kg.

Do not imagine that a rocket depends on the presence of air for its operation. It does not need air to "push against" and would actually work better in a vacuum. It is the molecules of ejected gas that push forward on the rocket. The gas is accelerated backward by a force exerted by the rocket, and Newton's third law tells us that the gas molecules must push forward on the rocket. It is this reaction force that propels the rocket. The idea of jet propulsion is not new. The first steam engine of recorded history was that of the Greek philosopher Hero (130 B.C.). Steam escaping through the nozzles of his device was the ancestor of the exhaust escaping from a jet plane. In modern times certain lawn sprinklers operate on the same principle. In some models a gear system is even provided to convert some of the rotational energy into translational kinetic energy, and the sprinkler "walks" across the lawn. The locomotion of the squid, a cephalopod, depends largely on the reaction force of ejected liquid, and the locomotion of an astronaut during a space walk depends on the reaction force of ejected gas.

5-3 Conservation of Momentum

To derive the law of conservation of momentum, we shall consider first an isolated system consisting of just two bodies. Suppose A and B, of mass m_1 and m_2 respectively, are traveling toward the right at velocities v_1 and v_2, with A overtaking and colliding with B (Fig. 5-2). When they collide, A exerts a force $+F$ on B, and B exerts an equal and opposite force $-F$ on A. The bodies receive different accelerations, since their masses

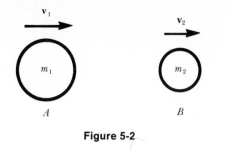

Figure 5-2

are different. Suppose that the duration of the collision is Δt. Then, as we have shown in the previous section, $+\mathbf{F}\Delta t$ equals the change in momentum of body B, and $-\mathbf{F}\Delta t$ equals the change in momentum of body A. These changes are numerically equal, since the forces are equal and opposite and the time of contact is the same. Thus the gain in momentum of one body exactly equals the loss of momentum of the other body; the total momentum of the system has remained constant during the interaction between the two bodies.

It is not hard to extend the proof to include a system containing any number of interacting bodies—for instance, a collection of 10^{23} molecules in a tank of compressed air. No part of the system can undergo a change of momentum unless another part undergoes an equal and opposite change; the justification for this statement is Newton's third law. In general, we state the *law of conservation of momentum* as follows:

The total linear momentum of an isolated system of bodies remains constant.

The importance of this law arises from its generality. It is true whether or not some mechanical energy has been transformed into some other form of energy such as heat or sound energy, as in an inelastic collision (see Sec. 6-9). It is true for any number of bodies making up a system. For instance, when a grenade explodes in mid-air, the fragments move in various directions with various speeds, but the vector sum of the momentum of all the fragments equals the

momentum of the original body. (See Fig. 4-41 on page 104.) The reason for the quite general validity of the law of conservation of momentum is that it follows from Newton's third law, which is always true.

To express mathematically the fact that the total momentum of the two bodies remains constant, we use $\mathbf{v}_1$ and $\mathbf{v}_2$ for the initial velocities (before collision) and $\mathbf{V}_1$ and $\mathbf{V}_2$ for the final velocities (after collision):

$$m_1\mathbf{v}_1 + m_2\mathbf{v}_2 = m_1\mathbf{V}_1 + m_2\mathbf{V}_2$$

If all motions take place along the same line, we can use $+$ and $-$ signs to designate directions; vector notation is not needed for straight-line collision problems.

Example 5-4

A car of mass 600 kg traveling at 20 m/s collides with a stationary truck of mass 1400 kg. The two vehicles are locked together after collision; what is their combined velocity?

After the collision the car and truck form a single object of combined mass 2000 kg.

Total momentum before collision

$$= \text{total momentum after collision}$$

$$m_C v_C + m_T v_T = m_{C+T} v_{C+T}$$

$$(600 \text{ kg})(20 \text{ m/s}) + (1400 \text{ kg})(0) = (2000 \text{ kg})(v_{C+T})$$

$$v_{C+T} = \boxed{6 \text{ m/s}}$$

Example 5-5

What is the recoil velocity of a 3-kg gun that fires a bullet of mass 0.060 kg at a velocity of 200 m/s? Let v stand for velocities before firing, and V for velocities after firing.

$$m_G v_G + m_B v_B = m_G V_G + m_B V_B$$

$$(3 \text{ kg})(0) + (0.06 \text{ kg})(0)$$

$$= (3 \text{ kg})(V_G) + (0.06 \text{ kg})(200 \text{ m/s})$$

$$V_G = \boxed{-4 \text{ m/s}}$$

On a very small scale, conservation of momentum is the basis for the *ballistocardiograph*, a diagnostic tool useful in the study of circulatory problems. During each heartbeat a certain mass of blood is given a forward momentum by the heart muscles; the heart muscles and the rest of the body recoil backward. For diagnostic purposes, the patient is strapped on a light table that is free to move back and forth along a line parallel to the length of the patient's body. The complex motion of the table is recorded during many heartbeat cycles and compared with motion recorded with a subject known to be in good health. Considered as a whole, the patient is an isolated body whose momentum is therefore constant (zero). The changes in momentum of blood and heart muscles are compensated by observable changes in the momentum of the body frame and the table to which it is attached. The heart muscles are only a small fraction of the total body mass, so the body recoil is very small. The blood itself, while considerable, flows on the whole simultaneously in both directions, and only the changes associated with changes in the capacity of the heart chambers during the heartbeat give rise to net changes in momentum of the blood.

Example 5-6

A normal young human adult of mass 60 kg is placed on a ballistocardiograph table of mass 2 kg. Data are taken that indicate an acceleration of the table of 0.05 m/s² during a certain 0.1-s portion of the heartbeat. Estimate (a) the displacement of the table; (b) the change in momentum during the 0.1-s interval; (c) the force exerted by the heart muscles.

(a) $s = \frac{1}{2}at^2$

$= \frac{1}{2}(0.05 \text{ m/s}^2)(0.1 \text{ s})^2 = \boxed{2.5 \times 10^{-4} \text{ m}}$

This is only 0.25 mm, indicating the need for sophisticated and sensitive recording techniques.

(b) Final velocity:

$$v = at = (0.05 \text{ m/s}^2)(0.1 \text{ s}) = 0.005 \text{ m/s}$$

Final momentum:

$$p = mv$$

$$= (62 \text{ kg})(0.005 \text{ m/s}) = \boxed{0.31 \text{ kg} \cdot \text{m/s}}$$

(c)   net $F = ma$

$$= (62 \text{ kg})(0.05 \text{ m/s}^2) = \boxed{3.1 \text{ N}}$$

See References at the end of the chapter for sources of further information on ballistocardiography.

The law of conservation of momentum can be used as the basis of an experiment to measure mass. Consider a collision between two objects whose centers of mass* are moving along the same straight line. Let v_1 and v_2 be their respective velocities before collision, and let V_1 and V_2 be their velocities after collision.

$$m_1 v_1 + m_2 v_2 = m_1 V_1 + m_2 V_2$$
$$m_1(v_1 - V_1) = m_2(V_2 - v_2)$$

$$\frac{m_1}{m_2} = \frac{V_2 - v_2}{v_1 - V_1} \qquad (5\text{-}3)$$

The velocities v_1, v_2, V_1, and V_2 can be measured (for instance, with a movie camera). Hence, we have found one way of measuring the mass m_2 in terms of some "standard" mass m_1 such as the standard kilogram. As we said in Sec. 3-3, *to measure a property is to define it*; we may say

---

* As shown in Sec. 4-9, the center of mass (c.m.) of an extended body or system is located at the same point as the center of gravity (c.g.). Newton's second law then takes the form

$$\mathbf{a}_{\text{c.m.}} = (\Sigma\mathbf{F})/(\Sigma m)$$

where $\Sigma\mathbf{F}$ is the sum of all forces acting on the body and Σm is the total mass of the body. In this sense, the mass of a body can be considered to be located at the center of mass.

that the law of conservation of momentum provides a way of defining mass. Applying these ideas to the straight-line collision described in Example 5-4, we have $v_1 = 20$ m/s, $v_2 = 0$, $V_1 = 6$ m/s, $V_2 = 6$ m/s. Therefore,

$$\frac{m_1}{m_2} = \frac{6-0}{20-6} = \frac{6}{14}$$

If an observer had measured only the velocities, knowing nothing at the start about the masses, she would have been able to deduce the ratio of the two masses, and she would have found 6:14, as is actually the case (600 kg:1400 kg).

5-4   Collisions at an Angle

To illustrate the vector nature of momentum, we shall analyze a glancing collision (one that is not head-on) in the xy plane. The vector sum of all momenta before collision $\mathbf{p}_1$ must be equal to the vector sum of all momenta after collision $\mathbf{p}_2$. For two vectors to be equal in magnitude and direction, their x components must be equal, and their y components must also be equal. [You can construct a proof of this by asking how

$(\mathbf{p}_1 - \mathbf{p}_2)$ can be zero. See the first footnote on page 75.]

Example 5-7

A puck A on an air table has mass 0.2 kg and an initial velocity of 1.60 m/s at an angle 45° to the x axis (Fig. 5-3). It makes a glancing collision with a stationary puck B of mass 0.4 kg, and after collision A is observed to be moving at 1.00 m/s at 20°. What are the magnitude and direction of B's final velocity?

This collision is best approached by the component method. The initial momentum $\mathbf{p}$ of puck A has magnitude $(0.2 \text{ kg})(1.6 \text{ m/s}) = 0.320$ kg·m/s. Similar units apply to all masses, velocities, and momenta, and will be omitted for simplicity. The components of $\mathbf{p}$ are found and entered in a table: $p_x = 0.320 \cos 45° = +0.226$; $p_y = -0.320 \sin 45° = -0.226$. We also enter the components of momentum after collision, using p_x' and p_y' for the components of B's momentum after collision.

Before collision

	p	p_x	p_y
puck A	$(0.2)(1.6) = 0.320$	$+0.226$	-0.226
puck B	$(0.4)(0) = 0$	0	0
totals before collision		$+0.226$	-0.226

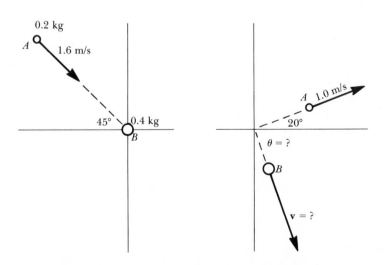

Figure 5-3   *Left*: before collision. *Right*: after collision.

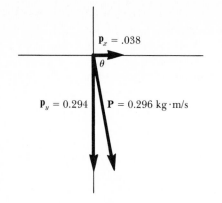

$$P_x = .038$$

$$\theta$$

$$P_y = 0.294 \qquad P = 0.296 \text{ kg} \cdot \text{m/s}$$

Figure 5-4

After collision

	p	p_x	p_y
puck A	$(0.2)(1.0) = 0.200$	$+0.188$	$+0.068$
puck B	$(0.4)(v') = ?$	p_x'	p_y'
totals after collision		0.188	0.068
		$+p_x'$	$+p_y'$

For the x components:

$$0.226 = 0.188 + p_x'; \qquad p_x' = 0.038 \text{ kg} \cdot \text{m/s}$$

For the y components:

$$-0.226 = 0.068 + p_y'; \qquad p_y' = -0.294 \text{ kg} \cdot \text{m/s}$$

The final momentum $\mathbf{p'}$ is the vector sum of its components (Fig. 5-4):

$$p' = \sqrt{0.038^2 + 0.294^2} = 0.296$$

$$\tan \theta = \frac{0.294}{0.038} = 7.73; \qquad \theta = 83°$$

The velocity of B after collision is

$$v = \frac{p'}{m} = \frac{0.296 \text{ kg} \cdot \text{m/s}}{0.40 \text{ kg}} = 0.74 \text{ m/s}$$

Thus the final velocity of puck B is

> 0.74 m/s, 83° below the x axis

5-5 Frames of Reference

To study the motion of a billiard ball, a player uses the table top as his frame of reference. In physics, a *frame of reference* is a set of three mutually perpendicular axes, such as the x, y, and z axes, relative to which positions in space can be measured. The space coordinates of a body depend on the location of the origin of the frame of reference and on the orientation of the axes. They also depend on the time t if the body is in motion; until the start of the 20th century it was assumed that the rate of flow of time is the same for observers in different frames of reference even if these frames are in relative motion.

The relative motion of two frames of reference is illustrated by a numerical example (Fig. 5-5).

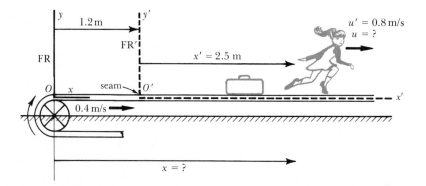
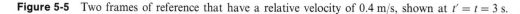

Figure 5-5 Two frames of reference that have a relative velocity of 0.4 m/s, shown at $t' = t = 3$ s.

At an airport, a conveyor belt is moving horizontally with a uniform velocity $v = 0.4$ m/s relative to the earth, which we call frame of reference FR. A seam in the belt is considered to be the origin O' of the moving frame of reference FR'. The origin of the stationary frame of reference* is a point O just above the pulley. We use primed symbols x', u' and t' to refer to measurements in frame FR', and unprimed symbols x, u, and t for measurements in frame FR. At $t = t' = 0$, the two frames coincided; at $t = t' = 3$ s, the origin O' has reached the position shown in Fig. 5-5. A child is running along the conveyor belt with velocity $u' = 0.8$ m/s, and at $t' = 3$ s the child is 2.5 m from the seam. It is clear from the figure that

$$x = x' + vt'$$
$$3.7 \text{ m} = 2.5 \text{ m} + (0.4 \text{ m/s})(3 \text{ s})$$

We use Eq. 2-15 (page 37) to find the child's velocity relative to frame FR:

$$u = u' + v \qquad (5\text{-}4)$$
$$1.2 \text{ m/s} = 0.8 \text{ m/s} + 0.4 \text{ m/s}$$

Note that t (in frame FR) and t' (in frame FR') are assumed to be equal to each other.

This example illustrates *Galilean relativity*—the way in which Galileo and Newton dealt with relative motion. We have taken the relative motion to be along the x axis (also along the x' axis), so $y = y'$ and $z = z'$. Collecting all these equations together, we have what is known as the *Galilean transformation*:

$$\left.\begin{array}{l} x = x' + vt' \\ y = y' \\ z = z' \\ t = t' \end{array}\right\} \qquad (5\text{-}5)$$

Since u is the rate of change of x, and u' is the rate of change of x', the first equation also implies

---

* Of course, it is only relative motion that is important; we could equally well call the belt stationary and the floor moving.

that $u = u' + v$, which is Eq. 5-4. The Galilean transformation is valid for relative velocities that are small compared with c, the speed of light. In our study of Einsteinian relativity in Chap. 28, we shall see that a modification is needed (the Lorentz transformation) that includes the Galilean transformation in the limit if $v \ll c$.

5-6 Inertial and Noninertial Frames

We have stated earlier that Newton's laws apply to any system or any part of a system of bodies. An *inertial frame of reference* is a frame relative to which Newton's laws are valid. It turns out that any other frame in uniform motion relative to an inertial frame is also an inertial frame. This powerful and simplifying result arises from the fact that Newton's second law is concerned not with velocity as such, but with acceleration, which is the rate of change of velocity. No accelerations of any space coordinates are introduced by constant relative motion of two frames of reference. In our example, if the earth is an inertial frame, then so is the conveyor belt.

We can say that the laws of Newtonian mechanics are *invariant* with respect to a change from one inertial frame to another.[†] To illustrate this invariance, let's use the numerical results of Example 5-4 to calculate the total momentum of the system in two frames of reference. In the earth frame FR_{earth} the 1400-kg truck is stationary (Fig. 5-6). We check the law of conservation of momentum as follows (masses in kg, velocities

---

[†] That is, the laws of mechanics are invariant with respect to the Galilean transformation.

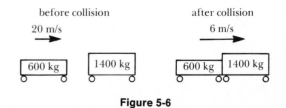

before collision   after collision

20 m/s   6 m/s

600 kg   1400 kg   600 kg   1400 kg

Figure 5-6

in m/s, momenta in kg·m/s):

$$(600)(20) + (1400)(0) \overset{?}{=} (2000)(6)$$
$$12{,}000 = 12{,}000$$

Now consider the same collision as viewed in a frame FR′ that is moving at 20 m/s toward the right. In this frame the car is originally stationary, and the truck comes to meet it at −20 m/s (to the left in frame FR′). After collision, the combined mass moves at 6 m/s − 20 m/s = −14 m/s.

$$(600)(0) + (1400)(-20) \overset{?}{=} (2000)(-14)$$
$$-28{,}000 = -28{,}000$$

Momentum is conserved in each frame. The *form* of the law is invariant, even though the numerical value of the total momentum is different in the two frames.

Not all frames are inertial; for instance, a frame attached to the rotating earth is, strictly speaking, not an inertial frame of reference. This is illustrated by the circular motion of winds in a hurricane. A large air mass veers toward the east as it moves northward from the equatorial region (Fig. 5-7). This violates Newton's first law, as viewed by someone whose frame of reference is attached to the earth, for the air molecules are subject to no net force, and yet they fail to move with uniform velocity in a straight line. However, an observer from outer space, whose frame of reference might be the "fixed" stars, would see no conflict, because he would see that the earth is turning and that the eastward velocity of the earth at the equator (about 1700 km/h) is greater than the eastward velocity (about 1100 km/h) at the latitude of the target, Newfoundland. Any air mass leaving the West Indies has a greater eastward component of velocity than does the target, so the air gets ahead of the target and moves toward the right. We see that there is an "unexplained" or "fictitious" force on the air, from the point of view of the observer whose frame of reference is attached to the rotating earth, and therefore his frame is not an inertial one. Similarly, air moving southward veers toward the west because the target is moving eastward faster than the air. The resulting pattern of winds is counterclockwise in the Northern Hemisphere; the same argument shows that hurricanes in the Southern Hemisphere have clockwise patterns (Fig. 5-8).

Fictitious forces arising from the rotation of the earth are often called *Coriolis forces*, after the French physicist Gaspard Coriolis who, in 1835, first studied them experimentally. Water in the Gulf Stream, like the air in Fig. 5-7, veers toward the right as it moves northward, giving the British Isles a vastly warmer climate than that of Labrador, which is at the same latitude. Coriolis forces also cause the line of oscillation of a Foucault pendulum, common in science museums, to move a bit toward the right during each swing. In daily life the effects of Coriolis forces are small, and we can consider the earth to be an inertial frame of reference for most purposes. Only when large distances and times are involved do the small fictitious forces due to the rotation of the earth produce significant effects. Fictitious forces are also called *inertial forces* or pseudoforces. In the next section we shall see that inertial forces of considerable magnitude can arise when a frame of reference has a large acceleration.

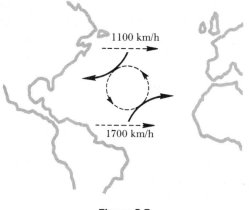

1100 km/h

1700 km/h

Figure 5-7

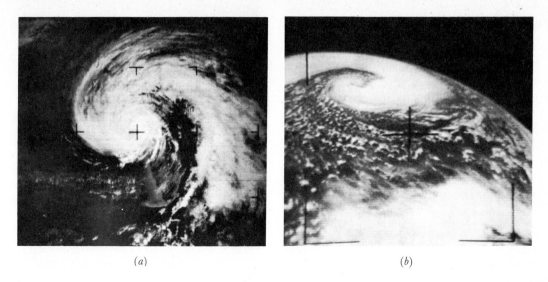

(a)   (b)

Figure 5-8 Satellite photographs showing hurricane wind patterns: (*a*) counterclockwise in the Northern Hemisphere and (*b*) clockwise in the Southern Hemisphere. In (*a*) the coastline of Florida is seen south of the eye of Hurricane Alma, June 9, 1966. The storms "wind up" as viewed from above.

5-7   *Weight, Weightlessness, and Artificial Gravity*

The weight of a body whose mass is *m* is the downward force of gravitation on it, given by Newton's law of gravitation. We call this the *true weight* of the body because it does not depend on the acceleration of our frame of reference.

Although the true weight of a body depends only on its position relative to other gravitating bodies, such as the earth, the moon, and the sun, the *apparent weight*—the sensation of weight—can be radically affected by the acceleration of the observer's frame of reference. For example, an astronaut whose true weight at the earth's surface is 800 N (180 lb) is sitting on a seat cushion in his capsule prior to launch. He is in equilibrium, and the downward 800 N of gravitational force (his true weight) is balanced by an upward 800 N caused by the compression of the seat cushion. The apparent weight is the actual force between the astronaut and the cushion that supports him; at this moment his apparent weight equals his true weight. Now suppose a malfunction occurs during lift-off and the astronaut is ejected upward, still strapped in his seat. The astronaut and seat are now projectiles having acceleration **g**, where **g** is a downward vector. There is no force between the astronaut and the seat; he has no sensation of weight and he is "weightless." He regains his apparent weight as soon as his parachute opens; the lift of the chute (which we imagine to be attached to the seat) causes the seat cushion to be compressed against him, giving the astronaut the sensation of weight. Of course, the sensation of weightlessness includes effects within the body; while he is weightless the astronaut's internal organs no longer press against each other in the normal fashion. Physiological changes due to prolonged weightlessness are studied during orbital flights.

For a more familiar example of the change in apparent weight caused by the acceleration of a frame of reference, we consider a man in an elevator.

Example 5-8

A man of mass 100 kg stands on a scale in an elevator that has an upward acceleration of 2 m/s². What is his apparent weight?

The apparent weight of the man equals the push P of the platform of the scale against the soles of his feet (Fig. 5-9a). The man's true weight, mg, is $100 \times 9.8 = 980$ N.

From Newton's second law applied to the man,

$$\text{net } F = ma$$

$$P - 980 \text{ N} = (100 \text{ kg})(+2 \text{ m/s}^2)$$

$$P = 980 \text{ N} + 200 \text{ N} = \boxed{1180 \text{ N}}$$

The man apparently weighs 200 N more than when at rest—a gain of about $\frac{1}{5}$ of his true weight. The reading of the scale changes from 980 N to 1180 N during the acceleration.

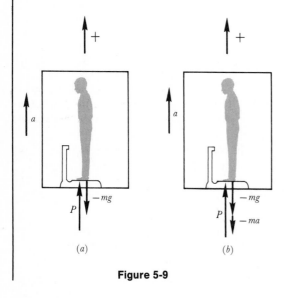

(a)   (b)

Figure 5-9

5-8 Inertial Forces

There are two equivalent ways of analyzing the forces in Example 5-8:

(a) To an outside observer the acceleration is known to exist (relative to an inertial frame), and the problem is solved by applying Newton's second law: $P + (-mg) = ma$; $P = m(g + a)$. These forces are shown in Fig. 5-9a. For an upward acceleration $a > 0$, $P > mg$, so the floor (scale platform) pushes upward with a force greater than the man's weight. According to Newton's third law, the man pushes downward on the scale with a force of magnitude P, and therefore the scale registers a force greater than mg. Similarly, when $a < 0$ (downward acceleration), the apparent weight is less than mg, giving rise to a "sinking feeling" in the pit of the stomach.

(b) If the man does not know that the elevator is accelerating, he considers himself to be in equilibrium under the action of two forces (Fig. 5-9b); the upward push of the scale platform P, and a downward "gravitational" force $-m(g + a)$. The fictitious inertial force $-ma$ that has arisen because of the (unknown to him) acceleration of his frame of reference is in every respect equivalent to a gravitational force. The man is at liberty to say either "someone accelerated the elevator upward" or "someone turned on an extra downward gravitational force."

We can generalize to say that an inertial force arises whenever a frame of reference is accelerated. In fact, if the frame's acceleration is **a**, an inertial (fictitious) force arises, given by $-m\mathbf{a}$, in a direction opposite to the acceleration of the frame of reference. We can understand the origin of inertial force by writing Newton's second law with $\Sigma\mathbf{F}_{\text{ext}}$ representing the externally applied forces, if any:

$$\Sigma\mathbf{F}_{\text{ext}} = m\mathbf{a}$$

Rearranging gives

$$\Sigma\mathbf{F}_{\text{ext}} - m\mathbf{a} = 0 \qquad (5\text{-}6)$$

Relative to himself, an observer has no acceleration and therefore he is always in equilibrium *in his own frame of reference*. The condition of equilibrium is $\Sigma\mathbf{F} = 0$, so we interpret Eq. 5-6 to mean that $\Sigma\mathbf{F}_{\text{ext}} - m\mathbf{a}$ is the effective force that acts on the observer in his own frame of

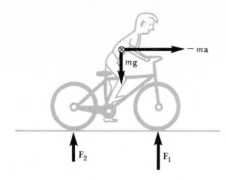

Figure 5-10 A bicyclist puts on the brakes and his acceleration **a** is toward the left. The inertial force −m**a** acts at the cyclist's center of mass.

reference. The extra force, caused by the acceleration of the frame of reference, is the inertial force, given by −m**a**.

The inertial force, −m**a**, acts at the center of mass of the body and is an "artificial gravity" to a person in the accelerated frame; it is in no way distinguishable from "real gravity." Albert Einstein made this equivalence a cornerstone of his theory of gravitation in the general theory of relativity.

The inertial force −m**a** always acts at the center of mass of the body that is in the acceler-

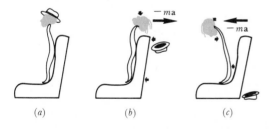

(a)   (b)   (c)

Figure 5-11 Effects of a rear-end collision on the spinal column. (a) Normal condition. (b) The car is struck from behind; **a** is toward the front of car. Inertial force −m**a** acts on the head, causing the neck to be hyperextended. (c) The car comes to rest, **a** is toward the rear of the car, and the inertial force −m**a** acts in a forward direction. (Adapted from Paul D. Cantor, ed., *Traumatic Medicine and Surgery for the Attorney*, Vol. 2, p. 390, by permission.)

ated frame of reference. When a bicyclist (Fig. 5-10) applies the brakes, his acceleration is negative, so −m**a** is positive. This force has a lever arm about an axis through the point of contact of the front wheels with the road, and therefore the inertial force −m**a** exerts a clockwise torque on the system and the cyclist tends to pitch forward while stopping. Similarly, in a rear-end automobile collision, whiplash injury occurs when a large inertial force acts on the head of the victim (Fig. 5-11).

Example 5-9

In a lawsuit concerning whiplash injury, a physicist testified that when a stationary car was hit from the rear by a similar car moving at 11 m/s (25 mi/h), the victim's head, of mass 3.4 kg, was subjected to an inertial force of 3500 N (800 lb) directed toward the rear of the car. Assume that the cars stuck together after the collision and that the accelerations involved were approximately constant in magnitude. (a) Calculate the magnitude of the acceleration during collision. (*Note:* A value of about 80g is considered an upper limit for a "safe" collision at this speed.) (b) Calculate the time required for the two cars to achieve their common final velocity in this inelastic collision.

(a) The inertial force (−m**a**) was stated by the physicist to have magnitude 3500 N. Hence the magnitude of the acceleration of the frame of reference FR_{car} was

$$\frac{\text{net } F}{m} = \frac{3500 \text{ N}}{3.4 \text{ kg}} = \boxed{1029 \text{ m/s}^2}$$

This is about 105g. The victim had a case.

(b) Starting from rest, the head acquired a backward velocity (in FR_{car}) of 5.5 m/s (recall that the cars were similar; conservation of momentum shows that in FR_{earth} the struck car received a velocity of $\frac{1}{2} \times$ 11 m/s). In FR_{car}, the impulse $F \Delta t$ on the head equals its change in momentum, $m \Delta v$.

$$(\text{net } F)(\Delta t) = m \Delta v$$

$$\Delta t = \frac{m \Delta v}{\text{net } F} = \frac{(3.4 \text{ kg})(5.5 \text{ m/s})}{3500 \text{ N}} = \boxed{0.00534 \text{ s}}$$

Summary   The momentum of a body is defined as the product of its mass and its velocity. Newton's second law may be restated in terms of change of momentum: The net force acting on a body equals the rate of change of its momentum. The product of net force and the time during which the force acts, defined as impulse, is equal to the total change of momentum.

Newton's third law may be used to derive the law of conservation of momentum: The total momentum of an isolated system remains constant. Since momentum is a vector quantity, the components of momentum are separately conserved during any collision or other interaction between bodies.

Rockets are propelled by the reaction force of the expelled exhaust gases against the body of the missile and do not require a medium such as air for their operation. In a vector sense, the total momentum of the system (rocket + exhaust gases) remains zero.

An inertial frame of reference is one in which Newton's laws of motion hold true; an accelerated frame of reference is not an inertial frame. Inertial forces are fictitious forces that arise because of the acceleration of the frame of reference. The inertial force on a body is opposite in direction to the acceleration of the frame of reference, and it acts at the body's center of mass. In an accelerated frame the apparent weight of a body differs from its true weight.

In the classical physics of Galileo and Newton, the forms of the laws of mechanics are the same in all inertial frames, unaffected by the uniform relative velocity of the observer and the frame in which the experiment is performed.

Check List

momentum
impulse

$$\Sigma \mathbf{F} = \frac{\Delta(m\mathbf{v})}{\Delta t} = \frac{\Delta \mathbf{p}}{\Delta t}$$

law of conservation of momentum
center of mass

inertial frame of reference
Coriolis force
true weight
apparent weight
inertial force $= -m\mathbf{a}$
Galilean relativity
Galilean transformation

Questions   5-1   When an apple falls to the ground and strikes the earth without rebound, what "becomes of" the momentum of the apple?

5-2   Can a woman in a rowboat "create" momentum by her unaided efforts?

5-3   A prospector carrying a bag of valuable uranium ore is trapped on absolutely smooth ice on a frozen lake in northern Canada. Is there any way in which he can move across the ice to safety, or must he freeze to death?

5-4   Is it possible for a rocket to attain a speed greater than the velocity with which the exhaust gases leave it?

5-5   A man stands on a wooden plank, which in turn rests on a concrete floor. If he hammers one end of the plank with a heavy mallet, can he make the "bumpmobile" move? If so, explain how. (See Fig. 5-12.)

Figure 5-12

5-6 An open freight car is coasting along on a frictionless track, and rain starts to fall, thereby increasing the mass of the car. Does the velocity of the car remain constant?

5-7 If a heavy ball is dropped from a great height, the rotation of the earth causes it to miss, by a small amount, the spot directly underneath the point of release. Does the ball land to the north, south, east, or west of the expected point? In which direction is the fictitious force on the ball?

5-8 You are a passenger on an interstate bus; the bus is well lighted inside, but it is a very dark night outside. A child at the rear of the bus rolls a steel ball down the center of the aisle. As you watch the motion of the ball, you observe it to move in a straight line until it is just past your seat, then it swerves to the left as it moves forward and strikes a mysterious black suitcase resting at the side of the seat of another passenger. Make two equivalent statements as to the possible cause of the ball's behavior.

5-9 During an earthquake in California, the acceleration of the ground is erratic and is of considerable magnitude. Using the concept of inertial force, explain how this causes damage to buildings.

MULTIPLE CHOICE

5-10 The dimensions of impulse are (a) $[\mathrm{MLT}^{-1}]$; (b) $[\mathrm{MLT}^{-2}]$; (c) $[\mathrm{ML}^2\mathrm{T}^{-2}]$.

5-11 Change in momentum equals (a) force; (b) acceleration; (c) impulse.

5-12 A body of mass 1 kg that has an initial speed of 1 m/s is acted on by a force of 1 N for 1 s. The increase of momentum is (a) 0.5 kg·m/s; (b) 1.0 kg·m/s; (c) 2.0 kg·m/s.

5-13 The law of conservation of momentum is most closely related to (a) Newton's first law; (b) Newton's second law; (c) Newton's third law.

5-14 Newton's first law is strictly valid in (a) any frame of reference; (b) any inertial frame; (c) any frame attached to the earth.

5-15 A Coriolis force is experienced by a body that is on the earth's surface and is (a) at rest relative to the surface; (b) moving relative to the surface; (c) both of these.

Problems **5-A1** Compute the momentum, in kg·m/s, of a fullback who has a mass of 120 kg and is moving at 2 m/s.

5-A2 Compute the momentum of a golf ball that has a mass of 60 g and is moving with a velocity of 70 m/s.

5-A3 If in Prob. 5-A2 the impact between the golf club and the ball lasted for 2×10^{-4} s, what was the rate of change of momentum? What force acted on the ball? What force acted on the club?

5-A4 A boy holds a 2-kg air rifle loosely and fires a bullet that weighs 0.049 N. The muzzle velocity of the bullet is 300 m/s. What is the recoil velocity of the gun?

5-A5 If the boy holds the rifle of Prob. 5-A4 tightly against his body, the recoil is less. Explain. Calculate the new recoil velocity if the boy's mass is 38 kg.

5-A6 In a freight yard a train is being made up. An empty freight car, coasting at 8 m/s, strikes a loaded car that is stationary, and the cars couple together. Each of the cars has a mass of 3000 kg when empty, and the loaded car contains 14,000 kg of bottled soft drinks. With what speed does the combined mass start to move?

5-A7 An astronaut of mass 60 kg carries an empty oxygen tank of mass 12 kg. She throws the tank away from herself with a speed of 2 m/s (measured relative to the fixed stars). With what velocity does the astronaut start to move through space?

5-B1 At 4:40 P.M. the momentum of a speedboat was measured as 20,000 kg·m/s, and at 4:42 P.M. it was 50,000 kg·m/s. Assuming that the boat was moving along a straight course and that its speed increased steadily (constant force), what forward force was exerted on the boat? By what was this force exerted? (Do not find the acceleration.)

5-B2 What force, acting for 0.001 s, will change the velocity of a 100-g baseball from 30 m/s eastward to 40 m/s westward? (Do not find the acceleration.)

5-B3 What forward force, applied for 1.25 s, is needed to give a 7-kg bowling ball a speed of 6 m/s? (Do not find the acceleration.)

5-B4 For how long a time must a tow truck pull with a force of 500 N on a stalled 1200-kg car to give it a forward velocity of 2 m/s? (Do not find the acceleration.)

5-B5 A fire hose sends 900 kg of water every minute against a burning building. The water strikes the building with velocity $+20$ m/s and does not bounce back. (a) What is the rate of change of momentum of the water? (b) What force does the building exert on the water? (c) What force does the water exert on the building? (*Note:* Be careful to affix proper signs, $+$ or $-$, to your answers to this problem.)

5-B6 A ventilator fan moves 80 m³ of air per minute, and the blast of air strikes a wall at a speed of 20 m/s. What force is exerted on the wall by the moving air? (See Table 1-3 for the density of air.) Assume that the air molecules make inelastic impact at the wall.

5-B7 An 80-kg fullback is moving at 5 m/s; he is tackled head-on by a 100-kg linebacker who is approaching him at 3 m/s. With what speed, and in which direction, do the pair of players move after the tackle?

5-B8 A 4-kg duck sitting on a pole 8 m tall is struck by a 20-g bullet that is moving horizontally at 200 m/s. The bullet remains embedded in the bird. (a) What horizontal velocity does the bird acquire? (b) How far from the base of the pole does the duck strike the ground?

5-B9 One way of measuring the muzzle velocity of a bullet is to fire it horizontally into a massive block of wood placed on a cart. Assuming no friction, we then measure the velocity with which the wood (containing the bullet) starts to move. In one experiment the bullet had a mass of 50 g, and the wood and its cart had a mass of 18 kg. After the shot, a stopwatch was used to determine that the cart, wood, and bullet moved at constant velocity, traveling 8 m in 0.40 s. What was the original speed of the bullet?

5-B10 A life raft of mass 200 kg carries two swimmers of mass 50 kg and 70 kg, respectively. The raft is initially floating at rest; then the swimmers simultaneously dive off from the midpoints of opposite sides of the raft, each with a horizontal velocity of 4 m/s. With what speed and in what direction does the raft start to move?

5-B11 A basketball player jumps into the air with both arms extended vertically overhead. Just as he reaches maximum height, he pulls one arm down so that his c.g. is moved to a point 4 cm closer to his toes. Do the fingertips of the other hand go higher or lower as a result of this maneuver? How much?

5-B12 In Fig. 5-5 the suitcase is 1.4 m from the seam. What are the values of x' and x for the suitcase at $t' = t = 5$ s?

5-B13 A 55-kg student is in an elevator that is accelerating downward at 2 m/s². (*a*) What is the net force on the student? (*b*) What is her apparent weight (that is, the upward force of the floor on her)?

5-B14 What upward acceleration of an elevator will cause a passenger's apparent weight to increase from 700 N to 850 N?

5-B15 A crate of glassware of mass 80 kg is on the smooth, flat, horizontal rear deck of a pickup truck that is traveling westward at 16 m/s. The truck approaches a traffic light and slows down to 2 m/s in 3.5 s. What is the magnitude and direction of the fictitious inertial force on the crate?

5-C1 An astronaut of mass 80 kg (which includes his back pack) is on a trajectory far above the moon's surface, moving 10 m/s upward at an angle of 60° with the horizontal. A short burst of compressed nitrogen is ejected horizontally at 100 m/s (all velocities are relative to the moon's surface). After this maneuver the astronaut is moving vertically upward. (*a*) What mass of gas was ejected? (*b*) What are the astronaut's new mass and speed? (*Hint:* Horizontal and vertical components of momentum are each conserved.)

5-C2 The horizontal component of initial velocity of a projectile is v_{x0}, the projectile reaches a maximum height H, and the range is R. If, when the projectile is at its maximum altitude, it explodes into two equal fragments, with one fragment falling vertically downward, prove that the other fragment strikes the ground at a distance $\frac{3}{2}R$ from the gun.

5-C3 What is the magnitude and direction of the force exerted by the stream of water of Example 5-1 if the water strikes the roof horizontally and is deflected without loss of speed to be 30° from horizontal, parallel to the roof surface? (*Hint:* Consider the horizontal and vertical components of momentum separately.)

5-C4 Consider the phenomenon of "weight transfer" that occurs due to inertial force when the truck of Fig. 5-13 is being brought to a stop. The total mass of the loaded truck is 6 metric tons, and the center of mass is 2 m from the front axle, 2 m from the rear axle, and 1 m above the road surface. How much weight does each axle bear (*a*) when the

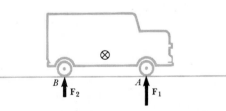

Figure 5-13

truck is moving at constant velocity; (*b*) when the truck is being braked at 5 m/s^2? (*Hint:* The c.m. and the c.g. are at the same point; see Sec. 4-9. Apply the torque condition in turn about each axis at *A* and *B*, perpendicular to the length of the truck. The inertial force acts at the c.m.)

5-C5 A book 22 cm tall, 18 cm wide, and 5 cm thick is standing upright on a rough table top. The table is given a forward acceleration in a direction perpendicular to the flat vertical cover of the book. (*a*) Does the book tend to fall forward or backward? (*b*) What acceleration of the table gives the book just enough torque (due to the combined action of gravity and inertial force) to topple over, rotating about a horizontal axis on the table that is the lower edge of one of the book's covers?

References

1. F. I. Ordway, "Principles of Rocket Engines," *Sky and Telescope* **14**, 48 (1954).
2. H. W. Lewis, "Ballistocardiography," *Sci. American* **198**(2), 89 (Feb. 1958). Conservation of momentum in the study of heart action. The subject lies on a movable platform, and as the heart ejects blood, the platform recoils. Additional references on ballistocardiography are found in biologically oriented textbooks: R. W. Stacy, D. T. Williams, R. E. Worden, and R. O. McMorris, *Essentials of Biological and Medical Physics* (McGraw-Hill, New York, 1955) pp. 382–383; S. G. G. MacDonald and D. M. Burns, *Physics for the Life and Health Sciences* (Addison-Wesley, Reading, Mass., 1975), pp. 85–87; H. J. Metcalf, *Topics in Classical Biophysics* (Prentice-Hall, Englewood Cliffs, N.J., 1980), pp. 84–86.
3. J. E. McDonald, "The Coriolis Effect," *Sci. American* **186**(5), 72 (May 1952).
4. "The Amateur Scientist," *Sci. American* **182**(4), 183 (Apr. 1960). A discussion of Coriolis forces.
5. R. P. Bauman, "Visualization of the Coriolis Force," *Am. J. Phys.* **38**, 390 (1970).
6. A. Shapiro, *The Bathtub Vortex* (film). Coriolis force arising from the earth's rotation causes counterclockwise rotation when the drain valve of a tank of water is opened.
7. F. Miller, Jr., *Inertial Forces—Translational Acceleration* (film).
8. J. Stull, *Conservation of Momentum—Inelastic Collisions; Conservation of Elastic Collisions* (films).

6

Conservation of Energy

6-1 Work

Whatever the literary or biological usages of the term may be, in a physical sense *work* is done only if there is a displacement **s** of a particle or other object on which a force **F** acts. It is only when an object is moved by a force that has a component acting in the direction of motion that work is done. If θ is the angle between the directions of **F** and **s**, the magnitude of the component of **F** parallel to **s** is $F \cos \theta$. The work done during a displacement is defined as a product:

Work = (force component)(displacement)

$$W = (F \cos \theta)(s) \qquad (6\text{-}1)$$

If, as is often the case, the force is in the same direction as the displacement, then $\theta = 0$, and $\cos \theta = 1$. In this special case we can use the simpler formula

$$W = Fs$$

Although work is a scalar quantity, it is calculated as the product of two vectors. This type of product is called the *dot product*:

$$W = \mathbf{F} \cdot \mathbf{s}$$

The "dot" multiplication symbol includes the factor $\cos \theta$. The equation $W = \mathbf{F} \cdot \mathbf{s}$ reduces to $W = Fs$ if **F** and **s** are parallel.

In any system of units, work units are products of the force unit and the length unit; in SI certain work units have received special names. In the mks system, work and energy are measured in newton·meters (joules). One *joule* (J) is the work done when a body is moved one meter against an opposing force of one newton. Similarly, in the cgs system, one *erg* (one dyne·centimeter) is the work done when a body is moved one centimeter against an opposing force of one dyne. As with all mks and cgs units, there is a simple relationship (involving only powers of 10) between the joule and the erg:

$$
\begin{aligned}
1 \text{ J} &= (1 \text{ N})(1 \text{ m}) \\
&= (10^5 \text{ dyn})(10^2 \text{ cm}) \\
&= 10^7 \text{ dyn·cm} = 10^7 \text{ ergs}
\end{aligned}
$$

Conservation of energy at the bowling alley: Some of the kinetic energy of the ball has been transformed into potential energy of the pins that have been raised up. (American Machine and Foundry Company)

A joule is thus 10 million ergs. The joule and its multiples, such as kJ, MJ, and GJ, are the preferred SI units for work and energy.

Whatever the system of units, the dimensions of work are the same: $[\text{work}] = [\text{F}][\text{L}]$, or $[\text{work}] = [\text{MLT}^{-2}][\text{L}] = [\text{ML}^2\text{T}^{-2}]$. Note that the mass or weight of the object moved is of no consequence for the definition of work, unless it so happens that the force of gravity is the force against which the work is being done. In general, the only requirement for work to be done is that a force be exerted and something or some point be moved in a direction parallel to the force or some component of the force. Newton's third law is important here: whenever work is done by a force, something exerts an equal and opposite force on the agent that does the work. Thus we can equally well think of work as being done *by* an applied force or *against* an opposing force. The two viewpoints are equivalent.

Example 6-1

How much work is done by gravity on a sled weighing 200 N that slides 40 m down a hill (measured along the road) whose angle with the horizontal is 30°?

The angle between **F** and the direction of the displacement is 60° (Fig. 6-1). Therefore the component of force in the direction of the displacement is

$$F \cos \theta = (200 \text{ N})(\cos 60°)$$
$$= (200 \text{ N})(0.500) = 100 \text{ N}$$

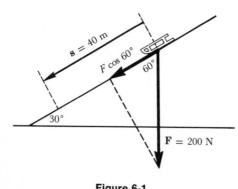

Figure 6-1

The work done by gravity while the sled slides 40 m is

$$W = (100 \text{ N})(40 \text{ m}) = \boxed{4000 \text{ J}}$$

6-2 Work Done Against a Variable Force

In defining work by the equation $W = (F \cos \theta)s$, we have tacitly assumed that the force **F** is constant in magnitude and direction throughout the displacement **s**. This is far from being true in many actual situations, even when $\cos \theta = 1$. For instance, when an astronaut is lifted to an altitude of 200 km, the force of gravity against which work is done decreases during the ascent, and so the work to lift him cannot be computed from $(F \cos \theta)s$. The value of F changes during the displacement. More prosaically, we can ask how much work is done in pushing a thumbtack into a bulletin board if the opposing force due to the board is not constant.

We can use a graphical interpretation to approach this problem. Suppose the force varies in some fashion (Fig. 6-2) as the displacement increases. (For simplicity we assume that the force and displacement are parallel to each other.) During any short displacement interval Δs_1, the force is practically constant in magnitude, and the work done is approximately $F_1 \Delta s_1$. This

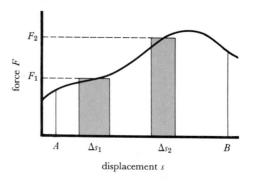

Figure 6-2 Work done during displacement against a variable force.

work is numerically equal to the area of the shaded rectangle shown on the graph. For another short interval Δs_2, the work equals the area $F_2 \Delta s_2$. For the whole motion from $s = A$ to $s = B$, the total work is approximately equal to the sum of the areas of all the rectangles. You will recognize that this involves much the same technique as was used in Sec. 2-5 to find the displacement of a car that has a variable velocity (see Fig. 2-5). In the limit, as each Δs approaches zero and the number of rectangles approaches infinity, the sum of the rectangular areas approaches the area under the curve. Thus *the work done is equal to the area under the force-displacement curve.*

We can say that the work to move a body from $s = A$ to $s = B$ is the definite integral of F with respect to s between the limits A and B:

$$W = \lim_{\substack{n \to \infty \\ \Delta s \to 0}} \sum_{i=1}^{n} F_i \Delta s_i$$

or

$$W = \int_A^B F \, ds$$

_____ **Example 6-2**

A car is pushed by a forward force that varies according to the graph in Fig. 6-3, where the curved portion when plotted to the scale shown is one-quarter of a circle with center at D. What work is done in moving the car 7 m, starting from $s = 0$?

To calculate the work, we use simple geometry to find the area under the force curve. The area of a

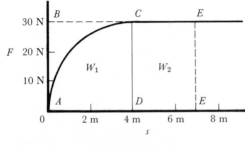

Figure 6-3

circle of radius R is πR^2, which is $\frac{1}{4}\pi$ times the area of a circumscribing square. Hence the area W_1 under the quadrant AC is $\frac{1}{4}\pi$ times the area $ABCD$. The area W_2 under the section CE is the area of a rectangle.

$$W = W_1 + W_2$$
$$= \int_{s=0}^{s=4} F \, ds + \int_{s=4}^{s=7} F \, ds$$
$$= \tfrac{1}{4}\pi(30 \text{ N})(4 \text{ m}) + (30 \text{ N})(3 \text{ m})$$
$$= 94 \text{ N} \cdot \text{m} + 90 \text{ N} \cdot \text{m} = \boxed{184 \text{ J}}$$

6-3   Energy

Whenever work is done on a body, it gains *energy.* Thus if 7 J of work is done on a body, the body has gained 7 J of energy. We may speak of the energy of a body as "stored work." It is possible to distinguish between two kinds of mechanical energy, depending on how the work was done on the body.

Potential energy (PE) is the energy of a system of bodies due to the relative position of the parts of the system. We often say that potential energy is "due to position," but some examples will make it clear that it takes at least two bodies (that is, a system) in order to have PE. When a bricklayer gives a brick PE by lifting it, she does work against gravity. The bricklayer performs work on a system, and two bodies are involved—brick and earth. When you wind your watch, you change the relative position of the various atoms of iron in the watch spring. The spring (a system of atoms) is distorted, and you have done work against elastic forces.

Chemical and electrical phenomena also illustrate the idea that PE involves a system of bodies. A lump of coal is said to have PE; we can burn coal in a steam engine and obtain some mechanical work from it. Note, however, that the chemical PE is that of a system: carbon atoms and oxygen atoms. The chemical equation

$$C + O_2 = CO_2 + \text{thermal energy}$$

is symmetrical with respect to carbon and oxygen, and it shows that carbon by itself has no more chemical PE than does oxygen by itself. The *system* has PE because the atoms are separated from each other, and the energy is released when the atoms, under the influence of electric forces of attraction, rearrange their positions to form CO_2. When a parallel-plate capacitor is charged and stores energy, the PE of the charges is due to their relative positions. Some free electrons have been moved from one plate to the other, so that there is an excess of negative charge on the other. There are many ways in which a system may have PE due to the relative positions of its parts; mechanical PE is, however, confined to work done against gravitational or elastic forces.

Kinetic energy (KE) is the energy a body has because of its motion. To give velocity to a body, it must be accelerated. By Newton's second law this requires a force, and by Newton's third law an equal and opposite force acts back on whatever applies the forward force. We call this force, which is exerted *by* the body on which we push or pull, the *reaction force*. Therefore, to give a body a velocity, work must be done to move it through some displacement against the reaction force. All bodies in motion possess kinetic energy. The energy of the winds and the seas is kinetic; work done to speed up the molecules is stored as KE.

It is important to understand that no work is done if there is no motion, no matter how much force is applied. If a 10-ton boulder rests for 100,000 years on top of another rock, a force is exerted, but no work is done. From a physical viewpoint no work is required to hold a 15-N book at arm's length, since there is no displacement.* Also, no work is done if the force is applied perpendicularly to the direction of motion.

Thus if a stationary sled weighing 200 N is on frictionless level ice ($\mu_k = 0$) and a child pulls with a horizontal force of 3 N, the sled starts to move. By Newton's third law, the sled pulls back with a force of 3 N against the child, and if the child maintains a steady 3-N force until the sled moves 4 m, the work done by the child on the sled is (3 N)(4 m) = 12 J. The sled therefore acquires 12 J of KE. No work is done against gravity, however, since the 200 N force of gravity acts downward, perpendicularly to the motion; in fact, the weight of the sled has no bearing on the situation so far. If now the sled comes to a level stretch of ice having $\mu_k = 0.015$, the frictional force is $\mu_k F_n = (0.015)(200 \text{ N}) = 3$ N, and the child's forward force is balanced by the force of friction. The sled is in equilibrium, and its already acquired velocity remains constant. As the sled moves another 4 m, the child, still exerting 3 N of force, does another 12 J of work. This time the work is done against the force of friction, and the ground and sled become slightly warmer. Finally, work *will* be done against gravity if the sled enters on a gently sloping frictionless hill that rises 1.5 m for every 100 m measured along the ground (Fig. 6-4). The downhill component of the force of gravity on the sled is 3 N, and the sled is in equilibrium

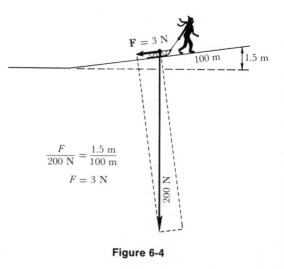

$$\frac{F}{200 \text{ N}} = \frac{1.5 \text{ m}}{100 \text{ m}}$$

$$F = 3 \text{ N}$$

Figure 6-4

---

* No work is done on the book; however, work is done within the muscles. It is impossible for a normal muscle to "lock in" and become rigid. Muscles contract slightly, then release, then contract again, and so on indefinitely as long as force is applied by the muscle. Work, in a physical sense, is done, since these displacements are in the direction of the force. This energy is dissipated within the muscle.

(hence the velocity remains constant). As the child pulls the sled 4 m, he again does 12 J of work, this time against gravity. The sled gains 12 J of PE. From start to finish the child has done 36 J of work; at the end of the trip the sled has 12 J of KE and 12 J of PE, and 12 J of heat has been distributed between the sled and the ground.

6-4 Calculation of Mechanical Energy

A professional engineer's handbook contains many formulas, which cover applications too numerous and specialized to be worth remembering. This is as it should be, but our use of formulas is quite different. We derive (and use) only a few formulas, chosen to throw into prominence the essential physical factors on which energy, for instance, depends. Formulas such as we are about to derive are basic and are used over and over again.

If a body of mass m is raised vertically a height h, the force of gravity is mg; hence the work done against gravity is mgh. (We assume that the body remains near the surface of the earth, so that $\mathbf{g}$ is constant in magnitude and direction.) The formula is true in any system of units, but in discussing gravitational PE we prefer to emphasize weight rather than mass. Therefore we write the force of gravity as W, the weight of the body, and the work to raise it a vertical distance h simply as Wh.* Thus we have the formula

$$\text{Gravitational PE} = mgh = Wh \qquad (6\text{-}2)$$

$$J = (N)(m)$$

$$\text{erg} = (\text{dyn})(\text{cm})$$

Whether or not the object is lifted straight up, the factor h in Eq. 6-2 refers to the *vertical* rise measured from some reference level where the gravitational PE is assumed to be zero. In any practical problem it is always *change* in PE that

_____
* It is perhaps confusing that the same symbol W is used for both work and weight. The context will make clear which is intended.

is required, so the reference level for zero PE can be quite arbitrary—sea level, the floor of a room, or the top of a table.

_____ **Example 6-3**

A roller coaster car of mass 3000 kg proceeds from point A (Fig. 6-5) to point B and then to C. What is its PE at C? How much PE did the car lose in going from A to C?

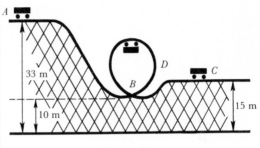

Figure 6-5

First we note that the weight of the car is 3000×9.8 N. At C the PE (relative to the ground level) is

$$\text{PE} = (3000 \text{ kg})(9.8 \text{ m/s}^2)(15 \text{ m})$$

$$= 4.3 \times 10^5 \text{ N·m} = \boxed{4.3 \times 10^5 \text{ J}}$$

In going from A to C, the car falls 18 m.

$$\Delta\text{PE} = W(h_2 - h_1)$$

$$= (3000 \text{ kg})(9.8 \text{ m/s}^2)(18 \text{ m})$$

$$= 5.3 \times 10^5 \text{ N·m} = \boxed{5.3 \times 10^5 \text{ J}}$$

Note that the exact nature of the car's path (hills, loop, and so on) as it goes from A to C has no bearing at all on the net change in PE during the overall motion.

_____ **Example 6-4**

A butterfly raises its four wings from horizontal to vertical positions; each wing is approximately rectangular, 3 cm tall by 4 cm long, of mass 0.020 g. What is the increase in PE?

The c.g. of each wing is raised by 1.5 cm.

$$PE = 4mgh = 4(0.020 \text{ g})(980 \text{ cm/s}^2)(1.5 \text{ cm})$$

$$= 118 \text{ dyn} \cdot \text{cm} = \boxed{118 \text{ erg}}$$

Kinetic energy is given to a body when it speeds up. When a body is at rest, it has no KE. Let us apply a force **F** to a body of mass m that is initially at rest ($v_0 = 0$), and let us allow this force to operate for a sufficient time for the body to acquire a velocity v. Let s be the displacement, the distance through which the body moves. Here ΣF is simply **F**, since all other forces on the body are balanced. We calculate the displacement by first finding the acceleration:

$$a = F/m$$
$$v^2 = v_0{}^2 + 2as$$
$$s = \frac{v^2}{2a} = \frac{v^2}{2(F/m)} = \frac{v^2 m}{2F}$$

By Newton's third law, while we are applying the force **F**, the body reacts on us with an equal and opposite force $-$**F**. We do work in moving the body against this reaction force, and the work done by us is the product of the force and the distance moved. By definition, work done against this reaction force (we may call it "work done against inertia") is the KE stored in the body.

Work done against inertia

$$= \text{(reaction force)(displacement)}$$

$$KE = (F)\left(\frac{v^2 m}{2F}\right)$$

or

$$KE = \tfrac{1}{2}mv^2 \qquad (6\text{-}3)$$
$$J = (\text{kg})(\text{m/s})^2$$
$$\text{erg} = (\text{g})(\text{cm/s})^2$$

It may be objected that our derivation of $KE = \tfrac{1}{2}mv^2$ is limited to a uniformly accelerated body acted on by a constant force. Fortunately, it can be shown by the use of calculus that the work done against inertia is given by $\tfrac{1}{2}mv^2$ regardless of the way in which the body reached its final velocity (see the footnote following Eq. 28-11 on page 683).

Note that the displacement and the force applied do not appear in the formula for the kinetic energy. If a smaller force had been applied, the acceleration would have been less, and the body would have had to move farther in order to attain the same speed v. The product Fs would have remained the same. Our two formulas for gravitational PE and KE, Eqs. 6-2 and 6-3, are easily remembered, and they serve to emphasize the significant physical quantities: weight and vertical rise for gravitational PE, mass and velocity for KE.

Example 6-5

A boy on a motorbike is scooting along at 20 m/s. If the boy's mass is 50 kg, what is his KE?

$$KE = \tfrac{1}{2}mv^2$$
$$= \tfrac{1}{2}(50 \text{ kg})(20 \text{ m/s})^2$$
$$= 10,000 \text{ kg} \cdot \text{m}^2/\text{s}^2 = \boxed{10,000 \text{ J}}$$

Example 6-6

What is the KE of a truck of weight 40,000 N (4.5 tons) moving down a turnpike at 90 km/h? How does this compare with the PE that would be stored if the truck were lifted 30 m straight up?

First, we find the velocity in m/s:

$$v = 90 \frac{\text{km}}{\text{h}}\left(\frac{1000 \text{ m}}{1 \text{ km}}\right)\left(\frac{1 \text{ h}}{3600 \text{ s}}\right) = 25 \text{ m/s}$$

In this example, we use W/g for the mass of the truck (Eq. 3-8 on page 62).

$$KE = \tfrac{1}{2}mv^2 = \tfrac{1}{2}(W/g)v^2$$

$$= \frac{1}{2}\left(\frac{40,000 \text{ N}}{9.8 \text{ m/s}^2}\right)(25 \text{ m/s})^2$$

$$= \boxed{1,280,000 \text{ J}}$$

If the truck were lifted 30 m straight up, its PE would be

$$(40{,}000 \text{ N})(30 \text{ m}) = \boxed{1{,}200{,}000 \text{ J}}$$

6-5   Conservation of Energy

The concept that energy can neither be created nor destroyed represents one of the great generalizations of the nineteenth century. Before the work of Mayer, Rumford, Helmholtz, and especially the English physicist James Joule (1818–1889), the possibility of conversion of energy to heat and vice versa was known, but the *quantitative* equivalence of thermal energy and other forms of energy was obscure. Joule showed experimentally that whenever a certain amount of heat (1 calorie) disappears or appears, the same amount of mechanical work (4.19 J) appears or disappears. This *law of conservation of energy* was extended by Joule and others to include all the then-known forms of energy (including electrical and chemical), and it was first stated somewhat as follows: *The total amount of mechanical, thermal, chemical, electrical, and other energy in any isolated system remains constant.*

In 1905 Einstein broadened the law still further to include the equivalence of mass and energy. This will be discussed in Chap. 28. In everyday life, however, we usually need to consider only mechanical, chemical, thermal, or electrical energies. The law of conservation of energy, even in this limited form, is a profound generalization, and we shall see, in Sec. 6-6, how to use it in solving problems.

The logical significance of the law of conservation of energy has been variously interpreted by authors. Some feel that we are using the law when we say that kinetic energy equals reaction force times displacement, for otherwise, without such a postulate, how can we "know" that the work is stored in the body? Our viewpoint is that it is fairly obvious that work is done against a reaction force, and it is merely a matter of definition to say that this work is stored as kinetic energy. We consider the law of conservation of energy to refer to something far broader than mechanical processes. The generalization that *all* forms of energy (including thermal energy, electrical energy, mass energy, and so forth) are quantitatively exchangeable with mechanical energy seems to us to be the essence of the energy principle that we call the law of conservation of energy.

6-6   The Energy Principle in Solution of Problems

It is often possible to solve mechanical problems by use of the law of conservation of energy. This is scarcely surprising, since our formula for kinetic energy (Eq. 6-3) came directly from Newton's second law. When we use this formula in a problem or a derivation, we are really only using Newton's second law in a "predigested" way that has already taken care of the kinematical part of the problem. In the following examples, energy changes other than changes in mechanical energy are assumed to be negligible. We call this limited use of the law of conservation of energy the *energy principle.*

Example 6-7

In the problem of the anxious space-station worker (see Example 3-6 on page 63), we can use the energy principle to find how far the moving section travels before stopping. The woman is pushed back by the section, and work is done *on* her by the section.

$$\begin{pmatrix} \text{Work done on} \\ \text{woman by section} \end{pmatrix} = \begin{pmatrix} \text{loss of KE of} \\ \text{moving section} \end{pmatrix}$$

$$(600 \text{ N})(s) = \frac{1}{2} \left(\frac{70{,}000 \text{ N}}{9.8 \text{ m/s}^2} \right)(0.3 \text{ m/s})^2$$

$$s = \boxed{0.54 \text{ m}}$$

This is the same answer that was obtained in Chap. 3, but here we got the result without computing the acceleration and without using the equations of uniformly accelerated motion.

_____ Example 6-8

A bowling ball is returned to the bowler by a mechanism that places it on a return ramp 100 cm above the alley level (Fig. 6-6). If the ball is given an initial speed of 0.6 m/s at point A, what is its speed when it reaches the bowler on the short level stretch DE? Ignore friction; ignore the rotational KE of the ball.

Figure 6-6

To have a consistent set of mks units, we first convert the heights from centimeters to meters. Then we apply the energy principle:

$$KE_1 + PE_1 = KE_2 + PE_2$$

$$\tfrac{1}{2}(m \text{ kg})(0.6 \text{ m/s})^2 + (m \text{ kg})(9.8 \text{ m/s}^2)(1.0 \text{ m})$$
$$= \tfrac{1}{2}(m \text{ kg})v^2 + (m \text{ kg})(9.8 \text{ m/s}^2)(0.4 \text{ m})$$

The mass cancels out, and the solution of the equation gives

$$v = \boxed{3.48 \text{ m/s}}$$

A direct solution of this problem would be more complicated. You would have to use a force diagram to find the acceleration of the ball in the slope AB, and the final velocity (at B) would become the initial velocity for a second problem to cover the interval BC. The energy method skips some of the steps and gives an answer for the final velocity. Note, however, that the energy method cannot be used to find times or accelerations.

Example 6-9

A stone of mass 2 kg is thrown at 5 m/s upward at a 30° angle from a cliff 20 m high (Fig. 6-7). How fast will the stone be moving when it strikes the ground? (Ignore air resistance.)

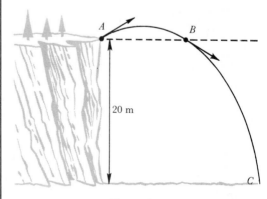

Figure 6-7

The direct method would make this into a projectile problem (Sec. 2-11), for which you would have to find the horizontal and vertical components of the velocity at C. Combining these components by the parallelogram rule would give the magnitude of the velocity at C. The energy method skips the intermediate steps. The weight of the stone is 19.6 N.

$$KE_1 + PE_1 = KE_2 + PE_2$$

$$\tfrac{1}{2}(2 \text{ kg})(5 \text{ m/s})^2 + (19.6 \text{ N})(20 \text{ m}) = \tfrac{1}{2}(2 \text{ kg})v^2 + 0$$
$$25 \text{ J} + 392 \text{ J} = (1 \text{ kg})v^2 + 0$$

$$v = \sqrt{417 \text{ J}/1 \text{ kg}} = \boxed{20.4 \text{ m/s}}$$

The angle of throw has nothing to do with this particular problem, although of course the angle *does* affect the maximum height of rise, time in the air, and distance out from the cliff. If the angle had been 20° instead of 30°, these latter quantities would have been different, but the final speed would still have been 20.4 m/s, as is indicated by the fact that all the quantities in the energy equation would have been the same.

The mass m of the stone is also immaterial. Both KE and PE are proportional to m, so the mass would cancel out of the energy equation.

6-7 Equilibrium and Potential Energy

It is a consequence of the law of conservation of energy that a system, left to itself, will tend toward a configuration of least potential energy. To prove this, assume that a body is initially at rest, with KE = 0. Since PE + KE = constant, the PE cannot spontaneously increase, for there is no way the KE can decrease ($\frac{1}{2}mv^2$ can never be negative). Thus the body's PE will decrease to the lowest value allowed by the mechanical construction of the system.

Let us interpret the three types of equilibrium in the light of energy changes (refer back to Figures 4-22, 4-23, and 4-24 on pages 90–91). The pendulum, in *stable* equilibrium, has a minimum PE: a small displacement in either direction causes the c.g. to rise, and the gravitational PE increases. The pole vaulter, in *unstable* equilibrium, has a maximum PE: a small displacement in either direction causes the c.g. to fall, and the gravitational PE decreases. The wheel, in *neutral* equilibrium, has constant PE: a slight rotation in either direction does not change the c.g., and the PE is constant. We conclude that a body is in stable, unstable, or neutral equilibrium according to whether the gravitational PE is a minimum, a maximum, or a constant.

6-8 Power; Efficiency

James Watt (1736–1819) in his pioneering investigations assumed that an exceptionally powerful dray horse could lift 550 lb a distance of 1 ft in 1 s and could perform work at this rate more or less steadily during a working day. Horses of lesser quality required a longer time to do the same amount of work. The quantity that interested Watt was *power*, defined as the rate of doing work, and Watt gave the name *horsepower* (hp) to this "standard" rate of 550 ft·lb/s. The SI unit of power is the *watt* (W), a rate of doing

work equal to one joule per second:

$$1 \text{ watt} = \frac{1 \text{ joule}}{1 \text{ second}}$$

Most people think of the watt as a unit of electric power, and it is true that electric or any other power can be measured in watts. However, James Watt was a mechanical engineer, and the watt is fundamentally a mechanical unit. It appears in electricity only because the electrical unit "volt" is an mks unit defined in terms of the mechanical work unit, the joule. (One reason for the popularity of mks units is their tie-in with basic electrical units.) It is sometimes useful to know that 746 W = 1 hp.

Power is the rate of doing work and is expressed as the time derivative of work. Thus

$$P = \lim_{\Delta t \to 0} \frac{\Delta W}{\Delta t}$$

or

$$P = \frac{dW}{dt}$$

A simple special case arises when work is done by a constant force that moves a body against the force of friction. The total work during a displacement s is $\mathbf{f} \cdot \mathbf{s}$; frictional force is always parallel to the displacement, so the dot product becomes simply fs.

$$W = fs$$

$$\frac{dW}{dt} = f\frac{ds}{dt}$$

or

$$P = fv$$

This equation gives the power required to move a body at speed v against a constant opposing force f.

Example 6-10

What is the power P of an engine that pulls a 500,000-kg train at a steady speed of 40 m/s along a horizontal track for which the coefficient of friction is 0.02?

The force of friction is given by

$$f = (500{,}000 \text{ kg})(9.8 \text{ m/s}^2)(0.02) = 98{,}000 \text{ N}$$
$$P = fv = (98{,}000 \text{ N})(40 \text{ m/s})$$
$$= 3.92 \times 10^6 \text{ N·m/s} = 3.92 \times 10^6 \text{ J/s}$$
$$= 3.92 \times 10^6 \text{ W} = \boxed{3920 \text{ kW}}$$

It is often useful to consider the *efficiency* of a device, defined as the work output divided by the work input:

$$\text{Efficiency} = \frac{\text{work output}}{\text{work input}}$$

If, as is usual, the input and output take place during the same time interval, the efficiency is also the *power* output divided by the *power* input. Thus an 80-W fluorescent lamp may have an output of visible energy at the rate of 8 W; its efficiency is $8 \text{ W}/80 \text{ W} = 10\%$. The efficiency of some types of simple machines is discussed in Sec. 6-11.

6-9 Energy Changes in Collisions

Let us look closely at Example 5-4, in which we studied the collision of a car and a truck. Before collision, all the KE of the system was in the car:

$$\tfrac{1}{2}mv^2 = \tfrac{1}{2}(600 \text{ kg})(20 \text{ m/s})^2 = 120{,}000 \text{ J}$$

After collision, the system had less kinetic energy:

$$\tfrac{1}{2}mv^2 = \tfrac{1}{2}(2000 \text{ kg})(6 \text{ m/s})^2 = 36{,}000 \text{ J}$$

At first glance it seems that the law of conservation of energy was violated. There was no obvious increase of potential energy, since all the motions were horizontal. Where, then, did the other 84,000 J of energy go?

We can account for the "missing" energy if we realize that some force must have been applied to the car in order to slow it down. At first, work was done against the force of the bumpers and other parts of the car. Then the metal crumpled, and work was done against internal friction. The point is that regardless of the exact process, some force had to be exerted to slow the car down, and this force must have been exerted through some distance, however short. The car's velocity could not decrease instantaneously unless an infinite force acted on it. If this force had not been present, the car would have sailed right through the truck and there would have been no collision! This hidden energy may properly be called PE if it is recoverable, but generally if KE is lost in a collision it is transformed mostly into thermal energy as the colliding bodies are permanently deformed. A minute fraction of the KE may go into, say, sound energy or light energy (if sparks fly).

We define an *inelastic impact* as one in which some KE is "lost" by transformation into other forms of energy. There are various degrees of inelastic impact; a collision such as the one discussed in Example 5-4, in which the bodies stick together, is said to be completely inelastic. The essential feature of a completely inelastic or a partially elastic impact is that some of the KE of the system is transformed during the collision.

Turning now to the other extreme, we define an *elastic impact* as one in which no KE is lost. A bouncing ball may be tremendously deformed during impact, but if the elastic limit (Sec. 9-1) has not been exceeded, the ball returns exactly to its original shape. Momentarily the ball has much PE, but this is returned to the form of KE as the ball separates from the ground. No energy remains in the ball as either elastic PE or heat.

A familiar example of elastic impact is the head-on collision of two billiard balls of equal mass (Fig. 6-8). To derive a general result, we work with symbols rather than with specific

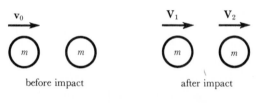

before impact   after impact

Figure 6-8

numbers. Let v_0 be the initial velocity of a moving ball of mass m heading straight for a stationary second ball with the same mass. We have two unknowns: the velocities V_1 and V_2 of the two balls after collision. We set up two equations, one for the conservation of momentum (always true) and one for the conservation of KE (true for elastic impact).

Conservation of momentum:

$$mv_0 + m(0) = mV_1 + mV_2$$

Elastic impact:

$$\tfrac{1}{2}mv_0^2 + \tfrac{1}{2}m(0)^2 = \tfrac{1}{2}mV_1^2 + \tfrac{1}{2}mV_2^2$$

The first equation gives $V_2 = v_0 - V_1$. Substituting this value of V_2 into the second equation, we obtain, after simplifying,

$$V_1^2 - V_1 v_0 = 0$$

or

$$V_1(V_1 - v_0) = 0$$

One root* of this quadratic equation is $V_1 = 0$; whence $V_2 = v_0$. This means that the first ball stops dead, and the second ball takes up the entire velocity v_0 of the first ball. This result is well known to pool players, although in practice rotational and frictional effects must also be considered.

On the microscopic level, conservation of momentum is well illustrated by *Brownian motion* (Fig. 6-9). This is often demonstrated by viewing particles of cigarette smoke through a microscope; the rapid zigzag motion of the smoke particles is the result of incessant random bombardment by fast-moving molecules of nitrogen, oxygen, and other components of air. The molecules themselves are, of course, far too small to be seen with even the most powerful microscope, but as a result of statistical fluctuations in the average effect of many collisions per second, the momentum of a smoke particle

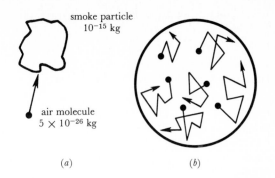

smoke particle
10^{-15} kg

air molecule
5×10^{-26} kg

(a)   (b)

Figure 6-9 Brownian motion. (a) A single collision. (b) Schematic view of smoke particles viewed through a microscope.

changes perceptibly in an erratic fashion. It is somewhat as if a battleship were bombarded by BB shot—one could deduce the existence of the BB shot by observing the motion of the ship, in principle, even though the recoil velocity of the ship is small because of its great mass. The Brownian motion (discovered by botanist Robert Brown in 1827) gives visual evidence for the existence of air molecules. More precisely, a billiard-ball model of a molecule is suggested by these observations.[†]

Even on the sub-submicroscopic level of smallness involved in nuclear collisions, a billiard-ball model of a nucleus is good enough to predict behavior with some degree of accuracy. (See also Sec. 1-7, where the relatively sharp boundary for nuclear matter is discussed.) The basic reaction in a nuclear reactor is the capture of a neutron by a ^{235}U nucleus, which results in the highly unstable ^{236}U nucleus. The ^{236}U divides by fission into two nearly equal fragments, plus several neutrons. One of these neutrons can keep the reaction going if it is captured by another ^{235}U nucleus. The technological problem lies in the fact that the emitted

---

* The other root $V_1 = v_0$ merely says that it is possible for the first ball to retain its original velocity (it could miss the second ball completely).

[†] The mathematical theory of the Brownian motion was worked out by Einstein in a series of five papers published in the first decade of the 20th century.

neutrons are fast (say 10^6 m/s), whereas only slow neutrons (say 10^3 m/s) are effectively captured. Hence the use of a *moderator*, a substance added to the pile to slow down the neutrons by successive collisions. Among commonly used moderators are deuterons (heavy hydrogen nuclei in heavy water) and carbon (in the form of graphite). Which of these is more effective? The relative masses are neutron 1, deuteron 2, and carbon nucleus 12. If a neutron traveling at a known velocity v_0 collides elastically head-on with a stationary deuteron, it is a routine problem in conservation of momentum to find the velocities of neutron V_N and deuteron V_D after collision.

Conservation of momentum:

$$1v_0 + 2(0) = 1V_N + 2V_D$$

Elastic impact:

$$\tfrac{1}{2}(1)v_0{}^2 + \tfrac{1}{2}(2)(0)^2 = \tfrac{1}{2}(1)V_N{}^2 + \tfrac{1}{2}(2)V_D{}^2$$

Combining these equations, we obtain

$$V_N = -\tfrac{1}{3}v_0 \quad \text{and} \quad V_D = +\tfrac{2}{3}v_0$$

That is, the neutron rebounds with $\tfrac{1}{3}$ of its original velocity. After two collisions it will have $\tfrac{1}{3} \times \tfrac{1}{3}$ or $\tfrac{1}{9}$ of its original velocity, and if $v_0 = 10^6$ m/s, after only 7 collisions the velocity will be less than 10^3 m/s, as desired.

Repeating the problem with carbon as the target, we obtain $-\tfrac{11}{13}v_0$ for the neutron's velocity after one collision, not very much less than its original velocity. Obviously, heavy water,* with its deuterons, is a more effective moderator than graphite, and heavy-water reactors are smaller and more compact than those using graphite. In the economics of reactor design, other factors must also be considered. Heavy water is expensive, but even the carbon is relatively costly, since it must be painstakingly purified to avoid capture of neutrons by "impurity atoms" before they have a chance to hit another ^{235}U nucleus. One might think that the protons in ordinary water would be still more effective, for, as in the case of the two equal billiard balls, the neutron would be completely stopped[†] at the first collision. Unfortunately, the proton has a strong affinity for capturing a neutron; in fact, a proton plus a neutron makes a deuteron, according to the equation $^1_1\text{H} + ^1_0\text{n} = ^2_1\text{H}$. Thus with ordinary water many of the neutrons are absorbed and are not available for continuing the chain reaction with ^{235}U. Reactors using ordinary water are less efficient than those using heavy water (see Sec. 33-2).

_____ **Example 6-11**

A ball of mass 2 kg slides down a frictionless chute (Fig. 6-10) and then strikes a stationary 3-kg block on a horizontal table. The coefficient of kinetic friction between the block and the table is 0.1. If the ball and the block stick together, how far do they slide before coming to rest?

We solve the problem in several steps. The weight of the ball is $(2 \text{ kg})(9.8 \text{ m/s}^2) = 19.6$ N; the weight of the block is $(3 \text{ kg})(9.8 \text{ m/s}^2) = 29.4$ N.

(*a*) Conservation of energy for path *AB*: The PE at *A* is Wh; the KE at *B* is $\tfrac{1}{2}mv^2$.

$$KE + PE = KE + PE$$

$$0 + (19.6 \text{ N})(2.5 \text{ m}) = \tfrac{1}{2}(2 \text{ kg})v^2 + 0$$

$$v = \boxed{7 \text{ m/s}}$$

This is the speed of the ball just before impact.

---

* It was Germany's all-out collection of heavy water from occupied Norway in the early 1940s that apprised the scientists of other countries that the Germans were attempting to release nuclear energy from uranium. Norwegian scientists had been studying heavy water, which is produced as a by-product of the great hydroelectric installations of that country.

---

† This assumes a head-on collision with a stationary proton. In practice, not all collisions would be head-on; the target protons would also be in motion at some 10^3 m/s due to thermal agitation. For both these reasons, after many collisions the average KE of the scattered neutrons will become equal to that of the protons among which they move.

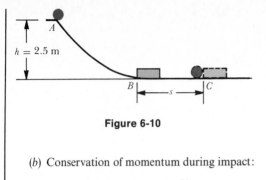

$h = 2.5 \text{ m}$

Figure 6-10

(b) Conservation of momentum during impact:

$$m_{ball}v_{ball} = m_{ball+block} V_{ball+block}$$
$$(2 \text{ kg})(7 \text{ m/s}) = (5 \text{ kg}) V$$

$$V = \boxed{2.8 \text{ m/s}}$$

(c) Finally we apply the law of conservation of energy to the path BC: the KE of the system just after impact is dissipated as work is done against friction. The force of friction is

$$f = \mu_k F_n = 0.1(5 \text{ kg})(9.8 \text{ m/s}^2) = 4.9 \text{ N}$$

$$\text{KE at } B = \text{work against friction}$$

$$\tfrac{1}{2}(5 \text{ kg})(2.8 \text{ m/s})^2 = (4.9 \text{ N})s$$

$$s = \boxed{4 \text{ m}}$$

See the For Further Study Sec. 6-12 for a discussion of collisions that are intermediate between perfectly elastic and completely inelastic.

6-10 The Significance of the Conservation Laws

We have by now come across two important laws of physics, both of them expressed in the form of conservation laws. Taken together, the laws of conservation of mechanical energy and conservation of momentum are adequate to give answers to most of the questions that might be asked about the mechanics of physical objects. These conservation laws are equivalent to Newton's laws; as a matter of fact, we derived them from Newton's laws of motion in the first place (Secs. 5-3 and 6-4). It would be entirely possible

to do it the other way around: assume the conservation laws and derive Newton's laws from them. Let us therefore ask which is more fundamental—Newton's second law in the form net force = mass × acceleration, or Newton's second law in the form net force = (change of momentum)/(time)? So far, we have seen no reason to prefer one form of the law to the other. Situations arise, however, in which the second way (involving change of momentum) is distinctly preferable. What if the mass of a body changes while it is being acted on by a force? There could then be a change of momentum even if velocity were constant. Acceleration would be zero; the equation net $F = ma$ would not apply.

_____ **Example 6-12**

A train of empty coal cars is being pulled along a straight, level piece of track at 20 m/s. If rain starts to fall and the cars receive 300 kg of water per minute, how much additional force must the engine supply to keep the train moving at a constant speed of 20 m/s?

Here the momentum changes because the mass changes. The amount of momentum that is added in each minute is $(300 \text{ kg})(20 \text{ m/s}) = 6000 \text{ kg·m/s}$. We now apply Newton's second law in the momentum form:

$$F = \frac{\Delta(mv)}{\Delta t}$$

$$= \frac{6000 \text{ kg·m/s}}{60 \text{ s}} = \boxed{100 \text{ N}}$$

A rather small force, to be sure, but note that the equation net $F = ma$ would have been inadequate to cope with the problem without a number of simplifying assumptions.

Rate of change of momentum is taken as a definition of force in realms far beyond Newton's experience. We shall see in Sec. 28-5 how the law of conservation of momentum is the basis for a more general concept of mass in Einstein's theory of special relativity. In relativity, as in all physics, the laws of conservation of momentum and energy are of fundamental significance.

Summary   Work is done on a body whenever it is moved against an opposing force. Work equals displacement times the component of force in the line of motion ($W = Fs$). Energy is stored work. Metric units for work and energy are the joule and the erg. Potential energy (PE) is due to the relative position of the parts of a system of bodies, and this includes gravitational, elastic, chemical, and some electrical energy, as well as nuclear energy. Kinetic energy (KE) is due to motion, and this includes much of what is known as thermal energy. Some useful formulas for mechanical energy are

$$\text{Gravitational PE} = Wh = mgh \qquad \text{KE} = \tfrac{1}{2}mv^2$$

The energy of an isolated system remains constant. The laws of conservation of momentum and of energy include the content of Newton's laws and are general enough to be useful in many non-Newtonian aspects of contemporary physics. The statement that net force equals the rate of change of momentum serves as a general definition of force.

In collisions, momentum is always conserved, but kinetic energy is conserved only in elastic collisions. In an inelastic collision, some KE is transformed to heat as the colliding bodies are permanently deformed, or else some of the KE is stored as PE in one or more of the colliding bodies.

Power is the rate of doing work. One watt is 1 joule per second.

Check List

work	$\text{KE} = \tfrac{1}{2}mv^2$
$W = Fs$	law of conservation of energy
joule	power
erg	watt
energy	$P = fv$
PE	efficiency
KE	inelastic impact
gravitational PE $= Wh = mgh$	elastic impact

Questions

6-1   A car is coasting on a level stretch of road. If the driver turns on the headlights, where does the energy come from   (a) if the headlights are connected to a battery;   (b) if they are connected directly to a generator geared to the wheels? Does the car slow down in either case?

6-2   Is there any source of energy available to us on earth that does not eventually trace back to the sun's energy? What about muscular energy? What is the source of the sun's energy?

6-3   When an artificial satellite is moving in an orbit around the earth, it has more energy than it had when it was on earth before launching. Where did this energy come from? Into what forms is this energy transformed when the satellite is circling the earth? What will be the eventual fate of this energy when the satellite spirals toward the earth and burns up?

6-4   Is some of the sun's energy used up in causing the earth to move in an orbit (assumed circular for convenience)?

6-5   The tides in some parts of Norway rise 15 m or more, and when the tide goes out, the water dashes against the rocks and beach and flows slowly out to sea, ready to be raised again by the moon in $12\tfrac{1}{2}$ hours as the earth turns. What becomes of the PE that has been stored in the water? If the water were trapped in a dam, the PE could be used by man; tidal energy stations have actually been constructed. What is the ultimate source of this energy?

6-6 What is wrong with this sentence? "The motor did 12,000 watts of work while pumping 0.1 m^3 of water from the well."

6-7 Show that the kilowatt·hour (kW·h) is a unit of energy. How many joules equal 1 kW·h? (*Note:* This product of units is usually abbreviated kWh.)

6-8 In Example 6-9, is the speed at A the same as that at B? Is the velocity at A the same as that at B?

6-9 A flash bulb is carefully weighed on a good analytical balance; then it is "fired," and after cooling off it is weighed again. Will the two observed weights be the same?

6-10 The dimensions of both torque and work are $[ML^2T^{-2}]$. Does this mean that torque and work are the same?

6-11 When a jet plane lands on the deck of a carrier, is this an elastic or an inelastic "collision"? What becomes of the momentum of the jet? What becomes of its KE?

MULTIPLE CHOICE

6-12 A body of mass 1 kg that has an initial speed of 1 m/s is acted on by a net force of 1 N for 1 s. The increase in KE is (*a*) 1.0 J; (*b*) 1.5 J; (*c*) 2.0 J.

6-13 A body of mass m has kinetic energy E. Its momentum is (*a*) $4E^2/m$; (*b*) $2E/\sqrt{m}$; (*c*) $\sqrt{2mE}$.

6-14 A body of mass m has momentum p. Its kinetic energy is (*a*) p/m; (*b*) $p^2/2m$; (*c*) $\sqrt{2mp}$.

6-15 For a freely falling body, the following quantity is constant: (*a*) $mgh + mv^2$; (*b*) $gh + \frac{1}{2}v^2$; (*c*) $mgh - \frac{1}{2}mv^2$.

6-16 During a collision between two bodies, momentum is conserved (*a*) only if KE is conserved; (*b*) only if the bodies are not permanently deformed; (*c*) always.

6-17 During a collision between two bodies, kinetic energy is conserved (*a*) only if the collision is perfectly elastic; (*b*) whenever the bodies do not stick together; (*c*) always.

Problems

6-A1 (*a*) Compare the momentum of a car of mass 1 metric ton moving at 1.5 m/s with that of a 25-g bullet moving at 1000 m/s. (*b*) Compare the KE's of the same objects.

6-A2 Use the energy principle, in conjunction with the definition of elastic impact, to prove that if a falling body makes *elastic* collision with the ground, it bounces back to the same height from which it was originally dropped.

6-A3 A child who weighs 150 N is sitting on a horizontal floor. (*a*) How much work is required to move the child 0.6 m against an opposing force of friction equal to 40 N? (*b*) How much work is required to lift the child 0.6 m straight up?

6-A4 How much work must be done to push a thumbtack 0.8 cm into a bulletin board against an opposing force of 30,000 dyn?

6-A5 A ladder 5 m long weighs 200 N, and its c.g. is at the center. What is the increase of PE of the ladder when it is raised from a horizontal position to a vertical position?

6-A6 How much work is done by a train's engine that pulls a train of mass 6×10^6 kg 1 km over a level track if the coefficient of rolling friction is 0.01?

6-A7 When a girl who weighs 400 N walks up a hill that is 130 m long, her PE increases by 30 kJ. What is the vertical rise of the hill?

6-A8 A student who weighs 800 N sleeps in a bed 76 cm above the floor. The student's c.g. is 90 cm from the soles of his feet. (*a*) What is the increase in his PE when he gets up in the morning? (*b*) Where does this energy come from?

6-A9 While pulling down a classroom window shade, a teacher exerted a downward force increasing steadily from 7.4 N to 17.4 N, and the length of the shade was extended by 1.5 m. (*a*) What average force opposed the motion? (*b*) What was the shade's increase of PE?

6-A10 A wild goose of mass 4.5 kg is flying at a speed of 4 m/s. What is the bird's KE?

6-A11 What is the KE of an 8-metric-ton truck moving at 10 m/s up a hill that makes an angle of 20° with the horizontal?

6-A12 What is the KE of a neutron that has a mass of 1.67×10^{-27} kg and is moving at 2000 m/s?

6-A13 Calculate the KE of an 800-kg sports car moving at 4 m/s.

6-A14 How fast is a sprinter who weighs 560 N running if her KE is 2800 J?

6-A15 A boy pushes with a steady force of 60 N on another boy who is on frictionless roller skates. How far must he push in order to give his friend 390 J of KE?

6-A16 A fly takes off horizontally with KE equal to 150 ergs. What force was exerted by the fly's legs if the force was exerted through a distance of 2 mm?

6-A17 The fly of Prob. 6-A16 weighs 29.4 dyn. What is its mass? What is its velocity just after takeoff?

6-B1 Eight books, each 4 cm thick and weighing 20 N, are lying horizontally on a table. What work is required to construct from them a stack of books 32 cm high?

6-B2 A .22-caliber bullet of mass 0.0026 kg is shot into a massive wooden target at 320 m/s. It penetrates 7 cm into the target; what is the average force exerted on the bullet by the wood? (Use the energy principle; do not find the acceleration.)

6-B3 An arrow of mass 0.1 kg was fired directly upward by an archer who exerted an average force of 75 N on a bowstring that was pulled back 0.6 m. After it came down, the arrow penetrated 30 cm vertically into the earth. Use the energy principle to answer the following questions: (*a*) With what speed did the arrow leave the bow? (*b*) How high did the arrow go? (*c*) What average upward force did the earth exert on the arrow while bringing it to rest?

6-B4 A 70-kg boy, starting from rest, slides down a 30° hill that is 60 m long. He arrives at the bottom with a speed of 10 m/s. How much thermal energy has been shared between the surface of the hill and the seat of his pants? (Use the energy principle; do not find the acceleration.)

6-B5 A dolphin of mass 70 kg starting from rest reached a speed of 8 m/s by expenditure of 3200 J of muscular energy. How much work was done to overcome the fluid friction (viscous drag) of the water?

6-B6 A Ping-Pong ball of mass *m* rolls off the edge of a table 0.8 m high. When the ball strikes the floor, its speed is 5 m/s. How fast was it rolling when it left the table? (Use the energy principle.)

6-B7 A stone of mass *m* is thrown at some angle at 14.7 m/s from a cliff 19.6 m high. With what speed does the stone strike the ground below the cliff? (Use the energy principle.)

6-B8 A soccer player gives a ball of mass 1 kg an initial horizontal speed of 10 m/s. The ball slides 16 m along the turf against an average force of 2.0 N. What is the final speed of the ball? (Use the energy principle.)

6-B9 A 20-kg sled is coasting with constant velocity at 5 m/s over perfectly smooth, level ice. It enters a rough stretch of ice 20 m long in which the force of friction is 8 N. With what speed does the sled emerge from the rough stretch? (This is the same as Prob. 3-B18. This time, use the energy principle; do not find the acceleration.)

6-B10 (a) Use the energy principle to find with what speed the roller coaster car of Example 6-3 reaches point B starting from rest (ignore friction). (b) Find the speed at point C.

6-B11 A motorcycle daredevil rider zoomed down a ramp, starting from point A 12 m above the ground, and took off at an angle from point B, 2 m above the ground (Fig. 6-11). Between A and B the rider and his cycle, of total mass 200 kg, received 250 kJ of energy from the engine and lost 98 kJ to friction. To what height above ground level did the rider ascend if his horizontal velocity at C was 40 m/s?

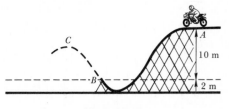

Figure 6-11

6-B12 A sack of mail of mass 40 kg slides down an inclined post-office chute 5 m long that has a vertical drop of 2.0 m. The force of friction (parallel to the chute) is 120 N. (a) What is the loss of PE of the bag? (b) How much work is done against friction? (c) With what speed does the bag reach the bottom of the chute if it is given an initial speed of 3 m/s? (Use the energy principle.)

6-B13 A small ball of mass 2 kg rolls down a frictionless chute (as in Fig. 6-10) that has a vertical fall of 1.8 m. The ball then strikes a stationary 3-kg block on a horizontal rough floor. The coefficient of kinetic friction between the block and the floor is 0.4. If the ball and the block stick together, how far do they slide before coming to rest? (Use conservation of momentum and the energy principle.)

6-B14 An egg weighing 0.49 N is dropped from a building 15 m high. (a) Find the egg's KE just as it strikes the pavement. (b) The target is a thick sheet of foam rubber, and the egg penetrates 30 cm into the rubber while stopping. What elastic upward force, assumed constant, is exerted by the rubber on the egg?

6-B15 A boy weighing 400 N hangs on the end of a rope 6 m long. How much PE does he gain when a friend pulls him aside so that the rope makes an angle of 30° with the vertical?

6-B16 A girl of mass m swings back and forth on the end of a rope 7 m long that is attached to the ceiling of a gymnasium. If she approaches within 3 m of the ceiling during each cycle, what is her speed as she passes through the lowest point of her swing? (Use the energy principle.)

6-B17 Masses of 3 kg and 4 kg are hanging at opposite ends of a cord that passes over a frictionless pulley. The system is held stationary for a while, then released. What will be the speed of the masses when the 4-kg mass has descended 0.5 m below its starting point? (Use the energy principle.)

6-B18 A car of mass m is moving with speed v along a level road; the coefficient of static friction between road and tire is μ_s. Use the energy principle to derive a formula for the minimum stopping distance without skidding, in terms of v, μ_s, and g. Why is the coefficient of *static* friction appropriate here?

6-B19 A heavy chain 140 cm long is stretched out on a frictionless table top, with one end of the chain just at the edge and the length of the chain perpendicular to the edge. The chain is displaced so that a tiny bit projects over the edge; the chain starts to pick up speed and slithers off the table. What is its speed when the last link of the chain is just leaving the table? (*Hint:* Assume the mass of the chain to be m.)

6-B20 A certain machine performs 60 J of work when a force of 5 N is applied through a distance of 15 m. What is the efficiency of the machine?

6-B21 Out of a total of 400 J of work put into a machine, an amount equal to 120 J produced heat in overcoming friction. What is the efficiency of the machine?

6-B22 How much energy must be supplied to a machine whose efficiency is 75% if 1200 J of useful work is to be obtained?

6-B23 A storage battery lights up a 100-W lamp and also runs a motor that is 90% efficient. The useful power output of the motor is 1 kW. What is the battery's loss of chemical PE in 10 min? (Express your answer in kJ.)

6-B24 What is the power of an engine that keeps a boat moving at a steady speed of 25 m/s if water resistance exerts a constant drag of 8800 N on the boat?

6-B25 What must be the input of electrical power, in watts, to a motor that is 85% efficient and delivers useful work at the rate of 600 W (0.8 horsepower)?

6-B26 Three men, each of whom develops 200 W (0.27 horsepower), lift a 3000-N piano to a penthouse 30 m above the street. If the pulley system is 70% efficient, how much time is required for the job?

6-B27 The pilot of the first successful human-powered, heavier-than-air flight could generate an average of 0.32 kW (0.43 horsepower) during the flight. The aircraft covered 2.4 km at an average speed of 4.1 m/s. How much energy was expended during the test flight?

6-B28 A rocket of mass 50 metric tons reaches a speed of 2000 m/s in 1 min after launching. (*a*) What is the rocket's KE? (*b*) Neglecting work done against gravity and against air resistance, what average power (in horsepower) is supplied by the rocket engine?

6-B29 Lindbergh flew the Atlantic in 1927 in 33 h, 30 min, 30 s, with a 230-hp motor in his single-engine plane. (*a*) Use Appendix Table 4 to convert the historical unit of power, hp, to an SI unit, the kilowatt. (*b*) Compute the total work in joules performed by the engine. What assumption did you make in solving this problem? (*c*) What force of air resistance opposed Lindbergh's plane at a time when his engine was causing the plane to move at 45 m/s?

6-B30 Have someone time you as you run up a flight of stairs as fast as possible. (*a*) Using your weight and the vertical rise of the stairs, calculate the power you developed. Express your answer both in watts and in horsepower. (*b*) Why didn't James Watt use human students of college physics instead of the exceptionally powerful dray horses mentioned on page 133?

6-B31 A motor whose useful output is 10 kW is used to lift an elevator and its load (weighing altogether 18,000 N) to a height of 10 m. The motor is 60% efficient. (*a*) How much work is done? (*b*) How much energy must be supplied to the motor? (*c*) How much time will be required for the trip?

6-B32 According to James Watt, a horse can keep a weight of 150 lb moving upward at a constant speed of $2\frac{1}{2}$ mi/h. Show that this statement is in agreement with the definition of 550 ft·lb/s for the horsepower. (One mile = 5280 ft.)

6-B33 Energy is stored by pumping 3×10^6 kg of water into a reservoir that is a shallow lake 40 m above a river. Calculate the power of a pump that can fill the reservoir in 12 h.

6-B34 Make a detailed energy check for the elastic collision of a neutron and a deuteron considered on pages 135–136. The total KE before impact is that of the proton, $\frac{1}{2}(1)v_0{}^2$ units. Using the final velocities calculated on page 136, show that the total KE of the system after impact is also $\frac{1}{2}(1)v_0{}^2$ units.

6-B35 How much KE is lost during the inelastic collision at point B in Fig. 6-10?

6-B36 A neutron of mass 1 unit, having velocity v_0, makes a head-on elastic collision with a stationary oxygen nucleus of mass 16 units. (*a*) What forward velocity does the oxygen nucleus receive? (*b*) What fraction of the KE of the neutron is transferred to the oxygen nucleus?

6-B37 A ball of mass 1 kg moving at 8 m/s westward overtakes and collides with a ball of mass 3 kg moving at 4 m/s westward. The balls stick together after impact. (*a*) Calculate the magnitude and direction of the velocity of the combined mass after impact. (*b*) Calculate the total KE of the system before and after impact, and find how many joules of KE were lost during the collision.

6-B38 A ball of mass 6 kg, moving at 2 m/s eastward, strikes head-on a ball of mass 2 kg that is moving at 2 m/s westward. The balls stick together after the impact. (*a*) What is the magnitude and direction of the velocity of the combined mass after collision? (*b*) Compute the total KE of the system before and after impact, and find how many joules of KE were lost during collision.

6-B39 Solve Prob. 6-B38 if the collision is perfectly elastic. Find the magnitude and direction of each ball's velocity after the collision.

6-B40 Repeat the derivation leading up to Eq. 6-3, but do not assume that $v_0 = 0$. Show that the work done against the reaction force (inertial force) is $\frac{1}{2}mv^2 - \frac{1}{2}mv_0{}^2$. Interpret your answer physically.

6-C1 Two objects of mass m_A and m_B are held together by a strong, light thread and are also acted on by a light spring that is compressed as shown in Fig. 6-12. When the restraining thread is burned, the two objects fly apart with velocities $-v_A$ and $+v_B$. Use the law of conservation of momentum to solve for the ratio of the velocities, v_A/v_B. Then compute the ratio of the kinetic energies, KE_A/KE_B, and prove that the ratio of the KE's of the two objects is the reciprocal of the ratio of their masses.

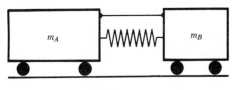

Figure 6-12

6-C2 The arrangement for measuring the speed of a bullet described in Prob. 5-B9 is not very practical, since friction would prevent the cart from rolling at constant velocity,

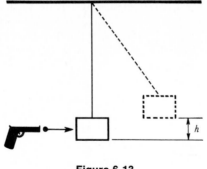

Figure 6-13

and the measurement of the short interval of time might prove difficult. A more precise arrangement is the ballistic pendulum (Fig. 6-13). After the bullet lodges in the wooden block, the block swings over and the height of rise is measured. (*a*) In one experiment, the string was 5 m long and the rise was 8 cm. The mass of the bullet was 9 g, and the mass of the block of wood was 4 kg. What was the forward velocity of the block at its lowest position, just after impact? (*Hint:* Use the energy principle.) (*b*) Using the velocity found in (*a*), apply the law of conservation of momentum to find the original velocity of the bullet.

6-C3 A glider is on a frictionless airtrack that is inclined 8° with the horizontal. The glider is released from a point 100 cm from a bumper at the bottom of the track, and it rebounds after losing 20% of its kinetic energy during the collision. (*a*) What is the glider's speed just after hitting the bumper? (*b*) How far from the bumper does the glider travel on the rebound?

6-C4 A roll of heavy but perfectly flexible paper is initially at rest at the top of a rough, inclined plane. The roll is released, and it unwinds to form a single long strip of paper lying at rest stretched out along the plane. Show that there is a loss of PE for the paper. What becomes of this PE? After thorough discussion of this problem, you may wish to refer to Ref. 7 at the end of the chapter.

6-C5 A 4-kg ball makes a perfectly elastic head-on collision with a stationary ball and re-bounds with $\frac{1}{5}$ of its original speed. What is the mass of the struck ball?

6-C6 A ball of mass m makes a perfectly elastic head-on collision with a stationary ball of mass M and rebounds with $1/n$ of its original speed. Derive a formula for M in terms of m and n. (This is generalization of Prob. 6-C5.)

For Further Study

6-11 Machines

Typical problems of modern industrial life in-volve the use of machines that allow us to exert large forces by means of the application of much smaller forces. As in the use of a lever to pry loose a boulder, the applied force often is much less than the force exerted on the load. Figure 6-14 illustrates such a simple machine. The lever arms are assumed to be in a 3:1 ratio; therefore, point A moves down three times as far as point B moves up. The force at A is $\frac{1}{3}$ that at B (from the

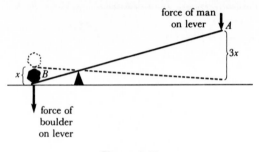

force of man on lever

A

3x

x

B

force of boulder on lever

Figure 6-14

torque condition). The work done by the lever on the stone equals the work done on the lever by the man, if it is assumed that no energy is transformed into heat due to friction.

In general, however, because of friction the mechanical work done by the machine is less than the work done on the machine. We must now define some terms: The *actual mechanical advantage* (AMA) is the ratio of the output force to the input force. The *ideal mechanical advantage* (IMA) is the ratio of the input distance (the distance through which the input force moves) to the output distance (the distance moved by the load). The *efficiency* of the machine is the ratio of the work output to the work input.

A typical machine is illustrated schematically by the box in Fig. 6-15, which contains a collection of gears, pulleys, and levers the exact construction of which we do not need to know. Since

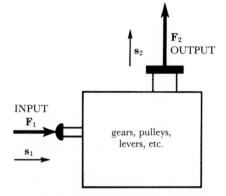

Figure 6-15 A simple machine (schematic).

F_1 and s_1 are parallel, the work input is $F_1 s_1$; similarly, the work output is $F_2 s_2$. Suppose we apply a force of 2 N and push the input button inward a distance of 0.02 m, and suppose that as a result the output platform moves 0.002 m, exerting a force of 15 N on the load.

$$\text{Work input} = F_1 s_1$$
$$= (2\ \text{N})(0.02\ \text{m}) = 0.04\ \text{J}$$
$$\text{Work output} = F_2 s_2$$
$$= (15\ \text{N})(0.002\ \text{m}) = 0.03\ \text{J}$$

$$\text{Efficiency} = \frac{\text{work output}}{\text{work input}}$$
$$= \frac{0.03\ \text{J}}{0.04\ \text{J}} = 0.75 = 75\%$$

$$\text{AMA} = \frac{\text{force output}}{\text{force input}} = \frac{15\ \text{N}}{2\ \text{N}} = 7.5$$

$$\text{IMA} = \frac{\text{input distance}}{\text{output distance}} = \frac{0.02\ \text{m}}{0.002\ \text{m}} = 10$$

Note that

$$\text{Efficiency} = \frac{\text{work output}}{\text{work input}}$$
$$= \frac{(\text{force output})(\text{output distance})}{(\text{force input})(\text{input distance})}$$
$$= \frac{\text{force output}}{\text{force input}} \bigg/ \frac{\text{input distance}}{\text{output distance}}$$
$$= \text{AMA/IMA}$$

This is illustrated in the above example:

$$\frac{\text{AMA}}{\text{IMA}} = \frac{7.5}{10} = 75\%$$

At the most (100% efficiency), the distance ratio equals the force ratio and the AMA equals the IMA.

Example 6-13

A tennis net is tightened by means of a crank (Fig. 6-16). The rope is wound around a spindle of radius 1.25 cm, and the crank handle moves in a circle of

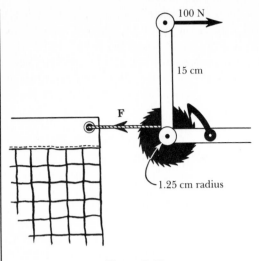

Figure 6-16

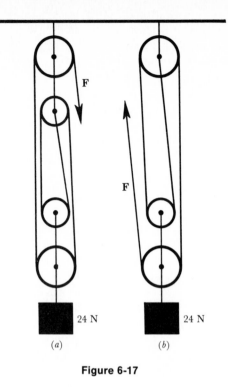

Figure 6-17

radius 15 cm. If the efficiency is 60%, what is the force in the rope when a player exerts a force of 100 N on the handle in the most effective direction?

The distances moved by the crank and by the spindle are found by considering one complete revolution of the crank:

$$\text{IMA} = \text{distance ratio} = \frac{2\pi(15\text{ cm})}{2\pi(1.25\text{ cm})} = 12$$

Since efficiency = AMA/IMA,

$$0.60 = \text{AMA}/12$$

$$\text{AMA} = 7.2$$

But AMA = force ratio, so

$$7.2 = F/(100\text{ N})$$

$$F = \boxed{720\text{ N}}$$

Example 6-14

In the pulley system shown in Fig. 6-17a, the weight of the load is borne equally by four ropes. (This is strictly true only if all the ropes are vertical. The diagram has been slightly distorted for clarity.) The downward rope on the right in (a) doesn't count, because it merely changes the direction of the force **F**; see the equivalent diagram in (b). The distance ratio is 4:1, which can be verified experimentally or by a little geometrical reasoning. (We know that the distance ratio must be 4:1, since the load is shared among four ropes and hence IMA = 4.) If a load of 24 N is lifted by application of 10 N,

$$\text{Efficiency} = \frac{\text{AMA}}{\text{IMA}} = \frac{24\text{ N}/10\text{ N}}{4} = \frac{24}{40}$$

$$= \boxed{60\%}$$

For a machine that is designed to convert small motions into large ones, the IMA and AMA are less than 1. A bicycle pedal is an example; the foot of the rider moves about 6 cm for every 30 cm moved by the rim of the wheel. Here the applied force must be greater than the output force, even for 100% efficiency. Convenience, rather than magnification of force, is the goal.

An inclined plane, while it is not usually thought of as a machine, does have a mechanical advantage: by means of a plane a heavy weight may be raised by application of a smaller force. Indeed, the erection of the Pyramids, a prodigious feat accomplished through use of human labor, was made possible through ingenious use of ramps and inclined planes.

Example 6-15

A team of movers uses an inclined plane to get a piano that weighs 3600 N over a doorstep (Fig. 6-18). The plane is 3.0 m long and the step is 0.27 m high. If the movers apply a force of 500 N, what are the IMA, AMA, and efficiency?

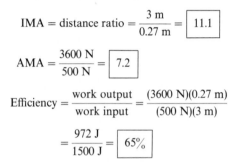

$$\text{IMA} = \text{distance ratio} = \frac{3 \text{ m}}{0.27 \text{ m}} = \boxed{11.1}$$

$$\text{AMA} = \frac{3600 \text{ N}}{500 \text{ N}} = \boxed{7.2}$$

$$\text{Efficiency} = \frac{\text{work output}}{\text{work input}} = \frac{(3600 \text{ N})(0.27 \text{ m})}{(500 \text{ N})(3 \text{ m})}$$

$$= \frac{972 \text{ J}}{1500 \text{ J}} = \boxed{65\%}$$

The efficiency could also be found by dividing the AMA by the IMA: 7.2/11.1 = 65%.

The work done against friction = (input of mechanical work) − (useful output of mechanical work):

$$W = 1500 \text{ J} - 972 \text{ J} = 528 \text{ J}$$

This work against friction becomes heat, and the plane and piano get slightly warmer.

Within the system (men + piano + plane + earth), energy has been conserved. We start with chemical PE (assuming the men ate a good nourishing breakfast). After the operation, this chemical PE has become partly gravitational PE, as the piano was lifted (that is, as the piano and the earth were separated from each other), and partly thermal KE, as the molecules of plane and piano were speeded up and as heat was released in the muscles of the men during their effort.

6-12 Partially Elastic Collisions

Our study of collisions in Sec. 6-9 was limited to two rather special cases—completely inelastic impact and perfectly elastic impact. To deal with intermediate cases, we define a constant known as the *coefficient of restitution e*, which depends on the nature of the colliding bodies:

$$e = \frac{\text{relative velocity of separation}}{\text{relative velocity of approach}}$$

For a completely inelastic impact, in which the bodies stick together, the relative velocity of separation is 0 and $e = 0$. In Sec. 6-9 we calculated the loss of KE in a completely inelastic collision.

For a perfectly elastic collision, $e = 1$ and the bodies separate as fast as they come together. The proof, for a head-on collision, is based on the definition of elastic collision (no loss of KE). Let the masses of the bodies be m_1 and m_2, their

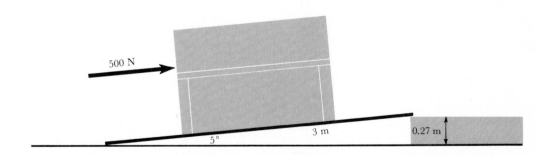

Figure 6-18

initial velocities v_1 and v_2, and their velocities after collision V_1 and V_2. Then the relative velocity of approach before collision is $v_1 - v_2$. The relative velocity of approach after collision is $V_1 - V_2$; hence the relative velocity of *separation* after collision is $V_2 - V_1$.

Elastic impact:

$$\tfrac{1}{2}m_1v_1{}^2 + \tfrac{1}{2}m_2v_2{}^2 = \tfrac{1}{2}m_1V_1{}^2 + \tfrac{1}{2}m_2V_2{}^2$$

Conservation of momentum:

$$m_1v_1 + m_2v_2 = m_1V_1 + m_2V_2$$

H. E. Edgerton, Massachusetts Institute of Technology

Figure 6-19 Partially elastic impact of a bouncing tennis ball, photographed at intervals of $\tfrac{1}{60}$ s. Relative velocity of separation is less than relative velocity of approach. Note the momentary storage of potential energy as the ball is deformed during impact.

Rearrangement of these equations gives

$$m_1(v_1{}^2 - V_1{}^2) = m_2(V_2{}^2 - v_2{}^2)$$
$$m_1(v_1 - V_1) = m_2(V_2 - v_2)$$

Dividing one equation by the other, we obtain

$$v_1 + V_1 = V_2 + v_2$$

or

$$v_1 - v_2 = V_2 - V_1$$

Thus we have proved that the relative velocity of separation equals the relative velocity of approach, and hence $e = 1$ for a perfectly elastic collision. For a good grade of steel ball bearing bouncing on a steel plate, e may be as high as 0.9; for an orange bouncing on the floor, e may be as low as 0.1 or less. Most collisions are partially elastic (or partially inelastic, if you perfer); this means that some mechanical energy is lost as the bodies are permanently deformed or otherwise changed (Fig. 6-19).

Example 6-16

What is the coefficient of restitution in the collision of Prob. 2-B18?

The relative velocity of approach is the velocity with which the ball struck the ground; call it v_0 (the detailed solution gives 28 m/s for this velocity, but the value is not needed). By the terms of the problem, the ball lost one-fifth of its speed in the collision, so the relative velocity of separation was $\tfrac{4}{5}v_0$:

$$e = \frac{\tfrac{4}{5}v_0}{v_0} = \boxed{0.80}$$

In a partially inelastic collision, as in a completely inelastic collision, some mechanical energy is lost, as seen in the following example.

Example 6-17

A car of mass 600 kg overtakes and makes a rear-end collision with a truck of mass 3000 kg. Before impact the car is moving at 25 m/s, and the truck at

10 m/s. (a) If the coefficient of restitution is 0.6, what are the velocities of the car and truck after collision? (b) How much KE has been lost during the collision?

(a) We let V_C be the final velocity of the car, and V_T be the final velocity of the truck. Thus the velocity of separation of car and truck is $V_T - V_C$. There are two equations and two unknowns. One equation is the expression of the law of conservation of momentum, and the other equation is the expression of the definition of the coefficient of restitution.

Conservation of momentum:

$$(600 \text{ kg})(25 \text{ m/s}) + (3000 \text{ kg})(10 \text{ m/s})$$
$$= (600 \text{ kg})V_C + (3000 \text{ kg})V_T$$

Coefficient of restitution:

$$0.6 = \frac{V_T - V_C}{25 \text{ m/s} - 10 \text{ m/s}}$$

Simplifying these two equations, we obtain

$$45,000 = 3000\, V_T + 600\, V_C$$
$$9 = V_T - V_C$$

whence

$$V_C = \boxed{+5 \text{ m/s}}$$

$$V_T = \boxed{+14 \text{ m/s}}$$

Both velocities are in the forward direction after the collision.

(b) To find the KE's before and after collision, we use the original velocities and the new velocities just computed.

Before collision:

$$KE_C = \tfrac{1}{2}(600 \text{ kg})(25 \text{ m/s})^2 = 187,500 \text{ J}$$
$$KE_T = \tfrac{1}{2}(3000 \text{ kg})(10 \text{ m/s})^2 = 150,000 \text{ J}$$
$$KE_{C+T} = 337,500 \text{ J}$$

After collision:

$$KE_C = \tfrac{1}{2}(600 \text{ kg})(5 \text{ m/s})^2 = 7500 \text{ J}$$
$$KE_T = \tfrac{1}{2}(3000 \text{ kg})(14 \text{ m/s})^2 = 294,000 \text{ J}$$
$$KE_{C+T} = 301,500 \text{ J}$$

We see that the car lost 180,000 J of KE and the truck gained 144,000 J. The net loss of KE was

$$\boxed{36,000 \text{ J}}$$

According to the law of conservation of energy, the KE in Example 6-17 was not "lost" but was transformed into other forms of energy, mostly thermal energy (that is, internal energy that contributed to a slight rise in the temperature of the colliding bodies). Considering the process in detail, the solid parts of the truck and car were distorted during the collision, and thus elastic PE was momentarily stored in the bumpers and elsewhere. However, after the vehicles separated, the metallic parts probably vibrated, perhaps invisibly, and as these vibrations died away due to internal molecular friction, the elastic PE was transformed into heat. A minute amount of the energy of the collision was transformed into sound energy and possibly also into radiant energy (light).

Problems

6-C7 A windlass whose efficiency is 80% is used to raise a bucket of water from a well. The radius of the axle is 4 cm, and the radius of the crank handle is 30 cm. The bucket weighs 20 N empty, and it holds 38 liters (10 gallons) of water. (a) What is the IMA? (b) What is the AMA? (c) What force, applied in the most advantageous way, must be applied to the crank handle in order to raise the bucketful of water?

6-C8 A chain hoist in a garage has pulleys arranged so that the load rises 1 cm for every 50 cm moved by the worker's hand. When an engine block weighing 1000 N is hoisted 80 cm, the mechanic expends 1600 J of energy. Calculate (a) the IMA; (b) the efficiency; (c) the AMA; (d) the force exerted by the worker.

6-C9 A trunk weighing 900 N is pulled up an inclined plane at a steady speed by a force of 100 N parallel to the plane. The vertical rise is 0.16 m, and the plane is 4.00 m long. Calculate (a) the IMA; (b) the AMA; (c) the efficiency.

6-C10 What is the IMA of the lever system consisting of the radius bone pictured in Fig. 4-18 (page 88) if the load is vertical as in the figure?

6-C11 Make some measurements on Fig. 6-19, and determine the coefficient of restitution for the bouncing tennis ball.

6-C12 A billiard ball of mass m moving at $+2$ m/s collides head-on with an equal, stationary ball. The coefficient of restitution is 0.7. Compute the velocity of each ball after the collision.

6-C13 What fraction of the original KE of the system is lost in the collision of Prob. 6-C12?

6-C14 What is the coefficient of restitution of a golf ball that when dropped from a height of 100 cm, rebounds to a point 9.75 cm below the starting point?

6-C15 A 2-kg body slides along a horizontal surface for which the coefficient of kinetic friction is 0.4. When its speed is 6 m/s, it strikes a stationary 3-kg body in a partially elastic collision. The second body slides 1.0 m before coming to rest. Calculate the coefficient of restitution for the impact.

6-C16 The inflation of the official basketball is standardized in the United States so that when the ball is dropped from a distance of 6 ft (measured from the bottom of the ball) it will rebound to a height of 49 to 54 in., measured from the top of the ball. The ball is 10 in. in diameter. What is the coefficient of restitution of a ball that is of the maximum permissible "liveness"?

6-C17 Reconsider Prob. 6-B38, if the coefficient of restitution is 0.8.

6-C18 A ball is dropped from a tower of height h at the same instant that another ball of equal mass is projected upward from the bottom of the tower, with a velocity just enough to raise it to the top of the tower. Show that if the balls collide head-on, the falling ball will rise, on the rebound, to a height that is $(3 + e^2)(h/4)$ above the ground, where e is the coefficient of restitution. Discuss the answer when $e = 1$.

References

1. J. R. Mayer, "The Conservation of Energy," *Phil. Mag.* **24**, 371 (1862). Translated in W. F. Magie, *A Source Book in Physics* (McGraw-Hill, New York, 1935), pp. 196–203.
2. J. P. Joule, "On the Mechanical Value of Heat," *Phil. Mag.*, various memoirs in the period 1843–50. Excerpts in Magie, *A Source Book in Physics*, pp. 203–11.
3. V. V. Raman, "Where Credit is Due—The Energy Conservation Principle," *Phys. Teach.* **13**, 80 (1975). Historical background; states that Mayer's initial interest in the subject was triggered by observations on the oxidation of blood. See also comment by P. M. Rinard, "Mayer and Energy," *Phys. Teach.* **13**, 196 (1975), and book review by S. G. Brush in *Am. J. Phys.* **42**, 920 (1974).
4. R. S. Mackay, "On the Strength of Insects," *Am. J. Phys.* **26**, 499 (1958).
5. E. S. Ferguson, "The Measurement of the 'Man-Day,'" *Sci. American* **255**(4), 96 (Oct. 1971).
6. B. Heinrich, "The Energetics of the Bumblebee," *Sci. American* **288**(4), 97 (Apr. 1973).
7. I. M. Freeman, "The Dynamics of a Roll of Tape," *Am. J. Phys.* **14**, 124 (1946).
8. M. S. Feld, R. E. McNair, and S. R. Wilk, "The Physics of Karate," *Sci. American* **240**(4), 150 (Apr. 1979).
9. J. Walker, "Judo and aikido applications of the physics of force," in Amateur Scientist, *Sci. American* **243**(1), 150 (July 1980).
10. J. Stull, *Conservation of Energy* (film).

7

Rotation

It has been said that nature is in unending motion. In Chap. 2 we studied the kinematics of linear motion—a description of uniformly accelerated motion. Looking at the motions characteristic of our civilization, as distinct from those occurring naturally, one is struck by the increased emphasis on motions that are *periodic*, that is, motions in which a particle or other object repeatedly returns to its original position. The progress of civilization involves our increasing control of energy, and control generally means keeping flywheels, pistons, piano strings, and pendulums in motion and yet at a stationary location. This is not to say that there are no periodic motions in uncivilized nature. On the contrary, the revolution of the earth about the sun and the vibration of a meadowlark's vocal cords are eloquent reminders of the periodicities that have existed since before humans appeared on the planet. The vibrations and rotations of molecules are an important part of the atomic model of the structure and behavior of matter. In this chapter and the next we shall study the simplest form of rotation (uniform circular motion) and the simplest form of vibration (simple harmonic motion).

The force of the rails supplies part of the downward (centripetal) force as these riders are at the top of a vertical curve at an amusement park.

7-1 Angular Quantities

In spite of a somewhat forbidding notation, there is nothing particularly difficult about the equations that describe circular motion, for they are strictly analogous to the familiar equations for linear motion (Sec. 2-6). In fact, the chief difficulty—and one that can be easily overcome—may well be the use of the Greek alphabet. It is useful to employ separate symbols to keep linear and angular concepts distinct yet comparable.

A particle in circular motion (Fig. 7-1) moves through an angle θ when it moves a distance s measured around the circumference of the circle.

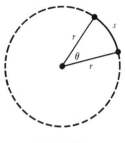

Figure 7-1

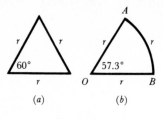

Figure 7-2

We call the arc *s* the *linear displacement** and θ the *angular displacement*. We shall find it convenient to measure angles (and angular displacement) in a new unit, the radian. A *radian* (rad) is the angle whose arc equals the radius of the circle; it is equal to about 57.3°. [Bulge out one side of an equilateral triangle (Fig. 7-2a) so that it becomes an arc of radius *r* (Fig. 7-2b). In doing so, you decrease the opposite angle. Thus, angle *AOB*, which is 1 rad by definition, is a little less than 60°.] There are 2π radians in a complete circle, since the circumference is 2π times the radius, and 2π arcs each equal to the radius can be fitted around the circumference. Thus, the size of a radian is $(1/2\pi)$ times a complete circle: $(1/2\pi) \times 360° = 57.3°$. If the subtended arc is $\frac{1}{2}r$, then the angle is $\frac{1}{2}$ rad, and so forth. In general,

$$\theta = \frac{s}{r}$$

$$s = r\theta \qquad (\theta \text{ in radians})$$

Angular velocity (denoted by ω, the Greek letter omega) is the rate of change of angular displacement:

$$\omega = \lim_{\Delta t \to 0}\left(\frac{\Delta\theta}{\Delta t}\right) = \frac{d\theta}{dt}$$

(Compare Eq. 2–4, which gives an analogous definition of linear velocity as the time rate of change of linear displacement.) Angular velocity can be measured in such units as revolutions per minute and degrees per second. However, if radian measure is used, linear velocity along the arc is related to angular velocity in the same

---

* Even though the path is curved, it is a "linear" displacement since the arc is measured along the circle in linear units such as meters or centimeters.

way that linear displacement is related to angular displacement:

$$v = r\omega \qquad (\omega \text{ in rad/s})$$

Angular acceleration (denoted by α, the Greek letter alpha) is the rate of change of angular velocity:

$$\alpha = \lim_{\Delta t \to 0}\left(\frac{\Delta\omega}{\Delta t}\right) = \frac{d\omega}{dt}$$

(Compare Eq. 2-10 on page 31, which gives an analogous definition of linear acceleration as the time rate of change of linear velocity.) Angular acceleration can be measured in such units as revolutions per minute per second and degrees per minute per hour. If radian measure is used, linear and angular acceleration are related in a simple fashion:

$$a = r\alpha \qquad (\alpha \text{ in rad/s}^2)$$

Note that *s*, *v*, and *a* are measured along the tangent to the circle.

With only the change of Latin symbols to corresponding Greek symbols, all the equations of Sec. 2-6 are equally valid in angular measure. These similarities are summarized below:

$$s = r\theta \qquad (7\text{-}1)$$

$$v = r\omega \qquad (7\text{-}2)$$

$$a = r\alpha \qquad (7\text{-}3)$$

Linear	Angular	
$v = v_0 + at$	$\omega = \omega_0 + \alpha t$	(7-4)
$s = \dfrac{v_0 + v}{2}t$	$\theta = \dfrac{\omega_0 + \omega}{2}t$	(7-5)
$v^2 = v_0{}^2 + 2as$	$\omega^2 = \omega_0{}^2 + 2\alpha\theta$	(7-6)
$s = v_0 t + \frac{1}{2}at^2$	$\theta = \omega_0 t + \frac{1}{2}\alpha t^2$	(7-7)

Equations 7-1 to 7-3 *require* use of radian measure, since linear and angular quantities appear in the same equation. However, Eqs. 7-4 through 7-7 are valid in any consistent set of angular units.

Example 7-1

What is the angular size, in radians, of a 2-m football player viewed by a spectator 100 m away (Fig. 7-3)?

Figure 7-3

We know that the angle is small, and so we make the simplifying assumption that the arc s is equal to the player's height:

$$\theta = \frac{s}{r} = \frac{2 \text{ m}}{100 \text{ m}} = \boxed{0.02 \text{ rad}}$$

Note that an angle in radians is a pure number (length/length) and has no dimensions.

Example 7-2

A ventilator fan is turning at 600 rev/min when the power is cut off, and it turns through 1000 rev while coasting to a stop. Calculate the angular acceleration and the time required to stop.

Just as we did earlier for problems in linear motion, we make a box with knowns and unknowns and select a suitable equation from our list.

$$
\boxed{
\begin{array}{l}
\omega_0 = 600 \text{ rev/min} \\
\omega = 0 \\
\alpha = ? \\
\theta = 1000 \text{ rev}
\end{array}
}
\qquad
\begin{array}{l}
\omega^2 = \omega_0{}^2 + 2\alpha\theta \\[1em]
\alpha = \dfrac{\omega^2 - \omega_0{}^2}{2\theta} \\[1em]
= \dfrac{0 - (600 \text{ rev/min})^2}{2(1000 \text{ rev})} \\[1em]
= \boxed{-180 \text{ rev/min}^2}
\end{array}
$$

To find the time,

$$\omega = \omega_0 + \alpha t$$

$$t = \frac{\omega - \omega_0}{\alpha} = \frac{0 - 600 \text{ rev/min}}{-180 \text{ rev/min}^2}$$

$$= \boxed{3.33 \text{ min}}$$

Note that in this problem we did not have to use radian measure, since no linear and angular quantities appeared together in the same equation. For convenience, we used *revolutions* as the unit for angles.

Example 7-3

A boy on a bicycle is traveling at 10.0 m/s. What is the angular speed of a point on the tire of the bicycle if the tire's radius is 34.0 cm?

Since the tire is always in contact with the road, the linear velocity relative to the axle of any point on the tire equals the forward velocity of the bicycle. We must use radian measure, since both linear and angular quantities are involved.

$$v = r\omega$$

$$\omega = \frac{v}{r} = \frac{10.0 \text{ m/s}}{0.340 \text{ m}} = 29.4 \frac{(\ )}{\text{s}}$$

$$= \boxed{29.4 \text{ rad/s}}$$

The dimensionless radian has been filled in to make a complete angular velocity unit. Recall that use of $v = r\omega$ ensures that the angular velocity will be in rad/s.

7-2 Uniform Circular Motion

Consider the motion of a speck of dust on the rim of a phonograph record turning at $33\frac{1}{3}$ rev/min. We cannot call this "uniform motion," for the *direction* of the motion is continuously changing. Velocity is a vector, and the velocity of the dust

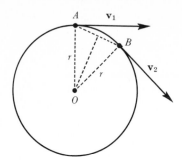

Figure 7-4 Change of velocity during a finite time interval.

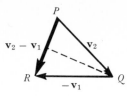

Figure 7-5 The vector difference $\mathbf{v}_2 - \mathbf{v}_1$ is found by adding $\mathbf{v}_2$ and $-\mathbf{v}_1$.

speck is not constant since only the magnitude (speed) is constant. We see that in spite of the apparent simplicity and constancy of the motion, the particle is actually accelerated, since the velocity changes.

A useful formula for the acceleration of a particle in uniform circular motion can easily be derived. In Fig. 7-4 the linear velocity of the particle at a certain instant is $\mathbf{v}_1$; after a short time interval Δt the particle has moved from A to B and the velocity is $\mathbf{v}_2$ (numerically the same as $\mathbf{v}_1$, but directed differently). As always, the acceleration is the rate of change of velocity—that is, the change of velocity divided by the time required for the change. We use the vector equation for average acceleration during an interval Δt,

$$\bar{\mathbf{a}} = \frac{\mathbf{v}_2 - \mathbf{v}_1}{\Delta t}$$

Even though the magnitude of the velocity has not changed, the *vector* difference $\mathbf{v}_2 - \mathbf{v}_1$ is not zero. We have not studied vector subtraction, but we can write $\mathbf{v}_2 - \mathbf{v}_1 = \mathbf{v}_2 + (-\mathbf{v}_1)$, and use ordinary vector addition. Since $-\mathbf{v}_1$ is the same as $\mathbf{v}_1$, but oppositely directed, the vector diagram is constructed by laying off vectors parallel to $\mathbf{v}_2$ and $\mathbf{v}_1$ of Fig. 7-4. We find that $\mathbf{v}_2 - \mathbf{v}_1$ is a vector perpendicular to the dashed line of Fig. 7-5, which is parallel to the average direction of $\mathbf{v}$.* Hence the average acceleration (proportional

---

* The altitude of an isosceles triangle bisects the vertex angle.

to the change in velocity $\mathbf{v}_2 - \mathbf{v}_1$) is directed toward the center of the circle; it is called a *centripetal acceleration.* (*Centripetal* means "seeking the center.") The acceleration is not constant any more than the velocity is constant. The direction of each is continuously changing, with the average acceleration always perpendicular to the average velocity.

To find the instantaneous acceleration, we must use a very short time interval. In the limit, as $\Delta t \to 0$, the *direction* of the instantaneous acceleration is perpendicular to the direction of the instantaneous velocity $\mathbf{v}$. The *magnitude* of the acceleration is found from $a =$ limit of $\Delta v/\Delta t$ as $\Delta t \to 0$. For small angles, the triangles AOB and PQR in Figs. 7-4 and 7-5 may be replaced by sectors (Fig. 7-6). The vertex angle $\angle AOB$ equals the angular speed times the time interval, $\omega \Delta t$. Then $\angle PQR$ also equals $\omega \Delta t$ (similar triangles, with the sides mutually perpendicular). The velocities $\mathbf{v}_1$ and $\mathbf{v}_2$ each have magnitude v, so we label PQ and RQ equal to v, and the magnitude of the change in v is PR, which we now call Δv. Finally, from the sector

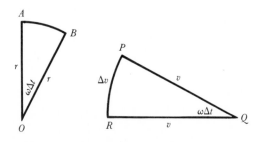

Figure 7-6 Instantaneous acceleration is found by letting Δt approach zero.

PQR we have the approximate equation

$$\Delta v/v \approx \omega\, \Delta t$$

or

$$\Delta v/\Delta t \approx \omega v$$

Passing to the limit, as $\Delta t \to 0$, we see that the magnitude of the centripetal acceleration, denoted by a_c, is

$$a_c = \omega v \qquad (7\text{-}8)$$

This is a most interesting equation for the centripetal acceleration, involving as it does both angular velocity and linear velocity. We can interpret it as follows: The factor v tells us how much velocity *magnitude* there is to change, and the factor ω tells us how rapidly the *direction* of v is changing. For computational purposes, it is convenient to use either ω or v, not both. Thus, since $v = r\omega$,

$$a_c = \omega v = \omega(r\omega) = \omega^2 r \qquad (7\text{-}9)$$

$$a_c = \omega v = \left(\frac{v}{r}\right)v = \frac{v^2}{r} \qquad (7\text{-}10)$$

Other formulas for centripetal acceleration involve the *period T* (time for one revolution, s/rev) and the *frequency f* (number of revolutions per second, rev/s).

$$a_c = \omega v = \frac{2\pi}{T}\left(\frac{2\pi r}{T}\right) = \frac{4\pi^2 r}{T^2} \qquad (7\text{-}11)$$

or, since f is the reciprocal of T,

$$a_c = 4\pi^2 rf^2 \qquad (7\text{-}12)$$

These five formulas for centripetal acceleration are entirely equivalent.

In Eq. 7-11 we have used several relations that will prove useful in later work. In uniform circular motion the particle moves through 2π radians (one revolution) in T seconds (one period), and hence the angular speed is

$$\omega = \frac{2\pi}{T}$$

Also, the particle moves a distance of $2\pi r$ (one circumference) in T seconds (one period), and hence the linear speed around the circle is

$$v = \frac{2\pi r}{T}$$

Example 7-4

Calculate the centripetal acceleration of a speck of dust on the rim of a phonograph record 30 cm in diameter turning at $33\frac{1}{3}$ rev/min. The radius is 15 cm, and the circumference is $2\pi(15\text{ cm}) = 94.2$ cm.

First method: The angular velocity is

$$\omega = \left(33.3\,\frac{\text{rev}}{\text{min}}\right)\left(\frac{2\pi\text{ rad}}{1\text{ rev}}\right)\left(\frac{1\text{ min}}{60\text{ s}}\right) = 3.49\,\frac{\text{rad}}{\text{s}}$$

$$a_c = \omega^2 r = (3.49\text{ rad/s})^2(15\text{ cm})$$

$$= \boxed{182\text{ cm/s}^2}$$

Second method: The linear velocity of the dust particle is

$$v = \left(\frac{94.2\text{ cm}}{1\text{ rev}}\right)\left(33.3\,\frac{\text{rev}}{\text{min}}\right)\left(\frac{1\text{ min}}{60\text{ s}}\right) = 52.3\,\frac{\text{cm}}{\text{s}}$$

$$a_c = \frac{v^2}{r} = \frac{(52.3\text{ cm/s})^2}{15\text{ cm}} = \boxed{182\text{ cm/s}^2}$$

The centripetal acceleration in this example is about $\frac{1}{5}g$.

If the acceleration of the earth (for instance) is toward the sun, then why doesn't the earth eventually fall into the sun? The truth of the matter is that it *does* fall toward the sun—but it never gets any closer. In the absence of the centripetal acceleration, the earth would continue moving in a straight line—that is, it would fly off on a tangent (Fig. 7-7), and after a time t it would have arrived at some point such as C. The earth actually arrives at B, and the distance

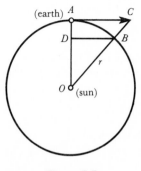

Figure 7-7

CB represents a fall toward the sun. In fact, it may be verified that the laws of falling bodies apply, and $CB \approx \frac{1}{2}at^2$, where *a* is the centripetal acceleration previously derived, equal to v^2/r. (See Prob. 7-C6.)

A particle in nonuniform circular motion has two accelerations: the centripetal acceleration $\mathbf{a}_c$, which is directed along the radius, and the tangential acceleration $\mathbf{a}_t$, whose magnitude is given by Eq. 7-3, $a = r\alpha$. The vector sum of $\mathbf{a}_c$ and $\mathbf{a}_t$ is the resultant acceleration, directed at some angle relative to the radius.

7-3 Centripetal Force

Having studied the kinematics (description) of uniform circular motion, we are now able to examine the causes of such motion (the problem of dynamics). Circular motion is no exception to the general rule that forces are needed to cause accelerations of bodies having mass. *Centripetal force*, which causes centripetal acceleration, acts on the body in motion and is directed toward the center of the circle. Any type of force can serve as a centripetal force. In the following examples we illustrate how elastic, frictional, gravitational, and magnetic forces can cause uniform circular motion of a body having mass.

horizontal circle of radius 2 m at an angular speed of 100 rev/min?

First we find the magnitude of the centripetal acceleration:

$$a_c = \omega^2 r$$

$$= \left[\left(100 \, \frac{\text{rev}}{\text{min}} \right) \left(\frac{1 \, \text{min}}{60 \, \text{s}} \right) \left(\frac{2\pi \, \text{rad}}{1 \, \text{rev}} \right) \right]^2 (2 \, \text{m})$$

$$= 219 \, \text{m/s}^2$$

The block is acted on by three forces: its weight, the upward force of the surface, and the tension in the rope (an elastic force). The vertical forces balance, and the net force equals the tension in the rope:

net $F = ma$

$$T = (10 \, \text{kg})(219 \, \text{m/s}^2) = \boxed{2190 \, \text{N}}$$

The direction of *T* is the same as that of a_c, namely, toward the center of the circular path. We have found the force *on* the block; by Newton's third law the block exerts an equal and opposite (centrifugal) force on the pivot.

Frictional forces often supply the necessary centripetal force, as for an automobile going around a level curve.

Example 7-5

A 10-kg block rests on a smooth surface and is attached to a vertical peg by a rope (Fig. 7-8). What is the tension *T* in the rope if the block moves in a

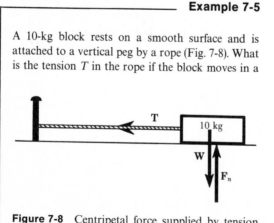

Figure 7-8 Centripetal force supplied by tension in a rope.

Example 7-6

What is the maximum speed at which a car of mass *m* can go around an unbanked curve of radius 40 m, if the coefficient of static friction between tires and road is 0.7 (Fig. 7-9)?

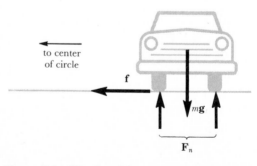

Figure 7-9 Centripetal force supplied by friction.

As in the preceding example, the weight of the car mg is balanced by the upward push F_n of the road. The net force is that due to friction, and $f = 0.7F_n = 0.7mg$, since the car is presumably about to skid and the maximum force of friction is being obtained.

Here it is convenient to use $a = v^2/r$ for the centripetal acceleration.

$$\text{net } F = ma = m\frac{v^2}{r}$$

$$(0.7)(m \text{ kg})(9.8 \text{ m/s}^2) = (m \text{ kg})\left(\frac{v^2}{40 \text{ m}}\right)$$

$$v = \boxed{16.6 \text{ m/s}}$$

If this speed is exceeded, there will be insufficient force to cause the car to maintain the acceleration, unless v^2/r is kept constant by an increase in r. The car skids in an arc of greater radius of curvature. (Note that the mass of the car does not appear in the answer.)

To illustrate centripetal acceleration caused by gravitational force, consider a car going over the top of a hill.

_____ **Example 7-7**

What is the force of a 700-kg sports car on the road surface, as it goes at 10 m/s over the crest of a hill having a radius of curvature of 35 m measured in a vertical plane (Fig. 7-10)?

We are asked for the force exerted by the car, but Newton's second law must be applied to the motion of the car, and so we consider the forces *on* the car. The weight of the car is 700 kg $\times$ 9.8 m/s^2 = 6860 N.

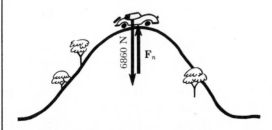

Figure 7-10 Centripetal force supplied by combined action of gravitation and force of the road.

$$\text{net } F = ma = m\frac{v^2}{r}$$

$$6860 \text{ N} - F_n = (700 \text{ kg})\frac{(10 \text{ m/s})^2}{35 \text{ m}}$$

$$6860 \text{ N} - F_n = 2000 \text{ N}$$

$$F_n = 4860 \text{ N} \quad \text{(upward)}$$

This is the upward push of the road on the car. Newton's third law tells us that the force of the car on the road is equal and opposite, that is,

$$\boxed{4860 \text{ N (downward)}}$$

As the car goes over the crest of the hill, it tends to "take off" and presses less against the road than when on a level stretch.

Magnetic force on a moving charge is responsible for the circular motion of a charged particle in a particle accelerator.

_____ **Example 7-8**

In a cyclotron a beam of protons (each of mass 1.67×10^{-27} kg) is moving in a circle of radius 80 cm. If an electromagnet supplies a force of 8.00×10^{-13} N directed toward the center of revolution, what is the velocity of the protons? What is the KE of each proton?

$$\text{net } F = ma = m\frac{v^2}{r}$$

$$v = \sqrt{\frac{Fr}{m}} = \sqrt{\frac{(8.00 \times 10^{-13} \text{ N})(0.80 \text{ m})}{1.67 \times 10^{-27} \text{ kg}}}$$

$$= 1.96 \times 10^7 \text{ m/s}$$

$$= \boxed{19{,}600 \text{ km/s}}$$

$$\text{KE} = \tfrac{1}{2}mv^2$$

$$= \tfrac{1}{2}(1.67 \times 10^{-27} \text{ kg})(1.96 \times 10^7 \text{ m/s})^2$$

$$= \boxed{3.20 \times 10^{-13} \text{ J}}$$

Although 3.20×10^{-13} J may seem to be too small to bother with, it is actually, as we shall see in Chap. 32, a great deal of energy for a single

Figure 7-11 Bending magnets in the interior of the main accelerator ring at the Fermi National Accelerator Laboratory in Batavia, Illinois. The magnets supply centripetal force on protons moving in a circle of radius 1.00 km. By adding additional superconducting magnets in the space below the original magnets, a much larger centripetal force was obtained in 1981.

atomic particle to have. Of course, if spread out among the 10^{23} or so atoms that make up a marble or a bullet, 3.20×10^{-13} J would be insignificant indeed.

The circular path of particles in a large accelerator is shown in Fig. 7-11.

7-4 *Centrifugal Force*

If there is a force directed *toward* the center exerted *on* a rotating body, by Newton's third law there must be a force directed *away* from the center exerted *by* the rotating body. This reaction force, exerted on the axis or whatever other body supplies the inward force, is equal and opposite to the centripetal force. For instance, in Example 7-5, the block pulls outward on the peg with a force of 2190 N. In Example 7-6 the car

pushes outward against the road with a horizontal force of 0.7 times its weight.

The term *centrifugal force* (*centrifugal* meaning "fleeing the center") is reserved for a very special situation—one in which the observer is moving along with the rotating body. The observer is then in an accelerated frame of reference, with the acceleration **a** of the frame directed toward the center. As we saw in Sec. 5-8, in such an accelerated frame of reference an inertial force equal to $-m\mathbf{a}$ acts on a body of mass m. This inertial force is directed opposite to the acceleration of the frame of reference, and for circular motion it is therefore directed away from the center. Only to an observer in the rotating frame of reference does the term *centrifugal force* have any significance.

Usually we take the point of view of an outside observer, for whom centrifugal force is a meaningless term. Sometimes, however, it is helpful to use the rotating frame of reference. Thus it is stated that in a clothes dryer, "centrifugal force throws the water outward" as the drum containing wet clothes rotates. This is a correct statement, but only from the point of view of an "insider"—the point of view of a drop of water in its rotating frame of reference. From the point of view of an outside observer—the person washing the clothes—this statement is incorrect. In the stationary frame of reference, we describe what happens as follows: "The molecular forces between water and fiber supply the necessary centripetal force to keep the water drops moving in a circle. When the speed is increased, the required force is increased, and sooner or later the molecular forces become insufficient. The water then flies off on a tangent and moves away from the center of the drum because of lack of sufficient force." In other words, in the stationary frame centrifugal force does not *cause* the water to fly outward. The water will do that anyway, due to its inertia, since its natural tendency is to move in a straight line (Newton's first law) *unless* acted on by a centripetal (or other) force.

Other examples of centrifugal force (the inertial force exerted on a body in a rotating frame of reference) are common. We on the earth (a rotating frame of reference) can talk about centrifugal force causing a decrease in the weight of a body at the equator. The upward force is a fictitious or inertial force arising from acceleration of the frame of reference; in this respect it is analogous to a Coriolis force (Sec. 5-6). Corpuscles can be separated from blood serum in a centrifuge. A cream separator is a type of centrifuge in which the heavier liquid, milk, is collected at the rim of the rotating dish. The effect is the same as turning on a strong gravitational force directed horizontally outward. In the ultracentrifuge, rotational speeds up to 600,000 rev/min are used to obtain separation of complex protein molecules that would require thousands of years to separate under the action of ordinary gravity.

7-5   Banking of Curves

Let us consider the predicament of a bicyclist going around a level (unbanked) curve. If he remains vertical, the forces acting on him are (1) his weight $\mathbf{W}$ ($= m\mathbf{g}$); (2) the normal force $\mathbf{F}_n$ of the road; and (3) a frictional force $\mathbf{f}$ acting toward the center of the circle. These forces are not in equilibrium, and the resultant force (which is simply the frictional force, since the others bal-

ance) is the centripetal force that causes the centripetal acceleration. Without sufficient friction, the cyclist is unable to go around the curve and will skid in a curve of greater radius, as described in Example 7-6.

Even when the coefficient of friction is adequate, the cyclist will topple over unless he leans inward (Fig. 7-12) while going around a curve. The vector sum of $\mathbf{f}$ and $\mathbf{F}_n$ represents the force $\mathbf{R}$ of the ground on the cycle. If the cyclist does not lean over, $\mathbf{R}$ does not pass through the center of gravity, and the result is a counterclockwise torque tending to push the wheels out from under the rider. Leaning at just the proper angle causes $\mathbf{R}$ to act along the frame of the bicycle, and no instability results.

Example 7-9

The combined mass of a motorcycle and its rider is 100 kg. What is the necessary force of friction if the cyclist is to go around a curve of 80 m radius at 20 m/s (Fig. 7-13)? If the coefficient of friction is 0.6, will the cyclist negotiate the curve successfully? At

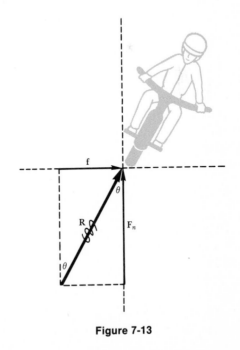

Figure 7-13

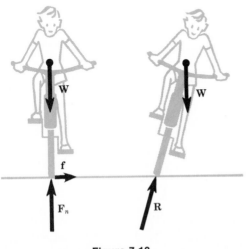

Figure 7-12

what angle should he lean over to avoid a spill?

$$\text{Centripetal force} = m\frac{v^2}{r}$$

$$= (100\ \text{kg})\frac{(20\ \text{m/s})^2}{80\ \text{m}}$$

$$= \boxed{500\ \text{N}}$$

The weight of the motorcycle and the cyclist is $100\ \text{kg} \times 9.8\ \text{m/s}^2 = 980\ \text{N}$.

$$\text{Maximum force of friction} = \mu_s F_n$$

$$= (0.6)(980\ \text{N})$$

$$= \boxed{588\ \text{N}}$$

Since the required force is less than the maximum possible force of friction, the cyclist will be able to make it.

Now determine the cyclist's angle with the vertical:

$$\tan\theta = \frac{f}{F_n} = \frac{500\ \text{N}}{980\ \text{N}} = 0.510$$

$$\theta = \boxed{27°}$$

It is instructive to look at this situation from the cyclist's point of view. Since the cycle is an accelerated frame of reference, *in that frame* there is an inertial force $-m\mathbf{a}$ (Sec. 5-8). (The magnitude of $-m\mathbf{a}$ is just equal to the magnitude of the force of friction $\mathbf{f}$, as in the previous example.) Figure 7-14a shows schematically the four forces acting on the cyclist, including the inertial force. If the rider does not lean inward, there is a torque (b) even though $\Sigma\mathbf{F} = 0$ in the accelerated frame. If the rider leans inward (c), the forces act

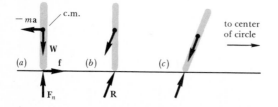

Figure 7-14 Forces acting on the cyclist, as observed in his own frame of reference. The inertial force is $-m\mathbf{a}$.

in the same line and the cycle is stable against tipping over.

Curves on highways are banked, computed to be safe at some particular speed. The normal force exerted by the roadbed has a horizontal component that serves as the centripetal force, and the vertical component of the normal force serves to support the weight. Since the road itself furnishes the centripetal force as one component of the compressional force (the normal force), no frictional force (parallel to the surface) is needed. A car can negotiate a properly banked curve even on glare ice for which $\mu_s \approx 0$.

The force diagram (Fig. 7-15) shows only two external forces acting on the car. They are not balanced, of course, and the resultant force is horizontal, directed toward the center of the circle. This net F serves as the centripetal force. From the diagram, $(\text{net } F)/W = \tan\theta$, and so net $F = W\tan\theta = mg\tan\theta$. From Newton's second law,

$$\text{net } F = ma$$

$$mg\tan\theta = m\frac{v^2}{r}$$

$$\tan\theta = \frac{v^2}{rg}$$

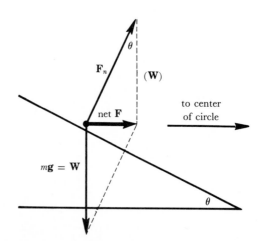
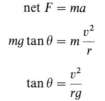

Figure 7-15 Banking of a frictionless curve. Centripetal force is supplied by a component of the road's normal force $\mathbf{F}_n$.

This formula shows that the angle of banking depends on the speed and on the radius of the curve. The absence of m or W from the formula shows that all cars and trucks (and bicycles) require the same banking on the same curve at any given speed, regardless of their weights.

Of course, if the car goes around a banked curve too fast, some frictional force will be needed to supplement the horizontal component of the normal force. If it goes around the curve too slowly, frictional force is again needed to oppose some of the horizontal component of the normal force, to keep the car from sliding downhill into the ditch. On a very slippery road, it is as bad to go too slowly as too fast around a banked curve. At the speed for which the curve is designed, friction is not needed, and road conditions are immaterial.

7-6 *Rotational Inertia*

Just as the tendency of a body to resist change in its linear velocity is called its linear inertia, so its tendency to resist change in its angular velocity is called its *rotational inertia*. To apply Newton's laws to rotational acceleration, we need quantities corresponding to force and mass; we look for an equation of the form net ( ) = ( )α, which will correspond to net $F =ma$ in linear motion.

Bodies having rotational inertia, such as a propeller or a door on its hinges, are accelerated only when a torque is applied. Force must be applied, but there must also be an effective lever arm if there is to be rotational acceleration. We see intuitively that torque takes the place of force when rotational motions are considered. Consider a body consisting of a single small mass m at a distance r from an axis, acted on by a single force F as shown in Fig. 7-16. The torque on the body is Fr. We use Newton's second law, remembering that the tangential acceleration is given by $a = r\alpha$:

$$\text{net } F = ma$$
$$F = mr\alpha$$
$$\tau = Fr = mr^2\alpha$$

Thus we infer that the quantity mr^2 is for rotation what the mass m is for linear motion.

If the body contained many particles, with various masses located at various distances from the chosen axis, the total torque would be the sum of the various torques, each computed as above. Then

$$\text{net } \tau = (\Sigma mr^2)\alpha$$

In this equation, Σmr^2 represents the sum of the masses making up the body, each multiplied by the square of its distance from the axis of rotation. For convenience, we call this sum the *moment of inertia* of the body and denote it by the symbol I. The dimensions of moment of inertia are $[ML^2]$; moments of inertia are expressed in units such as $\text{kg}\cdot\text{m}^2$ and $\text{g}\cdot\text{cm}^2$. We may now write

$$\text{net } \tau = I\alpha \qquad (7\text{-}13)$$

which is Newton's second law for rotation. Note how similar this equation is to the familiar linear form net $F = ma$.

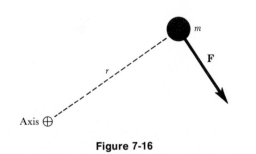

Figure 7-16

Example 7-10

A fisherman starts his outboard motor by applying a steady tangential force of 100 N by means of a starter rope. The flywheel has a moment of inertia of 0.05 $\text{kg}\cdot\text{m}^2$ and a diameter of 0.2 m. Ignoring the effects of friction and the opposing torque due to the compression of the motor, calculate the angular speed of the flywheel after it has turned through half a revolution starting from rest.

We first find the angular acceleration; then we use the equations of uniformly accelerated angular motion (Eqs. 7-4 to 7-7) to find the final speed.

$$\text{net } \tau = I\alpha$$

$$(100 \text{ N})(0.1 \text{ m}) = (0.05 \text{ kg} \cdot \text{m}^2)\alpha$$

$$\alpha = 200 \text{ rad/s}^2$$

$$\omega^2 = \omega_0{}^2 + 2\alpha\theta$$

$$= 0 + 2(200 \text{ rad/s}^2)(\pi \text{ rad})$$

$$= 1260 \text{ rad}^2/\text{s}^2$$

$$\omega = 35.4 \text{ rad/s}$$

$$= (35.4 \text{ rad/s})(1 \text{ rev}/2\pi \text{ rad})$$

$$= \boxed{5.64 \text{ rev/s}}$$

The moment of inertia of a body depends on how the masses are distributed relative to the axis of rotation. There is no unique moment of inertia of a body; the value depends on the choice of the axis of rotation. A simple example will make this clear.

Example 7-11

A thin, light, wooden meter stick (Fig. 7-17) is loaded with two small masses as follows: 3.0 kg at the 20-cm mark and 5.0 kg at the 70-cm mark. What is the moment of inertia about an axis through the zero end of the stick?

$$I = \Sigma mr^2$$

$$= (3 \text{ kg})(0.2 \text{ m})^2 + (5 \text{ kg})(0.7 \text{ m})^2$$

$$= 0.12 \text{ kg} \cdot \text{m}^2 + 2.45 \text{ kg} \cdot \text{m}^2$$

$$= \boxed{2.57 \text{ kg} \cdot \text{m}^2}$$

What is the moment of inertia of the same system about an axis through the 100-cm end of the meter stick?

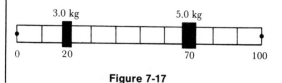

Figure 7-17

Table 7-1   Moments of Inertia for Some Simple Regular Bodies

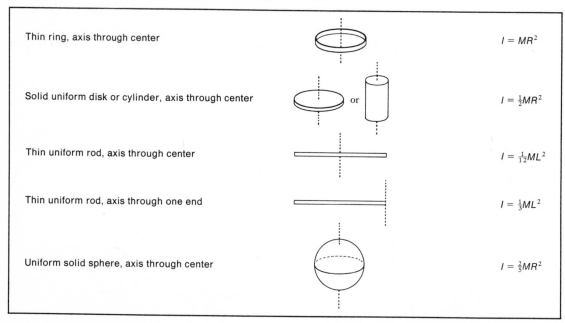

Thin ring, axis through center	$I = MR^2$
Solid uniform disk or cylinder, axis through center	$I = \frac{1}{2}MR^2$
Thin uniform rod, axis through center	$I = \frac{1}{12}ML^2$
Thin uniform rod, axis through one end	$I = \frac{1}{3}ML^2$
Uniform solid sphere, axis through center	$I = \frac{2}{5}MR^2$

$$I = \Sigma mr^2$$
$$= (3 \text{ kg})(0.8 \text{ m})^2 + (5 \text{ kg})(0.3 \text{ m})^2$$
$$= 1.92 \text{ kg} \cdot \text{m}^2 + 0.45 \text{ kg} \cdot \text{m}^2$$
$$= \boxed{2.37 \text{ kg} \cdot \text{m}^2}$$

The second moment of inertia is less than the first one. It can be proved that for all possible axes parallel to each other, the least moment of inertia is obtained if the axis passes through the center of mass.

If mass is distributed continuously, as in the case of most real bodies, it is a problem of integral calculus to perform the summation indicated in Σmr^2. Results for some simple regular bodies are summarized in Table 7-1. In each case, the dimensions of the formula for moment of inertia are $[ML^2]$, as they should be to agree with the dimensions of Σmr^2.

7-7 Linear and Rotational Analogies

From what has been said about Newton's second law, it may be suspected that most, if not all, of the concepts and laws of linear motion may be carried over into rotational motion. This is indeed the case. We pass over the proofs of these theorems and state a few results. In all cases, to obtain an angular form of some linear law or formula, replace *force* by *torque, mass* by *moment of inertia,* and of course *linear* displacement, velocity, or acceleration by *angular* displacement, velocity, or acceleration. For example, *rotational KE* is given by $\frac{1}{2}I\omega^2$ by analogy with the linear form $\frac{1}{2}mv^2$; *angular momentum,* denoted by L, is given by $I\omega$ by analogy with the linear form mv. *Work* is done whenever a body is rotated against an opposing torque (compare the linear statement on page 125), and hence

$$\text{Work} = (\text{torque})(\text{angular displacement}) = \tau\theta$$

by analogy with the linear formula

$$\text{Work} = (\text{force})(\text{linear displacement}) = Fs$$

Another analogy concerns the statement of Newton's second law. If mass is not constant, we saw in Chap. 5 that $F = ma$ must be replaced by the more general form $F = d(mv)/dt$—force equals the rate of change of linear momentum. Similarly, the rotational equation $\tau = I\alpha$ (which assumes constant I) must, in case I varies, be replaced by $\tau = d(I\omega)/dt$—torque equals rate of change of angular momentum. Change of moment of inertia is by no means as unusual as change of mass, as we shall see in Secs. 7-8 and 7-9.

We summarize linear and rotational analogies in a brief table. Recall that in these equations radian measure *must* be used for θ, ω, and α.

Linear formulas	Angular formulas
net $F = ma$	net $\tau = I\alpha$
$KE = \frac{1}{2}mv^2$	$KE = \frac{1}{2}I\omega^2$
$W = Fs$	$W = \tau\theta$
linear momentum	angular momentum
$p = mv$	$L = I\omega$

7-8 Conservation of Angular Momentum

Angular momentum, like linear momentum, is a vector quantity:

$$\mathbf{L} = I\omega \qquad (7\text{-}14)$$

The angular momentum vector $\mathbf{L}$ is parallel to the angular velocity vector ω; these vectors are along the axis of rotation in the direction that an ordinary (right-hand) screw would advance if turned along with ω. Thus for a phonograph record turning in the usual direction, ω and $\mathbf{L}$ are downward-pointing vectors.

The law of conservation of angular momentum may be stated as follows:

The total angular momentum of an isolated system remains constant.

For instance, a flywheel rotating at constant angular speed has constant angular momentum

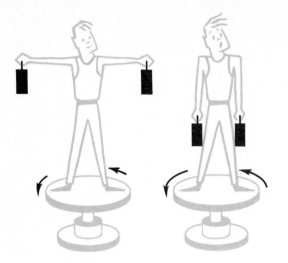

Figure 7-18   Conservation of angular momentum.

unless it is acted on by an external torque, such as might be caused by a motor or by friction in the bearings. The product $I\omega$ remains constant.

Applications of the law of conservation of angular momentum are among the most interesting in the field of mechanics. Some of the conclusions are unexpected, because of the possibility of changing moment of inertia without changing mass, by a redistribution of mass. A student standing on a frictionless rotating table (Fig. 7-18) can change his own speed of rotation by changing his moment of inertia. If he draws in his arms, his moment of inertia decreases, and hence, since $I\omega$ remains constant, his angular velocity ω increases. The effect is more pronounced (sometimes too much so for safety) if he carries a mass in each hand to make a larger change in moment of inertia.

Example 7-12

The turntable and student in Fig. 7-18 have total moment of inertia I_0, and each mass has moment of inertia $\frac{1}{2}I_0$ in the extended position at radius r. The system's initial angular velocity is ω_0. If the student draws the masses in to a radius of $\frac{1}{2}r$, what is the new angular velocity ω'?

The masses are small enough to be considered as "points," with $I = mr^2$. Since $I \propto r^2$, the new moment of inertia of each mass is $(\frac{1}{2})^2$ as much as before, that is, $\frac{1}{8}I_0$. Now we apply the law of conservation of angular momentum:

Total L before $=$ total L after

$$(I_0 + \tfrac{1}{2}I_0 + \tfrac{1}{2}I_0)\omega_0 = (I_0 + \tfrac{1}{8}I_0 + \tfrac{1}{8}I_0)\omega'$$

$$\boxed{\omega' = 1.6\omega_0}$$

If the student holds an electric drill vertically while stationary on a nonrotating platform, the system has zero angular momentum until the current is turned on. When he turns the drill on, giving its rotor an angular momentum $+I\omega$, the total $I\omega$ must still be zero, so the student's angular momentum must become $-I\omega$. He starts to rotate in a direction opposite to that of the drill. Since the student's moment of inertia is so much larger than that of the moving part of the drill, his angular velocity is correspondingly less, in order for his $I\omega$ to be numerically equal to (but opposite in direction to) the drill's $I\omega$.

Much the same phenomenon may account for minute variations in the length of the day. Due to internal rearrangements of the earth's matter hundreds or thousands of miles below the surface, its moment of inertia varies. The variations in I, although quite large, are only a small fraction of the total I for the earth. Nevertheless, present-day clocks have been developed to such a degree of perfection that reliable indications of erratic variations in the earth's rotation have been obtained. For instance, during the 20-year period from 1910 to 1930, the length of the day changed by 0.0047 s, a rather abrupt change on a geological time scale. It is because of such variations in the period of the earth's rotation that the unit of time, the second, is no longer defined as 1/86,400 of a mean solar day, as it was for many centuries. Because of these changes in the period of the earth's rotation, from time to time a "leap second" is added to or subtracted from a day. Thus there were 61 seconds in the last minute of June 30, 1981.

Figure 7-19 The small propeller at the tail of the helicopter keeps the ship from rotating about a vertical axis.

A diver or acrobat makes good use of the conservation of angular momentum. If the diver wishes to execute a double somersault, she pulls in her arms and doubles up her legs, thus decreasing her moment of inertia. According to the law of conservation of angular momentum, her angular velocity increases correspondingly, enabling her to make the turns in the short time available. Even a cat seems to know about conservation of angular momentum; it is through intricate changes of moment of inertia (chiefly through maneuvering the legs and tail) that a falling cat arranges to land feet first.

On a much grander scale, astronomers have discovered *pulsars*—collapsed stars only a few kilometers in diameter that rotate incredibly rapidly, in many cases faster than 1 rev/s.* These are believed to be dying stars that have collapsed to the universal nuclear density of some 10^{14} g/cm^3 (see Prob. 1-B3). Their rapid rotation is a consequence of conservation of angular momentum, reminiscent of the way a spinning ice skater gains angular speed as the arms are

drawn in toward the axis of rotation. $I\omega$ remains constant, so ω increases dramatically as I decreases during the collapse of the star. Some of the finer details are of interest: Pulsars radiate energy (mostly in the form of radio waves, with a few young and bright pulsars emitting visible light), and this energy comes basically from the rotational KE ($\frac{1}{2}I\omega^2$) of the star. As is to be expected, pulsars are observed to slow down as their content of rotational KE decreases. Occasionally a pulsar suffers a tiny but abrupt change in period; presumably, this is the result of a starquake in which the mass of the star is redistributed, changing the moment of inertia.

Finally, the body of a helicopter with one set of rotors would tend to rotate in a direction opposite to that of the rotors, to keep the total angular momentum of the system zero. To counteract this, some helicopters have two sets of lift rotors that revolve in opposite directions, so that the total $I\omega = 0$ without requiring rotation of the body of the helicopter. More commonly, a small auxiliary propeller at the tail supplies an auxiliary horizontal force, to give a torque that opposes the reaction torque of the main rotors (Fig. 7-19). We have here an illustration of

---

* For comparison, our sun has a diameter of about 10^9 km and spins on its axis with a period of about 26 days.

Newton's third law for rotation: The forward torque on the rotors is supplied by the body of the helicopter through the motor, and therefore the rotors supply an equal and opposite reaction torque on the ship.

The "automatic pilot" of an airplane illustrates the vector nature of angular momentum. A wheel with a considerable moment of inertia is mounted in bearings so that its axis is free to point in any direction, with the torque due to gravity balanced out. The wheel is kept rotating by a small motor, and its axis remains pointing in the same direction in space, since no external torque is applied. If the plane's course changes, the wheel's axis turns relative to the plane, initiating an action to return the plane to its course. Thus a reliable control is obtained, based essentially on Newton's first law. Another illustration of the vector nature of angular momentum is afforded by the gyroscope, discussed in the For Further Study Sec. 7-9.

Summary   The kinematics of uniformly accelerated rotational motion is quite analogous to the kinematics of uniformly accelerated translational motion; the same equations are used, with appropriate changes in symbols. Radian measure is not needed except where linear and angular quantities appear in the same equation. A complete circle contains 2π radians.

A particle moving in a circular path has a linear acceleration toward the center, since the linear velocity is changing in a vector sense due to change in direction. This centripetal acceleration is given by $a_c = \omega v$, where v is the linear velocity and ω is the angular velocity; equivalent forms of this equation are $a_c = v^2/r$, $a_c = \omega^2 r$, and $a_c = 4\pi^2 r/T^2$. The necessary centripetal force to cause the change in direction of a body moving in a circle is given by $F_c = m\omega v = mv^2/r = m\omega^2 r = 4\pi^2 rm/T^2$.

Centrifugal force is the outward inertial force exerted on a moving body when the observer is in the rotating frame of reference moving with the body. From the point of view of an outside observer, centrifugal force is a meaningless concept.

Curves are banked so that the normal force of the road, due to compression, has an inward horizontal component that can serve as the centripetal force. In this way the need for frictional force is lessened or eliminated.

In rotation, moment of inertia takes the place of mass, and torque takes the place of force. Newton's second law for rotation reads net $\tau = I\alpha$. Angular kinetic energy and work are computed by formulas analogous to those used for similar linear quantities: $KE = \frac{1}{2}I\omega^2$ and $W = \tau\theta$. The angular momentum $I\omega$ of an isolated system remains constant in magnitude and direction.

Check List   angular displacement θ
radian
angular velocity ω
angular acceleration α
centripetal acceleration a_c
period T
frequency f
centripetal force
centrifugal force

moment of inertia
angular momentum
law of conservation of angular
    momentum

$$\omega = \frac{2\pi}{T}$$

$$v = \frac{2\pi r}{T}$$

$$a_c = \frac{v^2}{r} = \frac{4\pi^2 r}{T^2} = 4\pi^2 r f^2 = \omega^2 r$$

$$I = \Sigma m r^2$$

LINEAR FORMULAS
net $F = ma$
$KE = \frac{1}{2}mv^2$
$W = Fs$
linear momentum $= mv$

ANGULAR FORMULAS
net $\tau = I\alpha$
$KE = \frac{1}{2}I\omega^2$
$W = \tau\theta$
angular momentum $= I\omega$

Questions

7-1 A fan rotates through 7π rad. After this rotation, what is the position of a blade that was originally pointing straight up?

7-2 A penny is laid on the rough surface of a phonograph turntable, near the outer edge, and the motor is turned on. The penny does not slip, and it turns with the turntable with a constant speed of $33\frac{1}{3}$ rev/min. Is the velocity constant? Is the penny in equilibrium?

7-3 Is is possible for a body to be accelerated if its speed is constant?

7-4 Consider an atom of aluminum near the rim of a phonograph turntable turning at 45 rev/min. Does any centripetal force act on the atom? If so, what is this force caused by?

7-5 What is the source of the centripetal force on the pilot of a plane that is executing a vertical loop-the-loop, when the plane is at the bottom of the loop, curving upward?

7-6 When an eagle flying westward starts to swerve to the left, what is the source of the centripetal force on the bird? What is the direction of the force?

7-7 "The earth doesn't fall into the sun because centrifugal force pushes it outward." Comment on this sentence.

7-8 A heavy fan with its blades attached to a horizontal axle is resting on a table, supported by four symmetrically placed, short, vertical legs, each leg bearing one-fourth the weight of the fan. If the fan is turned on, are the forces in the legs still equal? When the fan has reached constant speed, are the forces in the legs equal?

7-9 In a game of marbles played on a perfectly smooth, level surface, is it possible for a marble to roll in an arc of a circle?

7-10 In the example of the student holding the drill (Sec. 7-8), what will happen if the student turns the drill (while it is running) so that it points down instead of up?

7-11 What happens to the length of the day when a sprinter starts running in an easterly direction? What happens to it when he stops running? What happens to it when an office worker takes an elevator to the top floor? Explain why these changes in the earth's angular speed are too small to observe.

MULTIPLE CHOICE

7-12 When a motor turns through an integral number of revolutions, the angle in radians is an even multiple of (a) $\pi/2$; (b) π; (c) 2π.

7-13 In applying the equation for motion with uniform angular acceleration, $\omega = \omega_0 + \alpha t$, radian measure for ω and α (a) must be used; (b) may be used; (c) cannot be used.

7-14 Centrifugal force is an inertial force when considered by (a) an observer at the center of the circular motion; (b) an outside observer; (c) an observer who is moving with the particle that experiences the force.

7-15 The dimensions of angular momentum are (a) $[ML^2T^{-1}]$; (b) $[MLT^{-2}]$; (c) $[ML^2T^{-2}]$.

7-16 A rotating fan blade slows down from 300 rads/s to 150 rad/s. Compared with the initial value, the blade's KE is now (a) $\frac{1}{2}$ as much; (b) $\frac{1}{4}$ as much; (c) $\frac{1}{16}\pi$ as much.

7-17 If the angular momentum of a body is increased to twice its original value, (a) the KE is twice its original value; (b) the KE is 4 times its original value; (c) the angular velocity is 4 times its original value.

Problems

7-A1 Express the following angles in radians: (a) $180°$; (b) $30°$; (c) $90°$; (d) $50°$; (e) 0.30 rev.

7-A2 Express the following angles in degrees: (a) 1.40 rad; (b) 0.33 rad; (c) $\pi/6$ rad; (d) 0.50 rev; (e) 2π rad.

7-A3 What is the linear velocity of a point on the rim of a flywheel 2 m in diameter if the wheel is turning at an angular velocity of 5 rad/s?

7-A4 What is the angular velocity in rad/s of a $33\frac{1}{3}$-rev/min phonograph turntable?

7-A5 A record of diameter 30.48 cm (12 in.) is on a turntable that rotates at $33\frac{1}{3}$ rev/min for 4 min. Through what distance, measured along the arc, has a point on the rim of the record moved?

7-A6 The moon has a diameter of 3480 km (2160 mi) and is 384,000 km (240,000 mi) from the earth. Calculate its angular size (a) in radians; (b) in degrees.

7-A7 What is the angular size, in radians, of a football player whose image, viewed 3 m from a television screen, is 5 cm high?

7-A8 In aviation, a "standard turn" for level flight of a propeller-type plane is one in which the plane makes a complete circular turn in 2 min. If the speed of the plane is 180 m/s, (a) what is the radius of the circle? (b) what is the centripetal acceleration of the plane?

7-A9 (a) The earth turns on its axis with a period of 23 hours, 56.1 minutes. Compute its angular velocity, in rad/s. (b) Compute the magnitude of the centripetal acceleration of a point on the earth's equator (radius of earth is 6.37×10^6 m).

7-A10 A car goes around a curve at 20 m/s. The radius of the curve is 50 m. Calculate the centripetal acceleration of the car.

7-A11 (a) Calculate the angular velocity (in rad/s) of the minute hand of a wall clock. (b) Calculate the magnitude of the centripetal acceleration of a paint speck on the minute hand, 30 cm from the axis of rotation.

7-A12 Calculate the moment of inertia about its axis of a bicycle wheel of mass 4 kg and radius 0.30 m. Assume the wheel's mass to be concentrated at the rim.

7-A13 In Sec. 6-8 we saw that the power required to keep a body moving at velocity v against an opposing force f is given by $P = fv$. Use the method of analogy (Sec. 7-7) to write down the corresponding rotational equation. State the meaning of each symbol appearing in your equation.

7-A14 A small ball of mass 12 kg is swinging on the end of a light wire 5 m long. What is the moment of inertia of the ball? (You may assume all the mass of the ball to be concentrated at one point.)

7-A15 Calculate the angular momentum of a body of moment of inertia 15 kg·m² that is turning at 4 rev/s.

7-A16 What are the dimensions of angular momentum? of torque?

7-B1 The specifications of a certain hi-fi turntable call for it to reach its final angular velocity of $33\frac{1}{3}$ rev/min while turning through $\frac{1}{5}$ rev, starting from rest. What is the angular acceleration of such a table? (Express your answer in rev/s^2 and in rad/s^2.)

7-B2 A fan blade turning at 1750 rev/min slowed down to 850 rev/min in 0.3 min   (a) Calculate the angular acceleration of the fan blade, in rev/min^2.   (b) How many revolutions did it make during this time interval?

7-B3 An amusement park ride consists of a flat turntable that rotates about a vertical axis while people try to sit on it. The turntable reaches its final speed of 0.6 rev/s in 25 s. Assuming the angular acceleration to be constant, calculate   (a) the value of the angular acceleration in rev/s^2;   (b) the number of revolutions turned during the 25 s.

7-B4 As a Yo-Yo is let fall, its angular speed increases steadily with an angular acceleration of 20 rad/s^2. Through what angle has it turned when its final angular speed is 60 rad/s?

7-B5 A fly of mass 0.6 g is sunning itself on a phonograph turntable at a location 6 cm from the axis. The turntable is turned on and rotates at 45 rev/min. Calculate the inward (centripetal) force needed to keep the fly from slipping.

7-B6 A car of mass m goes around an unbanked curve at 20 m/s. If the radius of the curve is 50 m, what is the least coefficient of friction that will allow the car to negotiate the curve without skidding?

7-B7 A car goes around a level unbanked curve of radius R. The coefficient of static friction is μ_s. Derive a formula for the maximum speed, expressing v in terms of μ_s, g, and R. Does the safe speed depend on the mass of the car?

7-B8 What centripetal force must act on a truck of weight 15 kN that rounds a curve of radius 60 m at 20 m/s?

7-B9 Calculate the centripetal force on a car of mass 800 kg traveling at 24 m/s around a corner of radius 64 m.

7-B10 A 50-kg boy is swinging on a rope 6 m long. He passes through the lowest position with a speed of 3 m/s. What is then the tension in the rope? (*Hint:* Use net $F = ma$. The net force is the resultant of the weight and the tension; see Example 7-7.)

7-B11 In the Rotor Ride at an amusement park, riders are pressed against the inside vertical wall of a rotating drum 2.5 m in radius that has an angular speed of 27 rev/min about a vertical axis.   (a) Viewed from a stationary frame of reference outside the drum, what are the magnitude and direction of the force of the wall on a rider whose mass is 60 kg? The rider's weight is borne by the floor.   (b) While the drum is rotating, the floor is removed. If the coefficient of static friction between the rider and the wall is 0.6, will she slip or will she be held pressed against the vertical wall? (See Ref. 7 at the end of this chapter for a film depicting this motion.)

7-B12 A car of mass 2000 kg goes over the crest of a hill whose radius of curvature is 40 m measured in a vertical plane.   (a) What is the force of the car on the road surface if the car's speed is 15 m/s?   (b) Calculate the magnitude and direction of the necessary force between car and road if the speed at the top of the crest is 20 m/s. Interpret your answer.

7-B13 A plane comes out of a power dive, turning upward in a curve whose center of curvature is 1500 m above the plane. The plane's speed is 300 m/s.   (a) Calculate the upward force of the seat cushion on the 80-kg pilot of the plane.   (b) Calculate the upward internal force on a 0.08-kg sample of blood in the pilot's brain. What happens if the heart muscles are unable to supply this force?

7-B14 A 50-kg skier starting from rest coasts down a frictionless 30° hill that is 40 m long, enters a frictionless horizontal stretch, then turns sharply in a horizontal circle of radius 30 m. What horizontal force is exerted (by what?) on the skier during the turn?

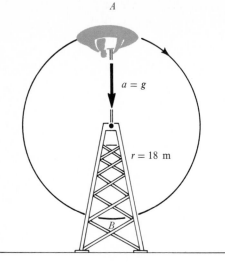

A

$a = g$

$r = 18$ m

B

Figure 7-20

Figure 7-21

7-B15 A carnival ride has cars that move at constant speed in a vertical circle of radius 18 m. (*a*) What must the angular velocity be so that a 100-kg passenger will be "weightless" when at the highest point A (Fig. 7-20)? (*b*) What will the passenger's apparent weight be at the bottom point B?

7-B16 At what angle should the well-known cyclist pictured in Fig. 7-21 lean inward if he is negotiating a curve of radius 20 m at a speed of 7 m/s?

7-B17 At what angle should a level curve of radius 170 m be banked for cars and trucks traveling at 24.6 m/s (55 mi/h)?

7-B18 A racetrack curves with a radius of 1000 m. At what angle should it be banked for cars traveling at 322 km/h (200 mi/h)?

7-B19 What is the proper speed for a car to go around a slippery curve of radius 50 m if the road is banked at an angle of 25°?

7-B20 Compute the moment of inertia of a pole vaulter's pole that is vertical and is pivoted about an axis on the ground through its lowest point. The pole is 5 m long and has mass 3 kg.

7-B21 Compute the moment of inertia of the loaded meter stick of Example 7-11 about an axis through the 30-cm mark on the stick.

7-B22 A playground carousel is in the form of a solid, uniform, horizontal disk of radius 1.4 m and mass 100 kg. If a child applies a force of 84 N for 2.5 s, tangent to the rim, what will be the final angular velocity of the carousel, starting from rest?

7-B23 A farmer, fleeing from an enraged bull, seeks safety by going through a gate that is standing ajar at an angle of 90° with the fence and closing it behind him. The gate has a moment of inertia of 2000 kg·m², and the farmer exerts a force of 180 N at a point 4 m from the hinge. As the gate swings shut, the farmer adjusts the direction in which he pulls, always applying force in the most advantageous way. How long will it take him to shut the gate?

7-B24 What is the rotational KE of a drum majorette's baton that is rotating about an axis through its center at an angular speed of 3 rev/s? Assume the baton to be a uniform rod 0.6 m long, of mass 0.5 kg. Would the light rubber ball at one end have much effect?

7-B25 A man on a platform (Fig. 7-18) has a small 2-kg object in each hand. Initially, his arms are extended, each object being 1 m from the axis of rotation, and he is rotating at 8 rad/s. The platform and man's body are assumed to have a constant moment of inertia of 1 kg·m². (*a*) Calculate the total angular momentum of the system (man plus platform plus objects). (*b*) The man now draws his arms in until each object is 0.3 m from the axis. What is the new angular velocity of the system? (*c*) Calculate the total KE of the system in part (*a*) and also in part (*b*). How do you account for the increase of KE? Has there been any change in PE?

7-B26 When a truck is brought to rest by conventional brakes, its KE is transformed into heat in the brake system plus the energy of tearing loose some asbestos fibers from the brake linings and forming some steel dust in the wear and tear of the brake drums. It is proposed to save some of this energy by storing it in a flywheel on board the truck, to be reused as the truck starts up again. Suppose that a truck of mass 10^4 kg moving at 25 m/s stores 50% of its KE in a flywheel that is a solid disk of mass 16 kg and radius 0.50 m. (*a*) What is the flywheel's final speed, in rev/min? (*b*) If the flywheel material will fly apart (due to internal stress) if it stores more than 14×10^6 J of rotational KE, is the answer to (*a*) a safe speed?

7-B27 A uniform solid grindstone wheel has mass 8 kg, and its radius is 10 cm. (*a*) Calculate the rotational KE of the grindstone when it is turning at 1800 rev/min. (*b*) After the grindstone's motor is turned off, a knife blade is pressed against the outer edge of the grindstone with a perpendicular force of 6 N. The coefficient of kinetic friction is 0.85. What will be the angular speed of the grindstone 5 s later?

7-B28 A roller coaster car (Fig. 6-5) of weight W starts 24 m above the bottom of a loop. If the loop is 15 m in diameter, show that the downward force of the rails on the car when it is upside down at the top of the loop is $\frac{7}{5}W$. (*Hint:* First use the energy principle to find the value of v^2 for the car at this point; use W/g for m.)

7-B29 In a certain ultracentrifuge a G-field of 300,000 times gravity is realized (this means that the acceleration is 300,000g). If the rotor is 12 cm in radius, what must the rotational speed be?

7-B30 A centrifuge used to separate uranium isotopes has a rotating chamber 7.5 cm in diameter and turns at 80,000 rev/min. What is the acceleration of a point on the circumference of the rotor?

7-C1 A hockey puck at rest on smooth ice is tethered by a light string, constrained to move in a horizontal circle of radius 5 m. An identical puck glides across the ice at 6.28 m/s, in a direction perpendicular to the initial direction of the string, and strikes the tethered puck head-on. There is a perfectly elastic collision, then (later) another perfectly elastic collision. How far has the second puck moved, 7 s after the first impact?

7-C2 Verify (partially) the statement at the end of Example 7-11. To do this, first find the c.m. of the loaded stick (see Secs. 4-6 and 4-9), and then compute the moment of inertia about an axis through the c.m. Your answer should turn out to be less than either of the answers to Example 7-11 and less than that to Prob. 7-B21.

7-C3 The door of a bank vault has a mass of 900 kg and may be considered to be many thin rods stacked one above the other, each pivoted frictionlessly at one end. The door is 2.0 m wide and 3 m high. (*a*) What is the moment of inertia of the door? (*b*) If a clerk pulls with a force of 120 N in the most effective way, through what angle will the door move in 2 s, starting from rest?

7-C4 A pole vaulter releases his pole without giving it a sideways shove, and the pole, starting from an upright position, topples over and falls to earth. The pole has length L and mass M. Show that the linear velocity with which the end of the pole strikes the ground is given by $\sqrt{3gL}$. (*Hint:* Use the energy principle; first find the angular velocity as the pole strikes the ground.)

7-C5 Refer to Prob. 4-C8. Prove that the tension in the string when the ball passes through its lowest position is the same as the tension in the string in position A before the thread is burned. (*Hint:* Use the energy principle to find the ball's velocity at its lowest point.)

7-C6 In Fig. 7-7, prove that CB is approximately equal to $\frac{1}{2}at^2$, where $a = v^2/r$, thus making a connection between uniform circular motion and the kinematics of falling bodies. [*Hint:* First use plane geometry to show that if $\angle AOB$ is small, $(DB)^2 \approx (AD)2r$. Then make use of the fact that $CB \approx AD$ and $DB \approx AB$.]

7-C7 A basket of ripe tomatoes of mass 20 kg is being hoisted by a windlass. The rope is wrapped around an axle that is a solid cylinder of wood having radius 0.1 m and mass 10 kg. The mass of the crank handle is negligible. The operator lets go of the handle when the basket is 5 m above the ground, and the rope unwinds. With what linear speed does the basket strike the ground? (*Hint:* Use the energy principle. The angular velocity of the cylinder is at all times given by $\omega = v/r$, where v is the velocity of the basket.)

7-C8 A solid cylindrical can of frozen juice has mass m and radius r. The can starts from rest and rolls down a plane that is inclined at an angle θ with the horizontal. After the can has rolled through n complete revolutions, what is the velocity v of its center of mass, parallel to the plane? (*Hint:* Use the energy principle. The angular velocity of the cylinder is at all times given by $\omega = v/r$.)

7-C9 A solid sphere of mass M and radius R rolls without slipping down a plane of length L and height H. (*a*) What is the linear velocity of its center of mass when the ball reaches the bottom of the plane? (*Hint:* The KE at the bottom is partially translational and partially rotational. At all times, $v = r\omega$. Use the energy principle, and solve for v.) (*b*) What would the speed be if the sphere slides, without rolling, down a smooth plane of length L and height H? (*Hint:* In this case the energy is all translational.)

7-C10 The planet Venus is unusual in having practically no spin angular momentum about its own axis (rotation on axis once in 243 days; the earth turns on its axis once in 1 day). It has been suggested that Venus originally was spinning like the earth and its angular momentum was canceled when it captured a small backward-moving planet that later fell to the surface. Calculate the mass of the supposed planet if it moved at 10 km/s tangent to the surface of Venus before capture. (Data: Assume Venus and earth to be identical uniform solid spheres, each of mass 6×10^{24} kg and radius 6×10^6 m. The angular momentum of the incoming planet would be negative; consider that for it $I\omega$ is the same as $mr^2\omega$, or mrv.)

7-C11 An air-driven miniature ultracentrifuge spins at 100,000 rev/min to give a centripetal acceleration of $160,000g$. (*a*) What is the radius of the rotor? (*b*) If the rotor is a disk made of steel with the height equal to the radius, calculate the moment of inertia of the rotor. (*c*) What force tangent to the rim of the rotor must be supplied by the air jets to bring the rotor up to speed in 50 s?

For Further Study

7-9 Gyroscopes and Tops

Small boys and girls have played with tops for centuries. A top is a symmetrical body that spins about an axis; it is observed that if the point of support is not at the c.g., the direction in which the axis points slowly changes (Fig. 7-22). This rotational motion of the axis is called *precession*, a phenomenon that can hardly have escaped your notice if you have ever observed a spinning top. If the top spins on a sharp point, and if its angular speed is constant, its axis maintains a constant angle with the vertical, and the top end of the axis describes a circle. In the usual cases the precessional motion is relatively slow compared with the spin motion of the top itself, and we shall assume this to be so in what follows.

The questions we have to answer are these: (1) Why doesn't the top fall down? and (2) What factors determine the direction and magnitude of the precessional motion of the axis? Let us fix our attention upon a particularly simple top, consisting of a bicycle wheel mounted on an axle. The wheel has a moment of inertia I about its axis, and its spin velocity is ω rad/s. We wish to apply Newton's second law to the motion of this system. First, we have to consider the vector interpretation of angular velocity and torque. In

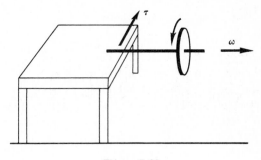

Figure 7-23

Fig. 7-23, by convention the spin velocity ω is represented by a vector pointing along the axis to the right. This is the direction that an ordinary (right-handed) screw would advance if a screwdriver were turned in the direction of spin. (If the wheel were spinning in the other direction, the vector ω would be directed along the axis to the left.) Angular accelerations are also represented by vectors according to the right-hand screw rule. Similarly, a torque vector is directed in the same direction as the acceleration that the torque tends to cause. Disregarding the spin for the moment, in Fig. 7-23 the force of gravity would pull the wheel down to the right, that is, rotate it around the edge of the table. The right-hand screw rule tells us that the vector τ due to gravity is along the edge of the table in the direction shown.

Suppose now that the torque due to gravity acts for a very short time interval Δt, giving rise to an angular impulse $\tau \Delta t$. By Newton's second law, in the momentum form (by analogy with Eq. 5-2 on page 109),

$$\tau \Delta t = \Delta(I\omega) \qquad (7\text{-}15)$$

In other words, the torque causes a change in angular momentum. After the time Δt has

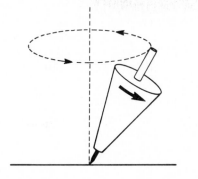

Figure 7-22 Precession of the axis of a top.

elapsed, the final angular momentum of the wheel is different from its initial value. The final value equals the initial value plus the change:

$$(I\omega)_2 = (I\omega)_1 + \Delta(I\omega)$$

Substituting from Eq. 7-15, we get

$$(I\omega)_2 = (I\omega)_1 + \tau\,\Delta t \qquad (7\text{-}16)$$

This is a vector equation. Let us apply it to the top for two situations: (1) The wheel is not spinning, and (2) the wheel is spinning. Even though the *change* in $I\omega$ is the same in both cases, the final $I\omega$'s are different because the initial $I\omega$'s are different.

1. *Wheel is not spinning.* In this case, $(I\omega)_1 = 0$; hence, Eq. 7-16 gives $(I\omega)_2 = \tau\,\Delta t$. This says that the wheel's final angular momentum is a vector parallel to the torque vector τ. In other words, the whole system rotates clockwise around the table edge as an axis, and falls down, just as would have been expected.

2. *Wheel is spinning.* In this case, the wheel has an initial angular momentum $(I\omega)_1$, and Eq. 7-16 tells us that the final angular momentum is to be found by vector addition (Fig. 7-24). The final angular momentum $(I\omega)_2$ turns out to be a horizontal vector, practically equal in magnitude to $(I\omega)_1$. The direction of the axis changes, but the axis remains horizontal since both $(I\omega)_1$ and $\tau\,\Delta t$ are vectors in a horizontal plane. The wheel does not fall over but precesses at a slow rate. The magnitude of the precessional angular

velocity can easily be derived using Fig. 7-24. For small $\Delta\theta$,

$$(I\omega)_1 \approx (I\omega)_2 \approx I\omega,$$

and

$$\Delta\theta = \frac{AB}{(I\omega)_1} = \frac{\tau\,\Delta t}{I\omega}$$

Thus the precessional angular velocity Ω is given by

$$\Omega = \lim_{\Delta\theta \to 0}\left(\frac{\Delta\theta}{\Delta t}\right) = \frac{\tau}{I\omega} \qquad (7\text{-}17)$$

It is impossible to go into the numerous ramifications of the theory of tops and gyroscopes,* even with the simplifying assumption of slow precession. We content ourselves with two applications.

First, it is well known that the earth is not a perfect sphere, but bulges slightly at the equator owing to the centrifugal force that acted while the earth was still molten. The gravitational attraction of the moon (and to a lesser extent, the sun) tends to "straighten out" the axis of the earth because of the unequal forces on the two parts of the bulge. At any instant, the force acting on the bulge on one side of the earth is greater than that on the bulge on the other side, since the two bulges are at different distances from the moon. Thus a net torque tends to "straighten up" the earth's axis (Fig. 7-25). However, since the earth is spinning, the torque causes a precession to take place instead of decreasing the angle θ. In Eq. 7-17, τ is small and I is large, so the precessional motion is very slow, about 26,000 years for one complete cycle. Still, the effect of this motion was actually observed by ancient Greek astronomers. The axis of the earth now points approximately to Polaris, the pole star; by 14,000 A. D. the axis will point

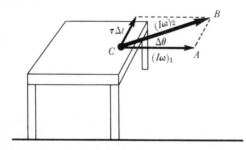

Figure 7-24

---

* A gyroscope is, technically, a top that is mounted in such a way as to be subject to no external torque.

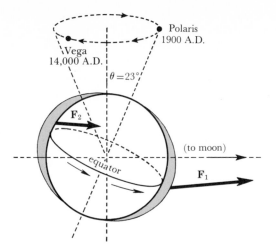

Figure 7-25 Precession of the earth's axis.

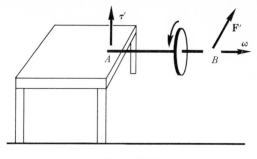

Figure 7-26

more or less toward the star Vega, which is now 46° from the pole.

Second, if while a wheel is spinning as shown in Fig. 7-26, you attempt to increase the precessional motion by applying a force $\mathbf{F}'$ to the axle, this force gives rise to a torque τ' directed upward (right-hand screw rule). Upward torque means that $\Delta(I\omega)$ is upward, and the axle rises at the B end. Therefore a *horizontal* force gives rise to an *upward* motion! The spinning wheel "reacts" in a most peculiar way. The heavy flywheel of a racing car has a distinct gyroscopic action of this sort (Fig. 7-27). When the driver turns the car toward the left, forces $\mathbf{F}$ and $\mathbf{F}'$ are applied by the frame of the car. The rear end of

the shaft, at B, moves upward, and the front end, A, moves downward. The car "digs in" as it rounds the curve. During a turn toward the right, the front end of the car "takes off" and tends to rise. It is to prevent this that speedway races are run with the cars continually turning toward the left rather than toward the right. All this assumes that the flywheel turns counterclockwise as viewed from the rear of the car, as is the case for almost all cars in the United States.

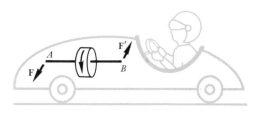

Figure 7-27

Problems

7-C12 What is the direction of the vector ω representing the angular velocity of the spinning flywheel in Fig. 7-27? (*a*) upward; (*b*) downward; (*c*) from A to B; (*d*) from B to A.

7-C13 The flywheel of an airplane spins counterclockwise as viewed from behind. When the plane makes a right turn, does the plane tend to nose up or nose down? Would a jet plane be subject to this effect?

7-C14 Calculate the rate of precession of the bicycle wheel in Fig. 7-23. The mass of the wheel is 10 kg, with practically all the mass concentrated in the rim. The wheel has a radius of 35 cm, and it is spinning at 5 rev/s = 31.4 rad/s. The axle is supported on the table at a point 30 cm from the c.g. of the wheel.

7-C15 Calculate the torque (in N·m), furnished by forces $\mathbf{F}_1$ and $\mathbf{F}_2$ in Fig. 7-25, that causes the precessional motion of the earth's axis at the rate of 1 revolution in 26,000 years. Use astronomical data from Appendix Table 7, and assume that the earth can be approximately described by a solid sphere of uniform density.

7-C16 The top of Fig. 7-22 is a solid cone with radius $R = 3$ cm and height along the axis $H = 6$ cm. The axis is inclined at 30° from the vertical. For a solid cone the center of gravity is $\frac{3}{4}H$ from the tip, and the moment of inertia is $\frac{3}{10}MR^2$ (the mass M will cancel out). What is the spin angular velocity when the top precesses around once in 2 s?

References

1. G. W. Gray, "The Ultracentrifuge," *Sci. American* **184**(6), 42 (June 1951).
2. J. W. Beams, "Ultrahigh-Speed Rotation," *Sci. American* **204**(4), 134 (Apr. 1961).
3. S. F. Singer, "Satellites for Physicists," *Physics Today* **9**(4), 21 (Apr. 1956). Describes the precession of an artificial satellite due to the earth's bulge.
4. R. C. Eastman, "Painless precession," *Am. J. Phys.* **43**, 365 (1975).
5. C. Frohlich, "The Physics of Somersaulting and Tumbling," *Sci. American* **242**(3), 154 (Mar. 1980).
6. J. G. Kreifeldt and M-C. Chuang, "Moment of Inertia: Psychophysical Study of an Overlooked Sensation," *Science* **206**, 588 (1979). Discusses a largely unrecognized contribution to an object's "feel."
7. F. Miller, Jr., *Inertial Forces: Centripetal Acceleration* (film). The rotor ride.
8. F. L. Friedman, *Velocity and Acceleration in Circular Motion* (film).

8

Gravitation and Planetary Motions

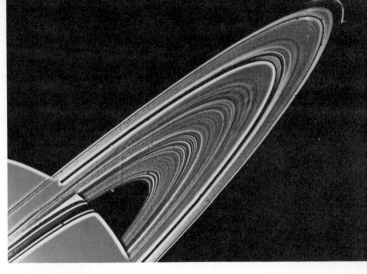

In our study of physics, we have made free use of the rather obvious fact that material bodies have a weight that is somehow related to the gravitational pull of the earth. It is time now to examine the gravitational force in greater detail. We shall see how the great physicist Isaac Newton made a connection between gravitational force and the motions of the planets in their orbits around the sun.

8-1   *Gravitation*

In 1798 Henry Cavendish completed an experiment that demonstrated directly the gravitational force of attraction between small bodies a few centimeters apart. Two lead balls about 5 cm in diameter were attached to the ends of a light rod suspended in a horizontal position by a long, fine wire about 100 cm in length. (Figure 8-1 shows a modern, compact version of Cavendish's apparatus.) Two larger lead balls, each about 20 cm in diameter, were placed almost touching the small balls. The gravitational attraction between the balls caused the moving

system to rotate to a new equilibrium position determined by the stiffness of the wire. By carefully shielding the apparatus from air currents, Cavendish was able to measure the tiny force of gravitation (amounting to less than $\frac{1}{20}$ dyne). Without specially designed apparatus of this sort, the gravitational forces between ordinary objects are masked by frictional forces of one kind or another, or by the tremendously larger force of gravitation due to the earth.

Cavendish's experiment, the first terrestrial measurement of gravitational force, came more than a century after Newton had announced his *law of universal gravitation* in 1687:

> **Every particle in the universe attracts every other particle with a force that is directly proportional to the product of the masses of the two particles and inversely proportional to the square of the distance between them.**

It was a triumph of inductive reasoning that Newton was able to arrive at this law from considering the application of his laws of motion to

This computer-assembled, two-image mosaic of Saturn's rings was taken by NASA's Voyager I in 1980 at a range 8 × 10^6 km. The rings are loose assemblages of many small chunks of matter; each chunk is in orbit around Saturn.

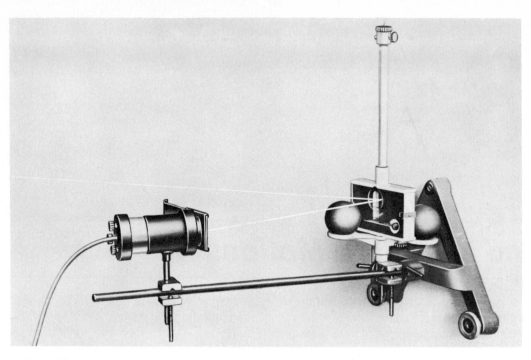

Figure 8-1 Modern form of the Cavendish apparatus. Two small balls (one visible in photo) are on a bar suspended by a fine vertical metal wire. When the massive stationary lead balls are shifted, the gravitational force on the small balls changes. The resulting motion is made evident by the shift of a beam of light reflected from a small mirror attached to the moving system.

the paths of the moon and the planets, as described by the German astronomer Johannes Kepler (1571—1630). Specifically, Kepler had formulated three laws of planetary motion. These laws are kinematical—a *description* of motion—and are based on data collected during a lifetime of observations, chiefly of Mars, by his teacher Tycho Brahe.

Kepler's laws:

   1. The orbit of any planet around the sun is an ellipse, with the sun at one focus of the ellipse.
   2. The line joining any planet to the sun sweeps out equal areas in equal times.
   3. For any two planets in the solar system, the squares of the periods of revolution are in the same proportion as the cubes of their average distances from the sun.

We shall discuss the significance of Kepler's laws more fully in Sec. 8-3. For the present, we note that two of Kepler's kinematical laws, combined with Newton's own laws of motion, lead to facets of Newton's law of gravitation; knowledge of falling bodies is also needed. The evolution of the law of gravitation may be summarized schematically as follows:

Kepler's second law + Newton's laws of motion:
   Force between planet and sun is either an attraction or a repulsion, directed along the line joining planet and sun.*

---

* This is not as obvious as it seems. The force might well have been tangent to the orbit, or perpendicular to the orbit, or oriented in some other direction.

Kepler's first law + Newton's laws of motion:
Force between planet and sun is an attraction that varies inversely as the square of the distance.

Acceleration g of a falling body independent of mass of body + Newton's laws of motion:
Force between planet and sun is proportional to the product of the masses of planet and sun.

Putting all these statements together, we write Newton's law of universal gravitation as

$$F = G\frac{mm'}{r^2} \qquad (8\text{-}1)$$

where G is a constant of proportionality, and m and m' are the masses of any two bodies separated by a distance r. It is assumed that the bodies are "particles," that is, their sizes are small compared with their separation r.

The Cavendish experiment dealt with a law of nature already known to be true from astronomical observations. The significance of the experiment was in the determination of the constant G ("Newton's constant"). In Eq. 8-1, Cavendish used his measured values of F, m, m', and r to compute G, obtaining the value 6.7×10^{-8} dyn·cm²/g². (The peculiar unit for G is necessary to make Eq. 8-1 dimensionally correct. This is an example of a constant that has units. The velocity of light, the acceleration due

to gravity, and other well-known constants also have units.) In the mks system

$$G = 6.7 \times 10^{-11}\ \text{N·m}^2/\text{kg}^2$$

Cavendish then proceeded to use his delicate measurement of a force less than the weight of the ink required to print the word "gravitation" to determine the mass of the earth itself. Consider a 1-kg object at the surface of the earth. Even though the earth is not a "particle," the gravitational pull of the earth acts as if all its matter were concentrated at the center,* so r in Eq. 8-1 is the radius of the earth, 6.4×10^6 m. The force on the object is its weight, which equals 9.80 N. We now know everything in Eq. 8-1 but the mass of the earth, which we call m'.

$$F = G\frac{mm'}{r^2}$$

$$9.8\ \text{N} = \left(6.7 \times 10^{-11}\ \frac{\text{N·m}^2}{\text{kg}^2}\right)\frac{(1\ \text{kg})(m')}{(6.4 \times 10^6\ \text{m})^2}$$

$$m' = 6.0 \times 10^{24}\ \text{kg}$$

Finally, Cavendish concluded that the average density of the earth is

$$\frac{6.0 \times 10^{24}\ \text{kg}}{\frac{4}{3}\pi(6.4 \times 10^6\ \text{m})^3} = 5500\ \text{kg/m}^3$$

or 5.5 times the density of water. Since the density of the upper crust is known to be only about 2.5 times the density of water, it is evident that the earth must have a dense core, to bring the average density up to the calculated value.

If we consider only the earth's gravitational pull and ignore the moon, the sun, and the other planets, a body's true weight can never be zero except at the center of the earth. At the earth's center, the gravitational force due to an off-center element of mass Δm is balanced by an equal and opposite gravitational force due to an equal mass $\Delta m'$ located symmetrically opposite to Δm (see Fig. 8-2). Assuming the earth to be of

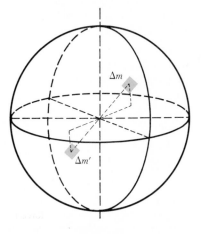

Figure 8-2

---

* Newton delayed publication of his law of gravitation for more than a decade while he worked out a satisfactory proof of this statement, which is true for a *spherical* earth whose density at any point depends only on the distance of that point from the center of the earth.

uniform density,* as a body moves through a hypothetical tunnel up toward the surface of the earth, its weight increases steadily and reaches a maximum value at the surface. This maximum weight is mg and also equals GmM/R^2, where M is the mass of the earth and R is the radius of the earth. Above the earth's surface the weight decreases and is given by GmM/r^2 at points where $r > R$. It is sometimes stated that a space probe can get far enough away from the earth to "escape the earth's gravitational pull," but this statement is wrong. At any finite distance r there is *some* gravitational pull; only as $r \to \infty$ does the true weight of a body approach zero.

Example 8-1

The mass of the moon is 7.4×10^{22} kg, and its radius is 1740 km (about 1080 mi). Calculate the magnitude of the acceleration due to gravity (a) on the moon's surface and (b) 1000 km above the moon's surface. Ignore the gravitational pull of the earth.

(a) The weight of a test body of mass m is mg.

$$mg = G\frac{mM}{r^2}$$

$$g = \left(6.7 \times 10^{-11}\ \frac{\text{N} \cdot \text{m}^2}{\text{kg}^2}\right)\frac{(7.4 \times 10^{22}\ \text{kg})}{(1.74 \times 10^6\ \text{m})^2}$$

$$= 1.6\ \text{N/kg} = \boxed{1.6\ \text{m/s}^2}$$

This is about $\frac{1}{6}$ of the value of g at the earth's surface.

(b) At an altitude of 1000 km, the distance from the moon's center is 2740 km. Since the force on m is inversely proportional to the square of the distance, we use a proportion:

$$\frac{m(g')}{m(1.6\ \text{m/s}^2)} = \frac{(1740\ \text{km})^2}{(2740\ \text{km})^2}$$

$$g' = \boxed{0.65\ \text{m/s}^2}$$

There is a fundamental question about mass that must be answered. Are we talking about the same thing when we speak of "mass" as inertia (Newton's first and second laws) and when we speak of "mass" as capable of exerting force at a distance on another such mass (Newton's law of gravitation)? Not necessarily. It is a matter for experiment to show whether a body that has 3 times as much weight as another also has exactly 3 times as much inertia. Galileo's famous (but probably apocryphal) experiment at the Leaning Tower of Pisa is such an experiment. We are all familiar with the bare facts: A heavy ball and a light ball dropped at the same time reached the ground at (almost) the same instant; therefore the accelerations were the same. The onlookers had expected the heavier ball to strike first, since more force (its weight) acted on it. Evidently, the heavier ball also had more inertia, in just the right proportion. We can use Newton's second law: $a = (\text{net }F)/m = W/m$, and since the accelerations were observed to be equal, $W_1/m_1 = W_2/m_2$ for the two balls; that is, *weight is proportional to mass*. Galileo's experiment was inaccurate, because of air resistance, but this fundamental and crucial experiment has been repeated in modified form by Newton, Bessel, Eötvös, and others, with the result that weight and mass are known to be in strict proportion, to within 3 parts in 10^{11} or better.[†] The proportionality between *gravitational mass* and *inertial mass* was accepted by Albert Einstein (1879–1955) as a basic axiom of the general theory of relativity.

Because of this proportionality, the same unit can be used for each kind of mass. Thus a body whose inertial mass is 1 kg also has a gravitational mass of 1 kg. Almost invariably "masses" are measured by comparing their weights on a balance. Thus it is *gravitational* mass that is measured. However, the inertial mass of an ob-

---

* This is not in fact true, and the weight of a body in a mine can only be calculated accurately by taking account of the dense core of the earth.

[†] This means that if the mass of one body is, say, 1.00000000000 times that of another body, the ratio of their weights is also 1.00000000000, not 0.99999999997 or 1.00000000003. See Ref. 2 at the end of this chapter.

ject can be measured directly, without any use of gravitation or weight. One such method is the inertia balance pictured in Fig. 9-16 on page 205.

8-2 Gravitational Field

Let us for a moment imagine that we are exploring the solar system with the help of a small unmanned space probe of known mass. We call this probe a *test body*. At any given point (for instance, halfway between Mars and Jupiter), we can measure the net gravitational force $\mathbf{F}_{grav}$ acting on the test body. This would be a difficult task, but in principle we could deduce $\mathbf{F}_{grav}$ by observing the acceleration $\mathbf{a}$ (relative to the fixed stars) of the test body whose mass is known. This force, or weight, is the resultant of all gravitational forces due to the sun, Mars, Jupiter, and all other astronomical bodies everywhere in the universe. If we measure the forces on various test bodies, we find, of course, that the force of gravitation $\mathbf{F}_{grav}$ is proportional to the mass of the test body that we use. We conclude that a vector can be drawn at this point in space whose magnitude is F_{grav}/m and whose direction is the direction of the resultant gravitational force on the test body. This ratio, force per unit mass, is called the *gravitational field strength* at the point in question.

Now we send our probe to a different point in space and measure the field strength there; it will no doubt have a different magnitude and direction and be represented by a different vector. The *gravitational field* is a collection of vectors, one located at every point in space, that give the magnitude and direction of the gravitational force per unit mass. Gravitational field strength is measured in newtons per kilogram:

$$\mathbf{g} = \frac{\mathbf{F}_{grav}}{m} \qquad (8\text{-}2)$$

We use the symbol $\mathbf{g}$ because gravitational field strength has dimensions of acceleration (since force has dimensions of mass × acceleration) and is, in fact, the acceleration of a test body placed at the point in question.

Example 8-2

What is the weight at the earth's surface of a body whose mass is 4 kg?

$$W = mg$$

$$= (4 \text{ kg})(9.8 \text{ N/kg}) = \boxed{39.2 \text{ N}}$$

In this example, we use N/kg for the unit of gravitational field strength.

If we know the location of all external bodies that exert a significantly large gravitational force on the test body, we can calculate the field using Newton's law of gravitation. The mass of the test body cancels out. Thus, in the simple case of a test body of mass m in the gravitational field of a single other body of mass m_1 at a distance r,

$$F_{grav} = \frac{Gmm_1}{r^2}$$

The magnitude of $\mathbf{g}$ at the location of the test body is

$$g = \frac{F_{grav}}{m} = \frac{Gm_1}{r^2}$$

and the direction of $\mathbf{g}$ is toward the body whose mass is m_1.

We may, if we wish, interpret a field as a property of space itself. Thus we can say that the region of space where P is located has somehow been changed by the presence of distant gravitating bodies like the earth and the moon; this region of space now has the property that a test body placed there experiences a force. This viewpoint about fields may or may not be helpful. However, from any viewpoint, gravitational field is an observable quantity, operationally defined in terms of the procedure used for measuring it.

In later chapters we shall discuss other kinds of fields. All field strengths are defined as force per unit "something," and in the case of gravitational field that "something" is mass. Other test bodies are used to explore and measure electric and magnetic fields.

Example 8-3

Find the resultant gravitational field at the location of a spacecraft that is at a point P (Fig. 8-3) between the earth and the moon. It is new moon, with the moon on a line between earth and the sun; the sun is 1.5×10^{11} m from point P. Masses (from Appendix Table 7) are as follows: $m_{earth} = 6.0 \times 10^{24}$ kg; $m_{moon} = 7.4 \times 10^{22}$ kg; $m_{sun} = 2.0 \times 10^{30}$ kg.

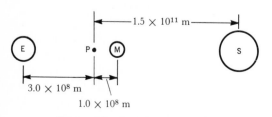

Figure 8-3 (not drawn to scale)

The field is the resultant of three fields:

$$g_{sun} = \frac{Gm_{sun}}{r_1^2}$$

$$= \frac{(6.7 \times 10^{-11} \text{ N·m}^2/\text{kg}^2)(2.0 \times 10^{30} \text{ kg})}{(1.5 \times 10^{11} \text{ m})^2}$$

$$= 6.0 \times 10^{-3} \text{ N/kg} \quad \text{(toward sun)}$$

$$g_{earth} = \frac{Gm_{earth}}{r_2^2}$$

$$= \frac{(6.7 \times 10^{-11} \text{ N·m}^2/\text{kg}^2)(6.0 \times 10^{24} \text{ kg})}{(3.0 \times 10^8 \text{ m})^2}$$

$$= 4.5 \times 10^{-3} \text{ N/kg} \quad \text{(toward earth)}$$

$$g_{moon} = \frac{Gm_{moon}}{r_3^2}$$

$$= \frac{(6.7 \times 10^{-11} \text{ N·m}^2/\text{kg}^2)(7.4 \times 10^{22} \text{ kg})}{(1.0 \times 10^8 \text{ m})^2}$$

$$= 0.5 \times 10^{-3} \text{ N/kg} \quad \text{(toward moon)}$$

The resultant field is a vector sum, with the direction of the field toward the sun taken as positive:

$$\mathbf{g} = \mathbf{g}_{sun} + \mathbf{g}_{earth} + \mathbf{g}_{moon}$$

$$= (6.0 - 4.5 + 0.5) \times 10^{-3} \text{ N/kg}$$

$$= 2.0 \times 10^{-3} \text{ N/kg} \quad \text{(toward sun)}$$

The weak gravitational fields found in Example 8-3 are to be compared with the much stronger gravitational field at the surface of the earth, which is $\mathbf{g} = 9.8$ N/kg (toward the center of the earth).

8-3 Dynamics of Planetary Motions—The Newtonian Synthesis

To Isaac Newton (1642–1727) we owe the extension of mechanics to astronomical phenomena, a synthesis that was both a culmination of the work of many scientists and the foundation for much to come during the two centuries following 1700. The dynamical content of Newton's first two laws, including $F = ma$, was known to Galileo (1564–1642), and the kinematical description of planetary motions had been found by Kepler (1571–1630). By 1665, there was a widespread belief among scientists that some unifying principle was awaiting discovery. In the introduction to his *Principia*, Newton wrote that his intention had been to induce the "forces of nature" (the laws) from the "phenomena of motion" (experimental observations), and from these laws to "deduce the motions of the planets, the comets, the moon, and the sea." It was his genius to have done just this, at about 22 or 23 years of age, using information that was known to many of his scientific contemporaries, including Christopher Wren, Robert Hooke, and Edmund Halley.

At the outset, the fact that planets move in curved paths rather than in straight lines means that they are not in equilibrium. Therefore, some net force acts on a planet at every point in its orbit to give it a centripetal acceleration. Newton's first law, the law of inertia, led him to look for the source (or sources) of such a force. Earlier workers, especially Descartes (1596–1650), proposed that a planet is caught up in vortices (whirlpools) in an invisible fluid; but no detailed agreement with experiment was possible. (Newton later worked out a mechanical theory of fluids, showing that a vortex theory is untenable.) Kepler, not realizing the nature of

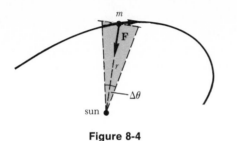

Figure 8-4

inertia, thought that a "magnetic" force from the sun might keep a planet moving forward in its orbit. Although his mechanical ideas were wrong, Kepler's precise description of planetary motions gave Newton the necessary clues from which he developed the law of gravitation. Let us see how Kepler's laws are a necessary consequence of Newton's laws of motion and his law of gravitation.

Kepler's second law (the law of areas) states that the line joining any planet to the sun sweeps out equal areas in equal times. This law is a consequence of the fact that the force on the planet is always directed toward a center (the sun). Such a force is called a *central force*. The exact nature of the central force (an inverse-square gravitational force in this case) is not involved in the proof of the law of areas. In Fig. 8-4, a body of mass m is moving in an orbit, subject to a force directed toward the sun. The force **F** has no lever arm relative to the sun, hence the torque on m is always zero. Newton's second law in its general form says that the net torque equals the rate of change of angular momentum.*

$$\text{net } \tau = \frac{d}{dt}(I\omega)$$

$$0 = \frac{d}{dt}(mr^2\omega)$$

---

* The moment of inertia of the planet, relative to the sun, is $I = mr^2$. Net $\tau = I\alpha$ is correct at any instant, but this form of Newton's second law is not very useful here, because I is not constant (r varies).

The rate of change of $mr^2\omega$ is zero, hence $mr^2\omega$ is a constant. But m is constant (at nonrelativistic speeds—see Chap. 28), and therefore

$$r^2\omega = \text{constant} \qquad (8\text{-}3)$$

Both r and ω may change as the planet moves, but the product $r^2\omega$ remains constant. Next we calculate the rate of change of area. From Fig. 8-4 we see that $\Delta A \approx \frac{1}{2}(\text{base}) \times (\text{altitude}) \approx \frac{1}{2}(r\,\Delta\theta)(r)$. (This approximation becomes better as $\Delta\theta \to 0$.) Thus,

$$\Delta A \approx \tfrac{1}{2}r^2\,\Delta\theta$$

$$\frac{\Delta A}{\Delta t} \approx \tfrac{1}{2}r^2\frac{\Delta\theta}{\Delta t}$$

and in the limit, as $\Delta t \to 0$ and $\Delta\theta/\Delta t \to \omega$,

$$\frac{dA}{dt} = \tfrac{1}{2}r^2\omega$$

Using Eq. 8-3 we obtain

$$\frac{dA}{dt} = \text{constant} \qquad (8\text{-}4)$$

This is Kepler's second law. Note that the force could equally well have been a repulsive one. Such a motion occurs when a positively charged helium nucleus moves past a positively charged target nucleus, Fig. 8-5 (see Rutherford's scattering experiment, Sec. 30-1). The law of areas holds for this motion, even though the moving nucleus is not in a closed orbit; what is important here is that the force on the moving body is *central*—that is, it is directed toward or away from a center.

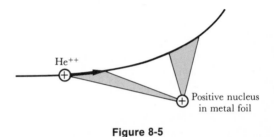

He^{++}

Positive nucleus in metal foil

Figure 8-5

Our analytical proof of Kepler's law of areas is essentially based on the first two of Newton's laws of motion. Newton himself gave an equivalent geometrical proof. The *magnitude* of the central force is not specified by the argument so far; it might vary inversely as r ($F \propto 1/r$), be proportional to $1/r^2$ or $1/r^3$, or vary in some other way. The law of areas shows a central force, directed toward the sun, of some as yet unspecified magnitude.

Kepler's first law (the law of orbits) states that the orbit of any planet around the sun is an ellipse, with the sun at one focus of the ellipse. The proof of this law requires that we assume a special form of central force—in fact, we must assume the inverse-square law of gravitation. Newton's proof is omitted, since it involves some formal calculus. There are only a few force laws that give a bounded orbit* in which the planet stays in the solar system for an infinitely long time. Newton investigated several such laws, including the inverse-square force law that was conjectured (without proof) by his close friend Halley and by others. Curiously, Newton found that $F \propto 1/r^5$ also gives a closed orbit, in fact a circle, with the center of force (the sun) *on* the orbit, rather than at the focus (center of the circle) as observed. (A planet in such an orbit would also be disastrously hot once a year as it passed through the sun!)

Our discussion of the Newtonian synthesis is now almost complete: We have seen that the law of areas and the law of elliptical orbits, coupled with Newton's second law of motion, are consistent with an inverse-square law of gravitation, in which $F \propto 1/r^2$ and F is directed toward the sun.

The role of mass in the law of gravitation that Newton developed is simple: The force is proportional to the product of the masses of the interacting bodies. Galileo had found that the acceleration of any falling body at a given point on the earth is the same, independent of the mass m of the body. Newton devised experiments to substantiate this to a high degree of precision. Since g is a constant (by experiment), $F = mg$ means that $F \propto m$—the force is proportional to the mass of the falling body. The *earth* is also "falling" toward the center of mass of the (earth + body) system, and the force F' on the earth is proportional to *its* mass m'; thus $F' \propto m'$. Now by Newton's third law, $F = F'$, and so F is proportional to m' as well as to m. The mutual force between two interacting bodies is therefore proportional to both masses. Thus, finally, we have arrived at Newton's law of gravitation:

$$F = G\frac{mm'}{r^2}$$

Kepler's third law (the law of periods) states that for any two planets in the solar system, the squares of their periods of revolution are in the same proportion as the cubes of their average distances from the sun. It was a gratifying check on his work that Newton found the law of periods to be a consequence of the law of gravitation that he had derived using only Kepler's first two laws. We can easily derive Kepler's third law for the special case of *circular* orbits (a circle is a special case of an ellipse, with the two foci coinciding at the center of the circle). The planet's centripetal acceleration is caused by the force of gravitation of the sun's mass M acting on the mass m_1 of the planet, which is at a distance r_1 from the sun (see Eq. 8-1). We apply Newton's second law to planet 1:

$$\text{net } F = ma$$

$$G\frac{Mm_1}{r_1{}^2} = m_1\frac{v_1{}^2}{r_1}$$

We now introduce the period T, which is the time for one revolution of the planet around the sun. The orbital velocity for a circular orbit is $v_1 = \text{circumference/period} = (2\pi r_1)/T_1$. Thus,

$$G\frac{Mm_1}{r_1{}^2} = \frac{m_1}{r_1}\frac{4\pi^2 r_1{}^2}{T_1{}^2}$$

---

* *Any* law in which the force depends only on r allows the trivial case of an orbit with the sun at the center of a circle; in this case the speed and centripetal force are constant.

Rearranging, we get

$$\frac{T_1{}^2}{r_1{}^3} = \frac{4\pi^2}{GM} \qquad (8\text{-}5)$$

For another planet of period T_2 at a radius r_2, we obtain by the same reasoning

$$\frac{T_2{}^2}{r_2{}^3} = \frac{4\pi^2}{GM}$$

The constant $4\pi^2/GM$ is the same for all planets revolving around the same sun, and so we conclude that

$$\frac{T_1{}^2}{r_1{}^3} = \frac{T_2{}^2}{r_2{}^3}$$

and hence

$$\frac{T_1{}^2}{T_2{}^2} = \frac{r_1{}^3}{r_2{}^3} \qquad (8\text{-}6)$$

for any two planets revolving around the sun. This is Kepler's third law.* Note that the mass of the planet canceled out. This means that even a tiny, marble-sized "planet" 149,000,000 km (93,000,000 miles) from the sun would have a period of revolution equal to 365 days, the same as the earth or any other planet at the same distance from the sun.

Kepler's third law (Eq. 8-6) is valid for an elliptical orbit if r represents the *mean distance* from planet to sun, defined as the average of the closest approach (at perihelion) and farthest distance (at aphelion).

Example 8-4

The asteroid Icarus is a small planet, several hundred meters in radius, whose orbit is oriented in such a way that a close approach to the earth is possible. The perihelion distance is 28.0×10^6 km,

---

* For an elementary extension of this discussion to take account of the motion of the sun itself, as well as the elliptical nature of orbits, see Ref. 3 at the end of this chapter. See also Sec. 8-4.

and the aphelion distance is 294.6×10^6 km. Calculate the period of revolution of Icarus.

First we find the mean distance r_1 by averaging the perihelion and aphelion distances:

$$r_1 = \tfrac{1}{2}(28.0 + 294.6) \times 10^6 \text{ km}$$
$$= 161.3 \times 10^6 \text{ km}$$

The earth's mean distance is $r_2 = 149.6 \times 10^6$ km (Appendix Table 7), and the earth's period is $T_2 = 365$ days. By Kepler's third law,

$$\frac{T_1{}^2}{(365 \text{ days})^2} = \frac{(161.3 \times 10^6 \text{ km})^3}{(149.6 \times 10^6 \text{ km})^3}$$

$$T_1 = \boxed{409 \text{ days}}$$

The Newtonian synthesis combined the terrestrial laws formulated by Galileo and Newton with the astronomical laws of Kepler into one general framework that included the radical and powerful law of universal gravitation. Newton worked out many consequences of his gravitational theory. He gave a quantitative explanation of the tides, which are caused by the gravitational attraction of the moon and the sun. He showed that comets are members of the solar system, having elliptic, parabolic, or hyperbolic orbits with the sun at one focus. The moon's motion is very complicated because of the gravitational attraction of bodies such as the sun and Jupiter; Newton gave the first reliable predictions of the moon's motion, of great practical value for navigation. Since Newton's time, the law of gravitation has been extended far beyond the solar system—for example, to the mutual revolution of a pair of stars about their center of mass. With all these successes, Newton never thought he had "explained" gravitation, because he had no experimental evidence for how the force can be transmitted through space between, say, the earth and the sun. Only in recent decades are testable theories for the transmission of gravitational force being proposed (see Secs. 19-3 and 33-9), but experiments are extremely difficult.

In some ways Newtonian mechanics is incomplete. It is precisely applicable to large-scale bodies moving at low speeds. The extension of Newtonian mechanics to the realm of the very small, known as quantum mechanics, is of overriding importance on the atomic scale of things. The theory of relativity (Chap. 28) extends Newtonian mechanics to the realm of very high speeds.

At low speeds, for large bodies, both quantum mechanics and relativity theory reduce to "ordinary" Newtonian mechanics with an incomparably high degree of precision. In fact, as far as planetary motions are concerned, only one difference between the predictions of the Newtonian gravitation and Einsteinian general relativity is large enough to be observed. Close study of the motion of the innermost planet Mercury reveals that its orbit is not a simple ellipse. Instead, there is a slow *advance of perihelion* (Fig. 8-6). The perihelion point P (the closest approach to the sun) moves at the slow rate of about 574 seconds of arc (574″, or 0.16°) per century. The orbit does not close up on itself. Such an advance of perihelion is predicted by Newtonian gravitational theory, because the other planets (Venus, Earth, and especially Jupiter) exert small forces called *perturbations* that are not in the same direction as the main force exerted by the sun. Classical (Newtonian) mechanics, taking account of all known sources of perturbation (including frictional drag of interplanetary matter), predicted an advance of 531″ per century. This is, however, 43″ less than the observed value. It is most encouraging that the theory of general relativity gives the required 43″ excess rate of advance of perihelion. In fact, possible inadequacies in general relativity can be tested if the observed 43″ excess cannot precisely be explained within observational limits of error. This is an active field of current research, of tremendous theoretical importance, but of essentially no practical importance for observable everyday mechanics. The effect on Mercury's orbit is large enough to be noticed because the sun's gravitational field, proportional to $1/r^2$, is strongest there for any natural planet. A solar probe, circling the sun much closer than Mercury, would be most helpful in the further study of gravitational theories.

8-4   The Nature of Scientific Theories

We have illustrated how Kepler's laws led Newton to the law of gravitation. In the detailed working out of the theory, Newton did not obtain Kepler's third law in exactly the form in which it was stated by Kepler. Newton obtained the following proportion (written for the case in which the two planets compared are Mars and the earth):

$$\frac{T^2_{\text{Mars}}\left(1 + \dfrac{m_{\text{Mars}}}{m_{\text{sun}}}\right)}{T^2_{\text{earth}}\left(1 + \dfrac{m_{\text{earth}}}{m_{\text{sun}}}\right)} = \frac{d^3_{\text{Mars}}}{d^3_{\text{earth}}} \qquad (8\text{-}7)$$

Here T is the period, the time for one revolution of the planet around the sun, and d is the average distance from the sun. Kepler's form of the proportion Eq. 8-6 omits the factors in parentheses involving the masses of the planets and the sun. Does this mean that Kepler was "wrong" and Newton "right"? Not at all. Kepler used astronomical data to construct a theory that agreed with the experimental facts within the limit of

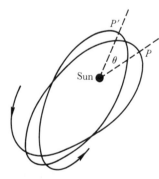

Figure 8-6 Advance of perihelion of Mercury's orbit (greatly exaggerated for clarity; actual angle θ is about 10^{-8} degree for perihelia of successive orbits).

observational error. Other theories might agree with observation equally well, and such theories would be equally "correct."

It is to be expected that at any time our most cherished laws of physics may suffer the same fate that Kepler's third law suffered. But note that the new law that replaces the old one always must contain the old law as a special case. For example, Newton's form of Kepler's third law (Eq. 8-7) reduces to the Kepler form if the factors in brackets are set equal to 1, and this is very nearly true in view of the special circumstance that the sun is so much more massive than any of the planets. We say that Kepler's form is a first approximation to the law of planetary periods, and Newton's form is a second approximation. Are there third, fourth, ... nth approximations? We face the prospect with equanimity, knowing that this is the way science progresses, each new step including all that went before.

As a matter of fact, Newton's law of gravitation has already received similar treatment at the hands of Einstein, and the Einstein law of gravitation includes Newton's form as a special case. Almost every prediction of the two laws of gravitation is the same, within experimental error. As discussed in Sec. 8-3, Mercury's orbit around the sun, however, is just noticeably different on the basis of the two laws, and the Einstein law agrees with observation while the Newton law does not. This does not mean that we have "repealed" Newton's law of gravitation. It is very useful and almost exact first approximation, valid for the situations in which we apply it.

Another example of the changing validity of a "law" of physics is supplied by Newton's second law itself. The 'breakdown" of Newton's second law became evident in the early 20th century as atomic phenomena became more clearly understood. The finishing touch was given in 1926 by Schrödinger, Born, Heisenberg, and others, who developed a new mechanics (quantum mechanics) to apply to a single atom.* This does not detract in the least from the validity of Newton's laws as we use them in this text and as engineers use them for the most complicated calculations. We always deal with large-scale objects (marbles, cars, even dust particles) that consist of many billions of atoms grouped closely together. We do not expect a political poll to shed any light on a specific individual's voting preference; yet the poll does reflect the average behavior of a large group. Similarly, we should not expect a description of the average behavior of a group of atoms (Newton's laws) to predict the behavior of a single electron or atom (quantum mechanics). A single atomic particle's behavior is, of course, basic, but it is nonclassical in the sense that not all of the familiar (classical) large-scale laws of Newtonian mechanics are applicable. We shall return to this subject briefly in Secs. 16-3 and 16-4 and in Chaps. 29–31, where both the classical and nonclassical properties of electrons are studied in some detail.

---

* For a single particle, the quantum-mechanical version of Newton's second law is quite different from net $F = ma$, but when it is applied to the motion of a large collection of particles, this quantum-mechanical law reduces to the familiar net $F = ma$, just as Newton's form of Kepler's third law reduces to Kepler's form.

Summary   Newton's law of universal gravitation was derived by the application of the laws of motion to an explanation of Kepler's laws of planetary motion. The gravitational force of one body on another depends on the product of the two masses and is also inversely proportional to the square of the distance between the bodies. This kind of mass is called gravitational mass. Experiment shows that gravitational mass and inertial mass for any body are proportional; since this is so, the same units (kg, g) are used for these two kinds of mass. The constant in the law of gravitation was first measured by Cavendish.

Gravitational field strength is defined as force per unit mass, and in the mks system it is measured in newtons per kilogram.

Kepler's second law, the law of areas, is true for any orbit that is described under the action of an attractive or repulsive central force; Kepler's law of orbits requires that the central force be an attractive inverse-square force, such as that of gravitation. The Newtonian synthesis extended the range of physical laws to the realm of astronomical phenomena.

Laws of physics are based on experiment and may be superseded by other, more general laws as new experimental facts become known. The new laws must include the old laws as special cases.

Check List

Newton's law of universal gravitation
Kepler's laws
inertial mass
gravitational mass

gravitational field strength
central force
mean distance
perturbation

Questions

8-1 How would you go about determining whether two bodies had the same weight? How would you go about determining whether two bodies had the same inertial mass? the same gravitational mass?

8-2 Two small ball bearings are placed 1 cm apart on a smooth, level metal surface. Will they accelerate toward each other because of gravitational attraction, and, if so, will they eventually touch each other? Answer this question for two conditions: (*a*) on earth; (*b*) in interstellar space, where there is no force pressing the balls against the surface and hence no friction.

8-3 If a body of mass 36 kg were taken to the moon, where the acceleration due to gravity is $\frac{1}{6}$ as much as on the earth, what would be its weight? What would be its mass?

8-4 A spaceship is far away from the sun, moving with practically constant velocity in an orbit that is practically a straight line. In Fig. 8-7, *AB* and *CD* represent motions that each take place in one day's time. Use plane geometry to show that Kepler's second law is true in this case.

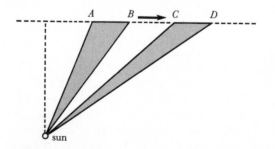

Figure 8-7 Kepler's law of areas for the special case of uniform motion in a straight line.

8-5 The mass of an astronaut on the moon, relative to his mass on the earth, is   (*a*) greater;   (*b*) the same;   (*c*) less.

8-6 In the cgs system, the unit for gravitational field strength is   (*a*) dyne/cm;   (*b*) g/cm;   (*c*) dyne/g.

8-7 Gravitational field strength can be expressed in the following units: (*a*) m/s^2;   (*b*) N/kg;   (*c*) both of these.

8-8 Planet *A* has radius *r* and density *d*; planet *B* has radius 2*r* and density 2*d*. Relative to *A*, the strength of the gravitational field at the surface of planet *B* is   (*a*) greater;   (*b*) the same;   (*c*) less.

8-9 For a star with a single planet, a precisely circular orbit for the planet is   (*a*) always the case;   (*b*) possible, but very unlikely;   (*c*) impossible.

8-10 Kepler's law of areas is true   (*a*) only for an inverse-square attractive force law;   (*b*) for a force law in which force varies directly as the distance, always directed toward a central point;   (*c*) for any force law consistent with the law of conservation of energy.

Problems

8-A1 What is the gravitational force between a teacher of mass 100 kg and a student of mass 60 kg when the student sits in the front row of the classroom, 2 m from the teacher?

8-A2 A space explorer 3 billion km away from a certain star observes that the gravitational force of attraction exerted on the spaceship by the star is 200 N. What will the force become when the ship has approached a position half a billion km from the star?

8-A3 The radius of the moon is 1730 km (1080 mi). What is the downward force (due to the moon's gravitational attraction) on an astronaut hovering in an orbit 1730 km above the moon's surface if her weight on the surface of the moon is 88 N (20 lb)?

8-A4 How far apart are two objects each of mass 1 metric ton if the gravitational force between them is 1 nN (nanonewton)?

8-A5 Suppose an astronaut goes into a circular orbit about the sun with an orbital radius of 149×10^6 km (93×10^6 mi). How long would be required for the astronaut to make one revolution around the sun?

8-B1 Calculate the gravitational force of attraction between 10 cm^3 of iron and 10 cm^3 of aluminum separated by 100 cm.

8-B2 Compute the gravitational force between a proton and an electron in the $^1_1$H atom, using the following data: mass of proton, 1.67×10^{-27} kg; mass of electron, 9.11×10^{-31} kg; radius of electron orbit, 5.29×10^{-11} m.

8-B3 What is the force of gravitation (in dynes) between two touching spherical iron balls, each 8 cm in radius?

8-B4 A small research rocket of mass 120 kg is released from rest at a point one earth-radius above the earth's surface.   (*a*) What is the rocket's weight at this point?   (*b*) With what initial acceleration does it start to fall?

8-B5 (*a*) How fast must a satellite of mass *m* be launched horizontally at the surface of the earth if its orbit is to be a circle just grazing the highest mountains? (Assume no air resistance.)   (*b*) What time would elapse between successive passes of this imaginary lowest satellite? (*Hint:* Use astronomical data from Appendix Table 7. Equate the centripetal force to the gravitational pull of the earth. The mass of the satellite cancels out.)

8-B6 (a) Calculate the mass of the sun, given the length of the year to be 365.3 days, mean radius of the earth's orbit 1.50×10^8 km, and Newton's gravitational constant $G = 6.67 \times 10^{-11}$ N·m²/kg². (*Hint:* Solve first in symbols; the mass of the earth cancels out. Check your answer by referring to Appendix Table 7.) (b) The sun's radius is 6.95×10^8 m; calculate its average specific gravity.

8-B7 (a) Give two reasons for expecting the acceleration due to gravity (relative to the earth) to be less at the equator than at the poles. (b) If an Olympic athlete could throw the shot 18.31 m at Helsinki (latitude 60.2° N), where g is 9.8191 m/s², how much farther could he throw it in Los Angeles (latitude 34.1° N), where g is 9.7969 m/s²? (*Hint:* See Sec. 2-11.)

8-B8 Prove that the mass m_P of a planet is given by $m_P = r_P{}^2 g_P/G$, where r_P is the planet's radius and g_P is the acceleration due to gravity on the planet's surface.

8-B9 Using data from Example 8-3, calculate the magnitude and direction of the resultant gravitational field when the explorer's spacecraft is halfway around the moon, 1.0×10^8 m from the moon, observing the hidden side of the moon's surface from a point on the extension of the line joining earth and moon. As in Example 8-3, it is new moon.

8-B10 What is the dimensional formula for the gravitational constant G, in terms of $[M]$, $[L]$, and $[T]$?

8-B11 The acceleration of a falling body near the earth's surface, at a distance R from the earth's center, is 9.80 m/s². (a) Use a suitable proportion to calculate the acceleration toward the earth of a falling body that is $60R$ from the earth's center. (b) The moon is in an orbit of radius $60R$, with a period of revolution 27.3 days. Show, as Newton did, that the centripetal acceleration of the moon toward the earth agrees with the answer to part (a). For this part you need to know that the earth's radius R is 6.38×10^6 m.

8-B12 Make a calculation to show that the gravitational force of the sun on the moon is always greater than the force of the earth on the moon. Thus at new moon the net force is away from the earth. This being so, how is it that the moon goes around the earth and not around the sun? Or does it? (Use astronomical data from Appendix Table 7.)

8-B13 A certain small planet (asteroid) revolves about the sun with a period of 8 years. What is its average distance from the sun? [Express your answer in astronomical units; 1 a.u. = 147×10^6 km (93×10^6 mi), the average distance of the earth from the sun.]

8-B14 A planet's closest distance from the sun is a, and its farthest distance is b (Fig. 8-8). Use Kepler's second law to prove that $v_A/v_B = b/a$.

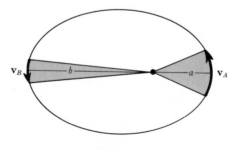

Figure 8-8

8-B15 When a satellite revolves in an orbit around the earth, the centripetal force equals the gravitational attraction. (*a*) Derive a formula for the period T (time for one revolution) of a satellite in a circular orbit. (*b*) Using astronomical data from Appendix Table 7, calculate the period of a satellite in a circular orbit of radius 7200 km.

8-B16 Use data in Appendix Table 7 to calculate the ratio $\dfrac{m_{earth}}{m_{sun}}$. What is the value of the factor $\left(1 + \dfrac{m_{earth}}{m_{sun}}\right)$ that occurs in Eq. 8-7, the Newtonian form of Kepler's third law? (Express in decimal form.) Is this factor sufficiently close to unity to be neglected if observations are accurate to 1 part in a million?

References

1. H. Cavendish, "The Density of the Earth," *Phil. Trans.* **17** (1798). The original account of the measurement of *G*. Reprinted in W. F. Magie, *A Source Book in Physics* (McGraw-Hill, New York, 1935), pp. 105–111.
2. R. H. Dicke, "The Eötvös Experiment," *Sci. American* **205**(6), 84 (Dec. 1961). See also P. G. Roll, R. Krotkov, and R. H. Dicke, "The Equivalence of Inertial and Passive Gravitational Mass," *Ann. Phys.* **26**, 442 (1964).
3. F. Miller, Jr., "Kepler's Third Law and the Mass of the Moon," *Am. J. Phys.* **34**, 53 (1966).
4. T. C. Van Flandern, "Is Gravity Getting Weaker?" *Sci. American* **234**(2), 44 (Feb. 1976).
5. G. Gamow, "Gravity," *Sci. American* **204**(3), 94 (Mar. 1961).
6. J. W. Beams, "Finding a Better Value for *G*," *Physics Today* **24**(5), 34 (May 1971).
7. F. Miller, Jr., *Measurement of "G"—The Cavendish Experiment* (film).

9
Elasticity and Vibration

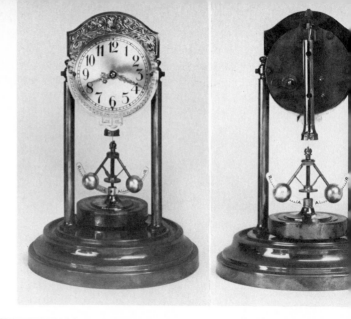

The role of elasticity in everyday affairs is usually greatly underestimated. The short-range interatomic electric forces we call elastic forces are all around us, often taken for granted. Suppose we use a rubber band to hold together a bundle of letters. We are accustomed to calling the force exerted by the rubber band an elastic force because we can easily see the distortion produced in the rubber (a stretch in this instance) and can feel the reaction to the force that we apply to cause the distortion. If we wrap the package with string, applying the same force, we tend to ignore the stretch of the string because it is not readily visible. However, the two situations differ only in degree. The string must stretch, be it ever so slightly, in order to give rise to the contractile force it exerts. A gnat alights on the flight deck of a 60,000-ton carrier, and the deck buckles (slightly) in order to supply the upward force necessary to support the gnat's weight. When you pry the cap off a bottle of soda, the opener must be slightly distorted to supply

the necessary force on the bottle cap. In Sec. 4-4 we made a model for static and kinetic friction based on the ideas of elasticity. Another example of "hidden" elastic force is afforded by the force of the rails beneath the wheels of a freight car. The car sinks into the rails just far enough so that the upward elastic force exerted by the rails is just enough to balance the weight of the car. If the upward force were insufficient, the car would experience a net downward force, and it would move downward until the forces were balanced again. Except for the minuteness of the motion, this behavior is in no way different from the behavior exhibited when a letter is placed on a spring-operated postage scale. The platform of the scale sinks down just far enough.

Short-range electric forces, to be discussed in Chap. 31, are of several kinds. Consider a solid such as the iron crystal shown in Fig. 1-1 on page 14. To increase the separation of the atoms requires a force to overcome *cohesive* electric forces; to push the atoms together one must

The vibrating system of this 400-day clock is a disk suspended by a fine, flat wire; the vibration period is determined by the disk's moment of inertia and the wire's stiffness. Two adjustable balls allow fine control of the moment of inertia. The letters A and R refer to the French words avance (fast) and retard (slow). The clock, from the Horolavar collection, is owned by Charles Terwilliger.

overcome *repulsive* electric forces. Each atom in the solid is in equilibrium under the influence of the opposite effects of these two rather different aspects of electric force.

9-1 Elastic Constants—Hooke's Law

One of the most important reasons for studying elasticity is that it gives us an approach to the study of vibrational motion, such as the up-and-down motion of a massive object suspended by a spring or a wire. First, however, we shall consider how the stretch of a wire depends on the material of which it is made, as well as on its length and cross-sectional area.

Robert Hooke (1635–1703) found that the stretch ΔL of an iron wire was proportional to the applied force ΔF, up to some maximum force. This proportionality between force and stretch is known as *Hooke's law:**

$$\Delta F = k\,\Delta L$$

where k, given by $\Delta F/\Delta L$, is the force constant of the wire.

Just as specific gravity is a constant for a material, not depending on the particular size and shape of the object that has been fashioned from the material, so each substance has a number of elastic constants that do not depend on the particular size and shape involved. All other things being equal, it is evident that a fine iron wire will stretch more than a thick iron wire, and yet the substance, iron, is the same in each case. It is possible to define an elastic property of the material itself, called a *modulus of elasticity,* and there are as many elastic moduli as there are types of distortion. We shall consider three simple types of distortion, and define a modulus for each.

---
* The term "law" is somewhat misleading. Hooke's law is an empirical observation that very closely correlates observed variables; it is not a law in the sense that Newton's laws are always true, under all circumstances.

Stretch Modulus. Change in length ΔL of a wire or rod is caused by a change in stretching force ΔF (see Fig. 9-1). The ratio $\Delta F/\Delta L$ is constant for any given wire, but of course it varies from wire to wire depending not only on the elastic modulus of the material, but also on the length L and the cross section A. When we consider wires of the same substance, but of different cross-sectional area and length, experiment shows that the amount of stretch is proportional to the original length of the wire. A wire 5 m long stretches twice as much as a wire 2.5 m long, other things being equal. Also, experiment shows that to produce a given stretch in wires of equal length, the required force is proportional to the cross-sectional area. All this is summarized in one equation:

$$\frac{\Delta F}{A} = E\,\frac{\Delta L}{L}$$

Stress = modulus × strain

where *stress* is the force per unit cross-sectional area and *strain* is the stretch or compression per unit length. Stress has the dimensions of pressure, being force/area; strain is a pure number, being length/length. E is the *stretch modulus*— also called *Young's modulus*—for the material.

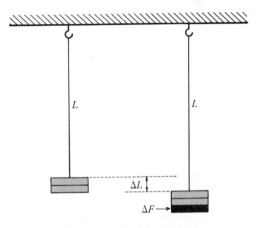

Figure 9-1 Stretch of a wire.

$$\text{Stretch modulus} = \frac{\text{stretching stress}}{\text{stretching strain}}$$

$$E = \frac{\Delta F / A}{\Delta L / L}$$

The stretch modulus, like all moduli, has the dimensions of force/area. The SI unit for pressure, N/m^2, is called the pascal (Pa).

$$1 \text{ pascal} = 1 \text{ Pa} = 1 \text{ N/m}^2$$

Example 9-1

A telephone wire 125.00 m long and 1.00 mm in radius is stretched to a length of 125.25 m when a force of 800 N is applied. What is the stretch modulus for the material?

The cross-sectional area of the wire is

$$\pi(10^{-3} \text{ m})^2 = 3.14 \times 10^{-6} \text{ m}^2$$

$$\text{Stress} = \frac{\Delta F}{A} = \frac{800 \text{ N}}{3.14 \times 10^{-6} \text{ m}^2}$$

$$= 2.54 \times 10^8 \text{ N/m}^2$$

$$\text{Strain} = \frac{\Delta L}{L} = \frac{0.25 \text{ m}}{125 \text{ m}}$$

$$= 0.002 = 2 \times 10^{-3}$$

(The stretch is 0.2% of the original length.)

$$E = \frac{\text{stress}}{\text{strain}} = \frac{2.54 \times 10^8 \text{ N/m}^2}{2 \times 10^{-3}}$$

$$= 1.27 \times 10^{11} \text{ N/m}^2 = \boxed{12.7 \times 10^{10} \text{ N/m}^2}$$

Using SI terminology, this modulus would be expressed as 127 GPa.

To interpret this experimental behavior, let us make a model of the wire (Fig. 9-2). Consider a stretched wire; the force arises from the cohesion between the atoms. That is, the atoms in the crystals are so close together that short-range electric forces are considerable. We can imagine the wire to consist of many parallel planes of atoms stacked on top of each other (Fig 1-1 on page 14). Due to a load, each pair of neighboring

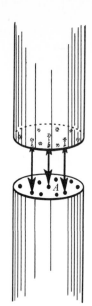

Figure 9-2 Separation of atomic planes during stretch of a wire.

planes separates slightly. The total stretch is the sum of the individual stretches. For a longer wire, more planes are separated, each pair by the same small amount; therefore, we expect ΔL, the total stretch, to be proportional to L, as experiment shows. Likewise, for a larger cross section A, more atoms act together at each layer. Hence we expect the required force to be proportional to A. We see that the model accounts satisfactorily for the main experimental facts.

Our simple model does not account for some of the details of the stretching process. As we know, if sufficient force is applied to a wire, it no longer remains in equilibrium but stretches a great deal and breaks. Before this happens, however, a more subtle discrepancy appears. If the applied stress is somewhat greater than some well-defined amount called the *elastic limit*, the wire does not break, but neither does it return to its original length when the stress is removed. The wire acquires a permanent distortion. In the design of structures, the elastic limit must never be exceeded. In fact, architects and engineers usually allow a "safety factor" and design for not more than $\frac{1}{5}$ to $\frac{1}{3}$ the elastic limit, especially in machinery where stress is repeatedly applied and removed.

Technically speaking, *elasticity* is the ability of a material to recover its original length (or shape or volume) after the stress is removed. In this sense, steel and glass are highly elastic materials, but rubber is not very elastic, although it is easily stretched. When the elastic limit of a biological material such as bone is exceeded, it tends to become plastic and "creeps" with a slow extension during a period of many hours or even days. This must be allowed for when a patient is in traction.

The *ultimate tensile strength* of a material is the stress required to break a wire or rod by pulling on it. This is, of course, greater than the elastic limit. By way of illustration, tables give the following values for a certain type of steel alloy:

Stretch modulus	20×10^{10} N/m^2
Elastic limit	3×10^8 N/m^2
Ultimate tensile strength	5×10^8 N/m^2

For bone, the stress is usually a compression or (during fracture) a shear. For crumbling due to compression, the ultimate stress is about 2×10^7 N/m^2.

Example 9-2

A man of weight 800 N stands on the ball of one foot. His tibia is 36 cm long; other dimensions are given in Fig. 9-3. Calculate (a) the stress, (b) the strain, and   (c) the change in length of his tibia.

First we must solve a problem in statics. We assume that the Achilles tendon acts vertically upward on the heel with a force **T**; the floor pushes up on the ball of the foot with a force of 800 N, equal to the man's weight.

Applying the torque condition about an axis through A gives

$$T(5 \text{ cm}) = (800 \text{ N})(14 \text{ cm})$$
$$T = 2240 \text{ N}$$

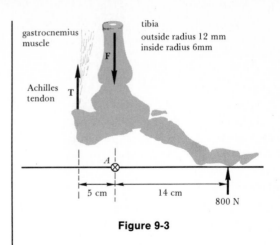

tibia
outside radius 12 mm
inside radius 6mm

gastrocnemius muscle

Achilles tendon T

Figure 9-3

To balance the vertical components of force we have

$$F = T + 800 \text{ N} = 3040 \text{ N}$$

The compression force in the tibia is almost 4 times the man's weight! The tibia is not solid. From the dimensions given we calculate the area of cross section to be $\pi(12 \text{ mm})^2 - \pi(6 \text{ mm})^2 = 339$ mm^2, or

$$A = 339 \text{ mm}^2 \left(\frac{1 \text{ m}}{10^3 \text{ mm}} \right)^2 = 3.4 \times 10^{-4} \text{ m}^2$$

Now we are ready for the problem in elasticity.

(a) $$\text{Stress} = \frac{\text{force}}{\text{area}} = \frac{3040 \text{ N}}{3.4 \times 10^{-4} \text{ m}^2}$$
$$= \boxed{9.0 \times 10^6 \text{ N/m}^2}$$

This is about half the ultimate stress for crumbling of bone.

(b) From the definition of stretch modulus we have

$$\text{Strain} = \frac{\text{stress}}{\text{modulus}}$$

$$\text{Strain} = \frac{\Delta L}{L} = \frac{9.0 \times 10^6 \text{ N/m}^2}{2 \times 10^{10} \text{ N/m}^2} = \boxed{4.5 \times 10^{-4}}$$

The bone compresses by 0.00045, about $\frac{1}{20}$ of 1%.

(c) $$\Delta L = (L)(\text{strain}) = (36 \text{ cm})(4.5 \times 10^{-4})$$
$$= \boxed{0.016 \text{ cm}}$$

The compression is about $\frac{1}{6}$ mm.

Note that if the problem refers to a basketball player who comes down on one foot, there is an additional inertial force $-m\mathbf{a}$ (see Sec. 5-8) also acting to compress the tibia.

Shear Modulus—Rigidity. We have considered the stretch modulus in some detail because the basic ideas are applicable to any elastic modulus, including a second form of distortion called *shear*. There is no change of any dimension during shear, but the *shape* of the body changes. For instance, as shown in Fig. 9-4, a book lying on a table is given a sideways shear by means of a force applied to the top cover (and an equal and opposite force of friction applied by the table to the bottom cover). Each page moves relative to its neighbor. In this case the stress is the applied force divided by the area of one of the pages; the strain is the horizontal distance moved divided by the height of the book.

$$\text{Shear modulus} = \frac{\text{shearing stress}}{\text{shearing strain}}$$

$$n = \frac{\Delta F/A}{\Delta x/h}$$

A model based on the mutual cohesion of adjacent layers of atoms (Fig. 9-5) will give a qualitative interpretation of the experimental facts: the amount of shear is inversely proportional to the *tangential* area A, and it is also directly proportional to the height of the book (that is, to the number of parallel planes, each of which is sheared the same tiny amount).

The shear modulus is also called the *rigidity modulus*, for it is this property of a substance that

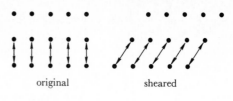

Figure 9-5 Separation of atomic planes during shear of a solid body.

determines how well a body will retain its shape when a shearing stress is applied. Indeed, if it were not for the property that the shear modulus describes, a solid block of steel would collapse under its own weight like a block of melting butter. The action of a common machine screw illustrates nicely several stresses (Fig. 9-6). When properly tightened, A and B must withstand a sideways shearing stress; also, as the shaft is compressed its length decreases, so the essential factor determining the force with which the surfaces A and B are pressed together is the stretch modulus.

Bulk Modulus. For our third example of a simple elastic distortion, we consider a solid block of iron on the end of a long wire. When the block is lowered to a great depth in the ocean, the pressure of the water causes the volume (bulk) of the block to diminish.

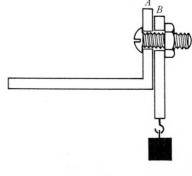

Figure 9-6

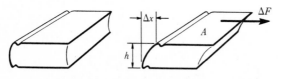

Figure 9-4 Shear of a solid body.

$$\text{Bulk modulus} = \frac{\text{volume stress}}{\text{volume strain}}$$

$$= \frac{\text{increase in pressure}}{\dfrac{\text{decrease in volume}}{\text{original volume}}}$$

$$B = \frac{\Delta P}{\dfrac{-\Delta V}{V}} = -\frac{V\,\Delta P}{\Delta V}$$

Values of the bulk modulus are positive; the minus sign in the equation compensates for the fact that ΔV (a decrease in volume) is negative.

The well-known fact that most solids are not so easily compressed as liquids is borne out by the figures in Table 9-1. The pressure required to produce a given small relative change in volume is about 70 times as much for steel as for water. The only elastic modulus that applies to a liquid or gas is the bulk modulus, since liquids

Table 9-1   Elastic Constants*

Substance	Stretch Modulus E, N/m²	Shear Modulus n, N/m²	Bulk Modulus B, N/m²
Solids			
steel, annealed	20 $\times 10^{10}$	8.0 $\times 10^{10}$	14 $\times 10^{10}$
iron, wrought	19	7.5	14
iron, cast	12	4.6	9.0
copper	10	3.9	11
aluminum	7.0	2.5	7.5
magnesium	4.1	1.6	3.1
lead	1.5	0.6	3.4
brass	9.0	3.4	8.3
glass	5.8	2.4	4.5
diamond	†	†	62
fused quartz	5.6	2.5	2.7
bone (approx.)	2		
hair (approx.)	0.2		
Liquids			
water			0.20 $\times 10^{10}$
mercury			2.5
ethyl alcohol			0.10
glycerin			0.45
sea water			0.21
Gases‡			
air			1.01 $\times 10^5$
hydrogen			1.01
helium			1.01
carbon dioxide			1.01

In tables such as this one, where magnitudes do not differ greatly, it is customary in handbooks to give a power of 10 at the head of each column, understood to apply to each entry. Note that one could write 2.0×10^{11} and 2×10^9 for the first and last entries in the column for E, but using the same power for all entries makes the relative values more readily apparent.

* Typical values; depend on heat treatment, cold work, and other conditions.
† Diamond is not listed because of the difficulty of tensile tests on diamond specimens.
‡ All gases at 101 kPa (1 atm) pressure with no temperature change during compression or expansion.

and gases have no definite length or shape. Table 9-1 gives values of the elastic constants for various substances in mks units of newtons per square meter. These moduli can be expressed in cgs or British units by making use of the fact that $1 \text{ N/m}^2 = 1 \text{ Pa} = 10 \text{ dyn/cm}^2 = 1.45 \times 10^{-4} \text{ lb/in.}^2$. There are elastic limits for distortions of the shear and volume types just as for the stretch type.

Our discussion of elasticity has been limited to highly idealized cases of "pure" stretch, "pure" shear, and so on. Actually, many elastic distortions are complex. A wire that stretches must also change shape, since it is now a cylinder of greater length and smaller diameter. The volume usually increases slightly, too. Exact treatment of elasticity is highly mathematical, but our aim has been rather to point out some main features of elasticity and to indicate how widespread the applications are.

It may seem that we have really explained nothing about elasticity, since we have put all the credit (or blame) on the mysterious interatomic forces that we call short-range electric forces. But actually "gravitational force" is an equally noncommittal term, and this phenomenon is equally unexplainable and equally fundamental to the nature of things. Only our great practical familiarity with gravitational force (weight) leads us to think of it as better "ex-

plained" than the forces of elasticity. We are equally familiar with elastic forces, to be sure, and yet they seem more mysterious because no simple numerical formulas can be written comparable to Newton's elegant law of gravitation. Whether or not you are satisfied with this state of affairs is probably only a matter of degree of sophistication.

9-2 Simple Harmonic Motion

A *vibration* is a to-and-fro motion, generally along a straight line (linear vibration) or along an arc of a circle (angular vibration). Typical examples of linear vibration are the vibration of a heavy load placed on the end of a spring or rubber band and the motion of an automobile engine's piston (Fig. 9-7a). A pendulum and the balance wheel of a watch illustrate angular vibratory motion (Fig. 9-7b). A brief look around us reveals many vibrations in nature. A raindrop strikes a leaf on a tree, and the leaf gyrates up and down; the tree itself sways in the breeze. The vocal cords of a bull moose vibrate many times per second, causing periodic compression of the air, which in turn causes vibratory motion of his mate's eardrum. The ticking of a grandfather clock, the clang of a dropped piece of metal, the twang of a guitar string—all these are among

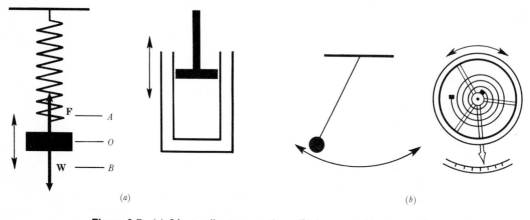

(a)   (b)

Figure 9-7 (a) Linear vibratory motion. (b) Angular vibratory motion.

the sounds that result from countless types of vibratory motion. Because of the importance of vibrations in music, the term "harmonic motion" has come to mean the same as "vibratory motion." One particularly simple form of harmonic motion occurs frequently both in the large-scale world and at the atomic level, and so we proceed to define and study *simple harmonic motion* (SHM), especially in its relation to Hooke's law for elasticity.

First of all, it is obvious that a body that is in vibratory motion has a somewhat specialized acceleration. Suppose the body has some initial velocity toward the right as it passes through the equilibrium position O (Fig. 9-8). The body must slow down, eventually stop, and return toward O; otherwise it would be lost forever and the motion could not repeat itself. On the way out, the velocity v is positive but decreasing in magnitude—a negative acceleration. On the way back, v is negative and increasing in magnitude—also a negative acceleration. Thus whenever the displacement is positive, the acceleration is negative. Similarly, when s is negative (and the body is between O and P'), the acceleration is positive. Thus for *any* harmonic motion, the acceleration varies in magnitude and is opposite in direction to the displacement.

"Simple" harmonic motion is so called because the acceleration varies as smoothly as possible. The very simplest equation for a variable acceleration is a direct proportion. Thus we define SHM as motion in which the acceleration is directed toward the center of the motion and is directly proportional to the displacement from the center.

$$\mathbf{a} = (\text{constant})(-\mathbf{s}) \qquad (9\text{-}1)$$

The force needed to cause SHM can be found from Newton's second law. From Eq. 9-1,

$$m\mathbf{a} = (\text{a different constant})(-\mathbf{s})$$

or

$$\mathbf{F} = k(-\mathbf{s})$$

We define the *force constant k* by

$$k = \frac{|\mathbf{F}|}{|\mathbf{s}|}$$

To avoid cumbersome notation, we write

$$k = \frac{F}{s} \qquad (9\text{-}2)$$

where it is understood that F is the magnitude of the *restoring* force exerted on the body when the displacement is s.

Hooke's law supplies just the right sort of force, with $F \propto s$. Any spring, wire, bar, or other system that obeys Hooke's law has a force constant and therefore will give rise to SHM.

Example 9-3

A certain spring is 30.0 cm long when a load of 40.2 N hangs from it and 37.0 cm long when the load is 45.2 N. What is the force constant of the spring?

The added force is 5.0 N, causing a stretch of 0.070 m:

$$k = \frac{F}{s} = \frac{5.0\ \text{N}}{0.070\ \text{m}} = \boxed{71\ \text{N/m}}$$

Some commonly occurring vibratory motions are only approximately SHM. A mass on a rubber band fails the test because the rubber band's length depends to some extent on its previous history; also, the displacement is not strictly proportional to the force. The piston in the

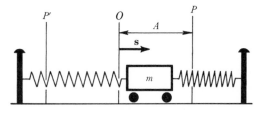

Figure 9-8

cylinder of an automotive engine moves in only approximate SHM because of the finite length of the connecting rod.

9-3 Period of Simple Harmonic Motion—The Reference Circle

In our study of SHM, we consider a cart of mass m that is supported by frictionless wheels as it moves to and fro horizontally (refer back to Fig. 9-8). The net force on the cart is that caused by the springs, since its weight is balanced by the upward force of the table. The force constant of the spring system is k.

Several technical terms are used in the description of SHM. A *cycle* is one complete to-and-fro vibration (P to P' and back again), and the *period* T is the time required for this to take place. The frequency f is the number of vibrations per unit time; this is the reciprocal of the period. Thus, if the period is 0.01 s, the body makes 100 vibrations per second. The *displacement* s is the vector measured from the equilibrium point to the position at any time t, and the *amplitude* A is the maximum value of the displacement. The body moves a total distance of $4A$ in one cycle, but of course the velocity is by no means constant. The *phase* of the vibration is a measure of the position of the particle in its motion (Fig. 9-9). Two vibrations are in phase with each other if both particles pass through their equilibrium points at the same time, going in the same direction.

To derive a formula for the period of a body in SHM, we use a *reference circle* to make clear

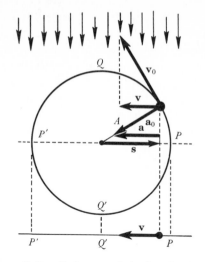

Figure 9-10   Reference circle for simple harmonic motion.

the relation of SHM to uniform circular motion. We have seen that, by definition, any motion for which a/s is a constant is SHM. It so happens that another simple motion has this same property. Consider a particle having uniform circular motion with velocity v_0 in a vertical circle of radius A. As in Fig. 9-10, let light shine vertically on the particle in this reference circle. The shadow (or projection) of the *reference particle* moves back and forth along a line parallel to the horizontal diameter. The shadow has maximum velocity at the midpoint of its motion, for then the reference particle is at Q (or Q'), and v, the horizontal component of v_0, equals v_0 itself (Fig. 9-11a). At the extreme points P and P', the horizontal component of v_0 is zero (Fig. 9-11b), which agrees with the fact that a vibrating particle is momentarily stationary as it changes direction of motion (its acceleration is not zero even though its velocity is zero; compare the motion of a vertically thrown baseball, which has a downward acceleration of 9.8 m/s² even when its instantaneous velocity is zero at the highest point).

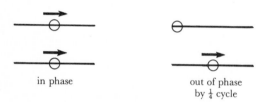

Figure 9-9

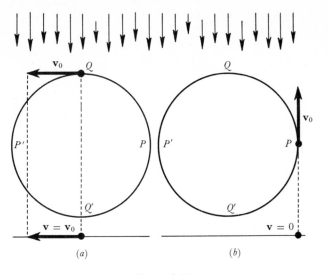

Figure 9-11

We now prove that the motion of the shadow is SHM. Let v_0 be the constant speed of the reference particle. The centripetal acceleration a_0 of the reference particle is constant in magnitude (equal to $v_0{}^2/A$) and directed along the radius. The acceleration a of the shadow is the horizontal component of a_0. By similar triangles, $a/s = a_0/A = $ constant. Hence the shadow moves in SHM, since the constancy of the ratio of acceleration to displacement is the definition of SHM. We also note that A, the radius of the reference circle, equals the amplitude of the SHM of the shadow. The time T for one complete vibration of the shadow is the same as the time for one complete revolution of the reference particle. Hence,

$$T = \frac{\text{circumference of reference circle}}{\text{speed around the circle}} = \frac{2\pi A}{v_0}$$

$$v_0 = \frac{2\pi A}{T} \qquad (9\text{-}3)$$

Use of this equation for v_0 allows an easy calculation of the maximum velocity and maximum acceleration in any SHM.

Example 9-4

A prong of a tuning fork vibrates in SHM, at a frequency of 500 vib/s. The prong moves through 2.00 mm, 1.00 mm on either side of center. What is its maximum velocity? What is its maximum acceleration?

The period is $\frac{1}{500}$ s = 0.002 s, and the amplitude is 1.00 mm = 1.00×10^{-3} m. At the midpoint of the motion,

$$v_0 = \frac{2\pi A}{T} = \frac{2\pi(1.00 \times 10^{-3}\text{ m})}{2 \times 10^{-3}\text{ s}}$$

$$= \boxed{3.14 \text{ m/s}}$$

At either endpoint,

$$a_0 = \frac{v_0{}^2}{A} = \frac{(3.14 \text{ m/s})^2}{1.00 \times 10^{-3}\text{ m}} = \boxed{9870 \text{ m/s}^2}$$

The fact that the projection of uniform circular motion is SHM can be demonstrated as shown in Fig. 9-12.

To derive a formula for the period of the SHM, we use the energy principle. During the SHM of a body, there is a continual transformation of

Figure 9-12 The projection of uniform circular motion is simple harmonic motion. The shadow of one ball attached to a vertical rotating turntable is cast on a screen, and alongside it is the shadow of another ball vibrating vertically on the end of a spring. When the period and phase of the turntable are properly adjusted, the two shadows move side by side at all times.

energy from KE to PE and back again. Referring to Fig. 9-13, suppose the body has been moved by some outside agent from its equilibrium position (2) through a distance A to some new position (1), thus compressing the spring. The work done is stored in the spring as elastic PE. The work was done against a variable force, which was 0 at position (2) and kA at position (1). As we saw in Sec. 6-2, work done against a variable force is represented by the area under the force-displacement curve. In this case, for a Hooke's law force, the area of the triangle of Fig. 9-14 is $\frac{1}{2}(kA)A$, or $\frac{1}{2}kA^2$. The work, and hence the PE stored, is

$$\text{PE} = \tfrac{1}{2}kA^2$$

If the body is now released, it moves back to position (2), and by the energy principle the total energy at (2) equals the total energy at (1):

$$\text{KE}_2 + \text{PE}_2 = \text{KE}_1 + \text{PE}_1$$
$$\tfrac{1}{2}mv_0{}^2 + 0 = 0 + \tfrac{1}{2}kA^2$$

Hence,

$$\frac{A^2}{v_0{}^2} = \frac{m}{k}$$

$$\frac{A}{v_0} = \sqrt{\frac{m}{k}}$$

and since $T = 2\pi A/v_0$,

$$T = 2\pi \sqrt{\frac{m}{k}} \qquad (9\text{-}4)$$

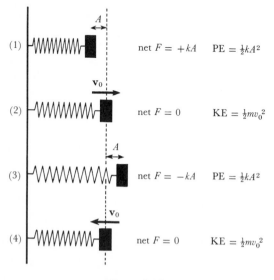

(1)	net $F = +kA$	PE $= \tfrac{1}{2}kA^2$
(2)	net $F = 0$	KE $= \tfrac{1}{2}mv_0{}^2$
(3)	net $F = -kA$	PE $= \tfrac{1}{2}kA^2$
(4)	net $F = 0$	KE $= \tfrac{1}{2}mv_0{}^2$

Figure 9-13

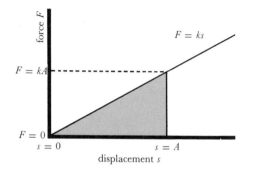

Figure 9-14 The shaded area represents the work done during displacement against a Hooke's law force.

This is the basic equation for the period of any SHM, and it shows how the period is related to the essential physical attributes of the vibrating system. If friction is small, as it usually is, *mechanical* energy is conserved, and our application of the energy principle is correct.

Note that the amplitude A cancels out, and the period of vibration depends only on the inertia of the object (represented by m) and the stiffness of the spring (represented by k). The period is independent of the amplitude of vibration. Equation 9-4 shows that for any given load, a stiff spring (large k, i.e., large ratio F/s) has a shorter period T, which agrees with experience. The equation also shows that for any given spring, a "heavy"* load vibrates more slowly; that is, T is larger. In Eq. 9-4, it is assumed that the force constant k is really constant. If the ratio F/s is not a constant, then the motion is not SHM in the first place, and Eq. 9-4 does not apply.

In solving problems in SHM, it will be sufficient to remember the definition of force constant, the idea of the reference circle (especially $v_0 = 2\pi A/T$), and the equation $T = 2\pi\sqrt{m/k}$.

Example 9-5

The springs of a 1000-kg car compress vertically 7.0 mm when a 100-kg man steps in. With the man in the car, how many vibrations per second does the body of the car make after being jarred while going over a bump?

First we find the force constant:

$$k = \frac{F}{s} = \frac{mg}{s} = \frac{(100 \text{ kg})(9.8 \text{ m/s}^2)}{0.0070 \text{ m}}$$

$$= 1.40 \times 10^5 \text{ N/m}$$

$$T = 2\pi\sqrt{\frac{m}{k}}$$

$$= 2\pi\sqrt{\frac{1100 \text{ kg}}{1.40 \times 10^5 \text{ N/m}}} = 0.557 \text{ s}$$

---

* It is the mass (inertia), not the weight, that is important here.

The frequency of vertical vibration is the reciprocal of the period:

$$f = \frac{1}{T} = \frac{1}{0.557 \text{ s}} = \boxed{1.80 \text{ s}^{-1}}$$

The car body bounces up and down 1.80 times per second.

Actually, cars are built with shock absorbers that provide frictional damping, and the vibration may last only one cycle or less.

Example 9-6

A spring having force constant 250 N/m is loaded with a mass of 10 kg. (*a*) Find the period of vibration. (*b*) If the mass is displaced 30 cm and then released, find the velocity with which it passes through the equilibrium position. (*c*) What is the total energy of the vibrating mass? (*d*) How much time is required for the mass to move the 30 cm from the endpoint of the motion to the equilibrium position?

(*a*)
$$T = 2\pi\sqrt{\frac{m}{k}}$$

$$= 2\pi\sqrt{\frac{10 \text{ kg}}{250 \text{ N/m}}} = \boxed{1.256 \text{ s}}$$

(*b*)
$$v_0 = \frac{2\pi A}{T}$$

$$= \frac{2\pi(0.30 \text{ m})}{1.256 \text{ s}} = \boxed{1.50 \text{ m/s}}$$

(*c*) The total energy of the vibrating mass is the same at any point of the motion. There are two points at which calculation of this energy is simple. First, as the mass passes through the equilibrium position, the energy is all KE:

$$\text{Maximum KE} = \tfrac{1}{2}mv_0^2$$

$$= \tfrac{1}{2}(10 \text{ kg})(1.50 \text{ m/s})^2$$

$$= \boxed{11.3 \text{ J}}$$

Second, at the endpoints, the energy is all PE:

$$\text{Maximum PE} = \tfrac{1}{2}kA^2$$

$$= \tfrac{1}{2}(250 \text{ N/m})(0.30 \text{ m})^2$$

$$= 11.3 \text{ N·m} = \boxed{11.3 \text{ J}}$$

(d) The time required to move the 30 cm is $\frac{1}{4}$ of the complete period, since the reference particle has moved through 90°, or $\frac{1}{4}$ of a complete circle. Hence,

$$t = \tfrac{1}{4}T = \tfrac{1}{4}(1.256 \text{ s}) = \boxed{0.314 \text{ s}}$$

It would be incorrect to try to use the laws of uniformly accelerated motion (such as $s = \frac{1}{2}at^2$) to find the time, since acceleration is *not* constant. The very idea of SHM is one of *variable* acceleration, according to the proportion $a \propto s$. *The equations of uniformly accelerated motion cannot be used for SHM problems.*

An equation for the displacement y of a body in SHM can be derived using the reference circle. From Fig. 9-15 we see that $y = A \cos \theta$, where A, the radius of the reference circle, is the amplitude of vibration. The phase angle θ, which is the steadily increasing angular displacement of the reference particle, is given by $\theta = \omega t$, where t is the time. The reference particle sweeps out 2π radians in one period T, so ω, in rad/s, is given by $2\pi/T$ or by $2\pi f$. Therefore, the equation $y = A \cos \theta$ can be written as

$$y = A \cos \frac{2\pi t}{T} = A \cos 2\pi f t \qquad (9\text{-}5)$$

In Eq. 9-5, it is assumed that time is measured from a point of maximum displacement (if $t = 0$, $\cos 0° = 1$, and $y = +A$). The significance of the period T in the denominator is as follows: imagine that t increases from 0 to T; then $2\pi t/T$ increases from 0 to 2π rad (360°), and the cosine function goes through one complete cycle of its variation.

Several equivalent equations for SHM can be written down, depending on the initial conditions. For instance, if we choose $t = 0$ to be an instant when the vibrating particle is at the center of its vibration, moving upward, a suitable equation would be

$$y = A \sin \frac{2\pi t}{T} = A \sin 2\pi f t \qquad (9\text{-}6)$$

since this makes $y = 0$ when $t = 0$. Also, the motion could be horizontal along the x axis, as in the following example.

Example 9-7

A particle is vibrating horizontally, to and fro along the x axis, making 30 vib/min. The total extent of its travel, from one extreme to the other, is 2.80 m. At $t = 0$, the particle is released from a point that is farthest to the left of center. (a) Write an equation for the motion. (b) What is the displacement of the particle 1.00 s after it is released?

(a) The amplitude is 1.40 m. The period T is 2 s/vib, and therefore $2\pi/T = (2\pi \text{ rad})/(2 \text{ s}) = \pi \text{ rad/s}$.

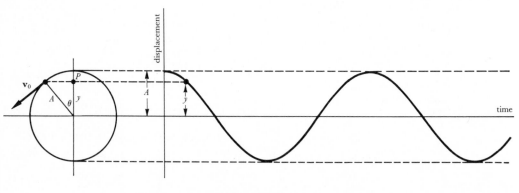

Figure 9-15

Since the particle has a maximum negative displacement at $t = 0$, we choose a negative cosine function. The motion is given by

$$x = -1.40 \cos(\pi t)$$

(b) At $t = 1.00$ s,

$$x = -1.40 \cos[(\pi \text{ rad/s})(1.00 \text{ s})]$$
$$= -1.40 \cos(\pi \text{ rad}) = -1.40 \cos 180°$$
$$= -1.40(-1) = \boxed{+1.40 \text{ m}}$$

In general, we say that both the sine function and the cosine function are "sinusoidal"—their graphs are identical, except for a phase difference of $\frac{1}{4}$ cycle. Thus we conclude that a body that is acted on by a Hooke's law force moves in SHM, with its displacement a sinusoidal function of time.

In Sec. 5-3 (page 110) we saw how the law of conservation of momentum could be used to measure the ratio of two masses by measuring the various velocities appearing in Eq. 5-3. The equation for the period of SHM gives another way of comparing masses without making use of gravitational forces. One form of "mass balance" is shown in Fig. 9-16. The unknown mass is placed on the platform, which vibrates from side to side with a period determined by the stiffness of the flat springs and the inertia of the system. Weight has nothing to do with this balance, which would work equally well in interstellar space.

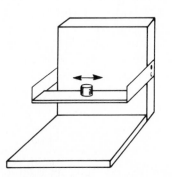

Figure 9-16   A mass balance.

Example 9-8

When a standard mass of 1.00 kg is placed on the platform of a mass balance, the vibration rate is 125 vib/min. What is the mass of an unknown object for which the vibration rate is 243 vib/min? Neglect the mass of the moving platform.

$$T = 2\pi \sqrt{\frac{m}{k}}$$

Squaring both sides gives

$$T^2 = \frac{4\pi^2 m}{k}$$

$$T^2 = \text{constant} \times m$$

since k is constant for any particular spring system. Hence,

$$\frac{T_1{}^2}{T_2{}^2} = \frac{m_1}{m_2}$$

Then, since the frequencies f are the reciprocals of the periods T,

$$\frac{f_2{}^2}{f_1{}^2} = \frac{m_1}{m_2}$$

This equation can now be used to find the unknown mass:

$$\frac{(243 \text{ vib/min})^2}{(125 \text{ vib/min})^2} = \frac{1.00 \text{ kg}}{m}$$

$$m = \boxed{0.265 \text{ kg}}$$

In actual practice, the mass of the supporting platform would not be negligible. However, the mass of the platform can be found by timing the vibration with two known masses.

Example 9-9

When a standard mass of 1.00 kg is placed on the platform of a mass balance, the frequency of vibration is 120 vib/min. When two standard 1.00 kg masses are placed side by side on the platform, the rate is 90 vib/min. What is the mass of the platform?

We call the unknown mass of the platform m'.

$$\frac{(120 \text{ vib/min})^2}{(90 \text{ vib/min})^2} = \frac{m' + 2.00 \text{ kg}}{m' + 1.00 \text{ kg}}$$

$$m' = \boxed{0.286 \text{ kg}}$$

Knowing the mass of the platform, we could now proceed to measure the mass of any unknown object, as in the previous example.

We have now found two ways of measuring (or defining) inertial mass. Are we really defining the same thing? The momentum method is based on the law of conservation of momentum, and the SHM method just discussed is based on the energy principle. Ultimately, both the conservation of momentum and the energy principle are derived from Newton's laws of motion, as we have shown in earlier chapters, and so the two methods necessarily give identical results for the measurement of mass. Our conviction that this is so is a measure of our faith in the general validity of Newton's laws. The two ways of measuring inertial mass are entirely equivalent to each other.

9-4 The Simple Pendulum

When a small, heavy pendulum bob is swinging at the end of a light rod or string, the pendulum is said to be "simple," since we assume that all moving parts that have inertia are concentrated at one place, all moving parts have the same speed, and the restoring force acts at a single point. These assumptions are never quite true, since even the lightest string has some mass, even the smallest bob has some distribution of mass, and all parts of the pendulum cannot be moving at the same speed. A real or "physical" pendulum (see Sec. 9-7) can be studied by methods similar to those we are about to use for the idealized *simple pendulum*.

The bob of the pendulum moves from P to Q to P' and back again (Fig. 9-17) with many of the characteristics of simple harmonic motion: The

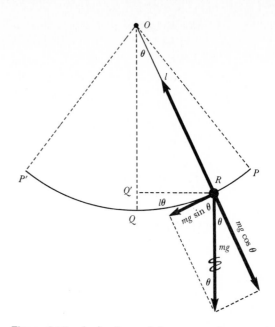

Figure 9-17 A simple pendulum moves in approximate SHM.

speed is maximum at Q; the acceleration is greatest at P and P' (when the velocity is momentarily zero, changing from $+$ to $-$ or from $-$ to $+$). To find out whether a pendulum does actually move in SHM, we ask the usual question: Is the restoring force proportional to the displacement? In other words, *is* there a force constant?

The outward component of the weight, $mg \cos \theta$, is more than balanced by the tension in the string; there is a net force along RO to supply the necessary centripetal force on the mass. The component of the weight tangent to the arc is also unbalanced, and this unbalanced force (equal to $mg \sin \theta$) is the restoring force directed along the arc. The displacement is the length of arc, $l\theta$. If there is a force constant, then it is defined by

$$k = \frac{\text{restoring force}}{\text{displacement}} = \frac{mg \sin \theta}{l\theta} \qquad (9\text{-}7)$$

This is not quite constant, since $\sin \theta$ is not the same as θ. In the triangle ROQ', $\sin \theta = $ (chord

$RQ')/l$, whereas the angle θ (in radians) is (arc $RQ)/l$. As the angle θ decreases, however, the arc and the chord become practically equal, and the ratio $(\sin\theta)/\theta$ approaches 1. Hence, if the angular displacement of a simple pendulum is small, the force constant is

$$k = \frac{F}{s} = \frac{mg}{l}\left(\frac{\sin\theta}{\theta}\right) \approx \frac{mg}{l} \qquad (9\text{-}8)$$

and to the extent to which k is a constant, the motion is SHM. The period of the pendulum is

$$T = 2\pi\sqrt{\frac{m}{k}} \approx 2\pi\sqrt{\frac{m}{\frac{mg}{l}}} \qquad (9\text{-}9)$$

The mass m cancels out, and

$$T \approx 2\pi\sqrt{\frac{l}{g}} \qquad (9\text{-}10)$$

It is important to realize that this simple formula for the period of a simple pendulum is an approximation, but a very good one if the angle of swing is small.* It is also important to note that the symbol m canceled out, because weight (mg) is proportional to mass (m). This being so, the period of a pendulum is independent of the mass of the bob, depending only on the length of the string.† On the other hand, since g appears in the formula, we have a means of determining the value of the acceleration due to gravity at any place.

_____ **Example 9-10**

What is the acceleration due to gravity on a planet where a space explorer's simple pendulum 40.0 cm long vibrates 100 times in 240 s?

_____
* The error is 0.1% if the angle θ_0 ($\angle POQ$) is 7°, and only 1% if θ_0 is as large as 23°.
† This property of the period of a pendulum was used by Bessel in his investigation of the proportionality of inertial mass and gravitational mass referred to on page 180.

The period is $(240 \text{ s})/100 = 2.40$ s.

$$T \approx 2\pi\sqrt{\frac{l}{g}}$$

$$T^2 = \frac{4\pi^2 l}{g}$$

$$g = \frac{4\pi^2 l}{T^2} = \frac{4\pi^2(40.0 \text{ cm})}{(2.40 \text{ s})^2} = \boxed{274 \text{ cm/s}^2}$$

Note that $T \approx 2\pi\sqrt{l/g}$ is a special formula, valid for a special case (a simple pendulum vibrating with not too large an amplitude), whereas the similar-appearing formula $T = 2\pi\sqrt{m/k}$ is a general formula for the period of *any* simple harmonic motion.

9-5 Angular Simple Harmonic Motion

The ideas we have developed may be carried over into the study of angular SHM by using the analogies discussed in Sec. 7-7 (page 163). We need only substitute *angular displacement* θ for *linear displacement*, *moment of inertia* I for *mass*, and *torque* τ for *force*. Instead of force constant $k = F/s$, we use torsion constant $k' = \tau/\theta$, where τ is the torque required to cause a rotation of θ radians. By analogy with the linear formula $T = 2\pi\sqrt{\dfrac{m}{F/s}}$, the period of oscillation is given by

$$T = 2\pi\sqrt{\frac{I}{\tau/\theta}} = 2\pi\sqrt{\frac{I}{k'}} \qquad (9\text{-}11)$$

_____ **Example 9-11**

A flat disk of radius 20 cm and mass 4 kg is supported by a thin vertical rod for which the torsion constant τ/θ is 0.9 N·m/rad (Fig. 9-18). The disk is twisted a few degrees in its own plane, then released. What is the period of torsional oscillation?

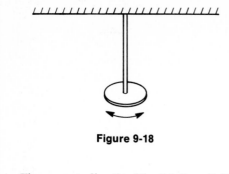

Figure 9-18

The moment of inertia of the disk (from Table 7-1) is

$$I = \tfrac{1}{2}MR^2 = \tfrac{1}{2}(4 \text{ kg})(0.20 \text{ m})^2 = 0.08 \text{ kg}\cdot\text{m}^2$$

$$T = 2\pi\sqrt{\frac{I}{\tau/\theta}} = 2\pi\sqrt{\frac{0.08 \text{ kg}\cdot\text{m}^2}{0.9 \text{ N}\cdot\text{m}}} = \boxed{1.87 \text{ s}}$$

Unless the elastic limit of the supporting rod or wire is exceeded, the motion is true SHM for any angular amplitude. The moving system of an anniversary clock (which is wound once a year) has two or four masses suspended from a thin, flat wire (see the photograph on page 192). The amplitude of angular vibration is several complete revolutions, and the period is, say, 15 s, independent of amplitude. The period is controlled by adjusting the positions of the masses, thus varying I. The torsion constant τ/θ depends

on the length and diameter of the wire and the shear modulus of the material of the wire.

9-6 Non-simple Harmonic Motion

The relative ease with which our study of SHM gave us useful formulas is impressive. We are also impressed by the ubiquitous occurence of SHM, depending as it does only on the ever-present elastic behavior of bodies as expressed by Hooke's law. However, not all vibrations are "simple," since the acceleration need not be proportional to the displacement as in SHM. Consider a ball bearing bouncing up and down on a highly elastic steel plate, always returning to the same height from which it fell. This vibration is less simple than that of a mass on the end of a spring, because the ball bearing's velocity changes abruptly (during impact) rather than smoothly. The two graphs in Fig. 9-19 compare SHM with the motion of a bouncing ball. The force acting in each case is represented by a dotted line. Note that for the bouncing ball, the force is a constant downward (negative) force equal to the weight of the ball, except for a very short fraction of each period, say 0.001 s, during which the ball is in contact with the base plate. During contact, the ball is acted on by a large upward (positive) force. Point P represents the

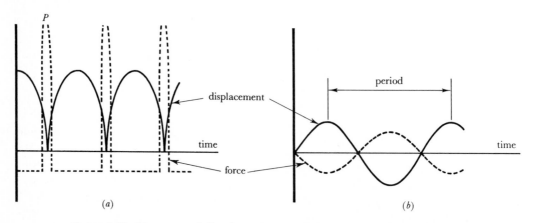

Figure 9-19 Two types of vibration. (*a*) Bouncing ball. (*b*) Mass on a spring.

instant of maximum distortion, corresponding to maximum upward force. Contrast this complex behavior of the bouncing ball with that of the mass on a spring. In SHM the displacement changes smoothly, and the force (and acceleration) are also free from sudden wild variations. In SHM the force is always proportional to the displacement, but oppositely directed. Both force and displacement are zero at the same time, and the restoring force is negative when the displacement is positive, and vice versa.

Other vibrations that are not SHM occur because of the difference between static and kinetic coefficients of friction (Sec. 4-4). When a piece of chalk squeals as it is drawn across the blackboard, the vibration is due to the "stick-slip" phenomenon. The chalk sticks because of static friction until the applied force causes it to slip. Since kinetic friction, at the start, is less than static friction, the chalk's speed increases rapidly. Now the dependence of friction on speed comes into play. As the speed increases, eventually the chalk "sticks" and the process starts all over again. The frequency of this process depends in a complicated way on the properties of matter. The bouncing ball, the squeal of a car's brakes, the passage of a violin bow across the string, even the vibrations of human vocal cords—all illustrate so-called *relaxation vibrations* in which motions change abruptly.

Summary   Elastic behavior in all kinds of distortion is described by Hooke's law, which states that stress is proportional to strain. Hooke's law is not valid above the elastic limit. Stretch modulus, shear modulus, and bulk modulus are names for the ratio of stress to strain in special kinds of distortion. For liquids and gases the only modulus that is not zero is the bulk modulus.

Simple harmonic motion occurs whenever the acceleration of a body is proportional to the displacement of the body from its equilibrium position. A Hooke's law restoring force gives rise to such an acceleration, and SHM occurs widely in nature. The ratio of restoring force to the displacement is the force constant: $k = F/s$. The period T of a SHM is the time for one complete vibration; the frequency f is given by $f = 1/T$.

The projection of uniform circular motion on any straight line is SHM. The radius of the reference circle equals the amplitude A of the SHM. The speed v_0 of the reference particle equals the maximum velocity of the SHM particle as it passes through center: $v_0 = 2\pi A/T$. In SHM, the displacement of a vibrating particle is a sinusoidal function of time.

During a vibration, KE is continually being transformed to PE (usually elastic), and vice versa. The motion eventually dies out as the (KE + PE) total is transformed into internal energy of the system.

The period of any SHM is given by $T = 2\pi\sqrt{m/k}$. As a special case, the motion of a simple pendulum turns out to be approximately SHM, if the amplitude of swing is not too great. The simple pendulum's period is given by $T \approx 2\pi\sqrt{l/g}$, an equation that can be used to measure the acceleration due to gravity.

In relaxation vibrations the restoring force varies abruptly. No general formula exists to give the period of all types of relaxation vibrations. One illustration is furnished by a bouncing ball; other relaxation vibrations are a result of the "stick-slip" phenomenon of friction.

Check List   Hooke's law   harmonic motion   torsion constant
                   stress   simple harmonic motion   relaxation vibration
                   strain   force constant $k = F/s$
                   modulus of elasticity   period
                   elastic limit   frequency $v_0 = \dfrac{2\pi A}{T}$
                   ultimate tensile strength   displacement
                   stretch modulus   amplitude $T = 2\pi\sqrt{m/k}$
                   Young's modulus   reference circle
                   shear modulus   phase angle $T \approx 2\pi\sqrt{l/g}$
                   bulk modulus   simple pendulum $T = 2\pi\sqrt{I/k'}$

Questions   **9-1**   Does the springiness of a telephone dial obey Hooke's law? Make a rough trial yourself, and try to tell whether it takes more force to hold the dial in the "0" position (all the way around) than to hold it with only a small displacement, such as when dialing "1" or "2."

9-2   Does a roller window shade obey Hooke's law?

9-3   Which of the three elastic moduli (stretch, shear, or bulk) is of chief importance in each of the following situations?   (*a*) A child stands on a long board that sags, with the ends remaining "square." Which modulus of wood is involved?   (*b*) A helical spring made of steel, such as those shown in Fig. 9-8, is deformed by pulling on one end; each loop spreads slightly. Which modulus of the steel is involved?   (*c*) A bicyclist pumps up his tire. He builds up a pressure by pressing down the handle of the pump, which forces the piston of the pump down. Which modulus of the air is involved?   (*d*) A movie star gets a permanent wave. Which modulus of the hair is involved?   (*e*) A worker uses a screwdriver to tighten a screw, and she twists the handle of the screwdriver while the blade of the tool is firmly seated in the slot of the screw. Which modulus of the shaft material of the screwdriver is involved?

9-4   In Example 9-2 the torque condition was applied to a group of small bones collectively called "the foot." Why was it permissible to ignore the weight of these foot bones in the calculation?

9-5   When a body is vibrating in linear SHM, is its acceleration zero at any point in the motion?

9-6   An airplane loops the loop in a vertical circle, picking up speed as it approaches the bottom of the loop and losing speed as it approaches the top of the loop. Does the shadow of the plane move across the ground in SHM?

9-7   The presence of oil beneath the earth is often indicated by a salt dome over the oil-bearing rock. The salt dome has a smaller specific gravity than the surroundings. Explain how a geophysicist on the surface can determine the probable location of oil by observing the stretch of a delicate spring on which a mass hangs.

9-8   A mass on the end of a string is swinging back and forth as a pendulum. Is it in equilibrium at any point of the motion?

9-9   A pendulum clock keeps perfect time when in Los Angeles. Will it gain or lose when it is taken to a summer resort in the mountains near the city?

9-10   In what respect does the motion of a pendulum fail to be exactly SHM? That is, does it pick up speed too quickly or too slowly just after being released from an elevated position? (*Hint:* Is $\sin\theta$ greater or less than θ?)

9-11 Give an example of SHM other than those mentioned in the text. Give an example of relaxation vibration other than those mentioned in the text.

9-12 Which of the following units is not suitable for an elastic modulus?   (a) dyne/cm^2; (b) kg/mm^2;   (c) N/m^2.

9-13 A wire stretches a certain amount under load. If the wire's length and diameter are both doubled, the stretch caused by the same load is   (a) greater;   (b) the same;   (c) less.

9-14 The dimensions of force constant are   (a) $[MT^{-2}]$;   (b) $[MLT^{-1}]$;   (c) $[MLT^{-2}]$.

9-15 At an instant when a body vibrating in linear SHM is farthest from the midpoint of its motion,   (a) its acceleration is a maximum;   (b) its velocity is zero;   (c) both of these.

9-16 The PE of a stretched spring is proportional to   (a) the square of the force constant; (b) the square of the amount of stretch;   (c) both of these.

9-17 A mass m on a string of length L is swinging as a pendulum. If the mass is made $\frac{1}{2}m$ and the length made $2L$, the period becomes   (a) greater;   (b) the same;   (c) less.

Problems

Note: Recall that the SI unit pascal (Pa) is the same as the N/m^2.

9-A1 What is the stress in a steel wire that is 5 m long and 0.04 cm^2 in cross section if the wire bears a load of 20 kg? Express your answer in pascals.

9-A2 A rubber band originally 40 cm long is stretched to a length of 42 cm by a certain load. What is the strain?

9-A3 What is the strain in a wire cable of original length 40 m whose length increases by 4 cm when a load is lifted?

9-A4 A piece cut from a bicycle inner tube is 50 cm long when it carries a load weighing 20 N. The force constant is 200 N/m. What will be the approximate length of the piece of rubber when the load is 30 N?

9-A5 The pan of a postal scale goes down 15 mm when a 0.28-N (1-oz) letter is weighed. What is the force constant of the spring, in N/m?

9-A6 Two identical springs, each of length L and force constant k, are connected together to form a new spring of length $2L$. What is the force constant of the new spring?

9-A7 A pendulum has a period of 1.25 s   (a) What is the frequency?   (b) How many vibrations does it make in 50 s?

9-A8 A vibrating reed in a harmonica makes 200 vib/s. What is the period?

9-A9 What is the time required for 100 vibrations of a loaded spring if the frequency is 4 vib/s?

9-A10 The piston of an engine moves a total distance of 9 cm from one extreme point to the other.   (a) What is the amplitude?   (b) What is the displacement when the piston is 4.5 cm from one end of its stroke?

9-A11 A pendulum is pulled aside +8 cm from its equilibrium position and allowed to vibrate.   (a) What is the amplitude?   (b) How far is the pendulum from its extreme position when the displacement is 0 cm?

9-A12 What are the dimensions of   (a) period?   (b) frequency?   (c) force constant?

9-A13 Compute the time for one vibration of a spring whose force constant is 4 N/m if the load has a mass of 1 kg.

9-A14 Compute the length of a clock pendulum that ticks once each second. (*Hint:* It ticks twice during each complete vibration.)

9-A15 A sailor being rescued at sea dangles from a helicopter on the end of a line 30 m long. What is his period of vibration?

9-B1 A crate weighing 8000 N is hoisted by a ship's derrick that has a wire cable 10 m long. The cable consists of 40 strands of steel, each 5×10^{-6} m^2 in cross-sectional area. By how much does the cable stretch?

9-B2 An experimenter studying the stretch modulus of human bone reported in a journal that a load 1600 N produced a stretch of 0.5 mm in a sample of femur 15 cm long. Can you deduce from these data the cross-sectional area of the bone specimen? (See Table 9-1.)

9-B3 A piece of beef tendon 25.00 cm long and of cross section 7.8 mm × 13 mm stretched to a length 25.19 cm when it supported vertically a load of 0.96 kg. Calculate the stretch modulus for the tendon material.

9-B4 A steel elevator cable 30 m long must supply an upward tension of 15,000 N. If the permissible stretch of the cable is 5 mm, what must be the total cross-sectional area of all the strands of the cable? Express your answer in cm^2.

9-B5 A wire made of the steel described on page 195 is stretched until it breaks. What is the strain in the wire just before it breaks?

9-B6 Show that it is impossible to give a 1% stretch to a wire made of the steel described on page 195.

9-B7 A uniform pole of weight 4000 N is lying horizontally on the ground, pivoted at one end. A nylon rope of length 20 m and diameter 1.6 cm is used to apply a vertical force at the free end of the pole. Young's modulus for nylon is about 4×10^8 Pa. How much does the rope stretch before the pole starts to move?

9-B8 A truck of mass 10 metric tons on a lift in a service station is supported by a vertical steel column that is a solid cylinder 3 m long and 10 cm in radius. By how much is the length of the column changed by the load?

9-B9 A 10-kg object is whirled in a horizontal circle on the end of a wire. The wire is 0.75 m long, has cross-sectional area 10^{-6} m^2, and is made of a material whose ultimate tensile strength is 4.8×10^8 Pa. Calculate the greatest angular speed the object can have.

9-B10 A pendulum consists of a 10-kg mass hanging from a steel wire of unstretched length 2 m and cross-sectional area 2 mm^2. (*a*) What is the stretch of the wire when the load is hanging motionless at its lowest position? (*b*) What is the stretch when the load is swinging through the lowest point after having been released from an angular displacement of 60°?

9-B11 A camper is being pulled by a car (Fig. 9-20), and the coupling is required to transmit horizontal forces up to 7000 N. The pin is 3 cm in diameter, 10 cm long, and is made of steel for which the maximum shearing stress is 8×10^7 Pa. Is this arrangement safe? (*Note:* A safety factor of 5 is generally allowed by design engineers.)

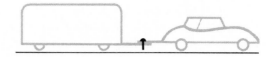

Figure 9-20

9-B12 A rubber eraser 2.5 cm × 1.0 cm × 8.0 cm is clamped at one end, with the 8-cm edge vertical. A horizontal force of 1.2 N is applied at the free end, perpendicular to the 1-cm edge, and the top of the eraser is displaced 2.4 mm horizontally. Calculate the shearing stress, the shearing strain, and the shear modulus.

9-B13 A jug contains 4000.0 cm^3 of kerosene at atmospheric pressure (101 kPa). When the cork is pushed in slightly, the pressure on the liquid increases to 690 kPa, and the volume of the liquid is 3998.5 cm^3. Calculate the bulk modulus of kerosene.

9-B14 What increase in pressure is needed to cause the volume of a block of aluminum to decrease by 1%?

9-B15 A cast-iron anchor of mass 1460 kg is heaved overboard and settles to the bottom of the ocean, where the pressure is 1.8×10^6 Pa greater than at the surface. Calculate the original volume of the anchor, and find the change in volume, in cm^3, caused by the increased pressure.

9-B16 Artificial diamond crystals have been made by subjecting carbon in the form of graphite to a pressure of 1.5×10^{10} Pa at a high temperature. Assuming that natural diamonds were formed at similar high pressures within the earth and that the bulk modulus and density at high temperature are roughly those given in Tables 9-1 and 1-3, what must have been the original volume of the fabulous Kohinoor diamond, whose mass before cutting was about 160 g?

9-B17 An object of mass 2 kg is moving with SHM of frequency 120 vib/min and amplitude 0.8 m. What is the restoring force when the object is at one of the endpoints of its motion?

9-B18 Assume that the piston of Prob. 9-A10 moves in SHM and makes 3 vibs/s. How fast is it moving when passing through the midpoint of its path?

9-B19 A point on a horizontal vibrating piano string is moving up and down in SHM with a frequency of 150 vib/s. The maximum excursion of the point below its normal position is 1.0 mm. (*a*) What is the maximum velocity of the particle? (*b*) What is the maximum acceleration of the particle?

9-B20 A load of mass 1.20 kg hangs from a long, light spring. When pulled down 0.16 m below its equilibrium position and released, it vibrates with a period of 2 s. (*a*) What is its velocity as it passes through the equilibrium position? (*b*) What is the force constant? (*c*) By how much will the spring shorten if the load is allowed to come to rest and is then removed?

9-B21 A load of mass 800 g is hanging from a light spring whose force constant is 20 N/m. The load is pulled down 10 cm from its equilibrium position and released. (*a*) How long is required for the load to reach the equilibrium position again? (*b*) What is then its velocity?

9-B22 A raft of mass 400 kg is floating in a pond. When an 80-kg swimmer climbs on board, the raft sinks 10 cm deeper into the water to a new equilibrium position. If the swimmer jumps off, how many vertical vibrations does the empty raft make in 5 s?

9-B23 A grocer weighs 2 kg of bananas on a scale whose platform has mass 1 kg. The system has a force constant of 30 N/m. What is the period of vibration while the scale is coming to its final reading?

9-B24 Write an equation for the vertical motion of a body of mass 4 kg on a spring of force constant 100 N/m if the motion has amplitude 3 m and the body passes downward through its equilibrium position at $t = 0$.

9-B25 In Example 9-6, what is the KE of the vibrating mass at an instant when its PE is 8.2 J?

9-B26 How much work is done to stretch a spring whose force constant is 800 N/m from an original unstretched length of 35 cm to a length of 40 cm?

9-B27 What energy must be given to a 10-kg mass suspended from a spring so that it will oscillate in SHM with a period of 0.5 s and amplitude 2 cm?

9-B28 A 4-kg mass is vibrating horizontally in SHM of amplitude 2 m with a period of 3.14 s. (*a*) Calculate the force constant. (*b*) Calculate the PE of the mass when it is at an extreme position. (*c*) Check your answer to part (*b*) by calculating the KE (from $\frac{1}{2}mv^2$) as the mass passes through its equilibrium position.

9-B29 A slingshot consists of a light leather cup containing a stone that is pulled back against the elastic force of two rubber bands. It takes a force of 30 N to stretch the bands 1 cm. (*a*) What is the PE of a 0.050-kg stone that is pulled back 20 cm from the equilibrium position? (*b*) With what speed does the stone leave the slingshot?

9-B30 In using the mass balance of Fig. 9-16, an experimenter finds the following data for the time required for 10 vibrations: with 20-g load, 4 s; with 120-g load, 6 s; with unknown load, 18 s. What is the mass of the unknown load? (*Hint:* First find the mass of the platform.)

9-B31 The platform of a mass balance (Fig. 9-16) has negligible mass. On earth, the period of vibration is 0.1 s when a 1-kg load is on the platform. What would be the period if the apparatus were in a space lab one earth-radius above the earth's surface, with an astronaut of mass 64 kg on the platform?

9-B32 A small computer placed on board a satellite must withstand a maximum acceleration of $10g$. To test this, the computer is attached to a horizontal table that is driven from side to side in SHM at 25 vib/s. What must be the amplitude of vibration of this shake-test apparatus in order to give a maximum acceleration of $10g$?

9-B33 A package is on a platform that vibrates in vertical SHM with period 0.3 s. (*a*) At what point in the motion is the package most likely to lose contact with the platform? (*b*) What is the maximum amplitude of SHM for which the package always remains in contact with the platform?

9-B34 Derive a formula for the maximum KE of a particle in SHM, in terms of the mass m, the amplitude A, and the period T. Check your answer for dimensional consistency.

9-B35 The value of gravitational acceleration on the moon is about $\frac{1}{6}$ that on the earth. What is the period on the moon of a simple pendulum whose period on the earth is 1.5 s?

9-B36 A perfectly elastic steel ball bearing is dropped from a height of 19.6 cm and bounces up and down on a steel plate. What is the frequency of this relaxation oscillation?

9-C1 A steel elevator cable must carry a total load of 1000 kg and accelerate it upward at 2.20 m/s². Taking a safe working stress for steel to be 1.0×10^8 Pa, calculate the necessary diameter of the cable. Express your answer in cm.

9-C2 A steel wire 25 m long has a cross sectional area of 0.5 mm². Young's modulus for steel is 20×10^{10} Pa. (*a*) Compute the force constant of the wire. (*b*) If a 10-kg load is hung on the wire and is then pulled down a short distance below its equilibrium position, what will be the frequency of vertical vibration when the load is released?

9-C3 A wire of length L and radius r is made of a material whose stretch modulus is E. Derive a formula for the force constant k of this wire, in terms of L, r, and E. Make a dimensional check for your formula.

9-C4 An object of mass m is supported by a wire of length L and radius r. When set into small vertical oscillation, the period is T. (*a*) Derive a formula for the stretch modulus of the wire, in terms of m, L, r, and T. (*b*) Make a dimensional check for your formula.

9-C5 A 10-kg load suspended by a brass wire 10 m long is observed to vibrate vertically in SHM at a frequency of 10 vib/s. What is the cross-sectional area of the wire?

9-C6 A particle vibrates in SHM with period 3 s and amplitude 10 cm. (*a*) How long is required for the particle to move from one end of its path to the nearest point 5 cm from the equilibrium position? (*b*) What is the particle's speed at this point?

9-C7 A 2-kg mass is in static equilibrium, hanging at the bottom end of a long, light, vertical spring. When pulled down 0.30 m below its equilibrium position and released, it vibrates with a period of 2 s. What force does the spring exert on the mass when the mass is (*a*) at its lowest point; (*b*) passing through the midpoint of its motion; (*c*) at its highest point?

9-C8 An ancient military harbor was protected by an underwater reef that was 4.00 m out of water at low tide and 1.20 m under water at high tide. Friendly boats whose keels were 1.00 m under water crossed the reef during the "clear" intervals of time when the water was deep enough. What was the duration of each clear interval? Assume the tides to be SHM, of period 12.5 h.

9-C9 A steel ball *A* of mass 3*m* supported on a light cord 3 m long is held out at a 60° angle from the vertical, as shown in Fig. 9-21. A second steel ball *B* of mass *m* rests just at the edge of the table, which is 1 m high. Ball *A* is released and strikes ball *B* head-on in a perfectly elastic collision. (*a*) Approximately how long after *A* is released does *B* strike the floor? (*b*) How fast is *A* moving when it strikes *B*? (*c*) How far out from the base of the table does *B* strike?

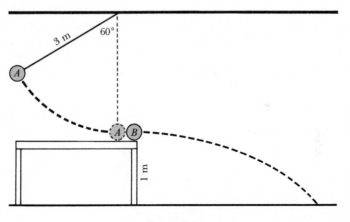

Figure 9-21

9-C10 A light spring is hanging loosely from the ceiling. A load is attached to the free end of the spring, and it is observed that the load, when released, moves downward a distance of 50 cm before starting to rise again. What is the period of this SHM? (*Hint:* The mass is not given. Assume it to be *m*, and hope that is cancels out of your final equation for the period.)

9-C11 The moving system of an anniversary clock (pages 192 and 208) may be approximated by four small balls, each of mass 250 g, with the center of each ball 10 cm from the vertical axis of rotation. The stiffness of the wire is such that a torque of 2000 cm·dyn causes the moving system to twist through an angle of 0.4 rad. What is the time required for one complete angular vibration of the system of balls?

9-C12 A flat disk 15 cm in radius, of mass 8 kg, is suspended in a horizontal plane by a vertical wire attached to its c.g. (Fig. 9-18). It is found that a force of 4 N, applied tangentially at the circumference of the disk, rotates it through 45°. Calculate   (a) the torsion constant in N·m/rad, and   (b) the time for the disk to make 40 torsional vibrations.

9-C13 (a) Estimate the PE delivered by a practitioner of karate whose hand comes down on the center of a piece of wood (supported at both ends), causing the board to sag at the middle. The top layers of the distorted board are compressed and the bottom layers are stretched. Assume the stored PE, just before the break, equals the work required to cause an average length change of 1 mm in a piece of wood 40 cm long and 2 cm × 8 cm in cross section. A typical value of Young's modulus for wood, along the grain, is 3×10^9 Pa. (*Hint:* First find the average force constant.) (b) Is your answer consistent with the statement that a karate blow can deliver several kilowatts of power over several milliseconds (see Ref. 8 on page 150 of Chap. 6)?   (c) The same reference states that the hand has a peak velocity of 12 to 14 m/s and can deliver a force of over 3000 N. Is this a reasonable value of the force for a hand of mass 0.5 kg that moves at peak velocity just before a blow that has a duration of 2 ms?

For Further Study

9-7   The Physical Pendulum

We have already considered the simple pendulum in Sec. 9-4. No real pendulum is "simple," of course, for the entire bob cannot be at one point but must be spread out over a finite volume. Such a real pendulum is called a *physical pendulum*. To develop the theory of the physical pendulum, we follow almost exactly the derivation made in Sec. 9-4 for the simple pendulum. It is only necessary to deal with angular SHM instead of linear SHM (Sec. 9-5).*

Consider a body of any shape supported on an axis that is at a distance h from the center of gravity (Fig. 9-22). If the body is given an angular displacement θ away from its equilibrium position, there is an unbalanced torque equal to the force times the lever arm. The weight acts at a lever arm $h \sin \theta$, so that the restoring torque is $\tau = mg(h \sin \theta)$. The torsion constant τ/θ is

$$\frac{\tau}{\theta} = \frac{mgh \sin \theta}{\theta} \qquad (9\text{-}7')$$

which is not quite constant since $\sin \theta$ is not the same as θ. As the angle θ decreases, the ratio $(\sin \theta)/\theta$ approaches 1. Hence, if the angular displacement of the physical pendulum is small, the torsion constant is

$$\frac{\tau}{\theta} = mgh\left(\frac{\sin \theta}{\theta}\right) \approx mgh \qquad (9\text{-}8')$$

and to the extent to which τ/θ is a constant, the motion is SHM. The period of the pendulum is

$$T = 2\pi \sqrt{\frac{I}{\tau/\theta}} \approx 2\pi \sqrt{\frac{I}{mgh}} \qquad (9\text{-}10')$$

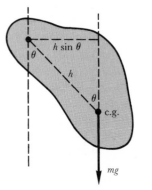

Figure 9-22   A physical pendulum.

---

* It is instructive to compare the equations we are about to derive with the similarly numbered equations in Sec. 9-4 for the *linear* SHM of the simple pendulum.

This approximate formula has the same limitations as the corresponding formula $T \approx 2\pi\sqrt{l/g}$ for a simple pendulum. It is a very good approximation if the angle of swing is small.

Although the period of a pendulum depends on the amplitude of swing, this does not affect the accuracy of a clock. It is the function of the mainspring, or of falling weights, to supply energy to the vibrating system to compensate for frictional losses. The amplitude of swing is thus held constant, and the period is constant, even though its value is slightly greater than that given by Eq. 9-10′.

9-8 Simple Harmonic Motion from the Viewpoint of Formal Calculus

In our discussion of SHM in Sec. 9-3, we used a graphical interpretation based on the reference circle. This is a valid approach. However, you can perhaps gain a greater insight into SHM through an analytical approach using some elementary calculus. We need to use the derivatives of the sine and cosine, and the chain rule (Appendix E on pages 838–840). We consider motion along the y axis; without using the reference circle, we will verify that a sinusoidal motion (such as shown in Fig. 9-15) is SHM, and derive the equation for the period.

We assume that the displacement of a particle is given by

$$y = A\cos\left(\frac{2\pi t}{T} - \phi\right) \qquad (9\text{-}12)$$

Compare Eq. 9-5; here we have inserted an arbitrary phase angle ϕ. In particular, if $\phi =$

$90° = \pi/2$ rad, $A\cos\left(\dfrac{2\pi t}{T} - 90°\right)$ is the same as $A\sin\dfrac{2\pi t}{T}$, which is the alternate form given in Eq. 9-6.

Now differentiate Eq. 9-12 twice with respect to time.

$$v = \frac{dy}{dt} = -A\frac{2\pi}{T}\sin\left(\frac{2\pi t}{T} - \phi\right)$$

$$a = \frac{dv}{dt} = \frac{d^2 y}{dt^2} = -A\frac{4\pi^2}{T^2}\cos\left(\frac{2\pi t}{T} - \phi\right)$$

or, in view of Eq. 9-12,

$$a = \frac{d^2 y}{dt^2} = -\frac{4\pi^2}{T^2}y \qquad (9\text{-}13)$$

But this is the definition of SHM: acceleration is proportional to displacement, and in the opposite direction. We see that Eq. 9-12 does indeed describe SHM. Next we multiply both sides of Eq. 9-13 by the mass m of the particle, and recall the definition of force constant: $k = |F|/|y|$.

$$F = ma = -m\frac{4\pi^2}{T^2}y$$

$$k = \frac{|F|}{|y|} = m\frac{4\pi^2}{T^2}$$

whence

$$T = 2\pi\sqrt{\frac{m}{k}}$$

We have derived Eq. 9-4, the formula for the period of a SHM.

Problems

9-C14 Show that the formula for the period of a physical pendulum contains, as a special case, the formula for the period of a simple pendulum. To show this, assume a *small* mass at the end of a *light* string of length l, and remember the definition of moment of inertia.

9-C15 What is the period of oscillation for a uniform meter stick suspended on a nail passing through a hole at one end? (Use Table 7-1 to find the moment of inertia of the stick.)

9-C16 A uniform rod of mass M and length L is supported by a horizontal axis passing through its center. A small object, also of mass M, is attached to the lower end of the rod. (a) Where is the c.g. of the system? (b) Show that the period of oscillation of this pendulum is given by $T = 2\pi\sqrt{2L/3g}$.

9-C17 Compute the period of oscillation of the loaded meter stick of Example 7-11 when the stick is suspended vertically about a horizontal axis passing through the 100-cm mark. (*Hint:* You must find the c.g.; the moment of inertia about the given axis has already been found in Example 7-11.)

9-C18 A particle moving in SHM along the x axis has some initial speed as it passes through $x = 0$ at time $t = 0$. The particle reaches $x = -2$ m at time $t = 3$ s, then turns around and heads for the origin. (a) Calculate the period. (b) Calculate the maximum velocity. (c) Write an equation for x as a function of t. (d) Write an equation for v as a function of t. (e) How much time is required for the particle to move from $x = -0.5$ m to $x = -1.0$ m?

9-C19 The motion of a particle is described by the equation $y = 2\cos(10t - 0.5)$, where y is in m, t is in s, and the numerical constants have appropriate units. The phase angle is in radians. (a) Calculate the period of vibration. (b) Calculate the displacement when $t = 0$. (c) Calculate the maximum speed of the particle. (d) Calculate the magnitude and direction of the velocity when $t = 0.12$ s. (e) Calculate the acceleration when $t = 0$. (f) What is the first time after $t = 0$ that the displacement is 0?

References

1. W. F. Magie, *A Source Book in Physics* (McGraw-Hill, New York, 1935), pp. 17–19. Galileo's discussion of the pendulum.
2. F. L. Friedman, *Velocity in Circular and Simple Harmonic Motion; Velocity and Acceleration in Simple Harmonic Motion* (films).
3. J. Stull, *Simple Harmonic Motion: The Stringless Pendulum* (film).

10

Wave Motion

We make contact with our surroundings through the five or more senses that we, like other sentient creatures, possess. One biology textbook* lists six major groups of specialized sense receptors—those sensitive to light, sound, touch and pressure, gravity and motion, taste and smell, and temperature. The function of a sense organ is to receive information about the outside world and transmit the information to the nervous system in order that the organism may respond in a coordinated fashion to the stimuli of its environment. From the point of view of physics, this process always involves transfer of energy. It is noteworthy that the two senses considered most useful to higher organisms such as man—sight and sound—require the transmission of energy from source to receptor through a region of space that may range from a few centimeters up to many kilometers. Another point of similarity between sight and sound is that neither is instantaneous; we never see or hear anything at the very moment that it occurs. Thus we seek to

describe and understand the flow of energy from one place to another at a finite speed.

A stream of machine-gun bullets or a blast of air molecules from a fan can carry energy; this *corpuscular* transfer of energy, called *convection*, will be discussed in Chap. 14. We turn our attention now to *wave motion*, the other important method by which energy moves from place to place.

10-1 Waves and Disturbances

In the future we shall frequently use the concept of *waves*. Electromagnetic waves include radio waves, infrared radiation, visible light, ultraviolet radiation, x rays, and gamma rays. All of these travel at a speed of 3.00×10^8 m/s (186,000 mi/s) in a vacuum. We all have had some first-hand experience with other types of waves—sound waves, water waves, and waves on stretched strings, as in a guitar or piano. Less familiar are earth waves (earthquakes), temperature waves (so-called heat waves or cold waves of popular meteorology), probability waves, and

---

* G. G. Simpson and W. S. Beck, *Life: An Introduction to Biology*, 2nd ed., Harcourt Brace Jovanovich, 1965, p. 362.

This surfer at Kauai, one of the Hawaiian Islands, is riding just ahead of the crest of an incoming wave.

even matter waves. Perhaps enough has been said to indicate the value of studying wave motion as such. What are the abstract properties of waves in general? What is common to the description of all these seemingly different types of waves? How is energy transmitted by a wave?

To most students the word "wave" conjures up a picture of water waves moving across a lake, or perhaps of a flag undulating in the breeze. These are *periodic waves*, in which motions are repeated at regular intervals. But a wave can also consist of only a single disturbance—never repeated at any given place. Such *pulses*, or *shock waves*, are in evidence when a high-speed bullet sets air molecules into rapid one-way motion, followed by a relatively slow diffusion of the molecules back to their original positions.

The key word is discussing waves is "disturbance." The reason there are so many types of waves is that there are so many ways of disturbing the physical state of a body or a substance. In water waves, it is the change of position of molecules that is the disturbance; the surface rises and falls. In sound waves, we can consider changes of pressure to be the disturbance; pressure does not remain constant at any given point. In general, a wave may be defined as the *propagation of a disturbance*. It is our purpose to find out what sorts of disturbances commonly exist, and how these disturbances are "handed on" from point to point as the wave travels through a medium.

We illustrate "disturbance" first by describing a *wave of enthusiasm* that passes through a crowd of devotees waiting outside a concert hall for a glimpse of their favorite performers. Word that they will appear soon reaches the first few rows of the crowd, and then the disturbance spreads at a more or less definite rate toward the rear echelons. Here is a nonphysical wave; the disturbance is an emotional state rather than pressure, height, temperature, or some other physical state. The crowd is the "medium" through which the wave moves, and the mechanism by which the wave moves is rumor. As with most waves,

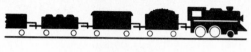

Figure 10-1

there are (1) a disturbance, (2) a medium* that can be disturbed, and (3) a connection between neighboring points of the medium—some means by which neighboring points can influence one another. All these features are present in the rather esoteric wave of enthusiasm (a pulse) that we have described.

10-2 Compression Pulses

On a molecular scale, a compression wave is started whenever a solid object is struck. For instance, when a hammer strikes the head of a vertical iron nail, the molecules of iron at the top surface are subject to a net force caused by the hammer. According to Newton's second law, this force causes an acceleration, and the molecules start to move downward. They, in turn, push on neighboring molecules, and so eventually a pulse is transmitted to the tip of the nail. It is important to note that as the wave travels along the nail, a pressure is momentarily built up wherever the molecules are closer together than is normal. Indeed, it is this squeezing together of the molecules—a distortion—that gives rise to the elastic force that pushes the next molecules along.

On a larger scale, a model of such compression waves can be easily observed when a stationary freight train backs up (Fig. 10-1). Before the engine applies the initial force, the couplings are loose, with perhaps a few centimeters of slack between each pair of cars. When the engine starts moving backward with a steady speed, one

---

* We can do without the medium if we can think of some property of space that can exist in a vacuum, such as gravitational, magnetic, or electric fields. Since these disturbances can exist in a vacuum, they permit waves that travel through a vacuum and need no medium. Light and other electromagnetic waves are of this type.

Figure 10-2

car after another is set into motion, and you can hear a "clank, clank, clank, ..." as the wave travels the length of the train. If in a 100-car train the pulse has reached as far as the 30th car, this means that the couplings between the first 30 cars are tight, and these cars, moving as a rigid unit, are in the process of taking up the slack of the 31st coupling. The final 70 cars are still stationary at this instant. The velocity with which the pulse travels down the train is called the *wave velocity*. This is much faster than the speed of the engine; it will usually take only 5 or 10 s for such a pulse to travel the length of a 100-car train. We have described a wave of compression; in similar fashion, if the engine starts forward, a wave of tension will be initiated.

If we replaced the cars by small heavy balls, and the loose couplings by stretched springs (Fig. 10-2), then our model would more closely resemble the structure of a solid body in which compression waves can travel.

10-3   *Velocity of Compression Waves*

We can make a shrewd guess as to the factors on which the velocity of a compression wave should depend, using the freight-train model. Each car must, in turn, be accelerated to a velocity equal to that of the engine. This requires time. The more heavily loaded the cars are, the greater their mass (inertia) and, by Newton's second law, the less the acceleration produced by a given force and so the longer the time required. Thus we expect the wave velocity to be *less* for a loaded train. A similar argument holds for the system of balls in Fig. 10-2. Considering a set of cars (or balls) held together by stiffer springs, we would expect a *greater* wave velocity, since there would be less "give" to the springs and thus less lost motion.

Passing to the molecular scale of things, we then expect the wave velocity to depend somehow on the density of the material and its modulus of elasticity. A detailed analysis is possible using the energy principle, which in turn is based on Newton's second law (Sec. 10-9). However, we can obtain a valuable insight by the method of *dimensional analysis*. We *assume* that the wave velocity v_w depends only on the density d and an elastic modulus E according to some simple formula such as $v_w = Ed$; $v_w = E/d$; $v_w = \sqrt{E/d}$; or $v_w = E/d^2$. To decide among the various possibilities, we write

$$v_w = KE^x d^y \qquad (10\text{-}1)$$

where x and y are two as yet unknown exponents and K is an unknown dimensionless constant. If Eq. 10-1 is true, it must be true dimensionally. Now the dimensions of velocity are

$$[v] = [LT^{-1}]$$

The dimensions of E are those of force/area (Sec. 9-1); hence,

$$[E] = [FL^{-2}] = [MLT^{-2}][L^{-2}]$$
$$= [ML^{-1}T^{-2}]$$

Also, density is mass/volume, and the dimensions of d are

$$[d] = [ML^{-3}]$$

The constant K has been assumed to have no dimensions. Putting all these dimensions into Eq. 10-1, we have

$$[v] = [E]^x[d]^y$$
$$[LT^{-1}] = [ML^{-1}T^{-2}]^x[ML^{-3}]^y$$
$$[LT^{-1}] = [M^x L^{-x} T^{-2x}][M^y L^{-3y}]$$
$$[LT^{-1}] = [M^{x+y} L^{-x-3y} T^{-2x}] \qquad (10\text{-}2)$$

For this equation to be true, the exponents of L must be the same on both sides, and the exponents of T must be equal. This gives two equations:

Exponents of $[L]$: $\qquad 1 = -x - 3y$
Exponents of $[T]$: $\qquad -1 = -2x$

Solving these equations simultaneously for x and y gives the values $x = \frac{1}{2}$ and $y = -\frac{1}{2}$.* Therefore,

$$v_w = KE^{1/2}d^{-1/2}$$

or

$$v_w = K\sqrt{\frac{E}{d}} \qquad (10\text{-}3)$$

This sort of dimensional analysis is far from being hocus-pocus; in the hands of experts it is a valuable tool. Without prior knowledge, and without detailed use of any physical laws, we have been able to find a plausible form of a physical law. What we had to work with was a "reasonable guess" as to which variables we expected to be significant. To be sure, we have not determined the constant K, but this could be found by analysis based on Newton's laws. In any event, the constant K could be found by experiment, now that we know the form of the law to be $K\sqrt{E/d}$ and not $KE/\sqrt{d}$ or KEd^2 or some other collection of exponents. For many uses, the value of K is immaterial.

Example 10-1

If the stretch modulus of an iron rod is reduced by heat treatment to 81% of its original value, how will the speed of a compression wave in the iron be altered?

We use our formula for wave velocity, and both the density and the unknown K cancel out.

$$\frac{v_w{}'}{v_w} = \frac{K\sqrt{E_2/d}}{K\sqrt{E_1/d}} = \sqrt{\frac{E_2}{E_1}} = \sqrt{\frac{0.81\,E_1}{E_1}}$$

$$= \sqrt{0.81}$$

$$= \boxed{0.90}$$

The new speed is 90% of the original value.

---

* It is gratifying to see that these values of x and y also make the exponents of [M] balance. The left side of Eq. 10-2 can be written $[\text{M}^0\text{LT}^{-1}]$, and for $x = \frac{1}{2}$, $y = -\frac{1}{2}$, we have $x + y = 0$; the right side of Eq. 10-2 also contains $[\text{M}^0]$, a pure number. If this has *not* checked out, it would have meant that our initial assumption as to the form of the law was wrong, and the method would have failed in this particular application.

As a matter of fact, analysis (Sec. 10-9) shows that the constant K is equal to 1, and so the formula can be written

velocity of compression wave $\qquad v_w = \sqrt{\dfrac{\text{elastic modulus}}{\text{density}}}$

For a long solid rod in which the pulse travels along the length of the rod, the proper elastic modulus to use is the stretch modulus E, and we can write

velocity of compression wave in solid rod $\qquad v_w = \sqrt{\dfrac{E}{d}}$

For a large body of liquid or gas, where the pulse spreads out in all directions, the proper elastic modulus to use is the bulk modulus B,[†] and we can write

velocity of compression wave in liquid or gas $\qquad v_w = \sqrt{\dfrac{B}{d}}$

In a liquid or gas, compression waves are called *sound waves* because as compression reaches the eardrum, a complicated mechanical and neurological process is initiated that results in the sensation we know as sound. *Sound waves are compression waves.* Strictly speaking, only compression waves occurring in air should be called sound waves, unless the eardrum is actually in contact with another medium such as water (as for a skin diver). However, the term *sound* is often used for any compression wave in any medium. We speak of the passage of sound waves through a plaster wall, or of the propagation of sound waves along a steel railroad track.

Example 10-2

A railroad worker strikes a steel rail with a sledge hammer, and another worker hears the shock wave, which comes to her through the rail, 0.20 s after she sees the blow. How far apart are the workers?

---

† The values of B tabulated for gases in Table 9-1 are for constant temperature. These values of B are not suitable for calculating the speed of sound in a gas, because temperature changes occur during rapid compression and expansion of the gas. See Sec. 16-7 for further discussion of the speed of sound in a gas.

To solve the problem we must first find the speed of sound in steel. From Table 1-3, we find that the density of steel is 7.8 g/cm^3 or 7800 kg/m^3. From Table 9-1, the stretch modulus of steel is found to be 20×10^{10} N/m^2.

$$v_w = \sqrt{\frac{E}{d}} = \sqrt{\frac{20 \times 10^{10} \text{ N/m}^2}{7800 \text{ kg/m}^3}}$$

$$= 5.06 \times 10^3 \text{ m/s}$$

The distance s between the workers, measured along the rail, is

$$s = v_w t$$
$$= (5.06 \times 10^3 \text{ m/s})(0.20 \text{ s})$$
$$= 1.01 \times 10^3 \text{ m}$$
$$= \boxed{1.01 \text{ km}}$$

(We have assumed that the speed of light waves is so great that the time for the second worker to receive the message via light is negligibly small. A brief calculation, using 3×10^8 m/s for the speed of light, will justify this conclusion.)

Our next example involves compression waves in a liquid, and so we use the bulk modulus in computing the wave velocity.

──────────── **Example 10-3**

A sailor strikes the hull of his ship with a hammer, and 0.54 s later he hears the echo, reflected from the ocean bottom directly below the ship. How far below him is the ocean bottom?

We must first find the speed of sound in sea water. The density of sea water is 1.025 g/cm^3 or 1025 kg/m^3 (Table 1-3). Using the value for the bulk modulus of sea water from Table 9-1 gives

$$v_w = \sqrt{\frac{B}{d}}$$

$$= \sqrt{\frac{2.1 \times 10^9 \text{ N/m}^2}{1.025 \times 10^3 \text{ kg/m}^3}} = 1430 \text{ m/s}$$

The round trip is a distance of $2h$; hence,

$$2h = (1430 \text{ m/s})(0.54 \text{ s})$$
$$h = \boxed{390 \text{ m}}$$

Speed of Sound in Air. Most familiar of all compression waves are sound waves in air. Experiment shows that their wave velocity is 331 m/s at 0°C and 343 m/s at 20°C. (These speeds are about 1100 ft/s, or 750 mi/h.) It is often sufficiently precise to use

$$v_w = 340 \text{ m/s}$$

for the speed of sound in dry air. Near room temperature, we can use the convenient fact that the speed of sound increases by 0.6 m/s for every 1°C rise in temperature. Thus, at 25°C,

$$v_w = 331 + (0.6)(25) = 346 \text{ m/s}$$

or, by an alternative calculation,

$$v_w = 343 + (0.6)(5) = 346 \text{ m/s}$$

This handy shortcut applies only to *air* at temperatures not too far from room temperature. A method for calculating the speed of sound in any gas at any temperature is discussed in Sec. 16-7.

10-4 Longitudinal and Transverse Waves

Many (but not all) waves involve disturbances that have a direction. In the freight-train model (Sec. 10-2), each coupling is stretched parallel to the train's motion and each car is displaced from the neighboring cars, the displacement being parallel to the direction of motion of the pulse; we call such a wave *longitudinal*. In a *transverse* wave the disturbances are directed at right angles to the direction of propagation of the wave. The most familiar longitudinal waves are sound waves, and the most familiar transverse waves are waves on strings. Water waves are described as a combination of transverse and longitudinal waves. Any individual molecule of water swishes back and forth while it bobs up and down. In the rest of this chapter we shall often use water waves to illustrate transverse waves, partly because

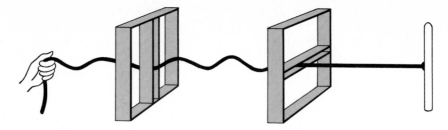

Figure 10-3

they move slowly and their transverse (up and down) disturbances are easily visualized.

If the disturbances of all particles in the medium are in the same plane, then the wave is said to be *plane polarized*. The polarization of transverse waves on a rope is shown in Fig. 10-3. It is evident that the plane of vibration (a vertical plane in the illustration) can be determined with the aid of a slot through which the rope passes. A longitudinal wave, in head-on view, would show no preferred plane, and the orientation of a slot or other analyzing device would have no effect on the transmission or reflection of a longitudinal wave. In other words, polarization has no meaning for longitudinal waves. Historically, the observation of polarization effects for x rays showed them to be transverse, not longitudinal.

10-5 Graphical Representation of Waves

Let us now consider water waves on the surface of a lake. The medium consists of water molecules, and the disturbance is y, which is the displacement of the surface above or below the normal level of the water. To show the difference between "wave" and "vibration," we construct several graphs of the disturbance y. Let us suppose that a swimmer dives into the lake and a short *wave train* of rather definite shape spreads out toward a fisherman, one of whose lines enters the water at A (Fig. 10-4). We start our stop watch when the swimmer enters the water; $t = 0$ at this instant. As the wave train moves toward the left, it eventually moves past point A, and

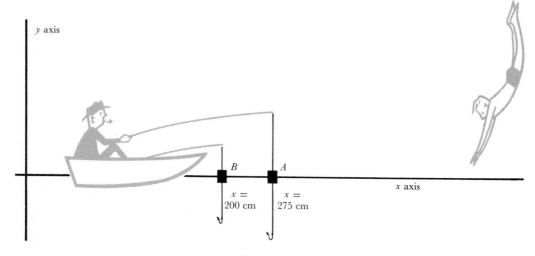

Figure 10-4

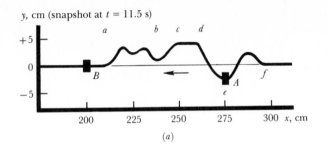

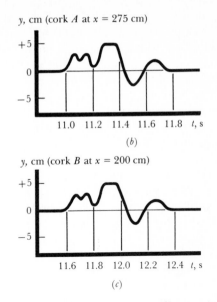

Figure 10-5 (*a*) Wave-form graph at $t = 11.5$ s; y as a function of x, at a given time. (*b*) Vibration graph for cork A at $x = 275$ cm; y as a function of t, at a given point. (*c*) Vibration graph for cork B at $x = 200$ cm.

the cork A on the fishline starts to bob up and down. Two (related) graphs can be constructed: y as a function of x, at some given time, and y as a function of t, at some given point.

Figure 10-5*a* is a graph of y as a function of x, at the particular time $t = 11.5$ s. In other words, Fig. 10-5*a* might be a picture (a snapshot) of the wave train, "freezing" the motion of the wave at $t = 11.5$ s. Such a graph is called a *wave-form* graph. As shown, for $x = 275$ cm (which is the position of the cork A), the value of the displacement at this instant is $y = -3$ cm; for other places (other corks, such as B, at other values of x) the displacement may be quite different at this same time.

If the fisherman pays attention only to his cork, not knowing that a swimmer has dived, he observes his cork bobbing up and down. Figure 10-5*b* is a graph of y as a function of t, at the point A ($x = 275$ cm); such a graph is called a *vibration graph* of the motion at a point. The cork remains stationary for 11.0 s until the leading edge (*a*) of the wave front hits it. The cork rises up a bit, wiggles up and down twice, then remains at the high level for 0.1 s until the plateau (*cd*) has passed it. The cork then bobs down to a negative value of y as the trough (*e*) passes

by. Finally, after a momentary $+$ value of y, the cork reaches zero level and remains at $y = 0$ as time increases indefinitely.

For cork B at some neighboring point (at $x = 200$ cm), the same crazy motion is observed, but starting at a later time. The graph of B's motion is shown in Fig. 10-5*c*; this graph is the same as Fig. 10-5*b*, except for a different time axis.

Now it is immediately seen that the *shape* of Fig. 10-5*a* is exactly the same as the *shape* of Fig. 10-5*b*, even though these graphs have different meanings. A little reflection will show that it must always be so—if the snapshot reveals some characteristic kink in the wave form, that same kink will show up in the time graph of the motion at any point.

We see here the difference between a vibration and a wave. A *vibration* is a motion or disturbance of a single particle of the medium, whereas a *wave* is an interrelated set of vibrations of many neighboring particles. We studied vibrations in Chap. 9 and found that some periodic vibrations, such as simple harmonic motion, are simpler to describe than others, such as relaxation vibrations. In a general sense, even a single nonperiodic pulse may be called a vibration. But the term "wave" must be applied only to the totality of many vibrations. No single graph suffices to describe a wave. The wave-form graph moves in its entirety, so to speak, and a complete description of the wave would involve a series of instantaneous snapshots—a motion picture, actually—or else a number of recorders making time records at a sequence of points spaced along the path of the wave.

In mathematical terms, a vibration is described by y as a function of t; a wave form is described by y as a function of x; and a wave is described by y as a function of both t and x. See Sec. 11-8 for a mathematical discussion of wave motion.

The *wave velocity* is the rate at which any given characteristic of the wave progresses from place to place. Thus we can fix our attention upon the trough (e) of the wave form, and we can deduce the wave velocity from the data given in the graphs. From Fig. 10-5b, the trough reached cork A at $t = 11.5$ s. (The same information is contained in Fig. 10-5a.) As shown in Fig. 10-5c, the trough reached cork B at $t = 12.1$ s. Since A and B are 75 cm apart, and the trough required 0.6 s to go from A to B, the wave velocity is 75 cm/0.6 s = 125 cm/s.

We have given a kinematical discussion of the water waves—a description of the motion, rather than a dynamical study of the causes in terms of force and mass. For large-scale water waves, such as those on the surface of the ocean, the connection between neighboring water molecules arises from gravity and from the elastic forces that are evidenced by the relative incompressibility of water. The force of gravity tends to "level out" any wave crest, but the inertia of the molecules keeps the water moving past the equilibrium point, and so a crest becomes a trough. We thus expect the density of the water and the acceleration due to gravity to enter into an expression for the velocity of such water waves. For small ripples the surface of the water is highly curved, and surface tension forces enter into the picture. We will not pursue further the dynamical study of water waves, since the mathematical complexities are considerable.

Nothing prevents us from making a graphical study of compression waves (sound waves) in the same way we have studied transverse waves. The disturbance is the pressure P rather than the displacement along the y axis. We need only plot P versus x, and P versus t, using pressure as the ordinates for the pair of graphs. Figure 10-6 represents a shock wave, or pulse, in air, resulting from a sudden compression in a limited region of space where a test bomb was located; the air pressure before and after the pulse was 101 kPa. (Note the analogy with the swimmer who created a localized disturbance where he dived into the water.) As always, the vibration graph has the same shape as the wave-form graph. The fact we

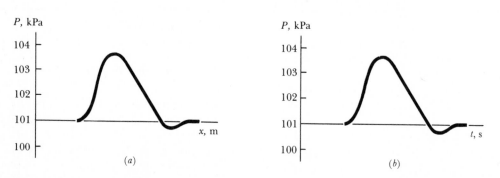

Figure 10-6 (*a*) Wave-form graph for a pulse; P as a function of x, at a given time. (*b*) Vibration graph for same shock wave; P as a function of t, at a given point.

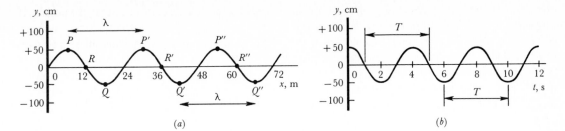

Figure 10-7 (a) Wave-form graph of a periodic water wave; snapshot at $t = 3.0$ s. (b) Vibration graph at one point of same periodic water wave; record of the motion at the point R' at $x = 36$ m.

wish to emphasize is that Fig. 10-6a is a *graph* of the pressure wave form and in no way represents any wave form in two dimensions that could be photographed, as was the water wave.

10-6 Periodic Waves

We have deliberately avoided mention of periodic, sustained waves, since the general ideas apply equally well to pulses (shock waves). However, if the source of disturbance varies regularly, the wave form has additional symmetry, and the wave is said to possess a wavelength and a frequency.

Let us redraw a and b of Fig. 10-5, this time assuming that instead of a diver and a single (complex) splash, some mechanism is periodically creating a $+$ and $-$ disturbance of the y coordinate of the water surface at the point where previously we imagined the dive taking place. Assuming, for simplicity, that the disturbance is one of SHM, the two graphs will be sinusoidal in appearance. The wave-form graph (Fig. 10-7a) and the vibration graph (Fig. 10-7b) are identical in appearance, as we have seen must always be the case. Successive particles P, P', P'', ..., are all at their highest points at the same instant (at $t = 3.0$ s, according to the legend accompanying the graph) and are momentarily at rest. Particles R, R', R'', ..., are all at their normal positions at this instant and are all moving

in the same direction. (See Prob. 10-B3.) Particles that have the same displacement and are moving in the same direction with the same velocity are said to have the same *phase* of vibration. For instance, in Fig. 10-7a, particles P and P' are "in phase" with each other, as are particles P and P'', R and R', R and R'', Q and Q', Q and Q''. Particles P' and Q' are exactly "out of phase"—sometimes referred to as "180° out of phase."*

To describe a periodic wave, we make use of five quantities, some of which we have already encountered in our study of vibration: (1) The *period* T is the time for one complete vibration of any given particle of the medium. (2) The *frequency* f is the number of vibrations of any particle per unit time. As in any SHM, $f = 1/T$. (3) The *amplitude* A is the maximum displacement of any particle, measured from its equilibrium position. (4) The *wavelength* λ (lambda) is the distance between any two successive particles that have the same phase. (5) The *wave velocity* v_w is the velocity with which any specific phase of motion (for instance, crest, trough, or compression) is propagated through the medium.

An important relationship is true for periodic waves of all kinds. Study of Fig. 10-7 shows that

---

* If the SHM's of these two particles were represented with the aid of a single reference circle, the reference particles would be at opposite ends of a diameter of the circle, 180° apart.

during the time T (one period), point P' moves down and up and makes one complete vibration; meanwhile, the wave crest has traveled from P' to P'', and the crest that was at P has moved over to P'. That is, the crest moves one wavelength in a time equal to the period of vibration of any individual particle. Thus,

$$v_w = \frac{s}{t} = \frac{\lambda}{T}$$

For any periodic wave, $1/T = f$, and hence,

$$v_w = f\lambda \qquad (10\text{-}4)$$

Wave velocity = (frequency)(wavelength)

Example 10-4

The wavelength of sound waves emitted by a whistle is 1.7 m. If the velocity of sound in air is 340 m/s, how many times per second does the pressure reach its maximum value above normal (atmospheric) pressure?

$$v_w = f\lambda.$$

$$f = \frac{v_w}{\lambda} = \frac{340 \text{ m/s}}{1.7 \text{ m}} = 200 \text{ s}^{-1}$$

$$= \boxed{200 \text{ vib/s}}$$

We interpret 200 s^{-1} as 200 "times" per second, or 200 vib/s. "Vibration" is simply a count, having no dimensions. One vibration is a "cycle," and frequency can also be expressed in *cycles per second*, or in *hertz* (Hz). These units for frequency are equivalent; we shall use them interchangeably:

$$1 \text{ vib/s} = 1 \text{ c/s} = 1 \text{ s}^{-1} = 1 \text{ Hz}$$

The hertz* is coming into increasing use, especially for frequencies of electromagnetic and acoustic waves. It is the SI unit for frequency.

---

* Named for Heinrich Rudolf Hertz (1857–1894), a pioneer experimental investigator of electromagnetic waves.

Example 10-5

What is the distance between adjacent regions of minimum electric field in an electromagnetic wave whose velocity in glass is 2.0×10^8 m/s and whose frequency is 10^{15} Hz?

The disturbance is evidently an entity known as "electric field," about which we need know nothing further at this time. Regions of minimum electric field are in the same phase of disturbance, and so the distance between adjacent regions of minimum electric field is equal to one wavelength.

$$v_w = f\lambda$$

$$\lambda = \frac{v_w}{f} = \frac{2.0 \times 10^8 \text{ m/s}}{10^{15} \text{ s}^{-1}} = \boxed{2.0 \times 10^{-7} \text{ m}}$$

The pictorial representation of a periodic compression wave is not easy. We might imagine a large number of quick-acting pressure gauges spaced at close intervals along a pipe in which normal atmospheric pressure is 100.00 kPa (10^5 N/m^2) (Fig. 10-8). A loudspeaker creates periodic disturbances at one end of the pipe. As the wave progresses down the pipe, the reading of each gauge goes through a sequence of values, fluctuating between 100.01 kPa in a *condensation* and 99.99 kPa in a *rarefaction*. We have drawn a wave whose period is 0.004 s, so the gauges would have to be quick-acting indeed to follow such rapid changes.[†] Figure 10-8 represents four snapshots of the row of gauges, taken at intervals 0.001 s apart as shown by the timer. Gauges a and e are in phase, always reading the same at any given instant. The wavelength is the distance between any two adjacent in-phase pressure gauges, such as a and e, b and f, or c and g. As indicated in the figure, the pressure amplitude of the wave is 0.01 kPa, since this is the maximum excursion of the pressure from the normal value.

---

† It is entirely practical to measure rapid pressure changes using crystal transducers. Such crystals generate a small piezoelectric voltage proportional to the pressure, and when used as microphones, they respond in the audio-frequency range and higher.

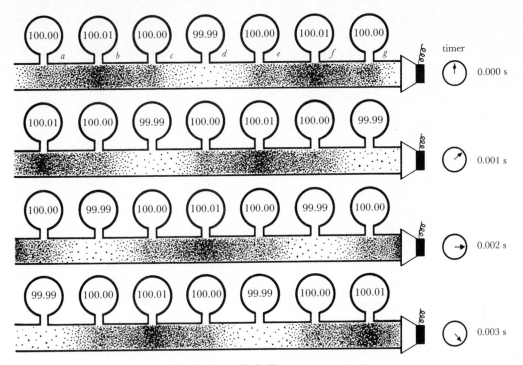

Figure 10-8 Periodic pressure wave.

10-7 The Superposition Principle—Beats

In our study of projectile motion (Sec. 2-11 on page 40), we used the superposition principle when we considered the resultant motion to be the vector sum of the horizontal and vertical motions. The principle can also be applied to vibrations and wave motions.

If a compression wave A acting alone would create a certain disturbance at a point, and if wave B would create a (perhaps different) disturbance at the same point, the *superposition principle* says that the resultant disturbance is the algebraic sum of the disturbances due to A and B.* Stated in this way, the theorem seems trivial, but we should realize that we are making

---

* For a transverse wave, the resultant is the *vector* sum of the disturbances.

a definite assumption about a medium when we invoke the superposition principle for waves of some sort in the medium. Let us illustrate the superposition theorem by reference to compression waves in an aluminum rod. Wave A might be capable of imparting a certain pressure P_A at some reference point. Wave B, acting alone, would be able to impart a pressure P_B. If both waves act simultaneously, will the pressure be $P_A + P_B$? The answer is "yes," provided we do not go beyond the elastic limit of aluminum. For compression waves having a very large pressure amplitude, each wave being almost enough to exceed the elastic limit, the superposition of the two waves might well cause the rod to deform permanently or even fracture. In such a case the total disturbance certainly is not equal to the sum of the component disturbances.

Fortunately, in ordinary sound waves in air, the pressure amplitudes are small [± 0.01 kPa (0.0001 atmosphere) for a very intense sound].

Under these circumstances we are justified in using the superposition principle. Waves on strings (say, for a violin string being bowed) also obey the superposition principle, unless the amplitude of vibration is excessively large.

As an illustration of the usefulness of the superposition principle, consider the effect of two waves of different frequencies. We assume that both waves act on the same point, and we plot, in Fig. 10-9, the vibration graphs of the two waves and their superposition. The vibration graph of the resultant disturbance is the sum of the vibration graphs of the two waves. Figure 10-9 shows a 0.2-s portion of the vibration graphs of A (50 vib/s) and B (60 vib/s). If the waves are in phase at $t = 0$ s, the less rapidly vibrating wave (A) lags behind until at $t = 0.05$ s, A and B are exactly out of phase. They are in phase again at $t = 0.10$ s, out of phase again at $t = 0.15$ s, and so on. The amplitude of the resultant wave (C) rises and falls once during each 0.1-s interval. These periodic changes in amplitude are called *beats*. The number of beats per second equals the difference in the frequencies of the two combining waves. In our example, waves of frequencies 50 and 60 vib/s combine; they must be out of phase and cancel each other 10 times per second. It is equally true to say that they reinforce each other 10 times per second. The resultant amplitude fluctuates from twice that of either wave to zero and back again, 10 times

per second. The phenomenon of beats extends to all types of waves. For instance, the "squeals" heard on some radios are due to beats (also called heterodyning) between electrical oscillations of slightly different frequencies.

It is not difficult to derive an equation for the beat frequency. As in Fig. 10-9, let the vibrations be in phase at $t = 0$, and let the periods be T_A for the slow vibration and T_B for the fast vibration ($T_A > T_B$). When the vibrations are next in phase, the elapsed time is one beat period T_{beat}; there have been n peaks of the fast vibration and $n - 1$ peaks of the slow vibration. This gives

$$(n - 1)T_A = nT_B = T_{\text{beat}}$$

which can be solved to give

$$n = \frac{T_A}{T_A - T_B}$$

Then the beat period is nT_B; that is,

$$T_{\text{beat}} = \frac{T_A T_B}{T_A - T_B}$$

The beat frequency is the reciprocal of the beat period, so

$$f_{\text{beat}} = \frac{T_A - T_B}{T_A T_B} = \frac{1}{T_B} - \frac{1}{T_A}$$

or

$$f_{\text{beat}} = f_B - f_A$$

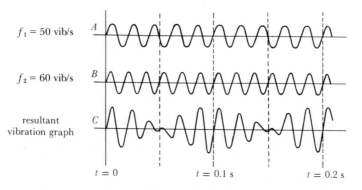

$f_1 = 50$ vib/s A

$f_2 = 60$ vib/s B

resultant vibration graph C

$t = 0$ $t = 0.1$ s $t = 0.2$ s

Figure 10-9

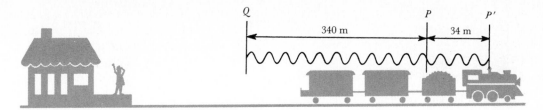

Figure 10-10 Doppler effect, source in motion. The sinusoidal graph represents pressure variations.

This shows that the beat frequency is the difference between the two superposed frequencies. See Sec. 10-10 for a proof, using elementary trigonometry, that also gives the additional result that the perceived frequency is the average of f_A and f_B.

10-8 The Doppler Effect

A phenomenon exhibited by all types of wave motion is the *Doppler effect*,* which has to do with the apparent change in frequency of a wave due to the relative motion of the source and the observer. Whether it is the source or the observer that is moving, the effect of relative motion is to make the apparent frequency greater if the source and observer are approaching each other. On the other hand, the apparent frequency is lowered if the source and observer are moving away from each other. In our study of the Doppler effect for sound waves, we shall see that it makes a difference whether it is the source or the observer that moves. The interested student can derive general formulas, or (preferably) proceed directly from first principles in each problem, as illustrated in the following examples.

Example 10-6

A through train is receding from a station platform at 34 m/s, blowing a whistle whose true frequency is 300 vib/s. The speed of sound in air is 340 m/s. What

---

* The effect was first calculated in 1842 by Christian Doppler, who analyzed experiments involving musicians on moving railroad cars.

is the apparent frequency, as judged by an observer on the platform?

Since it is the train that is moving, a pulse, once emitted, travels through the air with the usual velocity of 340 m/s. However, the wavelength is increased (Fig. 10-10). In 1 s, the whistle emits 300 pulses, but the train has moved 34 m during this 1 s. The last pulse is emitted at P' at the same instant that the first of the 300 pulses has reached the point Q. The wavelength is "stretched out"; we have 300 pulses in (340 + 34) m = 374 m, and so

$$\lambda' = \frac{374 \text{ m}}{300} = 1.247 \text{ m}$$

The apparent frequency is

$$f' = \frac{v_w}{\lambda'} = \frac{340 \text{ m/s}}{1.247 \text{ m}} = \boxed{272.7 \text{ s}^{-1}}$$

Example 10-6 shows that if the *source* is moving, the wavelength is altered, but the wave velocity remains the same. On the other hand, if the *observer* moves, the wave velocity relative to him is altered, but the wavelength remains the same since the source is stationary.

Example 10-7

Solve Example 10-6, assuming that the whistle is stationary and the observer is moving away from it at 34 m/s.

Here the wavelength is unaltered, but since the observer is moving away from the source (Fig. 10-11), the condensations are not catching up to him as rapidly as if he were stationary. The wavelength is

$$\lambda = \frac{340 \text{ m/s}}{300 \text{ vib/s}} = 1.133 \text{ m}$$

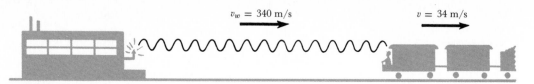

$v_w = 340$ m/s $v = 34$ m/s

Figure 10-11 Doppler effect, observer in motion. The sinusoidal graph represents pressure variations.

However, the apparent wave velocity is

$$v_w' = (340 - 34) \text{ m/s} = 306 \text{ m/s}$$

Hence,

$$f' = \frac{v_w'}{\lambda} = \frac{306 \text{ m/s}}{1.133 \text{ m}} = \boxed{270.0 \text{ s}^{-1}}$$

We can summarize the Doppler effect for sound waves in a simple way. The observed changes depend on v/v_w, the ratio of the relative velocity to the wave velocity. If the source moves, the *wavelength* changes by a relative amount equal to v/v_w; if the observer moves, the *frequency* changes by a relative amount equal to v/v_w. If we follow the methods of Examples 10-6 and 10-7, we can derive equations for the apparent frequency. The results are as follows:

sound waves
source in motion $$f' = \frac{f}{1 - v/v_w} \qquad (10\text{-}5)$$

where v is the velocity of the source, and

sound waves
observer in motion $$f' = f(1 + v/v_w) \qquad (10\text{-}6)$$

where v is the velocity of the observer. In these equations, v is considered positive if the observer and source are in relative motion *toward* each other.*

---

* Equations 10-5 and 10-6 can be combined into a general equation, in which we also include the effect of the velocity v_{air} of the air relative to earth (a possible wind).

$$f' = f \frac{v_w + v_{air} + v_{obs}}{v_w + v_{air} - v_{source}}$$

Equation 10-6 can be rearranged to give

$$\frac{f' - f}{f} = \frac{v}{v_w}$$

or

$$\frac{\Delta f}{f} = \frac{v}{v_w} \qquad (10\text{-}7)$$

This means, for example, that a 2% change in frequency will be heard if the relative velocity is 2% of the wave velocity. It can be shown (Prob. 10-C2) that the other equation for the Doppler effect reduces to the form of Eq. 10-7 when v/v_w is small compared with 1.[†] Thus Eq. 10-7 serves as an approximate formula for any Doppler effect calculation if the relative velocity of source and observer is small compared with the wave velocity.

Example 10-8

A certain spectrum line of hydrogen has a frequency 4.56571×10^{14} Hz as measured in a laboratory on the earth. What is the relative velocity of the earth and a star if the spectroscope shows that the same hydrogen line in the light reaching the earth from the star has a frequency 4.56711×10^{14} Hz? The wave velocity v_w is the speed of light, 3.00×10^8 m/s.

[†] The answers to Examples 10-6 and 10-7 differ slightly. We expected a difference, since the ratio v/v_w is $34/340 = \frac{1}{10}$, certainly not "small" compared with 1.

Here the relative velocity is very small compared with the wave velocity, since the relative change in frequency is so small.

$$\frac{\Delta f}{f} = \frac{0.00140 \times 10^{14} \text{ Hz}}{4.56571 \times 10^{14} \text{ Hz}} = 3.07 \times 10^{-4}$$

We can now use the approximate equation:

$$\frac{v}{c} = \frac{\Delta f}{f}; \qquad v = \frac{\Delta f}{f} c$$

$$v = (3.07 \times 10^{-4})(3.00 \times 10^8 \text{ m/s})$$

$$= 9.2 \times 10^4 \text{ m/s} = \boxed{92 \text{ km/s}}$$

Since the frequency is *raised*, we know that the star and the earth are *approaching* each other at 92 km/s. It is impossible to say which of the two is "really" in motion; the theory of relativity (Chap. 28) tells us that only the relative motion can be observed by this or any other means.

Summary A wave is the propagation of a disturbance. The nature of the disturbance may be mechanical, electrical, thermal, or even nonphysical. The wave velocity is the speed with which the disturbance is propagated. If a disturbance can exist only in a medium, the neighboring particles of the medium must somehow react on one another for the disturbance to be propagated. In compression waves the elastic forces between molecules supply the connection between neighboring particles. The wave velocity of a compression wave depends on the elastic modulus and on the density of the medium. For waves in solids, the stretch modulus is used; for waves in liquids and gases, the bulk modulus is used.

In a longitudinal wave the disturbance is parallel to the direction of propagation. Sound waves and other compression waves are longitudinal waves. In a transverse wave the disturbance is in a direction at right angles to the direction of propagation. Waves on strings are examples of transverse waves, and waves on the surface of a liquid are approximately transverse. A transverse wave is said to be plane polarized if all the disturbances are parallel to each other. Longitudinal waves cannot be polarized.

A wave is distinguished from a vibration by the fact that a wave consists of many vibrations of many particles, with definite phase relationships between the vibrations of neighboring particles. Two graphs are required to represent a wave: the vibration graph of any given particle (disturbance versus time), and the wave-form graph of the whole wave, at any given time (disturbance versus position). These two graphs necessarily have the same shape, although they are plotted with different variables.

Periodic waves are caused by a regularly repeated series of disturbances. The wavelength is the shortest distance between two particles having the same phase of vibration. For all periodic waves, of whatever kind, wave velocity = frequency × wavelength. The amplitude of a wave is the maximum value of the disturbance.

The superposition principle states that the combined effect of two waves is the sum of the effects of the separate waves. The superposition of two periodic waves gives rise to beats as the resultant amplitude rises and falls. The beat frequency is equal to the difference between the frequencies of the two combining waves.

The Doppler effect is the apparent change in frequency of a wave due to the relative motion of the source and the observer.

Questions

10-1 Give three examples of disturbances that can be propagated as a wave motion.

10-2 Does every wave have a wavelength? Does every wave have a wave velocity?

10-3 While a compression wave is moving past a point in a gas, is there any change in the density of the gas at this point?

10-4 The bulk modulus of air does not depend on temperature, but the density does (the latter decreases when the temperature rises). How would you expect the speed of sound in air to depend on the temperature?

10-5 Which elastic modulus enters into the formula for the speed of a compression pulse traveling through a liquid?

10-6 A fisherman is fishing from the end of a pier in a dense fog, and all that he can see is the motion of the cork on his own fishline. The cork starts to bob up and down; can the fisherman be sure that a wave is moving past the cork's position?

10-7 In what sort of unit might you measure the amplitude of the wave of enthusiasm mentioned in Sec. 10-1? In what unit would you measure the wave velocity of this wave? How would you produce a periodic wave of enthusiasm? Could such a wave have a wavelength?

10-8 When a flag waves in the breeze, is this a wave in the strict physical sense of Sec. 10-5? If so, what is the disturbance? What is the medium? Approximately what is the frequency?

10-9 Two waves of equal frequency, wavelength, velocity, and amplitude are traveling in the same direction through a medium. Is the amplitude of the resultant wave, found by the superposition principle, necessarily twice that of the individual waves?

10-10 What is the difference between a vibration and a wave?

10-11 Is the Doppler effect "real," or is it a physiological phenomenon in the listener's ear? Would a microphone plugged into a tape recorder in a moving car approaching a source of sound actually record an altered frequency?

10-12 An observer and a source of sound waves are fixed relative to each other. If a strong wind blows toward the observer, (*a*) is the wavelength changed? (*b*) is the wave velocity changed? (*c*) is the frequency changed?

MULTIPLE CHOICE

10-13 All waves, including shock waves and compression pulses, have (*a*) a wavelength; (*b*) a wave velocity; (*c*) both of these.

10-14 Longitudinal waves cannot (*a*) have a frequency; (*b*) transmit energy; (*c*) be polarized.

10-15 Sound waves travel faster in water than in air because water has a greater (*a*) density; (*b*) bulk modulus; (*c*) both of these.

10-16 If the wave velocity and frequency of a wave are both doubled, the wavelength is (a) half as large; (b) the same; (c) twice as large.

10-17 The formation of beats is a phenomenon of (a) transverse waves; (b) longitudinal waves; (c) both of these.

10-18 The driver of an automobile traveling at $\frac{1}{10}$ the speed of sound blows his horn. To an observer in a car that is ahead of the first one, also moving at $\frac{1}{10}$ the speed of sound, the apparent frequency of the horn is (a) raised by about 20%; (b) lowered by about 20%; (c) unaltered.

Problems

Note: Use data from tables in Chaps. 1 and 9 where needed.

10-A1 Two metal rods have the same density, but the stretch modulus of one metal is four times that of the other. What is the ratio of the speed of compression waves in the two metals?

10-A2 Two liquids have the same density, but the volume modulus of one liquid is one-fourth that of the other. What is the ratio of the speeds of compression waves in the two liquids?

10-A3 Calculate the speed of sound in a copper rod.

10-A4 Calculate the speed of sound in an aluminum rod.

10-A5 What is the speed of sound in water, (a) in m/s? (b) in ft/s?

10-A6 Calculate the speed of sound in glycerin.

10-A7 An unidentified metal fragment is analyzed by measuring the speed of sound in a rod made out of the metal. The rod has mass 6.10 kg and volume 2.04×10^{-3} m^3. The wave velocity in the rod is 4200 m/s. Calculate the stretch modulus of the metal.

10-A8 (a) At what time does cork A (Fig. 10-5a) start moving? (b) At what time does cork B start moving? (c) In what direction (up or down) is cork A moving at $t = 11.4$ s?

10-A9 (a) What is the frequency, in vib/min, of the wave of Fig. 10-7? (b) In which direction (to the right or the left) is the wave moving? (c) What is the wave velocity?

10-A10 What is the wavelength of the compression waves shown in Fig. 10-8 if the pressure gauges are spaced 1.6 m apart?

10-A11 What is the frequency of the compression wave shown in Fig. 10-8?

10-A12 How often does a compression pass an observer who is listening to a sound of wavelength 1.7 m and speed 340 m/s?

10-A13 What is the distance between rarefactions in a sound wave whose velocity is 345 m/s and whose frequency is 230 Hz?

10-A14 What is the shortest wavelength of sound that can be heard by a human if the frequencies of audible sound lie between 20 Hz and 20,000 Hz?

10-A15 Find the wavelength of an ultrasonic wave used by a bat for navigation purposes if the frequency is 40,000 Hz and the speed of sound in air is 340 m/s.

10-A16 What is the wavelength (in meters) of waves broadcast by a TV station whose frequency is 50 MHz (megahertz)?

10-A17 Two whistles having frequencies 408 vib/s and 414 vib/s are sounded simultaneously. How many beats are heard per minute?

10-A18 How fast would an observer have to move toward a sound source in order to hear the frequency of an emitted sound as 2% higher than normal?

Note: Use speed of sound in air = 340 m/s, unless otherwise specified.

10-B1 A 90-car train standing on a siding is started in motion by an engine. If there is 6 cm of slack between cars and the engine moves at a constant speed of 30 cm/s, how much time is required for the pulse to travel the length of the train?

10-B2 In Example 10-3 the combination of units $\sqrt{\dfrac{\text{N/m}^2}{\text{kg/m}^3}}$ is stated to yield m/s. Prove this statement.

10-B3 The points R, R', R'' in Fig. 10-7a are all in phase. Are they moving upward or downward at $t = 3.0$ s?

10-B4 What is the velocity of the compression wave of Fig. 10-8 if the pressure gauges are spaced 1.6 m apart?

10-B5 A sound wave of velocity 600 m/s has a wavelength of 240 cm. At a certain instant one of the molecules of the medium is at its normal position. How long will it be before this same molecule is again at its normal position?

10-B6 Find the wavelength of an ultrasonic compression wave in a magnesium rod if the frequency is 23 kHz.

10-B7 Calculate the frequency of compression waves in an aluminum rod if the wavelength is 20 cm.

10-B8 On a cold day when the speed of sound in air is 330 m/s, a blind man taps his cane on the sidewalk, and 0.060 s later he hears an echo from a building. How far is the man from the building?

10-B9 A boy standing on a steel bridge sees a construction worker strike a rivet, and 0.08 s later he feels the compression pulse at his feet. (*a*) How far away is the worker? (*b*) How long after feeling the pulse does the boy hear the sound through the air?

10-B10 Periodic SHM waves spread out over the surface of a lake where two women are fishing 21 m apart. Each woman's cork bobs up and down 20 times per minute. At a time when F's cork is at its crest, G's cork is at its lowest point, and there is one additional crest somewhere between the two women. What is the speed of the water waves?

10-B11 On a day when the speed of sound in air is 336 m/s, two whistles are blown simultaneously. The wavelengths of the sounds emitted are 7.00 m and 8.00 m, respectively. How many beats per second are heard?

10-B12 Two sources of sound are vibrating simultaneously with frequencies 2000 Hz and 2010 Hz. On a certain day the speed of sound is 340 m/s. (*a*) How many beats are heard? (*b*) At any instant, how far apart in space are the regions of maximum amplitude of vibration?

10-B13 A campus radio station is broadcasting on a frequency of 91.700 MHz. (*a*) What is the wavelength of the electromagnetic waves? (*b*) If another station simultaneously broadcasts at 91.702 MHz, how many beats per second are heard?

10-B14 A train approaches a station at a constant speed of 20 m/s sounding a 700-Hz whistle. (*a*) What frequency is perceived by an observer on the platform of the station? (*b*) After the train passes the station platform, what frequency is perceived by the observer as the train moves away?

10-B15 A factory whistle near a roadside emits a sound whose most intense component is at 500 Hz. (*a*) What is the apparent frequency heard by a passenger in a car that is approaching the factory at 25 m/s? (*b*) What is the apparent frequency after the car has passed the factory and is moving away from the whistle?

10-B16 A child walks at 50 cm/s toward a stationary man. If the child is whistling a note of frequency 600 Hz, what is the apparent frequency heard by the man?

10-B17 (*a*) With what speed should an observer in a plane fly toward a stationary source of sound of frequency 900 Hz so that the apparent frequency becomes 1500 Hz? (Assume that at this altitude and temperature the speed of sound is 300 m/s.) (*b*) If the source of 900 Hz is on the plane, with what speed should the plane fly toward a stationary observer so that the apparent frequency becomes 1500 Hz?

10-C1 An observer in a mountain town hears a train whistle and 4 s later hears the start of an echo from a cliff. The echo's frequency is 0.88 that of the sound heard directly. (*a*) How far is the train from the cliff? (*b*) How fast, and in what direction, is the train moving?

10-C2 Show that if v/v_w is small compared with 1, Eq. 10-5 reduces to $f' = f(1 + v/v_w + \cdots + \cdots)$, which is similar to Eq. 10-6 except for negligible terms. [*Hint:* For convenience, let $v/v_w = x$, and use the binomial theorem, which can be written as

$$(1 + x)^n = 1 + nx + \frac{n(n-1)}{2 \cdot 1} x^2 + \frac{n(n-1)(n-2)}{3 \cdot 2 \cdot 1} x^3 + \cdots.]$$

10-C3 During a lunar landing, it is necessary to measure the speed of descent v. A beam of electromagnetic waves of frequency 2.40 GHz is emitted vertically downward from the spacecraft module. The waves are reflected from the lunar surface; the beam returns as if it were emitted from a source moving toward the module at $2v$. The reflected waves are superposed on the emitted wave, and an on-board computer uses the beat frequency to calculate the speed. (*a*) What is the wavelength of the radar waves emitted by the module? (*b*) If the beat frequency is 1260 Hz, calculate the speed of descent.

10-C4 Two points A and B on the earth are at the same longitude and 60° apart in latitude (see Appendix Table 7 for the earth's radius). An earthquake at point A sends two waves toward B. A transverse wave travels along the surface of the earth at 4.6 km/s, and a longitudinal wave travels through the body of the earth at 7.7 km/s. (*a*) Which wave arrives at B first? (*b*) What is the time difference between the arrival of the two waves at B's seismograph?

10-C5 The transmitter of a radar-equipped patrol car sends out radio waves of frequency 10.5 GHz. When reflected by an approaching automobile (serving as a virtual source moving at $2v$), the reflected beam makes 2160 beats/s with the outgoing beam. (*a*) What is the automobile's speed? (*b*) Was it above the legal limit?

10-C6 Nolan Ryan's fast ball (Prob. 2-B4) had a velocity component 29.2 m/s in the line of sight from press box to home plate. What frequency shift, in MHz, was given when infrared radiation of wavelength 10.59 μm (from a CO_2 laser) was reflected by the ball?

10-C7 Use values in Tables 9-1 and 1-3 to calculate the speed of a compression wave in air. Can you explain why your answer is about 20% lower than the observed value? (*Hint:* See footnote to Table 9-1. This will be studied in Sec. 16-7.)

10-9 Derivation of the Equation for Velocity of Compression Waves

By the method of dimensional analysis, we have shown in Sec. 10-3 that the velocity of a compression wave along a solid rod is $K\sqrt{E/d}$, where E is the stretch modulus $\dfrac{\Delta F/A}{\Delta L/L}$. The dimensionless constant K could not be found by dimensional analysis; we stated without proof that its value is 1. There follows a derivation of the equation for the velocity of a compression wave based on Newton's second law.

We shall use the expression of Newton's second law that is summarized in the energy principle. You will recall that we successfully applied Newton's laws in this form when we derived the period of the simple harmonic motion of a body on a spring (Sec. 9-3). To derive the speed of a compression wave, we shall use the railroad-train model described in Sec. 10-2.

Consider a long rod of metal (Fig. 10-12) of density d. The cross section of the rod is A. With the rod initially at rest and clamped at one end, we suddenly apply a constant force ΔF at the other end of the rod, moving that end to the right with a steady velocity u. After a time Δt has elapsed, the compression pulse has reached a point c. During this time the metal to the left of c has been compressed by a small amount $ab = u\,\Delta t$; this much metal is moving at velocity u and is about to set into motion some additional metal beyond c. We let v_w be the wave velocity, and

hence $ac = v_w\,\Delta t$. To apply the energy principle, we need to know the PE of the compressed metal. This PE was stored in the rod because of the compression by an amount $u\,\Delta t$. The work done to compress the rod, against a variable force that increases steadily from 0 to ΔF, is the average force times the displacement (see Fig. 9-14 on page 202). Therefore the PE of the compressed portion of the rod is

$$\frac{(0 + \Delta F)}{2}(u\,\Delta t) \quad \text{or} \quad \tfrac{1}{2}\Delta F(u\,\Delta t)$$

The compressed part of the rod, bc, is in motion, and so it has some KE given by $\tfrac{1}{2}mu^2$. Its mass is $Vd = ALd\ (=Av_w d\,\Delta t)$, and the velocity is u. According to the energy principle, the work done by the applied force ΔF equals the gain in PE plus the gain in KE:

$$\Delta F(u\,\Delta t) = \tfrac{1}{2}\Delta F(u\,\Delta t) + \tfrac{1}{2}(Av_w d\,\Delta t)(u^2) \quad (10\text{-}8)$$

To simplify this we need to express u in terms of v_w. We have

$$u\,\Delta t = \Delta L$$

and

$$v_w\,\Delta t = L$$

Therefore $u/v_w = \Delta L/L$, and

$$u = v_w\,\frac{\Delta L}{L}$$

Putting this into Eq. 10-8 gives, after some algebra,

$$\Delta F = Av_w{}^2 d\,\frac{\Delta L}{L}$$

Solving for the wave velocity v_w and rearranging terms gives

$$v_w = \sqrt{\frac{(\Delta F/A)/(\Delta L/L)}{d}}$$

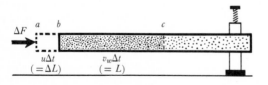

Figure 10-12

The numerator is the stretch modulus E, as defined in Sec. 9-1. Hence,

$$v_w = \sqrt{\frac{E}{d}}$$

10-10 A Mathematical Description of Beats

The resultant of two vibrations of equal amplitudes and different frequencies is given by combining two vibrations of the form given by Eq. 9-6 (page 204):

$$y = A \sin 2\pi f_1 t + A \sin 2\pi f_2 t$$

Now we use the trigonometric identity

$$\sin a + \sin b \equiv 2 \cos \frac{a-b}{2} \sin \frac{a+b}{2}$$

which leads to

$$y = \left[2A \cos 2\pi \left(\frac{f_1 - f_2}{2} \right) t \right] \sin 2\pi \left(\frac{f_1 + f_2}{2} \right) t$$

(10-9)

This is the equation for a superposition such as curve C in Fig. 10-9. The sine term represents a vibration having the average frequency $\frac{1}{2}(f_1 + f_2)$. The term in square brackets represents a slowly varying resultant amplitude that fluctuates between 0 and $2A$ with a frequency $\frac{1}{2}(f_1 - f_2)$. The number of beats per second is twice this value, $f_1 - f_2$, because the cosine function goes through zero twice during each complete cycle. Thus Eq. 10-9 tells us that the beat frequency is $f_1 - f_2$, the difference between the frequencies of the two superposed vibrations. The equation also shows that the perceived frequency is the *average* of the two combining frequencies.

References   1. N. Isenor and K. Woolner, "Chaucer's Theory of Sound," *Phys. Today* 33(3), 115 (Mar. 1980). Contains a 54-line quotation from *House of Fame*, with commentary. Chaucer had a good insight into sound as a wave motion in air.
2. O. Laporte, "Shock Waves," *Sci. American* 181(5), 14 (Nov. 1949).
3. J. Strickland, *Doppler Effect; Formation of Shock Waves* (films).
4. F. Miller, Jr., *Nonrecurrent Wavefronts* (film).

11

Interference
and Stationary Waves

We have studied wave motion in general and have used sound and water waves to illustrate the propagation of a disturbance through a medium. So far, we have tacitly assumed that once a disturbance is created, it spreads out indefinitely, with no hindrance or absorption. In practice, we are often faced with the likelihood that two or more waves will be simultaneously passing through a region of space. Sound waves are reflected at a wall; water waves from several boats may reach a floating log at about the same time; waves on a stretched string sooner or later reach the end of the string and must be reflected or absorbed. We use the term *interference* to describe the result of the superposition of several waves traveling through the same region of space. Interfering waves do not always tend to cancel or oppose one another, as might be expected from the nontechnical use of the word; there can be constructive as well as destructive interference, and intermediate cases of many sorts. The interference of two waves traveling in opposite directions, having the same wavelength, frequency, and speed, gives rise to the phenomenon of *stationary waves*, which is of utmost

importance in the production of musical and other sounds. The phenomenon is most pronounced if the waves also have identical amplitude.

11-1 Interference

Graphical methods are adequate for a description of what happens when two identical waves interfere. Consider first transverse waves on a long string (Fig. 11-1). The wave represented by the dotted line is moving toward the right, and the dashed-line wave is moving toward the left. At the instant shown in the top graph (for which we have taken $t = 0$), the two waves happen to be exactly out of phase. At the point P, for instance, the dotted-line wave tends to move the string downward, and the dashed-line wave tends to move the same particle of string upward. The net effect is zero displacement. A similar argument holds for every particle of the string, and therefore at $t = 0$ the entire string is straight—in its natural, undistorted state. After a brief interval of time equal to $\frac{1}{4}$ period, each

Indian musicians of the Bolivian highlands.
Principles of resonance and tuning of pipes are known in many cultures.

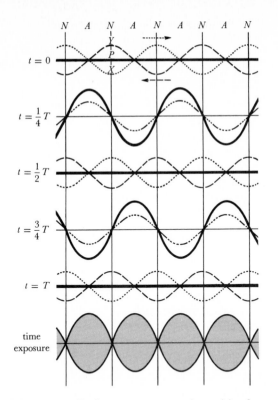

Figure 11-1 Stationary wave on a string, arising from interference of two identical waves traveling in opposite directions.

bance, are called *antinodes*. From the graphs, it is evident that *nodes are half a wavelength apart*, and also that *antinodes are half a wavelength apart*. We call the whole phenomenon by the name *stationary wave*—but the two waves that interfere are not at all stationary. It is the location of nodes and antinodes that is stationary. This is shown in the time exposure of Fig. 11-1, in which the nodes are located at fixed points along the string. If the amplitudes of the interfering waves are not equal, then a stationary pattern is obtained in which the nodes are minima, but are not of zero displacement.

Although we have used waves on strings for our example, periodic waves of any sort can combine to give stationary wave patterns. The discussion is general; it is not limited to waves on strings but applies to waves of compression, electromagnetic waves, and all other periodic waves. For instance, the acoustic properties of concert halls are strongly influenced by stationary wave patterns formed by the interference of waves reflected from the walls and ceiling. For a discussion of the mathematical representation of wave motion and stationary waves, see the For Further Study Sec. 11-8 (pages 262–264).

wave has moved $\frac{1}{4}$ wavelength. The dotted-line wave has taken the trough X to the right, and the dashed-line wave has taken the crest Y to the left. The two waves are now exactly in phase; they cooperate to make a king-sized wave form. (We are here using the superposition principle of Sec. 10-7.) We can continue this graphical procedure, $\frac{1}{4}$ period at a time, and obtain resultant wave forms as shown in Fig. 11-1. Thus at certain times (such as $t = 0$, $t = \frac{1}{2}T$, and $t = T$) the string is undistorted, with no net disturbance anywhere along the string. At other times (such as $t = \frac{1}{4}T$, $t = \frac{3}{4}T$, and $t = \frac{5}{4}T$) the string is violently disturbed.

We note in Fig. 11-1 that some points, marked N, never move from their normal positions. Such points of no disturbance are called *nodes*. The points marked A, which have maximum distur-

11-2 Vibrating Strings

We can now discuss the vibration of a string that is fixed at both ends. Such a string (or wire) might be found in a piano, on a guitar or violin, or between two telegraph poles along a railroad right of way. If any particle of the string is moved aside and released, it vibrates back and forth, perpendicular to the length of the string, and this disturbance spreads out along the string in both directions. The wave velocity depends on the tension in the string and on the string's mass per unit length. It can be shown (Sec. 11-9) that the velocity of a transverse wave along a string or wire is

$$v_w = \sqrt{\frac{F}{m/L}} \qquad (11\text{-}1)$$

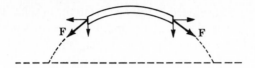

Figure 11-2 The net force on a short segment of string is the resultant of the two force vectors; the resultant is directed toward the equilibrium position.

where F is the tension and m/L is the mass per unit length. This equation is a reasonable one. If the string is a "heavy" rope, the mass (inertia) of 1 m of the rope is greater than the mass of 1 m of a "light" string or thread. The greater inertia of the molecules causes neighboring particles to resist being set into motion, and the disturbance is handed on at a slower rate. This checks with the presence of m/L in the denominator of our formula. On the other hand, if the tension is greater, there is more net force acting to accelerate any segment of the string (Fig. 11-2), and we expect larger accelerations. The neighboring particles are disturbed sooner, and the wave travels along the string faster. This checks with the presence of F in the numerator of our formula.

Example 11-1

A clothesline 20 m long has a mass of 0.80 kg and is stretched between two fixed posts with a tension of 400 N. If the rope is struck a sharp blow at one end, how much time will be required for the pulse to travel to the opposite post and back again?

First we find the wave velocity. The mass per unit length is

$$\frac{m}{L} = \frac{0.80 \text{ kg}}{20 \text{ m}} = 0.04 \text{ kg/m}$$

The wave velocity is

$$v_w = \sqrt{\frac{F}{m/L}} = \sqrt{\frac{400 \text{ N}}{0.04 \text{ kg/m}}}$$

$$= \sqrt{10,000 \text{ N} \cdot \text{m/kg}} = 100 \text{ m/s}$$

Now the time for the round trip can be found:

$$t = \frac{s}{v_w} = \frac{40 \text{ m}}{100 \text{ m/s}} = \boxed{0.4 \text{ s}}$$

The pulse of Example 11-1 will be reflected at each post and will travel back and forth, making one complete trip every 0.4 s, until finally it dies out because of internal molecular friction in the rope. Some energy will be transferred to the posts if they are not absolutely rigidly set into the ground.

An important concept has been introduced in Example 11-1. When the clothesline is struck near one end, the pulse returns again and again, with a definite frequency—2.5 vib/s in this case. This frequency of vibration does not depend on the exact nature of the blow and is characteristic of the string. A *natural frequency* of a vibrating body is one that is characteristic of the body; we have just seen in Example 11-1 how one of the natural frequencies of a stretched rope depends on the length, mass, and tension of the rope.

Passing now to *periodic* waves traveling back and forth along a string, let us imagine a piano string on which sinusoidal displacement waves travel back and forth. Because of the reflections of the waves at the fixed ends, we will have interference between two identical waves traveling in opposite directions. That is, the conditions for obtaining stationary waves will be fulfilled. As a consequence, nodes and antinodes will be set up, spaced $\frac{1}{2}\lambda$ apart. One condition must be fulfilled: nodes *must* exist at each end of the string, for the string is surely never disturbed at these two points. There may or may not be nodes between the ends, and the string may vibrate in 1 segment, 2 segments, 3 segments, or any other integral number of segments. These are called *modes* of vibration. We can summarize by saying that for each different mode of vibration there is a different distribution of nodes, but, in all modes, nodes must exist at each end of the string. The possible frequencies of vibration are related to the number of segments, as shown in our next

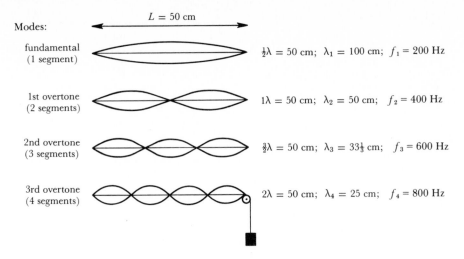

Modes:

fundamental (1 segment) $\frac{1}{2}\lambda = 50$ cm; $\lambda_1 = 100$ cm; $f_1 = 200$ Hz

1st overtone (2 segments) $1\lambda = 50$ cm; $\lambda_2 = 50$ cm; $f_2 = 400$ Hz

2nd overtone (3 segments) $\frac{3}{2}\lambda = 50$ cm; $\lambda_3 = 33\frac{1}{3}$ cm; $f_3 = 600$ Hz

3rd overtone (4 segments) $2\lambda = 50$ cm; $\lambda_4 = 25$ cm; $f_4 = 800$ Hz

$L = 50$ cm

Figure 11-3 Modes of vibration of a string.

example. The lowest natural frequency is called the *fundamental* frequency, and the other frequencies are called *overtones*.

_____ **Example 11-2**

A guitar string 50 cm long has a mass of 50 g and is under tension of 4×10^8 dyn. What are the wavelengths and frequencies of the first four modes of vibration?

The wavelengths are found directly from diagrams (Fig. 11-3) that show the locations of the possible nodes. We make use of the fact that nodes are spaced $\frac{1}{2}\lambda$ apart. For instance, when the string is vibrating in 3 segments, with each segment $\frac{1}{2}\lambda$ in length, we have $3(\frac{1}{2}\lambda) = 50$ cm, whence $\lambda = \frac{100}{3}$ cm $= 33\frac{1}{3}$ cm. The final column is filled in from knowledge of the wave velocity:

$$v_w = \sqrt{\frac{F}{m/L}} = \sqrt{\frac{4 \times 10^8 \text{ dyn}}{50 \text{ g}/50 \text{ cm}}}$$

$$= 2 \times 10^4 \text{ cm/s} = 20{,}000 \text{ cm/s}$$

Knowing the wave velocity, we can find the corresponding natural frequencies from $v_w = f\lambda$. The fundamental frequency is

$$f_1 = \frac{v_w}{\lambda_1} = \frac{20{,}000 \text{ cm/s}}{100 \text{ cm}} = 200 \text{ Hz}$$

Similarly, the first overtone's frequency is

$$f_2 = \frac{v_w}{\lambda_2} = \frac{20{,}000 \text{ cm/s}}{50 \text{ cm}} = 400 \text{ Hz}$$

and so forth for the frequency of any overtone.

It is evident from Example 11-2 that the natural frequencies of the allowed vibrations of a string are all integral multiples of the fundamental frequency. Such remarkable simplicity is a result of the fact that nodes on a string are equally spaced. If, as in this case, the overtones have frequencies that are integral multiples of the fundamental frequency, they are called *harmonics*.* For any string, the natural frequencies are in the ratios $1:2:3:4:\ldots$ and hence are harmonics as well as overtones.

_____ **Example 11-3**

A horizontal string 1.20 m long has a mass of 562 mg. It is attached at one end to a tuning fork producing 100 vib/s; the other end of the string runs over a pulley and supports a light pan. What load should

_____
* So called because musical scales and harmony are based on integral ratios of frequencies.

be placed in the pan to cause the string to vibrate in 4 segments?

The distance between nodes is $\frac{1}{4}$ the length of the string, or 0.30 m. This means that $\frac{1}{2}\lambda = 0.30$ m, or $\lambda = 0.60$ m. We next find the wave velocity from the frequency and wavelength:

$$v_w = f\lambda = (100 \text{ s}^{-1})(0.60 \text{ m}) = 60 \text{ m/s}$$

Finally, we find the tension from the velocity formula that applies to strings. The mass of the string is 562×10^{-3} g, or 562×10^{-6} kg.

$$v_w = \sqrt{\frac{F}{m/L}} = \sqrt{\frac{FL}{m}}; \qquad v_w{}^2 = \frac{FL}{m}$$

$$F = \frac{v_w{}^2 m}{L} = \frac{(60 \text{ m/s})^2(562 \times 10^{-6} \text{ kg})}{1.20 \text{ m}}$$

$$= 1.69 \text{ N}$$

Using $m = W/g$, we conclude that the mass of the load in the pan is given by

$$\frac{1.69 \text{ N}}{9.80 \text{ m/s}^2} = \boxed{0.172 \text{ kg}} \text{ or } \boxed{172 \text{ g}}$$

A type of laboratory balance makes use of the fact that the natural frequency of a vibrating wire depends on the wave speed, which in turn depends on the tension (Eq. 11-1). Two vibrating wires are subject to forces due to the weight of the unknown load and that of a standard mass. The ratio of the two frequencies is measured electronically and processed to give the unknown mass, which is displayed in a built-in calculator readout.

11-3   Vibrating Air Columns in Pipes

The production of certain natural frequencies in a vibrating column of air was known to the ancients, who fashioned shepherds' pipes and flutes on this principle. It was found that the frequency of vibration depended on the effective length of the air column—hence the strategically placed air holes along the length of a flute or recorder. We can analyze the modes of vibration of pipes in much the same way as we discussed modes of vibration of strings. As before, the location of nodes is the key to understanding what happens. When a condensation approaches a closed end

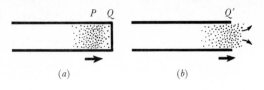

Figure 11-4   (a) Closed pipe. (b) Open pipe.

of a pipe (Fig. 11-4a), the molecules at P are free to move and swish back and forth as pressure builds up or as a partial vacuum is created. However, molecules at Q are in contact with the solid end of the pipe. Hence they cannot move, and so Q must be a node of displacement (the displacement remains 0). This is also called a motion node. We conclude that *the closed end of a pipe is a motion node*. On the other hand, if a condensation approaches an open end of the pipe, as at Q' (Fig. 11-4b), the molecules moving to the right have inertia and are carried out into the open air, leaving a partial vacuum behind. There is a maximum amount of to-and-fro motion at the open end of a pipe. In other words, *the open end of a pipe is a motion antinode*.

A vibrating column of air also has nodes and antinodes of pressure. A motion node is a pressure antinode, and a motion antinode is a pressure node. At a closed end (Q in Fig. 11-4a), the pressure fluctuates a great deal as molecules are squeezed together and then relax; point Q is a pressure antinode. At an open end (Q' in Fig. 11-4b), the pressure remains constant, equal to atmospheric pressure, as molecules move freely in and out of the pipe; point Q' is a pressure node.

To study the natural frequencies of open and closed pipes, we use our knowledge of the locations of nodes and antinodes. In what follows, the terms "nodes" and "antinodes" refer to *motion* nodes and antinodes, unless otherwise specified.

For a pipe that is open at both ends (Fig. 11-5) there must be antinodes at each end. Remembering that antinodes are spaced $\frac{1}{2}\lambda$ apart, we can make diagrams of the various modes of vibration and determine the wavelength for any given

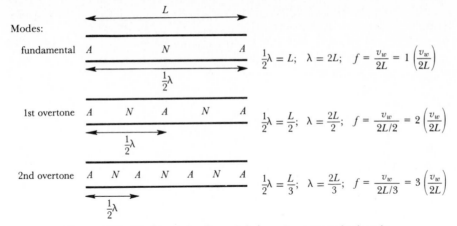

Figure 11-5 Modes of vibration of air in a pipe open at both ends.

mode by inspecting the diagrams. The calculations summarized in Fig. 11-5 show that for an open pipe the natural frequencies are integral multiples of the fundamental frequency. This is exactly the same result we found for vibrating strings—the natural frequencies are in the ratio 1:2:3:4:.... The reason for this similarity is not hard to find. Comparing Fig. 11-5 with Fig. 11-3, we see that the vibrating body (string or air column, as the case may be) is bounded at each end by similar conditions—either both ends are nodes or both ends are antinodes. Replacing every N in Fig. 11-5 by A and every A by N would give a distribution of nodes and antinodes just as in Fig. 11-3. In other words, the symmetry properties of strings and open pipes are essentially the same, and so the modes of vibration, and therefore the natural frequencies, would be expected to be similar.

If a pipe is closed at one end, there must be an antinode at the open end and a node at the closed end. We can prepare in the usual way a table of values for the natural frequencies of a closed pipe (see Fig. 11-6), remembering that from node to

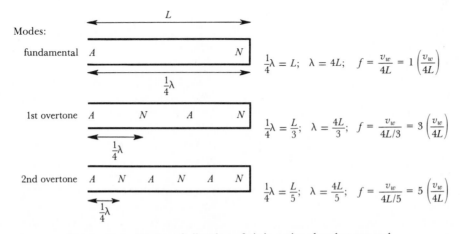

Figure 11-6 Modes of vibration of air in a pipe closed at one end.

antinode is $\frac{1}{4}\lambda$. We see that for a closed pipe the possible frequencies are *odd* multiples of the fundamental frequency; even multiples of the fundamental frequency are not possible. The symmetry properties of a closed pipe differ from those of an open pipe or a string because the situation at one end is different from the situation at the other end. The natural frequencies for a closed pipe are in the ratio $1:3:5:7:\ldots$. Only odd harmonics occur; the 1st overtone is the 3rd harmonic, the 2nd overtone is the 5th harmonic, and so on.

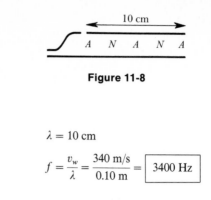

Figure 11-8

$\lambda = 10$ cm

$$f = \frac{v_w}{\lambda} = \frac{340 \text{ m/s}}{0.10 \text{ m}} = \boxed{3400 \text{ Hz}}$$

Example 11-4

What is the fundamental frequency of a whistle that is 10 cm long from blowhole to open end? Assume the speed of sound in air to be 340 m/s.

We first draw a diagram of the pipe (Fig. 11-7), putting antinodes at each end, since the pipe is open at both ends. According to the statement of the problem, we want the fundamental frequency, that is, the mode that has the fewest possible nodes and antinodes. This means that there is a single node, at the center of the pipe. From the diagram, $\frac{1}{2}\lambda = 10$ cm; $\lambda = 20$ cm. Now, since the speed of sound is 340 m/s,

$$f = \frac{v_w}{\lambda} = \frac{340 \text{ m/s}}{0.20 \text{ m}} = \boxed{1700 \text{ Hz}}$$

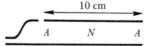

Figure 11-7

Example 11-5

By blowing harder, the whistler can cause the pipe to vibrate in its first overtone. What is the frequency of this overtone?

For the desired mode of vibration, the location of nodes and antinodes is as shown in Fig. 11-8.

Example 11-6

What is the lowest frequency that can be produced in the pipe of Example 11-4 if the end of the pipe is sealed with a close-fitting plug?

The pipe is now a closed pipe, and the diagram of nodes and antinodes is shown in Fig. 11-9.

$$\tfrac{1}{4}\lambda = 10 \text{ cm}; \qquad \lambda = 40 \text{ cm}$$

$$f = \frac{v_w}{\lambda} = \frac{340 \text{ m/s}}{0.40 \text{ m}} = \boxed{850 \text{ Hz}}$$

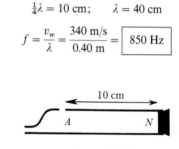

Figure 11-9

The speed of sound in a gas depends on the density, which in turn depends on the temperature. As we shall see in Sec. 16-7, v_w is proportional to the square root of the absolute temperature ($v_w \propto \sqrt{T}$), which implies that for air near room temperature, v_w changes by about 0.60 m/s for every 1 C° change in temperature. For this reason the natural frequencies of any pipe depend on the temperature. Every player of a wind instrument knows how important it is to "warm up" the instrument. Temperature differences between the pipes of an organ may cause the organ to be badly out of tune. A change of frequency of as little as 1% would be distressingly noticeable, and this could occur if some pipes were affected more than others by uneven temperatures in an organ loft.

Example 11-7

A closed organ pipe is 0.750 m long and vibrating in its fundamental mode. (*a*) What is the frequency on a day when the temperature is 15°C and the speed of sound in air is 340 m/s? (*b*) If the temperature rises to 25°C, what does the frequency become?

(*a*) The fundamental wavelength of the closed pipe is 4 times its length, or

$$\lambda = 4 \times 0.750 \text{ m} = 3.00 \text{ m}$$

Hence

$$f_1 = \frac{v_w}{\lambda} = \frac{340 \text{ m/s}}{3.00 \text{ m}} = \boxed{113.3 \text{ Hz}}$$

(*b*) At 25°C the speed of sound in air is $[340 + (0.60 \times 10)]$ m/s = 346 m/s. The wavelength is still 3.00 m (ignoring the negligible expansion of the pipe itself), since the locations of nodes and antinodes have not changed. The fundamental frequency is now

$$f_1' = \frac{v_w}{\lambda} = \frac{346 \text{ m/s}}{3.00 \text{ m}} = \boxed{115.3 \text{ Hz}}$$

11-4 Resonance

Our study of stationary waves on strings and in air columns has shown that each vibrating body has a certain set of natural frequencies, characteristic of the size and shape of the body and the nature of the medium. We have met with natural frequencies before: a simple pendulum has a single natural frequency, the reciprocal of its period. A mass on the end of a spring has a natural frequency of vibration. These examples from Chap. 9 are simpler than those discussed in this chapter, for the pendulum and the vibrating object on the spring each have only a single natural frequency, unlike the string and the pipe, which have an infinite sequence of natural frequencies.

A child sitting on a swing can be set into vibration by a friend who applies impulses at the proper times. Experience tells us that even a small child can give a large amplitude of vibration to a heavy playmate by applying repeated small impulses at just the right times. *Resonance* is the building up of a large vibration by repeated application of small impulses whose frequency equals one of the natural frequencies of the resonating body. In the example of the child on the swing, a large amplitude is built up only if the impulses are applied in synchronism with the natural period of vibration of the swing. A slight mismatch of frequency would result in little or no vibration.

Acoustic resonance occurs in many musical instruments (Sec. 11-7) and in auditoriums. Resonance is not always helpful. For instance, if a loudspeaker enclosure has a natural frequency of 200 Hz, the sounds of this frequency will be "amplified" and the reproduced music will have an unpleasant boomy sound. Bridges have been known to collapse because of mechanical resonance as large vibrations were built up from gusts of wind (see Ref. 29 at the end of this chapter). On the other hand, electrical resonance (Secs. 23-16 and 24-3) is helpful in radio receivers, which must respond with large amplitude to weak signals at the frequency of the desired station, but must reject (that is, respond very little to) the signals of other stations that broadcast at other frequencies. A certain "rattle" in the family car may be noticed only at 53 km/h, and not at 43 or 63 km/h. The loose part evidently has a natural frequency that is "excited" by the vibration of the motor; the excitation is effective only when the motor supplies impulses at the resonant frequency of the vibrating part.

Resonance is related to physical limitation of a vibrating body or medium. A pipe of infinite length has no natural frequency of vibration. However, if a loudspeaker is placed in front of a pipe that is terminated by either a closed end or an open end, it will cause a large disturbance of the air column if its frequency (the exciting frequency) closely matches one of the natural

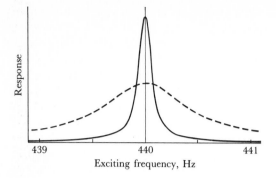

Figure 11-10 Resonance curves for two bodies, each with natural frequency 440.0 Hz. Solid curve: sharp resonance; dashed curve: broad resonance.

frequencies of the pipe. Thus every natural frequency of a body is also a resonant frequency.

A resonance is said to be "sharp" if the response is large only for exciting frequencies that are very close to the natural frequency (Fig. 11-10). The sharpness of a resonance depends on the rate at which the vibration would die away if left to itself. In general terms, a resonance is sharp if the energy stored in the vibrating body is large compared with the energy that is dissipated per cycle. Let us illustrate this by discussing a simple experiment with the A string of a piano, which has a natural frequency of 440.0 Hz. If a singer emits a 440.0-Hz tone, the string can be heard to vibrate after the sung tone ceases. The response to 440.1 Hz is less than to 440.0 Hz, and the response to 441.0 Hz is so small as to be inaudible. This rather sharp resonance can be predicted, for we know that the vibration of a struck piano string endures for many seconds. In, say, 25 s, the string vibrates 11,000 times, and the energy lost per cycle is evidently a small fraction of 1%. The rattle of a part of a car, mentioned earlier, is a broader resonance; if the rattle is most pronounced at a frequency corresponding to 53 km/h, it can also be heard if the car's speed is 52 or 54 km/h. This broad resonance is correlated with the fact that the vibration, once started, would die out in less than a second if it were not continuously excited by the outside source of energy.

11-5 Musical Sounds and Their Sources

Some of the most interesting and familiar applications of stationary waves and resonance involve musical instruments such as the piano, violin, trombone, saxophone, and even the musical saw. The primary distinction between music and noise is, of course, a cultural one. We are accustomed to calling certain sounds musical because they are "pleasing." Sharp differences of opinion exist as to whether certain sounds or combinations of sound are music or noise, and the opinions of one generation or cultural group are not necessarily the same as those of another. However, in broad outline, the general attributes of the component sounds that make up music are not controversial. Musical sounds are characterized by pitch, loudness, and quality. These are psychological terms denoting sensations; we shall see that each of these three attributes is largely (but not entirely) dependent on one of three physical attributes: frequency, intensity, and overtone structure, respectively.

Psychological Sensations	Physical Attributes
pitch	frequency
loudness	intensity
quality	overtone structure

Pitch and Frequency. Musical sounds are periodic waves in air, and are characterized by wavelength and frequency. The sensation of *pitch* is closely related to the *frequency* of a sound wave; high-pitched sounds are of high frequency, and low-pitched sounds are of low frequency. Hence our first correlation is pitch (a sensation) and frequency (a physically measurable attribute of a wave). In these days of high fidelity and hearing aids, we are familiar with the fact that the human ear has frequency limits. The sensation of sound is not produced in the brain unless the frequency lies between limits that are roughly 20 Hz and 20,000 Hz. Even this range is not available to all persons throughout their lives; for most, the process of aging involves a loss of frequencies above, say, 10,000 Hz or even less. For some, a hearing defect may involve relative loss of low-frequency sounds.

Compression waves of frequencies above about 20,000 Hz are called *ultrasonic waves*. Although they do not affect the human ear, and hence have no pitch, such waves are in every other way similar to the sound waves we have discussed. For example, stationary ultrasonic waves are easily produced. Because of their high frequency (and short wavelength), ultrasonic waves produce some effects in matter that are little, if at all, evident at audible frequencies. Medical diagnosis is made possible by the fact that the reflection and absorption of ultrasonic waves depend on tissue density. Thus images of soft internal structures can be produced (Fig. 11-11). At the other extreme of inaudible sounds, infrasonic waves of, say, 1 to 10 Hz are of great importance in engineering, for they can lead to destructively large vibrations if one of a structure's natural frequencies can be excited by a resonance type of interaction (Sec. 11-4).

As a matter of terminology, two sounds are said to be one *octave* apart if their frequencies are in the ratio 2:1. Thus an interval of 7 octaves, representing almost the complete span of a piano, from lowest C to highest C, is a ratio $2 \times 2 \times 2 \times 2 \times 2 \times 2 \times 2 (= 2^7)$, or 128:1. The lowest C on the piano has frequency 32.70 Hz, and the highest C, 7 octaves higher, is 128×32.70, or 4186 Hz. It is seen that the human ear's sensitivity extends well beyond the piano frequencies, in the direction of both higher and lower frequencies. Another interval that is worth knowing is called a *fifth*. This interval corresponds to a frequency ratio 3:2. Another simple frequency ratio, 5:4, is called a *major third*, and 6:5 is a *minor third*. Note that the pleasing chord "do-mi-sol," exemplified by the tones C-E-G, contains a fifth (C-G), a major third (C-E), and a minor third (E-G). The frequencies in the chord are in the simple ratios 4:5:6. Further information on musical scales can be found in the References at the end of the chapter.

Loudness and Intensity. The *loudness* of a sound depends on the rate at which acoustic energy enters the ear of the listener. When a sound wave travels through the air, the individual gas molecules oscillate back and forth but do not move far from their equilibrium positions. Nevertheless, energy is handed on from one molecule to the next, and this flow of energy can be measured by physical means. One direct method of measuring the energy flow is to place a sound-absorbing body in the path of the sound wave. As energy is absorbed, the temperature of the absorber rises at a steady rate. If we know the thermal characteristics of the absorber, we can compute the rate of energy flow. Of course, such measurements are extremely delicate, and the increases in temperature are very small. In actual practice, it would be more convenient to use a microphone (connected to an amplifier and a meter of some sort) for routine measurements, and calibrate the microphone once and for all by the more difficult, but more direct, thermal method. The rate of flow of energy in a wave, power per unit cross-sectional area, is called the *intensity I* of the wave.

$$\text{Intensity} = \frac{\text{power}}{\text{area}}$$

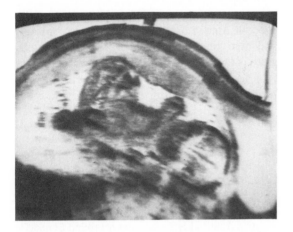

Figure 11-11 Ultrasonic scan, made at 2.25 MHz, showing fetus about 20 days before delivery. Because of possible damage to the fetus, such a scan is used only when needed to plan a safe delivery.

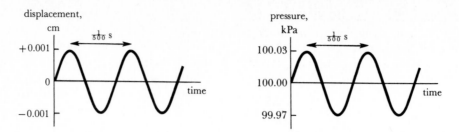

Figure 11-12 Two representations of a sound wave of frequency 500 Hz, having the maximum intensity that can be tolerated by the ear.

where the area is oriented perpendicular to the incoming wave. The SI unit for intensity is J/s·m², or W/m².

Motions and pressures in a sound wave are surprisingly small. Even for the noise of a military jet plane, which is close to the threshold of pain, gas molecules are displaced only about 0.001 cm at the most (assuming a wave whose frequency is about 500 Hz). In other words, the displacement amplitude in such a wave is 0.001 cm. In the same wave, in the condensations where the molecules are crowded together, the pressure is only 0.03 kPa (0.0003 atm) above the normal atmospheric pressure of 1 atm. That is, the pressure amplitude is only 0.03 kPa (Fig. 11-12). These relatively small changes in pressure are enough to drive the eardrum beyond a safe limit. The sound wave described in Fig. 11-12 has an intensity of about 1 W/m²; at the other extreme, the faintest detectable sound* has an intensity of about 10^{-12} W/m².

The psychological sensation of loudness is almost entirely determined by the physical attribute, intensity, other factors being equal.

The *intensity level* β of a sound wave is measured in *bels*. The bel is defined in terms of the ratio of two intensities: two sounds differ by one bel (named for Alexander Graham Bell) if their intensities are in the ratio 10:1. Usually a smaller unit, the *decibel* (dB) is used (1 dB = $\frac{1}{10}$ bel). A

---

* The faintest sound that can be heard has a pressure amplitude of only 0.00002 Pa. Since atmospheric pressure is about 1×10^5 Pa (Sec. 12-2), this means that the ear responds to changes of $2 \times 10^{-5}/10^5$, or 1 part in 5 billion!

logarithmic scale is used to measure the intensity level in decibels:

$$\beta = 10 \log \frac{I}{I_0} \qquad (11\text{-}2)$$

where I_0 is some arbitrary reference level.

To find the difference between the intensity levels of two sounds whose intensities are I_1 and I_2, we can use the equation (see Prob. 11-B26)

$$\Delta\beta = \beta_2 - \beta_1 = 10 \log \frac{I_2}{I_1} \qquad (11\text{-}3)$$

The decibel is approximately the threshold of discrimination of the human ear under the most favorable circumstances. For instance, sounds differing by only 0.2 dB are indistinguishable in loudness.

Usually, when comparing two sounds, the effective area A of the listener's ear is constant, and then, since $I = P/A$, the ratio of the intensities is the same as the ratio of powers. We can thus use a power ratio instead of an intensity ratio when comparing two sounds:

$$\Delta\beta = 10 \log \frac{P_2}{P_1} \qquad (11\text{-}4)$$

Example 11-8

What is the difference in intensity level between the sound from 6 cheerleaders and the sound from 6000 cheering fans?

The power ratio is 6000:6, so $P_2/P_1 = 1000$.

$$\Delta\beta = 10 \log 1000 = 10(3.0) = \boxed{30 \text{ dB}}$$

Example 11-9

A hi-fi's power output is 40 W at some audible mid-frequency such as 500 Hz. The output is 3 dB less if the frequency is decreased to 20 Hz. What is the power output at 20 Hz?

We place the larger power in the numerator of the expression P_2/P_1, so that the ratio is greater than 1 and its logarithm is positive.

$$\Delta\beta = 3 = 10\log\frac{40\text{ W}}{P_1}$$

$$0.3 = \log\frac{40\text{ W}}{P_1}$$

Using a log table or a pocket calculator, we find that $0.301 = \log 2.00$. Hence,

$$\log 2.0 = \log\frac{40\text{ W}}{P_1}$$

$$2.0 = \frac{40\text{ W}}{P_1}; \qquad P_1 = \boxed{20\text{ W}}$$

Strictly speaking, the decibel is not an absolute unit, because it gives the *ratio* of two intensities or two loudnesses. In many applications "0 dB" is defined as 10^{-12} W/m^2, the faintest detectable sound for the human ear. A sound level of 53 dB is therefore a sound that is 53 dB louder than the threshold of hearing. A calculation similar to that of Example 11-9 shows that such a sound has an intensity of 2×10^{-7} W/m^2.

Quality and Overtone Structure.
Two sounds of equal loudness and pitch are often

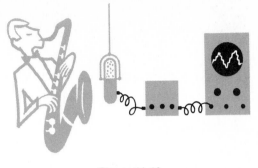

Figure 11-13

distinguishable from each other. A violin tone is said to have a *quality* that is different from that of a saxophone. What physical attribute is correlated with such obvious differences between sounds? Figure 11-13 shows an experimental setup for measuring the wave form of the pressure variations in a sound wave. As the wave spreads out, the pressure at the microphone varies in some fashion. The microphone, amplifier, and oscilloscope together perform the function of displaying on the screen a graph of pressure vs. time. Strictly speaking, the scope displays the vibration graph at a fixed point (the microphone). However, we know from Sec. 10-5 that the wave-form graph has the same shape as the vibration graph. We immediately notice that the wave forms of violin, trumpet, and clarinet (Fig. 11-14) are different, and so we correlate *quality* (a sensation) with *wave form* (a physically measurable characteristic).

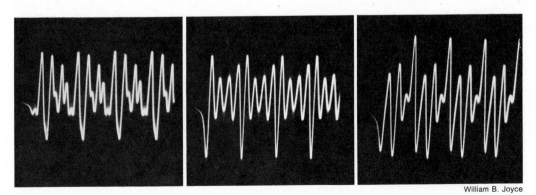

William B. Joyce

Figure 11-14 Wave forms of musical sounds. Left: violin; center: trumpet; right: clarinet.

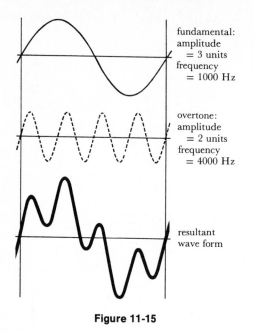

fundamental:
amplitude
= 3 units
frequency
= 1000 Hz

overtone:
amplitude
= 2 units
frequency
= 4000 Hz

resultant
wave form

Figure 11-15

The measurement of wave form in quantitative terms involves the superposition principle. It was shown by the French mathematician and physicist Joseph Fourier (1768–1830) that a complex wave form can be represented as a sum, or superposition, of simple harmonic motion wave forms. Thus the complex wave form of Fig. 11-15 can be represented as the super-

position of two simple waves: a fundamental vibration (light line) of amplitude 3 units and frequency 1000 Hz, and an overtone (dashed line) of amplitude 2 units and frequency 4000 Hz. The superposition of these two simple wave forms (Fig. 11-15) gives the observed wave form (heavy line). In actual practice, more than two components are often needed to represent a complex wave form. To represent a wave form, therefore, we construct a table showing the frequencies and relative intensities of the various overtones that together make up a complex wave.* Typical results for two sources of musical sounds are shown in Table 11-1. We call this table a table of *overtone structure* for each musical sound, and we summarize by saying that the sensation *quality* is closely related to the physical attribute *overtone structure*.

Sounds emitted by living animals contain a wide and continuous range of frequencies rather than the simple discrete overtones of a musical instrument. The relative distribution of frequencies, as well as the overall intensity, vary with time, as in the *acoustic spectrogram* shown in Fig. 11-16. Three variables are correlated: time (measured on the horizontal axis), frequency (measured on the vertical axis), and intensity at any given frequency and time (shown by the density or blackness of the recording).

Any psychological sensation of sound is not determined entirely by the corresponding physical attribute. A typical interrelation is that between pitch and loudness. A loud foghorn sounds "lower" in pitch than an exactly similar sound of the same frequency that is not so intense. In this case frequency is inadequate, by

Table 11-1   Typical Overtone Structures for Two Musical Sounds

CLARINET		FRENCH HORN	
Frequency, Hz	Relative Intensity, %	Frequency, Hz	Relative Intensity, %
400	36	100	3
800	0	200	22
1200	34	300	24
1600	9	400	44
2000	17	500	3
⋮	⋮	600	2
4000	3	700	1

---

* If the same two simple waves shown in Fig. 11-15 were added, but with the wave of frequency 4000 Hz shifted in phase by $\frac{1}{2}$ cycle so that it starts down at the instant the 1000 Hz wave starts up, an entirely different wave form would be obtained. In spite of this, the ear would hear little if any difference in the quality. This physiological fact (the relative insensitivity of the ear to phase differences between components of a complex tone) is the reason we prefer overtone structure to wave form as a means of describing the quality of a musical sound.

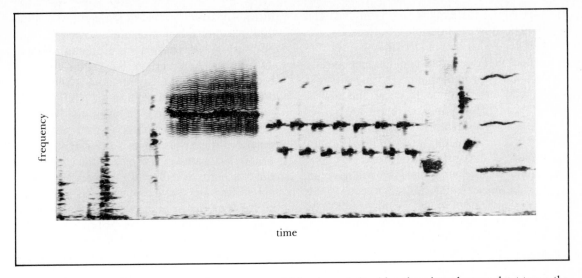

Figure 11-16 Acoustic spectrogram of song sparrow, *Melospiza melodia*. After three introductory short tones, the bird gave a sustained "whirring" sound with a broad band of frequencies, followed by seven chirps and a closing sequence. The recordist gave the following phrase, sung with appropriate pitch changes: "Madge, Madge, Madge, PUT the tea kettle on the stove, do it NOW." (Recording by Robert D. Burns.)

itself, to describe and predict the sensation of pitch that will be observed. As another example, consider two sounds of equal intensity (measured in W/m^2). If sound *A* is of frequency 5000 Hz and sound *B* is of frequency 50,000 Hz, then *B* has much less loudness—zero loudness, in fact—since its frequency is beyond the range of audible sound. Quality and pitch are also interrelated, to a slight extent.

11-6 The Human Ear

A convenient way to summarize data about the ear's sensitivity to sounds of different frequency is through an audibility graph (Fig. 11-17). The *average* ear can just hear a sound whose frequency is 2000 Hz if the intensity level is about 10 dB (relative to the standard 0-dB level of 10^{-12} W/m^2); a few persons (about 10% of the population) can hear a 0-dB sound at this frequency. Much more intensity is required to hear

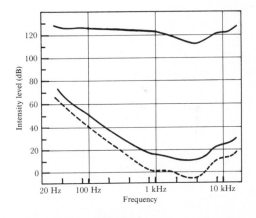

Figure 11-17 Audibility of sound as a function of frequency. Solid curve: average human ear. Dashed curve: limit of audibility for person with exceptionally acute hearing. Top curve: threshold of pain.

sounds of frequency closer to the limits of audibility; for example, the graph shows that an intensity level of at least 50 dB is needed to detect a 100-Hz sound. By Eq. 11-2, this is 10^5 times the intensity of the faintest audible sound. The graph also shows a threshold of discomfort (or pain) at somewhat above 120 dB.

The fact that the sensation of loudness is only approximately proportional to the intensity of a sound indicates that the physical and neural processes in the ear and the brain are rather complicated. The dependence of pitch on frequency is similarly complex. A full description of the ear's remarkable anatomy and functioning would require a treatise; we will give an application of elementary physics to each of the three major divisions of the ear (Fig. 11-18).

The **outer ear,** open to the atmosphere, contains the ear canal, about 2.7 cm long, terminated by the tympanum (eardrum). Consider the canal to be a closed pipe vibrating in its fundamental mode; $\frac{1}{4}\lambda = 0.027$ m, from which $\lambda = 0.108$ m, and for $v_w = 350$ m/s the resonant frequency is about 3200 Hz. The ear canal thus enhances the pressure at the closed end (tympanum) for sounds in the mid-range of audible frequencies, 1 to 5 kHz. The resonance is fairly broad.

The **middle ear** transmits sound from the tympanum to the oval window that closes off one end of the inner ear. The coupling is through a delicate and intricate lever system made up of

three tiny bones (ossicles). If they give a force multiplication (ideal mechanical advantage), it is not very great. The magnification of *pressure* (force/area), however, is large—up to 22—because the area of the tympanum is about that much greater than the area of the oval window. Thus a pressure wave is set up in the lymph (a liquid similar to spinal fluid) in the inner ear. A general physical principle is involved here. At an interface between two media, the energy of a wave is partially reflected and partially transmitted, and here the useful energy is that transmitted from the air in the outer ear to the liquid in the inner ear. High efficiency of energy transfer depends on the near equality for the two media of the product $v_w d$ of the wave velocity and the density (this product is called the acoustic impedance of the medium). For lymph versus air this ratio is as large as 3100; the ossicles, loosely bound by small tendons, serve the important purpose of matching the acoustic impedances of air and lymph, thus increasing the energy transfer into the cochlea.

The **inner ear** contains the semicircular canals (used for detection of orientation and acceleration) and the cochlea, a snail-shaped cavity divided along its length by the basilar membrane, which supports the organ of Corti. Here is where sound energy is converted into electric energy, to go to the brain as nerve impulses; here, too, is where frequency discrimination takes place. In the organ of Corti, nerve fibers are somehow connected to about 30,000 sensory cells called hair cells, each of which has about 50 stereocilia (hairs) about 1 nm long. When the basilar membrane is distorted, nerve endings are stimulated. It would be an attractive theory (but not altogether correct) to suppose that the basilar membrane contains transverse stretched entities like piano strings, one tuned to each observable frequency. For one thing, in the membrane the tension and mass per unit length do not vary enough to account for the wide range (almost 1000:1) of audible frequencies. It is generally accepted that for a pure tone of, say, 1000 Hz, stationary waves in the lymph and in

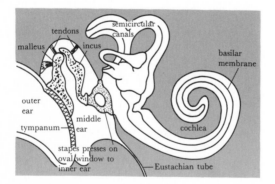

Figure 11-18 Middle and inner ear, approximately twice life size. The ossicles include the malleus, incus, and stapes. Adapted from *Life: An Introduction to Biology*, George Gaylord Simpson, et al., © 1957 by Harcourt Brace Jovanovich.

the basilar membrane interact to give a broad antinode extending over a region of perhaps 50 hair cells. The combined effect of nerve impulses from these adjacent sensory cells is decoded in the brain to give the pitch sensation. This is primarily a "place" theory of pitch discrimination rather than wholly a "resonance" or a "telephone" theory. The basilar membrane is thinnest and under greatest tension at the fixed end nearest the oval window; the free end, at the apex of the cochlear spiral, is heavier and under less tension. The wave velocity along the membrane, given by $v_w = \sqrt{T/(m/L)}$, is largest near the oval window. Analysis is more difficult than our earlier study of a stretched wire's nodes and antinodes, since v_w is not constant. Nevertheless, scale models and theory both show that regions of greatest distortion occur along the basilar membrane closer to the oval window as the frequency is increased. In Fig. 11-19 is shown the destruction of a rather sharply defined region of the organ of Corti in an ear that had been subjected to prolonged intense sound of frequency 125 Hz.

The ear is an extremely sensitive detector of acoustic radiation. A motion of the tympanum of only 10^{-11} m (less than the size of a single atom) can give rise to a noticeable sensation.

Figure 11-19 Complete degeneration (lower center of photograph) of part of guinea pig organ of Corti due to exposure to pure tone of 125 Hz at 148 dB for 4 hours ($\times 80$). [G. Bredberg and coworkers, *Science*, **170**, 863 (1970).]

11-7 Musical Instruments

We now discuss briefly one of many musical instruments—the violin—not so much for its intrinsic importance as for the review and illustration of the ideas of this chapter that it affords.

In general, a musical instrument consists of a *generator* of sound and a *resonator*, which selects and accentuates certain of the overtones. In the violin the generator is a string of steel, nylon, or gut, "fixed" at the bridge and at the nut (Fig. 11-20). The string is under considerable tension, supplied by the pegs, which are held firmly in their holes by frictional forces. The mass per unit length is characteristic of the type of string used; the strings of lower pitch are wound with fine aluminum or silver wire to give them added mass per unit length without sacrifice of flexibility. Wave velocity along the string is given by the formula $v_w = \sqrt{F/(m/L)}$, and stationary waves can exist on the string with a node at the bridge and a node at the nut. Such waves are set up by the bow, which pulls the string aside a few millimeters until the force of static friction becomes inadequate and the string slips back. The bow repeatedly picks up and releases the string, and so the primary motion of the string is a relaxation oscillation with a more or less triangular wave form (Fig. 11-21). Such a wave form is not "simple"; it contains many overtones, all of which are among the natural frequencies of the string. The string vibrates simultaneously in many modes, each of which has a node at each end. The vibrations of the string are transmitted through the bridge to the belly of the violin,

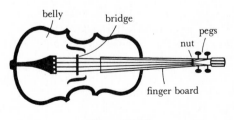

Figure 11-20

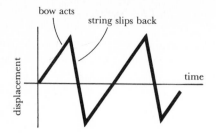

Figure 11-21 Vibration graph for a point on a violin string, idealized as a triangular relaxation oscillation.

Table 11-2 Frequencies of Fundamentals and Overtones in Violin Strings

A String, Hz	E String in Tune, Hz	E String Slightly out of Tune, Hz
440	660	661
880	**1320**	**1322**
1320	1980	1983
1760	**2640**	**2644**
2200	3300	3305
2640	3960	3966
⋮	⋮	⋮

which in turn sets into vibration the enclosed air column. The sound wave spreads out in the room from the belly and the back of the violin, as well as from the vibrating air column enclosed in the body of the violin.

The making of music is an art. The violin maker contributes to the art by constructing the resonator in such a way that certain overtones are accentuated and others are weakened or suppressed. The performer contributes to the art by drawing the bow across the string at the right place, with the right pressure and speed, to further modify the overtone structure of the resulting sound. Although perhaps neither the maker nor the artist can describe his or her procedure in analytical terms, the goal is the efficient transfer of sound energy from the string into the surrounding air, and the modification of the relaxation oscillation of the generator, by means of the resonator, into a wave form having pleasing overtone structure. To play a melody, the artist changes the length of the string by pressing a finger down somewhere on the finger board, between the nut and the bridge. The distance between the finger and the bridge is one-half the fundamental wavelength. As the string is shortened, the fundamental frequency and all the overtones are raised in frequency, but the overtone structure remains about the same.

The four strings are tuned in fifths (Sec. 11-5), with each string's fundamental frequency three-halves that of the next lower string. The musician makes good use of the phenomenon of beats in tuning her instrument. First she tunes her A string to 440 Hz by comparison with a tone of standard pitch, played perhaps on a piano or on the oboe of a symphony orchestra. The violinist adjusts the tension of her string, by means of the peg, until she hears no beats between her A and that of the standard when both are sounded together. To tune her E string, she also uses beats, this time between overtones. The desired frequency is $\frac{3}{2} \times 440 = 660$ Hz. As seen from Table 11-2, the out-of-tune E produces 2 beats/s between its 1322-Hz overtone and the 1320-Hz overtone of the A; other overtones produce 4 beats/s, 6 beats/s, and so on. This makes a sensitive way of adjusting the two strings to an exact 3:2 ratio of frequencies. The table also illustrates the basis of harmony: if the two tones are in a simple ratio (3:2 in this case), the lack of audible beats* between their overtones is very pleasing.

Among musical instruments are those with generators consisting of strings, air columns, vibrating bars, reeds, membranes, and plates. Resonators are usually air columns, although in the piano the resonator is a large, thin, wooden surface. The resonators may be untuned—as in the violin or piano, where no one frequency

---

* Beats between widely separated overtones in Table 11-2, such as 880 Hz and 660 Hz, are too rapid to be heard as distinct fluctuations in loudness.

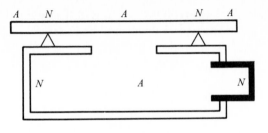

Figure 11-22

quency of the bar. As seen from the following table, only the fundamental frequency of the bar is in resonance with any of the natural frequencies of the pipe; hence the total effect is a pure tone of the single frequency 400 Hz. We may think of the resonator as being a selective amplifier, reinforcing and building up just one of the many frequencies present in the complex vibration of the bar.

should be emphasized—or tuned—as for the dinner gong pictured in Fig. 11-22. In this gong the generator is a flat bar, supported at two nodal points near, but not at, the ends. For a solid bar, the distance from node to antinode is not constant, which indicates that the frequencies of the overtones are not integral multiples of the fundamental. That is, the overtones are not harmonics (Sec. 11-2); this lack of simple relationship between the overtones of a bar of metal accounts for the "clanging" sound when such a bar is dropped on a concrete floor. In order to remove the inharmonious overtones, the air chamber (a closed pipe) is adjusted in length (tuned) so that the fundamental frequency of the pipe resonates with the fundamental fre-

Natural frequencies of bar, Hz	Natural frequencies of closed pipe, Hz
400	400
1104	1200
2172	2000
.	.
.	.

We have discussed only a relatively few aspects of wave motion and sound, preferring to go deeply into fundamentals rather than superficially into applications. We hope that with this discussion of basic ideas you will be able to read further, according to your interests, in the fields of acoustics, musical sounds, speech and hearing, and ultrasonics.

Summary   When two identical waves, having the same frequency, wavelength, and speed, travel in opposite directions through a medium, the resultant disturbance can be found by the superposition principle. The combined effect of the two waves is to produce a stationary wave. Certain points in the stationary wave, called motion nodes, remain at rest, and other points, called motion antinodes, undergo vibration with maximum amplitude. The distance from node to node is $\frac{1}{2}\lambda$; from antinode to antinode it is $\frac{1}{2}\lambda$; from node to antinode it is $\frac{1}{4}\lambda$. Pressure nodes (antinodes) occur wherever there are motion antinodes (nodes).

For a vibrating string, the wave velocity depends on the tension and the mass per unit length. The string can vibrate in 1, 2, 3, 4, . . . segments, always with a node at each end. A vibrating body has many possible modes of vibration. The lowest frequency of vibration is called the fundamental frequency, and the higher frequencies are called overtones. An overtone frequency that is an integral multiple of the fundamental frequency is called a harmonic. The natural frequencies of vibration for a string are 1, 2, 3, 4, . . . times the fundamental frequency. The modes of vibration of a column of air in a pipe are such that a closed end is a node and an open end is an antinode. If the pipe is open at both ends, the natural frequencies are 1, 2, 3, 4, ... times the fundamental frequency. If the pipe is closed at one end, the natural frequencies are 1, 3, 5, 7, ... times the fundamental frequency.

Resonance is the building up of a large vibration by repeated application of small impulses whose frequency equals one of the natural frequencies of the resonating body.

Musical sounds are characterized by the psychological sensations of pitch, loudness, and quality. Although there is some interaction between these sensations, to a large extent they depend on frequency, intensity, and overtone structure in that order. Intensity is defined as power per unit area; differences of intensity level are measured in decibels. A musical instrument consists of a generator and a resonator. Most instruments vibrate simultaneously in many modes, leading to a complex overtone structure that is characteristic of the instrument. Two tones differ in pitch by an octave if their frequencies are in the ratio 2:1; they differ by a fifth if their frequency ratio is 3:2.

Check List

interference	fundamental frequency	intensity
stationary wave	overtone	decibel
node	harmonic	overtone structure
antinode	resonance	octave
$v_w = \sqrt{F/(m/L)}$	pitch	fifth
natural frequency	loudness	major third
mode of vibration	quality	minor third

Questions

11-1 What are the conditions necessary for the production of stationary waves with well-defined nodes?

11-2 A glass of water is on a kitchen table, and the floor of the room vibrates because of a clothes dryer in action. For certain speeds of rotation of the dryer, stationary concentric ripples are observed on the surface of the water in the glass. Explain.

11-3 Give an example of constructive interference.

11-4 A high-speed motion picture is taken of the string of a double bass that is being played loudly. Would you expect any frame of the sequence of photographs to show the string absolutely straight?

11-5 Observe the construction of piano or guitar strings, and try to decide whether the mass per unit length for a "low" string is the same as for a "high" string. Why is the "low" string not of solid construction?

11-6 State three factors on which the fundamental frequency of a string depends.

11-7 Are overtones of all musical sounds harmonics?

11-8 In what physical respects does a loud, high flute tone differ from a soft, low violin tone?

11-9 Does a clarinet behave like an open pipe or a closed pipe? (See Table 11-1.) What about a French horn?

11-10 Describe a musical instrument in which resonance is desirable. Describe a musical instrument in which resonance is undesirable.

11-11 For the human voice, considered as a musical instrument, what is the generator? What is the resonator? How can one distinguish between the voices of a soprano and a tenor?

11-12 Why does a column of marching soldiers "break step" when crossing a bridge?

11-13 Make the following experiments on a piano: Gently press down the key of middle A without sounding the tone, thus releasing the damper for this string only.   (a) While holding the key down, strike vigorously an A that is one or two octaves lower, and explain why the higher string starts to vibrate.   (b) Repeat the experiment, still holding the A key down and striking a lower A♭ or A♯; explain your result.   (c) What if the lower A key is held down and the upper A key is struck?

11-14 Which of the 14 "syllables" sung by the sparrow (Fig. 11-16) is most like a tone from a musical instrument?

MULTIPLE CHOICE

11-15 In the sequence of "snapshots" of Fig. 11-1, the energy of the string is entirely kinetic at   (a) $t = 0$;   (b) $t = \frac{1}{4}T$;   (c) both of these times.

11-16 In a stationary wave system, the distance between adjacent antinodes is   (a) $\frac{1}{4}\lambda$;   (b) $\frac{1}{2}\lambda$;   (c) λ.

11-17 For a certain pipe, two out of the many overtones produced are separated by an octave. The pipe could be   (a) an open pipe;   (b) a closed pipe;   (c) either of these.

11-18 A possible unit for intensity of a sound wave is   (a) mW;   (b) mW/m²;   (c) J/m².

11-19 The intensity of sound from 10,000 cheering fans at a football game is louder than the sound of one fan by   (a) 20 dB;   (b) 40 dB;   (c) 100 dB.

11-20 The musical interval between the first overtone and the second overtone of a vibrating string is   (a) an octave;   (b) a fifth;   (c) a fourth.

Problems

11-A1 What is the velocity of transverse waves on a string of length 5 m that has a mass of 0.015 kg and is stretched with a tension of 25 N?

11-A2 Calculate the speed of transverse waves on a rope 12 m long, stretched with a tension of 40 N. The mass of the rope is 2 kg.

11-A3 Stationary waves are produced on a string for which the velocity of transverse waves is 180 m/s. The frequency of vibration is 600 Hz. How far apart are the nodes?

11-A4 A source vibrating with frequency 60 Hz sets up stationary waves on a string. The nodes are 25 cm apart. What is the wave velocity?

11-A5 What are the first three natural frequencies of a string whose fundamental frequency is 300 Hz?

11-A6 A cello A string 80 cm long is stretched with sufficient tension so that the fundamental frequency is 220 Hz. What is the velocity of transverse waves on this string?

11-A7 What is the fundamental frequency of an 86-cm pipe, open at both ends, in which the speed of sound is 344 m/s?

11-A8 The fundamental frequency of a pipe closed at one end is 430 Hz, and the speed of sound is 344 m/s. How long is the pipe?

11-A9 What is the frequency of the next-to-the-lowest natural frequency of a pipe 42 cm long and closed at one end on a day when the speed of sound is 336 m/s?

11-A10 (a) A pipe open at both ends is 86 cm long, and the speed of sound is 344 m/s. What is the fundamental frequency of the pipe?   (b) If the pipe is now closed at one end, what does the fundamental frequency become?

11-A11 What are the first three natural frequencies of the string of Prob. 11-A6?

11-A12 What are the first three natural frequencies of the pipe of Prob. 11-A7?

11-A13 What are the first three natural frequencies of the pipe of Prob. 11-A8?

11-A14 The lowest tone of a viola has a frequency of 131 Hz. What is the frequency of a tone that is three octaves higher?

11-A15 Use Fig. 11-17 to determine the frequency below which the average ear cannot hear a sound whose intensity level is 40 dB above that of the most easily heard sound.

11-B1 Make a dimensional check for Eq. 11-1.

11-B2 A wire of length L has circular cross section of radius r, and the density of the material is d. The wire is stretched with a force F. (a) Derive a formula for the velocity of transverse waves on the wire. (b) Derive a formula for the frequency of the nth natural frequency of this wire. (c) Make dimensional checks for your answers to parts (a) and (b).

11-B3 A clothesline of total mass 0.8 kg is stretched between posts 10 m apart. It is observed that when one pole is struck by a lawnmower, the transverse pulse reaches the other pole in 0.25 s. What is the tension in the rope?

11-B4 A construction engineer wishes to determine the tension in an aluminum power cable, of cross section 7 cm^2, which is stretched between two towers. The length of cable is 200 m. A worker strikes the wire a sharp blow at one end and feels the return of the transverse wave pulse 5.00 s later. (a) What is the tension in the cable? (b) How much did the wire stretch while it was being hung? (Assume constant tension throughout; use data from tables in Chaps. 1 and 9.)

11-B5 To what tension should a violinist adjust his A string in order to tune its fundamental frequency to 440 Hz? The distance from bridge to nut is 32.5 cm, and the string's mass is 2 g.

11-B6 A string 125 cm long has mass 500 mg and is stretched with a force of 4 N. (a) Draw a diagram of the locations of nodes and antinodes for the string when it is vibrating in its next-to-the-lowest natural frequency. (b) Calculate the frequency of this mode of vibration.

11-B7 When the tension is 27 N, a string 200 cm long has a fundamental frequency of 150 Hz. (a) What is the mass of the string? (b) With what tension must the string be stretched so that it vibrates in three segments at 150 Hz?

11-B8 How far, and in which direction, should a cellist move his finger to adjust an A string from an out-of-tune 449 Hz to an in-tune 440 Hz? The string is 68 cm long, and the finger is 18 cm from the nut.

11-B9 A string under tension 1080 N, of mass per unit length 0.00300 kg/m, has many resonant frequencies. One such frequency is 450 Hz, and the next higher frequency is 600 Hz. How long is the string?

11-B10 In the mid-range of a piano each note is sounded by several identical strings that are struck simultaneously by a single hammer. Two A strings are vibrating at 440 Hz. If one string's tension is increased by 1%, how many beats per second are heard?

11-B11 Two identical mandolin strings under tension 150 N are sounding tones of frequency 523 Hz (middle C). The peg of one string slips slightly, and the tension becomes 146 N. How many beats are heard?

11-B12 On a day when the velocity of sound is 342 m/s, a clarinetist adjusts the keys of her instrument so that the distance from reed to first open hole is 57 cm. In operation, sound waves are reflected from the reed as from a fixed surface so that the clarinet behaves like a pipe closed at one end. What are the frequencies of the fundamental and first two overtones?

11-B13 A bugler, by adjusting his lips correctly and blowing with the proper pressure, can cause his instrument to produce a sequence of tones, among which are the following— ... 440, 660, 880, 1100, ... Hz—all without changing the length of the air column. (*a*) Does the bugle behave like an open pipe or a pipe closed at one end? (*b*) What is the effective length of this bugle? (Use 350 m/s for the speed of sound.)

11-B14 Two identical organ pipes, both at 27°C, are sounding tones of frequency 602 Hz. The temperature of one pipe goes up to 31°C; how many beats per second are heard? (See pages 246–247.)

11-B15 Open organ pipes, 60.00 cm and 60.70 cm long, are sounded together on a day when the speed of sound is 340 m/s. How many beats per second are heard?

11-B16 A space explorer who does a little anthropology on the side observes Ronald Qrxxt, a native of a distant planet, who emits a mating call of frequency 300 Hz. This tone resonates with the fundamental frequency of the future Mrs. Qrxxt's intercranial cavity, which is a tube 60 cm long and closed at one end, located midway between two of her heads. (*a*) Calculate the speed of sound in the planet's atmosphere. (*b*) If Qrxxt is successful on a day when the speed of sound on the planet is as calculated in (*a*), what must be the speed of sound on a warmer day if his rival, Donald Prxxg, is to succeed in capturing the attentions of the lady in question? Prxxg's mating call is at 320 Hz.

11-B17 A metal rod 2.25 m long is clamped at its center and set into longitudinal vibration. (*a*) Make a diagram of the locations of motion nodes and motion antinodes for the first three modes of vibration. (*b*) Are the natural frequencies in the ratio $1:2:3:4:...$ or $1:3:5:7:...$? (*c*) When the rod is vibrating in its fundamental mode, the tone produced is in tune with a whistle of frequency 1100 Hz. What is the wave velocity of the compression waves in the rod?

11-B18 (*a*) Make a diagram of the locations of motion nodes and motion antinodes for the fundamental mode of vibration of the resonant air column in the gong pictured in Fig. 11-22, and explain with the aid of the diagram why the frequency 800 Hz does not occur in the table on page 257. (*b*) Using 344 m/s for the velocity of sound in air, calculate the total length of the resonant air column, assuming it to be vibrating in its fundamental mode.

11-B19 A brass rod 0.60 m long is clamped at its center and stroked longitudinally so that its fundamental mode is in tune with a loudspeaker vibrating at 2740 Hz. (*a*) Sketch the locations of motion nodes and motion antinodes for the vibrating rod. (*b*) Calculate the speed of sound in brass. (*c*) Using data from Table 1-3, calculate the stretch modulus of brass. (Your answer should be within 10% of the value in Table 9-1.)

11-B20 A wire whose mass per unit length is 10^{-2} kg/m is stretched with a tension of 900 N. The wire's fundamental frequency is in tune with the first overtone of a pipe 1.5 m long that is closed at one end. The speed of sound in air is 340 m/s. How long is the wire?

11-B21 A jet airplane inside a hangar is making a considerable noise (see page 250), and sound energy is flowing out through a door of dimensions 9×2 m. If the energy flowing through the door were converted efficiently into electrical energy, could a 40-W lamp be operated at full brilliance?

11-B22 One sound is 400 times as intense as another sound. What is the difference in intensity level of these two sounds, in dB?

11-B23 By soundproofing a living room near an airport, a noise reduction of 37 dB is obtained. What ratio of intensities does this correspond to?

11-B24 Verify the statement on page 251 that a sound level of 53 dB has an intensity of 2×10^{-7} W/m^2.

11-B25 The total effective area of the auricles of a listener is about 40 cm^2. (a) How much power (in watts) enters the ears of a person enjoying a concert if the intensity level is 60 dB above the threshold of hearing? (b) How long would be required for a microjoule of energy to enter the listener's ears?

11-B26 Derive Eq. 11-3 from Eq. 11-2.

11-B27 Under favorable conditions of loudness and frequency, the ear can just determine two sounds to be of different intensity if one is 0.6 dB louder than the other. Prove that to be perceptibly louder, a sound's intensity must be increased by about 15%. (*Hint:* Calculate the ratio I_2/I_1.)

11-C1 A tuning fork of frequency 500 Hz is held above a vertical tube partly filled with water. For what lengths of air column will resonance occur? (Take the velocity of sound to be 340 m/s.)

11-C2 The fundamental frequency of vibration of a string depends on its mass, length, and tension according to the equation $f = Km^x L^y F^z$. (a) Determine the form of the equation, finding x, y, and z by the method of dimensional analysis (Sec. 10-3). (b) Derive the equation, using the known formula for the velocity of a transverse wave on a string, and thus find the value of the dimensionless constant K.

11-C3 A wire 0.80 m long, of total mass 40 g, is stretched between two points, and a pipe 1.50 m long and closed at one end is placed nearby. The speed of sound in air is 340 m/s. What should be the tension in the wire in order for the 4th overtone of the wire to be in resonance with the 1st overtone of the pipe?

11-C4 A sudden handclap takes place in front of the opening of a tube of length L that is closed at the far end. The pulse travels back and forth in the pipe, sometimes changing from condensation to rarefaction and vice versa. As implied by the discussion in the first paragraph of Sec. 11-3, such changes of phase take place at the open end of the pipe but not at the closed end. Show that after the pulse has made two round trips, it is again starting down the pipe as a condensation. Thus prove that $\lambda = 4L$ for a closed pipe in its fundamental mode.

11-C5 A guitar string under tension 200 N, vibrating in its fundamental mode, gives 6 beats/s with a tuning fork. The player increases the string tension to 242 N and again gets 6 beats/s. What is the frequency of the tuning fork?

11-C6 (a) What must be the stress in a stretched steel wire in order for the speed of longitudinal waves to be equal to 80 times the speed of transverse waves? (*Hint:* Solve the problem first in symbols.) (b) Is this possible without exceeding the ultimate tensile strength of the steel described on page 195?

For Further Study

11-8 The Mathematics of Wave Motion

For simplicity, we will consider a transverse wave moving along the x axis with wave velocity v_w. In Sec. 10-5 (page 224) we noted that such a wave is represented by y as a function of both x and t. Suppose that at $t = 0$ a pulse $y = f(x)$ has a distinguishing feature such as a spike at a certain value of x (Fig. 11-23), and at a later time t the wave has moved a distance $v_w t$ to the right.

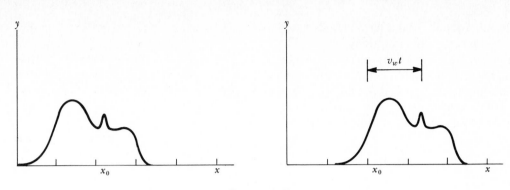

Figure 11-23

To shift a function of x to the right by an amount $v_w t$, we subtract $v_w t$ from x. The equation of the displacement at any time t is therefore

$$y = f(x - v_w t) \qquad (11\text{-}5)$$

This equation agrees with the qualitative discussion of Sec. 10-5. It gives y as a function of both x and t. If x is held constant, y is a function of time t; that is, y versus t is a *vibration* graph. If t is held constant, y is a function of x; that is, y versus x is a *wave-form* graph. The same function $f(\;\;)$ describes both curves, as discussed in connection with the water splash of Fig. 10-5 (page 225).

The wave of Eq. 11-5 is traveling in the positive x direction. This is shown by the fact that the argument $x - v_w t$ must remain unaltered to have $f(\;\;)$ represent the same spike. If t increases by Δt, the point of observation must *increase* by Δx. Then $x + \Delta x - v_w(t + \Delta t) = x - v_w t$, which can be solved to give $\Delta x / \Delta t = v_w$, as expected. By similar reasoning, the equation $y = f(x + v_w t)$ represents a wave traveling to the left along the x axis. So far, the form of the function $f(x \pm v_w t)$ is perfectly arbitrary.

Now let's consider a *periodic* wave having a definite wavelength and frequency. Such a wave is sinusoidal:

$$y = A \sin \frac{2\pi}{\lambda}(x \pm v_w t)$$

From $v_w = f\lambda = \lambda/T$, this can be written as

$$y = A \sin 2\pi \left(\frac{x}{\lambda} \pm \frac{t}{T} \right) \qquad (11\text{-}6)$$

At a fixed time t, y repeats in a distance of one wavelength λ, because then $2\pi(x/\lambda)$ goes from 0 to 2π and the sine function is back to its original value. Similarly, at a fixed position x, y repeats in a time of one period T, because then $2\pi(t/T)$ goes from 0 to 2π. In Eq. 11-6 the $+$ sign is for a wave traveling to the left, and the $-$ sign is for a wave traveling to the right.

We are ready now to consider the stationary wave pattern arising from the interference of two identical waves traveling in opposite directions. By the superposition principle, the resultant disturbance is

$$y = A \sin 2\pi \left(\frac{x}{\lambda} + \frac{t}{T} \right) + A \sin 2\pi \left(\frac{x}{\lambda} - \frac{t}{T} \right)$$

We make use of the trigonometric identity $\sin a + \sin b \equiv 2 \sin \left(\dfrac{a+b}{2} \right) \cos \left(\dfrac{a-b}{2} \right)$, with $a = 2\pi(x/\lambda + t/T)$ and $b = 2\pi(x/\lambda - t/T)$. Then $(a+b)/2 = 2\pi(x/\lambda)$, and $(a-b)/2 = 2\pi(t/T)$. The resultant disturbance is

$$y = \left[2A \sin \frac{2\pi x}{\lambda} \right] \cos \frac{2\pi t}{T}$$

This is the mathematical formulation of a stationary wave pattern. At any given value of x,

y is a function of t; the cosine term can be written as $\cos 2\pi ft$ where f, the frequency, is $1/T$. The amplitude of vibration at any point is given by $2A\sin(2\pi x/\lambda)$. This amplitude is 0 at a series of x values (the nodes), such that $\sin(2\pi x/\lambda) = 0$. Since $\sin(n\pi) = 0$ where n is any integer, we can locate the points where amplitude is zero.

$$2A\sin\left(\frac{2\pi x}{\lambda}\right) = 2A\sin(n\pi)$$

$$\frac{2\pi x}{\lambda} = n\pi$$

$$x = n\left(\frac{\lambda}{2}\right)$$

The nodes are separated by $\lambda/2$, as we saw graphically in Fig. 11-1.

11-9 Derivation of Equation for Velocity of Transverse Wave on a String

For a transverse wave on a string, we have stated in Sec. 11-2 that the wave velocity is $v_w = \sqrt{F/(m/L)}$, where F is the tension and m/L is the mass per unit length. To prove this we consider a wave crest that is an arc of a circle. Of course, no part of the string itself is moving in a circular path. Individual particles of the string move up and down, transversely to the direction of propagation of the wave. To derive the wave velocity formula, we find it convenient to change our frame of reference. Instead of letting the crest move past us along a stationary string, let us climb aboard an observation car that has the same velocity as the wave (Fig. 11-24). Now, as we ride along, the crest is always just abreast of us, seemingly stationary, and the string is moving backward with a velocity v_w relative to us, like a snake going around a tree. The advantage of this new viewpoint is that now we have a physical body (string particles) moving in an arc of a circle whose center is fixed. We can apply the equations of uniform circular motion to the particles of the string.

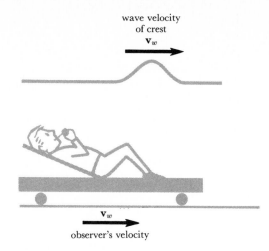

Figure 11-24 Wave on a string, viewed from a moving frame of reference.

Consider a short segment of the string having length ΔL (Fig. 11-25). The mass of the segment is (mass per unit length) $\times$ (length of segment) $= (m/L)\Delta L$. The net force on the segment is caused by the curvature of the string. At point A, the force $\mathbf{F}$ has a downward component that can be calculated by use of similar triangles:

$$\frac{F_y}{F} = \frac{\frac{1}{2}\Delta L}{R}$$

An important approximation has been made here: we assume that the string is not sharply

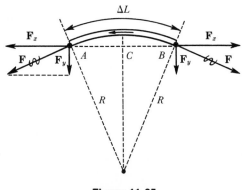

Figure 11-25

curved, so that the chord $\overline{AB}$ is equal to the arc $\overset{\frown}{AB}$, and hence AC is approximately equal to $\frac{1}{2}\Delta L$. Solving for F_y gives $F_y = F(\Delta L/2R)$ for the downward component of force acting at each end of the segment. The total net downward force is therefore twice this amount, or $(F\,\Delta L)/R$. The horizontal components of the forces at the ends of the segment cancel each other. We now apply Newton's second law and solve for the wave velocity v_w. The centripetal acceleration is v_w^2/R, so

$$\text{net } F = ma$$

$$\frac{F\,\Delta L}{R} = \left(\frac{m}{L}\Delta L\right)\left(\frac{v_w^2}{R}\right)$$

$$v_w = \sqrt{\frac{F}{m/L}}$$

Our proof has several interesting features. First, we are limited to disturbances that are not too radically curved. This is not a serious limitation. In a typical case, a guitar string 50 cm long might be pulled aside 0.5 cm at the center and then released. The string is never very much disturbed from its normal (straight) condition. Second, we note that in a complex wave form different parts of the string have different radii of curvature. However, since the radius of curvature cancels out and does not enter into the final result, we see that all parts of a disturbance travel at the same velocity. This means that a complex wave form is preserved in shape as it moves along the string. Another way of stating this same conclusion is to say that the wave velocity is independent of the wavelength of the disturbance.

References

1. J. Bernstein, "Tsunamis," *Sci. American* **191**(2), 60 (Aug. 1954). A description of water waves in the Pacific Ocean having wavelengths of from 150 to 1000 km and periods of vibration from $\frac{1}{4}$ to 1 h. See also the letter by W. G. Van Dorn on page 2 of the Jan. 1955 issue of *Sci. American*. A related phenomenon, the exceptionally high tides in the Bay of Fundy, is discussed briefly in *National Geographic*, Aug. 1957, p. 156.
2. G. von Békésy, "The Ear," *Sci. American* **197**(2), 66 (Aug. 1957). An article of particular interest to biologists. Physical mechanisms are discussed.
3. G. Oster, "Auditory Beats in the Brain," *Sci. American* **229**(4), 94 (Oct. 1973).
4. E. A. G. Shaw, "Noise pollution—what can be done?" *Physics Today* **28**(1), 46 (Jan. 1975).
5. A. H. Benade, *Fundamentals of Musical Acoustics* (Oxford University Press, Fair Lawn, N.J., 1976).
6. T. D. Rossing, "Musical Acoustics," (Resource Letter MA-1) *Am. J. Phys.* **43**, 944 (1975). Gives many journal and book references to all aspects of musical acoustics.
7. T. D. Rossing, "Physics and psychophysics of high-fidelity sound," *Phys. Teach.* **17**, 563 (1979); **18**, 278 (1980); **18**, 426 (1980).
8. H. Fletcher, "The Pitch, Loudness, and Quality of Musical Tones," *Am. J. Phys.* **14**, 215 (1946).
9. F. A. Saunders, "Physics and Music," *Sci. American* **179**(1), 33 (July 1948).
10. C. F. Hagenow, "The Equal Tempered Musical Scale," *Am. J. Phys.* **2**, 81 (1934).
11. C. Williamson, "Intonation in Musical Performance," *Am. J. Phys.* **10**, 171 (1942).
12. B. Patterson, "Musical Dynamics," *Sci. American* **231**(5), 78 (Nov. 1974).
13. S. E. Stickney and T. J. Englert, "The Ghost Flute," *Phys. Teach.* **13**, 518 (1975).
14. J. W. Coltman, "Acoustics of the Flute," *Physics Today* **21**(11), 25 (Nov. 1968).
15. W. L. Klein and H. J. Gerritsen, "Comparison Between a Musical and a Mathematical Description of Tone Quality in a Boehm Flute," *Am. J. Phys.* **43**, 736 (1975).
16. R. Herman, "Observations on the Acoustical Characteristics of the English Flute," *Am. J. Phys.* **27**, 22 (1959). The English flute is a recorder.
17. A. H. Benade, "The Physics of Wood Winds," *Sci. American* **203**(4), 145 (Oct. 1960).

18. J. C. Schelling, "The Physics of the Bowed String," *Sci. American* **230**(1), 87 (Jan. 1974).
19. C. M. Hutchins, "The Physics of Violins," *Sci. American* **207**(5), 79 (Nov. 1972).
20. E. D. Blackman, "The Physics of the Piano," *Sci. American* **213**(6), 88 (Dec. 1965).
21. T. D. Rossing, "Acoustics of percussion instruments," *Phys. Teach.* **14**, 546 (1976); **15**, 278 (1977).
22. J. Sundberg, "The Acoustics of the Singing Voice," *Sci. American* **236**(3), 82 (Mar. 1977).
23. F. S. Crawford, "Singing Corrugated Pipes," *Am. J. Phys.* **42**, 278 (1974).
24. W. C. Walker, "Demonstrating resonance by shattering glass with sound," *Phys. Teach.* **15**, 294 (1977). See also G. E. Jones and W. P. Gordon, "Apparatus for a shattering experience," *Am. J. Phys.* **47**, 828 (1979).
25. E. R. Smith and P. D. Loly, "The great beer bottle experiment," *Am. J. Phys.* **47**, 515 (1979). Speed of sound by resonance of air cavities. Parts I–III are recommended.
26. D. R. Griffin, "How Bats Guide Their Flight by Supersonic Echoes," *Am. J. Phys.* **12**, 343 (1944).
27. G. Neuweiler, "How bats detect flying insects," *Phys. Today* **33**(8), 34 (Aug. 1980). Fairly technical. There is a sharp acoustical resonance at 83 kHz in the structure of the hearing organ of a studied bat species.
28. V. O. Knudsen, "Architectural Acoustics," *Sci. American* **209**(5), 78 (Nov. 1963).
29. F. Miller, Jr., *Tacoma Narrows Bridge Collapse* (film). Large-scale destructive mechanical resonance.

12

Fluids

Some of the most spectacular achievements of recent years have been concerned with the exploration of the upper atmosphere of the earth. Even at low altitudes, airplanes and missiles travel through a resisting air at speeds undreamed of a few decades ago, propelled by jets and rockets that make use of high-speed gases. In the ocean, advanced types of submarines use the best possible streamlining as they cruise or remain at rest, securely suspended by the buoyant force of the water. The study of the ocean itself is proceeding at an increased pace, as scientists chart the flow of undersea "rivers" and study effects of waves and tides. All these and many other areas of applied science are concerned with fluids; in this chapter we study some basic facts about fluids at rest and in motion.

12-1 Definition of a Fluid

The general term *fluid* includes any substance that has no rigidity; thus *liquids* and *gases* are fluids. The bulk modulus (Sec. 9-1) of a fluid correlates change in volume with change in pressure. A fluid cannot sustain a static (steady-state) shearing stress, and the shear modulus is zero for any fluid. Thus a fluid does not have a definite length or shape. We can make a distinction between liquids and gases based on the common observation that a liquid has a surface. In other words, the cohesive forces between the molecules of a liquid are sufficient to give a definite volume to a quantity of material. Gases have no such "natural" volume, but expand to fill completely the container in which they are placed. Both liquids and gases are more easily compressed than solids, as shown by the smaller bulk modulus values in Table 9-1. In addition, as a result of gravitational forces, the density of a given sample of fluid is not strictly constant. It is well known that the air is more dense at sea level than at 3 km altitude. The bulk modulus of air is so small that the weight of the air above any given level compresses the air beneath it appreciably and so increases its density. To a much smaller degree, the density of sea water at 3 km depth is greater than at the surface.

A light ring pulled up out of a water surface carries a thin film of water with it.
Can you explain how the container can be filled to more than its geometrical volume?

One of the earliest discoveries about fluids is Archimedes' principle, which describes the phenomenon of buoyancy, or "lift." Archimedes' principle is applicable to both liquids (ships at sea) and gases (balloons in the air). To study it in detail, however, we need to use the concept of pressure and its relationship to depth.

12-2 Pressure

Pressure is defined as force per unit area, the force understood to be perpendicular to the area:

$$P = \frac{F}{A}$$

Thus an object weighing very little can exert a tremendous pressure if the force acts only on a small surface area. On the other hand, the weight of 5×10^{15} metric tons of air creates only a relatively small pressure on the earth, since the force is spread out over the entire surface of the globe. Pressure units are force units divided by area units, such as N/m^2, dyn/cm^2, lb/ft^2, and $lb/in.^2$.*

The SI unit for pressure is the *pascal* (Pa), equivalent to $1 N/m^2$. The pascal is a rather small pressure unit; the kilopascal (10^3 Pa, or 1 kPa) is often more convenient. For practical purposes, pressures are often expressed as so many atmospheres, where one *atmosphere* (atm) is normal atmospheric pressure,[†] which is approximately $1.013 \times 10^5 N/m^2$ or 101.3 kPa or $1.013 \times 10^6 dyn/cm^2$ or $14.7 lb/in.^2$. Another common unit of pressure, used especially in meteorology, is the *bar*, which is 100 kPa. A millibar (mbar) is 100 Pa. The *millimeter of mercury* (mm Hg) is used in many gas-law problems and

is defined as $\frac{1}{760}$ atm. It is the pressure due to a column of mercury 1 mm high. The mm Hg is also known as the *torr*, named for Evangelista Torricelli, who invented the barometer in 1643. The *inch of water* is used in engineering for measuring relatively small deviations from normal atmospheric pressure, as in a household vacuum cleaner. Pressure units can be readily manipulated with the aid of a few conversion factors:

$$
\begin{aligned}
1 \text{ atm} &= 101.3 \text{ kPa} \\
&= 1.013 \times 10^5 \text{ N/m}^2 \\
&= 1.013 \times 10^6 \text{ dyn/cm}^2 \\
&= 1.013 \text{ bar} \\
&= 76 \text{ cm Hg} \\
&= 760 \text{ mm Hg} \\
&= 760 \text{ torr} \\
&= 14.7 \text{ lb/in.}^2
\end{aligned}
$$

Example 12-1

What is the pressure on the pavement if a 10-metric-ton truck's weight is supported by 6 wheels, each having 0.01 m² of surface in contact with the concrete?

$$P = \frac{F}{A} = \frac{(10 \times 10^3 \text{ kg})(9.8 \text{ N/kg})}{6 \times 0.01 \text{ m}^2}$$

$$= \boxed{1.63 \times 10^6 \text{ Pa}}$$

Example 12-2

Compute the pressure exerted by the stylus on a phonograph record if the groove has a width of 0.001 in. (one mil) and the recommended stylus force is the weight of a 1-g object.

We assume that contact is made over a circular area of radius 0.0005 in. (0.00127 cm or 1.27×10^{-5} m). The area of contact is $\pi(1.27 \times 10^{-5} \text{ m})^2 = 5.07 \times 10^{-10} \text{ m}^2$.

$$P = \frac{F}{A} = \frac{(10^{-3} \text{ kg})(9.8 \text{ N/kg})}{5.07 \times 10^{-10} \text{ m}^2} = \boxed{19 \times 10^6 \text{ Pa}}$$

---

* The dimensional equation for pressure is [pressure] = $[FL^{-2}]$ or $[ML^{-1}T^{-2}]$.

[†] Normal atmospheric pressure is defined as the pressure at the bottom of a column of mercury exactly 76 cm high, if the mercury is at such a temperature that its density is 13.5951 g/cm³ and if it is located at a point where $g = 980.665$ cm/s².

Comparing these examples, we see that the pressure due to the stylus is 19,000 kPa, almost 200 atm, and more than 10 times the pressure due to the truck!

12-3 Pressure in Liquids at Rest

The pressure at the bottom of a regular container in which there is some liquid is easy to compute. The volume of liquid is found from the base area A and the height h (Fig. 12-1). If the density of the liquid is d, the weight of the column of liquid is

$$W = mg = (Ahd)g = Ahdg$$

The pressure is force per unit area, or $\dfrac{Ahdg}{A}$, whence

$$P = hdg \qquad (12\text{-}1)$$
$$N/m^2 = (m)(kg/m^3)(m/s^2)$$
$$dyn/cm^2 = (cm)(g/cm^3)(cm/s^2)$$

Thus, since g is essentially constant, the pressure at any point in a uniformly dense liquid at rest* depends only on the height and density of the

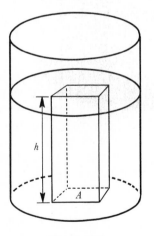

Figure 12-1

---

* If the liquid is moving, then Bernoulli's equation (Sec. 12-8) must be applied.

liquid above the point. Suppose two different liquids are used to measure the same pressure. Then $h_1 d_1 g = h_2 d_2 g$, or

$$\frac{h_1}{h_2} = \frac{d_2}{d_1}$$

The height of liquid corresponding to a given pressure is therefore inversely proportional to the density of the liquid. Mercury has a density 13.6 times that of water (Table 1-3); hence 1 cm of mercury is equivalent to 13.6 cm of water.

Equation 12-1 for pressure can be used for the pressure at the bottom of a short column of gas, provided the density of the gas is sufficiently constant. This is *not* true, of course, for the pressure due to the atmosphere, as the following example shows.

Example 12-3

If the earth's atmosphere were everywhere of the same density as at the surface of the earth (1.290 kg/m³), how high would the atmosphere have to extend in order to give rise to the observed atmospheric pressure of 1.013×10^5 N/m²?

From $P = hdg$, we have

$$h = \frac{P}{dg} = \frac{1.013 \times 10^5 \text{ N/m}^2}{(1.290 \text{ kg/m}^3)(9.80 \text{ m/s}^2)}$$

$$= 8.0 \times 10^3 \text{ m} = \boxed{8.0 \text{ km}}$$

An atmosphere only 8 km high would not even cover the top of Mt. Everest, and yet the observed sea-level pressure would be obtained from such an atmosphere if its density were uniform and equal to the sea-level value. We conclude that the density of the atmosphere must taper off at higher altitudes; it is incorrect to use our simple formula $P = hdg$ here.

Example 12-4

During a research project, deep-sea photographs were made at a depth of 7.50 km (24,600 ft). (a) What is the pressure at this depth? (b) What is the force on the plane surface of the window of a camera enclosure that measures 12 cm × 15 cm?

(a) We assume that water is sufficiently incompressible so that its density is approximately constant. The specific gravity of sea water is 1.025 (Table 1-3); therefore its density is 1025 kg/m³.

$$P = hdg$$
$$= (7.50 \times 10^3 \text{ m})(1025 \text{ kg/m}^3)(9.8 \text{ N/kg})$$
$$= 7.53 \times 10^7 \text{ N/m}^2 = \boxed{75{,}300 \text{ kPa}}$$

This is a pressure of 743 atm.
(b) The area of the window is 0.018 m².

$$\text{Force} = (\text{pressure})(\text{area})$$
$$= (7.53 \times 10^7 \text{ N/m}^2)(0.018 \text{ m}^2)$$
$$= \boxed{1.36 \times 10^6 \text{ N}}$$

The force on the window is about 10^6 N, or 300,000 lb.

Although pressure is a scalar quantity and has no direction, the *force* exerted on a surface immersed in a fluid is a vector, just like any force. If the fluid is not moving, the direction of the force is perpendicular to the surface of the object. This can be demonstrated in an indirect way. Consider two small rectangular sections of the bottom of a container, both at depth h below the surface of a liquid (Fig. 12-2). If the pressure at A were greater than that at B, liquid would flow from A to B, and the liquid would not be motionless. Hence, since the liquid is assumed to be at rest, the pressures at A and B are equal

(given by hdg at each point). Next, assume that the force on B is *not* perpendicular to the surface. This would mean that there would be a component $\mathbf{F}_t$ parallel to the surface, as well as a component $\mathbf{F}_n$ normal (perpendicular) to the surface. Now we have defined a fluid as a substance that has no rigidity, which means that it is unable to withstand a shearing stress. Thus the component $\mathbf{F}_t$ would cause a flow of liquid parallel to the surface. But the liquid was assumed to be at rest; therefore there can be no component of force $\mathbf{F}_t$, and only a normal component $\mathbf{F}_n$ can exist. Since $P = F/A$, the force is given by $F = F_n = PA$, acting perpendicularly to the surface in every case. Suppose a small, thin sheet of metal is suspended at an angle under the surface; the fluid exerts two forces on it, $\mathbf{F}_1$ and $\mathbf{F}_2$, each equal in magnitude to PA. The vector sum of these forces is zero, and the sheet is in equilibrium. These conclusions are equally valid for all fluids at rest, gases as well as liquids.

12-4 Pascal's Principle

When the photographic equipment was submerged at great depth in Example 12-4, the tremendous pressure that we calculated was due to the weight of the water above it. In addition, there is a relatively small pressure on a submerged object due to the weight of air above the water. If a change of weather causes an increase in atmospheric pressure, this change, however slight, is transmitted through the water and acts on the submerged camera. This transmission of changes in pressure was studied by Blaise Pascal (1623–1662), a French philosopher and scientist, whose experimental facts may be summarized in *Pascal's principle*:

> **Change of pressure exerted at any point in a confined fluid is transmitted undiminished in all directions to all points in the fluid.**

The hydraulic lift in a garage is a machine (a device for doing work) based on Pascal's prin-

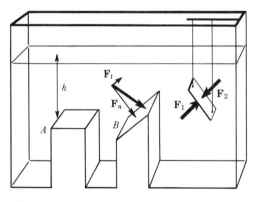

Figure 12-2

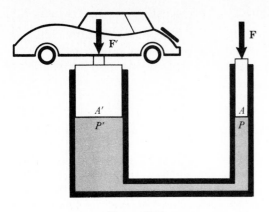

Figure 12-3

ciple. In the lift (schematically illustrated in Fig. 12-3), the weight of the car $\mathbf{F}'$ is balanced by a small force $\mathbf{F}$ applied to a piston of small area. The basic idea underlying the operation of such devices is that any increase in pressure at one piston is transmitted undiminished to the other piston. Usually, differences in pressure due to differences in level [which could be found from $\Delta P = (h_2 - h_1)dg$] are negligible compared with the larger pressures built up by the elastic forces of repulsion between the molecules of the fluid. In any case, if the two pistons are at the same level, we can say that $P' = P$; thus,

$$\frac{F'}{A'} = \frac{F}{A}$$

and hence

$$F' = \frac{A'}{A} F \qquad (12\text{-}2)$$

If now A' is much greater than A, the force is magnified considerably. Indeed, only the strength and resistance to leakage of the piston chambers limit the forces that can be applied in this manner. P. W. Bridgman (1882–1961) succeeded in designing a leak-proof cylindrical chamber in which the higher the pressure, the greater was the efficiency of sealing and the less the leakage. Pressures up to 140 gigapascals (140 GPa $= 140 \times 10^9$ Pa, about 20×10^6 lb/

in.^{2}) were achieved by Bridgman in this way.[*] We will use our next example, which deals with the hydraulic press, as an opportunity to review our earlier study of machines (Sec. 6-11).

Example 12-5

In the press shown schematically in Fig. 12-4, the small piston has a radius of 1.25 cm, and the large piston has a radius of 20 cm. The worker operates the press by means of a lever, with lever arms as shown in the diagram. If she applies a force of 100 N (22 lb) on the lever, what force is exerted on the bale of scrap paper, assuming 100% efficiency?

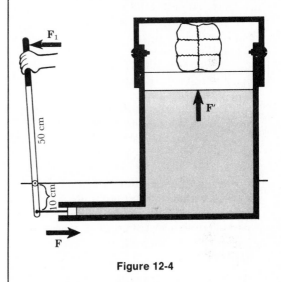

Figure 12-4

The output force of the lever supplies the input force for the press. We have two machines "in series," and the overall mechanical advantage is the product of two force ratios:

$$\frac{F'}{F_1} = \left(\frac{F'}{F}\right)\left(\frac{F}{F_1}\right)$$

---

[*] Magnitude in any field is spectacular, but it is important to note that the 1947 Nobel Prize in physics went to Bridgman not just for exceeding the previous record for pressure. It was the careful planning of experiments and the masterful interpretation of his data in terms of fundamental molecular phenomena that won for the Harvard professor of physics the highest award in his field.

From Eq. 12-2, the mechanical advantage of the press is

$$\frac{F'}{F} = \frac{A'}{A} = \frac{\pi(20 \text{ cm})^2}{\pi(1.25 \text{ cm})^2} = 256$$

The mechanical advantage of the lever is

$$\frac{F}{F_1} = \frac{50 \text{ cm}}{10 \text{ cm}} = 5$$

$$\frac{F'}{F_1'} = \left(\frac{F'}{F}\right)\left(\frac{F}{F_1}\right) = (256)(5) = 1280$$

The overall mechanical advantage is 1280.

$$F' = (1280)(100 \text{ N})$$

$$= \boxed{1.28 \times 10^5 \text{ N}}$$

This force is about equal to the weight of an object of mass 13 metric tons.

A worker can exert this large force by applying a force of only 100 N. The catch is that the worker will have to move the lever many centimeters to cause even 1 mm of motion of the big piston. Energy is conserved, so that if the big piston goes up 10^{-3} m, the work done on the bale is

$$(1.28 \times 10^5 \text{ N})(10^{-3} \text{ m}) = 128 \text{ J}$$

In order to do this work by applying a force of 100 N, the worker must move her end of the lever 128 J/100 N = 1.28 m. This disparity of distances is, of course, the essential feature of all machines that magnify force.

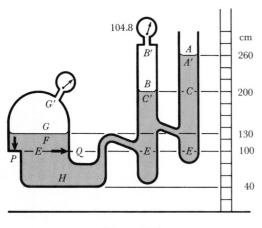

Figure 12-5

As a further illustration of Pascal's principle and the pressure-depth formula (Eq. 12-1), we calculate the pressure at various points in the somewhat complicated-looking apparatus of Fig. 12-5.

Example 12-6

Calculate the pressure at points C and B' in the apparatus of Fig. 12-5. The liquid is alcohol of specific gravity 0.816 (density 816 kg/m³); the enclosed volumes contain air. Atmospheric pressure on the day of the experiment is 100 kPa.

The pressure at A' is the same as that at A (no change in level). For the liquid column $A'C$ we have

$$\Delta P = hdg = (0.60 \text{ m})(816 \text{ kg/m}^3)(9.8 \text{ N/kg})$$

$$= 4800 \text{ Pa} = 4.8 \text{ kPa}$$

Hence the pressure at C is

$$P_C = P_{A'} + \Delta P$$

$$= 100 \text{ kPa} + 4.8 \text{ kPa}$$

$$= \boxed{104.8 \text{ kPa}}$$

The pressure at B' is the same as that at B (the difference in level causes a negligible ΔP because d in hdg is negligibly small for air). The pressures at B, C', and C are all equal since these points are at the same level. Hence

$$P_{B'} = P_B = P_{C'} = P_C$$

$$= \boxed{104.8 \text{ kPa}}$$

12-5 Measurement of Pressure

Many common devices for measurement of pressure make use of a combination of Pascal's principle and the pressure-depth relationship, as described for Fig. 12-5. The liquid-filled *manometer* may take various forms. For instance, in measuring the heating value of fuel gas, it is important to know the pressure at which the gas is passing through the flowmeter. Since the pressure is not very different from atmospheric pressure, an open-tube water manometer is often

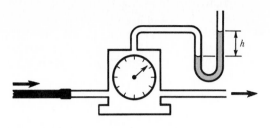

Figure 12-6 Open-tube manometer; pressure nearly equal to atmospheric pressure.

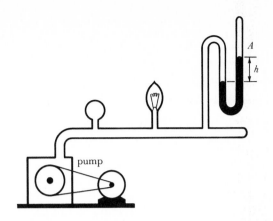

Figure 12-7 Closed-tube manometer; pressure nearly zero. The sealed part of the manometer tube (*A*) is evacuated.

used. As drawn in Fig. 12-6, the pressure is in excess of atmospheric pressure by an amount equal to *hdg*. The *sphygmomanometer**—the device the doctor uses to determine your blood pressure—uses two mercury columns to measure the pressure applied to an artery by means of a close-fitting, hollow jacket that expands as air is pumped in. When the external pressure equals or exceeds the maximum pressure in the artery (which occurs during the part of the heartbeat cycle called systole), the doctor hears no rhythmic heartbeat. This pressure is expressed in mm Hg, as read from the difference in level of the two columns of mercury. Maximum pressures usually range from 120 to 180 mm Hg (120 to 180 torr), depending on age, sex, weight, and many other factors.

The following examples of pressure-measuring devices illustrate the great variety of such instruments:

1. For measuring pressures that differ only slightly from an absolute vacuum, the *closed-tube manometer* is used, usually with mercury as the indicating liquid. As drawn in Fig. 12-7, the manometer is indicating a small pressure; the levels would be equal if the experimental system were completely evacuated.

2. The *mercury barometer* (Fig. 12-8) measures atmospheric pressure. Although it is simple in principle—essentially a closed-tube manometer—a user must take many precautions if readings are to be precise and repeatable.

Temperature increases cause the mercury to expand, so that its density decreases. Thus, according to $P = hdg$, a longer column of mercury is supported by a given atmospheric pressure, and normal atmospheric pressure would seem to be more than 760 mm Hg unless a

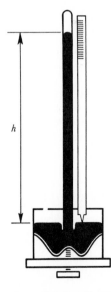

Figure 12-8 Mercury barometer. The space above the mercury column is evacuated.

---

* From the Greek word *sphygmos*, meaning pulse.

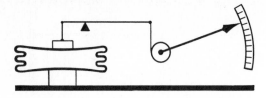

Figure 12-9   Aneroid barometer.

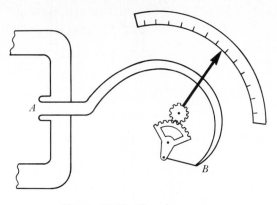

Figure 12-10   Bourdon gauge.

temperature correction is made. The expansion of the scale used to measure heights (usually of brass) must also be taken into account, as well as the curvature of the top surface of the mercury column due to surface tension (Sec. 12-7).

3. The *aneroid barometer* (Fig. 12-9) depends on the elastic properties of a sealed, evacuated can acted on by external (atmospheric) pressure. A sensitive lever arrangement magnifies the motion and on some instruments moves a pen point up and down a slowly rotating drum. The aneroid barometer can be made so sensitive that a measurable decrease in reading results from merely lifting the instrument from the floor to a desk top.

4. Water or steam pressure is usually measured by a *Bourdon gauge* (Fig. 12-10), in which the sealed chamber AB uncoils slightly, actuating a pointer through a system of gears and levers. This is somewhat like a more rugged version of the aneroid barometer.

Any physical property that depends on pressure could be used for measurement. Many properties such as density and electrical resistivity depend slightly on pressure and so can be used in instruments for measuring high pressures.

The pressure we have been discussing is *absolute pressure*—force per unit area. In some applications, the useful quantity is the *gauge pressure*, defined as the absolute pressure minus the atmospheric pressure.

Example 12-7

What is the fractional change in pressure registered by an aneroid barometer when it is raised vertically 1.00 m?

The density of the air is essentially constant for a change of altitude of only 1 m, so the pressure-depth equation can be used in this problem.

The density of air is 1.29×10^{-3} g/cm^3.

$$P = hdg$$
$$= (100 \text{ cm})(1.29 \times 10^{-3} \text{ g/cm}^3)(980 \text{ cm/s}^2)$$
$$= 126 \text{ dyn/cm}^2$$

Since normal atmospheric pressure is about 1.013×10^6 dyn/cm^2, the fractional change is

$$\frac{126 \text{ dyn/cm}^2}{1.013 \times 10^6 \text{ dyn/cm}^2} = \boxed{0.00012}$$

This is about $\frac{1}{80}$ of 1%!

Example 12-8

An automobile tire is pumped up to an absolute pressure of 300 kPa (44 lb/in.^{2}). What is the gauge pressure?

Tire pressure is usually measured by a pressure gauge that is calibrated to read 0 when open to the atmosphere. If atmospheric pressure is 100 kPa (14.5 lb/in.^{2}),

$$\text{Gauge pressure} = 300 \text{ kPa} - 100 \text{ kPa}$$
$$= \boxed{200 \text{ kPa}}$$

The gauge reads 200 kPa (29 lb/in.²), which is a useful quantity—the excess pressure within the tire that allows it to function.

In this book, the term "pressure" refers to absolute pressure unless otherwise noted.

12-6 Archimedes' Principle

We are now in a position to make a close study of Archimedes' principle, which has numerous applications. When a body is wholly or partially immersed in a fluid, the body seems to weigh less. Archimedes' principle states that *the loss of weight equals the weight of the fluid displaced by the body.* This apparent loss of weight may be thought of as an upward *buoyant force* (BF) supplied by the fluid. A body *floats* if its weight exactly equals the weight of the displaced fluid, for then the net force on the body is zero. A body *sinks* if the BF on it is less than its weight. However, there is still a BF on a body even if it is not floating. For instance, victims of polio, multiple sclerosis, or arthritis exercise in pools where the BF of the water helps make it possible for weakened muscles to move the limbs. In precise weighing, it is necessary to correct for the buoyancy of the air displaced by the volume of the body being weighed.

A geometrical proof for Archimedes' principle for a certain special case can be based on the differences between pressures at various depths. Consider the solid block shown in Fig. 12-11, immersed in fluid of uniform density d. The pressure at depth h_1 is $h_1 dg$, and the downward force on the top of the block is $F_1 = Ah_1 dg$. Likewise, the upward force on the bottom of the block is $F_2 = Ah_2 dg$, and because h_2 is greater than h_1, F_2 is greater than F_1. The net upward force is $F_2 - F_1 = A(h_2 - h_1)dg$, but since $A(h_2 - h_1)$ is the volume of the block, we see that $A(h_2 - h_1)dg$ is the weight of the fluid that would have occupied the space now filled by the block. Hence the BF equals the weight of fluid displaced.

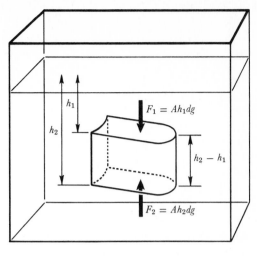

Figure 12-11

This proof is far from being a general one, but it can be improved and made satisfactory. For instance, we could extend it to a body of irregular shape by dividing the body into a large number of small vertical cylinders, each of very small cross-sectional area. Also, we assumed the fluid's density to be uniform; the proof for a fluid of variable density can be carried out using advanced mathematical methods. However, Archimedes himself doubtless was led to his principle by physical reasoning that is elegant in its simplicity and yet correct and quite general. Consider the irregularly shaped body in Fig. 12-12*a* immersed in a fluid whose density is less than the body's density. The forces acting on the surface of the body are due to the surrounding fluid; these forces would be the same whether the body were replaced by an identically shaped one of iron, lead, stone, or whatever. In particular, let us imagine the immersed body to be replaced by a body A' of exactly the same size and shape as A, composed of the same material as the surrounding fluid. The buoyant force is just sufficient to support A', since Fig. 12-12*b* represents merely a uniform bowl of fluid in equilibrium. Hence the vector sum of all the forces of the fluid on A (the same forces as on A') must be just sufficient to

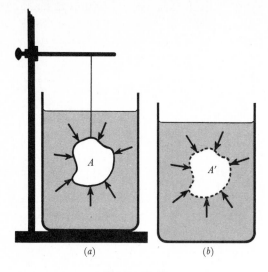

(a)   (b)

Figure 12-12

balance the fluid that has been displaced. This is Archimedes' principle.

It is said that Archimedes (who died in 212 B.C.) came upon the principle while in his bath, cogitating on how he might detect a fraudulent crown.* Gold is one of the densest metals (specific gravity 19.3; see Table 1-3), and any substance such as copper or silver that might be used as an alloy would necessarily make the specific gravity of the crown lower than that of pure gold. In modern terminology, the problem resolved itself into finding the density of an irregularly shaped object (a crown) whose volume could not easily be found by measurement.

Example 12-9

A crown of mass 4280 g weighs 4280×980 dyn in air, and when immersed in water it weighs 4000×980 dyn. What is the specific gravity of the material?

Let V = the volume of the crown. By Archimedes' principle,

$$BF = \text{weight of fluid displaced}$$
$$(4280 - 4000)(980)\ \text{dyn} = V(1.00\ \text{g/cm}^3)(980\ \text{cm/s}^2)$$
$$V = 280\ \text{cm}^3$$

$$d = \frac{\text{mass}}{\text{volume}}$$

$$= \frac{4280\ \text{g}}{280\ \text{cm}^3} = \boxed{15.3\ \text{g/cm}^3}$$

Hence the specific gravity of the material is 15.3. Since pure gold has a specific gravity of 19.3, the crown has been alloyed with a lighter substance such as silver (specific gravity 10.5) or copper (specific gravity 8.9).[†]

Example 12-10

An irregular object weighs 200×980 dyn in air, 160×980 dyn when immersed in water, and 170×980 dyn when immersed in oil. Calculate (a) the density of the object; (b) the density of the oil.

(a) The loss of weight in water is 40×980 dyn. By Archimedes' principle this is the weight of the displaced water. The mass of the displaced water is thus 40 g. The volume of displaced water is

$$V_{\text{water}} = \frac{40\ \text{g}}{1.00\ \text{g/cm}^3} = 40\ \text{cm}^3$$

Since this is also the volume of the irregular object, the object's density is

$$d = \frac{m}{V} = \frac{200\ \text{g}}{40\ \text{cm}^3} = \boxed{5.00\ \text{g/cm}^3}$$

(b) The weight of displaced oil is the BF, given by $(200 - 170)980$ dyn; the mass of displaced oil is 30 g.

$$d_{\text{oil}} = \frac{m}{V} = \frac{30\ \text{g}}{40\ \text{cm}^3} = \boxed{0.75\ \text{g/cm}^3}$$

---

* Perhaps he observed the apparent partial loss of weight of his arms and legs while in the bath. Whether he actually ran down the streets of Syracuse clad only in a towel and shouting "Eureka!" ("I have found it!") is open to question.

[†] Pure gold is rather soft, and some degree of alloying is necessary for jewelry, coins, and other objects. Gold coinage, consisting of 90% gold and 10% copper, has a specific gravity of 17.2.

When a body floats in a liquid, it sinks into the liquid just far enough so that the buoyant force equals its own weight. It is then in equilibrium, and net $\mathbf{F} = 0$. If a floating log is pushed down into the water slightly, the buoyant force is increased because more water is displaced. The net force is now upward, and the log accelerates upward (by Newton's second law) and returns to its equilibrium position. In fact, if a floating block has uniform cross section, the restoring force is proportional to the distance below or above the equilibrium position, and hence the motion is SHM (Sec. 9-2). The familiar battery tester (*hydrometer*) measures specific gravity by means of a float whose equilibrium position indicates the specific gravity of the liquid (Fig. 12-13). Since the specific gravity of the acid in a fully charged lead battery is about 1.30, whereas that of the acid in a "dead" battery is only about 1.10, the condition of the battery can be seen at a glance.

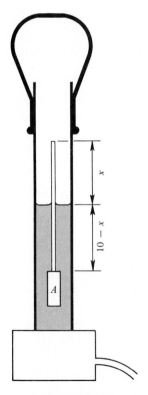

Figure 12-13

_____ **Example 12-11**

The float in Fig. 12-13 is constructed as follows: A portion A of volume 3.00 cm^3 is totally immersed, and a stem 10 cm long with cross section 0.200 cm^2 is partially immersed. The mass of the whole float is 4.80 g. When floating in a liquid of specific gravity 1.20, how much of the stem projects above the liquid surface?

Let x = the length above the surface. The volume below the surface in cm^3 is $3.00 + (10 - x)(0.200)$. Hence the weight of displaced fluid is found from the volume and the density of the fluid:

$$\{[3.00 + (10 - x)(0.200)] \text{ cm}^3\}$$

$$\times \left(1.20 \, \frac{\text{g}}{\text{cm}^3}\right)\left(980 \, \frac{\text{cm}}{\text{s}^2}\right)$$

The BF equals the entire weight of the float, since it is floating and is in equilibrium. We now apply Archimedes' principle, expressing both sides of the equation in dynes:

$$\text{BF} = \text{weight of fluid displaced}$$
$$(4.80)(980) = [3.00 + (10 - x)(0.200)](1.20)(980)$$
$$x = \boxed{5.00 \text{ cm}}$$

A balloon floating in air presents a slightly different problem. The entire balloon is immersed in the air, so the BF is constant (ignoring slight changes in density of air with altitude). The airship can be kept in equilibrium only by adjusting its weight so that the net force is zero. The downward force consists of the weight of the balloon plus the gas inside it plus the payload plus the ballast, if any. The upward force is the weight of an equal volume of air.

_____ **Example 12-12**

A blimp of volume 3000 m^3 is filled with helium (density 0.178 kg/m^3). If the blimp bag and framework weigh 3500 N and the payload, which includes personnel and instruments, is 23,000 N, how much ballast should be carried to maintain equilibrium? (The density of air is 1.293 kg/m³.)

The kilonewton (kN) is a good-sized force unit for this problem. The gravitational field strength is 9.8 N/kg.

$$BF = \text{weight of air displaced}$$
$$= (3000 \text{ m}^3)(1.293 \text{ kg/m}^3)(9.8 \text{ N/kg})$$
$$= 38.0 \text{ kN}$$

There are three downward forces:

1. Weight of helium in blimp:

    $(3000 \text{ m}^3)(0.178 \text{ kg/m}^3)(9.8 \text{ N/kg}) = $ 5.2 kN

2. Weight of blimp: $= $ 3.5 kN

3. Payload: $= $ 23.0 kN

    Total downward force $= \overline{31.7 \text{ kN}}$

Hence, to secure equilibrium an additional downward force of

$$38.0 \text{ kN} - 31.7 \text{ kN} = 6.3 \text{ kN}$$

must be supplied by the ballast. The mass of the ballast is

$$\frac{6300 \text{ N}}{9.8 \text{ N/kg}} = \boxed{643 \text{ kg}}$$

12-7 Surface Tension

Before leaving the subject of fluids at rest we will discuss the phenomena that take place at the surface of a liquid. As we have repeatedly pointed out, "cohesive forces" exist between atoms and between molecules; these forces are not gravitational, but are ultimately electrical in nature. A water molecule, although electrically neutral, is not symmetric and gives rise to enough electric field to distort the charge distribution of a nearby molecule. The small attractive force between these distorted charge distributions is called a Van der Waals force. The cohesive forces act between neighboring molecules and fall off rapidly as the distance between molecules increases. For this reason they are designated as "short-range" forces.

It is evident that there must also be repulsive forces between molecules that are extremely close together, for it is an observed fact that a large force must be applied in order to compress a liquid or a solid. These repulsive forces are also electric in nature, but in a subtle way. They are related to the Pauli exclusion principle (to be discussed in Chap. 31), according to which the electron clouds of neighboring atoms cannot freely interpenetrate. In popular terminology, two bodies cannot occupy the same region of space. The repulsive forces are effective at very short range, less than the range of the cohesive Van der Waals forces. For our study of liquids, we do not need to know the details of the origins of the short-range forces of cohesion and repulsion.

In the liquid state the cohesive forces are insufficient to give the structure rigidity, but the molecules are close enough to each other that there is sufficient cohesion to make possible a definite surface. To study surface phenomena more quantitatively, let us consider the energy changes that accompany the formation of a liquid surface. In order to increase the surface area of a given amount of liquid, molecules must be brought from deep* inside the liquid to the surface. To open up a space for an additional molecule, some molecules in the surface must be moved apart; thus work must be done against the cohesive intermolecular forces. After the molecule has become part of the (enlarged) surface area, it is in equilibrium under the action of two sets of intermolecular forces. Cohesive forces are balanced by repulsive forces. An analogous situation is the equilibrium of a book resting on a table top.

We see that a molecule at or very near the surface has potential energy, in much the same way that molecules of a stretched rubber band have elastic PE or a satellite has gravitational PE. The "surface PE" of a liquid therefore in-

---

* "Deep" here means several molecular diameters—about 10^{-9} to 10^{-8} m. At greater depths, the short-range electric forces are of negligible importance.

creases whenever the total surface area increases. When left to itself, a liquid tends to assume the shape that has the least surface area for a given volume, for in this way the potential energy is least. A falling raindrop would be spherical if it were not for its weight and for air resistance, since a sphere has minimum surface area for a given volume. We define the *coefficient of surface tension* of a liquid as potential energy per unit surface area; this constant is denoted by γ, the lower-case Greek gamma. A typical value of γ, for a water surface in contact with air, is 0.073 J/m^2 or, in cgs units, 73 erg/cm^2. (This is one of the few places in physics in which the erg is an energy unit of convenient size.)

An alternative way of looking at surface tension is to consider the force that must be applied when a surface is "stretched out" to a larger area. If a force **F**, applied parallel to the surface, moves an imaginary line of length L through a distance Δs perpendicular to the direction of the line, the increase in PE is the work done, equal to $F \Delta s$; the area created is $L \Delta s$. Hence, from the definition of γ,

$$\gamma = \frac{F \Delta s}{L \Delta s} = \frac{F}{L} \qquad (12\text{-}3)$$

We therefore interpret γ as a contractile force per unit length—a definition that is equivalent to our earlier definition of γ as surface energy per unit area. Since a joule is a newton·meter, the two units J/m^2 and N/m are equivalent. Similarly, an erg/cm^2 and a dyn/cm are equivalent units.

One of the standard methods of measuring the coefficient of surface tension of a liquid is by means of a Du Nouy torsion balance. A light ring of platinum wire, 4.00 cm in circumference, is suspended at one end of a light horizontal rod; the other end of the rod is attached perpendicularly to a fine horizontal steel wire that is fixed at one end and attached to a knob at the other end (see the photograph on page 267). As the knob is turned, the end of the rod tends to move upward, but it can be kept stationary by application of a small downward force. Such a

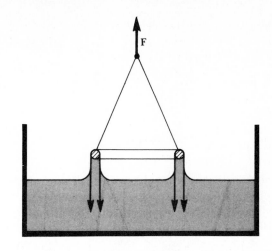

Figure 12-14

torsion balance is capable of measuring a force of 0.1 dyn, and, with a fine quartz fiber instead of a steel wire, it can be made sensitive enough to measure forces as low as the weight of a 0.001-μg object (approximately 10^{-6} dyn). As the ring is pulled out of the liquid, it carries a thin film of liquid with it (Fig. 12-14). When just about to break loose, the liquid film is vertical, and there is a downward force of contraction, represented in Fig. 12-14 by the downward force vectors. The contractile force is proportional to the circumference of the ring; the length of surface along which the force acts is approximately twice the circumference of the ring, since the sheet of liquid that is being stretched has two sides.* The working equation for the instrument is therefore $\gamma = F/2L$, where F is the magnitude of the measured force just as the liquid film breaks and L is the circumference of the ring.

The coefficient of surface tension can also be determined by observing the rise in a capillary tube. Let us take water in a clean glass tube as an example (Fig. 12-15). A molecule at P is acted on by three forces: a force of *cohesion* **C** exerted by

---

* Strictly, the total length is the sum of the inside circumference and the outside circumference of the ring.

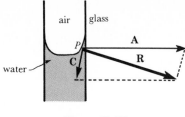

air   glass

P

C

A

R

water

Figure 12-15

adjacent water molecules; a force of *adhesion* **A** due to attraction by the glass molecules; and a slight force of adhesion caused by the attraction of the air molecules (not shown in the diagram).* The reason that water "wets" glass is that the adhesive forces are so much greater than the cohesive forces that the resultant inward force **R** is practically perpendicular to the surface of the glass. The liquid adjusts itself until its surface is perpendicular to the resultant force; if it did not, there would be a force component parallel to the surface, and the liquid, having no rigidity, would flow.† Near the glass, the resultant force on a water molecule is directed inward, toward the glass, and hence the liquid rises until the surface is perpendicular to the force. If the inside radius of the capillary tube is small, the top surface of the water in the tube is a hemisphere, with the edge of the liquid surface forming a horizontal circle of radius r. We now consider the equilibrium of the cylinder of water, of height h (Fig. 12-16). The upward force of surface tension (equal to γ dynes for every centimeter of length) is exerted on the circular ring where the surface ends, and the force is γ times the length of the ring. The downward force on the cylinder of water is the weight of the liquid, equal to its volume $\pi r^2 h$ times its weight per

---

* The term *cohesion* refers to forces between like molecules, and *adhesion* to forces between unlike molecules.
† The same phenomenon is illustrated by the fact that the surface of the water in a lake is horizontal, perpendicular to the apparent force of gravity (the resultant of the gravitational attraction of the earth and the outward inertial force due to the earth's rotation).

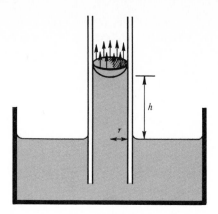

h

r

Figure 12-16   Rise of water in a capillary tube.

unit volume, dg. For equilibrium,

$$\text{Upward force} = \text{downward force}$$
$$\gamma(2\pi r) = (\pi r^2 h)dg$$

Solving this for γ, we get

$$\gamma = \frac{hrdg}{2}$$

The same equation shows that

$$h = \frac{2\gamma}{rdg} \qquad (12\text{-}4)$$

It is evident that the height of rise can be very great if r is small enough.

Example 12-13

How high does methyl alcohol rise in a glass tube 0.1 mm in diameter? The surface tension is 23 dyn/cm, and the specific gravity is 0.8.

$$h = \frac{2\gamma}{rdg}$$

$$= \frac{2(23 \text{ dyn/cm})}{(5 \times 10^{-3} \text{ cm})(0.8 \text{ g/cm}^3)(980 \text{ cm/s}^2)}$$

$$= \boxed{12 \text{ cm}}$$

If the cohesive forces are sufficiently large in comparison to the adhesive forces, the liquid will

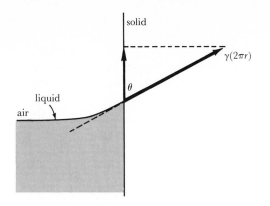

Figure 12-17

Table 12-1 Surface Constants

Interface	°C	Coefficient of Surface Tension γ erg/cm^2 or dyn/cm
water-air	0	75.6
water-air	20	72.8
water-air	100	58.9
ethyl alcohol-air	20	22.3
mercury-air	15	487.
NaCl (molten)-N$_2$	803	114.
nitrogen (liq.)-N$_2$	−203	10.5
water-benzene	20	35.0

Interface	Angle of Contact
water-clean glass	0°
ethyl alcohol-glass	0°
mercury-glass	140°
water-silver	90°
water-paraffin	107°
kerosene-glass	26°

not wet the glass or other container, and the rise is less. For any given liquid-solid combination the *angle of contact* θ (Fig. 12-17) is 0 only if the liquid wets the surface. The upward component of force is $\gamma(2\pi r)\cos\theta$, and Eq. 12-4 is replaced by

$$h = \frac{2\gamma\cos\theta}{rdg}$$

In extreme cases, the level near the glass may actually be depressed, as in the case of mercury in a glass tube (Fig. 12-18). The resultant force **R** can be perpendicular to the surface only if the mercury surface curves downward as shown. Here $\theta = 140°$, $\cos\theta = -.766$, and h is negative.

Many common phenomena depend on surface tension. For example, absorbency of a blotter or a towel is due to a capillary action in the tiny

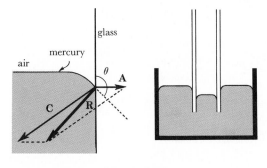

Figure 12-18

tubes formed by the crisscross of the fibers of paper or cloth. The dissolving of "instant" powdered coffee depends on the wettability of the granules. As shown in Table 12-1, water has a relatively high coefficient of surface tension—about 72 dyn/cm—as compared with other common liquids such as alcohol (23 dyn/cm) or carbon tetrachloride (20 dyn/cm). A commercial (nonalcoholic) antiseptic trademarked ST-37 has a value of only 37 dyn/cm. This minimizes the formation of drops that might block the entrance to tiny cracks in the skin or wound. A not-too-clean needle or razor blade can "float" on water due to the combined action of surface tension and Archimedes' principle. The water does not wet the metal (which may have a thin oily film), and as the needle sinks, the water's surface is deformed, giving rise to an upward component of force that may partially balance the weight if the diameter of the needle is not too large.

12-8   Fluids in Motion—
        Bernoulli's Equation

Turning now to the subject of fluids in motion, or hydrodynamics, let us first note that the mathematical difficulties are formidable. Some of the finest mathematicians and physicists of past and present have found the subject worthy of their best efforts, and only with the advent of electronic computers has there been substantial progress in the field.

We can distinguish between two broad classifications of fluid flow. In *streamline flow*, the molecules of fluid move from point to point without any rotational motion or turbulence. There may, however, be some energy losses due to fluid friction (viscosity). *Turbulent flow* takes place above a certain critical speed, which depends on the fluid and the shape of the pipe or on the shape of the stationary body past which the fluid is flowing. Little eddies or whirlpools

absorb energy, and the frictional drag of the fluid increases sharply (Fig. 12-19).

When a liquid flows at speed v through a tube of cross section A, an amount of liquid contained in a volume $(v \, \Delta t)A$ of the tube passes any point during a time Δt. Thus, $\Delta V = vA \, \Delta t$, and we find that the product vA is the volume rate of flow:

$$vA = \frac{\Delta V}{\Delta t}$$

$$\left(\frac{m}{s}\right)(m^2) = \frac{m^3}{s}$$

We see that vA is constant if the volume does not change—that is, if the fluid is incompressible. The speed is large wherever the cross section is small. As an illustration, imagine incompressible cars, moving in the same direction bumper to bumper in both lanes of a highway, coming to a construction area in which one lane is blocked off. If streamline flow is maintained, as assumed,

Figure 12-19   Streamline flow and turbulence around a model of an airfoil.

the cars move twice as fast through the constricted area, accelerated perhaps by bumps from behind if the drivers are not alert. Another illustration of the constancy of vA is found in the shower stall. Water must flow through many small holes in the shower head, so v increases, giving a stream of fast-flowing jets of water, each of small cross section.

Example 12-14

During a resting condition, the heart pumps blood at the rate of 3.6×10^3 cm^3/min through an aorta of cross section 0.8 cm^2. (*a*) What is the average blood velocity in the aorta? (*b*) The blood spreads out into a capillary network that is equivalent to about 5×10^6 fine tubes, each of diameter 8×10^{-4} cm (the diameter of a red corpuscle). Calculate the average blood velocity in a capillary.

The volume flow is $(3.6 \times 10^3)/60 = 60$ cm^3/s.

(*a*) $\quad v = \dfrac{\text{volume flow}}{\text{area}} = \dfrac{60 \text{ cm}^3/\text{s}}{0.8 \text{ cm}^2} = \boxed{75 \text{ cm/s}}$

(*b*) The cross section of each capillary is $\pi(4 \times 10^{-4}$ cm$)^2 = 5.0 \times 10^{-7}$ cm^2. Taken together, the total area of all the capillaries is

$$(5 \times 10^6)(5.0 \times 10^{-7} \text{ cm}^2) = 2.5 \text{ cm}^2$$
$$A_1 v_1 = A_2 v_2$$
$$(0.8 \text{ cm}^2)(75 \text{ cm/s}) = (2.5 \text{ cm}^2)v_2$$
$$v_2 = \boxed{24 \text{ cm/s}}$$

Blood flows only about $\frac{1}{3}$ as fast in the capillaries as in the aorta.

We can easily calculate the work done when the volume of a fluid changes by ΔV, the pressure P remaining constant. (If the pressure is not constant, the methods of integral calculus can be used to evaluate the sum of all the ΔW values.)

$$\text{Work} = \text{force} \times \text{displacement}$$

$$\Delta W = F(\Delta x) = (PA)(\Delta x) = P(A\,\Delta x)$$

or

$$\Delta W = P\,\Delta V \tag{12-5}$$

This useful equation gives the work done when a constant pressure P causes a change in volume ΔV.

Example 12-15

Assume that the left ventricle of the heart is analogous to a simple piston and chamber, and that 60 cm^3 of blood (density 1.0 g/cm^3) is ejected into the aorta during each stroke of the piston against an average pressure of 105 torr (105 mm Hg). (*a*) Calculate the work done by this ventricle muscle during a single contraction. (*b*) Calculate the power in watts, assuming 72 heartbeats per minute.

(*a*) $\qquad P = (105 \text{ torr})\left(\dfrac{101.3 \text{ kPa}}{760 \text{ torr}}\right)$

$\qquad\qquad = 14 \text{ kPa} = 1.40 \times 10^4 \text{ N/m}^2$

$\Delta V = (60 \text{ cm}^3)\left(\dfrac{1 \text{ m}}{10^2 \text{ cm}}\right)^3 = 6.0 \times 10^{-5} \text{ m}^3$

$\Delta W = P\Delta V = \left(1.40 \times 10^4 \dfrac{\text{N}}{\text{m}^2}\right)(6.0 \times 10^{-5} \text{ m}^3)$

$\qquad = \boxed{0.84 \text{ J}}$

(*b*) $\quad \text{Power} = \dfrac{\Delta W}{\Delta t} = \left(0.84 \dfrac{\text{J}}{\text{beat}}\right)\left(\dfrac{72 \text{ beats}}{60 \text{ sec}}\right)$

$\qquad = \boxed{1.0 \text{ W}}$

Note: The right ventricle also does work against hydrostatic pressure, but this power is much smaller, only about 0.1 W, since the pulmonary pressure against which work is done is much less than the aortic pressure.

The work calculated above is that performed by a resting heart. During a lifetime an enormous amount of work is done by the heart muscle. (See P ob. 12-B32.)

Many of the simpler phenomena of fluids can be correlated by means of a fundamental equation derived by Bernoulli.* The state of a fluid at any point can be characterized by four quantities: the velocity $\mathbf{v}$, the density d, the pressure P,

---

* Daniel Bernoulli (1700–1782), whose father and uncle were also famous Swiss mathematicians and physicists.

and the height h above some reference level. Bernoulli related these quantities to each other through a general principle:

If an incompressible fluid is in streamline flow, quantity ($\frac{1}{2}dv^2 + hdg + P$) is constant at every point in the fluid.

We can express this principle in a form known as *Bernoulli's equation*:

$$\tfrac{1}{2}dv_1{}^2 + h_1 dg + P_1 = \tfrac{1}{2}dv_2{}^2 + h_2 dg + P_2 \quad (12\text{-}6)$$

where the subscript 1 refers to an arbitrary point in the fluid and the subscript 2 refers to any other point. A proof of Bernoulli's equation for an incompressible fluid in steady, streamline flow is given in Sec. 12-10; we note here that the proof depends on two of the conservation laws: the law of conservation of mass and the law of conservation of energy.

Bernoulli's equation can be simplified in many special cases, if P, v, or h is constant, as often happens. A few examples illustrate varied applications of Bernoulli's equation.

Case 1 Velocity constant. The equation becomes $hdg + P =$ constant. For instance, the pressure in a tank of water decreases as the height above the bottom level increases. Thus $h_1 d_1 g + P_1 = h_2 d_2 g + P_2$; if the fluid is incompressible, $d_1 = d_2 = d$, and the equation can be rearranged to give the familiar formula $P_1 - P_2 = (h_2 - h_1)dg$, relating the magnitudes

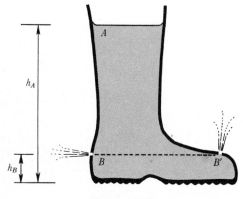

Figure 12-20

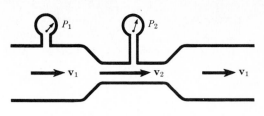

Figure 12-21 Venturi flow meter.

of the change in pressure and the change in height (consistent with Eq. 12-1).

Case 2 Pressure constant. A rubber boot is partially filled with water, and two pinholes are made at the same distance below the surface of the water (Fig. 12-20). (The density is considered to be constant.) The pressure at A is atmospheric pressure, and so is the pressure at hole B, since the water at this point is open to the atmosphere. Hence, canceling out the pressure from each side of Bernoulli's equation, we see that $\frac{1}{2}dv^2 + hdg =$ constant. The speed of the water at A is practically zero, but the speed at B is greater than that at A, since the value of h at B is less than at A. The speeds of the two streams at B and B' are the same, although the water emerges in different directions, since the force due to the pressure is always perpendicular to the surface on which it acts. The term hdg is just the gravitational PE of 1 cm^3 of the liquid, and the term $\frac{1}{2}dv^2$ is the KE of 1 cm^3. The total mechanical energy* remains constant, in accordance with the energy principle.

Case 3 Height constant. The equation becomes $\frac{1}{2}dv^2 + P =$ constant. The pressure decreases wherever the velocity increases. Among the applications of this aspect of Bernoulli's equation are the Venturi meter for measuring rate of flow of a liquid or gas; the curve of a spinning baseball; and the lift of the wings of a plane.

In a Venturi meter the rate of flow of a fluid (usually a liquid) is shown by the pressure decrease as the fluid is forced through a constriction in the pipe and consequently speeds up (Fig. 12-21). Such meters are used by water departments, which encounter large flow rates.

---

* The energy principle (Sec. 6-6) represents the conservation of *mechanical* energy and applies only when there are no frictional losses. It is for this reason that Bernoulli's principle is restricted to streamline flow.

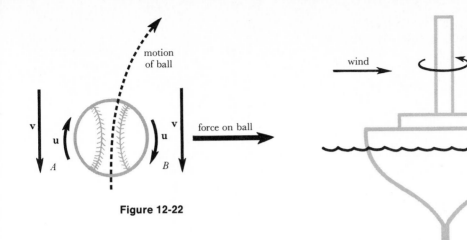

Figure 12-22

Figure 12-23

When a baseball is thrown forward, the air rushes past it at a velocity **v.** If, however, the ball is spinning, it drags some air with it (especially if the surface is rough), with a velocity **u** (Fig. 12-22). Hence the resultant speed of air at B is $v + u$, while at A the resultant speed of the air is $v - u$. By Bernoulli's equation the pressure is greatest at A, where the air speed is least, and the ball is pushed sideways, to the right, as shown. The whole phenomenon, called the *Magnus effect*, depends on the viscous drag of the fluid as well as on Bernoulli's principle. The ability of a baseball pitcher to throw a curve is by no means an optical illusion.

On the same principle, the Flettner rotor ship in the 1920s took advantage of the winds without using sails (Fig. 12-23). A tall, rough, cylindrical shaft was kept rotating by a small auxiliary motor; if a wind was blowing, the Bernoulli force acted on the rotor and hence on the ship. The power supplied by the motor was needed only to overcome friction in the bearings; the power to drive the ship came from the wind. The ship did not prove a commercial success, however.

As a final example of Bernoulli's equation, we consider the "lift" of an airfoil. If the shape of the leading edge of the airfoil is right, streamline flow is maintained except for some turbulence just above the wing (refer back to Fig. 12-19). Since flow lines are crowded together above the wing, the air speed is greater, just as if the air were flowing through a constriction in a pipe. Greater speed means less pressure, and so the downward force caused by the air pressure on the upper surface is less than the upward force on the lower surface. The lift also depends on the angle that the wing makes with the horizontal. Newton's third law

is illustrated here. The underside of the wing pushes air downward (called down-draft), so the air pushes the wing upward. The wing cannot be tilted too much, for then turbulence becomes excessive and the plane "stalls."

12-9 Viscosity

Fluids in motion exhibit a certain resistance to motion that is called *viscosity*—a sort of internal molecular friction. The viscous drag is caused in liquids by short-range molecular cohesive forces, and in gases by collisions between fast-moving molecules. In both liquids and gases, the drag is proportional to the speed, as long as the speed is slow enough so that streamline flow takes place. However, in turbulent flow the viscous drag increases rapidly, more nearly proportional to the square or cube of the velocity. Streamlining of cars and planes means just what it says—designing them to permit streamline flow at high velocities, thus reducing friction. There is no simple law giving the relationship between force and velocity (see Fig. 12-24 for typical experimental results). Whatever the law,

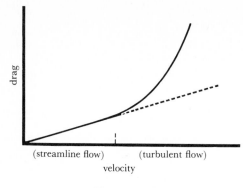

(streamline flow)   (turbulent flow)

velocity

Figure 12-24

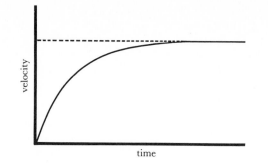

Figure 12-25 Velocity-time graph for a falling body, illustrating terminal velocity.

one thing is certain: viscous drag increases as velocity increases.

The dependence of the force of viscosity on velocity is nicely illustrated by the *terminal velocity* reached by a falling body. As a parachutist falls, his or her velocity increases, and hence the upward viscous force of the air increases. Sooner or later the upward drag equals the downward force (the parachutist's weight), and then the net force on the person is zero. The parachutist has reached a constant terminal velocity that depends on his or her weight and the size of the parachute. A typical value of terminal velocity for a person-parachute combination is about 40 km/h (25 mi/h). While the system is picking up speed, the viscous drag is less than the person's weight. Using Newton's second law, we have

$$\text{net } F = ma$$

or

$$\text{Weight} - \text{viscous drag} = ma$$

The weight is constant, but the viscous drag is increasing, so the net force is not constant and hence the acceleration is not constant. The velocity increases asymptotically toward the terminal velocity, as shown in Fig. 12-25. Without a parachute, the area exposed to the air stream is much smaller, and the balance of forces occurs at a higher speed. If the parachutist falls far enough before opening the chute, a terminal velocity of about 200 km/h (about 125 mi/h) is attained, regardless of the height from which the fall

began. (Records exist of a few cases of survival after a fall of many kilometers without a parachute.)

In liquids, terminal velocity is illustrated by the motion of a steamship. The forward thrust of the propellers causes the velocity to increase. As the velocity increases, so does the viscous drag of the water. Eventually, equilibrium (a condition of no acceleration) is reached when the forward thrust equals the viscous drag, and then the ship's velocity remains constant.

Viscosity depends on temperature, but in a different fashion for liquids and gases. The viscosity of most liquids decreases as temperature increases—molasses in June is less viscous than molasses in January. On the other hand, at high temperatures the viscosity of gases increases.*

Plastic flow, the permanent distortion of an object by an applied force, may be thought of roughly as an exaggerated case of viscous flow, with a very high coefficient of viscosity. The technical definition of plastic flow includes the provision that the substance flows only after

---

* It is relatively easy to explain the increased viscosity of gases as being due to the fact that at high temperatures the molecules move faster and collide more often with each other, giving rise to increased internal friction. The decrease of viscosity of a liquid is harder to explain, and it probably depends on the fact that at higher temperatures the molecules are farther apart, and the cohesive forces of internal molecular friction are therefore less effective.

some minimum stress is exceeded; lead, soap, and tallow candles are plastic. On the other hand, tar, sealing wax, some glues, and most glasses are highly viscous liquids. Crystalline quartz generally fractures before it flows appreciably and is not considered a liquid. The detailed study of liquids and plastic solids is complex and difficult. For instance, the apparent elasticity and viscosity of protoplasm depend on many factors, such as the nature of the protein lattice structure within the cells. Many cooks have experienced the sudden change in viscosity of a bowl of fudge that is being beaten. These examples show that the terms "liquid" and "solid" are only useful first approximations to the behavior of real substances. Much remains to be understood about the ways in which molecules interact with each other at close quarters in the no-man's-land between solid and liquid.

Summary   A fluid is a substance, such as a liquid or gas, that has no rigidity. Liquids are distinguished from gases by the presence of a surface.

Pressure is force per unit area. The difference in pressure between two points in a liquid at rest equals the vertical height separating the two points times the weight per unit volume of the liquid. A similar statement is true for gases, if the gas can be considered to have uniform density. Changes in pressure exerted at any point in a confined fluid are transmitted undiminished in all directions to all parts of the fluid (Pascal's principle).

The loss of weight of a body wholly or partially immersed in a fluid equals the weight of the fluid displaced (Archimedes' principle). The volume of an irregular solid can be measured using Archimedes' principle, thus leading to a way of determining density.

The molecules at the surface of a liquid are in a state of contractile tension due to cohesive forces between the molecules. The coefficient of surface tension represents the potential energy per unit area of surface; it is numerically equal to the contractile force per unit length of boundary. The angle of contact is the angle between a liquid surface and an adjacent solid surface.

According to Bernoulli's equation, the quantity $(\frac{1}{2}dv^2 + hdg + P)$ is constant throughout an incompressible fluid that is at rest or in streamline flow. At a given level, the pressure is least where the velocity is greatest.

Viscosity is fluid friction, caused (in a liquid) by cohesive forces between molecules and (in a gas) by collisions between molecules. The viscosity of liquids decreases as temperature rises, but the opposite is true for gases.

Check List

fluid	mm Hg	streamline flow
liquid	torr	$\Delta V/\Delta t = vA$
gas	$P = F/A$	$\Delta W = P\Delta V$
pressure	$P = hdg$	turbulent flow
absolute pressure	Pascal's principle	Bernoulli's equation
gauge pressure	Archimedes' principle	$\frac{1}{2}dv^2 + hdg + P$
pascal (pressure unit)	coefficient of surface	$\quad$ = constant
atmosphere (pressure	$\quad$ tension	viscosity
$\quad$ unit)	angle of contact	terminal velocity
bar		

12-1 At ordinary temperatures, which of the following are not fluids? Air; diamond; glass; taffy; kerosene; mercury; iron.

12-2 Suppose the undersea camera of Example 12-4 is first pointed horizontally and then turned downward to photograph the ocean bottom. Compare the magnitudes of the total force on the glass window in these two orientations.

12-3 Explain why "water seeks its own level."

12-4 Is the buoyant force on a submerged submarine lying on the floor of the ocean exactly the same as when the submarine is only 5 m below the surface? Why?

12-5 What factors determine the maximum height to which a balloon can rise?

12-6 In Fig. 12-26 the thin, hollow, glass ball is empty and is sealed off from the surrounding atmosphere. It is exactly counterbalanced by a small, solid, brass counterweight. If the air is pumped out of the bell jar, does the glass ball rise or fall?

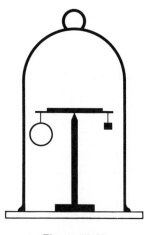

Figure 12-26

12-7 A solid cube of ice is floating in a glass of water, with the water level just even with the top of the glass. When the ice melts, will the water level be the same, will it fall below the edge of the glass, or will some water overflow?

12-8 A bucket of water is on a spring balance that reads 100 N. If a 1-kg live trout is placed in the bucket and the trout floats, will the reading of the balance be changed?

12-9 Which of the following units could *not* be used for the coefficient of surface tension? erg/cm^2; dyn/cm; N/m^2; N/m; J/m^2; lb/in.^{2}; g/cm.

12-10 A boy sends a Frisbee traveling horizontally through the air, giving it a spin as he throws it forward. As viewed from above, the disk is spinning clockwise. Does the Frisbee veer to the boy's right or left?

12-11 Is the ship in Fig. 12-23 coming toward the observer or moving away?

12-12 What's wrong with the proposed perpetual-motion machine in Fig. 12-27? The capillary tube is of such an inside diameter that water would rise 10 cm in it. The tube is cut off to be only 8 cm long, and it is expected that water will continually rush out in a fountain.

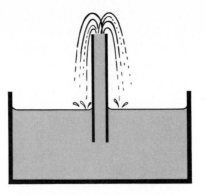

Figure 12-27 Perpetual motion?

12-13 Collisions have occurred between ships that were steered too close together in parallel paths. Explain.

12-14 The dimensions of pressure are (a) $[\mathrm{ML^2T^{-2}}]$; (b) $[\mathrm{MLT^{-1}}]$; (c) $[\mathrm{ML^{-1}T^{-2}}]$.

12-15 During a tornado, air pressure is 8×10^4 Pa inside a house and 1.0×10^5 Pa outdoors. The net force on a window of area $0.5\ \mathrm{m^2}$ is (a) 1.0×10^4 N; (b) 4.0×10^4 N; (c) 5.0×10^4 N.

12-16 The pressure at the bottom of a column of mercury 7.6 cm tall is (a) 7.6 torr; (b) 0.10 atm; (c) both of these.

12-17 An object weighs 5 N in air and 2 N when immersed in a liquid. The buoyant force is (a) 2 N; (b) 3 N; (c) 5 N.

12-18 The dimensions of the coefficient of surface tension are (a) $[\mathrm{MT^{-2}}]$; (b) $[\mathrm{MLT^{-2}}]$; (c) $[\mathrm{ML^2T^{-1}}]$.

12-19 For water in a clean glass capillary tube, the angle of contact is (a) exactly $0°$; (b) less than $1°$, but not exactly $0°$; (c) $90°$.

Problems

12-A1 What is the pressure on the ground beneath one of the spikes of a 100-kg football player whose shoes contain six spikes each, the cross section of the spikes at the ground being $0.4\ \mathrm{cm^2}$ for each spike? Express your answer in $\mathrm{N/m^2}$ and in atm.

12-A2 What is the pressure due to wind on the wall of a building 30×5.5 m if the horizontal force of the wind on the wall is 11 kN? Express your answer in $\mathrm{N/m^2}$ (Pa) and in atm.

12-A3 What is the total downward force of the atmosphere on a horizontal table top that is 50 cm $\times$ 160 cm? What is the total upward force on the underside of the same table top?

12-A4 Make the following conversions: $44.1\ \mathrm{lb/in.^2} = ?$ atm; $44.1\ \mathrm{lb/in.^2} = ?$ mm Hg; 20 atm $= ?$ kPa; 190 mm Hg $= ?$ torr; 1000 atm $= ?\ \mathrm{N/m^2}$.

12-A5 Make the following conversions: 110 kPa $= ?\ \mathrm{N/m^2}$; 30 atm $= ?\ \mathrm{lb/in.^2}$; 76 mm Hg $= ?\ \mathrm{lb/in.^2}$; 0.1 atm $= ?$ mm Hg; 10^{-10} atm $= ?$ torr.

12-A6 A household vacuum cleaner produces a partial vacuum that is measured by an open-tube manometer (Fig. 12-6) to be 60 cm $\mathrm{H_2O}$ below atmospheric pressure.

(*a*) What is the gauge pressure, in torr? (*b*) What is the absolute pressure, in torr? (*c*) What total force, in N, is available if the nozzle is 2.50 cm in diameter?

12-A7 In the normal adult human, the venous pressure just before the blood enters the heart is slightly less than 1 atm. The difference is so small that it is usually measured in millimeters of water. Express a typical value of 38 mm H_2O as (*a*) torr; (*b*) dyn/cm^2; (*c*) millibars (mbar).

12-A8 The specific gravity of mercury is 13.6. Calculate the pressure in N/m^2 and in lb/in.^{2} at the bottom of a flask of mercury that is 8 cm deep.

12-A9 Compute the pressure at the lowest point of a tank car full of gasoline (specific gravity = 0.7). The tank is in the form of a closed horizontal cylinder 220 cm in diameter and 10 m long.

12-A10 A student is given a sample of one of the liquids in Table 2-1. He finds that the pressure at the bottom of a 50-cm column of the liquid is 6.17×10^4 dyn/cm^2 greater than atmospheric pressure. Identify the liquid.

12-A11 A homeowner wishes to adjust a newly purchased barometer. The home is on a hill that is 300 m above the level of an airport where barometric pressure is reported to be 99.1 kPa. (*a*) What is the pressure difference, in kPa, between the home and the airport? (*b*) What is the correct setting of the home barometer?

12-A12 A stone of volume 0.05 m^3 is lowered to the bottom of a lake. How much weight does the stone lose?

12-A13 Calculate the buoyant force on an ice cube of volume 16 cm^3 that is immersed in alcohol of specific gravity 0.8.

12-A14 Suppose that, as a fog clears, 125 tiny drops of water coalesce into one larger drop. What is the ratio of the total surface energy of the tiny drops to the surface energy of the large drop?

12-B1 A 0.3-kg spherical grapefruit is resting on a table. If the material of the fruit yields until the pressure is reduced to 3000 N/m^2, what is the radius of the flat spot on the bottom of the grapefruit?

12-B2 A swimming pool measures 10 m $\times$ 25 m $\times$ 4 m deep. (*a*) What is the force on the bottom due to the water? (*b*) What is the force on the vertical wall at one end of the pool? (*Hint:* Use the average pressure, which equals the pressure at the average depth.)

12-B3 Crew members attempt to escape from a sunken submarine that is lying on its side in 100 m of water. What force must they apply on the escape hatch, which is a door 1.2 m $\times$ 0.6 m? The pressure inside the submarine is 1 atm.

12-B4 (*a*) Calculate the force, in dynes, on a horizontal surface of area 3 cm^2 at *P* in Fig. 12-5. (*b*) Calculate the force on a 3-cm^2 vertical surface at point *Q*.

12-B5 Calculate the pressure, in kPa, read by the gauge at *G'* in Fig. 12-5.

12-B6 In Fig. 12-5, what is the pressure at a point in the liquid 80 cm above the level of the table that supports the meter stick?

12-B7 The density of blood is about 1.1 g/cm^3. If there were no structures in a giraffe's neck to give partial support to the blood, what would be the difference in pressure, in torr, between the blood in a giraffe's brain and that in its heart? The brain is 2.2 m above the heart. See Ref. 2 at the end of the chapter.

12-B8 In an experiment on the elasticity of living organs, a researcher used boric acid solution as the indicating liquid in a manometer. She found that when the pressure changed from 6 cm of boric acid solution to 18 cm of boric acid solution, the volume of a urinary bladder changed from 300 cm^3 to 200 cm^3. The specific gravity of boric acid solution is 1.08. (a) Express the pressure change in dyn/cm^2; in mm Hg. (b) If the bladder were a solid body obeying Hooke's law, what would be the bulk modulus of the organ? (How do you account for the greatly different order of magnitude of your answer from the values listed for other materials in Table 9-1?)

12-B9 A hydraulic lift in a service station is to raise a truck of mass 6 metric tons by compressed air at a pressure of 2000 kPa. The large cylinder of the lift has a radius of 10 cm. Is the air pressure sufficient for the job?

12-B10 In a solid-waste recycling plant, the small piston of a hydraulic press has radius 3 cm, and the large piston has radius 30 cm. Compressed air is available at a gauge pressure of 500 kPa. What force can be applied to a load of used metal beverage cans?

12-B11 A cube of aluminum 4 cm on an edge floats in a beaker of mercury. (a) What volume of aluminum is immersed in the mercury? (b) What volume of lead, placed on top of the cube, would cause the cube to be just immersed in the mercury? (Use densities from Table 1-3.)

12-B12 A geologist analyzing an irregular piece of ore finds that the sample is balanced by 98 g when weighed in air and 72 g when weighed in water. What is the specific gravity of the ore?

12-B13 A moon rock weighs 5 × 980 dyn in air, and 3 × 980 dyn when immersed in alcohol of specific gravity 0.8. (a) What is the rock's volume? (b) What is its specific gravity?

12-B14 A stone of volume 0.06 m^3 weighs 1400 N in air. What fraction of its weight would it lose if it were submerged at the bottom of a fresh-water lake?

12-B15 A piece of cork weighs 15 × 980 dyn in air. To find its density, a student hangs the cork from one arm of a balance and hangs an iron sinker from the cork. When the sinker is immersed in water, the apparent weight of sinker plus cork is 127.5 × 980 dyn; when both sinker and cork are immersed, their total apparent weight is 65 × 980 dyn. From these data, calculate the density of cork.

12-B16 A 40-kg child is just floating in water, with negligible volume in the air. What is the child's volume?

12-B17 A swimmer who can float in a pool with only a negligible volume (the nostrils) exposed stands on a spring scale at poolside, which reads 88 kg. When standing on the same scale in hip-deep water, the reading is 60 kg. Calculate (a) the swimmer's body volume and (b) the volume of each of the swimmer's legs. (c) Estimate the volume of a human nose to see if it is really "negligible" in this context.

12-B18 The cross section (at the water level) of a sea-going tanker is 800 m^2. How far will the tanker sink into the water if 3000 m^3 of crude oil of specific gravity 0.9 is pumped into it?

12-B19 A weather balloon made of light, flexible plastic weighs 6 N when empty; it carries a load of 90 N of apparatus when filled with helium near the earth's surface. What is the volume of the balloon when full?

12-B20 If the hydrometer of Example 12-11 were placed in a dead battery where the specific gravity is 1.10, what length of stem (x in Fig. 12-13) would project above the liquid surface?

12-B21 It is desired to measure the specific gravity of olive oil by Archimedes' principle. An irregularly shaped piece of metal, of unknown composition, is available. The following data are taken: weight of metal in air = 135.0 × 980 dyn; weight of metal in water $= 85.0 \times 980$ dyn; weight of metal in oil $= 89.0 \times 980$ dyn. (*a*) Calculate the specific gravity of the metal. (*b*) Calculate the specific gravity of the oil.

12-B22 The coefficient of surface tension of soap solution is 25 erg/cm². (*a*) What is the increase in total surface energy when a soap bubble is blown up from a radius of 1 cm to a radius of 3 cm? (*b*) What is the source of this energy?

12-B23 The coefficient of surface tension of a mixture of alcohol and water is measured with the apparatus shown in Fig. 12-14. It is observed that the film breaks when an upward force of 600 dyn is applied. The average radius of the ring is 1.40 cm. Calculate the coefficient of surface tension of the liquid.

12-B24 What upward force is needed to break the vertical film of water adhering to a thin horizontal wire ring of circumference 4 cm (Fig. 12-14)? The coefficient of surface tension for water is 73 dyn/cm.

12-B25 A pencil 8 mm in diameter is held vertically with its lower end immersed in a beaker of ethyl alcohol, for which the coefficient of surface tension is 22 dyn/cm. The pencil is wet by the liquid. What are the magnitude and direction of the force exerted by the liquid on the pencil?

12-B26 A toy boat of white pine (specific gravity 0.45) is floating on the surface of a calm lake. The wood is 8 cm wide, 30 cm long, and 2 cm thick. (*a*) If, to the water near the right-hand 8-cm side, is added a detergent that makes the coefficient of surface tension 40% of its normal value, what is the direction of the net force on the boat? (*b*) Ignoring viscous drag of the water, what will be the velocity of the boat 5 s after the detergent is added to the water?

12-B27 The coefficient of surface tension for water is about 72 dyn/cm. (*a*) What is the total upward force exerted on the pavement during a rainstorm due to the boundary of a hemispherical water bubble on the pavement? The bubble is 2 cm in diameter. (*Hint:* The water film has two surfaces, each exerting upward contractile force.) (*b*) The upward force found in part (*a*), exerted over a circular area of pavement, means that the pressure inside the bubble is greater than atmospheric pressure. Calculate this excess pressure, in dyn/cm². (*c*) Solve the problem in symbols, showing that for a spherical bubble of radius R the excess pressure is $\Delta P = 4\gamma/R$.

12-B28 How high will water rise in a glass capillary tube 0.02 mm in diameter?

12-B29 What is the diameter of a capillary tube in which glycerin rises 5 cm? The glycerin wets the glass, and its coefficient of surface tension is 63 dyn/cm.

12-B30 Pure water is placed in a clean silver cup. (Refer to Table 12-1 for the angle of contact.) (*a*) Construct a force diagram similar to Fig. 12-15, showing the vectors **A, C,** and **R** (ignore the effect of air molecules above the liquid surface). What is the direction of the resultant force **R**? (*b*) For a water-silver interface, what is the ratio C/A of the cohesive force to the adhesive force? (*c*) What would you expect if a capillary tube of silver were placed in a beaker of water?

12-B31 In an effort to get an indestructible standard of length depending on the properties of one substance, Sir Humphry Davy (1778–1829) suggested that the natural standard of length be equal to the diameter of a tube, wet by water, in which the capillary rise of water, at the melting point of solid water (0°C), equals the tube's diameter. (*a*) Derive a formula for this length. (*b*) Check the dimensions of your formula. (*c*) Using 75.6 erg/cm² for the coefficient of surface tension of water at 0°C, calculate the length of Sir Humphry's unit.

12-B32 (*a*) Calculate the total work done by the muscle of the resting heart of Example 12-15 during an 80-year lifetime (of relative inactivity). (*b*) Through what vertical height could this amount of work lift the Great Pyramid of Cheops (see Prob. 1-A8)?

12-B33 The gauge pressures in a Venturi meter (Fig. 12-21) are $P_1 = 100$ kPa and $P_2 = 80$ kPa. The liquid is water. (a) If the velocity of the stream at point 1 is 3.00 m/s, what is the velocity at point 2? (b) What is the ratio of areas at points 1 and 2? (c) What is the ratio of diameters at points 1 and 2?

12-B34 In Fig. 12-21, pressure gauge P_1 reads 190 kPa and P_2 reads 140 kPa. If the velocity of the stream at point 1 is 3.00 m/s, what is the velocity at point 2? The liquid is gasoline of specific gravity 0.66.

12-B35 In Fig. 12-20, the hole at B is 40 cm below the water level and 3.6 cm above the floor. Assuming no loss of speed due to friction at the hole, how far out from the boot does the stream of water strike the floor? (Use Bernoulli's equation; the density of the liquid cancels out.)

12-B36 A storage tank is filled with water to a depth of 3.8 m, and a short hose is connected to a faucet at the bottom of the tank. Neglect friction. (a) With what speed does water leave the nozzle of the hose? (b) If the hose is pointed upward, how high does the stream of water rise?

12-B37 Air flows past the upper surface of a horizontal streamlined airplane wing at 260 m/s and past the lower surface of the wing at 200 m/s. The density of air is 1.0 kg/m^3 at the flight altitude, and the area of the wing is 20 m^2. (a) What is the pressure difference between the two surfaces of the wing? (b) Calculate the net lift on the wing.

12-B38 Calculate the surface area of the raft of Prob. 9-B22 (page 213).

12-C1 In Example 12-4, sea water was assumed to be of uniform density. Check this assumption, using data from Table 9-1. (a) Find the fractional change in volume of a given mass of sea water lowered from the surface to a depth of 7.50 km, where the pressure was calculated to be 7.54×10^7 N/m^2. (b) Calculate the specific gravity of sea water at 7.50 km if the sea-level value is 1.03.

12-C2 A 300-kg wooden log is floating with 25% of its volume above the water level. (a) What is the volume of an iron object that must be attached to the underside of the log in order to completely submerge it? (b) What is the mass of the iron object?

12-C3 A skin diver's average specific gravity is 0.98, and her mass is 60 kg. For good maneuverability while walking on a submerged ship's deck in fresh water, the diver desires an apparent weight 0.08 times her weight in air. (a) What volume of iron weights should she attach to her feet? (b) How much does this amount of iron weigh in air?

12-C4 The concept of isostasy in geology implies that mountains "float" on a denser substratum, the earth's mantle, with the tallest mountains penetrating deepest into the mantle. Suppose a cube of side L and density d' floats in a liquid of density d (Fig. 12-28)

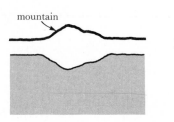
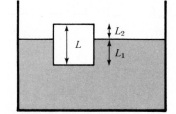

Figure 12-28

with a length L_1 beneath the liquid and L_2 above the surface ($L_1 + L_2 = L$). Derive formulas for L_1 and L_2 in terms of L, d, and d', and show that both L_1 and L_2 increase as L increases, thus illustrating isostasy.

12-C5 Consider the KE of the blood flowing as described in Examples 12-14 and 12-15. Since the pumping action occupies only a small fraction of the heart cycle, the *maximum* blood velocity is larger than the average value, and it can be shown that the correct effective velocity for calculating KE is about 1.87 times the average velocity. (a) If 60 cm^3 of blood is given a velocity of 75 cm/s, as in Example 12-14, this KE is supplied 72 times per minute. Show that the power required is only about $0.07 \times 2 = 0.14$ W (considering both ventricles) and is thus negligible in the *resting* heart. (b) For the vigorously active heart, during exercise, the blood flow is 10 times as great as for the resting heart, that is, 3.6×10^4 cm^3/min. Show that during exercise the power associated with KE is about 70 W for each ventricle, for a total of about 140 W. (*Hint:* In 1 s, 10 times as much blood is given 10 times as much speed, as compared with the resting heart.) (c) Using $W = P\Delta V$, explain how the power associated with hydrostatic pressure is about 10 W for the active heart, still only a fraction of the total power.

For Further Study

12-10 Derivation of Bernoulli's Equation

To derive Bernoulli's equation, we use the law of conservation of mass and the law of conservation of energy. Consider a tube of variable cross section through which an incompressible fluid moves in streamline flow (Fig. 12-29). In a time Δt, the volume of fluid that enters at end 1 is $(v_1 \Delta t)A_1$, and the mass of fluid that enters is $v_1 \Delta t A_1 d$. In this same time, the mass of fluid that leaves end 2 is $v_2 \Delta t A_2 d$. The law of conservation of mass tells us that no mass is created or absorbed within the tube. Since we assume that the flow is steady and the fluid is incompressible, of density d throughout, this means that the mass that enters end 1 in time Δt equals the mass that leaves end 2 during the same interval. That is,

$$v_1 \Delta t A_1 d = v_2 \Delta t A_2 d$$
$$v_1 A_1 = v_2 A_2 \qquad (12\text{-}7)$$

From this we see that the speed v is large where the cross section is small.*

To force the fluid through the tube, some external force $P_1 A_1$ acts at end 1, doing work *on* the fluid, and work is performed *by* the fluid as

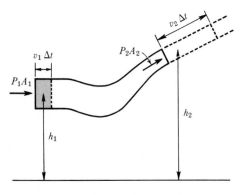

Figure 12-29

---

* It may be asked what forces act to accelerate the molecules of fluid, which speed up and slow down according to Eq. 12-7. The key word in this connection is that the fluid is "incompressible," which means that short-range repulsive forces act whenever the molecules come too close together. These forces, essentially electric in nature, serve to cause the necessary changes in momentum of the molecules.

it exerts a force P_2A_2 at end 2. Thus at end 1, the system gains energy $P_1A_1(v_1\,\Delta t)$, and at end 2 it loses energy $P_2A_2(v_2\,\Delta t)$. The net work done on the system is

$$P_1A_1(v_1\,\Delta t) - P_2A_2(v_2\,\Delta t)$$

Other energy changes are associated with PE and KE: the net increase in KE is

$$\tfrac{1}{2}(v_2\,\Delta tA_2d)v_2{}^2 - \tfrac{1}{2}(v_1\,\Delta tA_1d)v_1{}^2$$

and the net increase in PE is

$$(v_2\,\Delta tA_2d)gh_2 - (v_1\,\Delta tA_1d)gh_1$$

Putting all these results together, we apply the law of conservation of energy. The work done on the system equals the increase in KE and PE:

$$P_1A_1(v_1\,\Delta t) - P_2A_2(v_2\,\Delta t)$$
$$= \tfrac{1}{2}(v_2\,\Delta tA_2d)v_2{}^2 - \tfrac{1}{2}(v_1\,\Delta tA_1d)v_1{}^2$$
$$+ (v_2\,\Delta tA_2d)gh_2 - (v_1\,\Delta tA_1d)gh_1$$

Remembering that $v_1A_1 = v_2A_2$, we simplify and rearrange, obtaining Bernoulli's equation:

$$\tfrac{1}{2}dv_1{}^2 + h_1dg + P_1 = \tfrac{1}{2}dv_2{}^2 + h_2dg + P_2$$

References

1. V. A. Greulach, "The Rise of Water in Plants," *Sci. American* **187**(4), 78 (Oct. 1952).
2. J. V. Warren, "The Physiology of the Giraffe," *Sci. American* **231**(5), 96 (Nov. 1974).
3. R. Markus, "The Teapot Effect," *Physics Today* **9**(9), 16 (Sept. 1956). An interesting and not completely understood example of liquid flow. I. Reba, "Applications of the Coanda Effect," *Sci. American* **214**(6), 84 (June 1966). More about the teapot effect.
4. D. L. Webster, "What Shall We Say About Airplanes?" *Am. J. Phys.* **15**, 228 (1947). Physical problems involved in flight.
5. D. C. Hazen and R. F. Lehnert, "Low Speed Flight," *Sci. American* **194**(4), 46 (Apr. 1956).
6. J. E. McDonald, "The Shape of Raindrops," *Sci. American* **190**(2), 64 (Feb. 1954).
7. G. Mili et al., "Baseball's Curve Balls—Are They Optical Illusions?" *Life*, Sept. 15, 1941, p. 83. F. L. Verwiebe, "Does a Baseball Curve?" *Am. J. Phys.* **10**, 119 (1942). R. M. Sutton, "Baseballs Do Curve and Drop," *Am. J. Phys.* **10**, 201 (1942).
8. N. Smith, "Bernoulli and Newton in Fluid Mechanics," *Phys. Teach.* **10**, 451 (1972).
9. F. J. Almgren, Jr., and J. E. Taylor, "The Geometry of Soap Films and Soap Bubbles," *Sci. American* **235**(1), 82 (July 1976).
10. L. Trefethen, *Examples of Surface Tension; Surface Tension and Contact Angles; Formation of Bubbles; Surface Tension and Curved Surfaces; Breakup of Liquid Drops* (films).
11. A. Shapiro, *The Magnus Effect* (film).

13

Temperature and Expansion

In our study of mechanics we assumed, without much quantitative evidence, that work done against friction becomes internal energy, shared in some fashion among the bodies that slide or rub against each other. We may, if we wish, consider much of the next five chapters as an elaboration of the law of conservation of energy, interpreted to include heat as a form of energy flow. It is our intention to develop the theory of heat to the point at which we can state two broad generalizations, the first and second laws of thermodynamics. In order to do this, however, we must first find out how to measure quantitatively temperature, internal energy, and heat.

13-1   Temperature

The sensations of hot and cold are among the very first experiences of a newborn baby. Every known primitive language has words for hot, warm, cool, and cold. These descriptive terms are applied to the concept of *temperature*, a concept with which most of us have at least a nodding acquaintance. Yet upon careful consideration, it becomes evident that if we are to use

the idea of temperature in our work, we need to have a way of measuring it. For instance, exactly what is meant when we speak of a temperature of 37°C? For that matter, what is meant by a mass of 37 kilograms? Looking back, you will realize that our ability to use the concept of mass depends on the existence of some fixed or reference mass (the international prototype kilogram), and a means of comparing masses by acceleration or momentum experiments (Sec. 5-3). In a similar fashion we deal with temperature by defining fixed temperatures on a scale and adopting a *thermometer* as a means of making comparisons between temperatures. Actually there are two kinds of mass: inertial mass (defined by acceleration experiments) and gravitational mass (defined by weighing experiments). The *operation*, or experimental procedure, really defines what is meant by the concept of mass. Just so, the thermometer really defines the concept of temperature, and we shall see that there are as many different kinds of temperature as there are types of thermometers. Fortunately, most common thermometers yield Celsius temperature scales that are in close agreement with each other, and for many purposes it is unneces-

Joints in rigid pavements allow for thermal expansion and contraction of roads and sidewalks.
This highway buckled when the temperature exceeded the expectations of the designer.

sary to specify the exact type of thermometer used.

When using any type of thermometer, an assumption is made that is so fundamental that for a long time it was not even stated in words. Suppose we place an upright tumbler containing cold milk (system M) in a saucepan containing warm water (system W). Eventually the two systems come to *thermal equilibrium*, which means that there are no further changes in any of the measurable quantities associated with either system (for example, the density). We make the reasonable assertion that these systems have the same temperature if they are in thermal equilibrium. Now we allow a third system—a thermometer of some sort (system T)—to come to thermal equilibrium first with W, then with M. We find that the thermometer gives the same reading in each case. This is an example of the

Zeroth law of thermodynamics:

If a system is in thermal equilibrium with two other systems, then these other systems are in thermal equilibrium with each other.

It is called the zeroth law because it was only after the first and second laws of thermodynamics had been formulated that it was recognized as a separate and independent law of nature. In effect, the zeroth law of thermodynamics supplies the framework by which the concept of temperature can be established; without the zeroth law, the first and second laws would have no meaning.

It it tempting to think of molecular motion as a fundamental property by which we "ought" to define temperature, for most of the phenomena of heat are associated with the random motions of atoms or molecules. For instance, when a car remains in direct sunlight for a while, the iron atoms in the fender gain kinetic energy and move faster than before. The faster-moving atoms, acting against the attractive and repulsive forces of cohesion, vibrate about their equilibrium positions with slightly greater amplitudes. If an observer touches the hot metal, KE is transferred to his finger by a collision process, and the sensation of warmth is produced in his brain. The pressure exerted by the air in a tire is due to the impact of molecules on the walls of the tire; at higher temperatures, the molecules move faster, and the pressure is greater.

However, we do not define temperature in terms of molecular motion. For one thing, the process of radiation would not easily fit into such a scheme: How is the heat from the sun transmitted through 149 million kilometers of vacuum to reach the earth? What is the temperature of empty space? And, as we shall see in our discussion of thermodynamics (Sec. 17-5), mere absence of random molecular motion is no guarantee that a body possesses no internal energy or is at a temperature of absolute zero. There may be other forms of available energy (such as potential energy) remaining in the atoms, even though at ordinary temperatures these other forms of energy are of negligible importance in comparison with the random kinetic energies of the molecules of a solid, liquid, or gas.

Let us emphasize once more our reason for bringing up at this time these disquieting suggestions about the definition of temperature. As a student, you will (understandably) want a nice, packaged definition of temperature that will stand up in later work. We are unable to give such a definition until Chap. 16. For the present, we merely point out that many definitions are *possible*, because many types of thermometers have been invented. For ordinary work, it is our good fortune that commonly used thermometers agree well enough with each other (and with the thermodynamic scale to be discussed later) that we are spared the necessity of choosing any particular thermometer as standard.

13-2 Temperature Scales

The human senses are not precise enough to determine small changes in temperature. An anxious father, feeling his child's brow, can distinguish between high fever and no fever at all, but such measurements are not quantitative. The science of medical diagnosis was tremendously advanced by the invention in about 1593 by Galileo of the "thermoscope," based on the expansion of gases or liquids, and the later development of thermometers that quantitatively indicated temperature.* Before going into constructional details of thermometers, let us first discuss temperature scales and the fixed reference temperatures used to define the various scales.

Of the many temperature scales used in various places and times, only two are still in common use. Both use as fixed points the melting point of ice and the boiling point of water. For scientific work, the *Celsius scale* is almost universally used. On it, ice melts at 0°C and water boils at 100°C. Because of this 100° range, the Celsius scale was formerly called the centigrade scale.† We will use the Celsius scale in this book. However, it is useful to be able to convert temperatures to and from the *Fahrenheit scale*, still in widespread use in English-speaking countries. On the Fahrenheit scale, ice melts at 32°F and water boils at 212°F. Since an interval of 100 Celsius degrees equals an interval of 180 Fahrenheit degrees, we see that a change of one Fahrenheit degree (1 °F) is only $\frac{100}{180}$, or $\frac{5}{9}$, as large as a change of one Celsius degree (1 °C). To visualize the relation between temperatures on the Celsius and Fahrenheit scales, a diagram similar to Fig. 13-1 may prove useful. Formulas can be derived that relate °C to °F, but these are actually unnecessary and a burden on the memory; it is preferable to proceed directly from first principles, as in the following examples.

---

* Early thermometers indicated only *changes* of temperature, and a fever thermometer had to be individually calibrated before each use by making a mark at the point indicated when the bulb was placed in the mouth of a person known to be in good health.

† The choice of the ice point and the steam point as fixed points, with the interval divided into 100 degrees, was due to Anders Celsius (1701–1744), a Swedish astronomer. (Celsius originally called the ice point 100°C and the steam point 0°C!)

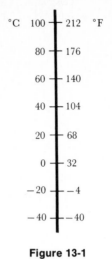

Figure 13-1

Example 13-1

A mountain climber finds that water boils at 80°C. What is the Fahrenheit temperature of this boiling water?

We consider *changes* in temperature from some standard temperature, such as an assumed original temperature of 100°C (which we know to be 212°F). The new temperature is 20°C less than the normal boiling temperature. This is a decrease of $\frac{9}{5}(20)$, or 36, °F. (Remember that 1 °F is smaller than 1 °C, so that it takes more of them to represent a given change.) Hence the new temperature is

$$212° - 36° = \boxed{176°F}$$

Example 13-2

Research has indicated that humans can stand an air temperature of 900°F for over a minute if insulated with clothing about 1 cm thick. Express this air temperature on the Celsius scale.

We can work from a reference temperature of 32°F (which we know to be 0°C). The new air temperature is 868°F above the melting point of ice. This interval is $\frac{5}{9}(868°) = 482.2°C$. Since the melting point of ice is at 0°C, the air temperature is

$$0° + 482.2° = \boxed{482.2°C}$$

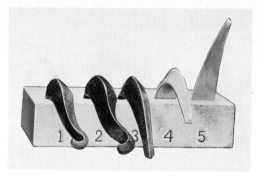

Figure 13-2 A series of pyrometric cones of different ceramic compositions that have all had the same heat treatment. The degree of softening depends primarily on the kiln temperature.

The Celsius scale we have been discussing is the International Practical Scale, which is defined to have 100 degrees between the ice point and the steam point.* The thermodynamic or Kelvin scale, to be discussed in later chapters, differs from the Celsius scale only by a constant: $T_K = T_C + 273.15$. Thus any temperature difference ΔT, or $T_2 - T_1$, is the same on either scale. We will use T for temperature, adding subscripts as in T_C or T_K only occasionally, where needed for clarity.

Any property of a substance that depends uniquely on temperature may be used as a temperature indicator. Such a property is called a *thermometric property* of the substance. For example, a yellow-hot piece of iron is hotter than a red-hot piece; color is therefore a thermometric property, and thermometers can be made that depend on color changes (radiation pyrometers, discussed in Sec. 14-7). The pressure of an enclosed gas also depends on temperature, and gas thermometers can be constructed in which change in temperature is indicated by change

---

in pressure (Sec. 13-5). The electrical resistivity of most metals increases as their temperature goes up (Sec. 20-6), and so we have the resistance thermometer, consisting of a coil of platinum wire whose resistance indicates its temperature. An interesting thermometric property is used to determine (roughly) the temperature in a furnace: A series of ceramic cones of different compositions are placed in the furnace. Each cone softens at a different temperature, and so the furnace temperature is determined (to within a few degrees) by inspection of the row of cones (Fig. 13-2). Enough has been said to indicate that many properties of substances are usable as thermometric properties. However, thermometers based on the thermal expansion of solids and liquids are often used for reasons of convenience, ease of reading, and cost.

13-3 Thermal Expansion of Solids

Almost all solids expand upon heating; no matter which common thermometer is used, the expansion is sufficiently uniform so that for practical purposes we can say that length is a linear function of the temperature. Figure 13-3 represents the change in length of a certain brass rod, whose length increases by a small amount ΔL for each small temperature increase ΔT.

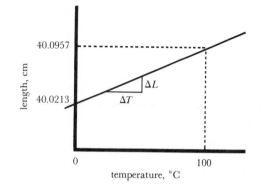

Figure 13-3 Uniform expansion of a certain brass rod.

However, we expect the change in length of a long rod of brass to be more for a given change in temperature than that of a short rod, for the longer rod has more small crystals, each of which expands the same amount. Thus the change in length is proportional not only to the change in temperature, but also to the original length L. We therefore write the basic equation for the expansion of a solid as follows:

$$\Delta L = L\alpha\,\Delta T \qquad (13\text{-}1)$$

where α (the Greek letter alpha) is the *coefficient of linear expansion* for the given material. To find the final length, the amount ΔL is added to or subtracted from the original length, according to whether the temperature increases or decreases. Values of α for various solids are shown in Table 13-1.*

Example 13-3

Compute the coefficient of linear expansion for brass, using the data given in Fig. 13-3.

From the diagram

$$\Delta L = 40.0957 \text{ cm} - 40.0213 \text{ cm} = 0.0744 \text{ cm}$$

Solving Eq. 13-1 for α, we get

$$\alpha = \frac{\Delta L}{L\,\Delta T} = \frac{0.0744 \text{ cm}}{(40.0 \text{ cm})(100°\text{C})}$$

$$= \boxed{18.6 \times 10^{-6}\ (°\text{C})^{-1}}$$

In the denominator we used 40.0 for the length L, consistent with our knowledge of ΔL to only 3-figure precision. The expansion is so small that either the initial or the final length would give the same value of α to 3-figure precision.

The unit for α is simply $1/°\text{C}$ or $(°\text{C})^{-1}$, a fractional change in length per degree. In words, a coefficient of $18.6 \times 10^{-6}\ (°\text{C})^{-1}$ means that any given length of the material changes by 18.6 parts per million of its original length for every Celsius degree change in temperature.

---
* For a few substances, over a limited range of temperature, α is negative; such substances shrink upon heating.

Table 13-1   Coefficients of Expansion

Substance	Coefficient of Linear Expansion α, $(°\text{C})^{-1}$	Coefficient of Volume Expansion β, $(°\text{C})^{-1}$
Solids		
iron or steel	11×10^{-6}*	33×10^{-6}
aluminum	26	77
brass	19	56
ordinary glass	8.5	26
Pyrex glass	3.3	10
fused quartz	0.4	1
platinum	9.0	27
concrete	12	36
lead	29	87
Liquids		
methyl alcohol		1134×10^{-6}
carbon tetrachloride		581
glycerin		485
mercury		182
turpentine		900
gasoline		960
Gases[†]		
air		3670×10^{-6}
carbon dioxide		3740
hydrogen		3660
helium		3665

* See page 197 for a note on powers of 10 in tables.
[†] At constant pressure.

Example 13-4

A piece of steel railroad track is exactly 20 m long on a winter day when the temperature is $-10°\text{C}$. What is its length on a hot summer day when the temperature is $35°\text{C}$?

$$\Delta L = L\alpha\,\Delta T$$
$$= (20 \text{ m})[11 \times 10^{-6}\ (°\text{C})^{-1}](45°\text{C})$$
$$= 0.010 \text{ m}$$

We know that steel expands when it is heated, so we *add* ΔL to the original length. The new length is

$$L' = 20.000 \text{ m} + 0.010 \text{ m} = \boxed{20.010 \text{ m}}$$

In railroad construction, rail sections as long as several kilometers are prevented from buckling by careful anchoring to the ground. Similarly, concrete roads are poured in slabs to allow for expansion (see photograph on page 296). The forces generated by thermal expansion are large, as illustrated by the following example.

Example 13-5

A steel rail exactly 20 m long at $-10°C$ is firmly clamped between two fixed supports exactly 20 m apart. What is the force of compression in the rail if its temperature rises to 35°C? The rail's cross-sectional area is 50 cm^2 ($= 50 \times 10^{-4}$ m^2).

The force is the same as would be required to compress the rail back to its original length. For this we use Young's modulus (Sec. 9-1). The change in length ΔL is 0.010 m, as computed in Example 13-4. According to Table 9-1, the stretch modulus for steel is 20×10^{10} N/m^2 (200 GPa).

$$\frac{\Delta F}{A} = \frac{E \, \Delta L}{L}$$

$$\Delta F = \frac{E A \, \Delta L}{L}$$

$$= \frac{(20 \times 10^{10} \text{ N/m}^2)(50 \times 10^{-4} \text{ m}^2)(0.010 \text{ m})}{20 \text{ m}}$$

$$= \boxed{5.0 \times 10^5 \text{ N}}$$

This is more than 100,000 lb of force.

Considered as a thermometric property, linear expansion of solids is exploited in thermostats and inexpensive thermometers. To gain sensitivity, thin strips of different metals are welded or riveted together along their lengths and clamped at one end. When the temperature changes, the combination must bend in an arc, as shown in the schematic diagram of a thermostat (Fig.

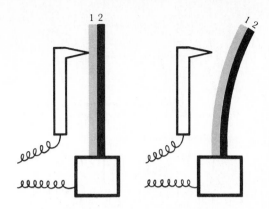

Figure 13-4 Thermostatic relay.

13-4), where metal 1 has a greater coefficient of expansion than metal 2. As the temperature rises, the electric contact is opened, turning off the furnace. In a thermometer of this type, a more sensitive indication is achieved by forming the bimetallic strip into a compact spiral that would be many centimeters long if unrolled.

The advantages of a low coefficient of expansion are illustrated by the use of Pyrex brand glass for ovenware. Glass is a poor conductor of heat, and so when placed in an oven, the inner and outer surfaces of a dish become hot before the glass in the interior has a chance to warm up. Because of thermal expansion, stresses are set up that may shatter the glass. One way to avoid this is to heat the glass dish very slowly, to allow time for equalization of temperature throughout. Even rapid heating is possible if the glass has a low coefficient of expansion, as Pyrex has, for then the stresses will be small (see Table 13-1). Fused quartz has an even lower coefficient of linear expansion; a white-hot piece of quartz at 1500°C may be removed from a flame and plunged into liquid air ($-192°C$) without shattering.

Crystalline substances such as calcite (Iceland spar, $CaCO_3$) or blue vitriol (ordinary copper sulfate, $Cu_2SO_4 \cdot 5H_2O$) are held together by cohesive forces between atoms. Unless the crystal is highly symmetrical, the forces are likely

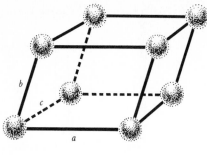

Figure 13-5

to be different along different directions, both because of differing interatomic distances, such as *a*, *b*, and *c* in Fig. 13-5, and because of different strengths of forces between atoms of different kinds. We expect, therefore, that the coefficients of linear expansion will be different along the various directions in a crystal. Such is actually the case. For example, the linear expansion coefficient of a single crystal of bismuth metal varies from 15.4×10^{-6} (°C)$^{-1}$ in a certain direction to 10.8×10^{-6} (°C)$^{-1}$ in another direction at right angles to the first one. Ordinary bismuth metal expands equally in all directions, however, since it is made up of millions of tiny crystals oriented at random.

Both the elastic properties of a substance and its thermal expansion are intimately dependent on the same basic phenomenon—cohesive forces between molecules. This is why substances that are easily compressed also usually have high expansion coefficients and low melting points. All these attributes of a soft metal such as lead are related to the relatively weak cohesive forces between atoms in the crystal.

13-4 *Thermal Expansion of Liquids*

In the case of liquids, the cohesive forces are so weak that a rigid crystal is not possible. The KE of molecular motion is large enough to prevent "locking in" of the molecules. According to these ideas, we can understand why liquids generally

are more easily compressible than solids (Table 9-1) and why their thermal expansion is greater (Table 13-1). The *coefficient of volume expansion* is defined as the fractional change in volume per degree change in temperature. Thus an equation similar to Eq. 13-1 can be used:

$$\Delta V = V\beta \Delta T \qquad (13\text{-}2)$$

where ΔV is the change in volume, V the original volume, ΔT the change in temperature, and β (Greek letter beta) the coefficient of volume expansion. To find the final volume, the amount ΔV is added or subtracted, according to whether the temperature has increased or decreased.

Example 13-6

A 2000-cm^3 quartz beaker is brimful of turpentine at a room temperature of 20°C. If the beaker and contents are put in a refrigerator so that the temperature becomes 5°C, what is the volume of the turpentine?

We can ignore the expansion of the quartz beaker, since its coefficient is so small. In this problem, ΔT is 15°C. From Table 13-1, β is 900×10^{-6} (°C)$^{-1}$. Hence,

$$\begin{aligned} \Delta V &= V\beta \, \Delta T \\ &= (2000 \text{ cm}^3)[900 \times 10^{-6} \text{ (°C)}^{-1}](15\text{°C}) \\ &= 27 \text{ cm}^3 \end{aligned}$$

We subtract this amount from the original volume, since the temperature has decreased. Hence the final volume of the turpentine is

$$2000 \text{ cm}^3 - 27 \text{ cm}^3 = \boxed{1973 \text{ cm}^3}$$

The liquid-in-glass thermometer consists of a bulb and an indicating tube of small inside diameter, or *bore*. Mercury is commonly used as a thermometric substance down to −40°C, its freezing point; alcohol or propane can be used down to temperatures below −100°C. If the bulb is of large volume and the indicating tube is of very small bore, such a thermometer can be made sensitive enough to indicate changes of 0.001° or less.

Figure 13-7 Mercury-in-glass thermometers.

Example 13-7

A mercury thermometer (Fig. 13-6) has a quartz bulb of volume 0.300 cm³ and a stem of bore 0.0100 cm. How far does the indicating thread of mercury move when the temperature changes from 30.0°C to 40.0°C? Ignore the expansion of the quartz bulb.

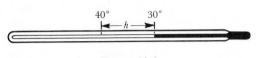

Figure 13-6

From Table 13-1, the coefficient of volume expansion for mercury is 182×10^{-6} (°C)$^{-1}$. Hence,

$$\Delta V = V\beta\Delta T$$
$$= (0.300 \text{ cm}^3)[182 \times 10^{-6}(°C)^{-1}](10.0°C)$$
$$= 5.5 \times 10^{-4} \text{ cm}^3$$

Since the additional volume of the cylinder of mercury in the indicating tube is equal to the change in volume of the mercury, and since volume = (cross-sectional area)(height), we find the height of rise to be

$$h = \frac{\text{volume}}{\text{area}} = \frac{5.5 \times 10^{-4} \text{ cm}^3}{\pi(0.005 \text{ cm})^2} = \boxed{7.0 \text{ cm}}$$

In practice, the bulb is made of ordinary glass, and its expansion would be appreciable. The bulb is a solid; since a solid expands in each of three independent directions, its volume coefficient is approximately three times its linear coefficient. (Note the relation between α and β for the solids in Table 13-1.) The volume coefficient of Pyrex is

$$3 \times 3.3 \times 10^{-6} (°C)^{-1} = 10 \times 10^{-6} (°C)^{-1}$$

The *relative* or *apparent* volume coefficient of expansion for mercury in Pyrex is then

$$(182 - 10) \times 10^{-6} (°C)^{-1} = 172 \times 10^{-6} (°C)^{-1}$$

During its construction, the liquid-in-glass thermometer is sealed off at the end of the capillary tube to prevent loss of liquid by evaporation. The upper part is evacuated (Fig. 13-7a) in order to prevent building up a pressure as the mercury expands. Sometimes the space above the liquid is filled with an inert gas, and a little cavity at the very top prevents excessive buildup of pressure (Fig. 13-7b). The clinical thermometer (Fig. 13-7c) has a constriction just above the bulb so that the mercury thread is broken as the temperature falls. Surface tension keeps the mercury in the stem from flowing back through the narrow opening. In this way the patient's temperature can be read after the thermometer has been removed.

Most liquids expand more or less uniformly, as does mercury (Fig. 13-8). However, the expansion of water shows a curious behavior; the

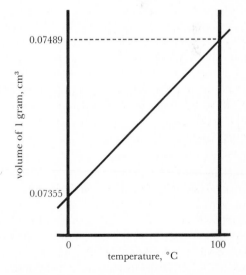

Figure 13-8 Uniform expansion of mercury.

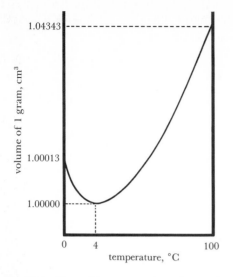

Figure 13-9 Nonuniform expansion of water (scales exaggerated for clearness).

volume of 1 g of pure water varies with temperature, as shown in Fig. 13-9. In the region 0 to 4.0°C, water *contracts* as the temperature increases, and the volume coefficient is negative. At 4.0°C, water has the least volume and maximum density, and the coefficient of expansion is zero. When water is used as a reference material in determining the specific gravity of other substances, the temperature of 4.0°C is often used, since variations in the temperature at which measurements are made will then have only a slight effect on the density of water near this point.

Biological life cycles are profoundly affected by the expansion characteristics of water. In a fresh-water lake, as the surface water becomes chilled to 4°C, it becomes denser and sinks, thereby pushing up warmer water from below, which in turn is chilled at the surface. A continual mixing takes place, and the water deep in the lake remains at about 4°C until the colder (and less dense) water at the surface freezes. This is why lakes freeze from the surface down—a fortunate fact for the survival of living organisms in fresh water. The situation is complicated by the fact that the temperature of maximum densi-

ty decreases with increasing pressure. This leads to incipient instability and allows wind-induced mixing to become effective in cooling deep water below 4°C.

13-5 Thermal Expansion of Gases

In contrast to those in solids and liquids, the cohesive forces between the widely separated molecules of a gas are so small as to be almost negligible. The thermal expansion of a confined gas is intimately related to its pressure, as we shall see in Chap. 15. For the present, let us imagine the pressure on a confined mass of gas to be held constant. Common experience tells us that as the temperature increases, we must not allow the pressure to build up; otherwise the volume change will be meaningless, or even zero. Therefore we specify that the coefficient of volume expansion for a gas is measured at *constant pressure*.

A glance at Table 13-1 shows that the coefficients for gases are much larger than those for liquids or solids. It is also evident that the coefficients for all gases are practically the same. This is to be expected, since the only differences between the various gases would be in their cohesive forces, which are so small as to have only a minor effect, or in the variable sizes of the molecules, which in a gas are tiny compared with the distances between molecules. We shall consider the gas laws in detail in Chaps. 15 and 16.

In applying the standard expansion equation, $\Delta V = V\beta\Delta T$, to gases, we must be cautious in our choice of initial volume V. Since, in contrast to solids or liquids, the change in volume ΔV may be a large fraction of the volume, it is necessary to make a definite statement about the measurement of V itself. The values of β listed in Table 13-1 are *based on the original volume at 0°C*. Thus we may say that hydrogen expands by 0.00366 of its 0° volume for every degree rise in temperature, if the pressure remains constant.

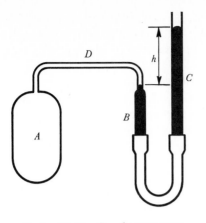

Figure 13-10 Gas thermometer.

The behavior of hydrogen and helium, for instance, can be allowed for, since the necessary corrections are small and well known. The hydrogen gas scale, so corrected, is called the ideal gas scale; it has considerable theoretical importance, being identical to the Kelvin scale of absolute temperatures. In actual use, the gas thermometer is complicated and bulky and does not respond quickly to changes of temperature.

Example 13-8

In a constant-pressure gas thermometer the volume of the gas is 546 cm³ at 0°C and 746 cm³ at 100°C. What is the temperature when the volume is 628 cm³?

The volume increased by 200 cm³ for a 100°C change in temperature; each increase of 1 cm³ corresponds to a temperature increase of 0.50°C. The observed volume is 82 cm³ greater than the volume at 0°C. Hence the temperature is

$$T = (82 \text{ cm}^3)\left(\frac{0.50°C}{1 \text{ cm}^3}\right) = \boxed{41°C}$$

An alternative solution uses a proportion:

$$\frac{100°C - 0°C}{746 \text{ cm}^3 - 546 \text{ cm}^3} = \frac{T - 0°C}{628 \text{ cm}^3 - 546 \text{ cm}^3}$$

$$T = \boxed{41°C}$$

The gas thermometer (Fig. 13-10) is capable of high precision, and it is a standard with which other types of thermometers may be compared and calibrated. If the gas container A is heated, the gas expands and pushes the mercury in arm B down and that in arm C upward. Thus the height h changes, and so there is a pressure change as well as a volume change. The flexible connecting tubing allows arm C to be lowered, to bring the pressure back to the original value. When so used, the thermometric property is the *volume* of the gas, and the device is a *constant-pressure gas thermometer*. The connecting neck D is of small volume compared with A, but even so, correction must be made because this portion of the gas is not at the same temperature as that in A. The gas thermometer cannot, of course, be used below the temperature at which the gas liquefies.

The chief advantage of a gas as the thermometric substance is that the readings are relatively independent of the exact composition of the gas. What slight differences remain between

It is also possible and more convenient to use the gas thermometer in such a way that the *volume* remains constant; the thermometric property is then the *pressure* exerted by the gas. If the necessary (small) corrections are made, the calculated temperatures agree with each other for the constant-pressure gas thermometer and the constant-volume gas thermometer.

Summary Temperature is a measure of how hot or cold a body is, as determined by some type of thermometer. The fixed points on which the Celsius scale is based are the melting point of ice, 0°C, and the boiling point of water, 100°C. On the Fahrenheit scale these points are at 32°F and 212°F. A change of 1 Celsius degree is the same as a change of $\frac{9}{5}$ Fahrenheit degrees. Various thermometric properties of matter may be used to define temperature scales. Commonly used thermometers make use of the expansion of gases and liquids and also of changes in electrical properties; at ordinary temperatures, such thermometers are in close, though not exact, agreement with each other.

The change in length of a solid is proportional to the original length and to the change in temperature, according to the equation $\Delta L = L\alpha\,\Delta T$. The factor α is the coefficient of linear expansion. Similarly, the change in volume of a solid, liquid, or gas is given by $\Delta V = V\beta\,\Delta T$, where β is the coefficient of volume expansion. For a solid, β is approximately 3α.

Check List

thermal equilibrium
zeroth law of thermodynamics
thermometric property
Celsius temperature scale
Fahrenheit temperature scale

coefficient of linear expansion
coefficient of volume expansion
constant-pressure gas thermometer
constant-volume gas thermometer

Questions

13-1 Name several thermometric properties of matter.

13-2 In the manufacture of the first light bulbs, the electrical leads were made of platinum, which has a coefficient of expansion nearly equal to that of ordinary glass. Explain how this helps make a firm glass-metal joint.

13-3 If the stem of the thermometer described in Example 13-7 were engraved with markings every 0.2°C, how far apart would the markings be?

13-4 Analyze the act of shaking down a fever thermometer, in terms of mass, forces, and acceleration.

13-5 Describe qualitatively the construction of a mercury-in-glass thermometer that indicates temperatures between 42°C and 46°C on a scale 20 cm long.

13-6 Why are the concrete slabs on a highway separated from each other by narrow cracks filled with pitch?

13-7 In using the equation $\Delta L = L\alpha\,\Delta T$, which length L should be used—the initial length or the final length? Does it really matter which length is used?

13-8 Why is it incorrect to say that hydrogen expands by 0.00366 of its original volume for every Celsius degree rise in temperature?

13-9 If the room temperature goes up, barometric pressure remaining the same, how will the reading of a mercury barometer be affected? Will it seem to indicate a pressure that is too high or too low? (*Hint:* The density of the mercury changes. See Sec. 12-3.)

13-10 Why doesn't an aluminum cup shatter when it is suddenly filled with boiling water?

13-11 A hole is cut in a flat metal plate. When the plate is heated, will the hole become larger or smaller?

13-12 Explain how the metal cap on a bottle of catsup can be loosened by holding it briefly in boiling water.

MULTIPLE CHOICE

13-13 A temperature of 5°C is equal to   (*a*) 9°F;   (*b*) 37°F;   (*c*) 41°F.

13-14 A temperature of 68°F is equal to   (*a*) 20°C;   (*b*) 34°C;   (*c*) 36°C.

13-15 A circular metal hoop is sawed through at one point, leaving a gap of 1.4 mm. If the hoop is then put into an oven and heated, the width of the gap will   (a) increase;   (b) decrease;   (c) remain the same.

13-16 A bar made of iron for which $\alpha = 11 \times 10^{-6} \ (°C)^{-1}$ is 10.000 cm long at 20°C. At 19°C the length is   (a) 11×10^{-6} cm greater;   (b) 11×10^{-6} cm less; (c) 11×10^{-5} cm less.

13-17 A unit for volume coefficient of expansion could be   (a) $m^3 \cdot (°C)^{-1}$;   (b) $(°C)^{-1}/m^3$;   (c) $(°C)^{-1}$.

13-18 At about 4°C, a gram of water has maximum   (a) density;   (b) energy;   (c) volume.

Problems   **13-A1**   A typhoid-fever patient's temperature ranged from 96.8°F to 104.0°F during the course of the illness. Express these temperatures in °C.

13-A2   What is the Fahrenheit temperature at which pure water has its maximum density?

13-A3   Express in °C: 68°F; 50°F; $-328°F$; 98.6°F.

13-A4   Express in °C: 0°F; $-40°F$; 752°F; 77°F.

13-A5   Express in °F: 20°C; 500°C; $-40°C$; $-273°C$.

13-A6   Express in °F: 30°C; $-10°C$; 5000°C; $-78°C$.

13-A7   A TV weather announcer said that the current temperature was "33 degrees Fahrenheit—that's 0 degrees Celsius." For what small range of temperatures could this be a correct statement, assuming that the thermometers were correct and that the readings were correctly rounded off?

13-A8   A patient's temperature decreased by 2.7 °F overnight. Express this change in °C.

13-A9   Steel rails 15 m long are laid on a winter day when the temperature is $-10°C$. How much space must be left between the rails to allow for expansion at a summer temperature of 40°C?

13-A10   A bridge expands 3 mm for every Fahrenheit-degree temperature change. How much does it expand for a change of 1°C?

13-A11   A radiator in the attic of a house is connected to the furnace by a vertical iron pipe 12.0 m long. If the radiator rests snugly on the floor when the pipe is at 5°C, how high off the floor will it be lifted when the pipe contains steam at 100°C?

13-A12   The steel main span of the Golden Gate Bridge is 1280 m long. What allowance must be made for expansion if the temperature varies between $-10°C$ and 55°C?

13-A13   A metal rod, observed to be 102.503 cm long at 20°C, is 102.639 cm long at 90°C. Out of which of the metals listed in Table 13-1 might the rod be made?

13-A14   What is the volume coefficient of expansion of a metal whose linear coefficient is $12 \times 10^{-6} \ (°C)^{-1}$?

13-A15   A motorist purchases 30 liters of gasoline on a cold day when the temperature is $-20°C$. What is the volume of this gasoline after it has warmed up to $+20°C$?

13-A16   A block of metal is 5.000 cm × 8.000 cm × 10.000 cm at 25°C. The linear coefficient of expansion is $25 \times 10^{-6} \ (°C)^{-1}$. What is the volume of the block if it is immersed in an ice bath?

13-B1   At what temperature is the reading of a Fahrenheit thermometer twice that of a Celsius thermometer?

13-B2   Newton used a thermometer filled with linseed oil. The ice point was 0°N, and body temperature was 12°N. What is the temperature on this scale of an ice-salt mixture that is at $-18°C$? Suggest one or more disadvantages of this scale.

13-B3 In assembling a steel bridge, a steel rivet 3.803 cm in diameter is to be placed in a hole 3.800 cm in diameter. Would it be possible to do the job by first cooling the rivet in Dry Ice ($-78°C$)? Air temperature is $22°C$.

13-B4 In precise surveying, a steel tape is used that is calibrated at $20°C$ and is at a specified tension. (a) If the tension is maintained constant, but the temperature increases to $29°C$, what is the percent error in the reading of the tape? (b) Does the tape read too high or too low? (c) What would be the error, in cm, when measuring a lot 15.00 m wide?

13-B5 A brass cannon has a bore 9.000 cm in diameter at $20°C$. A steel liner, in the form of a thin cylinder of outside diameter 9.025 cm at $20°C$, is to be inserted. To what temperature should the cannon be heated so that the liner (still at $20°C$) can be inserted?

13-B6 To what temperature should both the cannon and liner of Prob. 13-B5 be heated so that the liner becomes loose and can be extracted?

13-B7 An iron ball 3.000 cm in diameter is 0.001 cm too large to go through a hole in a brass plate when the ball and plate are at $20°C$. At what temperature, the same for both ball and plate, will there be a snug fit?

13-B8 An iron rim is to be placed around a wagon wheel whose circumference at $15°C$ is 270.662 cm. The rim is in the form of a band having cross section 2.50 cm^2 and inside circumference at $15°C$ of 270.020 cm. (a) To what temperature must the band be heated to fit snugly on the wheel? (b) What tension will exist in the band after it cools to $15°C$?

13-B9 A brass rod of length L and cross section 3 cm^2 is cooled from $100°C$ to $20°C$. What force is needed to stretch the rod back to its original length?

13-B10 During a Young's modulus experiment, weights were stretching an iron wire of length L and cross section $2.00 \times 10^{-7} \text{ m}^2$. Room temperature dropped by $10°C$; what added force was needed to restore the wire to its original length?

13-B11 A lead rod and an aluminum rod, each 40.00 cm long at $0°C$, are clamped together at one end with their free ends coinciding. Compute the separation of the free ends of the rods when the system is placed in a steam bath.

13-B12 An aluminum cup is brimful, containing 250 cm^3 of glycerin at $20°C$. What volume of glycerin overflows when the cup is placed in an oven that is at $150°C$?

13-B13 An open steel drum of volume 200.0 liters (50 gallons) is just filled with turpentine when the temperature is $15°C$. The drum and contents are heated to $40°C$ and then allowed to cool to $15°C$ again. How much turpentine remains in the drum?

13-B14 An aluminum block of volume V is heated from $10°C$ to $25°C$. What pressure must be applied to keep the block from expanding?

13-B15 What is the apparent coefficient of volume expansion of glycerin in an aluminum container?

13-B16 A fused quartz container is completely filled with mercury and sealed off at $20.00°C$. What is the pressure, in kPa and in atmospheres, inside the container if the temperature becomes $20.40°C$?

13-B17 Starting at $0°C$, through what temperature change would hydrogen have to be heated in order to cause a 7.32% change in volume?

13-B18 In a constant-pressure gas thermometer, the volume of the gas is 400 cm^3 at $20°C$ and 450 cm^3 at $100°C$. What temperature is indicated when the volume is 540 cm^3?

13-B19 A rod has original length L_0. When the temperature changes by ΔT, its final length is L_T. The coefficient of linear expansion is α. Prove that $L_T = L_0(1 + \alpha \Delta T)$.

13-B20 A body has original volume V_0. When the temperature changes by ΔT, its new volume is V_T. The coefficient of volume expansion is β. Prove that $V_T = V_0(1 + \beta \Delta T)$.

13-C1 A grandfather clock has a period of exactly 2 s at 15°C. The pendulum consists of a small, heavy bob on a thin, brass rod.   (a) What will the period be if the temperature rises to 32°C?   (b) At the end of a day, how many seconds will the clock have gained or lost?

13-C2 At 20°C, a steel rod of length 40.000 cm and a brass rod of length 30.000 cm, both of the same diameter, are placed end to end between two rigid supports 70.000 cm apart, with no initial stress in the rods. The temperature of the rods is now raised to 60°C.   (a) What is the stress (the same in both rods) at 60°C?   (b) How far from its old position is the new junction between the rods? [*Hint:* First find the total expansion if the rods were free; then find the stress, from the fact that the sum of the two compressions (due to stress) equals the sum of the two expansions (due to temperature). Finally, find the contraction of each temperature-expanded rod due to the stress.]

13-C3 At 20°C, the specific gravity of aluminum is 2.699 and that of glycerin is 1.248. A block of aluminum whose volume is 100.0 cm^3 at 20°C is immersed in glycerin at 20°C.   (a) What is the apparent weight of the aluminum block?   (b) What is the apparent weight of the block if the system is heated to 70°C?

References   1. E. E. Jones, "Fahrenheit and Celsius, a history," *Phys. Teach.* **18**, 594 (1980).
2. C. B. Boyer, "Early Principles in the Calibration of Thermometers," *Am. J. Phys.* **10**, 176 (1942). Two-point calibration; one-point calibration; Fahrenheit, Celsius, Réaumur, and Kelvin scales.
3. R. E. Wilson, "Standards of Temperature," *Physics Today* **6**(1), 10 (Jan. 1953).
4. C. H. Anderson and W. H. Morewood, "Long rails," *Phys. Teach.* **17**, 601 (1979).

14

Heat
and Heat Transfer

Now that we have studied the measurement of temperature, we are able to discuss the quantitative phenomena of heat that involve the interactions of material bodies.

14-1 The Caloric Theory and Its Overthrow

If a kilogram of lead and a kilogram of iron are placed side by side in front of a brisk fire burning in a fireplace, and if the blocks of metal are allowed to stand for, say, 10 minutes, it is observed that the lead becomes warmer than the equal mass of iron. Experiments of this sort were made in the late 1700s by Black, Rumford, and others in an effort to test the theory that heat is a fluidlike substance, called "caloric," that flows from a hot body into a cold one. The experimental facts (which are as true today as 200 years ago) are as follows: Every material object, such as a lead ball or an aluminum saucepan, has a "thermal capacity," which is a measure of the quantity of heat required to make it one degree warmer. Some early measurements were indeed made with fires in fireplaces serving as the source

of heat, and it was assumed that a fire, burning steadily, produced in 40 minutes twice as much "quantity of heat" as it did in 20 minutes. It was soon found that different iron objects had different thermal capacities, in proportion to their masses, and so the concept of "specific heat" was introduced to represent the thermal capacity per unit mass. We shall define these quantities in Sec. 14-3 and give units for them.

Nothing in the above paragraph would contradict the idea that heat is a material fluid. As a matter of fact, scientists still speak of the "flow" of heat, but no longer is it supposed that an actual physical substance (the caloric of the old theory) flows. The existence of caloric would mean that heat should have weight and mass, and that a body should gain weight and mass when it is heated. That this does not happen* was demonstrated about 1760 by Joseph Black

---

* Since the molecules of a body are, on the average, moving faster when the body is hot than when it is cold, they have slightly greater mass, according to the Einstein equation $\Delta E = (\Delta m)c^2$ (Sec. 28-6). To this extent, heat does have mass. But this increase in mass is so slight as to be, so far, entirely beyond detection.

A blackbody is a good emitter and absorber of radiation. The word "BLACK" was painted on a porcelain dish and a flame directed at the inside surface, behind the letter A. The heated black surface emits more radiation than the surrounding porcelain, which is at the same temperature.

(1728–1799) and is evidence against the existence of caloric.

The investigators could still "save the phenomenon" by imagining caloric to be a *weightless* fluidlike substance, intimately associated with the material particles of a substance. In 1797 Count Rumford (1753–1814) designed a simple but decisive experiment to test whether a fluid (weightless or not) resides in a heated body. One of the most colorful natural philosophers of his time, Rumford was born Benjamin Thompson in North Woburn, Massachusetts. After a stay in England, he became minister of war in Bavaria, where he received his title. While supervising the boring of cannon at the military arsenal at Munich, he noticed the great rise in temperature of the shavings left in the cannon by the boring tool. Unlike the dozens or hundreds of his predecessors in many countries who must have observed this heat, Rumford had some scientific training and an inquiring mind that was interested in discovering basic scientific facts even if they were not immediately useful.

Rumford was finally led to press a *blunt* boring tool against the inside of a cannon that was being turned by horses. The stationary tool exerted a considerable force—about 45,000 N (10,000 lb)—on the 51-kg (113-lb) cannon, and after half an hour of turning, the cannon's temperature had risen to 54°C (130°F)—strong evidence of the production of a large quantity of heat by friction. Only about 50 g (2 oz) of gun metal had been scraped loose by the action of the borer, and Rumford reasoned that it was quite unlikely that so much caloric could come from this minute amount of metallic dust. (He had previously determined that metal powder or scrapings had the same thermal behavior as did bulk metal.) In another experiment, Rumford caused water to boil by the heat generated by friction: "It would be difficult to describe the surprise and astonishment expressed in the countenances of the bystanders, on seeing so large a quantity of cold water heated, and actually made to boil without any fire." And Rumford confessed to "a degree of childish pleasure" at the result of the experiment.

If the cannon were turned for a long enough time, an amount of caloric approaching infinity could be obtained from a finite amount of gun metal, which itself suffered no apparent change. Rumford rejected the idea of caloric for this reason and concluded that heat is a form of motion.

Sir Humphry Davy (1778–1829) also did some crucial experiments at about the same time. In one carefully designed experiment, Davy managed to melt some blocks of ice by rubbing them against each other on a day when the temperature was below 0°C. In another experiment, performed by a clockwork mechanism that rested on a cake of ice in an evacuated chamber, some wax was melted by friction.

The caloric theory was ingenious and, in its day, explained all the then known phenomena; it was "true" in the same sense that any physical theory is "true" (see Sec. 8-4). The theory was overthrown when well-thought-out experiments gave new results not explainable by the existence of caloric. This is a fascinating chapter in the history of science. Rumford's own description of his motives, his experiments, and his conclusions should be read by everyone interested in the development of scientific methods. (See references at the end of this chapter.)

14-2 Internal Energy, Heat, and Work— The First Law of Thermodynamics

In our earlier study of mechanics, we saw that the conservation of energy is a powerful concept that successfully describes changes in gravitational potential energy, elastic potential energy, and kinetic energy. Yet there are situations where the concept seems to fail. For example, suppose we measure the velocities of two hockey pucks before and after a collision. Although we would find that momentum is conserved, we would probably find that kinetic energy is not conserved. However, if we could precisely measure the temperature of each puck before and after the collision, we would find that their temperatures are higher immediately after the collision. Could this change in temperature reflect a change in some new form of energy, thus saving the concept of energy conservation? The answer is yes. One of the triumphs of 19th-century theoretical physics was the proposal that an object possesses *internal energy*, and that the object's temperature reflects the amount of its internal energy. The symbol U is used for internal energy.

One form of internal energy is the *random kinetic energy* of the molecules of a system. The molecules of a hockey puck have random KE as they vibrate around their average positions in the puck. This random KE is internal energy, and it is greater at higher temperatures. Thus we can say that an increase in temperature implies an increase ΔU of internal energy. The reverse is not true, however: there can be an increase ΔU of internal energy without a change of temperature—if the molecules of a body gain random *potential* energy.

Returning to our gliding hockey puck, in addition to internal energy U it also has a macroscopic, large-scale KE, which can be calculated from the external forces that set it into motion. The molecules are moving in parallel paths (except for their random vibrations) with some amount of *ordered* KE, given by $\frac{1}{2}Mv^2$, where M is the total mass of the puck and v is the common translational velocity of all the molecules. This ordered KE is not part of the puck's internal energy.

The theoretical proposal for internal energy describes two ways in which the internal energy of a system can change.* First, mechanical work may be performed *by* the system, and this would tend to *decrease* its internal energy, or work may be performed *on* the system, which would tend to *increase* its internal energy. Second, the internal energy of a system may increase or decrease

---

* Here, and elsewhere, by *system* we mean a definite amount of matter not necessarily all in one chunk—a single hockey puck, a pair of pucks, a glass of water in which an ice cube is floating.

due to interaction with another system at a different temperature. This transfer of energy due solely to a temperature difference is called *heat*. Thus the law of conservation of energy can be restated as the

First law of thermodynamics:

All the heat added to a system can be accounted for as mechanical work, an increase in internal energy, or both.

We can express the first law of thermodynamics in the form of an equation:

$$Q = W + \Delta U \qquad (14\text{-}1)$$

where Q = heat added to a system from the surroundings

W = net mechanical work performed by the system on the surroundings

ΔU = increase of internal energy of the system

Rearranging Eq. 14-1 gives

$$\Delta U = Q - W$$

This equation can be considered the *definition* of internal energy (or, strictly, the definition of *change* in internal energy, since internal energy, like potential energy, can be measured from any arbitrary reference condition).

Example 14-1

A vertical rod supports a load of weight 200 N. The rod and load are in an oven, which can be heated electrically (Fig. 14-1). When 400.0 J of heat is absorbed, the rod expands to become 1 mm longer. What is the change in internal energy of the rod?

The work done by the rod is the increase of PE of the load, given by $\Delta PE = mgh = (200 \text{ N})(10^{-3} \text{ m}) = 0.2 \text{ J}$.

$$\Delta U = Q - W$$

$$= 400.0 \text{ J} - 0.2 \text{ J} = \boxed{399.8 \text{ J}}$$

In this case, the external work performed by the rod is only a small fraction of the heat input.

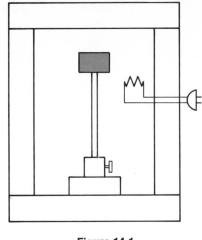

Figure 14-1

An *isolated system* is defined as one that has no interaction with its surroundings. No work is done by or on such a system ($W = 0$). Also, the system is thermally insulated, so there is no heat flow to or from the surroundings ($Q = 0$). In such a system then, $\Delta U = Q - W = 0 - 0 = 0$. This leads to another statement of the first law of thermodynamics:

The internal energy of an isolated system remains constant.

The law of conservation of energy, as developed and used in our earlier work, is seen to be a special case of the first law of thermodynamics, in which only mechanical energy and work were considered.

Note carefully that one always speaks of heat in a context of *change* or *flow* of energy. The "total heat energy" or the "heat content" of a body or a system is a meaningless concept and cannot be used. However, "heat gained" and "heat lost" are perfectly acceptable and useful ideas. An example will illustrate this point. Suppose a book is moved steadily 0.5 m horizontally along a rough table by an applied force of 10 N that is just equal to the force of kinetic friction. The work done on the book is (10 N)(0.5 m) =

5 J; the value of W is negative, since work done *by* the book is defined to be positive. The heat flow is essentially zero, because the cloth cover of the book is a good insulator. Now the first law gives $\Delta U = Q - W = 0 - (-5 \text{ J}) = +5 \text{ J}$. The internal energy increases, and the book's temperature increases slightly. We could also have reached the same final condition or state by simply placing the book on a hotplate for a short time. Then Q would be positive and W would be 0, and now $\Delta U = Q + 0$. If the temperature of the hotplate and the time of contact are properly chosen, ΔU by this process can be made to be the same as when the book is pushed along the table. If we can reach the same final internal energy with or without heat flow, then there is no unique value possible for the "heat content" of the book. It would be equally impossible to give meaning to the "work content" of the book. Heat and work are *processes* by which energy can be transferred, and only the resulting change of internal energy has meaning.

Quantity of heat refers to the amount of energy transferred to or from a body because of a temperature difference between the body and its surroundings. Thus quantity of heat Q can be measured in usual energy units, such as the joule. However, before this was fully understood, special units for quantity of heat were used, and they have persisted to this day. A *calorie* was originally defined in thermal terms as the heat required to raise the temperature of one gram of water one Celsius degree.* Likewise, the *Bristish thermal unit* (Btu), used by engineers and environmentalists, is the heat required to raise the temperature of one pound of water one Fahrenheit degree. It is useful to know that 1 Btu is about equal to 1054 J or 252 cal.

Much effort and ingenious experimentation lie behind the statement that exactly the same temperature rise of a system is produced by the

---

* The *Calorie*, with a capital C, used by dieticians, is the kilocalorie (kcal). It is the heat required to raise the temperature of one kilogram of water one Celsius degree. This Calorie is 1000 calories.

absorption of a given amount of heat (measured in calories) as by the performance of an equivalent amount of work (measured in joules). This ratio, 4.186 J/cal, is called the *mechanical equivalent of heat.*

The first law of thermodynamics did not spring into being fully formed, but rather it evolved from the speculations and experiments of many workers. As long ago as 1797, Rumford had an inkling of the first law when he reflected on the production of heat in his cannon-boring experiment. He wrote, "In a case of necessity, the Heat thus produced might be used in cooking victuals. But no circumstances can be imagined, in which this method of procuring Heat would not be disadvantageous; for, more Heat might be obtained by using the fodder necessary for the support of a horse, as fuel." Here he anticipated the energy relationships of metabolism, an aspect of biology that he was not equipped to explore. Julius Mayer (1814–1878), a German physician, published in 1842 a paper in which he stated that "the warming of a given weight of water from 0°C to 1°C corresponds to the fall of an equal weight from the height of about 365 meters." We see from this that Mayer had a clear idea of the transformation of energy between mechanical and thermal forms.

James Joule (1818–1889) published a series of papers from 1843 to 1850 in which he measured the mechanical equivalent of heat in many ways: He churned water by a paddle system; he stirred mercury with iron paddles; he heated water by friction, passing it through a thick piston perforated with many small holes; he rubbed iron rings together under mercury; he measured the heat produced electrically in a wire through which electrons flowed. Using all these methods, Joule obtained results agreeing within 5%, truly a remarkable achievement. It is because of the variety of his methods that Joule's work is of such great significance. The German physiologist and physicist Hermann von Helmholtz* (1821–1894) presented in 1847 a great paper that con-

solidated the work of others and gave the first systematic publication of the law of conservation of energy with convincing mathematical arguments.

We see that the decade 1840–1850 was remarkably fruitful, and it is perhaps pointless to try to assign "credit" to Mayer, Joule, or Helmholtz for the first law of thermodynamics. Our brief excursion into the history of science shows how a major discovery often develops gradually, with a breakthrough coming as a result of independent but simultaneous efforts on the part of a number of workers. Present-day physics is in this tradition.

14-3 Specific Heat Capacity

As we noted in Sec. 14-1, substances differ markedly in the temperature rise caused by the absorption of a given amount of heat. The *heat capacity* of a body (measured in J/kg) is the heat absorbed to raise its temperature one degree. To describe a property of matter, we define the *specific heat capacity* of a substance as the heat capacity per unit mass. According to this definition, the specific heat capacity c (measured in $J/kg \cdot °C$) is the heat absorbed per unit mass per degree rise in temperature.

At room temperature, the specific heat capacity of water is found experimentally to be close to 4186 J/kg·°C, or 4.186 J/g·°C. As seen from Table 14-1, this is by far the largest specific heat capacity of any common substance. In view of the definition of the calorie, the measured specific heat capacity of water is very close to 1.000 cal/g·°C. This non-SI unit is convenient for problems involving water.[†]

---

* A direct descendant, on his mother's side, of William Penn.

[†] The term "specific heat" is used in several ways. Some authors use it as equivalent to specific heat capacity. Others use "specific heat" as the dimensionless ratio of the specific heat capacity of a substance relative to that of water; this is similar to the use of "specific gravity" as the density of a substance relative to that of water (Sec. 1-6). In this usage, specific heat is numerically equal to specific heat capacity, if the latter is measured in $cal/g \cdot °C$.

Table 14-1 Specific Heat Capacities of Some Solids, Liquids, and Gases

Substance	Temperature or Temperature Range, °C	Specific Heat Capacity, J/kg·°C	cal/g·°C
Solids			
iron or steel	20	460	0.11
copper	15–100	390	0.093
aluminum	15–100	920	0.22
glass	20–100	840	0.20
lead	20	128	0.0306
sodium chloride	20	880	0.21
ice	−3	2100	0.50
wood	—	1700	0.4
asbestos	20–98	820	0.195
silver	20	240	0.056
diamond	20	500	0.12
Liquids			
water	15	4186	1.000
mercury	20	138	0.033
glycerin	50	2500	0.60
methyl alcohol	20	2500	0.60
benzene	20	1720	0.41
Gases*			
steam	110	2010	0.48
air	50	1050	0.25
helium	20	5190	1.24
neon	20	1050	0.25

* At constant pressure.

How much heat is required to raise the temperature of 400 g of aluminum from 40°C to 45°C? First we note that if we had 400 g of *water*, the heat required would be 400 cal for every degree, or 400(5) = 2000 cal in all. But aluminum requires only 0.22 as much heat as water requires, since the specific heat capacity of aluminum is 0.22 cal/g·°C (Table 14-1). Therefore the required heat is 2000 cal × 0.22 = 440 cal. If 400 g of aluminum were to cool from 45°C to 40°C, 440 cal of heat would "flow out" of the aluminum.

In general, if the temperature of a body changes by a small amount ΔT, the heat ΔQ that flows in or out of it is given by

$$\Delta Q = mc\,\Delta T$$

where m is its mass and c is its specific heat capacity. If the temperature changes by a finite amount, from T_1 to T_2, the sum of all the ΔT's is $T_2 - T_1$, and the total heat flow Q is the sum of all the ΔQ's. Thus

$$Q = mc(T_2 - T_1) \qquad (14\text{-}2)$$

The product mc is the heat capacity, or *water equivalent*, of the body. The latter term is more descriptive; for instance, in the above discussion, the aluminum could have been replaced by 88 g of water insofar as its thermal behavior is concerned. Then specific heat capacity would no longer enter into the problem; the heat flow would be given by

(Water equivalent)(temperature change)

$$= (88 \text{ cal/°C})(5°C) = 440 \text{ cal}$$

As is true of most properties of matter, specific heat capacity varies slightly with temperature. Some of the values given in Table 14-1 are average values, calculated from

$$c = (1/m)[Q/(T_2 - T_1)]$$

for the temperature ranges shown. Values given at some definite temperature, such as 20°C, are found from the small heat flow ΔQ needed to cause a very small temperature difference ΔT, centered at the desired temperature. The variation of specific heat capacity with temperature is ordinarily rather small, but at very low temperatures the specific heat capacity of a solid is usually strongly dependent on the temperature.

In using Eq. 14-2, we are making two assumptions: (1) External work done by or on the system is zero or negligibly small ($W = 0$, hence $Q = \Delta U$), and (2) all the change ΔU of internal energy is manifested as temperature change, with no changes in internal PE. These assumptions are reasonable for a liquid or solid that is not going through a change of phase, such as melting or boiling.

In many cases, where a system of bodies is isolated and no mechanical work is done on or by the system, conservation of energy means conservation of *internal* energy. The heat gained by one part of the system equals that lost by another part of the system, as illustrated below.

Example 14-2

An automobile radiator contains 20 liters of water. If 200,000 cal of heat is given to the cooling system by the motor, what is the rise in temperature?

First we find the mass of the water; 20 liters = 20,000 cm³.

$$\text{Mass} = (\text{density})(\text{volume})$$
$$= (1 \text{ g/cm}^3)(20,000 \text{ cm}^3) = 20,000 \text{ g}$$

Now consider the isolated system

$$[(\text{water}) + (\text{cylinder block of motor})]$$

Heat gained by water

$$= \text{heat lost by cylinder block}$$
$$mc(T_2 - T_1) = 200,000 \text{ cal}$$
$$(20,000 \text{ g})(1 \text{ cal/g} \cdot °\text{C})(T_2 - T_1) = 200,000 \text{ cal}$$

$$T_2 - T_1 = \frac{200,000 \text{ cal}}{20,000 \text{ cal/°C}} = \boxed{10°\text{C}}$$

Example 14-3

Repeat Example 14-2, assuming that the radiator is filled with glycerin instead of water.

From Table 1-3, the density of glycerin is 1.26 g/cm³. Twenty liters of glycerin therefore has a mass of 25,200 g.

Heat gained by glycerin

$$= \text{heat lost by cylinder block}$$
$$(25,200 \text{ g})(0.60 \text{ cal/g} \cdot °\text{C})(T_2 - T_1) = 200,000 \text{ cal}$$

$$T_2 - T_1 = \frac{200,000 \text{ cal}}{15,120 \text{ cal/°C}} = \boxed{13.2°\text{C}}$$

It is evident that glycerin is not as efficient as water in cooling a car engine. In fact, if water cost $100 a gallon, it would still be the most desirable coolant, since a given flow of heat would cause a smaller temperature rise in water than in any other common liquid.

The *method of mixtures* is often used to measure specific heat capacity in the laboratory, as in the following example.

Example 14-4

400 g of lead BB shot is heated to 100°C and dropped into a 100-g glass beaker containing 200 g of water at 20°C. The specific heat capacity of glass is 0.20 cal/g·°C. The mixture is stirred until temperature equilibrium is reached; the final temperature is 24.2°C. What is the specific heat of lead?

$$\binom{\text{heat gained}}{\text{by water}} + \binom{\text{heat gained}}{\text{by beaker}} = \binom{\text{heat lost}}{\text{by lead}}$$

$$[(200 \text{ g})(1 \text{ cal/g} \cdot °\text{C})(24.2° - 20.0°)]$$
$$+ [(100 \text{ g})(0.20 \text{ cal/g} \cdot °\text{C})(24.2° - 20.0°)]$$
$$= [(400 \text{ g})(c)(100° - 24.2°)]$$
$$840 \text{ cal} + 84 \text{ cal} = (400 \text{ g})(c)(75.8°\text{C})$$

$$c = \frac{924 \text{ cal}}{(400 \text{ g})(75.8°\text{C})} = \boxed{0.0305 \text{ cal/g} \cdot °\text{C}}$$

The heat capacity of the beaker is its mass times its specific heat capacity, or

$$(100 \text{ g})(0.20 \text{ cal/g} \cdot °\text{C}) = 20 \text{ cal/°C}$$

In other words, it is equivalent to 20 g of water, and we could have added this amount of "water equivalent" to the original 200 g of water for a simpler solution of the problem.

Another example will perhaps clarify the energy principle as exemplified by the first law of thermodynamics. Many mechanical problems are highly simplified by neglecting friction. Let us now take friction into account, using the first law of thermodynamics (that is, the law of conservation of energy).

Example 14-5

A 30-kg boy on a 5-kg sled takes a running start from the top of a slippery hill, moving at 4.00 m/s. The hill is 60 m long and drops 2.00 m for every 10 m traversed (Fig. 14-2). The boy and sled reach the bottom of the hill, moving at 10.0 m/s. (a) How much mechanical energy has been transformed into internal energy of the sled and the snow? (b) If the steel runners of the sled have a mass of 1.500 kg, and if we assume that half the transformed energy remains in the runners, by how much does the temperature of the steel runners rise during the ride downhill?

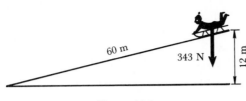

60 m

343 N

12 m

Figure 14-2

(a) The top of the hill is 12 m above the bottom. The weight of the boy and sled together is (35 kg) $(9.8 \text{ m/s}^2) = 343$ N. As the sled and boy slide down, some (but not all) of their initial PE is transformed into KE. The rest of the energy becomes internal energy. We now apply the first law of thermodynamics to the system (sled + snow), assuming that no heat flows to or from the system during the quick ride downhill.

Gain in internal energy

$$= \text{loss of mechanical energy}$$

$$\Delta U = (PE_1 + KE_1) - (PE_2 + KE_2)$$

$$= (343 \text{ N})(12 \text{ m}) + \tfrac{1}{2}(35 \text{ kg})(4.00 \text{ m/s})^2$$

$$- [0 + \tfrac{1}{2}(35 \text{ kg})(10.0 \text{ m/s})^2]$$

$$= 4120 \text{ J} + 280 \text{ J} - 0 - 1750 \text{ J}$$

$$\Delta U = \boxed{2650 \text{ J}}$$

This is the amount of mechanical energy "dissipated" in heat. Of course it has not disappeared; it has merely been transformed into internal energy, a less useful form.

(b) It is assumed that half the energy—1325 J— remains in the runners. To find the temperature rise

of the steel runners is a simple problem involving the specific heat capacity of steel (Table 14-1):

$$\Delta U = mc(T_2 - T_1)$$

$$1325 \text{ J} = (1.500 \text{ kg})(460 \text{ J/kg} \cdot {}^\circ C)(T_2 - T_1)$$

$$T_2 - T_1 = \boxed{1.92 {}^\circ C}$$

Under some circumstances the rise in temperature due to friction can be considerable. A piece of sandpaper becomes noticeably warm while being used. When a speeding train is brought to a halt, its iron brake shoes may become red hot. In other cases the temperature rise might be unobservably small, but it must be present, according to the first law. The stylus and the surface of a phonograph record must become slightly warmer as the record spins.

In some cases it is not so easy to see that "disappearing" energy is really being transformed into internal energy because of friction. When an object on the end of a spring vibrates up and down, there is a continual transformation of energy from PE to KE and back to PE (Fig. 9-13). However, the vibration dies out because there are "frictional" forces between atoms that to a slight extent oppose the motion of the spring. To move against these forces requires work, regardless of whether the spring is stretching or contracting. There is no other source for this work, unless the total mechanical energy of the system is to be gradually used up. The spring becomes warmer. *Random* KE of the atoms increases, as the *ordered* KE of the whole body decreases and becomes zero. Energy is conserved.

14-4 Change of Phase; Heats of Fusion and Vaporization

Matter exists in three states, known as *phases*: solid, liquid, and gaseous. Let us review the behavior of a common substance, H_2O. Let us place 1 g of it in the form of ice at, say, $-20{}^\circ C$ in a thin metal cup on a hotplate maintained at $130{}^\circ C$. According to Table 14-1, the specific heat capacity of ice is 0.50 cal/g·°C, so a certain quantity of heat is required to raise it to 0°C. Further heat is then absorbed, but it is found that instead of the ice becoming warmer, the temperature remains 0°C while the ice melts. Experiment shows that about 80 cal of heat is required to melt 1 g of ice, the temperature remaining constant at 0°C the entire time. Only when all the ice has become water at 0°C does the temperature again begin to rise. In the early days it was expected that *any* absorption of heat would result in a temperature rise, and so the behavior of a melting piece of ice was a puzzle. The 80 cal is called the *heat of fusion*,* since its effect is to cause melting (fusion) rather than even the tiniest rise in temperature.

As we continue to supply heat, the H_2O, in the liquid phase now, warms up until 100°C is reached, at which point a similar behavior is observed. The water remains at 100°C until *all* of it is converted to water vapor (steam),† and then (if this gas is confined) the temperature begins to rise again. The *heat of vaporization*‡ of water at 100°C is considerably greater than its heat of fusion—about 540 cal/g is absorbed without any rise in temperature. Finally, if water vapor in a sealed box is placed in a furnace, it can be heated as much as desired without any further latent heats, until the molecules (absorbing *heat of dissociation*) are separated into hydrogen and oxygen atoms, and H_2O molecules cease to exist as such.

We interpret change of phase as a reorganization of the molecular arrangements. In a solid the molecules are so strongly bound to each other by cohesive forces that a definite crystal structure is possible. When ice melts, the molecules in liquid water are at about the same distance from each other as they were in the ice crystals, but they are no longer rigidly bound to each other. To disrupt the structure requires work (force × displacement). In much the same way, work must be done to break up a large rock into an equal amount of sand, even though the fragments have about the same total volume as the rock. In the vapor state the molecules are actually separated to a considerable distance, and the required work is larger.

The volume of 1 g of H_2O is 1.1 cm^3 in the form of ice, 1.0 cm^3 in the form of water, and 1700 cm^3 in the form of water vapor at 100°C. We know that there are cohesive forces acting to hold the molecules of a liquid together, as shown by the phenomenon of surface tension. Most (about 92%) of the heat required to vaporize 1 g of water at 100°C is used up in separating the molecules—doing work against internal cohesive forces and causing an increase in the internal energy of the molecules. About 8% of the total energy is used in expanding the gas—doing work on the surrounding molecules that are pushing in with atmospheric pressure. In terms of the first law of thermodynamics, to vaporize 1 g of water requires 2256 J (540 cal):

$$Q = W + \Delta U$$
$$2256 \text{ J} = 169 \text{ J} + 2087 \text{ J}$$

From what has been said, it is evident that heat may be transformed into internal energy in two ways: it may become random KE of the molecules, in which case the temperature increases, or it may become internal PE, in which case the temperature remains constant and the energy is stored in the substance and its surroundings.§ We can consider these two transformations as we take the gram of water through a reverse cycle, starting with water vapor at

* Also called *latent* heat of fusion, from the Latin word *latere*, to lie hidden or concealed.

† Technically, "steam" is the same as "water vapor" and is invisible. The white visible cloud that is commonly miscalled steam is in reality a collection of many tiny droplets of liquid water, similar to a mist or fog.

‡ Also called *latent* heat of vaporization.

§ Recall that PE is a joint property of several bodies (Sec. 6-3). When water boils away in an open dish, there are two systems in which PE can be stored. Gravitational PE is stored in the system (earth + H_2O molecules) as the molecules rise. More important by far is the PE stored in the system (liquid molecules + vapor molecules) as work is done against the short-range electric forces of attraction between molecules.

Table 14-2 Heats of Fusion and Vaporization

Substance	Melting Point, °C	Heat of Fusion, kJ/kg	cal/g	Boiling Point, °C	Heat of Vaporization, kJ/kg	cal/g
oxygen	−218	13.8	3.30	−183	210	51
ammonia	−75	453	108	−33	1370	327
water	0	334	80	100	2256	540
mercury	−39	11.8	2.8	357	290	70
lead	327	25	5.9	1753	862	206
aluminum	660	380	90	2450	11400	2720
copper	1083	134	32	2300	5070	1211
uranium	1133	84	20	3900	1900	454
tungsten	3410	184	44	5900	4810	1150

130°C. After the temperature falls to 100°C, liquid drops of water form from the vapor, and the temperature remains at 100°C until all the vapor is condensed. The heat of vaporization—540 cal/g—is *released* during this process. As the liquid H_2O cools down to 0°C, more heat is given off (as the random KE of the molecules decreases), and then, when the H_2O changes from liquid to solid, 80 cal/g of heat is given off while the temperature remains at 0°C (the PE decreases and the molecules lock into position in tiny crystals).

The heats of fusion and vaporization of substances vary widely; see Table 14-2, where these heats and the corresponding transition temperatures (at 1 atm pressure) are given. It is important to note that the heats of fusion and vaporization (denoted by Δh_f and Δh_v) involve no temperature changes and are expressed in SI units J/kg or kJ/kg, or in the non-SI unit cal/g, rather than in J/kg · °C or cal/g · °C, as are specific heat capacities. Heats of fusion or vaporization can be measured by the method of mixtures. In the following example, the calorimeter cup is stated to be "well insulated," so there is no flow of heat into or out of the system.

Example 14-6

A well-insulated copper calorimeter cup of mass 100 g contains 400 g of water at 5°C. When 40 g of ice at −8°C is added, it is found that 23.6 g of the ice melts. What is the heat of fusion of ice?

Since not all the ice melts, we know that the final temperature is 0°C, with 16.4 g of ice floating in 423.6 g of water at 0°C. For the specific heat capacities of ice and copper we use the values from Table 14-1, which are 0.50 and 0.093 cal/g·°C, respectively.

$$\begin{pmatrix} \text{Heat absorbed to} \\ \text{raise 40 g of ice to} \\ 0°C \end{pmatrix} + \begin{pmatrix} \text{heat absorbed} \\ \text{to melt} \\ \text{23.6 g of ice} \end{pmatrix}$$

$$= \begin{pmatrix} \text{heat given off} \\ \text{by water cooling} \\ \text{to } 0°C \end{pmatrix} + \begin{pmatrix} \text{heat given off} \\ \text{by copper cup} \\ \text{cooling to } 0°C \end{pmatrix}$$

$$(40 \text{ g})(0.50 \text{ cal/g} \cdot °C)[0° - (-8°)] + (23.6 \text{ g})(\Delta h_f)$$
$$= (400 \text{ g})(1 \text{ cal/g} \cdot °C)(5° - 0°)$$
$$+ (100 \text{ g})(0.093 \text{ cal/g} \cdot °C)(5° - 0°)$$
$$160 \text{ cal} + (23.6 \text{ g})(\Delta h_f) = 2000 \text{ cal} + 46 \text{ cal}$$

$$\Delta h_f = \boxed{79.9 \text{ cal/g}}$$

Example 14-7

What is the result of adding 10 g of steam at 100°C to 50 g of ice at 0°C?

We have to feel our way through this problem in order to find out whether all the ice melts. Then we can set up an equation to find the final temperature. During condensation, 10 g of steam *can* give up as much as

$$(540 \text{ cal/g})(10 \text{ g}) = 5400 \text{ cal}$$

To melt all the ice requires (80 cal/g)(50 g) = 4000 cal. Hence all the ice *will* melt, heat will be left over to warm up the ice water formed from the ice, and the final temperature will lie between 0°C and 100°C. Call this final temperature T.

$$\text{Heat given off} = \text{heat absorbed}$$

$$\left(\begin{array}{c}\text{condensation}\\\text{of steam}\end{array}\right) + \left(\begin{array}{c}\text{cooling of}\\\text{hot water}\end{array}\right)$$

$$= \left(\begin{array}{c}\text{melting}\\\text{of ice}\end{array}\right) + \left(\begin{array}{c}\text{warming up of}\\\text{cold water}\end{array}\right)$$

$$(10 \text{ g})(540 \text{ cal/g}) + (10 \text{ g})(1 \text{ cal/g} \cdot {}^\circ\text{C})(100^\circ - T)$$
$$= (50 \text{ g})(80 \text{ cal/g}) + (50 \text{ g})(1 \text{ cal/g} \cdot {}^\circ\text{C})(T - 0^\circ)$$
$$5400 + 1000 - 10T = 4000 + 50T$$
$$T = 40.0^\circ\text{C}$$

The result is

$$\boxed{60 \text{ g of water at } 40.0^\circ\text{C}}$$

The heat given off when water freezes is put to good use on a farm when produce is stored in an unheated shed. Large tubs of water are set in the shed, and as the outside temperature goes below 0°C, the water freezes. As this happens, the heat of fusion is given off to the surroundings, keeping the shed and contents at about 0°C until all the water freezes. A similar action takes place when dew forms on the leaves of a plant. As the water vapor in the air condenses and becomes liquid water, the heat of vaporization is given out, and this heat may well help prevent plant damage due to freezing.

14-5   Heat of Combustion

Thermal energy is usually obtained by burning some kind of fossil fuel.* For example, we saw in Sec. 6-3 that during the combustion of carbon (coal is about 80% carbon), the *system* C + 2 O changes configuration to give a product CO_2 that has less PE than the original reactants had.

---

* "Burning" and "combustion" are both terms describing rapid oxidation.

Table 14-3   Heats of Combustion

Substance	MJ/kg	kcal/kg
coal (anthracite)	33	8000
wood (approx.)	15	3500
gasoline	48	11500
furnace oil	44	10600
methyl alcohol	22	5300
fuel alcohol (denatured)	27	6500
carbohydrates	17	4000
proteins	17	4000
fats	40	9500

For anthracite coal, it has been found that when 1 kg of coal combines with enough oxygen to ensure complete combustion, the thermal energy released is about 33×10^6 J (33 MJ), which is the same as 8000 kcal. Thus the *heat of combustion* is measured in MJ/kg, kWh/kg, kcal/kg, kcal/mol, or (by engineers and enviromentalists) Btu/lb. Some typical values are given in Table 14-3.

The heat of combustion of gases is usually given for a unit volume. Thus, methane, the major constituent of natural gas, can supply about 40 MJ/m^3. At 0°C and 1 atm, methane has a density of 0.715 kg/m^3, so the heat of combustion of natural gas is actually about 56 MJ/kg.

Some global figures are of interest. The known worldwide supply of fossil fuel (chiefly coal) has an energy content of about 2×10^{23} J, and lower-grade sources such as shale oil may possibly account for another 3×10^{23} J, making a total known supply of about 5×10^{23} J. This can be compared with an estimate of the maximum possible fossil fuel based on the amount of free oxygen in the atmosphere of the earth. We assume that all, or nearly all, of this oxygen is the result of photosynthesis going back as far as 4×10^8 years ago, when the surface of the earth was covered with the extensive vegetation that is now mostly fossil fuel. The overall reaction is

$$\text{Solar energy} + CO_2 \rightarrow C + O_2$$

Details of the photosynthesis process are complex, but only the end result is necessary for our argument. It is much easier to measure oxygen in the air than to guess at buried fossil fuel deposits. By this line of reasoning the total initial supply of fossil fuel cannot have exceeded about 1×10^{25} J. The *recoverable* energy that can be utilized economically is surely much less than the estimated maximum possible primeval fossil fuel, so the two figures are in substantial agreement.

14-6 Transfer of Heat by Convection and Conduction

With the understanding that heat is not a material substance, we can talk about the flow of heat, by which we mean transfer of heat from one place to another. There are three fundamentally distinct ways in which heat may be transferred: convection, conduction, and radiation. Each of these methods of heat transfer requires a temperature difference.

The *convection* of heat is well illustrated by the blower of a space heater in a large gymnasium. Hot air is moved bodily from the point of origin to the spectators in the bleachers. Kinetic energy of the air molecules is transported by simply moving the molecules en masse to the desired point. Convection currents in gases and liquids are often caused by variations in density due to temperature changes. Hot air, which is less dense than cold air, rises from a point above a radiator and strikes the ceiling, spreading out through the room. The force causing this motion arises from pressure exerted by nearby cooler (and denser) air. In this way, a continuous flow takes place. A similar convection phenomenon keeps the temperature of a lake at about $4°C$ (the temperature of the densest water) before any of it cools to freezing (Sec. 13-4). Most of the changes in air temperature that appear on a weather map are due to convection currents set up as masses of warm or cold air move over land and water.

The *conduction* of heat is a more intimate process, in which KE is handed on from one molecule to an adjacent molecule by collisions.

Table 14-4   Thermal Conductivity of Common Materials

Substance	Approximate Thermal Conductivity K, cal·cm/cm²·s·°C
Masonite	0.00011
white pine	0.0002
building brick	0.0015
glass	0.0025
concrete	0.002
steel	0.11
copper	0.92

When a spoon is placed in a cup of hot coffee, the vibrating silver molecules in the spoon bump each other, but they do not migrate away from their average positions. The slow-moving molecules in the handle eventually receive KE through a chain of events, and so the handle becomes warm. As one would expect, heat flows by conduction only if there is a temperature difference, for only then (on the average) will fast molecules be striking slow molecules and be able to give up energy.

Substances differ markedly in their *thermal conductivity* and may be classified as insulators (such as asbestos or cork) and conductors (such as aluminum or iron). Metals are especially good conductors of heat (Table 14-4), since a metal has free electrons not attached to any particular nucleus. These free electrons can drift through the metal and transfer KE by their collisions with each other and with more massive ions and nuclei in the crystal lattice. (Such a drift partakes of the nature of convection as well as that of conduction.) The same free electrons are responsible for the conduction of electric currents; it is no accident that silver and copper, which are the best conductors of heat, are also the best conductors of electricity.

The amount of heat that flows through a flat slab of material (Fig. 14-3) in time t is proportional to the time, to the area A of the slab, and

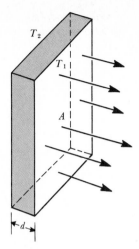

Figure 14-3 Heat flow through a slab of area A. The temperature gradient is $(T_2 - T_1)/d$.

to the *temperature gradient* $\Delta T/\Delta x$, where ΔT is the temperature difference between the two faces of the slab, which are separated by a thickness Δx. In Fig. 14-3 the temperature gradient is $(T_2 - T_1)/d$, where one face is at temperature T_2, the other face is at temperature T_1, and the thickness of the slab is d. The constant of proportionality is called the *thermal conductivity K* of the material. Thus, in a time t the heat flow is given by

$$Q = KAt \frac{T_2 - T_1}{d} \qquad (14\text{-}3)$$

$$\text{cal} = \left(\frac{\text{cal} \cdot \text{cm}}{\text{cm}^2 \cdot \text{s} \cdot {}^\circ\text{C}}\right)(\text{cm}^2)(\text{s})\left(\frac{{}^\circ\text{C}}{\text{cm}}\right)$$

14-7 Transfer of Heat by Radiation

A third method heat transfer, *radiation*, requires no physical medium and is essentially electrical in nature. Whenever electric charge undergoes acceleration, an electromagnetic wave is formed that sends energy through space, in somewhat the same way as waves spread out over the water

when a stone is dropped into a pond. Electromagnetic waves, which we shall study in Chap. 24, are called radio waves when they are vibrating slowly enough to cause a response in a radio receiver. They may be generated when electrons move up and down in the tower of a transmitting antenna. Electromagnetic waves of various higher frequency ranges are known by other names: heat, infrared radiation, visible light, ultraviolet radiation, x rays, and gamma rays. They all travel through a vacuum at 3×10^8 m/s (186,000 mi/s). Heat waves *are* radio waves and can be generated electrically, as in the diathermy machine or the microwave oven. In solids it is far more commonly the vibratory motion of electrically charged particles (ions) that gives rise to the electromagnetic radiation we call "heat radiation" or "infrared radiation." As shown in Table 14-5, the various electromagnetic radiations are characterized by a wide variety of mechanisms for production and recep-

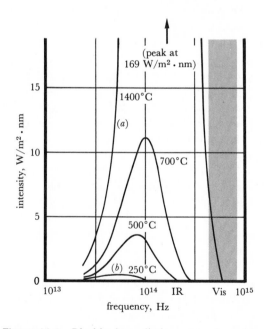

Figure 14-4 Blackbody radiation at various temperatures. The shaded region includes the frequencies of visible light. (*a*) White-hot piece of coal, at 1400°C. (*b*) Blackberry pie, at 250°C.

Table 14-5   Electromagnetic Radiations

Type of Radiation	Usual Source	Usual Detector	Approximate Frequency Range, Hz (cycles/s)	Approximate Wavelength Range, meters*
Radio waves	electric circuits	electric circuits		
lowest frequency used in communications			10^4	3×10^4
AM broadcast band			$0.55 \times 10^6 - 1.60 \times 10^6$	545–188
FM broadcast band			$88 \times 10^6 - 108 \times 10^6$	3.40–2.78
TV broadcast bands			$54 \times 10^6 - 890 \times 10^6$	5.55–0.34
radar and microwaves			$10^9 - 10^{11}$	$3 \times 10^{-1} - 3 \times 10^{-3}$
highest experimental frequency			10^{12}	3×10^{-4}
Infrared radiation	hot bodies	thermocouples thermometers skin	$10^{11} - 3.8 \times 10^{14}$	$3 \times 10^{-3} - 8 \times 10^{-7}$
Visible light	hot bodies electric arcs	eye photocell photographic plate	$3.8 \times 10^{14} - 7.5 \times 10^{14}$	$8 \times 10^{-7} - 4 \times 10^{-7}$
Ultraviolet radiation	electric arcs	photographic plate photocell Geiger counter	$7.5 \times 10^{14} - 3 \times 10^{17}$	$4 \times 10^{-7} - 10^{-9}$
X rays	impact of electrons on target	photographic plate Geiger counter ionization chamber	$3 \times 10^{17} - 3 \times 10^{19}$	$10^{-9} - 10^{-11}$
Gamma rays	radioactive nuclei	Geiger counter ionization chamber scintillation counter	$> 3 \times 10^{19}$	$< 10^{-11}$

Note: The boundaries between the various types of electromagnetic radiation are not sharp. For instance, radiation in the neighborhood of 3×10^{17} Hz might equally well be called "extremely high-frequency ultraviolet radiation" or "extremely low-frequency x rays." Likewise, the range from 10^{11} to 10^{12} Hz can be produced either by electrical means (radio) or by thermal means (infrared sources).

* Wavelength = velocity/frequency. Short wavelengths are often expressed in angstrom units, microns, or nanometers: 1 angstrom (1 Å) = 10^{-10} m; 1 micrometer (1 μm) = 1 micron (1 μ) = 10^{-6} m; 1 nanometer (1 nm) = 10^{-9} m.

tion; the unifying property of all these radiations is their *wave velocity*.

The essential thing for us now is the fact that every body radiates not just one type of radiation, but some of each. Consider two of the graphs in Fig. 14-4. Curve *a* is for a white-hot piece of coal in a fireplace, and it shows that the most intense radiation is in the infrared (IR) region, with some visible light and a small amount of ultraviolet (UV) radiation (too weak to be seen on the graph). There is even *some* radiation in the radio band, although the amount is relatively so small that we don't have to worry about TV interference from a fireplace! When a blackberry pie (at 250°C) is set outdoors to cool, the total radiation is much less, and also the character of the radiation is different (curve b). The peak radiation is now at about 5×10^{13} Hz (in the infrared region), and there are relatively more of the infrared and radio waves. Visible light from the pie is present, but far too weak to detect. The total radiation (all frequencies combined) is, of course, much less for the pie than for the glowing coal.*

The only way that energy can traverse a vacuum is by electromagnetic radiation. In fact, "empty" space between the stars is crisscrossed by many electromagnetic radiations, coming from the stars and galaxies, and it has been estimated that 4×10^{-14} J of energy (on the average)—mostly in the infrared, visible, and ultraviolet regions—is contained in each cubic meter of "empty" space. In addition, the earth is bathed in radiation with a continuous spectrum, similar to Fig. 14-4, with a peak corresponding to a temperature of only 3° above absolute zero. This peak, found in the microwave region, has a frequency of 3×10^{11} Hz (far off the scale at the left of Fig. 14-4), and its wavelength is about 1 mm. This microwave radiation is very nearly isotropic (of the same intensity from all directions of space) and is the remnant or relic of the Big Bang, which took place some 2×10^{10} years ago.

Practically all the radiation received by the earth comes from the sun; at a place where the sun is directly overhead, 8.13 J (1.94 cal) would reach each square centimeter of the earth's surface every minute if the atmosphere were perfectly transparent. This has been going on for

several billion years; why, then, has the earth's temperature not increased to some fantastic figure? The answer lies in the radiation *from* the earth to the surrounding space. The earth has become just hot enough so that the daily loss of heat by radiation (which depends on the earth's temperature) just balances the daily gain in energy from the sun. The sun itself is a gigantic H-bomb, continually converting rest mass into energy (Sec. 28-6). Most of the radiation from the sun does not reach the earth, but passes on out into space.

Not all surfaces are equally effective in emitting or absorbing radiation, even though their temperatures are the same. A white or shiny body is a poor absorber of radiation, since some of the incoming energy is reflected back away from the interior. By the same token, a white or shiny body is a poor *emitter* of radiation, since radiant energy within the body is reflected back. The surface looks shiny from the inside as well as from the outside! A perfect absorber is called a *blackbody*, and such a body absorbs all radiations, not merely visible light. A blackbody is also a perfect emitter. The photograph on page 310 shows a porcelain dish on which the word "black" has been applied with black paint and baked in; the central part of the marking is heated by a gas flame. It is evident that the same region that is a good absorber (when cool) is also a good emitter (when hot). According to these ideas, a living-room "radiator" should be painted black for efficient radiation of heat. A light-colored paint job might be decorative, but is hardly in accord with the laws of physics. However, most of the heat transfer from a "radiator" is by direct contact of air molecules with faster-moving iron atoms or paint molecules; such a process is conduction and does not depend greatly on the blackness of the surface.

At a given temperature, most bodies are good emitters, look much alike, and emit the same distribution of frequencies of radiant energy. In a fireplace, the "red-hot" iron poker is almost indistinguishable from the "red-hot" carbon

---

* Precise statement of the laws of radiation makes use of absolute temperature (Sec. 15-6). Stefan's law says that the total radiation, all frequencies included, is proportional to the fourth power of the absolute temperature. Wien's law says that the frequency of the most intense radiation (the peak of the curve) is directly proportional to the absolute temperature.

(coal or ash) that is at the same temperature. The *radiation pyrometer* is a device to determine the temperature of a furnace by comparing the color of the contents with that of an electrically heated wire at a known temperature. To the extent to which all substances are "black," or nearly so, it is immaterial just what substance is in the furnace. The same radiation pyrometer can be used to measure the temperature of molten iron or molten glass.

Summary

Heat is the transfer of energy due solely to a temperature difference between a body and its surroundings. Internal energy of a body or a system can be changed in two ways: mechanical work can be done on or by the body, and energy can be transferred to or from a body by a flow of heat. The first law of thermodynamics is an extension of the law of conservation of energy, taking into account both of the ways in which internal energy can change. The first law states that the internal energy of an isolated system remains constant.

The heat capacity, or water equivalent, of a body is the heat absorbed or given off per unit change in temperature. The specific heat capacity of a substance is the heat absorbed or given off per unit mass per unit change in temperature. Specific heat capacity can be measured in joules per kilogram per degree Celsius, or calories per gram per degrees Celsius. When heat flows into or out of a body, the quantity of heat Q that flows to cause a temperature change $T_2 - T_1$ is given by $Q = mc(T_2 - T_1)$, where m is the mass, c is the specific heat capacity, and Q is the heat that flows.

When heat flows into a body, the internal energy of the body increases in one or both of two ways. If the random KE of the molecules increases, the temperature increases. If the PE of the molecules increases, as in the separation of the molecules against cohesive forces during fusion or vaporization, the temperature does not increase, and the heat absorbed is a latent heat. Both processes are reversible: heat is given out when a body cools down (KE decreases) or when it condenses or solidifies (PE decreases).

The heat of combustion is the thermal energy released per unit mass when a substance combines with enough oxygen to ensure complete combustion.

Transfer of heat takes place in three ways. Convection is the motion of heated matter, in bulk. Conduction is due to transfer of KE from molecule to neighboring molecule by collision. In metals, heat conduction and electrical conduction are both largely due to a cloud of free electrons within the metal. Radiation includes all types of electromagnetic waves, from radio waves, which are of lowest frequency, to gamma rays, which are of highest frequency. All electromagnetic waves travel with the same velocity (3×10^8 m/s) through a vacuum. A hot body emits some of each type of radiation, but the frequency of the most intense radiation depends on the temperature of the body. The total radiation, all frequencies combined, increases rapidly with temperature. A good absorber is also a good emitter of radiation. A blackbody has a surface that absorbs and emits radiation of all frequencies without hindrance; the distribution of frequencies and the total amount depend only on the temperature.

heat
internal energy
$Q = W + \Delta U$
calorie
Calorie
heat capacity

specific heat capacity
water equivalent
heat of fusion
heat of vaporization
heat of combustion
convection

conduction
thermal conductivity
radiation
types of electromagnetic
  waves
blackbody

Questions

14-1 What is the difference between "heat" and "temperature"?

14-2 Correct the following statements: (a) "The heat of the sun's surface is 6000°C." (b) "Body heat of a whale is 36°C." (c) "The heat required to melt lead is 5.86 cal/g·°C." (d) "The specific heat capacity of copper is 0.093 cal."

14-3 Which of the following are *not* units of energy: joule, calorie, Calorie, foot · pound, watt, horsepower, erg, Btu?

14-4 What is the difference between specific heat and specific heat capacity? What is the SI unit for each of these quantities?

14-5 If two buckets of equal size are filled, one with water and one with glycerin, both at 50°C, which will cool down to room temperature first? Why? Suppose one bucket were painted black and the other white. Would this affect your answer?

14-6 A book lying near the edge of a table is pushed over and allowed to fall to the floor. When the book strikes the floor, its KE disappears. What becomes of this KE?

14-7 How much heat is needed to convert 1 kg of ice at 0°C to 1 kg of water vapor at 100°C?

14-8 An Eskimo places a stone near a fireplace, then takes the warm stone to bed with him and rests his feet on the stone. Explain how convection, conduction, and radiation are involved in this custom.

14-9 Why does a metal seem colder to the touch than wood outdoors on a cold winter day?

14-10 Is any light of visible frequency emitted by the surface of a soot-covered teakettle filled with boiling water?

14-11 It is known that the interior of the earth is maintained at a temperature of several thousand degrees Celsius by a source of heat due to radioactive disintegration. This being the case, does the earth absorb the same amount of heat from the sun (Sec. 14-7) as it radiates?

MULTIPLE CHOICE

14-12 The first law of thermodynamics is concerned with the conservation of (a) energy; (b) momentum; (c) both of these.

14-13 When a liquid is heated but does not change its state, its molecules gain (a) PE; (b) KE; (c) both of these.

14-14 Heat of fusion of a substance is called a "latent" heat because, when melting, its molecules (a) gain PE; (b) gain KE; (c) become cooler.

14-15 If 1 g of steam at 100°C loses 560 cal of heat, the final temperature is (a) 20°C; (b) 80°C; (c) 100°C.

14-16 A unit for heat of combustion could be (a) J/kg; (b) the same as a unit for heat of vaporization; (c) both of these.

14-17 An object that is a perfect absorber of radiation is (a) a perfect emitter of radiation; (b) a blackbody; (c) both of these.

Problems

14-A1 How much heat is required to raise the temperature of 300 g of water from 50°C to 70°C? Express your answer in calories.

14-A2 How much heat is wasted when a steamship engine pours 400 kg of water at 80°C into a lake that is at 10°C? Express your answer in joules.

14-A3 In a laboratory experiment, 900 J of work was done against friction. How much heat, in calories, was generated?

14-A4 What is the specific heat capacity of water, in kJ/kg·°C?

14-A5 It is found that 240 cal of heat is needed to heat 100 g of a substance from 6°C to 14°C. What is the specific heat capacity of the substance?

14-A6 A metal block of mass 500 g is warmed from 20°C to 35°C when it absorbs 2000 cal of heat. Calculate the specific heat capacity of the metal.

14-A7 How much energy is needed to heat 40 kg of asbestos from 10°C to 50°C?

14-A8 What is the water equivalent of 3 kg of lead?

14-A9 How much heat, in calories, is given off by 300 g of water as it cools from 100°C to 20°C?

14-A10 How much heat is needed to bring 5 kg of water from 20°C to within 1°C of the boiling temperature?

14-A11 How much heat must be supplied to a 15-g ice cube, originally at 0°C, in order to convert it to 15 g of water at 20°C?

14-A12 How much heat is needed to melt 2 kg of lead, originally at 0°C?

14-A13 While condensing from a vapor into a liquid at 61.5°C, 30 g of chloroform liberated 1770 cal of heat. What is the heat of vaporization of chloroform?

14-A14 The heat of vaporization of liquid helium, at its normal boiling point of −269°C, is 6 cal/g. How much helium is evaporated when an electric heater coil supplies 100 J of energy to a flask of liquid helium?

14-A15 What is the wavelength of the center frequency of a TV station that sends out programs on Channel 4 (frequency range 66–72 MHz)?

14-A16 A certain type of x-ray radiation from copper atoms has wavelength 1.54×10^{-10} m. What is the frequency of this radiation?

14-A17 A certain infrared radiation has wavelength 10^{-7} m. Express this wavelength in angstrom units; in micrometers; in nanometers.

14-A18 What is the frequency, in GHz, of radar waves whose wavelength is 2.00 cm?

14-A19 What is the wavelength, in angstroms, of the most intense radiation from an electric heater whose temperature is 700°C? (See Fig. 14-4.)

14-B1 When 500 g of lead at 100°C is placed in an aluminum calorimeter cup of mass 120 g that contains 200 g of olive oil at 20°C, the final temperature of the mixture is 29.0°C. What is the specific heat capacity of olive oil?

14-B2 In a laboratory experiment to determine the heat of fusion of ice, 59.1 g of ice at 0°C was placed in a calorimeter cup that contained 320 g of warm water. The water equivalent of the cup was 20.1 g. All the ice melted, and the cup and contents were cooled from 34.3°C to 17.0°C. Calculate the heat of fusion of ice from these data. (*Note:* The lack of agreement between your answer and the accepted value is typical of the degree of precision commonly obtained in heat measurements in elementary laboratories.)

14-B3 An aluminum pan is placed on a stove at 4:15 P.M. The pan has a mass of 680 g and contains 300 g of water at 20°C. The flame supplies 100 cal of heat to the system each second. (*a*) When will the water begin to boil? (*b*) When will the temperature of the pan begin to rise above 100°C?

14-B4 What is the result of adding 10 g of steam at 100°C to 25 g of ice at 0°C?

14-B5 What is the result of adding 10 g of steam at 100°C to 50 g of ice at 0°C?

14-B6 What is the result of adding 10 g of steam at 100°C to 100 g of ice at 0°C?

14-B7 To measure the temperature of a flask of liquid air, a researcher let a 500-g chunk of aluminum come to thermal equilibrium with the liquid air, then quickly dropped the chilled aluminum into a 200-g copper cup containing 400 g of water. The cup and the water, both originally at 80°C, came to a final temperature of 31°C. What was the temperature of the liquid air? The average specific heat capacity of aluminum for the temperature range used is 0.182 cal/g·°C.

14-B8 A glass beaker whose water equivalent is 60 g contains 200 cm³ of glycerin at 20°C. What will be the final temperature if 20 cm³ of copper at 100°C is placed in the beaker?

14-B9 How much steam at 100°C would have to be added to 90 g of ice at 0°C so that the entire system becomes water at 40°C?

14-B10 An ice cube of mass 40 g is in an aluminum cup of mass 100 g. The system is originally at −35°C. What will be the result of adding 8000 cal to the system?

14-B11 Steam is allowed to bubble through water contained in an aluminum cup. Some of the steam condenses into water, giving off its heat of vaporization, and then, as water, cools down. The following data are taken: initial temperature, 10°C; final temperature, 40°C; mass of empty cup, 100 g; mass of cup plus cold water, 218 g; mass of cup plus warm water plus condensed steam, 225 g. Calculate the heat of vaporization of steam.

14-B12 A closed, unheated storage room at 1°C contains sacks of vegetables whose total water equivalent is 450 kg. As a result of a sudden drop in the outside temperature, heat leaves the room at the rate of 600 kcal/h. (*a*) How long will it take for the temperature of the room to go from 1°C to −1°C? (*Note:* The vegetables do not freeze, because the starch and sugar in them lower the freezing point to a few degrees below −1°C.) (*b*) How long will be required for the same temperature drop if a tank containing 200 kg of water is in the storage room along with the vegetables?

14-B13 About how much water is evaporated in an hour from a lake 1 km² in area if 0.1% of the sun's radiation gets through the atmosphere and is absorbed by the water? The sun's rays strike the water at an angle of 60° with the vertical. Assume negligible temperature change of the large body of water. The heat of vaporization of water at this temperature is about 580 kcal/kg.

14-B14 An artisan hammers out a silver ashtray whose mass is 90 g. If the hammer head has a mass of 450 g and it strikes the tray with a speed of 10 m/s, what is the rise in temperature of the tray after 30 strokes of the hammer? (*Note:* Silver is much softer than the iron hammer head. Hence it is reasonable to assume that the hammer rebounds with negligible speed. Assume also that 50% of the heat generated remains in the ashtray.)

14-B15 A truck of mass 10 metric tons is traveling at 70 km/h. How much heat, in MJ and in kcal, is dissipated in the brake system when the truck is brought to rest?

14-B16 An ice skater of mass 60 kg starts to glide at 5 m/s and comes to rest after traveling 20 m. How much ice is melted? (Assume that half the heat generated by friction is absorbed by the ice, and also assume that the ice beneath the skates is at 0°C.)

14-B17 A lead bullet of mass 25 g is fired at 400 m/s at a tree trunk, and the bullet emerges on the other side with a speed of 300 m/s. (a) What is the loss of KE? (b) Assuming that 60% of the loss of mechanical energy is stored as internal energy in the bullet, and also assuming the specific heat capacity to remain equal to the value given in Table 14-1, calculate the bullet's rise in temperature. Will any of the bullet melt?

14-B18 What is the ratio of the energy needed to evaporate some water in the ocean to the energy needed to lift the same water to form a cloud 2.0 km above the ocean?

14-B19 An overweight student attempts to reduce by lifting two 4-kg masses vertically through 0.8 m. How many times must this be done to "work off" the energy equivalent of one 70-kcal slice of bread? Is your answer reasonable? Discuss.

14-B20 The basal metabolic rate (BMR) is the rate of energy production when in a resting state. For an adult male human the BMR is about 90 W. What would be the minimum intake of energy from food sources to maintain bodily functions at the minimum level? Express your answer in kcal/day and compare with your own caloric intake, which is probably greater than 2500 kcal/day.

14-B21 An office worker prepares for a coffee break by heating a cup of water with a 200-W electric immersion heater placed directly in the cup. The cup is of glass and has mass 200 g; it holds $\frac{1}{4}$ liter and is $\frac{4}{5}$ full of water at 15°C. How much time is required to bring the temperature of the water to 60°C?

14-B22 In a physiological experiment, it was found that when the surrounding air temperature was in the range 19°C to 31°C, a human subject lost about 100 kcal of heat per hour due to all causes. Of this loss, about 20% was due to evaporation from an invisibly thin layer of moisture on the skin. The heat of vaporization of water at skin temperature is 580 cal/g. (a) Taking the surface area of the subject's body to be 1.0 m^2, what would be the thickness of a layer of sweat that would be completely evaporated in 15 min? (b) Can you explain qualitatively why the heat of vaporization of water at body temperature is higher than the value quoted in the text for 100°C? (c) Suggest two ways in which the other 80% of the heat loss might take place.

14-B23 Suppose a person who is exercising strenuously produces a power that is 25 W greater than his basal metabolic rate (when resting; see Prob. 14-B20). Estimate the mass of water (in the form of sweat) that would have to be evaporated per hour to dissipate this additional power.

14-B24 Could an adult in a famine area exist at minimum level (basal metabolic rate 90 W) if the only food available were 400 g of carbohydrate and protein per day?

14-B25 A lead ball of mass m kg is dropped onto a pavement from a height of 40 m. Assuming that 60% of the heat generated remains in the lead, calculate the temperature rise of the ball.

14-B26 Using the modern value of the mechanical equivalent of heat (4.19 J/cal), calculate the height of fall that Mayer stated to be "about 365 meters" (Sec. 14-2).

14-B27 A frozen water pipe in a house is 5 m long, and its inside diameter is 1 cm. The pipe is iron, and its mass is 2 kg. To thaw the ice, an electric current is set up in the pipe by a transformer that can supply electric energy at the rate of 1500 W, of which 60% is useful in warming up the pipe and ice. The original temperature of pipe and ice is −15°C. (a) How much heat is needed to melt the ice? (b) How much time is required for the job?

14-B28 How much coal must be burned in a furnace to heat (one time) the air in a medium-sized, single-story house (about 700 kg) from −5°C to 20°C? Assume that 75% of the fuel value of the coal can be utilized in heating the air.

14-B29 The skin temperature of an adult is 34°C, maintained by a flow of heat from a central core at 37°C (blood temperature). On this model, energy is supplied by metabolism to replace heat lost by conduction through a layer of subcutaneous tissue 3 cm thick. The surface area of an adult is 1.8 m², and the thermal conductivity of tissue is 0.0012 cal/cm·s·°C. (*a*) Estimate the basal metabolic rate, in watts. (*b*) To treat severe burns, it is proposed to immerse the patient in a cold-water bath to bring the skin temperature to 25°C. To maintain the core at a life-sustaining 37°C, an additional heat source must be provided, perhaps by diverting some blood through a heat exchanger. Estimate the rate at which heat must be supplied over and above the basal rate calculated in part (*a*).

14-B30 One wall of a brick house is 5 m × 10 m × 20 cm thick. How much heat per day flows through the wall if the inside temperature is 20°C and the outside temperature is −10°C? Assume that there are no insulating layers of stagnant air near the inner or outer brick surfaces. Express your answer in kcal/day and in kWh/day.

14-B31 A lamp bulb inside an ornamental spherical glass shell 15 cm in radius and 0.5 cm thick radiates 100 W of thermal power. What is the difference in temperature between the inner and outer surfaces of the glass?

14-B32 Because of radioactivity in the crust of the earth (see Sec. 32-1), the temperature gradient in nonvolcanic regions is about 3°C for each 100 m change in depth below the surface. (*a*) Assuming the thermal conductivity of granitic crust material to be about the same as for concrete, calculate the heat flow (in μcal/s) through 1 cm² of the earth's surface. (*b*) Estimate the heat flow, in J/day, through the entire surface of the earth.

14-B33 An ice-cube container is in the form of a cube of outside edge 30 cm, made of plastic insulating material whose thermal conductivity is 9×10^{-5} cal·cm/cm²·s·°C. The base, walls, and lid are 5 cm thick. If 6.0 kg of ice at 0°C is placed in the container and taken on a picnic on a day when the temperature is 35°C, how long will it be before all the ice is melted?

14-C1 The temperature of a sample of molten lead that is near its temperature of solidification is falling at the rate 9.0°C/min. (*a*) If the lead continues to lose heat at the rate indicated by this value and solidifies completely 20 min after the temperature becomes constant, what is the heat of fusion for lead indicated by these data? (*Hint:* Let *m* be the mass of the sample.) (*b*) What assumption have you made about the specific heat capacity of lead?

14-C2 An aluminum rod has a diameter of 0.50 cm. What will be the change in length of the rod if it absorbs 15 kJ of heat? (*Hint:* Solve in symbols first. Use tables from Chaps. 1, 13, and 14.)

14-C3 The ends of a cast-iron rod are held clamped between two fixed posts 2 m apart with a tension of 100 N. How much heat must be absorbed by the rod to reduce the tension to zero? (*Hint:* Solve in symbols first. Use tables from Chaps. 1, 9, 13, and 14.)

14-C4 A torque of 0.025 m·N is needed to turn the paddles of a cake mixer at 300 rev/min. (*a*) At what rate is heat being generated in the batter? (*b*) What temperature rise would occur in 500 g of batter as a result of mixing it for 4 min? (Assume specific heat capacity of the batter to be 0.9 cal/g·°C.)

14-C5 A solar house is designed to store a million kcal of heat. The storage facilities consist of sealed drums in the attic containing Glauber's salt ($Na_2SO_4 \cdot 10 H_2O$), which is warmed up from 25°C to 40°C during the day. (*a*) What mass of salt is needed? (*b*) What should be the volume of the drums? [Data for Glauber's salt, in the impure form used: specific heat capacity (solid), 0.46 cal/g·°C; specific heat capacity (liquid), 0.68 cal/g·°C; melting point, 32°C; heat of fusion, 60 cal/g; specific gravity, 1.46.]

14-C6 The waste heat from a nuclear power plant is 5×10^5 kcal/s. If it could be recovered and converted into electric energy with 15% efficiency, what would be the value of the energy wasted in 1 day? The wholesale cost of 1 kWh (3.6×10^6 J) is about 2¢.

14-C7 A fossil-fuel power plant that generates 1000 MW (10^9 W) of electric power discharges 1600 MJ per second as waste heat into a river. (*a*) At what rate (in MW) is thermal energy obtained from the burning of coal? (*b*) What is the thermal efficiency (power output/heat input) of the plant? (*c*) If the river flow is at 5×10^8 kg of water per hour, what is the temperature rise of the river due to this thermal pollution?

14-C8 Charles Coulomb (better known for the law in electrostatics that bears his name) made an early estimate of the sustained power that can be developed by a human. He knew that a party of French marines had climbed a 2923-m mountain in the Canary Islands in 8 h. Assume that each marine had a mass of 70 kg. (*a*) Calculate the mechanical power (in watts) developed by each marine during the climb. (*b*) Marine mess sergeants know that a dietary intake of about 10,000 kcal/day would be required for the marines if they were to perform this feat on a daily basis. Bearing in mind that the basal metabolic rate for energy expenditure during rest is about 90 W, calculate the dietary intake chargeable to the normal or resting state, and then calculate the efficiency of the marines (ratio of work output to net dietary intake).

14-C9 A factory roof measures 30 m × 50 m. If the sun's radiation strikes the roof broad-side, and if the energy could be converted into mechanical work with 10% efficiency, what power, in kilowatts, would be available for use in the factory?

14-C10 Suppose that beginning now only 98% of the solar radiation of 1.94 cal/cm²·min (see page 324) is reradiated, the balance being trapped by increased CO_2 in the atmosphere arising from the burning of fossil fuels. Assume also that 10% of the *trapped* radiation goes to melt polar ice caps. How much would the ocean level rise in 10 years? Data: radius of earth, 6.4×10^6 m; area of ocean surface, 4.0×10^{14} m². (*Hint:* Use the projected earth surface πr^2 in calculating thermal input.)

14-C11 Global average annual rainfall is about 1.0 m. (*a*) How much energy (in J) is released worldwide during a year by condensation of water vapor into liquid water? (*b*) Where did this energy come from?

14-C12 Architects in the United States often need to know the heat in Btu that flows each hour through a sheet of area 1 ft² due to a temperature gradient of 1°F/in. By what factor should the coefficient K in the cgs unit cal·cm/cm²·s·°C be multiplied to obtain K in Btu·in./ft²·h·°F?

References

1. D. Roller, *The Early Development of the Concepts of Temperature and Heat: The Decline of the Caloric Theory* (Harvard Case Histories in Experimental Science No. 3), (Harvard Univ. Press, Cambridge, 1950). See especially Rumford's investigation of weight ascribed to heat (pp. 47–61); his experiments on heat produced by friction (pp. 61–81); and Davy's experiments (pp. 81–89). *Caution:* The early writers sometimes used the words "heat" or "degree of heat" where we would today use "temperature." The context will tell whether temperature or quantity of heat is meant.

2. W. F. Magie, *A Source Book in Physics* (McGraw-Hill, New York, 1935). Selections from the writings of Newton, Amontons, Fahrenheit, Taylor, Black, Rumford, Davy, Gay-Lussac, and others. See especially the work of Black on latent heat (pp. 139–44); Rumford

on heat produced by friction (pp. 151–61); and Davy's experiments on heat produced by friction (pp. 161–65).

3. S. C. Brown, "Count Rumford's Concept of Heat," *Am. J. Phys.* **20**, 331 (1952).
4. M. S. Powell, "Count Rumford: Soldier, Statesman, and Scientist," *Am. J. Phys.* **3**, 161 (1935). More about his personal life than about his physics.
5. M. Wilson, "Count Rumford," *Sci. American* **203**(4), 42 (Oct. 1960).
6. S. C. Brown, "Benjamin Thompson, Count Rumford," *Phys. Teach.* **14**, 270 (1976).
7. F. J. Dyson, "What Is Heat?" *Sci. American* **191**(3), 58 (Sept. 1954).
8. H. Weinstock, "Thermodynamics of cooling a (live) body," *Am. J. Phys.* **48**, 339 (1980).

15

Thermal Behavior of Gases

Now that we have developed some familiarity with the concepts of temperature and of heat as it is related to energy, we proceed to the study of the thermal behavior of gases and vapors. One of the most useful generalizations about gases is the concept of an ideal gas, whose behavior can be described by rather simple equations.

15-1   Ideal Gases

In spite of their much greater familiarity in daily life with matter in the solid and liquid states, physicists understand the behavior of gases better than that of either solids or liquids. The gaseous state is actually far simpler than the liquid or solid state, because the molecules in a gas are so far apart that the cohesive forces between them are usually small. Granted that the cohesive forces between molecules of O_2 gas are probably different from those between molecules of CO_2 gas, the practical effect on gas behavior in either case is so small as to be negligible in ordinary work. Again, it is natural to suppose that the behavior of these two gases

would be somewhat affected by the sizes of the individual molecules. However, as a practical matter, the volume actually occupied by the molecules is usually a tiny fraction of the total volume of the container in which the gas is confined. An *ideal gas* is one in which these disturbing effects of cohesive forces and molecular volume are so small as to be negligible. We thus expect the laws of gases to be simple and universal, the same for all ideal gases, regardless of their chemical composition. No real gas is, of course, an ideal gas, but we shall find that under ordinary conditions of temperature and pressure we shall not be very wrong if we consider all gases ideal.

15-2   Boyle's Law

How shall we describe the behavior of a gas? Four measurable quantities are of interest: the pressure, volume, mass, and temperature of a given sample of gas. Together, these quantities determine the *state* of the sample of gas. The first experimental results were obtained by the

Hoarfrost has been deposited on a wire fence. Water vapor in the air has gone directly from the vapor phase to the solid phase without passing through an intermediate liquid phase.

English investigator Robert Boyle (1627–1691), who studied the changes in volume of a gas as the pressure was varied. In order to arrive at a physically significant law, Boyle simplified the problem by doing his experiments under controlled conditions. He kept the mass m of gas constant (by having no leaks in the container), and he kept the temperature T constant. Under such circumstances, the relation between pressure P and volume V has the simple form known as *Boyle's law*:

$$\left(\begin{array}{c}\textbf{temperature and}\\ \textbf{mass constant}\end{array}\right) \quad PV = \text{constant} \quad (15\text{-}1)$$

or

$$P = \frac{\text{constant}}{V}$$

In other words, the pressure of a given mass of ideal gas is inversely proportional to the volume if the temperature remains constant. The graph of P versus V is a hyperbola (Fig. 15-1), and as we increase the pressure, passing from state 1 to state 2, Eq. 15-1 tells us that the product PV is constant. Boyle's law is often written as

$$\left(\begin{array}{c}\textbf{temperature and}\\ \textbf{mass constant}\end{array}\right) \quad P_1V_1 = P_2V_2$$

As elsewhere, the subscript 1 refers to the initial state, and the subscript 2 refers to the final state.

In applying Boyle's law we have considerable freedom in our choice of units for pressure and volume. For convenience, we often take atmospheric pressure to be 10^5 N/m^2, or 100 kPa

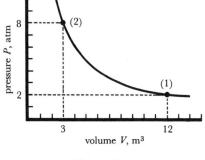

Figure 15-1

(kilopascals); a more precise conversion is 1 atm = 101.3 kPa (Sec. 12-2 on page 268).

Example 15-1

An oxygen tank of volume 0.06 m^3 is filled to a pressure of 20,000 kPa. What was the original volume of the oxygen, when it was at a pressure of 1 atm? (Assume no temperature change.)

In using Boyle's law, it is necessary to use similar units on each side of the equation. Since 1 atm is about 100 kPa, the final pressure is

$$P_2 = (20{,}000 \text{ kPa})\left(\frac{1 \text{ atm}}{100 \text{ kPa}}\right) = 200 \text{ atm}$$

Now we apply Boyle's law:

$$P_1V_1 = P_2V_2$$
$$(1.00 \text{ atm})V_1 = (200 \text{ atm})(0.06 \text{ m}^3)$$
$$V_1 = \boxed{12 \text{ m}^3}$$

Solving the problem in another way will show that it is unimportant just which units are chosen for P (or V), provided they are the same (for similar quantities) on both sides of the equation:

$$P_1V_1 = P_2V_2$$
$$(100 \text{ kPa})V_1 = (20{,}000 \text{ kPa})(0.06 \text{ m}^3)$$
$$V_1 = \boxed{12 \text{ m}^3}$$

It is essential to note that the mass of oxygen has not changed. The same mass is involved both at 20,000 kPa and at 100 kPa.

The value of the constant in Boyle's law depends on several factors. Before considering the effects of mass and temperature, we must consider some of the consequences of the molecular nature of matter.

15-3 The Masses of Individual Atoms— Avogadro's Number

So far, we have described atoms in relation to each other. When we say that ^{238}U has an atomic weight of 238 units and ^{56}Fe has an atomic weight of 56 units, we are merely saying that

$$\frac{\text{Weight of one } ^{238}\text{U atom}}{\text{Weight of one } ^{56}\text{Fe atom}} = \frac{238}{56}$$

This knowledge of the *ratios* of atomic weights comes from the data of chemistry—from careful experiments in which compounds containing uranium and iron are weighed. Now, bodies that have more inertia also weigh more, in strict proportion. Therefore, the chemical data allow us to say that

$$\frac{\text{Mass of one } ^{238}\text{U atom}}{\text{Mass of one } ^{56}\text{Fe atom}} = \frac{238}{56}$$

In this way the *ratios* of atomic masses were known many years ago. It is quite a different thing, however, to find out the mass of an individual atom.

Suppose we have 56 g of iron (^{56}Fe) in one lump and 238 g of uranium (^{238}U) in another. These two chunks of matter are each said to contain 1 *mole* of the metal; this term simply means that in each case the mass in grams numerically equals the atomic weight. We also know that *each chunk contains the same number of atoms*. To see that this statement is true, let us imagine two large sealed boxes A and B, with box A containing golf balls and box B containing marbles. The total mass of box A is measured to be 238 kg, and the total mass of box B is 56 kg. Suppose we reach in and remove one golf ball from box A and find its mass to be 47.6 g. Then we pull out one marble from box B and find that its mass is 11.2 g. The ratio of the two masses is 47.6 g:11.2 g, which may also be written as 238:56. We conclude that there are as many golf balls in A as there are marbles in B, the greater total mass of box A being exactly accounted for by the greater mass of the individual balls. Similarly, there must be some definite number of atoms in a mole of any element. This number is called *Avogadro's number*, N_A.

For a chemical compound, the word *mole* is used to refer to an amount whose mass in grams

equals the *molecular* weight.* A mole of molecules contains N_A molecules, just as a mole of atoms contains N_A atoms. For instance, the atomic weight of carbon is 12, and that of oxygen is 16. The molecular weight of CO_2 is $12 + 16 + 16 = 44$. Therefore 1 mole, or 44 g, of CO_2 contains N_A molecules; there are N_A atoms of C and $2N_A$ atoms of O, a total of $3N_A$ atoms in the N_A molecules of CO_2.

Although the *existence* of Avogadro's number was known as soon as the existence of identical atoms was postulated, the *measurement* of the number is difficult, and Avogadro himself had no clear idea of its value. One early method was to let a drop of oil spread out on the surface of water, forming a thin film. As the film spreads, it gets thinner, until finally it becomes a *monolayer*, a layer one molecule thick. We know the original volume of the drop (from its mass and density) and the area of the film, so the thickness of the monolayer can be found. This gives the diameter of a single molecule, and hence the volume of a single molecule can be estimated. In this way the mass of a single molecule can be estimated, and therefore the number of molecules in a known mass of the oil can be found. Modern measurements, more precise but less direct, give for Avogadro's number

$$N_A = 6.02 \times 10^{23} \text{ molecules per mole}$$

This is also the number of atoms per mole of an element.[†]

The mass m of a single iron atom can be found by a proportion:

$$\frac{56 \text{ g}}{6.02 \times 10^{23} \text{ atoms}} = \frac{m}{1 \text{ atom}}$$

$$m = 9.3 \times 10^{-23} \text{ g}$$

---

* Molecular weight is relative molecular mass and has no units (see Sec. 1-7). Thus the molecular weight of H_2 is 2.0, and the mass of a mole of H_2 is 2.0 g. The mole (for which the symbol is mol) is one of the seven base units in SI.
[†] A more precise value of Avogadro's number is $N_A = 6.022094 \times 10^{23}$ atoms per mole, with some uncertainty in the last digit.

The cgs system is commonly used for such calculations, since the mass of a mole is usually expressed in grams.

Example 15-2

How many atoms are contained in a speck of colloidal silver that can just be seen under a microscope? Assume the speck to be a cube 10^{-6} cm on an edge. The atomic weight of silver is 107, and its specific gravity is 10.5.

First we find the mass of the silver from its volume and density.

$$\text{Volume} = (10^{-6} \text{ cm})^3 = 10^{-18} \text{ cm}^3$$

$$\text{Mass} = (\text{density})(\text{volume})$$

$$= (10.5 \text{ g/cm}^3)(10^{-18} \text{ cm}^3)$$

$$= 10.5 \times 10^{-18} \text{ g}$$

We make a proportion, knowing that one mole (107 g) has 6.02×10^{23} atoms:

$$\frac{6.02 \times 10^{23} \text{ atoms}}{107 \text{ g}} = \frac{N \text{ atoms}}{10.5 \times 10^{-18} \text{ g}}$$

$$N = \left(\frac{10.5 \times 10^{-18} \text{ g}}{107 \text{ g}}\right)(6.02 \times 10^{23} \text{ atoms})$$

$$= 0.59 \times 10^5 \text{ atoms} = \boxed{5.9 \times 10^4 \text{ atoms}}$$

This tiny piece of silver contains 59,000 atoms

15-4   The Effect of Mass

The value of the "constant" in Boyle's law depends on several factors. First, let us imagine that the quantity of gas changes, temperature and volume remaining fixed. The pressure is caused by the impact of molecules as they strike the walls of the container. It is reasonable to expect that doubling the mass (that is, doubling the number of gas molecules) will cause the number of impacts per second to double, and hence the pressure will be doubled. Experiment verifies that the product PV is proportional to the mass of gas. If m is the mass of gas, we may

say that

$$\left(\begin{array}{c}\text{at constant} \\ \text{temperature}\end{array}\right) \qquad \frac{P_1 V_1}{m_1} = \frac{P_2 V_2}{m_2} \qquad (15\text{-}2)$$

In other words, if m is increased, then so is PV; PV/m remains the same.

Example 15-3

A compressed-air storage tank whose volume is 112 liters contains 3.00 kg of air at a pressure of 18 atm. How much air would have to be forced into the tank to increase the pressure to 21 atm, assuming no change in temperature?

$$\frac{P_1 V_1}{m_1} = \frac{P_2 V_2}{m_2}$$

$$\frac{(18 \text{ atm})(112 \text{ liters})}{3.00 \text{ kg}} = \frac{(21 \text{ atm})(112 \text{ liters})}{m_2}$$

$$m_2 = 3.50 \text{ kg}$$

The mass of air that must be forced in is

$$3.50 \text{ kg} - 3.00 \text{ kg} = \boxed{0.50 \text{ kg}}$$

When dealing with different kinds of gas—for example, hydrogen as compared with oxygen—experiment shows that we must use equal *numbers of molecules* rather than equal masses. For instance, there are actually more molecules in 8 g of H_2 (whose molecular weight is 2.0) than in 80 g of O_2 (whose molecular weight is 32). The 8 g of H_2 represents $(8 \text{ g})(1 \text{ mole}/2 \text{ g}) = 4$ moles, and hence has $4 \times 6.02 \times 10^{23}$ molecules. Similarly, 80 g of O_2 is only $80/32 = 2.5$ moles, and contains only $2.5 \times 6.02 \times 10^{23}$ molecules. All else being equal, we expect that the pressure exerted by the 8 g of H_2 would be greater, in the ratio of 4 to 2.5. Thus we can say that

$$\left(\begin{array}{c}\text{at constant} \\ \text{temperature}\end{array}\right) \qquad \frac{PV}{n} = \text{constant} \qquad (15\text{-}3)$$

where n is the number of moles of gas present.

Example 15-4

A certain tank contains 6.4 kg of oxygen at a pressure of 4.20 atm. What would be the pressure if the oxygen were pumped out and replaced by 6.6 kg of carbon dioxide (CO_2) at the same temperature?

The molecular weight of O_2 is $16 + 16 = 32$; the molecular weight of CO_2 is $12 + 2(16) = 44$.

$$6.4 \text{ kg of } O_2 = (6400 \text{ g})\left(\frac{1 \text{ mol}}{32 \text{ g}}\right)$$

$$= 200 \text{ mol}$$

$$6.6 \text{ kg of } CO_2 = (6600 \text{ g})\left(\frac{1 \text{ mol}}{44 \text{ g}}\right)$$

$$= 150 \text{ mol}$$

$$\frac{P_1 V_1}{n_1} = \frac{P_2 V_2}{n_2}$$

Here $V_1 = V_2$, since it is the same tank in both cases.

$$\frac{(4.20 \text{ atm})(V_1)}{200 \text{ mol}} = \frac{(P_2)(V_2)}{150 \text{ moles}}$$

$$P_2 = \boxed{3.15 \text{ atm}}$$

What we are really saying here is that at a given temperature and volume, the pressure exerted by any single molecule does not depend on the type of molecule involved. This remarkable fact illustrates the essential simplicity of gases as compared with liquids or solids. We shall see reasons for this simplicity in Chap. 16, "The Kinetic Theory of Gases."

15-5 The Effect of Temperature

Finally, in our approach to a general gas law, we ask, what are the effects of temperature changes on the pressure exerted by a confined gas, the volume and mass remaining constant? Experiment shows (Fig. 15-2) that the pressure increases *uniformly* as the temperature increases (with certain exceptions to be noted later). The graph of pressure versus temperature is thus a straight line. Note that in Fig. 15-2 the graph of

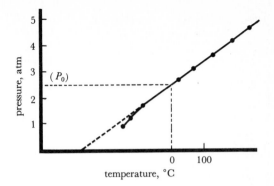

Figure 15-2 Pressure of 100 g of CO_2 gas in a container of volume 20 liters.

P versus T is shown as a dashed line below a certain temperature. By this we mean to indicate that the pressure would become zero at some temperature *if it continued to decrease at the same rate as it does near room temperature.* Any ideal gas is assumed to behave in this ideal way, but in actual practice, before the gas pressure becomes zero, the gas liquefies at some temperature and then, of course, no longer even *is* a gas.

As the temperature of the gas decreases, the pressure deviates somewhat from the straight-line relationship. We expect such behavior on the basis of what we know about molecules. The slower-moving molecules are more affected by mutual cohesive forces, and the pressure exerted by them drops off a little. Finally, when the temperature is low enough and the molecules slow enough, the cohesive forces cause the molecules to stick together as a liquid.

15-6 The Absolute Temperature Scale

The mathematical equation of the straight line in Fig. 15-2 may be written as

$$P_T = P_0(1 + bT_C) \qquad (15\text{-}4)$$

where P_T is the pressure at a Celsius temperature T_C, P_0 is the pressure at $0°C$, and b is the

Table 15-1   Pressure Coefficients*
for Some Gases, in $(°C)^{-1}$

air	0.00366
carbon dioxide	0.00371
hydrogen	0.00366
oxygen	0.00367
helium	0.00367
ethane	0.00375
nitrogen	0.00367

* Measured at room temperature with original pressure
about 1 atm.

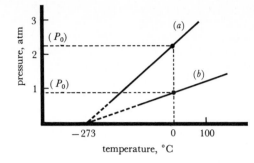

Figure 15-3   At absolute zero, the pressure exerted
by any ideal gas would be zero. (a) 4 kg of H_2 in
20 m^3. (b) 36 kg of O_2 in 30 m^3.

pressure coefficient for the gas. Experiment
shows that the constant b is practically the
same for all gases, being about 0.00366 $(°C)^{-1}$
for such gases as hydrogen that are most nearly
"ideal" (Table 15-1). The differences that do
exist are minor and show that no gas is strictly
ideal. This being so, we can rework Eq. 15-4
into a form that makes calculations very simple.
Using the numerical value of 0.00366 $(°C)^{-1}$ for
b (the same for all ideal gases),

$$P_T = P_0(1 + 0.00366T_C) \qquad (15\text{-}5)$$

At what temperature would the pressure be
zero if the straight line continued to be a true
description of the course of events? From Eq.
15-5, it is evident that to make $P_T = 0$, we must
have $(1 + 0.00366T_C) = 0$. That is, the required
temperature is

$$T_C = -\frac{1}{0.00366 \; (°C)^{-1}} = -273°C$$

Thus there is a certain temperature, *the same
for all ideal gases*, at which the pressure would
become zero *if* the gas behaved "ideally" all the
way down. We call this temperature *absolute
zero*. In Fig. 15-3 the different masses of different
gases (*a* and *b*) are enclosed in containers of
different volume, but the graphs each extend
toward the same point, zero pressure at $-273°C$.
By measuring the pressure of a gas at two
temperatures (which may be 0°C and 100°C if

desired) and extrapolating the graph down to
zero pressure, it is possible to determine absolute zero in the laboratory without any risk
of frostbite. Careful experiments have given
$-273.15°C$ as the value of this important constant, but we shall usually round the value off
to $-273°C$.

Absolute temperature is given by the Celsius
temperature (T_C) plus 273° and is denoted by
T_K, or more simply by T. Absolute temperature,
measured in kelvins (K), represents the number
of degrees that a body is above absolute zero.
On the absolute scale, the size of the degree is
the same as on the Celsius scale, and on both
scales there are 100 degrees between the ice
point and the steam point.

The symbol K is not written with a degree
mark, because the kelvin is a unit of measure.
It makes sense to say that a body at 300 K is
twice as hot as a body at 150 K, just as a trip
of 300 km is twice as long as a trip of 150 km.
A similar statement that a body at 10°C is twice
as hot as a body at 5°C would be meaningless,
since the zero of the Celsius scale is arbitrary.
The kelvin is one of the seven base units in SI.

The concept of absolute zero has deep significance, as we shall see in Chap. 17. However,
for the present we are defining absolute zero
merely in order to get a simpler equation
describing the effect of temperature on the
pressure of an ideal gas. A little algebra is

needed here. Since $b = 1/273$, Eq. 15-5 can be written as

$$P_T = P_0\left(1 + \frac{1}{273}T_C\right) \qquad (15\text{-}6)$$

$$P_T = P_0\left(\frac{273 + T_C}{273}\right) = \frac{P_0}{273}(273 + T_C)$$

Calling $0°C$ (273 K) equal to T_0, and remembering that $273° + T_C = T$, we can write

$$P_T = \left(\frac{P_0}{T_0}\right)T \qquad (15\text{-}7)$$

$$\frac{P_T}{T} = \frac{P_0}{T_0} \qquad (15\text{-}8)$$

We see that whereas the pressure *depends on* the Celsius temperature T_C (Eq. 15-6), it is actually *proportional to* the absolute temperature T (Eq. 15-7). This is the reason for using the absolute temperature in the gas laws—it allows a simple proportion.

The proportionality between P_T and T can be clearly seen directly from Fig. 15-4. Since the pressure versus temperature graph for any particular amount of gas in a given volume is a straight line, triangles *ABC* and *AED* are similar. Hence $P_T/T = P_0/T_0$. This proportion is not true unless absolute temperatures are used and, of course, is strictly true only for ideal gases.

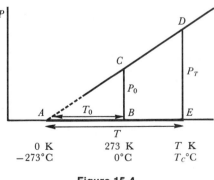

Figure 15-4

Example 15-5

The pressure gauge of a truck tire reads 200 kPa (about 30 lb/in.^{2}) on a cold day when the thermometer stands at $-3°C$. What will the pressure reading be after the tire has stood for a while in a heated garage at $27°C$?

First, we take account of atmospheric pressure, which is about 100 kPa. If the gauge read zero, the air in the tire would still be at atmospheric pressure; at a gauge reading of 200 kPa, the total (absolute) pressure in the tire is 200 kPa + 100 kPa = 300 kPa. Next, we convert both Celsius temperatures to absolute temperatures by adding $273°$: $-3°C$ is 270 K, and $27°C$ is 300 K.

Since the mass and volume of the air remain constant, the pressure is proportional to the absolute temperature.

$$\frac{P_2}{T_2} = \frac{P_1}{T_1}$$

$$\frac{P_2}{300\text{ K}} = \frac{300\text{ kPa}}{270\text{ K}}$$

$$P_2 = 333\text{ kPa}$$

This is the total pressure; the gauge will read

$$333\text{ kPa} - 100\text{ kPa} = \boxed{233\text{ kPa}}$$

We have used Eq. 15-6 to describe the effect of temperature on the *pressure* of a gas (at constant volume). A very similar equation describes the effect of temperature on the *volume* of a gas (at constant pressure). As the temperature decreases, both the pressure and the volume of an ideal gas must tend toward zero at the *same* temperature (absolute zero). Thus,

$$V_T = V_0(1 + \tfrac{1}{273}T_C) \qquad (15\text{-}9)$$

The same factor, $(1 + \tfrac{1}{273}T_C)$, applies both to pressure changes and to volume changes of an ideal gas. We see now that it is no accident that the coefficients of volume expansion for most gases (Table 13-1) are almost equal to 0.00366 $(°C)^{-1}$ $(\tfrac{1}{273}\text{ K}^{-1})$, the same as the pressure coefficients in Table 15-1.

15-7 The General Gas Law

We have seen (Eq. 15-3) that PV/n is a constant for any given temperature, but the constant depends on the temperature. We have also seen (Eqs. 15-7 and 15-8) that the pressure is proportional to the absolute temperature, other things being equal. Combining these ideas, we can write the *general gas law*

$$\frac{PV}{nT} = R \qquad (15\text{-}10)$$

where P is the pressure, V is the volume, n is the number of moles, T is the absolute temperature, and R is a new constant. Here at last we have a constant that is really constant—not dependent on the mass of gas, its chemical formula, or the temperature. We denote this *universal gas constant* by the symbol R. If P is in N/m^2 or Pa, V in m^3, and T in K, then

$$R = 8.314 \ \text{J/mol} \cdot \text{K}$$

This is the SI value of the universal gas constant.*

For many purposes we shall not actually have to use *any* number for R if we remember one combination of P, n, T, and the corresponding experimental value of V. The following set of values is called *standard temperature and pressure* (STP) for 1 atm pressure and 0°C (273 K):

$$
\begin{aligned}
P_0 &= 1 \text{ atm} \\
n_0 &= 1 \text{ mol} \\
T_0 &= 273 \text{ K} \\
V_0 &= 22.4 \text{ liters}
\end{aligned}
$$

Several examples will make clear how to use the general gas law in various systems of units. In all cases, *absolute* temperatures are used.

---

* If P is in atm, V in liters, and T in K, then R has the value 0.0821 liter·atm/mol·K.

Example 15-6

If 4 moles of gas exerts a pressure of P_1 atm when confined in a volume of 40 m^3 at 300°C, what would be the pressure of 200 moles of the same gas placed in a 100-m^3 tank at 600°C?

$$\frac{P_1 V_1}{n_1 T_1} = \frac{P_2 V_2}{n_2 T_2}$$

$$\frac{(P_1 \text{ atm})(40 \text{ m}^3)}{(4 \text{ mol})(573 \text{ K})} = \frac{(P_2)(100 \text{ m}^3)}{(200 \text{ mol})(873 \text{ K})}$$

$$P_2 = \boxed{30.5 P_1 \text{ atm}}$$

Example 15-7

How much helium gas (molecular weight 4.00) is contained in a high-altitude balloon whose volume is 5.00 m^3, if the temperature is -23°C and the pressure is 30 cm Hg?

Here we use the set of values for standard conditions to fill in one side of the gas-law equation, remembering that 1 atm is 76 cm Hg pressure and 22.4 liters = 0.0224 m^3. The unknown is n_1, the number of moles of helium.

$$\frac{P_0 V_0}{n_0 T_0} = \frac{P_1 V_1}{n_1 T_1}$$

$$\frac{(76 \text{ cm Hg})(0.0224 \text{ m}^3)}{(1 \text{ mol})(273 \text{ K})} = \frac{(30 \text{ cm Hg})(5.00 \text{ m}^3)}{(n_1)(250 \text{ K})}$$

$$n_1 = 96.2 \text{ moles}$$

The mass of helium is

$$(96.2 \text{ mol})\left(\frac{4.00 \text{ g}}{1 \text{ mol}}\right) = \boxed{385 \text{ g}}$$

Example 15-8

Solve the preceding example, assuming the balloon to be filled with hydrogen instead of helium. The molecular weight of hydrogen (H_2) is 2.02.

Just as before, we calculate that there are 96.2 moles of gas in the balloon. However, since the molecular weight is different, the mass of gas is also different.

The mass of hydrogen is

$$(96.2 \text{ mol})\left(\frac{2.02 \text{ g}}{1 \text{ mol}}\right) = \boxed{194 \text{ g}}$$

The *ratio method* of using the gas law is easily applied and appeals to common sense. Starting with

$$\frac{P_1 V_1}{n_1 T_1} = \frac{P_2 V_2}{n_2 T_2} \qquad (15\text{-}11)$$

we solve algebraically for the desired unknown. For instance, if we wish to compute the new pressure, the equation becomes

$$P_2 = P_1 \left(\frac{V_1}{V_2}\right)\left(\frac{T_2}{T_1}\right)\left(\frac{n_2}{n_1}\right) \qquad (15\text{-}12)$$

We think of the new pressure P_2 being related to the old pressure P_1 through a series of "ratio factors" such as V_1/V_2. Each ratio factor is dimensionless if similar units are used in numerator and denominator. It is unnecessary to derive the formula each time; common sense will tell you whether to use V_1/V_2 or V_2/V_1, and similarly for the other ratio factors, as the following examples show.

_____ **Example 15-9**

A balloon containing 2 m³ of hydrogen at 10°C and 1 atm pressure rises to an altitude where the temperature is $-30°C$ and the pressure is 0.2 atm. Assuming that the balloon bag is free to expand, what is its new volume? We first write $V_2 =$

$(2 \text{ m}^3)\left(\underline{}\right)\left(\underline{}\right)$ where the blank spaces are

to be filled in from the given data. We know that decreasing the temperature makes the volume *less*, so the temperature ratio factor must be less than 1. Therefore we write $\left(\dfrac{243 \text{ K}}{283 \text{ K}}\right)$ rather than $\left(\dfrac{283 \text{ K}}{243 \text{ K}}\right)$. Similarly, we know that decreasing the pressure causes the volume to *increase*, so the pressure ratio is greater than 1. Hence we use $\left(\dfrac{1 \text{ atm}}{0.2 \text{ atm}}\right)$ for the

pressure ratio. We now fill in the ratios and solve:

$$V_2 = (2 \text{ m}^3)\left(\frac{243 \text{ K}}{283 \text{ K}}\right)\left(\frac{1 \text{ atm}}{0.2 \text{ atm}}\right) = \boxed{8.59 \text{ m}^3}$$

_____ **Example 15-10**

What pressure is exerted by 64 g of oxygen (O_2) confined at 40°C in a flask of volume 5 liters?

If we had 32 g of O_2 (1 mol) in 22.4 liters at 0°C, the pressure would be 1 atm, or 101.3 kPa. However, we actually have 64 g of O_2, so this mass ratio factor would itself double the pressure. Also, the temperature and volume are not standard, and each affects the pressure in its own way. We use a succession of ratio factors:

$$P = (1 \text{ atm})\left(\frac{64 \text{ g}}{32 \text{ g}}\right)\left(\frac{313 \text{ K}}{273 \text{ K}}\right)\left(\frac{22.4 \text{ liters}}{5 \text{ liters}}\right)$$

$$= \boxed{10.3 \text{ atm}}$$

It is a matter of personal preference whether you use the gas law in the form of a proportion (Eq. 15-11) or in some ratio-factor form (such as Eq. 15-12). The two methods are completely equivalent. In neither case is it necessary to use a numerical value for the gas constant R if you remember the standard volume, 22.4 liters for 1 mole of ideal gas at 0°C and 1 atm pressure.

When two or more different gases are in the same container, a *partial pressure* can be calculated for each gas; this is the pressure that each gas would exert if it alone occupied the whole container. Dalton's *law of partial pressures* states that the total pressure is the sum of the partial pressures of the component gases. This behavior of a mixture of gases is in accord with the model of an ideal gas discussed in Sec. 15-1.

15-8 Saturated Vapor Pressure

Suppose we have an open dish of water in a room at 20°C. As we shall see in Chap. 16, the average KE of the molecules depends on the temperature, being greater at 20°C than at 10°C, for instance. Because of incessant collisions, head-on as well as glancing, the molecules continually exchange energy. At *any* temperature, however, not all molecules have the same speed. If the motions are in random directions, there may be chance collisions resulting in very high or very low speeds. For instance, two molecules might "gang up on" a third molecule to give it a larger-than-average speed. At any instant there must be some molecules that are moving slowly or not at all and others that have acquired exceptionally high speeds. The faster molecules are able to escape from the liquid, overcoming the cohesive forces pulling them back to the surface, in much the same way that a spacecraft leaves the earth when it is fired upward at sufficient speed. In each case the initial KE must be sufficient to supply the PE gained by the molecule (or spacecraft) as it moves against the cohesive molecular force (or gravitational force, in the case of the spacecraft). In this way we account for the evaporation that takes place when a bowl of water is left standing in the open. We can also explain why evaporation is a cooling process: the molecules that leave are the faster ones, and the *average* KE of those left behind decreases. In arid regions, desert travelers carry drinking water in porous bags so that it will be cooled by evaporation. Looking at it from another viewpoint, we may say that the heat of vaporization required to change the liquid to a vapor is supplied by the water in the bag. Since heat flows out of the water without any work being done, the water's internal energy decreases and its temperature decreases. Evaporation can also take place from a solid substance, although the cohesive forces are much stronger than in liquids, and fewer molecules leave the surface.

Let us next imagine uncovering our bowl of water inside a cylinder that has been sealed off after all air molecules have been removed by a vacuum pump (Fig. 15-5a). After a short while, which may be less than 0.001 s, the empty space becomes filled with water molecules crisscrossing and bouncing off the walls of the enclosure.

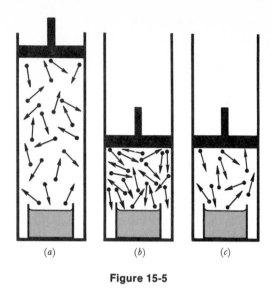

(a)   (b)   (c)

Figure 15-5

Some of these molecules will find their way back into the liquid; the process builds up until the number leaving per second just equals the number bouncing back in. The space is then said to be *saturated* with water vapor. The molecules in the cylinder exert a pressure that is called the *saturated vapor pressure.*

If the volume of the cylinder is now suddenly reduced (Fig. 15-5b), the pressure exerted by the gas is momentarily increased. The vapor is now *supersaturated*, but this is not a stable condition. The number of molecules leaving the surface per second is the same as before (since the temperature has not changed, and the average molecular KE is the same). However, the number coming back into the liquid per second has been increased because of the increased density of vapor. There is a net flow of molecules back into the liquid, and the vapor pressure returns to the original saturated value, at which there is a balance between molecules leaving and molecules returning (Fig. 15-5c). Our conclusion is that saturated vapor pressure and density depend only on the temperature of the liquid, and not at all on the volume of the container. In Table 15-2 are listed values of the saturated vapor pressure for water at various tempera-

Table 15-2 Saturated Vapor Pressure and Density of Water Vapor

Temperature, °C	Saturated Vapor Pressure, torr	Saturated Density of Water Vapor, g/m³
0	4.58	4.85
10	9.21	9.40
20	17.55	17.30
30	31.86	30.37
⋮	⋮	⋮
90	526	424
99	733	579
100	760	598
101	788	618
110	1,074	827
200	11,650	7,840
300	64,400	45,600
374	166,000	318,000

tures; pressures are expressed in torr (1 torr = 1 mm Hg = $\frac{1}{760}$ atm).

In actual practice, we are more likely to deal with a cup of water on a dining-room table, with the added complication that air molecules fill up most of the space in the room. Collisions with air molecules make the diffusion and return of water molecules to the cup much slower; equilibrium is reached only after several hours, rather than practically instantaneously as in the previous illustration. If a cup of water at 20°C is placed on the table in a sealed room full of dry air, and if the pressure of the air is originally 760.0 torr (760.0 mm Hg), then in due course of time the total pressure in the room will rise to

$$760.0 + 17.6 = 777.6 \text{ torr}$$

What is more probable is that the room is not airtight. If outside atmospheric pressure is 760 torr, there will be a slight excess of pressure in the room, which causes a flow of air and water vapor through chinks and cracks around the doors and windows. Eventually, in the steady condition, the pressure in the room will be a total of 760 torr, of which about 742 torr is due to air molecules and 18 torr is due to water molecules (see Table 15-2). The air in the room will be practically* saturated with water vapor as long as any water at all remains in the dish.

15-9 Relative Humidity

Relative humidity is the ratio of the density of the water vapor in the air at any temperature to the density that would exist if the air were saturated with water vapor at that temperature. Relative humidity has great biological importance. A warm summer day when the relative humidity is 80% or more is likely to be uncomfortable. The human skin is normally cooled by evaporation, and if the surrounding air is nearly saturated, there will be only a slight net evaporation, and hence the skin will remain warm. On the other hand, if the air is not saturated, moist tissues will tend to lose more water molecules than they receive. Generally, a relative humidity of at least 40% is desirable. Some simple calculations will show that many liters of water might have to be evaporated to "humidify" a large room in winter.

Example 15-11

An auditorium measures 10 m × 20 m × 30 m and is heated to 20°C with air that has been drawn in from the outside at 0°C. Assuming the most favorable case—that the outside air is completely saturated—what is the relative humidity in the auditorium?

At 0°C, the incoming air, which is saturated, contains 4.85 g of water in each cubic meter (Table 15-2). After it has been heated to 20°C, we have the same 4.85 g in a slightly larger volume. This volume of the heated air, which was 1 m³ at 0°C, is found from the gas law, with P and n remaining constant.

---
* Not quite saturated; after all, slightly more water molecules must be leaving the cup than returning, since a few of those leaving the cup are being lost through the chinks and crevices to the outside world.

$$\frac{P_1 V_1}{n_1 T_1} = \frac{P_2 V_2}{n_2 T_2}$$

$$\frac{1 \text{ m}^3}{273 \text{ K}} = \frac{V_2}{293 \text{ K}}$$

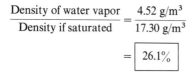

$$V_2 = (1 \text{ m}^3)\left(\frac{293 \text{ K}}{273 \text{ K}}\right) = 1.07 \text{ m}^3$$

Therefore the density of water vapor in the auditorium is 4.85 g/1.07 m³ = 4.52 g/m³. At 20°C, the saturated density is 17.30 g/m³. Hence the relative humidity is

$$\frac{\text{Density of water vapor}}{\text{Density if saturated}} = \frac{4.52 \text{ g/m}^3}{17.30 \text{ g/m}^3}$$

$$= \boxed{26.1\%}$$

Example 15-12

How much water would have to be evaporated in the auditorium of Example 15-11 in order to bring the relative humidity up to 40%?

At 20°C, a 40% relative humidity would correspond to a water-vapor density that is 40% of the saturated value, and each cubic meter would contain 0.40(17.30 g) = 6.92 g. We already have 4.52 g of water vapor in each cubic meter brought in from the outside, so we must add 2.40 g to each cubic meter. The volume of the room is 6000 m³, so the total amount of water that would have to be evaporated into the air is

$$\text{Mass of water} = (6000 \text{ m}^3)(2.40 \text{ g/m}^3)$$

$$= 14,400 \text{ g} = \boxed{14.4 \text{ kg}}$$

This is 14.4 liters of water!

To add such amounts of water to the air and to replace it several times an hour as the air is circulated is a formidable engineering problem. There is also the practical difficulty of dew formation ("sweating"). The humidified air that finds its way to a cold windowpane or a cold wall might at the lower temperature be supersaturated, and dew would form; that is, water vapor would leave the air and become liquid.

15-10   The Vapor-Pressure Curve and Change of Phase

We have discussed rather fully one type of change of phase—that from liquid to vapor (*evaporation*) and vice versa (*condensation*). Much the same sort of process is involved when a solid changes to a liquid (*melting*) or a liquid changes to a solid (*freezing*). A third process, known as *sublimation*, takes place when molecules of a solid go directly into vapor form. The "wasting away" of Dry Ice or a moth ball are examples of sublimation. The opposite process, when molecules go from vapor phase to solid phase, is called *deposition* (shown in the photograph on page 333).

To get a unified view of changes of phase, let us plot a graph of saturated vapor pressure versus temperature, using the substance H_2O as an example (Fig. 15-6). The curve AB represents the same data on vapor pressure as those presented in Table 15-2; the other segments of the graph, AC and AD, are also based on experiment. At any point such as q, e, b, or m on curve AB, liquid and saturated vapor are in dynamical

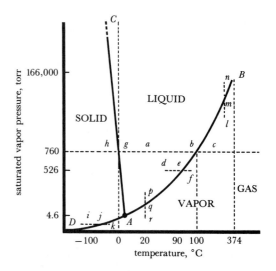

Figure 15-6   Phase diagram for H_2O; saturated vapor pressure as a function of temperature (scales exaggerated for clearness). A is the triple point; B is the critical point.

equilibrium with each other and can exist together, with a continual exchange of molecules. Likewise, curve AC gives the pressure and temperature at which solid ice and liquid water can exist together (in equilibrium), and AD gives the conditions for steady coexistence of solid ice and saturated water vapor. Point A is the *triple point*, which for H_2O is at $T = +0.01°C$ and $P = 4.58$ torr. At the triple point, solid, liquid, and gaseous H_2O can exist together, freely exchanging molecules but having no tendency to accumulate in any one of the three phases.

To illustrate the use of the graph, let us make some simple changes in temperature or pressure (one at a time).

1. At a pressure of 760 torr (760 mm Hg, standard atmospheric pressure) we start at room temperature, say 20°C. Point a on the graph corresponds to these conditions, and the substance is liquid. As we increase the temperature (keeping pressure constant all the while), we move along the line abc. The saturated vapor pressure increases until finally it reaches 760 torr when the temperature reaches 100°C. Any small bubble in the water at 20°C has by now expanded greatly (Fig. 15-7).* Bubbles rise rapidly and break the surface, and the water boils. In general, the *boiling point* of a liquid is the temperature at which the saturated vapor pressure of the liquid equals the surrounding atmospheric

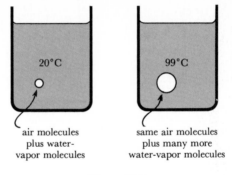

air molecules   same air molecules
plus water-   plus many more
vapor molecules water-vapor molecules

Figure 15-7

pressure. At point b we have a change of phase from liquid to vapor. Continuing to increase the temperature, we reach a point (such as c) where the substance is a vapor. With care, it is possible to heat liquid water at 760 torr to 110°C or 115°C before it violently changes to the vapor state.

2. If we heat some water on a mountaintop where atmospheric pressure is 526 torr, we will be on a line such as def. The water will boil at 90°C (point e), at which temperature the saturated vapor pressure equals atmospheric pressure. The curve AB therefore represents a boiling-point curve; the boiling point at any pressure can be read off the curve or taken from Table 15-2. Pressure cookers are a necessity at high altitudes, and even at sea level, cooking can be hastened in a pressure cooker, in which the boiling point may be as high as 125°C.

3. If we move from point a toward lower temperatures, the water freezes when we reach 0°C (point g), and we enter the region marked "solid." At point h the ice is at a temperature of, say, $-10°C$, and is unable to remain in equilibrium with liquid water if the pressure is 760 torr. Point g is the *melting point* of water at standard atmospheric pressure of 760 torr (760 mm Hg); in general, the melting point is that temperature at which the solid and liquid can exist together, with molecules leaving the solid and returning in equal numbers. Curve AC (practically a straight line) is the melting-point curve for H_2O. According to the figure, the melting point of ice *decreases* as the pressure increases, since the line

---

* The space within the bubble is saturated with water vapor at any time, since it is a closed volume in contact with liquid. At 20°C the water molecules inside a bubble exert a pressure of 17.55 torr (Table 15-2), and so by subtraction we find that the air molecules in the bubble exert a pressure of 742.45 torr, to make up the total pressure of 760 torr. As the temperature rises, the saturated vapor pressure of the water molecules increases greatly as more water molecules leave the liquid and enter the bubble. The pressure of the *air* molecules must decrease greatly, and according to the gas law this can be done only if the volume of the bubble expands greatly. The total pressure inside the bubble remains 760 torr. The pressures at the two temperatures are divided as follows: at 20°C: (17.55 torr due to water vapor) + (742.45 torr due to air); at 99°C: (733 torr due to water vapor) + (27 torr due to air).

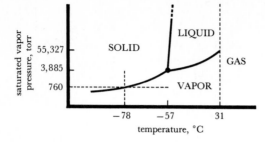

Figure 15-8 Saturated vapor pressure of CO_2 as a function of temperature (scales exaggerated for clearness).

AC slopes upward to the left. This characteristic is related to the fact that water expands when it freezes.

4. Finally, sublimation is possible along a line such as ijk, provided the temperature is low enough and the surrounding vapor pressure is low. This is not usually the case for H_2O. Carbon dioxide (CO_2) sublimes at ordinary atmospheric pressure. It is this fact that makes Dry Ice "dry." Study of the vapor-pressure diagram for CO_2 (Fig. 15-8) shows that solid CO_2 becomes vapor at $-78°C$ if the pressure is 1 atm. In fact, CO_2 can be a liquid only if the pressure is 3885 torr or more. In general, the *sublimation point* is that temperature at which the solid and vapor can exist together, with molecules leaving the solid and returning in equal numbers.

The liquid-vapor curve AB does not continue indefinitely; at a certain temperature called the *critical temperature* (374°C for water) the curve ends. Below 374°C, application of sufficient pressure will cause the vapor to liquefy, the change taking place along a line such as lmn. However, if the temperature is above 374°C, no amount of pressure is sufficient to cause formation of liquid. This is understandable, since at high temperatures the molecules move so fast that the cohesive forces are unable to form a surface. *The critical temperature is the temperature at or above which no amount of pressure, however great, will cause a gas to liquefy.* At 400°C, it would be possible to compress 1 g of water vapor to a volume so small that the density would be greater than that of lead, but it would still be a gas, and no liquid surface would form. Sometimes a substance is called a *gas* above the critical temperature and a *vapor* below the critical temperature. From this point of view, nitrogen and oxygen are gases at ordinary temperatures. Before the discovery of critical temperature, scientists tried (and failed) to make liquid oxygen by use of tremendous pressures. If oxygen is first cooled to a temperature below $-119°C$, its critical temperature, then a very moderate pressure of 50 atm is sufficient to liquefy it.*

In our discussion of vapors, we have dealt mostly with a familiar substance, H_2O, in solid, liquid, and gaseous form. All of these ideas are equally true for other substances, such as CO_2 or oxygen, but their behaviors may appear entirely different because the numerical values of critical temperature and triple point are so different. However, these are differences of degree only.

---

* If highly compressed oxygen is allowed to expand suddenly, the cohesive forces are no longer negligible, since the molecules are close together. To expand the gas requires that work be done against the cohesive forces, and as a result the gas may well be cooled below its critical temperature as its internal random KE is converted to internal random PE. The same phenomenon occurs on a lesser scale if a tank of compressed air is opened. As the air expands and rushes out, it is often noted that the valve becomes cold enough so that water vapor from the air around the tank condenses on the valve and freezes. (See photo on page 371.)

Summary If the molecules of a gas exert no appreciable cohesive forces on one another, and if the molecules are small compared with the space between them, then the gas is an ideal gas. To a large extent these conditions are fulfilled by ordinary gases at ordinary temperatures, with the result that all gases obey the same gas law regardless of chemical composition. The form of the gas law is simplified by using absolute temperature, which is equal to the Celsius temperature plus 273°. The pressure, volume, and temperature of an ideal gas are related by the general gas law, $PV/nT = R$, where the amount of gas is given by n, the number of moles, and R is the general gas constant. Boyle's law is a special case of the general gas law, with temperature held fixed.

Any gas that is below its critical temperature is called a vapor and can be liquefied if the pressure is sufficiently great. When the number of molecules that leave a liquid surface each second equals the number returning to the surface, the space adjacent to the surface is said to be filled with saturated vapor. Under these conditions, the pressure exerted by the molecules is called the saturated vapor pressure; the value of this pressure does not depend on the volume of the container, but only on the temperature. Saturated vapor pressure increases rapidly as temperature increases, and when the saturated vapor pressure equals the surrounding atmospheric pressure, the liquid boils.

Relative humidity is given by the ratio of the density of water vapor in the air to the density of water vapor that would be contained if the air were saturated.

The saturated vapor-pressure diagram for a substance gives much information about changes of phase from liquid to vapor (boiling), solid to liquid (melting), and solid to vapor (sublimation). At the triple point the three phases can coexist and remain in equilibrium with each other. If the temperature is above the critical temperature, no amount of pressure will cause a gas to form a liquid surface.

Check List

ideal gas	$PV/nT = R$	melting point
state	partial pressure	sublimation point
Boyle's law	saturated vapor	triple point
mole	saturated vapor pressure	critical temperature
Avogadro's number	relative humidity	vapor
absolute temperature	boiling point	gas
$T = 273 + T_C$		

Questions

15-1 Molecules of different gases have different sizes. Explain why this fact is of little practical importance as far as the gas laws are concerned.

15-2 How many molecules are there in a nanomole of carbon dioxide?

15-3 Which has the greater number of atoms—a kilogram of lead or a kilogram of iron? (See Periodic Table, Appendix Table 6, for the atomic weights.)

15-4 Starting with the general gas law (Eq. 15-10), write a simpler law for the relation of volume and absolute temperature, when mass and pressure remain constant. (This is called the law of Charles and Gay-Lussac.) The same formula can be derived starting with Eq. 15-9.

15-5 Explain why the saturated vapor pressure of water is so much greater at 100°C than at 20°C (about a 43:1 ratio—see Table 15-2), whereas the ratio of absolute temperatures is 373:293 (only 1.27:1). In your explanation, solve the general gas law for P, and tell which factor has increased so much.

15-6 Why is evaporation a cooling process?

15-7 How is it possible to have boiling water at room temperature (20°C)? Would a flask containing such boiling water be hot to the touch?

15-8 Use Fig. 15-6 to explain what would happen to ice at $-5°C$ if sufficient pressure were applied.

15-9 Explain in words what happens when water at 20°C undergoes a change from a state p to a state r along a line such as pqr in Fig. 15-6.

15-10 Two thermometers are hung side by side in a room that is not at 100% relative humidity. One thermometer has a moist cloth wrapped around its bulb. Are the readings of the two thermometers identical?

15-11 Carbon dioxide is usually shipped in tanks at a pressure of about 65 atm at room temperature. Is the material in the tank solid, liquid, or gas?

15-12 What is the condition (solid, liquid, vapor, or gas) of CO_2 at   (a) 0°C and 760 torr; (b) 20°C and 3000 torr;   (c) 20°C and 50,000 torr?   (d) Is there any temperature at which it would be possible to have liquid CO_2 in a beaker open to the atmosphere?

15-13 What is the difference between a gas and a vapor?

15-14 What is meant by the statement: "The boiling point of liquid hydrogen is $-253°C$"?

15-15 A biological preparation is to be dried by boiling off the water in a tissue sample, but to avoid damaging the specimen the temperature should not be above 30°C. How can this be done?

15-16 Air is (principally) a mixture of nitrogen (boiling point $-196°C$) and oxygen (boiling point $-183°C$). When first made, liquid air contains both of these substances. Is a flask of liquid air at the temperature of liquid nitrogen or at the temperature of liquid oxygen (assume that it is boiling)? After standing for a while, one of the materials boils away, leaving the other substance in almost a pure state. Which of the two substances is contained in a flask of liquid air that has stood for a while?

MULTIPLE CHOICE

15-17 Boyle's law describes the behavior of an ideal gas   (a) only if the mass is constant; (b) only if the temperature is constant;   (c) both of these.

15-18 A tire pressure gauge reads 186 kPa on a day when atmospheric pressure is 100 kPa. The absolute pressure in the tire is   (a) 86 kPa;   (b) 186 kPa;   (c) 286 kPa.

15-19 At absolute zero, an ideal gas would have   (a) no pressure;   (b) no volume;   (c) both of these.

15-20 The universal gas constant R can be found from the pressure P, temperature T, number of moles n, and volume V according to the relation   (a) $R = PVT/n$;   (b) $R = PV/nT$; (c) $R = PT/nV$.

15-21 The number of atoms in a mole of CO_2 is about   (a) 6×10^{23};   (b) 12×10^{23}; (c) 18×10^{23}.

15-22 The saturated vapor density in the space above a confined liquid depends on (a) temperature;   (b) volume;   (c) both of these.

Problems

[*Note:* Assume atmospheric pressure to be 100 kPa (100 kN/m²) unless otherwise stated.]

15-A1 The cork of a popgun is inserted so tightly that a pressure of 4 atm is required to dislodge it. Air is admitted through a hole at A, which is 24 cm from the cork at B

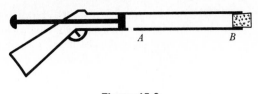

Figure 15-9

(Fig. 15-9). How far from A is the piston when the cork pops out, assuming no temperature change?

15-A2 Water is pumped into the bottom of a sealed, vertical storage tank whose volume is 24 m³. When the pressure gauge on the tank shows that the air pressure in the tank is 300 kPa above atmospheric pressure, what is the volume of water in the tank?

15-A3 A large research balloon whose volume is 400 m³ is to be filled with helium at atmospheric pressure. The helium is stored in cylinders of volume 0.05 m³ at a pressure of 20 atm. How many cylinders are required?

15-A4 How many moles of H_2O are there in a glass that contains 180 g of water?

15-A5 How many moles of H_2O are there in a piece of ice that has a mass of 180 g?

15-A6 (a) How many atoms of hydrogen are contained in 180 g of water, H_2O? (b) How many atoms of oxygen are contained in this same amount of water?

15-A7 A tank containing 0.1 m³ of nitrogen at room temperature and a pressure of 6×10^6 N/m² is connected through a valve to an empty tank of volume 0.4 m³. The valve is opened and the nitrogen allowed to expand. After the system returns to room temperature, what is the pressure in the tank?

15-A8 Convert to K: 40°C; −60°C; 68°F; 212°F.

15-A9 Given that a certain mass of ideal gas occupies a volume of 24 liters at 300 K and 1 atm, compute the volume of the same mass of the same gas at 1500 K and 3 atm.

15-A10 A certain amount of an ideal gas occupies 320 liters at 200 K and 2 atm. What pressure will the same mass of gas exert if confined in 80 liters at 500 K?

15-A11 A sealed tank contains air at 27°C and a pressure of 4×10^5 N/m². What will be the pressure if the temperature goes up to 77°C?

15-A12 A truck tire contains air at 5°C. A gauge reads 200 kPa above atmospheric pressure. What will the gauge read if the temperature rises to 30°C?

15-A13 On a hot, steamy day, the temperature is 30°C and there is 28 g of water vapor in each cubic meter of air. What is the relative humidity?

15-A14 How much water vapor is contained in each cubic meter of air if the temperature is 20°C and the relative humidity is 30%?

15-A15 At what temperature does water boil if atmospheric pressure is 526 torr?

15-A16 What is the atmospheric pressure on a mountaintop where water boils at 99°C?

15-B1 The atomic weight of $_1^1H$ (ordinary hydrogen) is 1.008. Use Avogadro's number to compute the mass of a proton.

15-B2 Compute the mass of a billion atoms of the gold isotope $_{79}^{197}Au$.

15-B3 How many atoms are in a steel (iron) paper clip of volume 0.02 cm³?

15-B4 A bottle contains 500 cm³ of ethyl alcohol (CH_3CH_2OH). (a) What is the mass of the liquid? (b) How many molecules of alcohol are in the bottle?

15-B5 At a symposium in 415 B.C., Socrates appeased the gods by ritualistically pouring a libation of wine onto the ground. The libation contained 50 g of water, and by now these water molecules have been perfectly mixed with all the water (10^{21} kg) in the oceans, rivers, and clouds of the earth. The average human body has about 40 kg of water. (*a*) How many water molecules are in a human body? (*b*) About how many of the water molecules in Socrates' libation are now in your own body?

15-B6 As a result of burning coal (which contains small concentrations of nickel and other metals), the air above some cities contains about 0.06 mg of nickel per cubic meter of air. How many nickel atoms are taken into the body when a person takes a breath and inhales 450 cm^3 of air?

15-B7 It has been found that only 8×10^{-20} g of the powerful sex attractant bombykol ($C_{16}H_{30}O$) will cause a response in the silkworm moth *Bombyx mori*. How many molecules are contained in this amount of the material?

15-B8 How many objects are in box *A* and in box *B* (page 335)?

15-B9 A vertical cylindrical tank 42 cm tall and 2.00 cm in radius is open at the top. Atmospheric pressure is 1.013×10^5 N/m^2. A close-fitting cylindrical plug of mass 6.00 kg is inserted at the top and falls inside. If the temperature of the trapped air does not change, how far from the bottom of the cylinder is the base of the plug when it comes to rest?

15-B10 Writing Boyle's law in the form $PV = k$, show by dimensional analysis that the constant k has the same dimensions as work.

15-B11 A mercury barometer column (Fig. 12-8) has an air bubble of radius 1.00 mm, 60 cm below the top of the column. How far below the top of the column will the bubble be when it has risen to a point where its radius is 2.00 mm? Assume constant temperature, and ignore effects due to surface tension.

15-B12 A vertical glass capillary tube is held with its sealed end down; the upper end is open to the atmosphere. A column of air 10.0 cm long is confined in the tube by a thread of mercury that is 12.0 cm long. When the tube is inverted so that the open end is down, the column of confined air is 13.7 cm long. Assuming that there is no temperature change, calculate atmospheric pressure, in cm Hg and in torr.

15-B13 A cylindrical diving bell 4 m tall, with an open bottom, is lowered to the bottom of a lake on a day when atmospheric pressure is 100 kPa. It is observed that water rises inside the bell to within 1 m of the top. (*a*) What is the pressure of the air trapped in the diving bell? (*b*) How deep is the lake? (Assume constant temperature.)

15-B14 On a day when atmospheric pressure is 760 torr (760 mm Hg), a defective barometer tube of uniform inside diameter contains mercury to a height of 740 mm (*h* in Fig. 12-8); the 60 mm of clear space above the mercury contains a small amount of air. What is the true atmospheric pressure on another day when this barometer reads 725 mm?

15-B15 In a vessel of 1 m^3 capacity are placed the following: (1) as much hydrogen (H_2) as would occupy 1 m^3 at 0°C at atmospheric pressure; (2) as much nitrogen (N_2) as would occupy 2 m^3 at 0°C at 1.5 atm pressure; and (3) as much oxygen (O_2) as would occupy 3 m^3 at 0°C at 2 atm pressure. The mixture is allowed to come to 0°C. (*a*) Calculate the partial pressures exerted by each gas. (*b*) Calculate the total pressure of the mixture. (*c*) Calculate the mass of the mixture.

15-B16 A tank contains 5 liters of oxygen under a pressure of 2560 torr. A second tank, cubical in shape and 30 cm on an edge, contains hydrogen at 320 torr pressure. The two tanks are connected and their contents allowed to mix. Assuming that the temperature does not change, calculate the resultant pressure, the same in each tank.

15-B17 A scientist stores 45 g of a certain gas at a pressure of 900 torr. Overnight, the container develops a slight leak and the pressure drops to 870 torr. What mass of the gas has escaped?

15-B18 A certain amount of gas in a volume 2 m^3 exerts a pressure 400 kPa at 200°C. (a) What will the pressure become if the gas is compressed to a volume of 1 m^3 at a temperature of 100°C? (b) How many moles of gas are involved?

15-B19 (a) How many moles of propane gas (C_3H_8) is contained in a tank of volume 30 liters if the temperature is 50°C and the pressure is 8 atm? (b) What is the mass of this gas? (c) How many molecules of propane are in the tank?

15-B20 What pressure, in atm, is exerted by 1 kg of methane (CH_4) confined in a tank of volume 100 liters at 50°C?

15-B21 Calculate the pressure, in atm, exerted by 100 moles of carbon dioxide gas (CO_2) in a tank of volume 300 liters at a temperature of 100°C.

15-B22 (a) How many kilomoles of nitrogen gas (N_2) at −50°C occupy a volume of 5 m^3 at a pressure of 8.2 atm? (b) What is the mass of this gas?

15-B23 Calculate the density of ozone (O_3) at 520 torr and 60°C.

15-B24 A flask of volume 80 liters contains nitroxyl fluoride (NO_2F) at 760 torr and 27°C. What will the pressure become if 20 g of the gas leaks out?

15-B25 Calculate the density of neon (a monatomic gas whose atomic weight equals its molecular weight; see Periodic Table, Appendix Table 6) when the pressure is 15 atm and the temperature is 300°C.

15-B26 One of the few gaseous compounds of uranium is uranium hexafluoride (UF_6). Calculate the ratio of the density of the gas UF_6 at 1200 K to that of N_2 gas at 300 K, both gases being under the same pressure.

15-B27 A balloon is filled with helium at 0°C and 1 atm. The weight of the balloon is 0.80 N, and it contains 25 mol of helium. (a) What is the volume of the balloon? (b) What is the useful payload that the balloon can lift at the earth's surface? (Molecular weight of helium, 4.00; average molecular weight of air, 29.)

15-B28 A soccer ball of constant volume 2.24 liters is pumped up with air at 20°C so that a pressure gauge reads 86 kPa above atmospheric pressure. (a) How much air is in the ball? (b) During the game, the temperature rises to 30°C. How much air must be allowed to escape to bring the gauge pressure back to its original value? (The molecular weight of air may be taken to be approximately 29.)

15-B29 At what temperature are the readings on the Kelvin scale and the Fahrenheit scale the same? (*Hint:* Use the precise value of absolute zero.)

15-B30 From the data given in Example 15-6, calculate the value of the pressure P_1.

15-B31 Dew has formed on the outside of a glass of cold water that is on a table in a well-ventilated room at 30°C. As the glass and its contents warm up, it is observed that the dew disappears when the temperature of the water reaches 20°C. What is the relative humidity in the room?

15-B32 In a damp basement, the relative humidity is 80% at 10°C. How many grams of water could be evaporated into each cubic meter of the basement before moisture would start to condense on the walls?

15-B33 Consider water vapor at its critical temperature and pressure. Express the critical pressure in Pa; in atm; in lb/in.^{2}. Express the density of water vapor at its critical point in g/cm^3.

15-B34 What is the vapor pressure of water in a greenhouse where the temperature is 30°C and the relative humidity is 70%?

15-B35 A classroom measures 2 m × 4 m × 6 m and is supplied with air drawn from the outside at 0°C and heated to 20°C. The relative humidity outdoors is 70%. (*a*) What was the volume of the air when it was outdoors, before being heated and placed in the room? (*b*) What mass of water vapor is in the room? (*c*) What is the relative humidity in the room? (*d*) Is the relative humidity in a satisfactory range for comfort and health?

15-C1 In order to measure the saturated vapor pressure of water at 50°C, an experimenter started with a sealed container containing some liquid water, some water vapor, and some air. At 20°C the total pressure was 628.0 torr, and at 50°C the total pressure was 765.5 torr. Using data from Table 15-2, calculate the saturated vapor pressure of water at 50°C.

15-C2 A teaspoonful of an organic oil (volume 4.00 cm^3) dropped on the surface of a quiet lake spreads out to cover an area of 4400 m^2, about an acre. (*a*) What is the thickness of the oil film? (*b*) Explain why this monolayer (one molecule thick) is thicker than the atomic plane separations of Fig. 1-1. (*c*) Estimate the order of magnitude of Avogadro's number, assuming the oil to have molecular weight 300 and specific gravity 0.9. [*Hint:* Find the number of molecules by assuming each molecule to be a cube whose edge has been found in part (*a*).]

15-C3 If the volume of the bubble in Fig. 15-7 is 2.00 mm^3 at 20°C, calculate the volume at 99°C.

15-C4 (*a*) Do you expect water vapor at 100°C and 1 atm to be an ideal gas? (*b*) To test whether the ideal gas law is even approximately true for water vapor at 100°C and 1 atm, calculate the density of this substance from its molecular weight, the temperature, and the pressure, assuming the ideal gas volume of 22.4 liters for 1 mol at STP. Compare with the experimental density listed in Table 15-2.

15-C5 In a damp basement 5 m × 10 m × 3 m, the relative humidity at 30°C is 85%. How much water would have to be removed by a dehumidifier in order to bring the relative humidity down to 20%? Would a 4-liter (1-gallon) jug hold this much water?

15-C6 In a chemistry experiment the measured volume of hydrogen gas collected over water is 380 cm^3 at 20°C; the barometer stands at 740 torr, and the pressure inside the collecting chamber is 6.8 cm of water greater than atmospheric pressure. (*a*) What mass of H$_2$ gas, in moles, is in the chamber? (*b*) What would be the dry volume of the gas if corrected to 0°C and 760 torr?

15-C7 Calculate the density of water vapor at the critical temperature and pressure, assuming the vapor to obey the ideal gas law. Why is your calculated value less than the experimental value given in Table 15-2?

References 1. J. B. Conant, *Robert Boyle's Experiments in Pneumatics* (Harvard Case Histories in Experimental Science No. 1), (Harvard University Press, Cambridge, 1950). See especially the foreword; the introduction (pp. 11–19); and Boyle's law (pp. 57–67).

2. P. W. Bridgman, "Synthetic Diamonds," *Sci. American* **193**(5), 42 (Nov. 1955). See the phase diagram for the substance carbon (p. 46). Is the diamond phase stable at room temperature, according to the phase diagram?

3. H. P. Burstyn and A. A. Bartlett, "Critical point drying: Application of the physics of the *PVT* surface to electron microscopy," *Am. J. Phys.* **43**, 414 (1975). Describes a way of avoiding disruptive surface-tension effects when drying a biological specimen.

4. J. L. Gaines, "The Dunking Duck," *Am. J. Phys.* **27**, 190 (1959).

5. "A Robot Shadoof—New Waterbird for Egypt," *Saturday Review* (June 3, 1967). Introduction by John Lear (p. 49) and technical article by R. B. Murrow (p. 51). Proposes a "dunking bird" for river-valley irrigation.

6. F. Miller, Jr., *Critical Temperature* (film).

16

The Kinetic
Theory of Gases

In this chapter, we shall continue our study of the mechanical and thermal behavior of gases. To be sure, we have already used many of the concepts about to be discussed, such as temperature, molecular motions, and specific heat capacity. However, it is time now to look at these aspects of gases more closely. We shall build on Newton's second law for some of our theory (the pressure exerted by moving molecules, Sec. 16-2), but we shall also see how the non-Newtonian quantum theory is indicated by measurements of specific heat capacities (Sec. 16-3). The study of gases has been, in more than one way, a stepping-stone for the advance of physical knowledge.

16-1 Heat and Temperature

First of all, let us make a sharp distinction between heat and temperature. Recall that heat is a flow of energy, due to temperature difference, to or from a large collection of molecules that can be called a body or a system. When a gram of water at 20°C absorbs 10 cal of heat, the

random KE of the 3×10^{22} molecules of H_2O increases by a total of 10 cal (42 J); the molecules speed up, and the system gains internal energy. Heat flow can also increase the random PE of a system, as, for example, when a gram of water is being vaporized. Also, heat flow may cause external work to be done on or by the system. In all these cases, "heat flow" describes what happens to an ensemble of many molecules—a system. On the other hand, temperature difference is what causes heat flow. In the next section we shall prove that *absolute temperature in a gas or liquid is proportional to the average translational KE per molecule due to random motions.* As an illustration, consider a swimming pool full of cold water at 5°C and a teacup full of hot water at 90°C. There is much more total internal KE in the pool, since there are many molecules, but the pool's temperature is lower than the teacup's because the average KE

---

* In a solid the molecules are in contact but are not free to "drift" and can have no random translational KE. In solids the *vibrational energy* is proportional to the absolute temperature.

A quick push on the plunger of a bicycle pump causes adiabatic compression of the air in the pump and in the tire.

per molecule is less than the average per molecule in the cup.

Each word in our statement about absolute temperature has definite significance: "proportional," because the factor relating absolute temperature to average KE depends on the choice of temperature scale (we shall see later what the factor of proportionality is); "average," because not all the molecules are moving at the same speed, and if they were, collisions would soon destroy equality; "translational," since rotational or vibrational types of KE are without effect in causing collisions in a gas and hence cannot cause heat transfer (a basic effect of temperature difference); "kinetic" energy, since increases of PE such as a latent heat take place without rise in temperature; "per molecule," since we are distinguishing between temperature and heat; due to "random" motions, since the "ordered" motions of all the molecules of a body are considered to be mechanical energy. The molecules of a baseball are given ordered motions when the ball is thrown. These "organized" motions of translation and rotation are not in themselves indicative of a temperature rise, but are classed as KE of the ball as a whole. When the ball strikes a catcher's mitt and is stopped, its KE is transformed into an increase in random KE of molecules of ball and mitt, and the ball's temperature rises.

16-2 Molecular Pressure

The interpretation of temperature we have just given is based on our understanding of nature, as summarized by Newton's laws of motion and the general gas law. The first step in the analysis is to use Newton's second law to derive a formula for the pressure P exerted by the molecules in a closed box of volume V. We assume that there are N molecules, each of mass m, moving with equal speeds* v in the x direction toward the face

---

* The assumption of equal speeds is for simplicity; a derivation of Eq. 16-2 not making this assumption is given in Sec. 16-6.

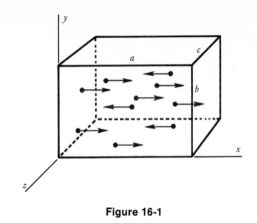

Figure 16-1

bc of the box (Fig. 16-1). The problem is to find the force on the box face bc due to impacts of molecules as they strike and bounce back. This problem is similar to the problem of the fire hose sending water against a burning building (Example 5-1). Now, as then, we find it convenient to use Newton's second law in the form "net force = rate of change of momentum." A molecule has a momentum $+ mv$ before collision with the wall and a momentum $- mv$ after collision, assuming that it bounces back without loss of energy. The change in momentum is

$$(+mv) - (-mv) = 2mv$$

at each collision. In a time t, a molecule moves vt meters, and since the molecule has to travel a distance of $2a$ meters between successive collisions at the same surface, the number of collisions in time t is

$$\frac{(vt \text{ meters})}{(2a \text{ meters/collision})} = \frac{vt}{2a} \text{ collisions}$$

Now we apply the second law:

$$\text{net force} = \frac{\begin{pmatrix} \text{change in} \\ \text{momentum} \\ \text{per collision} \end{pmatrix} \begin{pmatrix} \text{total} \\ \text{number of} \\ \text{collisions} \end{pmatrix}}{\text{time}}$$

$$= \frac{(2mv)(vt/2a)}{t} = \frac{mv^2}{a}$$

Thus the force exerted by each molecule is mv^2/a. There are N molecules, so the total force is Nmv^2/a, and the pressure on face bc is force/area:

$$P = \frac{Nmv^2}{(a)(bc)} = \frac{Nmv^2}{V}$$

where $V = abc$ = volume of the box. To make our derivation more realistic, we must take account of the fact that the molecules are moving in random directions, not just parallel to the x axis. This introduces a factor of $\frac{1}{3}$, since on the average a molecule has a component of velocity along any one of three axes. Only $\frac{1}{3}$ of the momentum changes are "effective" in any one direction. Our final result is, therefore,

$$P = \frac{1}{3}\frac{Nmv^2}{V} \qquad (16\text{-}1)$$

or

$$PV = \tfrac{1}{3}Nmv^2 \qquad (16\text{-}2)$$

We can use this equation to compute the speeds of molecules in gases.

Example 16-1

What is the speed of the molecules in a flask of oxygen (O_2) at $0°C$ and 1 atm pressure?

If there is 1 mole of oxygen present, the volume is 22.4 liters $= 2.24 \times 10^{-2}$ m³. The pressure of 1 atm is equal to 1.01×10^5 N/m² (see Sec. 12-2). First, we solve Eq. 16-2 for the velocity:

$$3PV = Nmv^2$$

$$v = \sqrt{\frac{3PV}{Nm}}$$

Since N is the number of molecules and m is the mass of each, the product Nm is the total mass of gas, 32 g $= 3.2 \times 10^{-2}$ kg in this case, since we are assuming that 1 mole of oxygen is present.

$$v = \sqrt{\frac{3(1.01 \times 10^5 \text{ N/m}^2)(2.24 \times 10^{-2}\text{ m}^3)}{3.2 \times 10^{-2}\text{ kg}}}$$

$$= \sqrt{21.2 \times 10^4\text{ m}^2/\text{s}^2} = 4.60 \times 10^2\text{ m/s}$$

$$= \boxed{460 \text{ m/s}}$$

This is an average speed, about $\frac{1}{4}$ mi/s in the example. Individual O_2 molecules may have speeds much greater or smaller than this value, which is called the root-mean-square (rms) speed.*

Let us look at Eq. 16-2 more closely. We recognize mv^2 as just twice the translational KE per molecule. Therefore, replacing mv^2 by 2KE, we get

$$PV = (\tfrac{1}{3}N)(2\text{KE per molecule})$$

or

$$PV = (\tfrac{2}{3}N)(\text{KE per molecule}) \qquad (16\text{-}3)$$

This, then, is the result of applying Newton's second law to calculate the pressure due to moving molecules.

We now observe that the general gas law, based on experiments with gases at various temperatures and pressures, also makes a statement about the product PV for an ideal gas:

$$PV = nRT$$

or

$$PV = \frac{N}{N_A}RT \qquad (16\text{-}4)$$

where n, the number of moles of gas, is given by the total number of molecules N divided by Avogadro's number N_A (6.02×10^{23} molecules/mol). We can write Eqs. 16-3 and 16-4 in words as follows:

16-3: pressure times volume is proportional to number of molecules times KE per molecule

16-4: pressure times volume is proportional to number of molecules times absolute temperature

The conclusion is immediate and far-reaching: *absolute temperature is proportional to KE per*

---

* The more general analysis of Sec. 16-6 takes account of the fact that the molecules do not all have the same speed. Then Eq. 16-2 is replaced by $PV = \frac{1}{3}Nm\overline{v^2}$, where $\overline{v^2}$ is the mean (average) of the squares of the speeds. The square root of this quantity, $\sqrt{\overline{v^2}}$, is the rms speed.

molecule. This is what we set out to prove. In so doing, we have not only strengthened our understanding of temperature; we have also strengthened our belief in the adequacy of Newton's laws, and we have made more plausible our model of molecules as hard, elastic particles like billiard balls or machine-gun bullets.

To find the KE per molecule, or per mole, we equate Eqs. 16-3 and 16-4.

$$\left(\frac{2}{3} N\right)(\text{KE per molecule}) = \frac{N}{N_A} RT$$

or

$$\text{KE per molecule} = \frac{3}{2}\left(\frac{R}{N_A}\right) T$$

Here R is the gas constant per mole, and R/N_A is the gas constant per molecule. This latter quantity is such a basic and universal constant that it is given a special name and symbol:

$$Boltzmann\text{'s constant} = k = \frac{R}{N_A}$$

The value of Boltzmann's constant is

$$k = \frac{8.31 \text{ J/mol}\cdot\text{K}}{6.02 \times 10^{23} \text{ molecules/mol}}$$
$$= 1.38 \times 10^{-23} \text{ J/molecule}\cdot\text{K}$$

So, our final results are

$$\text{KE per molecule} = \tfrac{3}{2}kT \qquad (16\text{-}5)$$
$$\text{KE per mole} = \tfrac{3}{2}RT \qquad (16\text{-}6)$$

The derivation has been made for an ideal gas. The fact that most real gases deviate slightly from the general gas law can be accounted for by a detailed analysis including factors we have neglected, such as the weak cohesive forces between molecules and the volume of the molecules themselves.

Although our derivation has been made for gases, the final conclusion is also true for liquids and solids. Imagine an aluminum cup full of water sitting on a table at 20°C. The water molecules are being continually bombarded by air molecules, and there would be a tendency for

an exchange of energy (heat flow) unless the average KE's were the same for air and water molecules. We know that if the temperatures of air and water are the same, there is no heat flow. Hence we conclude that the average KE's are the same in air and water at the same temperature. A similar argument shows that the average KE of the vibrating aluminum atoms has the same value as that of the air and the water molecules.

One illustration will show the power of the kinetic theory of gases in correlating and illuminating earlier observations and hypotheses. In 1811, Avogadro was led to suggest that at any temperature and pressure, the number of molecules per unit volume is the same for all gases. The kinetic theory, which came later, gives a direct proof of Avogadro's brilliant hypothesis, as follows: Solving Eq. 16-1 for the number of molecules per unit volume, we get $N/V = 3P/mv^2$. Now the product mv^2 is proportional to the absolute temperature T. Thus at any given temperature and pressure, both P and mv^2 are fixed. Therefore N/V is completely determined by the temperature and the pressure, and this ratio is independent of anything else, such as the identity of the gas. We now know that if the temperature is 0°C and the pressure is 1 atm, the ratio N/V equals $(6.02 \times 10^{23} \text{ molecules})/(22.4 \text{ liters})$ for any ideal gas.

16-3 Specific Heat Capacities of Gases

We are already familiar with specific heat capacity, a property of matter that tells how the temperature of a substance changes upon the addition of heat (Sec. 14-3). Specific heat capacity is measured in joules per kilogram per degree or calories per gram per degree. In this section we shall see that close study of specific heat capacity will help us understand better the nature of temperature and heat. We shall find in this study some discrepancies of fundamental significance, which are related to the quantum theory of molecular vibration and rotation.

First, we attempt to predict the specific heat capacity of a gas from the kinetic theory, using helium as an example. Suppose we have 1 mole of helium (4.00 g) at 0°C (273 K), and we heat it up by 1° to a temperature of 1°C (274 K). How much heat must be added to give the additional KE to the molecules? We recall that the KE of n moles of gas is $\frac{3}{2}nRT$, and we assume the value of the gas constant to be exactly 8.315 J/mol·K.

$$KE_{274\,K} = \tfrac{3}{2}(1\text{ mol})(8.315\text{ J/mol·K})(274.00\text{ K})$$
$$KE_{273\,K} = \tfrac{3}{2}(1\text{ mol})(8.315\text{ J/mol·K})(273.00\text{ K})$$

Subtracting, we get

$$\Delta KE = \tfrac{3}{2}(1\text{ mol})(8.315\text{ J/mol·K})(1.00\text{ K})$$
$$= 12.47\text{ J}$$

$$= (12.47\text{ J})\left(\frac{1\text{ cal}}{4.18\text{ J}}\right) = 2.98\text{ cal}$$

Hence the specific heat capacity of helium is expected to be

$$\frac{2.98\text{ cal/}°C}{4.00\text{ g}} = 0.745\text{ cal/g·}°C$$

Measurement of the specific heat capacity of helium gives a value of 0.75 cal/g·°C, in excellent agreement with our predicted value.

Calculation of the specific heat capacity of neon (molecular weight 20.2) proceeds in much the same way. Just as for helium, we expect 2.98 cal to be required to heat 1 mole of neon 1°C; then the specific heat capacity (per gram) would be (2.98 cal/°C)/(20.2 g) = 0.148 cal/g·°C. The experimental value is 0.15 cal/g·°C. So far, so good.

Things are different if we calculate the specific heat capacity of H_2, O_2, or N_2. Experiment shows that to heat 1 mole of one of these diatomic gases 1°C requires not 2.98 cal, but about 4.9 cal. However, we can understand in a general way where the extra energy goes. A basic assumption is made, called the *law of equipartition of energy:*

Energy given to an assemblage of molecules is, on the average, equally shared among all possible modes of energy absorption.

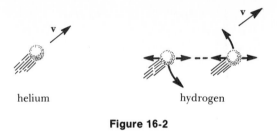

Figure 16-2

The hydrogen molecule can rotate on its axis and have rotational KE ($\frac{1}{2}I\omega^2$) in addition to translational KE ($\frac{1}{2}mv^2$). The atoms can also have KE and PE as they vibrate to and fro along the line joining them in the molecule (Fig. 16-2). The helium and neon molecules, on the other hand, consist of single atoms, have no appreciable rotational inertia, and hence can have no appreciable rotational energy. To raise the temperature of hydrogen 1°C requires more heat than for helium, since that portion of the internal energy that is stored as rotational or vibrational energy is not effective in raising the temperature. This illustrates our statement that temperature is proportional to the *translational* KE.

Another prediction of the theory is borne out by experiment. It is found that if a gas is allowed to expand while being heated (thereby preventing a buildup of pressure), the specific heat capacity is greater than if the volume is kept constant. Suppose a mole of any gas is confined within a container having a piston (Fig. 16-3). As heat is supplied and the piston is allowed to move, mechanical work is done against an opposing force PA, where P is the pressure both inside and outside the container, and A is the area of the piston. This energy has to come from somewhere. The source of heat must supply not only the increased KE of the molecules but also the external work done as the piston moves. The specific heat capacity is greater, by an amount that is almost the same for all gases, since all gases obey the same gas law. Experiment confirms this fully. At constant pressure the specific heat capacity c_p of H_2 or O_2 is 40% greater than the specific heat capacity c_v at constant volume.

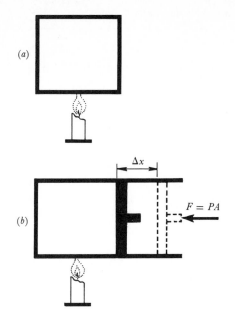

(a)

(b)

Figure 16-3 When the same amount of heat is supplied, which gas becomes warmer? Container *a* is at constant volume, and container *b* is at constant pressure.

For a discussion of the ratio of the specific heat capacities of a gas, see Sec. 16-4.

In view of all these successes, it came as a distinct shock, in the early 1900s, to find that some predictions of the kinetic theory of gases were not at all true. Consider the case of hydrogen, for which we expect the specific heat capacity at constant volume to be 4.9 cal/mole·°C. This is not far from the observed value if we observe a 1° change near room temperature, say from 293 K to 294 K. However, to change 1 mole of H_2 from 93 K to 94 K, only about 2.98 cal is required, the same as for 1 mole of helium. Evidently, at low temperatures the hydrogen molecules are unable to absorb rotational energy! Physicists spoke of the "breakdown of the law of equipartition of energy"—a most disturbing prospect.

In Sec. 31-8 we shall discuss molecular rotations and vibrations as revealed by the spectroscope. Not all rotational speeds are possible; a molecule may possess 0, 1, 2, 3, ... quanta of

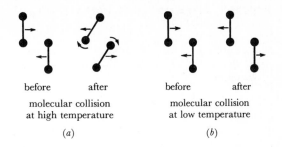

before   after
molecular collision
at high temperature
(a)

before   after
molecular collision
at low temperature
(b)

Figure 16-4 Collisions between molecules of a diatomic gas.

rotational KE, but it is impossible for a molecule to rotate very slowly, with just $\frac{1}{2}$ or $\frac{1}{5}$ or $\frac{1}{1000}$ of a quantum of KE. Here is an explanation of the curious behavior of H_2 gas at low temperatures (Fig. 16-4). At 93 K, the average translational KE of the molecules is much less than even one quantum of rotational KE. As a result, the collisions take place without angular recoil, as in Fig. 16-4*b*. To all intents and purposes, the molecules cannot rotate, and all the heat is effective in raising temperature, just as in the case of helium. At room temperature the collisions are energetic enough to allow more or less free rotation, with 5 or 10 quanta, and the specific heat capacity has its expected higher value because extra energy may now be stored as rotational KE.

It is interesting to note that this "freezing out" of rotational motion at low temperatures provided some early evidence for the quantum of mechanical energy. The kinetic theory has served two purposes: to give us a model for the cause of pressure as due to molecular impacts, and to give us an indication that the model is insufficient to fully describe reality.

16-4 The Ratio of the Specific Heat Capacities of a Gas

In Sec. 16-3 we saw that the specific heat capacity at constant pressure should be greater than that at constant volume, since at constant pressure some extra heat must be supplied in order to do work as the gas expands against external pressure.

Table 16-1 Measured Specific Heat Capacities of Gases at Room Temperature*

Type of Gas	Chemical Formula	C_p	C_v	$C_p - C_v$	$\gamma = C_p/C_v$
Monatomic	He	4.97	2.98	1.99	1.67
	Ne	4.97	2.98	1.99	1.67
Diatomic	H_2	6.87	4.88	1.99	1.41
	O_2	7.03	5.03	2.00	1.40
	N_2	6.95	4.95	2.00	1.40
Polyatomic	CO_2	8.83	6.80	2.03	1.32
	H_2O	8.20	6.20	2.00	1.32
	C_2H_6	12.35	10.30	2.05	1.20
	$C_4H_{10}O$	32.5	31.	1.5	1.05

*In cal/mol·K.

At *constant volume,* for an ideal monatomic gas, we can carry through a calculation in symbols similar to that in Sec. 16-3 for helium. For n moles,

At $T_1 + 1°$: $KE = \frac{3}{2}nR(T_1 + 1°)$

At T_1: $KE = \frac{3}{2}nRT_1$

Since the volume does not change and no heat is absorbed as rotational or vibrational energy, all the heat required to raise the gas 1 K becomes extra translational KE of the molecules. Subtracting, we obtain

$$\Delta KE = \frac{3}{2}nR$$

and the specific heat capacity per mole is given by $\frac{3}{2}R$ ($= 3.0$ cal/mol·K, since $R = 2.0$ cal/mol·K).

At *constant pressure,* the gas must expand against the surroundings. As seen from Fig. 16-3, the work required is

Work = (force)(distance)

$$\Delta W = (PA)(\Delta x)$$

$$\Delta W = P \Delta V \qquad (16\text{-}7)$$

since $A \Delta x$ is ΔV, the change in volume. From the gas law,

$$PV = nRT$$

Hence,

$$\Delta W = P \Delta V = nR \Delta T$$

and the extra work, for a 1-K rise in temperature ($\Delta T = 1$ K) is

$$\Delta W = nR \quad \text{(for } n \text{ moles)}$$

or

$$\Delta W = R \text{ per mole} = 2.0 \text{ cal/mol·K}$$

Thus the *difference* between the specific heat capacity* at constant pressure C_p and the specific heat capacity at constant volume C_v is expected to be equal to R—that is, 2.0 cal/mol·K for all ideal gases, regardless of composition. Table 16-1 shows this difference to be approximately 2.0 cal/mol·K for many gases.

The *ratio* of the specific heat capacities is denoted by the symbol γ (gamma) and is a dimensionless number. This constant enters into the formula for the speed of sound in a gas (Sec 16-7) and is usually measured by acoustical experiments.

For helium and other monatomic gases,

$$\gamma = \frac{C_p}{C_v} = \frac{\frac{3}{2}R + R}{\frac{3}{2}R} = \frac{\frac{3}{2} + 1}{\frac{3}{2}}$$

$$= \frac{5}{3} = 1.67$$

---

* We use C_p and C_v for specific heat capacity per mole, and c_p and c_v for specific heat capacity per gram.

We interpret terms in the expression for γ with the help of the first law of thermodynamics, $Q = \Delta U + W$. In the numerator, at constant pressure, ΔU is $\frac{3}{2}R$ per mole, an increase of internal energy, and W is R per mole, work done to expand the gas. Thus for C_p, $Q = \frac{3}{2}R + R = \frac{5}{2}R$. In the denominator, at constant volume, we have $W = 0$, hence $Q = \Delta U = \frac{3}{2}R$.

For H_2, O_2, N_2, and other diatomic gases, the possibility exists that energy may be absorbed by molecular rotation. These molecules can rotate in two significant ways, not three (rotation about the molecular axis absorbs no appreciable energy, since the moment of inertia about this axis is assumed to be zero); thus we say that a diatomic molecule has two *degrees of freedom*, as far as rotation is concerned. On the other hand, these molecules (as well as all other molecules) have three translational degrees of freedom, as the molecule can move along the x, y, or z axis. It seems very reasonable that any energy absorbed by the molecules is shared among these various degrees of freedom; for instance, energy would not be absorbed in such a way that *all* the molecules would gain translational energy along the y axis only. This share-the-energy behavior is a fundamental prediction of classical mechanics, the law of equipartition of energy (Sec. 16-3). A monatomic gas, with 3 degrees of freedom, has $C_v = \frac{3}{2}R$, so we see that each degree of freedom contributes $\frac{1}{2}R$ to the value of C_v. In the case of H_2, there is a total of 5 degrees of freedom: $\frac{3}{5}$ of the energy absorbed goes into translation, and $\frac{2}{5}$ goes into rotation. Translational energy is to rotational energy as $3:2$, and our ratio of specific heat capacities becomes

$$\frac{C_p}{C_v} = \frac{\text{(translational)} + \text{(rotational)} + \text{(work)}}{\text{(translational)} + \text{(rotational)}}$$

so that for a diatomic gas with molecules free to rotate,

$$\gamma = \frac{\frac{3}{2}R + \frac{2}{2}R + R}{\frac{3}{2}R + \frac{2}{2}R} \qquad (16\text{-}8)$$

$$= \frac{7}{5} = 1.40$$

It is evident that if other ways of absorbing energy, such as vibration, exist, both the numerator and the denominator of the fraction increase by the same amount, and the ratio becomes closer to 1. Table 16-1 shows how these ideas are verified by experiment. Note particularly how the complex vibrations of ethane (C_2H_6) and ether ($C_4H_{10}O$) are indicated by the large values of their specific heat capacities, and by values of γ approaching 1. The fact that $C_p - C_v$ for ether departs from the predicted value of 2.0 cal/mol·K merely indicates that ether is not an ideal gas at the temperature of these measurements.

Other departures are attributed to the quantum effects mentioned at the end of Sec. 16-3. At very low temperatures, collisions are not energetic enough to give the molecules even one quantum of rotational energy. As a result, the rotational terms ($\frac{2}{2}R$) are missing from both numerator and denominator in Eq. 16-8, and γ for H_2, O_2, and N_2 becomes $\frac{5}{3}$ (1.67), the same as for monatomic gas.

We can illustrate some of the ideas of this section by a small digression into the theory of solids. The atoms in solids absorb energy primarily through increased amplitude of vibration. In 3 dimensions, each atom has 3 degrees of freedom for KE (for example, $\frac{1}{2}mv_x^2$, $\frac{1}{2}mv_y^2$, and $\frac{1}{2}mv_z^2$), and also 3 degrees of freedom for vibrational PE, as for a mass on a spring (for example, $\frac{1}{2}k_1x^2$, $\frac{1}{2}k_2y^2$, and $\frac{1}{2}k_3z^2$). With a total of 6 degrees of freedom, we expect C_v to be $6(\frac{1}{2}R) = 3R = 5.94$ cal/mol·°C. Usually a solid is kept at atmospheric pressure and allowed to expand while the specific heat capacity is measured, but since a solid does not expand upon heating nearly as much as a gas, the external work is small, and $C_p \approx C_v$.

It has been known since the early 1800s that the specific heat capacities per mole of many simple solids, such as elements, are indeed about 6 cal/mol·°C; this is called the "law" of Dulong and Petit. Some solids, such as diamond, however, have abnormally low specific heat capacity at room temperature. This is correlated with the

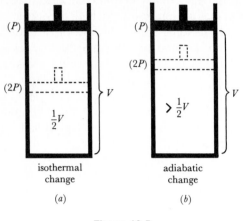

isothermal change
(a)

adiabatic change
(b)

Figure 16-5

hardness of the diamond crystal, indicating strong binding forces between atoms of carbon and also indicating a correspondingly large value for a quantum of vibrational energy. Even at room temperature, the vibrations are partially "frozen out," so it is only at very high temperatures that the specific heat capacity of diamond approaches 6 cal/mol·°C. For iron, aluminum, and copper, room temperature is already high enough to allow full vibrational equipartition, free of quantum effects.

16-5   The Adiabatic Gas Law

An *isothermal** process is one in which the temperature does not change. Boyle's law, $P_1V_1 = P_2V_2$, describes such a process (Fig. 16-5a). An *adiabatic*[†] process is one in which no heat is transferred. For example, a piston is suddenly pushed down, doubling the pressure on an enclosed mass of gas (Fig. 16-5b). The process is adiabatic, since there is insufficient time for heat to flow out of the gas. Even a slow compression would be adiabatic if the cylinder and piston were thermally insulated from the surroundings. In an adiabatic compression, $Q = 0$ (by defini-

---

* From Greek roots meaning "equal temperatures."
† From Greek roots meaning "not going through."

tion); hence the first law of thermodynamics gives $\Delta U = -W$. The work W is negative, since we are doing work *on* the gas. Therefore ΔU is positive: the internal energy increases, and the temperature increases. This temperature rise causes an expansion of the gas, so to speak; the pressure is doubled, but the volume is still more than half its original value. It is reasonable to expect that the specific heat capacities of the gas will be involved in this process, since we must relate the change in internal energy to the change in temperature. The result of analysis is

adiabatic process for ideal gas $$P_1V_1{}^\gamma = P_2V_2{}^\gamma$$

where γ is the ratio of specific heat capacities discussed in Sec. 16-4. The final temperature can be found from the gas law, as usual, since $PV = nRT$ at all times.

Example 16-2

A cylinder contains 10 liters of air at 3 atm and 300 K. (a) If the pressure is suddenly doubled, what are the new volume and temperature? (b) If the pressure is slowly doubled, what are the new volume and temperature?

(a) For the sudden change, the adiabatic gas law is used. Air consists mostly of N_2 and O_2, both diatomic molecules, and so we may take γ for air as 1.40.

$$P_1V_1{}^\gamma = P_2V_2{}^\gamma$$

$$(3 \text{ atm})(10 \text{ liters})^{1.40} = (6 \text{ atm})(V_2)^{1.40}$$

$$\left(\frac{10 \text{ liters}}{V_2}\right)^{1.40} = \frac{6 \text{ atm}}{3 \text{ atm}} = 2$$

$$\frac{10 \text{ liters}}{V_2} = 2^{1/1.40} = 2^{0.714} = 1.64‡$$

$$V_2 = \frac{10 \text{ liters}}{1.64} = \boxed{6.10 \text{ liters}}$$

---

‡ The use of logarithms to find $2^{0.714}$ goes as follows (the logarithms can be looked up in a log table): $\log 2 = 0.301$; $(0.714)\log 2 = 0.215$; $2^{0.714} = $ antilog $0.215 = 1.64$. Many pocket calculators give y^x in one operation.

Now we need to find the new temperature. The mass of gas has not changed, and so we use

$$\frac{P_1 V_1}{T_1} = \frac{P_2 V_2}{T_2}$$

$$T_2 = (300 \text{ K})\left(\frac{6.10 \text{ liters}}{10 \text{ liters}}\right)\left(\frac{6 \text{ atm}}{3 \text{ atm}}\right)$$

$$= \boxed{366 \text{ K}}$$

(b) For a slow change, heat flows out of the cylinder as the gas is compressed, and the temperature remains constant. Boyle's law applies, and we easily obtain

$$V_2 = \boxed{5 \text{ liters}}$$

$$T_2 = T_1 = \boxed{300 \text{ K}}$$

If a gas is in a well-insulated container, no heat can flow, and even a slow change is adiabatic under such circumstances. We should point out that it is perfectly possible to have changes that are neither isothermal nor adiabatic, so that the division we have made is to some extent arbitrary, dictated by ease of solving problems.

Summary Absolute temperature of an ideal gas is proportional to the average random translational KE per molecule. The pressure exerted by the molecules of a gas can be computed from Newton's second law, and this allows determination of molecular speeds from observed data on temperature, pressure, and volume. At any given temperature, molecular speeds are less for molecules of greater mass, since the average KE's of molecules of all gases are the same at any temperature.

The specific heat capacity per mole of an ideal gas at constant pressure is greater than the specific heat capacity per mole at constant volume by an amount equal to the gas constant R; the ratio of the specific heat capacities depends on the structure of the gas molecules. At ordinary temperatures, when a diatomic molecule such as H_2 absorbs heat, both the rotational KE and the translational KE increase, the relative amounts being in accordance with the law of equipartition of energy. At very low temperatures, rotation is no longer possible, since the average energy of a colliding molecule is much less than a single quantum of rotational KE. At such temperatures, gases with complex molecules behave like monatomic gases, for which all the internal energy is in the form of translational KE.

Check List

$PV = \frac{1}{3}Nmv^2$
translational KE of n moles $= \frac{3}{2}nRT$
translational KE per molecule $= \frac{3}{2}kT$
Boltzmann's constant
specific heat capacity per mole of a gas

specific heat capacity at constant volume
specific heat capacity at constant pressure
law of equipartition of energy
adiabatic gas law

Questions 16-1 What is the difference between "heat" and "temperature"?

16-2 The molecules in a flask of helium have a total random translational KE of 10 μJ; a flask of oxygen contains twice as many molecules and has a total random translational KE of 14 μJ and internal rotational energy of 8 μJ. Which flask is hotter?

16-3 Isotopes of gases are sometimes separated by diffusion, depending on the differences of the speeds of the molecules. Neon consists of a mixture of two stable isotopes of molecular weight 20 and 22, respectively. Which neon molecules are moving faster if both isotopes have the same temperature? Which would you expect to diffuse more readily through a porous membrane?

16-4 Give two reasons, having to do with the molecules of a gas, why the general gas law might fail to represent the behavior of that gas.

16-5 If the molecules of a gas were not tiny points but had a finite (though small) diameter, would they make more or fewer collisions per second with the walls of the box in Fig. 16-1? (Assume the same temperature, and thus the same speed.) Would the pressure be greater or less than that computed from Eq. 16-1?

16-6 In deriving Eq. 16-1, we ignored the collisions that must be taking place between molecules in the box. Explain why this is justified. (*Hint:* See Sec. 5-3.)

16-7 A complex molecule such as alcohol or ether is able to absorb energy in the form of internal vibrations in many ways, as well as by rotation of the molecule as a whole and by linear motion of the center of mass of the molecule itself. Would you expect the specific heat capacity per molecule of alcohol vapor to be greater or less than that of oxygen, which does not have so many ways in which to vibrate?

16-8 The plunger of a bicycle pump is suddenly pushed down, compressing the air within the cylinder. Explain why the air becomes warm.

16-9 In a real gas there are small cohesive forces between molecules. If the valve of a tank of compressed air is opened and the gas is allowed to expand suddenly, the metal parts of the valve become cold. Explain, on the basis of the kinetic theory of gases.

MULTIPLE CHOICE

16-10 Deviations from the ideal gas law take place at low temperatures because   (*a*) molecular collisions are inelastic;   (*b*) molecular cohesive forces become less;   (*c*) molecular volumes are not negligible.

16-11 If the average speed of molecules in a gas is tripled, the temperature is multiplied by a factor of   (*a*) 3;   (*b*) 6;   (*c*) 9.

16-12 If the average speed of molecules of an ideal gas is v, the pressure caused by these molecules is proportional to (*a*) $\sqrt{v}$;   (*b*) v;   (*c*) v^2.

16-13 A wire wrapped around a long metal bar carries electric current for a certain length of time, supplying a definite amount of heat to the bar. If the ends of the bar were rigidly clamped, so that the bar were prevented from expanding as its temperature increased, the final temperature (relative to an unclamped condition) would be   (*a*) less;   (*b*) the same;   (*c*) greater.

16-14 A container of ideal gas has volume V under a pressure P. If the pressure is suddenly doubled, the new volume is   (*a*) less than $V/2$;   (*b*) equal to $V/2$;   (*c*) greater than $V/2$.

16-15 The room-temperature value of γ for hydrogen is about 1.40. At very high temperatures, the value of γ is   (*a*) greater;   (*b*) the same;   (*c*) less.

Problems   **16-A1**   The total random translational KE of the water molecules in an 18-liter can is 4.4×10^6 J; the total random translational KE of the water molecules in a 360,000-liter swimming pool is 8×10^{10} J. Which body of water—in the can or in the pool—is at the higher temperature?

16-A2 A 1-liter bottle contains liquid mercury at a temperature such that the total random translational KE of the molecules is 3.5×10^5 J. What is the total random translational KE of the molecules in a beaker containing 300 cm^3 of the same material at an absolute temperature 1.3 times that of the bottle?

16-A3 An oxygen molecule (O_2) bounces back and forth between two walls of a cubical container 40 cm on an edge and makes 1000 round trips each second. What is its speed?

16-B1 A machine gun fires 20 bullets per second. Each bullet has mass 100 g, is moving at 300 m/s, and bounces back from the target with its speed unaltered. What is the average force on the target? (See Sec. 5-2.)

16-B2 Calculate the rms speed of the molecules in a tank of H_2 gas (molecular weight 2) if the temperature is 546°C and the pressure is 3 atm.

16-B3 A total mass of 6.0 kg of a certain gas is confined in a tank of volume 8 m^3, and the pressure is 24 kPa. What is the rms speed of the molecules?

16-B4 Suppose that the pressure in Example 16-1 were P instead of 1 atm, the temperature still being 0°C. Show that Eq. 16-2 will give the same velocity. Thus, show that the molecular speed for a given gas depends only on temperature and not at all on pressure.

16-B5 (a) Calculate the average KE of a single O_2 molecule in Example 16-1. (b) Calculate the gravitational PE increase when a molecule moves from the bottom to the top of a tank 3 m on an edge. (c) Are gravitational effects significant in this tank?

16-B6 Make a dimensional check for Eq. 16-2 ($PV = \frac{1}{3}Nmv^2$).

16-B7 (a) Calculate the rms speed of the molecules in a tank of volume 8 m^3 that contains 50 kg of gas at a pressure of 5 atm (505 kPa). (b) If the temperature of the gas is 40°C, what is the molecular weight?

16-B8 At 50°C the rms speed of certain molecules is v. What will the rms speed of these same molecules become if the temperature rises to 300°C?

16-B9 The rms speed of atoms in a certain star's atmosphere at 18,000 K is v. What is the rms speed of the same kind of atoms higher up in the star's atmosphere where the temperature is 2000 K?

16-B10 Calculate the temperature of the water in the can and in the pool in Prob. 16-A1.

16-B11 Calculate the temperature of the bottle in Prob. 16-A2.

16-B12 What would be the temperature of an ensemble of molecules similar to the oxygen molecule in Prob. 16-A3?

16-B13 Calculate the total translational KE of 4 moles of an ideal gas at 600 K.

16-B14 The best modern vacuum pumps are able to reduce the pressure in a container to about 1 picotorr (10^{-12} torr). (a) At such pressure, how many molecules are in 1 cm^3 at 0°C? (b) Assuming each molecule to be at the corner of a cube x cm on an edge, calculate x, the approximate average distance between molecules in the gas.

16-B15 "Absolute temperature is proportional to KE per molecule." This means that KE per molecule = (constant)(absolute temperature). Find the SI value of this constant.

16-B16 The specific heat capacity at constant pressure of a certain ideal gas is 0.220 cal/g·°C. The molecular weight of the gas is 32. (a) What is the specific heat capacity at constant volume, in cal/g·°C? (b) What is γ for this gas? (c) What can you tell about the molecular structure of this gas?

16-B17 For N_2O gas, $\gamma = 1.303$, and $C_p - C_v = 2.05$ cal/mol·K. Calculate C_p and C_v.

16-B18 Reconsider Prob. 15-A1, taking account of the fact that the temperature of the air would inevitably be increased during the sudden compression while the popgun is being fired. The process is described by the adiabatic gas law instead of by Boyle's law. (a) How far does the piston move before the cork pops out? (b) What is the temperature of the air just before the cork pops out (assumed to be at an initial temperature of 20°C)?

16-B19 A cylinder contains 300 cm³ of neon at 17°C and 100 kPa pressure. The gas is suddenly compressed to a pressure of 120 kPa. Calculate (a) the final volume and (b) the final temperature of the neon.

16-B20 Consider the metals iron, copper, aluminum, and lead. Calculate the specific heat capacity per cubic centimeter for each metal. Also calculate the specific heat capacity per mole for each metal. Arrange your results in the form of a neat table. Discuss any regularities in the table.

16-B21 How much heat is required to increase the temperature of 3 moles of ethane (C_2H_6) by 20 K at constant volume? (See Table 16-1.)

16-B22 Starting at 0°C, how much heat must be slowly added to 2 moles of oxygen (a) to double the volume, at constant pressure? (b) to double the pressure, at constant volume?

16-C1 Starting with Eq. 16-2, prove that the rms speed of molecules of a gas of molecular weight W at absolute temperature T is given by the equation $v = \sqrt{3RT/W}$. Use the equation to check the result of Example 16-1.

16-C2 Prove that Eq. 16-2 is also equivalent to $P = \frac{1}{3}dv^2$, where d is the density of the gas.

16-C3 Make dimensional checks for the equations derived in Probs. 16-C1 and 16-C2.

16-C4 Measurements on sodium atoms (Na) are made using a beam that escapes through a small hole in an oven heated to 500 K. With what speed do atoms leave the oven?

16-C5 Isotopes of uranium are separated by a thermal diffusion process that depends on a slight difference of rms speeds for two gaseous UF_6 molecules. Calculate the ratio of the rms speed of $^{235}U^{19}F_6$ to that of $^{238}U^{19}F_6$ (*Hint*: See Prob. 16-C1.)

For Further Study

16-6 The Distribution of Molecular Speeds

We derived in Sec. 16-2 an expression for the pressure exerted by molecules moving in a box: $P = \frac{1}{3}Nmv^2/V$. The derivation is marred by several weaknesses. For one thing, we assumed that all molecules have the same speed v. Also, the factor $\frac{1}{3}$ was introduced in a way that appealed to intuition but was not rigorous. We now proceed to a more detailed study of molecular motions, leading to the same result we derived earlier with the aid of simplifying assumptions.

We assume identical molecules, each of mass m, but suppose that they have different speeds

$v_1, v_2, \ldots, v_N$. We also suppose that the molecules are moving through the box in various directions, not all parallel to the x axis. Four such molecules are shown in Fig. 16-6. Molecule 1 has velocity $\mathbf{v}_1$ (a vector), and the components of its velocity have magnitudes denoted by v_{1x}, v_{1y}, and v_{1z}. Molecule 2 has components v_{2x}, v_{2y}, v_{2z} (for clearness, only v_{2x} is shown), and so forth for molecules 3, 4, ..., N. In particular, molecule 3 is moving along the y axis, so $v_{3x} = 0$.

Consider the impacts of molecule 1 on the face bc of the box. Molecule 1 approaches this face with a velocity v_{1x}, and the x component of the change of momentum at each collision is $2mv_{1x}$. Molecule 1 makes $v_{1x}/2a$ collisions per second

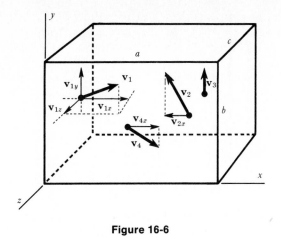

Figure 16-6

with face bc. By Newton's second law, the force on bc due to repeated impacts of molecule 1 is the rate of change of momentum:

$$F_1 = (2mv_{1x})\left(\frac{v_{1x}}{2a}\right)$$

$$= \frac{mv_{1x}^2}{a}$$

(The components of velocity v_{1y} and v_{1z} do not contribute at all to the force on bc.) Similarly, molecule 2 causes a force on bc equal to

$$F_2 = \frac{mv_{2x}^2}{a}$$

The total force of all N molecules on face bc is the sum of the forces caused by the separate molecules.

$$F = \frac{mv_{1x}^2}{a} + \frac{mv_{2x}^2}{a} + \cdots + \frac{mv_{Nx}^2}{a}$$

$$= \frac{m}{a}\left(v_{1x}^2 + v_{2x}^2 + \cdots + v_{Nx}^2\right)$$

$$= \frac{m}{a}\left(\Sigma v_x^2\right) \qquad (16\text{-}9)$$

To evaluate the sum, we recall that the magnitude of a vector is found by applying the Pythagorean theorem to the perpendicular com-

ponents. This theorem is general and applies to vectors in three dimensions as well as two. Thus, $v_1^2 = v_{1x}^2 + v_{1y}^2 + v_{1z}^2$, and similarly for v_2, $v_3, \ldots, v_N$. Let us make a table and form the sum of each column:

$$v_1^2 = v_{1x}^2 + v_{1y}^2 + v_{1z}^2$$
$$v_2^2 = v_{2x}^2 + v_{2y}^2 + v_{2z}^2$$
$$v_3^2 = v_{3x}^2 + v_{3y}^2 + v_{3z}^2$$
$$\vdots \qquad \vdots \qquad \vdots \qquad \vdots$$
$$v_N^2 = v_{Nx}^2 + v_{Ny}^2 + v_{Nz}^2$$
$$\overline{\Sigma v^2 = \Sigma v_x^2 + \Sigma v_y^2 + \Sigma v_z^2} \qquad (16\text{-}10)$$

For example, suppose molecule 1 has a large velocity, more or less diagonally through the box, with components $(3, 2, 3)$ in some system of units. Suppose molecule 2 is moving upward and to the left, with components $(-1, 2, 0)$. If molecule 3 is moving upward, its components might be $(0, 3, 0)$. The table of values would look like this:

$$v_1^2 = 9 + 4 + 9$$
$$v_2^2 = 1 + 4 + 0$$
$$v_3^2 = 0 + 9 + 0$$
$$\overline{\Sigma v^2 = 10 + 17 + 9}$$

We now make the assumption that the molecules are moving in *random directions*. That is, there is no breeze blowing through the box, and there is no preferred direction of motion for the molecules. In our example with only three molecules, the y direction is distinctly favored; but in the case of a measurable quantity of gas* we can surely say that if the molecules are in random motion,

$$\Sigma v_x^2 = \Sigma v_y^2 = \Sigma v_z^2$$

Therefore Eq. 16-10 becomes

$$\Sigma v^2 = \Sigma v_x^2 + \Sigma v_x^2 + \Sigma v_x^2$$

or

$$\Sigma v_x^2 = \tfrac{1}{3}\Sigma v^2 \qquad (16\text{-}11)$$

---

* Even a millionth of a mole of gas contains 6×10^{17} molecules!

Substituting into Eq. 16-9, we obtain

$$F = \frac{m}{a}\left(\frac{1}{3}\Sigma v^2\right) \qquad (16\text{-}12)$$

Since there is no speed common to all the molecules, our next step is to introduce some sort of average speed. Guided by Eq. 16-12, we define the mean-square speed as

$$\overline{v^2} = \frac{v_1{}^2 + v_2{}^2 + \cdots + v_N{}^2}{N} = \frac{\Sigma v^2}{N}$$

It is the average (or "mean") of the squares of the speeds of the N molecules. Multiplying by N, we see that

$$\Sigma v^2 = N\overline{v^2}$$

and hence

$$F = \frac{1}{3}\frac{m}{a}N\overline{v^2}$$

The rest of the proof goes much as in Sec. 16-2:

$$P = \frac{\text{force}}{\text{area}} = \frac{1}{3}\frac{m}{abc}N\overline{v^2}$$

$$PV = \tfrac{1}{3}Nm\overline{v^2} \qquad (16\text{-}13)$$

This means that PV is proportional to the average of the squares of all the speeds of the molecules, rather than to the square of any one speed.

In Example 16-1 we found that $v = \sqrt{3PV/Nm}$. How shall we interpret speed computed in this way? From Eq. 16-13 we see that $\overline{v^2} = 3PV/Nm$, so $\sqrt{\overline{v^2}} = \sqrt{3PV/Nm}$. Thus we are calculating the *root-mean-square speed*—the square root of the mean of the squares of the speeds. Without further specification, it is the rms speed ($\sqrt{\overline{v^2}}$ or v_{rms}) that is meant when we speak of the "speed" of the molecules in a gas. There are several other ways of describing the distribution of molecular speeds. For instance, we could simply average the speeds, which would result in the *mean speed* (v_m). Another index of molecular speeds that is sometimes useful is the *most probable speed* (v_p), the speed that the greatest number of molecules have. The distribution of speeds in a gas has been studied by James Clerk Maxwell in Cambridge,

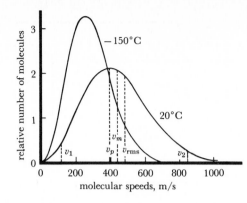

Figure 16-7 Maxwell distribution curve for molecular speeds at two temperatures. The ordinate is in percent, for a 10 m/s range of speeds. As an example, at 20°C, about 2% of the molecules have speeds between 395 and 405 m/s.

Ludwig Boltzmann in Vienna, and others. The *Maxwell distribution curve* (Fig. 16-7) shows that at any instant some molecules have very small speeds such as v_1 (a few are nearly at rest), and others have speeds such as v_2, far above the average value. For a large number of molecules in random motion, the rms speed is about 8% greater than the mean speed.*

In view of these results, it ordinarily makes little difference which of the various molecular average speeds is used to describe a gas, as long as we remember that there are a considerable number of molecules in the assemblage whose speeds differ by a substantial amount from the average value. Since the total KE of a collection of molecules equals $\tfrac{1}{2}M\overline{v^2}$ (Prob. 16-C6), and since KE is so importantly related to temperature, the rms speed is usually preferred as an all-around indicator of molecular speeds.

16-7   *The Speed of Sound in a Gas*

The ratio of specific heat capacities of a gas appears in the formula for the speed of sound, and values of γ are usually found by measuring

---

* This is to be expected, since the rms method of averaging gives greater weight, through squaring, to the larger speeds.

the speed of sound. In Sec. 10-3 we saw that the speed of sound in a fluid is

$$v_w = \sqrt{\frac{B}{d}}$$

where B is the bulk modulus and d is the density. We will use the gas laws to calculate d and B for an ideal gas (a gas that obeys the gas laws). The density is

$$d = \left(\frac{W}{V_0}\right)\left(\frac{P}{P_0}\right)\left(\frac{T_0}{T}\right)$$

where W is the molecular weight (numerically equal to the mass of 1 mole) and V_0, P_0, T_0 refer to standard temperature and pressure. For 1 mole of any ideal gas at any temperature and pressure,

$$\frac{PV}{T} = \frac{P_0 V_0}{T_0} = R$$

This gives

$$d = \frac{WP}{RT} \qquad (16\text{-}14)$$

The bulk modulus was defined in Sec. 9-1 as

$$B = \frac{\Delta P}{-\Delta V/V} = -\frac{V\,\Delta P}{\Delta V}$$

or, in the limit of a small pressure change,

$$B = \lim_{\Delta V \to 0}\left(-\frac{V\,\Delta P}{\Delta V}\right) = -V\frac{dP}{dV}$$

(The $-$ sign indicates that P decreases when V increases.)

To calculate dP/dV we use the gas law; but which shall we use, the isothermal law $P_1 V_1 = P_2 V_2$, or the adiabatic law $P_1 V_1^{\gamma} = P_2 V_2^{\gamma}$? We must recognize that in a sound wave the molecules of the gas undergo temperature changes; the gas becomes momentarily warmer wherever a condensation exists. A gas is a poor conductor of heat. At audible frequencies of, say, 1000 Hz, there is insufficient time for heat to flow out of the compressed region before the pressure relaxes and the condensation becomes a rarefaction, cooling the gas by expansion. The tendency

now is for heat to flow into the region, but before this can happen the region becomes a condensation and warms up again. The whole process is an adiabatic one (no flow of heat) rather than an isothermal one (no temperature change). Therefore, to find the bulk modulus, we differentiate the adiabatic gas law.

$$PV^{\gamma} = P_0 V_0^{\gamma}$$
$$P = (P_0 V_0^{\gamma})V^{-\gamma}$$
$$\frac{dP}{dV} = (P_0 V_0^{\gamma})(-\gamma V^{-\gamma-1})$$
$$= (PV^{\gamma})(-\gamma V^{-\gamma-1})$$
$$= -\gamma P V^{-1}$$

The bulk modulus is, then,

$$B = -V\frac{dP}{dV} = -V(-\gamma P V^{-1})$$
$$B = \gamma P \qquad (16\text{-}15)$$

We have proved that the adiabatic volume modulus of an ideal gas is γ times the pressure.

Finally, we calculate the speed of sound:

$$v_w = \sqrt{\frac{B}{d}} = \sqrt{\frac{\gamma P}{d}} = \sqrt{\frac{\gamma P}{WP/RT}}$$

or

$$v_w = \sqrt{\frac{\gamma RT}{W}} \qquad (16\text{-}16)$$

The velocity is proportional to the square root of the absolute temperature and is independent of pressure.

Example 16-3

Calculate the speed of sound in CO_2 gas at 50°C if the speed at 0°C is 258 m/s.

$$\frac{(v_w)_2}{(v_w)_1} = \sqrt{\frac{T_2}{T_1}}$$

$$(v_w)_2 = (v_w)_1 \sqrt{\frac{T_2}{T_1}} = (258 \text{ m/s})\sqrt{\frac{323 \text{ K}}{273 \text{ K}}}$$

$$= \boxed{281 \text{ m/s}}$$

16-C6 Prove that the total translational KE of N molecules is given by $\frac{1}{2}$(total mass)(v_{rms}^2), and hence, insofar as KE and absolute temperature are concerned, the molecules behave as if they all had the same speed, equal to the rms value.

16-C7 Ten molecules have speeds as follows, in arbitrary units: 2, 3, 0, 2, 2, 3, 3, 6, 1, 2. Calculate   (a) the most probable speed;   (b) the mean speed;   (c) the rms speed.

16-C8 Explain in words why the pressure has no effect on the speed of sound in a gas.

16-C9 Calculate the speed of sound in CO_2 gas at 20°C. (*Hint:* See Table 16-1.)

16-C10 If the speed of sound in a certain gas is 320 m/s at 200°C, what is the speed in the same gas at 400°C?

16-C11 At what temperature would the speed of sound in air be 300 m/s, as assumed in Prob. 10-B17 (page 237)?

16-C12 A pipe 60.0 cm long, closed at one end, resonates in its fundamental mode with frequency 140.0 Hz when filled with ethylene vapor (C_2H_4) at 27°C. Calculate the value of γ for ethylene at this temperature.

16-C13 In Prob. 11-B16*a* Qrxxt was successful on a day when the temperature was 333 K. At what temperature was Prxxg successful?

Reference

1. N. H. Fletcher, "Adiabatic Assumption for Wave Propagation," *Am. J. Phys.* **42**, 487 (1974); **44**, 486 (1976). A mathematical discussion of the use of the adiabatic bulk modulus in calculating the speed of sound in air.

17

The Second Law of Thermodynamics

The science of thermodynamics provides powerful methods of analyzing changes in pressure, temperature, and volume (for solids and liquids as well as gases) and the relations between these variables and quantities of heat and mechanical work. We have already used the first law of thermodynamics (Sec. 14-2); we repeat it for the sake of completeness:

All the heat added to a system can be accounted for as mechanical work, an increase in internal energy, or both.

In symbols

$$Q = W + \Delta U \quad \text{or} \quad \Delta U = Q - W$$

The first law is a statement of the law of conservation of energy, with heat flow being included along with mechanical work as a way of changing the internal energy of a system. Recall that for an ideal gas, internal energy is random translational KE of the molecules, and in this case $U = \frac{3}{2}nRT$ for n moles of gas. Thus for an ideal gas, U is proportional to the absolute temperature.

17-1 The Second Law of Thermodynamics

It is difficult to give a brief and all-inclusive statement of the second law, partly because there are so many ways of stating it, all of them equivalent.

The second law deals with a *trend* in nature. An event does not necessarily happen just because the conservation of energy (first law of thermodynamics) allows it. For example, a book slides across a rough table and comes to rest. We realize that according to the first law, the table and book have become slightly warmer; the increased random KE of their molecules has come from the original *ordered* translational KE of the book as a whole. Later, without violating the first law, the table and book *could* cool down, and the book *could* absorb the table's molecular KE and start to move. But we know that this won't happen. Another example: You have, no doubt, seen a motion picture of a diver going off a springboard into a pool. When projected backward, the result is amusing precisely because it is so certainly outside the realm of possibility.

Cohesive forces between molecules must be overcome as air expands. The work required to separate the molecules comes from the kinetic energy of the gas molecules, and therefore the gas cools during expansion. Atmospheric water vapor has condensed as frost on the low-pressure side of the system.

Yet it would not violate the law of conservation of energy (the first law of thermodynamics) for the churning water molecules to all get together after the dive and push the diver back up into the air. It is a fact of life, however, that the events described above are *irreversible*, as are *all* events on the macroscopic scale of things (when friction is taken into account). We are dealing here with what has been called "time's arrow." We cannot unscramble an egg; plants and animals grow older, not younger.

The trend is from order to disorder, from structure to randomness. The measure of the disorder of a system is given the name *entropy*. Thus we can state the second law of thermodynamics in terms of entropy:

The entropy of the universe never decreases; during any process, the entropy either remains constant or else increases.

Of course, the entropy of a *portion* of the universe can decrease, as when a human sorts out a deck of playing cards. In fact, entropy decreases during most activities of humans, animals, and plants. A living organism is a highly structured system, a pocket of low entropy. However, although human activity may decrease entropy in a small region of the universe, we can be sure that elsewhere, usually in the near environment of the human, entropy has increased and there has been a *net* increase of entropy in the universe.

The entropy of a system of molecules (such as a book or a beaker of water) is just as real and measurable as its energy. See Sec. 17-6 for a quantitative discussion of entropy changes.

Another statement of the second law of thermodynamics, equivalent to the one just given, applies to any closed system:

It is impossible to have a process whose only result is to transfer energy from a cold place to a warm place.

In other words, heat tends to flow from a hot place to a cold place. If we mix 1 kg of boiling water at 100°C with 1 kg of ice-cold water at 0°C, the result is 2 kg of water at 50°C; such calculations were made in Chap. 14. Now the total energy of the system has remained constant—as much heat was lost by the hot water as was gained by the cold water (first law of thermodynamics). Nothing in the first law would prohibit the reverse process from happening: during the mixing process, heat *might* leave the cold water (thereby freezing it) and flow into the hot water (thereby causing it to boil). For that matter, the first law would not be violated if the molecules of the 2 kg of 50° water were to sort themselves out into two groups, with the faster ones in one corner of the bucket and the slower ones in another corner, setting up a temperature difference again. But all these unlikely happenings would violate the second law, for we would have heat flowing spontaneously "from cold to hot." The second law says that it is highly unlikely that heat will flow from cold to hot—so unlikely that we can safely assert it to be an impossibility in everyday life.

We can make a connection between the two statements of the second law by noting that disorder is always more probable than order.* Heat tends to flow from a hot place to a cold place because the original situation (with molecules separated into fast and slow groups) is less probable than the final situation (in which molecular energies have, on the average, been equalized by many collisions). Thus the trend toward the *most probable configuration* is the basis for both statements of the second law: increase in disorder, and equalization of temperature differences.

The whole discussion of the second law, of entropy, and of time's arrow revolves around two facts: the atomicity of matter, and the extremely large number of particles involved in

---

* Just think of the 6.35×10^{11} possible different bridge hands, all equally probable, that can be dealt from a mere 52 cards—yet there are only 4 arrangements in which you, the player, get 13 cards all of one suit (a highly ordered result).

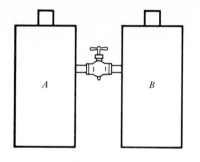

Figure 17-1

everyday events. On a submicroscopic scale—say, a dozen molecules in a small region of a container of gas—random collisions might act to make little "pockets" of hot spots (faster molecules) and cold spots (slower molecules). Averaged out, though, there is no lasting observable effect. On a larger scale of things—the macroscopic scale—imagine tanks of air, A and B, connected by a pipe, both tanks containing about a mole of gas (6×10^{23} molecules) at the same temperature (Fig. 17-1). For a significant amount of heat to flow "spontaneously" (that is, without external influence) from tank A to tank B, the molecules in A that happen to be moving faster would have to strike the pipe in such a way as to move into tank B, and likewise the slower molecules in B would have to move into tank A. It would be like tossing a coin "heads" 10^{23} times in a row—possible, but only one chance in $2^{10^{23}}$, which is about $10^{10^{22}}$. This is an unbelievably large number. If your friendly computer can print 10,000 zeros each second, it would take 300 billion years to print out $10^{10^{22}}$—longer than the age of the universe! Under such circumstances, "highly improbable" means "practically impossible." Such considerations of probability are at the very heart of the second law. It is in this probabilistic sense that we speak of the "impossibility" of the spontaneous setting up of a temperature difference or the spontaneous decrease of the entropy of the universe.

Since the entropy of the universe never decreases (second law of thermodynamics), it must tend toward some maximum value.* Eventually, when the entropy of the universe has reached its maximum value, everything will be at the same temperature (estimated to be a few degrees above absolute zero), and there will be no way to convert the thermal energy of the universe into useful mechanical work. This "heat death" is far in the future.

17-2 Heat Engines

A *heat engine* is a device for getting useful mechanical work out of the heat of combustion of fuel. Our civilization depends on heat engines—not only for transportation (locomotives, aircraft, automobiles, and trucks) but also for supplying mechanical energy to be converted into electrical energy by the generators in power plants using fossil fuels or nuclear fuels. In every case, a fuel is burned to supply random KE, which must be converted to useful (ordered) energy in the form of mechanical work. The key problem here is the unavailability of energy. There is a tremendous amount of internal energy in the oceans, for instance, but it is not available for doing work unless there is a region of lower temperature to which heat can flow.[†] Energy is "available" when it can be converted into mechanical or "useful" work.

In a heat engine a fixed amount of working substance, usually a gas, starts at some state given by a pressure, volume, and temperature. Some or all of P, V, and T change during expansion or compression, heating or cooling, or change of phase, until finally P, V, and T have

---

* Rudolf Julius Emanuel Clausius (1822–1888) gave a rallying cry for a generation of physicists and chemists when he stated the laws of thermodynamics in a terse couplet:

   Die Energie der Welt ist constant.

   Die Entropie der Welt strebt einem Maximum zu.

† Such a difference, no more than 25°C, is available at different depths in some tropical oceans. Engineering feasibility studies are being made of this possible energy source.

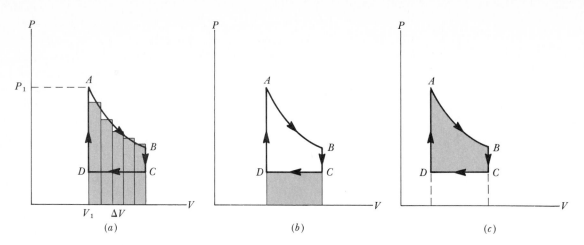

Figure 17-2 (*a*) Work done *by* the gas during expansion from *A* to *B*. (*b*) Work done *on* the gas during compression from *C* to *D*. (*c*) The shaded area is the net work done by the gas during one complete cycle.

their original values again and the working substance is ready for another cycle. The work output of a heat engine is most easily found by representing the cycle on a *P-V* diagram such as Fig. 17-2. (Information about temperature change and heat flow are not explicitly shown in such a diagram.) Suppose the working substance is an ideal gas, initially in state *A*, and suppose the gas first expands isothermally to state *B*, where the volume is $2V_1$ and the pressure is $P_1/2$ (Boyle's law). The work done *by* the gas during a small volume change is given by Eq. 16-7 on page 360:

$$\Delta W = P \, \Delta V$$

and the total work from *A* to *B* is approximately equal to the sum of the areas of many narrow rectangles (Fig. 17.2*a*). In the limit, as each $\Delta V \rightarrow$ 0, the work done becomes equal to the area under the curve. During path *BC*, the pressure decreases while the volume is held constant; no work is done since $\Delta V = 0$. During *CD*, work is done *on* the gas, as some external pressure compresses it; this work is the area under the horizontal line (Fig. 17-2*b*). Finally, the work from *D* to *A* is zero ($\Delta V = 0$). Thus the *net* work output during the cycle is the area enclosed within the closed curve *ABCDA* (Fig. 17-2*c*).

This is a general result and is true for a *P-V* cycle of any shape.

Now what about temperature changes and heat flow during the various segments of this particular cycle? During *AB*, heat must flow into the gas in order to keep the temperature constant as assumed (in terms of the first law, $\Delta U = 0$ for the isothermal change, so $Q = W$, and we have already seen that $W > 0$ for this part of the cycle). During *BC*, $W = 0$, hence $Q = \Delta U$. From $PV = nRT$, we see that *T* decreases since *P* decreases; hence ΔU and *Q* are negative. The temperature at *C* is less than at *A* or *B*, and heat flows out of the system during *BC*. . . . We could go on, but enough has been said to indicate that during a complete cycle, net work can be performed by or on the system, and heat can flow into or out of the system.

The best heat engine is one that gives the maximum output of useful work for a given heat input. The *thermal efficiency*, also called the *first-law efficiency*, is defined as

$$e = \frac{\text{work output}}{\text{heat input}}$$

The overall efficiency will be less than this, because of friction and other irreversible processes. We may represent a heat engine by the schematic

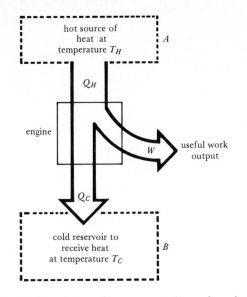

Figure 17-3 Schematic representation of a heat engine.

diagram of Fig. 17-3. The source of heat, A, according to the second law, must be at a higher temperature T_H than the reservoir B if heat is to flow from A to B. The heat Q_C rejected by the engine to the "cold" reservoir is less than the heat input Q_H received by the engine from the "hot" source; the remainder of the heat energy is converted into useful work W. By the first law, $Q_H = Q_C + W$; hence,

$$W = Q_H - Q_C$$

$$\text{Efficiency} = \frac{\text{work output}}{\text{heat input}} = \frac{W}{Q_H}$$

$$e = \frac{Q_H - Q_C}{Q_H}$$

The science of thermodynamics had its origin in the studies of N. L. Sadi Carnot (1796–1832), a young French engineer and physicist who tried to improve the efficiency of steam engines. His conclusions were general and apply to steam engines, diesel engines, gas turbines, and all other devices that convert heat into work. Carnot considered engines all working between two

temperatures T_H and T_C and came to three conclusions: (1) He proved that all *reversible** engines, of whatever construction, have the same efficiency when working between the same two temperatures. (2) He proved that any nonreversible engine has a lower efficiency than a reversible engine working between the same two temperatures. (3) He derived a formula for the efficiency of one particularly simple reversible engine, one that uses the *Carnot cycle* (illustrated in Fig. 17-4). For a Carnot engine, it turned out that

$$\frac{Q_H - Q_C}{Q_H} = \frac{T_H - T_C}{T_H} \qquad (17\text{-}1)$$

and therefore we can write

$$\text{Maximum efficiency} = \frac{T_H - T_C}{T_H} = \frac{\Delta T}{T_H} \qquad (17\text{-}2)$$

where temperatures T_H and T_C are the absolute (Kelvin) temperatures of source and reservoir, respectively.[†] Note that we are able to call this the *maximum* efficiency of *any* engine operating between temperatures T_H and T_C because of Carnot's proofs mentioned in (1) and (2) above. Not all engines have this efficiency, but Carnot's analysis showed that at best the efficiency of an engine has a theoretical limit.

It can be seen from Eq. 17-2 that the only way to get 100% efficiency from an engine would be to have the cold reservoir at absolute zero; if $T_C = 0$, then the efficiency would be T_H/T_H, or 100%. This would mean that $Q_C = 0$; an engine that was 100% efficient would reject no heat and would convert *all* the heat input into useful

---

* A reversible engine can be run in either direction; when the arrows in Fig. 17-3 are reversed, work is put into the system and we have a heat pump (Sec. 17-4). The presence of friction would cause an engine to be nonreversible, because work is done *against* friction whether the engine is run forward or backward.

† From Eq. 17-1 it follows that $T_C/T_H = Q_C/Q_H$. This is Kelvin's definition of the absolute temperature scale in terms of a Carnot engine (see Sec. 17-5).

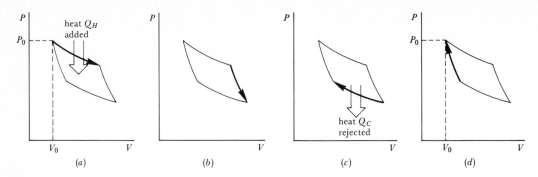

Figure 17-4 Carnot cycle, in which a fixed amount of working substance (an ideal gas) is originally at pressure P_0 and volume V_0. (a) Isothermal expansion—work W_1 is done by the gas as it expands; heat Q_H is supplied to keep the temperature constant at T_H. (b) Adiabatic expansion—gas is isolated from surroundings, allowed to expand and cool to T_C; no heat is added or removed. (c) Isothermal compression—heat Q_C is removed to keep the temperature constant at T_C while work W_2 is done on the gas. (d) Adiabatic compression back to original P_0 and V_0—no heat is added or removed. The overall result is that $Q_H - Q_C$ = net input of heat during cycle, and $W_1 - W_2$ = net output of work done by gas during cycle.

work. But this would violate the second law of thermodynamics, since disorder (random KE) would be converted entirely into order (mechanical work) and the entropy of a closed system would decrease. This cannot happen. So we have still another statement of the second law of thermodynamics:

> **No engine or other cyclic process can have as its sole result the conversion of heat entirely into work.**

In other words, it is impossible to have an engine with 100% efficiency. From this discussion, we can also infer that since T_C cannot be zero, the absolute zero of temperature can never be reached.

Actual heat engines are thus less than 100% efficient, even in the absence of frictional losses or thermal losses due to heat conduction. The maximum possible efficiency, given by $(T_H - T_C)/T_H$, is called the *Carnot efficiency* of the engine. Most real engines, such as a steam engine or a gas turbine, cannot be designed to have the Carnot efficiency, even in theory. Fric-

tional losses and conduction of heat through the metal parts of the engine lower the efficiency still more.

To illustrate these ideas, in a simple type of steam engine, the heat input might be at 200°C, the temperature of superheated steam at a pressure of 11,650 torr (Table 15-2). The steam is allowed to expand, pushing a piston, and the working substance, H_2O, condenses into water at 100°C. According to Carnot's theorem, the greatest possible efficiency for *any* engine operating between these temperatures is

$$\frac{473 \text{ K} - 373 \text{ K}}{473 \text{ K}} = 21\%$$

In other words, at least 79% of the heat input is unavailable for doing work, and this 79% is carried away in the hot water that is the end result of the cycle. In practice, frictional and other losses cut down the overall efficiency of a piston type of steam engine from 21% to about 10% or 12%. The efficiency can be increased by operating at higher temperatures and by improving mechanical design, as illustrated in the next example.

Example 17-1

The most efficient engine yet constructed is a supercharged, spark-ignited engine of the piston type, developed by Cooper-Bessemer, a division of Cooper Energy Services. The working substance, a mixture of natural gas and air, operates between temperatures of 1870°C in the firing chamber and 430°C in the exhaust. The overall efficiency, including all frictional losses, is such that in 1 hour, an input of 6.85×10^9 cal of heat produces 1.20×10^{10} J of useful mechanical energy. Calculate (*a*) the Carnot efficiency, (*b*) the actual efficiency, and (*c*) the power of this engine.

(*a*) For this engine, with $T_H = 2143$ K and $T_C = 703$ K, the Carnot efficiency is

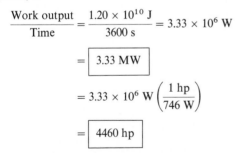

$$\frac{2143 \text{ K} - 703 \text{ K}}{2143 \text{ K}} = \boxed{67\%}$$

(*b*) The efficiency of the engine as actually constructed is

$$\frac{\text{Work output}}{\text{Heat input}} = \frac{1.20 \times 10^{10} \text{ J}}{(6.85 \times 10^9 \text{ cal})(4.18 \text{ J/1 cal})}$$

$$= \boxed{42\%}$$

(*c*) The power output, in megawatts and horsepower, is

$$\frac{\text{Work output}}{\text{Time}} = \frac{1.20 \times 10^{10} \text{ J}}{3600 \text{ s}} = 3.33 \times 10^6 \text{ W}$$

$$= \boxed{3.33 \text{ MW}}$$

$$= 3.33 \times 10^6 \text{ W} \left(\frac{1 \text{ hp}}{746 \text{ W}}\right)$$

$$= \boxed{4460 \text{ hp}}$$

Our study of heat engines shows that energy is "available" only when there is a reservoir at lower temperature to which heat can be rejected. This means that entropy, which is a measure of disorder, is also a measure of the unavailability of energy. In the example of mixing 1 kg of hot water and 1 kg of cold water, *some* energy was available at the start of the operation, since a heat engine could (conceivably) have been operated to make use of the temperature difference between the two bodies of water. However, after the mixing, this energy is no longer available because there is no temperature difference. The energy of the system has remained constant (first law), but the entropy of the system has increased (second law). More energy is unavailable than was unavailable at the start of the process.

17-3 Available Work and Thermal Pollution

We have seen how the second law of thermodynamics limits the efficiency of a heat engine to some maximum value, which we called the Carnot efficiency. Thus, not all the energy in a source such as a kilogram of coal can be used in a heat engine to do the desired task (such as turning a shaft).

Available work is the maximum work that can be provided by a fuel or by a system, such as an engine, by *any* process consistent with the laws of thermodynamics. The heat of combustion of a fuel is almost entirely "available" work, because (conceivably) it could operate a heat engine whose cold reservoir would be nearly at absolute zero. (Remember that the definition includes "any" process.) In this case, the Carnot efficiency would be nearly 100%, since T_H and ΔT would be nearly equal. As another example, the PE of a stretched spring is available work.

The *first-law efficiency* (*e*) has already been defined as the ratio of work output to heat input:

$$e = \frac{\text{work usefully performed}}{\text{heat input}}$$

In ordinary usage, "efficiency" means first-law efficiency, also called thermal efficiency.

The *second-law efficiency* (*ε*) is a measure of the fraction of the available work that is actually done (or heat transferred) by the device or process considered.

$$\varepsilon = \frac{\text{heat or work usefully transferred}}{\substack{\text{maximum possible heat or work that can}\\ \text{be usefully transferred using the same}\\ \text{energy input}}}$$

The value of ε can never exceed 1. Note that energy or work, as such, is never "used up"—it is merely transformed from one form to another (first law). But *available* work *is* used up or consumed; entropy increases, and the unavailability of work increases (second law). In terms of available work, the second-law efficiency for a process or task can be written as

$$\varepsilon = \frac{\substack{\text{minimum available work that could}\\ \text{perform a task}}}{\substack{\text{available work actually used to}\\ \text{perform the same task}}}$$

The laws of thermodynamics have been known for a century and a half, but their social consequences have only recently come into public awareness. Although we think of electric energy as "clean energy," we must realize that the generating plant itself is not necessarily clean. Such a plant is usually a large steam engine that takes in heat (from fossil fuel or nuclear fuel) and delivers mechanical power to the axles of large electric generators. A major portion of the input of heat becomes unavailable—it is *waste heat*.

Example 17-2

A steam-electric plant supplies a city with 612 MW of electric power, using 4500 metric tons of coal each day. The steam is at 277°C, and the exhaust water is at 40°C. The electric generators have 90% efficiency. (*a*) Calculate the first-law efficiency of the plant and compare it with the Carnot efficiency. (*b*) How much thermal energy is wasted each day?

In solving this problem we use the fact that a metric ton is 10^3 kg.

(*a*) The mechanical power input to the generator is 612 MW/0.90 = 680 MW. In 1 day the mechanical work done is

$$(680 \times 10^6 \text{ J/s})(8.64 \times 10^4 \text{ s}) = 5.88 \times 10^{13} \text{ J}$$

The heat of combustion of coal (from Table 14-3) is 33×10^6 J/kg, so the heat input in one day is

$$(3.3 \times 10^7 \text{ J/kg})(4.50 \times 10^6 \text{ kg}) = 1.48 \times 10^{14} \text{ J}$$

$$e = \frac{\text{work output}}{\text{heat input}}$$

$$= \frac{5.88 \times 10^{13} \text{ J}}{1.48 \times 10^{14} \text{ J}} = 0.40 = \boxed{40\%}$$

The Carnot efficiency (the maximum possible at these temperatures) would be

$$\frac{\Delta T}{T_H} = \frac{550 \text{ K} - 313 \text{ K}}{550 \text{ K}} = \boxed{43.1\%}$$

(*b*) The energy wasted is given by

Heat output = heat input − work output

$$= (14.8 \times 10^{13} - 5.88 \times 10^{13}) \text{ J}$$

$$= \boxed{8.9 \times 10^{13} \text{ J}}$$

This is 60% of the heat supplied by the coal.

The results of this example look pretty bad— more than half the energy of the coal is wasted. But, in fact, the hypothetical power plant does almost as well as possible by operating at nearly the Carnot efficiency. The wasted heat, while large in amount, is at a low temperature, and a still colder reservoir would be needed to convert this heat into mechanical energy. A real improvement would be to use this waste heat ("low-temperature heat") for household heating or industrial processes.

Sometimes waste heat is dissipated into the atmosphere. Often, water from a river or lake is circulated through the engine's condenser on a "once-through" basis. This can result in significant heating of the river. As the next example shows, a rather large river is needed.

Example 17-3

Calculate the temperature rise of the Hudson River (average flow rate 1.6×10^5 kg/s) if all the waste heat of the previous example were discharged into the river.

Power output $= 680 \times 10^6$ J/s

$$\text{Rate of heat input} = \frac{680 \times 10^6 \text{ J/s}}{0.40}$$

$$= 1700 \times 10^6 \text{ J/s}$$

Rate of heat rejection $= (1700 - 680) \times 10^6$ J/s

$$= 1020 \times 10^6 \text{ J/s}$$

The specific heat capacity of water is 4186 J/kg·°C, and hence the temperature rise is

$$\Delta T = \frac{Q}{mc}$$

$$= \frac{1020 \times 10^6 \text{ J/s}}{(1.6 \times 10^5 \text{ kg/s})(4186 \text{ J/kg·°C})} = \boxed{1.5\text{°C}}$$

The temperature rise would be 2.5°C in a season when the flow rate is 0.6 of the average rate. In practice, only a fraction of the water would be diverted through the power plant, with a correspondingly larger ΔT, but the net result would be the same after the hot exhaust water was thoroughly mixed with cold water downstream from the plant.

From the last example it appears that *one* medium-sized power plant could probably be tolerated, ecologically, but the problem will become critical as more and more power plants compete for available cooling water in the few rivers large enough to be useful. Thus, five plants like the one in Example 17-3 on the upper Hudson River would cause a temperature rise of 12.5°C during low-flow seasons, an intolerable prospect. In addition to ecological damage to the river environment, waste heat from power plants (whether fossil-fueled or nuclear-fueled) may cause subtle but perhaps important climatic changes.

Solar energy, in the present state of the art, involves use of collectors on a roof or side of a building to absorb "low-temperature heat," with correspondingly low Carnot efficiency (T_H is only slightly greater than T_C). Low efficiency is, however, of secondary importance, since the heat input costs nothing. Thermal pollution is also unimportant if the aim is to heat a building or warm up some water. It is quite a different problem to use solar energy to turn an electrical generator; here the low Carnot efficiency would lead to vast quantities of rejected heat for which we would have to find some use if thermal pollution were to be kept within reasonable limits.

17-4 The Heat Pump and the Refrigerator

The second law of thermodynamics tells us that heat will not *of itself* flow from a cold place to a hot place. However, if we are willing to expend some energy and do some work, it *is* possible to cause heat to flow "uphill." The *refrigerator* is a good example: heat is caused to flow from a cold place (the ice cubes) to a warm place (the kitchen). But this is done only by virtue of the fact that a motor is turning, expending electrical energy as the price to be paid for the backward transfer of heat. Some houses are built with reversible *heat pumps*. In summer they are air conditioners, moving heat from a cool place (the rooms of the house) to a warm place (the outside). In winter the heat is moved from outside to inside—once more from a cool place to a warm place. Such a "furnace" burns no fuel and has no chimney. The source of heat is free—usually from pipes buried in the ground—but the owner pays for comfort by paying for electric power to keep the motor of the heat pump turning. In general, any device that moves heat energy from a cold place to a warm place is a heat pump.

There are many ways to heat a house—furnace, solar heater, use of geothermal energy from a nearby hot spring (if any), or a heat pump run by an electric motor or driven by a heat engine, such as a diesel engine. To focus our ideas, we will discuss just one way of heating a house, the heat pump.

For house heating, we must move heat "uphill" from the cold outdoors to the warm interior. The heat pump is a heat engine operated in reverse; Fig. 17-5 is similar to Fig. 17-3

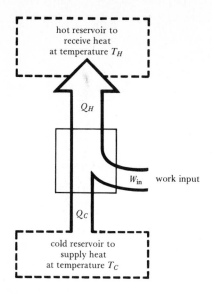

Figure 17-5 Schematic representation of a heat pump.

for a heat engine, but the arrows are reversed. If the device were run as an engine, its efficiency, which is W/Q_H, would have its maximum (Carnot) value if

$$\frac{W_{\text{max}}}{Q_H} = \frac{\Delta T}{T_H} \quad \text{or} \quad W_{\text{max}} = Q_H \frac{\Delta T}{T_H} \quad (17\text{-}3)$$

where $\Delta T = T_H - T_C$. When the device is run backward, as a heat pump, we are interested in knowing how much work is needed to move a quantity of heat Q_H into the house. The *coefficient of performance* (COP) is the measure of the pump's effectiveness, defined as

$$\text{COP} = \frac{Q_H}{W_{\text{in}}} = \frac{\text{heat moved}}{\text{work input}} \quad (17\text{-}4)$$

If the heat pump operates as efficiently as possible (limited by the second law), we can solve Eq. 17-3 for Q_H/W and find

$$\text{COP}_{\text{max}} = \frac{T_H}{\Delta T} \quad (17\text{-}5)$$

Since $T_H > \Delta T$, the value of COP_{max} is always >1. In actual practice, because of friction and other irreversible processes, the COP would be less than the ideal or maximum value given by Eq. 17-5. The second-law efficiency for the actual heat pump is, therefore,

$$\frac{(Q_H)_{\text{actual}}}{(Q_H)_{\text{max}}}$$

which can be written as

$$\frac{(Q_H/W_{\text{in}})_{\text{actual}}}{(Q_H/W_{\text{in}})_{\text{max}}} \quad \text{or} \quad \varepsilon = \frac{\text{COP}_{\text{actual}}}{\text{COP}_{\text{max}}}$$

Using Eq. 17-5 for the maximum COP, we find for the second-law efficiency of a heat pump

$$\varepsilon = \text{COP}_{\text{actual}} \left(\frac{\Delta T}{T_H}\right)$$

Example 17-4

A heat pump powered by an electric motor brings heat from outdoors (at 0°C) to air at 43°C, measured at the hot-air register in the room. (*a*) Calculate the maximum COP. (*b*) The heat pump has a COP of 2.20; calculate the second-law efficiency.

(*a*) $\quad \text{COP}_{\text{max}} = \dfrac{T_H}{\Delta T} = \dfrac{316 \text{ K}}{43 \text{ K}} = \boxed{7.35}$

(*b*) $\quad \varepsilon = \dfrac{\text{COP}_{\text{actual}}}{\text{COP}_{\text{max}}} = \dfrac{2.20}{7.35} = \boxed{0.30}$

This leaves plenty of room for improvement. For a given energy input (in this case electric energy), only 30% as much heat is usefully transferred to the place we want it as could be transferred if the heat pump had the maximum possible coefficient of performance.

Although the electric motor itself is better than 90% efficient in transforming electric energy into mechanical work, there is a large loss of available energy at the power plant, where a heat engine is turning the shaft of a generator. Usually about $\frac{2}{3}$ of the heat of combustion of the fuel is rejected, unused, to the "cold" reservoir of this heat engine, and the second-law efficiency of a typical power plant is only about $\frac{1}{3}$. In the above example, the overall second-law efficiency is the product $\varepsilon_1\varepsilon_2 = (0.30)(\frac{1}{3}) = 0.10$. There is a way to improve this situation. If the heat pump is run by a heat engine (such as a small diesel engine on the premises) instead of by an electric motor, the rejected heat of the engine could be reclaimed locally and used to help heat the house, instead of being wasted at the power plant. This strategy is probably most useful for larger buildings such as an apartment complex.

Finally, we should note that the COP of a heat pump can be significantly increased if a source of heat is available that is closer to the desired house temperature. In most areas, ground water a few meters below the surface of the earth is at a temperature of about $13°C$ throughout the year.

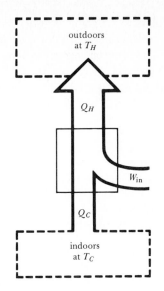

Figure 17-6 Schematic representation of an air conditioner.

Example 17-5

Suppose the heat pump of Example 17-4 uses ground water at $13°C$ instead of the atmosphere at $0°C$. Calculate the COP.

Now the temperature difference ΔT is only $43°C - 13°C = 30°C = 30$ K.

$$COP_{max} = \frac{T_H}{\Delta T} = \frac{316 \text{ K}}{30 \text{ K}} = 10.5$$

and if the second-law efficiency is still 0.30, the actual COP is now

$$COP_{actual} = (10.5)(0.30) = \boxed{3.2}$$

This COP is almost 50% better than that of Example 17-4. For a given expenditure of work (electric energy), almost 50% more heat can be transferred into the house if the ground water is available. Of course, there must be sufficient natural circulation of the ground water (not much is required) so that the source will not cool off appreciably as heat is pumped from the aquifer into the house.

The *air conditioner* or *refrigerator* is a heat pump that moves heat uphill from a cold place (the room) to a warm place (outdoors). The work is usually supplied by an electric motor (Fig. 17-6).

$$COP = \frac{\text{heat moved*}}{\text{work required}} = \frac{Q_C}{W_{in}}$$

From the first law, $W_{in} + Q_C = Q_H$; hence $W_{in} = Q_H - Q_C$, and

$$COP = \frac{Q_C}{W_{in}} = \frac{Q_C}{Q_H - Q_C} \qquad (17\text{-}6)$$

If the machine were run backwards, as an engine, and had the maximum (Carnot) efficiency, we would have (by algebraic manipulation of Eq. 17-1)

$$\frac{Q_H}{Q_C} = \frac{T_H}{T_C}$$

Equation 17-6 thus leads to

$$COP_{max} = \frac{T_C}{T_H - T_C} = \frac{T_C}{\Delta T} \qquad (17\text{-}7)$$

The usual room air conditioner has an actual COP of about 1.8 to 2.0. The temperature at the

---

* The numerator is now Q_C instead of Q_H, since the performance is judged by the quantity of heat that leaves the room or the ice-cube tray.

cooling coil of the machine is about 13°C (55°F) to allow for dehumidifying action, an important comfort consideration. (Moist air in the room becomes supersaturated when it is cooled below the dew point near the cooling coil, and liquid water condenses out and is drained off.)

Example 17-6

A house air conditioner moves heat from a coil at 13°C (55°F) to the outside environment at 32°C (90°F). (*a*) Calculate the maximum possible COP. (*b*) Calculate the overall second-law efficiency if the actual COP is 2.0 and the heat engine at the power station operates with a second-law efficiency of 0.33. (*c*) If the air conditioner moves 2.5×10^7 J of heat out of the house in 1 h, what must be the power rating of the air conditioner's motor?

(*a*) $\qquad \text{COP}_{\text{max}} = \dfrac{T_C}{\Delta T} = \dfrac{286 \text{ K}}{19 \text{ K}} = \boxed{15.1}$

(*b*) For the air conditioner itself,

$$\varepsilon = \frac{\text{COP}_{\text{actual}}}{\text{COP}_{\text{max}}} = \frac{2.0}{15.1} = 0.13$$

Taking account of the inability of the power plant to convert all the available work in the fuel (approximately equal to the heat of combustion) into electric energy, we have an additional factor of 0.33, making the overall second-law efficiency for the task

$$\varepsilon = (0.13)(0.33) = \boxed{0.043}$$

(*c*) $\quad W_{\text{in}} = \dfrac{\text{heat moved}}{\text{COP}} = \dfrac{2.5 \times 10^7 \text{ J}}{2.0}$

$$= 1.25 \times 10^7 \text{ J}$$

$$\text{Power} = \frac{\text{work}}{\text{time}} = \frac{1.25 \times 10^7 \text{ J}}{3600 \text{ s}} = \boxed{3500 \text{ W}}$$

Since 1 horsepower = 746 watts, a motor of about 5 horsepower would be used to air-condition this typical house.

It is worthy of note that the air conditioner rejects to the surroundings the *sum* of the heat moved and the work input (see Fig. 17-6). There is a significant rise in the air temperature of a large city where many air conditioners are being used.

In closing our discussion of engines and refrigerators, we return to the two kinds of efficiency already defined. The first-law efficiency is *device*-oriented; a furnace delivers to a house a fraction of the heat of combustion of the fuel, or a heat engine converts a fraction of the heat input into mechanical work. The second-law efficiency is *task*-oriented and will depend on the specific device or machine used to perform the desired task. We illustrate by a familiar task: heating a house with a furnace.

Example 17-7

In a certain furnace, 60% of the heat of combustion of the fuel is converted into thermal energy of hot air at 43°C. The outside temperature is 0°C. Calculate (*a*) the first-law efficiency of the furnace; (*b*) the second-law efficiency for the task.

(*a*) The first-law efficiency is

$$e = \frac{\text{heat output}}{\text{heat of combustion}} = \boxed{0.60}$$

From the point of view of the first law (conservation of energy), this furnace could be improved but not very much. To deliver heat Q to the air, the heat input (from combustion) would be $Q/0.60 = 1.67Q$. The heat of combustion represents an actual using up of available work of $1.67Q$ (see the second paragraph of Sec. 17-3).

(*b*) Consider the *task* of heating the house. A heat pump *could* do the job, with a maximum possible coefficient of performance (Eq. 17-5) of

$$\text{COP}_{\text{max}} = \frac{T_H}{\Delta T} = \frac{316 \text{ K}}{43 \text{ K}} = 7.35$$

Since

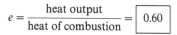

we have

$$W_{\text{min}} = \frac{Q}{\text{COP}_{\text{max}}} = \frac{Q}{7.35} = 0.136Q$$

The second-law efficiency is

$$\varepsilon = \frac{\text{minimum available work used to perform task}}{\text{available work actually used}}$$

$$= \frac{0.136Q}{1.67Q} = \boxed{0.082}$$

We see from this example that the task of heating a house is performed with a second-law efficiency of only 8.2% if the given furnace is used. Other ways of heating a house can be developed—the use of a heat pump, for example—that approach more closely the ideal of 1.00 for the maximum second-law efficiency.

The commercial aspects of thermodynamics are interwoven with ecology, economics, and sociology in such a complex way that any further discussion in this book would have to be greatly oversimplified. See the references at the end of the chapter for additional readings.

17-5 Absolute Zero

So far, we have used absolute zero simply as a convenient tool in the description of the behavior of ideal gases. The kinetic theory showed that for an ideal gas the KE of the molecules is zero at absolute zero. At temperatures of 5 or 10 degrees above absolute zero, however, most gases are far from ideal, if indeed they have not yet condensed into liquid or solid. It is questionable, therefore, whether our definition of absolute zero has much meaning, since the gas laws that were used to define absolute zero no longer are working well near absolute zero. We now have a better way of defining absolute zero, based on Carnot's work with heat engines.

We have seen (Sec. 17-2) that the efficiency of a Carnot engine would be 100% if the temperature of the cold reservoir were at absolute zero. For 100% efficiency, *all* the heat input Q_1 would be converted into useful work, so that no heat would be rejected to the reservoir. This leads to

Lord Kelvin's definition of absolute zero:

Absolute zero is the temperature of a reservoir to which a Carnot engine would reject no heat.

This is the only definition that will stand up under close scrutiny. A Carnot engine* is not restricted to any particular substance. Lord Kelvin, extending these ideas, developed the *absolute temperature scale*, which does not depend on the properties of any substance, and so there is no problem of deciding which thermometric property and which substance should be used. (This was the problem raised in Sec. 13-1.) Kelvin also showed that the ideal-gas thermometer scale is identical with the absolute thermodynamic scale, which is one reason we have stressed the properties of ideal gases in our work. Unlike the Celsius scale, the Kelvin scale makes use of just *one* reproducible temperature. The size of the Kelvin degree is fixed by *defining* the triple point of H_2O (Sec. 15-10) to be exactly 273.16 K. Since 0°C is 0.01°C below the triple point of water, this makes absolute zero equal to −273.15°C.

Absolute zero is sometimes defined (erroneously) as the temperature at which all random molecular motion stops. If there is random molecular motion *that can be utilized*, then surely we are not at absolute zero. But this is not the whole story. Even if the molecules were devoid of random linear, rotational, or vibrational KE, there might still be sources of available energy, such as magnetic energy or electric energy or some as yet unknown form of internal energy. Such molecules (in theory, at least) could run an engine and so would *not* be at absolute zero, since there would then be a lower temperature to serve as a reservoir.

Of course, the residual magnetic and other energies are so small that they are usually of no significance whatever. At ordinary temperatures,

---

* Remember that a "heat engine" is any device for transforming heat into mechanical energy. This process may be carried out entirely without moving parts, and the working substance may be solid, liquid, or gas.

the molecular KE so far overshadows these smaller energies that it may well be said that absolute temperature is proportional to the KE of the molecular motion. At very low temperatures, below 1 K, magnetic and other energies are the major source of available energy, since molecular KE is, by comparison, already zero. The quest for ever lower temperatures is one of the active fields of research in physics. At present, the lowest attainable temperature for bulk matter in the solid or liquid state is about 10^{-4} K. On an atomic scale, temperatures of less than 10^{-7} K have been achieved for spinning nuclei, but such temperatures are not useful since no larger body such as a crystal lattice can be put into thermal equilibrium with the nuclei for which the temperature is defined. The properties of matter are radically different in some respects at very low temperatures (for instance, note the phenomenon of superconductivity in some metals—Sec. 20-6), and low-temperature research is bringing us a better understanding of the properties of matter and the basic laws of physics.

Summary The second law of thermodynamics may be stated in several equivalent ways. Some of the consequences of the second law are as follows: (*a*) Heat tends to flow from a hot place to a cold place; (*b*) no heat engine can have 100% efficiency; (*c*) internal energy tends to become more and more unavailable for transformation into mechanical work; (*d*) the disorder of the universe tends to increase; (*e*) the entropy of the universe tends toward a maximum.

Heat engines are devices for extracting useful work from a source of heat; it is impossible to convert all the heat that flows into an engine into work unless the cold reservoir is at absolute zero. A heat engine operated in reverse is a heat pump or a refrigerator, in which heat flows from a cold place to a hot place; an external source of mechanical work is required to perform this feat.

Absolute zero is the temperature of the cold reservoir of a heat engine to which no heat would be rejected. At absolute zero, random molecular or atomic motions would have the lowest possible values permitted by quantum theory; in addition, all other forms of available energy would have to be removed from the atoms or molecules.

Check List first law of thermodynamics
second law of thermodynamics
entropy
heat engine
heat pump
refrigerator
first-law efficiency of a heat engine
thermal efficiency of a heat engine

$$e = \frac{Q_H - Q_C}{Q_H}$$
Carnot engine
$$\text{Carnot efficiency} = \frac{T_H - T_C}{T_H}$$
available work
second-law efficiency of a process
absolute zero

Questions **17-1** Mercury can be made to boil (under pressure) at a temperature as high as 473°C. Why would mercury be a more efficient working substance for a heat engine than water? (Mercury boilers are used in some high-efficiency power plant installations.)

17-2 In some parts of the Indian Ocean, the surface water is 23°C warmer than the water 100 m below the surface. Would it be possible to operate a steam engine continuously, using this temperature difference? Would some working substance in the engine other than H_2O make this possible, or is the whole idea theoretically unsound?

17-3 Would it be possible to cool a room by placing an electric refrigerator with its door open in the room and allowing a fan to blow cool air from the ice-cube compartment out into the room?

17-4 Why is it not advisable to place the back of a household refrigerator snugly against the kitchen wall?

17-5 Quantum theory predicts that a vibrating molecule can have only certain specified values of vibrational energy and that there is a least amount of vibrational energy that is possible. In other words, it is impossible for a molecule to have zero vibrational energy. The molecule has this "zero-point" energy even at absolute zero (insofar as a model is possible). Is such energy "available"? Does possession of this energy conflict with Kelvin's definition of absolute zero?

MULTIPLE CHOICE

17-6 When water in a well-insulated container was stirred for a while, the temperature was observed to rise. During this process (a) heat flowed into the water; (b) work was done on the water; (c) both of these.

17-7 The thermal efficiency of a Carnot engine is (a) 1; (b) equal to the ratio of heat output to heat input; (c) the maximum possible value for any engine operating between the same two temperatures.

17-8 The coefficient of performance of a heat pump (a) is always less than 1; (b) would be 1 if there is no frictional loss; (c) can be greater than 1.

17-9 For a closed system that is isolated from its surroundings, (a) the total energy is conserved; (b) the total entropy is conserved; (c) both of these.

17-10 When ice cubes are made in a household refrigerator, the entropy of the water (a) increases; (b) remains the same; (c) decreases.

17-11 At absolute zero a molecule could have (a) translational KE; (b) some unavailable vibrational energy; (c) neither of these.

Problems   **17-A1** A heat engine absorbs 5 cal from the source of heat and does 8.37 J of mechanical work. What is its efficiency?

17-A2 A heat engine takes in 100 kJ of energy from a source at 800 K and rejects 40 kJ of heat to a reservoir at 300 K. What is the efficiency of the engine?

17-A3 What is the maximum efficiency of an engine that operates between fixed temperatures of 600 K and 420 K?

17-A4 A heat engine performs 2000 J of work and at the same time rejects 6000 J of heat to the cold reservoir. What is the efficiency of the engine?

17-A5 An engine whose efficiency is 10% does 220 J of work. How much heat does it take in from the hot reservoir?

17-A6 An engine whose maximum (Carnot) efficiency is 30% obtains energy from a hot reservoir at 600 K. What is the temperature of the cold reservoir?

17-B1 An engine whose efficiency is half the Carnot efficiency operates between temperatures of 100°C and 500°C and rejects heat to the cold reservoir at the rate of 20 kW. What is the rate of heat input to the engine, in kW?

17-B2 A certain automobile obtains useful (mechanical) power from the engine at the rate of 136 kW when moving at a certain speed. The thermal efficiency is 30%, and the gas mileage at this speed is 6.4 km/liter. What is the gas mileage at the same speed if an air conditioner is turned on that requires an extra 3.7 kW of mechanical power?

17-B3 A steam power plant has a boiler that can safely withstand a pressure of 84.74 atm. The steam is rejected at 100°C, and 20% of the mechanical output is used to overcome friction. (*a*) What is the maximum temperature of the boiler? (See Table 15-2.) (*b*) What is the Carnot efficiency of the engine? (*c*) What is the maximum possible overall efficiency of the power plant?

17-B4 A Carnot engine performs work at the rate of 1600 kW. The source of heat is at 727°C, and there is a difference of temperature of 400°C between the input heat source and the output heat reservoir. How much heat is being "lost" each hour?

17-B5 A Carnot engine takes in 25 kcal of heat from a reservoir at 1000 K and performs 2×10^4 J of work. What is the temperature of the cold reservoir?

17-B6 An engine working at maximum efficiency between temperatures of 300 K and 500 K rejects 1800 J of heat during a certain time. How much work is performed during this period?

17-B7 The useful power output of an engine is 25 kW, and the frictional losses within the engine are 5 kW. The engine operates between temperatures of 227°C and 77°C. (*a*) What is the Carnot efficiency of the engine? (*b*) If the thermal efficiency has the maximum possible value, how many joules of thermal energy are taken in per hour by the engine? (*c*) What is the overall efficiency of the engine?

17-B8 In a research program to utilize the earth's heat for electric power generation, an exploratory hole is being drilled in Idaho to a depth (about 2000 m) where the rock temperature is about 130°C. The surface temperature is about 20°C. (*a*) Calculate the maximum efficiency of such a generating plant. (*b*) Explain how even such a low efficiency as calculated in part (*a*) could be worth utilizing.

17-B9 A floating power plant using oceanic temperature differences has been proposed. The surface water is at 22°C and the deep seawater is at 2°C. (*a*) Calculate the maximum efficiency of such a plant. (*b*) If the useful output is 80 MW, at what rate is thermal energy being rejected to the surrounding water by the plant? (*c*) Explain why even such a low efficiency as calculated in part (*a*) could be worth utilizing. What are some of the foreseeable difficulties?

17-B10 A fossil-fueled electric generating plant has an output of 900 MW of power, using steam at 227°C and exhaust water at 47°C. (*a*) If the system operates with maximum (Carnot) efficiency, at what rate (in MW) is heat rejected to the environment? (*b*) What temperature rise would the waste heat cause in a river in which the flow rate is 2×10^5 kg/s?

17-B11 A typical house heat pump is designed to deliver air at 37°C at the duct (warmer than room temperature to avoid the sensation of chilliness). For efficient exchange of heat to the air, the "hot" temperature of the heat exchanger should be 8°C warmer than the air, that is, 45°C. The temperature outdoors is assumed to be 5°C. (*a*) Calculate the maximum COP of the heat pump operating between 5°C and 45°C. (*b*) Calculate the actual COP if the second-law efficiency is 0.33. (*c*) What power must be supplied by the motor if heat is supplied to the house at the rate of 24 kW?

17-B12 A household refrigerator is a heat pump with the hot source at room temperature (say, 27°C) and the cold reservoir at the temperature of the freezing compartment (−15°C). (*a*) Calculate the maximum COP. (*b*) Calculate the actual COP if the second-law efficiency is 0.30.   (c) If the cost of mechanical energy (as realized by an electric motor) is 5¢/kWh (1 kWh = 3.6 MJ), what would be the cost of making 1 kg of ice cubes by changing water at 27°C to ice at −15°C?

17-B13 (a) What power in watts would a Carnot refrigerator require in order to pump heat at the rate of 41.4 W from an ice-cube tray at −1°C to a room at 21.7°C?   (b) An actual refrigerator (see Reference 11 at the end of the chapter) requires 31.6 W to perform the task studied in part (a). What were the actual COP and the Carnot COP for this refrigerator?

17-C1 A run-down watch is dropped into a beaker of acid and is dissolved completely, releasing a considerable amount of heat in a chemical reaction.   (a) Would the amount of heat released be less, or would it be greater, if the watch were first wound before it was dropped into the acid?   (b) Make a rough calculation of the temperature difference to be expected due to the winding of the watch. Assume that the acid and watch have a total water equivalent of 130 g, and that the owner of the watch, while winding it, turned the stem through 10 revolutions, exerting a force of 1.2 N at a lever arm of 3 mm.

17-C2 In the second part of the Carnot cycle (Fig. 17-4b), work is done by the ideal gas as it expands adiabatically.   (a) Using the first law, prove that this work depends only on temperatures T_H and T_C.   (b) Prove that an equal amount of work is done on the gas in part d of the cycle. In this way, verify the statement in the caption to the figure that the net output of work during the entire cycle is just $W_1 - W_2$.

17-C3 (a) In the Carnot cycle (Fig. 17-4), is there any change in the internal energy of the gas during part a or part c of the cycle?   (b) Using the result of Problem 17-C2, prove that the efficiency of the Carnot cycle is given by $(Q_H - Q_C)/Q_H$.

For Further Study

17-6   Calculation of Entropy Changes

Before discussing entropy, let us recall some of the properties of internal energy U. Only *changes* of internal energy are defined by the equation $\Delta U = Q - W$; increase of internal energy equals the difference between the heat input to a system and the work done by the system. A system in a given state (such as 1 kg of water at 100°C) has a definite amount of internal energy (relative to some arbitrarily assumed reference energy), no matter how the system reached this state. The same cannot be said about Q or W. Work W and heat or heat flow Q are things you do to a system; internal energy is something a system has.

Entropy, like internal energy, is something a system has when it is in any given state, no matter how it reached that state. Entropy is denoted by the symbol S, and like internal energy, only *changes* of entropy are defined. When a small quantity of heat ΔQ flows into or out of a system, the change in entropy of the system is given by

$$\Delta S = \frac{\Delta Q}{T} \qquad (17\text{-}8)$$

where T is the absolute temperature of the system. Units of entropy include J/K, cal/K, and kcal/K.

Example 17-8

When 80 cal of heat is supplied to an ice cube at 0°C, 1 g of ice becomes water at 0°C. What is the entropy change?

$$\Delta S = \frac{80 \text{ cal}}{273 \text{ K}} = +0.183 \text{ cal/K}$$

The change is positive because ΔQ is positive when heat flows into the system.

Let us calculate the entropy change when 1 kg of water at 100°C is mixed with 1 kg of water at 0°C to produce 2 kg of water at 50°C. Here the temperature is hardly constant enough to apply Eq. 17-8 as it stands, but we can make an approximate calculation using many small steps. First consider a small change as 1 kg cools from 100°C to 99°C. During this step, 1000 cal flows out of the hot water.

$$\Delta S = \frac{\Delta Q}{T} \approx \frac{-1000 \text{ cal}}{372.5 \text{ K}} \approx -2.685 \text{ cal/K}$$

where 372.5 K is the average of the initial temperature (373 K) and the new temperature (372 K). Let us consider that this 1000 cal is used to heat the cold kilogram from 0°C to 1°C. During this step the cold kilogram gains entropy:

$$\Delta S = \frac{\Delta Q}{T} \approx \frac{+1000 \text{ cal}}{273.5 \text{ K}} \approx +3.656 \text{ cal/K}$$

So far, during this initial step in the approach to equilibrium, the internal energy of the system has remained constant:

$$\Delta U = -1000 \text{ cal} + 1000 \text{ cal} = 0$$

but the entropy of the system has increased:

$$\Delta S = -2.685 \text{ cal/K} + 3.656 \text{ cal/K}$$
$$= +0.971 \text{ cal/K}$$

This is in accord with the second law of thermodynamics, as stated in terms of entropy on page 372. The original temperature difference of 100 K is now only 98 K, and a Carnot engine working between 99°C and 1°C would be less efficient than one working between 100°C and 0°C. Thus energy has become less available, in accordance with our statement that entropy is a measure of the unavailability of energy (page 373). Also, there is less order in the system after the temperatures of the two masses of water have been partially equalized, thus confirming that increase of entropy implies an increase of disorder (page 372).

We could continue the calculation of ΔS for each 1° step until finally 1 kg of water at 51°C mixes with 1 kg at 49°C to produce the final state. The change of entropy for the final 1° step works out to be +0.0098 cal, smaller than that for the first step, but still positive. To a good approximation, the total entropy change is the sum of the values of ΔS. Carrying this through for the mixing process considered, we get

$$S_2 - S_1 = \Sigma \Delta S$$
$$= +0.971 + \cdots + 0.0098 \approx +24 \text{ cal/K}$$

A precise evaluation of $S_2 - S_1$ can be made using integral calculus. We use the specific heat capacity of water to get dQ in terms of dT. The change of entropy of the hot water is

$$S_2 - S_1 = \int \frac{dQ}{T} = \int_{373}^{323} \frac{mc \, dT}{T}$$
$$= (1000 \text{ g}) \left(1 \frac{\text{cal}}{\text{g·K}}\right) \int_{373}^{323} \frac{dT}{T}$$
$$= 1000 \frac{\text{cal}}{\text{K}} \Big[\ln T\Big]_{373}^{323}$$
$$= 1000 \frac{\text{cal}}{\text{K}} (\ln 323 - \ln 373)$$
$$= 1000 \frac{\text{cal}}{\text{K}} (5.778 - 5.922)$$
$$= -143.9 \frac{\text{cal}}{\text{K}}$$

Similarly, the change of entropy of the cold water is

$$S_2 - S_1 = 1000 \frac{\text{cal}}{\text{K}} \int_{273}^{323} \frac{dT}{T}$$
$$= 1000 \frac{\text{cal}}{\text{K}} (\ln 323 - \ln 273)$$
$$= 1000 \frac{\text{cal}}{\text{K}} (5.778 - 5.609)$$
$$= +168.2 \frac{\text{cal}}{\text{K}}$$

The net change of entropy of the system is

$$-143.9 \frac{\text{cal}}{\text{K}} + 168.2 \frac{\text{cal}}{\text{K}} = +24.3 \frac{\text{cal}}{\text{K}}$$

Problems

17-C4 Calculate the change of entropy when 400 g of water at 0°C is converted to 400 g of ice cubes at 0°C in a household refrigerator.

17-C5 A 0.4-kg block of aluminum is placed in an oven and heated from 27°C to 227°C. (*a*) What is the increase in the block's internal energy?   (*b*) What is the increase in the block's entropy?

17-C6 Calculate the net change of entropy when 40 g of water at 0°C is mixed with 100 g of glycerin at 50°C.

17-C7 Which of the following are suitable units for entropy? N/K; J/K; N·m/kg·K; cal/g·K; cal/K; kWh/K; kg·m/K.

17-C8 When a hot bowl of soup is allowed to cool to room temperature, its entropy decreases. Explain how this does not violate the entropy form of the second law of thermodynamics.

References

1. C. M. Summers, "The Conversion of Energy," *Sci. American* **225**(3), 148 (Sept. 1971).
2. S. W. Angrist, "Perpetual Motion Machines," *Sci. American* **218**(1), 115 (Jan. 1968).
3. W. Carnahan et al., *Efficient Use of Energy: A Physics Perspective.* A report of a study group sponsored by the American Institute of Physics, 1974. (Pub. by National Technical Information Service, U.S. Dept. of Commerce, Springfield, Va. 22151.)
4. R. H. Socolow, ed., "Efficient use of energy," *Physics Today* **28**(8), 23 (Aug. 1975). A summary of the material in Reference 3.
5. A. B. Meinel and M. P. Meinel, "Physics looks at solar energy," *Physics Today* **25**(2), 44 (Feb. 1972).
6. "Solar Energy Technologies," a series of 13 articles in *Bull. Atomic Scientists* **31–33**. See that publication's indexes for 1975–1977 for a complete listing. The following are of special interest:

   F. von Hippel and R. H. Williams, "Solar Technologies," **31**(11), 25 (Nov. 1975). An overview.

   C. Zener, "Solar Sea Power," **32**(1), 17 (Jan. 1976).

   M. J. Antal, Jr., "Tower Power," **32**(5), 58 (May 1976).

   M. H. Ross and R. H. Williams, "Energy Efficiency: Our Most Underrated Energy Resource," **32**(11), 30 (Nov. 1976).

   R. S. Caputo, "Solar Power Plants: Dark Horse in the Energy Stable," **33**(5), 46 (May 1977).

   F. von Hippel and R. H. Williams, "Toward a Solar Civilization," **33**(10), 12 (Oct. 1977). Final article in the series.
7. Rubber under tension contracts when thermal energy is absorbed. Simple rubber-band engines are described in C. L. Stong's "Amateur Scientist" department of *Scientific American* by R. Hayward, **194**(5), 149 (May 1956), and by P. B. Archibald, **224**(4), 118 (Apr. 1971).
8. G. Walker, "The Stirling Engine," *Sci. American* **229**(2), 80 (Aug. 1973). A low-noise, low-pollution external combustion engine.
9. G. Waring, "Energy and the automobile," *Phys. Teach.* **18**, 494 (1980).
10. H. S. Leff and W. D. Teeters, "EER, COP, and the second law efficiency for air conditioners," *Am. J. Phys.* **46**, 19 (1978).
11. A. A. Bartlett, "Introductory experiment to determine the thermodynamic efficiency of a household refrigerator," *Am. J. Phys.* **44**, 555 (1976). Describes a method of finding the coefficient of performance of a refrigerator.

18

Electric Charge

As we have seen in Sec. 1-5, one of the distinguishing attributes of matter is the existence of electric forces of several kinds. We are now prepared to study these forces in more detail in the next six chapters, and to learn how the application of fundamental ideas has led to so many useful developments in science, industry, and the home. With very few exceptions, we shall deal only with the long-range aspect of electric forces. These are complex enough, to be sure. One of the difficulties in making an orderly presentation of the subject is the high degree of interrelationship between the concepts. At the start, you will be asked to take certain facts for granted and even to use freely some terms that have not yet been precisely defined. However, before you finish the course, you will (we hope) gain for yourself an inclusive view of electric phenomena.

18-1 Electric and Magnetic Forces

Our study of the electrical structure of matter (Sec. 1-7) showed that each atom has a heavy, positively charged central nucleus, surrounded by a cloud of small, negatively charged electrons.

A "charged body" can be produced by creating an excess or a deficiency of electrons, so that a body can contain a net negative charge or a net positive charge.

The force between *stationary* charged bodies is called *Coulomb force* and will be discussed in Sec. 18-5. In addition to the Coulomb forces that are always present, there are forces that arise only when two charged bodies are in motion relative to each other. These forces due to *motion* of charges are called *magnetic forces;* they are discussed in detail in Chap. 22. Magnetism is thus a subdivision of electricity, dealing with the effects of charges in motion.

Another way of looking at this division is to note the distinction between charge and current. The rate of flow of charge is called *current*. Using this terminology, we can say that Coulomb forces are caused by electric charges, as such, whereas magnetic forces are caused by electric currents.

18-2 Electrification of Bodies

It has long been known that small bits of paper, fluff, and so on, can be "picked up" by a rod of

Negative charge on the rubbed plastic pen has repelled an excess of negative charge (electrons) to the leaves of the aluminum-foil electroscope.

glass, hard rubber, amber, or plastic that has been rubbed with fur, silk, or cloth. This phenomenon was observed by the Greeks, whose word for amber was "elektron." In Elizabethan times a piece of amber, glass, or other material that could exert a force on a small test body such as a scrap of paper or a silk thread was said to be "electrified."* These long-range electric forces could be observed at distances of more than a meter.

Not until the early 19th century did investigators come to the conclusion that there are two, and only two, kinds of electricity. For instance, let body A be a glass rod that has been rubbed with silk, and let B be a rubber rod that has been rubbed with fur. A force of attraction is observed between body A and body B. However, if C is a glass rod rubbed with silk (C is thus identical with A), then it is found that A is repelled by C, and B is attracted to C. If A and B are oppositely charged, it is impossible to find a third kind of charge that will be attracted by both A and B or repelled by both A and B.

Historically, there has been a great deal of needless confusion related to the naming of the two kinds of charge. The kind of electricity usually found on rubbed glass was called "vitreous," and the kind usually found on rubbed amber was called "resinous." Charles Du Fay (1698–1739) advocated a "two-fluid" concept of electricity, embracing vitreous and resinous electricity on equal footing. Benjamin Franklin (1706–1790) proposed a "one-fluid" theory in which vitreous electricity was an excess of positive electricity, and resinous electricity a deficiency of positive electricity—hence called negative electricity by Franklin. As we now know, there *are* two kinds of electricity, and so the two-fluid theory is actually "correct"; yet the one-fluid theory has some merit, since in most large-scale applications only one type of electricity is free to "flow." Franklin's choice of the terms "positive" and "negative" was unfortunate, since the free electrons that

move through wires and light our lamps and turn our motors are of the negative sort. This negative electricity actually exists and is not merely a deficiency of positive electricity.

Let us look briefly at those features of these two electric theories that made them so difficult to accept. In the one-fluid theory, the fluid was "electricity," and it repelled itself. Since oppositely charged bodies attract each other, and since a negative body was characterized by absence of the electric fluid, it became necessary to assume that electric fluid attracts ordinary matter. Now, however, arises the problem of repulsion between two negatively charged bodies, which were supposed to have no electric particles of any sort. Franklin himself worried about this; his follower the German Aepinus (1724–1802) boldly assumed that particles of ordinary matter repel each other! In the ordinary, unelectrified state, all matter was assumed to have just enough electric fluid so that the force of attraction on the particles of matter exerted by the electric particles was exactly balanced by the repulsion exerted by the particles of matter on each other. A cumbersome, artificial system of this kind could not endure long except as a last resort.

The proponents of the two-fluid theory had their troubles, too. A major problem was posed by the idea that uncharged matter contained equal quantities of positive and negative electric particles, ready to be separated by friction and yet not producing any observable effect. The production of "something" out of "nothing" is always open to question. From the vantage point of the 20th century we see no difficulty, for we are accustomed to the idea of a nuclear atom with electrons and positive charges already in existence, separated by a small distance. Let us realize, however, that the efforts of Du Fay, Franklin, and others in the 18th century to construct models are in the best modern tradition. Our own models of atomic structure and the nature of electricity will no doubt seem equally confused to a generation of physicists 200 years hence; we can only hope that in the present

---

* The earliest use of the term "electrification" included magnetic as well as electrostatic effects.

generation we do not suffer unduly from model-itis (see footnote, page 81).

Our belief in the atomic nature of electricity rests on a firm experimental foundation, which we shall study in due course. Among such experiments are the Rutherford scattering experiment (Sec. 30-1), which showed the existence of a positively charged nucleus, and the Millikan oil-drop experiment (Sec. 29-1), which showed that charges on bodies are multiples of a unit charge, the charge of one electron. In a solid body, the nuclei are fixed in position except for the slight to-and-fro vibrations that are the manifestation of thermal energy. At higher temperatures, the nuclei vibrate with greater amplitude. However, unless the solid melts, nuclei so rarely leave their places in the crystal structure that they take little part in observable electric phenomena in solids. On the other hand, in certain substances some of the orbital electrons are rather easily detached from the electron cloud surrounding a nucleus.

Electrification by friction is really a process of electrification by contact. If a glass rod is touched momentarily to a piece of silk, electrons have a tendency to leave the glass and attach themselves to the silk. This leaves an excess positive charge on the glass at the point of contact. It would be most surprising if such a tendency did not exist when dissimilar substances are brought into intimate contact. Rubbing the glass with silk is a more effective way of electrifying it than merely touching it with silk, because more parts of the surfaces are brought into contact in this way. The primary mechanism is still one of contact, and surface layers of excess charge are created—a negative layer on the silk (excess of electrons) and a positive layer on the glass (deficiency of electrons). The process of electrification by friction is as yet very imperfectly understood.

18-3 Conductors and Insulators

Metals, unlike other solids, contain *free electrons*. For instance, in iron each nucleus has 26 protons, and there are 26 electrons per atom,

making iron as a whole electrically neutral (refer back to Fig. 1-2). However, not all the 26 electrons are bound to one nucleus; roughly 1 to 3 of them are free to move through the crystal, attached to no particular nucleus. In a small piece of iron of mass 56 g (1 mole) there are 6×10^{23} atoms* arranged in a crystal structure, shown schematically in Fig. 1-1*a*. There are $26 \times 6 \times 10^{23}$ electrons in the piece of iron, of which perhaps $24 \times 6 \times 10^{23}$ are bound to their respective nuclei, and $2 \times 6 \times 10^{23}$ are free to move throughout the metal. The large number of free electrons in iron make it a good *conductor* of electricity. In a good *insulator*, such as glass, all electrons are bound to one or another of the nuclei, and only the relatively few electrons right at the surface can move under the influence of a closely approaching dissimilar atom or molecule.

We have described two extreme cases, the good conductor and the good insulator. Intermediate behavior is possible, and all substances may be classified in a sequence ranging from "good conductors" to "poor conductors" (which are also called "good insulators"). Silver and copper are the best conductors of electricity, but all metals are good conductors. Glass, quartz, sulfur, and paraffin are typical insulators. In Chap. 20 we shall see how the conductivity of a substance can be measured quantitatively.

18-4 Electrostatic Experiments— Conservation of Charge

Very simple instrumentation will suffice for the qualitative study of electric forces and the motions of electric charges. The *electroscope* (page 390) has a central rod and a flexible leaf of gold or aluminum foil that is free to move. The outer case may be connected to the earth, although this is not necessary for all experiments. The central rod assembly is insulated from the case by a support made of rubber, sulfur, or some

---

* This is Avogadro's number. See Sec. 15-3.

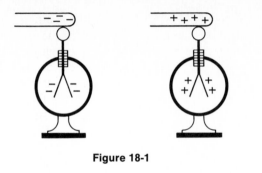

Figure 18-1

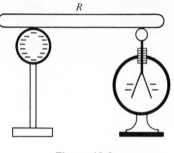

Figure 18-2

other insulating material. Sometimes an electroscope is constructed with two movable leaves. In one modern form, rugged enough to be used by mineral prospectors in the field, the moving conductor is a fine quartz fiber coated with gold to make it is good conductor of electricity. With the aid of an electroscope, several experiments can be performed that give important information about charges and charge distribution.

1. *Like charges repel each other and unlike charges attract each other.* The repulsion of like charges is shown by the fact that the leaves of the electroscope diverge when a charged rod of either sign is touched to the knob (Fig. 18-1). Some of the surface charge on the rod is shared among the rod, the knob, and the leaf structure. If the rod is removed, the leaves remain somewhat spread apart, indicating that some charge has remained on the leaves. The repulsion of like charges spreads the leaves apart until mechanical equilibrium is reached, when Coulomb forces are balanced by gravitational forces.

2. *Substances differ markedly in their electrical conductivity.* In Fig. 18-2, a rod R is placed in contact with a charged sphere and an electroscope that was originally uncharged. The leaves begin to diverge, but at a rate that depends on the rod. If R is of copper, iron, aluminum, or some other metal, the leaves diverge almost instantly; however, if R is of a poorly conducting material, such as glass or rubber, the leaves diverge slowly—perhaps imperceptibly. We can distinguish between conductors and insulators

in this fashion and get a rough idea of their relative conductivities. The electroscope itself makes good use of these differences. The leaves must be good conductors so that charges can flow; hence the use of a metal foil. The supporting plug at the top of the container must be a good insulator so that charge applied to the knob will not "leak" to the surrounding case.

3. *Electric charges are of two kinds.* This can be shown by first charging the electroscope negatively (Fig. 18-3). If a negatively charged rod is now brought near the knob, not touching it, the leaves diverge a bit. We interpret this as a repulsion of electrons driving additional excess negative charge to the leaves. If a positively charged rod is brought up, the leaves collapse a little. We interpret this as an attraction of unlike charges; the rod pulls up some of the excess electrons that were on the leaves.

4. *Electric forces act at a distance.* This fact is shown by the experiments just described to illustrate item 3. The quantitative measurement of

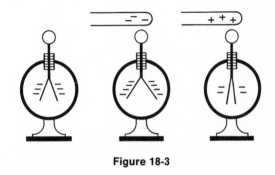

Figure 18-3

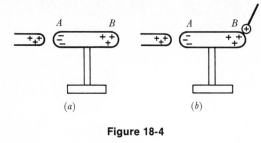

(a)   (b)

Figure 18-4

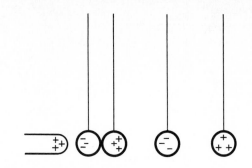

Figure 18-5   Separation of charge by induction.

how electric forces depend on distance will be discussed in Sec. 18-5.

5. *A body can be charged by induction.* If a positively charged rod is brought near a neutral metal body mounted on an insulating stand (Fig. 18-4*a*), free electrons in the metal are attracted toward the rod, leaving an excess positive charge at the opposite end of the body. We say that a charge has been *induced* at each end of the metal body. Of course, no charge has been created; it has merely been separated into two equal and opposite amounts. We can test this with a "proof plane," which might be a small metal disk or ball on a long insulating handle. Touching the proof plane to end *B* allows a sharing of charge (Fig. 18-4*b*), and the proof plane receives a small positive charge, a sample of the charge at *B*. The sign of the charge of the proof plane can be tested with the aid of an electroscope, as in item 3. If two metal spheres, initially uncharged, are hung from silk threads and are touching each other, they can be charged by induction as in Fig. 18-5. The spheres can be separated while the charge is held bound by the rod, and a considerable amount of positive and negative charge can be isolated in this way.*

Still another induction experiment consists of "grounding" one end of a body while the negative charge is held bound at the other end by the charged rod's action-at-a-distance (see Fig.

18-6*a*). Electrons flow from the earth to end *B* and neutralize the positive charge at this end (Fig. 18-6*b*). Upon removing the earth connection, the negative charge is still bound at end *A* (Fig. 18-6*c*). Finally, when the charged rod *R* is removed, the negative charge spreads out, and we have a body with a net excess of negative charge (Fig. 18-6*d*).

Induced charges are produced even in good insulators, but the separation of such charges is impossible because no flow can take place. Nevertheless, the effect is noticeable, as when a fountain pen is rubbed on clothing and held near a small piece of paper. The molecules of the paper are distorted, and a layer of positive charges appears at the surface. If a positively charged rod had been used, the induced surface layer would have been negative. In either case, a force of attraction between unlike charges would cause the pen or rod to "pick up" the paper.

The formation of thunderclouds and lightning strokes illustrates both electrification by friction (water droplets move through air) and induction (charged cloud attracts opposite charges from earth). Similarly, a gasoline truck moving on the highway becomes charged as the tires separate from the pavement. To prevent hazardous "lightning strokes" to the earth, a dangling chain provides a conduction path to the ground and prevents an accumulation of charge.

6. *When charges are separated by contact, equal amounts of positive and negative charge are produced.* A small piece of fur glued to an

---

* This sort of operation, done in a mass-production fashion, is the basis for electrostatic generators delivering large quantities of charge at high voltage suitable for making artificial lightning or doing experiments in high-energy physics (see Sec. 33-6).

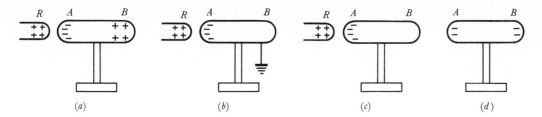

Figure 18-6 Charging a metallic body by induction.

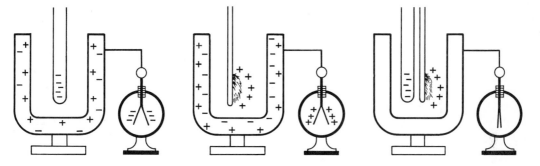

Figure 18-7 Positive and negative charge is produced in equal amounts.

insulating handle is stroked by a piece of hard rubber. It is easy to show qualitatively that the rubber gains a net negative charge and the fur gains a net positive charge. For a quantitative test of the equality of these charges, we use an induction method: A metal cup* on an insulating stand is connected by a wire to an electroscope (Fig. 18-7). Originally, the cup and electroscope are uncharged. If the negatively charged piece of rubber is held in the cup, not touching the metal, a separation of charge takes place by induction, and the outer surface of the cup becomes negatively charged. The electroscope leaves share this charge, and they diverge a certain amount. If the rubber rod is removed and the fur is inserted, the leaves diverge the same distance as before, but this time the electroscope is positively charged. As a final test of the equality of the charges, the

rubber and the fur are both inserted, not touching each other or the metal cup. No induction takes place, and the electroscope leaves do not diverge. By experiments of this sort, Faraday was led to the idea of the indestructibility of electric charge.

7. *The charge on a solid or hollow conducting body resides entirely on the surface of the body.* The electrons in the body of a metal conductor are free to move, and since like charges repel, any excess charge moves to the surface.[†] This can be tested by a proof plane that acquires a charge when touched to point X (Fig. 18-8) but acquires no charge when touched to an interior point such as Y.

In summary, we can express the facts revealed by these and other experiments in the form of a

---

* Michael Faraday (1791–1867) used an ordinary pewter ice pail for his historic researches in 1843. Faraday was an English physicist noted for his experimental investigations in many fields of electricity and magnetism.

[†] This would be true only for an inverse-square law of electric force, and the absence of charge on the inner surface of a hollow conductor is a sensitive proof of Coulomb's inverse-square law. See Sec. 18-7.

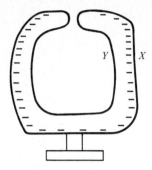

Figure 18-8 Any excess charge is distributed over the *outside* surface of a hollow conductor.

law of conservation of electric charge:

The algebraic sum of the electric charge in any closed system remains constant.

In other words, charge can be neither created nor destroyed, although it can be moved from place to place. We accept this as a fundamental conservation law, with the same validity as the other conservation laws dealing with mass, energy, and momentum. We shall see in Chap. 28 how the facts of relativity require us to qualify the laws of conservation of energy and mass, and thus we are understandably cautious about proclaiming any conservation law as absolute and final. The validity of the law of conservation of charge in the realm of nuclear particles is still under discussion, and it may well be that some modification may eventually have to be made.* This possibility need not concern us, for whatever law is developed in the future must include the familiar conservation law as a special case. This is the way in which scientific advances come about.

---

* For example, when a positive electron (Sec. 32-5) and a negative electron come together, their rest masses are converted into radiant energy, according to the equation $\Delta E = (\Delta m_0)c^2$. The equal and opposite charges cancel out, to be sure, so in a sense the law of conservation of charge is valid. In another sense, we are not merely rearranging charges; we are *destroying* charges that originally added up to zero.

18-5 Coulomb's Law

The law of force between two charged bodies[†] was investigated by the French engineer and scientist Charles-Augustin Coulomb (1736–1806), who also made important studies of sliding friction and the twisting of wires. To measure the small forces involved, Coulomb in 1788 used a torsion balance very similar in principle to that used 10 years later by Cavendish in his measurement of the force of gravitation (described in Sec. 8-1). A small, charged, metal sphere was placed at a known distance from another similar small, charged sphere that was part of the moving system, and the force was calculated from the observed twist of the fiber. Although he experienced some difficulty with forces due to undesired induced charges, Coulomb was able to conclude that an inverse-square law of electric force accounted for his results. *Coulomb's law* states:

The force between two charges at rest is directly proportional to the product of the magnitudes of the charges and inversely proportional to the square of the distance between them.

In symbols,

$$F = k\frac{QQ'}{r^2} \qquad (18\text{-}1)$$

where F is the magnitude of the force, Q and Q' are the magnitudes of the charges, r is the distance[‡] between them, and k is an as yet undefined constant, to be determined by experiment.

[†] The law was surmised by Daniel Bernoulli in 1760. Joseph Priestley in 1767, John Robison in 1769, and Henry Cavendish in about 1775 all made experiments to test an inverse-square law. See Sec. 18-7 for a discussion of scientific priority, as illustrated by the investigations of the law of force for electric charges.

[‡] The charges in Coulomb's law are considered to be point charges, which is an idealization. The distance between the centers of *spherically symmetric* distributions of charge can also be used as r in Coulomb's law; we omit proof of this fact. A similar situation holds for gravitational force, which is also an inverse-square force (refer to the footnote on page 179).

Note that Coulomb's law contains an implicit definition of magnitude of charge: the ratio of the magnitudes of two charges equals F_1/F_2, the ratio of the magnitudes of the forces they would exert on a third "test charge" at the same distance r. This definition is made plausible by the experimental fact that the total force exerted by n equal charges all at one point is n times the force exerted by any one of them placed at the same point. The situation is quite analogous to the way the ratio of two gravitational masses can be defined (F_1/F_2 for attraction by a given test body such as the earth). Similar operational definitions have been made in Sec. 3-3 for the ratio of two forces (a_1/a_2 for acceleration of a given test body), and the ratio of two inertial masses (F_1/F_2 for the force necessary to cause a given acceleration).

You will immediately be impressed by the similarity between Coulomb's inverse-square law and Newton's inverse-square law for gravitational forces, $F = Gmm'/r^2$. Many scientists believe that this similarity is more than accidental, but the precise connection, if any, has eluded theoretical physicists of the highest caliber. We will consider Coulomb's law as an independent law of physics, based on experimental evidence.

The constant k in Coulomb's law can be determined as soon as we adopt a system of units for charge, distance, and force. In this book, *only SI units are used when dealing with electrical quantities*. Indeed, one reason for introducing SI units such as the newton and joule in our study of mechanics was to lay the foundation for our use of these same units in electricity. The SI unit of charge is the *coulomb* (C), which is based on the SI unit of current called the *ampere* (A). We shall define the ampere in Chap. 22 in terms of the magnetic force between two currents. In other words, in defining the ampere, we place emphasis on the forces between moving charges. For the present we need only say that current I measures the rate of flow of charge. A coulomb is the quantity of charge carried past a given point in one second by a current of one ampere.

Symbolically, we can say

$$Q = It \qquad (18\text{-}2)$$
$$\text{coulombs} = (\text{amperes})(\text{seconds})$$

We could write 1 A = 1 C/1 s = 1 C/s, but this statement, though true, is not a definition. Equation 18-2 emphasizes our choice of the ampere as the *basic* electric unit for this and future chapters.* The coulomb is a derived unit, in the same sense that the joule is a derived unit in mechanics.

Example 18-1

How much charge flows through the bulb in a student's desk lamp in 1 h if the current through the bulb is 0.5 A?

$$Q = It$$
$$= (0.5 \text{ C/s})(3600 \text{ s}) = \boxed{1800 \text{ C}}$$

A problem in language arises; strictly speaking it is charge that flows, and current *exists* in a wire. "Current" means "rate of flow of charge." Thus the sentence "A current of 0.5 A flows through the bulb" really says (ungrammatically) "A rate of flow of charge of 0.5 C per second flows through the bulb." We shall avoid such inexact statements as "Current flows through a wire."

On the atomic scale, a "natural" unit of charge immediately suggests itself—the charge e of the electron. The most fundamental of all atomic constants, the charge of the electron has been the subject of many brilliant investigations since about 1897, with the result that all electrons are now known to be identical, having a charge equal to 1.6020×10^{-19} C (Sec. 29-1). Accurate to three significant figures, the charge of the electron is

$$e = 1.60 \times 10^{-19} \text{ C/electron}$$

---

* The ampere is one of the seven base units of the Système International (SI) units (page 7). The dimensions of charge are $[Q] = [IT]$.

We could therefore define one coulomb in terms of the charge of the electron. The reciprocal of e is

$$\frac{1}{1.6020 \times 10^{-19} \text{ C/electron}}$$

$$= 6.2422 \times 10^{18} \text{ electrons/C}$$

In other words, a coulomb might be *defined* as the charge carried by this number of electrons. Such a definition would be a permanent one, based on atomic standards. However, the present-day precision of such a standard of charge would fall far short of the precision of the atomic standards of length and time mentioned in Sec. 1-4.

Example 18-2

How many electrons flow through the wire filament in a flashlight bulb in 1 ms (10^{-3} s) if the current is 50 mA (50×10^{-3} A)?

$$Q = It$$
$$= (50 \times 10^{-3} \text{ A})(10^{-3} \text{ s}) = 50 \times 10^{-6} \text{ C}$$

Number of electrons

$$= (50 \times 10^{-6} \text{ C})\left(\frac{1 \text{ electron}}{1.60 \times 10^{-19} \text{ C}}\right)$$

$$= \boxed{3.12 \times 10^{14} \text{ electrons}}$$

In 0.001 s, 312 trillion electrons flow through the filament, carrying a total charge of 50 μC.

Having defined the coulomb as a unit of charge, we are able to ask what is the value of the constant k in Coulomb's law. In principle, k can be found by experiments such as Coulomb performed, in which known charges are separated by a measured distance. However, the most precise values for k are found indirectly, by analysis of experiments that are in turn based on Coulomb's law. Such experiments give the value of k as

$$k = 9.0 \times 10^9 \text{ N} \cdot \text{m}^2/\text{C}^2 \qquad (18\text{-}3)$$

It is significant that k is numerically equal to 10^{-7} times the square of the speed of light; this is a consequence of the electrical nature of light and the other electromagnetic waves. We use k to avoid fractions and a factor 4π. Coulomb's law is often written in the form $F = (1/4\pi\epsilon_0)QQ'/r^2$, where $1/4\pi\epsilon_0 \equiv k$.

Example 18-3

What is the force of attraction between a positive charge of 4 microcoulombs (μC) and a negative charge of 5 μC, separated by 30 cm?

$$F = \frac{kQQ'}{r^2}$$

$$= \frac{\left(9 \times 10^9 \dfrac{\text{N} \cdot \text{m}^2}{\text{C}^2}\right)(4 \times 10^{-6} \text{ C})(5 \times 10^{-6} \text{ C})}{(0.30 \text{ m})^2}$$

$$= \boxed{2.00 \text{ N}}$$

If a charge is acted on by two or more other charges, the resultant force is found by vector addition, as is true for all forces.

Example 18-4

In the x-y plane, a charge of -48 μC is on the x axis at $x = 0.6$ m, and a charge of $+8$ μC is on the y axis at $y = 0.3$ m (Fig. 18-9). Calculate the magnitude and direction of the resultant force on a charge of $+2$ μC that is at the origin.

Using Coulomb's law, we find

$$F_1 = \frac{(9 \times 10^9 \text{ N} \cdot \text{m}^2/\text{C}^2)(2 \text{ }\mu\text{C})(48 \text{ }\mu\text{C})}{(0.6 \text{ m})^2} = 2.4 \text{ N}$$

$$F_2 = \frac{(9 \times 10^9 \text{ N} \cdot \text{m}^2/\text{C}^2)(2 \text{ }\mu\text{C})(8 \text{ }\mu\text{C})}{(0.3 \text{ m})^2} = 1.6 \text{ N}$$

The magnitude of the resultant force is

$$F = \sqrt{(2.4 \text{ N})^2 + (1.6 \text{ N})^2} = \boxed{2.88 \text{ N}}$$

$$\tan \theta = \frac{1.6 \text{ N}}{2.4 \text{ N}} = 0.667; \quad \theta = \boxed{33.7°}$$

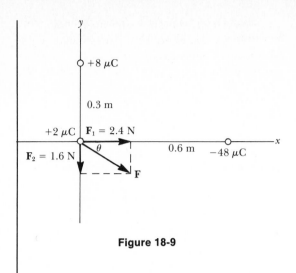

Figure 18-9

The resultant force on the $+2 \ \mu C$ charge is 2.88 N, at an angle $33.7°$ below the x axis.

18-6   Electrolysis

Many liquids such as oil, alcohol, or pure water are poor conductors of electricity. However, water solutions of many salts are good conductors. Such solutions, known as *electrolytes*, differ from nonconducting solutions (such as sugar in water) in that the molecules of solute are dissociated into charged fragments called *ions*. In a typical case, silver nitrate ($AgNO_3$) breaks up in water into silver ions, Ag^+, and nitrate ions, NO_3^-. Each silver atom has given one electron to an NO_3^- ion, and hence each Ag^+ ion consists of a nucleus with 47 positive charges (its atomic number) surrounded by a cloud of only 46 electrons. Neutral NO_3 would have 31 electrons: 7 electrons for the N atom (of atomic number 7) and 24 electrons for the three O atoms (each of atomic number 8). After receiving one electron from the Ag atom, each NO_3^- ion has 32 electrons in its cloud instead of the 31 that would make it neutral.

If a solution of silver nitrate is placed in a cup with two metallic electrodes A and B, and a charge is caused to flow, a movement of ions

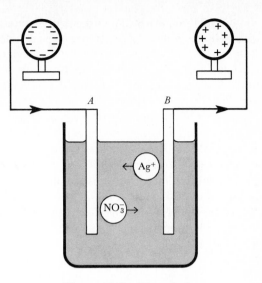

Figure 18-10   Electrolysis.

takes place in the liquid. We do not at this time have to specify just why the charge flows; it would be sufficient to connect the electrodes to spheres (Fig. 18-10) that had been charged by electrostatic means.* For a continuous effect, we could connect the electrodes to a battery or a generator, which could cause a steady flow of charge through the circuit. We are interested at present in what takes place in the electrolytic cell. The positive Ag ions are attracted to the negative electrode A. Each Ag^+ ion receives a free electron when it touches the metal electrode, thus increasing the number of electrons in the cloud from 46 to the normal 47. The silver ion thus becomes an atom of silver metal and sticks to the electrode. In this way, electrode A becomes silver-plated and gains mass. At electrode B, electrons leave the solution and become free electrons in the metal. However, the prediction of just where these electrons come from requires considerable knowledge of chemistry. It turns out that if B is an inert metal such as platinum, electrons leave the OH^- ions that are in the

---

* Electrolysis was first observed in this way by Giovanni Beccaria in 1758, some 22 years before the discovery of electric current and the invention of electric batteries.

solution, and water and oxygen gas are produced at B. If B is silver metal, silver atoms leave the metal and go into solution as Ag^+ ions, leaving free electrons behind in the metal electrode; these free electrons flow through the external metallic circuit. To summarize, in the solution the carriers of charge are positive and negative ions moving in opposite directions; in the metal the carriers of charge are free electrons. The reactions at the electrodes cannot always be easily predicted, although it is usually safe to say that the positive ions will accept free electrons from a negative metal electrode and will become neutralized and "plate out" at the negative electrode.

It is an easy matter to calculate the mass of substance that is deposited during electrolysis. We need to know the atomic weight of the ion, its excess charge,* and the total quantity of charge that flows. Many ions can exist in several states of charge. For instance, copper has an atomic number 29, and the neutral atom has an electron cloud of 29 electrons. The Cu^+ ion has 28 electrons in the cloud, and the Cu^{2+} ion has only 27 electrons in the cloud. To plate out each Cu^{2+} ion requires that two electrons flow through the external circuit in order to neutralize the doubly charged ion.

Since each atom requires 2 electrons for neutralization, the number of atoms of Cu is

$$\frac{3.00 \times 10^{22} \text{ electrons}}{2 \text{ electrons/atom}} = 1.50 \times 10^{22} \text{ atoms}$$

Finally, we use Avogadro's number and the mass of a mole of copper:

Mass of Cu

$$= (1.50 \times 10^{22} \text{ atoms})\left(\frac{63.6 \text{ g}}{6.02 \times 10^{23} \text{ atoms}}\right)$$

$$= \boxed{1.58 \text{ g}}$$

The total charge of a mole of electrons is called the faraday, Q_F. Thus

$$Q_F = N_A e$$
$$= (6.023 \times 10^{23})(1.602 \times 10^{-19} \text{ C})$$
$$= 96,500 \text{ C}$$

We use the faraday in a concise summary of the laws of electrolysis:

In electrolysis, one faraday of charge (96,500 coulombs) is required to deposit or liberate one mole of any monovalent substance.

Example 18-6

Solve Example 18-5, using the faraday.

As before, the charge passing through the cell is 4800 C. The valence of Cu^{2+} is 2, so a charge $2Q_F$ will plate out 1 mole (63.6 g).

We use a proportion to find the mass m that is deposited.

$$\frac{m}{4800 \text{ C}} = \frac{63.6 \text{ g}}{2(96,500 \text{ C})}$$

$$m = \boxed{1.58 \text{ g}}$$

The study of electrolysis brings out the close connection between the atomic nature of matter and the atomic nature of electricity. Several decades before the discovery of the electron, it was clearly realized that if matter is atomic, the

---

Example 18-5

A current of 8 A is maintained for 10 min through an electrolytic cell containing Cu^{2+} ions. How much copper is deposited?

$$Q = It = (8 \text{ A})(600 \text{ s})$$
$$= 4800 \text{ C}$$

If n electrons flow through the cell,

$$n = (4800 \text{ C})\left(\frac{1 \text{ electron}}{1.60 \times 10^{-19} \text{ C}}\right)$$

$$= 3.00 \times 10^{22} \text{ electrons}$$

---

* In chemistry, the excess charge on an ion is often called its valence n; Cu^{2+} is a divalent copper ion, with $n = 2$.

facts of electrolysis require the existence of "atoms of electricity" having a charge of the order of magnitude of 10^{-19} C. The equation $Q_F = N_A e$ has been used in several ways. The faraday is experimentally known with great precision (the latest value is written as 96,484.6 $\pm$ 0.3 C); hence N_A can be found if e is known, and vice versa. The first rough estimates of e were made in this way, using approximate values for Avogadro's number. It is customary now to invert the process and use the precise modern value of e (Sec. 29-1) to find Avogadro's number.

Summary   Electric forces between charges at rest are called Coulomb forces, and electric forces between charges in relative motion are called magnetic forces. When two dissimilar substances are placed in close contact, electrons tend to leave one substance and go to the other. If the substances are insulators, the electrons are bound to the nuclei, and transfer of charge takes place only at the points of contact. Electrification of insulators by friction is effective because contact is made at many points. In metallic conductors, free electrons can flow through the metal and take part in the conduction process.

There are two kinds of charge, positive and negative, and the law of conservation of charge states that the algebraic sum of the electric charge in any closed system remains constant. A neutral, or uncharged, body has equal amounts of positive and negative charge. Like charges repel each other, and unlike charges attract each other. The law of force between stationary electric charges is Coulomb's law, which states that the force is directly proportional to the product of the magnitudes of the charges and inversely proportional to the square of the distance between them. The SI unit of charge is the coulomb, which is the charge carried past a given point in one second due to a current of one ampere.

In electrolytes, electric current is carried by ions that are produced by dissociation of molecules. An ion has an excess or deficiency of electrons, and the net charge of an ion is a small multiple of the electron charge. When an ion reaches a metal electrode, its charge may be neutralized by the flow of free electrons to or from the electrode. In electrolysis, the mass of substance deposited or liberated is proportional to the charge that flows. One faraday is that amount of charge that will deposit or liberate one mole of a monovalent substance. The numerical value of the faraday is 96,500 C, which is the charge of one mole of electrons.

Check List   Coulomb forces
magnetic forces
positive electricity
negative electricity
conductor

insulator
electric induction
Coulomb's law
coulomb

electrolyte
ion
valence
faraday

Questions   **18-1**   Why is magnetism considered a subdivision of electricity?

18-2   In what respects is modern electric theory a two-fluid theory? In what respects is it a one-fluid theory?

18-3 Do protons contribute to the flow of electricity through a metal?

18-4 Why is a plastic pen electrified more strongly when it is rubbed vigorously on a coat sleeve than when it is merely touched to the cloth?

18-5 Reword item 3 of Sec. 18-4 to apply to a situation in which the electroscope was originally charged positively (that is, had a deficiency of electrons), and rods of either sign are brought up.

18-6 Explain the action of a rubbed comb that picks up bits of paper. Why do the pieces eventually fly off the comb with an initial velocity? Why is it difficult to pick up pieces of paper on a damp day?

18-7 Suggest a method by which Coulomb might have placed exactly half as much charge on one of two identical metal balls as on the other.

18-8 Correct the following statement: "The charge on the raindrop was 463 μA."

18-9 Why should an airplane be connected to ground before a refueling operation begins?

18-10 A small battery has fallen out of a transistor radio, and it is found that the $+$ and $-$ labels on the terminals are illegible. Describe how the principles of electrostatics can be used to determine the polarities of the battery terminals.

18-11 How many electrons are required to neutralize the charge of a mole of a monovalent element?

18-12 How many electrons are in the electron cloud of a Pb^{4+} ion? (*Hint:* Use the Periodic Table, Appendix Table 6, to find the atomic number of lead.)

MULTIPLE CHOICE

18-13 The leaves of a charged electroscope diverge just enough so that the repulsive Coulomb forces are balanced by   (*a*) magnetic forces;   (*b*) gravitational forces;   (*c*) nuclear forces.

18-14 In SI, the base unit in electricity is the   (*a*) volt;   (*b*) ampere;   (*c*) coulomb.

18-15 When a current of 10 A is maintained for 2 s, the charge that flows is   (*a*) 5 C;   (*b*) 10 C;   (*c*) 20 C.

18-16 Two equal point charges 2 m apart repel each other with a force of 2×10^{-8} N. The magnitude of each charge is about   (*a*) 2.2×10^{-9} C;   (*b*) 3.0×10^{-9} C;   (*c*) 6.3×10^{-9} C.

18-17 A charge of 96,500 C passes through an electroplating bath containing Ni^{2+} ions. The amount of nickel plated out is   (*a*) 0.5 mole;   (*b*) 1 mole;   (*c*) 2 moles.

18-18 If there is a current of 1 A for 10 s through an electrolytic cell containing Ag^+ ions (each of mass 1.8×10^{-25} kg), the mass of silver deposited is about   (*a*) 2×10^{-24} kg;   (*b*) 6×10^{-24} kg;   (*c*) 10^{-5} kg.

Problems

18-A1 In 5 min, 3600 C of free electrons enter one end of a conductor, and 3600 C move out the other end. What is the current through the conductor?

18-A2 A service station charges a battery for 5 h, using a current of 30 A. How much charge passes through the battery?

18-A3 How many electrons are needed to carry a nanocoulomb of charge?

18-A4 In a type of memory cell on a semiconductor chip in a computer, one bit of information is represented by an excess charge of 0.8 pC. A single high-energy particle, arising from radioactive contamination in the encapsulating shell of the chip, can release the equivalent of 6×10^6 electrons in the semiconductor material. Can computer error result from these randomly occurring events?

18-A5 What current in picoamperes is equivalent to the flow of ten million electrons per second?

18-A6 What quantity of charge passes through the wire filament in a lamp bulb in 15 min if the current is 0.3 A?

18-A7 Two charged bodies exert a force on each other of 16 mN. What will be the force between the same two bodies if the distance between them is halved?

18-A8 What force does a charge of 10^{-10} C exert on a charge of 4×10^{-8} C that is 3 m away?

18-A9 How far apart must two charges, each of 2 μC, be placed so that force between them is 2 μN?

18-A10 (*a*) What charge is needed to neutralize 8 moles of a trivalent ion? (*b*) How many electrons does this require?

18-A11 How many electrons are needed to neutralize the charge of 5 moles of a divalent ion?

18-A12 A certain ion has 3 excess electrons. How many moles of this ion will be deposited by 24 faradays of charge?

18-B1 Assuming that the electron in a hydrogen atom is 0.53×10^{-10} m (0.53 Å) from the nucleus (which consists of a proton), (*a*) compute the force of electrical attraction between the electron and the proton; (*b*) compute the gravitational attraction between the electron and proton; (*c*) show that the ratio of these two forces is approximately 10^{39}. (Use Appendix tables for necessary data.)

18-B2 What would be the force of repulsion between a milligram of electrons on the moon separated by 3.84×10^5 km from another milligram of electrons on the earth? (Use Appendix tables for necessary data.)

18-B3 How far apart are two electrons if the Coulomb force between them equals the weight (on earth) of an electron? (Use Appendix tables for necessary data.)

18-B4 Two helium nuclei (alpha particles), each of charge $+2e$, approach each other to within a distance of 10^{-12} m in vacuum. What is the force of repulsion?

18-B5 A cosmic-ray particle as shown in Fig. 33-12 (page 810) consists of a bare $_{26}$Fe nucleus (an iron atom stripped of all its orbital electrons). Calculate the Coulomb force exerted on an electron 2 nm from the particle in a photographic emulsion.

18-B6 Calculate the repulsive Coulomb force between two protons 4×10^{-15} m apart inside a nucleus. Why doesn't a force of such magnitude, acting on a proton of mass about 10^{-24} kg, cause the nucleus to fly apart? (See Sec. 1-5.)

18-B7 Calculate the force of attraction between an electron, of charge $-e$, and a copper nucleus, of charge $+29e$, at a separation of 10^{-10} m.

18-B8 A small plastic sphere coated with a thin metallized surface has mass 0.05 g and carries a charge of $+8$ nC. It is suspended by a light insulating thread at a point 3 cm below the center of a small fixed conducting sphere carrying -5 nC. When the thread is cut, with what acceleration does the plastic sphere start to fall?

18-B9 A charge $Q_1 = +10$ μC is on the x axis at $x = 0$, and a second charge $Q_2 = +3$ μC is on the x axis at $x = 5$ m. A third charge $Q_3 = -10$ μC is placed on the x axis at $x = 15$ m. Calculate the magnitude and direction of the force on Q_2.

18-B10 A triangle ABC, marked out on a flat surface, has sides of the following lengths: $AB = 4$ m, $BC = 5$ m, and $AC = 3$ m. At the corners are the following charges: -40 μC at A, -160 μC at B, and $+90$ μC at C. What are the magnitude and direction of the net force on the charge at A?

18-B11 Three charges, each of $+80$ μC, are equally spaced along a straight line, successive charges being 6 m apart. Calculate   (a) the force on one of the end charges;   (b) the force on the central charge.

18-B12 Equal charges of $+15$ μC are placed at the four corners of a square 0.3 m on a side. Calculate the magnitude and direction of the force on one of the charges.

18-B13 Equal charges of $+8$ μC are placed at the three corners of an equilateral triangle 2 m on a side. Calculate the magnitude and direction of the force on one of the charges.

18-B14 In 1 μs, how many electrons flow through a sensitive research galvanometer that reads 72 pA?

18-B15 Iron articles are sometimes cadmium-plated to increase rust resistance. Calculate the mass of cadmium deposited by a current of 50 A during an 8-h working day if the cadmium is present in the form of Cd^{2+} ions.

18-B16 The electrochemical equivalent is defined as the mass of substance deposited by 1 C of charge. Calculate the electrochemical equivalent for Cr^{3+}.

18-B17 A student performs a laboratory experiment in which she plates copper from a solution containing copper ions (which may be Cu^+ or Cu^{2+}). The initial mass of the electrode was 21.031 g. After a current of 0.300 A was passed through the solution for 25 min, the plate was removed and dried, and its mass was found to be 21.329 g. What can the student conclude about the charge of the copper ions in the solution?

18-B18 In the production of hydrogen gas, a current of 100 A is used. Charge flows for 12 h through a cell containing water. H_2 gas is released at one electrode, and O_2 gas at the other electrode. The ions involved are H^+ and O^{2-}.   (a) What mass of hydrogen is produced?   (b) What volume is occupied by the released hydrogen, measured at 27°C and 760 torr?   (c) What volume of oxygen is produced?

18-B19 Chlorine is often produced by electrolysis from a solution containing Cl^- ions. (a) What is the current through an electrolytic cell in which 15 kg of Cl_2 gas is produced every 24 h?   (b) What is the volume of this gas, stored as Cl_2 at 0°C and 100 atm?

18-C1 Two Ping-Pong balls painted with aluminum paint are suspended from the same point by threads 50 cm long. The mass of each ball is 20 g. When equal charges are given to the two balls, they come to rest in an equilibrium position in which their centers are 60 cm apart. Calculate the charge on each ball.

18-C2 A small ball of mass 0.01 kg carries charge $+8$ μC and is suspended by a light thread 50 cm long that is attached at the intersection of a wall and the ceiling. What positive charge, fixed on the wall 40 cm below the ceiling, will cause the ball to be in equilibrium 30 cm out from the wall?

18-C3 Each of two small spheres is charged positively, the combined charge totaling 6×10^{-9} C. When they are placed with their centers 3 cm apart, the force of repulsion is 5×10^{-5} N. What is the magnitude of each charge?

18-C4 Two small charges Q_1 and Q_2 are situated on the x axis at points having the following coordinates: $Q_1 = -4$ μC at $x = -3$ m; $Q_2 = +1$ μC at $x = 2$ m. Find a point on the x axis at which a third charge Q_3 could be placed and experience no net force.

18-C5 Three charges are located as follows: $+5\ \mu\text{C}$ at the origin of coordinates; $-3\ \mu\text{C}$ on the x axis at $x = +2$ m; and $+8\ \mu\text{C}$ on the y axis at $y = +2$ m. Calculate the magnitude and direction of the net force on the $+8\text{-}\mu\text{C}$ charge.

18-C6 A small metal ball (A) of mass 20 g carrying a charge $+0.07\ \mu\text{C}$ is electrically insulated from a light vertical spring that supports it in equilibrium 12 cm above an insulating platform. An identical metal ball (B), initially uncharged, is firmly attached to the platform, directly beneath ball A. When a charge $-0.07\ \mu\text{C}$ is given to B, A's new equilibrium position is 10 cm above B. Now the platform is slowly raised until the balls come into contact. (*a*) Explain what happens, and (*b*) calculate the period of oscillation of ball A.

18-C7 A spoon has surface area 50 cm² and is to be silver-plated by passing a current of 300 mA through a solution containing Ag^+ ions. What thickness of silver is deposited in 40 min? (The density of silver is 10.5 g/cm³.)

18-C8 It is desired to gold-plate a metal object to a thickness of 0.020 mm. The object has a total surface area (all sides) of 50 cm² and is in an electrolytic bath containing Au^{3+} ions. If a current of 3 A is used, how much time will be required for the job?

18-C9 What current should be used in an electrolytic bath containing Ag^+ ions to silver-plate in 3 min both sides of a coin 3 cm in diameter with silver 100 μm thick? (The density of silver is 10.5 g/cm³.)

18-C10 A lead-covered telephone cable was buried in moist earth in 1940, and the average leakage current from a 3-m section of the cable into the earth was 15 mA. The reaction at the sheath involves Pb^{4+} ions. The lead sheath was 5 cm in diameter and 2 mm thick. Was the cable still serviceable in 1982?

18-C11 In a physiological experiment, each heartbeat of a long-distance runner caused 0.00410 C of charge to flow through a small electrolytic cell strapped to his chest. When he ran a 5000-m race in 13 min, 20 s, a total of 9.05 mg of silver was deposited on the cathode from a solution containing Ag^+ ions. Calculate the average rate, in min^{-1}, of the athlete's heartbeat during the race.

For Further Study

18-7 Scientific Priority and Coulomb's Law

Over the years, undignified squabbles over priority have taken place between workers who each claimed to have first discovered some physical law or to have invented some device. Robert Hooke (1635–1703), a versatile English scientist who was a bit more sensitive to these matters than some of his contemporaries, found it expedient to announce his law of elasticity in ana-gram form as "ceiiinossssttuu" in 1676. He thus guarded his priority, and unraveled the anagram two years later as *Ut tensio sic uis*,* which can be translated "As the extension, so the force." Christian Huygens (1629–1695) used a similar Latin anagram to "pin down" his priority in the matter of the true shape of Saturn's rings.

---
* In the 20th century a Latin phrase might still seem to many to be a "hidden message," but in 1678 Latin was the accepted language for communicating learned discoveries.

It is always a legitimate question to ask who is to be credited for a major discovery. To illuminate this matter a little, let us look closely at the unfolding of the inverse-square law for electric forces, known as Coulomb's law. Action at a distance had been introduced to physics in the shape of Newton's law of universal gravitation, published in 1687. When it became clear that electric forces decrease rapidly as distance increases, a natural assumption was made that $F \propto 1/r^n$, where n is some exponent. This assumption that the force depends on some *power* of r was justified by experiment. It was generally known by 1760 that the charge on the surface of an irregularly shaped body (Fig. 18-11) is distributed irregularly, but for similarly shaped bodies the pattern of charge distribution is the same. This observation, plus some geometrical reasoning, is sufficient to show that the force law must be of the form $F \propto 1/r^n$, and not, for instance, of the form

$$F \propto a^{-kr} \quad \text{or} \quad F \propto 1/r^n + 1/r^m$$

The problem, then, was to determine the value of the exponent. Daniel Bernoulli, in 1760, guessed that $n = 2$, but he had no proof for his guess.

Joseph Priestley (1733–1804), the discoverer of oxygen, was informed in 1766 by his friend Benjamin Franklin that cork balls inside a metal cup were totally unaffected by the electrification of the cup. Now from Newton's work it was known that if the earth were a hollow spherical shell, a body anywhere within the shell would experience no force of gravitation. This is a consequence of the fact that gravitation follows an inverse-*square* law. Hence Priestley drew the correct conclusion that since the cork balls experienced no force, the law of electric force must also be an inverse-square law.

To prove this statement, consider a point P somewhere inside a uniformly charged hollow sphere (Fig. 18-12), and construct a small double cone through P, as shown, intersecting the surface in areas A_1 and A_2. If ϕ is small, the surface segment is almost plane, and the projection of the segment s_1 (perpendicular to the line of sight) is given by $s_1 \cos \theta_1$. Similarly, the projection of the other arc is $s_2 \cos \theta_2$. By similar triangles,

$$\frac{s_1 \cos \theta_1}{r_1} = \frac{s_2 \cos \theta_2}{r_2}$$

But $\theta_1 = \theta_2$ (they are exterior base angles of an isosceles triangle whose vertex is at O). Hence, after squaring, we get

$$\frac{s_1{}^2}{r_1{}^2} = \frac{s_2{}^2}{r_2{}^2}$$

The area of each portion of the surface is proportional to the square of its linear dimension, so we have

$$\frac{\text{Area } A_1}{r_1{}^2} = \frac{\text{Area } A_2}{r_2{}^2}$$

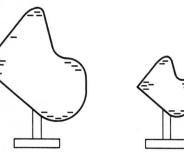

Figure 18-11

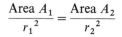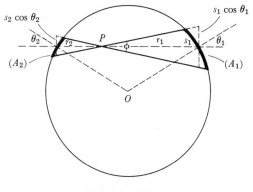

Figure 18-12

Now the shell is uniformly charged; therefore the charges on areas A_1 and A_2 are proportional to the areas; thus

$$\frac{Q_1}{{r_1}^2} = \frac{Q_2}{{r_2}^2} \qquad (18\text{-}4)$$

If the inverse-square law is true, a test charge Q_0 placed at P will experience two opposing forces $kQ_1Q_0/{r_1}^2$ and $kQ_2Q_0/{r_2}^2$, and these forces are equal by Eq. 18-4. The net force on the test charge due to both charged areas is zero at any arbitrary point P; the greater charge on area A_1 has been exactly compensated for by the greater distance r_1. The entire sphere can be divided up in this way by small cones passing through P, and so the net force on a charge anywhere inside the spherical shell is zero. To complete the proof, one must also show that no exponent other than 2 will give this result. Priestley apparently did not take account of the fact that the electrified cup was not a sphere.*

Priestley published his conclusion in London in 1767. For some reason the paper seems to have escaped the notice of other scientists of his day, or perhaps he himself failed to present his case forcefully enough. At any rate, in 1769 John Robison of Edinburgh attempted a direct experimental proof of the law, measuring the forces between small charged bodies. He arrived at the exponent 2.06—satisfactory enough in view of the difficulty of the experiment.

A few years later, Henry Cavendish (1731–1810), who was to become known for his work on Newton's law of gravitation a decade later, turned his attention to the problem and chose to expand on Priestley's indirect method. He placed a metal sphere of diameter 31 cm within two close-fitting hemispheres 34 cm in diameter. The inner sphere was completely enclosed but nowhere touched the outer hemispheres except by means of a removable wire (Fig. 18-13). Cavendish now electrified the system as strongly as

Figure 18-13 Cavendish's experiment to test the inverse-square law for electrostatic force.

possible, allowing every opportunity for charge to flow to the inner sphere by means of the connecting wire. He then withdrew the wire, isolating the inner sphere, and finally separated and removed the hemispheres. He tested the inner sphere by touching it with a pith ball and found it to be electrically neutral. His conclusion was similar to Priestley's—that the absence of any flow of charge to the inner sphere showed that electric forces obey an inverse-square law.

Cavendish went further: he tested the sensitivity of his apparatus by deliberately charging the inner sphere to the least detectable amount. In this way he placed limits of accuracy on his conclusion, which allowed him to state that the exponent must lie between 1.96 and 2.04. This important experiment of Cavendish's, as well as his gravitational experiment (Sec. 8-1), deserves to be studied carefully by students interested in the development of scientific methods. Cavendish's results were not published during his lifetime.

The next investigator of the law of electric force was, of course, Coulomb, who published his researches in a series of papers during the years 1788 and 1789. Coulomb no doubt expected to get an inverse-square law, as did all the others. He had already invented the torsion balance, independently of the Rev. John Michell (1724–1793) (whose torsion balance was applied by Cavendish to the gravitational problem), and he had studied the torsion of wires. It was a

---

* Analysis by Karl Friedrich Gauss (1777–1855) extended the proof to include any *closed* metallic cup of whatever shape.

Table 18-1 Coulomb's Experiment

Observed Force	Observed Distance	Distance Calculated from Inverse-Square Law
36 units	36 units	36 units
144	18	18
576	$8\frac{1}{2}$	9

natural step for Coulomb to apply the torsion balance to the measurement of the exceedingly small forces of electrical repulsion and attraction.* Using figures from his first paper, we can construct a table (Table 18-1) relating distance between charges and the force of repulsion (expressed in arbitrary units). Within experimental error, the data indicate an inverse-square law of force.

Following Coulomb, other workers repeated Cavendish's experiment with more sensitive detectors of charge. Faraday, in 1838, actually put himself inside a 4-m wire cage and had the cage charged to an extent that artificial lightning would jump to the walls of the room. He felt no effect on himself, and a sensitive electroscope inside the cage showed not the slightest indication of charge inside the cage. In 1870, a century after Cavendish, Maxwell used a delicate electrometer and found no effect, leading to a value of the exponent between 1.9999 and 2.0001. In 1971 it was shown by E. R. Williams, J. E. Faller, and H. A. Hill that if $F \propto 1/r^{2+x}$, then the deviation x from an exact inverse-square force law is less than 3×10^{-16}.

In view of all this activity, why then has the law become known as Coulomb's law when Coulomb's experiment was neither the first nor the most precise? Scientific priority is based on *publication* of results in a form suitable for the use of others. For this reason, Cavendish lost out

because he never published his method or his results, even though he kept an excellent record of his experiments in his private notebooks. It was Maxwell who brought Cavendish's notes to light and edited and published them for the first time in 1879, a century too late. It is hard to explain Cavendish's eccentricity in this respect, since he did publish some of his researches in the usual channels, including the gravitational experiment for which he is famous. He was best known in his lifetime as a chemist and is one of those to whom the discovery of the composition of water is attributed. He inherited great wealth and retired from much of public life and from society to devote himself to scientific work. As Maxwell said,

> Cavendish cared more for investigation than for publication. He would undertake the most laborious researches in order to clear up a difficulty which no one but himself could appreciate, or was even aware of, and we cannot doubt that the result of his enquiries, when successful, gave him a certain degree of satisfaction. But it did not excite in him that desire to communicate the discovery to others which, in the case of ordinary men of science, generally ensures publication of their results. How completely these researches of Cavendish remained unknown ... is shown by the external history of electricity.[†]

Coulomb, on the other hand, contributed to his subject through a series of well-organized and comprehensive papers. He described his apparatus in detail, discussed the possible sources of error, and, most important, checked his result by an entirely different experiment. His first method used the torsion balance to measure repulsive forces between two small, similarly charged spheres. His second method involved attraction instead of repulsion and was dynamic instead of static (he timed oscillations of a small charged body placed near a charged sphere several feet in diameter). Having obtained an

---

* Forces as small as 0.0001 grain (about 10^{-7} N, less than a millionth of an ounce) could be determined with considerable accuracy by Coulomb.

[†] *The Electrical Researches of the Hon. Henry Cavendish*, ed. by J. Clerk Maxwell, 1879. Quoted in W. C. D. Whetham, *The Theory of Experimental Electricity* (Cambridge University Press, 1912), p. 15.

inverse-square law by two different methods, he felt able to announce the law as proved.

The subsequent experiments of Faraday, Maxwell, and others contribute nothing new but merely extend the precision of the value of the exponent. Such experiments must be made, of course, for if they failed to give an exponent of exactly 2 (within experimental error), the consequences would be disastrous for much related electrical theory. But for its scientific importance, which includes publication and comprehensiveness, Coulomb's work stands out, and we justifiably speak of the law as Coulomb's law.

References

1. W. F. Magie, *A Source Book in Physics* (McGraw-Hill, New York, 1935), pp. 400–403, 408–420. Selections from the original papers of Franklin and Coulomb.

2. D. Roller and D. H. D. Roller, *The Development of the Concept of Electric Charge* (Harvard Case History in Experimental Science No. 8), (Harvard University Press, Cambridge, Mass, 1954). See especially Priestley's work on the inverse-square law (pp. 69–72) and Coulomb's experiments (pp. 72–80).

3. A. E. Moyer, "Benjamin Franklin: Let the experiment be made," *Am. J. Phys.* **14**, 536 (1976). A 10-page description of Franklin's electrical experiments, with many old illustrations.

4. S. Devons, "Benjamin Franklin as experimental philosopher," *Am. J. Phys.* **45**, 1148 (1977).

5. O. Jemifenko and D. K. Walker, "Electrostatic Motors," *Phys. Teach.* **9**, 121 (1971).

6. W. C. D. Whetham, *The Theory of Experimental Electricity* (Cambridge University Press, 1912), pp. 1–22. Cavendish's experiments are described on pp. 15–20.

19

Electric Field

19-1   *The Concept of Electric Field*

What can we say about the mechanism of electric force? The theory of action at a distance described in Chap. 18 is sufficiently noncommittal to be accepted without much critical thought. Electric forces exist; these forces act at a distance. Yet this is almost no theory at all. To *describe* what happens is often useful, but most of us wish to *explain* a phenomenon in terms of familiar laws and facts of physics. In other words, we intuitively seek a model for electric forces. To do so is permissible, but only if we have been inoculated against the scourge of modelitis. If we keep in mind the limitations of any model and remember that a model is not the same thing as reality, we can gain much help from a well-constructed model.

Let us discuss several models that have been proposed as descriptions of electric force. A useful concept, that of electric field, was introduced by Michael Faraday (1791–1867). Consider the space surrounding a positively charged sphere (Fig. 19-1*a*). We explore the region, using

an imaginary test charge, and we find that no matter where we put the test charge, it experiences a force such as $\mathbf{F}_1$ or $\mathbf{F}_2$. The force $\mathbf{F}_3$ is larger than $\mathbf{F}_2$, since at m the test charge is closer to the sphere than at n. At any point in space a force vector can be drawn; such a collection of infinitely many vectors is called a *vector field*, or simply a *field*. The direction of the field is defined as the direction of the force on a *positive* test charge. The strength or intensity $\mathbf{E}$ of the electric field at a point is defined as the force on a test charge at that point divided by the magnitude of the test charge. Thus,

$$\mathbf{E} = \frac{\mathbf{F}}{Q} \qquad (19\text{-}1)$$

In this definition, $\mathbf{E}$ and $\mathbf{F}$ are parallel vectors. The magnitude of $\mathbf{E}$ is given by

$$E = \frac{F}{Q} \qquad (19\text{-}2)$$

Electric field strength equals force per unit charge, and E is measured in newtons per

During a thunderstorm, large electric fields are built up in the region between clouds and the earth, as well as in the region between the clouds themselves.

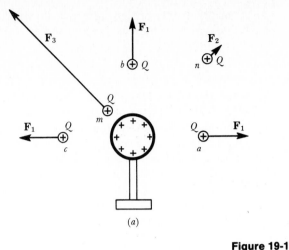

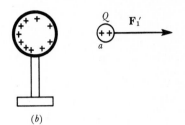

(a)

(b)

Figure 19-1

coulomb. It is always assumed that the test charge itself is so small that it does not appreciably alter the distribution of the charges that cause the field. For instance, if a very small test charge Q is placed at a, b, or c near a uniformly charged metal sphere (Fig. 19-1a), the force vectors labeled $\mathbf{F}_1$ are all equal in magnitude if a, b, and c are equidistant from the sphere's center. However, if the test charge is larger, the charges on the sphere will be redistributed, as in Fig. 19-1b, and the force on the (larger) test charge at a will be $\mathbf{F}'_1$. The ratio $\mathbf{F}'_1/Q$ will be equal to the magnitude of the field strength at point a, to be sure; but this ratio is the field strength due to the redistributed charge and is not the same as the original field strength we wished to measure. It is not unusual in physics for a measuring device (the test charge, in this case) to have an effect on what is being measured. It is like trying to explore the surface of the moon on foot—the explorer's own footprints are part of what he sees. To overcome this difficulty, we imagine using a test charge so small that making it still smaller would cause no appreciable change in the ratio F/Q. In other words, the ratio F/Q approaches a limit as Q (and F) approach zero, and this limit is E.

Example 19-1

Calculate the magnitude and direction of the electric field at a point 50 cm directly above a charge $Q' = -2 \times 10^{-6}$ C.

Since by definition the field strength equals the force on a unit *positive* charge, and a positive charge would be attracted by the given negative charge, we see at once that the direction of the field at the point in question is downward. The magnitude of the force on a test charge Q is found from Coulomb's law:

$$E = \frac{\text{force}}{\text{charge}} = \frac{kQQ'/r^2}{Q} = \frac{kQ'}{r^2}$$

$$= \frac{(9 \times 10^9 \text{ N·m}^2/\text{C}^2)(2 \times 10^{-6} \text{ C})}{(0.50 \text{ m})^2}$$

$$= \boxed{7.2 \times 10^4 \text{ N/C}}$$

It is not always as easy to calculate the magnitude and direction of the electric field vector $\mathbf{E}$ as it is in the example just given. If there are many charges, perhaps on several conductors, it might be a difficult task. We would have to use vector addition, with one vector contributed by each of the charges (Fig. 19-2). The point we wish to

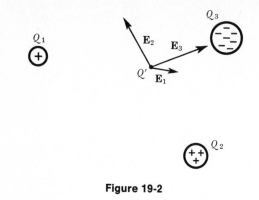

Figure 19-2

consider a river of varying depth, flowing in a curved path, perhaps having here and there some swirling eddies or stagnant places. The flow pattern can be defined as a velocity field, for at every point in the river a velocity could be measured in both magnitude and direction. For an example of a scalar field, we could measure the density of the sea at various latitudes, longitudes, and depths. Density is a scalar quantity, and so this collection of density values, one for every point in space, is a scalar field.

19-2   Lines of Force

As an aid in visualizing electric field, Faraday and his successors drew *lines of force* that everywhere have the direction of the electric field. In Fig. 19-3 we show lines of force that represent the electric field surrounding a positive point charge. The lines are directed away from the + charge, since like charges repel each other and the imaginary test charge is a + one. The lines extend all the way (!) to infinity, since there is *some* Coulomb force on the test charge no matter how far away it is. Only a representative number of lines are drawn; a line could equally well be drawn through *any* point, such as *P*. For this special case of a single isolated point charge, the lines of force are straight lines; in general, when the direction of **E** is calculated at various points, the field lines turn out to be

emphasize is that, regardless of the origin of a field, we can *measure* the field by use of a test charge. Here, as elsewhere in physics, to specify a measurement procedure is to define a concept.* We have set up a procedure (measuring the force per unit charge) that *defines* what we mean by electric field. In general, Example 19-1 shows the magnitude of the electric field of a point charge *Q* to be given by

$$E = kQ/r^2$$

We have met with a vector field before—in Sec. 8-2, where we found $g = GM/r^2$ for the magnitude of the gravitational field of a point mass *M*. In principle, we can explore the gravitational field in any region of space by measuring the magnitude and direction of the force on a small test mass *m*. The nature of any force field is defined by the kind of test body we use: a small test charge *Q* to explore an electric field, or a small test mass *m* to explore a gravitational field. In our study of the magnetic field, we shall use still another kind of test body to explore the field.

The concept of field is not limited to force fields such as we have been discussing. For an example of a vector field that is not a force field,

---

* For other examples of this "operational" viewpoint, see pages 112–113 and 205, where operational definitions of inertial mass are discussed. Also, in Chap. 17, our definition of absolute zero was an operational one made in terms of a procedure involving Carnot engines.

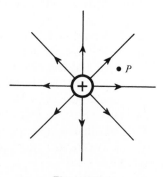

Figure 19-3

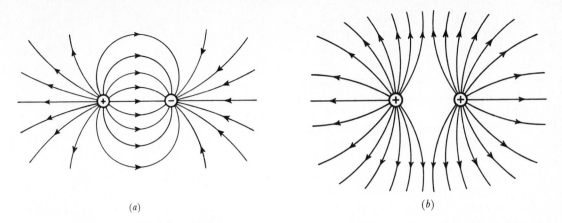

Figure 19-4 (*a*) Electric field in the neighborhood of a pair of unlike charges of equal magnitude. (*b*) Electric field in the neighborhood of a pair of like charges of equal magnitude.

curved. The field near a pair of equal but opposite charges is shown in Fig. 19-4*a*, and the field near a pair of equal and like charges is shown in Fig. 19-4*b*.

Lines of force originate on + charges and terminate on − charges; no line starts in midair, so to speak. Another descriptive property of the lines of force is their "density"; the field is strongest where the lines are closest together.

19-3 Models for Electric Field

Nowadays, we consider that lines of force have no objective reality but are merely a convenient and useful *representation* of a field. Faraday went much farther; he made a model of electric forces according to which the lines of force were considered to be the bounding edges of what he called tubes of force (Fig. 19-5). Faraday, like most scientists of his time, imagined all space to be filled with an invisible weightless "ether" that had certain elastic properties. The tubes of force, being made of ether, were like stretched rubber bands or, better yet, like the springy wires used to give shape to the Victorian era's bustle and corset. In addition, Faraday advanced the hypothesis that adjacent tubes of force repelled

each other. Thus in Fig. 19-4*a*, the repulsion between adjacent tubes causes the lines of force to bulge outward, and since the lines of force tend to straighten out, the force of attraction between unlike charges is explained. This ingenious model was refined and developed by Maxwell and others of the "Cambridge school" of English physicists to include a description of electromagnetic forces as well as electrostatic forces.

The difference between the action-at-a-distance model and the lines-of-force model can be illustrated by considering the nature of charge according to the two models. The action-at-a-distance model considers charges to be located at certain places, attracting or repelling each other

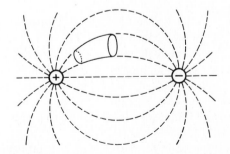

Figure 19-5 A short section of a tube of force.

much as the earth attracts the moon. The electric field is then just a mathematical or geometrical description of the Coulomb forces. In the Faraday-Maxwell model, the emphasis is on the field—that is, on the space between the charges. This space is filled with elastic tubes of force (of varying cross sections), and the charges are thought of as having existence only as the representation of a sliced-off "raw end" of a tube of force.

Physicists today do not adhere in detail to Faraday's concept of the electric field, in which the forces are a sort of by-product of properties of a medium. For one thing, the medium (ether) would have to be present even in a vacuum, and its mechanical properties would have to be miraculous indeed. The Michelson-Morley experiment (Sec. 28-2) in 1887 showed that the velocity of light is the same in a north–south direction as in an east–west direction, even though the rotation of the earth is continually taking us in an easterly direction; other motions are surely present as the earth revolves around the sun, the sun moves through our galaxy, and our galaxy moves through space. This experiment showed that there is no "ether wind" blowing past a moving observer. All other attempts to prove the existence of the ether have also failed. The Faraday-Maxwell model also lost ground with the discovery of the free electron about 1900. We are accustomed now to think of charged particles moving through cyclotrons, radio tubes, and transistors, and even bombarding us from outer space in the form of cosmic rays. For such "free" particles, the cumbersome apparatus of tubes of force seems to be of little help.

To Faraday goes the credit for introducing and emphasizing the concept of field; even if we reject the idea of an ether, we can still think of electric field as a property of space itself, requiring no medium of any kind. This viewpoint pervades scientific thought even now, and we recognize Faraday's contribution as a major one in the history of physcis.

A third model for describing forces that act at a distance has been developed in the last several

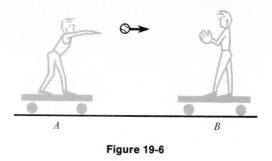

Figure 19-6

decades. Imagine two boys standing on frictionless carts (Fig. 19-6), playing a game of catch. Boy A recoils to the left when he throws the ball; boy B receives an impulse to the right when he catches the ball. We can look on this as an example of action at a distance, with the force being "transmitted" by the ball. If the boys used a heavier (that is, "massier") ball, or if they exchanged it more often or threw it with greater speed, there would be a greater force of repulsion between the boys. The simple model we have considered is incapable of describing an attractive force. (See Ques. 19-10 at the end of this chapter.) However, working along these lines, theoretical physicists have constructed a satisfactory model for electromagnetic forces of attraction as well as repulsion, including Coulomb forces, magnetic forces, and the interatomic forces that are responsible for elasticity and molecular cohesion. The theory is complex and highly mathematical. The "particles" that are, in a sense, bandied back and forth are called *photons*. Under suitable circumstances, photons can have a "real" existence and are emitted as light from an atom that is electrically excited, as in a neon sign. We see here how intimately related are the theories of electricity and light; the very existence of electric forces is postulated to be due to particles of light.

In Chap. 1 we stated that there are three types of force between matter—electric, nuclear, and gravitational. For each type of force, there is a field theory according to which particles are emitted and absorbed at a very rapid rate. In the model of strong nuclear forces, the field particles are called *mesons*; it is for this reason that mesons are sometimes spoken of as "nuclear glue."* For example, the continual exchange of mesons between the proton and the neutron is thought to keep the two nucleons bound together in the $^2_1$H nucleus. Even gravitational forces are believed to have their field particles, called *gravitons*. Under suitable circumstances, free photons can be formed (light is emitted), and free mesons can be formed (the particles materialize in high-energy particle accelerators). The possibility of detecting free gravitons, if they exist, seems to be beyond existing experimental technique.

We have discussed three models of electric field: action at a distance, with no further explanation; the Faraday-Maxwell theory, in which lines and tubes of force represent a strain in the ether; and the photon-field theory, which is action at a distance with an added model for transmission of force by impact of particles. In our future work, we shall use the first model of field, drawing lines of force as a representation but without giving them any mechanical properties.

19-4   Potential Difference

When we explore the electric field in the region of space around a charged body or near the terminals of a battery or generator, we imagine that a small, positive test charge is moved from point to point. For simplicity, we consider first a uniform electric field directed from point A to point B. If we move the test charge from B to A and electric forces act on it, we must do work on the charge, and its PE will increase. The change in PE per unit charge is called the *electric potential difference* between the points in question; the mks unit for potential is the *volt* (V). *Two points differ in potential by 1 volt if 1 joule of work is*

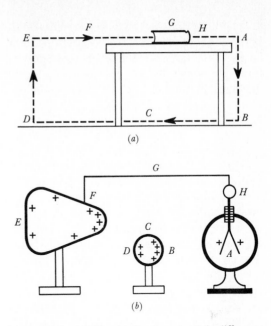

Figure 19-7   (a) Gravitational potential difference. (b) Electric potential difference.

required to move 1 coulomb of charge from one point to the other. In symbols,

$$V_{AB} = \frac{W_{B \to A}}{Q} \tag{19-3}$$

$$\text{volts} = \frac{\text{joules}}{\text{coulombs}}$$

where V_{AB} is the difference in potential of point A relative to point B, and $W_{B \to A}$ is the work done against electric forces to move a test charge Q from B to A.

A gravitational analogy will prove helpful in our study of electric potential.[†] When a test mass m, such as the book in Fig. 19-7a, is moved around on a horizontal table, no work is done against gravity, and there is no change of gravitational potential. All points on the table top are

---

* The mesons themselves are believed to be composed of quarks, which are bound together by the exchange of field particles called gluons. See Sec. 33-10 for further details.

[†] The analogy is not perfect, however. There are two kinds of electric charges, positive and negative; but, as far as we know, negative masses do not exist outside the pages of science fiction.

at the same potential; similarly, all points on the floor are at the same (but lower) potential. If the height of the table is h, the difference of gravitational PE is mgh, and the gravitational potential difference between the two levels is

$$(V_{AB})_{\text{grav}} = \frac{W_{B \to A}}{m} = \frac{mgh}{m} = gh$$

Gravitational potential is expressed in joules per kilogram, a unit for which no special name has been invented. In mechanical problems a body may be assigned zero PE at any convenient level—for instance, when it is at sea level, or at the level of the floor, or on the table top. There is thus no absolute value of gravitational potential for any given location. It is potential *difference* between two levels that is useful. The energy principle is stated in terms of *changes* of energy, and the choice of "reference level" of zero PE is arbitrary.

These ideas are carried over into our discussion of electric potential. *Potential difference* is the useful quantity. Just as in the gravitational case, "the potential at a point" has meaning only if some point of zero potential is specified. Often the earth (the "ground" of a circuit) is taken to be at zero electric potential, but other choices are possible.

Examples of potential difference (PD) are not hard to find. Let us discuss a few representative situations in detail.

(*a*) Two metal spheres are charged, with A having an excess of positive charge and B an excess of negative charge (Fig. 19-8). Since the test charge is defined as a positive one, we see that work must be done on it to bring the unit positive test charge from Q to P. Therefore P is at higher potential than Q.

(*b*) Water tends to flow downhill. The usual explanation is from the point of view that the force of gravity has a component parallel to the hill, and this force component "pulls the water down the hill." A more sophisticated way of looking at this same fact is to consider energy

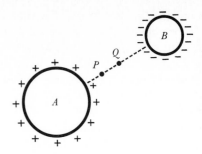

Figure 19-8 Potential difference; P is at a higher potential than Q.

changes. Water flows downhill because its gravitational PE is less in the lower position. A heavy particle, subjected to no other force than its own weight, always seeks the lowest attainable level, where its gravitational PE is least. Similarly, *positive* electric charge tends to flow from points of high potential to points of low potential—"downhill," so to speak. We do not have to consider electric forces in detail to know that their net result is to push or pull charges into a position of least PE. This means that if the charges on a metallic conductor are at rest, all points of the conductor are at the same potential. In Fig. 19-7b, the pear-shaped object, the connecting wire, and the leaves of the electroscope are all at the same potential, for if this were not so, the free electrons in the metal would readjust their distribution, flowing away from any region in which they had a higher PE into a surrounding region in which they had a lower PE. All parts of the small metallic sphere are also at the same potential; indeed, the distribution of its charge is self-adjusting to make this so. Points E, F, G, H, and A are all at one potential; points B, C, and D are at another potential. The analogy between points in parts (*a*) and (*b*) of Fig. 19-7 is complete, letter for letter.

(*c*) An ordinary dry cell (Fig. 19-9a) consists essentially of a carbon rod and a zinc can, separated by a moist paste containing ammonium chloride and manganese dioxide, which can be considered an electrolytic solution. When the cell is on the shelf or on a laboratory table, with

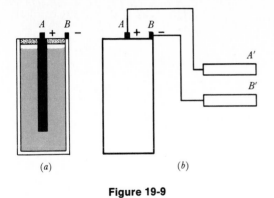

(a)   *(b)*

Figure 19-9

nothing connected to it, experiment shows that there is a slight excess of positive charge on the carbon rod* and a slight excess of negative charge on the zinc can. To move a charge of $+1$ C from B to A through the air requires 1.5 J of work, and the PD is 1.5 J/C, or 1.5 V. If the terminals are connected by wires to a pair of plates (Fig. 19-9b), the plates become charged (this process takes a small fraction of a second), and then the PD between A' and B' is also 1.5 V.

Example 19-2

What is the PD between two points if 200 J of work is required to move 40 C from one point to the other?

$$V_{AB} = \frac{W_{B \to A}}{Q} = \frac{200 \text{ J}}{40 \text{ C}} = 5 \text{ J/C}$$

$$= \boxed{5 \text{ V}}$$

Example 19-3

How much work is required to move an electron between the two terminals of a particle accelerator whose PD is 4 million volts?

---

* Remember that a deficiency of electrons is entirely equivalent to an excess of positive charge.

$$W_{B \to A} = V_{AB}Q$$
$$= (4 \times 10^6 \text{ V})(1.60 \times 10^{-19} \text{ C})$$

$$= \boxed{6.40 \times 10^{-13} \text{ J}}$$

Example 19-4

In a laboratory experiment, an electron leaves the heated cathode K of an evacuated tube (Fig. 19-10) with negligible initial velocity and is accelerated through an applied PD of 500 V. What is the velocity of the electron as it approaches the accelerating electrode A?

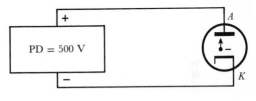

Figure 19-10

While the electron is being repelled by the cathode K and attracted by the anode A, it loses PE, just like a baseball that falls toward the earth. The change in PE equals the change in potential times the charge. Now we apply the energy principle, using 9.11×10^{-31} kg for the electron mass:

$$\text{Loss of PE} = \text{gain of KE}$$
$$(1.60 \times 10^{-19} \text{ C})(500 \text{ V}) = \tfrac{1}{2}(9.11 \times 10^{-31} \text{ kg})(v^2)$$
$$v = \sqrt{\frac{2(1.60 \times 10^{-19} \text{ C})(500 \text{ V})}{9.11 \times 10^{-31} \text{ kg}}}$$
$$= \sqrt{1.77 \times 10^{14} \text{ C} \cdot \text{V/kg}} = \boxed{1.33 \times 10^7 \text{ m/s}}$$

A note on units: In this example, we arrived at velocity $= \sqrt{\text{C} \cdot \text{V/kg}}$. This fine collection of units can be shown to be equal to m/s, as follows:

$$\sqrt{\frac{\text{C} \cdot \text{V}}{\text{kg}}} = \sqrt{\frac{\text{C} \cdot (\text{J/C})}{\text{kg}}} = \sqrt{\frac{\text{J}}{\text{kg}}}$$
$$= \sqrt{\frac{\text{N} \cdot \text{m}}{\text{kg}}} = \sqrt{\frac{(\text{kg} \cdot \text{m/s}^2)\text{m}}{\text{kg}}}$$
$$= \sqrt{\frac{\text{m}^2}{\text{s}^2}} = \text{m/s}$$

19-5 Equipotential Surfaces

An *equipotential surface* is defined as a surface all points of which are at the same potential. The PD between any two points on the surface is zero. Since PD is work per unit charge, our definition implies that no work is done when a test charge is moved from one point to another on an equipotential surface. As an example, consider a point charge $+Q$ from which lines of force radiate outward (Fig. 19-11). We can easily prove that the equipotential surface designated by V_1 is a sphere surrounding the charge. To take a test charge from A to B along a path that lies in the spherical surface would require no work, since the motion is at all times perpendicular to the direction of the force. Hence A and B are at the same potential, and the sphere is an equipotential surface. The larger sphere, labeled V_2, is also an equipotential surface, and there is a PD equal to $(V_1 - V_2)$, or ΔV, between any point on the first sphere and any point on the second sphere. A more complicated example is shown in Fig. 19-12, which represents the field surrounding a pair of equal and like charges. This is the same as Fig. 19-4b, with the addition of dashed lines to represent some of the equipotential surfaces; they are no longer spherical in shape.

An important relationship between lines of force and equipotential surfaces is seen in both figures. The equipotential surfaces are every-

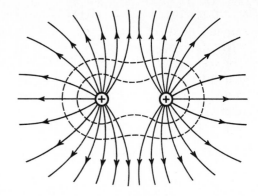

Figure 19-12 Electric field in the neighborhood of a pair of like charges. The equipotential surfaces are shown as dashed curves.

where perpendicular to the lines of force. This must be true in general, for if the lines of force were *not* perpendicular to an equipotential surface, then a test charge moving along the equipotential surface would experience a component of force parallel to its motion, and work would be done. This means that a PD would exist between two points on the same equipotential surface—which is contrary to definition.

The close relationship between field and potential leads to a new interpretation of electric field strength. Since

$$\text{Work} = (\text{force})(\text{displacement})$$

where the displacement is assumed to be parallel to the force, we can write

$$\frac{\text{Work per unit charge}}{\text{Distance moved}} = \text{force per unit charge}$$

or

$$\frac{\text{Potential difference}}{\text{Distance moved}} = \text{field strength}$$

$$-\frac{\Delta V}{\Delta s} = E \qquad (19\text{-}4)$$

A positive test charge tends to move from a region of high potential to a region of low poten-

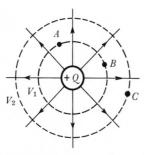

Figure 19-11 Lines of force (solid lines) and equipotential surfaces (dashed curves) for an isolated point charge.

tial—"downhill," in the direction of *decreasing* potential. The − sign is used in Eq. 19-4 so that E is positive when $\Delta V/\Delta s$ is negative. The change in potential per unit distance* is called the *potential gradient*; we see that the magnitude of the field strength equals the negative of the potential gradient. Potential gradients are measured in volts per meter.

Example 19-5

A fine iron wire 25 cm long is connected between the terminals of a battery whose terminal voltage is 6 V. What are the magnitude and direction of the electric field inside the wire?

$$E = -\frac{\Delta V}{\Delta s} = -\frac{6\text{ V}}{0.25\text{ m}} = \boxed{-24\text{ V/m}}$$

The direction of the field is toward the negative terminal of the battery, since a positive test charge inside the wire would move in this direction. The direction of field is always in the direction of *decreasing* potential.

Example 19-6

In Example 19-5, what is the PD between point P of the wire, which is 5 cm from the + terminal of the battery, and point Q, which is 15 cm from the + terminal?

The field inside the wire is uniform, equal to −24 V/m. Hence,

$$\Delta V = -E\,\Delta s = -(-24\text{ V/m})(0.10\text{ m})$$

$$= \boxed{2.4\text{ V}}$$

Point P is 2.4 V higher in potential than point Q, since P is closer to the + terminal.

---

* The distance moved must be measured along the direction of the lines of force; that is, one must proceed from one equipotential surface to a neighboring one along the shortest (perpendicular) path. To obtain the field strength "at a point," we must take the limit of Eq. 19-4 as $\Delta s \to 0$; in other words, $E = -dV/ds$.

We now have two mks units for field strength: newtons per coulomb (force per unit charge) and volts per meter (potential gradient). We can show that these units are equivalent, using the definition of the volt:

$$E = \frac{\text{V}}{\text{m}} = \frac{\text{J/C}}{\text{m}} = \frac{\text{J}}{\text{C}\cdot\text{m}} = \frac{\text{N}\cdot\text{m}}{\text{C}\cdot\text{m}} = \frac{\text{N}}{\text{C}}$$

In most engineering applications, the volt per meter or the volt per centimeter is used to measure electric field. For instance, tables show that the field strength required to cause a spark in dry air is about 10^6 V/m. The electric field near the surface of the earth (due to an excess of negative charge on the earth) is normally about 100 V/m, but this figure may increase to 600 V/m or more during a thunderstorm. Since a volt per meter is equal to a newton per coulomb, this means that if a coulomb of positive charge could be collected together and held at one point near the earth's surface, it would experience a downward force of about 100 N (approximately 23 lb).

The potential, relative to infinity, of an isolated point charge or a uniformly charged sphere can be found using integral calculus. This is done in Sec. 19-9. The result is that the potential at a point distance r from a point charge Q is given by

$$V = \frac{kQ}{r} \qquad (19\text{-}5)$$

The difference in potential V_{AB} between two points A and B can be calculated from Eq. 19-5 and used to find the work required to move a test charge from one point to the other.

Example 19-7

In Fig. 19-11, points A and B are 0.9 m from a charge of $+8 \times 10^{-9}$ C, and C is 1.6 m from the charge. (*a*) Calculate the values of the potentials V_1 and V_2 on the two equipotential surfaces shown in the figure. (*b*) How much work is required to move a test charge of $+4\ \mu$C from C to B? (*c*) How much work is required to move the test charge from B to A?

$$(a) \quad V_1 = \frac{kQ}{r_1} = \frac{(9 \times 10^9 \text{ N} \cdot \text{m}^2/\text{C}^2)(8 \times 10^{-9} \text{ C})}{0.9 \text{ m}}$$

$$= 80 \frac{\text{N} \cdot \text{m}}{\text{C}} = 80 \text{ J/C} = \boxed{80 \text{ V}}$$

$$V_2 = \frac{kQ}{r_2} = \frac{(9 \times 10^9 \text{ N} \cdot \text{m}^2/\text{C}^2)(8 \times 10^{-9} \text{ C})}{1.6 \text{ m}}$$

$$= 45 \frac{\text{N} \cdot \text{m}}{\text{C}} = 45 \text{ J/C} = \boxed{45 \text{ V}}$$

(b) The potential difference between B and C is 80 V − 45 V = 35 V. Work must be done *on* the test charge to bring it closer to $+Q$.

$$W_{C \to B} = (V_{BC})(Q)$$

$$= (35 \text{ V})(4 \times 10^{-6} \text{ C}) = \boxed{1.4 \times 10^{-4} \text{ J}}$$

(c) B and A are at the same potential, so

$$W_{B \to A} = \boxed{0}$$

Since electric potential is related to work, which is a scalar quantity, it is much easier to add potentials than to add fields. The potential V due to a number of charges is simply the algebraic sum of the potentials due to each charge. Finding the resultant electric field $\mathbf{E}$ is more difficult, since vector addition must be used to add the contributions of the various charges.

The gravitational analogy for Eq. 19-5 is

$$V_{\text{grav}} = \frac{GM}{r}$$

which is based on Newton's inverse-square law of gravitation in the same way that $V_{\text{elec}} = kQ/r$ is based on Coulomb's inverse-square law of electric force.

19-6   Capacitance

If two conductors are placed near each other, the combination is called a *capacitor*. Usually, but not necessarily, the conductors are in the form of parallel plates. In Fig. 19-13 we show two arbitrarily shaped conductors, with some +

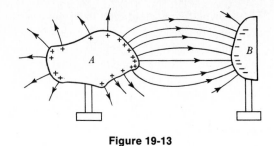

Figure 19-13

charge on one body and an equal − charge on the other. Of one thing we can be sure: if the charges are at rest, there is no electric field at any point within the metal of either conductor. If this were not so, the free electrons in the metal would move under the influence of the field, and so would not be at rest. By similar reasoning, at any point on the surface of a conductor there can be no component of electric field parallel to the surface, for if there were, free electrons would flow along the surface. All points of conductor A must therefore be at the same potential, since the potential gradient along the surface is zero. We conclude that *if the charges are at rest, the surface of a conductor is an equipotential surface*, and all points in the interior have the same potential as the surface. The piling up of charge at the pointed end of conductor A is explained by saying that if the charges were *not* so distributed, the conductor would not be an equipotential surface.* The charges automatically adjust themselves so that the surface of A is an equipotential surface. The surface of B is also an equipotential surface, at some different potential. There is a single PD between body A and body B. In Fig. 19-13, A is at a higher potential than B, since it would require work to move a unit test + charge from B to A.

We now prove that the PD between the plates of any capacitor is directly proportional to the charge on either plate. We assume equal and

---

* Incidentally, as we have seen in Sec. 18-4, the charges all reside on the surface of a metallic conductor, whether the body is solid or hollow.

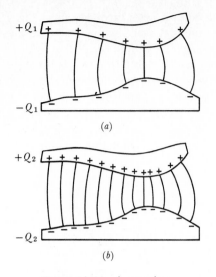

$+Q_1$

$-Q_1$

(a)

$+Q_2$

$-Q_2$

(b)

Figure 19-14 A capacitor.

opposite charges on the plates.* If the charge on the plates is increased, the strength of the field between the plates is increased in proportion to the charge. The relative distribution of lines of force remains the same, but if there are twice as many charges on the plates, the lines are twice as close together and the field is twice as great. This is illustrated in Fig. 19-14, where the same capacitor is charged with a small charge Q_1 in (a) and with a large charge Q_2 in (b). The PD is defined as work per unit test charge. If a test charge is moved from the bottom plate to the top plate along any of the lines of force, the work is greater in (b), since the force is greater while the displacement is the same. Therefore, the potential difference V is proportional to the charge. For any pair of conductors, the ratio Q/V would be expected to be a constant. This constant is called the *capacitance* C of the capacitor.

---

* It is possible, of course, to place unequal charges on two metallic plates. However, if the plates are close together and of large area, the charges on the facing portions of the plates will be equal, and the excess charge of the more highly charged plate will lie on the outer surface of that plate. Lines of force from these excess charges will terminate on opposite charges on some neighboring conductor or conductors.

$$\text{Capacitance} = \frac{\text{charge on either plate}}{\text{PD between the plates}}$$

$$C = \frac{Q}{V} \tag{19-6}$$

$$\text{farads} = \frac{\text{coulombs}}{\text{volts}}$$

The mks unit for capacitance is the *farad* (F); a capacitor has a capacitance of one farad if one coulomb of charge causes a PD of one volt. The farad is an inconveniently large unit. For most applications, the microfarad ($1~\mu\text{F} = 10^{-6}$ F) or the picofarad ($1~\text{pF} = 10^{-12}$ F) is used. Circuit-diagram symbols for a capacitor are —||— or —||— or —||—, the last one representing a variable capacitor.

For the special case of a pair of parallel plates close together, it can be shown that

$$C = \frac{A}{4\pi k d} \tag{19-7}$$

where A is the area of each plate, d is their separation, and k is the constant ($= 9 \times 10^9~\text{N} \cdot \text{m}^2/\text{C}^2$) that enters into Coulomb's law.

A check of the units of this equation for capacitance is instructive.

$$C = \frac{A}{4\pi k d} = \frac{\text{m}^2}{(\text{N} \cdot \text{m}^2/\text{C}^2)(\text{m})} = \frac{\text{C}^2}{\text{N} \cdot \text{m}}$$

$$= \frac{\text{C}^2}{\text{J}} = \frac{\text{C}}{\text{J/C}} = \frac{\text{C}}{\text{V}} = \text{F}$$

The check of units is successful.

One way of charging a capacitor is to connect the two plates to a source of PD such as a battery. In Fig. 19-15 the battery (represented by the symbol —|i|—) causes a momentary flow of electrons from the top plate through the battery to the bottom plate. If we arbitrarily "ground" the bottom plate and call its potential zero, then the top plate will be at a potential of $+45$ V. The PD between A and B is 45 V. Just enough charge (given by $Q = CV$) flows so that the top plate is an equipotential surface at a potential of $+45$ V and the bottom plate is an equipotential surface

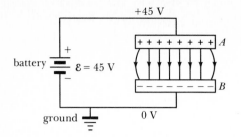

+45 V

battery $\mathcal{E} = 45$ V

A
B

ground $\quad$ 0 V

Figure 19-15 A parallel-plate capacitor, showing uniform field between the plates.

at a potential of 0 V. The lines of force are perpendicular to the equipotential surfaces, as must always be true. For this special case of a parallel-plate capacitor with plates close together, the lines of force are parallel, and the field between the plates is *uniform* in magnitude and direction. Work is done when a capacitor is charged; see Sec. 19-10 for a discussion of energy storage in capacitors.

The process of "grounding" a conductor or a point in a circuit, as in Fig. 19-15, deserves a brief comment. The earth is such a large body that charges can flow into or out of it without appreciably changing its potential relative to infinity (or relative to the moon or the sun). This means that the capacitance of the earth is so large that ΔV (given by $\Delta V = \Delta Q/C$) is negligible for any charge flow ΔQ encountered in the laboratory or even in large-scale engineering.* The potential of the earth is, therefore, a useful reference potential, and grounding a circuit by connecting a point to earth gives that point a fixed potential that is designated 0 V. In circuit diagrams, "ground" is represented by $\equiv$ or $\overline{}_{\!/\!/\!/\!/}$.

19-7 Dielectrics

So far in our discussion of capacitors we have tacitly assumed the two plates to be separated by a vacuum. If this is not so, the capacitance is

---

* A good analogy is the water in the oceans; sea level does not change appreciably even if many cubic meters of water are added to or removed from the ocean.

increased by a factor that depends on the electrical nature of the insulating medium between the plates. Such a medium is called a *dielectric medium* or simply a *dielectric*. We define the *dielectric constant K* of a medium as a ratio of capacitances: if C_{med} is the capacitance of a capacitor whose plates are separated by the medium, and C_{vac} is the capacitance of the same pair of plates separated by a vacuum, then

$$K = \frac{C_{\text{med}}}{C_{\text{vac}}} \qquad (19\text{-}8)$$

Equation 19-8 is an operational definition of dielectric constant, since it specifies a way of measuring K. Let us now try to interpret the dielectric behavior of matter by a molecular model.

The molecules of many substances are said to be *polar*, which means that the positive and negative charges of the molecules are separated by a small distance. For instance, research has shown that H_2O is a bent molecule (see Fig. 19-16), with the negative oxygen ion not coinciding with the point midway between the two positive hydrogen ions. This permanent polar structure of the water molecule is equivalent to a pair of equal and opposite charges separated by a small distance. Such an arrangement of charges is called a *dipole*. The dipoles in water give rise to electric forces within the liquid that partially cancel the Coulomb forces between other charged bodies immersed in the water.

Coulomb's law is always true, whether or not there is a dielectric medium. That is, there is only

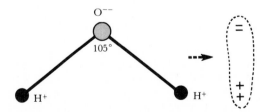

Figure 19-16 A water molecule is equivalent to a dipole.

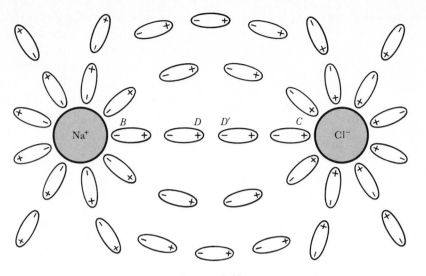

Figure 19-17

one force law, namely,

$$F = \frac{kQQ'}{r^2}$$

for the force between two stationary electric charges, wherever they may be. No new law of force is needed to explain the reduced force between charges immersed in a dielectric.

To illustrate this point for a liquid, let us analyze the forces acting on the sodium ion when sodium chloride is dissolved in water. The water molecules, which are polar, cluster around the Na^+ and Cl^- ions, as shown in Fig. 19-17. The net electric force on the Na^+ ion is the vector sum of three forces caused by (a) the Cl^- ion, (b) the ring of negative charges such as that at B, and (c) the ring of positive charges such as that at C. All other charges, such as those at D and D', have no effect, since they occur in equal and opposite pairs close together. Each of the three forces (a), (b), and (c) can be computed (in principle, at least) by application of Coulomb's law: (a) is the force on the Na^+ ion, directed toward the right, exerted by the Cl^- ion; (b) is zero, since the forces exerted on Na^+ by this symmetrical ring of negative charges have a vector sum equal to zero; (c) is directed toward the left and tends to oppose

(a). The *net* force on Na^+ is therefore less than it would have been if the charges had been in a vacuum. It is for this reason that water is such a good solvent for NaCl and so many other diverse chemical compounds; the attraction between the component parts of the molecules is weakened. Even symmetrical nonpolar molecules such as those of benzene are distorted slightly by neighboring charged bodies, and so there is a slight dielectric effect, but not nearly so strong an effect as in the case of polar molecules, where dipoles are already permanently present.

Consider now a charged capacitor first with plates separated by a vacuum, then (Q remaining the same) with the plates immersed in a dielectric fluid such as water, benzene, or air. According to the argument of the preceding paragraphs, the effect of the dipoles in the fluid is to reduce the net force on a test charge immersed in the medium, and hence less work is required to take a test charge from one plate to the other. Since potential difference is work per unit charge, we see that V has been decreased by the presence of the dielectric medium, and therefore $C(=Q/V)$ has been increased. The ratio of the two capacitances is defined as the dielectric constant of the medium, as in Eq. 19-8.

Table 19-1　Dielectric Constants*

vacuum	1
dry air at 1 atm	1.0006
water	80
carbon tetrachloride	2.24
benzene	2.28
castor oil	4.67
methyl alcohol	33.1
glass	4–7
amber	2.65
wax	2.25
mica	2.5–7

* All at 20°C.

It is difficult to imagine an experiment to measure the force on a test charge immersed in a solid dielectric medium; for this reason we have limited our discussion of dielectric constant to fluids. Nevertheless, it should be evident that dipoles *are* formed in a solid insulator when an electric field is applied, and the capacitance increases when a solid slab of a material like glass or mica is inserted between a pair of plates. Even though the molecular interpretation of solid dielectrics is complex, Eq. 19-8 still serves as a definition of dielectric constant for a solid and provides a procedure for its measurement. Note that K is the ratio of two capacitances, and hence the dielectric constant of a substance is a dimensionless quantity. Dielectric constants for some substances are given in Table 19-1.

For any parallel-plate capacitor, the definition of K requires that Eq. 19-7 be modified to read

$$C = \frac{KA}{4\pi k d} \qquad (19\text{-}9)$$

If SI units are used, $C = 8.85 \times 10^{-12}\dfrac{KA}{d}$.

19-8　The Use of Capacitors in Circuits

In electric circuits, capacitors are used to store charge, somewhat as a bucket can be used to store a quantity of water. Do not imagine, how-

ever, that the capacitance of a capacitor is the "amount of charge it can hold." Since $Q = CV$, the charge Q can be increased to any desired value, with a corresponding increase in the potential difference V. The limit is reached only when the insulation between the plates breaks down and a spark jumps across the gap. The water analogy would be closer if we were to think of a cylindrical storage tank of very great height. As water is pumped in, the pressure rises, and the "capacity" of the tank is limited only by the ability of the seams and joints at the bottom of the tank to withstand the pressure of the water. At any given time, the amount of water stored in the tank is proportional to the pressure P, just as the charge stored in a capacitor is proportional to the potential difference V.

Example 19-8

A capacitor whose capacitance is 20 μF is momentarily connected to a dry cell that maintains a potential difference of 1.5 V between its terminals. It is then disconnected. How much charge is stored in the capacitor?

The resistance of the connecting wires and the internal resistance of the cell are of no consequence except to determine how rapidly the capacitor becomes charged. In a typical case, after 10^{-6} s or 10^{-8} s or some similar very short time, the capacitor is fully charged and the electrons have ceased flowing. The plates have reached a PD equal to the PD of the cell.

Using Eq. 19-6, we have

$$Q = CV = (20 \times 10^{-6}\ \text{F})(1.5\ \text{V})$$

$$= 30 \times 10^{-6}\ \text{C} = \boxed{30\ \mu\text{C}}$$

Example 19-9

In measuring the charge of the electron by the Millikan oil-drop method (Sec. 29-1), a uniform electric field is obtained by connecting a pair of parallel circular plates to a source of PD, as in Fig. 19-15. If the plates are 10 cm in radius and are separated by 3 cm, what charge on the plates will give a field of 400 V/cm?

If the field is to be 400 V/cm, the PD between the plates must be

$$V = (400 \text{ V/cm})(3 \text{ cm}) = 1200 \text{ V}$$

We next compute the capacitance of the capacitor formed by the plates, which are separated by air, whose dielectric constant is very nearly 1.

$$C = \frac{KA}{4\pi kd}$$

$$= \frac{(1)(\pi)(1 \times 10^{-1} \text{ m})^2}{4\pi(9 \times 10^9 \text{ N·m}^2/\text{C}^2)(3 \times 10^{-2} \text{ m})}$$

$$= 9.26 \times 10^{-12} \text{ F}$$

Now we can compute the charge on the plates:

$$Q = CV = (9.26 \times 10^{-12} \text{ F})(1.20 \times 10^3 \text{ V})$$

$$\boxed{= 1.11 \times 10^{-8} \text{ C}}$$

The physical construction of capacitors depends on the desired capacitance and the PD that must be sustained without rupture of the dielectric. In the formula

$$C = \frac{KA}{4\pi kd}$$

K, A, or d can be varied. For low values of C (such as those needed in the tuning stage of a radio receiver), parallel plates separated by air are used (Fig. 19-18a). The capacitance is changed by rotating one set of plates to give a variable effective area of overlapping A. Small "trimmer" capacitors used for fine adjustment (Fig. 19-18b) have mica dielectric, and C is changed by changing the separation d. Capacitors of fixed value are often constructed of two rolled-up aluminum foils separated by waxed paper (Fig. 19-18c). An essential compromise must be made between high capacitance (which requires small separation d) and the ability to withstand high voltages without rupture (which requires a large d). Capacitors are therefore rated by "working voltage" as well as capacitance. For about a dollar, one can buy 1000 μF rated at 6 V, 40 μF rated at 450 V, or 0.0005 μF rated at 10,000 V. Some large oil-filled capacitors are designed to store a small charge at a PD of a million volts or more, for use in artificial lightning experiments.

As circuit elements, capacitors are notable for the fact that no steady flow of electrons can take place. Instead, a charge builds up, and the dielectric material or vacuum between the plates is subject to an electric field.

Capacitors can be connected together in several ways, and the equivalent capacitance of the combination can be calculated by certain

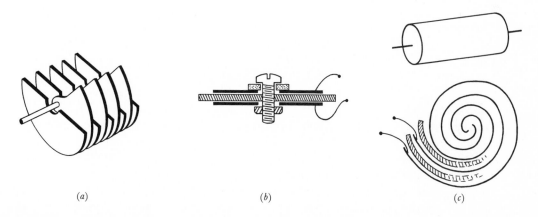

(a) (b) (c)

Figure 19-18 Typical capacitors. (a) 30–300 pF, rated at 300 V. (b) 5–40 pF, rated at 100 V. (c) Tubular fixed capacitor, 0.05 μF, rated at 600 V.

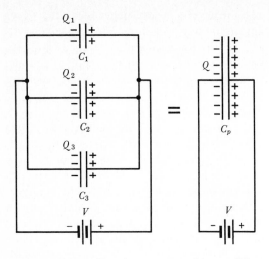

Figure 19-19 Capacitors connected in parallel.

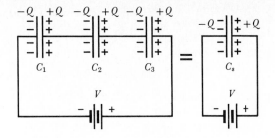

Figure 19-20 Capacitors connected in series.

formulas. When capacitors are connected in parallel as in Fig. 19-19, charge flows from the battery or other source and is shared among the various plates that are connected together. If a PD of V is applied, each capacitor receives a different quantity of charge, given by $Q = CV$ in each case. The total charge that leaves the battery is

$$Q = Q_1 + Q_2 + Q_3 + \cdots$$

Hence,

$$C_p V = C_1 V + C_2 V + C_3 V + \cdots$$

Canceling V, we get

$$C_p = C_1 + C_2 + C_3 + \cdots$$

(capacitors in parallel)   (19-10)

If a number of capacitors are connected in series, as in Fig. 19-20, a certain charge Q leaves the battery and flows on one plate of C_1. This repels an equal charge $-Q$ to the left-hand plate of C_2, leaving a charge $+Q$ on the right-hand plate of C_1. Similarly, the plates of C_3 acquire charges $-Q$ and $+Q$. The total PD from A to B is the sum of the PD's of the individual capacitors, but the charges on all the capacitors are the same, equal to the charge that flows from the battery.

$$V = V_1 + V_2 + V_3 + \cdots$$

$$\frac{Q}{C_s} = \frac{Q}{C_1} + \frac{Q}{C_2} + \frac{Q}{C_3} + \cdots$$

Canceling Q, which is the same for each capacitor, gives

$$\frac{1}{C_s} = \frac{1}{C_1} + \frac{1}{C_2} + \frac{1}{C_3} + \cdots$$

(capacitors in series)   (19-11)

Summary An electric field exists in any region of space where a small test charge would experience an electric force. The magnitude, or intensity, of the field equals the force per unit charge, and the direction of the field is the direction of the force on a small positive test charge. Another example of a force field is a gravitational field, whose magnitude is defined as force per unit mass. In general, a field may be thought of as a collection of vectors or scalars defined throughout a region of space. Lines of force are used to represent the direction of an electric field. Various models have been proposed for electric, nuclear, and gravitational fields.

Electric potential is the electric potential energy per unit charge. The potential difference between two points is measured in joules per coulomb, or volts. The PD is 1 V if 1 C of charge gains or loses 1 J of electric PE when moved from one point to another. An equipotential surface is a surface all points of which have the same potential. If the free charges in a conductor are at rest, they reside on the surface and are distributed in such a way that the surface is an equipotential surface. Lines of force "originate" on positive charges and "terminate" on equal negative charges, and equipotential surfaces are always perpendicular to the lines of force. The electric field strength can also be measured in terms of potential gradient, which is the rate of change of potential with respect to distance measured along a line of force.

Capacitance is defined as the stored charge per unit PD between the plates of a capacitor. For a parallel-plate capacitor, the capacitance depends on the area and separation of the plates and on the dielectric material between the plates. The dielectric constant of a medium is the factor by which the capacitance of a capacitor is increased when the capacitor is immersed in the medium. Formulas exist for the combined capacitance of capacitors in series and in parallel.

Check List

electric field strength
gravitational field
  strength
lines of force
photons, mesons,
  gravitons
potential difference
volt
equipotential surface

potential at a point
potential gradient
capacitance
farad
dielectric constant

$V_{AB} = \dfrac{W_{B \rightarrow A}}{Q}$

$V = kQ/r$

$C = KA/4\pi kd$

$C_p = C_1 + C_2 + C_3 + \cdots$

$\dfrac{1}{C_s} = \dfrac{1}{C_1} + \dfrac{1}{C_2} + \dfrac{1}{C_3} + \cdots$

Questions

19-1 Is electric field strength a vector quantity or a scalar quantity? Is electric potential a vector quantity or a scalar quantity?

19-2 Why must a *small* test charge be used when measuring an electric field?

19-3 Draw in some representative equipotential lines in Fig. 19-4a.

19-4 Figure 19-21a represents a small part of the surface of the moon, assumed to be horizontal, and the space above the moon. Draw in a few representative lines of force (solid) and a few equipotential surfaces (dotted) for the gravitational field above the lunar surface. Repeat, for the *mascon* (mass concentration) of Fig. 19-21b, in which a

(*a*) Normal lunar surface

(*b*) Lunar surface near a mascon

Figure 19-21

small deposit of heavy material is buried below the surface. How can the existence of a mascon be detected?

19-5 Why are the lines of force in Fig. 19-13 perpendicular to the surfaces at the points where the lines and surfaces meet?

19-6 A diver is in midair during a dive. What is the direction of the gravitational potential gradient at the diver's position?

19-7 What is the gravitational field strength at the center of the earth?

19-8 A spacecraft moves around Mars in an orbit that is the intersection of a gravitational equipotential surface and a plane. Describe the orbit. (Ignore the gravitational attraction of the sun and the other planets.)

19-9 On the weather map of Fig. 19-22, wind velocities and temperatures are shown at various points. Explain how the map illustrates the following concepts: (a) vector field; (b) scalar field; (c) temperature gradient. Where is the temperature gradient the greatest?

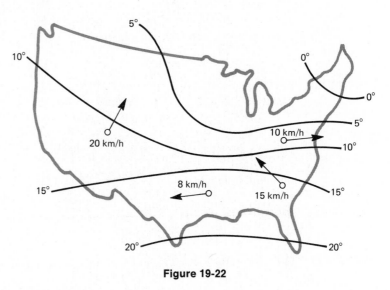

Figure 19-22

19-10 Construct a model by which two boys on frictionless carts, as in Fig. 19-6, can exert an attractive force on each other by throwing and catching a boomerang rather than a ball. (See also Ref. 39 at the end of Chap. 33.)

19-11 What is the difference between potential difference and difference of potential energy?

19-12 Explain why all parts of a conductor are at the same potential if there is no flow of electrons.

MULTIPLE CHOICE

19-13 In Fig. 19-7b, suppose the charge on the small sphere were moved around to give a uniform charge distribution instead of being concentrated at B, and all other charges in the diagram remained as shown. The potential energy of the sphere would (a) increase; (b) remain the same; (c) decrease.

19-14 Two plates of metal that have fixed charges $+Q$ and $-Q$ are immersed in a tank of oil. If the oil is pumped out, the electric field at a point midway between the plates (a) increases; (b) remains the same; (c) decreases.

19-15 Ten equal capacitors, each of capacitance C, are connected in series. The resultant capacitance of the combination is given by (a) $0.1C$; (b) $10C$; (c) $100C$.

19-16 Ten equal capacitors, each of capacitance C, are connected in parallel. The resultant capacitance of the combination is given by (a) $0.1C$; (b) $10C$; (c) $100C$.

19-17 In SI, the dimensions of electric field strength are (a) $[MLT^{-2}I^{-2}]$; (b) $[ML^2T^{-3}I^{-1}]$; (c) $[MLT^{-3}I^{-1}]$.

19-18 In SI, the dimensions of capacitance are (a) $[ML^2T^2I^2]$; (b) $[M^{-1}L^{-2}T^4I^2]$; (c) $[M^{-1}L^{-1}T^{-3}I^{-1}]$.

Problems

19-A1 The force on a charge of -3×10^{-7} C is measured to be 0.60 N, in an upward direction. What are the magnitude and direction of the electric field at this point?

19-A2 What electric force is exerted on a charge of 2×10^{-8} C that is placed in a field whose strength is 6×10^4 N/C?

19-A3 What are the magnitude and direction of the electric field at a point where an electron experiences an upward force of 3.2×10^{-16} N?

19-A4 Calculate the force on a charge of 30 μC that is in a field whose strength is 125 N/C.

19-A5 What is the PD between two points if 12 J of work is required to move 15 C from one point to the other?

19-A6 How much work is needed to move a charge of $+3$ C from A to B if the potential of A relative to ground is $+100$ V and the potential of B relative to ground is $+118$ V?

19-A7 Three points A, B, and C are on a straight line, at intervals of 20 cm. The potentials of the points are $+20$ V, $+80$ V, and $+140$ V, respectively. What is the magnitude of the potential gradient?

19-A8 (a) What is the potential gradient in the space between the parallel plates of a capacitor that are separated by 2.5 mm if the PD between the plates is 2000 V? (b) What is the electric field in the space between the plates?

19-A9 A charge of $+5000$ μC is on one plate of a capacitor, and -5000 μC is on the other plate. The PD between the plates is 400 V. What is the capacitance of the capacitor?

19-A10 What charge flows from a 9-V battery when a 40-μF capacitor is connected to it?

19-A11 How much charge can be placed on the plates of a 40-μF capacitor whose maximum (breakdown) PD is 450 V?

19-A12 Calculate the capacitance (in pF) of a capacitor that acquires a PD of 800 V when 10^{12} electrons are taken from one plate and placed on the other plate.

19-B1 Two charges of value $Q_1 = +96$ μC and $Q_2 = -40$ μC are 7 m apart in air. At a point P, which is on the line between the charges, 4 m from Q_1, calculate (a) the field due to Q_1; (b) the field due to Q_2; (c) the resultant field.

19-B2 Two charges of $+100$ μC and $+60$ μC are 4 m apart. Calculate the magnitude and direction of the electric field at a point midway between the two charges.

19-B3 Calculate the magnitude and direction of the electric field at the center of a square 2 m on an edge if charges of $+6$ μC each are at three of the corners of the square.

19-B4 The moon has, in round numbers, $\frac{1}{81}$ as much mass as the earth. Show that on a journey from moon to earth, the resultant gravitational field is zero when a spacecraft is $\frac{1}{10}$ of the way to the earth. (*Hint:* Let M = mass of earth, x = distance from earth, and y = distance from moon.)

19-B5 An electron is located at a point at which the electric field is 3000 V/m. (*a*) What is the force on the electron? (*b*) What is the acceleration of the electron?

19-B6 A drop of water in a fog has a net charge of 200 electrons. What is the radius of the drop if it is suspended motionless on a day when the earth's electric field is 300 V/m?

19-B7 It is found that 3.2 J of work is needed to move a charge of 0.4 mC from one insulated conductor to another insulated conductor that is 20 cm away. What is the average field strength in the region between the conductors?

19-B8 Starting from rest, a charge of -2×10^{-8} C is moved from P to Q in Fig. 19-8 by a muscle that does 11×10^{-6} J of work. The test particle arrives at Q with 3×10^{-6} J of KE. What is the PD between P and Q?

19-B9 If a proton gains 9.6×10^{-12} J of electric PE while being moved from X to Y, (*a*) what is the PD between X and Y, and (*b*) which point is at the higher potential?

19-B10 A proton moving horizontally southward at 2×10^5 m/s enters a region of electric field. What is the magnitude and direction of the field that will cause the proton to come to rest while traveling 5 cm? (See Appendix for mass of proton.)

19-B11 What speed will an electron acquire if it is accelerated through a PD of 220 V?

19-B12 A storm cloud may have a PD relative to a tree of, say, 80 MV. If during a lightning stroke, 50 C of charge is transferred through this PD and 1% of the energy is absorbed by the tree, how much water (sap in the tree) can be boiled away, starting at $30°$ C?

19-B13 A parallel-plate capacitor has square plates 6 cm $\times$ 6 cm, separated by 1.5 mm of glass for which the dielectric constant is 5; calculate the capacitance of the capacitor, in pF.

19-B14 A parallel-plate capacitor of capacitance 0.04 μF has a charge of 16 μC. The plates are separated by 0.5 mm of air. What is the magnitude of the electric field at a point between the plates?

19-B15 It requires 1.2×10^{-16} J to move an electron from one plate of a charged capacitor to the other. The plates are 5.0 mm apart. Calculate the field strength at a point between the plates.

19-B16 During an electroshock treatment, a 20-μF capacitor is charged to a PD of 1.5 V and then discharged through the patient's heart. The process is repeated 5 times per second; what is the average current?

19-B17 A 100-μF capacitor is charged to a PD of 40 V, and then connected to a circuit consisting of a switch and an electrolytic cell in series. The electrolytic cell contains Cu^{2+} ions. When the switch is closed, the capacitor discharges through the cell. The operation is repeated 2000 times. What mass of copper is plated out?

19-B18 A parallel-plate capacitor has a slab of glass (of dielectric constant 4) separating the plates. The capacitor is first connected to a 12-V battery and then disconnected from the battery. The PD between the plates is observed by connecting them to a meter, which reads 12 V. The slab of glass is now pulled out. What is the new reading of the meter?

19-B19 A parallel-plate capacitor with plates separated by air acquires 3 μC of charge when connected to a battery of 900 V. The plates, still connected to the battery, are then immersed in castor oil. How much charge flows from the battery?

19-B20 Three capacitors having capacitances 1 μF, 2 μF, and 6 μF are available. (a) What is the maximum value of the capacitance that can be formed from these three? (b) What is the minimum value? Draw circuit diagrams showing the connections for each case.

19-B21 A capacitor of 6 μF and one of 3 μF are connected in parallel and charged by connecting the combination to a battery of 30 V. Calculate (a) the PD of each capacitor; (b) the charge on each capacitor.

19-B22 Capacitors of 6 μF and 3 μF are connected in series, and the combination is charged by connecting it to a battery of 30 V. Calculate (a) the charge on each capacitor; (b) the PD of each capacitor.

19-C1 A proton is placed in an electric field of strength 10 kV/m. Calculate (a) the acceleration; (b) the displacement after 0.1 μs, starting from rest.

19-C2 Solve Prob. 19-B3 if one charge next to the vacant corner is changed to -6 μC.

19-C3 In a small television picture tube, an electron moves horizontally at 6×10^7 m/s through the space between two 3 cm $\times$ 3 cm horizontal deflection plates separated by 1 cm. The PD between the plates is 4000 V. (a) How long is the electron in the deflecting electric field? (b) Calculate the vertical impulse (see Sec. 5-2) given to the electron. (c) What is the vertical component of the electron's velocity after it leaves the deflection plates?

19-C4 A capacitor of 5 μF is connected in parallel with the combination of a 6-μF capacitor and a 3-μF capacitor that are in series. Find (a) the net capacitance of the entire combination; (b) the PD across the 3 μF capacitor when 12 V is maintained across the 5-μF capacitor.

19-C5 A capacitor is designed to have a capacitance of 0.0001 μF and is to operate safely at a PD of 1000 V. The dielectric is mica, for which the dielectric constant is 3 and the breakdown field is 100 kV/mm. What must the area of each plate be, allowing a safety factor of 5 (it must withstand 5 times the design voltage)?

19-C6 A model of a red blood cell describes the cell as a capacitor: a positively charged liquid of surface area A separated by a membrane of thickness t from the surrounding negatively charged fluid. Tiny electrodes introduced into the interior of the cell show a PD of 100 mV across the membrane. The membrane thickness was estimated to be 100 nm and its dielectric constant to be 5. (a) From the mass given in Table 1-2 (page 9), estimate the volume of the cell (assumed spherical) and thus find its surface area. (b) Calculate the capacitance of the cell. (c) Calculate the charge on the surface of the membrane. How many electronic charges does this represent?

19-C7 If the breakdown field strength for dry air is 3×10^6 V/m, what is the maximum charge that can be stored in a parallel-plate air capacitor whose plates are each 20 cm $\times$ 30 cm? (Hint: The spacing of the plates is not given. Call it d, carry d through your calculation, and trust that it will cancel out of your final answer.)

19-C8 (a) Derive a formula correlating the charge density (Q/A) on the plates of an air-filled parallel-plate capacitor with field strength E in the space between the plates. (b) Assuming (correctly) that your formula is also valid for the space near the surface of a charged metal sphere, calculate the radius of the smallest sphere that can hold a charge of 0.05 C in air if the breakdown field strength of dry air is 3×10^6 V/m.

For Further Study

19-9 Potential Due to a Point Charge

The electric potential V at a point P is, by definition, the work per unit charge required to bring a test charge from infinity to the point P. We calculate the work required to go from $r = \infty$ to $r = r$ as the area under the force graph (Fig. 19-23). Here the force is given by Coulomb's law, so

$$W = \lim_{\Delta r \to 0} \sum \Delta W$$

$$= \lim_{\Delta r \to 0} \sum_{r=r}^{r=\infty} \frac{kQQ_1}{r^2} \Delta r$$

$$= kQQ_1 \lim_{\Delta r \to 0} \sum r^{-2} \Delta r$$

The limit of the sum is a definite integral:

$$W = kQQ_1 \int_r^{\infty} r^{-2} \, dr$$

The table of integrals in the Appendix tells us that $\int x^n \, dx = (x^{n+1})/(n+1)$; here x is replaced by r and $n = -2$. Thus,

$$W = kQQ_1 \left[\frac{r^{-1}}{-1} \right]_r^{\infty} = kQQ_1 \left[\frac{\infty^{-1}}{-1} - \frac{r^{-1}}{-1} \right]$$

or

$$W = \frac{kQQ_1}{r}$$

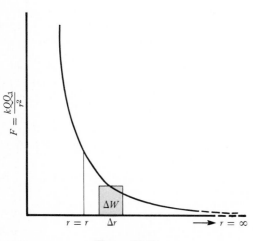

$$r = r \quad \Delta r \qquad \qquad \longrightarrow r = \infty$$

Figure 19-23

Finally, we divide by Q_1 to get the work per unit charge, which is the potential at point P, obtaining

$$V = \frac{kQ}{r}$$

This is Eq. 19-5. We shall find use for this result in our study of the Bohr atom (Sec. 30-3).

19-10 Energy Storage in Capacitors

The act of forcing charge into a capacitor is analogous to the stretching of a spring. At the start, when the capacitor is "empty," it is easy to move a few electrons from one plate to the other through an external circuit. However, as the capacitor becomes charged, an electron approaching the negative plate is repelled by the charges already on that plate, and an electron leaving the positive plate is attracted back by the excess of positive charge on that plate. Thus, the greater the charge on a capacitor, the more work is required to place additional charge on the plates. In a similar fashion, the more a spring is already stretched, the more work is required to stretch it an additional small amount. In Sec. 9-3 (page 202) we found that the PE stored in a stretched spring is the area under the force-displacement graph; this PE turned out to be $\frac{1}{2}kA^2$, where k is the force constant or stiffness of the spring and A is the final (maximum) displacement. The area under the curve was found graphically.

Exactly similar reasoning is used to find the work done in charging a capacitor to some final charge Q_f. The work ΔW required to move a charge of ΔQ coulombs against an opposing PD of V volts is given by $\Delta W = V \Delta Q$ joules. While the capacitor is being charged, its PD varies from 0 to its final value V_f in such a way that V is at all times proportional to Q (Fig. 19-24)—just as F is always proportional to x as a spring

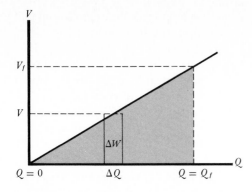

Figure 19-24 The PE of a charged capacitor is $\frac{1}{2}Q_fV_f$.

is stretched. (The factor of proportionality between V and Q is $1/C$, since Eq. 19-6 can be rearranged to read $V = Q/C$.) Hence the total work done in charging the capacitor is the area under the curve, and from Fig. 19-24 we obtain

$$PE = \tfrac{1}{2}V_fQ_f$$

We can drop the subscripts, which have now served their purpose, using Q and V for the final charge and PD, respectively, and we can also use the relation $C = Q/V$ to obtain three equivalent equations for the PE of a charged capacitor:

$$PE = \tfrac{1}{2}VQ = \tfrac{1}{2}CV^2 = \tfrac{1}{2}\left(\frac{1}{C}\right)Q^2 \quad (19\text{-}12)$$

The same result can be obtained by an integration:

$$PE = W = \lim_{\Delta Q \to 0} \sum \Delta W = \lim_{\Delta Q \to 0} \sum \frac{1}{C} Q\,\Delta Q$$

$$= \frac{1}{C}\int_{Q=0}^{Q=Q_f} Q^1\,dQ = \frac{1}{C}\left[\frac{Q^2}{2}\right]_0^{Q_f}$$

$$= \frac{1}{C}\left(\frac{Q_f^2}{2}\right) - \frac{1}{C}\left(\frac{0^2}{2}\right)$$

$$= \tfrac{1}{2}\left(\frac{1}{C}\right)Q_f^2$$

Let us note the close analogy with the formula for the PE of a spring.

$$PE = \tfrac{1}{2}\left(\frac{1}{C}\right)Q^2 \quad \text{(PE of a charged capacitor)}$$

$$PE = \tfrac{1}{2}(k)A^2 \quad \text{(PE of a stretched spring)}$$

The reciprocal of the capacitance is a measure of the "stiffness" of a capacitor, just as the force constant k is a measure of the stiffness of a spring. If C is very large, the capacitor absorbs a great deal of charge with only a little rise in its PD, and it might be said to be not very stiff. Mechanical analogies of this sort are useful in discussing electrical oscillations (Sec. 24-3).

Just *where* is the energy stored in a charged capacitor? The answer to this question depends to a large extent on which model of electric field we adopt. If we like the action-at-a-distance model, we probably think of the PE as being stored at or on the charges themselves. After all, we did mechanical work on the charges while we were charging up the capacitor. If we like the Faraday-Maxwell field theory, we think of the energy as stored in the space between the plates, somehow related to an elastic stress of the lines of force. This is a superficially attractive model, for we know that the molecules of a dielectric medium such as glass or mica *are* distorted and under stress while in a field. However, the model fails to explain how the energy is stored in a vacuum, where there are no molecules to be polarized.

Whatever model of electric field we use, it is easy enough to write an expression for the energy density in the space between the plates of a capacitor. Using symbols already defined, we have

$$\text{Stored energy} = \tfrac{1}{2}CV^2 = \tfrac{1}{2}\left(\frac{KA}{4\pi kd}\right)V^2$$

$$\text{Volume of dielectric} = Ad$$

$$\text{Energy density} = \frac{\text{energy}}{\text{volume}}$$

$$= \frac{\tfrac{1}{2}(KA/4\pi kd)V^2}{Ad}$$

$$= \frac{K}{8\pi k}\left(\frac{V}{d}\right)^2$$

But the field strength E is the potential gradient, or V/d. Hence our final result is

$$\text{Energy density} = \frac{K}{8\pi k} E^2 \qquad (19\text{-}13)$$

This formula says that even in a vacuum, where $K = 1$, energy is stored in any region of space where a field exists. Indeed, energy density may well be the most fundamental aspect of electric fields.

The present-day photon model gives a simple interpretation of energy storage in a field. It will be recalled that electric forces are considered to be caused by many photons (particles of light) being exchanged between the charges. Energy storage in the space between the charges is a natural result of this model, since each photon has a certain amount of energy. In an exactly similar fashion, a juggler who maintains a group of oranges moving from hand to hand stores energy "in space." The concept of energy storage can be extended to the gravitational field, for which the model postulates the exchange of gravitons (Sec. 19-3) between two attracting masses.

Problems

19-C9 A 50-μF capacitor is charged to a PD of 300 V. How much electric PE is stored in this capacitor?

19-C10 A capacitor has parallel plates separated by 1.5 mm of air. When 4×10^{15} electrons are moved from one plate to the other, a field of 50 kV/m is set up between the plates. (*a*) What is the PE of the charged capacitor? (*b*) How much work would be required to move one additional electron from one plate to the other?

19-C11 A charge of $+2$ μC is at $x = 0$ and a charge of -18 μC is at $x = 10$ m. (*a*) At what point or points on the x axis is the electric field equal to 0? (*b*) At what point or points on the x axis is the electric potential equal to 0?

19-C12 Charges of $+40$ μC are at each corner of a square 2 m on a side. How much work must be performed to move one of the charges from its corner position to the center of the square?

19-C13 A photographer's flashbulb outfit contains a capacitor that is charged for 20 s by a current of 30 mA and then discharged quickly through a xenon-filled glow tube. The tube requires 100 J of energy for a successful discharge. What must the capacitance of the capacitor be?

19-C14 In a severe storm on July 19, 1976, at Kennedy Space Center, a cloud 6 km above a weather tower contained -39 C of charge in a relatively small volume. During a lightning flash, three strokes hit the tower a few tens of milliseconds apart. (*a*) What was the potential of the cloud relative to earth just before the first stroke? (*b*) During the first stroke a charge of 25 C was neutralized. A peak current of 200 kA was observed. If the average current was $\frac{1}{10}$ the peak value, what was the duration of the stroke?

19-C15 A 10,900-μF capacitor was charged to 40.0 V then discharged through a resistor immersed in a well-insulated cup containing 37.4 g of water at 18.0°C. The process was repeated 100 times, and the final temperature of the water was 23.5°C. (*a*) What value for the specific heat capacity of water is given by these data? (*Hint:* One calorie is defined as 4.186 J.) (*b*) If the temperatures are known to within ± 0.05°C, within what range of values does the calculated specific heat capacity lie? Is this a satisfactory result?

19-C16 On a certain day the electric field surrounding the earth was about 150 V/m in the air near the surface. How much electric energy was stored in a layer of air 2 m thick over a football field that measures 91.4 m $\times$ 48.8 m?

19-C17 A 3-μF capacitor is charged to 30 V, and a 5-μF capacitor is charged to 6 V. The batteries are removed, and the capacitors are then connected together, + to + and − to −. (*a*) Compute the new PD, the same for each capacitor. (*b*) Compute the new charge on each capacitor. (*Hint:* Total charge is conserved, algebraically, during the redistribution process.) (*c*) What is the loss of electric PE during this process? What becomes of this PE?

References

1. H. Kondo, "Michael Faraday," *Sci. American* **189**(4), 90 (Oct. 1953).
2. L. B. Loeb, "The Mechanism of Lightning," *Sci. American* **180**(1), 22 (Jan. 1949).
3. R. E. Orville, "The Lightning Discharge," *Phys. Teach.* **14**, 7 (1976).
4. A. A. Few, "Thunder," *Sci. American* **233**(1), 80 (July 1975). A good description of the lightning process and thunder.
5. M. A. Uman et al., "An Unusual Lightning Flash at Kennedy Space Center," *Science* **201**, 9 (1978). A report co-authored by 15 scientists describing the lightning flash that is the basis for Prob. 19-C14.
6. B. N. Turman, "Is a swimmer safe in a lightning storm?" *Phys. Teach.* **18**, 388 (1980).
7. A. Gemant, "Electrets," *Physics Today* **2**(3), 8 (1949). Discusses permanently "frozen in" electric dipoles.
8. Methods of making electrets are described in C. L. Stong's "Amateur Scientist" department of *Scientific American* by G. O. Smith, **203**(5), 202 (Nov. 1960), and by S. L. Khanna, **219**(1), 122 (July 1968).
9. J. E. McDonald, "The Earth's Electricity," *Sci. American* **188**(4), 32 (Apr. 1953).
10. A. D. Moore, "Electrostatics," *Sci. American* **226**(3), 47 (Mar. 1972). Describes modern applications of electrostatic phenomena.

20

Electric Energy

20-1 The Energy Method in Electricity

Our preliminary survey of electric charge in Chap. 18 revealed a general law of force—Coulomb's law. One would expect that the motions of electrons, protons, and other charged particles could be predicted by a straightforward use of Newton's laws of motion in conjunction with Coulomb's law. In like fashion, one would expect to be able to use Newton's laws to discuss the complex motion of a speeding airplane (and it *is* a complex motion, if we seek to take account of turbulence, air resistance, production of shock waves, and other factors). Electric forces are even more complex, if we wish a detailed picture of what happens. In mechanical problems, we have several times used the energy principle for an easy solution of problems for which the use of Newton's laws would have been too difficult. One example of this procedure is our derivation of the period of a simple harmonic motion (SHM)(Sec. 9-3) by the energy principle. Another example is afforded by our use of Bernoulli's equation (Sec. 12-8), which is based on the energy principle. In applying Bernoulli's equation we talk about pressure, velocity, and height at "point A" and "point B," but we have no detailed picture of forces and accelerations, and we do not ask what happens between points A and B, or how it happens. The energy method is easier than a complete dynamical solution, but, as payment for this ease, we get less information. For instance, the energy method is unable to give us an equation for position as a function of time, in SHM; for this information we used the reference circle to derive Eq. 9-5.

Fortunately, energy considerations are sufficient to give us much useful information about the behavior of electric charge. In Chap. 19 we introduced the concept of potential difference, which is work per unit charge. In this chapter we will continue to place emphasis on energy changes of charges, rather than on the electric fields that would give a more detailed description than we usually need. For the time being, we do not concern ourselves with how the force on a charge varies as it is moved from point to point. When using the energy method, we pass over the details and deal only with the total work done by or on the charge.

An immersion heater is connected to a source of emf, and water is made to boil as electric energy is converted to thermal energy.

20-2 Electromotive Force

Much of the usefulness of electricity in science and technology and in our daily lives arises from the possibility of continuous flow of charge through an electric circuit. A simple circuit is shown in Fig. 20-1, in which a dry cell is connected to a flashlight bulb. We know that the bulb becomes warm and lights up, and hence we conclude that some agency is at work forcing charges to move in an endless path around the circuit. This agency is the cell; we say that the cell is the *seat* of an *electromotive force*.

As the electrons move through the filament of the lamp, they lose electric PE. This energy is

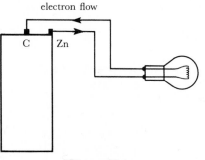

Figure 20-1

somehow transferred to the atoms of the tungsten filament of the bulb, and the atoms vibrate with greater amplitude and hence become warmer. If we wish to make a crude model for this process, we can think of free electrons rubbing the ions as they drift through the metal of the filament on their way toward the positive terminal of the cell. According to this model, the heat is produced as a sort of frictional effect. In a more detailed model based on quantum mechanics, the energy is lost when the electron waves are scattered by imperfections in the small metal crystals of the filament. After an electron reaches the + terminal of the cell, it moves through the cell from + to −, gaining electric PE at the expense of some of the cell's chemical PE. (It *gains* PE because it is moving *against* the internal electric field inside the cell.) When our wandering electron finally reaches the zinc cup again, it is ready to start on another round trip. We have here a mechanism for the steady flow of charge, as long as the chemical PE of the cell holds out. The essential action of the cell is to transform energy from chemical PE to electric PE. A similar transformation is shown in Fig. 20-2a, where a headlamp is supplied with energy from a car battery.

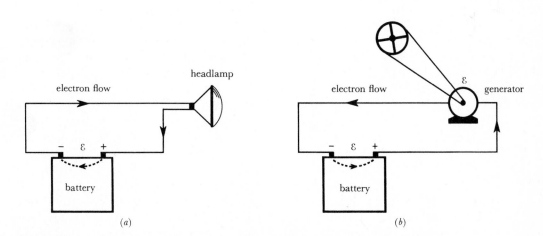

(a)   (b)

Figure 20-2 Seats of emf. (*a*) Chemical energy is transformed into electric energy in the battery. (*b*) Electric energy, which was transformed from mechanical energy in the generator, is transformed into chemical energy in the battery.

The reverse process is also possible; if charges are forced backward through a cell, electric energy is transformed into chemical energy. This is what happens when a car's storage battery is "charged up" (Fig. 20-2b). The transformation of energy is said to be *reversible*. Enough energy can be stored in recently developed types of reversible cells to make electric automobiles an economic possibility.

Our discussion of the electric cell can be paralleled by a discussion of other seats of electromotive force. In an electric generator, it is mechanical energy that is transformed into electric energy. A thermocouple transforms thermal energy into electric energy. These devices will be discussed in Chap. 23.

With this background, we now define a seat of electromotive force.

A seat of electromotive force is a device within which nonelectric energy can be reversibly transformed into electric energy.

The numerical value of an electromotive force (emf) is denoted by ε and is measured in joules per coulomb, or volts. Thus in a battery whose emf is 6 V, each coulomb of charge passing through the battery gains or loses 6 J of electric PE, depending on the direction of flow. The transformation of energy takes place at the rate of 6 J for every coulomb involved. The emf of a battery depends only on its physical and chemical construction and the temperature; it is not dependent on the amount of current through it.

The concepts of potential difference and electromotive force are easily confused, since both are measured in terms of work per unit charge. The distinguishing feature of a seat of emf is the *transformation* of nonelectric energy into electric energy. To illustrate this point, let us imagine connecting a wire between two oppositely charged bodies such as A and B in Fig. 19-8 (page 416). There is a PD between the bodies, but no emf. When the connection is made, electrons will momentarily flow from B to A, to be sure; but no continuous current is possible unless another wire is added to provide a return path from A to B. However, negatively charged electrons would not flow along this return wire back to the negatively charged conductor B, since like charges repel each other. Therefore some source of electric energy, a seat of emf, would have to be added to the circuit in order to obtain a continuous flow. It is not hard to extend this type of reasoning to the mechanical system of Fig. 19-7a (page 415). The book would fall to the floor "on its own," and its KE would become internal energy during the inelastic collision with the floor (Sec. 6-9). Even if the table and floor were frictionless, continuous motion around the path could take place only if a source of mechanical energy were "in the circuit" (for instance, a muscle helping the book back up from D to E on each trip).

20-3 Resistors, Resistance, and Ohm's Law

There must be at least one seat of emf in any circuit in which continuous current exists. The electric energy supplied by the seat of emf can, of course, be converted into mechanical or chemical energy. In addition, a certain amount of the electric energy is transformed into heat, as we noted for the light bulb in the last section. The way in which electric energy is utilized in a motor is fundamentally different from the way in which it is utilized in a lamp bulb. The motor is (electrically speaking) reversible, and if it were turned by a crank powered by a muscle or a waterfall, it would act as a generator to send electric energy into a circuit (see Sec. 23-2). On the other hand, the lamp bulb is not reversible; it is impossible to generate electric current in a wire by building a fire under a lamp bulb that is connected to the two ends of a wire. This is why we say that a lamp bulb "dissipates" electric energy; it is a one-way street for electric energy. Such a conductor, in which electric energy is dissipated no matter in which direction the charges flow, is called a *resistor*. Every conductor is a resistor, in the sense that the flow of electrons is impeded to a greater or lesser extent.

In 1827 the German physicist Georg Simon Ohm (1787–1854) studied the current through resistors that were in the form of long metallic wires. Ohm found that the current through a resistor is proportional to the potential difference between the ends of the wire. That is, the ratio of PD to current is constant for any given resistor. **Ohm's law** is given in symbols as

$$R = \frac{V}{I} \qquad (20\text{-}1)$$

where V is the potential difference (PD), I is the current, and the constant ratio R is defined as the *resistance* of the wire. In words,

The ratio of potential difference between any two points in a circuit to the current is constant and equals the resistance between the two points.

Ohm's law is valid only for metallic resistors at a definite temperature and for steady currents, but within these limits it is remarkably precise. The law is approximately valid for other types of resistors but fails completely for semiconductors and for gases. Unless specifically stated otherwise, all resistors in circuits discussed in this book are assumed to obey Ohm's law.

If a resistor does not obey Ohm's law, the resistance R is still defined and measured as V/I; the only difference is that R is not a constant independent of V and I.

Ohm's law is based on experiment. The constancy of R is a definite statement about the behavior of a certain class of substances (metals), and it is possible to derive Ohm's law from fundamental considerations about the drift of free electrons through metals. Ohm's law has about the same status as Hooke's law, which is related to the short-range electric forces between atoms. Neither Hooke's "law" nor Ohm's "law" is exact, nor are they of the same logical importance as Newton's laws of motion or Coulomb's law. Nevertheless, Ohm's law is an important

generalization that is the basis for much circuit theory.

The SI unit of resistance is the *ohm*; a resistor has a resistance of 1 ohm if a PD of 1 volt maintains a steady current of 1 ampere. Rearranging Eq. 20-1 gives another form of Ohm's law:

$$V = IR \qquad (20\text{-}2)$$
$$\text{volts} = (\text{amperes})(\text{ohms})$$

The symbol Ω, the Greek capital omega, is often used to represent ohms.

Example 20-1

What is the resistance of a lamp bulb through which the current is 0.5 A when the applied PD is 120 V?

$$R = \frac{V}{I} = \frac{120 \text{ V}}{0.5 \text{ A}} = \boxed{240 \ \Omega}$$

20-4 *Conventional Current*

There are two equivalent ways of describing the same electric current: as a flow of negative charge in one direction (say, toward the right), or as an equal flow of positive charge in the opposite direction (toward the left). In metals and in vacuum tubes the carriers of charge are electrons, which are negative. The flow of positive charge is not entirely unknown, however. For example, charge carriers of both signs are involved in electrolysis (see Fig. 18-10, where Ag^+ ions move toward the left while NO_3^- ions move toward the right). Also, the current in many high-energy particle accelerators consists of a flow of positive charges (protons); and on a smaller scale, transistors of some types make use of the flow of "holes" that behave like positive charges. Since both positive and negative carriers of charge are known to be able to flow, our choice of words to describe the direction of electric current can be made quite arbitrarily.

In this book we define the direction of current to be the direction in which *positive* charges would have to move in order to give the same electric and magnetic effects as the flow of actual charges $(+, -,$ or both) through the circuit. This *conventional current,* the flow of an equivalent positive charge, is what we mean by the term "current." Our definition of the direction of current, which is in widespread use, fits in well with the use of a *positive* test charge for measuring electric potential difference. A positive test charge loses energy when it moves from a region of high potential to a region of low potential, much as a test mass loses gravitational PE when it flows downhill. Thus in Fig. 20-2a the conventional current is directed away from the + terminal of the battery, and the hypothetical positive charge flows "downhill" from the + terminal through the headlamp and returns to the battery at the − terminal. The direction of this conventional current is opposite to the direction of electron flow.

Consider a resistor *AB* (Fig. 20-3), whose resistance is *R*, through which the current is *I*. We are not concerned with *why* the charges flow; it is enough to know that they *do* flow, for some reason connected with one or more emf's somewhere in the circuit. We assume that the resistor itself is free from sources of energy. Since there is a flow of charge through the resistor and heat is produced, this charge is losing electric PE. In other words, there must be a potential difference between the terminals of the resistor; *A* is positive relative to *B*. *The mere existence of a current through a resistor means that the terminal at which conventional (positive) charge enters is positive relative to the other terminal.*

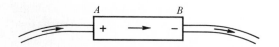

Figure 20-3 When charge flows through a resistor, the end at which positive charge enters the resistor becomes positive relative to the end at which the charge leaves. The arrows represent the direction of the conventional current through the resistor.

20-5   Electric Power

During his classical investigations of the mechanical and electrical equivalent of heat, Joule studied the production of heat in resistors. He found that the rate of production of heat in a metallic resistor is proportional to the square of the current. Now the rate of production of energy is power (*P*), measured in SI units as watts (= joules per second), and so we can state that $P = (\text{constant}) \times I^2$ for a given metallic resistor at a given temperature. To show that this result is a consequence of Ohm's law requires only a little algebra and the use of definitions already made:

PD ≡ work per unit charge: $V = \dfrac{W}{Q}$

Charge ≡ (current)(time): $V = \dfrac{W}{It}$

Power ≡ work per unit time: $V = \dfrac{P}{I}$

Ohm's law, $V = IR$: $IR = \dfrac{P}{I}$

Hence,

$$P = (IR)I$$

or

$$P = I^2R \qquad (20\text{-}3)$$

$$\text{watts} = (\text{amperes})^2(\text{ohms})$$

Example 20-2

An electric iron has a resistance of 10 Ω. How much heat is produced in 10 min if the current is 10 A?

$$P = I^2R = (10\text{ A})^2(10\text{ }\Omega) = 1000\text{ W}$$

Heat is being produced at the rate of 1000 J/s.

$$\text{Energy} = Pt = (1000\text{ J/s})(10\text{ min})\left(\frac{60\text{ s}}{1\text{ min}}\right)$$

$$= \boxed{6 \times 10^5\text{ J}}$$

Converting the energy to heat units, we obtain

$$6 \times 10^5 \text{ J} \left(\frac{1 \text{ cal}}{4.19 \text{ J}} \right) = \boxed{1.43 \times 10^5 \text{ cal}}$$

In Chap. 21 we shall study electric circuits in detail; we shall derive formulas for resistors in series and in parallel, and we shall consider emf's in series and in parallel. For the present, we consider energy and power in a very simple circuit (Fig. 20-4) in which a device D is connected to the two terminals of a "black box" B. Without knowing anything about the nature of B and D, we can say that if conventional charge flows from the positive terminal of B at a rate of I amperes, and if the PD across the terminals of B (and D) is V volts, then energy is being supplied by B, and absorbed by D, at the rate of P watts, where

$$P = VI \tag{20-4}$$

This follows directly from the definition of the volt as a joule per coulomb:

$$\text{watts} = \left(\frac{\text{joules}}{\text{coulomb}} \right) \left(\frac{\text{coulombs}}{\text{second}} \right)$$

In Eq. 20-4, we use V, the symbol for PD, since we do not know, or need to know, the *reason* for the potential difference between the terminals of the boxes. Similarly, if the box B contains an emf $\mathcal{E}$, and the current through the seat of emf is I, the rate at which nonelectric energy is being converted into electric energy by the seat of emf is

$$P = \mathcal{E}I \tag{20-5}$$

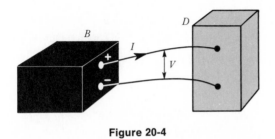

Figure 20-4

These power equations are general and do not depend on Ohm's law. $\mathcal{E}$ is used in Eq. 20-5 since we are concerned here with the transformation of energy in a seat of emf.

Example 20-3

Suppose that in Fig. 20-4 the device D consists of a motor that uses 400 W, and a lamp bulb. If $V = 100$ V and $I = 5$ A, how much electric power is received by the bulb?

Power input to D:

$$VI = (100 \text{ V})(5 \text{ A})$$
$$= 500 \text{ W}$$

Power to bulb:

$$500 \text{ W} - 400 \text{ W} = \boxed{100 \text{ W}}$$

Example 20-4

If the black box B of the preceding problem dissipates heat at the rate of 40 W, calculate the value of the emf inside the box.

The box is delivering electric energy at the rate of 500 W to device D, and electric energy is being transformed into heat inside the box at the rate of 40 W. Hence the total rate of production of electric energy from nonelectric forms must be 540 W.

$$P = \mathcal{E}I$$

$$\mathcal{E} = \frac{P}{I} = \frac{540 \text{ W}}{5 \text{ A}} = \boxed{108 \text{ V}}$$

Electric energy may be measured in joules (watt·seconds). Another common unit is the kilowatt·hour (kWh), which is the energy produced, transformed, or dissipated in 1 h if the power is 1000 W. One kilowatt·hour equals $(1000 \text{ J/s})(3600 \text{ s}) = 3.6 \times 10^6$ J. The commercial cost of electric "power" (really energy) is from 3¢ to 6¢ per kWh, depending on the amount purchased. Of course, the unit kilowatt·hour is not reserved for electric energy; it can equally well be used for thermal, mechanical, or other forms of energy.

20-6 Resistivity

To compare the resistances of various substances, we need to define a property that does not depend on the size and shape of the conductor. Other things being equal, we would expect the resistance to be directly proportional to the length of the conductor. We also expect that the electron flow will be easier if the conductor has a larger cross section, since the free

electrons can "spread out" more. Thus we define the *resistivity* ρ (Greek letter rho) as the constant in the equation

$$R = \frac{\rho L}{A} \qquad (20\text{-}6)$$

where R is the resistance, L is the length, and A is the cross section. The unit of resistivity is the ohm·meter, as seen by solving for ρ:

$$\rho = \frac{RA}{L} = \frac{\Omega \cdot m^2}{m} = \Omega \cdot m$$

Table 20-1 shows that silver has the smallest resistivity of any metal, with that of copper not much greater. Of all physical properties of matter, electrical resistivity perhaps shows the greatest range of values. Fused quartz, an excellent insulator, has a resistivity more than 10^{25} times that of silver. Our next examples show that the resistance of a conductor does not depend solely on the amount of metal used.

Table 20-1   Approximate* Resistivities ρ and Temperature Coefficients of Resistance α, All at 0°C

Substance	ρ, $\Omega \cdot m$	α, (°C)$^{-1}$
silver	1.47×10^{-8}	0.0038
copper	1.59×10^{-8}	0.0039
gold	2.27×10^{-8}	0.0034
aluminum	2.60×10^{-8}	0.0040
tungsten	5.0×10^{-8}	0.0046
iron	11.0×10^{-8}	0.0052
platinum	11.0×10^{-8}	0.00392
constantan (60 Cu, 40 Ni)	49×10^{-8}	0.00001
mercury	94×10^{-8}	0.0009
Nichrome (60 Ni, 24 Fe, 16 Cr)	100×10^{-8}	0.0002
carbon	4×10^{-5}	−0.0005
germanium	2×10^{0}	
silicon	3×10^{4}	
boron	1×10^{6}	
wood (maple)	3×10^{8}	
Celluloid	4×10^{12}	
glass	10^{11}–10^{13}	
amber	5×10^{14}	
sulfur	1×10^{15}	
mica (colorless)	2×10^{15}	
fused quartz	5×10^{17}	

* Exact values depend on purity, heat treatment, and other factors.

Example 20-5

In a power station, a "bus bar" designed to carry many amperes of current is in the form of a slab of copper 2 m long and 1 cm × 10 cm in cross section. (a) What is the resistance of the bar at 0°C? (b) What PD is needed to cause a current of 5000 A through the bar?

(a) The cross section is $(10^{-1}\,m)(10^{-2}\,m) = 10^{-3}\,m^2$. To find the resistance, we look up the resistivity of copper in the table and substitute into the formula:

$$R = \frac{\rho L}{A} = \frac{(1.59 \times 10^{-8}\,\Omega \cdot m)(2\,m)}{10^{-3}\,m^2}$$

$$= \boxed{3.18 \times 10^{-5}\,\Omega}$$

(b) To find the necessary PD, we use Ohm's law:

$$V = IR = (5 \times 10^{3}\,A)(3.18 \times 10^{-5}\,\Omega)$$

$$= 15.9 \times 10^{-2}\,V = \boxed{0.159\,V}$$

Such a small PD would be negligible in a power station, where the PD between the terminals of the generator might be several hundred volts.

Example 20-6

What will the resistance of the bar of Example 20-5 become if it is stretched out to form a long wire that is 1 mm × 1 mm in cross section?

The length of the wire equals the volume divided by the area of the cross section:

$$L = \frac{(2m)(10^{-1}\ m)(10^{-2}\ m)}{(10^{-3}\ m)(10^{-3}\ m)} = 2 \times 10^3\ m$$

The wire has the same amount of copper as the bar had, but it is now in the form of a wire 2000 m long (over a mile long), and its cross section is now only $10^{-6}\ m^2$.

$$R = \frac{\rho L}{A} = \frac{(1.59 \times 10^{-8}\ \Omega \cdot m)(2 \times 10^3\ m)}{10^{-6}\ m^2}$$

$$= \boxed{31.8\ \Omega}$$

Like most physical properties, resistivity depends on temperature. The resistivity of a pure metal increases with temperature, except for certain special alloys, and we describe this increase by the use of a *temperature coefficient* of *resistance* α (alpha), which is analogous to the temperature coefficient of linear expansion (Sec. 13-3). The defining equation, for temperature changes that are not too great, is

$$\rho_T = \rho_0(1 + \alpha T) \qquad (20\text{-}7)$$

where T is the temperature in degrees Celsius, ρ_T is the resistivity at $T°C$, ρ_0 is the resistivity at 0°C, and α is the temperature coefficient of resistance, in $(°C)^{-1}$, based on a reference temperature of 0°C. Values for α are tabulated in Table 20-1; these temperature coefficients are several hundred times as large as the linear coefficients of expansion for solids and are comparable to the volume coefficients of expansion for gases. It is noteworthy that for most common metals, α is about 0.004 $(°C)^{-1}$, which means that the resistivity would become zero at about $-1/0.004°C = -250°C$ if the linear relationship of Eq. 20-7 remained valid at low temperatures. Actually, the resistivities of all pure metals tend toward zero as absolute zero $(-273.2°C)$ is approached. If impurities are present, the resistivity tends to "level off" at some finite value as the temperature is lowered.

For any given conductor, the resistance R is proportional to the resistivity ρ, so Eq. 20-7 is equivalent to

$$R_T = R_0(1 + \alpha T) \qquad (20\text{-}8)$$

The variation of resistance with temperature affords one way of making precise temperature measurements. Platinum is often used for this purpose because of its high melting point and its freedom from corrosion and oxidation effects.

Example 20-7

The resistance of a platinum resistance thermometer is 200.0 Ω at 0°C and 257.6 Ω when the thermometer is immersed in a crucible containing melting $SbCl_3$. What is the melting point of this compound?

Solving Eq. 20-8 for the temperature T, and substituting, we find

$$T = \frac{R_T - R_0}{\alpha R_0} = \frac{257.6\ \Omega - 200.0\ \Omega}{[0.00392\ (°C)^{-1}](200.0\ \Omega)}$$

$$= \frac{57.6\ \Omega}{0.784\ \Omega \cdot (°C)^{-1}} = \boxed{73.5°C}$$

In 1911, the Dutch physicist Kamerlingh Onnes discovered that below 4.2 K the resistivity of mercury becomes essentially zero. Some (but not all) metals and their compounds show an abrupt decrease of resistivity at some finite temperature called the *transition temperature*, and the resistivity then remains exactly zero* for all lower temperatures. This phenomenon of *superconductivity* is an active subject of research. Once started, a current in a loop of superconducting material persists for hours or days, with no emf in the circuit, and can be detected by its magnetic field. Transition temperatures vary from 0.0002 K (for rhodium) to as high as 23 K (for the niobium-germanium compound Nb_3Ge).

---

* A more precise statement would be that the resistivity becomes unmeasurably small. Present techniques show that the resistivities of metals in this state are less than $4 \times 10^{-25}\ \Omega \cdot m$.

Placing a superconducting material in a magnetic field lowers its transition temperature; in a sufficiently strong field, characteristic of the substance, superconductivity is not possible even at absolute zero. Very large currents, in the kiloampere range, are possible in superconductors without excessive I^2R losses. As a result, among the areas now under active development are superstrong magnetic fields for research in particle physics; magnetic support (levitation) of high-speed trains; and distribution of huge amounts of electric power over resistanceless transmission lines. Although much research and development remains to be done, these examples and others show that superconductivity is no longer merely a laboratory curiosity.

Semiconductors are characterized by moderate resistivities (Table 20-1) and have negative temperature coefficients; germanium and silicon are typical examples. The model of a semiconductor attributes poor conduction either to a relatively few free electrons (*n*-type semiconductors) or to a relatively few vacant spots ("holes") in the crystal for which a large cloud of free electrons compete (*p*-type semiconductors). The model successfully explains the negative temperature coefficient. (At higher temperatures, there are more free electrons, but still far fewer than in a metal. Hence the conductivity goes up, and the resistivity goes down.) The model also explains the strong dependence on a minute concentration of impurity atoms. (The impurities may bring in the free electrons necessary to increase conduction.) The transistor (Sec. 23-10) is a semiconductor device, as is the "crystal" of the early crystal-set radio.

Summary

A seat of electromotive force is a device that is able to transform mechanical, chemical, thermal, or other nonelectric energy into electric energy, and vice versa. Emf's are measured in joules per coulomb, or volts. A device has an emf of 1 V if 1 J of energy is transformed during the passage of 1 C of charge. Common seats of emf are electric generators, cells, and thermocouples.

Ohm's law states that for a metallic conductor at any given temperature, the ratio of potential difference to current is a constant, and this constant is defined as the resistance. A conductor's resistance is 1 Ω if there is a current of 1 A when the PD is 1 V.

The energy principle in electricity focuses our attention on how a unit positive test charge gains or loses energy. Conventional current is therefore defined as the flow of positive charge. In metals, the direction of the conventional current is opposite to the direction of flow of the free electrons that are the charge carriers. The rate of production of heat in a conductor is proportional to the square of the current. The proportionality constant is the resistance of the conductor.

Resistivity is a property that depends only on the substance and its temperature; it is defined as ρ in the formula $R = \rho L/A$. Resistivity is measured in ohm·meters. There is a wide variation in resistivities of substances, ranging from metallic conductors through semiconductors to insulators. Resistivity and resistance depend on temperature, and the temperature coefficient of resistivity is defined as α in the formulas $\rho_T = \rho_0(1 + \alpha T)$ and $R_T = R_0(1 + \alpha T)$. For most metals, α is positive and resistance increases with temperature, but for semiconductors α is usually negative. At temperatures near absolute zero, some metals, alloys, and compounds become superconducting and have zero resistivity.

seat of electromotive
    force
resistor
Ohm's law
resistance
ohm
$V = IR$

conventional current
$P = I^2R$
$P = VI$
$P = \varepsilon I$
watt
kilowatt·hour
resistivity

$R = \rho L/A$
temperature coefficient
    of resistance
$R_T = R_0(1 + \alpha T)$
superconductivity
transition temperature
semiconductor

Questions

20-1 Electrons leave a dry cell and flow through a lamp bulb back to the cell. Which terminal, the + or the − terminal, is the one from which electrons leave the cell? In which direction is the conventional current?

20-2 When a dry cell is furnishing current through an external circuit, electrons inside the cell are moving toward the − terminal. Explain how they are able to do this.

20-3 Both PD and emf are measured in volts. What is the difference between these concepts?

20-4 Could you construct two wires of the same length, one of copper and one of iron, that would have the same resistance at the same temperature?

20-5 What is "constant" about the alloy constantan (Table 20-1)?

20-6 State two ways in which p-type semiconductors differ from ordinary metallic conductors such as copper.

MULTIPLE CHOICE

20-7 If a 1-MΩ resistor is connected for 1 min to a PD of 1 kV, the heat developed in the resistor is   (a) 3.6 J;   (b) 60 J;   (c) 1000 J.

20-8 A piece of wire is cut into four equal parts, and the pieces are bundled together side by side to form a thicker wire. Compared with that of the original wire, the resistance of the bundle is   (a) the same;   (b) $\frac{1}{4}$ as much;   (c) $\frac{1}{16}$ as much.

20-9 If the temperature does not change, the current I is proportional to the potential difference V for   (a) a metallic wire;   (b) a semiconductor;   (c) both of these.

20-10 If a wire that is already in a superconducting state is placed in a strong magnetic field, its resistance   (a) becomes much greater;   (b) remains the same;   (c) becomes much less.

20-11 If the temperature does not change, when power P is being developed in a wire, the current is proportional to   (a) $\sqrt{P}$;   (b) P;   (c) P^2.

20-12 A suitable unit for resistivity would be   (a) Ω/m^2;   (b) Ω/m;   (c) $\Omega \cdot m$.

Problems

20-A1 What PD is needed to cause a current of 2 mA through a resistance of 4 MΩ?

20-A2 Calculate the resistance of a transistor if the current is 0.05 A when the PD is 9 V.

20-A3 What is the current through a 300-Ω resistor if the PD between its terminals is 15 V?

20-A4 What is the power rating of a theater light in which a current of 12 A is caused by a PD of 120 V?

20-A5 The battery of a pocket calculator supplies 300 mA at a PD of 6 V. What is the power rating of the calculator?

20-A6 What is the current through a 1200-W electric heater if the resistance is 75 Ω?

20-A7 In a stereo system each speaker has an effective resistance of 4 Ω, and the system is rated at 60 W of power in each channel. There is a fuse in each speaker circuit that will blow at 4 A. Is this system adequately protected against overload?

20-A8 Calculate the resistance of a 1000-W heater in which the current is 5 A.

20-A9 Calculate the energy in kWh expended in running a 400-W motor for 20 min.

20-A10 A certain wire has a resistance of 100 Ω. Find the resistance of a wire of the same material, at the same temperature, that is 5 times as long and has a cross section 4 times as large.

20-B1 A current of 5 A through a battery is maintained for 30 s, and in this time 900 J of chemical energy is transformed into electric energy. (*a*) What is the emf of the battery? (*b*) How much electric power is available for Joule heating and other uses?

20-B2 If energy costs 5¢ per kWh, what does it cost to leave a 25-W porch light burning all night (average 10 h) every day for a year?

20-B3 In some television sets in the 1970s the picture-tube filament was kept warm at all times, using 10 W of power to provide "instant-on" viewing. If the set was idle 16 h/day, (*a*) calculate the dollar cost per year, at 5¢ per kWh, for this convenience; (*b*) calculate the ecological cost, in kg/y of bituminous coal (fuel value 30 MJ/kg) burned in a 35%-efficient power plant.

20-B4 At 5¢ per kWh, what does it cost to heat a 200-liter tank of water from 20°C to 60°C?

20-B5 What is the efficiency of a $\frac{1}{4}$-hp motor that draws a steady current of 1.67 A from a 120-V line? (See Appendix Table 4 for conversion of horsepower into watts.)

20-B6 An electric iron draws 9 A from a 120-V line. The iron's mass is 0.8 kg, and it is originally at 20°C. Forty percent of the heat is lost to the room by radiation. What will the temperature of the iron be 2.0 min after it is connected? Assume constant resistance.

20-B7 A car battery of emf 12 V is used to run a radio receiver through which the current is 1.5 A, and at the same time to operate the motor of a 20-W electric shaver. How much chemical energy is transformed into electrical energy in 5 min?

20-B8 What is the current from a 12-V automobile battery while it supplies 800 W (1.08 hp) to a starter motor?

20-B9 An x-ray tube used for therapy operates at 4 MV with a beam current of 20 mA striking the metal target. The power in this beam is transferred to a stream of water flowing through holes drilled in the target. What rate of flow, in kg/min, is needed if the temperature rise of the water is not to exceed 50°C?

20-B10 An average electron-beam current of 25 μA strikes a 3-kg copper target in a linear accelerator for which the equivalent PD is 20 GV. If the target were not cooled, how long would be required for its temperature to rise from 20°C to the melting point? For this temperature range, the average specific heat capacity of copper is about 0.11 cal/g·C.

20-B11 A coil of wire is immersed in a cup of water and is connected to a source that delivers 3 A. After the water starts to boil, it is found that 50 g of water is vaporized in 5 min. (*a*) What is the resistance of the coil? (*b*) Explain why the specific heat capacity of the cup does not enter into this problem.

20-B12 The 1990 annual consumption of electric energy in the United States is predicted (with some uncertainty) to be 13,000 kWh per person. About half this energy will be generated in coal-fired power plants having average efficiency 40%. (a) How much coal, with heat of combustion 30 MJ/kg, will be burned in 1990 to provide electric energy for a family of 4? (b) How much heat will be rejected to the environment?

20-B13 A standard test of a small cell used in an electric watch is for the cell to supply 104 μA into a load of 13 kΩ for 800 h. How much energy (in J and in kWh) is stored in a fresh cell?

20-B14 It is estimated that there is one electric clock per person in the United States (population 2.3×10^8), each using energy at the rate of 2.5 W. To supply this energy, about how many metric tons of coal is burned per hour in coal-fired electric generating plants that are on the average 25% efficient? (See Table 14-3 on page 320.)

20-B15 A bird whose feet are 5 cm apart perches on a power line that carries 765 A. If the resistance of the wire is 12 $\mu\Omega$/m, what is the PD between the bird's feet?

20-B16 A wire of resistance 2 Ω is drawn out to four times its original length. What is the new resistance of the wire?

20-B17 A metal wire of length 70 m and diameter 2 mm is connected to a source of PD of 1.56 V, and the current is found to be 2.69 A. Identify the metal.

20-B18 A telephone cable contains many insulated 22-gauge copper wires, each of diameter 0.64 mm. Calculate the total resistance at 0°C of one circuit pair from the central office to a subscriber 8.0 km away and back to the central office.

20-B19 The resistance of a coil of iron wire is measured with the coil immersed in an ice bath; then the coil is placed in an oil bath and heated. At what temperature will the resistance be double the original value?

20-B20 A tungsten filament in a lamp bulb is rated at 60 W when the bulb is connected to a 120-V source. The filament's resistance at 0°C is 18 Ω. (a) Calculate the hot resistance of the filament. (b) Estimate the temperature of the filament, assuming an average temperature coefficient of resistance, for the temperature range considered, to be 0.0049 (°C)$^{-1}$.

20-B21 A resistor is made by winding on a spool a 32.00-m length of constantan wire of diameter 0.800 mm. Calculate, to 3 significant figures, the resistance of the wire at (a) 0°C; (b) 50°C. Assume ρ at 0°C to be 49.00×10^{-8} Ω·m.

20-B22 A wire of uniform cross section is stretched along a meter stick, and a PD of 0.600 V is maintained between the 0-cm mark and the 100-cm mark. How far apart on the wire are two points that differ in potential by 3 mV?

20-C1 A researcher has available 2 g of gold and wishes to form it into a wire having resistance 100 Ω at 0°C. How long should the wire be?

20-C2 (a) Derive a formula for the resistance of a wire in terms of its radius, density, total mass, and resistivity. Check your formula for dimensional consistency. (b) Calculate the resistance at 0°C of an iron wire 1 mm in radius whose mass is 40 g.

20-C3 In one form of plethysmograph (a device for measuring volume), a rubber capillary tube of inside diameter 1.00 mm is filled with mercury at 20°C. The resistance of the mercury is measured with the aid of electrodes sealed into the ends of the tube. If 100.00 cm of the tube is wound in a spiral around an upper arm, the blood flow during a heartbeat causes the arm to expand, stretching the tube to a length of 100.03 cm. From this, assuming cylindrical symmetry, the change in volume can be found, which gives an indication of blood flow. (a) Calculate the resistance of the mercury. (*b*) Calculate the fractional change in resistance during the heartbeat. (*Hint*: The cross-sectional area of the mercury thread decreases by the same fraction that the length increases, since the volume of mercury is constant.)

20-C4 The copper wire of a magnet coil has resistance 6.00 Ω at 20°C. What is the resistance when the current is large enough to raise the temperature to 75°C? (*Hint*: First find the resistance at 0°C.)

20-C5 A motor connected to 120-V direct-current mains develops 400 W of useful mechanical power, and 32 W is lost through friction. The current from the mains is 4 A. (*a*) What electric power is being supplied to the motor? (*b*) What is the motor's efficiency? (*c*) What is the resistance of the motor?

For Further Study

20-7 Energy Storage

So many of the physical principles encountered in our study of physics are involved in energy storage that it is difficult to find a logical place to bring the ideas together. By now, having encountered mechanical, thermal, and electric energy, we are in a position to consider the importance of storing energy and some of the ways of doing so.

In daily life, the demand for energy is not steady. Lights are turned on in the evening, as are many TV sets, but by and large the greatest demand for electric energy is during the working hours when business and factories are in full swing. Some electric utilities offer a reduced rate (in cents per kWh) for energy used during the nighttime "off-peak" hours. This is because the fixed costs of a generating plant (such as labor cost, management salaries, and land taxes) go on whether or not the plant is actually supplying energy. Therefore it is most economical for the plant to generate and distribute power at a steady rate, 24 hours per day, and the effort to accomplish this equality of load (demand) is called *load-leveling* or *peak-shaving*.

Some load-leveling can be accomplished easily by scheduling. For example, an electroplating plant can work from 9 P.M. to 7 A.M.

when other demand for power is less. On the other hand, air-conditioning is usually demanded at just the worst time of day, contributes significantly to the peak demand, and helps cause brownouts. In an ingenious arrangement, an air-conditioned home can be cooled in daytime by blowing air over a large amount of solid substance that has been frozen by a conventional electric refrigerator during the night. In this way the *electric* use is transferred to the time of least demand and least cost. Similarly, small houses in England, Germany, and other Western European countries are heated in the daytime by blowing air over warm bricks in a well-insulated container. The bricks are heated electrically during the night when energy is more readily available and cheaper. Thermal energy stored in hot bricks is an example of *sensible heat*—the bricks get warm as perceived by the senses. In many cases, a greater amount of thermal energy can be stored or released as *latent heat* associated with change of state (Sec. 14-4). Thus, when ice melts at 0°C, it absorbs 80 kcal/kg, but the temperature does not rise; the freezing of ice releases this heat, which is latent rather than sensible.

An experimental solar building at the University of Delaware (Solar One) was a proving ground for testing the basic physical concepts in one of several full-scale applications of energy storage through latent heat. We will look at some of the details of this home heating plant.*

For solar heating, about 184 kWh of energy is stored as latent heat of fusion in 3200 kg of sodium thiosulfate pentahydrate (photographers' "hypo," $Na_2S_2O_3\cdot5H_2O$) contained in 300 sealed plastic pans around which air is circulated and warmed. The best working substance is one that has a low cost per kWh stored as latent heat of fusion; is not toxic, corrosive, combustible, or explosive; and can be stored in low-cost, nonleaking, sealed containers. In these terms, Glauber's salt ($Na_2SO_4\cdot10H_2O$) is least expensive, about \$0.35/kWh of latent heat. It is available commercially at about \$20 per ton in carload lots, from chemical plants as a by-product and from salt lakes and deposits (see Prob. 14-C5 on page 330). However, its melting point of 32°C is too low to give a comfortable, "warm" feeling to circulating air. $Na_2S_2O_3\cdot5H_2O$ is chosen because its melting point is in the range 48°C to 52°C as the water of hydration changes from $5H_2O$ to $2H_2O$ to anhydrous over this range. The total heat of fusion is 0.058 kWh/kg. (Incidentally, it is the water of hydration in both Glauber's salt and hypo that gives the large heats of fusion per mole.) Supercooling must be avoided if all the latent heat is to be given up at about 50°C ($\pm2°$); therefore, a small nucleating device is inserted into each container before sealing. As for cost, the hypo material is currently at about \$2.70/kWh (photographic grade), but it need not be so pure, so its cost can be brought down to less than \$0.75/kWh in quantity production of technical-grade sodium thiosulfate pentahydrate.

In the solar house, then, storage of 184 kWh as latent heat comes to an investment of \$138 for the salt, to which must be added about \$560 for the containers and loading cost. The cost of the solar energy collectors on the roof is about \$100/m². These are one-time costs. If the sensible heat evolved in cooling the crystals and liquid is included, the actual thermal energy increases

by about 10% to 205 kWh. In Prob. 20-C10 the solar input is considered; it turns out that in 9 hours (3 hours centered on solar noon on each of 3 "best clear days"), the unit can be recharged to its capacity of 205 kWh, and the expected winter household heating of 5.9 kW (average) can therefore be maintained for 34 hours, or $1\frac{1}{2}$ cold and cloudy days.

These numerical calculations are included here to give the flavor of the research and development required to implement a typical energy storage scheme. The realization of *any* practical energy storage scheme requires similar detailed calculations and study, and usually involves many disciplines of pure science and technology.

Coolness storage can also be used to achieve load-leveling and peak-shaving. For this purpose the working substance is a eutectic mixture of Glauber's salt, sodium chloride, and ammonium chloride ($Na_2SO_4\cdot10H_2O/NaCl/NH_4Cl$), which has a melting point of 13°C and a heat of fusion of 0.050 kWh/kg. An important purpose of coolness storage is to lessen the daytime power demands for house air-conditioners that may lead to power brownouts. The eutectic is liquefied as warm air is circulated over it during the daytime, and it is refrozen at night during the period from 10 P.M. to 6 A.M. Units of this type have been tested that have storage capacity in the range of 64 kWh.

Several modes of energy storage are employed in transportation. Chemical PE, in the form of gasoline or diesel fuel, is stored until used; electric energy is stored in the battery. Consider a 10,000-kg truck that is lifted through a vertical height of 100 m while going up a hill. This PE is not recovered when the truck goes down the other side of the hill; instead, the truck's mechanical PE is transformed mostly into thermal energy in the brake drums and slightly into the work of shredding a minute amount of brake lining and powdering a little steel. Similarly, when the truck comes to a stop on a level road, its KE is transformed mostly into thermal energy.

As we see, an important problem (or task) in transportation is to stop the vehicle without

---

* See Refs. 14 and 15 at the end of the chapter. Since these publications are not likely to be found in most physics libraries, we quote the data in some detail, with the translation of energy units from Btu to kWh, and mass units from "pounds" to kg. For reference, the costs are given as of the time of the research, in 1976. We are indebted to Dr. Maria Telkes for much of this information; she has spent more than 30 years in this development.

squandering available work. The second law of thermodynamics gives us an insight here. Ordinary braking dissipates the vehicle's KE (which is ordered motion) into low-temperature energy (which is random motion, disordered molecular energy of high entropy). The term *regenerative braking* applies to any process whereby the vehicle's KE is transformed into some form of mechanical or electric energy that is available to be used again or "regenerated" in starting up the vehicle after stopping. One way of storing mechanical energy is to speed up a pair of flywheels—rotating disks whose KE is given by $\frac{1}{2}I\omega^2$ (Sec. 7-7)—whose energy can be saved and given back to the truck as it climbs the next hill.* On a much larger scale, the flywheel storage of energy has been demonstrated in subway cars.

Example 20-8

A crowded subway car of mass 3×10^4 kg approaches a station at 20 m/s. There are four flywheels, each of mass 60 kg and effective radius 0.15 m. Half of the car's initial translational KE is transformed into additional rotational KE of the flywheel system; this energy is available later on for starting up from rest when leaving the station. If the initial angular velocity of the flywheels is 1000 rad/s (about 10,000 rev/min), what is their final angular velocity when the train is standing in the station?

For each disk, the moment of inertia is

$$I = M(R_{\text{effective}})^2 = (60 \text{ kg})(0.15 \text{ m})^2$$
$$= 1.35 \text{ kg} \cdot \text{m}^2$$

Total $I = 4 \times 1.35 \text{ kg} \cdot \text{m}^2 = 5.40 \text{ kg} \cdot \text{m}^2$

In the four flywheels, the initial stored energy is

$$\tfrac{1}{2}I\omega_1{}^2 = \tfrac{1}{2}(5.40 \text{ kg} \cdot \text{m}^2)(10^3 \text{ rad/s})^2$$
$$= 2.7 \times 10^6 \text{ J}$$
$$\text{KE of car} = \tfrac{1}{2}mv^2 = \tfrac{1}{2}(3 \times 10^4 \text{ kg})(20 \text{ m/s})^2$$
$$= 6.0 \times 10^6 \text{ J}$$

Half of this energy is to be stored as the flywheels' increased KE. By the law of conservation of energy, we have

$$\tfrac{1}{2}I\omega_2{}^2 = \tfrac{1}{2}I\omega_1{}^2 + 3.0 \times 10^6 \text{ J}$$
$$= 2.7 \times 10^6 \text{ J} + 3.0 \times 10^6 \text{ J} = 5.7 \times 10^6 \text{ J}$$

Hence,

$$\omega_2{}^2 = \frac{5.7 \times 10^6 \text{ J}}{(\tfrac{1}{2})(5.40 \text{ kg} \cdot \text{m}^2)}$$

$$\boxed{\omega_2 = 1450 \text{ rad/s}}$$

Thus the angular speed of the flywheels increases by 45%.

To avoid excessive loss of energy to air friction, the rotating flywheels are operated in a near-vacuum of only $\frac{1}{30}$ atm. The centripetal force required to keep the outer parts of a wheel rotating in its circular path must be supplied largely by the tensile strength of the material near the axle, where the strain is greatest. Thus the shape of the flywheels has been optimized, with a slight bulge near the axle, even though this is not the way to get maximum moment of inertia. The coupling of flywheel energy to and from the subway wheels is indirect: the flywheels are part of a device that serves as either a motor or a generator, as required (see Sec. 23-3). It is estimated that in a subway system the load-leveling described here would save 30% of the total energy now wasted, would decrease the peak load (during start-up) by 80%, and would reduce temperatures in uncomfortably warm subway tunnels, which at present serve to absorb the heat generated in the braking process.

City buses are being developed in which flywheel energy would last for perhaps 10 km before it would be necessary to reenergize the flywheels at an electric connection. However, unless some form of regenerative braking is used, the pollution problem would merely be shifted from the city streets to the electric generating plants (which would still be an advantage, since large plants are more efficient and can also afford large-scale pollution control devices).

---

* Two oppositely rotating flywheels would be needed so that the angular momentum would be zero; otherwise there would be serious gyroscopic effects, as in the racing car of Sec. 7-9.

In discussing various forms of load-leveling, we have been dealing with problems raised by uneven *demand* for energy. Equally, we must consider the uneven *production** of energy. Although gasoline, natural gas, coal, and oil can be stockpiled and used as needed, the same is not true of some of the alternative sources of energy on which we must certainly depend in the future. The sun shines only during the day; winds do not blow steadily; the tides rise and fall. One way of storing energy is gravitational. When power is available, water is pumped uphill into an artificial lake; later the water can flow down again, through turbines, to generate electricity. It is interesting that Niagara Falls is retained as a daytime tourist attraction only by the use of load-leveling. The visible Falls are diverted at night to a mere trickle while the main flow generates power that is used to pump water up to a higher level from which it can flow down during the day to give useful power. In this way, normal flow over the Falls is maintained during the day and the tourists are happy. In effect, Niagara Falls is an energy source that is available only during nighttime hours. In a host of similar circumstances we can suppose that the primary source (solar, wind, or tidal) can be converted to electric energy at the time it is available; this electric energy is in turn transformed into gravitational or other PE and recovered as electric energy when needed.

It is also possible to use gravitational energy storage as a form of load-leveling, as in the next example. Energy for pumping can be taken from the electric power lines during the night, when demand is low.

Example 20-9

At a hydroelectric plant, 2000 tons (2×10^6 kg) of water is pumped every second for 8 h to a lake 90 m above the turbines. The lake is 3 km long,

---

* Of course, energy is never "produced"—it is only transformed.

1.5 km wide, and 14 m deep. (*a*) How much PE can be stored? (*b*) If the stored energy is returned during an 8-h day, what is the average power available for load-leveling? (*c*) At what rate must energy be expended to fill the reservoir during the night in 8 h if the process has an efficiency of 67%?

(*a*)   Mass of water = (3000 m)(1500 m)(14 m)

$$\times (1000 \text{ kg/m}^3)$$

$$= 6.30 \times 10^{10} \text{ kg}$$

$$\text{PE} = mgh = (6.30 \times 10^{10} \text{ kg})(9.8 \text{ N/kg})(90 \text{ m})$$

$$= \boxed{5.56 \times 10^{13} \text{ J}}$$

(*b*)   Power = $(5.56 \times 10^{13} \text{ J})/(8 \times 3600 \text{ s})$

$$= 1.93 \times 10^9 \text{ W} = \boxed{1930 \text{ MW}}$$

Almost 2000 megawatts is available during the daytime peak-demand period.

(*c*)   Power input = (1930 MW)/(0.67)

$$= \boxed{2880 \text{ MW}}$$

Pumping large quantities of water is not likely to be economical when the primary electrical source is at low voltage, such as from photovoltaic cells exposed to solar radiation. A "hydrogen economy" has been proposed in which energy can be stored by the electrolysis of water into hydrogen and oxygen: $2H_2O + \text{energy} = 2H_2 + O_2$. If certain safety factors can be managed, the existing transcontinental pipeline system can be used, or modified, to distribute hydrogen gas (or some less difficult gas that can be manufactured from hydrogen). Hydrogen can also be stored by diffusion into some metals to form solid metal hydrides. The *fuel cell*, used in space vehicles and some small terrestrial vehicles, combines hydrogen and oxygen stored in on-board tanks to run an electrolytic cell backward, creating electric PE whose only pollutant is water: $2H_2 + O_2 = 2H_2O + \text{energy}$.

As we shall see in Sec. 24-9, energy is stored in any region of space where a magnetic field exists. Large magnets carrying large currents can store useful amounts of energy. In such a system it is essential to avoid the loss of energy through

Joule heating in the resistance of the magnet coils. This power loss, given by I^2R, can be avoided by reducing R to 0 through use of a superconducting material for the wires of the magnet coil. Low-temperature physics thus may play a significant role in energy technology.

In the present state of the art, the only commercially practical energy storage methods are hydroelectric storage for electric utilities and battery storage for transportation vehicles such as the electric car or truck. An important consideration is the *energy density* of the storage method, defined as energy stored per unit mass. The common lead-acid storage battery in a car can store about 0.1 MJ/kg.

Storage batteries now being studied are of exotic design. One such battery has a solid electrolytic material ($Na_2O \cdot 11Al_2O_3$) separating electrodes made of sodium and sulfur; it offers the promise of an energy density as great as 1.1 MJ/kg. Such a ten-fold improvement in energy density would make an electric automobile or truck economically and environmentally attractive for the user. However, if all vehicles ran on storage batteries, a major increase in the number of electric power plants would be needed. In effect, automobiles and trucks would be running on coal or uranium instead of petroleum. The environmental problems would be shifted from city streets to the locations of the power plants, and the economic and environmental costs of additional electric power transmission lines would also have to be considered.

We have only scratched the surface of the goals, technologies, and problems of energy storage. Our aim has been to look at the physical principles of a few of the available methods. If you are alert, you will increasingly hear of these and other methods of energy storage that are or will be crucial to the energy-intensive society in which we live.

Example 20-10

The specifications for a lead-acid storage battery of mass 18 kg state that it can deliver 41 A·h of charge at a PD of 12 V. Calculate the energy density of the stored electric PE.

$$PE = QV = (41 \text{ A})(3600 \text{ s})(12 \text{ J/C}) = 1.77 \text{ MJ}$$

$$\text{Energy density} = \frac{1.77 \text{ MJ}}{18 \text{ kg}} = \boxed{0.10 \text{ MJ/kg}}$$

Problems

20-C6 A suitable unit for energy density would be   (a) kW/m^3;   (b) kWh/m^2·s;   (c) J/kg.

20-C7 The concept of regenerative braking refers to   (a) storage of a car's KE for later use when starting up;   (b) generating energy from a car's fuel;   (c) using energy from a car's battery to power the brakes.

20-C8 Estimate the net volume of the solar-heat storage unit described in Sec. 20-7 using the given data (the specific gravity of hypo is 1.70). Add 10% for the volume of the plastic container material, and multiply the result by 1.5 to allow for the necessary air passages. Finally, find the edge of a cube into which all this will fit.

20-C9 A pile of bricks of mass 2000 kg is heated by an 8-Ω wire that passes around and through the pile. The bricks have a specific heat capacity of 0.20 cal/g·°C.   (a) How much sensible heat (in MJ) can be stored if a current of 15 A is maintained for an 8-h period?   (b) What is the energy density of this storage system?   (c) What is the temperature rise of the bricks?

20-C10 It is stated on page 324 that if the earth's atmosphere were perfectly transparent, solar radiation would reach the earth's surface at the rate of 8.13 J/cm^2 · min.   (a) Express this *solar constant* in kW/m^2.   (b) It is stated on page 449 that a solar house could recharge its thermal storage system with 205 kWh in 9 h. Calculate the required area of the collecting surface under the following assumptions: 67% of the radiation incident at the top of the atmosphere reaches the collector (the rest is lost through atmospheric absorption and reflection); 50% of the energy reaching the collector is utilized; the "best clear day" is clear for 90% of the time.

20-C11 (*a*) It is proposed to store tidal energy by using a barrier to dam up a bay at high tide and recover the PE of the trapped sea water at low tide, 6.25 h later. If the bay is 20 km × 3 km in area, and the difference between high tide and low tide is 1.8 m, calculate the stored energy, in MJ. (*Hint:* The water is raised through an average height of 0.9 m.) (*b*) If the stored energy is used during a period of 6.25 h, at 50% efficiency, what power, in MW, is thus available? (*c*) Compare with Example 20-9, and discuss the outlook for tidal energy storage as a significant method of load-leveling.

20-C12 Calculate the energy density in MJ/kg of the energy stored in the hydroelectric system of Example 20-9. Compare with the value found in Example 20-10 for the lead-acid battery, and with the value of 46 MJ/kg for the heat of combustion of gasoline.

References

1. W. F. Magie, *A Source Book in Physics* (McGraw-Hill, New York, 1935), pp. 420–431, early discoveries of Galvani and Volta about electric currents; pp. 465–472, Ohm's work.
2. L. G. Austin, "Fuel Cells," *Sci. American* **201**(4), 72 (Oct. 1959).
3. H. B. Steinbach, "Animal Electricity," *Sci. American* **182**(2), 40 (Feb. 1950).
4. R. T. Cox, "Electric Fish," *Am. J. Phys.* **11**, 13 (1943).
5. H. Grundfest, "Electric Fishes," *Sci. American* **203**(4), 115 (Oct. 1960).
6. C. L. Stong, in the "Amateur Scientist" department of *Sci. American* **206**(1), 145 (Jan. 1962). Report on an apparatus to demonstrate nerve and muscle potentials of aquatic animals.
7. R. F. Wheeler, in C. L. Stong's "Amateur Scientist" department, *Sci. American* **214**(2), 120 (Feb. 1966). Electrical signals from microscopic animals.
8. J. L. Harrison and A. A. Bartlett, "The power to tell time," *Phys. Teach.* **16**, 304 (1978). Background material for Probs. 20-B13 and 20-B14.
9. F. R. Kalhammer, "Energy Storage Systems," *Sci. American* **241**(6), 56 (Dec. 1979). Discusses pumped water, compressed air, batteries, and thermal storage.
10. R. F. Post and S. F. Post, "Flywheels," *Sci. American* **229**(6), 17 (Dec. 1973). Energy storage in flywheels. See also a rather unfavorable cost analysis of energy storage in flywheels by R. W. Moses, Jr., Letter to the Editor, *Physics Today* **28**(10), 15 (Oct. 1975).
11. D. S. Ward, "Advances in Solar Heating and Cooling Systems," *Phys. Teach.* **14**, 199 (1976).
12. M. Young, "Solar Energy," *Phys. Teach.* **12**, 243 (1974); **14**, 226 (1976).
13. A. R. Tamplin, "Solar Energy," *Environment* **15**(5), 16 (June 1973).
14. M. Telkes, "Solar Energy Storage," *ASHRAE Journal* (Sept. 1974), 38–44. Includes 25 references.
15. M. Telkes, "Thermal Energy Storage," *IECEC Record* (1975), 111–115.
16. W. A. Shurcliff, "Active-Type Solar Heating Systems for Houses: A Technology in Ferment," *Bull. Atomic Scientists* **32**(2), 30 (Feb. 1976).

21

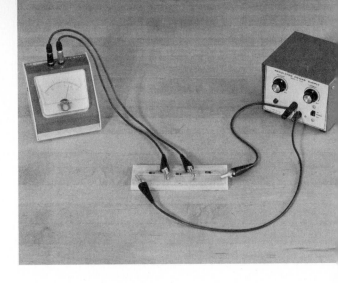

Electric Circuits

So far, we have discussed three circuit elements—capacitors, seats of emf, and resistors. A surprisingly large number of simple circuits involving only these circuit elements are in common use and can be discussed with the help of Ohm's law. Among such instruments are the Wheatstone bridge and the potentiometer, which are used routinely in the physics laboratory and in medical research. We shall discuss these and other circuits in this chapter, but we shall introduce no new theory, relying on our understanding of PD, emf, and Ohm's law to see us through.

In drawings of electric circuits, certain symbols are standard. A capacitor is represented by —||— or —|(— . A resistor is represented by a zigzag line —⋀⋁⋀—; a variable (adjustable) resistor, also called a rheostat, is represented by —⋀⋁⋀— or —⋀⋁⋀—— . A cell or a battery is represented by —⊣|⊢— , where the long line is the positive terminal and the short line is the negative terminal. A battery (several cells connected together in series) is sometimes represented thus: —⊣|ı|⊢— . Unless specifically stated otherwise, we shall assume all connecting wires to have zero resistance. The junction between two connected wires is indicated by a thick spot: ——⊢—— . A grounded point, connected to earth, is represented by ⏚ or ⏛.

21-1 Kirchhoff's Laws

For the systematic application of Ohm's laws to circuits, we use Kirchhoff's laws, which in turn are based on two of the conservation laws. Kirchhoff's first law is based on the law of conservation of charge.

Kirchhoff's first law:

The algebraic sum of the currents entering any junction point in a circuit is zero.

That is, in a given time, as much charge flows into any point as flows away from that point. If this were not true, there would be a net accumulation or diminution of charge at the point. Kirchhoff's first law is illustrated at each of the junctions in Fig. 21-1. At A, $(12 + 3)$ A enters and 15 A leaves. At C, $(6 + 4 + 2)$ A enters and

Three resistors are connected in series. The potential difference between the terminals of the middle resistor is indicated by a voltmeter connected in parallel with it.

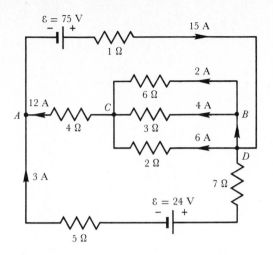

Figure 21-1 Kirchhoff's first law can be verified at each junction.

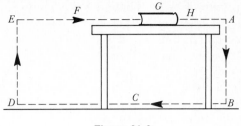

Figure 21-2

12 A leaves. You should verify Kirchhoff's first law at the other junctions, B and D.

Kirchhoff's second law (the loop theorem) is based on the law of conservation of energy.

Kirchhoff's second law:

The algebraic sum of the changes in potential around any closed path is zero.

If we remember that potential changes are measures of work done per unit charge, it should be evident that Kirchhoff's second law is a statement of the energy principle, in a form useful for electric circuits. A mechanical analogy may prove helpful here: Suppose a student slides a physics book to the edge of the table, lets it fall to the floor, slides it along the floor, then lifts it up and returns it to the starting point (Fig. 21-2; see also the discussion of Fig. 19-7). The total work done on the book against gravitational forces is zero, for the round trip $GHABCDEFG$. If this were not so, and the gravitational PE of the book at G after the trip were greater than the gravitational PE before the trip, then a perpetual-motion machine could be constructed (letting the book fall the "hard" way and lifting it up the "easy" way). The impossibility of con-

structing such a machine is asserted by the law of conservation of energy (more generally, by the first law of thermodynamics). Similarly, for electric charge, conservation of energy requires that no net work be done *against electric forces* in taking a test charge around any complete circuit. Therefore there can be no net change in potential after going completely around a circuit.

In a circuit, changes of potential can arise in several ways. If there is a current through a resistor, the PD across the resistor is given by Ohm's law, $V = IR$. The product IR is called the "IR drop" in the resistor. Another way in which potential may change is in a cell or battery. In a cell there is a PD equal to the emf; in addition, if the cell has an internal resistance r, there is a PD equal to Ir in the cell. Let us apply Kirchhoff's second law to several simple circuits.

(*a*) A dry cell of negligible internal resistance is connected to a load of resistance 0.5 Ω; the emf of the cell is 1.5 V. In Fig. 21-3 the arrows represent the direction of the conventional current away from the positive terminal of the cell and through the load resistor from A to B.

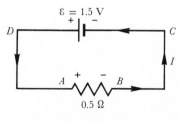

Figure 21-3

Whenever there is a current through a resistor, the end of the resistor at which charge enters becomes positive relative to the end from which it leaves. In this case, A is positive relative to B, and we indicate this fact by drawing a $+$ sign near A and a $-$ sign near B. We also label the two terminals of the cell $+$ and $-$. Note well the difference in method here. The $+$ and $-$ signs attached to $\mathcal{E}$ are determined by the physical structure of the cell, battery, generator, or other seat of emf. As an example, for a dry cell, the $-$ sign is at the zinc can, regardless of the direction of the current. On the other hand, the $+$ and $-$ signs by which we label the resistor depend entirely on the direction of the current (if any). A resistor has no inherent $+$ or $-$ terminals, as does a seat of emf.

Having labeled both the emf and the resistor in Fig. 21-3, we now apply Kirchhoff's second law. Traversing the circuit in the direction $ABCDA$, we have a *fall* of potential ($+$ to $-$) in AB equal to $-I(0.5\ \Omega)$, followed by a *rise* of potential ($-$ to $+$) in CD equal to $+1.5$ V. Kirchhoff's second law says that the net change of potential is zero:

$$-I(0.5\ \Omega) + 1.5\ \text{V} = 0$$

whence

$$I = \frac{1.5\ \text{V}}{0.5\ \Omega} = \boxed{3\ \text{A}}$$

We can with equal success traverse the circuit in the opposite direction, $ADCBA$. We have a fall ($+$ to $-$) in the emf followed by a rise ($-$ to $+$) as we go through the resistor from B to A:

$$-1.5\ \text{V} + I(0.5) = 0$$

whence

$$I = \frac{-1.5\ \text{V}}{-0.5\ \Omega} = \boxed{3\ \text{A}}$$

which is the same answer as before.

(b) To study a more complex circuit, let us redraw Fig. 21-1, this time filling in $+$ and $-$ signs for each resistor and each emf (assumed to have no internal resistance). We also fill in the

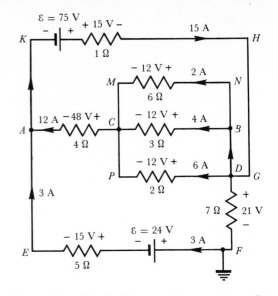

Figure 21-4 Kirchhoff's second law can be verified around any closed path.

values of the IR drops in the resistors, obtained in each case by Ohm's law (Fig. 21-4). In this diagram the currents are assumed known, and we wish to verify Kirchhoff's second law. Around path $ACMNBDGHKA$:

$$+48\ \text{V} + 12\ \text{V} + 15\ \text{V} - 75\ \text{V} \overset{?}{=} 0$$
$$0 = 0$$

Also, around path $AEFDPCA$:

$$+15\ \text{V} + 24\ \text{V} + 21\ \text{V} - 12\ \text{V} - 48\ \text{V} \overset{?}{=} 0$$
$$0 = 0$$

Similarly, the net change of potential is zero around *any* closed path in Fig. 21-4.

Example 21-1

What is the PD between points A and D in Fig. 21-4? Starting at A, we go to D along the path $ACMNBD$:

$$\Delta V = +48\ \text{V} + 12\ \text{V} = \boxed{+60\ \text{V}}$$

Hence D is 60 V higher in potential than A. As a check, we can use the path $AKHGD$:

$$\Delta V = +75 \text{ V} - 15 \text{ V} = \boxed{+60 \text{ V}}$$

When some point of a circuit such as F in Fig. 21-4 is "grounded," the potential of such a point may be considered to be zero.

Example 21-2

If point F (Fig. 21-4) is grounded, what is the potential of point K?

Here F is assumed to be at a potential of 0 V. Around path $FDGHK$:

$$\Delta V = +21 \text{ V} + 15 \text{ V} - 75 \text{ V} = -39 \text{ V}$$

Hence if F is grounded, the potential of K is

$$\boxed{-39 \text{ V}}$$

21-2 Terminal Voltage of a Cell

A cell or battery is no exception to the rule that irreversible "Joule" heat is produced by a current. This means that a cell has *internal resistance*, denoted by r. The internal resistance of a fresh "dry" cell is about 0.05 Ω, but as the moist electrolyte dries out, the internal resistance may increase to as much as 100 Ω or more. Also, hydrogen gas released by electrolysis during the passage of charge through the cell collects around the carbon rod; this too increases the resistance, since a gas is a poor conductor of electricity.* The cell is said to become *polarized* during use. The addition of the compound MnO_2

---

* The emf of a cell depends on the nature of the electrodes and the chemical composition of the electrolyte. Thus the emf of a dry cell also changes when it becomes polarized, since it is now to some extent a hydrogen-zinc cell instead of a carbon-zinc cell.

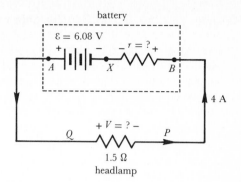

Figure 21-5 A car battery AB gives rise to a current through a headlamp PQ.

to the cell is helpful in two ways: the MnO_2 combines with the H_2 gas to form water, thereby not only removing the H_2 but also replenishing moisture that may have evaporated. From what has just been said, one would expect the internal resistance of a cell not to be constant but to depend on the age and past history of the cell and on the amount of current through it. Nevertheless, in problem solving we usually can consider the internal resistance to be a constant, at least to a first approximation.

Taking into account the internal resistance, we find that the PD between the terminals of a cell, the *terminal voltage* of the cell, is not always equal to the cell's emf. Consider a storage battery of emf 6.08 V that is delivering 4 A to a headlamp of resistance 1.5 Ω. In Fig. 21-5 we represent the internal resistance of the battery as a resistor r drawn next to the symbol; the battery as a whole is represented by a box that includes ε and r, with terminals A and B. In actuality, $\overset{\varepsilon,\,r}{\dashv\!\!\mid\!\!\text{\small I}\!\!\mid\!\!\vdash\!\!\!\wedge\!\!\wedge\!\!\wedge\!\!\!\vdash}$ might be a better symbol for a battery of emf ε and internal resistance r, but by drawing these symbols separately we make it easier to apply Kirchhoff's second law. We must not forget, however, that the junction point, such as X in Fig. 21-5, is never available to us, and we could never connect a voltmeter between A and X to measure the emf ε directly.

_____ **Example 21-3**

For the circuit in Fig. 21-5, calculate (a) the PD across the headlamp terminals PQ, (b) the terminal voltage of the battery, and (c) the internal resistance of the battery.

(a) By Ohm's law, across the headlamp

$$V = IR = (4.00 \text{ A})(1.50 \text{ } \Omega) = \boxed{6.00 \text{ V}}$$

(b) The terminal voltage of the battery is the PD between its terminals. We calculate this PD by going from B to A. Taking the bottom path $BPQA$, we see that the terminal voltage is

$$(4.00 \text{ A})(1.50 \text{ } \Omega) = \boxed{6.00 \text{ V}}$$

the same as the PD across the headlamp that is connected across the terminals of the battery.

(c) Since the current through r is from B to X, the point B is positive relative to X and is so marked in the figure. The internal Ir drop is $(4.00 \text{ A})(r)$. Now we apply Kirchhoff's second law around the path $AQPBXA$:

$$-6.00 \text{ V} - (4.00 \text{ A})(r) + 6.08 \text{ V} = 0$$

$$r = \frac{0.08 \text{ V}}{4.00 \text{ A}} = \boxed{0.02 \text{ } \Omega}$$

Calling the terminal voltage V, we see that in Example 21-3 the terminal voltage is less than the emf of the battery; in fact,

$$V = \varepsilon - Ir$$

where ε is the emf and r is the internal resistance

of the battery. It is also possible for the terminal voltage of a battery to be greater than its emf, if electrons are being forced through the battery in a direction opposite to that in which they would normally flow. This is what happens to a string of storage batteries being charged in a service station.

_____ **Example 21-4**

Two batteries, of emf 6.20 V and 12.45 V, are being charged by a battery charger that sends 20 A through the cells (Fig. 21-6). The internal resistances of the batteries are 0.01 Ω and 0.03 Ω, respectively. What is the terminal voltage of each battery?

We use Ohm's law to calculate the Ir drop inside each battery. In the 6.20-V battery, $Ir = (20 \text{ A})(0.01 \text{ } \Omega) = 0.20 \text{ V}$. In the 12.45-V battery, $Ir = (20 \text{ A})(0.03 \text{ } \Omega) = 0.60 \text{ V}$. These internal Ir drops are labeled $+$ to $-$ as shown, since current is being forced through the internal resistances of the batteries and the $+$ end of a resistor is always the one at which current enters. Now we find the PD from A to B as an algebraic sum of two potential changes, the emf and the internal Ir drop:

$$V = +6.20 \text{ V} + 0.20 \text{ V} = \boxed{6.40 \text{ V}}$$

Similarly, for the other battery, the PD from C to D is

$$V = +12.45 \text{ V} + 0.60 \text{ V} = \boxed{13.05 \text{ V}}$$

These terminal voltages are _greater_ than the emf's of the batteries. In this case, $V = \varepsilon + Ir$.

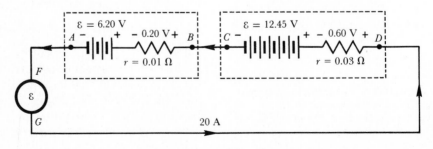

Figure 21-6

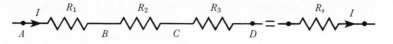

Figure 21-7 Resistors connected in series.

In general, the terminal voltage of a battery or other seat of emf is given by

$$V = \varepsilon \pm Ir \qquad (21\text{-}1)$$

Whether to use the $+$ or the $-$ sign depends on the direction of the current through the battery. As a special case, we see that if the current through a cell or battery or other seat of emf is zero, then $V = \varepsilon$ and the terminal voltage equals the emf. This points out one way to measure emf: we need only measure the terminal voltage under circumstances in which there is no current through the seat of emf. The potentiometer circuit (Sec. 21-8) is designed to do just this. An ordinary voltmeter is definitely ruled out for this purpose, since there must be *some* current through the voltmeter to produce a deflection of the needle by magnetic force.

21-3 Resistors in Series and in Parallel

Many (but not all) complex circuits can be simplified by mentally replacing a group of resistors by a single resistor equivalent to the group.

The resistors R_1, R_2, and R_3 in Fig. 21-7 are said to be connected *in series*, because the charges move through the resistors one after the other, with no short cuts. The three resistors as a unit have resistance R_s and carry a current I. The PD from A to D is V_s, so by Ohm's law $R_s = V_s/I$. Now in a series circuit, the PD from A to D equals the sum of the PD's across the individual resistors (the total work required to take a unit test charge from A to D equals the sum of the work from A to B plus that from B to C plus that from C to D).

$$V_s = V_1 + V_2 + V_3 + \cdots$$

$$R_s = \frac{V_s}{I} = \frac{V_1 + V_2 + V_3 + \cdots}{I}$$

$$= \frac{IR_1 + IR_2 + IR_3 + \cdots}{I}$$

Hence, for resistors in series,

$$R_s = R_1 + R_2 + R_3 + \cdots \qquad (21\text{-}2)$$

Our proof is general, for any number of resistors in series, although for simplicity we have shown only three resistors in Fig. 21-7. The essential features of the proof are (*a*) that the PD's are added for resistors in series, and (*b*) that the current is the same through each resistor in a series group.

The resistors in Fig. 21-8 are said to be connected *in parallel*. In a parallel connection, the incoming current divides, some going through each resistor. On the other hand, the PD is the same across each resistor, and is equal to the PD across the equivalent resistor R_p. (The work required to take a unit test charge from A to B is the same by any path, whether CD, EF, or GH.) By Kirchhoff's first law, the total current entering point E equals that leaving point E.

$$I = I_1 + I_2 + I_3 + \cdots$$

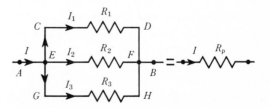

Figure 21-8 Resistors connected in parallel.

By Ohm's law, the equivalent resistance is

$$R_p = \frac{V}{I}$$

and therefore

$$\frac{1}{R_p} = \frac{I}{V} = \frac{I_1 + I_2 + I_3 + \cdots}{V}$$

$$\frac{1}{R_p} = \frac{I_1}{V} + \frac{I_2}{V} + \frac{I_3}{V} + \cdots$$

Hence, for resistors in parallel,

$$\frac{1}{R_p} = \frac{1}{R_1} + \frac{1}{R_2} + \frac{1}{R_3} + \cdots \qquad (21\text{-}3)$$

The essential features of this proof are (a) that the currents are added for resistors in parallel, and (b) that the same PD exists across each resistor in a parallel group.

Example 21-5

Resistors of resistance 6 Ω, 12 Ω, and 4 Ω are connected in parallel. What is the equivalent resistance of the combination?

$$\frac{1}{R_p} = \frac{1}{6} + \frac{1}{12} + \frac{1}{4} = \frac{2}{12} + \frac{1}{12} + \frac{3}{12} = \frac{6}{12}$$

$$R_p = \frac{12}{6} = \boxed{2\,\Omega}$$

It will be helpful to remember that the combined resistance of any number of resistors in parallel is always less than the smallest resistance in the combination. As an analogy from highway engineering, the opening of a parallel bypass, no matter how narrow, will always allow a greater flow of cars, and hence the "resistance" of the combination is less than that of either the road or the bypass taken separately.

A complicated circuit can often (but not always) be simplified by repeated application of the formulas for resistors in series and in parallel.

Example 21-6

What is the net resistance between A and F in Fig. 21-9a?

This circuit is a combination of series and parallel units. We start from the "inside" and use our two basic formulas.

Step 1: Replace the parallel combination BC by a single resistor.

$$\frac{1}{R} = \frac{1}{30} + \frac{1}{15} = \frac{1}{30} + \frac{2}{30} = \frac{3}{30}$$

$$R = \frac{30}{3} = 10\,\Omega$$

Substituting 10 Ω for the combination, we can redraw our circuit as in Fig. 21-9b.

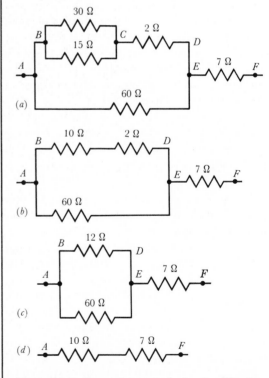

(a)

(b)

(c)

(d)

Figure 21-9 Successive stages in the simplification of a circuit diagram.

Step 2: Replace the series combination *BD* by a single resistor.

$$R = 10 + 2 = 12 \ \Omega$$

We are now able to redraw our circuit as in Fig. 21-9c.

Step 3: Replace the parallel combination *AE* by a single resistor.

$$\frac{1}{R} = \frac{1}{12} + \frac{1}{60} = \frac{5}{60} + \frac{1}{60} = \frac{6}{60}$$

$$R = \frac{60}{6} = 10 \ \Omega$$

This allows us to redraw the circuit as in Fig. 21-9d.

Step 4: Finally, using the series formula, we see that the resistance from *A* to *F* is

$$R = 10 + 7 = \boxed{17 \ \Omega}$$

Not all circuits can be simplified by this technique. Figure 21-10a shows a "simple" circuit whose net (equivalent) resistance *A* to *K* can be found by repeated use of the series and parallel formulas (the answer turns out to be 10 Ω). The

circuit of Fig. 21-10b, however, is actually a "non-simple" circuit that cannot be further simplified by elementary methods; it is impossible to use the method outlined above to find the resistance from *A* to *B*. The difficulty is that the 5-Ω resistor is neither in series nor in parallel with any other resistor. In this book we shall usually deal with simple circuits, except in special cases; see Sec. 21-9 for a method of finding currents in non-simple circuits.

21-4 Emf's in Series and in Parallel

When several emf's are connected in series, they can be replaced by a single emf equal to the algebraic sum of the individual emf's. This follows from the energy principle: the total rate of transformation of nonelectric energy into electric energy equals the algebraic sum of the rates of transformation in the various individual seats of emf. Thus, in Fig. 21-11 three batteries are shown; the net emf between *A* and *B* is − 10 V + 4 V + 13 V = +7 V, with *B* positive relative to *A*. Since the cells are in series, their internal resistances can also be added up, to give 4 Ω.

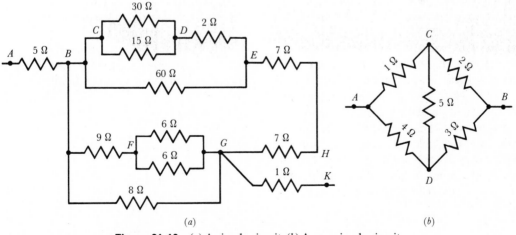

Figure 21-10 (*a*) A simple circuit. (*b*) A non-simple circuit.

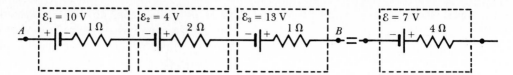

Figure 21-11 Batteries connected in series; equivalent $\mathcal{E}$ and r are found by addition.

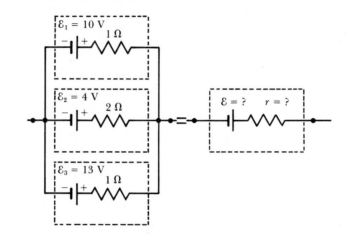

Figure 21-12 Batteries connected in parallel; equivalent $\mathcal{E}$ and r cannot be found in any simple way.

The situation is more complicated when batteries are connected in parallel. In Fig. 21-12, it is difficult to compute easily the effective $\mathcal{E}$ and r of the combination. Certain currents exist within the batteries, and all we know is that the terminal voltages of all three batteries will be equal since they are connected in parallel. In the specific example of Fig. 21-12, the 4-V battery is probably being charged by current from the other batteries, but the magnitude of the current through it cannot be easily found.* There is only one way in which we can make a simple treatment of cells in parallel: if they are identical in emf *and in internal resistance*, then the net emf equals the emf of each cell, and the net internal resistance equals the parallel resistance of all the cells (and hence is $1/n$ times the internal resistance of any one cell). Common dry cells and

batteries (Fig. 21-13) illustrate these points. The cylindrical flashlight cell is a true single cell of emf $1\frac{1}{2}$ V. The short, stubby, $1\frac{1}{2}$-V "cell" is actually a battery of four flashlight cells connected in parallel. The 9-V battery used in transistorized radios and calculators consists of six $1\frac{1}{2}$-V cells, connected in series to give a total emf of 9 V. A 12-V automobile storage battery consists of six 2-V batteries in series, and each 2-V battery may have up to several dozen individual cells in parallel with each other. The inter-

---

* Such questions can be answered by the systematic use of Kirchhoff's laws, as discussed in Sec. 21-9.

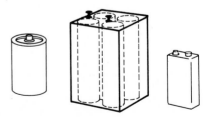

Figure 21-13 Cells and batteries.

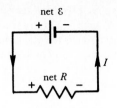

Figure 21-14

nal resistance of a car battery is thus very low (not only because of the use of cells in parallel but also because a liquid electrolyte is used instead of a paste). This allows large currents; in some cars the starter motor may require 50 A or more for a brief time.

21-5 Analysis of an Electric Circuit

We have now discussed the conditions under which a circuit can be replaced by the simplified equivalent form of a seat of emf connected to a single resistor. Kirchhoff's loop rule applied to the circuit of Fig. 21-14 then leads to a useful equation:

$$+(\text{net } \varepsilon) - I(\text{net } R) = 0$$

$$I = \frac{\text{net } \varepsilon}{\text{net } R} \qquad (21\text{-}4)$$

That is, the current through the circuit equals the net emf* divided by the net resistance of the circuit.

Example 21-7

What is the current in the 5-Ω resistor of Fig. 21-15a?

First we simplify the circuit and find the total current. For the combination of the 6-Ω and 3-Ω resistors, $1/R = \frac{1}{6} + \frac{1}{3}$, whence $R = 2\,\Omega$. We now redraw the original circuit as in 21-15b, remembering to draw in the internal resistance of the batteries. We now have a series combination, and the net resistance is given by $R = 1\,\Omega + 5\,\Omega + 1\,\Omega + 2\,\Omega = 9\,\Omega$. To find the net emf, we observe that the two batteries tend to send charges around the circuit in opposite directions, so that the net emf is 60 V − 6 V = 54 V. Now we apply Ohm's law to the entire circuit:

$$I = \frac{\text{net } \varepsilon}{\text{net } R} = \frac{54\text{ V}}{9\,\Omega} = \boxed{6\text{ A}}$$

This is the current through each battery and through the 5-Ω resistor.

---

* Sometimes the symbol Σε is used for "net ε," just as we use ΣF for "net F" in Newton's second law.

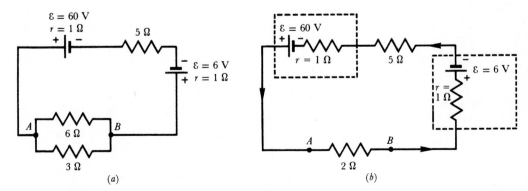

Figure 21-15

Example 21-8

What are the currents through the 3-Ω and 6-Ω resistors in Fig. 21-15a?

Here we apply Ohm's law to a portion of the circuit. The section AB is redrawn in Fig. 21-15b. Since 6 A flows between A and B, and the net resistance from A to B is 2 Ω, Ohm's law tells us that $V_{AB} = IR = (6 \text{ A})(2 \text{ } \Omega) = 12 \text{ V}$. This is the PD across each of the two resistors. The current through each resistor is found by further application of Ohm's law:

Through 6 Ω: $I_1 = \dfrac{12 \text{ V}}{6 \text{ } \Omega} = \boxed{2 \text{ A}}$

Through 3 Ω: $I_2 = \dfrac{12 \text{ V}}{3 \text{ } \Omega} = \boxed{4 \text{ A}}$

As a check, we find that I_1 and I_2 add up to 6 A, which is the total current entering point A. Thus Kirchhoff's first law is satisfied.

Example 21-9

Calculate the terminal voltage of each battery in Fig. 21-15b.

To find the terminal voltages, we use the solution of Example 21-7, in which we calculated the current through each battery to be 6 A. In Fig. 21-16 we fill in the Ir drops in the internal resistances of the batteries, paying due regard to signs. (Remember that the current is from $+$ to $-$ through a resistor.) The terminal voltage of the 60-V battery is found by going from E to F through the battery; a drop of 6 V

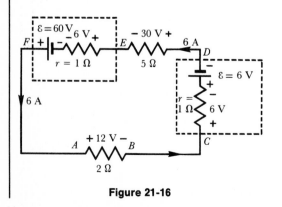

Figure 21-16

is followed by a rise of 60 V, so the net PD is $-6 \text{ V} + 60 \text{ V} = +54 \text{ V}$, with F positive relative to E. Similarly, in going through the other battery from D to C, we find a rise of 6 V followed by a rise of 6 V; the net PD is $+6 \text{ V} + 6 \text{ V} = +12 \text{ V}$, and the terminal voltage of this battery is 12 V, with C positive relative to D.

Suppose that we connect ideal voltmeters* across the various parts of our circuit. According to the calculations of the preceding examples, the voltmeters would read as shown in Fig. 21-17. Kirchhoff's second law is illustrated here, and the net change of potential around the circuit is zero, as it should be. For instance, around the path $ABCDEFA$, the potential changes indicated by the voltmeters add up to zero:

$$-12 \text{ V} - 12 \text{ V} - 30 \text{ V} + 54 \text{ V} = 0$$

We have not yet extracted all possible nourishment from our study of the circuit of Fig. 21-15. Let us make an energy check, or—what amounts to the same thing—a power check. In any resistor (including the internal resistances of the batteries), the Joule heating is at the rate of $P = I^2R$, where the power P is measured in watts if I is in amperes and R in ohms. The table below shows that Joule heat is produced at a total rate of 324 W. We can account for this by the action of the two seats of emf. The 60-V battery is delivering current in the "normal" direction, and chemical PE is being transformed into electric energy at the rate of $P = \mathcal{E}I = (60 \text{ V})(6 \text{ A}) = 360 \text{ W}$. The 6-V battery is being charged as charge carriers are forced "backward" through it. Here electric energy is being transformed into chemical PE at the rate of $P = \mathcal{E}I = (6 \text{ V})(6 \text{ A}) = 36 \text{ W}$. The *net* rate of production of electric energy is

$$360 \text{ W} - 36 \text{ W} = 324 \text{ W}.$$

We see from the table that electric energy is transformed into heat in the various resistors

---

* That is, voltmeters of very high resistance, which would not divert any appreciable current from the circuit (see Sec. 21-6).

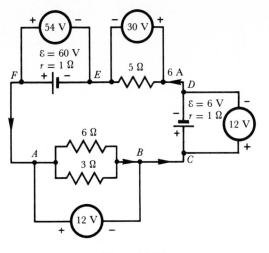

Figure 21-17

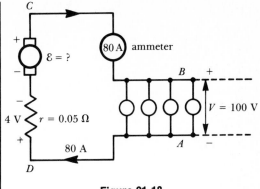

Figure 21-18

and in the batteries themselves at this same total rate of 324 W.

R, ohms	I, amperes	V, volts	P, watts
3	4	12	48
6	2	12	24
1	6	6	36
5	6	30	180
1	6	6	36
			324 W

Example 21-10

A portable generator is delivering 80 A to a string of lamps at a carnival (Fig. 21-18). The terminal voltage of the generator is 100 V, and its internal resistance is 0.05 Ω. (a) What is the emf of the generator? (b) Ignoring friction, what mechanical power must be supplied by the gasoline engine that turns the generator?

We represent the generator schematically as a seat of emf in series with 0.05 Ω.

(a) The Ir drop inside the generator is (80 A) (0.05 Ω) = 4 V. Applying Kirchhoff's second law around the circuit ABCDA:

$$+100 \text{ V} - \mathcal{E} + 4 \text{ V} = 0$$

$$\mathcal{E} = \boxed{104 \text{ V}}$$

(b) The rate of transformation of energy from mechanical to electrical form* is

$$P = \mathcal{E}I = (104 \text{ V})(80 \text{ A}) = 8320 \text{ W} = \boxed{8.32 \text{ kW}}$$

This is the rate at which the gasoline engine does work to supply the electrical energy. An engine of about 11 horsepower would be needed, since 1 hp = 746 W.

Example 21-11

In the preceding example, the load consists of a number of 50-W lamps connected in parallel. How many lamps are used?

In each lamp, $P = VI$; hence the current in each lamp is

$$I = \frac{P}{V} = \frac{50 \text{ W}}{100 \text{ V}} = 0.5 \text{ A}$$

Since the total current is 80 A, the number of lamps is

$$\frac{80 \text{ A}}{0.5 \text{ A/lamp}} = \boxed{160 \text{ lamps}}$$

Check: The resistance of each lamp bulb is 100 V/0.5 A = 200 Ω. Connecting 160 of these lamps in parallel gives a total resistance R_p, which we may

---

* We use $\mathcal{E}I$ rather than VI, since $\mathcal{E}$, an emf, represents the reversible transformation of energy.

calculate as follows:

$$\frac{1}{R_p} = \frac{1}{200} + \frac{1}{200} + \cdots + \frac{1}{200} = \frac{160}{200}$$

$$R_p = \frac{200}{160} = 1.25 \ \Omega$$

The total current through the system is

$$\frac{V}{\text{net } R} = \frac{100 \text{ V}}{1.25 \ \Omega} = 80 \text{ A}$$

which checks with the given data.

Kirchhoff's second law can be used to compute the PD across a capacitor, as in the following example.

_____ **Example 21-12**

Calculate the charge on the plates of the capacitor in the circuit of Fig. 21-19.

There is no steady current through a capacitor, although there *is* a PD across its terminals. Therefore we ignore the capacitor while we first calculate the current through the loop $ABCDEFA$. Note that net R includes the internal resistance of the battery.

$$I = \frac{\text{net } \varepsilon}{\text{net } R} = \frac{16 \text{ V}}{(2 + 5 + 1) \ \Omega} = 2 \text{ A}$$

The PD across C is found by taking a Kirchhoff loop around the path $ABYXA$: we have $+10 \text{ V} - V = 0$, whence $V = 10 \text{ V}$ across the capacitor. (This could

also be found by noting that the PD from X to Y is the same as from A to B, and the latter is equal to IR, or 10 V.) Finally, we calculate the charge on the capacitor from C and V:

$$Q = CV = (2 \times 10^{-6} \text{ F})(10 \text{ V})$$

$$= \boxed{20 \times 10^{-6} \text{ C}}$$

There is $+20 \ \mu C$ on plate Y, and $-20 \ \mu C$ on plate X.

21-6 The Ammeter and the Voltmeter

As their names indicate, the *ammeter* measures current (expressed in some multiple of the ampere), and the *voltmeter* measures potential difference (expressed in some multiple of the volt). Circuit connections for voltmeters and ammeters are quite different. An ammeter has *low resistance* and is connected *in series*, as in Fig. 21-18, where all the charges flowing through the generator also flow through the ammeter. On the other hand, a volmeter has *high resistance* and is connected *in parallel* with the part of the circuit whose PD is to be measured, as in Fig. 21-17.

The instruments in common use are both based on the same moving-coil *galvanometer*, which is fundamentally a current-indicating device depending on magnetic forces. We shall discuss the magnetic force on a wire carrying current in Chap. 22, and we shall study the moving-coil galvanometer in detail in Chap. 23. At this time we accept the galvanometer as a working instrument and inquire how to use it in practical measurement of current and potential difference.

Small currents can be measured by a galvanometer without further modification. For instance, a basic galvanometer movement might have a coil of resistance 100 Ω, and it might require $200 \times 10^{-6} \text{ A}$ (200 microamperes, or 200 μA) for a full-scale deflection. A current of $160 \times 10^{-6} \text{ A}$ would give a reading 80 % of full scale, and so forth. Usually, however, we need to measure larger currents—for instance, the current of 2 A through a lamp bulb or a small motor.

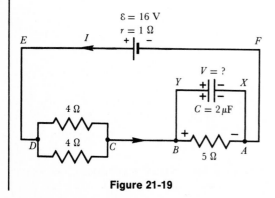

$\varepsilon = 16$ V
$r = 1 \ \Omega$

$V = ?$

$C = 2 \ \mu F$

$5 \ \Omega$

$4 \ \Omega$

$4 \ \Omega$

Figure 21-19

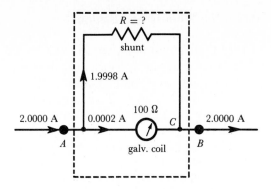

Figure 21-20 Conversion of a galvanometer into an ammeter, using a low-resistance shunt.

The unmodified galvanometer is far too sensitive for such a use.

To make a galvanometer less sensitive, we provide a low-resistance bypass to carry most of the current. Such a resistor, called a *shunt*, is connected in parallel with the galvanometer coil (Fig. 21-20). Let us compute the value of the shunt resistor needed to convert our galvanometer into an ammeter reading 2 A at full-scale deflection. Using a path ACB, we find that the PD from A to B is $(2 \times 10^{-4} \text{ A})(100 \text{ } \Omega) = 2 \times 10^{-2}$ V. Applying Kirchhoff's first law at A, we find that there is 1.9998 A through the shunt when the needle is deflected full-scale by the current of 0.0002 A through the coil. By Ohm's law,

$$R_{\text{shunt}} = \frac{2 \times 10^{-2} \text{ V}}{1.9998 \text{ A}} = 0.010 \text{ } \Omega$$

The shunt has a resistance of about 0.01 Ω, a very small resistance indeed. Another way of looking at the action of the shunt is to consider how the current divides. In Fig. 21-20, the shunt has a resistance about $\frac{1}{10,000}$ that of the coil. If the wire at A is carrying 2 A, the current splits in the ratio of 10,000 to 1; that is, $\frac{1}{10,001} \times 2$ A through the galvanometer coil, and $\frac{10,000}{10,001} \times 2$ A through the shunt. To three-figure accuracy, $\frac{1}{10,001} \times 2 = 2.00 \times 10^{-4}$ A, and the instrument reads full-scale, as desired. After adjustment of the shunt,

the final step in the conversion of our galvanometer into an ammeter is to mark off the scale with values ranging from 0 to 2 A, so that it will be direct-reading. The shunt is usually mounted inside the case, out of sight.

To construct a voltmeter, we modify a galvanometer in a somewhat different way. The original instrument could be used to measure a small PD: full-scale deflection requires 200×10^{-6} A through 100 Ω, and the required PD is given by

$$V = IR = (200 \times 10^{-6} \text{ A})(100 \text{ } \Omega) = 0.020 \text{ V}$$

The meter reads 20 mV full-scale, and its resistance is 100 Ω.

To measure a dry cell's terminal voltage, however, would require a larger range. This is made possible by a high-resistance resistor called a *multiplier*, connected in series with the galvanometer coil. In Fig. 21-21 we show the construction of a voltmeter that reads 2 V full-scale, starting with the same galvanometer used for the ammeter of Fig. 21-20. The galvanometer coil and the multiplier are in series, and at full-scale deflection each carries the current 2×10^{-4} A. By Ohm's law, the resistance of the circuit from A to B is

$$\Sigma R = \frac{2 \text{ V}}{2 \times 10^{-4} \text{ A}} = 10,000 \text{ } \Omega$$

Hence

$$100 \text{ } \Omega + R_{\text{multiplier}} = 10,000 \text{ } \Omega$$
$$R_{\text{multiplier}} = 9900 \text{ } \Omega$$

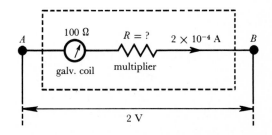

Figure 21-21 Conversion of a galvanometer into a voltmeter, using a high-resistance multiplier.

Thus we need to connect a resistor of 9900 Ω in series with the coil, in order to obtain a voltmeter of range 0 to 2 V. If the applied potential difference is reduced from 2 V to 1.6 V, the current through both coil and multiplier is reduced, and the deflection is 80% of full scale. We mark off the scale with values ranging from 0 to 2 V, so that the voltmeter is direct-reading. The multiplier is usually mounted inside the case, out of sight.

The voltmeter we have described has a high resistance (10,000 Ω total), and it diverts no more than 0.0002 A from the device to which it is connected. An *ideal voltmeter* is one that has an infinite resistance, and that therefore draws no current from the circuit whose PD is being measured. Voltmeters in common use in elementary physics laboratories have resistances such that they require about 0.001 to 0.01 A for full-scale deflection. Voltmeters used for research may need as little as 10^{-5} A; if even less disturbance of the circuit is required, transistorized voltmeters are available that require currents of 10^{-9} A or less.

To summarize, ammeters and voltmeters are modifications of the same moving-coil galvanometer. Ammeters have very low resistance and are connected in series with the circuit element for which the current is to be measured. Voltmeters have high resistance and are connected in parallel with the circuit element for which the PD is to be measured.

21-7 The Wheatstone Bridge

As final illustrations of electric circuits, we describe two instruments in which the galvanometer is used only as a "null" indicator, to show when the current is zero. The first of these devices is the *Wheatstone bridge** for measurement of resistance.

First let us closely examine the "ammeter-voltmeter" method of measuring resistance. If in Fig. 21-22a the voltmeter reads 10.0 V and the ammeter reads 2.00 A, it would seem that $R = (10.0 \text{ V})/(2.00 \text{ A}) = 5.00$ Ω. This would be true if the voltmeter were ideal. If, however, the voltmeter resistance is 1000 Ω, we know that the current through the voltmeter is

$$I = \frac{V}{R} = \frac{10.0 \text{ V}}{1000 \text{ Ω}} = 0.01 \text{ A}$$

and hence only 1.99 A goes through the unknown resistor R. The correct value of R is given by

$$R = \frac{V}{I} = \frac{10.0 \text{ V}}{1.99 \text{ A}} = 5.03 \text{ Ω}$$

We can obtain the correct value of R only by

* Invented by Sir Charles Wheatstone (1802–1875), who used the bridge to locate breaks and short circuits in telegraph lines.

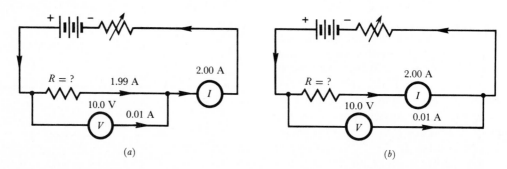

(a)   (b)

Figure 21-22 Ammeter-voltmeter method for measuring resistance. For either circuit, the unknown resistance is not given exactly by $R = V/I$.

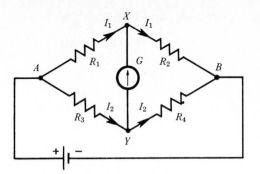

Figure 21-23 Wheatstone bridge circuit.

allowing for the effect of the voltmeter, that is, by knowing the resistance of the voltmeter.

A similar error arises in the connection of Fig. 21-22b; this time the current is correctly measured, but the 10.0-V reading of the voltmeter includes the IR drop in the ammeter as well as that in R. To find the correct value of R, we would have to know the resistance of the ammeter.

The Wheatstone bridge avoids these difficulties. The unknown resistance is determined in relation to three other resistances that are assumed known. If Fig. 21-23 the current entering at A splits in some proportion, with I_1 going through the top branch and I_2 through the bottom branch. If there is no current through the galvanometer G, what must be the relation between R_1, R_2, R_3, and R_4? Since (by hypothesis) there is no current between X and Y, all of I_1 goes on from R_1 to R_2; likewise, I_2 passes through both R_3 and R_4. Since X and Y are at the same potential,* the IR change from A to X must equal the change from A to Y. Therefore,

$$I_1 R_1 = I_2 R_3$$

Similarly,

$$I_1 R_2 = I_2 R_4$$

---
* If there were a PD between X and Y, there would be a current through the galvanometer.

Dividing equals by equals gives

$$\frac{I_1 R_1}{I_1 R_2} = \frac{I_2 R_3}{I_2 R_4}$$

whence

$$\frac{R_1}{R_2} = \frac{R_3}{R_4} \qquad (21\text{-}5)$$

This is the equation of balance for a Wheatstone bridge. Any unknown "arm" of the bridge can be found if the other three arms are known.

In a common form of Wheatstone bridge, R_3 and R_4 are segments of a uniform wire stretched along a meter sick (Fig. 21-24). According to Eq. 20-6, $R = \rho l/A$; if the cross section is uniform, R is proportional to l, and so R_3/R_4 equals l_3/l_4. This takes care of two of the resistances in the equation of balance—only the ratio need be known. R_2 is a standard resistance of known value, and R_x is the unknown resistance. The equation becomes

$$\frac{R_x}{R_2} = \frac{l_3}{l_4}$$

from which R_x is easily found.

Since the resistivity of a metal depends on temperature (Sec. 20-6), a Wheatstone bridge can be used to measure temperature. In this case R_x is a coil of wire, usually platinum or nickel, whose resistance is found by balancing the bridge. The temperature is then calculated from Eq. 20-7. One test for the purity of distilled water is measurement of the resistance of a standard-sized sample by means of a Wheatstone bridge.

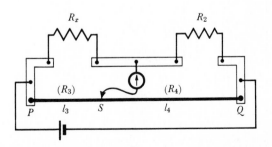

Figure 21-24 Slide-wire form of Wheatstone bridge.

The temperatures encountered in spacecraft are transmitted to earth by radio signals; the sensing device can be some form of Wheatstone bridge. Some lie detectors measure skin resistance by a Wheatstone bridge; it is presumed that a moist skin (indicated by lower electrical resistance) can be correlated with the subject's psychological state.

21-8 The Potentiometer

We have already pointed out that an ordinary voltmeter cannot be used to measure the emf of a cell. Such a voltmeter measures the terminal voltage correctly enough, but since the cell must supply some slight current through the voltmeter, the reading is given by $V = \varepsilon - Ir$, which is less than the emf ε.

Example 21-13

A voltmeter of resistance $1000\ \Omega$ is connected to an "old" dry cell whose emf is 1.500 V and whose internal resistance is $100\ \Omega$ (Fig. 21-25). What is the reading of the voltmeter?

The net resistance for the entire circuit is $1000\ \Omega + 100\ \Omega = 1100\ \Omega$. Hence,

$$I = \frac{\text{net } \varepsilon}{\text{net } R} = \frac{1.500 \text{ V}}{1100\ \Omega} = 0.001364 \text{ A}$$

Now we apply Ohm's law to the voltmeter alone:

$$V = IR = (0.001364 \text{ A})(1000\ \Omega)$$

$$= \boxed{1.364 \text{ V}}$$

Compared with the emf of 1.500 V, this represents a considerable error, due entirely to current through the voltmeter.

The *potentiometer* circuit* measures an emf by a null method that draws no current from the cell or other seat of emf. In the basic potentiometer circuit of Fig. 21-26, a "working battery" sets up a steady current through a slide wire BCD in the closed loop $ABCDEA$. The sliding contact C is adjusted so that there is no current through the galvanometer. Suppose that the unknown cell has an emf ε and internal resistance r, and let the galvanometer's resistance be R_g. There is no current through r or R_g when the potentiometer is in balance; therefore there are no IR drops in r and R_g, and no $+$ and $-$ signs are shown for them. We now apply Kirchhoff's second law to the path $BCGFB$:

$$-IR + (0 \text{ A})(R_g) + \varepsilon + (0 \text{ A})(r) = 0$$
$$\varepsilon = IR$$

---

* Invented by the German physicist Johann Poggendorff (1796–1877).

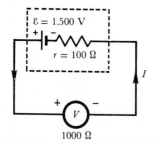

Figure 21-25 The voltmeter reads less than the emf of the cell.

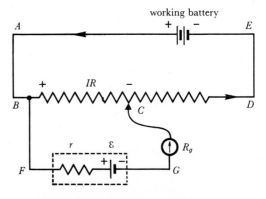

Figure 21-26 Potentiometer circuit.

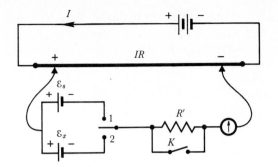

Figure 21-27 Comparison of two emf's, using a slide-wire potentiometer.

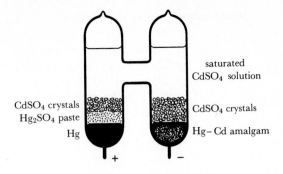

Figure 21-28 Weston standard cell.

When the potentiometer is in balance, the emf of the cell equals the IR drop in a portion of the slide wire. The internal resistance of the cell is of no consequence, since there is no current through it.

In practice, the unknown cell is compared with a *standard cell* whose emf $\mathcal{E}_s$ is known (Fig. 21-27). The potentiometer is first standardized by throwing the switch to position 1 and adjusting the slide wire for balance, giving a length l_1 (proportional to R_1). Then the potentiometer is balanced with the switch thrown to position 2, giving a length l_2 (proportional to R_2).

$$\frac{\mathcal{E}_x}{\mathcal{E}_s} = \frac{IR_2}{IR_1} = \frac{R_2}{R_1} = \frac{l_2}{l_1}$$

$$\mathcal{E}_x = \mathcal{E}_s\left(\frac{l_2}{l_1}\right)$$

Several forms of potentiometer have been developed for special purposes; many are made direct-reading (in volts) by proper adjustment of the working current I. One common type of standard cell is the Weston cell, which uses as its negative electrode a cadmium-mercury amalgam instead of the more common zinc. Such a cell has an emf that is reproducible and is relatively independent of temperature. At 20°C, $\mathcal{E}_s = 1.0183$ V for the saturated form of Weston cell (Fig. 21-28). These cells have high internal resistance, but this is allowable because the cur-

rent is zero when the potentiometer is balanced. To prevent undesired electrolysis within the cell, care must be taken not to draw more than about 0.0001 A from a standard cell. The protective resistor R' (which may be 5000 Ω or more) is in series with the standard cell while initial adjustments are being made. When the potentiometer is almost balanced, R' is shorted out by closing the key K, and the final critical adjustment of the slide wire is made to bring the galvanometer current to zero.

The potentiometer is a most valuable tool for chemists, biologists, and medical researchers. Since the emf of a cell depends on its electrodes and on the nature and concentration of the electrolyte, measurement of emf gives valuable information to the chemist. Also, numerous potentials exist in living tissue. For instance, there is a steady emf of about 50 mV between the inside and the outside of the giant axon nerve cell[*] of the squid *Loligo forbesi* when no nerve impulse is being transmitted. This emf rises sharply to about 95 or 100 mV during nerve action. These potential differences can be measured best with a potentiometer, since the potentiometer requires essentially no current from the source and the high internal resistance of biological tissue has no effect on the measurements.

---

[*] We must distinguish between a biological "cell" and an electric "cell." In this case, the nerve cell is also an electric cell in the same sense that a dry cell is a "cell."

Summary   Kirchhoff's first law is based on the law of conservation of charge, and it states that the algebraic sum of currents entering any point in a circuit is zero. Kirchhoff's second law is based on the law of conservation of energy, and it states that the algebraic sum of the changes in potential around any closed path is zero. These laws offer a systematic way of applying Ohm's law to a circuit.

The internal resistance of a cell depends on its physical construction and its past history. Because of internal resistance, the terminal voltage of a cell differs from its emf when there is current through the cell. If the cell is discharging, $V = \varepsilon - Ir$, and if the cell is being charged, $V = \varepsilon + Ir$.

Formulas for the equivalent resistance of resistors in series and in parallel can be used to reduce many circuits to a simple circuit. Not all circuits can be studied by such elementary means, however. Emf's in series can be added up, but there is no simple way to obtain the equivalent emf to represent a number of emf's connected in parallel, unless they are identical in emf and internal resistance. For a simple circuit for which a net emf and net resistance exist, Ohm's law implies that $I = $ net ε/net R.

The deflection of a moving-coil galvanometer is proportional to the current through its coil. An ammeter is a galvanometer with a low-resistance shunt in parallel with the coil. A voltmeter is a galvanometer with a high-resistance multiplier in series with the coil. A voltmeter will usually have some effect on the circuit to which it is connected, unless the resistance of the voltmeter is sufficiently large.

A Wheatstone bridge is used to make precise measurements of resistance. The method requires one standard resistor whose resistance is known, and two other resistors for which the ratio of resistances is known. This pair of resistors may be two segments of a uniform slide wire.

The potentiometer is used for measuring the value of an emf without drawing current from the seat of emf. The disturbing effect of a cell's internal resistance is thus eliminated.

Check List   Kirchhoff's first law
Kirchhoff's second law
terminal voltage
polarization of a cell
$V = \varepsilon \pm Ir$
$R_s = R_1 + R_2 + R_3 + \cdots$

$$\frac{1}{R_p} = \frac{1}{R_1} + \frac{1}{R_2} + \frac{1}{R_3} + \cdots$$

simple circuit
non-simple circuit

galvanometer
ammeter
voltmeter
shunt
multiplier
Wheatstone bridge

$$\frac{R_1}{R_2} = \frac{R_3}{R_4}$$

potentiometer

Questions   21-1   In Fig. 21-29, place $+$ and $-$ signs at the ends of the resistors A, B, C, D, and E to indicate the polarity of the PD across each resistor.

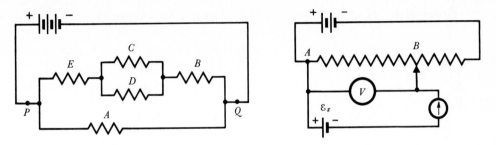

Figure 21-29

Figure 21-30

21-2 Check Kirchhoff's second law around the path $EFDGHKAE$ of Fig. 21-4.

21-3 Why is the internal resistance of a cell not a constant?

21-4 Under what circumstances is the terminal voltage of a battery greater than its emf?

21-5 Can the terminal voltage of a battery be zero?

21-6 Could you start a car by replacing a dead 12-V battery with eight flashlight cells connected in series, each with emf 1.5 V?

21-7 Is the mechanical construction of a voltmeter essentially different from that of an ammeter?

21-8 How does the electrical circuit of a voltmeter differ from that of an ammeter?

21-9 If the working battery of a Wheatstone bridge runs down a little so that its terminal voltage decreases, will this affect the value of R_x, calculated from Eq. 21-5?

21-10 In what way is a potentiometer superior to a voltmeter for measuring a PD? In what way is a voltmeter superior?

21-11 The circuit of Fig. 21-30 is a potentiometer that requires no standard cell. Show that if the sliding contact B is adjusted so that the galvanometer reads zero, the voltmeter reading equals the emf of the unknown cell ε_x.

21-12 Why would it ruin a standard cell to try to measure its emf with a low-resistance voltmeter?

MULTIPLE CHOICE

21-13 Kirchhoff's first law is based on the law of conservation of   (*a*) mass;   (*b*) energy;   (*c*) charge.

21-14 Kirchhoff's second law (the loop theorem) is based on the law of conservation of   (*a*) mass;   (*b*) energy;   (*c*) charge.

21-15 The terminal voltage of a battery   (*a*) is always equal to the emf;   (*b*) is always less than the emf;   (*c*) could be greater or less than the emf, depending on the direction of the current through the battery.

21-16 Eight equal resistors R are available. If they are connected in four parallel pairs and the pairs are connected in series, the total resistance is   (*a*) $2R$;   (*b*) $4R$;   (*c*) $8R$.

21-17 For best results, a voltmeter's resistance should be   (*a*) as large as possible;   (*b*) equal to the resistance of the circuit being measured;   (*c*) as small as possible.

21-18 To convert a galvanometer into an ammeter, you should add   (*a*) a low resistance in series;   (*b*) a high resistance in series;   (*c*) a low resistance in parallel.

Problems

21-A1 What is the PD between points F and P in Fig. 21-4?

21-A2 What is the PD between points E and N in Fig. 21-4?

21-A3 If point H in Fig. 21-4 is grounded instead of point F, what is the potential of point C?

21-A4 In Fig. 21-31, what is the PD across each of the three resistors?

21-A5 A 15-Ω resistor is connected in series with a parallel combination of 10 resistors, each of 200 Ω. What is the net resistance of the circuit?

21-A6 Five 20-Ω resistors are connected   (a) in series and   (b) in parallel. Compute the equivalent resistance in each case.

21-A7 Three equal resistors, each of resistance 6 Ω, can be connected in four different ways (Fig. 21-32). What is the equivalent resistance of each combination?

21-A8 A string of Christmas tree lights consists of ten bulbs in series, connected to a 120-V source of PD.   (a) What is the PD across each bulb?   (b) If one bulb is removed from its socket while the string is plugged in, what is now the PD across the socket terminals?   (c) While this bulb is removed, what is the PD across each of the remaining nine bulbs?

21-A9 The emf of an automobile storage battery is 12.20 V, and its internal resistance is 0.008 Ω. What is the terminal voltage of the battery when the starter motor is drawing 120 A from the battery?

21-A10 A test for a fresh dry cell, of emf 1.5 V, is to connect the terminals directly to the terminals of a heavy-duty ammeter. [*Warning:* Don't try this with an ammeter of insufficient range. (Why not?)] If the ammeter reads at least 30 A, the cell is considered fresh. What is the internal resistance of a fresh dry cell?

21-A11 In Fig. 21-23, $R_1 = 100\ \Omega$, $R_2 = 200\ \Omega$, $R_3 = 25.3\ \Omega$. For what value of R_4 will the bridge be balanced?

21-A12 In Fig. 21-24, $R_x = 36.0\ \Omega$ and $R_2 = 14.0\ \Omega$. The slide wire PQ is 100 cm long with the 0-cm mark at P and the 100-cm market at Q. When the bridge is balanced, what is the reading of the slider S, in centimeters?

21-A13 In Fig. 21-24, the slide wire PQ is 100 cm long, and the distance PS is 25 cm. The value of R_2 is 18 Ω; what is the value of R_x?

21-B1 What is the resistance between P and Q in Fig. 21-29 if $A = 20\ \Omega$, $B = 7\ \Omega$, $C = 15\ \Omega$, $D = 30\ \Omega$, and $E = 3\ \Omega$?

21-B2 What is the resistance between P and Q in Fig. 21-29 if $A = 10\ \Omega$, $B = 3\ \Omega$, $C = 6\ \Omega$, $D = 3\ \Omega$, and $E = 5\ \Omega$?

(a)

(b)

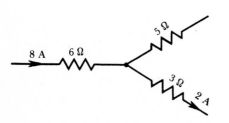

Figure 21-31

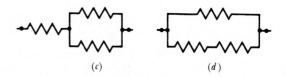

(c)   (d)

Figure 21-32

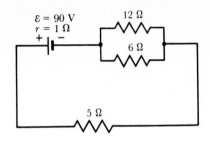

Figure 21-33

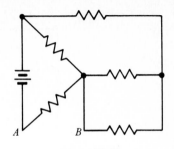

Figure 21-34

21-B3 Show that the resistance between A and K in Fig. 21-10a is 10 Ω, as stated in the text.

21-B4 What is the equivalent resistance between C and D in Fig. 21-10b?

21-B5 The battery in Fig. 21-29 has terminal voltage of 60 V. What is the current through the battery if $A = 30$ Ω, $B = 5$ Ω, $C = 6$ Ω, $D = 3$ Ω, and $E = 8$ Ω?

21-B6 (a) In Fig. 21-33, calculate the current through each resistor. (b) At what rate, in watts, is energy being transformed from chemical to electrical form in the battery? (c) Make a table of power dissipated in heat in the various circuit elements, and show that the sum of these powers agrees with your answer to part (b).

21-B7 Solve Prob. 21-B6, with the battery changed to one with $\varepsilon = 72$ V and $r = 3$ Ω.

21-B8 In Fig. 21-34 the battery has $\varepsilon = 36$ V and $r = 2$ Ω, and each resistor is 10 Ω. Calculate the current through the battery.

21-B9 In Fig. 21-35, (a) what is the current in the 5-Ω resistor? (b) what power is dissipated in the 6-Ω resistor? (c) what is the terminal voltage of each battery?

21-B10 In Fig. 21-36, what is the charge on the capacitor (a) when the switch is open and (b) when the switch is closed?

21-B11 In Fig. 21-37, the battery has emf 9 V and internal resistance 80 Ω. (a) What is the terminal voltage of the battery? (b) What is the potential of point A relative to the ground?

21-B12 (a) In Fig. 21-6, what is the PD between points F and G? (b) The battery charger has an emf of 20.5 V. What is its internal resistance?

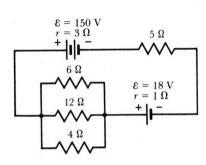

Figure 21-35

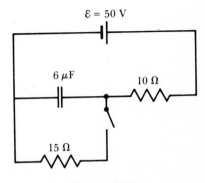

Figure 21-36

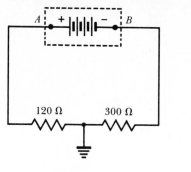

Figure 21-37

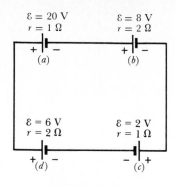

Figure 21-38

21-B13 Four batteries are connected in series, as shown in Fig. 21-38. Compute the terminal voltage of each battery, and verify Kirchhoff's second law by showing that the algebraic sum of the terminal voltages equals zero.

21-B14 Solve Prob. 21-B13 with the 8-V battery's polarity reversed.

21-B15 Two cells are connected in parallel, + to + and − to −. Cell 1 has $\varepsilon = 1.5$ V and $r = 1\ \Omega$; cell 2 has $\varepsilon = 2.1$ V and $r = 2\ \Omega$. What is the common terminal voltage of the cells when connected together in this way, with no external load?

21-B16 Two lamp bulbs, each designed to use 60 W when connected to a 120-V source, are connected in series across a 120-V source of PD. Assume that at the new (cooler) temperature each bulb's resistance is half of its value at the normal temperature. Calculate the total power used by the two bulbs when connected in series.

21-B17 A milliammeter reads 1 mA full-scale and has a resistance of 200 Ω. How would you convert this to a voltmeter that reads 10 V full-scale, by use of a single additional resistor? Draw the circuit and compute the value of the resistor that is needed.

21-B18 A biophysicist needs a voltmeter that reads 20 V full-scale and has available a milliammeter of resistance 50 Ω that requires 4 mA for full-scale deflection. Draw a circuit showing how a single additional resistor can be used to convert the milliammeter into the desired voltmeter. Calculate the value of the resistor that is needed.

21-B19 What is the value of a single resistor that would convert the milliammeter of Prob. 21-B17 into an ammeter reading 10 A full-scale? Draw the circuit and compute the value of the resistor that is needed.

21-B20 A milliammeter of resistance 100 Ω requires 3 mA for full-scale deflection. Draw a circuit showing how a single additional resistor can be used to convert the milliammeter into an ammeter reading 20 A full-scale. Calculate the value of the resistor that is needed.

21-B21 In Fig. 21-37, suppose the battery has emf 16 V and negligible internal resistance. When a voltmeter is connected in parallel with the 120-Ω resistor, it reads 4 V. (*a*) What is the resistance of the voltmeter? (*b*) What will the same voltmeter read if it is connected in parallel with 300-Ω resistor?

21-B22 (*a*) In the circuit of Fig. 21-22*a*, suppose the voltmeter (of resistance 2000 Ω) reads 50.0 V and the ammeter (of resistance 0.020 Ω) reads 1.000 A, giving an apparent value of $R = 50\ \Omega$. What is the true value of R? (*b*) Calculate the true R for the same meter readings, using the circuit of Fig. 21-22*b*. (*c*) Which method gives the best result?

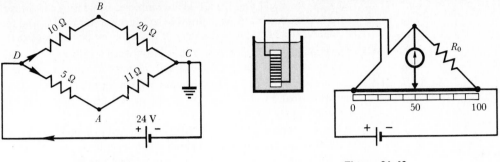

Figure 21-39   Figure 21-40

21-B23 Figure 21-39 shows a Wheatstone bridge that is almost balanced, with point C grounded. (*a*) Calculate the potential of point A. (*b*) Calculate the potential of point B. (*c*) If a galvanometer is connected between B and A, what is the direction of the current through it? (*d*) For what value of the resistor BC would the bridge be in balance?

21-B24 To what value should the 10-Ω resistor in Fig. 21-39 be changed in order to make the potential of point B equal to that of point A?

21-B25 A coil of iron wire is wound on an insulating support and immersed in an oil bath. The coil is made one arm of a Wheatstone bridge, and when the oil is at 0°C, the bridge is balanced, as shown in Fig. 21-40. If the oil bath is heated to 40°C, how far and in what direction must the sliding contact be moved? (*Hint:* See Table 20-1.)

21-B26 The slide wire BD of the potentiometer of Fig. 21-26 is 11.000 m long, and its resistance is 20 Ω. The working battery's terminal voltage is 1.5306 V. What length of wire BC is needed to balance the potentiometer when the cell FG is a standard cell of emf 1.0183 V and internal resistance 100 Ω?

21-C1 A 40-W, 120-V lamp bulb is connected in parallel with a 60-W, 120-V lamp bulb. What is their combined resistance?

21-C2 Compute the current through the 6-Ω resistor in Fig. 21-41.

21-C3 Compute the power dissipated in the 4-Ω resistor of Fig. 21-41.

21-C4 In Fig. 21-42, calculate the value of the unknown resistor R that makes the total resistance of the circuit from A to B also equal to R.

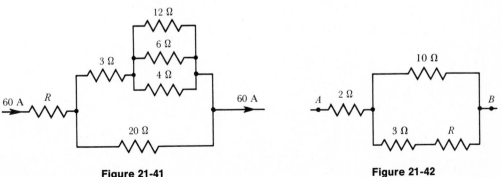

Figure 21-41   Figure 21-42

21-C5 (a) If points A and B in Fig. 21-34 are connected by a heavy wire of zero resistance, what will the current through this wire be? Each resistor is 10 Ω, and the battery has $\varepsilon = 36$ V, $r = 2$ Ω. (b) Make a power check for this circuit (page 464).

21-C6 The terminal voltage of a generator is 110 V when it delivers 10 A, and 104 V when it delivers 30 A. Calculate the emf and the internal resistance of the generator.

21-C7 A voltmeter of range 0 to 25 V requires 0.025 A for full-scale deflection. When connected to a battery of emf 12.18 V, the voltmeter reads 12.00 V. Calculate the internal resistance of the battery.

21-C8 When a voltmeter of resistance 200 Ω is connected to a dry cell, it reads 1.5200 V. If a 200-Ω resistor is connected in parallel with the cell and the voltmeter, the reading falls to 1.5160 V. Compute the emf and internal resistance of the cell. (*Hint:* Use simultaneous equations.)

21-C9 The reciprocal of resistance is conductance, denoted by G. The SI unit for conductance is the siemens (S), also known as the mho (read it backward!). (a) What is the conductance of a heater that develops 1800 W when connected to a PD of 220 V? (b) Derive a formula for the conductance G_p of a group of conductors $G_1, G_2, G_3, \ldots$ connected in parallel.

21-C10 Two identical resistors (R) and a cell (ε, r) are available. Currents through the cell are measured for various loads: I_1 for one resistor; I_2 for two resistors in series; I_3 for two resistors in parallel. Show that $2I_1I_2 + I_1I_3 = 3I_2I_3$.

21-C11 The student engineer of a campus radio station wishes to verify the effectiveness of the lightning rod on the antenna mast (Fig. 21-43). The unknown resistance R_x is between point C and point E. E is a "true ground" but is inaccessible to direct measurement since this stratum is several meters below the earth's surface. Two identical rods are driven into the ground at A and B, introducing unknown resistance R_y. The procedure is as follows: Use a Wheatstone bridge to measure R_1 between A and B, then connect A and B together with a heavy wire and measure R_2 between A and point C. (a) Derive a formula for R_x in terms of the observable resistances R_1 and R_2. (b) A satisfactory ground resistance would be $R_x < 2$ Ω. Is the grounding of the station adequate if measurements give $R_1 = 11$ Ω and $R_2 = 5$ Ω?

21-C12 An automobile battery has an emf of 12.60 V and internal resistance of 0.090 Ω. The headlights have total resistance of 5.00 Ω (assumed constant). What is the PD across the lamp bulbs (a) when they are the only load on the battery, and (b) when the starter motor is operated, taking an additional 30 A from the battery?

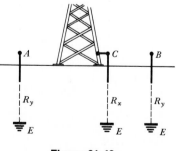

Figure 21-43

21-C13 A galvanometer reads 0.01 A full-scale and has resistance 60 Ω. It is desired to construct an ammeter reading 50 A full-scale by attaching a constantan shunt between the terminals of the meter. The terminals are 10.0 cm apart. What should be the cross-sectional area of the shunt, which is in the form of a straight bar?

For Further Study

21-9 Kirchhoff's Laws Applied to Non-Simple Circuits

We have seen that not all circuits can be simplified by use of the formulas for resistances in series and in parallel. To solve problems involving non-simple circuits, we use Kirchhoff's systematic formulation of Ohm's law. The method can best be explained by an example. We wish to find the current in each resistor of Fig. 21-44. Simple methods fail, since the two seats of emf are not in series (they do not carry the same current), nor are they exactly in parallel (their terminal voltages are not the same). There are three unknown currents, which we call I_1, I_2, and I_3; therefore we need three equations in this example. The general procedure is as follows:

(1) Assume arbitrary current directions in each conductor; if you choose wrongly, this will become apparent by a $-$ sign in the answer.

(2) Label each resistor and emf with $+$ and $-$ signs, the signs for the resistors being determined by the assumed directions of the currents.

(3) Write Kirchhoff's second law (the loop theorem, Sec. 21-1) for as many loops as possible, making sure that each loop contains at least one new circuit element not already used in another loop.

(4) Write Kirchhoff's first law for all but one of the junction points (nodes).

(5) You will find that you have exactly enough equations to solve simultaneously for the unknown currents.

In this example, we use two loops and one node.

$$ABEFA: -8I_3 - 2I_1 - 1I_1 + 7 = 0$$
$$BCDEB: +1I_2 - 36 + 3I_2 + 8I_3 = 0$$
$$\text{Node at } B: I_1 + I_2 - I_3 = 0$$

Upon simplification the equations become

$$\begin{cases} 3I_1 & + 8I_3 = 7 \\ & 4I_2 + 8I_3 = 36 \\ I_1 + I_2 - I_3 = 0 \end{cases}$$

Especially if a computer is available, these equations can easily be solved for the three unknown currents. The result is

$$I_1 = -3\text{A} \qquad I_2 = 5\text{ A} \qquad I_3 = 2\text{ A}$$

Since I_1 turned out to be negative, we know that our choice for its direction was wrong and I_1

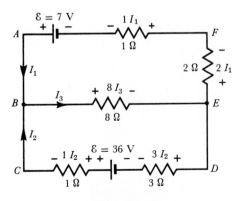

Figure 21-44

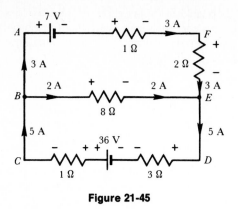

Figure 21-45

actually is from A to F. Our final solution is shown in Fig. 21-45. Note how Kirchhoff's first law is satisfied at B and at E. As a final check, the PD between E and B works out to be the same for any of the three paths, $EDCB$, EB, or $EFAB$ (Prob. 21-C14).

You should guard against writing down equations that are superfluous. Thus the loop $ACDFA$ correctly leads to

$$+1I_2 - 36 + 3I_2 - 2I_1 - 1I_1 + 7 = 0$$

but this equation is nothing new; it is simply the sum of the equations for the two loops already used. This is why each loop equation must have at least one new circuit element. Likewise, a node equation could be written for point E, but it would be algebraically equivalent to the node equation already written for B.

Problems

21-C14 Compute the PD from B to E in Fig. 21-45 by each of the three possible paths.

21-C15 Solve the circuit of Sec. 21-9, assuming other directions for I_1 and I_2.

21-C16 Make a power check for the circuit of Fig. 21-45. Show that the net rate of heat production is equal to the net rate of conversion of chemical energy into electric energy.

21-C17 (a) Compute the current through a 31-V battery, of negligible internal resistance, connected between points A and B of Fig. 21-10b. (*Hint:* With the battery $\mathcal{E}$ drawn in the circuit, there are three loops: $ABDA$; $CBDC$; $B\mathcal{E}ADB$. Use six simultaneous equations.) (b) From the current found in part (a), calculate the resistance between A and B.

21-C18 (a) Calculate the current through each battery in the circuit of Fig. 21-12. (b) Calculate the terminal voltage (the same for each battery).

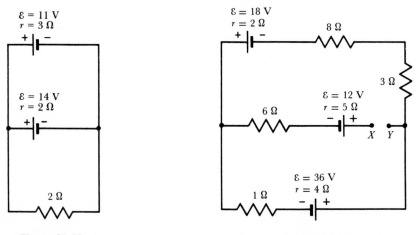

Figure 21-46   Figure 21-47

21-C19 (*a*) What is the current through the 2-Ω resistor in Fig. 21-46? (*b*) What is the terminal voltage of each battery?

21-C20 (*a*) Compute the PD between X and Y in Fig. 21-47, and tell which is at the higher potential. (*Hint*: This part of the problem can be solved without use of simultaneous equations.) (*b*) Points X and Y are now connected by a 10-Ω resistor. What is the current through this resistor?

21-C21 A battery of emf $\mathcal{E}$ and internal resistance r is connected to a load R. (*a*) Show that the power P dissipated in the load resistor is $\mathcal{E}^2 R(R + r)^{-2}$. (*b*) Set $dP/dR = 0$ to find the relation between R and r for maximum transfer of power from the battery to the load. (*Note*: This illustrates the technique of *impedance matching*.) (*c*) Show that at most 50% of the chemical energy in the cell can be delivered as thermal energy in the load resistor.

References

1. J. Busse, "Thermometry," in O. Glasser, *Medical Physics* (Year Book Publishers, Chicago, 1944), pp. 1561–1564. A discussion of the Wheatstone bridge and potentiometer.

2. P. H. Baker, "The Nerve Axon," *Sci. American* **214**(3), 74 (Mar. 1966).

3. E. D. Noll, "Determining the internal resistance of an energy source," *Phys. Teach.* **16**, 478 (1978).

22

Electromagnetism

22-1   *Electromagnetism—A Preview*

The ancients knew that a certain naturally occurring ore called magnetite (impure iron oxide, Fe_3O_4) could serve as a navigational aid, for a piece of the material pointed in a north–south direction when suspended by a thread. These "lodestones" were first found near Magnesia, an ancient city in Asia Minor, and became known as *magnets*. As early as A.D. 121 the Chinese knew that a piece of ordinary iron metal could be "magnetized" by bringing it near a lodestone, and navigation with the aid of the magnetic compass dates back ten centuries into the past. Without knowing it, the early natural philosophers who studied magnetism were dealing with electric forces of a kind that is different from the electrostatic (Coulomb) forces exerted by a rubbed piece of amber that is at rest. The early investigators studied *magnetic forces*, and, as we stated in Sec. 1-5, these forces are exerted by charges in motion. The lodestone fits into this

scheme; the magnetic force is that between certain spinning orbital electrons in the Fe_3O_4 and other spinning orbital electrons in the iron that it attracts. When a compass needle points north, it does so because of a magnetic force between some spinning orbital electrons in the iron atoms of the needle and (presumably) electrons, protons, or ions that move in circular paths deep within the earth.

Electromagnetism is perhaps not already as familiar to you as some other areas of physics. It is worthwhile to outline in advance our approach to the subject, and to state in capsule form some of the definitions and results that we shall develop in this chapter.

(1) The strength **B** of a magnetic field is *defined* and *measured* by force per unit test object. This test object can be a moving charge $Q\mathbf{v}$ or an equivalent current segment $I\,\Delta\mathbf{l}$. Field strength can also be measured by the torque on a current loop. Application: a motor.

Electrons flowing in flat, vertical coils give rise to a horizontal magnetic field directed away from the reader. At the center of the apparatus a beam of electrons is emitted upward from an electron gun in a partially evacuated tube. The magnetic force on the moving electrons bends the beam into a circle.

(2) Magnetic field is *caused* by moving charges—exactly the same entities that respond to a magnetic field. A source of magnetic field can be a moving charge, a current segment, or a current loop. Application: an electromagnet.

(3) On an atomic scale, spinning orbital electrons can give rise to the magnetism of iron and other ferromagnetic substances. Application: a permanent magnet.

(4) Magnetic and electric fields are intimately related: a changing B-field gives rise to an E-field, and a changing E-field gives rise to a B-field. Application: electromagnetic waves such as radio and light waves.

From this brief outline we see that the study of magnetism consists of the study of the forces exerted by and on moving charges.*

22-2   Magnetic Field

Somehow, even in a perfect vacuum, the space near a pole of a bar magnet or near a current-carrying wire is "different." We say that a *magnetic field* exists in such a region of space, and we ask how the magnitude and direction of a magnetic field can be defined and measured.

In Chapters 8 and 19 we described gravitational field and electric field; in each case the field is explored by measuring the force on a suitable test object. The magnitude of a gravitational field is force per unit mass (N/kg):

$$\mathbf{g} = \frac{\mathbf{F}_{grav}}{m}$$

This vector equation also shows that the direction of the gravitational field is the direction of the force on the unit test mass. Similarly, the magnitude of an electric field is force per unit charge (N/C):

$$\mathbf{E} = \frac{\mathbf{F}_{elec}}{Q}$$

where the vector equation shows that the direction of the field is the direction of the force on the unit test charge.

In similar fashion, we say that a *magnetic field* exists in any region of space where magnetic force would be exerted on a suitable test object. For our test object, we use a moving charge; an electron moving down the axis of a TV picture tube would serve. The magnetic forces are additional forces due to motion, over and above any Coulomb forces that would act on the charge even if it were stationary. Experiment shows that in a given field the magnetic force on a test charge depends both on the magnitude of the charge and on its velocity. Thus $F \propto Qv$, and we can use the ratio F/Qv (force per unit charge-velocity product) as a measure of the strength of the magnetic field at the point in question.

Now a complicating factor appears. Experiment shows that at a given point in space (say, the geometrical center of the room in which you are studying), the force on a moving charge depends on the direction of motion as well as on the product Qv. If the only magnetic field is that of the earth, and if the TV tube is aimed so that the electrons move downward and to the north (OP in Fig. 22-1), then the moving electrons experience no force; but if the stream of electrons is

* The term *electromagnetic force* is used to describe the totality of all electric forces, including electrostatic (Coulomb) forces and magnetic forces.

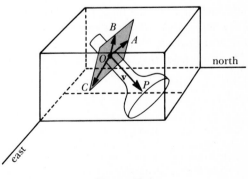

Figure 22-1

directed at right angles to the line OP (direction OA, OB, OC, or any other line in the tilted plane), then the force on the moving charge is a maximum.* The only unique direction in this situation is the direction OP, so we call this the direction of the magnetic field at O.

> **A moving charge experiences no force when moving parallel to a magnetic field; the force on a moving charge is a maximum when the motion is perpendicular to the magnetic field.**

Keeping these facts in mind, we represent a magnetic field at any point by a vector **B**, which is known as the *magnetic induction* or, more simply, the *strength of the B-field*. The *magnitude* of **B** is defined as

$$B = \frac{F_{mag}}{Qv} \qquad (22\text{-}1)$$

where F_{mag} is the magnetic force on a test charge Q moving with velocity v in such a direction as to give rise to the maximum force at the point in question. The *direction* of **B** is the direction in which the test charge moves when it experiences zero force. Like gravitational and electric fields, a magnetic field can be represented by lines of induction drawn through representative points, always in the direction of the vector **B**. The SI unit for B is the *tesla* (T); Eq. 22-1 gives

$$\text{tesla} = T = \frac{N}{C \cdot (m/s)} = \frac{N}{(C/s) \cdot m} = \frac{N}{A \cdot m}$$

The magnitude of a B-field is 1 T if a charge of 1 C moving at 1 m/s experiences a force of 1 N *due to its motion*. Other units for strength of a B-field are also used in the literature; see page 504.

---

* This is an idealized experiment. In actual practice even the maximum force on an electron moving through a TV tube would be too small to observe if the earth's magnetic field were the only field present. Much larger magnetic forces are exerted by the currents in the "deflection coils" in the set.

Example 22-1

An electron moving at 10^6 m/s experiences a force of 8×10^{-13} N when moving vertically upward ($\mathbf{v}_1$ in Fig. 22-2), and it experiences a force of 8×10^{-13} N when moving horizontally from north to south ($\mathbf{v}_2$) or from south to north ($\mathbf{v}_3$). For no other direction of motion is the force larger than 8×10^{-13} N. (*a*) What can you say about the direction of the magnetic field? (*b*) What is the magnetic induction?

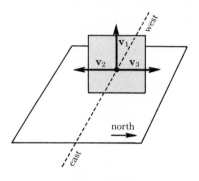

Figure 22-2

(*a*) The plane in which the electron moves to receive the maximum force is evidently a vertical north–south plane. The electron would experience no force if moving perpendicular to this plane; hence **B** is directed horizontally, either toward the east or toward the west. (Our present definitions do not allow us to decide between these two possibilities.)

(*b*) The magnitude of **B** is

$$B = \frac{F_{mag}}{Qv} = \frac{8 \times 10^{-13} \text{ N}}{(1.6 \times 10^{-19} \text{ C})(10^6 \text{ m/s})}$$

$$= 5 \text{ N per C} \cdot \text{m/s} = \boxed{5 \text{ teslas}}$$

This is a strong field; the magnetic induction B for the earth's field is about 6×10^{-5} T, and B is about 1 or 2 T for the field produced in the air gap between the highly magnetized iron pole pieces of a very strong electromagnet.

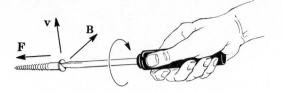

Figure 22-3   Illustration of $\mathbf{F} = Q\mathbf{v} \times \mathbf{B}$.

22-3   Direction of Magnetic Force on a Moving Charge

To specify the direction of magnetic force on a moving charge, we agree to define the direction of **B** so that the following *right-hand rule* is true:

> The direction of magnetic force on a moving positive charge is in the direction that an ordinary (right-hand) screw would advance when rotated from the direction of v toward the direction of **B**.

(For example, when the screw in Fig. 22-3 is turned so that vector **v** is rotated toward the direction of vector **B**, the screw advances in the direction of vector **F**.) This procedure is summarized by a vector symbolism called the *cross product*:

$$\mathbf{F} = Q\mathbf{v} \times \mathbf{B} \qquad (22\text{-}2)$$

The magnitude of **F** is

$$F = QvB \sin \theta \qquad (22\text{-}3)$$

The cross not only indicates the direction of **F**, but it also includes the factor $\sin \theta$, where θ is the angle between the directions of **v** and **B**.* The magnitude of the cross product is $QvB \sin \theta$, and the direction of $Q\mathbf{v} \times \mathbf{B}$ is perpendicular to **v** and **B**, according to the right-hand rule. [Note that

---

* Recall that the dot in the dot product (Sec. 6-1) symbolizes the *cosine* of the angle between two vectors. The dot product of two vectors gives a scalar, whereas the cross product of two vectors gives another vector.

for any two vectors, $\mathbf{M} \times \mathbf{N} = -(\mathbf{N} \times \mathbf{M})$, so it is important to preserve the order of the vectors in a cross product.] In Eq. 22-3 we can think of $B \sin \theta$ as the component of **B** perpendicular to **v**. Often **v** and **B** are perpendicular to each other, so that $\sin \theta = 1$, and then we have

$$F = QvB \qquad (22\text{-}4)$$

This equation, which is equivalent to Eq. 22-1, gives the magnitude of the maximum magnetic force on a charge Q moving with velocity **v** through a magnetic field whose induction is **B**. The direction of the force on a moving charge is perpendicular both to **v** and to **B**. This three-way perpendicularity of **F**, **v**, and **B** is illustrated in cyclotrons (Fig. 33-4). In circular accelerators such as those shown in Figs. 7-11, 33-5, and 33-8, protons move horizontally in a vertical magnetic field supplied by currents in iron-cored electromagnets. From $\mathbf{F} = Q\mathbf{v} \times \mathbf{B}$ we know that the electromagnetic force is perpendicular to both **v** and **B** and is directed horizontally, toward the center of the orbit. This force has magnitude QvB and serves as the centripetal force necessary to keep the protons moving in a circle. We shall discuss particle accelerators more fully in Chap. 33.

To repeat, the magnetic force on a moving charge is perpendicular to the direction of motion and to the direction of the magnetic field; for maximum effect, the motion should be perpendicular to the direction of the magnetic field.

Example 22-2

In a particle accelerator a proton moves horizontally toward the south through a region in which the magnetic induction B is 10 T in an upward direction. The proton's speed is 0.1 that of light. Find the magnitude and direction of the force on the proton.

The speed is

$$v = (0.1)(3 \times 10^8 \text{ m/s}) = 3 \times 10^7 \text{ m/s}$$

Since **v** is perpendicular to **B**, we use Eq. 22-4 to calculate the magnitude of the force on the proton.

$$F = QvB$$
$$= (1.60 \times 10^{-19} \text{ C})(3 \times 10^7 \text{ m/s})(10 \text{ T})$$
$$= \boxed{4.80 \times 10^{-11} \text{ N}}$$

Although this may seem to be a small force, the mass of a proton is so small that a very large acceleration is produced. The acceleration can be found by Newton's second law.

To find the direction of the magnetic force on the proton, we note that **v** is horizontal, to the south, and **B** is upward. By the right-hand rule, a screw rotating from the direction of **v** toward that of **B** would advance toward the west.

Hence **F** is $\boxed{\text{horizontal, to the west.}}$

Example 22-3

In Fig. 22-2, assume a positive charge to be moving upward with velocity $\mathbf{v}_1$, experiencing a force horizontally to the north. What is the direction of **B**?

We already found in Example 22-1 that **B** is horizontal, either toward the east or toward the west. We test both assumptions and find that only for **B** toward the east does the right-hand rule give a force toward the north as specified in the problem.

22-4 Force on a Current Segment in a Magnetic Field

Aside from charges in accelerators and in TV and radio tubes, most of the moving charges with which we are familiar are the free electrons that drift through a wire carrying a current. A simple geometrical proof allows us to modify the equations of Sec. 22-3 to forms useful for currents. (Corresponding equations are numbered similarly.) Consider a cylindrical segment of wire of length Δl (Fig. 22-4). Suppose there are N free electrons, each of charge q, in this segment of wire, and suppose each electron has a drift velocity $-\mathbf{v}$ toward the left, equivalent to a drift of N positive (conventional) charges with velocity **v** toward the right. The total charge in the segment is $Q = \Sigma q = Nq$. If the segment is placed in a

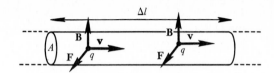

Figure 22-4 A current segment, illustrating $\mathbf{F} = q\mathbf{v} \times \mathbf{B}$ for the force on charges in a wire.

magnetic field of induction **B** directed at right angles to the wire, the total magnetic force on the N charges is

$$F = QvB = NqvB \qquad (22\text{-}5)$$

Now, current = charge/time, and the time required for any charge to travel a distance Δl is $\Delta l/v$. During this time, the cylinder will be emptied of the charge that is in it, and an entirely new set of charges will be in the cylinder. The charge that crosses one face of the cylinder in this time is Nq; hence

$$I = \frac{\text{charge}}{\text{time}} = \frac{Nq}{\Delta l/v} = \frac{Nqv}{\Delta l}$$

$$I\,\Delta l = Nqv$$

By putting this expression for Nqv into Eq. 22-5 we find that the magnitude of the maximum magnetic force on the moving electrons in the cylinder is given by

$$F_{\text{mag}} = I\,\Delta l\,B \qquad (22\text{-}4')$$

In a given magnetic field the magnetic force on a wire is not determined solely by the current. The force on a long wire is greater than that on a short wire, and the essential factor is the product $I\,\Delta l$, which is called a *current segment*. In view of Eq. 22-4', we can write

$$B = \frac{F_{\text{mag}}}{I\,\Delta l} \qquad (22\text{-}1')$$

which means that a current segment $I\,\Delta l$, instead of a moving charge Qv, could be used as a test body, and B could be measured as the force per unit current segment. According to Eq. 22-1', B in teslas can be measured in N per $(\text{A}\cdot\text{m})$ as well as in N per (C·m/s). These units are equivalent, since amperes equal coulombs per second. (In Sec. 22-11 we shall introduce still another unit equivalent to the tesla, the weber/m^2.) Just as for a single moving charge, directions are important. The force on a current segment is greatest if the current is perpendicular to the magnetic field. Also, the force on a current segment is perpendicular to the direction of the current and perpendicular to the direction of the magnetic field.

We represent a current segment by the vector $I\Delta\mathbf{l}$, where $\Delta\mathbf{l}$ has the direction of the conventional, or positive, current through the wire. In general, $I\Delta\mathbf{l}$ and $\mathbf{B}$ are not perpendicular. The vector equation analogous to Eq. 22-2 is

$$\mathbf{F} = I\Delta\mathbf{l} \times \mathbf{B} \qquad (22\text{-}2')$$

and the magnitude of $\mathbf{F}$ is given by

$$F = I\,\Delta l\,B \sin\theta \qquad (22\text{-}3')$$

where θ is the angle between the directions of $I\,\Delta\mathbf{l}$ and $\mathbf{B}$.

Example 22-4

A wire carrying 400 A is stretched horizontally in an east–west direction between two towers 60 m apart. The earth's magnetic field is downward and to the north, and the value of B at that location is 7×10^{-5} N/A·m. What is the magnitude of the magnetic force exerted on the wire?

After convincing ourselves that $\mathbf{B}$ is perpendicular to the current segments $I\Delta\mathbf{l}$, we use Eq. 22-4'.

$$F = I\,\Delta l\,B = (400\text{ A})(60\text{ m})(7 \times 10^{-5}\text{ N/A·m})$$

$$\boxed{= 1.68\text{ N}}$$

It may seem strange to consider a 60-m wire as a short segment. However, since the wire is straight and all parts of it are subjected to the same field, our procedure can be justified. Consider, for instance, dividing the wire into 6000 parts, each 0.01 m long. The force on each little segment would then be $\frac{1}{6000}$ of 1.68 N, and the total of all the forces on all 6000 segments would be 1.68 N, as we have already obtained by our short cut. If the wire were curved, or if $\mathbf{B}$ were not constant over the length of the wire,

then our simplified procedure would be incorrect, and we would use integral calculus to find the total force.

The magnetic force on a current segment is what makes an electric motor work. The moving-coil galvanometer and its modifications, the ammeter and the voltmeter, also depend on magnetic forces acting on current segments.

Example 22-5

A single wire in the rotor of a household motor may be 10 cm long and carry a current of 2 A. If the magnetic induction B is 0.6 N/A·m, what is the force on the wire?

$$F = I\,\Delta l\,B = (2\text{ A})(0.1\text{ m})(0.6\text{ N/A·m})$$

$$\boxed{= 0.12\text{ N}}$$

Since this is only about half an ounce, it is evident that in practical motors many current segments must be acted on simultaneously in order to obtain a useful force.

From the many equations of the last two sections, we can emphasize those that give magnetic force:

$$\mathbf{F} = Q\mathbf{v} \times \mathbf{B} \quad (22\text{-}2) \qquad\qquad \mathbf{F} = I\Delta\mathbf{l} \times \mathbf{B} \quad (22\text{-}2')$$

$$F_{\text{max}} = QvB \quad (22\text{-}4) \qquad\qquad F_{\text{max}} = I\Delta l\,B \quad (22\text{-}4')$$

Here F_{max} refers to the magnitude of the maximum magnetic force, obtained when the charge moves at right angles to $\mathbf{B}$.

22-5 Current Loops

We now have described two ways in which we might, in principle, determine the magnitude and direction of a magnetic field at any point. We can measure force per unit Qv or force per unit $I\,\Delta l$. Both methods are based on the fact that a charge experiences a force when it moves in a suitable direction in a magnetic field. However, the idealized current segment of strength $I\,\Delta l$ is hard to

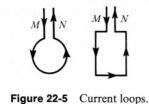

Figure 22-5 Current loops.

imagine, for there must be *some* conducting path by which the current enters and leaves the segment. For a more practical indicator of the direction of a magnetic field, we can use a circular or rectangular current loop (Fig. 22-5). Here we have a complete path for the current, but the net magnetic force on the two lead-in wires *M* and *N* is zero, since they are close together and carry equal currents in opposite directions. Such a loop tends to orient itself broadside to a magnetic field, for if it were at an angle (Fig. 22-6*a*) the two forces $\mathbf{F}_1$ and $\mathbf{F}_2$ would give rise to a torque, and the loop would rotate. When in the broadside position of Fig. 22-6*b*, the loop is in equilib-

rium, and the net force and net torque are both zero.* To obtain a larger effect, we can form a stiff wire into a long narrow coil (Fig. 22-7). When an outside source sets up a current through the wire, the coil tends to rotate until the axis AA' is parallel to the field **B**, with each loop broadside to the field. The coil points in the direction of the field, like a compass needle.

The most ancient device for indicating the direction of a magnetic field is, of course, the compass needle itself. There is more than accidental similarity between our long narrow coil and a magnetized iron nail hung by a thread (Fig. 22-8). In the model of an atom that we are using, electrons move in orbits around the nucleus. The

---

* Figure 22-6 illustrates a method for drawing vectors that are perpendicular to the plane of the paper. The current in the bottom wire is toward the reader and is represented by ⊙, which may be thought of as the head of an arrow just bursting through the paper. The symbol ⊗ represents the tail feathers of an arrow just disappearing into the paper. We shall use these symbols many times in the future.

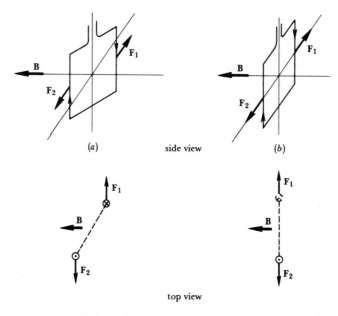

Figure 22-6 Rectangular current loop; the torque is zero when the loop is broadside to the magnetic field, as in (*b*).

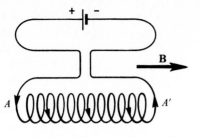

Figure 22-7 Long current-carrying coil becomes aligned with an applied magnetic field.

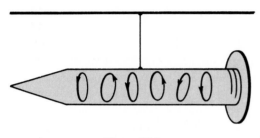

Figure 22-8

electrons also spin on their own axes. Both orbital motion and spin are, in a sense, analogous to current loops. We shall see in Sec. 22-9 that in iron and some other substances, groups of atoms are locked together in *domains* that give a far stronger magnetic effect than do single atoms in substances, such as aluminum, that do not form domains.

A magnetized compass needle is, therefore, basically a collection of current loops. The nee-

dle points northward because of the earth's magnetic force on the atomic currents of the domains. Another way of indicating the direction of a magnetic field is familiar to you from your high-school general science. You have undoubtedly sprinkled iron filings on a piece of paper that has been laid on top of one or more permanent magnets. Each little sliver of iron contains domains, and the magnetic force on the current loops that make up the domains acts to turn the sliver parallel to the field. This is an easy way of "mapping" the field. In the final analysis, we are still making use of magnetic forces on moving charges.

A magnetic field is considered known if the magnitude and direction of the induction are known at every point in space. Our test body is a moving charge, a current segment, or some equivalent current loop or coil. To help in visualizing the field, we draw lines of induction parallel to the direction of the field. (These are not called "lines of force," because in a magnetic field the *force* on the test body, given by $\mathbf{F} = Q\mathbf{v} \times \mathbf{B}$, is not parallel to the field.) Lines of induction near one end of a bar magnet are shown in Fig. 22-9. The diagrams show three different ways of exploring the field. A line of induction could, of course, be drawn through any point at which there is a field, such as *P*. However, only a representative number of lines can actually be drawn. Just as for an electric field, lines are closer together in regions of stronger field.

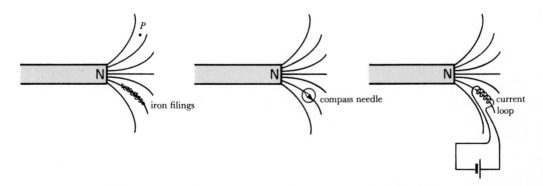

Figure 22-9 Three equivalent ways of exploring a magnetic field.

22-6 Measurement of the Strength of a Magnetic Field

We have rather glibly talked of using the force on a test object (such as a moving charge or a current segment) to "measure" the strength of a magnetic field. In actual practice, a magnetic field would be measured by a more convenient device that in turn would be calibrated at a laboratory such as the National Bureau of Standards. The primary standard of the strength of a B-field does indeed go back to the force on a current segment (see the definition of the ampere in Sec. 22-8). However, many physical properties depend to some extent on magnetic field and can thus be used as the basis of convenient secondary standards. For example, the electrical resistivity of a wire made of bismuth metal depends on the applied magnetic field. Thus, when great sensitivity is not needed, magnetic fields can be compared using a Wheatstone bridge to measure the resistance of a bismuth sample. Note that the exact theory behind the variation of resistance need not be known, provided only that the resistivity depends *reproducibly* on the applied magnetic field.

Another more modern instrument is the Hall effect gaussmeter.* A current of a few mA is maintained by an external emf connected to leads X and Y at the ends of a flat ribbon made of suitable material (Fig. 22-10). We need not concern ourselves with the composition of the material—many conductors and semiconductors show the effect. Nor do we need to concern ourselves with the *sign* of the charge carriers. They may be negative electrons, or they may be positive "holes" (Sec. 23-10). When the ribbon is placed broadside to the magnetic field that is to be measured (vertical $\mathbf{B}$ in Fig. 22-10), the magnetic force $Q\mathbf{v} \times \mathbf{B}$ is horizontal, pushing the charge carriers toward the electrodes at the sides of the ribbon. Thus a PD appears between C and D (the sign of the PD depends on the sign of the

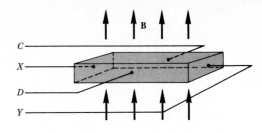

Figure 22-10 Hall effect probe for measuring magnetic field strength. X and Y are current leads; C and D are potential leads.

charge carriers). The PD is measured by a potentiometer, or even by a good high-resistance voltmeter. This probe, when calibrated in a known field, serves to measure an unknown field B that is proportional to the measured PD. The gaussmeter scale is calibrated to read the magnetic field strength directly in teslas, gauss, or kilogauss.

We have mentioned only two devices out of many that are routinely used where convenience and reproducibility are desired. What we have said about magnetic field measurement could apply to almost every physical measurement in the laboratory or the clinic. It is almost always too difficult to use the conceptually simple definition of the quantity. Thus, a secondary device (such as the Hall effect probe in our example) is calibrated directly or indirectly against a primary standard (usually by the manufacturer) and used for routine measurements.

22-7 Sources of Magnetic Field

We turn now to a second aspect of electromagnetism: how does a magnetic field arise? So far, we have discussed the measurement of the strength of a magnetic field without saying anything about how a magnetic field might be produced, except to intimate that moving charges (currents) interact with other moving charges (other currents). Since a moving charge $Q\mathbf{v}$ is equivalent to a current segment $I\,\Delta\mathbf{l}$, we will fix

---

* The term gaussmeter comes from the cgs unit for magnetic induction; 10^4 gauss (G) = 1 tesla (T).

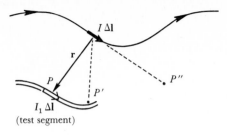

Figure 22-11 The magnitude of the B-field at point P, caused by $I\,\Delta\mathbf{l}$, is given by $\Delta B = k'(I\,\Delta l)/r^2$.

our attention on the magnitude and direction of the magnetic field caused by a current segment.

Experiment shows that the *magnitude* of the induction at point P (Fig. 22-11) is given by

$$\Delta B = k' \frac{I\,\Delta l}{r^2} \qquad (22\text{-}6)$$

with

$$k' = 10^{-7}\ \text{N/A}^2$$

We write this as ΔB rather than B, because it is caused by a *small* segment of length Δl. This equation is about as simple as it can possibly be. The field is proportional to the current I, proportional to the length Δl of the segment, and inversely proportional to the square of the distance r. The constant of proportionality is written as k', whose value is 10^{-7} N/A^2; the reason for this choice is discussed in the next section.

A current segment $I\,\Delta\mathbf{l}$ is a vector, having magnitude $I\,\Delta l$. Only the "broadside" component of $I\,\Delta\mathbf{l}$ is effective in producing a magnetic field. Thus the magnetic induction at P (Fig. 22-11) is given by $k'I\,\Delta l/r^2$; the induction at P'' is zero (head-on view of $I\,\Delta\mathbf{l}$); and the induction at P' could be computed by first resolving the vector $I\,\Delta\mathbf{l}$ into a broadside component and a head-on component, using only the broadside component to compute ΔB. Orientation is usually not a problem, since practical devices such as motors and galvanometers are (understandably) constructed to take advantage of proper orientation for maximum force.

The *direction* of $\Delta\mathbf{B}$ is perpendicular to both $I\,\Delta\mathbf{l}$ and $\mathbf{r}$; it is into the plane of the paper in Fig. 22-11.

The basic equation (22-6) gives the magnetic induction caused by a short current segment. In any practical situation we need to know the resultant or total value of B caused by all the segments of an entire wire that is part of a circuit. There are two cases of special interest to us.

Long Straight Wire. The magnetic field surrounding a long straight wire is represented by lines of induction that are concentric circles (Fig. 22-12). A simple right-hand rule is used to determine the direction of the field surrounding a wire

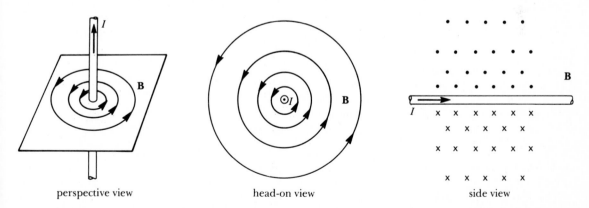

perspective view   head-on view   side view

Figure 22-12 Three views of the magnetic field surrounding a long straight wire. In each case, the fingers of the right hand would represent the direction of the field if the thumb pointed in the direction of the conventional (positive) current.

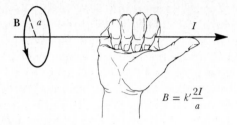

$$B = k' \frac{2I}{a}$$

Figure 22-13 Right-hand rule for field surrounding long straight wire.

that carries a current. Imagine grasping the wire with your right hand, your thumb pointing in the direction of the conventional (positive) current (Fig. 22-13). The fingers are curved around the wire and indicate the direction of the magnetic field.* This thumb rule for the direction of **B** and the right-hand rule for the direction of **F** will take care of all our future needs, including the study of motors, generators, and particle accelerators. It is shown in Sec. 22-14 that the magnitude of the induction at a point at a distance a from an infinitely long straight wire is

$$B = k' \frac{2I}{a} \qquad (22\text{-}7)$$

Note that the strength of a B-field decreases inversely as the first power of the distance from the wire.

Example 22-6

At an electroplating plant, a long horizontal wire connecting two buildings carries 500 A toward the east. What are the magnitude and direction of the magnetic induction at a point on the ground 10 m directly below the wire?

---

* A current of negative charges, such as a beam of electrons in a TV tube, creates a magnetic field just opposite to that of a positive current, such as a beam of protons in a particle accelerator; on such occasions we can use a left-hand rule to find the direction of **B**, or we can replace the negative current by a conventional (positive) current in the opposite direction.

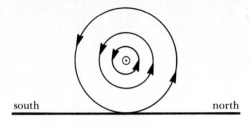

Figure 22-14

The *direction* of the field is found from the right-hand rule (Fig. 22-14). The field at ground level is toward the north.

The *magnitude* of the induction is found from Eq. 22-7:

$$B = k' \frac{2I}{a} = (10^{-7} \text{ N/A}^2)\left(\frac{2(500 \text{ A})}{10 \text{ m}}\right)$$

$$= 10^{-5} \text{ N/A·m} = 10^{-5} \text{ T}$$

Thus

$$\mathbf{B} = \boxed{10^{-5} \text{ T, toward the north}}$$

Two parallel wires that carry currents in the same direction attract each other. To see this, consider Fig. 22-15, which shows the two wires in cross section; the current in each wire is into the plane of the paper. Wire Y is in the field caused by wire X; applying the right-hand rule $\mathbf{F} = I \, \Delta \mathbf{l} \times \mathbf{B}$ to wire Y shows that $\mathbf{F}$ is to the left. A similar argument shows that the force on X is toward the right. (We would expect this by

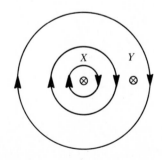

Figure 22-15 Force of attraction between two wires carrying current in the same direction.

Newton's third law—the force of X on Y is opposite to the force of Y on X.) In general, like currents attract each other, and unlike currents repel each other.

Circular Loop. At the center of a circular loop of radius R, the magnitude of B is given by

$$B = k'\frac{2\pi I}{R} \qquad (22\text{-}8)$$

and the direction is parallel to the axis of the loop (Fig. 22-16). The total field due to all the current segments that make up the loop is seen to be inversely proportional to the first power of the loop's radius. See Sec. 22-13 for a derivation of this formula.

Let us compare our two equations for magnetic induction B:

(point near a long straight wire) $\quad B = k'\dfrac{2I}{a}$

(center of a loop) $\quad B = k'\dfrac{2\pi I}{R}$

For each equation, the constant $k' = 10^{-7}\,\text{N/A}^2$. Thus the two equations give B in the same units (teslas):

$$T = \frac{N}{A^2}\left(\frac{A}{m}\right) \quad \text{or} \quad T = \frac{N}{A \cdot m}$$

The unit $N/A \cdot m$ agrees with the earlier definition of field as the force (measured in N) on a current segment (measured in $A \cdot m$). Also, we see that the field at a given distance from the wire is less (by a factor of π) than it is at the center of a loop. This is reasonable, since all portions of the loop are equally close to the point at which B is found (the center of the loop), whereas the field-producing current segments of the wire extend from $-\infty$ to $+\infty$, and the distant segments are less effective. See Sec. 22-14 for computational details.

22-8 Definition of the Ampere

We are now able to define the *ampere* (and therefore the coulomb)—something we have put off for more than four chapters. The discussion revolves around the constant k'. If we had already defined units of current, force, and length, then we would use $B = k'2I/a$ to find k' experimentally by substituting measured values of B, I, and a; the magnitude of B would be found by measuring the force on a known test segment $I_1\,\Delta l$ and using $B = F/I_1\,\Delta l$. You will recall that this is the approach used when we found the constant k in Coulomb's law (Sec. 18-5). However, since we have not yet defined the size of the ampere, we are free to select any numerical value for k' that we please. We choose the value of the constant k' to be $10^{-7}\,\text{N/A}^2$, for historical reasons,* and then Eq. 22-7 serves to define the unit of current that we call the ampere.

To see how the choice of k' leads to a definition of the ampere, let us compute the force per unit length on a long straight wire carrying current $I_1 = 1$ A, which is 1 m from a parallel wire also carrying a current $I_2 = 1$ A (Fig. 22-17). The magnitude of B at point P is given by

$$B = k'\frac{2I_2}{a} = \left(10^{-7}\frac{N}{A^2}\right)\frac{(2)(1\ A)}{1\ m}$$

$$= 2 \times 10^{-7}\frac{N}{A \cdot m}$$

---

* This choice of k' makes our ampere equal to the ampere that earlier workers had defined before SI units became popular.

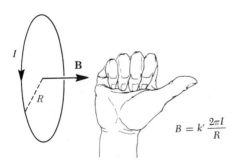

Figure 22-16 Right-hand rule for field at center of circular loop.

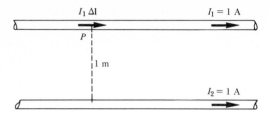

Figure 22-17   Definition of the ampere.

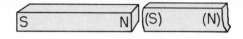

Figure 22-18

From Eq. 22-1′, the force on the current segment at P is given by $F = I_1 \,\Delta l \, B$. Hence,

$$\frac{F}{\Delta l} = BI_1 = (2 \times 10^{-7} \text{ N/A·m})(1 \text{ A})$$

or

$$\frac{F}{\Delta l} = 2 \times 10^{-7} \text{ N/m}$$

We see that making $k' = 10^{-7}$ N/A² leads to the following definition:

> **One ampere is the current in a long straight wire that exerts a force per unit length of exactly 2×10^{-7} N/m on a neighboring long parallel wire, 1 m distant, that carries an equal current.**

We use k' to avoid fractions and a factor 4π. Equation 22-7 is often written in the form $B = (\mu_0/4\pi)2I/a$, where $\mu_0/4\pi$ is the same as k'.

The National Bureau of Standards maintains a "current balance" that is based on Eq. 22-7 but uses coils instead of long wires. The ammeters you use in the laboratory are (indirectly) calibrated relative to this current balance by use of Eq. 22-7. In this way, our standards of current, charge, and PD can be reproduced in terms of the meter, kilogram, and second.

22-9 Magnets and Poles

In the early days, it was thought that a lodestone or a magnetized iron rod had "poles," and that the force of attraction or repulsion between two magnets was due to a sort of Coulomb law between magnetic poles of two kinds. The "north" pole of a bar magnet is the "north-seeking pole," which points northward in the earth's magnetic field. It was assumed that a compass needle had an excess of N poles at one end and an excess of S poles at the other end. This idea was reinforced by the phenomenon of magnetic induction (Fig. 22-18). If a piece of iron is brought near a permanent magnet, poles are induced in the iron in a way that reminds one of the separation of electric charge by induction (Sec. 18-4). Sometimes, if the right kind of iron is used, the piece becomes permanently magnetized. The effect is more pronounced if the iron bar is stroked or hammered while in the magnetizing field. One might say that magnetic poles have been separated "by induction."

Magnetic poles were once believed to have an objective existence of their own, separate from the atoms and molecules of the substance in which they were found. This concept of magnetic poles had to be abandoned, however, as additional facts were brought to light. For one thing, no *free* poles have ever been observed.* If you try to cut off a north pole from one end of a magnet (Fig. 22-19a), all you get is another short magnet, with a north pole at one end and a south pole at the other. No magnetic current has ever been observed, and there are no "conductors" of magnetic poles. The magnetic properties of a few substances, such as iron, cobalt, nickel, and certain alloys, are radically different from those of other metals, but the electrical properties of these same metals differ only slightly. For example, iron is a worse conductor of electricity than aluminum (Table 20-1), by a factor of 5, but iron

---

* Free poles are not prohibited by present-day complete electromagnetic theory. The hunt for the "magnetic monopole" is a field of active research, and the failure to observe it is something of a mystery.

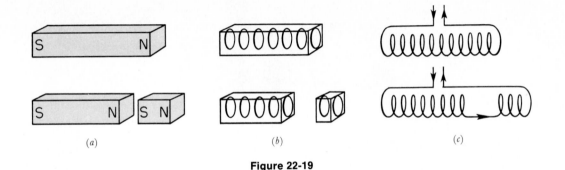

Figure 22-19

is from 500 to 1000 times as easily magnetized as aluminum.

In both iron and aluminum there are unpaired electron spins that do not cancel, and an external magnetic field can to some extent align the spin magnets in spite of thermal agitation. However, iron is very much more magnetic than aluminum because of a "cooperative" effect. Groups of 10^{15} or so iron atoms are locked together with all their electron-spin axes parallel. These groups, called *domains*, are roughly 0.01 to 0.1 mm on an edge in unmagnetized iron. During the magnetization process, some domains grow at the expense of others (Fig. 22-20). In union there is strength, and each domain exerts a force on its neighbor, helping to keep it oriented under the influence of the field. This is what is meant by the "cooperative" effect. In aluminum, where there are no domains, it is every atom for itself; aluminum is much less magnetic than iron. An aluminum rod will set itself parallel to a magnetic field, but the effect is noticeable only in the strongest fields.

It cannot be denied that the north end of a bar magnet attracts the south end of another bar magnet, and so we still speak of north poles and south poles of magnets, and we say that like poles repel and unlike poles attract one another. We now know that these magnetic forces are essentially electrical in nature; they arise from the interactions between the moving and spinning orbital electrons that are bound to the nuclei. We know that the radical differences between iron

and copper are due to the possibility of cooperation between oriented domains in the iron. We presume that *all* magnetic phenomena are ultimately to be described in terms of moving charges, and the free magnetic pole is only a useful fiction, a model that has probably outlived its usefulness. All statements containing the phrase "magnetic pole" can be interpreted in this way, the pole actually being the result of electronic motions and spins. Figures 22-19b and c, which represent a magnet in terms of atomic current loops or equivalent long coils, show how impossible it would be to produce free poles by cutting a bar magnet.

A long, thin, bar magnet offers the closest approximation to isolated poles. The magnetic field near one end of such a needle-like bar is shown in Fig. 22-21; almost all the lines of induction leave the immediate vicinity of the end of the needle. Under these conditions, it makes sense to think of poles located at the ends of the needle. It can even be shown that if a second long needle is brought near the first one, the magnetic force between the current loops represented by the poles should vary inversely as the square of the distance between the "poles." Only in this limited sense is there a Coulomb's law for magnetic poles.* Even for a bar magnet that is not a needle, the general direction of the field is away

---

* Coulomb made some measurements of this sort with his torsion balance, and established an inverse-square law for long, needle-shaped magnets.

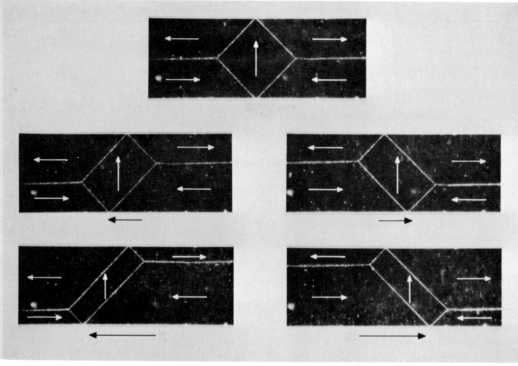

Courtesy of R. W. De Blois and C. D. Graham, Jr.

Figure 22-20 Growth of magnetic domains in a section of a single-crystal iron whisker 0.1 mm thick. The top photograph is for no applied magnetic field; domains oriented to the right and the left are of equal size. In the other photographs, arrows drawn below the whisker show the magnitude and direction of the applied field. It is seen that the process of magnetization consists of growth of some domains and shrinking of others. Domain boundaries are made visible by spreading a colloidal suspension of magnetite over the surface; strong magnetic fields in the region of the boundaries attract the magnetite to form the white lines visible in the photomicrographs.

from the region designated as a north pole and toward the region designated as a south pole. Permanent magnets are usually marked with polarities in accordance with this scheme; this does not mean that the poles are "real." The two situations in Fig. 22-22 are equivalent.

The net magnetic field caused by a coil or other current-carrying conductor is greatly

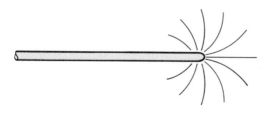

Figure 22-21

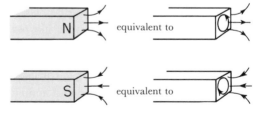

equivalent to

equivalent to

Figure 22-22

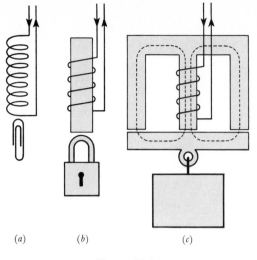

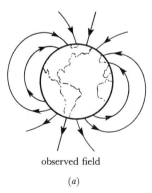

(a)   (b)   (c)

Figure 22-23

increased if a piece of iron or magnetic alloy is brought nearby. The oriented domains are equivalent to current loops inside the metal, and the magnetic field caused by these "atomic current loops" is added to the original field caused by the current in the external coil or wire. Three electromagnets are shown in Fig. 22-23; magnet b can support a much greater load than a, since much of the force is supplied by the domains within the iron core. The relatively weak magnetic field of the current in the winding has been sufficient to enlarge and align many domains in

the iron. Magnet c is still stronger, for the domains are head to tail for complete paths in the iron, thus helping keep the domains in alignment.

In most large electromagnets, the coil current causes Joule heating, which must be dissipated by a cooling system; the cost of the electric power (I^2R) wasted in heat is also a consideration. Modern efficient magnet coils use wires of a superconducting alloy (which is cooled below its transition temperature). For example, one commercially available magnet uses an alloy of 75% Nb, 25% Zr cooled to 4.2 K. Such a wire of only 0.2 mm diameter carries 25 A with no Joule heating whatever.

22-10   The Earth's Magnetism

As long ago as 1570, William Gilbert, court physician to Elizabeth I of England, constructed a permanent magnet in the form of a large lodestone sphere. Gilbert used a small compass needle to survey the magnetic field near the surface of the sphere and found that his model successfully represented the main features of the earth's magnetic field. At a point in the Northern Hemisphere, such as in England or the United States, the field is directed downward and to the north, as shown in Fig. 22-24a. These observations can be accounted for by assuming that a relatively short magnet, several hundred miles

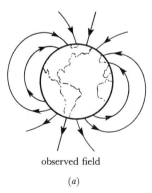

observed field

(a)

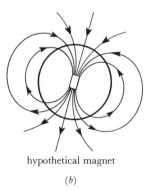

hypothetical magnet

(b)

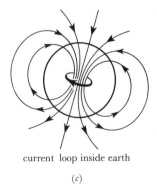

current loop inside earth

(c)

Figure 22-24 The earth's magnetism.

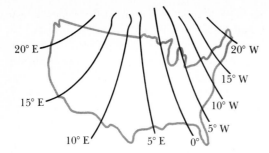

Figure 22-25 Lines of equal magnetic declination. In Maine, the compass points 20° W of true north.

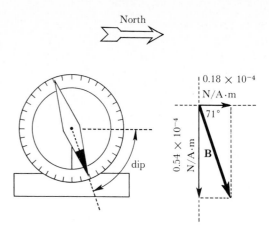

Figure 22-26 Dip needle.

long, is buried deep inside the earth (Fig. 22-24*b*). Since the lines of induction are directed downward for an observer in the Northern Hemisphere, we must assume that the "earth's magnet" has an S pole on the end that is beneath the north magnetic pole.

We know that permanent magnets are collections of current loops, and so we draw Fig. 22-24*c* as a modern equivalent of the outmoded permanent-magnet hypothesis. If the axis of the assumed current loop is more or less along the direction of the earth's axis of rotation, the magnetic North Pole will approximately coincide with the geographical North Pole, as it actually does. The magnetic North Pole is in northern Canada; this means that a compass needle does not point exactly north except at certain places. The difference between geographic north and magnetic north is called *magnetic declination;* this quantity varies from about 20°E to 20°W for different places in the United States, and also varies slowly from year to year. (See the map, Fig. 22-25.)

The ordinary compass needle responds only to the horizontal component of the earth's magnetic field, since it is pivoted in such a way as to prevent up-and-down motion. If a magnetized needle is mounted so as to be free to swing in a vertical north–south plane,* it is free to point in the direction of the field, and in the Northern Hemisphere it will assume a position in which it

---
* Magnetic north, not geographic north.

points downward and to the north. Such an instrument is called a *dip needle* (Fig. 22-26); the angle of dip is measured from the horizontal. Values of the horizontal and vertical components of **B** for the earth's field at Washington, D.C., are shown in Fig. 22-26; the vertical component is about three times the horizontal component at this location.

The ultimate explanation of the earth's magnetic field must somehow be found in circulating currents deep within the earth, or in the upper atmosphere, or both. The approximate coincidence of the magnetic poles and the geographic poles is surely significant, but the exact mechanism is still unknown. Perhaps a stream of charged particles (electrons, ions, or both) has been created by the high temperatures in the earth's core, and the motion of these charges is affected by the earth's rotation. The strength or location, or both, of the internal currents change gradually, for the angles of declination and dip change. For example, in 1580 the declination at London was measured by Gilbert to be 11°E; it was 0° in 1657, reached 25°W by 1820, and is decreasing again, with a value of 6°W in 1982. A second type of change is sudden and short-lived: "magnetic storms" are correlated with sunspot outbreaks and are a result of temporary currents of ions in the upper atmosphere. Finally, evidence points to radical changes in the magnetic axis of the earth in the geological past; during the last several million years (My) the magnetic axis reversed direction several times, in addition to undergoing irregular wandering. In the period from 2.5 My ago to 0.7 My ago, the "north" magnetic pole of the earth was in the Southern Hemisphere. This implies a change in the direction of the field-causing currents inside the earth.

The phenomena of terrestrial magnetism are complicated and variable, and not yet well understood. So far, however, physicists are confident that the earth's field originates in the motion of charges, as do all other magnetic fields.

22-11   Induced Emf and Magnetic Flux

In the 1820s Hans Christian Oersted in Denmark and André Marie Ampere in Paris found that a *magnetic* field exists in the neighborhood of a current segment $I \Delta l$ (or, equivalently, in the neighborhood of a moving charge $Q \mathbf{v}$). Scientists looked for the converse effect but found no *electric* field or emf in the neighborhood of a stationary magnet. In about 1832, Joseph Henry in the United States and Michael Faraday in England discovered that an emf is induced in a circuit only by a *changing* magnetic field. An important example is the emf induced in the secondary coil of a transformer, to be discussed in the next chapter.

To formulate a general law for *induced emf*, we first define *magnetic flux* Φ as a product:

$$\Phi = B \cos \phi \; A \qquad (22\text{-}9)$$

where $B \cos \phi$ is the component of $\mathbf{B}$ perpendicular to the plane of some complete circuit whose area is A (Fig. 22-27), and ϕ is the angle between $\mathbf{B}$ and the perpendicular to A.* In many cases $\mathbf{B}$ is broadside to the circuit, making $\phi = 0°$ and $\cos \phi = 1$, in which case

$$\Phi = BA \qquad (22\text{-}10)$$

The unit for flux is the *weber* (Wb). From Eq. 22-10 we obtain a new unit for magnetic induction:

$$B = \Phi/A$$
$$1 \text{ tesla} = 1 \text{ Wb/m}^2$$

Experiment shows that an emf ε appears (or is "induced") in any circuit through which the magnetic flux is changing. The induced emf equals the rate of change of magnetic flux:

$$\varepsilon = \lim_{\Delta t \to 0} \frac{\Delta \Phi}{\Delta t}$$

or

$$\varepsilon = \frac{d\Phi}{dt} \qquad (22\text{-}11)$$

$$1 \text{ V} = 1 \frac{\text{Wb}}{\text{s}}$$

---

* We assume that the circuit lies in a plane and that $\mathbf{B}$ is uniform in magnitude and direction at all points in the plane of the circuit. In a more general treatment, Φ is calculated as a limit of a sum of terms:

$$\Phi = B_1 \cos \phi_1 \Delta A_1 + B_2 \cos \phi_2 \Delta A_2 + \cdots$$

The limit of the sum is, of course, a definite integral, and hence $\Phi = \int B \cos \phi \, dA$. The vector notation for flux uses the dot product (Sec. 6-1): $\Phi = \int \mathbf{B} \cdot d\mathbf{A}$.

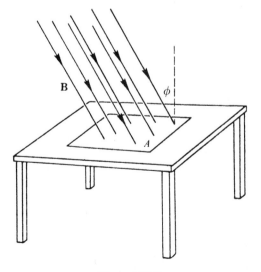

Figure 22-27

This equation is known as Faraday's law for induced emf.

The *direction* of an induced emf is given by

Lenz's law:*

An induced emf tends to set up a current whose action opposes the change that caused it.

A minus sign is sometimes used in writing Faraday's law ($\varepsilon = -d\Phi/dt$), indicating that the emf induced in a circuit equals the time rate of *decrease* of magnetic flux through the circuit. If a circuit or coil has N turns and the rate of change of flux is the same through each turn, then the magnitude of the total induced emf in the coil is found from

$$\varepsilon = N \frac{d\Phi}{dt} \qquad (22\text{-}12)$$

An induced emf arises whenever Φ changes. Since $\Phi = B \cos \phi\, A$, an emf can arise because of a change in B (transformers) or a change in $\cos \phi$ (generators). These devices will be discussed in Chap. 23. An emf associated with a change in area A is discussed in connection with Fig. 22-28.

Example 22-7

A square loop of wire 4 cm on an edge is lying on a horizontal table. An electromagnet above and to one side of the loop is turned on, causing a uniform magnetic field that is downward at an angle of 30° from the vertical, as in Fig. 22-27. The magnetic induction is 0.500 T, or 0.500 Wb/m². Calculate the average induced emf in the loop if the field increases from 0 to its final value in 200 ms.

The component of magnetic induction perpendicular to the plane of the loop is

$$B \cos \phi = (0.500\ \text{Wb/m}^2)(\cos 30°)$$
$$= 0.433\ \text{Wb/m}^2$$

The area of the loop is 1.6×10^{-3} m². Hence the flux is

$$\Phi = B \cos \phi\, A$$
$$= (0.433\ \text{Wb/m}^2)(1.6 \times 10^{-3}\ \text{m}^2)$$
$$= 6.93 \times 10^{-4}\ \text{Wb}$$

In finding the average induced emf we do *not* take a limit as $\Delta t \to 0$; hence we calculate $\Delta\Phi/\Delta t$, and we denote the average emf by a bar over the ε, as is usual for average values (Sec. 2-3). The magnitude of the average induced emf is

$$\bar{\varepsilon} = \frac{\Delta\Phi}{\Delta t} = \frac{6.93 \times 10^{-4}\ \text{Wb}}{200 \times 10^{-3}\ \text{s}}$$
$$= \boxed{3.46 \times 10^{-3}\ \text{V}}$$

The direction of ε is found from Lenz's law. Assume that ε is clockwise as viewed from above. By the right-hand rule, the induced magnetic field caused by this assumed ε would be downward; this would *aid* the downward field that is being established by the magnet. The induced emf must *oppose* the change that caused it; hence the assumed direction is wrong. We conclude that ε is directed counterclockwise around the loop during the time that the downward magnetic field is increasing.

Another, more general, way of looking at induced emf is to say that

Any changing magnetic field causes an electric field.

We see this illustrated in what we have just studied. A changing B means a changing flux Φ, which gives rise to an induced emf ε. This emf can serve as a source of electric energy when an electron moves around a circuit, so there is an electric force acting on the electron—that is, an electric field E (according to Eq. 19-1, $E = F_{\text{elec}}/Q$). The net result is that the changing B has given rise to E. Maxwell[†] extended this reasoning to empty space; neither a wire nor the presence of charges is needed for this relation between changing B-field and induced E-field.

---

* Developed by the German physicist Heinrich Lenz (1804–1865).

† James Clerk Maxwell (1831–1879), one of the great British physicists of the mid-19th century.

The *field* is there, whether or not a charge exists to be moved by the field.

Maxwell also showed that the converse is true:

Any changing electric field causes a magnetic field.

Thus there is a symmetry in electromagnetism, shown by this intimate connection between changing *B*-fields and changing *E*-fields. In Chap. 24 we shall discuss further how these ideas are related to the propagation of an electromagnetic wave, such as a radio or light wave.

22-12 *Motional Electromotive Force*

We have seen in Sec. 22-2 that, in general, a charged particle moving through a magnetic field experiences a magnetic force. If, then, a conductor moves in a suitable direction relative to a magnetic field, a magnetic force acts on the free electrons that are carried along with the conductor, and work is done.* We have here a mechanism for the transformation of mechanical energy into electric energy; in other words, the conductor becomes a seat of emf. We say that an *induced emf* is generated in a conductor that moves in a suitable direction through a magnetic field.

Until the 1830s, the only practical way of obtaining electric current was from cells and batteries of various types. It is instructive to contemplate the tremendous advance of technology that resulted from the discovery of induced emf by Henry and Faraday. The generators of the world's power stations transform mechanical energy into electric energy with high efficiency; more than 99% of the electric energy used per year in the United States is "generated" in this way, and less than 1% comes from energy transformations in other seats of emf, such as chemical cells.

---

*The work is done by whatever agency keeps the wire in motion. A magnetic field, as such, does no work on a moving charge since **F** is always perpendicular to **v**.

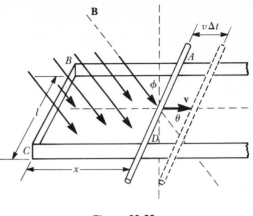

Figure 22-28

The original investigators found that an emf appears in a wire that moves relative to a magnetic field. The magnitude of the induced emf can be derived by a "thought experiment." Imagine a horizontal rod that rolls along rails with velocity **v** (Fig. 22-28). The flux Φ through the area of the circuit $BADC$ is increasing, because the area of this rectangular circuit is increasing. In a time Δt, the rod moves a distance $v \Delta t$ in the x direction, and the increased area is $\Delta A = l(v \Delta t)$. The increased flux is given by $B \Delta A \cos \phi$, or by $B \Delta A \sin \theta$ (note that $\sin \theta = \cos \phi$ since $\theta = 90° - \phi$). Thus

$$\Delta \Phi = Bl(v \Delta t) \sin \theta$$

The induced emf is given by $\Delta\Phi/\Delta t$; that is,

$$\mathcal{E} = \frac{Bl(v \Delta t) \sin \theta}{\Delta t}$$

or

$$\mathcal{E} = Blv \sin \theta \qquad (22\text{-}13)$$

where θ is the angle between the directions of **B** and **v**. If **v** is perpendicular to **B**, then $\sin \theta = 1$, and the maximum induced emf is obtained:

$$\mathcal{E} = Blv \qquad (22\text{-}14)$$

The induced emf equals the product of the only factors that might be expected to be significant: B is the magnitude of the magnetic induction, and l is the length of the wire that is moving with velocity v at an angle θ with the field. We interpret $v \sin\theta$ as the component of velocity perpendicular to **B**. To check units, note that

$$\varepsilon = Blv\sin\theta = (\text{N/A}\cdot\text{m})(\text{m})(\text{m/s})$$
$$= \text{N}\cdot\text{m/A}\cdot\text{s} = \text{J/C} = \text{V}$$

Example 22-8

The rod of Fig. 22-28 has resistance R, and the rails have negligible resistance. Calculate (a) the current through the rod; (b) the force on the rod; (c) the mechanical power supplied to the rod; (d) the electric power dissipated as Joule heating in the rod.

(a) $$I = \frac{\varepsilon}{R} = \frac{Blv}{R}$$

(b) $$F = IlB \qquad \text{(See Eq. 22-4.)}$$

(c) As shown in Sec. 6-8, the mechanical power is

$$P = Fv = \left(\frac{Blv}{R}\right)lBv = \frac{B^2l^2v^2}{R}$$

(d) Electric power $= I^2R = \left(\frac{Blv}{R}\right)^2 R = \frac{B^2l^2v^2}{R}$

The law of conservation of energy is illustrated by the agreement between the answers to parts c and d of Example 22-8. All of the work performed by the agent that pushes the rod turns up as Joule heating in the resistance of the rod.

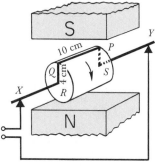

Figure 22-29

For maximum motional emf, a moving rod or wire should be at right angles to the field, and the velocity should be at right angles to both the field and the wire. Such a motion can be described as that of a conductor that is "cutting" lines of induction, as in Fig. 22-29.

Example 22-9

The rotating armature of a simple generator consists of a rectangular loop $SPQR$, to which connections are made by means of sliding contacts. The armature is turned at 1200 rev/min, and the magnetic induction is 0.5 N/A$\cdot$m. What is the magnitude of the induced emf between the terminals of the generator at the instant shown in Fig. 22-29?

The wire PQ is moving to the right, with a velocity given by the circumference of its circular path, multiplied by the number of revolutions per second:

$$v = 2\pi rf = 2\pi(0.04 \text{ m})\left(1200\,\frac{\text{rev}}{\text{min}}\right)\left(\frac{1 \text{ min}}{60 \text{ s}}\right)$$
$$= 5.03 \text{ m/s}$$

The magnetic field is directed vertically upward, from N pole to S pole. Since the wire is cutting perpendicularly across the lines of induction of the B-field, the induced emf is a maximum. Here $\sin\theta = 1$, and

$$\varepsilon = Blv = (0.5 \text{ N/A}\cdot\text{m})(0.10 \text{ m})(5.03 \text{ m/s})(1)$$
$$= \boxed{0.25 \text{ V}}$$

No emf's are induced in wires SP and RQ, because the magnetic force on free electrons in these wires (given by $\mathbf{F} = Q\mathbf{v} \times \mathbf{B}$) has no component parallel to the wires. Also, there is no emf in the lead-in wires, which are not in motion. Therefore the total emf between the terminals is just that due to the segment PQ, and $\varepsilon = 0.25$ V at this particular instant.

Let us apply Lenz's law to the generator of Example 22-9, in which the upper conductor moves to the right (shown in end view in Fig. 22-30). The best way to proceed is to *assume* a direction for ε and then apply Lenz's law to see if the assumption is correct. Let us assume that if there were a complete circuit, positive (conven-

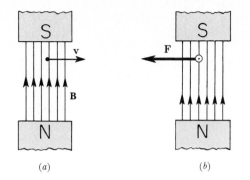

(a)   (b)

Figure 22-30 (a) Conductor moving to the right, cutting across a magnetic field. (b) Lenz's law: conductor moving to the right, giving rise to an induced current, which in turn causes a force to the left, opposing the motion.

tional) charge would flow toward us from P to Q and on through the circuit. Using the right-hand rule to get the force on this assumed current, we find that the force on the wire PQ, given by $I\,d\mathbf{l} \times \mathbf{v}$, would be toward the left (Fig. 22-30b). This force would indeed oppose the "change that caused it"; in this case the change is a motion to the right. Hence our assumed direction is correct. Charge would tend to flow from P to Q, out at X, through an external circuit, and back in at Y. The generator's positive terminal is at X, and its negative terminal is at Y at the instant shown.

Using a different approach, we can also find the direction of $\mathcal{E}$ by finding the direction of the force on a "conventional" + charge in the wire. For $\mathbf{v}$ to the right and $\mathbf{B}$ upward, $\mathbf{F} = Q\mathbf{v} \times \mathbf{B}$ gives a force *on the charge* that is toward us, showing that the induced current in the wire is toward us, as in Fig. 22-30b.

Lenz's law is a deduction from the law of conservation of energy. Thus in Fig. 22-30b, the force $\mathbf{F}$ is opposite to the velocity $\mathbf{v}$, and work must be done by an outside agent to move the wire to the right against the opposing force $\mathbf{F}$. Suppose the generator is connected by wires to a lamp. Mechanical energy is put into the system and transformed into Joule heat (I^2R) in the wires of the generator and in the lamp. However,

if the induced current were in the opposite direction ($\otimes$ instead of $\odot$), the wire would be doing mechanical work ($\mathbf{F}$ and $\mathbf{v}$ in the same direction), and *also* heat would be produced in the wires of the generator and in the lamp. This would clearly violate the law of conservation of energy, since there would be an expenditure of energy without any source of energy. We see that Lenz's law is a necessary consequence of the law of conservation of energy.*

Example 22-10

What emf is induced between the head and the toes of an athlete 1.7 m tall who is running at full speed of 10 m/s horizontally toward the east through the earth's magnetic field shown in Fig. 22-26? Is the head + or − relative to the toes?

The field is downward toward the north, 71° below horizontal. Only the component of $\mathbf{B}$ that is perpendicular both to the motion and to the vertical runner is effective in causing induced emf. This component is the horizontal component, given as 0.18×10^{-4} T in Fig. 22-26.

$$\mathcal{E} = Blv = (0.18 \times 10^{-4} \text{ T})(1.7 \text{ m})(10 \text{ m/s})$$

$$= \boxed{3.1 \times 10^{-4} \text{ V}}$$

Note that the factor $\sin \theta$ in Eq. 22-13 is taken care of in this case by using the horizontal component of $\mathbf{B}$, which is perpendicular to the runner's height. The direction of the induced emf is either upward or downward. Assume upward and see if that is correct. If so, conventional + charges move upward and, from $Q\mathbf{v} \times \mathbf{B}$, the force on them is toward the west (since $\mathbf{v}$ is upward and $\mathbf{B}$ is horizontal toward the north). This is exactly what is needed according to Lenz's law, for the effect of the induced motion of the charges must be such as to oppose the change (here an eastward running of the athlete) that caused the

---

* In chemistry, Le Chatelier's principle is similarly related to the law of conservation of energy: "If the equilibrium of a system is disturbed by a change in one or more of the determining factors (as temperature, pressure, or concentration) the system tends to adjust itself to a new equilibrium by counteracting as far as possible the effect of the change." (Webster's New International Dictionary, 3rd ed.)

emf. Thus we conclude that $\mathcal{E}$ is directed upward, and the head of the runner becomes + relative to her toes.

The emf calculated in this example is less than a millivolt and is of course not directly perceived by the runner. It is not entirely impossible that minute induced emf's of this sort may have something to do with "homing" or other subliminal responses of living organisms. Larger induced emf's on the ions in the bloodstream can be used to monitor the rate of flow of blood (see Prob. 22-B19).

Magnetic Units in the cgs System.

Some research workers in physics, chemistry, and biology still use cgs magnetic units that may have unfamiliar names. The cgs unit for magnetic induction B is the gauss (G), where 1 gauss $= 10^{-4}$ Wb/m$^2 = 10^{-4}$ tesla. The cgs unit for magnetic flux Φ is the maxwell, where one maxwell $= 10^{-8}$ Wb.

Summary
A magnetic field exists in any region of space where a moving electric charge would experience an electric force other than Coulomb force. Magnetic forces are caused by moving charges acting on other moving charges; we therefore speak of magnetism as a subdivision of electricity. A magnetic field can be explored by using a moving charge as a test body. The magnitude of the vector representing magnetic field is given by the force per unit Qv product. The force on a moving charge is always perpendicular to its motion and to the direction of the field. A moving charge experiences maximum force if its motion is perpendicular to the field. A current segment $I \, \Delta l$ can be used as a test body instead of a moving charge. If current segments are laid end to end, forming a loop or a coil, the coil tends to align its axis with the direction of magnetic field. The force on a current segment in a magnetic field equals the product of the normal component of magnetic induction and the strength of the current segment.

Permanent magnets such as a compass needle are essentially collections of current loops arising from the spins of the electrons in the electron clouds of the atoms. In iron and other strongly magnetic substances, the magnetic effect is strong because atoms are grouped in domains in which cooperation between neighboring atoms causes many orbits and spins to be in parallel alignment.

Magnetic fields are mapped by drawing representative lines of induction. In the space surrounding a wire through which charges flow, the magnetic field is represented by concentric circles. The right-hand rule gives the direction of the concentric lines of induction caused by a conventional current.

The earth's magnetic field is caused by circulating electric currents inside the earth, but the exact mechanism is not known. In the United States, the field is directed downward and to the north. Three quantities are used to describe the earth's field at any given point: the magnitude of the field, the declination, and the dip.

Magnetic flux is the product of the perpendicular component of magnetic induction and the area of a circuit. Induced emf equals the time rate of change of magnetic flux. The direction of the induced emf is given by Lenz's law, which states that an induced emf tends to oppose the change that caused it. When a wire moves across a magnetic field, an emf is induced in the wire.

magnetic induction
test bodies for
  measurement of
    strength of a *B*-field
current segment
tesla
$B = F/Qv$
$B = F/I \, \Delta l$
$\mathbf{F} = Q\mathbf{v} \times \mathbf{B}$
$\mathbf{F} = I \, \Delta l \times \mathbf{B}$

$$B = k' \frac{2I}{a}$$

$$B = k' \frac{2\pi I}{R}$$

right-hand rule
ampere
domain
magnetic declination
angle of dip

induced emf
Lenz's law
magnetic flux
$\Phi = B \cos \phi \, A$
weber

$$\varepsilon = N \frac{d\Phi}{dt}$$

$$\varepsilon = Blv \sin \theta$$

Questions

22-1 On what factors does the magnetic force on a moving charge depend?

22-2 How would you determine the direction of magnetic field at a given point, (*a*) using a compass, and (*b*) using a moving test charge?

22-3 Give two mks units for magnetic field strength, both equivalent to the tesla, and show that they are equivalent to each other.

22-4 What characteristic of iron causes it to concentrate magnetic lines of induction?

22-5 Is there any known way in which a magnetic field can be caused other than by the motion of electric charge?

22-6 When a bar magnet is cut in half along its length, what is the magnetic state of the two pieces? What is the result of cutting the same bar magnet in half by a perpendicular cut?

22-7 Lightning strikes a vertical metal flagpole, and a stream of electrons momentarily flows up the pole. What is the direction of the magnetic field at a point in the air just east of the center of the pole?

22-8 Suppose that a plastic phonograph record is rubbed along its edge with fur, charging the edge with an excess of electrons. When the charged record is turning in the usual direction, in which direction is the magnetic field at a point directly above the center of the record?

22-9 (*a*) In which direction (to the right or to the left) is the magnetic field at a point between the pole pieces of the magnet in Fig. 22-31? (*b*) In which direction (up or down) is

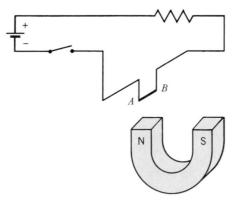

Figure 22-31

the magnetic field at a point above the N pole?   (c) In which direction (right or left) does the wire AB move when the switch is closed?

22-10 What is the direction of the force (up or down) on the current-carrying wire at P in Fig. 22-32?

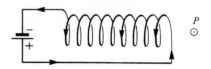

Figure 22-32

22-11 The general equation for the force on a charge Q is $\mathbf{F} = Q\mathbf{E} + Q\mathbf{v} \times \mathbf{B}$. Interpret each symbol in this equation.

22-12 Which is more important for a navigator to know—the dip or the declination?

22-13 Describe, in three dimensions, the general direction of the earth's magnetic field in Australia.

22-14 Without using iron or any other magnetic material, how could you construct a device that would serve as a compass?

22-15 Explain how magnetic storms are caused by eruption of charged particles from the sun.

22-16 On what factors does the induced emf in a conductor depend?

22-17 An iron girder in a railway bridge is in a vertical north–south plane, at an angle of 70° from the horizontal, with the bottom end of the girder north of its top end. During the years, the girder becomes magnetized as trains pound over the bridge. Explain.

22-18 What is the difference between magnetic induction and magnetic flux?

22-19 Suppose a horizontal metal rod, oriented perpendicularly to the B-field, is freely falling through the gap between the poles of the permanent magnet in Fig. 22-31. Does the induced emf cause free electrons in the rod to move toward or away from the reader? Explain, using Lenz's law.

22-20 An automobile is traveling south along a level highway in the United States, and the front axle is therefore cutting the earth's magnetic field.   (a) On which of the front hub caps do electrons tend to accumulate because of the induced emf?   (b) Would the effect be greater or less if the car were coasting southward down a hill 20° below the horizontal?

22-21 The magnetic fields needed for the operation of some small motors are supplied by permanent magnets. Will such a motor eventually lose its effectiveness as the magnetic energy of the permanent magnets is used up?

22-22 A significant factor in the design of high-current electromagnets is the mechanical force exerted on the wires in the inner layers of the winding, which are immersed in the strong magnetic field caused by the outer layers of the winding. Does this force tend to expand or contract the inner windings?

22-23 A small mass is supported in equilibrium on a vertical spring. When a battery is connected between the top and bottom ends of the spring, does the mass rise or fall?

MULTIPLE CHOICE

22-24 An electron moves horizontally to the west through a magnetic field that is downward. The force on the electron is toward the   (a) north;   (b) south;   (c) east.

22-25 There is an upward current in a vertical wire that is in a magnetic field directed horizontally to the south. The force on the wire is   (*a*) toward the east;   (*b*) toward the west;   (*c*) zero.

22-26 In an iron sample of mass 56 g, each magnetic domain contains   (*a*) about 10^4 atoms;   (*b*) about 10^{14} atoms;   (*c*) 6×10^{23} atoms.

22-27 The horizontal component of the earth's magnetic field in North America is   (*a*) toward the magnetic north;   (*b*) smaller in magnitude than the vertical component;   (*c*) both of these.

22-28 Which of the following is (are) a vector quantity?   (*a*) magnetic induction;   (*b*) magnetic flux;   (*c*) both of these.

22-29 The effect of an induced emf in a coil is always to oppose   (*a*) the current in the coil;   (*b*) the change of flux through the coil;   (*c*) the buildup of current in the coil.

Problems   **22-A1** A charge of 1.6×10^{-10} C is moving at 1 km/s through a magnetic field in such a direction that the magnetic force on it is a maximum. What is the magnetic induction if the maximum force on the charge is 4×10^{-8} N?

22-A2 What is the magnitude of the force on an electron that moves at 2% of the speed of light horizontally through a vertical magnetic field of 0.5 T?

22-A3 A rifle bullet having a net charge of 3×10^{-11} C moves at 200 m/s perpendicular to the earth's magnetic field. The strength of the earth's field is 6×10^{-5} T.   (*a*) What is the magnitude of the magnetic force on the bullet?   (*b*) Is this force observable?

22-A4 With the help of a diagram similar to Fig. 22-15, show that unlike currents repel each other.

22-A5 Calculate the magnetic force on a current segment 8 cm long, placed broadside to a magnetic field of strength 1.2 T. The current in the wire is 2500 A. Could a strong man supply enough force to hold the wire in equilibrium?

22-A6 What is the magnetic induction at a point 10 cm away from a long wire in which the current is 15 A?

22-A7 A vertical wire 30 cm long is in a horizontal magnetic field of 2.5 T. What current through the wire will cause a force of 0.8 N on the 30-cm segment?

22-A8 The earth's magnetic field at a certain place is 0.55×10^{-4} T, and the angle of dip is $70°$. Calculate the value of the horizontal component of the induction.

22-A9 Calculate the current in a circular loop 40 cm in radius that would cause a magnetic induction at the center of the loop equal to that of the earth's magnetic field, which is about 6×10^{-5} T.

22-A10 What emf is induced in a horizontal rod 4 cm long that is falling at 6 m/s in a region where there is a horizontal magnetic induction of 0.2 T perpendicular to the rod?

22-A11 A horizontal wire 6 cm long moves with a speed of 5 m/s across a magnetic field of induction 0.6 T. What emf is induced in the wire?

22-A12 What is the magnetic flux through a rectangular loop 40 cm tall and 50 cm wide if the coil is placed with its plane horizontal in the earth's magnetic field shown in Fig. 22-26?

22-A13 Calculate the emf induced in a coil of wire through which the magnetic flux changes from $+3$ Wb to -9 Wb in 0.400 s.

22-A14 Prove: 1 T = 1 V·s/m^2; 1 T = 1 Ω·C/m^2; 1 Wb = 1 Ω·C; 1 Wb = 1 J·s/C.

22-B1 (a) What is the magnetic force on a proton moving horizontally at (almost) the speed of light through a synchrotron (Fig. 7-11 on page 158) where there is a vertical magnetic induction of 2.5 T? (b) What is the ratio of the magnetic force calculated in part (a) to the weight of the proton when it is at rest?

22-B2 Calculate the strength of the magnetic field in the cyclotron of Example 7-8 (page 157).

22-B3 A deuterium nucleus $_1^2$H (a deuteron; see Sec. 1-7) has a speed of 2×10^7 m/s perpendicular to a magnetic field of induction 0.2 T. (a) What is the force on the deuteron? (b) What is the acceleration of the deuteron? (c) Calculate the radius of curvature of the orbit of the deuteron. (*Hint:* The magnetic force causes a centripetal acceleration.)

22-B4 An electron in a small cathode-ray tube in an oscilloscope is moving horizontally toward the viewer at 5×10^7 m/s and is deflected upward by a horizontal magnetic field of 6×10^{-4} T. (a) What is the direction of the B-field? (b) What is the magnetic force on the electron? (c) What is the radius of the circular path of the electron while in the magnetic field?

22-B5 An electron moves horizontally toward the east through a region where two fields exist: an upward electric field $E = 5000$ V/m, and a horizontal magnetic field, perpendicular to the electron's motion, of magnitude $B = 0.2$ T. (a) What is the speed of an electron that passes through this "velocity selector" with no resultant deflection? (b) In which direction is **B**? (c) Would a proton having the same speed as found for the electron in part (a) also pass through this same velocity selector?

22-B6 A current segment 1.00 cm long, located 2.00 m above the ground, carries a current of 20 A horizontally toward the west. What are the magnitude and direction of the B-field at a point that is 1.50 m west of the point on the ground directly below the current segment?

22-B7 A current in a circular path through a horizontal coil is counterclockwise as viewed from above. The coil, which has 50 turns, has a radius of 9 cm, and the current is 200 mA. (a) What is the direction of the magnetic field at the center of the loop? (b) What is the magnitude of the magnetic induction at the center? (c) Is this field stronger or weaker than the earth's magnetic field?

22-B8 Calculate the magnetic induction at the nucleus of an atom caused by the revolution of an electron in a circle of radius 0.5×10^{-10} m if the moving orbital electron is equivalent to a current of 10^{-3} A. (*Note:* These values are approximately correct for the motion of the electron in a hydrogen atom in its normal state.)

22-B9 There is an upward conventional current of 5 A in a long vertical wire; 24 cm east of the wire, 8×10^{20} electrons per second flow downward in a vertical current segment 5 cm long. What are the magnitude and direction of the force on the segment?

22-B10 (a) At a place where Fig. 22-26 is applicable, what should be the direction of a horizontal current-carrying wire if it is to experience a maximum force due to the earth's magnetic field (assume the declination to be 0°)? (b) What would this force be on a properly oriented current segment 2 cm long that carries a current of 50 A?

22-B11 Two long wires, 2 cm apart, carry currents of 200 A in opposite directions. (a) Is the magnetic force between them one of attraction or repulsion? Explain. (b) What is the magnetic force on a 2-cm length of one wire due to the field of the (very long) other wire?

22-B12 A long horizontal wire carrying 120 A toward the south is placed in a uniform magnetic field of 3×10^{-4} T that is directed horizontally toward the west. Describe the locus of points in space at which the resultant magnetic induction is zero.

22-B13 A copper wire 0.1 mm in radius is stretched between two points 4 m apart. What PD must be applied to the ends of the wire in order to obtain a magnetic induction of 0.2 milliteslas at a point near the center of the wire, 6 mm away from the wire?

22-B14 A horizontal rod 80 cm long is released from rest in a region where there is a uniform horizontal magnetic induction of 0.200 T perpendicular to the rod. How far has the rod fallen when the induced emf is 300 mV?

22-B15 During a space walk, two astronauts are separated by a connecting wire 12 m long that is moving horizontally, broadside to its length, at an orbital speed of 7.7 km/s over the north magnetic pole of the earth. The earth's magnetic induction at this point is 5×10^{-5} T. Calculate the induced emf, which is also the PD between the two astronauts.

22-B16 A flat coil has 200 turns of fine wire, each of radius 10 cm. The coil is first oriented broadside to a uniform magnetic field, and then, in 0.050 s, it is turned through 180° so that the plane of the coil is again broadside to the field. The average induced emf is found to be 0.250 V. Calculate the magnetic induction at the coil.

22-B17 A flat, circular wire loop of radius 4 cm and resistance 5 Ω is in a region of space where the magnetic field strength is 0.06 Wb/m², perpendicular to the plane of the loop. Calculate the current in the loop if the field decreases at a steady rate, becoming 0 after 200 μs.

22-B18 What emf is induced in a flat coil of 20 turns, each of radius 10 cm, that is broadside to a uniform magnetic field for which B is changing at the rate of 140 T/s?

22-B19 A heart surgeon monitors the rate of flow of blood through an artery using an electromagnetic flowmeter (Fig. 22-33) in which electrodes A and B are attached to the outer surface of a blood vessel of inside diameter 3 mm. Blood and vessel both have good electrical conductivity. (*a*) For a magnetic induction 0.04 T, an emf of 160 μV is developed. Calculate the velocity of the blood. (*b*) Verify that electrode A is +, as shown. Does the sign of the emf depend on the sign of the ions in the blood?

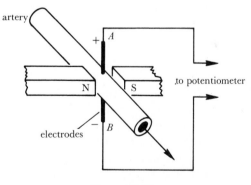

Figure 22-33

22-B20 Figure 22-33 can also represent an electromagnetic pump used to keep blood moving through an artery. It has the advantage of no moving parts and does not damage blood cells passing through. In place of the potentiometer, a source of PD is connected to electrodes A and B, giving rise to a vertical current through the blood. To give a force on the blood in the direction shown by the arrow, should A be made + or − relative to B?

22-C1 A coil of diameter 3 cm having 80 turns and resistance 6.0 mΩ is flat on a table, and a magnetic induction perpendicular to its plane changes at a steady rate from 0 to 0.8 T. How much charge flowed through the coil during the time required for the field to change?

22-C2 An aluminum wire 80 cm long has a square cross section 1 mm × 1 mm. The ends of the wire are welded together and the wire formed into a square loop. What is the current in the loop when a B-field perpendicular to the plane of the loop decreases steadily from 0.6 T to 0 in 0.002 s?

22-C3 To measure the momentary current surge in a guy wire at a tower during a lightning stroke, a strip of prerecorded magnetic tape was supported so that it stood out perpendicularly to the guy wire. After the stroke, it was found that the tape had been erased to a distance 12 cm from the guy wire. Previous calibration had shown that a magnetic induction of 0.03 T would erase the tape. What was the maximum current in the guy wire during the stroke?

22-C4 Suppose the apparatus of Fig. 22-28 is tilted so that the rails make an angle α with the horizontal, and suppose **B** is vertical ($\phi = 0$). Show that the frictionless rod (of resistance R) slides down the rails at a constant (terminal) speed given by $v = mgR$ $\sin \alpha / B^2 l^2 \cos^2 \alpha$. (*Hint*: Equate the magnetic force to the component of weight parallel to the rails.) What if $\alpha = 0$? What if **B** is reversed in direction, still vertical?

22-C5 A rod of resistance 0.002 Ω is pushed at a steady speed of 0.5 m/s along resistanceless rails spaced 0.3 m apart (*AD* in Fig. 22-28). The magnetic induction is 0.6 T downward at an angle of 30° from the vertical. (*a*) Calculate the emf generated in the rod and tell which end is at higher potential. (*b*) Calculate the current through the rod. (*c*) How much horizontal force is required to keep the rod moving? In which direction is this force? (*d*) How much mechanical power is expended by the muscles that move the rod? (*e*) At what rate is Joule heat being produced in the resistance of the rod? (*f*) Explain why the answers to parts (*d*) and (*e*) are the same.

For Further Study

22-13 Magnetic Field at the Center of a Current Loop

Using Eq. 22-6, we calculate the magnitude of the field at the center of a circular current loop of radius R (Fig. 22-34). All the little segments $I \Delta l$ are at a distance R from the center, and each segment $I \Delta l$ is oriented for maximum effect, broadside to the radius. The total field is

$$k' \frac{I \Delta l_1}{R^2} + k' \frac{I \Delta l_2}{R^2} + k' \frac{I \Delta l_3}{R^2} + \cdots$$

$$= k' \frac{I}{R^2} (\Delta l_1 + \Delta l_2 + \Delta l_3 + \cdots)$$

For a complete circular loop, the total of all the Δl's is the circumference, or $2\pi R$; hence

$$B = \frac{k'I}{R^2} (2\pi R)$$

$$= k' \frac{2\pi I}{R}$$

This is Eq. 22-8.

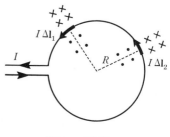

Figure 22-34

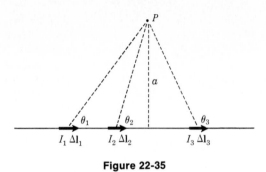

Figure 22-35

22-14 Magnetic Field Due to a Long Straight Wire

To calculate the magnitude of the field at a point near an infinitely long wire, we must find the resultant of infinitely many fields due to the infinitely many tiny current segments $I \Delta \mathbf{l}$, of which three are shown in Fig. 22-35. The segments differ in distance from P, and differ also in orientation relative to P. The magnetic induction $d\mathbf{B}$ at point P (Fig. 22-36) due to a small current element $I \, dx$ is out of the plane of the paper, perpendicular to both the current segment $I \, d\mathbf{x}$ and the radius vector $\mathbf{r}$. The field is caused by the component of $I \, d\mathbf{x}$ perpendicular to $\mathbf{r}$, which is given by $I \, dx \sin \phi$. Thus, from Eq. 22-6,

$$dB = k' \frac{I \, dx \sin \phi}{r^2} \qquad (22\text{-}15)$$

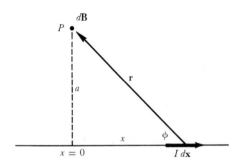

Figure 22-36

The total field at P due to half the wire is given by an integral between suitable limits:

$$B = \int_{x=0}^{x=\infty} dB \quad \text{or} \quad \int_{\phi=\pi/2}^{\phi=0} dB \quad \text{or} \quad \int_{r=a}^{r=\infty} dB$$

To carry out the integration, dB must be expressed in terms of one variable—x, ϕ, or r. If you like trigonometry, it is perhaps simplest to express everything in terms of ϕ. From the diagram, $\tan \phi = a/x$, or $x = a(\tan \phi)^{-1}$. Using the chain rule (Appendix E), we find

$$\frac{dx}{d\phi} = a(-1)(\tan \phi)^{-2}(\sec^2 \phi)$$

$$= -a \frac{\cos^2 \phi}{\sin^2 \phi} \frac{1}{\cos^2 \phi} = -\frac{a}{\sin^2 \phi}$$

Thus,

$$dx = -\frac{a \, d\phi}{\sin^2 \phi}$$

Also, $\sin \phi = a/r$, whence $1/r^2 = (\sin^2 \phi)/a^2$. Finally, we put the pieces together to get dB in terms of ϕ and $d\phi$:

$$dB = k'I \left(-\frac{a \, d\phi}{\sin^2 \phi} \right) (\sin \phi) \left(\frac{\sin^2 \phi}{a^2} \right)$$

The limits on ϕ for the positive half of the wire are $\phi = \pi/2$ (at $x = 0$) and $\phi = 0$ (at $x = \infty$). Thus the total B, due to both halves of the wire, is

$$B = 2 \int_{\phi=\pi/2}^{\phi=0} \left(-\frac{k'I}{a} \right) \sin \phi \, d\phi$$

$$= -\frac{2k'I}{a} \left[(-\cos 0) - \left(-\cos \frac{\pi}{2} \right) \right]$$

$$= -\frac{2k'I}{a} \left[(-1) - (0) \right]$$

or

$$B = k' \frac{2I}{a}$$

This is Eq. 22-7.

Problems

22-C6 A flat square loop of wire of length a on each side carries a current I. (a) Show that magnetic induction B at the center of the square is $11.31k'I/a$. (*Hint:* First calculate the induction caused by half of one side. The limits of integration are from $\phi = \pi/2$ to $\phi = \pi/4$.) (b) Check the answer to part (a) by showing that it lies between the value $12.56k'I/a$ for an inscribed circular loop (tangent to the square) and the value $8.88k'I/a$ for a circumscribed circular loop (touching the four corners of the square).

22-C7 A flat square coil 40 cm on an edge has 250 turns and is in a magnetic field perpendicular to its plane that changes with time t according to the following equation:

$$B = 0.5 \text{ T} + (0.01 \text{ T} \cdot \text{s}^{-1})t + (0.003 \text{ T} \cdot \text{s}^{-2})t^2 - (0.0001 \text{ T} \cdot \text{s}^{-3})t^3$$

(a) What is the magnitude of the emf in the coil at $t = 0$? (b) At what time does the magnitude of the emf have its maximum value? (c) What is the value of the emf at this time?

22-C8 Obtain Eq. 22-7 for the field due to a long straight wire by integrating Eq. 22-15 with respect to x. (*Hint:* Express $\sin \theta$ and r in terms of a and x. The resulting integral can be evaluated between $x = 0$ and $x = \infty$ with the aid of a table of integrals more extensive than that in this book.)

References

Note: In some references, the cgs unit for magnetic induction is used: 10^4 gauss = 1 Wb/m^2 = 1 T. For a vacuum, 1 oersted = 1 gauss.

1. W. F. Magie, *A Source Book in Physics* (McGraw-Hill, New York, 1935), pp. 437–444, Faraday; 473–489, Henry and Faraday on magnetic induction; 511–513, Lenz's law; 513–519, Henry's investigations.
2. M. Wilson, "Joseph Henry," *Sci. American* **191**(1), 72 (July 1954).
3. B. M. Swartz, "Joseph Henry—America's premier physics teacher," *Phys. Teach.* **16**, 348 (1978).
4. J. R. Nielsen, "Hans Christian Oersted—Scientist, Humanist and Teacher," *Am. J. Phys.* **7**, 10 (1939).
5. S. K. Runcorn, "The Earth's Magnetism," *Sci. American* **195**(3), 152 (Sept. 1956).
6. W.M. Elsasser, "The Earth as a Dynamo," *Sci. American* **198**(5), 44 (May 1958).
7. A. C. Cox, G. B. Dalrymple, and R. R. Doell, "Reversals of the Earth's Magnetic Field," *Sci. American* **216**(2), 44 (Feb. 1967). Evidence for nine reversals in 3.6 million years.
8. J. P. Dunn, M. Fuller, H. Ito, and V. A. Schmidt, "Paleomagnetic Study of a Reversal of the Earth's Magnetic Field," *Science* **172**, 840 (1971).
9. H. P. Furth, M.A. Levine, and R. W. Waniek, "Strong Magnetic Fields," *Sci. American* **198**(2), 54 (Feb. 1958).
10. D. Cohen, "Magnetic Fields of the Human Body," *Physics Today* **28**(9), 35 (1975).
11. B. M. Schwarzschild, "Weak magnetic fields in the human body," *Phys. Today* **32**(8), 18 (Aug. 1979).
12. J. E. Kunzler and M. Tannenbaum, "Superconducting Magnets," *Sci. American* **206**(6), 60 (June 1962).
13. G. Shiers, "The Induction Coil," *Sci. American* **224**(5), 80 (May 1971).
14. L. Kuhn and R. A. Myers, "Ink-Jet Printing," *Sci. American* **240**(4), 162 (Apr. 1979). Principles of electrostatics, electromagnetism, and surface tension are involved in high-speed printers under development.
15. F. Miller, Jr., *Paramagnetism of Liquid Oxygen; Ferromagnetic Domain Wall Motion* (films).

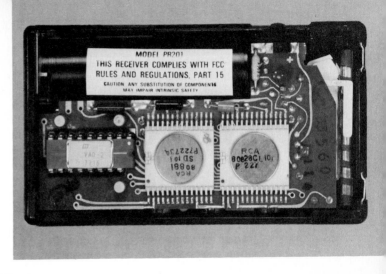

23

Applied Electricity

In this chapter we touch briefly on a representative few of the electrical machines and devices that play such a large part in the technical aspects of present-day culture. Our aim is to show these devices as logical applications of the basic principles we have already studied. First, we see that meters, motors, and generators have much in common.

23-1 Meters

At the heart of applied electricity is the direct-current meter. In Sec. 21-6 we saw how a galvanometer can be modified by a series resistor to make a voltmeter, or by a parallel shunt to make an ammeter. What factors determine the sensitivity of a galvanometer? In other words, what factors determine the deflection caused by a given current in the coil of the galvanometer? This is a problem in design. Let us study the galvanometer in detail, as an illustration of how physical principles are used in the design of apparatus.

The moving-coil galvanometer consists of a flat coil supported in the magnetic field of a permanent iron-alloy magnet (Fig. 23-1). As electrons move through the coil, they experience a magnetic force, which causes the coil to turn against the resisting torque of the springs S and S'. The "forward" torque is due to the magnetic

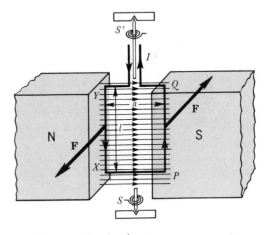

Figure 23-1 A galvanometer movement.

Paging receiver, shown full-size, contains 1400 transistors in two integrated circuits. Each circuit is in a square package with 28 connecting lugs (see Fig. 23-34 for an enlarged view of the chip). The rectangular integrated circuit with 17 lugs serves as an amplifier. The receiver emits a tone to summon the wearer to the telephone.(Receiver manufactured by Martin-Marietta.)

force on the vertical segments of the coil, PQ and XY. We calculate this torque τ, which equals force × lever arm, as follows: Using Eq. 22-4′, we find the force to be $F = I \Delta l B$; the total Δl for each segment is l, and the total force on each vertical wire is IBl. The lever arm is $a/2$; hence,

$$\tau = IBl\left(\frac{a}{2}\right)$$

The total torque acting on both wires is therefore $2(IBl)(a/2)$. Finally, if the coil consists of N turns of wire, the total torque τ is $NIBla$, or $NIBA$, since la is the area A of the coil.* This torque causes the coil to start to turn; as it does so, the springs wind up and give rise to a restoring torque. By Hooke's law, this restoring torque τ' is proportional to the angle of twist—say, $\tau' = k'\theta$, where k' is the "torsion constant" of the spring system. The coil comes to rest at such an angle that $\tau' = \tau$, for then the net torque $\tau' - \tau$ is zero, and there is no angular acceleration of the coil. Thus,

$$\tau' - \tau = 0$$
$$k'\theta - NIBA = 0$$
$$\theta = \left(\frac{NBA}{k'}\right)I \qquad (23\text{-}1)$$

If the design problem is to get the most deflection θ for a given current I, our "working equation" shows that we should strive for a strong magnetic induction B and a coil of large area A having many turns N. It also shows that we should use a spring system that is easily twisted—in other words, one that has a small k'.

Closer study reveals a defect in the design of our galvanometer movement. In Fig. 23-2a, where the galvanometer movement has turned through an angle θ, the force $\mathbf{F}$ is no longer fully effective, since the lever arm is less than $a/2$. In fact, if θ reached 90°, the magnetic force would produce no torque at all (Fig. 23-2b). To overcome this defect, the pole pieces are curved, and

---
* Although our derivation is for a rectangular coil, the result is valid if A is the area of a flat coil of any shape.

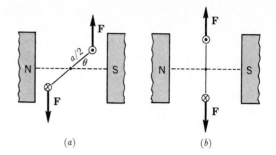

Figure 23-2 Top view of galvanometer coil. (a) Torque is reduced as coil starts to move. (b) Torque is zero when coil is broadside to the magnetic field.

a fixed cylindrical iron core is placed inside the coil to concentrate the lines of induction along radii (Fig. 23-3). The magnetic force on the current-carrying segments is the same as before, but now the lever arm is the full $a/2$, and the torque caused by the current does indeed depend only on N, B, A, and I, and not on the coil's position.

We have discussed one electrical device at some length, as an illustration of a typical engineering procedure. Its successful design follows from a full understanding of the underlying physical principles, as expressed in a working equation such as Eq. 23-1. The designer uses the equation to predict the effect of changes in the design or construction. As often happens, compromises must be made. For instance, for high sensitivity we need a large coil (large A), but then the large moment of inertia of the coil will make

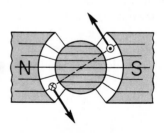

Figure 23-3 Shaping the pole pieces of a galvanometer and inserting a cylindrical iron core restores the lever arm to its full value.

the period of oscillation long,* and perhaps the meter will be too sluggish for convenient use. Again, high sensitivity can be obtained by using a very delicate spring system (small k'), but this would also lengthen the period, and in addition it would make the meter fragile and easily damaged. Looking at other factors in our working equation, we see that a strong magnetic induction (large B) is desirable. Modern alloys such as Alnico V (51% Fe, 24% Co, 14% Ni, 8% Al, 3% Cu) provide strong fields with small pole pieces; hence modern meters are considerably smaller than those of 90 years ago.

Permanent magnets are not "permanent," strictly speaking, for the domains always have a tendency to become disoriented. As a meter ages, the magnitude of the magnetic induction B tends to weaken, and so the sensitivity drops off. For this reason, in precise work a meter is calibrated at regular intervals. Finally, we must compromise on the number of turns of wire (N): for utmost sensitivity a coil should have many turns, but this would be physically awkward unless fine wire is used. In that event, the many turns of fine wire would have a relatively high resistance (Eq. 20-6), which is undesirable for a meter that is to measure small potential differences, such as the emf of a thermocouple.

---

* The period is given by $T = 2\pi\sqrt{I/k'}$, where k' is the torsion constant τ'/θ, and I is the moment of inertia (see Sec. 9-5).

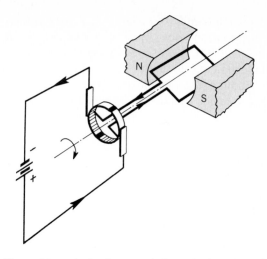

Figure 23-4 A simple motor (schematic; iron core of armature not shown).

Our analysis of the galvanometer movement is similar to the analysis that could be made for every instrument, machine, or device. This general procedure is typical of applied science, or engineering.

23-2 Motors and Generators

A direct-current (dc) motor is essentially a galvanometer movement that is allowed to turn continuously. In cross section (Figs. 23-4 and 23-5),

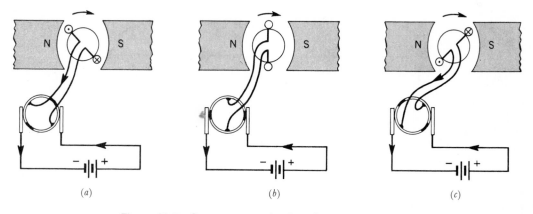

(a)   (b)   (c)

Figure 23-5 Commutator action in a direct-current motor.

we see the similarity: a motor has a coil (called the *armature*) wound on an iron core (not shown in Fig. 23-4), which concentrates the magnetic induction, just as in the galvanometer. To make continuous rotation possible, the armature leads are brought out to a pair of sliding contacts called a *commutator*. After the armature coasts through the "dead-center" position of Fig. 23-5*b*, the direction of current in any given wire is reversed as the brushes make contact with the other segment of the commutator. Note the reversal of current direction in the armature. Thus, whichever conductor is near the north pole is at that time carrying current in the proper direction to give rise to an upward force. In actual practice, an electromagnet is usually used to supply the magnetic field. In order to obtain a smooth-running motor, the armature may have a number of partial windings brought out to a many-segmented commutator.

The construction of a generator is basically the same as that of a motor. The armature is turned by some external torque (Fig. 23-6), and as the wires cut across the magnetic lines of force, an emf is induced according to $\varepsilon = Blv$ (Eq. 22-9).

For a given generator, B and l are constant, and the output emf is proportional to the speed with which the armature turns. Let us fix our attention on one conductor of the armature. In Fig. 23-7*a* it is moving upward across the lines of induction, and there is a maximum induced emf. When the armature reaches position *b*, the conductor is moving parallel to the field, and there is no induced emf. At *c*, this same conductor moves down across the lines of induction, and the induced current is directed opposite to that in *a*. If we plot a graph of the current in the lamp as a function of time, we get a sinusoidally varying current known as an *alternating current* (Fig. 23-7*d*). We see that it is not difficult to produce alternating currents. The frequency of an alternating current equals the number of times per second that the current changes from + to − and back again. Ordinary house current in the United States fluctuates at a frequency of 60 Hz. The study of alternating currents is highly technical and of great importance in engineering.

It is sometimes necessary to generate a steady, or *direct*, *current*—for instance, for the purpose of charging a storage battery in a car. The reversing action of a commutator can be used to make pulsating direct current out of alternating

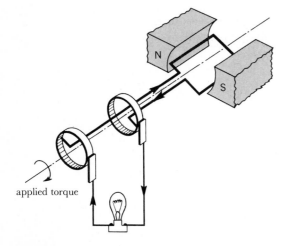

Figure 23-6 A simple generator. The armature is turned by a torque supplied by some source of mechanical power.

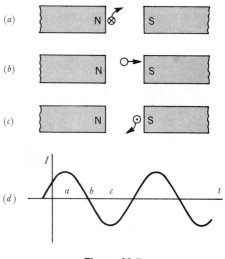

Figure 23-7

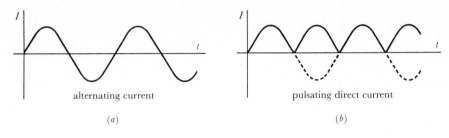

alternating current

pulsating direct current

(a)   (b)

Figure 23-8

current (Fig. 23-8). The commutator is a quick-acting switch that reverses the connections to the armature at just the right times to compensate for the reversals in current that would otherwise take place. In most cities, however, the direct-current motor and generator have disappeared from the scene, as far as public distribution of power is concerned. We shall see a reason for this when we study transformers in Sec. 23-5. Modern alternating-current (ac) motors do not need a commutator, and they are more compact and rugged than the direct-current motor we have described. Here, too, the full story is technical, and we will not pursue the matter further.

23-3 Back Emf and Back Torque

Since the physical construction of a motor is practically the same as that of a generator, it follows that if a dc generator is connected to a battery, it will run as a motor; and if a motor is turned by any external means, it will generate an emf.* This dualism is with us willy-nilly: whenever a motor is running, its generator action cannot be turned off. By Lenz's law (Sec. 22-11), the induced emf tends to oppose the change that causes it. In this case, the "cause" is the current through the armature, and so the induced emf tends to reduce this armature current. This in-

duced emf, which is the unavoidable result of generator action in a motor, is called *back emf*. When the motor is just starting up, its armature is not turning, and hence it is not generating a back emf. As the motor picks up speed, the back emf increases, and therefore the armature current decreases. This accounts for the fact that the current through a motor is larger while it is starting than while it is running at full speed.

Example 23-1

A motor has an armature of resistance 5 Ω. The magnetic field is supplied by field coils, which form an electromagnet of resistance 55 Ω connected in parallel with the armature. The motor and field coils are both connected to a 110-V dc line, and at full speed the back emf is 100 V. (a) What is the current through the motor when it first starts up? (b) What is the current when the motor is running at full speed?

(a) When starting, the circuit is a simple parallel circuit (Fig. 23-9a). The PD across AB is 110 V, and the PD across CD is also 110 V. We use Ohm's law to find the current in each branch.

Field coils: $I = \dfrac{V}{R} = \dfrac{110\,\text{V}}{55\,\Omega} = 2\,\text{A}$

Armature: $I = \dfrac{V}{R} = \dfrac{110\,\text{V}}{5\,\Omega} = 22\,\text{A}$

Total current when starting is

$$2\,\text{A} + 22\,\text{A} = \boxed{24\,\text{A}}$$

(b) As the motor picks up speed, the back emf $\mathcal{E}_b$ grows. We represent this emf by the symbol for a

---

* In some early automobiles, as in the 1924 Dodge, after the car was started, the starter motor was used as a generator (turned by the car's engine) to keep the storage battery charged.

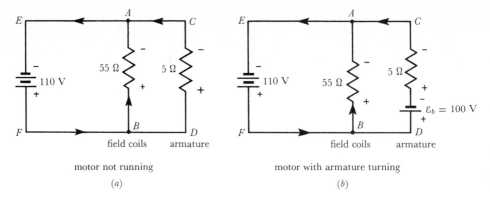

motor not running

(a)

motor with armature turning

(b)

Figure 23-9

battery; it is in every way equivalent to a battery connected in series with the armature (Fig. 23-9b).

Field coils: $\quad I = \dfrac{V}{R} = \dfrac{110 \text{ V}}{55 \text{ } \Omega} = 2 \text{ A}$

which is the same current as before.

Armature: We apply Kirchhoff's second law around the loop *CDBFEAC*, letting *I* represent the unknown current through the armature. The PD across the 5 Ω resistor is 5*I*, directed as shown. Hence, starting at *C* and going around the loop, we have

$$+5I + 100 \text{ V} - 110 \text{ V} = 0$$
$$5I = 10 \text{ V}$$
$$I = 2 \text{ A (instead of the}$$
$$\text{previous 22 A)}$$

The total current through the motor, while it is running, is

$$2 \text{ A} + 2 \text{ A} = \boxed{4 \text{ A}}$$

The large starting current of a motor is familiar in everyday life. A car's headlights may dim when the starter is operated; the house lights may momentarily dim when a refrigerator or furnace motor starts up. As we saw in Sec. 21-2, the terminal voltage of any source of current decreases when the current increases, due to the *IR* drop in the internal resistance of the source. When a refrigerator motor starts up, the

"source" consists of everything behind the floor socket, including the house wiring. Usually we can ignore the resistance of connecting wires, but in this case, an appreciable *IR* drop occurs in the house wiring because of the large value of *I* while the motor is starting. A lamp plugged into the same floor socket as the refrigerator becomes dim, since its terminal voltage always equals that of the refrigerator which it is connected in parallel.

Not only does a motor display generator action, but a generator behaves to a certain extent like a motor. If a generator is causing charges to flow through an external circuit, the charges must, of course, also flow through the armature. But these charges flowing in an armature that is in a magnetic field must experience a torque, and according to Lenz's law, this torque must oppose the applied torque by which the generator is being turned. The *back torque* exists only when there is a complete circuit for the flow of charges; a big generator turning at full speed could be wrecked by a sudden attempt to obtain a large current from it. We can consider the generator as a link in a chain by which mechanical energy is transformed into electric energy, and eventually into heat in a load such as a lamp bulb. The back torque is a necessary part of this chain, for it is only by pushing against *something* (the back torque) that the source of mechanical energy is able to do work.

In some electric cars and trucks, *regenerative braking* is used at several levels. When the accelerator pedal is released, the motor is reconnected as a generator supplying current that recharges the battery. The back torque gives a small amount of braking, roughly equivalent to "engine braking" in an internal combustion engine. When the brake pedal is pressed, the armature current is increased, giving enough back torque to cause a maximum of about -2 m/s^2 acceleration of the vehicle. Advantages of regenerative braking are threefold: First, wear of the mechanical brake system is reduced; second, battery life is increased by partial recharging during driving, and total electric power consumption is reduced; and third, the feel or "driveability" of the vehicle is improved.

23-4 Eddy Currents

We have seen (Sec. 22-12) that an emf is induced in any conductor that moves across magnetic lines of induction or in any conductor across which lines of induction move. The *relative* motion of conductor and magnetic field gives rise to an emf in a conductor. It often happens that there is no well-defined path for the induced current. For instance, a flat metal plate on the end of a rod is allowed to swing as a pendulum (Fig. 23-10), passing through a strong magnetic field at the lowest point of its swing. As the plate starts to enter the region of magnetic field (position 1), the induced emf causes free electrons in the metal to swirl around. Such currents, distributed throughout the body of a piece of metal, are called *eddy currents*. By Lenz's law, the direction of an eddy current must be such as to oppose the change that causes it. As the plate swings down, electrons flow clockwise, conventional current (shown in Fig. 23-10) is counterclockwise, and by the right-hand rule the front face of the plate becomes a north pole and the rear face becomes a south pole. These induced poles are repelled by the poles of the magnet that they are approaching. The eddy currents reverse direction

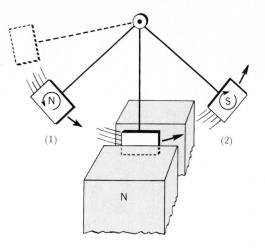

Figure 23-10 Eddy currents induced in the swinging plate are in such a direction as to produce magnetic poles that oppose the motion.

at the instant when the plate is passing through its lowest point. As the plate swings away (position 2), the eddy currents create poles that are attracted to the magnet—again opposing the motion. The currents reverse again when the plate is at its highest position and starts to move down again. The result of these "educated" eddy currents is always to oppose the motion of the plate, and so the plate soon comes to rest.

On a larger scale, some subway and rapid-transit cars have electromagnetic brakes that depend on eddy currents. An electromagnet hangs from the car near the steel rails (Fig. 23-11), and when the operator wishes to stop the

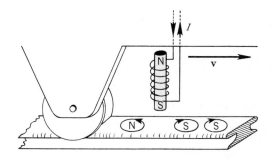

Figure 23-11 Eddy-current brake of a subway car (schematic).

car, he or she sets up a large current through the coils of the electromagnet. If the car is moving, eddy currents are induced in the rails, and the direction of the currents must be such as to oppose the motion of the car. Such brakes are free from sudden jerks, for as the car slows down, the eddy currents automatically die away. Thus the force decreases steadily to zero, and the car "eases in" to a quick but gentle stop. The current may be partially supplied by the car's motor; the operator throws a switch to disconnect the motor from the third rail or trolley line and connect it to the magnet. The motor now functions as a generator, and the back torque of the generator also helps stop the car. From an energy viewpoint, the overall result is that the KE of the car is transformed mostly into heat in the rails, as the eddy currents circulate in the rails. This is Joule heat, given by I^2R.

Another constructive use for eddy currents is in sorting municipal solid waste during the recycling process. Under development is a scheme in which waste is first shredded into small fragments. Then, after ferrous material is lifted out of the waste by conventional magnets, the residue is placed on a conveyor belt or ramp (Fig. 23-12). Embedded in the ramp are permanent magnets arranged in strips of alternating polarity, each strip making a 45° angle to the direction of motion of the particles. Eddy currents induced in nonferrous metals, such as aluminum, copper, brass, and lead, cause these particles to drift away from the straight-line path down the ramp; the residue of glass, plastic, and other nonmetals continues down the ramp, thus effecting a separation. (See Ref. 1 at the end of the chapter.)

Sometimes eddy currents represent a wasteful dissipation of energy. For instance, if the armature of a motor were wound on a solid iron core, eddy currents would be induced in the rotating iron, the armature would become warm, and the motor would not be very efficient. To reduce eddy currents, the core is "laminated," that is, constructed of many thin iron sheets rather than of one solid piece. The sheets are electrically

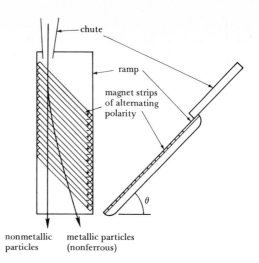

Figure 23-12 Schematic diagram of eddy-current metal separator (courtesy of E. Schlömann).

insulated from one another by lacquer or by thin surface layers of iron oxide or other compounds. As the armature turns, cutting across lines of induction, the induced emf is present just as before, but the induced *currents* are small on account of the high resistance.

23-5 *Transformers*

Electric power is the rate of doing work, and this rate depends on the product of the emf and the current in a circuit. The equation $P = \mathcal{E}I$ shows that the same power is obtained in a 12-V headlight bulb operating at 2 A as is obtained in a 120-V house-lighting circuit at a current of 0.2 A. For practical reasons, however, one or the other arrangement may be more suitable. The low-voltage circuit in the automobile is dictated by the low emf of the battery, freedom from shock hazard, and reliability of insulation. The high-voltage circuit in the home is desirable so that the current to any bulb or appliance will be relatively low; this avoids the expense of installing the heavy copper wire that would be necessary if a low-voltage, high-current circuit were used. The *transformer* is a device for

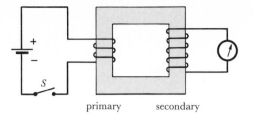

Figure 23-13

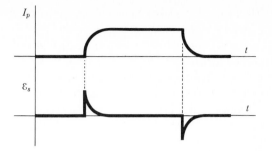

Figure 23-14 An emf $\mathcal{E}_s$ is induced in the secondary only when the current in the primary changes.

changing the effective value of the emf in a circuit. Indeed, the ease with which this can be done with alternating currents has led to the almost universal present-day use of ac power for home and industry.

A common type of transformer consists of an iron core on which are wound a primary coil and a secondary coil. Often these coils are called simply the *primary* and the *secondary*. Let us start by imagining the primary connected to a battery through a switch S (Fig. 23-13). When the switch is closed, a current is established in the primary, causing an increasing magnetic field that orients the magnetic domains in the entire iron ring. While the domains inside the secondary are being oriented, their lines of induction are cutting across the wires of that coil, and therefore a momentary emf is induced in the secondary. An equivalent way of describing the process would be to consider that the magnetic flux lines arising in the primary coil are confined to the iron core and also pass through the secondary coil; when the flux changes, the induced emf $\mathcal{E}$ is $d\Phi/dt$. As soon as the primary current reaches its steady value,* given by Ohm's law, there is no further change in the magnetization of the iron core, and the emf in the secondary coil becomes and remains zero. In other words, an emf *pulse* of short duration appears across the terminals of the secondary coil. If a galvanometer is connected to the secondary, there is a complete circuit, the induced emf is able to send a pulse of

charge through the galvanometer, and a momentary deflection is observed. If at some later time the switch S is opened, the primary current becomes zero, the domains return to their random orientation, and, while this is happening, lines of induction cut the secondary coil in a direction opposite to the cutting that previously took place while the domains were being oriented. Now the flux is decreasing, $d\Phi/dt$ is negative, and a momentary pulse is observed in the opposite sense to that previously observed (Fig. 23-14). Note that a *steady* current in the primary produces no effect in the secondary.

It should now be clear that a continually changing current in the primary will produce a continually changing current in the secondary. It can be shown that the ratio of primary emf $\mathcal{E}_p$ to secondary emf $\mathcal{E}_s$ equals the ratio of the number of turns in the primary N_p to the number of turns in the secondary N_s (Fig. 23-15):

$$\frac{\mathcal{E}_s}{\mathcal{E}_p} = \frac{N_s}{N_p} \qquad (23\text{-}2)$$

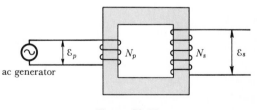

Figure 23-15

---

* This may take only a fraction of a second, or it may take several seconds, depending on the number of turns and the amount of iron.

This equation follows in part from the fact that if a certain emf is induced in each turn, the total emf is the sum of the emf's induced in the individual turns, and hence $\mathcal{E}_s$ is expected to be proportional to N_s. Thus, to get a large induced emf, we need many turns in the secondary coil.

Transformers can be either of the "step-up" or "step-down" variety, depending on whether the secondary emf is greater or less than the primary emf. The law of conservation of energy implies that the *power* in the primary equals the *power* in the secondary.* Hence, since $P = \mathcal{E}I$,

$$\mathcal{E}_p I_p = \mathcal{E}_s I_s \qquad (23\text{-}3)$$

or

$$\frac{I_s}{I_p} = \frac{\mathcal{E}_p}{\mathcal{E}_s} \qquad (23\text{-}4)$$

Thus for an ideal transformer, the ratio of the currents is the reciprocal of the ratio of the emf's.

Example 23-2

A certain ideal transformer's primary has 2000 turns, and its secondary has 50 turns. The primary is connected to a source having emf 120 V; the secondary is connected to a lamp bulb having resistance (when lit) of 0.6 Ω. (a) Calculate the emf of the secondary; (b) calculate the current through the bulb; (c) calculate the current in the primary; (d) calculate the power in the primary and in the secondary.

(a) This is a step-down transformer, since the secondary has fewer turns than the primary. The turns ratio is 2000 turns/50 turns = 40. Hence,

$$\mathcal{E}_s = (\tfrac{1}{40})(120 \text{ V}) = \boxed{3 \text{ V}}$$

(b) $$I_s = \frac{\mathcal{E}_s}{R_s} = \frac{3 \text{ V}}{0.6 \text{ Ω}} = \boxed{5 \text{ A}}$$

(c) $$\frac{I_s}{I_p} = \frac{\mathcal{E}_p}{\mathcal{E}_s}; \quad \frac{5 \text{ A}}{I_p} = \frac{120 \text{ V}}{3 \text{ V}}$$

$$I_p = \boxed{0.125 \text{ A}}$$

---

* This statement is true for an ideal transformer of 100% efficiency.

(d) Power in the primary:

$$\mathcal{E}_p I_p = (120 \text{ V})(0.125 \text{ A}) = \boxed{15 \text{ W}}$$

Power in the secondary:

$$\mathcal{E}_s I_s = (3 \text{ V})(5 \text{ A}) = \boxed{15 \text{ W}}$$

Modern transformers can be constructed to have efficiencies greater than 99%, and so our assumption of 100% efficiency in Example 23-2 is not far out of line.

23-6 Distribution of Electric Energy

Electric energy is transmitted through long distances at high voltage in order that, for a given power, the current in the line will be as small as possible. Any transmission line has some resistance, and the energy dissipated as Joule heat in the line, proportional to the square of the current, is wasted. Our next examples illustrate the advantage of using high voltages for transmitting electric energy.

Example 23-3

A transmission line between a power station and a factory has a resistance of 0.1 Ω in each of two wires. If 200 A is "delivered" at 100 V, (a) what useful power is delivered into the load? (b) What power is wasted in heating the line? (c) What total power must be supplied by the generator?

(a) Power into the load:

$$\mathcal{E}I = (100 \text{ V})(200 \text{ A}) = \boxed{20{,}000 \text{ W}}$$

(b) Power used in heating the line:

$$I^2 R = (200 \text{ A})^2 (0.2 \text{ Ω}) = \boxed{8000 \text{ W}}$$

(c) Power supplied by the generator:

$$20{,}000 \text{ W} + 8000 \text{ W} = \boxed{28{,}000 \text{ W}}$$

———————————— **Example 23-4**

Repeat Example 23-3 for the same line, assuming that the same useful power is delivered, but at a voltage of 1000 V instead of 100 V. Find (a) the current in in the line, (b) the power used in heating the line, (c) the total power supplied by the generator.

(a) Power into the load is 20,000 W, as before. Current in the line:

$$I = \frac{P}{\varepsilon} = \frac{20,000 \text{ W}}{1000 \text{ V}} = \boxed{20 \text{ A}}$$

(This is $\frac{1}{10}$ as much current as before.)

(b) Power used in heating the line:

$$I^2R = (20 \text{ A})^2(0.2 \text{ }\Omega) = \boxed{80 \text{ W}}$$

(c) Power supplied by the generator:

$$20,000 \text{ W} + 80 \text{ W} = \boxed{20,080 \text{ W}}$$

By using higher voltage, perhaps by employing a different generator design, the power used in heating the wires is reduced from 8000 W to only 80 W. This energy is lost, unless it possibly helps keep warm some birds perched on the wires.

At present, ac voltages as high as 765 kV are used in some long-distance intercity overhead transmission lines. This is perhaps an upper limit for several reasons, only a few of which are given here. The electric field near the wire approaches the value that permits breakdown of the insulating quality of the surrounding air, especially in damp weather. At ground level, 15–30 m below the wires, electric fields are suspected of causing adverse biological effects on crops, animals, and humans. Taller towers would be more costly, preempt more land, be unsightly, and be impossible in urban areas. Research and development are now directed toward increasing the current-carrying capacity of lower-voltage dc lines that could be buried underground and operate at about 200 kV. To do this, their resistance must be decreased, perhaps by cooling with liquid nitrogen or even by cooling to superconducting temperatures of less than 10 degrees Kelvin.

———————————— **Example 23-5**

A city of population 200,000 uses about 300 MW of power, which is supplied by an underground cable designed to operate at 230 kV. The city receives 98% of the power fed into the line at the generating plant. (a) How far from the plant can the city be if cable of round-trip resistance 0.1 Ω/km is available? (b) At what rate, in J/km·h, must thermal energy be dissipated by the earth in which the cable is buried?
Power loss is 2% × 300 MW = 6 MW.

The current is given by $I = \dfrac{300 \text{ MW}}{230 \text{ kV}} = 1300$ A.

The power loss is I^2R:

$$R = \frac{6 \times 10^6 \text{ W}}{(1.3 \times 10^3 \text{ A})^2} = 3.5 \text{ }\Omega$$

(a)   Distance $= \dfrac{3.5 \text{ }\Omega}{0.1 \text{ }\Omega/\text{km}} = \boxed{35 \text{ km}}$

(b) Power dissipated per km per hour is

$$\left(6 \times 10^6 \frac{\text{J}}{\text{s}}\right)\left(\frac{1}{35 \text{ km}}\right)\left(\frac{3600 \text{ s}}{1 \text{ h}}\right) = \boxed{6 \times 10^8 \frac{\text{J}}{\text{km}\cdot\text{h}}}$$

This is 150 kcal per meter per hour—a not inconsiderable amount.

Aside from the economic loss involved, heat dissipated in a buried cable can be a safety hazard. Cryogenic (low-temperature) cables offer a way of carrying more power without exceeding safety tolerances.

Of course, if the power is delivered to the city at high voltage, it must be transformed to a useful (lower) emf by use of a step-down transformer at the place where it is to be used. Distribution of electric energy has played a major role in the expanding economy of the past 80 years, and we should give the transformer full credit as the device that has made this possible. Its basic principles were discovered by Joseph Henry and Michael Faraday in the 1830s, and the engineering genius of Nikola Tesla and Charles P. Steinmetz in the late 19th century implemented the change from the dc system of Edison to our modern flexible and economical ac distribution system.* In each step-up or step-down transformer, no direct connection exists between the primary and secondary coils, and the insulation between the windings must be sufficient to withstand high voltages—up to hundreds of thousands of volts in some cases. The lines in the alleys and along the streets of a city are usually at 2300 V, serving areas of many blocks, and each house or group of several houses is served by its own step-down transformer whose secondary has a grounded center tap. This makes available 115 V for lamps and small appliances, and also 230 V for large appliances, such as stoves, furnaces, and dryers, in which the current would be too large if power had to be supplied at 115 V.

The primary winding of a pole transformer is of heavy wire, and its resistance may be only 0.1 Ω or less. This winding is permanently connected across a 2300-V line. At first glance, Ohm's law leads us to expect that this winding would continuously carry some 23,000 A. Actually, it carries only a very small fraction of an ampere, unless some lamp, motor, or other load is turned on in the secondary circuit inside the house. To explain this paradox, we need to study more closely just how a coil limits, or "impedes," alternating current.

23-7   Impedance of a Coil

Suppose we measure the current through a certain coil connected to a source of direct current (Fig. 23-16a). Using Ohm's law, we find the resistance of the coil to be

$$R = \frac{6 \text{ V}}{6 \text{ A}} = 1 \text{ }\Omega$$

However, repeating the experiment with a 6-V source of 60-Hz (60-cycle/s) alternating emf, represented by —Ⓥ—, we find a much smaller

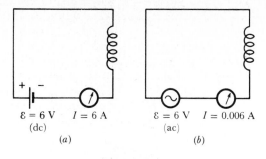

Figure 23-16

$\varepsilon = 6$ V $I = 6$ A
(dc)

(a)

$\varepsilon = 6$ V $I = 0.006$ A
(ac)

(b)

alternating current. The "apparent resistance" of the coil in Fig. 23-16b is

$$\left(\frac{6 \text{ V}}{0.006 \text{ A}} \right) = 1000 \text{ }\Omega$$

This ratio is not constant for alternating current of different frequencies, for if we repeated the experiment using a 120-Hz source, we would find that I was only 0.003 A, so that the ratio V/I for this coil would be 2000 Ω. The coil "impedes" the flow of charge, and we define *impedance Z* in the same way that resistance can be defined for a dc circuit:

$$\text{Impedance} = \frac{\text{alternating PD}}{\text{alternating current}}$$

$$Z = \frac{V}{I} \tag{23-5}$$

$$\text{ohms} = \frac{\text{volts}}{\text{amperes}}$$

For the 60-Hz experiment, the coil of Fig. 23-16 has a resistance of 1 Ω and an impedance of 1000 Ω.

The reason the current in Fig. 23-16b is so small is that the coil acts as a transformer, with the same winding serving as both primary and secondary. When there is a current through a coil, a magnetic field proportional to the current exists within the coil (Fig. 23-17). While the current is building up, lines of induction are cutting across the wires of the coil, and so an emf is induced in the coil by its own changing magnetic field. This phenomenon is called *self-*

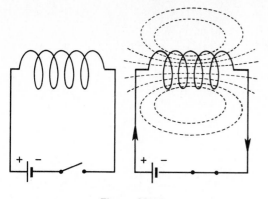

Figure 23-17

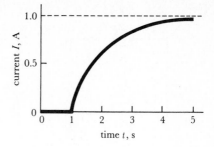

Figure 23-18 Rise of current through a coil; switch is closed at $t = 1$ s.

induction. The induced emf is a back emf, and by Lenz's law it tends to oppose the change. Thus the back emf keeps the current in a large iron-cored magnet from rising instantaneously to its final value (Fig. 23-18). For any coil, we can define a constant called its *self-inductance L*, measured in *henries* (H). The larger the self-inductance, the greater the back emf for a given rate of change of current.* Thus

$$\varepsilon_{\text{back}} = L\left(\frac{\Delta I}{\Delta t}\right)$$

It is shown in Sec. 23-14 that the reactance of a coil is given by $X_L = 2\pi f L$, where f is the frequency in Hz (cycles/s). If the ordinary resistance of a coil is negligible, its impedance Z is the same as its reactance X_L and therefore

$$Z \approx X_L = 2\pi f L \qquad (23\text{-}6)$$

We see now how the impedance of a coil can be so much larger than its resistance. If the induced back emf counteracts to a large extent the applied PD, then the net emf in the circuit is small and the current is small. As with any transformer action, the induced emf is greater if the alternating current reverses more rapidly. Also, the induced back emf in a coil will be greater if

---

* The *self-inductance* of a coil is defined as back emf divided by rate of change of current. A coil's self-inductance is 1 H if a back emf of 1 V is induced in the coil when the current through it changes at the rate of 1 A/s.

the coil has many turns, especially if it is wound on an iron core in which magnetic lines of induction are concentrated. The primary coil of a transformer fulfills all these conditions admirably, and so we expect the primary current to be small because of the large back emf.

Example 23-6

Calculate the self-inductance of a transformer primary winding if a 60-Hz generator whose emf is 115 V sends 0.1 A through the coil. Assume the coil to have negligible resistance.

$$X_L \approx \frac{V}{I} = \frac{115 \text{ V}}{0.1 \text{ A}} = 1150 \ \Omega$$

To find L, we use Eq. 23-6:

$$X_L = 2\pi f L$$

$$L = \frac{X_L}{2\pi f} = \frac{1150 \ \Omega}{2\pi(60 \text{ Hz})}$$

$$= \boxed{3.05 \text{ H}}$$

Finally, we seek a mechanism by which a current *does* exist in the primary coil when a circuit is closed in the secondary. For instance, when someone plugs in an electric iron, how does this cause the *primary* current to increase from its previously negligible value? The interaction is, of course, through the iron core of the transformer. The secondary current creates its own changing magnetic field, which in turn induces

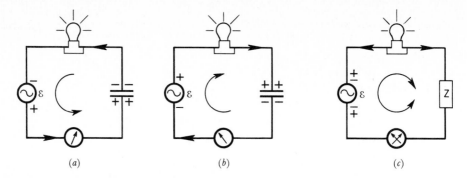

Figure 23-19 (a) and (b) Flow of charge in an alternating-current circuit containing a capacitor. (c) Equivalent circuit, replacing the capacitor by an impedance Z.

an emf in the primary coil—a sort of "back back emf." The primary current increases because the original back emf (which limited the current) is partially canceled by this third emf; the cancellation takes place only when there is current in the secondary coil. In this way, an increased load in the secondary circuit results in an increased current in the primary circuit—all without direct connection between the two coils.

23-8   Impedance of a Capacitor

The behavior of a capacitor in an ac circuit deserves a brief discussion, for there is a sharp contrast between its response to alternating and to direct currents. That is, the impedance of a capacitor is quite different from its resistance. When a capacitor is connected to a dc source of potential, such as a battery, in series with an ammeter and a lamp, there is a momentary rush of charge as the capacitor becomes charged. The ammeter needle swings over for an instant if its inertia is not too great, but it soon returns to zero and stays there. The lamp flashes momentarily. The situation is different if the source is an ac generator. The capacitor is alternately charged in one sense and then in the other, and charges are rushing to and fro through the meter and the lamp (Fig. 23-19a and b). No charge actually moves across the gap between the plates of the

capacitor,* but the lamp and meter don't know the difference. From the lamp's point of view, just as much heat and light are produced as if the charge really did flow in a complete circuit, as in Fig. 23-19c. To all intents and purposes, there is an alternating current *through* the capacitor as far as the rest of the circuit is concerned.

What limits the current through the capacitor in Fig. 23-19? In other words, what factors determine the impedance of a capacitor? The larger the capacitance, the more charge flows to and fro, since more charge can be momentarily stored on the plates (according to the equation $Q = CV$). The higher the frequency of the alternating current, the greater the rate of change of charge, and hence the greater the current. Thus the current is directly proportional both to C and to f, and the impedance, which is V/I, is inversely proportional both to C and to f. It is shown in Sec. 23-14 that the reactance of an ideal capacitor is given by $X_C = 1/(2\pi f C)$. If the leakage current through the capacitor is negligible, its impedance Z is the same as its reactance X_C, and therefore

$$Z \approx X_C = \frac{1}{2\pi f C} \tag{23-7}$$

_____

* We assume that the dielectric material between the plates is a perfect insulator; that is, the dc resistance of the capacitor is assumed to be infinite. This is a valid assumption for most practical capacitors.

Especially in audio amplifiers and radio circuits, where f is large, even a small capacitor's impedance is often low enough to be negligible. To this extent, a capacitor is a short circuit for alternating current but an open circuit for direct current.

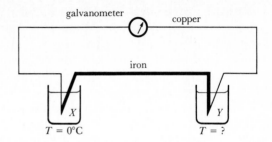

Figure 23-20 Thermocouple circuit; transformation of thermal energy into electric energy.

Example 23-7

Calculate the approximate impedance of a 0.02-μF capacitor in a circuit for which the frequency is (a) 60 Hz; (b) 6 MHz.

(a) $Z = \dfrac{1}{2\pi(60\ \text{Hz})(0.02 \times 10^{-6}\ \text{F})}$

$ = \boxed{1.33 \times 10^5\ \Omega} \quad$ (at 60 Hz)

(b) $Z = \dfrac{1}{2\pi(6 \times 10^6\ \text{Hz})(0.02 \times 10^{-6}\ \text{F})}$

$ = \boxed{1.33\ \Omega} \quad$ (at 6 MHz)

At 6 MHz (a radio frequency), the impedance is about 1 Ω, which is negligible in comparison with other impedances in a radio circuit.

See Secs. 23-14 through 23-17 for further study of ac circuits containing resistance, inductance, and capacitance.

23-9 Thermoelectricity

One way of measuring temperature electrically has already been discussed in Sec. 21-7. The resistance of a coil of wire depends on its temperature, and if the resistance is measured with a Wheatstone bridge, the temperature can be computed. Another common way of measuring temperature is by means of a *thermocouple*, which consists of a pair of junctions between two dissimilar metals. A simple thermocouple circuit, using iron and copper, is shown in Fig. 23-20. The two iron-copper junctions are X and Y, with X held at some reference temperature (0°C in the diagram). If X and Y are at different temper-

atures, there is a net emf in the circuit. For small temperature changes, the emf is proportional to the temperature *difference* between X and Y. Tables in handbooks give the value of this *thermal emf* for various pairs of metals; for iron-copper junctions near room temperature, the thermal emf is about 14×10^{-6} V for a temperature difference of 1°C (14 μV/°C). Such small emf's can be measured by observing the deflection of the galvanometer of Fig. 23-20. According to Ohm's law, $I = $ net $\mathcal{E}/$net R; in this circuit the net emf is the thermal emf due to the temperature difference, and the net resistance is the resistance of the junctions, the connecting wires, and the galvanometer coil. For the most precise work, thermal emf's are often measured with the help of a potentiometer (Sec. 21-8).

In order to discuss the theory of thermoelectricity, let us first consider two insulated blocks of metal that are initially uncharged (Fig. 23-21a). Iron and copper differ in crystal structure, density, conductivity, thermal expansion, and other properties, and so we expect that the number of

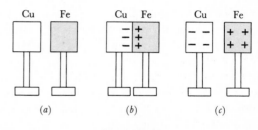

Figure 23-21

free electrons per unit volume will also be different in these dissimilar metals. If now we bring the blocks into contact, an electron that finds itself just at the boundary between iron and copper will experience a net force in one direction or the other, and electrons will tend to flow from one metal block to the other. Experiment shows that in the situation in Fig. 23-21b, an excess of electrons flows from iron to copper. A PD is set up between copper and iron merely because the metals are in contact; we call this the *contact potential difference* between iron and copper. If the blocks are separated again, as in (c), the excess charges become redistributed over their surfaces, as can be shown by using an electroscope and a proof plane (see Sec. 18-4). All this takes place with the two blocks at the same temperature.

In a complete circuit such as Fig. 23-22, if X and Y are at the same temperature, the contact PD at X exactly opposes that at Y, and there is no current. However, the contact PD between any pair of metals depends on temperature, just as almost every physical property does. If, then, X and Y are at different temperatures, the contact PDs will be different, a net PD will exist, and a steady current will exist. From this point of view, thermoelectricity is simply a manifestation of the fact that contact potential differences varies with temperature.

Why do we speak of thermal emf rather than thermal PD? Let us recall that a seat of emf is a place where some form of nonelectric energy is *reversibly* transformed into electric energy. The pair of junctions X and Y illustrate this transformation of energy. Heat is being converted into electric energy; in a similar fashion, chemical energy is transformed into electric energy in a battery, and mechanical energy is transformed into electric energy in a generator. The reversibility of the thermocouple (considered as a seat of emf) is illustrated by the following experiment. A battery sets up a steady electron current through an iron-copper circuit, as shown in Fig. 23-23. If both water baths are originally at 20°C, junction X becomes slightly warmer and junction Y becomes slightly cooler. While the electrons flow, electric energy is being transformed at junction X into heat in a way that is entirely different from the Joule heating in all resistors. At the same time, heat is being absorbed from the water at junction Y. If electrons are forced through the junctions in the other direction, the roles of X and Y are interchanged, and X absorbs heat instead of liberating heat. At a given junction, the transformation can go in either direction; this is what is meant by reversibility, a characteristic of all seats of emf. Thermoelectric refrigerators are based on this principle; they have no moving parts and can be operated as heaters simply by reversing the direction of the current.

Although thermal emf's are small, it is often possible to get a larger net emf by connecting several junctions in series to form a *thermopile*. Of course, the internal resistance of a thermopile is greater than that of a single pair of junctions, so the terminal voltage (given by $V = \varepsilon - Ir$) cannot be increased indefinitely. In practice, 10 to 50 thermocouples in series are about as many as can be used effectively.

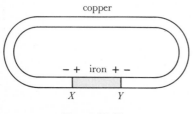

Figure 23-22

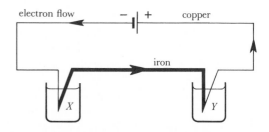

Figure 23-23 Thermoelectric heater (at X) and refrigerator (at Y).

23-10 Solid-State Devices— Diode and Transistor

No aspect of applied electricity is more characteristic of 20th-century technology than the field of electronics. Some of the end results, such as television and computers, are spectacular and are deeply ingrained in contemporary culture. We shall not touch even superficially on many of the ingenious devices and circuits of electronics. Instead, we shall illustrate in a general way some of the basic methods and circuits of electronics by study of two simple solid-state devices, the diode and the transistor. We shall only briefly mention their vacuum-tube counterparts, the vacuum diode and the triode, which are now used chiefly for special applications.

The carriers of charge in a semiconductor such as silicon or germanium may be either positive or negative. To see how this is possible we must consider semiconductor materials. Let us consider a crystal of silicon, which is a typical semiconductor (Table 20-1) with a resistivity of $3 \times 10^4 \ \Omega \cdot m$. Although a full understanding of semiconductors requires a knowledge of quantum mechanics that is beyond the scope of this book, it turns out that for pure silicon an electron can gain energy only if it receives more than some fixed amount of energy. There is an *energy gap*, and small increases of an electron's energy in a silicon crystal are ruled out. It is this energy gap that makes silicon crystals semiconductors; in a copper crystal there is no energy gap, and certain electrons are "free" to receive any amount of energy, no matter how small, and thus take part in the conduction process. The semiconductor's conductivity increases as temperature rises, as was mentioned in Sec. 20-6, because thermal agitation can help supply some of the energy required to get an electron across the gap. This is true for pure silicon.

Diodes. To make a *diode*, we create two different types of silicon by adding very small, carefully controlled amounts of impurity, measured in parts per million (Fig. 23-24). Silicon has valence 4, and in the crystal each Si^{4+} ion is surrounded by 4 electrons (shared with other Si^{4+} ions). If a few silicon atoms are replaced by donor atoms of valence 5, such as phosphorus, only 4 of the phosphorus valence electrons are tightly bound in the crystal, and one electron is loosely bound. The "doped" material has an excess of "almost free" negative electrons and is called *n*-type silicon. Similarly, *p*-type silicon can be made by replacing some silicon atoms by "acceptor" atoms of valence 3, such as boron; there are empty places to which electrons can

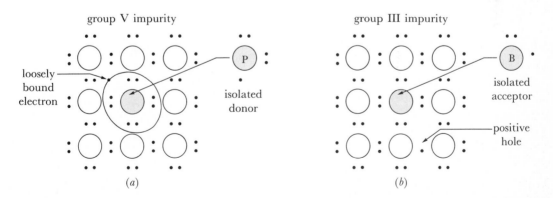

Figure 23-24 Doped semiconductor materials (schematic). The open circles represent Si^{4+} ions. (*a*) *n*-type silicon. (*b*) *p*-type silicon. (Adapted from Fig. 2-12 of *Principles of Solid-State Microelectronics* by S. N. Levine, © 1963 by Holt, Rinehart and Winston.)

move. These empty places, called *positive holes*, are equivalent to positive charges that are "almost free" to move. Another commonly used semiconductor material is germanium, with arsenic or aluminum as impurities.

A *p-n junction* is a tiny, single-crystal piece of silicon, germanium, or some other semiconductor that has been treated to make one side *p*-type and the other side *n*-type. The magnitude of the current through the semiconductor depends on the polarity of the battery or other source of emf in the external circuit; the determining factor is what happens at the junction. In the *junction diode* shown in Fig. 23-25*a*, the electrons in the *n* side are pushed across the junction by the − terminal of the battery (and are also attracted by the + terminal of the battery). Also, positive holes in the *p* side are aided in their travel across the junction by the battery's PD. There is a large current, and the diode is said to be *forward biased*. If the battery is connected as in Fig. 23-25*b*, the PD of the battery tends to assist elec-

trons in the *p* side to cross the junction; but there are very few free electrons in this material, so the current is very small. Likewise, the PD would help positive holes to flow from the *n* side to the *p* side, but there are very few positive holes in the *n*-type material. Under these circumstances the current is very small, and the diode is said to be *reverse biased*. We see that there is a large forward current and a very small reverse current (Fig. 23-25*c*). The diode is a one-way valve, allowing charges to flow only if the diode is forward biased. The value is essentially "off" when the diode is reverse biased. The symbol for a diode is ▸▮, where the arrowhead indicates the direction of "easy" flow of conventional (positive) charge when the diode is forward biased.

The diode is often used to make pulsating direct current when an ac source of potential is available, perhaps from a step-up or step-down transformer. Thus in Fig. 23-26, the secondary of the transformer makes *AB* a source of alternating PD. During half the cycle, when *A* is + relative to *B*, conventional + charges flow in the direction *AMNXYBA*; but during the other half of the cycle, when *A* is − relative to *B*, charges cannot flow and the current is zero. The graph of current *I* vs. time shows a pulsating direct current whose average value is given by the dashed line. The conventional current through the load resistor *R*

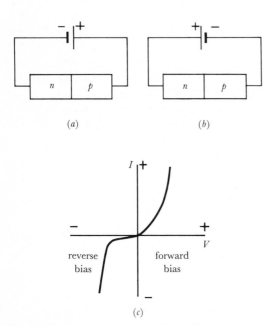

Figure 23-25 Junction diode. (*a*) Forward bias. (*b*) Reverse bias. (*c*) Graph of current vs. bias voltage.

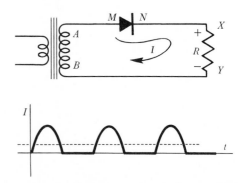

Figure 23-26 Rectifier circuit. The arrow represents the direction of conventional current.

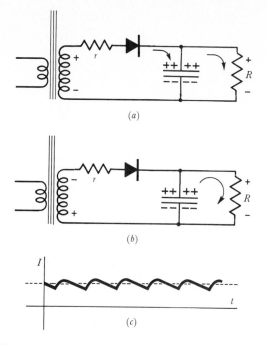

(a)

(b)

(c)

Figure 23-27 Solid-state power supply. If the diode accidentally becomes a short circuit, the small resistor r will overheat and burn out, serving as a fuse to protect the capacitor and load.

is from X to Y; hence X is $+$ relative to Y. To smooth out this current, a capacitor is used in parallel with the load resistor (Fig. 23-27). During the "rainy season" (a), while charges are flowing through the diode, some charges go through R, and in addition some flow onto the top plate of the capacitor, charging it as shown. During the "dry season" (b), when no charges flow through the diode, the capacitor discharges, sending charge through R in the same direction as before. The capacitor acts as a reservoir to smooth out the variations in current. With such an arrangement, called a *filter*, the current can be made practically constant (Fig. 23-27c). Also, the average value of the current is much larger than without the filter, and the PD between X and Y is practically constant. The whole circuit, called a *rectifier circuit*, is used in place of batteries to provide a steady PD. Every ac-operated radio or television set has a rectifier circuit of some sort, although other one-way devices, such as the vacuum-tube diode, can also be used, especially where very high dc voltages are involved.

Transistors. The *transistor* is a solid-state device that is essentially a current amplifier. That is, a relatively large current output is caused by a small change in the current input.

Bipolar junction transistors (BJT). A common type of transistor is formed of a single-crystal fragment of silicon doped with impurity atoms to make two *p-n* junctions back to back. For this reason it is called a *bipolar junction transistor* (BJT). In the *n-p-n* transistors shown in Fig. 23-28, the *emitter* and the *collector* regions are of *n*-type silicon with a heavy concentration of impurity atoms that are electron donors; the

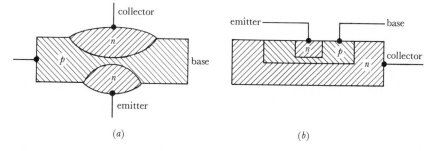

(a)

(b)

Figure 23-28 Physical construction of typical *n-p-n* junction transistors. The overall size is 1 mm or less; the space between emitter and collector is about 0.01 mm.

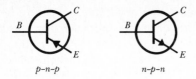

Figure 23-29 Symbols for transistors.

base is a very thin region with only a small concentration of *p*-type impurity atoms (electron acceptors, or holes). A *p-n-p* transistor is the mirror image of an *n-p-n* type; the emitter and collector are of *p*-type material, and the base is of *n*-type material. The symbol for a *p-n-p* transistor shows by the direction of the arrow on the emitter that conventional (positive) current is toward the base, while the opposite is true for the symbol for the *n-p-n* transistor (Fig. 23-29).

The operation of a simple transistor amplifier can best be explained by an example in which we use a *p-n-p* transistor; everything would be the same for an *n-p-n* transistor if the polarity of the battery were reversed and the words "electron" and "hole" were interchanged.

The emitter-base junction (*e-b* in Fig. 23-30*a*) is forward biased by the 15-V battery; therefore, a large number of positive holes flow from the emitter into the base. In the base, they are in a region where they can combine with the excess electrons in the base (which is of *n*-type material), but this is not very likely since the base is

very thin and is not very strongly doped with electrons. Typically, 95% to 99% of the emitter current goes on into the collector and on out of the transistor, since the negative terminal of the battery attracts the positive holes. We can apply Kirchhoff's first law to point *B* of Fig. 23-30*a*. The collector current is 1.0 mA, the input current is negative (0.030 mA out from the base), and hence 1.03 mA enters the emitter at *E*. In the transistor, the emitter current and the collector current are practically equal.

If there is no input current, electrons are lost from the *n*-type base region by two processes: They can flow across the forward-biased *e-b* junction toward the + terminal of the battery; also they can be lost by recombination with holes entering the base from the emitter. This loss of electrons from the base builds up a slight net positive charge in the base that tends to inhibit the flow of positive holes from the emitter. In fact, if these electrons in the base were not replenished, the current from the emitter into the base would stop entirely, and the collector current would be 0. By injecting electrons *into* the base (that is, by increasing the input current of conventional current *out* from the base), we to some extent cancel the inhibiting holes in the base and allow more holes to flow across the base-emitter junction and reach the collector. Thus, the flow of holes, and therefore the collector current, is controlled by the much smaller

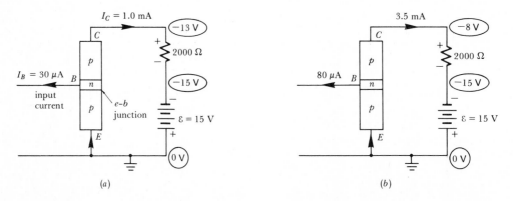

Figure 23-30 A transistor amplifier. The arrows represent the direction of conventional current.

base current. We then have *amplification of current*—control of a large current (I_C) by a much smaller current (I_B). Typical data for two values of the base current are shown in the two parts of Fig. 23-30. The current gain is

$$\frac{\text{Change in } I_C}{\text{Change in } I_B} = \frac{3.5 \text{ mA} - 1.0 \text{ mA}}{0.080 \text{ mA} - 0.030 \text{ mA}}$$

$$= \frac{2.5 \text{ mA}}{0.050 \text{ mA}} = 50$$

This amplifier has a current gain of 50.

The current amplifier described above could be used to build up any weak source of current. The photocell current in the exposure meter of a camera, originally in the microampere range, must be in the milliampere range to deflect the needle of the camera's built-in galvanometer. The detector of a spectrophotometer, so useful in analytical chemistry, may be a photocell. In most situations, however, it is a small *voltage* that must be amplified. We will not go into the ways in which various sources of really small voltage can be made to give small currents that are proportional to their PDs. Such sources include the output of the magnetic cartridge of a hi-fi system and the voltage pulse of a heart muscle (electrocardiograph) or of a brain wave (electroencephalograph).

The two circuit diagrams of Fig. 23-30 show how a voltage output can be obtained from the transistor current amplifier. The key element is the *load resistor* of 2000 Ω (2 kΩ) in series with the collector. Because of the *IR* drop in this resistor, the potential of the collector (point *C*) is less than the 15-V battery emf. When the collector current increases from 1.0 mA to 3.5 mA, the *IR* drop in the load resistor increases from 2.0 V to 7.0 V, and the potential of point *C* changes from -13 V to -8V. Thus, a change of 50 μA in base current results in a 5-V change in the collector voltage.

Field effect transistor (FET). The metal oxide semiconductor transistor (MOS) is fundamentally different from the bipolar junction transistor (BJT) already discussed. It makes use of a field effect and is one type of *field effect transistor* (FET, or MOSFET). Although the effect was known as long ago as 1930, only in the mid-1970s did the FET come into its own. As so often happens, advances in materials technology have determined the course of events, along with improved understanding of the basic physical phenomena in solids.

The FET transistor is voltage-controlled, unlike the junction transistor, which we have seen is a current-controlled device. The MOSFET is a thin single crystal of *p*-type silicon in which two shallow surface regions, called the *source* and the *drain*, are lightly doped with phosphorus to become *n*-type silicon. In one type of MOSFET, a very thin *channel* of *n*-type material lies under the *gate* electrode, and an insulating layer of silicon dioxide (quartz) separates the gate from this *n* channel. In Fig. 23-31 the drain is made a few volts positive relative to the source, which is at ground potential. The gate is initially set at a small negative voltage called the *bias voltage*. There is a small electron current from source to drain, through the channel. If the gate's potential is made slightly more negative, the electric field causes the channel to shrink and become narrower, the channel's resistance increases, and the electron current decreases. The larger the negative value of V_G, the narrower the

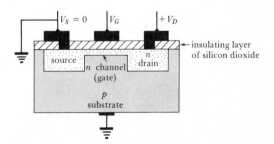

Figure 23-31 The solid black portions are metallic connections to the source, gate, drain, and substrate. The MOSFET is shown here with a small negative gate voltage, which gives rise to a narrow *n* channel beneath the gate. All voltages are measured relative to the source, which is grounded at $V_S = 0$.

channel; within limits, the decrease in drain current is proportional to the applied gate voltage. This allows amplification—gate voltage controls drain current.

In the circuit of Fig. 23-32 the 2000-Ω load resistor serves the same function as in the BJT circuit of Fig. 23-30. In each case, the *IR* drop in the load resistor is subtracted from the supply voltage to obtain the output voltage; in this way a current change is converted to a voltage change.

It would take us too far afield into technology to do more than mention the advantages of the MOSFET and other FETs. Electrically, they have the desirable property of drawing very little current from the input, since the gate is insulated and serves only to supply the controlling electric field. The MOSFET dissipates less power than the BJT, an important consideration when hundreds or thousands of transistors are packed into a small space.

But it is in fabrication techniques that the MOSFET has proved spectacularly successful. A complete radio would have up to a dozen or so transistors, with some associated capacitors and resistors. A pocket calculator requires perhaps 1000 transistors, intricately interconnected. The MOSFET lends itself admirably to the production of an *integrated circuit*, called an IC or a "chip." In fabrication of an IC, the many transistors, diodes, resistors, and capacitors are formed all at the same time on a single piece of doped silicon substrate by controlled evaporation of aluminum (for conductors), phosphorus and boron (for doping), silicon dioxide (for insulation), and other materials (for resistance paths). Masks are used to direct these substances into the desired pattern on the surface, often in several steps, with use of an acid at appropriate steps to remove silicon oxide from certain areas. The complex structure shown in the electron microscope photograph (Fig. 23-33) is only a very small portion of an IC chip formed by such an automatic and reliable process. A single IC chip measuring 6 mm × 6 mm can contain as many as 10,000 interconnected transistors, and 100 such chips can be made in one batch process and then sliced apart. Both BJT and MOS types of transistors lend themselves to fabrication of

Figure 23-33 Scanning electron micrograph of a small portion of a FET memory array. The vertical light-colored "bars" are aluminum interconnections about 2.3 μm wide.

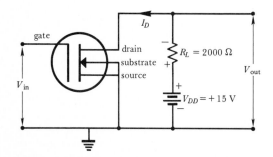

Figure 23-32 Voltage amplification by a MOSFET. The output voltage is $V_{DD} - I_D R_L$.

Figure 23-34 Left: IC chip from paging receiver (page 513) showing 28 connection points. Right: Close-up of the chip, which has 700 transistors in an area 3.3 mm × 3.6 mm.

ICs; the choice depends to some extent on economics, which favors the MOS for very large arrays or for very large markets. Most of the transistors in consumer goods such as radios, hi-fis, and calculators are of the BJT type. Up to 10^6 components per IC are now possible.

Figure 23-34 gives an idea of the scale of things. A physician's pocket-sized paging receiver emits a tone when the doctor is wanted on the telephone. The receiver is built around two ICs, each containing 700 transistors (see Fig. 23-34). A third IC is an operational amplifier (op-amp) that produces the signal to drive the loudspeaker. The complete receiver is shown on page 513. As the technology advances, we can expect a rapid proliferation of integrated circuits in many areas of daily life—automotive and medical diagnosis and monitoring, computers, games and toys, household appliances, wristwatches, and traffic control by computer, to name just a few. Needless to say, a very important use of ICs is in the largest computers, where up to a million transistors may be used in a single installation.

See the References at the end of the chapter for additional material on solid-state electronics and various applications of transistor technology.

23-11 Vacuum-Tube Devices— Diode and Triode

In contrast to the situation in solid-state electronics, where charge carriers can be either negative electrons or positive holes, in vacuum tubes the charge carriers are always negative electrons. We will discuss briefly two types of electron tubes.

The essential parts of a *vacuum-tube* **diode**, shown in Fig. 23-35a, are the *cathode K* and the *plate P*. These electrodes are sealed in an evacuated tube, with connections brought out to two of the base pins. In addition, a heater coil (*HH*) inside the cathode is provided so that the surface of the cathode can be brought to a dull-red heat. There is no direct electrical connection between the heater and the cathode.

We have seen that a distinguishing feature of a metal is the presence of a large number of free electrons that wander about inside the metal, attached to no particular nucleus. They are usually confined to the metal, but at any given temperature a certain number of electrons per second will be able to escape from the surface and fly off into space. This process is very similar to

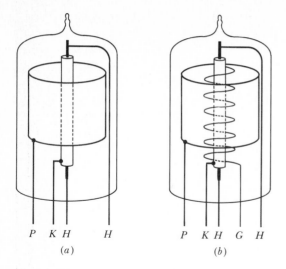

Figure 23-35 Vacuum tubes. (*a*) Diode. (*b*) Triode. *P*, plate; *K*, cathode; *HH*, heater; *G*, grid.

the evaporation of liquid molecules (Sec. 15-7); the analogy can be pushed far enough to include similarities to latent heat of vaporization and vapor pressure. The important thing for us is that this tendency for electrons to leave the surface increases greatly as the temperature of the metal goes up (compare the vapor-pressure curve of Fig. 15-6). When a diode's cathode is heated to a visible glow, electrons leave the surface at such a rate that the electron current from cathode to plate may easily become measurable. To obtain this current, the plate must be positive relative to the cathode so that electrons are attracted. If the plate is negative relative to the cathode, electrons are repelled and cannot reach the plate, and the current is zero. The diode is thus a one-way valve, allowing electrons to flow only if the plate is positive.*

Just as for the solid-state diode, a vacuum diode delivers pulsating direct current to a load resistor, and, as in Fig. 23-27, a capacitor can be

used to smooth out the variations in current. Vacuum diodes are used where large voltages are encountered, as in x-ray tubes and TV picture tubes. During the half-cycle when the diode is not conducting, it must withstand the peak value of the ac supply voltage without breakdown. Most solid-state diodes break down (become conducting) at inverse voltages of a few volts to a few thousand volts, depending on design and size. For special high-voltage applications, vacuum diodes are still the best solution.

As its name implies, the *vacuum triode* has three electrodes. In the form invented by Lee De Forest in 1906, an open-mesh *grid* is placed between the cathode and the plate (Fig. 23-35*b*). The grid (*G*) is made of fine wire so that it does not directly obstruct electrons from reaching the plate, but the electron flow can be controlled by changing the potential of the grid. In a suitable circuit, a relatively small change of grid potential can produce a large change in plate current, and therefore a large change in plate potential. In this way a triode is used to amplify small changes of potential. We see that the triode has many of the properties of the FET—it is a voltage-controlled device that does not draw much current from the input circuit to which it is connected. The cathode, grid, and plate of a triode are analogous to the source, gate, and drain of a FET.

Vacuum tubes are costly, bulky, and fragile; they dissipate a great deal of heat from their filaments; and they require complex circuit wiring. For these reasons, the solid-state integrated circuit has largely replaced the vacuum triode, except for certain specialized applications such as the final stages of a high-power radio transmitter.

There are energy considerations that apply to any amplifier, whether it is a solid-state or vacuum-tube amplifier. Let us contrast the usefulness of a 20:1 step-up transformer and that of an amplifier whose voltage gain is 20:1. Both devices magnify the input voltage the same amount. The transformer, if 100% efficient, delivers as much power as it receives. Certainly no

---

* It is interesting to note that Thomas Edison in 1883 observed a current through the vacuum to an electrode sealed into the side of an incandescent lamp bulb. This was before the discovery of the electron, and Edison missed the significance of his observation that there was no current unless the collector wire was positive.

increase of power can be obtained from a transformer; such a result would violate the law of conservation of energy. The amplifier, on the other hand, can deliver much more power than it receives. The ultimate source of the output power is the battery; the gate or the grid acts only to *control* the flow of electrons. In much the same way, an engineer at a hydroelectric installation can, by unaided muscular effort, open or shut a valve that may control the flow of many thousands of tons of water.

23-12   *The Cathode-Ray Tube*

Although vacuum tubes in radio receivers are a thing of the past, you are no doubt familiar with one specialized type of vacuum tube—the *cathode-ray tube*, which is used in oscilloscopes and television receivers, and for the display of computer output. The essential parts are shown in Fig. 23-36. Electrons emitted from an indirectly heated cathode K are accelerated to high speed as they move toward an anode that is at, say, $+20,000$ V relative to the cathode. Some electrons pass through small holes on the axis of the anode structure and emerge as a beam. The whole assembly is called an *electron gun*, and for best results and sharp focus, several electrodes, properly spaced, are used. The intensity of the beam is controlled by a negative grid within the electron gun. Next, the electron beam passes though a *deflection system*, which can be magnetic or, as shown in Fig. 23-36, electrostatic. A varying PD applied to the X plates supplies a horizontal E-field (force per unit charge), which causes the *horizontal* momentum component of the electron beam to change. Similarly, a varying PD across the Y plates causes the *vertical* component of momentum to change. After leaving the deflection system, the electron beam moves in a straight line toward the *fluorescent screen*, where the impact of the energetic electrons causes the emission of light from the material that coats the inside face of the tube.

The cathode-ray tube's usefulness is due to the very low mass of the electrons, which allows the beam to be deflected across the screen in a microsecond or less. Many ingenious circuits are used in the oscilloscope; the X plates are supplied with a steadily increasing voltage to give a uniform horizontal motion of the beam—a time base—followed by a quick return to the starting point. In the oscilloscope, vertical deflections are caused by the output of an amplifier having as its input the signal whose waveform is to be displayed. In the TV tube and in the computer-output scope, successive horizontal sweeps are displaced vertically a slight amount by the Y amplifier, and the intensity is varied during each sweep to form the picture or the displayed characters.

Many cathode-ray tubes, including all but the smallest TV tubes, use magnetic deflection instead of the electrostatic method. A coil of wire (Fig. 23-37) carrying current causes a B-field

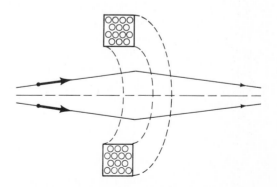

Figure 23-37   Magnetic lens.

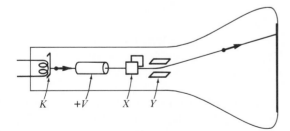

Figure 23-36   Cathode-ray tube.

through which the electron beam travels. If an electron's velocity makes an angle with the axis of the coil, there is a magnetic force, given by $\mathbf{F} = Q\mathbf{v} \times \mathbf{B}$, that changes the beam's direction. By careful design, coil systems can bring a beam of electrons to a sharp focus. Additional current-carrying coils placed beyond the focus coil are used to give deflections in the X and Y directions.

Images in color on the screen of a receiver or a computer display are formed by three fluorescent substances, each of which produces one color. These substances cover the screen's surface with an interlocking network of dots. When electron beams from three electron guns are guided to the proper places on the screen, the different dots of color fluoresce. At a distance, the tiny dots of color merge into a continuous colored image.

Summary

The sensitivity of a galvanometer depends on several design factors, and the solution of the design problem is one of compromise aided by study of the working equation for the instrument. A direct-current motor is basically a galvanometer with a commutator that automatically reverses the current at the proper times. A motor is a generator, and vice versa; this duality is illustrated by the phenomena of back emf in a motor and back torque in a generator.

Alternating currents are produced by a generator that has no commutator. The chief advantage of alternating currents is that transformers can be used to increase or decrease the emf in a circuit. No power is gained; in an ideal transformer the product of emf and current is the same in the primary coil as in the secondary.

Electric energy is distributed through long-distance transmission lines operating at high voltage in order to reduce I^2R losses in the wires.

Coils and capacitors behave differently in ac circuits and dc circuits. Impedance, a generalization of resistance, is defined as the ratio of PD to current for an ac circuit. The impedance of a coil is largely due to the back emf induced in its windings by its own changing current. The impedance of a capacitor is infinite for direct current, since there can be no steady current through a capacitor. Alternating current through a capacitor is possible, since charges flow back and forth in an external circuit as the plates are charged and discharged many times per second. The impedance of a coil goes up as frequency increases, while the impedance of a capacitor goes down as frequency increases.

A thermoelectric emf exists in a circuit where two junctions of dissimilar metals are at different temperatures. The effect is reversible, and a flow of charge through a junction generates or absorbs thermal energy.

Solid-state devices depend on the electrical properties of semiconductors with controlled amounts of substituted impurity atoms. Junction diodes are used to change alternating current into pulsating direct current. A filter capacitor can be used to smooth out the pulses. Transistors of various types are used to control the collector current by changes in the base current (BJT) or by changes in the gate voltage (FET).

In vacuum tubes, electrons are emitted by the heated cathode and flow through the tube to the positively charged plate. Vacuum diodes serve the same function as their solid-state counterparts. In the vacuum triode, the electron flow to the plate is controlled by the potential of the grid.

The cathode-ray tube is a vacuum tube that contains an electron gun, a deflection system, and a fluorescent screen. Magnetic or electrostatic fields are used to focus and deflect the electron beam.

Check List

$$\theta = \left(\frac{NBA}{k'}\right)I$$

armature
commutator
alternating current
direct current
back emf
back torque
eddy currents

$$\frac{\mathcal{E}_s}{\mathcal{E}_p} = \frac{N_s}{N_p}$$

impedance

$$Z = \frac{V}{I}$$

self-inductance
henry
$Z \approx X_L = 2\pi f L$
$Z \approx X_C = 1/2\pi f C$

thermocouple
thermopile
n-type semiconductor
p-type semiconductor
p-n junction
junction diode
bipolar junction transistor (BJT)
emitter
base
collector
field effect transistor (FET)
source
gate
drain
vacuum diode
vacuum triode
cathode
grid
plate
cathode-ray tube

Questions

23-1 Why is a fixed cylindrical iron core placed inside the moving coil of a galvanometer?

23-2 The output of a generator is, to start with, an alternating current. Describe two ways (one mechanical, one electrical) to convert alternating current to pulsating direct current.

23-3 Why has alternating current been so widely adopted in place of the earlier direct-current system prevalent some 90 years ago?

23-4 When a refrigerator motor starts up, the house lights dim momentarily. When an electric iron is plugged in, the lights stay dim for as long as the iron is turned on. Discuss the similarities and differences between these two situations.

23-5 Through a system of gears and belts, the armature of a car's generator is turned at a speed proportional to the engine speed. The headlights are connected directly to the armature. (a) Explain why the lights become brighter when the engine is "gunned" at high speed. (b) Considering the system (engine + generator + lamp filament), discuss the energy transformations, and tell which are reversible and which are not reversible.

23-6 Would the eddy-current brake (Fig. 23-11) operate on aluminum rails (which are nonmagnetic) as well as on steel rails?

23-7 What are some advantages and disadvantages of (a) using 12 V for distribution of electric power in an internal combustion automobile, and (b) using 120 V for distribution of electric power in a house?

23-8 Why is the core of a transformer made of many thin sheets of iron instead of one solid piece?

23-9 What can you say about the construction of a transformer for which the secondary emf equals the primary emf?

23-10 What would happen if the primary of a pole transformer outside a house were supplied with 2300 V dc instead of 2300 V ac?

23-11 Delegate a member of the class to find out from your local electric power company at what voltage the power is generated that eventually supplies the lamps in your classroom. Find out also how many transformers of all sorts there are between the generator and the lamps, and at what voltages the primary and secondary of each transformer operate.

23-12 Can a diode be used for amplifying a weak signal?

23-13 What is the difference between *n*-type germanium and *p*-type germanium?

23-14 Is the resistance of a *p-n* junction the same for forward bias as for reverse bias? If not, in which direction is the resistance greater?

23-15 What would happen if the battery of Fig. 23-30*a* were reversed? Would the device still function as a current amplifier?

MULTIPLE CHOICE

23-16 A motor connected to a 120-V source of PD runs at full speed when there is 10 A through a 3-Ω armature. The back emf is   (*a*) 30 V;   (*b*) 90 V;   (*c*) 120 V.

23-17 In a good transformer with 100 turns in the primary and 500 turns in the secondary, the efficiency is approximately   (*a*) 0.2;   (*b*) 1.0;   (*c*) 5.0.

23-18 The primary winding of a good transformer has a small (*a*) reactance; (*b*) impedance; (*c*) resistance.

23-19 Power lines between cities operate at high voltage in order to decrease   (*a*) current; (*b*) Joule losses in the wires;   (*c*) both of these.

23-20 The function of the capacitor in a rectifier circuit is to   (*a*) decrease the I^2R losses; (*b*) increase the peak voltage;   (*c*) smooth out the current through the load.

23-21 A *p-n-p* transistor consists of two junction diodes separated by thin *n*-type material in the base. In use,   (*a*) one diode is forward biased and the other is reverse biased; (*b*) both are forward biased;   (*c*) both are reverse biased when there is no base current.

Problems

23-A1 A certain galvanometer needle deflects 20° when the current through the coils is 6 μA. What deflection would be produced by the same current if the magnetic field were decreased to $\frac{4}{5}$ of its original value?

23-A2 The strength of the magnetic field of a "permanent" magnet of a certain moving-coil ammeter decreased to 96% of its original value in several decades. The meter was originally calibrated to read 5 A full-scale. What is now the reading caused by a current of 5 A?

23-A3 A generator furnishes an emf of 48 V when its armature turns at 400 rev/min. All other factors remaining the same, what will the emf be if the armature turns at 300 rev/min?

23-A4 The motor of an electric-powered delivery van uses 35 kW of power from a bank of batteries connected in series to give a terminal voltage of 216 V. What current is delivered from the batteries? Why is such a large voltage used?

23-A5 The back emf in a motor is 50 V when the motor is turning at 2000 rev/min. What is it when the motor turns at 3000 rev/min, if the magnetic field remains the same?

23-A6 A doorbell transformer has an 800-turn primary and a 40-turn secondary. When the primary is connected to a 60-Hz, 120-V line, what is the secondary emf?

23-A7 The transformer of Prob. 23-A6 delivers a current of 0.6 A to the bell. What is the current from the 120-V supply circuit?

23-A8 A transformer that powers a neon sign has 200 turns in the primary and 10,000 turns in the secondary. When the primary is connected to a 120-V ac source, the current through it is 60 mA. What are the emf and current in the secondary?

23-B1 A dc motor has field coils that are connected in parallel with the armature and with the supply line. The field coils have resistance 100 Ω, and the armature has resistance 4 Ω. When connected to a 120-V source of PD, the motor turns at such a speed that the back emf in the armature is 108 V. What is the total current through the motor?

23-B2 The armature of the motor of Prob. 23-B1 becomes "frozen" owing to improper lubrication and can no longer turn. If no fuses blow, what will the total current become?

23-B3 A motor connected as in Fig. 23-9 draws a total of 6 A from a 120-V dc line. The resistance of the armature is 2 Ω, and the back emf generated in it is 109 V. Calculate (*a*) the current through the armature; (*b*) the current through the field coil; (*c*) the resistance of the field coil.

23-B4 A motor such as that described in Fig. 23-9 is connected to a PD of 120 V. The resistance of the armature is 3 Ω and that of the field coils is 400 Ω. The back emf is 114 V. Calculate the electrical efficiency of this motor. (*Hint:* Subtract the Joule losses from the input power. Friction in the bearings, air resistance, and hysteresis would cause the efficiency to be somewhat less than calculated here.)

23-B5 Two flat coils of wire are lying on a table, with the small one (the secondary) completely inside the larger coil (the primary) (Fig. 23-38). (*a*) If the switch is closed, in which direction, clockwise or counterclockwise, will conventional charge flow in the primary? (*b*) What will be the direction of the momentary induced current in the secondary, just after the switch is closed?

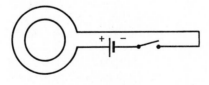

Figure 23-38

23-B6 If the switch in Fig. 23-38 has been closed for some time and is then opened, what will be the direction of the momentary current in the secondary?

23-B7 A transmission line between a substation and a shopping center consists of two wires, each of resistance 0.9 Ω; the current is 100 A and the PD at the input end of the line is 12 kV. Calculate (a) the power input; (b) the power loss in the line; (c) the power delivered to the load; (d) the efficiency of the line.

23-B8 Solve Prob. 23-B7 for the same input power to the same line, but with input end of the line at 3.0 kV instead of 12 kV. (*Hint:* First find the current.) Explain the much reduced efficiency of this arrangement compared with that of Prob. 23-B7.

23-B9 Energy is delivered to an industrial park at the rate of 1000 kW at a terminal voltage of 12 kV measured at the park. Each of two aluminum wires connecting the factory with the power station is 10 km long and 7 mm in radius. (a) What power is lost in the line? (b) What is the efficiency of the line? (c) What is the PD at the input end of the line?

23-B10 What is the impedance of a coil of self-inductance 15 mH, at frequency of 8 kHz, assuming the coil to have a negligible dc resistance?

23-B11 What back emf is induced in a coil of self-inductance 200 mH if the current changes at a steady rate from 12 A to 0 in 3 ms?

23-B12 A 0.01-H coil is made of heavy copper wire. At what frequency is its impedance 400 Ω?

23-B13 In the tone-control circuit of a hi-fi amplifier, a capacitor of 0.002 μF is in series with a resistor of 80 kΩ. (a) At what audio frequency is the impedance of the capacitor 15 times that of the resistor? (b) At what audio frequency is the impedance of the capacitor $\frac{1}{15}$ that of the resistor?

23-B14 The self-inductance of an iron-cored magnet is 8 H, and its resistance is negligible. What is the current through the magnet when it is connected to a 60-Hz alternating emf of 120 V?

23-B15 At what frequency does a well-insulated capacitor of value 0.01 μF have an impedance of 50 kΩ?

23-B16 What is the alternating current through a well-insulated 500-pF capacitor that is connected to a radio-frequency source whose frequency is 700 kHz and whose terminal voltage is 15 V?

23-B17 In a transistor circuit similar to Fig. 23-30, the emf of the battery is 9 V and the load resistor is 1500 Ω. When the base current is 40 μA, the collector current is 3.2 mA; when the base current is changed to 50 μA, the collector current is 3.9 mA. (a) What is the collector potential in each case? (b) What is the current gain of the amplifier?

23-C1 An electric motorcycle is powered by a permanent-magnet motor connected to a 24-V battery (actually, two 12-V batteries in series). The motor is rated at a maximum of 1.11 kW (1.5 hp), and the battery can deliver 90 A for a period of 1 h before recharging is necessary. (a) What is the efficiency of the motor? (b) How much electric energy is available from a single charge of the battery? (c) If the machine's range is 80 km (50 mi) on a single charge, what is the fuel cost per km? (Assume that electric energy costs 5¢ per kWh.) Compare with the fuel cost for a compact car that travels 50 km (30 mi) on 4 liters (1 gallon) of gasoline that costs 50¢ per liter ($1.89/gal). (d) What is the total cost per km if the $100 battery has a useful life of 1000 charge cycles?

23-C2 What will be the total current through the motor of Example 23-1 if the motor is loaded so as to run at only $\frac{4}{5}$ of full speed? (*Hint:* The back emf is proportional to the speed.)

23-C3 What total current will the motor of Prob. 23-B3 use if it is loaded down so that its speed is half the original speed? (*Hint:* The back emf is proportional to the speed.)

23-C4 What is the back emf in a motor that draws 7.1 A from a 220-V line if the field coils have a resistance of 200 Ω and the armature has a resistance of 2 Ω?

23-C5 The output of a generator is stepped up and rectified to give 800 kV at a power station on the Columbia River, and 3200 MW of power is delivered to Los Angeles, about 1350 km away. What must be the total cross-sectional area of aluminum wires in the transmission line if the power loss in the line is to be only 8% (that is, 256 MW)? (*Hint:* First find the current.)

23-C6 An iron-cored coil has self-inductance 4 H and resistance 1.2 Ω. What is the PD across the coil when the current is 30 A, increasing at the rate of 50 A/s?

For Further Study

23-13   The Emf of a Generator

An equation for the emf of a generator can be derived with the help of differential calculus. According to Eq. 22-11, the magnitude of an induced emf is equal to the rate of change of magnetic flux: $\varepsilon = -d\Phi/dt$.* In a generator the flux changes because the *orientation* of the plane of the loop changes. In Fig. 23-6, the component of **B** perpendicular to the plane of the armature is $B\cos\phi$, where ϕ measures the rotation of the loop (of area A) from its vertical position, and the flux through the loop is $B\cos\phi\ A$. If the generator is turning steadily at an angular speed ω rad/s, the angle ϕ increases according to $\phi = \omega t = 2\pi ft$, where f is the frequency in rev/s and t is the time. Hence

$$\Phi = B(\cos 2\pi ft)A$$

$$\varepsilon = -\frac{d\Phi}{dt} = -BA\frac{d}{dt}(\cos 2\pi ft)$$

To find the derivative of $\cos 2\pi ft$, we use the table of derivatives in Appendix E, where we find that if $y = \cos ax$, then $dy/dx = -a\sin ax$. Here a is $2\pi f$, x is t, and so we obtain

$$\varepsilon = 2\pi fBA\sin 2\pi ft$$

The generator's emf is a sinusoidal function of time. The magnitude of the emf is proportional to the frequency of rotation of the armature, as well as to the area of the loop and the strength of the magnetic induction.

23-14   Formulas for Reactance

That property of a coil or capacitor that limits the flow of *alternating* current is called its *reactance* (X_L or X_C, measured in ohms). The term *impedance* (Z, also measured in ohms) includes the effect of ordinary ohmic resistance as well as reactance. The impedance of a circuit or part of a circuit is defined as

$$Z = \frac{V_0}{I_0}$$

where V_0 is the maximum or peak PD across the circuit, and I_0 is the maximum or peak current through the circuit. The complicating factor is that V and I do not reach their peak values at the same time, unless the circuit is a pure resistance. In the next two sections we shall study some of the simpler aspects of ac circuits that contain both reactance and resistance.

Consider the current through a coil of self-inductance L connected to an ac generator whose

---

* The $-$ sign is used with $d\Phi/dt$ for the reasons discussed on page 500 in connection with Lenz's law.

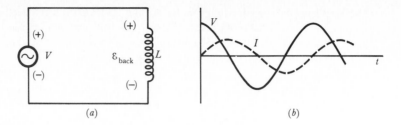

Figure 23-39 (a) Circuit containing an ac source and a coil. The (+) and (−) signs indicate the polarities of coil and generator at an instant when $I = 0$ and is increasing. (b) The current through the coil lags $\frac{1}{4}$ cycle behind the PD.

terminal voltage is V (Fig. 23-39). The resistance of the coil is assumed to be negligible. The frequency is f cycles/s, and the graph of I versus t is a sinusoidal curve whose equation may be written as $I = I_0 \sin 2\pi ft$. To avoid many factors of 2π in what follows, we use the *angular frequency* in rad/sec, given by $\omega = 2\pi f$. Thus we represent the current by the equation

$$I = I_0 \sin \omega t$$

As we saw in Sec. 23-7, the magnitude of the back emf in the coil depends on the rate of change of current and is given by $\mathcal{E}_{\text{back}} = L(dI/dt)$. The slope of the current graph is greatest when $t = 0$, so at this instant, when the current is *changing* most rapidly, the rate of change of flux is greatest and the back emf is greatest. Since the generator and the coil are in parallel, at any given instant their terminal voltages are equal. Hence,

$$V = \mathcal{E}_{\text{back}} = L \frac{dI}{dt}$$

Qualitatively, we see from the graphs that the current through the coil is not in phase with the PD across the coil. In fact, the current reaches its maximum value $\frac{1}{4}$ cycle *after* the applied PD is at its peak. This delay is, of course, caused by the back emf, which by Lenz's law opposes the change (an increase, in this case) of current. If the frequency is increased, the necessary terminal voltage of the generator must be larger, for then the time Δt is less for a given ΔI, the slope dI/dt is larger, and the back emf is larger.

To obtain a quantitative formula for the impedance of a coil of negligible resistance (that is, for its reactance), we use a formula for the derivative of the sine function in Appendix E that states that if $y = \sin ax$, then $dy/dx = a \cos ax$. Here the constant a is ω and x is t. Hence,

$$V = \mathcal{E}_{\text{back}} = L \frac{dI}{dt} = L \frac{d}{dt} (I_0 \sin \omega t)$$

$$= L I_0 \frac{d}{dt} (\sin \omega t)$$

$$= L I_0 \omega \cos \omega t$$

Thus

$$V = V_0 \cos \omega t$$

where V_0, the maximum value of V, is $\omega L I_0$. The reactance is

$$X_L = \frac{\text{max. value of applied PD}}{\text{max. value of current}} = \frac{\omega L I_0}{I_0}$$

or

$$X_L = \omega L$$

This is the formula stated without proof in Sec. 23-7. The $\frac{1}{4}$-cycle lag of the current through a coil is apparent from the fact that I is a sine function if V is a cosine function. Another way of stating this phase relationship is to say that *for an inductance, the applied voltage leads the current by $\frac{1}{4}$ cycle.*

The reactance of a capacitor can be calculated in much the same way. Consider the PD across the capacitor in Fig. 23-40, equal at all times to

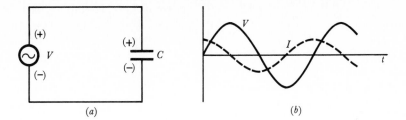

Figure 23-40 (a) Circuit containing an ac source and a capacitor. (b) The current through the capacitor leads the PD by $\frac{1}{4}$ cycle.

the generator's terminal voltage. If the origin of the time axis is chosen so that the graph of the PD is a sine curve, then the charge on the capacitor is also a sine curve, since $Q = CV$ at all times. The current is the rate of flow of charge, and the charges are able to flow into the capacitor most easily when the capacitor is uncharged and presents no opposing PD. This makes the current $\frac{1}{4}$ cycle out of phase with the applied PD; but in contrast to the behavior of the coil, the current in a capacitor reaches its peak *before* the PD does. We assume that the applied PD is given by $V_0 \sin \omega t$. Then,

$$I = \frac{dQ}{dt} = \frac{d}{dt}(CV) = C\frac{d}{dt}(V_0 \sin \omega t)$$

$$= CV_0 \omega \cos \omega t$$

or

$$I = I_0 \cos \omega t$$

where I_0, the maximum value of I, is ωCV_0. The reactance is

$$X_C = \frac{\text{max. value of applied PD}}{\text{max. value of current}} = \frac{V_0}{\omega CV_0}$$

or

$$X_C = \frac{1}{\omega C}$$

This is the formula stated without proof in Sec. 23-8. This $\frac{1}{4}$-cycle lead of the current through a capacitor is apparent from the fact that I is a cosine function if V is a sine function (Fig.

23-40). Another way of stating this phase relationship is to say that *for a capacitor*, the applied voltage *lags* the current by $\frac{1}{4}$ cycle.

23-15 Phasors

Let us fix our attention on the *voltage* across an inductor or a capacitor or a resistor. We have seen that voltage across a pure inductance *leads* the current by 90° ($\frac{1}{4}$ cycle); the voltage across a pure capacitance *lags* the current by 90°; and in a pure resistance the voltage is *in phase with* the current. To study more complex circuits we use a mathematical device somewhat reminiscent of the reference circle in simple harmonic motion (page 200). Varying voltages are represented by rotating arrows called *phasors*.* Later, we use phasors to represent reactances and impedances.

Consider the series circuit of Fig. 23-41a, in which there is an alternating current given by

$$I = I_0 \cos \omega t$$

through a resistor R and a coil whose reactance is X_L. (As shown in Sec. 23-14, $X_L = \omega L$.) We choose the *cosine* function so that I has its maximum value (I_0) at $t = 0$. The voltage of the

---

* Although designated by arrows having magnitude and direction, we do not call them vectors because they do not have spacelike properties. Like vectors, phasors can be "added" by the head-to-tail method or by the component method.

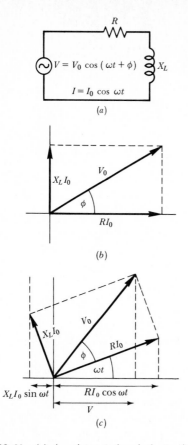

(a)

(b)

(c)

Figure 23-41 (a) A resistor and an inductor in series with a generator. (b) Voltages in the circuit, shown at a time of maximum current (at $t = 0$). (c) Voltages in the same circuit, a short time later than in part (b); all phasors have rotated through an angle ωt. Note that one right triangle shows that $V = V_0 \cos(\omega t + \phi)$, as assumed in Eq. 23-8.

generator varies sinusoidally at the same angular frequency ω as the current, but this voltage is not in phase with the current. Thus we write

$$V = V_0 \cos(\omega t + \phi) \qquad (23\text{-}8)$$

where ϕ is the *phase angle* of the circuit. We wish to find the relation between V_0 and I_0 and thus find the impedance of the combination from $Z = V_0/I_0$, and we also wish to find the phase

angle ϕ between the applied voltage and the current. If there were no resistance in the circuit, the voltage across X_L would lead the current by 90°, as in Fig. 23-39b, but here the phase angle is less than 90°.

The graphical interpretation, using phasors, is shown in Fig. 23-41. We set up two arrows of length RI_0 and $X_L I_0$ to represent the maximum or "peak" voltages across R and L, respectively. These arrows maintain their 90° orientation as both phasors rotate counterclockwise at ω rad/s. The projection of each phasor on the reference axis represents the instantaneous value of the PD across the resistor or inductor, as the case may be. At the moment shown in Fig. 23-41b, the current through the circuit is a maximum. In Fig. 23-41c, the PD across the resistor (RI) is $RI_0 \cos \omega t$, which is less than at $t = 0$, and the PD across the inductor $(X_L I)$, which was 0 at $t = 0$, is now $-X_L I_0 \sin \omega t$. The net PD across the combination is then

$$V = RI_0 \cos \omega t - X_L I_0 \sin \omega t$$

By Kirchhoff's loop rule, this net PD must equal V, the projection of V_0 on the reference axis. In making use of this construction, we use the fact that for phasors, as for vectors, the sum of the projections equals the projection of the sum.

We can make a new diagram involving only reactance, resistance, and impedance (all in Ω) by

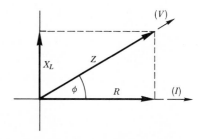

Figure 23-42 Phasor diagram for reactance and impedance in a series circuit. The arrow designated (I) indicates that the voltage across the resistance is in phase with the current.

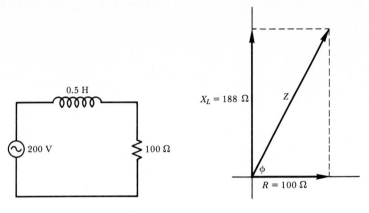

Figure 23-43 *R-L* circuit.

dividing each PD magnitude in Fig. 23-41b by I_0 (which is the same for each phasor, since in a series circuit the current is the same in each element). The division of V_0 by I_0 gives the impedance Z. Thus we obtain a useful form of the phasor diagram from which impedance and phase angle can be found (Fig. 23-42). From the diagram, we see that the current in the circuit lags the voltage by an angle ϕ, given by tan $\phi = X_L/R$. Thus, for the circuit considered,

$$Z = \sqrt{R^2 + X_L^2} = \sqrt{R^2 + (\omega L)^2}$$

Also,

$$\tan \phi = X_L/R \quad \text{or} \quad \tan \phi = \omega L/R$$

In actual practice, any real coil of wire has *some* resistance (unless it is superconducting near 0 K). Such a coil is drawn in a circuit diagram as a pure inductance in series with its own resistance.

The phasor diagram is shown in Fig. 23-43. Here $\omega = 2\pi f = 2\pi(60 \text{ Hz}) = 377 \text{ rad/s}$. Therefore,

$$X_L = \omega L = (377 \text{ rad/s})(0.5 \text{ H}) = 188 \ \Omega$$

$$Z = \sqrt{R^2 + X_L^2} = \sqrt{(100 \ \Omega)^2 + (188 \ \Omega)^2}$$
$$= 213 \ \Omega$$

$$I_0 = \frac{V_0}{Z} = \frac{200 \text{ V}}{213 \ \Omega} = \boxed{0.94 \text{ A}}$$

$$\tan \phi = \frac{X_L}{R} = \frac{188 \ \Omega}{100 \ \Omega} = 1.88$$

$$\phi = \boxed{62°}$$

The current lags 62° behind the applied PD.

To extend the analysis to a series circuit with capacitance, we use a *downward* phasor to represent X_C, since the voltage across a capacitor *lags* the current by 90°.

Example 23-8

A coil of inductance 0.50 H and resistance 100 Ω is connected to a 60-Hz source whose peak PD is 200 V. Calculate the peak current and the phase difference between the current and the applied voltage.

Example 23-9

In the circuit of Example 23-8, a capacitor of 20.0 μF is added. Calculate the maximum current and the phase angle for the circuit (Fig. 23-44).
    The reactance of the capacitor is

$$X_C = \frac{1}{\omega C} = \frac{1}{(377 \text{ Hz})(20 \times 10^{-6} \text{ F})} = 132.6 \ \Omega$$

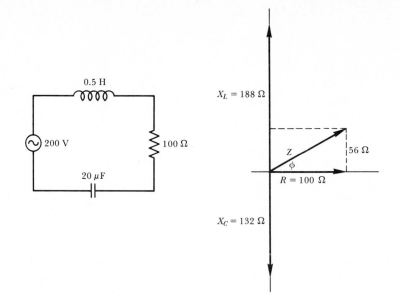

Figure 23-44 *R-L-C* circuit.

Therefore, the net reactance of the circuit is

$$X_L - X_C = 188.5\ \Omega - 132.6\ \Omega = 56\ \Omega$$

$$Z = \sqrt{R^2 + (X_L - X_C)^2}$$

$$= \sqrt{(100\ \Omega)^2 + (56\ \Omega)^2} = 115\ \Omega$$

$$I_0 = \frac{V_0}{Z} = \frac{200\ \text{V}}{115\ \Omega} = \boxed{1.75\ \text{A}}$$

$$\tan\phi = \frac{56\ \Omega}{100\ \Omega} = 0.56$$

$$\phi = \boxed{29°}$$

From these examples we see that the addition of the capacitor has to some extent canceled the reactance of the coil; Z is less, the current is larger, and the phase angle is closer to $0°$.

In general, for a series circuit, the impedance and phase angle are given by

$$Z = \sqrt{R^2 + (X_L - X_C)^2} \qquad (23\text{-}9)$$

$$\tan\phi = \frac{X_L - X_C}{R} \qquad (23\text{-}10)$$

If $\phi > 0$, the voltage leads the current or (equivalently) the current lags the voltage. Such a circuit is said to be "inductive" since $X_L > X_C$.

23-16   Series Resonance

For a given frequency, *resonance* occurs if the current through the series circuit is a maximum. For this to take place, the impedance Z must have its minimum value, which requires that $X_L = X_C$ and thus $\phi = 0°$. At resonance, the current is in phase with the voltage, and the circuit behaves like a pure resistance. Either L or C can be adjusted to "tune" the circuit for resonance at a given frequency.

Example 23-10

For what value of C in Fig. 23-45 would the current from the source be a maximum?

For resonance we must make $X_L = X_C$. Here,

$$X_L = \omega L = (1000 \times 2\pi\ \text{rad/s})(0.1\ \text{H}) = 628.3\ \Omega$$

Hence X_C must also be 628.3 Ω.

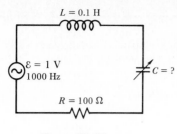

Figure 23-45

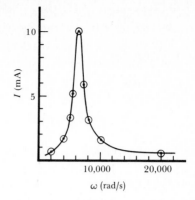

Figure 23-46   Resonance curve.

$$X_C = \frac{1}{\omega C}$$

$$C = \frac{1}{\omega X_C} = \frac{1}{(6283 \text{ rad/s})(628.3 \ \Omega)}$$

$$= 2.53 \times 10^{-7} \text{ F} = \boxed{0.253 \ \mu\text{F}}$$

When C is adjusted to this value, the circuit behaves like a pure resistance of 100 Ω, with the current given by

$$I = \frac{1 \text{ V}}{100 \ \Omega} = 0.010 \text{ A} = \boxed{10 \text{ mA}}$$

ω (rad/s)	X_L (Ω)	X_C (Ω)	Z (Ω)	I (mA)	ϕ
2000	200	1974	1776	0.56	$-87°$
4000	400	987	595	1.68	$-80°$
5000	500	790	306	3.26	$-71°$
5500	550	718	196	5.11	$-59°$
6283	628	628	100	10.00	$0°$
7000	700	564	169	5.92	$+54°$
8000	800	493	322	3.10	$+72°$
10000	1000	395	613	1.63	$+81°$
20000	2000	197	1805	0.55	$+87°$

The current in a circuit with fixed values of L and C changes as the frequency of the applied voltage changes. A few points on the resonance curve for the circuit of Fig. 23-45 are calculated using Eqs. 23-9 and 23-10, and plotted in Fig. 23-46. Note the current peak and the zero phase shift at the resonant frequency of 6283 rad/s (1000 Hz).

In general, the resonant frequency of an L-R-C circuit can be found by setting $X_L = X_C$. Then, from $\omega L = 1/\omega C$, we obtain

$$\omega_{\text{resonance}} = \sqrt{\frac{1}{LC}} \quad \text{(rad/s)} \qquad \text{(23-11)}$$

$$f_{\text{resonance}} = \frac{1}{2\pi} \sqrt{\frac{1}{LC}} \quad \text{(Hz)} \qquad \text{(23-12)}$$

The resonant frequency depends only on L and C. The value of the resistance R determines the maximum current and the sharpness of the resonance. If R is small compared to X_L or X_C at the resonant frequency, the resonance is "sharp," and there is a large current over only a small range of frequencies near the resonant frequency.

It is by the use of resonant circuits that a radio receiver is tuned to the frequency of a desired station. (See Sec. 24-3 for a different approach to the resonant frequency based on analogies with mechanical SHM.)

23-17   Power in an AC Circuit

Consider first a pure resistance R connected to a PD given by $V = V_0 \cos \omega t$. The instantaneous current is

$$I = \frac{V}{R} = \frac{V_0 \cos \omega t}{R} = I_0 \cos \omega t$$

which is exactly in phase with the instantaneous voltage V. At any instant, the power dissipated in the resistance is

$$P = RI^2 = RI_0{}^2 \cos^2 \omega t$$

We see that the power varies from 0 to a maximum value RI_0^2. Now the average value* of $\cos^2 \omega t$ is given by $\overline{\cos^2 \omega t} = \frac{1}{2}$, so we can write for the average power in a resistor

$$\bar{P} = RI_0^2 \overline{\cos^2 \omega t} = \frac{1}{2}RI_0^2$$

$$= R\left(\frac{I_0}{\sqrt{2}}\right)^2 = R(I_{\text{rms}})^2$$

where I_{rms}, the root-mean-square current, is $I_0/\sqrt{2} = 0.707 I_0$. The rms current is equal to the steady (dc) current that would produce the same rate of heating in the resistor. In this sense, the rms current is the "effective current"; it is equal to $1/\sqrt{2}$ times the peak or maximum current.

In a similar fashion, starting from

$$P = \frac{V^2}{R} = \frac{V_0^2 \cos^2 \omega t}{R}$$

we deduce that $V_{\text{rms}} = V_0/\sqrt{2}$. As a practical matter, the scale of an ordinary voltmeter or ammeter is calibrated to read the rms values. Unless specified otherwise, an ac voltage or current refers to the rms value.

Example 23-11

A capacitor is connected across a 120-V, 60-Hz power outlet. One side of the outlet is grounded, always at 0 V. Between what limits does the potential of the ungrounded plate of the capacitor fluctuate?

The rms voltage is 120 V, which implies that one terminal of the outlet is, at its peak, $120\sqrt{2}$ or 170 V above or below the other (grounded) terminal. Relative to the grounded terminal, the ungrounded plate fluctuates between $+170$ V and -170 V. An ac voltmeter connected to the plates would read 120 V.

---

* The bar over an expression represents the average value (see page 26). Since the graphs of $\cos^2 \omega t$ and $\sin^2 \omega t$ are identical in shape and differ only in phase—both are sinusoidal—we deduce that $\overline{\cos^2 \omega t}$ is the same as $\overline{\sin^2 \omega t}$. Each average equals $\frac{1}{2}$, since at any time $\cos^2 \omega t + \sin^2 \omega t = 1$.

Example 23-12

A 100-W lamp bulb is connected to a 60-V ac source. What is the peak value of the current through the bulb?

$P = VI$, so

$$I = \frac{P}{V} = \frac{60 \text{ W}}{120 \text{ V}} = 0.5 \text{ A}$$

This is the rms current and would be the reading of an ac ammeter connected in series with the bulb. The peak current is

$$I_0 = (0.5 \text{ A})\sqrt{2} = \boxed{0.71 \text{ A}}$$

For a circuit containing reactance, the current is not in phase with the voltage, and the power is less than $V_{\text{rms}}I_{\text{rms}}$. A little trigonometry is needed here. Let ϕ be the phase angle of the circuit.

$$
\begin{aligned}
P &= VI \\
&= [V_0 \cos(\omega t + \phi)] \cdot I_0 \cos \omega t \\
&= V_0 I_0 (\cos \omega t \cos \phi - \sin \omega t \sin \phi)(\cos \omega t) \\
&= V_0 I_0 \cos^2 \omega t \cos \phi - V_0 I_0 \sin \omega t \cos \omega t \sin \phi \\
&= V_0 I_0 \cos^2 \omega t \cos \phi - \frac{1}{2}V_0 I_0 \sin 2\omega t \sin \phi
\end{aligned}
$$

Now $\cos^2 \omega t$ is always positive and, as we have seen, averages out to $\overline{\cos^2 \omega t} = \frac{1}{2}$. On the other hand, $\sin 2\omega t$ is positive for half the cycle and negative for half the cycle, so $\overline{\sin 2\omega t} = 0$, and the second term drops out. Therefore, the average power is

$$\bar{P} = V_0 I_0 (\tfrac{1}{2}) \cos \phi$$

$$= \left(\frac{V_0}{\sqrt{2}}\right)\left(\frac{I_0}{\sqrt{2}}\right) \cos \phi \qquad (23\text{-}13)$$

or

$$\bar{P} = V_{\text{rms}} I_{\text{rms}} \cos \phi \qquad (23\text{-}14)$$

This is just like the formula for power in a dc circuit, with the additional factor $\cos \phi$, which is called the *power factor*. Usually the subscripts can be omitted, and we write

$$P = VI \cos \phi \qquad (23\text{-}15)$$

where it is understood that rms values are used for P, V, and I.

Recalling that $V_0 = I_0 Z$ (from the definition of Z), we can substitute into Eq. 23-13 to obtain

$$P = I^2 Z \cos \phi \qquad (23\text{-}16)$$

$$P = (V^2/Z) \cos \phi \qquad (23\text{-}17)$$

For a purely resistive circuit, Z is the resistance R, ϕ is $0°$, and these equations reduce to those derived from Ohm's law in Chap. 20.

--- **Example 23-13**

A motor draws 2.00 A from a 120-V, 60-Hz source, and its power factor is 0.80. What power is used by the motor?

$$P = VI \cos \phi = (120 \text{ V})(2.00 \text{ A})(0.80) = \boxed{192 \text{ W}}$$

In this example, note that the same 192 W of power could be delivered at 120 V to a purely resistive load (such as a lamp bulb) using only $192/120 = 1.60$ A. From the power company's viewpoint, the other 0.40 A is "wattless"; it is not paid for by the consumer, and it does no work on the consumer's premises. However, this "wattless" current does contribute to the Joule heating of the transmission line between the generator and the user; this is truly wasted power. The electrical systems of most factories are primarily inductive loads due to the magnet windings of their motors, with current lagging behind the voltage. A description of the use of capacitors, or their equivalents, to bring the power factor of a large industrial plant closer to the optimum value of 1.00 would take us too far afield.

A seeming paradox arises when a series circuit is at or near resonance. In Fig. 23-47 the re-

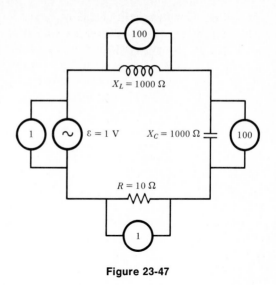

Figure 23-47

actances cancel, the total impedance is just 10 Ω, and the current is given by

$$I = \frac{\varepsilon}{Z} = \frac{1 \text{ V}}{10 \, \Omega} = 0.10 \text{ A}$$

Suppose we connect ideal voltmeters (of very high resistance) across each circuit element. The voltmeter across the inductance reads $V_L = IX_L = (0.1 \text{ A})(1000 \, \Omega) = 100$ V. All this with a 1-V generator! We must remember, however, that the meter readings are rms values, averaged over a complete cycle. V_L *leads* the current by 90°, whereas V_C *lags* the current by 90°. These voltages are thus 180° out of phase; at any instant $V_L = -V_C$, and Kirchhoff's loop theorem is satisfied at all times. Nevertheless, the capacitor's dielectric insulation must, twice each cycle, sustain a peak PD of as much as 141 V, even though the generator's PD is only 1 V.

Problems   23-C7   A flat coil of resistance 5 Ω has 500 turns, each 5 cm × 5 cm, and rotates at 3600 rev/min in a field of 0.05 T. The coil is connected to a 20-Ω load resistor. Calculate (a) the generated emf as a function of time;   (b) the maximum terminal voltage; (c) the rms terminal voltage.

23-C8   Calculate the resistance of a coil for which the impedance is 25 Ω and the reactance is 15 Ω.

23-C9   Calculate the impedance and the phase angle, at 1000 Hz, of a coil of inductance 30 mH and resistance 80 Ω.

23-C10 Calculate the impedance and phase angle, at 1000 Hz, of a circuit consisting of 0.01 μF in series with 20 kΩ.

23-C11 A 120-V bulb rated at 5 W is connected to a 120-V, 60-Hz source through a 20-μF capacitor. Show that the bulb is lit to practically full brilliance. (*Hint:* Compare the resistance of the bulb and the impedance of the series circuit.)

23-C12 Calculate the resonant frequency when a 0.002-μF capacitor is placed in series with a 500-μH coil of resistance 50 Ω.

23-C13 A capacitor C is in series with a coil of inductance 100 μH. For what value of C does the circuit resonate at 1 MHz?

23-C14 A 0.02-μF capacitance and a resistance R are connected in series across a 100-kHz source of potential. For what value of R is the PD across R equal to half the PD of the source?

23-C15 A coil of inductance 4 H and resistance 150 Ω is connected to a 120-V, 60-Hz PD. A capacitor is connected in series with the coil to bring the power factor up to 1.00. (*a*) What is then the current in the circuit? (*b*) What value of capacitance is required? (*c*) What peak PD must the capacitor withstand?

23-C16 A series circuit contains a 230-V, 60-Hz source of PD, a coil of 0.6 H, a resistor of 60 Ω, and a capacitor of 10 μF. Calculate (*a*) the current through the circuit; (*b*) the phase angle; (*c*) the power factor. (*d*) Is the circuit inductive or capacitive?

23-C17 When a small electromagnet is connected to a 120-V, 60-Hz source, an ammeter indicates 0.40 A through the magnet, and a wattmeter indicates the power used by the magnet to be 40 W. (*a*) Calculate the phase angle of the circuit. (*b*) Calculate the resistance of the magnet. (*c*) Calculate the inductance of the magnet.

23-C18 What is the peak PD between a wire and the ground for a 765-kV ac overhead transmission line?

References

1. E. Schlömann, "Recovery of Nonmagnetic Metals from Municipal Waste," *Phys. Teach.* **14**, 116 (1976). A constructive use of eddy currents.
2. T. H. Geballe and J. K. Hulm, "Superconductors in Electro-power Technology," *Sci. American* **243**(5), 138 (Nov. 1980). A lengthy and informative survey.
3. B. M. Schwarzschild, "High-capacity buried power lines," *Phys. Today* **32**(10), 20 (Oct. 1979). The purpose of superconducting cables will be to deliver large amounts of electrical energy compactly, at modest voltages.
4. G. Shiers, "The First Electron Tube," *Sci. American* **220**(3), 104 (Mar. 1969). The Edison effect; the Fleming valve (diode).
5. R. A. Chapman, "De Forest and the Triode Detector," *Sci. American* **212**(3), 92 (Mar. 1965).
6. W. S. Shockley, "Transistor Physics," *Am. Scientist* **42**, 41 (1954).
7. W. Brattain, "Genesis of the Transistor," *Phys. Teach.* **6**, 106 (1968).
8. R. E. Alley, Jr., "Semiconductors and Semiconductor Physics," *Phys. Teach.* **3**, 55 (1965).
9. W. C. Hittinger, "Metal-Oxide Semiconductor Technology," *Sci. American* **229**(2), 48 (Aug. 1973).
10. S. McDonald and E. Brown, "The transistor and attitude to change," *Am. J. Physics* **45**, 1061 (1977).
11. R. S. Mackay, in C. L. Stong's "Amateur Scientist" department, describes some interesting cyclic effects (very low-frequency oscillations) related to variations of impedance with current. *Sci. American* **205**(2), 143 (Aug. 1961).

24

Electromagnetic Waves

The type of electromagnetic radiation known as "light" is, of course, familiar to all of us, but the identification of light as an electromagnetic wave is a comparatively recent development. The wave nature of light has been known only since about 1800, and not much more than a century has elapsed since Maxwell proved (in 1864) that electromagnetic (e-m) waves are a necessary consequence of the laws of electricity and magnetism. As we pointed out in Chap. 14, not all electromagnetic radiation is visible, and in Table 14-5 (page 323) we listed various types of e-m radiation, which range from the low-frequency radio waves of communications to the high-frequency gamma rays of nuclear physics.

24-1 Models of Radiation

How shall we describe electromagnetic radiation? Certainly, any description (or model) of e-m radiation must include a means by which energy can be transported from place to place. The earth is warmed by infrared radiation from the sun; a photographic emulsion is "exposed"

when it absorbs light energy; radio waves transmit enough energy to control the flow of electrons in a television receiver. It is not hard to think of illustrations of energy transfer by other e-m radiations such as ultraviolet radiation, x rays, or gamma rays.

One simple model for radiation would be a stream of particles, or corpuscles; this view was held in the earliest times. By the time of Newton it was realized that energy can also be transmitted by means of waves. The pounding of the surf on the shore, the transmission of audible speech by sound waves—these illustrate the transmission of energy by wave motions. To decide between a corpuscular model and a wave model of light (and other e-m radiation), we must look for experiments that can be described by one model but not the other.

Here we come upon one of the most baffling chapters in the history of physics. Some experiments confirm the corpuscular model and seemingly cannot be explained by the wave model; other experiments are equally definite in causing us to reject the corpuscular model in favor of the wave model. In the end, we shall be led to accept

Two tank circuits are shown in this radio amateur's transmitter. The same variable capacitor is used; to change frequency bands, one of the two coils is selected by the switch at the right.

a new model, which combines features of both. We cannot describe this dual model of radiation at this point; first we must learn about the wave model and the corpuscular model separately before attempting to fuse them into a single acceptable model.

In Chaps. 24 through 27 we shall be concerned with those aspects of radiation that are best described by a wave model, chiefly the phenomena associated with the propagation of light, including reflection, refraction, interference, and diffraction. In Chaps. 29 through 33 we shall study those aspects of radiation that are best described by a corpuscular model, chiefly the phenomena associated with the emission and absorption of radiation. We cannot say that either model is "correct," nor can we say that it is "incorrect." Rest assured that we are not going to use either model unless it is "correct enough" to allow us to correlate observations and predict the results of new experiments. After all, such correlation and prediction are about all that one can ask of any model, however good.

24-2 Electric Oscillations

If energy is to be transmitted by some sort of electric wave, we must look for some type of electric oscillation that can be analogous to the vibration of a bell or string that sends out sound waves. We can learn a great deal by exploring the analogies between electric circuits and mechanical systems. Let us look at the characteristics of a vibrating body—say, an object of mass m on the end of a spring whose force constant is k (see Fig. 9-13). There is a continual interchange of energy from one form to another. The total energy of the system is sometimes stored as KE, sometimes as PE. In Sec. 9-3 we were able to use the energy principle to derive a formula for the period of oscillation T of the mass. The result was $T = 2\pi\sqrt{m/k}$. During this time the oscillating body goes through one cycle of its vibration. The period, which is independent of the amplitude of vibration, depends on the stiffness of the spring (a factor related to the force

constant k, and hence to the PE) and also on the inertia of the body (a factor related to the mass m, and hence to the KE). Such a vibrating mass has a *natural frequency*, and it can set into motion a medium such as air or water, thus giving rise to a wave. The frequency of the wave will be the frequency of the oscillating body; since frequency is the reciprocal of the period, we can write

$$f = \frac{1}{2\pi}\sqrt{\frac{k}{m}} \qquad (24\text{-}1)$$

In searching for an electrical analogy to mechanical oscillation, we recall that electric PE can be stored in a capacitor. In Sec. 19-10 the PE of a charged capacitor was derived:

$$PE = \tfrac{1}{2}(1/C)Q^2 \qquad (24\text{-}2)$$

where $1/C$ is analogous to the force constant k. (The PE of a stretched spring is given by $\tfrac{1}{2}kA^2$; see Sec. 9-3.) The electrical analog of KE must be sought for among magnetic effects, for KE is energy due to motion, and magnetism is caused by the motion of charges. Consider a coil whose self-inductance is L, through which a battery sends a steady current. If the resistance of the coil and battery are very small, only a negligible amount of Joule heat is being dissipated in the coil, and hence the seat of emf is transforming chemical energy into electric energy at a negligible rate. Not so during the time just after the switch was closed; while the magnetic field was being established, the magnetic flux through the coil was changing, a back emf was generated in the coil, and work had to be done to force charges through the coil against the opposing PD of the back emf. After the current reached its final value, the flux was constant and the back emf disappeared; but while the current and field were building up, a back emf existed and work had to be done. We can say that energy is stored in the magnetic field, and our analysis involving back emf gives us a mechanism by which work is done in establishing the field.

The energy stored in a current-carrying coil may be likened to KE, for it is due to the motion of charges. The situation is reminiscent of the storage of energy in a moving body such as an automobile. Disregarding friction, once the car gets up to speed, no further work need be done, and the KE remains constant. But while it is being accelerated, there is a continuous flow of energy from the fuel, and work is done. We will not here derive the formula for energy stored in a coil carrying a steady current, except to note that the back emf is proportional to the self-inductance L; hence we would expect the stored energy to be proportional to L. The result of a calculation given in Sec. 24-9 is that

$$KE = \tfrac{1}{2}LI^2 \qquad (24\text{-}3)$$

where L is the self-inductance (or simply the "inductance") of the coil, in henries, and I is the current, in amperes. The stored energy is measured in joules. There is a good analogy between this formula and the formula $KE = \tfrac{1}{2}mv^2$ for mechanical KE; L is analogous to m, and I^2 (related to motion of charge) is analogous to v^2 (related to motion of the body).

We are ready now to consider electric oscillations. To produce an oscillation we must arrange for the exchange of energy from KE to PE, and vice versa. This is accomplished in the circuit of Fig. 24-1. Suppose that the capacitor is initially charged (a), and then the switch is closed. Charge starts to flow through the coil (b), and some of the PE stored in the capacitor becomes KE as the magnetic field of the coil builds up. In (c), the capacitor is empty of charge, but charges are in motion in the coil, and all the energy of the system is in the form of KE. The process does not stop, however. The charges move on through the coil and begin to accumulate (d) on the bottom plate of the capacitor. Eventually (e) the capacitor is fully charged again (opposite to its original condition), and all the energy is again stored as PE, since the current is momentarily zero. The process could go on indefinitely except for the fact that the wires have some small resistance, so that the energy is dissipated in Joule heat and the oscillation dies out.

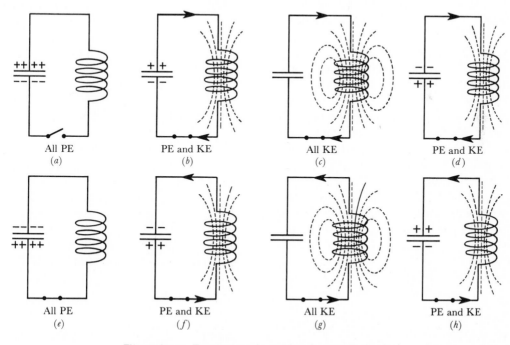

Figure 24-1 Energy transformations in an L-C circuit.

Note the point-by-point similarity of this process to the mechanical oscillation of a mass on a spring. There, too, the oscillation eventually dies out as work is done against internal friction in the spring and against the external air resistance. The ultimate fate of the mechanical energy is heat, just as in the electric oscillation. We can use the analogy between mechanical and electric oscillations to write down a formula for the frequency of an electric oscillation. If the force constant k is replaced by $1/C$, and if the mass m is replaced by L, the formula

$$f = \frac{1}{2\pi} \sqrt{\frac{k}{m}}$$

becomes

$$f = \frac{1}{2\pi} \sqrt{\frac{(1/C)}{L}}$$

Simplifying, we obtain

$$f = \frac{1}{2\pi} \sqrt{\frac{1}{LC}} \qquad (24\text{-}4)$$

This is the natural frequency of the simple circuit of Fig. 24-1. This equation for the resonant frequency was derived in Sec. 23-16 using the theory of alternating-current circuits.

In this section we have used the energy principle to find that $1/C$ is analogous to k, and L is analogous to m.

We could have proceeded directly from Kirchhoff's loop rule to find that $1/C$ is analogous to the force constant k, and L is analogous to the mass m. In Fig. 24-1, at any time, $V_{cap} + \mathcal{E}_{back} = 0$ by the loop rule; here V_{cap} is the PD across the capacitor, given by Q/C, and $\mathcal{E}_{back}$ is the back emf in the coil, given by $L \, \Delta I/\Delta t$ (see footnote on page 525). Therefore,

$$-V_{cap} = \mathcal{E}_{back}$$

$$-\frac{1}{C}(Q) = L \frac{\Delta I}{\Delta t} \qquad (24\text{-}5)$$

But since I is the rate of flow of charge, $\Delta I/\Delta t$ is analogous to the acceleration of charge. We see that Eq. 24-5 is analogous to the force equation

for a mass on a spring, which is

$$-kx = ma$$

provided we identify k with $1/C$, and L with m. This is the same result that we obtained by considering analogies to stored energy.

24-3 Electric Resonance

In Sec. 11-4 we saw that a natural frequency of a pendulum, string, pipe, or other vibrating object is also a resonant frequency. The mechanical system responds to an applied frequency that is in tune with a natural frequency of oscillation. Our simple electric circuit has only a single natural frequency,* given by $f = (1/2\pi)\sqrt{1/LC}$. If we insert an ac generator into our circuit, we find that the current is large only if the frequency of the generator equals the resonant frequency of the circuit. The resonance apparatus of Fig. 24-2

---

* Figure 24-2 illustrates *series resonance*. More complicated circuits can be worked out that have a series of natural frequencies similar to the series of frequencies of a string or a pipe.

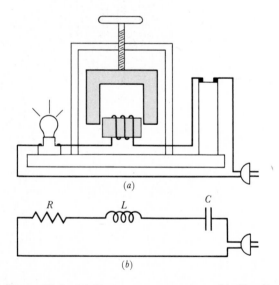

Figure 24-2 (a) Resonance apparatus. (b) Circuit diagram for resonance apparatus.

demonstrates this nicely. A resistance R (the lamp bulb) is included in the circuit; it does not alter the resonant frequency, and the brilliance of the bulb is an indication of the current. If the value of L is changed by raising or lowering the iron yoke, the lamp is brightest for the setting that makes L have just the right value in Eq. 24-4. In other words, we *tune* the circuit by adjusting its resonant frequency to match that of the source of energy.

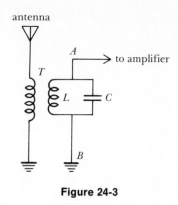

antenna

to amplifier

Figure 24-3

Example 24-1

For what value of L does the circuit of Fig. 24-2 resonate at 60 Hz, if $C = 10\ \mu F$ and $R = 100\ \Omega$?

The resistance of the bulb has no effect except to determine the value of the current at resonance. The condition for resonance is

$$f = \frac{1}{2\pi}\sqrt{\frac{1}{LC}}$$

$$L = \frac{1}{4\pi^2 f^2 C} = \frac{1}{4\pi^2 (60\ s^{-1})^2 (10 \times 10^{-6}\ F)}$$

$$= \boxed{0.704\ H}$$

The resonance of a tuned circuit is used in the "front end" of a radio receiver to select which of several stations will be heard. Schematically, Fig. 24-3 shows an antenna that receives e-m radiation of all frequencies—regular broadcast, television, aircraft, satellites, police calls, and so on. All these signals give rise to currents through the primary of the antenna transformer T. Emf's of many frequencies are induced in the secondary, but since the secondary is part of a resonant LC circuit,* the circulating current in the tank circuit is large only for that incoming signal whose frequency just matches that of the circuit. It is this resonance phenomenon that allows a listener to select the desired station. In radios, the tank circuits are usually tuned by changing the

---

* Picturesquely called a "tank circuit," on account of its energy storage.

capacitors; however, it is also possible to tune a receiver by varying L by means of ferrite slugs that are moved in and out of the coils. Radio frequencies are high enough that the required values of L and C are very small.

Example 24-2

Calculate the resonant frequency of a tank circuit in a television receiver for which the capacitance is 9 pF and the inductance in 4 μH.

$$f = \frac{1}{2\pi}\sqrt{\frac{1}{LC}}$$

$$= \frac{1}{2\pi}\sqrt{\frac{1}{(4 \times 10^{-6}\ H)(9 \times 10^{-12}\ F)}}$$

$$= \frac{10^9}{2\pi(6)} = 26.5 \times 10^6\ Hz$$

$$= \boxed{26.5\ MHz}$$

As viewed from terminals A and B, the LC circuit of Fig. 24-3 illustrates *parallel resonance*, for which the resonance frequency is given by the same equation $f = (1/2\pi)\sqrt{1/LC}$ that is used for series resonance (provided the resistance of the coil is negligibly small). There is a large circulating current in the tank circuit as energy storage fluctuates between the capacitor and the coil, but there is very little current between A and B. In other words, a parallel resonant tank circuit presents a large impedance to the surrounding circuit—infinitely large, if there are no resistive losses in L or in C.

24-4   Radiation

Electric oscillations differ from mechanical oscillations in one important respect: the electric oscillator loses energy not only because of Joule heat in the connecting wires, analogous to friction, but also by radiation. The close interaction between electric fields and magnetic fields is responsible for the radiation of energy, in the form of *electromagnetic waves* (for instance, radio waves, light waves, and x rays), that is broadcast by an oscillator.

We have seen in Chap. 22 that an induced emf appears when a magnetic field changes. For example, the induced emf in a generator or a transformer is caused by a changing magnetic induction as the magnetic flux through a circuit changes, or as magnetic lines of induction cut across a conductor. Now when an emf exists between two points in a wire, or in empty space, there must be an electric field (force per unit charge) in the region between the points. Thus when a wire is connected to the two terminals of a dry cell, the electric field set up in the wire accelerates the free electrons in the wire and causes a current. Even if there is no wire between the terminals, the electric field exists in the space between the terminals. Thus an emf implies an electric field.

As we saw in Sec. 22-11, *a changing magnetic field causes an induced electric field.* It is also true that *a changing electric field causes an induced magnetic field.* This symmetrical relationship between electric and magnetic fields provides the interaction needed to propagate a wave. Maxwell showed that when an e-m wave is passing through some point in space, both the electric field **E** and the magnetic induction **B** at that point fluctuate. **E** and **B** are both zero at the same time, and they reverse direction (together) twice each cycle. Another prediction of the

Maxwell theory of radiation is that **E** and **B** are perpendicular to each other and that both are perpendicular to the direction of propagation of the wave.

The velocity of an e-m wave (in a vacuum) is given by $\sqrt{k/k'}$, where k is the constant in Coulomb's law for charges (Sec. 18-5), and k' is the constant in the law of magnetic force (Sec. 22-7). It is natural that these electric and magnetic constants would enter into an expression for the speed of an electromagnetic wave. One of the early triumphs of the Maxwell theory was the numerical value of the predicted velocity of electromagnetic waves:

$$c = \sqrt{\frac{k}{k'}} = \sqrt{\frac{8.99 \times 10^9 \text{ N} \cdot \text{m}^2/\text{C}^2}{10^{-7} \text{ N}/\text{A}^2}}$$
$$= \sqrt{8.99 \times 10^{16} \text{ m}^2 \cdot \text{A}^2/\text{C}^2}$$
$$= 3.00 \times 10^8 \text{ m/sec}$$

Thus the constant $\sqrt{k/k'}$, an electrical quantity that can be measured by electrical means, works out to be equal to c, the speed of light in a vacuum. This is strong evidence for the fact that light waves are *electromagnetic* waves; the evidence becomes even stronger when it is realized that all matter, including sources of light, is composed of electric charges that might be expected to oscillate and hence broadcast e-m waves.

Before we describe the production of an e-m wave by an oscillating charge, let us first take a more detailed look at the description of an e-m wave.

24-5   Description of an Electromagnetic Wave

We can describe an electromagnetic (e-m) wave in the same way that we described a water wave in Sec. 10-5. We use the technique of a series of instantaneous snapshots taken at equal intervals of time. Suppose that the antenna tower of a radio station is located at the center of town. We station a number of observers at 75-m intervals

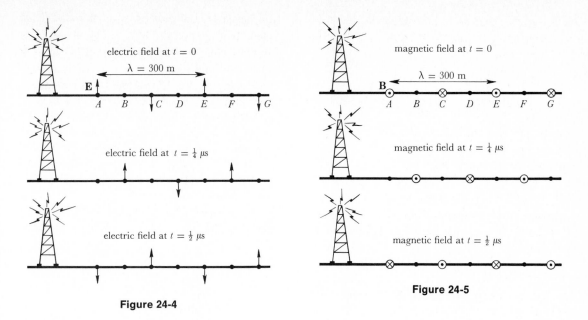

Figure 24-4

Figure 24-5

along a street leading north from the station and equip each observer with an inertialess, positively charged test body that can (in our imagination) respond to very rapid changes in electric field. For a test body we might use a pith ball attached to the end of a light, thin, springy glass fiber. If the charge on the pith ball is Q and it experiences a force **F**, then the direction of the field is the direction of that force. At twelve noon (which we call $t = 0$), the observers' pith balls all indicate electric fields that are either vertical or zero; as shown in Fig. 24-4, at this instant some balls are urged upward, others are urged downward, and still others indicate the field to be zero. A quarter of a microsecond later, the situation has changed, and A's pith ball registers zero field, B's is urged upward, and so on. At $t = \frac{1}{2}$ μs, the force on A's pith ball is downward, B's has returned to zero, and so on.

We have here all the attributes of a wave: At any given place, the electric field fluctuates (**E** is a function of t); and at any given time, the electric field varies along the street (**E** is a function of x). The physically measurable quantity that varies is electric field; hence we

could call our wave an "electric" wave. It is a transverse wave, since the quantity that varies is a vector perpendicular to the direction of propagation. The wave we have described is plane polarized, although other kinds of polarization can also be obtained. The wavelength λ of our wave is 300 m, the distance from, say, A to E, or B to F—always the shortest distance between two points in the same phase of vibration. The period is 1 μs, which is the time for one complete cycle of changes in **E** at any given point; the frequency f is $1/T = 10^6$ s^{-1} = 10^6 Hz = 1 MHz. The velocity of the wave is given as usual by the product of frequency and wavelength:

$$c = f\lambda = (10^6 \text{ Hz})(300 \text{ m})$$
$$= 3.00 \times 10^8 \text{ m/s}$$

While all this is going on, a second group of observers stationed at the same points are measuring the *magnetic* field. Each of these observers is equipped with an ideal, inertialess compass needle that points in the direction of the magnetic induction. The results of a series of rapid-fire measurements are shown in Fig. 24-5. At $t = 0$, the magnetic induction at point A is directed horizontally to the east (out of the plane of the

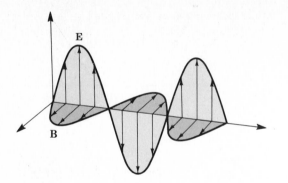

Figure 24-6 The **E** and **B** vectors in a plane electromagnetic wave, traveling along the positive x axis. The field vectors are mutually perpendicular and are in phase.

term "electromagnetic." A purely electric wave, shown in Fig. 24-4, is impossible; likewise, a purely magnetic wave, shown in Fig. 24-5, is impossible. In a sense, the two waves keep each other going by the mutual interaction between the changing E-field and the changing B-field.

Although our illustration was based on the long-wavelength e-m waves that we call radio waves, the wave model for visible light is exactly similar. Except for their short wavelengths (and high frequencies), light waves are in no way different from radio waves. They *are* radio waves, and their velocity, in a vacuum, is given by the electrical quantity $\sqrt{k/k'}$, which equals 3.00×10^8 m/s.

paper); but at this same instant, the magnetic induction at B is zero, and at C the magnetic induction is directed to the west, into the plane of the paper. Just as before, we interpret these measurements by means of a "magnetic" wave, transverse because the variable quantity (magnetic induction **B**) is a vector perpendicular to the direction of propagation.

We see that a radio wave is really two simultaneous waves traveling through the same region of space (Fig. 24-6). This is the significance of the

24-6 Production of Electromagnetic Waves

A simple generator of e-m waves is shown in Fig. 24-7, where an antenna is formed by two metal rods connected to some sort of electric oscillator. The exact nature of the oscillator need not concern us here; a tuned LC circuit might be used, but other circuits are also possible. The oscillator causes charges to flow back and forth along the antenna. The system is shown at several times. In

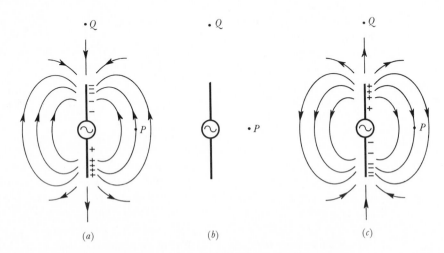

Figure 24-7 Electric field near a simple antenna.

(*a*), negative charge has been pushed to the end of the upper rod, and an excess of positive charge remains on the bottom rod. At a time that is $\frac{1}{4}$ of a cycle later (*b*), the charges have come together again, and the rods are momentarily uncharged. Still later, at (*c*), the oscillator has reversed polarity and the charges are reversed, with the top rod being positively charged. Considering the situation at any one instant, say (*a*), we see that the electric field surrounding the antenna is similar to that due to a pair of equal and opposite charges. At point *P* in (*a*), for instance, the electric field is upward; but half a cycle later, in (c), the field at *P* is downward. During the time between (*a*) and (*c*), positive charge is flowing up the antenna from bottom end to top end. This upward current produces a magnetic field, and the right-hand rule tells us that the lines of induction of the magnetic field are in the form of circles (Fig. 24-8). At point *P*, the magnetic induction is perpendicular to the plane of the paper, directed horizontally away from us. Later, while charges are flowing back down the antenna, the magnetic induction **B** reverses direction but is still horizontal.

We have had to overlook some details, but it is evident from our discussion that at point *P two* fields exist, and that the *E*-field and the *B*-field are perpendicular to each other. At *P*, the wave is propagated horizontally, broadside to the rods; at a point such as *Q*, the magnetic induction is zero (head-on view of the current; see Sec. 22-7),

and hence no wave is propagated along a direction parallel to the length of the antenna. There is a smooth variation in the intensity of the wave, from a maximum value if the antenna is viewed broadside to a zero value if viewed head-on. If, as is usually the case, the observer is far from the antenna (compared with its length), **E** and **B** are in phase, and both fields pass through their zero values at the same time.

At any instant, energy is stored in the space surrounding the antenna. Figure 24-9 shows the electric field (for simplicity, the accompanying magnetic field is not shown) for a given instant. The field in region *X* is upward, at *Y* it is zero, and in region *Z* it is downward. Electric PE is stored wherever there is an electric field; the diagram shows that the energy is (at this instant) localized in regions such as *X* and *Z*, separated by regions where no energy is stored. As the wave spreads out, the regions of energy concentration spread out. This is how the wave model explains the propagation of energy by radiation. The situation is analogous to the spreading out of a water wave. Gravitational PE is stored wherever there is a crest, and energy is propagated as the crest moves. One basic

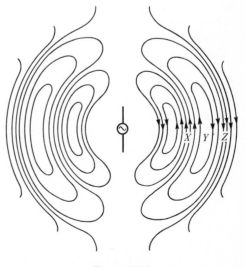

Figure 24-9

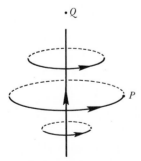

Figure 24-8 Magnetic field near a simple antenna.

difference should be noted: An electric or magnetic field can exist in a vacuum, and hence no medium is needed to transmit energy by an e-m wave. In a water wave, gravitational PE can exist only where there is water, and hence such a wave requires a medium (the water).

24-7   Radio

If a radio receiving antenna is placed parallel to the electric field vector of an e-m wave (Fig. 24-10), the free electrons in the metal will be urged back and forth as the field fluctuates. At any instant, the force on an electron (whose charge is Q) could be calculated from the equation $E = F/Q$. That is, $F = QE$. If the current in the antenna is great enough, a lamp bulb at the center can be lit to full brilliance by the charges that pulse back and forth. The effect is greatest if the antenna is parallel to the electric vector of the wave, since otherwise only a component of the force QE would be effective. An antenna whose length is equal to about half the wavelength of the radiation is much more efficient than a shorter or a longer one; this applies both to the transmitting antenna and to the receiving antenna.* Still another way to receive radio signals is to use a vertical loop as an antenna. As the e-m wave travels past the loop, the changing magnetic field induces a current if the plane of the loop is properly oriented. A loop can be used as a direction finder on a ship or plane to locate the direction of a source of e-m radiation, such as a commercial radio station or a radio beacon.

A radio transmitter and receiver together constitute a system by which information can be sent from one place to another. As we use the term, *information* is the opposite of randomness. For instance, if a 1000-Hz sound is being emitted by a tuning fork, information is being trans-

---

* This is a case of electric resonance; recall that an open pipe that is $\frac{1}{2}\lambda$ long responds well to an incident sound wave (Sec. 11-4).

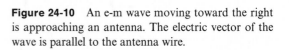

Figure 24-10   An e-m wave moving toward the right is approaching an antenna. The electric vector of the wave is parallel to the antenna wire.

mitted to the ear of the listener. A more complex type of information is transmitted by a speaking person, or by a symphony orchestra. To transmit the information of a 1000-Hz fork, it would seem natural to use a system somewhat like that in Fig. 24-11. A microphone would pick up the sound, producing feeble voltages that would be amplified, thus causing electrons to flow up and down in a vertical transmitting tower. The tower would radiate an e-m wave whose frequency is 1000 Hz, which would be received by a distant antenna. After further amplification, a pulsating current would be caused in the loudspeaker.

Unfortunately, the system of Fig. 24-11 is entirely impractical because of the long wavelength of the necessary e-m wave. Since the velocity is 3.00×10^8 m/s, we have

$$\lambda = \frac{\text{velocity}}{\text{frequency}} = \frac{3.00 \times 10^8 \text{ m/s}}{10^3 \text{ Hz}}$$

$$= 3.00 \times 10^5 \text{ m} = 300 \text{ km}$$

The wavelength of our 1000-Hz e-m wave would be 300 km (about 186 mi), and to obtain a useful amount of radiation, both the transmitting and receiving antenna would have to be at least $\frac{1}{4}\lambda$ in height—some 75 km high! While it is true that the antennas could be stretched out horizontally, they would have to be supported many kilometers above the surface of the earth in order to radiate such a long wave with useful intensity. Such dimensions are fantastically large. Radio waves of frequency 10^6 Hz ($\lambda = 300$ m)

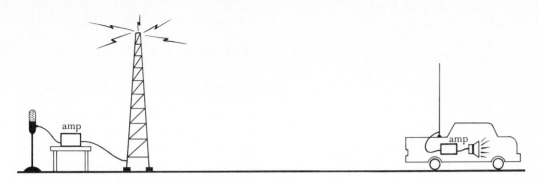

Figure 24-11 Hypothetical system of communication using unmodulated e-m waves, impractical because of long wavelengths required.

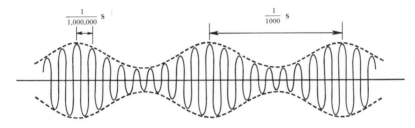

Figure 24-12 Amplitude modulation: a carrier wave of frequency 1 MHz is modulated by a tone of frequency 1000 Hz.

are practical enough—but the frequency of 1,000,000 Hz (1 MHz) is far above the audible range.

The problem of conveying audible information by radio waves was solved by the technique of *modulation*. To modulate a wave, some property of the wave is varied at a slow (or "audio") rate. Two types of modulation are in common use: *amplitude modulation* (AM) and *frequency modulation* (FM). If the amplitude of a wave is periodically changed, then information can be sent at a relatively slow rate, using the rapid, easily transmitted high-frequency wave as a *carrier*. For instance, if the amplitude of a 1-MHz wave varies as shown in Fig. 24-12, information is sent at 1000 Hz by a carrier wave whose frequency is 1,000,000 Hz. The very earliest communication by radio waves was by the use of dots and dashes (Fig. 24-13), a form of amplitude modulation. An on-off switch

(telegrapher's key) at the transmitter is sufficient to modulate a carrier wave in this simple fashion. We will not discuss the circuits by which a microphone produces a smoothly varying amplitude modulation of a carrier, as in Fig. 24-12.

In the receiver, the wave must be *demodulated* (or "detected") in order to recover the information. If the wave of Fig. 24-14a were applied directly to a loudspeaker, nothing would be heard. The speaker has far too much inertia to respond to variations as rapid as 10^6 Hz, and so it responds only to the average value of the

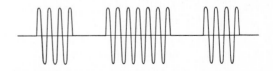

Figure 24-13 A simple form of amplitude modulation: letter F in Morse code.

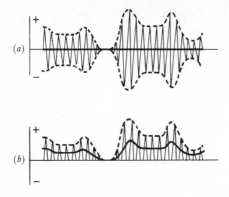

(a)

(b)

Figure 24-14 Demodulation of an AM signal. The heavy line represents the average loudspeaker current.

current (heavy line), which is zero. A simple way to demodulate an amplitude-modulated signal is to use a diode that passes current in only one direction (assumed to be the positive direction in Fig. 24-14). During the positive half of each cycle the diode conducts easily, but during the other half of each cycle, when the diode is reverse biased, it no longer conducts. After the negative portions of Fig. 24-14a are thus chopped off, the resulting wave (Fig. 24-14b) has a slowly varying average value (heavy line), and the loudspeaker can follow these variations to produce a sound wave that duplicates the original sound wave in the studio at the transmitter.

In frequency modulation the amplitude of the carrier wave remains constant while the frequency changes. In Fig. 24-15, which is typical of

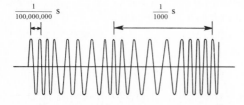

Figure 24-15 Frequency modulation: a carrier wave of frequency approximately 100 MHz is modulated by a tone of frequency 1000 Hz.

a commercial FM station whose "center frequency" is 99.9 MHz, the carrier frequency varies from 99.8 MHz to 100.0 MHz, and this variation takes place 1000 times per second if a flutist is playing a tone having audio frequency 1000 Hz. We will not discuss the means by which frequency modulation is produced at the transmitter, nor will we attempt to describe the circuits in a receiver that are used to demodulate an FM wave and thus recover the original (audio) information. One advantage of FM reception is that it is practically free from "static." The e-m radiation caused by lightning or ignition would show up as an amplitude change, and hence the FM receiver, which responds only to frequency changes, is unaffected.

A television receiver receives two kinds of information: picture information ("video") and sound information ("audio"). The video information is transmitted by AM, and the net result is to control the intensity of a beam of electrons in a cathode-ray tube (Sec. 23-12). The picture is built up as the electron beam is caused to sweep back and forth, striking the face of the picture tube in a succession of spots. The brightness of the spot of light changes rapidly, in proportion to the beam current, which is controlled by a grid, whose potential in turn is determined by the demodulated video signal. The sound part of the information is received by FM, with a demodulating circuit and an audio amplifier that feeds a loudspeaker. Typical carrier frequencies are 85 MHz for the center of TV channel 6, which extends from 82 to 88 MHz, and 98 MHz for the middle of the FM radio band, which extends from 88 to 108 MHz. Working out the wavelengths, we find that λ is about 3.3 m for both the TV carrier and the FM radio carrier. This has the practical advantage that it is possible to use antennas tuned for maximum efficiency—a structure of dimension $\frac{1}{2}\lambda$ is not too bulky to be placed on a rooftop or chimney. If you look carefully at TV or FM antennas, you will find that most of them have rods about a

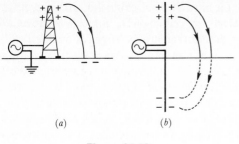

Figure 24-16

meter in length. What is more, you will note that the rods are horizontal, which shows that carrier waves for TV are polarized horizontally (the electric vector is horizontal and the magnetic vector is vertical). The antennas are set broadside to the direction of propagation of the wave. The carrier waves of FM stations are also polarized horizontally; hence a TV antenna serves admirably to receive FM broadcasts, especially if it is necessary to receive distant stations.

Standard-band AM stations require tall transmitting antennas, since $\frac{1}{2}\lambda$ is of the order of several hundred meters. One way to economize is to build the tower only $\frac{1}{4}\lambda$ high. Charges are continually rushing up and down the tower. The oscillator tends to cause charges to flow from the top, down the tower, through the oscillator, and out into the earth. Thus, when there is a net + charge at the top of the tower, there is a net − charge on the earth's surface, and lines of force can terminate on these charges, as shown in Fig. 24-16a. Half a cycle later, the signs of the net charges, both on tower and ground, are reversed. At a distance of approximately $\frac{1}{2}\lambda$ from the tower, some of the lines of force become detached and energy is radiated. The tower and ground together are equivalent to the longer doublet of Fig. 24-16b. Usually a set of heavy copper wires is buried in the ground, radiating out from the base of the tower, to facilitate the necessary to-and-fro flow of charges in the earth near the tower.

24-8 Sources of Radiation

If all e-m radiation is alike, except for wavelength and frequency, we naturally look for some similarity in the means of production of radiations. What general statement can we make that will apply to such diverse sources of radiation as radio transmitters, incandescent filaments in light bulbs, and x-ray tubes? The unifying concept here is that *electromagnetic radiation is produced when electric charge is accelerated.* We have seen this illustrated for radio waves. As charges rush back and forth along the antenna, they are undergoing rather rapid accelerations, especially at the two ends. (Recall that in SHM the acceleration is greatest at the turning points.) A stream of charges moving with a steady velocity emits no e-m wave, for then there is no acceleration of charge. True enough, a steady magnetic field surrounds a wire carrying a steady current, but an induced electric field is caused only by a *changing* magnetic field. Thus we see that the interaction between electric and magnetic fields requires a changing current— that is, an acceleration of charge—in order for an e-m wave to arise.

The sources of radio waves quite obviously involve acceleration of charge; free electrons drift back and forth through the metal of the wires and towers, driven by an electric oscillator. Going down the e-m spectrum to shorter wavelengths and higher frequencies, we next consider infrared radiation of typical wavelength 10^{-5} m, or 10^5 angstroms (1 Å = 10^{-10} m). Infrared radiation is emitted by the atoms of a solid or liquid that are in thermal agitation. Unless a solid is at absolute zero, the atoms are not rigidly locked in a crystal structure, but vibrate to and fro about their equilibrium positions. Since all atoms contain electric charges, it is reasonable to expect that the acceleration of these charges during thermal vibration should give rise to e-m waves.

Progressing now to still shorter wavelengths, we identify visible light and ultraviolet radiation as waves often related to accelerations of orbital electrons. In a general way, we would expect the orbital electrons in atoms to vibrate with high frequency because of the small mass of an electron as compared with that of a nucleus or an atom or molecule. Still higher frequencies are represented by x rays. When electrons are given a high speed by application of a large potential difference (say, 40,000 V), they strike a target and are suddenly stopped. This is a negative acceleration, and an e-m wave is emitted from the target. Here, too, we see that the acceleration of electric charge gives rise to the emission of electromagnetic radiation.

The wave model of e-m radiation is best for low-frequency radio waves, and the model becomes progressively worse as we go to higher and higher frequencies. As a sample of the trouble we get into, consider the motion of an orbital electron as it moves around the nucleus. We know that a particle in circular motion is accelerated toward the center of the circle. Why, then, does not the electron in a hydrogen atom, for instance, continually radiate an e-m wave? The energy of such a wave would have to come from the KE and electric PE of the electron. The situation is analogous to the frictional loss of energy by an artificial satellite, which spirals down to the earth as it loses energy. Calculation leads us to expect that all electrons would lose energy by radiation and spiral into the nucleus, with an increasing angular speed. The whole universe should go up in a blinding flash of radiation—the so-called "violet death." Such a catastrophe obviously has not taken place, and so we conclude that in this respect our wave model is inadequate. There is no doubt, however, that visible light *does* have wave properties, and so we shall continue to use the wave model whenever it is proper and fruitful to do so.

Summary A changing magnetic field gives rise to an induced electric field, and a changing electric field gives rise to an induced magnetic field. This interaction between electric and magnetic fields allows the propagation of electromagnetic waves, in which the electric and magnetic fields are perpendicular to each other and both are perpendicular to the direction of propagation. Electromagnetic waves are transverse and are often plane polarized; their velocity in a vacuum is $c = \sqrt{k/k'} = 3.00 \times 10^8$ m/s.

In general, electromagnetic waves are radiated whenever electric charge is accelerated. The complete electromagnetic spectrum extends from radio waves to gamma rays, but only for the relatively slow vibrations of radio waves can charges be set into continuous oscillation by purely electrical means. Many electric oscillators use tuned circuits whose frequency is given by $f = (1/2\pi)\sqrt{1/LC}$. Infrared, visible, ultraviolet, x-ray, and gamma-ray radiations are also connected with acceleration of charge, but the wave model is not entirely adequate to describe the emission of radiation of high frequency.

In order to communicate by use of radio waves, information is conveyed by some form of modulation. Either the amplitude or the frequency of the carrier wave can be varied.

Check List

$f = (1/2\pi)\sqrt{1/LC}$	modulation	carrier wave
electromagnetic wave	amplitude modulation	demodulation
information	frequency modulation	

24-1 Are radio waves a form of light?

24-2 Can e-m waves be propagated through a perfect vacuum?

24-3 What is the velocity of x rays in a vacuum?

24-4 Two radio stations having different frequencies are broadcasting in the same city at the same time. Why does a listener's radio receiver respond to only one of these stations at a time?

24-5 In an *L-C* circuit, why doesn't the capacitor simply lose its charge and remain in a discharged condition, as in Fig. 24-1*c*?

24-6 A police radio transmitter has a vertical antenna. Should the receiving antenna on a squad car be horizontal or vertical for the most efficient reception?

24-7 Explain, in terms of Eq. 22-11, why a loop antenna (page 562) picks up no induced emf when its plane is broadside to the transmitting radio beacon. (*Hint:* Imagine the antenna wire of Fig. 24-10 replaced by the loop.)

24-8 Why are the "rabbit ears" on a portable TV set about a meter long when fully extended?

24-9 What is the difference between amplitude modulation and frequency modulation?

24-10 A Boy Scout sends a message by blinking his flashlight on and off, using Morse code. This is an example of a modulated e-m wave. (*a*) What is the approximate frequency of the carrier? (*b*) Is the Boy Scout using amplitude modulation or frequency modulation?

24-11 A traffic signal changes from red to green. What sort of modulation is used to convey the information that it is safe to proceed?

24-12 Discuss the concept of information in relation to that of entropy (Sec. 17-1).

MULTIPLE CHOICE

24-13 The wavelength of visible light, compared with that of infrared radiation, is (*a*) longer; (*b*) shorter; (*c*) either longer or shorter, depending on the color of the light.

24-14 A patrol car sends out a radar beam of frequency 10.525 GHz. The wavelength of this radar beam is about (*a*) 3 cm; (*b*) 30 cm; (*c*) 300 cm.

24-15 The oscillation in a tank circuit such as Fig. 24-1 dies out because electric energy is converted to thermal energy and also to (*a*) mechanical energy; (*b*) nuclear energy; (*c*) radiation.

24-16 To increase the resonant frequency of a tank circuit, you should (*a*) increase the resistance; (*b*) increase the inductance; (*c*) decrease the capacitance.

24-17 In a beam of sunlight that strikes the earth, the *B*-field and the *E*-field are (*a*) in phase; (*b*) parallel; (*c*) neither of these.

24-18 Electromagnetic radiation is produced by electric charges that are (*a*) moving with constant velocity; (*b*) moving with changing velocity; (*c*) at rest, near some other stationary charge.

Problems 24-A1 What is the wavelength of a radar wave of frequency (*a*) 15,000 MHz? (*b*) 20 GHz?

24-A2 (*a*) What is the wavelength of a radio wave of frequency 108 MHz emitted by a transmitter in an artificial satellite? The antenna is a split dipole, as in Fig. 24-16*b*. (*b*) About how long should each section of the antenna be?

24-A3 Calculate the frequency in GHz of a radar wave whose wavelength is 2.5 cm.

24-A4 What is the wavelength of the interstellar hydrogen spectrum line whose frequency is 1.42×10^9 Hz?

24-A5 Compute the frequency of red light, whose wavelength in vacuum is 750 nm.

24-A6 Compute the wavelength of a standard-band AM station that broadcasts on an assigned frequency of 1200 kHz.

24-A7 Compute the wavelength of a campus FM radio station that broadcasts on an assigned frequency of 89.7 MHz.

24-A8 Express in angstroms: (*a*) 6.5×10^{-5} cm; (*b*) 1.54×10^{-10} m; (*c*) 546.1 nm; (*d*) 10.59 μm.

24-A9 Express in meters: (*a*) 6328 Å; (*b*) 0.56 Å; (*c*) 15 μm; (*d*) 593 nm.

24-A10 Make the following conversions: (*a*) 6000 Å = ? cm; (*b*) 0.71 Å = ? m; (*c*) 10^{-10} m = ? μm; (*d*) 5461 Å = ? nm; (*e*) 15 μm = ? Å.

24-B1 The current through a coil is 2 A, and 10 J of energy is stored in the magnetic field. How much energy will be stored if the current is 12 A?

24-B2 (*a*) How much energy is stored in a coil of inductance 10 H when the current through the coil is 800 mA? (*b*) How far would a 1-kg object have to fall to gain this much KE?

24-B3 The frequency of a tuned circuit is determined by a fixed inductance connected in series with a variable capacitor. The frequency is 1200 kHz when the capacitor is set at 200 pF; what is the frequency when the capacitor is increased to 800 pF?

24-B4 Calculate the resonant frequency of a tank circuit for which L is 200 mH and C is 400 pF.

24-B5 What is the resonant frequency of a tank circuit in which $C = 0.25$ μF and $L = 5$ H?

24-B6 Calculate the wavelength of an e-m wave whose frequency is determined by a tank circuit inductance of 100 μH and a capacitance of 120 pF.

24-B7 A diathermy machine used in physiotherapy generates e-m radiation that gives "deep heat" when absorbed in tissue. One assigned frequency for diathermy is 27.33 MHz. (*a*) What is the wavelength of this radiation? (*b*) What effective capacitance is needed for the tank circuit if the inductance is 1.6 μH?

24-B8 A standard broadcast band AM radio can be tuned by a variable capacitor through a range of 500 kHz to 1600 kHz. (*a*) What is the ratio of the maximum capacitance of the tuning capacitor to the minimum capacitance? (*b*) How is this change usually effected?

24-B9 A coil of inductance L and two capacitors, each of capacitance C, are available to form a tank circuit. What is the ratio of the maximum possible frequency to the minimum possible frequency?

24-B10 An amateur radio operator wishes to construct a receiver to operate in the 41-m band (that is, to receive waves of wavelength in the neighborhood of 41 m). She has a variable capacitor whose average value is 50 pF, and wishes to wind a coil to use in a tank circuit. What should be the inductance of the coil?

24-B11 In some radios, the tuned circuit consists of a fixed capacitor connected in parallel with a variable inductance. The radio is tuned to 88 MHz (low end of the FM band) when the inductance is 82 nH. (*a*) What must be the value of the inductance if it is desired to receive 108 MHz (high end of the FM band)? (*b*) What is the value of the fixed capacitance?

24-B12 In an oscillating tank circuit, prove that the maximum current through the coil is related to Q, the maximum charge on the capacitor, by the equation $I = 2\pi f Q$. (*Hint*: Use the energy method.)

24-C1 (a) Show that in an *L-C* circuit, at the resonant frequency the reactance of the coil equals that of the capacitor (see Eqs. 23-6 and 23-7). Derive a formula for this reactance in terms of *L* and *C*; check your answer dimensionally. (b) What is the value, in Ω, of the reactance of the coil and that of the capacitor in the circuit of Prob. 24-B6?

24-C2 The charge of the capacitor in a resonant circuit varies with time according to the equation $Q = Q_0 \sin 2\pi ft$, where Q_0 is the maximum value of the charge. By differentiation, derive the result stated in Prob. 24-B12.

24-C3 (a) What capacitance will resonate with a small, one-turn loop of inductance 100 pH to give a radar wave of wavelength 3 cm (see Table 14-5)? (b) If the capacitor has square parallel plates separated by 1 mm of air, what should the edge length of the plates be? (c) What is the common value, in Ω, of the reactance of the loop and that of the capacitor, at resonance?

For Further Study

24-9 Energy Storage in Coils

A coil of self-inductance *L* is connected to a battery of emf ε through a switch, as in Fig. 23-17. After the switch is closed, charges leave the battery and move through the coil; work is done on the charges because they move against the back emf $L(dI/dt)$ of the coil. This back emf is not constant (it becomes zero when the current through the coil reaches its final steady value), so we must use integration to find the total work done on the charges.

If in a time Δt a charge ΔQ moves through the coil, the average current is $\Delta Q/\Delta t$, and the back emf is $L(\Delta I/\Delta t)$. Hence,

$$\text{joules} = \left(\frac{\text{joules}}{\text{coulomb}}\right)(\text{coulombs})$$

$$\Delta W = \varepsilon_{\text{back}}\,\Delta Q = L\frac{\Delta I}{\Delta t}\,\Delta Q$$

Rearranging gives

$$\Delta W = L\frac{\Delta Q}{\Delta t}\,\Delta I$$

The total work done while the current is increasing from 0 to its final value *I* is found from a definite integral (see Appendix E):

$$W = \lim_{\substack{\Delta I \to 0 \\ \Delta t \to 0}} \sum L\frac{\Delta Q}{\Delta t}\,\Delta I$$

$$= \int_{I=0}^{I=I} LI\,dI = L\int_0^I I\,dI$$

$$= L\left(\frac{I^2}{2} - \frac{0^2}{2}\right) = \tfrac{1}{2}LI^2$$

This is Eq. 24-3. This work may be considered to be stored in the magnetic field that has been set up in the coil. It can also be interpreted as kinetic energy of the charges whose motion gives rise to the current in the coil.

References 1. W. F. Magie, *A Source Book in Physics* (McGraw-Hill, New York, 1935), pp. 528–538. Brief extracts from the writings of Maxwell and Rowland.
2. J. R. Newman, "James Clerk Maxwell," *Sci. American* **192**(6), 58 (June 1955).
3. P. Morrison and E. Morrison, "Heinrich Hertz," *Sci. American* **197**(6), 98 (Dec. 1957). An account of the work of the man who first obtained experimental evidence for the electromagnetic waves predicted by Maxwell's theory.

25

Geometrical Optics

The study of optics is the study of a small part of the electromagnetic spectrum. Visible light waves range in wavelength from about 380 nm (in the violet region) to about 760 nm (in the red region). The corresponding frequencies are high—in the neighborhood of 6×10^{14} Hz (600 THz) for green light of wavelength 500 nm. The approximate wavelength ranges of various colors are shown in Table 25-1. The preferred SI unit for wavelength is the nanometer (nm), which is 10^{-9} m. This is a useful unit, especially for

Table 25-1 Wavelengths of Visible Light

Color	Approximate Wavelength Range	
	(nm)	(Å)
violet	380–450	3800–4500
blue	450–490	4500–4900
green	490–560	4900–5600
yellow	560–590	5600–5900
orange	590–630	5900–6300
red	630–760	6300–7600

visible light, since the human eye can just about detect the difference in color of two sources that differ by 1 nm. You should also be familiar with the *angstrom unit* (Å), which is 10^{-10} m; thus, $1 \text{ Å} = 0.1$ nm. The angstrom is often used in dealing with atomic structure, since atoms are a few angstroms in size and are spaced a few angstroms apart in solid crystals (Fig. 1-1). In this book we will emphasize the nanometer for wavelengths of visible light.

In spite of the fact that visible light comprises only about one "octave" of the complete e-m spectrum, it is worthwhile to devote a considerable time to the study of optical phenomena. For one thing, of course, much of our knowledge of the physical world comes to us through optical instruments such as the eye, the telescope, and the microscope. Then, too, our study of visible light is to a certain extent the study of all e-m radiation. The wave-particle duality of radiation is perhaps best shown by visible light. Radio radiations (low frequencies) are practically "all wave" and show few or no particle properties; nuclear gamma rays (high frequencies) are practically "all particle," with wave properties ob-

Refraction at the faces of a parallel slab of glass; the emerging rays are parallel to the incoming rays. Also to be noted are the weak partial reflections at the top and bottom faces.

servable only with extreme difficulty. Visible light occupies a favored position in the e-m spectrum, where wave and particle aspects are about equally prominent. Since our aim is to explore and understand nature, we welcome the challenges as well as the practical uses of visible light. We shall find the wave model entirely adequate to explain how light often travels in a straight line, how it is reflected by a mirror and bent as it enters a prism or a lens, and how it spreads out into a diffraction pattern when it passes through a small opening.

Figure 25-1   Wave fronts on the surface of water.

25-1   *Huygens' Principle*

Christian Huygens* (1629–1695) was a versatile Dutch physicist and natural philosopher, born 14 years before Newton. His work in mechanics did much to pave the way for Newton's achievements, and he, along with Galileo, is credited with developing the pendulum clock, an invention of prime importance for the science of mechanics, where precise measurement of time intervals is necessary. Huygens was a firm believer in the wave theory of light, and he used a geometrical principle (without proof) to explain many of the properties of the transmission of light and other waves.

The concentric circles formed when a stone is dropped into a lake and water waves spread out (as in Fig. 25-1) are called wave fronts. The circle *a* represents a crest, as does the circle *b*. Between these circles is another circle *t*, which is a trough. In general, a *wave front* is defined as the locus of points having the same phase of vibration. All the points on *a* have the same phase, since all the molecules on this particular wave front are at their maximum displacement *above* the normal level. Likewise, *t* is a wave front on which all particles are at their maximum displacements *below* the normal level. Halfway between *t* and *b*

is another wave front where the particles are (momentarily) at the normal level but moving upward. This technical use of the word "front" is not unfamiliar; in meteorology a cold front is said to connect cities *X*, *Y*, and *Z*, which are simultaneously experiencing a sudden drop in temperature. As a wave spreads out, the wave fronts advance, and in the case of our water wave, the radius of each circular wave front continuously increases.

Huygens' principle, or Huygens' construction, tells us how to predict a new wave front when we know the position of an earlier one. The construction is as follows: Let every point on the wave front be considered the source of a small *wavelet* that spreads out in the forward direction (Fig. 25-2). The new wave front is the envelope of all the wavelets—that is, the line or surface tangent to all the little wavelets. If we now consider light waves spreading out in three dimensions, it is evident that the wavelets are small hemispheres, provided the speed of light is the same in all directions; hence we see that a spherical wave front will remain spherical and that the energy of the wave is carried away equally in all directions. Such directions of energy flow are called *rays*; the short arrows in Fig. 25-2 represent rays of light that diverge from the source *S*.

* The English pronunciation is Hy′genz.

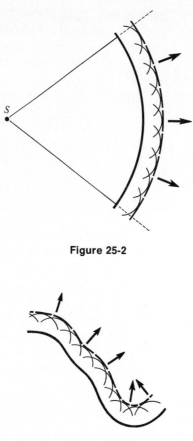

Figure 25-2

Figure 25-3

25-2 Straight-Line Propagation of Light

We are ready now to consider some of the well-known properties of light. Huygens' principle describes the propagation of light in a straight line, since wavelets from a plane wave front give rise to a new plane wave front (Fig. 25-4), and the corresponding rays are parallel to each other. Such a plane wave front can be thought of as a section of a very large spherical wave front; for instance, sunlight strikes the earth in wave fronts of radius 149 million kilometers. For all practical purposes, such a wave front is "plane," and the rays are "parallel" to each other.

In everyday experience, light travels in straight lines through openings, and the edge of an object casts a sharp shadow. Because of this, Newton never did accept a wave model for light. He felt (quite justifiably) that if light waves existed, they would naturally "bend around corners" and would *not* travel in straight lines through an opening or past the edge of an obstacle. So, on the basis of facts known to him, Newton rejected the wave model of light. Newton did not realize that the extreme smallness of the wavelength of light compared with the size of the usual opening or obstacle could account for the apparent straight-line propagation of light waves.

Let us illustrate this point by comparing the behavior of water waves and light waves as they pass through an opening. If *water* waves strike

Unless specifically stated otherwise, we shall always assume that light travels through a medium equally fast in all directions; such a medium is called an *isotropic* medium.* Glass, water, air, Lucite, and most other common substances are isotropic, but certain transparent crystals have enough structure to give rise to "easy" and "hard" directions for the propagation of light. Such a crystal is said to be *anisotropic*. In an isotropic medium, the wavelets are spherical, and the rays are always perpendicular to the wave fronts, as in Fig. 25-3. In an anisotropic medium, the wave fronts are ellipsoidal, as will be discussed in Sec. 26-11.

---

* From the Greek *iso*, equal, and *tropic*, turning or changing.

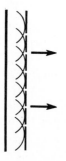

Figure 25-4

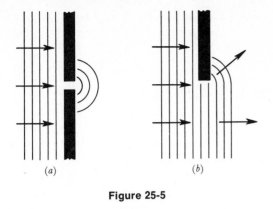

Figure 25-5

a gap a few centimeters wide in a long break-water (Fig. 25-5a), a disturbance spreads out in all directions on the other side. (Similarly, if water waves strike the end of the breakwater, as in Fig. 25-5b, they diffuse around the corner.) On the other hand, if *light* waves fall on an opening a few centimeters wide, the rays go straight through, with only a very slight "fuzziness" around the edges. We conclude that straight-line propagation is a relative thing. We shall prove in Sec. 26-5 that the determining factor is the size of the opening *as compared with the wavelength*. Water waves have wavelengths of the order of several centimeters to several meters, larger than the opening in the breakwater. But visible light waves have wavelengths of about 5×10^{-5} cm, which is small compared with an opening such as the pupil of the eye. Thus we anticipate that in the usual situation, light waves are propagated in practically straight lines.

All that we have said about the relative size of openings can be applied to the relative sizes of obstacles. A small rock sticking up above the surface of a lake hardly affects the passage of waves that strike it; the waves wrap around the rock and go on their way unimpeded. However, an obstacle even a few millimeters in size casts a sharp shadow of the light waves that are so much smaller in wavelength than the size of the obstacle.

These ideas are illustrated by sound waves. Many sounds have wavelengths of several meters or more (a 300-Hz blast from an automobile horn has wavelength of approximately 1.1 m); hence, street noises diffuse through an open window without any obvious straight-line propagation. Nevertheless, sound shadows and reflection (such as echoes) are possible even with long wavelengths. It is only necessary to have an obstacle that is large compared to the wavelength. This can be done in two ways: An echo can be received when an ordinary-sized (audible) wave is reflected from a mountainside or the wall of a large building, or very short-wavelength sound waves of ultrasonic frequency can be reflected by obstacles of ordinary size. The latter method is used in underwater range finding (sonar) and in navigation by bats. Light and other e-m waves also show a gradual transition from diffuse nature to straight-line propagation as wavelength is decreased. Standard-band radio waves ($f = 10^6$ Hz, $\lambda = 300$ m) spread out diffusely; TV broadcasts ($f = 10^8$ Hz, $\lambda = 3$ m) are easily blocked out by buildings or by the curvature of the earth; radar waves ($f = 10^{10}$ Hz, $\lambda = 0.03$ m) are small compared with a plane or even a man, and thus cast sharp shadows.

Geometrical optics, also called *ray optics*, is, according to Huygens' principle, the limiting case of wave optics that is valid when the obstacles or openings are large compared with the wavelength. For the mirrors, prisms, and lenses discussed in the rest of this chapter, ray optics is a very good approximation indeed.

25-3 Reflection

When a ray of light strikes a smooth surface such as a mirror, the reflected ray leaves the mirror in a definite direction, which is determined by two rules that can be called the *laws of reflection*: (1) The incident ray, the reflected ray, and the normal to the surface all lie in the same plane; (2) the *angle of incidence* equals the *angle of reflection*. These angles are measured from the

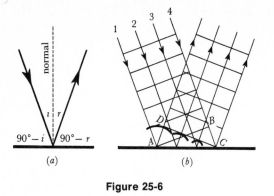

Figure 25-6

normal and are denoted by i and r in Fig. 25-6a. To prove that $i = r$, we use Huygens' principle. In Fig. 25-6b we show a wave front AB that is perpendicular to the incoming rays 1, 2, 3, and 4. While ray 4 is traveling from B to C, ray 1 spreads out from A in a sphere whose radius is AD. Hence $AD = BC$. Rays 2 and 3 strike the surface later than ray 1, and therefore their wavelets are correspondingly smaller. According to Huygens' principle, the reflected wave front is CD, which is tangent to all the wavelets. We now use plane geometry to prove that $i = r$. The right triangles ADC and ABC are congruent, since they have the same hypotenuse, and the leg AD of one triangle equals the leg CB of the other triangle. Hence $\angle BCA = \angle DAC$:

$$90° - i = 90° - r$$
$$i = r \qquad (25\text{-}1)$$

The *specular reflection** from a smooth surface such as we have just described is often a nuisance—for example, the glare from the glass in front of a framed watercolor. If a surface is not perfectly smooth, *diffuse reflection* takes place, with many reflected bundles of rays coming from small flat spots. Much of what we see is made visible by such a process. If the surface is so rough that there are no flat spots much larger than a wavelength, the light is *scattered* diffusely. We can think of scattering as the absorption and reemission of light by particles of a medium. The

---

* From the Latin *speculum*, mirror.

light of the sky reaches us after being scattered by air molecules, water droplets, or dust particles.

25-4 Refraction

The bending of light at the interface between two media is called *refraction*. The *angle of refraction* (θ_2 in Fig. 25-7) is the angle the ray in the second medium makes with the normal. Similarly to reflection, the angles of incidence and refraction lie in a plane that includes the normal to the surface. Refraction is caused by the fact that the speed of light is different in different media. This is in accordance with Maxwell's theory of electromagnetic waves, and advanced theory shows how to correlate the wave velocity with the electrical and magnetic properties of the medium. Most transparent substances are nonmagnetic, and the velocity of an e-m wave then works out to be $c/\sqrt{K}$, where K is the dielectric constant. For the very high frequencies of visible light, the dielectric constant is not strictly a constant but depends on frequency. Hence we expect the speed of light in a medium to depend somewhat on frequency, that is, on color. However, this effect is at most only a few percent over the whole range of visible light and can often be ignored.

The speed of light in a medium is denoted by c_n, and the speed of light in a vacuum is denoted

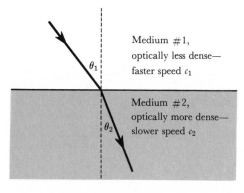

Medium #1, optically less dense— faster speed c_1

Medium #2, optically more dense— slower speed c_2

Figure 25-7

Table 25-2 Index of Refraction

Substance	n
crown glass (67% SiO_2, 12% Na_2O, 11% BaO, misc. 10%)	1.52
flint glass (39% SiO_2, 3% Na_2O 49% PbO, misc. 9%)	1.66
diamond	2.42
ice	1.31
water	1.333
benzene	1.50
air at 0°C and 1 atm	1.00029
hydrogen at 0°C and 1 atm	1.00013

by c ($=3.00 \times 10^8$ m/s). We now define a constant n, called the *index of refraction*:

$$n = \frac{\text{speed of light in vacuum}}{\text{speed of light in medium}} = \frac{c}{c_n} \qquad (25\text{-}2)$$

The index of refraction is a measure of *optical density*; for instance, the speed of light in glass is less than in air, and so glass is said to be optically denser than air. Some values of index of refraction are shown in Table 25-2; these are average values for visible light. The variation with color is minor; typical values of n for window glass might be 1.51 for red light, 1.52 for green light, and 1.53 for violet light. This means, of course, that in glass, red light travels faster than violet light. In a medium, the dependence of the speed of light on color (that is, on frequency) is called *dispersion*. Some of the consequences of dispersion, both useful and annoying, will be discussed more fully in Chap. 27, where we study optical instruments.

Example 25-1

Calculate the speed of light in diamond.
Since $n = c/c_n$,

$$c_n = \frac{c}{n} = \frac{3.00 \times 10^8 \text{ m/s}}{2.42}$$

$$= \boxed{1.24 \times 10^8 \text{ m/s}}$$

When a wave front (or a ray) enters a medium such as glass, the *frequency* of the light remains unaltered, while both the speed and the wavelength change. Think of a marching band slowing down as it passes a reviewing stand— the players get closer together, the "wavelength" decreases, but the same number per minute pass any point, otherwise they would pile up!

Example 25-2

The index of refraction of a certain glass is 1.50 for light whose wavelength in vacuum is 600 nm. What is the wavelength of this light as it passes through glass?

The frequency remains the same when light enters the glass. This frequency is

$$f = \frac{c}{\lambda} = \frac{3 \times 10^8 \text{ m/s}}{600 \times 10^{-9} \text{ m}}$$

$$= 5 \times 10^{14} \text{ Hz}$$

The speed in glass is given by

$$c_n = \frac{c}{n} = \frac{3 \times 10^8 \text{ m/s}}{1.50}$$

$$= 2 \times 10^8 \text{ m/s}$$

Hence the wavelength in glass is

$$\lambda_n = \frac{c_n}{f} = \frac{2 \times 10^8 \text{ m/s}}{5 \times 10^{14} \text{ Hz}}$$

$$= 4 \times 10^{-7} \text{ m}$$

$$= 400 \times 10^{-9} \text{ m}$$

$$= \boxed{400 \text{ nm}}$$

The wavelength in glass is only two-thirds as much as it is in vacuum.

We can generalize Example 25-2, working in symbols instead of with numbers. If λ_{vac} is the wavelength in vacuum and λ_n is the wavelength in the medium, we have

$$f = \frac{c}{\lambda_{vac}} = \frac{c_n}{\lambda_n}; \qquad \frac{c}{\lambda_{vac}} = \frac{c/n}{\lambda_n}; \qquad \lambda_n = \frac{\lambda_{vac}}{n}$$

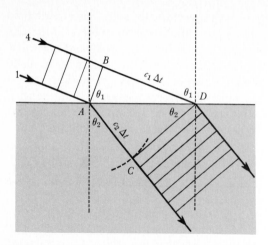

Figure 25-8 Derivation of Snell's law using Huygens' principle.

Thus the wavelength is reduced by a factor equal to the index of refraction.

Returning now to refraction, we apply Huygens' principle to derive a relationship between θ_1 and θ_2 (Fig. 25-7). The construction in Fig. 25-8 is much the same as for reflection; during the time interval Δt, ray 1 travels from A to C while ray 4 travels from B to D. The radius AC of the wavelet through C is given by velocity × time = $c_2 \Delta t$, whereas the distance BD is given by $c_1 \Delta t$. Using plane geometry, we identify $\angle BAD$ as equal to θ_1, and $\angle ADC$ as equal to θ_2. We now use the definition of the sine of an angle.

$$\sin \theta_1 = \frac{c_1 \Delta t}{AD}$$

$$\sin \theta_2 = \frac{c_2 \Delta t}{AD}$$

Dividing one equation by the other and simplifying, we get

$$\frac{\sin \theta_1}{\sin \theta_2} = \frac{c_1}{c_2}$$

But since $c_1 = c/n_1$ and $c_2 = c/n_2$, we can write

$$\frac{\sin \theta_1}{\sin \theta_2} = \frac{c/n_1}{c/n_2} = \frac{n_2}{n_1}$$

whence, finally,

$$n_1 \sin \theta_1 = n_2 \sin \theta_2 \qquad (25\text{-}3)$$

This is *Snell's law of refraction*, discovered experimentally in 1621 by the Dutch scientist Willebrord Snell and independently by his contemporary, the English investigator Thomas Hariot. In words, the product of the index of refraction for a given frequency of light and the sine of the angle measured from the normal is a constant, the same in each medium. Thus the angle is less in the medium whose index of refraction is greater. That is, light bends toward the normal when it enters an optically denser medium. Conversely, light bends away from the normal as it enters an optically less dense medium.

Example 25-3

A ray of light in air strikes a slab of glass, as shown in Fig. 25-9. Calculate the angle of refraction in the glass, which has an index of refraction of 1.5.

The index of refraction of air is 1.0003, and for most problems it is sufficiently accurate to call it exactly 1, the same as for a vacuum. Using Snell's law, we have

$$n_1 \sin \theta_1 = n_2 \sin \theta_2$$
$$(1)(\sin 60°) = (1.5)(\sin \theta_2)$$

$$\sin \theta_2 = \frac{(1)(0.866)}{1.5} = 0.577$$

$$\theta_2 = \boxed{35°}$$

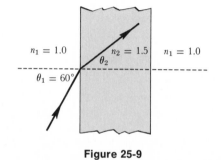

Figure 25-9

Example 25-4

An insulated lamp bulb is on the bottom of a swimming pool at a point 2.5 m from a wall. The pool is 2.5 m deep and filled with water to the top (Fig. 25-10). At what angle does the light leave the water at the edge of the pool?

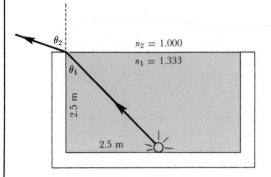

Figure 25-10

First we find θ_1, using simple trigonometry.

$$\tan \theta_1 = \frac{2.5 \text{ m}}{2.5 \text{ m}} = 1.00$$

$$\theta_1 = 45°$$

Next we apply Snell's law:

$$n_1 \sin \theta_1 = n_2 \sin \theta_2$$
$$(1.333)(\sin 45°) = (1.000)(\sin \theta_2)$$
$$(1.333)(0.707) = \sin \theta_2$$
$$\sin \theta_2 = 0.942$$
$$\theta_2 = \boxed{70°}$$

As expected, θ_2 is greater than θ_1, and the ray bends away from the normal as it enters the less dense medium (air).

One common application of Snell's law is shown in Fig. 25-11. Light strikes a slab of glass having parallel surfaces (for instance, a window pane), and the rays emerge without change of direction. Each ray is displaced slightly, parallel to itself.

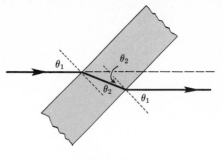

Figure 25-11

25-5 Total Reflection

It is always possible for an oblique ray to bend toward the normal as it enters a denser medium. If the ray travels exactly along the normal, it continues to do so in the second medium. However, a ray in a dense medium does not always emerge into a less dense medium. Suppose that $n_1 = 2$ and $n_2 = 1$. Then, according to Snell's law, $2 \sin \theta_1 = \sin \theta_2$. This is fine for values of $\sin \theta_1$ up to and including 0.5. But no angle exists for which $\sin \theta$ is greater than 1; the sine of 90° is 1, and sin 91° is the same as sin 89°, which is less than 1. We then ask what happens when θ_1, n_1, and n_2 have values that would make $\sin \theta_2$, as calculated from Snell's law, greater than one. We find experimentally that no refraction takes place; instead, the ray is completely reflected back into the first medium. This is called *total reflection*, and it occurs if the angle θ_1 in the denser medium is greater than a certain *critical angle*. We also find that *some* reflection takes place at any angle; as the angle in the denser medium is gradually increased (Fig. 25-12a, b, c, d),* the reflected ray becomes stronger and the refracted ray becomes weaker. In the limit, at the critical angle, the refracted ray just grazes the surface (e), but its intensity is zero. Beyond the critical angle, there is no refracted ray, and the

---

* The intensities have been calculated for a glass-air interface with $n = 1.50$, using the Maxwell theory of e-m waves.

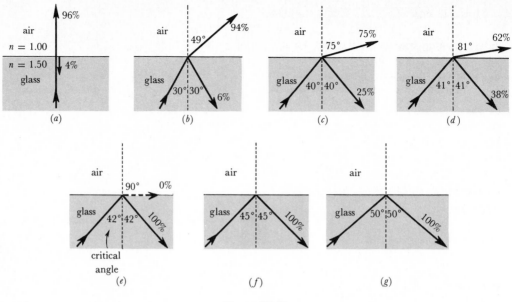

Figure 25-12

reflected ray's intensity exactly equals that of the incident ray (f, g). The critical angle can be easily computed, using Snell's law. For glass surrounded by air, the angle works out to be about 42°.

Example 25-5

Calculate the critical angle for a glass-air interface. The glass is a light crown glass of index of refraction 1.50 (Fig. 25-13).

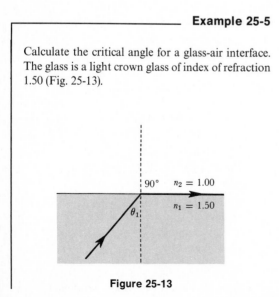

Figure 25-13

When θ_1 just equals the critical angle, the emerging ray just grazes the surface, and $\theta_2 = 90°$.

$$n_1 \sin \theta_1 = n_2 \sin \theta_2$$
$$(1.50)(\sin \theta_1) = (1)(\sin 90°)$$
$$1.50 \sin \theta_1 = (1)(1)$$
$$\sin \theta_1 = 1/1.50 = 0.667$$

$$\theta_1 = \boxed{42°}$$

Total reflection is used in many optical instruments, including binoculars (as we shall see in Fig. 27-27). In order to erect the image, as well as to obtain a longer path length between the objective lens and the eye lens, two 45° prisms are used. Silvered mirrors might be used, but the prisms serve the same purpose and give 100% reflection, whereas a conventional mirror might become tarnished. The angle in the glass is 45°, which is a few degrees greater than the critical angle for a glass-air interface. Total reflection takes place, and it is not necessary to silver the reflecting surface at all.

25-6 Thin Lenses

Lenses are used to concentrate or to disperse light and to form images. In this section we shall discuss the image-forming properties of simple lenses, with emphasis on the wave fronts and how they change.

A rudimentary lens could be formed by placing two prisms back to back, as in Fig. 25-14. Each ray is refracted toward the normal as it enters the glass, and is refracted away from the normal as it leaves the glass. Both refractions bend the ray away from the original (horizontal) direction. After passing through the "lens," the rays intersect in a zone AA, which might be called an image. A much better lens is shown in Fig. 25-15a, where the surfaces are curved. The various rays strike the surface at different angles, are refracted differently, and are brought to a much sharper focus F. If the lens surfaces are spherical (an easy shape to manufacture), the image is almost, but not quite, a mathematical point. For the time being, we shall consider that *thin* spherical lenses make perfectly sharp images, and we put off until Chap. 27 a discussion of the aberrations that cause fuzzy or color-fringed images.*

In Fig. 25-15a, the incident rays are parallel to the axis of the lens, and after passing through the lens the rays converge and meet at the *principal focus* F, also called the focal point. If the rays are parallel to each other but not to the axis, as in Fig. 25-15b, the image F' is formed at an off-axis point. The plane FF' is called the *focal plane*, and the distance from the lens to the focal plane is the *focal length* f. The focal length is usually measured from the center of the lens, but since the lens is assumed to be thin, the thickness is negligible compared with f, and the focal length can be measured from either surface without serious error.

---

* Our treatment of lenses is limited to thin lenses and to rays that do not make too large an angle with the axis of the lens. In the figures, the thickness of the lenses has been exaggerated for clarity.

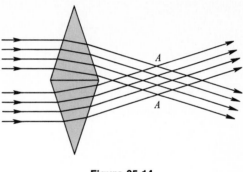

Figure 25-14

We have drawn a double-convex lens—one in which both surfaces bulge outward. Parallel rays striking such a lens produce a *real image*, defined as a point at which the rays actually intersect. A

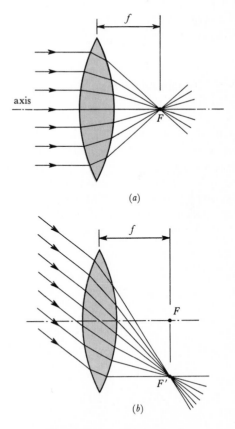

(a)

(b)

Figure 25-15 Converging lens.

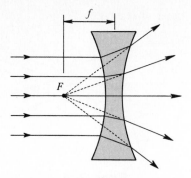

Figure 25-16 Diverging lens.

real image can be observed as a bright spot on a screen or photographic film. Another useful lens is the double-concave lens of Fig. 25-16. Refraction at the surfaces causes the parallel rays to diverge, giving rise to a *virtual image* at F, which is defined as the point from which the diverging rays seem to come.

25-7 Objects and Images— Ray Tracing

We usually represent an object by a small vertical arrow (*AB* in Fig. 25-17) to the left of the lens. To a good approximation, *all* the rays leaving B pass through the image point B'. We use the following symbols: p = object distance, q = image distance, f = focal length (all measured from the center of the lens); h = object size,

h' = image size. The *magnification m* is defined as

$$m = -\frac{h'}{h}$$

where the $-$ sign is used so that an *inverted* image as in Fig. 25-17 will give a *negative* ratio. Three rays that leave the head of the arrow are particularly easy to trace; any two of them would suffice to locate the image point B'. *Ray* 1, parallel to the axis, goes through the focal point F' (this follows from the definition of focal point). *Ray* 2, headed for the center of the lens, passes through without change of direction (Fig. 25-18), since the lens here is like a thin parallel-faced slab. *Ray* 3 passes through the other focal point F and emerges parallel to the axis. Any other ray leaving B will also pass through B' if the thin-lens approximation is valid.

Triangle *ABO* is similar to *A'B'O*, from which we obtain the magnification:

$$m = -\frac{h'}{h} = -\frac{q}{p} \qquad (25\text{-}4)$$

or, in words,

$$\frac{\text{Image size}}{\text{Object size}} = \frac{\text{image distance}}{\text{object distance}}$$

Also, triangle *FOC'* is similar to *FAB*, which yields

$$\frac{h'}{f} = \frac{h}{p - f}$$

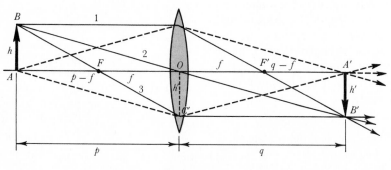

Figure 25-17

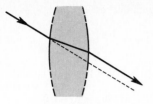

Figure 25-18

From Eq. 25-4, $h' = qh/p$, so

$$\frac{qh}{pf} = \frac{h}{p-f}$$

After canceling h, this equation can be rearranged to give the *lens equation*

$$\frac{1}{p} + \frac{1}{q} = \frac{1}{f} \qquad (25\text{-}5)$$

Note that the lens equation is consistent with the definition of focal length: For incident parallel rays, the object is at infinity. Then $1/\infty + 1/q = 1/f$, which gives $q = f$.

All rays leaving the base of the arrow (point A) converge to an image point A', which is also on the axis, at the base of the image. Three of these rays, including the central ray along the axis, are shown diverging from A in Fig. 25-17.

The same graphical method can be used to find the location and size of a virtual image

formed by a diverging lens (Fig. 25-19a). Rays leaving B appear to come from the virtual image point B'. Here, too,

$$m = -\frac{q}{p}$$

but since p is positive, we agree to call q *negative* for a virtual image, to preserve the form of the equation for m and still obtain a positive m for the erect image. A virtual image can also be formed by a converging lens (Fig. 25-19b). These diagrams also show that for any single lens a real image is inverted, whereas a virtual image is erect.

Two equivalent methods can be used to calculate the position of an image. The rest of this section is based on the lens equation. If you use the method of curvatures, to be discussed later, skip the rest of this section and proceed directly to Sec. 25-8.

The following *sign convention* is used when applying the lens equation:

p, q, and f are each positive for the "typical case" of a converging lens forming a real image of a real object.

Any change from the "typical" situation requires a minus sign. Thus, f is negative for a diverging lens; q is negative for a virtual image (formed to

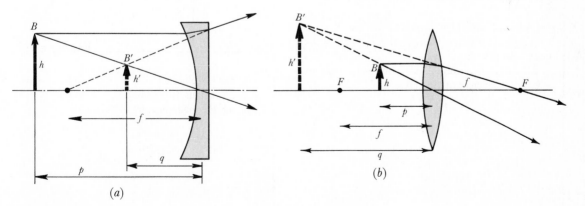

Figure 25-19

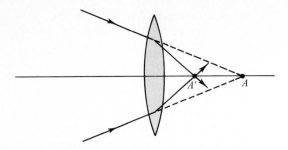

Figure 25-20   Virtual object at A; real image at A'.

the left of the lens if the rays travel from left to right); p is negative for a virtual object (located to the right of the lens).

	f	p	q
converging lens	$+$		
diverging lens	$-$		
real object		$+$	
virtual object		$-$	
real image			$+$
virtual image			$-$

In Fig. 25-20, A is a virtual object since rays striking the lens are coming toward A instead of diverging (as in the "typical case") from a real object to the left of the lens.

In the next few examples we illustrate the use of the lens equation and the sign conventions.

_____ **Example 25-6**

An object 2 mm high is placed 100 cm from a converging lens of focal length 20 cm. Where is the image and how large is it?

The lens is a converging one, so its focal length is $+20$ cm. The lens equation gives

$$\frac{1}{p} + \frac{1}{q} = \frac{1}{f}$$

$$\frac{1}{100 \text{ cm}} + \frac{1}{q} = \frac{1}{20 \text{ cm}}$$

from which

$$q = \boxed{+25 \text{ cm}}$$

The magnification is

$$m = -\frac{q}{p} = -\frac{25 \text{ cm}}{100 \text{ cm}} = -0.25$$

The image is real, it is inverted, and its size is

$$(0.25)(2 \text{ mm}) = \boxed{0.5 \text{ mm}}$$

_____ **Example 25-7**

How far from the lens of the previous example should a 3-mm object be placed in order to form an image on a screen that is 80 cm from the lens? How large will the image be?

Since the image is formed on a screen, we know that it is a real image.

$$\frac{1}{p} + \frac{1}{q} = \frac{1}{f}$$

$$\frac{1}{p} + \frac{1}{80 \text{ cm}} = \frac{1}{20 \text{ cm}}$$

from which

$$p = \boxed{26.7 \text{ cm}}$$

The magnification is

$$m = -\frac{q}{p} = -\frac{80 \text{ cm}}{26.7 \text{ cm}} = -3.00$$

The real image is inverted, and its height is

$$(3.00)(3 \text{ mm}) = \boxed{9 \text{ mm}}$$

_____ **Example 25-8**

An object is placed 15 cm from the same lens used in the previous two examples. Compute the position and the magnification of the image.

$$\frac{1}{15 \text{ cm}} + \frac{1}{q} = \frac{1}{20 \text{ cm}}$$

$$q = \boxed{-60 \text{ cm}}$$

$$m = -\frac{q}{p} = -\frac{-60 \text{ cm}}{15 \text{ cm}} = \boxed{+4}$$

The image is virtual, erect, and 4 times as large as the object. It is 60 cm from the lens, and since q is negative, the image is on the same side of the lens as the object.

In the next example, a virtual object is used with a diverging lens. This illustrates a common laboratory method for measuring the focal length of a diverging lens.

Example 25-9

A certain lens, about which we need know nothing, forms a real image at the 82-cm position on an optical bench (Fig. 25-21). A diverging lens is then placed at the 78-cm mark of the optical bench, and it is found that the real image is now at the 84-cm mark. What is the focal length of the diverging lens? What is the magnification of the diverging lens?

Ray 2 (extended as ray 2′) and ray 3 show the formation of the original image at 82 cm, before introduction of the negative lens. This image serves as a virtual object for the negative lens (virtual, since the rays don't actually pass through this point). The object distance is therefore negative, according to our sign convention for a virtual object.

$$\frac{1}{p} + \frac{1}{q} = \frac{1}{f}$$

$$\frac{1}{-4 \text{ cm}} + \frac{1}{6 \text{ cm}} = \frac{1}{f}$$

from which

$$f = \boxed{-12 \text{ cm}}$$

$$m = -\frac{q}{p} = -\frac{6 \text{ cm}}{-4 \text{ cm}} = \boxed{+1.5}$$

The new image is erect, and 1.5 times as tall as the original real image.

Note that the final direction of ray 2, extended backward, passes through the left-hand focal point of the negative lens, 12 cm back of the lens, at the 66-cm mark. Ray 3, coming from the original object, is undeviated as it passes through the center of the negative lens. Ray 1 is drawn emerging from the negative lens, parallel to the axis. You can use a

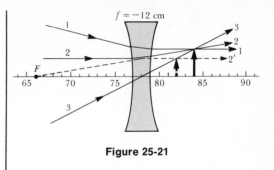

Figure 25-21

straight-edge to verify that before the negative lens was inserted, ray 1 would have gone to the right-hand focal point of the lens, at the 90-cm mark, and would have joined rays 2′ and 3 in forming the original image at the 82-cm mark.

25-8 Types of Lenses—The Diopter

To gain an intuitive understanding of what happens as light passes through a lens, let us consider wave fronts instead of rays. The *curvature* of a wave front is defined as the reciprocal of its radius; thus, a sphere of radius 0.2 m has a curvature of $1/(0.2 \text{ m}) = 5 \text{ m}^{-1}$. Wave fronts reaching the earth from the sun have practically no curvature ($\frac{1}{149,000,000}$ km^{-1} or 0.0000000067 km^{-1}). A plane wave, which is represented by parallel rays and plane wave fronts, has zero curvature. Spherical wave fronts are classified according to the behavior of the rays that are perpendicular to them (see Fig. 25-2). A point source of light, which might serve as an object for a lens, emits rays that travel outward in all directions; the wave fronts continually grow larger in radius. Such a wave front is called a *diverging wave front* and is said to have *negative curvature*. Similarly, if the rays associated with a spherical wave front are all heading for some point, called a real image, such a wave front is called a *converging wave front* and is said to have *positive curvature*.

We can redraw Fig. 25-15a, replacing the parallel rays by a plane wave AC, which strikes the double-convex lens (Fig. 25-22). During a short time interval Δt while the center of the

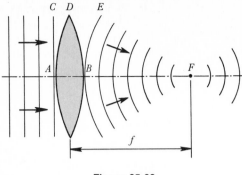

Figure 25-22

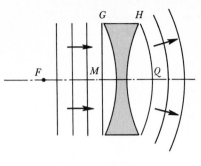

Figure 25-23

wave front is moving through glass from A to B, the outer edge of the wave front moves through air from C to E. Since the light travels faster in air than in glass, the wave front must curve around and become converging, as shown. The radius of curvature of the wave front BE is f,* but the wave front rapidly shrinks down to a point and then diverges. Fixing our attention on the two wave fronts AC and BE, we see that the lens has had a converging effect. The lens has acted on a wave front that had no curvature to start with, and it has produced a wave front that is converging with a positive curvature equal to $1/f$. For this reason a converging lens is called a *positive lens*. The exact shape of the lens is immaterial as far as the converging effect is concerned; any lens that is thicker in the center than at the edges is a converging, or positive, lens. Of course, a lens with a thicker bulge produces a greater effect and has a shorter focal length.

A *diverging lens* changes a plane wave front into a diverging one. Thus, in Fig. 25-23, the top of the wave front moves from G to H while the center of the same wave front (which has less glass to go through) moves from M to Q. After leaving the lens, the wave front continues to spread out, seeming to come from some virtual image at F. Any lens that is thinner in the middle than at the edges is a diverging, or negative, lens.

Several converging and diverging lenses are shown in Fig. 25-24.

The effect of a lens on the curvature of a wave front is called its *power*. That is, a powerful lens, one that has a short focal length, changes the curvature a great deal. We define the power of a lens as the reciprocal of its focal length. The power of a lens is often measured in *diopters*:

$$\text{Power in diopters} = \frac{1}{\text{focal length in meters}}$$

For any lens, the power depends on the curvatures of the surfaces and on the index of refraction of the glass according to the *lens maker's equation*, which is derived in Sec. 25-12:

$$\frac{1}{f} = (n-1)\left(\frac{1}{R_1} + \frac{1}{R_2}\right) \qquad (25\text{-}6)$$

where R_1 is the radius of the first surface of the lens and R_2 is the radius of the second surface. In this equation R_1 and R_2 are considered positive for convex surfaces; for a plane surface, $R = \infty$ and $1/R = 0$.

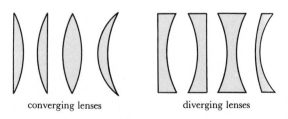

converging lenses   diverging lenses

Figure 25-24

---

* In the thin-lens approximation that we are using, $BF \approx f$.

Example 25-10

What is the power of a spectacle lens made of glass for which $n = 1.50$, if the outer surface is convex with radius 200 cm and the surface nearest the eye is concave with radius 100 cm?

This is a diverging lens, similar to the right-hand one in Fig. 25-24. We use Eq. 25-6 to calculate the power, first expressing the radii in meters.

$$\frac{1}{f} = (1.50 - 1)\left(\frac{1}{2.0\text{ m}} + \frac{1}{-1.0\text{ m}}\right)$$

$$= \boxed{-0.25\text{ m}^{-1}}$$

The power is -0.25 diopter. The $-$ sign indicates a diverging or negative lens. The shape of this spectacle lens, both surfaces bulging outward from the eye, allows room for eyelashes and the closing of the eyelids.

The power of the human eye (including both the lens and the cornea) is about $+60$ diopters, and the lenses ordinarily used for eyeglasses have powers of from $+5$ diopters to about -5 diopters, in steps of $\frac{1}{8}$ diopter. (See Sec. 27-2 for a more detailed study of the eye.)

25-9 Objects and Images— Method of Curvatures

We have used ray tracing to derive the lens equation (Eq. 25-5) that relates p, q, and f. Thus we can calculate the position of the image, as well as the magnification. Ray diagrams are use-ful but may become rather complicated, especially if several lenses are involved.

Another, more descriptive, way of looking at the behavior of a lens allows us to locate image positions using a cause-and-effect procedure. We rearrange the terms in the lens equation (Eq. 25-5) to obtain

$$-\frac{1}{p} + \frac{1}{f} = \frac{1}{q} \qquad (25\text{-}7)$$

Each term in this equation has a simple physical significance. We saw in Sec. 25-8 that the curvature* of a wave front is the reciprocal of its radius. Thus, $-1/p$ is the initial curvature, and $1/q$ is the final curvature. Also, we saw that $1/f$ is the power of the lens. Therefore, Eq. 25-7 has the following significance:

$$\begin{array}{c} \text{initial} \\ \text{curvature} \end{array} + \begin{array}{c} \text{power of} \\ \text{lens} \end{array} = \begin{array}{c} \text{final} \\ \text{curvature} \end{array}$$

You are already familiar with the only sign convention needed in using the method of curvatures:

A converging lens or wave front is $+$
A diverging lens or wave front is $-$

Let us interpret Fig. 25-25 with the help of our new form of the lens equation. The wave front that leaves the object A has a large curvature (small radius) when it is near A, but by the time the wave front reaches the lens it is flatter, and its

---

* Some authors use *vergence* as a general term for the curvature of a wave front.

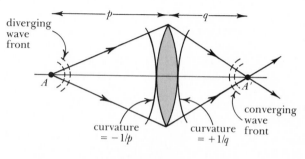

Figure 25-25

curvature is less. We see that $-1/p$ is the curvature of the wave front just as it reaches the lens; the curvature is negative because the wave front has *diverged* from the object A. The power of the lens (the effect of the lens on the wave front) is $1/f$, and $1/q$ is the curvature of the wave front just as it leaves the lens, before it shrinks down to form the image at A'. Our equation tells us how the lens has acted on the wave front to change its curvature. As a check, we can apply the equation to prove that $1/f$ is the power of the lens: If the incident wave front is plane, coming from a distant source, it has no curvature and $-1/p$ is $-1/\infty$, which is 0. Equation 25-7 becomes in this case $-1/\infty + 1/f = 1/f$, which shows that the change in curvature in this case is just equal to the power of the lens.

It is important to note that all the action represented by the equation takes place *at the lens*; p and q are involved in the problem only insofar as they are related to the curvatures of the wave fronts that are entering and leaving the lens.

To illustrate the method of curvatures, we shall solve some simple problems, some of which were also solved by ray-tracing methods in Sec. 25-8.* Note that in diagrams using the method of curvatures, wave fronts diverge from a point on the axis; we do not draw an extended object such as an arrow. We use the diagram chiefly to visualize changes in curvature. To obtain the magnification, we still use the simplest result of ray tracing, the fact that

$$\text{Magnification} = -\frac{\text{image distance}}{\text{object distance}}$$

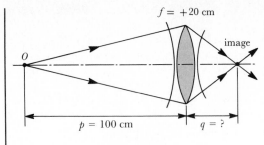

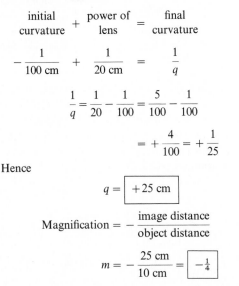

Figure 25-26

power. The rays diverge from the object at O, and by the time the wave front reaches the lens, the curvature is $-\frac{1}{100}$ cm^{-1}.

$$\begin{array}{ccccc} \text{initial} & & \text{power of} & & \text{final} \\ \text{curvature} & + & \text{lens} & = & \text{curvature} \end{array}$$

$$-\frac{1}{100 \text{ cm}} + \frac{1}{20 \text{ cm}} = \frac{1}{q}$$

$$\frac{1}{q} = \frac{1}{20} - \frac{1}{100} = \frac{5}{100} - \frac{1}{100}$$

$$= +\frac{4}{100} = +\frac{1}{25}$$

Hence

$$q = \boxed{+25 \text{ cm}}$$

$$\text{Magnification} = -\frac{\text{image distance}}{\text{object distance}}$$

$$m = -\frac{25 \text{ cm}}{10 \text{ cm}} = \boxed{-\frac{1}{4}}$$

This illustrates the fact that the magnification is *negative* whenever the curvature of the wave front changes sign while passing through the lens.

Example 25-11

An object (Fig. 25-26) is placed 100 cm from a lens of focal length $+20$ cm. Where is the image? What is the magnification?

The lens is a converging one, since it is thicker in the center than at the edges, and so it has positive

---

* Which form of the lens equation you use is a matter of personal preference.

Example 25-12

How far from the lens of the previous example should an object be placed in order to form an image on a screen that is 80 cm from the lens? (See Fig. 25-27.)

Since the image is formed on a screen, we know that it is formed by converging light; hence the curvature of the beam as it leaves the lens is positive, and equal to $+\frac{1}{80}$ cm^{-1}.

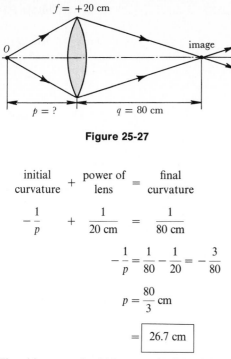

$$f = +20 \text{ cm}$$

O image

$$p = ? \qquad q = 80 \text{ cm}$$

Figure 25-27

$$\frac{\text{initial}}{\text{curvature}} + \frac{\text{power of}}{\text{lens}} = \frac{\text{final}}{\text{curvature}}$$

$$-\frac{1}{p} + \frac{1}{20 \text{ cm}} = \frac{1}{80 \text{ cm}}$$

$$-\frac{1}{p} = \frac{1}{80} - \frac{1}{20} = -\frac{3}{80}$$

$$p = \frac{80}{3} \text{ cm}$$

$$= \boxed{26.7 \text{ cm}}$$

The object must be 26.7 cm to the left of the lens. Note that p is positive, and the object is in the "typical" position discussed in Sec. 25-7.

Example 25-13

An object is placed 15 cm from the same lens used in the previous two examples. Compute the position of the image (Fig. 25-28).

$$\frac{\text{initial}}{\text{curvature}} + \frac{\text{power of}}{\text{lens}} = \frac{\text{final}}{\text{curvature}}$$

$$-\frac{1}{15 \text{ cm}} + \frac{1}{20 \text{ cm}} = \frac{1}{q}$$

$$\frac{1}{q} = -\frac{1}{15} + \frac{1}{20} = -\frac{4}{60} + \frac{3}{60} = -\frac{1}{60}$$

$$q = \boxed{-60 \text{ cm}}$$

The $-$ sign for the image distance indicates a virtual image. The lens is not strong enough to remove all the divergence of the original wave front; the rays still diverge (but to a lesser extent) after they leave the lens.

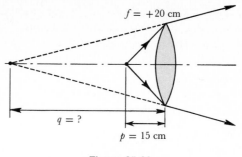

$$f = +20 \text{ cm}$$

$$q = ?$$

$$p = 15 \text{ cm}$$

Figure 25-28

A negative lens always gives a virtual image of a real object, since the original diverging wave front can only be made more diverging by the action of a negative lens.

Example 25-14

A negative lens of focal length 25 cm is placed 75 cm from an object. Where is the image? (See Fig. 25-29.)

$$\frac{\text{initial}}{\text{curvature}} + \frac{\text{power of}}{\text{lens}} = \frac{\text{final}}{\text{curvature}}$$

$$-\frac{1}{75 \text{ cm}} + \frac{1}{-25 \text{ cm}} = \frac{1}{q}$$

$$\frac{1}{q} = -\frac{1}{75} - \frac{1}{25} = -\frac{4}{75}$$

$$q = \boxed{-18.75 \text{ cm}}$$

The $-$ sign for q indicates a virtual image formed by diverging rays.

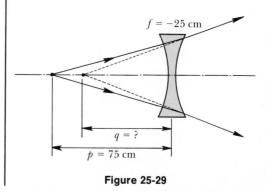

$$f = -25 \text{ cm}$$

$$q = ?$$

$$p = 75 \text{ cm}$$

Figure 25-29

The next example shows how a real image can be formed by a diverging lens—but only if the light has already been made converging before striking the lens.

Example 25-16

A slide projector has a converging lens of focal length 5 cm, and the screen is 3 m from the lens. (a) How far from the lens should the slide be placed? (b) What is the height (on the screen) of the image of a portion of the slide that is 4 mm high?

In this problem, the object distance is unknown, and the image distance is 300 cm.

$$(a) \qquad -\frac{1}{p} + \frac{1}{5 \text{ cm}} = \frac{1}{300 \text{ cm}}$$

$$\frac{1}{p} = \frac{1}{5} - \frac{1}{300} = \frac{59}{300}$$

$$p = \frac{300}{59} \text{ cm} = \boxed{5.08 \text{ cm}}$$

The slide should be placed just outside the principal focus of the lens.

$$(b) \qquad \text{Magnification} = -\frac{\text{image distance}}{\text{object distance}}$$

$$m = -\frac{300 \text{ cm}}{\frac{300}{59} \text{ cm}} = -\frac{(300)(59)}{300} = -59$$

The image size is -59 times the object size (the negative sign indicates an inverted image). Since the object is 4 mm high, the height of the image on the screen is

$$(4 \text{ mm})(59) = \boxed{236 \text{ mm}}$$

Our next example requires us to find the size of a virtual image.

Example 25-15

A certain lens, about which we need know nothing, forms a real image at the 82-cm position on an optical bench (Fig. 25-30). A diverging lens is then placed at the 78-cm mark of the optical bench, and it is found that the real image is now at the 84-cm mark. What is the power of the diverging lens? What is the magnification of this lens?

The incoming wave front, heading for the 82-cm mark, has 4 cm still to go to reach an image at $+82$ cm, so it has a positive curvature of $+\frac{1}{4}$ cm^{-1} when it strikes the lens surface. The lens decreases some (not all) of this curvature, to make the curvature $+\frac{1}{6}$ cm^{-1} as the wave front leaves the lens.

$$\begin{array}{c} \text{initial} \\ \text{curvature} \end{array} + \begin{array}{c} \text{power of} \\ \text{lens} \end{array} = \begin{array}{c} \text{final} \\ \text{curvature} \end{array}$$

$$+\frac{1}{4} \text{ cm}^{-1} + \frac{1}{f} = +\frac{1}{6} \text{ cm}^{-1}$$

$$f = -12 \text{ cm}$$

The power of the lens is

$$\frac{1}{f} = \frac{1}{-0.12 \text{ m}} = -8.3 \text{ m}^{-1} = \boxed{-8.3 \text{ diopters}}$$

$$\text{Magnification} = -\frac{\text{image distance}}{\text{object distance}}$$

$$= -\frac{+6 \text{ cm}}{-4 \text{ cm}} = \boxed{+1.5}$$

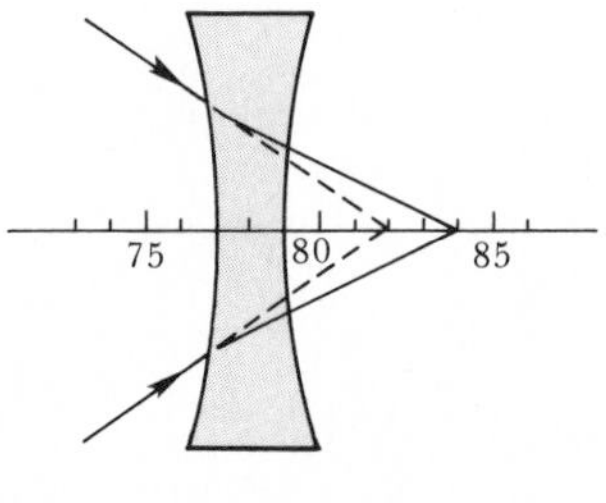

Figure 25-30

Example 25-17

A magnifying glass of power $+5$ diopters is held 16 cm from a page of fine print. How large is the image of a letter that is 2 mm high?

First we locate the *position* of the image, using the method of curvatures. The power of the lens is $+5$ diopters, so we know that it is a converging lens, with a focal length equal to $\frac{1}{5}$ m, or 20 cm. The object distance is 16 cm, and the image distance is unknown.

$$-\frac{1}{16 \text{ cm}} + \frac{1}{20 \text{ cm}} = \frac{1}{q}$$

$$q = -80 \text{ cm}$$

The negative sign indicates a virtual image, which is expected, since the object is inside the principal focus of the lens. We can now compute the magnification:

$$\text{Magnification} = -\frac{\text{image distance}}{\text{object distance}}$$

$$m = -\frac{-80 \text{ cm}}{16 \text{ cm}} = +5$$

The positive magnification indicates an erect image (see Fig. 25-19b). Finally, the image size is given by

$$(2 \text{ mm})(5) = \boxed{10 \text{ mm}}$$

Since the simple converging lens is used so often, it is worthwhile summarizing the object-image relationships for various cases. The results shown in Fig. 25-31 can be obtained by applying the method of curvatures as we have done in the examples. Two principal foci are shown, each at a distance f from the lens.

25-10 Mirrors

We have already used Huygens' principle to derive the law of reflection for plane mirrors. From a general point of view, a mirror is just

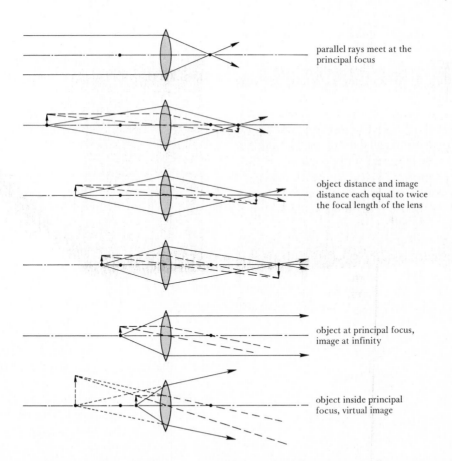

parallel rays meet at the principal focus

object distance and image distance each equal to twice the focal length of the lens

object at principal focus, image at infinity

object inside principal focus, virtual image

Figure 25-31 Object-image relationships for a converging lens. Solid lines represent divergence or convergence of wave fronts; dashed lines indicate formation of image by two intersecting rays.

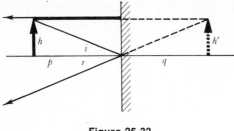

Figure 25-32

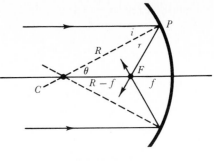

Figure 25-34

another surface that can modify the curvature of a wave front. The same ideas that we used for lenses can be applied to mirrors, with the added understanding that the wave front reverses its direction of motion while being reflected. First, consider a plane mirror. It is evident from Fig. 25-32 that the image distance q equals the object distance p and that $h' = h$. When you look at yourself in a plane mirror, the virtual image of your face is as far back of the surface as you are in front of it. The same result is found from the method of curvatures.

In Fig. 25-33, the wave front's curvature is not changed by reflection, and the original diverging wave front 1 becomes wave front 2, which is also diverging, seemingly from a virtual image I. Since $AB = CD$, wave fronts 1 and 2 have the same radius of curvature; hence the image distance equals the object distance.

When parallel rays strike a concave mirror, they converge to a focal point F (Fig. 25-34). We see that a concave mirror is a converging, or positive, mirror. It is fortunate that the same equation, $1/p + 1/q = 1/f$, that we used for

lenses can be applied to mirrors. First, however, we need to derive the mirror maker's formula, which is much simpler than the lens maker's formula. The focal length of a mirror is given by

$$f = \tfrac{1}{2}R \qquad (25\text{-}8)$$

where R is the radius of curvature of the mirror surface. The proof follows easily from Fig. 25-34, where C is the center of curvature. The angles i, r, and θ are all equal. Hence, if θ is small,* the two equal sides of the isosceles triangle CFP are approximately equal to half the base. This means that $R - f \approx \tfrac{1}{2}R$, whence $f \approx \tfrac{1}{2}R$. As for a lens, the power of a mirror is $1/f$, in diopters if f is in meters.

For a mirror of focal length $f\,(= \tfrac{1}{2}R)$, we use

$$\frac{1}{p} + \frac{1}{q} = \frac{1}{f}$$

with the following sign conventions: p, q, and f are all positive for the "typical case" of a concave (converging) mirror forming a real image of a real object. Thus f is negative for a convex (diverging) mirror; q is negative for a virtual image formed to the right of the mirror (behind the surface); p is negative for a virtual object (located to the right of the mirror, behind the surface). Just as for a lens, a real image is inverted and a virtual image is erect. The magnification is given, as for a lens, by $m = -q/p$. A converging

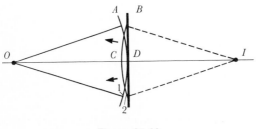

Figure 25-33

---

* This is the sort of approximation made for all of our work with lenses and mirrors. See the footnote on page 579.

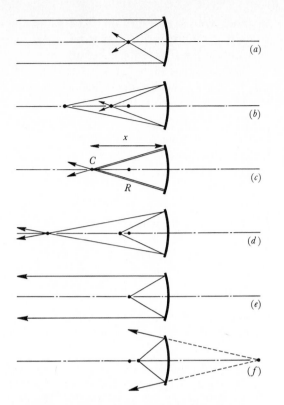

(a)

(b)

x

C

R

(c)

(d)

(e)

(f)

Figure 25-35 Object-image relationships for a converging mirror.

mirror, like a converging lens, forms a real, inverted image if the object is outside the principal focus, and it forms a virtual, erect image if the object is inside the principal focus. The six typical situations shown in Fig. 25-35 are analogous to those of Fig. 25-31, which are drawn for a converging lens.

The method of curvatures can easily be applied to mirrors. The same sign convention is used as for lenses (converging is +, diverging is −).

Example 25-18

An object of height 6 mm is placed 60 cm from a concave mirror of radius of curvature 30 cm. Where is the image, and how large is it?

The focal length is half the radius; hence $f = 15$ cm. The mirror is stated to be a concave one, so we know that it is a positive mirror.

$$\frac{1}{p} + \frac{1}{q} = \frac{1}{f}$$

$$\frac{1}{60 \text{ cm}} + \frac{1}{q} = \frac{1}{15 \text{ cm}}; \qquad q = \boxed{+20 \text{ cm}}$$

Since q is positive, the image is real. The magnification is

$$m = -\frac{q}{p} = -\frac{20 \text{ cm}}{60 \text{ cm}} = -\frac{1}{3}$$

The negative sign indicates an inverted image, and its size is

$$\tfrac{1}{3} \times 6 \text{ mm} = \boxed{2 \text{ mm}}$$

The image is 20 cm to the left of the mirror, real, inverted, and 2 mm tall.

The next example illustrates the solution of a mirror problem by the method of curvatures.

Example 25-19

A dentist holds a concave mirror of radius of curvature 50 mm at a distance of 20 mm from a cavity in a tooth (Fig. 25-36). Where is the image of the cavity, and how large is it?

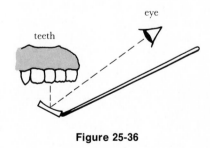

eye

teeth

Figure 25-36

The mirror's focal length is $f = \frac{1}{2}R = \frac{1}{2}(50 \text{ mm}) = 25$ mm; the power of this mirror is $\frac{1}{25}$ mm^{-1}

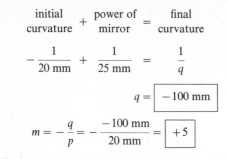

$$\underset{\text{curvature}}{\text{initial}} + \underset{\text{mirror}}{\text{power of}} = \underset{\text{curvature}}{\text{final}}$$

$$-\frac{1}{20\text{ mm}} + \frac{1}{25\text{ mm}} = \frac{1}{q}$$

$$q = \boxed{-100\text{ mm}}$$

$$m = -\frac{q}{p} = -\frac{-100\text{ mm}}{20\text{ mm}} = \boxed{+5}$$

The image is virtual, erect, 100 mm in back of the mirror, and 5 times as large as the cavity.

Convex mirrors are diverging, or negative. By a method similar to that which we used for concave mirrors, it can be shown that the focal length of a convex mirror is also equal to half its radius of curvature. Such a mirror always forms a reduced, upright, virtual image of a real object. Convex mirrors are used as safety mirrors, placed where aisles meet in factories (to avoid forklift truck collisions), or placed above strategic spots in stores (to reveal shoplifters).

25-11 Lens Combinations

When light passes through several lenses, the image formed by one lens becomes the object for a second lens, and so on for as many lenses as there may be. The overall magnification is the *product* of the magnifications of the separate lenses. One example will suffice to illustrate the method. We use the method of curvatures in order to visualize how light passes through the system of lenses. The problem can also be solved by several applications of the lens equation $1/p + 1/q = 1/f$, with proper attention to signs.

Example 25-20

Compute the position, size, and nature of the final image formed by the system shown in Fig. 25-37.

Lens A:

$$-\frac{1}{50} + \frac{1}{40} = \frac{1}{q}$$

$$q = +200\text{ cm}$$

$$m_1 = -\frac{+200\text{ cm}}{50\text{ cm}} = -4$$

The image is real, inverted, and 4 times as large as the object.

Lens B: The lenses are 220 cm apart, so the converging beam leaving lens A has a chance to reach a focus and diverge again. As the beam strikes B, it has a radius of curvature of 20 cm (= 220 cm − 200 cm) and is diverging.

$$-\frac{1}{20} + \frac{1}{10} = \frac{1}{q}$$

$$q = +20\text{ cm}$$

$$m_2 = -\frac{+20\text{ cm}}{20\text{ cm}} = -1$$

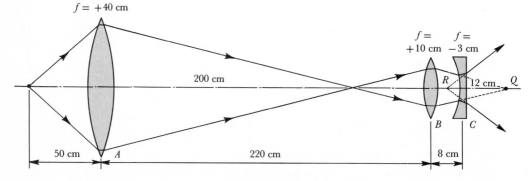

Figure 25-37

The image formed by lens B is real, inverted (relative to its object), and the same size as its object. The beam leaves lens B headed for a point Q that is 20 cm beyond the lens, that is, 12 cm beyond lens C.

Lens C: This is a tricky situation. The converging beam leaving lens B does not have a chance to reach its focus before lens C intervenes. The radius of curvature is shrinking and is equal to 12 cm when the surface of lens C is reached. Hence the initial curvature for lens C is positive.* The lens is negative because of its shape; such a lens tends to remove convergence and/or increase divergence. In this case, the lens actually succeeds in removing all the

---

* This is rare; the initial beam can be converging only if the beam has a past history and has already been acted on by one or more lenses or mirrors. This is an example of a virtual object. The "typical" real object sends out diverging rays, and so the initial curvature is usually negative.

convergence of the wave front and causes some divergence, as shown by the equation

$$+ \frac{1}{12 \text{ cm}} + \frac{1}{-3 \text{ cm}} = \frac{1}{q}$$

$$q = -4 \text{ cm}$$

$$m_3 = -\frac{-4 \text{ cm}}{-12 \text{ cm}} = -\frac{1}{3}$$

The image formed by lens C is virtual, 4 cm to the left of C, inverted relative to its object, and $\frac{1}{3}$ as large as its object.

The final image is at R, halfway between lens B and lens C. It is a virtual image. The overall magnification is

$$m_1 m_2 m_3 = (-4)(-1)(-\tfrac{1}{3}) = -\tfrac{4}{3}.$$

The three minus signs mean that three inversions have taken place, and hence the final image is virtual, inverted, and $\frac{4}{3}$ the size of the original object.

Summary   Huygens' principle states that as a wave propagates, the new wave front is the envelope of wavelets that spread out from all points of the old wave front. If no surface such as a mirror, prism, or lens intervenes, a plane wave remains plane, and a spherical wave spreads out spherically. These facts imply straight-line propagation of light waves; such is actually the case for waves whose wavelength is small compared with the sizes of obstacles or openings.

When light rays strike a smooth surface, both reflection and refraction may take place, in a plane determined by the incident ray and the normal to the surface. The angle of reflection equals the angle of incidence; the angle of refraction is determined by Snell's law, which states that $n_1 \sin \theta_1 = n_2 \sin \theta_2$. The index of refraction n is the ratio of the speed of light in vacuum to the speed in the medium. In a medium, the speed of light (and index of refraction) depend on the frequency (that is, the color) of the light. In most transparent solids and liquids, this variation, called dispersion, is relatively slight.

The critical angle is the angle of incidence, in the more optically dense medium, for which the light emerges into the less dense medium just tangent to the surface. If the critical angle is exceeded, total reflection takes place within the denser medium.

Images formed by converging rays are real, and images formed by diverging rays are virtual. Images can be located by ray tracing; image size is to object size as image distance is to object distance. Any lens that is thicker at its center than at its edges is a converging, or positive lens; a lens that is thinner at its center than at its edges is a diverging, or negative, lens. The power of a lens is the reciprocal of its focal length.

The curvature of a wave front is defined as the reciprocal of its radius. A diverging wave front has negative curvature, and a converging wave front has positive curvature. Images can be located by considering changes of curvature caused by the lens; initial curvature plus power of lens equals final curvature.

A plane mirror gives an erect virtual image having the same size as the object, and image distance equals object distance. Concave mirrors are positive, and convex mirrors are negative. The focal length of a mirror equals half its radius of curvature.

Check List

angstrom unit	Snell's law of refraction	curvature
Huygens' principle	critical angle	diverging wave front
wave front	principal focus	converging wave front
isotropic medium	focal length	positive lens or mirror
anisotropic medium	real image	diverging lens or mirror
angle of incidence	virtual image	power of a lens or mirror
angle of reflection	magnification	diopter
angle of refraction	lens equation	lens maker's equation
law of reflection	typical case	mirror maker's equation
index of refraction		

Questions

25-1 Would you expect Huygens' principle to apply to radio waves?

25-2 Do any radio waves travel in straight lines?

25-3 Give two examples of each of the following processes: (*a*) specular reflection; (*b*) diffuse reflection; (*c*) scattering.

25-4 A beam of white light containing all colors strikes a prism as shown in Fig. 25-38, and the rays of different colors are separated because their speeds in glass are different. From a general knowledge of the relative values of index of refraction, state whether the red rays or the violet rays are deviated the most from their original direction.

25-5 Can a ray in air be totally reflected at a glass surface if the angle of incidence (in air) and the index of refraction are just right?

25-6 Which of the lenses in Fig. 25-39 are positive, and which are negative?

25-7 Is it possible for a converging wave front to change into a diverging wave front without striking a surface of any kind?

25-8 Is it possible for a diverging wave front to change into a converging wave front without striking a surface of any kind?

25-9 A real image is to be formed by a plane mirror. What must be the nature of the initial wave front that strikes the mirror—converging or diverging?

25-10 Diamond has a higher index of refraction than glass. Is the critical angle for a diamond-air interface larger or smaller than for a glass-air interface? Does this have anything to do with the brilliance of a diamond gemstone?

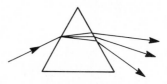

Figure 25-38   **Figure 25-39**

25-11 Sound waves of wavelength 15 cm are obstructed by a building that is 20 m wide and 15 m tall. Is the shadow "sharp"?

25-12 When hot air rises above a radiator, the wall behind the air stream seems to shimmer. Explain. Also explain why stars twinkle.

MULTIPLE CHOICE

25-13 Audible sound waves do not cast sharp shadows because their wavelength is   (a) variable;   (b) too small;   (c) too large.

25-14 When a beam of light traveling straight down in air strikes the surface of a horizontal flat glass plate, the beam is   (a) totally transmitted;   (b) totally reflected;   (c) partly reflected and partly transmitted.

25-15 If the critical angle for a certain liquid is 48°, the index of refraction is given by   (a) $n = \sin 48°$;   (b) $n = 1/\sin 48°$;   (c) $n = \tan 48°$.

25-16 If an object is placed 30 cm from a converging lens of focal length 10 cm, the image distance is   (a) 7.5 cm;   (b) 15 cm;   (c) 20 cm.

25-17 The focal length of a lens is 20 cm. The power of this lens is   (a) 0.05 diopter;   (b) 5 diopters;   (c) 20 diopters.

25-18 A virtual image is formed at a point P that is 80 cm in front of an observer's eye. This image can be   (a) seen by the observer if the eye is properly focused;   (b) recorded on a photographic film placed at P;   (c) neither of these.

Problems

25-A1 What is the speed of light in a clear plastic material whose index of refraction is 1.95?

25-A2 What is the index of refraction of a liquid in which the speed of light is 1.970×10^8 m/s?

25-A3 The speed of light in a certain kind of glass is five-eighths as much as in a vacuum. What is the index of refraction of the glass?

25-A4 Light of frequency 5×10^{14} Hz enters a clear plastic material whose index of refraction is 2.00. What is the wavelength of the light while it is in the plastic?

25-A5 What is the wavelength, in water, of blue light whose wavelength in air is 460 nm?

25-A6 What is the ratio of the speed of light in ice to the speed of light in diamond?

25-A7 In Fig. 25-40, a ray of light enters a liquid and is bent toward the normal as shown. Calculate the speed of light in the liquid.

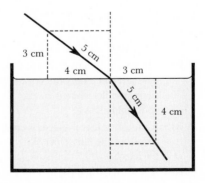

Figure 25-40

25-A8 (*a*) What is the focal length of a spectacle lens of power -8.0 diopters?   (*b*) Describe the lens; is it converging or diverging?

25-A9 (*a*) A converging lens has focal length 25 cm. Calculate the power of the lens, in diopters.   (*b*) Is it a positive or negative lens?

25-A10 What is the least height of a plane mirror on a wall in which a girl 1.60 m tall can just see her entire image from the top of her head to the soles of her shoes? (*Note:* Do not assume that the girl's eyes are at the top of her head.)

25-B1 The rays of the afternoon sun strike the surface of a lake at an angle of 70° with the vertical. At what angle, measured from the vertical, is the refracted ray in the water?

25-B2 A ray of light strikes a piece of glass ($n = 1.50$), making an angle of 50° with the surface. What angle does the refracted ray make with the surface?

25-B3 A ray of light starts from a point on the bottom of a dish of liquid 3 cm deep and strikes a point on the surface that is 4 cm from the point directly above the source. What is the least index of refraction for the liquid that will allow the ray to be totally reflected?

25-B4 A ray of light directed horizontally 30° N of E enters a vertical window pane that is in a N–S plane. If $n = 1.50$ for the glass, what is the direction of the ray in the glass?

25-B5 A flashlight on the bottom of a swimming pool 3 m deep sends a ray upward and at an angle such that the ray strikes the surface of the water 1.8 m from the point directly above the flashlight. What angle (in air) does the emerging ray make with the vertical?

25-B6 A ray of light from a boat's searchlight reaches a skin diver's eye at an angle of 43° from the vertical. What angle in air does the ray make with the *surface* of the water?

25-B7 Calculate the critical angle for an interface between flint glass and air.

25-B8 What is the index of refraction of a type of glass for which the critical angle at a glass-air interface is 40°?

25-B9 A lamp bulb is placed 400 cm from a spectacle lens of power $+0.45$ diopter. (*a*) What is the radius of curvature of the wave front just as the light strikes the lens? (*b*) What is the radius of curvature of the wave front just as the light leaves the lens?

25-B10 A point source is 6 m from a converging lens of focal length 2 m.   (*a*) Where is the image?   (*b*) What is the radius of curvature of the wave front just as the light leaves the lens?

25-B11 An object 6 mm high is 12 cm to the left of a diverging lens of focal length 4 cm. (*a*) Where is the image?   (*b*) Is it real or virtual?   (*c*) Is it erect or inverted? (*d*) How large is it?

25-B12 Repeat Prob. 25-B11 if the lens of focal length 4 cm is converging instead of diverging.

25-B13 A lens of power $+2.5$ diopters is held 30 cm above an object 4 mm wide. Describe the location, nature, and size of the image.

25-B14 A converging lens is 30 cm from a diverging lens of focal length 15 cm. A beam of parallel light enters the system from the left, as shown in Fig. 25-41, and the beam is again parallel when it emerges from the second lens. Calculate the focal length of the converging lens.

25-B15 A projector lens is placed 8 cm from a slide, and the real image is formed on a screen 6 m from the lens.   (*a*) What is the magnification?   (*b*) What is the focal length of the lens?

25-B16 A researcher in a space lab focuses a camera lens (of focal length 50 mm) to form an image of the earth. How far must she move the lens toward or away from the film (which?) in order to focus the camera on a fellow researcher 3.0 m away?

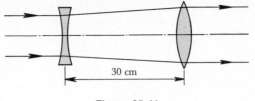

30 cm

Figure 25-41

25-B17 The moon's diameter is 3500 km, and it is 3.84×10^5 km from the earth. What is the size of the moon's image on the film in a camera whose lens has focal length 50 mm?

25-B18 A lens of power $+25$ diopters is used to form a real image of a postage stamp on a photographic film that is 5 cm from the lens. (*a*) How far from the lens is the stamp? (*b*) If the stamp is 3.0 cm tall, how tall is the image?

25-B19 How far from a converging lens of focal length 40 cm must an object be placed if the image is to be a real one, four times as large as the object? (*Hint:* Let $x =$ the unknown object distance; then the image distance is $4x$.)

25-B20 Repeat Prob. 25-B19, for the same converging lens of focal length 40 cm, if the image is to be a virtual one, four times as large as the object.

25-B21 In a laboratory experiment to measure the focal length of a diverging lens, first an image of an object is formed on a screen 45 cm to the right of a converging lens of unknown focal length. The diverging lens is then placed 25 cm to the right of the converging lens, and it is found that the image is again in focus if the screen is moved 10 cm (in which direction?). Calculate the focal length of the diverging lens.

25-B22 In each part of Fig. 25-19, only two of the three "easy" rays leaving the object are shown. Using a pencil, draw in the third ray. For part (*a*), the ray leaves B along a line directed toward the other focal point; how does it emerge from the lens? In part (*b*), the ray leaves B along a line directed away from the left-hand focal point; how does this ray emerge? In both parts, show that the emerging ray seems to come from the image of the head of the arrow.

25-B23 A bird views himself in a convex garden mirror that is a reflecting ball of diameter 40 cm. If the bird is 15 cm from the surface of the mirror, (*a*) where is the image? (*b*) what is the magnification?

25-B24 Prove that for a concave mirror, object distance p, image distance q, and focal length f are related by $1/p + 1/q = 1/f$. (*Hint:* Use a diagram similar to Fig. 25-17, but drawn for a mirror instead of a lens. Assume the position of F to be known.)

25-B25 An object 5 cm high is 40 cm from a concave mirror whose radius of curvature is 16 cm. Calculate (*a*) the focal length of the mirror; (*b*) the location of the image; (*c*) the nature of the image (real or virtual; erect or inverted; size of image).

25-B26 An object 2 cm high is 30 cm from a convex mirror of focal length 20 cm. Calculate (*a*) the location of the image; (*b*) the nature of the image (real or virtual; erect or inverted; size of image).

25-B27 An object 1 cm tall placed 12 cm from a spherical mirror gives a virtual image 4 cm tall. (*a*) Where is the image? (*b*) Is the mirror concave or convex? (*c*) What is the focal length of the mirror?

25-C1 (a) What is the critical angle for a glass-water interface? (b) In which medium must the incident ray be for total reflection? (Assume $n = 1.662$ for the glass used.)

25-C2 Three adjacent faces of a plastic cube of index of refraction n are painted black, with a clear spot at the painted corner serving as a source of diverging rays. Show that a ray from this corner to the center of a clear face is totally reflected if $n \geq \sqrt{3}$.

25-C3 A. H. Pfund's method for finding the index of refraction of glass is illustrated in Fig. 25-42. One face of a slab of thickness t is painted white, and a small hole scraped clear at point P serves as a source of diverging rays when illuminated from below. Ray PBB' strikes the clear surface at the critical angle and is totally reflected, as are rays such as PCC'. Rays such as PAA' emerge from the clear surface. On the painted surface there appears a dark circle of diameter D surrounded by an illuminated region or halo. (a) Derive a formula for n in terms of the measured quantities D and t. (b) What is the diameter of the dark circle if $n = 1.52$ for a slab 0.50 cm thick? (c) If white light is used, the critical angle depends on color due to dispersion (mentioned in Sec. 25-4). Is the inner edge of the white halo tinged with red or with violet light?

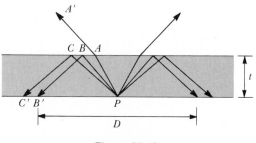

Figure 25-42

25-C4 A certain thin lens made of crown glass has a power of $+4$ diopters when in air. What is the power of the lens when it is immersed in water?

25-C5 A positive lens of power 5 diopters forms a real image 5 times as large as the object. How far apart are the object and the image?

25-C6 How far apart are the object and image if a lens of focal length 15 cm forms a real image 3 times as tall as the object?

25-C7 When an object is moved along the axis of a thin lens, the size of the image is 4 times the size of the object if the object is at a point A to the left of the lens, and also if the object is at a point B that is 30 cm farther from the lens. (a) Is the lens converging or diverging? (b) What is the focal length of the lens? (c) Where are points A and B?

25-C8 An object 6 mm high is placed 12 cm from a lens of focal length $+10$ cm. A second lens, of focal length $+30$ cm, is placed 45 cm beyond the first lens. (a) How far from the object is the final image? (b) What are the nature and size of the final image?

25-C9 An object is placed 30 cm to the left of a diverging lens of focal length 30 cm, and a converging lens of focal length 20 cm is placed 45 cm to the right of the first lens. (a) How far from the object is the final image? (b) What must be the size of the object if the final image is 2 mm tall?

25-C10 A lens of focal length $+15$ cm is 30 cm to the left of a lens of focal length -15 cm. (a) Where is the final image of an object 60 cm to the left of the converging lens? (b) What is the magnification of the combination?

25-C11 A dedicated sports car buff polishes the inside as well as the outside surface of a hub cap that is a section of a sphere. When he looks into one side of the hub cap, he sees an image of his face 42 cm in back of the hub cap. He then turns the hub cap over and sees another image of his face 14 cm in back of the cap. (a) How far is his face from the hub cap? (b) What is the radius of curvature of the hub cap?

25-C12 What is the radius of curvature of a makeup mirror that gives a twofold linear magnification of an eyebrow of an actress whose face is 24 cm in front of the mirror?

25-C13 A thin plano-convex lens made of glass of index of refraction n has a curved surface of radius of curvature R. (a) What is the power of the lens? (b) What is the power of a lens-mirror combination formed by giving a reflecting coating to the plane surface of the lens? (c) Repeat, if the reflecting coating is applied to the curved surface of the lens.

25-C14 Use the method of curvatures to prove that the power of two thin lenses that are touching each other is the sum of the powers of the two lenses.

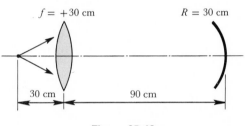

Figure 25-43

25-C15 (a) Calculate the position of the final image in Fig. 25-43. (Hint: The light passes through the lens twice.) (b) Describe the nature and magnification of the final image.

For Further Study

25-12 Derivation of the Lens Maker's Equation

The method of curvatures for thin lenses is proved by use of a theorem from plane geometry. According to the sagittal theorem,* the segment

---

* From the Latin *sagitta*, arrow (so called because the diagram resembles a bow and arrow). This is a special case of a theorem of geometry: for any two chords of a circle, the product of the segments of one chord equals the product of the segments of the other chord.

---

x in Fig. 25-44 is given by $x(2R - x) = h^2$. In cases of interest to us, x is negligible compared to $2R$; hence $x(2R) \approx h^2$, or

(sagittal theorem) $\quad x \approx \dfrac{h^2}{2R}$

Consider now the formation of an image by a lens. In Fig. 25-45 let the radii of curvature of the two lens surfaces be R_1 and R_2; the object distance is p, and the image distance is q. While the center of the wave front travels from P to Q in

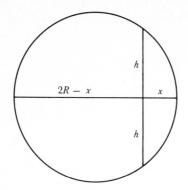

Figure 25-44

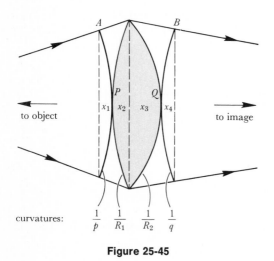

curvatures: $\quad \dfrac{1}{p} \quad \dfrac{1}{R_1} \quad \dfrac{1}{R_2} \quad \dfrac{1}{q}$

Figure 25-45

glass, the top of the same wave front travels from A to B in air. From Fig. 25-45 we have

$$AB - PQ \approx x_1 + x_4$$

Since the times for paths AB and PQ are equal,

$$\frac{AB}{c} = \frac{PQ}{c/n}$$

$$AB = n\,PQ$$

$$AB - PQ = n\,PQ - PQ$$
$$= (n-1)\,PQ$$

That is,

$$x_1 + x_4 = (n-1)(x_2 + x_3)$$

Now we use the saggital theorem. Substituting $x_1 = h^2/2p$, $x_2 = h^2/2R_1$, $x_3 = h^2/2R_2$, $x_4 = h^2/2q$, and canceling $h^2/2$, we obtain

$$\frac{1}{p} + \frac{1}{q} = (n-1)\left(\frac{1}{R_1} + \frac{1}{R_2}\right) \qquad (25\text{-}9)$$

The term on the right involves n, R_1, and R_2, which depend only on the shape of the lens and the index of refraction of the glass. To interpret this term, which is constant for a given lens, let the object be infinitely far from the lens ($p = \infty$); then the image is at the focal point ($q = f$). For this situation, $1/p = 0$, and $1/q = 1/f$, and from Eq. 25-9 we obtain

$$\frac{1}{f} = (n-1)\left(\frac{1}{R_1} + \frac{1}{R_2}\right) \qquad (25\text{-}6)$$

This is the *lens maker's equation* by which we can calculate the power ($1/f$) of a thin lens when we know the curvatures of its surfaces and the index of refraction of the glass. Note that R_1 and R_2 are considered to be $+$ for convex surfaces, as is true for both surfaces in Fig. 25-45.

Problems

25-C16 The index of refraction of the glass in a lens 8.00 cm in diameter is to be found in the laboratory. The focal length of the lens is $+22.9$ cm, and the lens is 0.562 cm thick at its center. Calculate the index of refraction. [*Hint*: Use the saggital theorem to find $(1/R_1 + 1/R_2)$.]

25-C17 To start a fire, an Arctic explorer makes a burning glass out of ice. The lens is plano-convex, 10 cm in diameter, and 1.5 cm thick at its center. Calculate the focal length and power of the lens in diopters. (*Hint*: Use the saggital theorem to find the radius of curvature.)

References   1. W. F. Magie, *A Source Book in Physics* (McGraw-Hill, New York, 1935), pp. 283–294, Huygens' principle; pp. 335–344, speed-of-light determinations by Römer, Bradley, Fizeau, and Foucault.

2. J. F. Mulligan and D. F. McDonald, "Recent Determinations of the Speed of Light," *Am. J. Phys.* **20**, 165 (1952); **25**, 180 (1957).

3. J. H. Rush, "The Speed of Light," *Sci. American* **193**(2), 62 (Aug. 1955).

4. J. W. Satterly, "An Early Determination of Snell's Law," *Am. J. Phys.* **19**, 507 (1951). The law of refraction was discovered (but not published) by Thomas Hariot and his friends, three years before Snell's work.

5. T. B. Greenslade, Jr., "Index of refraction by an updated Pfund's method," *Phys. Teach.* **17**, 394 (1979).

6. F. D. Smith, "How Images Are Formed," *Sci. American* **219**(3), 97 (Sept. 1968).

7. A. B. Fraser and W. H. Mach, "Mirages," *Sci. American* **234**(1), 102 (Jan. 1976).

8. N. J. Herrick, "Use of Frustrated Total Reflection to Measure Film Thicknesses and Surface Reliefs," *J. Applied Phys.* **33**, 2774 (1962). Includes an application of total reflection to fingerprinting.

9. J. Strickland, *Straight Wave Reflection from Straight Barriers; Circular Wave Reflection from Various Barriers; Reflection of Waves from Concave Barriers; Refraction of Waves* (films).

26

Wave Optics

According to the dual theory of light, which we are now approaching, light has wave properties, and also particle properties. More precisely, we can say that sometimes a wave model for light is useful, and at other times a corpuscular (particle) model is useful. Some phenomena can be interpreted by either model. For example, energy is transmitted by a beam of light. We know that a stream of particles can carry kinetic energy from one place to another; and we saw in Chap. 10 that energy can also be transmitted by wave motion. Without specifying the exact nature of light, we therefore define the *intensity* of a beam of light as the rate of flow of energy per unit cross-sectional area. This is the same definition we used for the intensity of a sound wave in Chap. 11. In general, however, neither the wave model nor the corpuscular model is, by itself, successful in describing the entire range of optical phenomena. The new "nonclassical" model is a dual model, combining portions of the older, or "classical," models with some new ideas. In this chapter we shall study the compelling evidence

for a wave model, bearing in mind that the evidence for a particle model is yet to come (in Chap. 29).

26-1 Newton's Corpuscular Theory of Light

It was not until about 1800 that clear experimental evidence for the wave nature of light was discovered by Thomas Young (1773–1829), a British physician and experimenter. Let us review the situation at that time. The *possibility* that light is a wave motion had been recognized by Huygens, Hooke, Newton, and others, and indeed the laws of reflection and refraction (geometrical optics) had been derived from Huygens' principle (Chap. 25). It is possible, however, to derive these same laws from a corpuscular theory. Anyone who has played billiards or tennis knows that the angle of incidence equals the angle of reflection, assuming no "English" or top spin on the ball. The law of reflection for per-

Circular interference fringes formed by the mercury green line of wavelength 546 nm. The complex fringes at the left are natural mercury; each isotope produces light of a slightly different wavelength. The sharp fringes at the right are from a single mercury isotope of mass number 198.

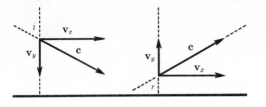

Figure 26-1

fectly elastic corpuscles was shown by Newton to be a direct consequence of the laws of motion. The corpuscle has components of velocity v_x and v_y as it approaches the surface (Fig. 26-1). As it leaves the surface, we find that v_x remains unaltered in magnitude and direction, since there is no horizontal force and the horizontal momentum of the system is conserved. The vertical component v_y remains unaltered in magnitude but reverses in direction; in this way the conditions for perfectly elastic collision are fulfilled. (Velocity of approach equals velocity of separation.) Applying these ideas to the reflection of light, the vector sum of v_x and v_y is the velocity of light c; the magnitude of c, and hence the kinetic energy of the bouncing corpuscle, is constant. From the diagram, it is evident that $i = r$, and hence the law of reflection is proved.

Newton also had a corpuscular theory for refraction. He postulated that a corpuscle of light is attracted toward a surface by some sort of force. Thus in Fig. 26-2, the force of the glass on the corpuscle would be directed along the normal. The normal component v_y increases, but the horizontal component v_x remains unaltered. The resultant velocity in glass is a vector that makes an angle θ_2 with the normal, and hence the ray of light bends toward the normal. If the supposed force of attraction is a very short-range force, the change in velocity will take place only during the time that the corpuscle is close to the surface, and the bending of the light ray will be abrupt. It is even possible to derive Snell's law from this model:

$$\frac{\sin \theta_1}{\sin \theta_2} = \frac{v_x/c_1}{v_x/c_2} = \frac{c_2}{c_1} = \text{constant}$$

The constant is the *reciprocal* of the constant c_1/c_2 derived in Sec. 25-4 by the wave model and Huygens' principle. However, the essence of Snell's law is that the ratio $\sin \theta_1/\sin \theta_2$ is a constant (any constant) for a given pair of media, and so we see that the corpuscular and wave models are equally useful in this case.

We should realize that Newton's corpuscular theory was, in its day, superior to the competing wave theory in certain respects, since his theory used the laws of mechanics that had already proved to be valid in other areas of physics. Newton emphasized the corpuscular aspects of his model on the basis of the then-known facts, especially the straight-line propagation of light. An important experiment to clarify the issue is obvious: one need only measure the speed of light in air and in glass. According to the wave theory, light should slow down as it enters glass (see Fig. 25-8); according to the corpuscular theory, it is attracted toward the surface and should speed up as it enters glass. In Newton's time the speed of light could not be measured in the laboratory, and so either assumption was possible. Not until 1850, over a century after the death of Newton, did Jean Foucault succeed in showing experimentally that the speed of light in

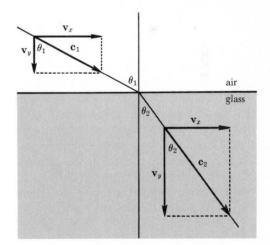

Figure 26-2 Newton's proof of Snell's law, using a corpuscular model for light.

water is less than in air; but this verification of the wave theory of light came as an anticlimax. In the early years of the 19th century the ingenious experiments of Young, Fresnel, and others had already provided direct evidence for the existence of light waves.

26-2 Interference

Just as in Sec. 11-1, we use the term *interference* to describe the result of the superposition of several waves traveling through the same region of space. The first experiment in which the interference of light waves was consciously obtained was that of Young (see Fig. 26-3). A narrow vertical slit S is illuminated by a source, and a wave front spreads out from S, striking two narrow slits P and Q that are close together.* Wavelets spread out from P and Q, and any given point on the screen receives light from both slits. It is found that the screen is not uniformly illuminated; some points, such as O, A, and B, are bright, and others, such as X and Y, are dark. Young interpreted this pattern as the result of constructive and destructive interference of light waves. Consider a point such as A. We wish to find the resultant illumination at A due to the two rays PA and QA. Since P and Q are on a wave front whose center is at S, the rays leaving P and Q are in phase with each other. Whether they are in phase when combined at A depends on the path difference. If QA is exactly one wavelength longer than PA, the rays will be in phase at the point A, and the resultant amplitude will be large.[†] In general, *the condition for a maximum is that the path difference is an integral number of wavelengths.* Thus, $QO - PO = 0\lambda$; $QA - PA = 1\lambda$; $QB - PB = 2\lambda$; and so on; in general, $QN - PN = n\lambda$ for the nth bright band. (The

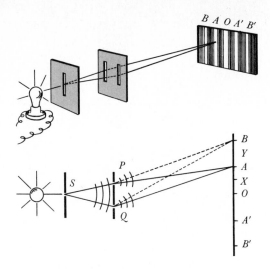

Figure 26-3 Young's experiment.

integer n is called the *order* of interference.) The central (undeviated) bright band is the 0th order, and there are two of each nth-order image, one on each side of center. Between the bright bands are dark bands, where the waves are completely out of phase; for these the path difference is $\frac{1}{2}\lambda$, $\frac{3}{2}\lambda$, $\frac{5}{2}\lambda$, and so on.

In order to calculate the positions of the bright bands, we simplify the problem by using parallel rays (plane wave fronts). According to Fig. 26-4, the path difference is DE, and therefore DE must equal $n\lambda$ if constructive interference is to take place. That is, $n\lambda/a = \sin\theta$, where a is the separation of the slits. Thus the general condition

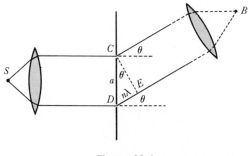

* The experiment can also be performed with pinholes instead of slits, with some sacrifice in intensity.
[†] Note the implicit use of the superposition principle here; see Sec. 10-7.

Figure 26-4

for a maximum (a bright region) of a double-slit pattern is

$$\left(\begin{array}{c}\text{double-slit}\\\text{maximun}\end{array}\right) \qquad n\lambda = a\sin\theta \qquad (26\text{-}1)$$

The same equation serves to locate the maxima for a grating consisting of many equally spaced slits.

Example 26-1

Parallel rays of wavelength 500 nm fall on a pair of slits that are 0.01 mm apart. The interference pattern is focused on a screen 2.30 m from the slits. Calculate the separation of the two 3rd-order bright bands, or images.

First we calculate the angular deviation for the 3rd order. The wavelength is 500 nm = 5×10^{-7} m, and a is 10^{-5} m.

$$n\lambda = a\sin\theta$$

$$\sin\theta = \frac{n\lambda}{a} = \frac{3(5\times 10^{-7}\text{ m})}{10^{-5}\text{ m}} = 0.150$$

$$\theta = 8.6°$$

The deviation y is found from the tangent of the angle (Fig. 26-5).

$$\tan 8.6° = \frac{y}{2.30\text{ m}}$$

$$y = (2.30\text{ m})(\tan 8.6°)$$
$$= (2.30\text{ m})(0.152) = 0.350\text{ m}$$

The 3rd-order images are each 0.350 m from the central bright image; hence the distance between them is

$$\boxed{2y = \quad 0.70\text{ m}}$$

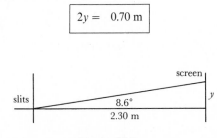

Figure 26-5

When the angular deviation is small, as in Example 26-1, the sine and the tangent are practically equal. In our next example we make use of this approximation.

Example 26-2

Two slits 0.03 mm apart are illuminated by parallel rays, and the 5th-order image is 14 cm from the central image, on a screen 2 m from the slits. Calculate the wavelength of the light used. From Fig. 26-6,

$$\tan\theta = \frac{14\text{ cm}}{200\text{ cm}} = 0.07$$

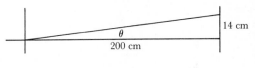

Figure 26-6

The angle is small, so $\sin\theta$ is also approximately 0.07.

$$n\lambda = a\sin\theta$$

$$\lambda = \frac{a\sin\theta}{n} = \frac{(0.03\times 10^{-3}\text{ m})(0.07)}{5}$$

$$= 4.2\times 10^{-7}\text{ m} = \boxed{420\text{ nm}}$$

This is violet light (Table 25-1).

Young's experiment does not produce a very bright interference pattern, since most of the light from the source is blocked out and only a small fraction passes through the slits. Two ways of getting a much brighter pattern make use of only one slit and a virtual image of that slit. In Fig. 26-7, a plane mirror M forms a virtual image S', and if the tilt of the mirror is small enough, S and S' are very close together and act as a double slit of small separation. The interference bands are visible in the region where the beams overlap. In the arrangement of Fig. 26-8, two virtual images of a slit are formed by two prisms placed

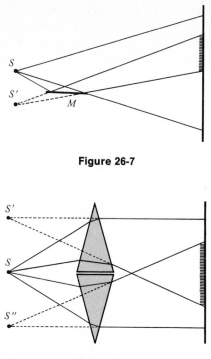

Figure 26-7

Figure 26-8

lengths of path difference will produce constructive interference.* These ideas apply to any wave motion, not merely to visible light. We are familiar with interference of radio waves, although perhaps unknowingly so. Normally, a television signal (of wavelength several meters) travels to the receiver by a direct line-of-sight path (Fig. 26-9a). If a plane flies past, a reflected beam may also reach the receiving antenna. If the situation is as in Fig. 26-9b, the two waves interfere destructively, but a few seconds later (c), as the plane moves, the path difference becomes an integral number of wavelengths, and constructive interference takes place. The resultant signal fluctuates in intensity while the plane is flying by, and the picture flutters. A similar effect can be caused at a slower rate by reflection from a drifting cloud or even by reflection from a trail of ions left by a meteor falling through the atmosphere.

Broadcasters make ingenious use of interference to beam their signals into a desired area. For instance, it would be uneconomical for a station near the seacoast to broadcast equally in all horizontal directions, as would be the case if a single vertical antenna were used. Instead, a pair of antennas spaced $\frac{1}{4}\lambda$ apart are fed from the same oscillator, but by electrical means the current in one antenna is caused to reach its peak just $\frac{1}{4}$ cycle later than that in the other antenna (Fig. 26-10). Point Y is $\frac{1}{4}\lambda$ farther from the antenna at A than from that at B. Also, A is caused to radiate $\frac{1}{4}$ cycle later than B, which is equivalent to another $\frac{1}{4}$ of path difference. These two quarter-wavelengths add up to $\frac{1}{2}\lambda$, and hence at Y the two beams interfere destructively, and no energy is radiated out to sea at Y. Just the opposite is

back to back. If the angles of the prisms are small, the separation between S' and S'' is small, and they act as a double slit. Here, too, the interference pattern is visible in the region where the beams overlap.

26-3 Interference of Radio, Water, and Sound Waves

The essential feature of interference is that *path difference is equivalent to phase difference.* Suppose that two beams reach a certain point in phase with each other—crest coinciding with crest and trough with trough. Stretching out the path of one beam by an extra half-wavelength is exactly equivalent to retarding its phase at the endpoint of the path by half a cycle of vibration. Thus, any odd number of half-wavelengths of path difference will produce destructive interference, while an even number of half-wave-

---

* In general, a phase change $\Delta\phi$ (in radians) is related to the path difference Δs by the equation

$$\Delta\phi = 2\pi\left(\frac{\Delta s}{\lambda}\right)$$

For example, for a path difference of $\frac{1}{2}\lambda$, the equation gives $\Delta\phi = 2\pi(\frac{1}{2}\lambda/\lambda) = \pi = 180°$, corresponding to a $\frac{1}{2}$-cycle change of phase.

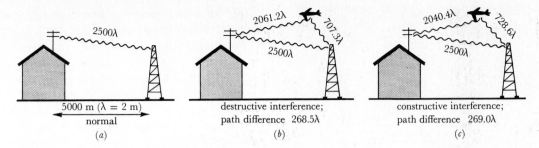

Figure 26-9 Constructive and destructive interference of radio waves. It is assumed that there is no phase change when the radio waves are reflected from the plane.

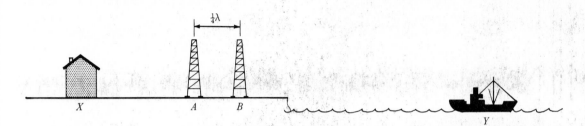

Figure 26-10 Tower A radiates $\frac{1}{4}$ cycle later than tower B; the radiation pattern is stronger at X than at Y (not drawn to scale).

true at X. The tower at A is $\frac{1}{4}\lambda$ closer to X than is B, but it is fed $\frac{1}{4}$ cycle later than B, so the waves reach X in phase, and the resultant amplitude is twice as much as from either antenna separately.

Considering interference from an energy viewpoint, we expect that the total energy radiated from the antennas must appear somewhere in the area surrounding the towers. Detailed calculation of such radiation patterns show that this is true. All the energy is accounted for, but the *distribution* of the energy depends on the number of antennas, their spacing, and their phasing. In a similar fashion, the total light energy falling on the two slits in Young's experiment is fully accounted for, even though certain places on the screen receive no light and other places are brighter than they would have been in the absence of the slit system. For visible light, as for radio waves, the amplitude of a wave is determined by the maximum magnitude E of the elec-

tric field vector (Fig. 24-4), and the intensity is proportional to E^2.

Interference is by no means confined to light, radio, and other e-m waves. In a ripple tank, a source of light shines through a shallow glass-bottomed tank filled with water, and the crests of ripples act as converging lenses to focus the light on a screen. The interference of two wavelets spreading out from adjacent sources is shown in Fig. 26-11; this is Young's experiment for water waves.

The constructive and destructive interference of sound waves is, of course, the basis of the stationary waves in pipes that were discussed in Sec. 11-3. One application of interference to direction finding is shown in Fig. 26-12. It is an interesting physiological fact that the brain can perform the function of mixing the output of the two ears, with the sensation depending on the resultant amplitude, which in turn depends on the phase difference between the sounds heard in

Figure 26-11 Constructive and destructive interference between water waves spreading out from two sources.

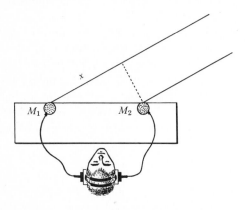

Figure 26-12 Sonic direction finder makes use of interference of sound waves.

the separate ears. If the path difference x is a half-wavelength, the outputs of the two microphones are out of phase, and the observer hears nothing. In use, the microphone assembly is ro-

tated until the sound is a maximum; the path difference is then zero, and the source is broadside to the line M_1M_2.*

In dealing with mechanical waves, we defined a motion node as a place where two vibrations combine to give zero net amplitude at all times. The nodes and antinodes in a pipe or on a string are caused by the interference of one wave with an equal reflected wave traveling in the opposite direction. In this chapter we have been studying interference of waves that are traveling in more or less parallel directions. When two such waves are brought together on a screen or on the retina of the eye, the places where they destructively or constructively interfere are also called "nodes" and "antinodes." Thus we consider an interference pattern on the screen or retina to be a pattern of nodes and antinodes. For light, the resultant electric vector of two interfering e-m waves is analogous to the resultant displacement vector of two interfering mechanical waves on a string or on water. In each case, nodes occur wherever the path difference between the two waves causes them to be permanently out of phase with each other.

26-4 The Grating

A series of many slits, equally spaced, is called a *grating*. Parallel rays falling on a grating form sharper and brighter maxima than they would with a pair of slits, but the basic principle is the same. In Fig. 26-13 we illustrate the formation of the 3rd-order maximum. The path length of each ray differs by $\pm 3\lambda$ from that of its neighboring ray. Thus, ray DD' is 3λ longer than ray CC', and it is 3λ shorter than ray EE'. Since the rays are

---

* While it is true that a maximum would also be heard if the path difference x were 1λ, 2λ, 3λ, . . . , in practical use this is no source of confusion. Most sounds contain components of many frequencies and have many wavelengths. The path difference x could not be equal to $n\lambda$ for all wavelengths at the same time, and no clear maximum would be heard except at the broadside position for which the path difference is 0λ. The condition $x = 0\lambda$ *can* be satisfied for all λ's simultaneously.

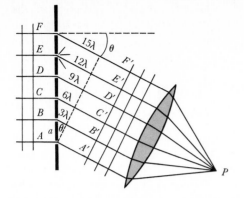

Figure 26-13 Grating with six slits. The lens might represent the eye lens, and P might be a point on the retina where the rays converge (the wave front shrinks to a point).

all in phase on the wave front $ABCDEF$, they will again be in phase on the wave front $A'B'C'D'E'F'$ if the angle θ is chosen just right. The condition for the 3rd-order maximum (a bright spot) is

$$3\lambda/a = \sin\theta \quad \text{or} \quad 3\lambda = a\sin\theta$$

where a is the space between adjacent slits, called the *grating space* or *grating constant*. In general, for the nth-order maximum,

$$n\lambda = a\sin\theta$$

This is the same equation as for a pair of slits. In fact, a pair of slits can be considered to be a rudimentary grating.

It should be noted that Fig. 26-13, and many others in this chapter, are simplified to show only those rays that are of interest. To be sure, wavelets *do* spread out from each slit. This means that rays diverge in all directions from each slit, as shown for the next-to-the-top slit. Our procedure is to pick a direction of observation, indicated by θ. This is done when we orient the lens and select a point P at which we observe the resultant amplitude (if any); this point might be on the retina of the eye or on a photographic plate. If it is objected "How do the rays know enough to bend at an angle θ?" we answer that we can later investigate the resultant amplitude in any other direction by the same method, but for the present we are concerned only with those rays that are eventually brought together at P. In thus drawing selected rays rather than wave fronts, we are making full use of the wave theory of light.

The sharpness of the "image" formed by a grating can be seen from the following argument: If there are 1001 slits and we are considering the 1st-order maximum, all the rays cooperate if θ is such that the top ray's extra path (MM') is exactly 1000λ (Fig. 26-14a). If we now increase the

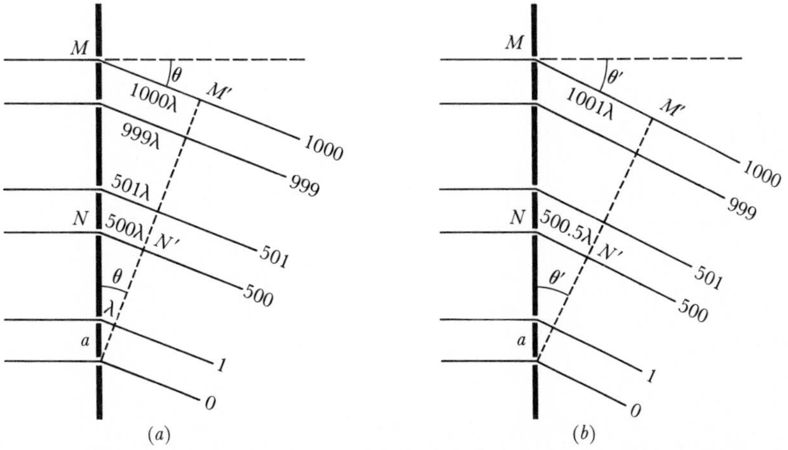

Figure 26-14 Grating with 1001 slits. (*a*) All rays in phase. (*b*) Rays cancel in pairs to give 0 resultant intensity.

angle very slightly to θ' (Fig. 26-14b), just enough to make $MM' = 1001\lambda$, the middle ray's extra path NN' will become 500.5λ. Hence ray 0 and ray 500 are 180° out of phase and cancel each other; likewise, ray 1 cancels ray 501, ray 2 cancels ray 502, and so on. Every ray from a slit in the bottom half of the grating cancels one in the top half. The resultant intensity is zero at the new angle, θ', which is very close to the former angle θ for which the intensity is a maximum. We can calculate the angles from the diagrams:

$$\sin\theta = \frac{1000\lambda}{1000a} = \frac{\lambda}{a}$$

$$\sin\theta' = \frac{1001\lambda}{1000a} = 1.001\frac{\lambda}{a}$$

The sines of these angles differ by only 0.1%. This means that the maximum is a sharp one. If we had only two slits, a similar small change of angle would cause the top ray to be only 0.001 cycle out of phase (Fig. 26-15), and it would still cooperate with the bottom ray to give a resultant intensity almost as large as at the peak. The qualitative difference between the interference pattern of a grating and that of a pair of very narrow slits is shown in Fig. 26-16.

We have so far tacitly assumed that the light striking a grating is *monochromatic*—that is, that it consists of a single color and has a single wavelength. If a mixture of colors falls on a grating, the equation $n\lambda = a\sin\theta$ tells us that to each different λ there corresponds a different θ. Since the colors are separated, a grating can serve the

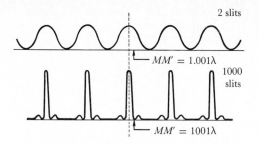

Figure 26-16 Comparison of interference pattern of two slits and that of a grating. Between the sharp maxima formed by the grating are many subsidiary maxima, too weak to be drawn to scale.

same function as a prism, though in an entirely different way. The angular separation between two colors can be quite large if the grating space a is small enough. We shall consider the grating spectroscope in the next chapter, as an example of an optical instrument.

26-5 The Single-Slit Diffraction Pattern

In our discussions of Young's experiment and the grating, we have always assumed *narrow* slits. In this way we could consider all the rays passing through a given slit to be in phase with each other; this is true enough if the slit width is small compared with the wavelength. In other words, rays 1, 2, and 3 in Fig. 26-17 are sufficiently in

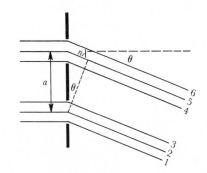

Figure 26-17

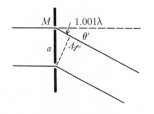

Figure 26-15

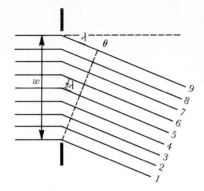

Figure 26-18 Single slit, 1st-order minimum.

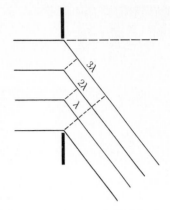

Figure 26-19 Single slit, 3rd-order minimum.

phase with each other to act as a single ray; and this "super ray" is in phase with the "super ray" composed of rays 4, 5, and 6 if $n\lambda = a\sin\theta$. Of course, rays 1 and 3 do differ by a slight (but negligible) amount for any real (but narrow) slit.

Turning now to a wider slit, we must take account of phase differences between rays passing through various parts of the slit. We have to deal with infinitely many rays, of which only nine are shown in Fig. 26-18. The term *diffraction* refers to the interference between the component rays of a single, broad wave front, as in this figure. If we consider the diffraction pattern formed by rays leaving the slit at an angle θ, and choose θ so that $\lambda = w\sin\theta$, where w is the width of the slit, then we can show that the resultant amplitude of all the rays is zero. Ray 1 is $180°$ out of phase with ray 5, ray 2 is $180°$ out of phase with ray 6, and so on. Thus, each ray in the bottom half of the slit cancels out a corresponding ray in the upper half. An exactly similar cancellation takes place for a grating at an angle slightly different from the angle giving an interference maximum (Fig. 26-14b). The only difference is that now we have a continuous group of infinitely many rays, instead of just 1000 or so.

The resultant amplitude from a single slit falls to zero also at larger angles. For instance, if the extreme path difference is 3λ, as in Fig. 26-19, we can divide the slit into three equal segments, each giving zero intensity as before. Thus, the general condition for a minimum of a single-slit diffraction pattern is

$$\left(\begin{array}{l}\text{single-slit}\\ \text{minimum, } n \neq 0\end{array}\right) \quad n\lambda = w\sin\theta \qquad (26\text{-}2)$$

Between these minima, the pattern has bright maxima (Figs. 26-20 and 26-21). A special case occurs if $n = 0$ and $\theta = 0$. No path differences exist, all of the many rays are in phase, and in this case they add up to give an intense central maximum. The photograph shows clearly that the central maximum of the diffraction pattern is twice as wide as the others.

An interesting paradox is illustrated in Fig. 26-22. By blocking out most of the single slit to leave only two narrow slits, the intensity in the direction shown (for which $n = 1$) is *increased*! Do not imagine that we get something for nothing. The total intensity from the single slit (area under the curve) is far greater than the total for two slits, as seen from Fig. 26-23, in which the two patterns are superposed.

Let us restate Eq. 26-2 in another way. Most of the energy of the wave lies in the central bright band, for which the 1st-order minimum is given

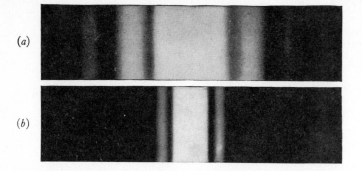

(a)

(b)

Figure 26-20 Diffraction pattern of a single slit. (a) Pattern of a narrow slit. (b) Pattern of a wide slit. (By permission from *Fundamentals of Optics*, 4th ed., by F. A. Jenkins and H. E. White. Copyright 1976. McGraw-Hill Book Co.)

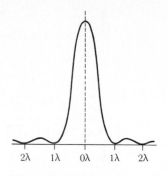

Figure 26-21 Intensity distribution for a single-slit diffraction pattern.

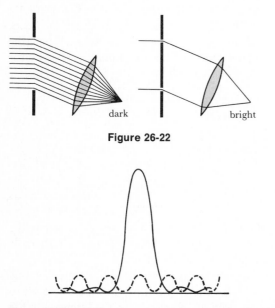

dark   bright

Figure 26-22

Figure 26-23 Diffraction pattern for a single slit of width w (solid curve) and pattern from a pair of narrow slits of separation w (dashed line).

by $\sin \theta = \lambda/w$. This leads to an important generalization:

> **The ratio of wavelength to size of opening determines to what extent light or any other wave fails to travel in a straight line.**

Similarly, the ratio λ/w determines the amount of diffraction when a wave front is limited by an opaque obstacle.

The diffraction patterns of some other apertures and obstacles are reproduced in Fig. 26-24. In each case, the pattern can be calculated by the methods of wave optics that we have illustrated for the single slit. The shadow of a straight edge is surrounded by a series of bands of uneven spacing; this is the "bending around corners" that is expected for wave motion. Such effects had actually been observed in the 17th century by Francesco Grimaldi, who failed to make the wave interpretation of his results. One difficulty in all such experiments is a geometrical one. Shadows are usually far from sharp because the source is broad. For instance, the sun subtends an angle of about $\frac{1}{2}°$ in the sky; different parts of the sun cast different shadows, which overlap to give a "fuzzy" edge. The shadow of a straight edge may well be diffuse enough to mask the diffraction effects unless special precautions are taken to use a small ("point") source. Figure 26-24b is especially interesting. The French physicist Siméon Poisson is said to have objected to the wave theory, arguing that the theory would predict a bright spot at the center of the shadow of a perfectly circular obstacle—a

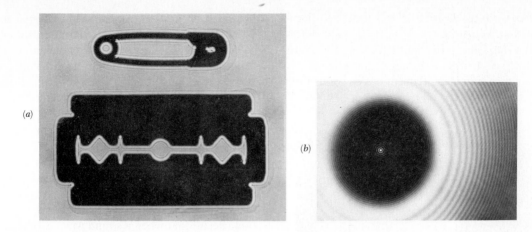

(b)

Figure 26-24 Diffraction of light. (*a*) Opaque objects were mounted on a glass plate 4 m from a point source. A photographic film was exposed at a position 2 m beyond the obstacles. No lenses were used. Note diffraction bands at all edges; no shadow is sharp. (*b*) Diffraction pattern due to a penny, courtesy of P. M. Rinard, *Am. J. Phys.* **44**, 70 (1976). The film was 20 m beyond the penny. The source of light, a He-Ne laser, was equivalent to a point source 20 m behind the penny. Note the bright spot at the center, the circular fringes near the edge of the shadow, and the complex diffraction pattern outside the geometric shadow.

seemingly absurd result.* When the experiment was performed by Dominique Arago in 1818, the bright spot was promptly found. No more dramatic proof of the wave nature of light can be found. A hypothesis (the wave theory) led to a new and hitherto meaningless experiment, with results contrary to "intuition" and yet in perfect agreement with prediction of the theory.

26-6 Applications of Interference

Some of the evidence for interference of light waves is familiar in everyday life and indeed was known to Newton and his contemporaries but inadequately interpreted by them. Interference fringes can be observed if light strikes an air wedge, as shown in Fig. 26-25. A fine wire or a sheet of tissue paper at *X* forms a thin wedge of

air between two flat pieces of glass. At any point such as *A* or *B*, rays are reflected both at the upper surface of the air wedge (ray 1) and at the lower surface of the wedge (ray 2). When these rays are brought together on the retina of the eye, the resultant amplitude is large or small depending on the path difference *PBR*. Since the path difference gradually increases as *B* moves out along the wedge, the observer sees a series of alternately bright and dark bands. For any two

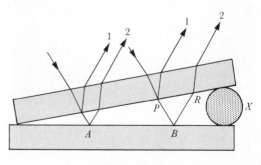

Figure 26-25

---

* The proof of this prediction is too advanced to be included in this book.

adjacent bands, the total (round-trip) path difference at B is just 1λ longer than the path difference at A. That is, the air wedge is $\frac{1}{2}\lambda$ thicker at B than at A. The wavelength of light can be found by a simple experiment of this sort. In the optical shop, a glass surface can be tested for flatness by laying it on top of a standard surface of known flatness known as an optical flat. If the fringes are wavy (Fig. 26-26a), the air film is of uneven thickness and the high spots of the surface under test can be polished until a parallel set of fringes is obtained (Fig. 26-26b).

A seeming paradox arises in air-wedge experiments. The resultant is dark where the pieces of glass are in contact, and yet according to the wave theory we would expect a bright band here since the path difference is zero. This illustrates a change of phase that takes place during reflection. Recall that when a sound wave is reflected at the end of a pipe, the result differs according to whether the reflection takes place at a closed end or an open end. At a closed end, a compression is reflected as a compression, which requires an abrupt reversal of the direction of motion of the air molecules. That is, there is a change of phase of half a cycle (180°) when reflection takes place at the closed end of a pipe. On the other hand, at an open end a compression is reflected as a rarefraction, and the air molecules move right on out into the open, with no change of phase. A similar situation exists when light waves are reflected at an optically "more dense" medium (Fig. 26-27a). Ray B is 180° out of phase with ray A, simply because it is reflected in air against a glass surface. However, in Fig. 26-27b, ray D is in phase with ray C, since this reflection

(a)

(b)

Figure 26-26 Testing an optical surface by interference. (a) Initial condition of surfaces, showing three high spots. (b) Surfaces polished flat to within a fraction of a wavelength.

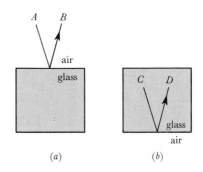

(a)   (b)

Figure 26-27 Light waves undergo a 180° phase change when reflected as in (a), but not when reflected as in (b).

is in glass against air—similar to the reflection of a sound wave at the open end of a pipe. We can now understand why the interference pattern is dark at the vertex of the air wedge. The path difference is zero, but one ray suffers the 180° phase change while the other does not. This is equivalent to an extra $\frac{1}{2}\lambda$ in the path of one ray, and destructive interference results.

Example 26-3

A plane glass plate 10 cm long is separated from another plane glass plate at one end by a piece of aluminum foil 0.004 cm thick. How far apart are the successive interference bands, viewed in light of wavelength 600 nm?

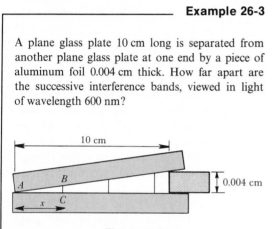

Figure 26-28

The foil creates an air wedge, and there is a dark band at the vertex A (Fig. 26-28), where the path difference is 0λ. The next dark band is at B, where the total path difference is 1λ. Hence,

$$BC = \tfrac{1}{2}\lambda = \tfrac{1}{2}(6 \times 10^{-7} \text{ m}) = 3 \times 10^{-7} \text{ m}$$

We find the distance x by a proportion:

$$\frac{x}{10 \text{ cm}} = \frac{3 \times 10^{-7} \text{ m}}{4 \times 10^{-5} \text{ m}}$$

$$x = 0.075 \text{ cm} = \boxed{0.75 \text{ mm}}$$

The bands are spaced 0.75 mm apart and can easily be seen with a magnifying glass. The total number of dark bands in the 10-cm length of glass is

$$\frac{10 \text{ cm}}{0.075 \text{ cm/band}} = 133 \text{ bands}$$

One way to obtain a thin air film of variable thickness is to place a long-focus lens on a plane glass surface. Figure 26-29 shows the arrangement, with the curvature much exaggerated for clarity. If the curved surface of the lens is almost flat, the air film varies from zero thickness (at A) to a dozen or so wavelengths (at the edge). When viewed by reflected light, a pattern of dark rings is observed; for any given circular ring, such as at B, the air film is of constant thickness. Ironically, Newton himself discovered these interference rings—they are called Newton's rings—but entirely missed their significance. He postulated that corpuscles of light had "fits of easy reflection and easy transmission" to explain the fact that light is partially reflected and partially transmitted at a glass-air surface. He had no consistent theory for the rings. After two and a half centuries it is easy for us to see that Newton's model was inadequate because it led nowhere and predicted no new results (compare the fruitfulness of the wave theory as exemplified by the Arago spot). The intellectual climate of Newton's day was not yet ready for a model in which invisibly small waves, 2 million of them per meter, behave like the water waves seen by the naked eye.

Other reflections from thin films are often observed. A soap bubble consists of a film of soap solution. Viewed in white light, a very thin bubble appears brilliantly colored. Suppose that for some particular color the thickness is exactly $\frac{1}{2}\lambda$. An extra 180° phase shift is introduced at the front surface but not at the back surface. The *effective* (round-trip) path difference is $1\lambda + \frac{1}{2}\lambda = 1\frac{1}{2}\lambda$, and light of this color is not reflected, since the two rays interfere destructively. (The light goes on through the film.) However, neighboring colors have longer or shorter wavelengths, and the film cannot be $\frac{1}{2}\lambda$ thick for all colors at once.

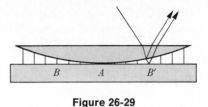

Figure 26-29

Hence the film reflects some colors more than others. As a result, a bubble is colored, and the colors change as the thickness of the film changes during evaporation. Eventually, the bubble's thickness is much smaller than any visible wavelength, the path difference is 0λ for all colors, and the bubble appears black (why not white?) just before it breaks.

Example 26-4

A soap bubble 500 nm thick is illuminated with white light. The index of refraction of the film is 1.35. What colors are *not* reflected?

The first wavelength that is not reflected is one for which 500 nm $= \frac{1}{2}\lambda$. Then the round trip of 1000 nm would be 1λ, and the phase change at the front surface would account for another $\frac{1}{2}\lambda$, making a total of $1\frac{1}{2}\lambda$. If $\lambda = 1000$ nm in film, the wavelength in air would be

$$(1000 \text{ nm})(1.35) = \boxed{1350 \text{ nm}}$$

This is in the infrared region.

Next, we consider a wavelength for which the round trip is 2λ. Then 1000 nm $= 2\lambda$, and $\lambda = 500$ nm in film, or, in air,

$$(500 \text{ nm})(1.35) = \boxed{675 \text{ nm}}$$

This is red light.

The next color that is completely removed by interference is one for which 1000 nm $= 3\lambda$; $\lambda = 333.3$ nm in film, or

$$\boxed{450 \text{ nm}}$$

in air. This is blue light.

Going to still shorter wavelengths, we find that the next destructive interference is for $\lambda = 250$ nm in film, or

$$\boxed{337.5 \text{ nm}}$$

in air. This is in the ultraviolet region.

We conclude that only two visible colors are totally canceled: red light of wavelength 675 nm and blue light of wavelength 450 nm. Other colors are reflected to a greater or lesser degree. Since red and blue are missing, the middle of the spectrum predominates, and the bubble appears greenish by reflected light.

The colors of oil films on pavements are caused in similar fashion by interference.

26-7   The Michelson Interferometer

We conclude our study of interference by describing the interferometer invented by a brilliant American physicist, A. A. Michelson (1852–1931). The instrument uses the air-wedge principle, but still allows the working area to be many centimeters in length instead of a fraction of a millimeter. In Figs. 26-30 and 26-31, light from a broad source S strikes a glass plate M, which is "half-silvered" to be partially reflecting, so that half the light is reflected to mirror X and half goes on through to the mirror Y. The beams reflected from X and Y emerge together and

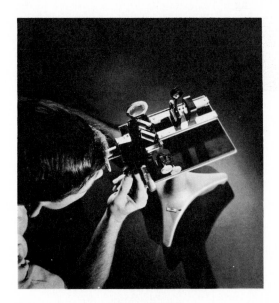

Figure 26-30

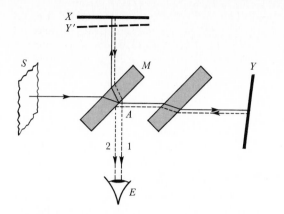

Figure 26-31 Michelson interferometer.

enter the eye E. Interference takes place between the beams, depending on the path difference.* Since M is a plane mirror, it forms a virtual image of Y at Y'. In effect, the two beams 1 and 2 are coming from two mirrors at X and Y'. The great advantage of the method is that the thickness of the air wedge between X and Y' can be adjusted, even to zero if desired, by moving X back and forth by means of a screw. The angle of the wedge can be adjusted by the screws that tilt mirror Y; thus a set of fringes of any desired spacing can be obtained. The fringes move across the field of view as X moves toward coincidence with the image Y'.

A typical use of the Michelson interferometer is to measure the wavelength of light, as described in the following example.

Example 26-5

Monochromatic green light is obtained by passing light from a mercury arc through a suitable filter. It is found that moving the mirror X through a dis-

---

* An extra glass plate is inserted in the path of beam 2 so that each beam traverses the same thickness of glass. This plate is not silvered.

tance of 0.273 mm causes a shift of 1000 fringes. Calculate the wavelength of the light.

The passage of 1000 fringes means that the total (round-trip) path has been changed by 1000 wavelengths. Thus,

$$1000\lambda = 2(0.273 \text{ mm})$$

$$\lambda = \frac{2(0.273 \text{ mm})}{1000} = 5.46 \times 10^{-4} \text{ mm}$$

$$= 5.46 \times 10^{-7} \text{ m}$$

$$= \boxed{546 \text{ nm}}$$

Michelson used his interferometer to determine the length of the standard meter in terms of the wavelength of a certain red line in the cadmium spectrum. He chose this line because it was the "purest" known at that time. By way of contrast, the sodium yellow "line" is really a close pair of lines, of wavelengths $\lambda_1 = 589.0$ nm and $\lambda_2 = 589.6$ nm. This is only a 0.1% difference in wavelength; but as a result, two separate interference patterns are obtained simultaneously. Simple arithmetic shows that 500 wavelengths of λ_1 equal $499\frac{1}{2}$ wavelengths of λ_2. Thus an air film of thickness $250\lambda_1$ is of proper thickness to cancel λ_1 [the round trip is $500\lambda_1$; an extra $\frac{1}{2}\lambda_1$ arises from the phase shift (Fig. 26-27b)]. This same air film is just right to reinforce λ_2, and due to this overlapping, no fringes would be seen for a film of this thickness. By pursuing this idea of visibility of fringes, Michelson was able to show the extreme purity of the cadmium red line he used. The mercury green line (546 nm) is a complex one, as shown by the interference pattern on page 602, which was obtained using an interferometer that makes sharp circular fringes. Each isotope of mercury produces a line of slightly different wavelength. In recent years it has proved possible to separate the isotopes of mercury, and it has been found that the interference fringes from the ^{198}Hg isotope are even "cleaner" than the cadmium fringes used by Michelson. Precise counting of fringes in an interferometer makes possible the definition of the meter in terms of the wavelength of a certain orange line of one of the krypton isotopes (see Sec. 1-4).

The Michelson interferometer was used in a famous experiment to study the effect of a mechanical "ether" in which light waves were supposed to vibrate. See Sec. 28-2 for a description of the Michelson-Morley experiment of 1887.

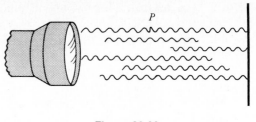

Figure 26-32

26-8   *Polarization of Light*

The electromagnetic waves described in Sec. 24-5 (page 558) are *plane-polarized waves*, since the variable quantity is an electric field, or a magnetic field, represented by vectors that lie in a plane. For the electric vector in Fig. 24-4, this is the vertical plane. We have also seen that the Maxwell theory predicts *transverse waves*, since the vectors are perpendicular to the direction of propagation. The fact that the portion of the electromagnetic spectrum that we call light can be polarized was known as long ago as 1669, and the full interpretation came two centuries later, based on the Maxwell theory of electromagnetic waves. In fact, one of the more subtle problems we have to discuss is the fact that most light in everyday life is *not* polarized. The question is not, how can light be polarized, but rather, why shouldn't it always be polarized? How can *un*-polarized light exist, when electromagnetic waves by their very nature are plane polarized?

Light waves are emitted by molecules or atoms that are excited by thermal or electrical means. In the classical model we are using, any given atom vibrates a few thousand or a few million times, sending out a relatively short burst of plane-polarized light called a *wave train*. In the usual source, such as the filament of a lamp bulb or the gas in a discharge tube, the vibrating atom makes many collisions per second with other atoms. If the atom vibrates freely for 10^6 cycles, it emits a wave train containing 10^6 wavelengths. For visible light of wavelength 5×10^{-7} m, this wave train would be emitted during less than 10^{-8} s, and it would be about half a meter long.

The wave train is plane polarized, with the electric vector in a plane determined by the chance orientation of the vibrating atom. If a beam of light from a flashlight bulb shines on a wall 10 m away, our model leads us to a picture somewhat like Fig. 26-32. Many wave trains fill the space between the flashlight and the wall; they start and stop at random, and the intensity at the wall averages out to a steady illumination. During an atomic collision, the radiation from a single atom might suffer an abrupt change in phase; this would cause a kink in the wave train, as at P. Looking head-on at a "point" source (actually composed of very many vibrating atoms), we see a superposition of many plane waves with all possible orientations of the electric vector. Each single wave train (such as XX in Fig. 26-33a) originates from one atom and is plane polarized with its electric vector perpendicular to the direction of propagation. The net effect averages out, and there is no preferred orientation of the resultant of all these randomly oriented vibra-

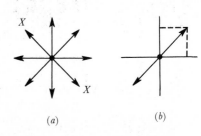

(a) (b)

Figure 26-33 (a) Unpolarized beam of light containing many superposed wave trains (head-on view). (b) A single plane-polarized wave train resolved into perpendicular components (head-on view).

tions. Thus we say that ordinary light is unpolarized but consists of components polarized in all possible transverse directions.

For purposes of calculation, any plane-polarized wave can be replaced by components vibrating along two arbitrary axes, as in Fig. 26-33b. Thus an unpolarized beam is equivalent to two equally intense beams polarized along any two arbitrary, mutually perpendicular axes. We represent polarized beams as in Fig. 26-34, the bars or dots indicating the direction of the electric vector. Unpolarized light is represented as containing equal proportions of two beams polarized at right angles to each other. It might seem that we could add up all the vectors of Fig. 26-33a to zero; however, remember that each arrow represents light from a different atom, and there is no *coherence* between the phases of their vibrations. The atoms in an ordinary light source vibrate independently, start and stop at random, and do not pass through zero at the same time. This difficulty has been overcome in the laser (Sec. 30-5), which gives a coherent beam from many cooperating atoms.

One way to determine experimentally the plane of vibration of a beam of light is to look at the source through a sheet of Polaroid film. This material consist of many tiny crystals of a quinine compound, oriented with their axes parallel to each other and spread out on a transparent base. The crystals act like one big crystal and have the unusual property of transmitting only light whose plane of vibration is properly oriented relative to the crystals. Light vibrating in the "wrong" direction is absorbed by the crystals.

Thus a sheet of Polaroid can be used as a *polarizer*, to make plane-polarized light out of ordinary light, and also as an *analyzer*, to determine whether and in what plane a beam of light is polarized.

The polarization of skylight offers a good example of the principles involved in the scattering of light. Unpolarized light from the sun is scattered by air molecules and accounts for the blue sky; simple tests with Polaroid show that skylight is often partially polarized. The effect is most pronounced if the observer looks through Polaroid at right angles to the sun's rays—for instance, if he looks straight up at sunset or sunrise (Fig. 26-35). We consider the two components of polarization separately. Half the light (A) can be considered to be polarized with the electric vector horizontal. The electric field exerts a horizontal driving force on electrons in the oxygen or nitrogen molecules, and, as a result, e-m waves are radiated by these accelerated electrons. The observer is broadside to the miniature antennas represented by the motion of the electrons, and hence he receives scattered light, polarized horizontally, of course, since the vibrating charges are moving horizontally. The other half of the original beam (B), polarized vertically, also sets electrons into motion, but the observer is looking at these motions head-on and therefore receives no scattered ray (Sec. 24-6).

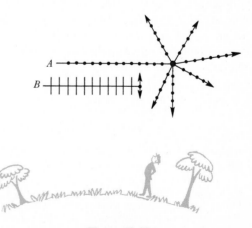

Figure 26-35

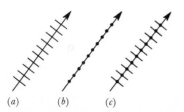

Figure 26-34 (a) and (b) Rays of polarized light. (c) Ray of unpolarized light.

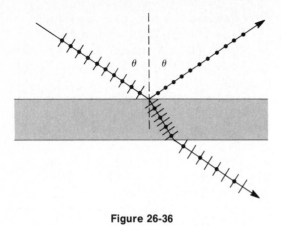

Figure 26-36

The overall result is that the scattered light reaching the observer is polarized.*

Reflection by an electrical insulator such as glass or water also is stronger if the electric vector of the rays is parallel to the surface. Detailed study by means of Maxwell's equations for e-m waves shows that an initially unpolarized beam becomes partially polarized by reflection, with the electric vector parallel to the surface (Fig. 26-36). The analysis is somewhat similar to the discussion of the polarization of scattered light in the previous paragraph. At *Brewster's angle*, given by $\tan \theta = n$, the polarization is strongest. We see here an indication that light is electrical in nature; light cannot be made plane polarized by reflection from a metal surface. The presence of free electrons in a conductor makes the process of reflection more complicated than for an insulator. Only with the advent of the e-m

---

* Skylight is blue because the efficiency of scattering is proportional to the 4th power of the frequency. Violet light has twice the frequency of red light and is scattered 2^4, or 16, times as readily. This dependence on frequency is true only if the scattering particle is small compared with λ. The condition is fulfilled by molecules of O_2 or N_2, which are about 2 Å in diameter. Even local condensations of a dozen or so air molecules, which occur at random in the air, are small compared with the wavelength and can act as scattering centers. Water drops, which are much larger than λ, scatter all wavelengths of light equally well; hence a cumulus cloud of water drops is white.

wave theory of light could a satisfactory theory of metallic reflection be worked out.

Much glare experienced during ordinary activities can be removed by wearing Polaroid glasses properly oriented. Light reflected from a lake or the ocean is partially or completely polarized, with the electric vector horizontal. The Polaroid glasses must be oriented so as to absorb light polarized in this way and to transmit the vertically polarized half of the diffusely scattered (unpolarized) light coming from solid bodies such as boats or fish. The discovery of polarization by reflection was made in 1808 by the French scientist Étienne Louis Malus (1775–1812).

The speed of light in certain transparent crystals—the anisotropic ones—depends on the plane of the electric vector. This is a consequence of the electrical nature of matter. For the crystal shown schematically in Fig. 26-37, it is natural to expect the electric forces along the axis AA' to be different from the forces along the axis BB'. Therefore, as an e-m wave passes through the crystal and secondary wavelets are set up at each atom, we would expect that the direction of the electric vector would be of importance. The wave velocity (and hence the index of refraction) depends on the direction of polarization of the electric vector. This gives rise to the phenomenon of double refraction. See Sec. 26-11 for further material on crystal optics.

It is worth noting that the polarization of light waves proves them to be *transverse*. Longitudinal disturbances such as sound waves show all the wave phenomena of interference and diffraction, but they cannot be polarized. It is not difficult to

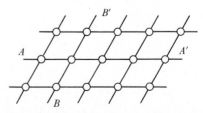

Figure 26-37 Anisotropic crystal (schematic).

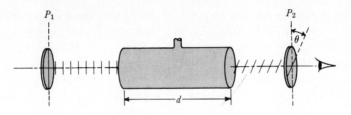

Figure 26-38 A polarimeter.

imagine a polarized beam of corpuscles—for instance, a stream of flat stones being "skipped" on the surface of a lake. The orientation of the stone is all-important for the success of the operation. We could in this way separate a beam of polarized stones from among the unpolarized (randomly oriented) stones striking the water. Some nuclear particles can be polarized by reflection in this way. We should not make the mistake of thinking that polarization, as such, proves the wave nature of light. Light is shown to be (1) transverse and (2) a wave motion by the *combined* evidence of polarization and interference.

26-9 Optical Activity

Certain substances have the property of rotating the plane of polarization of a beam of plane-polarized light. This property is called *optical activity*. Quartz, some sugars, and many other organic and inorganic solids and solutions are optically active, but, on the other hand, calcite is not active, in spite of its strong double refraction. In a typical experiment (Fig. 26-38), a beam is rendered plane polarized by a piece of Polaroid or a Nicol prism P_1, and a second Polaroid or Nicol prism P_2 is adjusted so that no light is transmitted. When the sample is inserted, light reappears, and the analyzer P_2 must be rotated through an angle θ in its own plane to make the intensity zero again. The angle of rotation depends on the material and the path length and is also strongly dependent on wavelength. If no wavelength is specified, the yellow sodium D-lines of average wavelength 589 nm are under-

stood. The specific rotation is measured in deg/mm for a solid, or deg/decimeter for a solution of concentration 1 g/cm^3. Typical values are 22 deg/mm for quartz and 77 deg/dm for a solution of nicotine in water.

Optical activity is related to a spiral structure of the individual molecules or of the crystal. Many substances, including quartz, can exist in either a right-handed or a left-handed form. The two sugars *d*-glucose (dextrose) and *l*-glucose are identical except for the direction of rotation. Since the rotation is proportional to the total number of molecules per unit length of path, the concentration of sugar in a commercial syrup can be tested by this means.

Glass and some transparent plastics become optically active when subjected to mechanical

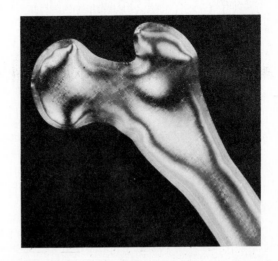

Figure 26-39 Photoelastic stress analysis of a model of human bone, using polarized light.

stress. Figure 26-39 is a photograph of a plastic model of a bone structure, loaded at one point. The model is placed between Polaroids, and the stress pattern is made visible by the corresponding rotation of the planes of polarization in the stressed regions of plastic. Some substances also become doubly refracting when stressed.

In this chapter we have given evidence supporting the wave model for light. Much additional evidence supports a corpuscular model, chiefly in connection with the emission and absorption of light (for instance, spectrum analysis and the photoelectric effect, to be discussed in Chaps. 29–31). Do not imagine, however, that Newton's original corpuscular model has thus been vindicated or reinstated. The simple mechanical "bullet" model of Newton is certainly false, and the newer corpuscular part of the dual model is quite different and more subtle, based on experimental facts unknown in Newton's day.

Summary   The early corpuscular theory of light, adopted by Newton, could explain reflection and refraction, on the assumption that light travels faster in a medium than in vacuum. This assumption later turned out to be wrong, but in the meanwhile the corpuscular theory was discredited by the discovery of interference. To obtain interference it is necessary to divide a wave front into two or more parts that come together again after traveling along paths of different lengths. Both constructive and destructive interference are possible, depending on the relative phases of the rays when they recombine. If the effective path difference between two rays is 0λ, 1λ, 2λ, ..., the resultant amplitude is a maximum, and the rays interfere constructively. If the path difference is $\frac{1}{2}\lambda$, $\frac{3}{2}\lambda$, $\frac{5}{2}\lambda$, ..., the resultant amplitude is zero and the rays interfere destructively.

If two rays strike perpendicularly a pair of narrow slits or a grating, the equation $n\lambda = a\sin\theta$ can be used to calculate the angles θ at which the maxima of intensity are found. For a single slit of width w, the equation $n\lambda = w\sin\theta$ gives the angles for zero intensity in the diffraction pattern. The basic concepts of interference and diffraction apply to all wave motions, including e-m waves, water waves, and sound waves.

Shadows are usually not perfectly sharp because the sources are usually of finite size. Even with a point source, the edge of a shadow shows diffraction bands caused by the wave nature of light. Wave effects are in general more pronounced for larger wavelengths, or for smaller obstacles, since then $\sin\theta\,(= n\lambda/w)$ is larger.

Interference from thin films of air, oil, water, and other substances can be applied to many laboratory measurements. Change of phase occurs when a ray is reflected by an optically more dense medium, but no change of phase takes place if a ray is reflected by an optically less dense medium. Certain reflections, therefore, in effect introduce an extra $\frac{1}{2}\lambda$ of path difference.

Interference indicates the wave nature of light, and polarization indicates the transverse nature of light. The electromagnetic wave theory of light predicts light to be transverse waves, and many phenomena, such as scattering and reflection, show the expected dependence on the direction of the electric vector of the e-m wave. Unpolarized light consists of many short wave trains vibrating in random orientation and phase. Such light is equivalent to equal proportions of two plane-polarized beams whose vibrations are parallel to any two arbitrary, mutually perpendicular directions. The passage of light through anisotropic crystals shows a marked dependence on polarization. Optical activity refers to the rotation of the plane of polarization as light passes through a medium.

Check List

interference	monochromatic light	colors of thin films
phase difference	diffraction	Michelson interferometer
nth order	single-slit pattern	coherence
double-slit pattern	$n\lambda = w\sin\theta$	polarized light
grating	phase change upon	Brewster's angle
$n\lambda = a\sin\theta$	reflection	optical activity

Questions

26-1 What additional experimental fact must be known before the refraction of light can be considered an adequate proof of the wave nature of light?

26-2 What is the path-length condition for the constructive interference of two rays of light?

26-3 Two radio antennas are emitting waves that are in phase at the towers. An observer is 20,000 m from one tower and 22,000 m from the other tower. The wavelength is 500 m. Is reception good at the observer's location?

26-4 What change, if any, takes place in the pattern on the screen of Fig. 26-3 if the entire apparatus is immersed in a transparent liquid of index of refraction n?

26-5 If white light shines on two slits as in the upper part of Fig. 26-3, what color (red or violet) appears closest to O in the 1st-order band of color?

26-6 Locate a point in Fig. 26-11 that is 5λ from the left-hand source and 6λ from the right-hand source. Locate a point that is $5\frac{1}{2}\lambda$ from the left-hand source and 6λ from the right-hand source.

26-7 What is the difference, if any, between the interference pattern of a pair of slits separated by 0.001 cm and the pattern of a grating having 1000 slits per cm?

26-8 If you look at a distant line source through a narrow slit, the source appears to become broader as the slit is narrowed. Explain.

26-9 Hold a straight-edge, such as a ruler or a sheet of paper, about 50 cm from the lamp bulb of a desk lamp, and look closely at the edge of the shadow cast on a surface held a few cm from the obstacle. Can you explain why the shadow is not perfectly sharp?

26-10 Instead of using the arrangement of Fig. 26-3, in which a pair of slits are illuminated by light from a single slit, an experimenter attempts to get an interference pattern by looking at a pair of sources, such as the headlights of a distant automobile. The path length for light from one headlight bulb is exactly 3 wavelengths longer than the path length from the other bulb. The experimenter expects to get an interference maximum but is disappointed. Explain why no interference takes place under these circumstances.

26-11 In Fig. 26-31, which surface of the diagonal mirror M is the half-silvered one?

26-12 As a soap bubble evaporates, just before it breaks its surface appears black when viewed by reflected light. Explain in detail why it is dark and not bright.

26-13 A few drops of crankcase oil leak out of a car onto a wet pavement (Fig. 26-40). The oil slick appears bright at its thinnest portion (the outer edge of the spot), where the

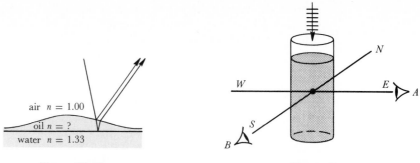

air $n = 1.00$

oil $n = ?$

water $n = 1.33$

Figure 26-40

Figure 26-41

thickness is much less than the wavelength of light. Are the two reflected rays in phase or out of phase? Is the index of refraction of used crankcase oil greater or less than 1.33?

26-14 What is the difference between interference and diffraction?

26-15 What fact about the nature of light is proved by the phenomenon of polarization?

26-16 A vertical beam of light polarized with the electric vector in an east–west plane travels through a dilute milky liquid and is scattered by small globules of fat (Fig. 26-41). Which observer, A or B, sees a more intense beam of scattered light? Is the observed beam polarized, and, if so, what is the direction of its electric vector?

26-17 Describe the use of Polaroid (*a*) as an analyzer, and (*b*) as a polarizer.

26-18 Refer to Fig. 10-3 (page 224), in which polarization of waves on a string is illustrated. Which slit is the analyzer and which is the polarizer?

26-19 Describe the use of a polarimeter to measure the concentration of a sugar solution.

MULTIPLE CHOICE

26-20 A path difference of $\frac{1}{4}\lambda$ is associated with (*a*) constructive interference; (*b*) destructive interference; (*c*) neither of these.

26-21 A single-slit diffraction pattern is produced with monochromatic light. The distance between adjacent minima can be increased by (*a*) changing the light to a lower frequency; (*b*) decreasing the width of the slit; (*c*) both of these.

26-22 When a light wave in air is reflected by a glass surface that is perpendicular to the direction of propagation of the wave, the wave undergoes a phase change of (*a*) $0°$; (*b*) $90°$; (*c*) $180°$.

26-23 The wave nature of light is best proved by the phenomenon of (*a*) polarization; (*b*) interference; (*c*) reflection.

26-24 When monochromatic light from a slit passes through a grating, the 3rd-order maximum is observed to be very sharp because (*a*) the grating has a large total number of slits; (*b*) the slits of the grating are very close together; (*c*) the wavelength is very short.

26-25 The coherence length of the wave train emitted as visible light due to the excitation of a single atom is in the range (*a*) 380 to 760 nm; (*b*) 1 to 100 cm; (*c*) 10 to 100 m.

Problems **26-A1** Two slits 1.8×10^{-6} m apart, illuminated perpendicularly by parallel rays, form the 2nd-order bright fringe at an angle of $30°$ from the direct beam. What is the wavelength of the light used?

26-A2 A coarse grating designed for use in the infrared region of the e-m spectrum has slits 0.5 mm apart. What is the angle of deviation of the 3rd-order image for radiation whose wavelength is 1050 nm?

26-A3 Light of wavelength 5×10^{-7} m falls on a pair of slits, and the 3rd-order bright fringe occurs at an angle of 30° from the direct beam. How far apart are the slits?

26-A4 (a) A slit 1.2×10^{-4} cm wide is illuminated perpendicularly with light of wavelength 480 nm. What is the angular deviation of the first *dark* band on either side of the central maximum? (b) What is the angular deviation of the first *bright* band on either side of the central maximum? (*Hint:* The bright band is approximately halfway between the 1st and 2nd dark bands.)

26-A5 When light of wavelength 500 nm falls perpendicularly on a single slit, the first two dark fringes (one on each side of the central bright fringe) are separated by 50°. How wide is the slit?

26-B1 A radio transmitter A operating at 60 MHz is 10 m from another similar transmitter B that is 180° out of phase with A. How far must an observer move from A toward B to reach the nearest point where the two beams are in phase?

26-B2 Radio waves of frequency 600 kHz are received at a location 10.0 km from the transmitter. The radio reception temporarily "fades" due to destructive interference between the direct beam and the beam that is reflected without phase change from a horizontal layer of charged particles (ions) formed in the atmosphere by a passing aircraft. Calculate the minimum height of the aircraft. (*Hint:* Reflection takes place from a point above the midpoint between the transmitter and the receiver.)

26-B3 A small ultrasonic transducer (loudspeaker) and a receiving transducer are 20 cm apart on a table top. A horizontal sheet of plywood 3 cm above the table serves as a reflector; reflection takes place at a point on the plywood directly above the midpoint between the two transducers. The speed of sound is 340 m/s. What are the three lowest ultrasonic frequencies, in kHz, for which destructive interference takes place? (*Note:* The receiving transducer responds to variations in pressure, for which there is no phase change upon reflection by the plywood.)

26-B4 Light from a flame in which sodium chloride has been placed has an average wavelength 589 nm. What is the 2nd-order angular deviation of a ray of this light that strikes perpendicularly a grating having 2000 slits per cm?

26-B5 An observer looks at a source through a grating having 1200 slits per cm. If λ is 600 nm, how many images of the source can be seen? (*Hint:* For what order would θ be $\pm 90°$?)

26-B6 A grating has 14,000 slits spread out over 4.0 cm. (a) For what wavelength of light, incident normally as in Fig. 26-14, will the angle between the two 3rd-order maxima be 90°? What color is this light? (b) What is the maximum wavelength observable in the 3rd order using this grating? Is this radiation visible?

26-B7 (a) Sunlight falls perpendicularly on a grating having 1000 slits per cm. The rays leaving the grating are focused on a screen 2.00 m from the grating. How far from the central image on the screen is the green region of the 3rd-order spectrum? (Use an average green wavelength from Table 25-1, page 570.) (b) Calculate the width of this green region on the screen. (Find the deviation for each of the two extreme green wavelengths in Table 25-1.)

26-B8 Parallel rays of red light of wavelength 633 nm strike perpendicularly a grating having 1500 slits per cm. The rays leaving the grating are focused by the eye lens on the retina, which is 2 cm from the eye lens. How far apart, measured on the retina, are the central image and one of the 3rd-order images?

26-B9 In the two parts of Fig. 26-14, $a = 1.000$ μm and $\lambda = 600.0$ nm. Show that the difference $\theta' - \theta$ is only about 3'. (*Hint:* Find θ and θ' separately.)

26-B10 Prove that if the wavelength is equal to or greater than the width of a slit, light striking the slit perpendicularly passes through without forming any dark interference bands.

26-B11 A laser emits light of wavelength 633 nm, which falls perpendicularly on a slit of width 1.20 μm. What is the width of the most intense central bright band on a screen 1.40 m from the slit? (*Hint:* Most of the energy lies between the two 1st-order minima.)

26-B12 Light of a certain color strikes perpendicularly a slit of width 2.2×10^{-6} m, and on a screen 1.5 m from the slit, the central maximum is 0.8 m wide. What are the wavelength and color of the light? (*Hint:* The central image is bounded by the two 1st-order dark fringes.)

26-B13 Sound waves of frequency 1700 Hz from a distant outside source enter a tall, narrow, open window perpendicularly. The window is 0.70 m wide and is in the center of a wall of an auditorium 20 m on an edge. The room is acoustically padded to eliminate reflections from the walls, and a number of observers are seated in a row along the wall opposite the window. How far from the corners of the room are the two observers who hear nothing? (The velocity of sound is 340 m/s.)

26-B14 An air wedge is formed by placing a thin blond hair between the edges of two plane glass plates that are in contact at the other end. When viewed by reflected light of wavelength 546 nm, 63 dark bands are observed, including one at the place where the plates are in contact. Calculate the diameter of the hair, in μm.

26-B15 Two glass plates 8 cm long are in contact at one edge and are separated at the other edge by a piece of aluminum foil 0.0015 cm thick. When the plates are illuminated perpendicularly by light of wavelength 600 nm, (*a*) what is the separation of adjacent fringes? (*b*) How many dark fringes are seen, including those at each end of the plates?

26-B16 Two glass surfaces are separated by an air film of uniform thickness 675 nm. If the film is illuminated perpendicularly by parallel light having wavelength 450 nm, explain why no light of this color is reflected by the air film.

26-B17 Repeat Prob. 26-B16, with the same light illuminating a thin *glass* film of the same thickness as before, namely 675 nm. The index of refraction of the glass film is 1.50. Do the reflected beams interfere constructively or destructively?

26-B18 A "nonreflecting" coated lens has a thin film of magnesium fluoride ($n = 1.25$) deposited on the surface of the glass ($n = 1.56$). What should be the thickness of the film in order that the reflected rays (Fig. 26-42) will be exactly out of phase for blue light of wavelength (in air) 450 nm? Express your answer in μm.

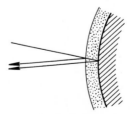

Figure 26-42

26-B19 If the fringes in Fig. 26-26b were photographed using light of wavelength 589 nm, estimate the thickness of the air film at the point diametrically opposite the point of contact. Express your answer in μm.

26-B20 A soap bubble of thickness 219 nm has index of refraction $\frac{4}{3}$. Make an analysis similar to that of Example 26-4 to show that the film appears purple (violet plus some red) when viewed perpendicularly in white light.

26-B21 Light of wavelength 633 nm is used to illuminate a Michelson interferometer. How many fringes pass the reference mark in the field of view when the movable mirror is shifted forward through a distance of 0.0600 mm?

26-B22 The index of refraction of a glass plate is 1.66. What is Brewster's angle when the plate is   (a) in air?   (b) in water?

26-B23 What is the angle of the sun above the horizon when light reflected from a calm lake is most strongly plane polarized?

26-C1 Light strikes a pair of slits at an angle θ_1 as in Fig. 26-43, and an interference maximum is formed at an angle θ_2. Derive a formula connecting θ_1, θ_2, a, and n for the nth-order maximum of constructive interference.

26-C2 (a) Derive a formula for the radius x of the kth dark ring in a Newton's rings experiment. The symbols are defined in Fig. 26-44. [Hints: Obtain a formula for y in terms of R and x; solve for x in terms of y (here you can assume that y^2 is negligible compared with Ry). For destructive interference, y must equal $k(\frac{1}{2}\lambda)$. Substitute this value of y into your formula for x.]   (b) Draw the first six rings to scale if $R = 200$ m and $\lambda = 5 \times 10^{-7}$ m.

26-C3 Using the formula derived in Prob. 26-C2, discuss the effect of replacing the air film by a film of liquid of index of refraction n, using the same glass surfaces and the same wavelength light. Could you find an unknown index of refraction for a liquid in this way?

26-C4 In a Newton's rings experiment with light of wavelength 600 nm (Fig. 26-44), the 10th dark circular fringe has a radius 2.00 cm when viewed from above.   (a) What is the radius of curvature of the lens? (Hint: Use the saggital theorem; see Sec. 25-12 or the result of Prob. 26-C2.)   (b) If the glass has index of refraction 1.50, what is the focal length of the lens?

26-C5 An experimeter wishes to determine the index of refraction of a piece of cellophane 0.0099 mm thick. The cellophane is placed broadside to the beam in one arm of a

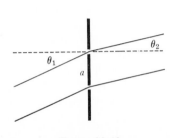

Figure 26-43

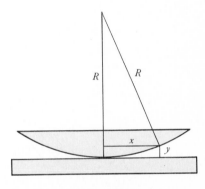

Figure 26-44

Michelson interferometer, and a shift of 20 fringes is observed. Light of wavelength 589 nm is used as a source. Calculate the index of refraction of cellophane. (*Hint*: There are more wavelengths in 0.0099 mm of cellophane than in the 0.0099 mm of air that it has replaced. For the round trip, the difference is 20λ.)

26-C6 Prove that when light strikes a glass plate at Brewster's angle, the refracted ray is perpendicular to the reflected ray.

26-C7 Light whose wavelength in vacuum is 546.1 nm falls perpendicularly on a biological specimen that is 1.50 μm thick. The light splits into two beams polarized at right angles, for which the indexes of refraction are 1.320 and 1.333, respectively. (*a*) Calculate the wavelength of each component of the light while it is traversing the specimen. (*b*) Calculate the phase difference between the two beams when they emerge from the specimen.

26-C8 A polarimeter tube 10 cm long (Fig. 26-38) contains sucrose dissolved in water at a concentration of 2.56 g/cm^3. The rotation θ is 170°. (*a*) Calculate the specific rotation for sucrose. (*b*) What is the concentration of a sucrose syrup that gives a rotation of 107° in the same apparatus?

26-C9 Can a micromole of quinine ($C_{20}H_{24}N_2O_2$) be detected in a polarimeter tube 2 mm in diameter and 20 cm long if a rotation of 0.1° can be measured? The specific rotation for quinine in a benzene solvent is 136 deg/dm for 1 g/cm^3.

For Further Study

26-10 Holography

Images can be formed without the use of lenses by a process known as *holography*. Unlike images that are formed by lenses or mirrors, a holographic image is truly three-dimensional; it is the result of the reconstruction of the wave fronts that would have reached the eye after an incident beam is scattered by points on the ob-

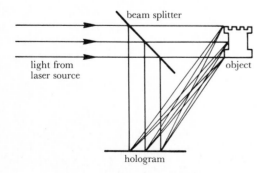

Figure 26-45   Production of a hologram.

ject. To form a hologram, the object (traditionally a chess piece!) is illuminated by a broad beam of monochromatic coherent light from a laser. Light scattered from the object strikes a photographic film (Fig. 26-45); every point on the film receives light from every point on the object. Simultaneously, a portion of the illuminating beam is removed by a beam splitter (such as a half-silvered mirror similar to that used in the Michelson interferometer) and passes directly to the film. The film, when developed, is called a hologram; the resultant density at any point of the hologram depends on both the amplitudes and the phases of all interfering wave fronts that reach that point.

To view the picture later, the image is reconstructed by placing the hologram in a laser beam; both a real and a virtual image are formed, in depth (Figs. 26-46 and 26-48). When the viewer's eye is moved from side to side, the "near" parts of the three-dimensional scene are seen to move*

---

* This relative motion is called parallax.

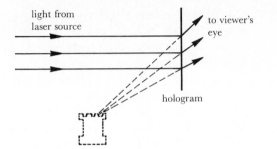

Figure 26-46 Viewing a holographic image.

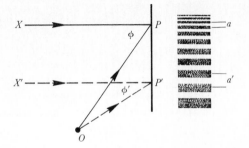

Figure 26-47 Production of hologram of a point object O that is not infinitely far from the photographic emulsion.

relative to the more distant parts. It is even possible to make holograms that allow the viewer to move in a 360° path around the image, seeing all sides of the image in a true 3-D representation.

The theory of holography is mathematically complicated. However, we can use our knowledge of interference to gain some insight into what takes place. In Fig. 26-47, XP and $X'P'$ represent parallel rays of a reference beam, and OP and OP' represent rays scattered by some point O on the object. The wave fronts interfere at P to produce a set of fringes; the exposure of the plate varies sinusoidally as in the upper part of Fig. 26-16, with the spacing between maxima dependent on the angle ϕ between the arriving rays. At another point P' closer to the object, the developed plate will also have fringes that are locally sinusoidal, but with a larger spacing between maxima. The hologram of this point source would be as shown at the right of Fig. 26-47.

Now what happens when a single broad reference beam falls on the hologram? Any small portion of the hologram is a grating, but only the 0th-order (straight-through beam) and the two 1st-order interference maxima are observed. This is because of the *gradual* (sinusoidal) variations of opacity. (Gratings with sharply bounded openings, as in Fig. 26-13, give *all* orders of interference, as in Fig. 26-16.) Parallel rays (a plane wave front) strike different parts of our hologram "grating," with results shown in Fig. 26-48. B and B' are the 0th-order maxima. A and A' are 1st-

order interference maxima, but since the grating space a' at P' is greater than the spacing a at P, the interference angle given by $\sin \theta' = \lambda/a'$ is smaller (see Eq. 26-1; here the order n is 1). Thus, $\theta' < \theta$. Rays A and A' form a virtual image that can be seen by an observer who simply looks through the hologram at a laser beam. The other 1st-order interference maxima C and C' form a real image that can be captured on a photographic plate or viewed with an eyepiece. If the hologram is cut into several pieces, each piece gives the same images as the entire hologram (although with less sharpness), for it is only the

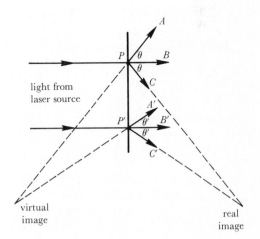

Figure 26-48 Formation of real and virtual images by 1st-order diffracted waves from a hologram.

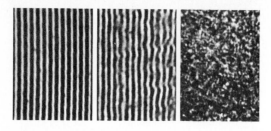

Figure 26-49 Enlarged portions of holograms of various objects. (*a*) Very distant point object. (*b*) Simple transparency as object. (*c*) Complex diffusely scattering object.

relative changes in grating space along the hologram that give rise to the images.

We have considered an object that is a single point source. Actual two- or three-dimensional objects consist of many such points; the resulting hologram is then a jumbled-up superposition of "gratings" (Fig. 26-49). Nevertheless, such a hologram, or part of one, can reconstruct the amplitude and phase of the light that originally fell on it from the object (Fig. 26-50). Thus, virtual and real images are formed of every point on an extended object.

In our example, a coherent source of light is needed so that the phase difference between two

rays arriving at P does not fluctuate randomly with time, thereby "washing out" the pattern on the film. The image can be enlarged simply by viewing the hologram in divergent light of wavelength longer than that with which it was made. If coherent x-ray beams can ever be achieved, high resolution can be obtained, and it will be possible to "see" (visually) biological or other structures much smaller than those it is now possible to see using visible light. Holography was first described in 1947 by Dennis Gabor, who received in 1971 Nobel Prize in physics for his pioneering work. With the advent of strong coherent light sources (lasers; see Sec. 30-5), the method has been applied in many areas of physics, chemistry, biology, and engineering. Many of these applications of holography are described in the references at the end of the chapter.

26-11 Double Refraction

The transparent crystal calcite (Iceland spar) is composed of molecules of calcium carbonate, $CaCO_3$, with the calcium atoms arranged in planes in a crystal lattice. Such crystals show a natural cleavage (Fig. 26-51), with the surfaces making definite angles with each other. By successive cleavages, a calcite crystal can be split into smaller crystals, not necessarily with sides of equal length, but always with the same angles as the larger crystal. This cleavage shows that the cohesive forces (short-range electric forces) between the atoms are not equal in all directions. In other words, the crystals are anisotropic. This

Figure 26-50 Reconstructed image from hologram similar to Fig. 26-49*c*.

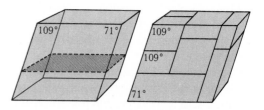

Figure 26-51

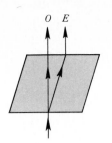

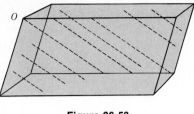

Figure 26-53

Figure 26-52 Passage of ordinary and extraordinary rays through a calcite crystal.

same structure leads to the optical phenomenon of double refraction. A single ray may (if properly oriented) split into two rays, as shown in Fig. 26-52. The *ordinary ray*, labeled *O*, behaves as we would expect it to; in the figure it is incident along the normal and travels straight through the crystal. The *extraordinary ray*, labeled *E*, behaves very peculiarly; it bends *away* from the normal as it enters the crystal (an optically denser medium), and bends *toward* the normal as it goes from crystal back into the air—exactly contrary to Snell's law. Experiment also shows that the two beams are plane polarized, with their electric vectors at right angles to each other.

These facts can be explained by Huygens' principle, with an additional assumption about the speed of propagation of light in the crystal. In calcite, the speed of light depends on the plane of vibration of the electric vector of the wave. A certain direction exists in the crystal, called the *optic axis*, such that *the speed is a maximum if the electric vector is parallel to the optic axis*. For calcite, the optic axis (O.A.) is the direction parallel to the line that makes equal angles with the three sides of the obtuse angles at *O* (Fig. 26-53). The O.A. is not just one line through a crystal; an optic axis can be drawn through *any* point in the crystal parallel to the axis through the corner. (Remember that any point inside the crystal could be made into a corner by proper cleavage.) A few of the optic axes of a certain calcite crystal are shown as dashed lines in Fig. 26-53.

The basic assumption is, now, that the speed of light depends on the orientation of the electric vector relative to the optic axis. Let us see how this assumption (which follows from the Maxwell theory of e-m waves) leads to double refraction. We resolve the incident (unpolarized) beam into two components polarized at right angles to each other and choose one component to have its electric vector in the plane determined by the optic axis and the incident ray (Fig. 26-54a). While the wave is traveling in air, Huygens' principle predicts straight-line propagation, assuming that the pinhole that limits the size of the beam is large enough compared with λ that diffraction can be ignored. When the beam reaches the calcite, each Huygens wavelet spreads out in an ellipsoid rather than a sphere: ray 6 travels fastest, since its electric vector is parallel to the O.A.; ray 2 travels slowest, since its electric vector is perpendicular to the O.A. A similar construction is made for another ray in the incident beam, and the new wave front is drawn tangent to the wavelets. By continued application of Huygens' principle, we see that the beam is propagated through the crystal as shown—still a plane wave, but with the ray not perpendicular to the wave front. Eventually the rays reach the top surface, and wavelets spread out into air. Now, in air, the rays travel equally fast in all directions, and the wave front moves upward, parallel to the original direction. We have described how the extraordinary beam passes through the crystal; as a by-product of the analysis, we are led to expect that the *E* beam in Fig. 26-54a is polarized parallel to the plane of the paper.

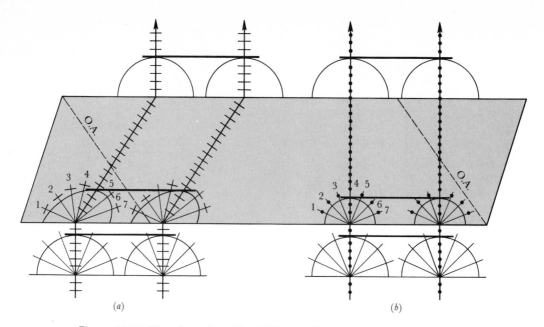

Figure 26-54 Wave fronts in calcite. (*a*) Extraordinary beam. (*b*) Ordinary beam.

The ordinary beam is represented by the other half of the incident beam, polarized as in Fig. 26-54*b*. In this case, each of the rays 1, 2, 3, 4, 5, 6, and 7 is vibrating perpendicular to the O.A., since each electric vector is perpendicular to the plane of the paper. Therefore, the wavelets in the crystal are spherical, and the wave front moves through in the ordinary fashion. To obtain the complete phenomenon, we must imagine (*a*) and (*b*) superposed. The original unpolarized beam splits into two beams, polarized at right angles to each other.

Two special cases are of interest. If the crystal is cut and polished so that light can enter along the O.A. (Fig. 26-55), then no separation of the beams takes place, and they travel through the crystal with the same speed. If the O.A. is parallel to the surface, we again find that no separation takes place, but in this case the *E* beam travels faster than the *O* beam. Proof of this statement is left as an exercise (Prob. 26-C10).

We have described double refraction in calcite, a typical "negative" crystal. In "positive" crystals, the O.A. is the direction of the electric vector for which the speed of a ray is *least* instead of *greatest.*[*] Still other crystals have two optic axes. Many crystals, such as NaCl, are isotropic, have no O.A., and are not doubly refracting. All the experimental results are rigorously predictable on the basis of the e-m wave model of light.

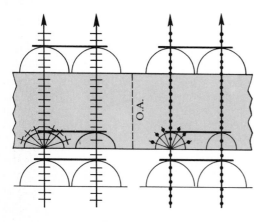

Figure 26-55

---

* Ice is such a crystal.

Problems
26-C10 Prove that if an unpolarized beam strikes the surface of a crystal whose O.A. is parallel to the surface, the O and the E beams do not separate, but one of them travels faster than the other. (*Hint:* Make a diagram similar to Fig. 26-55, but with the O.A. parallel to the surface.)

26-C11 For calcite the indexes of refraction are $n_O = 1.658$ and $n_E = 1.486$. (*a*) What is the ratio of the speed of the E ray to that of the O ray in calcite? (*b*) Which ray travels faster?

References
1. W. F. Magie, *A Source Book in Physics* (McGraw-Hill, New York, 1935), pp. 294–318. Selections from the original writings of Grimaldi, Newton, Young, and Malus.
2. F. Cajori, *A History of Physics*, rev. ed. (Macmillan, New York, 1929), pp. 89–97. Chiefly Newton's experiments with prisms; his rejection of the wave theory of light. See also pp. 148–171, where the development of the wave theory of light is discussed.
3. I. B. Cohen, "The First Explanation of Interference," *Am. J. Phys.* **8**, 99 (1940).
4. R. W. Pohl, "Discovery of Interference by Thomas Young," *Am. J. Phys.* **28**, 530 (1960).
5. E. B. Sparberg, "Misinterpretation of Theories of Light," *Am. J. Phys.* **34**, 377 (1966).
6. H. B. Lemon, "Albert Abraham Michelson: The Man and the Man of Science," *Am. Phys. Teach.* (*Am. J. Phys.*) **4**, 1 (1936).
7. R. S. Shankland, "Michelson and His Interferometer," *Physics Today* **27**(4), 37 (Apr. 1974).
8. A. G. Ingalls, "Ruling Engines," *Sci. American* **186**(6), 45 (June 1952). Construction of gratings of high quality.
9. P. Baumeister and G. Pincus, "Optical Interference Coatings," *Sci. American* **223**(6), 58 (Dec. 1970).
10. T. H. Waterman, "Polarized Light and Animal Navigation," *Sci. American* **193**(1), 88 (July 1955).
11. R. Wehner, "Polarized Light Navigation by Insects," *Sci. American* **235**(1), 106 (July 1976).
12. J. Walker, "Dazzling laser displays that shed light on light," in the "Amateur Scientist" department, *Sci. American* **243**(2), 158 (Aug. 1980).
13. E. N. Leith and J. Upatnieks, "Photography by Laser," *Sci. American* **212**(6), 24 (June 1965).
14. S. Heumann, "How to Make Holograms," in the "Amateur Scientist" department, *Sci. American* **216**(2), 122 (Feb. 1967).
15. A. G. Porter and S. George, "An Elementary Introduction to Practical Holography," *Am. J. Phys.* **43**, 954 (1975).
16. W. R. Schubert and C. R. Throckmorton, "Making a 360° Hologram," *Phys. Teach.* **13**, 310 (1975).
17. K. S. Pennington, "Advances in Holography," *Sci. American* **218**(2), 40 (Feb. 1968).
18. J. Strickland, *Superposition of Pulses; Interference of Waves; Effect of Phase Differences Between Sources; Single Slit Diffraction; Multiple Slit Diffractions; Diffraction and Scattering Around Obstacles* (films).
19. F. Miller, Jr., *Diffraction—Single Slit; Diffraction—Double Slit; Michelson Interferometer* (films).

27

Applied Optics

In this chapter we study the applications of the theory of light to a number of common optical instruments. All these instruments use lenses, in one way or another, and so we shall continually rely on our work in geometrical optics (Chap. 25) to find the positions of images (by either of two methods) and their sizes (by the ratio of image distance to object distance). We shall not ignore the disturbing effects of dispersion, nor shall we overlook the inevitable diffraction that follows from the wave nature of light.

27-1   *The Camera—Lens Aberrations*

The camera is simple in theory but is likely to be complicated in practice. In essence, a converging lens forms an inverted real image on a sensitized film or photographic plate; wherever light strikes, chemical PE is stored that can be released by proper chemical action during the process of developing. We do not wish to inquire further into the chemical and physical processes

by which the small grains of metallic silver are caused to appear wherever enough PE has been absorbed by the molecules of the emulsion. Our problem is to understand why the optical image may not be perfectly sharp, and how to improve it.

In our earlier work, we assumed (without proof) that a point source forms a point image. That is, all the rays that leave a source and pass through the lens come together at a single point. This is only an approximation to the truth, but one good enough for many purposes.* However, the precise way to study image formation is to trace the path of each ray, using Snell's law at each of the refracting surfaces. When this is done, a number of defects, called *aberrations*, of a simple camera lens appear. We shall discuss only three of the aberrations of a lens: spherical

---

* If you have studied Sec. 25-12, you will realize that the method of curvatures and the lens equation are based on the sagittal theorem, which is an approximation having greatest validity for thin lenses.

Two radio telescopes, each of diameter 27 m, operate as a variable-spacing interferometer with different baseline orientations and lengths up to 1 km. Wavelengths between 0.04 m and 0.05 m are used. The system is in a deep valley to minimize man-made electrical interference. (Photo by A. T. Moffet, Director of the Owens Valley Radio Observatory, California Institute of Technology.)

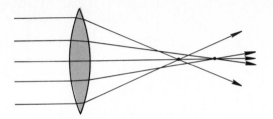

Figure 27-1 Spherical aberration.

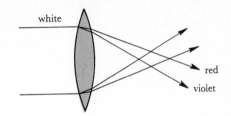

Figure 27-3 Chromatic aberration of a converging lens.

aberration, off-axis astigmatism, and chromatic aberration.

The surfaces of a simple lens are ordinarily sections of spheres; such surfaces are easy to construct by grinding and polishing. *Spherical aberration* is caused by the fact that such a lens has a longer focal length for rays near the center than for rays passing through an outer zone (Fig. 27-1). There is no one focal length, but rather a focal region. Most cameras are equipped with an adjustable stop, or diaphragm, which is primarily used to control the amount of light reaching the film. When the sharpest possible image is desired, the photographer reduces the effective diameter of the lens and compensates for the loss of light by a longer exposure time.

Off-axis astigmatism is a troublesome defect that is noticeable for objects not on the axis of the lens, as in Fig. 27-2. The image of point P is *two* short lines: a vertical line AB and a horizontal line CD closer to the lens. The "focus" is a "circle of least confusion" somewhere between the two line images. To reduce astigmatism, a photographer sees to it that the important part of the picture is centered in the field of view, so that the object points lie near the axis of the lens. For more expensive cameras, systems of lenses

have been designed to reduce off-axis astigmatism. The lens surfaces are symmetrical but not necessarily spherical.

A third major lens defect is *chromatic aberration*, which causes color-fringed images even for a lens of small opening and for an object and image on the axis. This is one of the effects of dispersion. The index of refraction of glass varies with wavelength, so that a point source of white light is spread out into a colored line. For simplicity, we consider in Fig. 27-3 rays from a distant source. Bearing in mind that refraction takes place at each surface of the lens, we see that the violet light, for which the index of refraction is greatest, is refracted the most. Thus the focal length is least for violet light and greatest for red light. Study of a diverging lens (Fig. 27-4) shows that the focal point for violet light is closest to the lens in this case too. Chromatic aberration can be corrected for two colors by using a combination of lenses as in Fig. 27-5, called an *achromatic doublet*. The converging lens tends to place the red image to the right of the violet one,

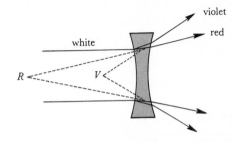

Figure 27-4 Chromatic aberration of a diverging lens.

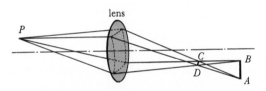

Figure 27-2 Off-axis astigmatism.

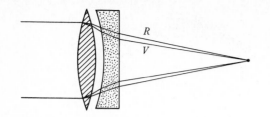

Figure 27-5 Achromatic doublet.

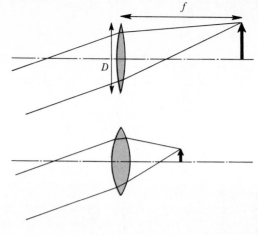

Figure 27-6

while the diverging lens has the opposite effect. With proper choice of the radii of curvature of the surfaces, and with the use of two different kinds of glass for the two lenses, a perfect match can be obtained for any two colors. For visual work, in which the eye rather than a photographic plate is used to receive the image, it is customary to correct the lens for yellow and blue. This will give nearly perfect correction also for green light, which lies between yellow and blue, but there may be some residual color at the extreme violet or red ends of the spectrum. An achromatic doublet is essential for a camera of high quality; in many cameras three colors are corrected by a triplet of three lenses.

Another important characteristic of a camera lens is its ability to form a bright image. The brightness or intensity* of an image depends on the rate at which energy reaches the film. Other things being equal, the intensity of an image of given size is proportional to the area of the lens opening. That is, intensity $\propto D^2$, where D is the diameter of the lens. The focal length of the lens also affects the intensity. Consider two lenses, each forming an image of the same distant object (Fig. 27-6). We have

$$\frac{\text{Image height}}{\text{Object height}} = \frac{\text{image distance}}{\text{object distance}}$$

$$= \frac{f}{\text{object distance}}$$

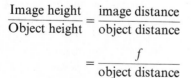

The image distance in each case is the corresponding focal length, and so the image height is proportional to the focal length f, and the image *area* is proportional to f^2. The smaller image will be brighter, since the same *total* energy per second is shared by fewer molecules of the film. Putting these results together, we see that the intensity I is proportional to D^2/f^2. The important factor is the ratio of focal length to diameter (f/D), which is called the f-*number of the lens*. A "fast" lens has a large diameter and a short focal length. A simple fixed-focus camera's lens is about $f/16$ or $f/22$, where $f/16$ means that $f/D = 16$. The smaller the f-number, the faster is the lens, and the more expensive because of the necessity of more elaborate corrections for various aberrations. A modern camera lens system may have an effective f-number of $f/1.6$ or less.

27-2 The Human Eye

The human eye is not much different in principle from a camera. A converging lens forms a real image on the retina (Fig. 27-7), and sensations are transmitted to the brain by a complex neural mechanism. The retina takes the place of the sensitized film. The camera can focus on objects

---

* Just as for sound waves, the intensity of light is defined as the rate of flow of energy per unit cross-sectional area (Sec. 11-5).

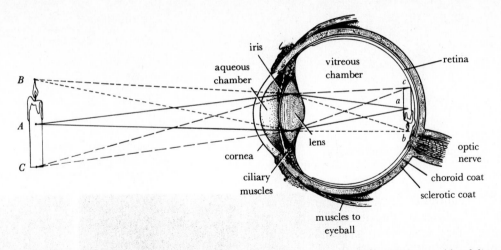

Figure 27-7 A diagram of the eye, somewhat idealized; the optic nerve is actually a bit to one side of the vertical plane through the eye. (From G. G. Simpson and W. S. Beck, *Life: An Introduction to Biology*, 2nd ed., Harcourt Brace Jovanovich, 1965.)

at different distances from the film; this is accomplished by moving the lens back and forth to change the image distance as necessary. In the eye, however, this focusing adjustment, called *accommodation*, is managed somewhat differently. The lens is fixed in position, with its center about 1.7 cm from the retina. Since it is made of a transparent, flexible tissue, its focal length can be changed by a set of small muscles that change the curvatures of the lens surfaces. In this way the equation $-1/p + 1/f = 1/q$ can be satisfied for any object (within limits), with q being about 1.7 cm for a "normal" eye. The accommodation of the lens for various object distances is shown *schematically* in Fig. 27-8. The power of the lens must be greater in (*b*), since the incoming wave front has greater divergence that must be changed into convergence.

Figure 27-8 is oversimplified by omission of the cornea (Fig. 27-7), which is the hard, transparent protective layer that the light strikes first. Actually, most of the ability of the eye to form an image is due to the outer surface of the cornea, which is in contact with air. (You will recall that it is the *relative* index of refraction that determines the ability of an interface to change the direction of a ray.) The other surface of the cornea, and both surfaces of the lens, are much less effective than the front surface. In a typical eye the power of the cornea is about $+40$ diopters, and the other surfaces, including the adjustable lens, vary from a total of $+20$ to $+24$ diopters. The overall power of the eye is thus $+60$ to $+64$ diopters. This variation is sufficient for most purposes, provided the object is not too close to the eye. If the lens is completely removed, as in an operation for cataract, the patient can still see, although an auxiliary spectacle lens of power $+10$ diopters or so must be used, to compensate for the loss of the eye lens, and no accommodation is possible.

Closer study of the eye, considered as an optical instrument, reveals that a number of serious defects are possible. One common defect of

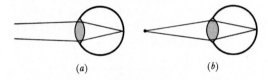

(*a*)   (*b*)

Figure 27-8

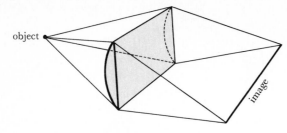

Figure 27-9 Cylindrical lens.

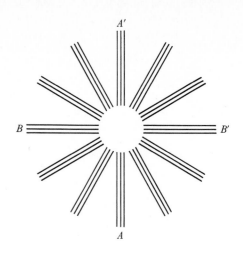

Figure 27-10 Chart for testing astigmatism of the eye.

vision is *astigmatism*, in which a point source forms a line image on the retina. The result is the same as for astigmatism of a camera lens, but the cause is different. A cylindrical lens has positive convergence for rays in only one plane, as shown in Fig. 27-9, since it is curved in only one dimension. A cylindrical lens is completely astigmatic*; in certain applications this is useful, but not in the camera or eye. Astigmatism of the eye may be caused by imperfectly spherical surfaces of the cornea or lens. A quick test for astigmatism can be made with the help of the chart shown in Fig. 27-10. If, for instance, each point source makes a vertical line image on the retina, the figure will remain sharp in direction AA', but will be blurred in the direction of BB'. To correct for astigmatism, spectacle lenses are used that have a compensating cylindrical curvature, properly oriented. Off-axis astigmatism in the eye is not a problem because mammals and some other vertebrates have adapted to it by scanning the field of view. The eyes are rotated in their sockets to make the fovea (intersection of the visual axis of the eye and the retina) coincide with the image of any object to be examined in detail.

The other major defects that the eye is subjected to are an inability to focus clearly on a very distant object (*myopia*, or nearsightedness) or on a very close object (*hypermetropia*, or farsightedness). To see how these defects can be corrected with the aid of spectacles, let us use the method of curvatures. (The lens equation

---

* From the Greek *a*, no, and *stigma*, point.

could also be used, with proper attention to the sign convention.)

We have seen that the power of the lens-cornea combination can adjust from $+60$ to $+64$ diopters in the "normal" eye. When accommodated for distant vision (to view an incoming wave front having curvature of 0 diopters), the lens is almost completely relaxed, and the power of the eye is the least, namely, $+60$ diopters, or $+60 \text{ m}^{-1}$. The curvature equation reads

$$\begin{array}{ccccc} \text{initial} & + & \text{power} & = & \text{final} \\ \text{curvature} & & \text{of eye} & & \text{curvature} \\ 0 & + & 60 \text{ m}^{-1} & = & 1/q \end{array}$$

This says that the final curvature $1/q$ is $+60 \text{ m}^{-1}$; the radius of curvature is $\frac{1}{60}$ m, or about 1.67 cm, just right to converge to the retina. How close can an object approach the normal eye and still be focused on the retina? As the object approaches, the muscles operate to increase the power of the lens-cornea combination from $+60$ to $+64$ diopters. While this happens, the curvature of the incident wave front must change from 0 to -4 diopters (Fig. 27-11). At the *near point*, the equation reads

$$-4 + 64 = +60$$

and the final curvature is the same as before (as it

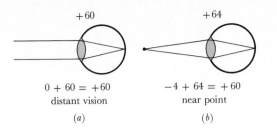

$$0 + 60 = +60$$
distant vision

(a)

$$-4 + 64 = +60$$
near point

(b)

Figure 27-11 Normal eye.

must be if the wave front is to converge on the retina, which is still $\frac{1}{60}$ m from the lens). We see that the near point is about $\frac{1}{4}$ m (25 cm) from the normal eye. (Try it.)

Consider now the myopic eye of a nearsighted girl, for which the eyeball is longer than usual. Even with the lens completely relaxed (Fig. 27-12a), the image is in front of the retina. To obtain a sharp image on the retina, the final curvature must be *less* than $+60$ m^{-1} (longer radius of curvature)—say, $+55$ m^{-1}. The required spectacle lens is negative, and the curvature equations in Fig. 27-12b show that a lens of power -5 diopters will allow sharp vision of a distant object.* If the maximum power of the eye is 64 diopters, without glasses this particular myopic eye has a near point that is calculated in

---

* Throughout this section we assume that the spectacle lens is close to the eye, so that there is no appreciable change of curvature as the wave front travels between the two lenses.

Fig. 27-12c; $-1/p = -9$ m^{-1}, and $p = 0.11$ m instead of the normal 0.25 m. This nearsighted person can see distinctly an object that is only 11 cm from her eye.

Hypermetropia can be analyzed in a similar fashion. The eyeball is too short, and a positive lens is used to help the eye lens. In Fig. 27-13 it is assumed that the lens is 1.59 cm from the retina ($\frac{1}{63}$ m), and so the final curvature must be $+63$ diopters for a sharp image. When the near point is calculated as in Fig. 27-13c, the eye lens is made as bulgy as possible, but p works out to be 1 m instead of the normal 0.25 m. This hypermetropic person cannot see an object clearly without glasses unless it is at least 1 m from the eye.

In middle age, most persons gradually lose some of their power of accommodation as the eye muscles become weaker. We have stated that the "normal" eye can add 4 diopters to its power by strong muscular effort. However, this ability depends on age; it is not unusual for a child of 10 to be able to add 15 or 20 diopters, whereas at age 75 the muscles may be able to add only 1 diopter, or even less. Thus, people tend to become farsighted as they grow older—a condition known as *presbyopia*. The remedy is the same as for hypermetropia, which is to help out the eye's own lens by adding a positive spectacle lens. When the ability to accommodate is lost, bifocals are worn so that the total power of the system can be altered as required for near and

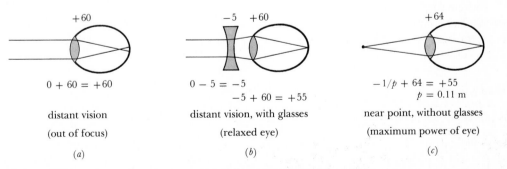

$$0 + 60 = +60$$

distant vision

(out of focus)

(a)

$$0 - 5 = -5$$
$$-5 + 60 = +55$$

distant vision, with glasses

(relaxed eye)

(b)

$$-1/p + 64 = +55$$
$$p = 0.11 \text{ m}$$

near point, without glasses

(maximum power of eye)

(c)

Figure 27-12 Myopic eye. In each figure the length of the eyeball is $\frac{1}{55}$ m (1.82 cm), and (for an in-focus image) the curvature of the wave front just after leaving the eye lens must be $+55$ diopters.

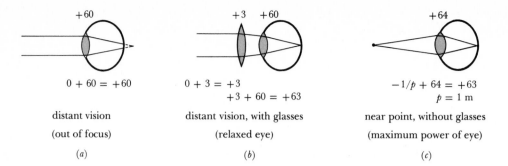

$$0 + 60 = +60$$

distant vision

(out of focus)

(a)

$$0 + 3 = +3$$
$$+3 + 60 = +63$$

distant vision, with glasses

(relaxed eye)

(b)

$$-1/p + 64 = +63$$
$$p = 1 \text{ m}$$

near point, without glasses

(maximum power of eye)

(c)

Figure 27-13 Hypermetropic eye. In each figure the length of the eyeball is $\frac{1}{63}$ m (1.59 cm), and (for an in-focus image) the curvature of the wave front just after leaving the eye lens must be $+63$ diopters.

distant vision.* In an inexpensive camera, the lens is of fixed power, and a positive "portrait attachment" can be placed in front of the camera lens if it is desired to photograph a nearby subject. This lens serves the same function as do spectacles for a human who has lost the power of accommodation. Television and movie cameras use zoom lenses of variable focal length, in which the power is changed continuously by changing the spacing between the elements of the lens system.

To summarize, myopia is corrected by using a negative lens; hypermetropia and presbyopia are corrected by using a positive lens.

Pursuing the analogy of the eye and the camera, we note the presence in the eye of a diaphragm, the iris, which is controlled by a set of involuntary muscles as the intensity of the light changes. Chromatic aberration is, fortunately, a minor problem in the eye, although it can be observed. The image on the retina is inverted, as in a camera, but the brain has learned to present a "right-side-up" picture to the consciousness.[†]

A simple experiment to prove this is described as follows: Use the point of a pencil to punch a small hole through a piece of thin cardboard or stiff paper. Holding the paper about 10 cm from the eye, look through the hole at a distant lamp or window. Place the pencil point as close to the eye as possible (within a few centimeters), and move it up and down. If you wear glasses, the pencil can rest on the spectacle lens. No lens action is possible, since the object is too close to the eye. When you move the pencil upward, the shadow on the retina also moves upward, but the brain interprets this as a downward motion. Looking through the hole, you will see the shadow appear to move downward.

All swimmers are aware that when the eyes are open under water, distant objects seem blurred and out of focus. Figure 27-14 illustrates the

---

*Trifocal glasses are by no means uncommon; the upper part of the lens is used for distant vision, the middle part for middle distances, and the lower part for reading.
[†] A relatively short time is required for the brain to reorganize its "switchboard." A subject who wears glasses that present everything upside down begins to see the world right side up in a week or so. If, after several weeks, the inverting glasses are removed, everything looks upside down again for another week or more.

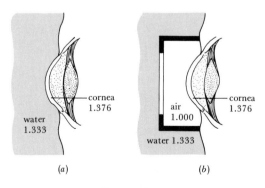

(a)   (b)

Figure 27-14

cause of this problem and its cure through the use of goggles. The average cornea has an index of refraction 1.376, which is considerably greater than that of air, which is 1.000. Therefore, in air, the front surface of the cornea has a significant converging power. In fact, almost all of the +40 diopters of the cornea is due to the *outer* surface of the cornea, where the ratio of indexes of refraction is 1.376/1.000. The inner surface of the cornea has only slight power because it is in contact with the liquid in the aqueous chamber (Fig. 27-7), which has a very similar index. For the swimmer under water, however, the ratio at the outer surface of the cornea is only 1.376/1.333, or 1.032. Thus, the overall power of the cornea-lens system in the eye is much weakened for an underwater swimmer. The remedy is to wear goggles (Fig. 27-14b). A thin sheet of glass has no power in itself, but now the cornea is in air and retains its full power. Reference 8 at the end of the chapter is an interesting study of vision in the dolphin, which does not wear underwater goggles but has about equal visual acuity both in and out of the water.

27-3   The Magnifier

The actual size of an object is of small consequence in determining how big it appears to be. For instance, a 2-cm penny held at arm's length (0.60 m) looms as large as a 20-cm pie plate at 6 m, and much larger than a 3500-km full moon at 384,000 km. The *angular size* of the object is what counts, and we can use the ratio h/d as a measure of angular size. (This is really the tangent of the angle θ in Fig. 27-15.) Let us compute the angular size of the three objects mentioned above.

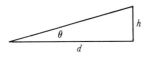

Figure 27-15

$$\text{Penny:} \quad \frac{h}{d} = \frac{2 \text{ cm}}{60 \text{ cm}} = \frac{1}{30} = 0.033 \text{ rad}$$

$$\text{Pie plate:} \quad \frac{h}{d} = \frac{20 \text{ cm}}{600 \text{ cm}} = \frac{1}{30} = 0.033 \text{ rad}$$

$$\text{Full moon:} \quad \frac{h}{d} = \frac{3500 \text{ km}}{384,000 \text{ km}} = \frac{1}{110} = 0.009 \text{ rad}$$

The penny and the pie plate have the same angular size and are almost 4 times as "large" as the moon (try it).

The usefulness of a magnifier, a microscope, or a telescope lies in its ability to magnify the *angular* size of a viewed object. In Chap. 25 we considered linear magnification m, defined as the ratio (image height)/(object height). Now we define *angular magnification* and denote it by M:

$$M = \text{angular magnification}$$
$$= \frac{\text{angular size of image}}{\text{angular size of object}}$$

The angular magnification produced by an optical instrument is often called its *magnifying power*.

The obvious way to make a small object appear larger is to increase its angular size by bringing it close to the eye. If a stamp collector places a stamp 2 cm from her eye, it seems large enough, to be sure, but the image is hopelessly blurred (try it). This is because the near point of a normal eye is about 25 cm. To see the stamp better, the collector uses a magnifying glass, which is a single converging lens. When she places the stamp just inside the focal point of the lens, a virtual image is formed at least 25 cm from the eye (Fig. 27-16). As seen from the ray diagram, the object and image have the same angular size as seen by the eye. The lens, in effect, moves the stamp back to a distance of 25 cm or more, without at the same time decreasing its angular size.

To compute the magnifying power of the lens, we compare the angular sizes with and without the lens. We first use the method of curvatures to find the object distance. We know that the image

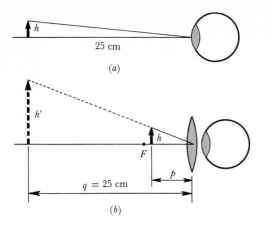

Figure 27-16 (a) Without aid, the object is of small angular size. (b) With the aid of a magnifier, the same object has a much larger angular size.

is a virtual one, 25 cm behind the lens, and the final curvature is $-\frac{1}{25}$ cm^{-1}.*

$$\begin{array}{ccccc}
\text{initial} & + & \text{power} & = & \text{final} \\
\text{curvature} & & \text{of lens} & & \text{curvature} \\
-\dfrac{1}{p} & + & \dfrac{1}{f} & = & -\dfrac{1}{25}
\end{array}$$

$$\frac{1}{p} = \frac{1}{25} + \frac{1}{f} = \frac{f+25}{25f}$$

which gives us

$$p = \frac{25f}{f+25} = \frac{f}{1+f/25}$$

As expected, the magnitude of p is just a bit less than f, since the denominator is slightly greater than 1. To find the image size we multiply the object size h by the linear magnification m. According to Eq. 25-4, $m = -q(1/p)$, and here $q = -25$ cm for the virtual image.

---

* In this and similar problems, distances are in centimeters unless otherwise specified, and the near point is assumed to be 25 cm from the eye. The lens equation could also be used, with due attention to the sign convention.

$$m = -q\left(\frac{1}{p}\right) = -(-25)\left(\frac{f+25}{25f}\right)$$

$$= \frac{f+25}{f} = 1 + \frac{25}{f}$$

This gives

$$h' = mh$$

or

$$h' = \left(1 + \frac{25}{f}\right)h \qquad (27\text{-}1)$$

The magnifying power of a magnifier is given by

$$M = \frac{\text{greatest possible angular size (aided)}}{\text{greatest possible angular size (unaided)}}$$

Now we *could* have moved the stamp to within 25 cm of our unaided eye, but no closer (for then we could not have seen it clearly). The unaided maximum angular size is therefore $h/25$. Using the magnifier, we have a larger stamp of size h', also at 25 cm, and so the new angular size is $h'/25$.

$$M = \frac{h'/25}{h/25} = \frac{h'}{h}$$

and so we see that in this case the angular magnification is the same as the linear magnification. Using Eq. 27-1 gives

$$M = \frac{h'}{h} = \frac{\left(1 + \dfrac{25}{f}\right)h}{h} = 1 + \frac{25}{f}$$

Thus, finally, we have found a formula for the greatest angular magnification, or magnifying power, of the lens. If we choose to relax the eye and move the virtual image back to infinity (by placing the object at the focal point of the lens), the image becomes larger, but the angular magnification decreases slightly to $25/f$ in the limit. A numerical example will serve to illustrate these principles, using the lens equation in the "curvature" form $-1/p + 1/f = 1/q$. The "ray" form of the lens equation, $1/p + 1/q = 1/f$, could also be used.

Example 27-1

A bug 3 mm tall is viewed by a person with normal vision using a magnifier of focal length $+2$ cm. Calculate the angular magnification (*a*) when the lens is adjusted for maximum magnification, (*b*) when the virtual image is 50 cm from the lens, and (*c*) when the virtual image is at infinity.

(*a*) At maximum magnification, the virtual image is at the near point, -25 cm.

$$-\frac{1}{p} + \frac{1}{2} = -\frac{1}{25}$$

$$p = \frac{50}{27} = 1.85 \text{ cm}$$

The bug must be 1.85 cm from the lens, which is just inside the focal point. We next compute the linear size of the image from $h' = h(-q/p)$:

$$h' = h\frac{-(-25)}{\left(\frac{50}{27}\right)} = h(25)(\tfrac{27}{50}) = h(13.5)$$

The image is $(3 \text{ mm})(13.5) = 40.5$ mm high. Without using the lens, we could have placed a 3-mm bug 25 cm from the eye; with the lens, we have a 40.5-mm bug, also 25 cm from the eye. The angular magnification is

$$M = \frac{40.5 \text{ mm}/25 \text{ cm}}{3 \text{ mm}/25 \text{ cm}} = \boxed{13.5}$$

As a check, we use the formula for M:

$$M = 1 + \frac{25}{f} = 1 + \frac{25}{2}$$

$$= 1 + 12.5 = 13.5$$

(*b*) If we relax the eye muscles somewhat, to focus on a virtual image 50 cm from the eye, the object distance must be recalculated.

$$-\frac{1}{p} + \frac{1}{2} = -\frac{1}{50}$$

$$p = \frac{100}{52} = 1.92 \text{ cm}$$

Under these circumstances,

$$h' = h\frac{-(-50)}{\left(\frac{100}{52}\right)} = h(50)(\tfrac{52}{100}) = h(26)$$

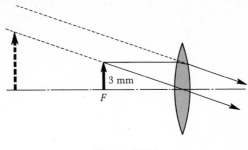

Figure 27-17

The height of the image is now $(3 \text{ mm})(26) = 78$ mm. However, this larger image is farther away.

$$M = \frac{78 \text{ mm}/50 \text{ cm}}{3 \text{ mm}/25 \text{ cm}} = \boxed{13.0}$$

(*c*) If the object is *at* the focal point, the image is at infinity and is infinitely large. The angular size is not infinite (Fig. 27-17), but exactly equals that of a 3-mm object at *F*.

$$M = \frac{3 \text{ mm}/2 \text{ cm}}{3 \text{ mm}/25 \text{ cm}} = \frac{25}{2} = \boxed{12.5}$$

This illustrates how the angular magnification approaches $25/f$ as the virtual image recedes to infinity, for a person with "normal" vision.

It was formerly believed that *minimum eyestrain* is achieved if a magnifier is used in such a way that the emerging rays are parallel, as in Fig. 27-17, for then the rays seem to come from infinity and the eye lens is most relaxed. Indeed, many optical instruments are designed with the implicit assumption that a "relaxed" eye is focused at infinity, and we use this assumption in this book. However, recent research has shown that when viewing a featureless area, such as the empty sky, the normal eye accommodates so that its "resting focus" is on a point about 1 m distant (not infinity). This is also probably the cause of the well-known phenomenon of "instrumental myopia"—one does not, and should not, focus a microscope to give a final virtual image at infinity. (See Ref. 7 at the end of the chapter.)

In most cases a virtual image at infinity would be only slightly different in angular size than if it

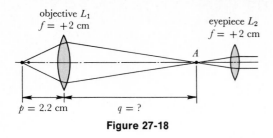

objective L_1
$f = +2$ cm

eyepiece L_2
$f = +2$ cm

A

$p = 2.2$ cm $q = ?$

Figure 27-18

were at 1 m; so in our work, placing the "minimum eyestrain" image at infinity is justifiable because of computational ease. We use

$$M = \frac{25}{f}$$

for the magnifying power of a magnifier used by a person with normal vision.

27-4 The Compound Microscope

To obtain higher magnification than is possible with one lens, we use a *compound microscope.* The object is placed near an *objective lens* L_1 of short focal length, and a real image A is viewed through an *eyepiece,* or *ocular,* L_2, which is a simple magnifier (Fig. 27-18). To form the real image at A, the object must be just beyond the focal point of L_1, and since the ratio (image distance)/(object distance) is large, the linear size of the real image A is considerably larger than that of the object. Further angular magnification is obtained when the real image at A serves as an object and is viewed through the eyepiece.

Example 27-2

The objective and the eyepiece of a microscope each have focal length 2 cm. If an object is placed 2.2 cm from the objective, (a) calculate the distance between the lenses when the microscope is adjusted for minimum eyestrain; (b) calculate the magnification.

(a) First we calculate the position of the real image A (Fig. 27-18).

$$-\frac{1}{2.2} + \frac{1}{2.0} = \frac{1}{q}$$

whence $q = 22$ cm. For minimum eyestrain, the eyepiece is adjusted so that parallel rays emerge, and hence A is at the focal point of the eyepiece. The total separation of the lenses is

$$22 \text{ cm} + 2 \text{ cm} = \boxed{24 \text{ cm}}$$

(b) If the object size is h, the real image's size is

$$h' = h \frac{22 \text{ cm}}{2.2 \text{ cm}} = 10h$$

Since the eyepiece is adjusted for minimum eyestrain, its magnifying power is given by $25/f = 25/2 = 12.5$. The overall magnifying power of the instrument is the product of the two magnifications:

$$M = (10)(12.5) = \boxed{125}$$

For many purposes, the conventional compound microscope has been replaced by specialized types of optical microscopes, such as the interference microscope, the phase contrast microscope, and the polarizing microscope. We shall not discuss the technical details of these instruments.

27-5 Resolving Power

In theory, it would seem possible to use a series of lenses to make a real image as large as desired. For instance, the real image A of Fig. 27-18 could have been viewed through a compound microscope instead of a simple magnifying glass. However, the wave nature of light sets an upper limit to the usefulness of any image.

The ability of a lens to form sharp images is called its *resolving power.* When a lens forms an image of a point source, the lens serves somewhat as a single slit, and so the image is surrounded by a diffraction pattern whose angular size is *roughly* $\theta = \lambda/D$, where D is the diameter of the lens. This follows from Eq. 26-2 (page 611), which states that the 1st-order minimum comes at an angle given by $\sin \theta = \lambda/w$ (remember that

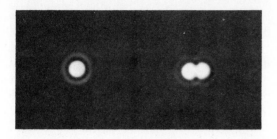

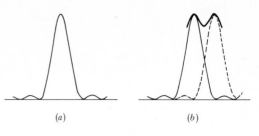

Figure 27-20

Figure 27-19 (*a*) Telescopic image of a point source, showing the diffraction disk (the Airy disk) and circular fringes. (*b*) Two point sources, barely resolved. (Reproduced by permission of Blackie and Son, Ltd., from *Light* by R. W. Ditchburn.)

for small angles $\sin \theta$ is approximately equal to θ). A circular opening such as a lens does not have a uniform width, however; on the average, its width is somewhat narrower than its diameter. Detailed analysis shows that a circular opening of diameter D is equivalent to a slit of width $0.82D$. The angular radius of the central part of the circular diffraction pattern is, therefore, given by

$$\theta = \frac{\lambda}{0.82D} \tag{27-2}$$

Of course, for visible light, λ is very small compared with the diameter of a microscope lens, but nevertheless there is an irreducible diffraction pattern surrounding each point of the image. For an objective of a given diameter, each point source makes a blurred image (Fig. 27-19), and two neighboring points of an object would be indistinguishable if their diffraction patterns overlapped.

The single-slit diffraction pattern of a point source was studied in Sec. 26-5. For the image formed by a lens of diameter D, analysis shows that 84% of the light lies within the central circle defined by $\theta = \lambda/0.82D$. This bright circle, due entirely to the wave nature of light, is called the *Airy disk** (Fig. 27-19a). Two point sources are

said to be resolved if the center of the Airy disk of one image coincides with the first minimum of the adjacent source's pattern (Fig. 27-20). This criterion, due to Rayleigh,[†] ensures that the resultant intensity shows a slight dip between the centers of the image of adjacent point sources. For an extended object, such as a biological cell, the "point sources" are adjacent points in the fine structure of the object.

We see that the resolving power of a microscope is limited by the wave properties of light. Mere magnification is worthless beyond a certain point, for the user would only be magnifying an inherently blurred image. There are two ways to improve the resolving power of a microscope or telescope, and both methods make λ/D smaller: (1) The wavelength can be decreased (some microscopes use ultraviolet light partly for this reason), or (2) the diameter can be increased (this is one reason for large-diameter telescopes).

Our discussion of resolving power has implied that the various lens aberrations have been made so small that diffraction effects are the only remaining source of fuzziness. This is actually true for microscopes, as a result of careful design of the objective lenses. Figure 27-21 shows the objective lens of a modern microscope; it consists of no less than 10 components, highly corrected for many aberrations, and has an effective focal length of a few millimeters. To seek further improvement in the resolving power

---

* Sir George B. Airy (1801–1892), British astronomer and physicist.

[†] John William Strutt (1842–1919), third Baron Rayleigh, an English physicist and Nobel Prize winner for his discovery of the noble gas argon.

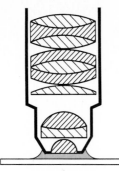

Figure 27-21 Microscope objective (schematic).

of a microscope using visible light would be comparable to trying to increase the efficiency of a heat engine beyond the Carnot value, or trying to move a body at a speed greater than that of light. It cannot be done; here the nature of light limits the sharpness of the image.

In principle, any form of wave motion can be used in a microscope. We are not limited to visible electromagnetic radiation (light). The *acoustic microscope* has been developed to the point where the resolving power is only slightly less than that of a microscope using light waves. Details down to about 1 μm (10,000 Å) have been resolved with acoustic waves of frequency 900 MHz. The specimen is in water, where the speed of compression waves is about 1500 m/s (Sec. 10-3), and the wavelength is about 1 μm:

$$\lambda = \frac{\text{wave velocity}}{\text{frequency}} = \frac{1500 \text{ m} \cdot \text{s}^{-1}}{900 \times 10^6 \text{ s}^{-1}}$$

$$= 1.7 \times 10^{-6} \text{ m}$$

One advantage of using acoustic waves is that the method is sensitive to *elastic* properties of the specimen rather than to *optical* properties such as index of refraction or opacity. Thus a whole new dimension of analysis is being opened up. Modern technology is being applied to the improvement of the acoustic microscope.

The *x-ray microscope* is being developed for certain uses. Since λ for x rays is orders of magnitude less than for visible light (by a factor of 10^3),

high resolving power is theoretically possible because λ/D can be made small. However, conventional lenses are impossible for x rays, and ingenious mirror systems are used.

The *electron microscope* gives us the highest resolving power of all instruments. See Sec. 29-6 for a discussion of the small value of λ associated with the electrons that are used in the electron microscopes described in Sec. 29-7.

The resolving power of the eye is limited by the wave nature of light. We assume green light of wavelength 555 nm (in air), for which the wavelength inside the eye is about 555 nm/1.33 = 416 nm. We also assume the diameter D of the pupil to be about 3 mm. For a point source the angular radius of the Airy disk is given by Eq. 27-2:

$$\theta = \frac{\lambda}{0.82D} = \frac{416 \times 10^{-9} \text{ m}}{0.82(3 \times 10^{-3} \text{ m})}$$

$$= 1.7 \times 10^{-4} \text{ rad} = 0.6'$$

The diameter of the Airy disk is, therefore, about 1.2'. Experiment shows that under favorable circumstances the eye can resolve two point sources separated by about 1', in good agreement with the calculation based on the diffraction of light.

The mammalian eye has evolved to the point where the size of the receptors (cones on the surface of the retina) is also consistent with the wave nature of light. At the fovea centralis (region of most distinct vision), the cones are about 2 μm in diameter. The angle subtended by a receptor, at a distance of about 2 cm from the lens, is given by

$$\theta = \frac{2 \times 10^{-6} \text{ m}}{2 \times 10^{-2} \text{ m}} = 10^{-4} \text{ rad} = 0.3'$$

The diameter of a receptor is thus about $\frac{1}{4}$ that of the Airy disk of a point source. No further improvement in visual acuity would result from a further decrease in the size of the receptors; in fact, smaller receptors would be less sensitive to light and less adapted to survival of the organism.

27-6 The Astronomical Telescope

The difference between an *astronomical telescope* and a microscope is one of degree only. In each case a real image is formed that, in turn, is viewed through an eyepiece. The microscope of Fig. 27-22a forms at A a real image that is larger than the object, whereas the real image in a telescope is smaller than the object (Fig. 27-22b). The moon is several thousand kilometers in diameter, but its real image may be only a few centimeters in diameter. However, the (smaller) real image has the great virtue that it is where we can get close to it; hence its *angular* size can be enormous. Usually a telescope is used to view a distant object (Fig. 27-22c), so the incident rays are practically parallel, and the real image A is at the focal point of the objective lens and also at the focal point of the eyepiece (when adjusted for minimum eyestrain).

To derive a formula for the magnifying power of a telescope, let us consider rays that come from one end P of the object, pass through the center of the lens, and form a real image P' in the focal plane of the objective lens whose focal length is F (Fig. 27-23a). The object is of height h, at some large distance p, and so its angular size as seen with the unaided eye is h/p. When

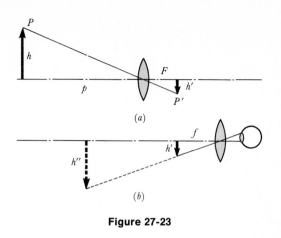

(a)

(b)

Figure 27-23

the image h' is viewed with the eyepiece of focal length f (Fig. 27-23b), the angular size of h'' remains the same as that of h' and is given by h'/f. Thus, for the telescope as a whole,

$$M = \frac{\text{angular size of image}}{\text{angular size of object without aid}} = \frac{\dfrac{h'}{f}}{\dfrac{h}{p}}$$

But, by similar triangles, $\dfrac{h}{p} = \dfrac{h'}{F}$, and hence

$$M = \frac{\dfrac{h'}{f}}{\dfrac{h'}{F}} = \left(\frac{h'}{f}\right)\left(\frac{F}{h'}\right) = \frac{F}{f} \qquad (27\text{-}3)$$

When adjusted for minimum eyestrain for viewing a distant object, the magnifying power of a telescope is the ratio of the focal lengths of the two lenses.

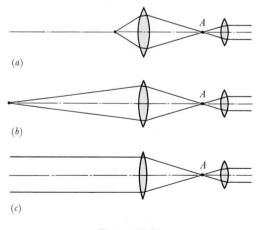

(a)

(b)

(c)

Figure 27-22

Example 27-3

A comet seeker's telescope has an objective lens of focal length 40 cm and an eyepiece of power 50 diopters. What is the magnifying power of the telescope?

First we find the two focal lengths, in the same units:

$$F = 40 \text{ cm} = 0.40 \text{ m}$$

$$1/f = 50 \text{ diopters} = 50 \text{ m}^{-1}$$

$$f = \tfrac{1}{50} \text{ m} = 0.02 \text{ m}$$

$$M = \frac{F}{f} = \frac{0.40 \text{ m}}{0.02 \text{ m}} = \boxed{20}$$

This is a low-power telescope, which is desirable when a large area of the sky is to be scanned.

A concave mirror can be used instead of an objective lens to form the real image, as in Fig. 27-24. All the largest telescopes, such as the 5.1-m (200-in.) Hale reflector at Mt. Palomar in California, are constructed in this way. For one thing, it would be impossible to obtain an optically uniform piece of glass to make a lens of this size—almost 17 ft in diameter! If supported at the edges, such a lens would sag under its own weight. A mirror need not be transparent, for only the front surface counts, and it is made reflecting by coating with a thin film of silver or aluminum. A large mirror is usually cast in Pyrex or quartz, to take advantage of the low expansion coefficients of these substances; the back is ribbed to provide rigidity. Figure 27-25 shows some of the ways in which the real image can be brought out to an eyepiece. For most astronomical work, a photographic plate or photocell receives the real image, and no eyepiece is used. In the very largest telescopes, the observer can be stationed at the focal point

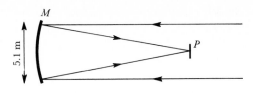

Figure 27-24   Reflecting telescope, with the curvature of the mirror exaggerated for clearness. P represents a photographic plate.

without obstructing an appreciable fraction of the incoming light (Fig. 27-25c).

The reflecting telescope has the great advantage that its objective is entirely free of chromatic aberration, since no light travels through glass. Spherical aberration is eliminated by using a parabolic rather than a spherical mirror; this works well for rays coming in along the axis, but introduces other aberrations for off-axis rays.

The objective diameters are made large for two reasons. First, a large diameter makes λ/D smaller and hence reduces the size of the central disk of the diffraction pattern. However, the "geometrical" image of almost all stars is much smaller than the diffraction disk, and hence unobservable. Just as for a microscope, the sharpness of an image is limited by diffraction. This sets a practical limit to the useful magnifying power of any telescope. Through careful design and workmanship, the theoretical limit can actually be reached for small telescopes, but the uneven idex of refraction of the air, caused by turbulence in the atmosphere, is a serious limi-

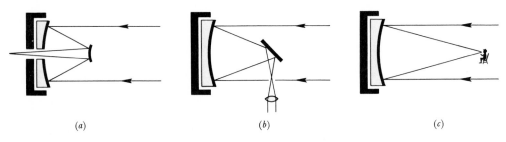

Figure 27-25   (a) Cassegrain focus.   (b) Newtonian focus.   (c) Prime focus.

tation for large instruments.* (The 5-m telescope at Mt. Palomar has never been used at greater than $\frac{1}{10}$ the theoretical resolving power, through no fault of the instrument itself.)

Telescopes of large diameter are used for a second reason: they gather more light and make a brighter image. The light-gathering power of a telescope is proportional to the surface area of the objective lens, given by the formula $A = (\pi/4)D^2$. The 5-m Mt. Palomar telescope of 1951 thus gathers 4 times as much light from a given source as the 2.5-m Mt. Wilson telescope of 1921. Since the brightness of a distant object is inversely proportional to the square of the distance, this means that the larger telescope can "see" twice as far—the fourfold increase due to the increased area of the mirror compensates for the fourfold decrease in brightness due to distance. In this way, the volume of space explored by telescopes using visible light, proportional to the cube of the distance, was increased eightfold in three decades.

Radio telescopes use e-m radiation of 10^{-3} m or more, instead of the e-m radiation of wavelength 5×10^{-7} m that we call visible light. Two dishes of diameter 27 m are shown on page 634; in each, incoming e-m radiation is focused on a small receiving antenna at the focal point of the parabolic mirror. A radio telescope must be turned to compensate for the earth's rotation if a fixed "radio star" is to be tracked. It is found that strong radiations are received from certain points in the sky, very seldom associated with bright visible stars. For radio telescopes, as for optical telescopes, resolving power depends on the ratio (λ/D) of the wavelength to the diameter of the mirror. For e-m radiation of wavelength 0.05 m, a single dish of diameter 27 m has $\lambda/D = 0.002$, which is far larger than the corresponding ratio for a small optical telescope or

---

* Only in 1974 was the observation of the surface of a large star (Betelgeuse) made possible by computer-assisted analysis of the image, which was greatly blurred by atmospheric turbulence.

even the human eye. In fact, the Airy disk for a point source having wavelength 0.10 m would appear as large as the full moon, and two such sources 0.25° apart would barely be resolved. A dramatic improvement in resolving power has been made possible by interferometry. By using a long baseline (which may even be of intercontinental dimensions) between two or more dishes, the effective aperture can be increased to such an extent that the resolving power for radio sources exceeds that for the largest optical telescopes. (See Refs. 14–16 at the end of the chapter.) Radio astronomy gives information that optical astronomy cannot give, because e-m radiation having wavelength 0.5 m (for example) easily penetrates distant dust clouds that would completely absorb visible e-m radiation. The energy-gathering power of large radio-telescope installations gives astronomers their farthest view of distant space.

27-7 The Galilean Telescope

The astronomical telescope suffers from two limitations: it gives an inverted image, and it is rather long and unwieldy. These are not serious faults for astronomical work, but for other uses the original form used by Galileo is often preferable. In the *Galilean telescope*, an objective lens gathers light (Fig. 27-26a), but the converging rays are not allowed to reach their focus. Instead, a strong diverging lens changes the curvature to -4 m^{-1}, and a virtual image is formed at B, 25 cm from the eye. Such a telescope is compact and gives an erect image, but it has a rather small field of view. When it is adjusted for minimum eyestrain, the rays emerge parallel to each other, as in Fig. 27-26b, and the virtual image is at infinity. The Galilean telescope is used for low-power instruments such as opera glasses.

In all telescopes, the magnifying power is greater for an objective of long focal length. This leads to the design of prism binoculars, where

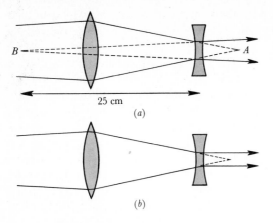

(a)

25 cm

(b)

Figure 27-26 Galilean telescope.

totally reflecting prisms are used instead of mirrors to give a longer path and allow the objective lens to have longer focal length (Fig. 27-27). The main function of the prisms is, however, to reinvert the image by multiple reflection. Even an inexpensive pair of binoculars makes use of achromatic lens combinations for objective and

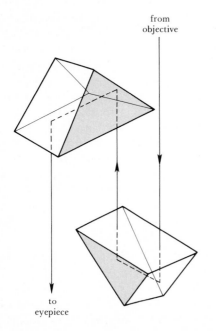

from objective

to eyepiece

Figure 27-27 Prism combination used in binoculars.

eyepiece. In the "spyglass," the inverted image is made erect by means of an additional lens, which requires a tube of unwieldy length.

27-8 The Prism Spectroscope

A spectroscope is used to analyze a beam of light by breaking it up into its component colors. Newton experimented with prisms while a student at Cambridge University. He found that white light can be broken up into the colors of the rainbow, and furthermore that any given spectrum color, red for instance, is "pure" and cannot be further modified by a second prism. Light of the various spectral colors, when recombined, gives white light. Newton did not follow up his important discovery, which might well have led to the spectroscope decades in advance of the work of others.

In Chap. 25 we gave Snell's law and stated that the index of refraction of glass is greater for violet light than for red. Thus we can "explain" the action of a prism in separating the rays of different colors. This is no explanation at all, of course, but the full theory of dispersion is too advanced for inclusion in this text. Accepting the *fact* of *dispersion* (that is, the dependence of n on λ), we now proceed to describe the prism spectroscope, which is illustrated in Fig. 27-28.

A slit S is illuminated by the source, and a *collimator* lens C makes the light parallel. This means that S is at the focal point of C. In the better instruments, C is an achromatic combination, so that the focal length does not depend on color. The parallel rays strike the prism, are bent variously, and emerge, still in parallel bundles. An objective lens L_1 forms real images at A, and the eye views these real images of the slit, one in each color, through an eyepiece L_2. The combination L_1 and L_2 is, of course, a telescope, and the observer uses the telescope to view the slit S, which has, in effect, been removed to infinity by the action of the collimator lens. For photographic work, the eyepiece is not used, and the film or plate is placed at A.

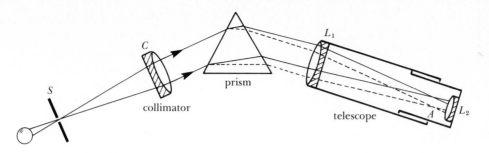

Figure 27-28 Prism spectroscope. Solid rays, red light; dashed rays, violet light.

We often speak of spectrum "lines." Thus, the strong green line of the mercury spectrum has a wavelength of 546 nm. The images are lines only because the slit is tall and narrow. A thin curved slit would give curved real images at A, one in each color (Fig. 27-29). For ease of measurement, a thin straight slit is preferable. If the slit is made wider, *each* of the lines in Fig. 27-29*a* becomes wider, since they are all images of the same slit.

One criterion of a good spectroscope is that it be able to show as separate lines two colors that have almost the same wavelength. For a spectro-

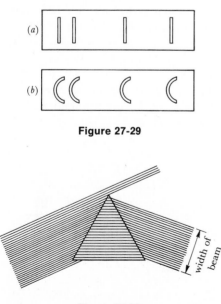

Figure 27-29

Figure 27-30

scope, *resolving power* is measured by the thinness of the line image of a very narrow slit. Just as for a microscope or telescope, the diffraction of light sets the limit. The size of the prism is usually the controlling factor (Fig. 27-30), since the prism limits the width of the beam in the same way that a single slit would. Two adjacent lines are resolved if the width of each line due to diffraction is less than their separation due to the difference between the two indexes of refraction for the two neighboring colors.

In use, a spectroscope is calibrated by light of known wavelengths. The photo on page 712 shows the sun's spectrum (center), flanked on each side by lines of the iron spectrum that are emitted by an incandescent electric arc between two iron electrodes. In this photograph, the wavelengths are in the visible region. To study the ultraviolet region, for which glass is opaque, quartz prisms can be used down to about 185 nm, fluorite to 125 nm, and LiF to 110 nm. At the other extreme, glass becomes opaque to infrared radiation at about 2200 nm (2.2 μm). Workers studying the far infrared region of the spectrum use prisms of rock salt, sylvite (KCl), and other substances that are transparent at the wavelength used.

27-9 The Grating Spectroscope

We saw in Sec. 26-4 how a grating forms a series of bright images for any given wavelength. The angle of deviation for the nth-order interference

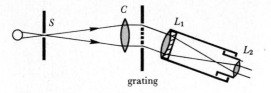

Figure 27-31 Grating spectroscope.

We can tabulate the results as follows:

Line	sin θ	θ
violet, 1st order	0.16	9°
red, 1st order	0.30	17°
violet, 3rd order	0.48	29°
red, 3rd order	0.90	64°

The angular separation between red and violet in the 3rd order is $64° - 29° = 35°$

maximum depends on the wavelength, according to the equation $n\lambda = a \sin \theta$. This means, of course, that a grating can replace a prism in a spectroscope. A simple grating spectroscope is shown in Fig. 27-31; the slit S, collimator C, and telescope $L_1 L_2$ are the same as for a prism spectroscope. Since $\sin \theta = n\lambda/a$, in any given order violet light (having the shortest λ) is deviated least, and red light is deviated most. Thus the grating spectroscope presents the colors in a sequence opposite to that of a prism spectroscope. The grating spectroscope furnishes several complete spectra, each line occurring in several orders.* The separation between two colors can be considerable, as shown in our next example.

If we compare the grating spectroscope and the prism spectroscope, we find that each has its advantages and disadvantages. As seen from the preceding example, the dispersion (ability to spread out the colors) of a grating spectroscope can be large, especially in the higher orders. The dispersion of a prism spectroscope depends on the rather small variation of index of refraction, and a spread of 5° between violet and red lines is considered excellent. The dispersion of a grating depends on λ and the grating space a, which can be made as small as desired. The resolving power of a grating can also be made much higher than that of a prism. As shown in Sec. 26-4, the sharpness of the line depends on the total number of rulings and on the order. In the 2nd order, a well-constructed grating 10 cm long having 5000 slits per centimeter (a total of 50,000 slits) can separate two colors that differ in wavelength by only one part in 100,000! Still another advantage of the grating is that the lens can be dispensed with entirely, in order to study the extreme infrared or ultraviolet spectra, which would be absorbed by a glass or quartz lens. The spectrograph of Fig. 27-32 has a grating ruled on

---

Example 27-4

A grating of 4000 slits per centimeter is used to form a spectrum of a source that consists of two lines, a violet one of wavelength $\lambda_1 = 400$ nm, and a red one of wavelength $\lambda_2 = 750$ nm. Calculate $\sin \theta$ for the 1st- and 3rd-order images of each line. What is the angular separation between red and violet in the 3rd order?

The grating space is $\frac{1}{4000}$ cm $= 2.5 \times 10^{-4}$ cm $= 2500$ nm. We compute $\sin \theta$ for each line, using $\sin \theta = n\lambda/a$. For instance, for the red line in the 3rd order:

$$\sin \theta = \frac{3(750 \text{ nm})}{2500 \text{ nm}} = \frac{2250 \text{ nm}}{2500 \text{ nm}} = 0.900$$

$$\theta = 64°$$

---

* An interesting phenomenon known as overlapping of orders is considered in Prob. 27-B24.

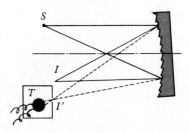

Figure 27-32 Infrared spectrograph.

a concave mirror. Without the rulings the mirror would form a real image of the source at I. The 1st- or 2nd-order interference maximum I' is focused on a blackened thermocouple T, which serves to indicate the presence of infrared e-m radiation.

On the other hand, the grating spectroscope is wasteful of light, since usually much of the available energy goes into the central (undeviated) image.* For this reason, prisms are used when weak sources must be examined, or when the very highest resolving power or dispersion is not needed. Figure 27-33 shows the spectra of several stars, formed in one exposure on a plate placed at the focal point of an astronomical telescope. In (a) each line is the image of a star; the telescope was purposely driven slightly too slowly, so that the star images were drawn out into lines by the rotation of the earth. For (b), a prism was placed in front of the objective lens and properly oriented, with the result that each line image was spread out into a tiny spectrum. In astronomy, grating spectra are used for studying strong sources: the sun and the brighter stars.

---

* The grooves of modern reflection gratings are often shaped to reflect a large fraction of the light in the approximate direction of the desired interference maximum.

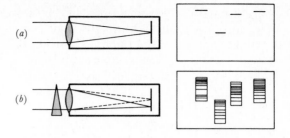

Figure 27-33 The spectra of many stars can be obtained with one exposure by placing a prism in front of the objective lens of a telescope.

In this chapter we have not attempted to describe all of the many optical instruments in use. The six or seven instruments we have studied are interesting and useful. More important for our study of physics, we have been able to use wave optics (and the wave-front interpretation of geometrical optics) as adequate models for predicting the performance of these common optical instruments. On the basis of our studies so far, the wave theory of light has carried the day. We shall see in Chap. 29, however, that for a complete interpretation of optical phenomena, the photon (corpuscular) model of light must also be used.

Summary The camera and the human eye each form a real image on a sensitive surface. They differ in their means of achieving accommodation; in the camera the lens is moved, whereas in the eye the power of the lens is changed. An ideal lens gives a sharp, undistorted image of the object, and any departure from this condition is called an aberration. Spherical aberration can be corrected by using a small lens opening. Chromatic aberration can be corrected by using combinations of lenses made of different kinds of glass. Astigmatism causes a line image of a point source. In the eye, such a defect is usually due to a cylindrical component of the curvature of the cornea and lens. A spherical camera lens suffers from astigmatism of a different sort, for rays that are not along the axis of the lens. Approximate correction of astigmatism is possible, using combinations of lenses that are not necessarily spherical.

The normal eye has a power that adjusts from about $+60$ diopters to $+64$ diopters, and the near point is about $\frac{1}{4}$ meter, or 25 cm, from the eye. Negative spectacle lenses are used to correct for myopia, and positive lenses to correct for hypermetropia and presbyopia. Cylindrical lenses are used to correct for astigmatism.

The simple magnifier forms a virtual image whose distance from the eye is at least 25 cm and may be as much as infinity. The angular magnification, or magnifying power, lies between $1 + 25/f$ and $25/f$, the latter value being for minimum eyestrain when the virtual image is at infinity.

The compound microscope and the telescope are similar in construction; the objective forms a real image of the object, and this image is received on a photographic film or is viewed with a magnifier. In the microscope the real image is larger than the object, whereas in the telescope the real image is smaller than the object. When a telescope is adjusted for viewing a distant object with minimum eyestrain, its magnifying power is the ratio of the focal lengths of the two lenses.

The resolving power of a microscope or telescope is its ability to show sharp detail; this is not the same as magnifying power, which is the ability to form a large image on the retina of the eye. Resolving power is limited by the diameter of the objective lens, which acts in much the same manner as a single slit does to form a blurred image of a point object. The essential factor that determines the resolving power is λ/D, the ratio of wavelength to lens diameter. In order to obtain the full theoretical resolving power, the lens aberrations must, of course, be negligibly small. Microscopes can use acoustic, x-ray, and electron waves to form images.

Most large astronomical telescopes use a concave mirror to form the real image directly on a photographic plate, photocell, thermocouple, radio antenna, or other detector of e-m radiation. The intensity of the image is proportional to the area of the objective mirror or lens.

A spectroscope usually consists of a slit, collimator, prism or grating, and telescope. Except for a lack of sensitivity, the grating spectroscope is superior to the prism spectroscope, for it has greater resolving power and dispersion, and it is adaptable to radiations such as infrared or ultraviolet for which absorption in prisms or lenses would be excessive.

Check List

aberration
spherical aberration
off-axis astigmatism
chromatic aberration
f-number of a lens
accommodation
near point
myopia
hypermetropia

presbyopia
angular magnification
$M = 1 + 25/f$
$M = 25/f$
minimum eyestrain
compound microscope
objective lens
eyepiece

resolving power of a lens
$M = F/f$
Galilean telescope
collimator
dispersion of
  a spectroscope
resolving power of
  a spectroscope

Questions

27-1   Describe three aberrations of a simple spherical lens. Does the lens of the human eye have these aberrations?

27-2   Why does a camera give a sharper image on the film if the lens is "stopped down" to have a smaller diameter?

27-3 Hypermetropia and presbyopia are different conditions, each corrected by a positive spectacle lens. Explain.

27-4 If you shut your eyes for a while in a dark room and then press gently (through the eyelid) on the extreme left part of the left eyeball, you "see" flashes of light at the *right* side of the field of view. Explain.

27-5 Why does a young person who wears glasses usually not require bifocals?

27-6 For any given person, does the magnifying power of a simple magnifier depend very much on where it is held?

27-7 Compare and contrast the compound microscope and the astronomical telescope.

27-8 Why cannot a microscope, using visible light, be constructed that will give a clear view of a sugar molecule?

27-9 Why is a radio telescope useful for research, in spite of the fact that the resolving power of even a large one is less than that of an optical telescope?

27-10 In what ways is a telescope using an objective mirror superior to one using an objective lens? What are the disadvantages, if any, of the reflecting telescope?

27-11 What are two advantages of the Galilean form of telescope?

27-12 Why are spectrum colors usually referred to as "lines"?

27-13 Explain the difference between the dispersion and the resolving power of a spectroscope.

27-14 What are some of the advantages and disadvantages of a grating spectroscope as compared with a prism spectroscope?

MULTIPLE CHOICE

27-15 An achromatic doublet is a combination of two lenses   (*a*) that are made of different kinds of glass;   (*b*) one of which is positive and one negative;   (*c*) both of these.

27-16 Relative to the lens of the eye, the power of the cornea is   (*a*) greater;   (*b*) about the same;   (*c*) less.

27-17 Astigmatism in the human eye can be corrected by a spectacle lens that is   (*a*) negative;   (*b*) cylindrical;   (*c*) spherical.

27-18 The resolving power of the human eye depends primarily on   (*a*) the focal length of the eye lens;   (*b*) the diameter of the pupil;   (*c*) the ratio of focal length to diameter of the pupil.

27-19 A large radio telescope is superior to a large visible-light telescope in   (*a*) ability to detect a weak source;   (*b*) resolving power;   (*c*) both of these.

27-20 A reflecting telescope is free of   (*a*) chromatic aberration;   (*b*) off-axis astigmatism;   (*c*) spherical aberration for an off-axis object.

Problems   **27-A1**   Which has the greater angular size: a football player 2 m tall, viewed from a stadium seat 100 m away, or the 5-cm image of the player, viewed by a spectator 3 m from his TV screen?

27-A2   Which has larger angular size for a viewer on earth: a satellite 16 m in diameter, 1200 km above the earth, or the planet Saturn, which is 112,000 km in diameter and averages 1.425×10^9 km from the earth?

27-A3   A camera lens in a television studio has a diameter of 5 cm and focal length of 12 cm. What is the *f*-number of this lens? Is it a "fast" lens?

27-A4   What is the focal length of an *f*/5.6 lens whose diameter is 4 cm?

27-A5 In 1675 the Dutch biologist Antonie van Leeuwenhoek, using a single lens probably of focal length 1.25 mm, discovered bacteria. What was the magnifying power of this early simple microscope?

27-A6 What is the magnifying power of a lens of focal length $+3.0$ cm when used by a person with normal vision?

27-A7 A microscope's objective lens forms a real image that is magnified 20 times; the eyepiece has an angular magnification 8. What is the overall magnification of the microscope?

27-A8 What is the angular magnification of an astronomical telescope that has an objective of focal length 2.10 m and an eyepiece of focal length 3 cm?

27-A9 An astronomical telescope has an objective lens of focal length 60 cm. What magnifying powers can be obtained if two eye lenses, with focal lengths 2 cm and 0.75 cm, are available?

27-A10 In James Thurber's *Many Moons* (Harcourt Brace Jovanovich, 1943), a little princess who wanted the moon settled for a gold moon the size of her thumbnail that covered the real one. Determine the angular size of your thumbnail, held at arm's length. Is it larger or smaller than that of the moon? (See Sec. 27-3.)

27-B1 A successful picture is taken using an exposure of $\frac{1}{200}$ s and a camera lens of speed $f/8$. The photographer wishes to repeat the shot, with the lens stopped down to $f/16$. What should the exposure time be?

27-B2 What is the approximate f-number of a human eye if the open diameter of the iris is 4 mm and the eye, focused on a distant object, has a power of $+60$ diopters?

27-B3 What is the distance from lens to retina in the myopic eye illustrated in Fig. 27-12a?

27-B4 A certain young girl can adjust the power of her eye's lens-cornea combination between limits of $+56$ diopters and $+66$ diopters. With the lens relaxed, she can see a distant star clearly. (*a*) How far is this girl's near point from her eye? (*b*) How far is her retina from her eye lens?

27-B5 A man can see clearly only objects that lie between 40 cm and 125 cm from his eye. Calculate the prescription of bifocal glasses that will enable him to see distant objects (through the top half) and also read a book at a distance of 20 cm (through the lower half). Express the powers of the lenses in diopters.

27-B6 A myopic student wears glasses of power -9 diopters to see distant objects, and the distance from eye lens to retina is 1.82 cm. (*a*) What is the power of the student's eye, when relaxed? (*b*) If the student can give an extra $+5$ diopters to her eye lens by muscular effort, where is the near point without glasses?

27-B7 A retired bank president can easily read the fine print of the financial page when the newspaper is held at arm's length (50 cm) from his eye. What should be the focal length of a spectacle lens that will allow him to read at a comfortable distance of 30 cm?

27-B8 The near point for a certain child is 8 cm; the far point (with eye lens relaxed) is 100 cm. The eye lens is 2.00 cm from the retina. (*a*) Between what limits, measured in diopters, does the power of this child's lens-cornea combination vary? (*b*) Calculate the power of the spectacle lens this child should use for relaxed distant vision. Is the lens converging or diverging?

27-B9 What is the magnifying power of a magnifying glass of strength 32 diopters when used by a person having normal vision? The lens is adjusted for minimum eyestrain.

27-B10 What is the magnification of a magnifier of focal length $+5$ cm if the lens is adjusted to give a virtual image 100 cm from the lens?

27-B11 Carry through an analysis, similar to that in Example 27-1, for a person whose near point is at 40 cm instead of 25 cm, using the same magnifier of focal length $+2$ cm. Calculate the angular magnification when the lens is adjusted for maximum magnification.

27-B12 What is the power, in diopters, of a magnifying glass that has a magnifying power of 6 when used by a person having normal vision? The lens position is adjusted to give maximum magnifying power.

27-B13 A slide projector's lens is $f/2.0$ and is 4 cm in diameter. What would be the maximum magnifying power of this lens if used as a magnifying glass by a person with normal vision?

27-B14 What is the magnifying power of a concave makeup mirror of radius of curvature 100 cm, used by a model whose face is 30 cm from the mirror? (*Hint:* For the angular size without aid, assume the model views his image in a plane mirror that is 30 cm from his face.)

27-B15 The objective lens of a microscope of focal length 2.9 mm is placed 3.0 mm from a specimen. The image is viewed with an eyepiece of focal length 25 mm, adjusted for minimum eyestrain. (*a*) How far apart are the lenses? (*b*) What is the overall magnification of the instrument?

27-B16 The objective lens of a device has focal length $+2$ cm, and the eyepiece has focal length $+3$ cm. When adjusted for minimum eyestrain, the lenses are 15 cm apart. (*a*) Is the device a microscope or a telescope? (*b*) What is the object distance? (*c*) What is the magnifying power?

27-B17 An astronomical telescope has an objective lens of power 0.80 diopters, and the eyepiece is 127.5 cm from the objective when adjusted for minimum eyestrain. (*a*) Calculate the magnifying power of this telescope. (*b*) How close would a person a kilometer away seem to be when viewed through the telescope?

27-B18 The objective lens of a device has focal length $+20$ cm, and the eyepiece has focal length $+2$ cm. When adjusted for minimum eyestrain, the lenses are 23 cm apart. (*a*) Is the device a microscope or a telescope? (*b*) What is the object distance? (*c*) What is the magnifying power?

27-B19 A pocket microscope used for viewing the stylus of a phonograph cartridge has an objective of focal length $+0.5$ cm and an eyepiece of focal length $+2$ cm. When adjusted for minimum eyestrain, the lenses are 6 cm apart. Calculate (*a*) the objective distance for the objective lens, and (*b*) the overall magnification of the microscope.

27-B20 An amateur astronomer makes a reflecting telescope whose mirror has a radius of curvature 1.80 m and whose eyepiece lens has focal length 1.0 cm. (*a*) What is the approximate overall length of the telescope when adjusted for minimum eyestrain? (*b*) What is the magnifying power? (*c*) How close would the moon seem when viewed through the telescope if the actual distance is 384,000 km?

27-B21 An elderly sailor is shipwrecked on a desert island but manages to save his spectacles. The lens for one eye has a power of $+1.25$ diopters, and the other lens has a power of $+8$ diopters. (*a*) What is the magnifying power of the telescope he can construct with these lenses? (*b*) How far apart are the lenses when the telescope is adjusted for minimum eyestrain?

27-B22 In the prism binoculars shown in Fig. 27-27, the optical path from objective to eyepiece is 40 cm, but the length of the instrument is much less than this because of the use of prisms to fold the path while also reinverting the image. (*a*) The eyepiece has focal length 5 cm; what is the magnifying power when adjusted for minimum eyestrain? (*b*) Design a terrestrial (noninverting) straight-tube telescope using the same lenses as the binoculars, with an additional "erecting" lens of focal length +9 cm inserted in the tube beyond the focal point of the objective at such a place that it forms a real image (with $m = -1$) at the focal point of the eyepiece. What is the overall length of such a telescope?

27-B23 A spotting scope in the form of a Galilean telescope has two lenses 15 cm apart when adjusted for viewing a distant object with minimum eyestrain. If the eyepiece has a power of -25 diopters, what is the focal length of the objective?

27-B24 A source emits three lines: a violet line of wavelength 400 nm, a green line of wavelength 550 nm, and a red line of wavelength 700 nm. Images of these lines are formed by a grating whose grating space is 2150 nm. Calculate $\sin \theta$ for all observable orders of each line, and arrange the images (11 in all) in sequence according to their $\sin \theta$ values, thus: $V_1, G_1, R_1, V_2, \ldots$. The resulting sequence illustrates a phenomenon known as the "overlapping of orders" in a grating spectrum of visible light.

27-B25 A grating having 5000 lines per centimeter forms a spectrum of a source of white light. Calculate the angular separation of the yellow and blue regions of the spectrum, in the 3rd order. (Use $\lambda = 580$ nm for yellow light and $\lambda = 470$ nm for blue light.)

27-C1 A compound microscope has an objective lens of focal length 1.20 cm, and an eyepiece of focal length 1.40 cm. The object is a bug 0.2 mm in diameter on a slide that is 1.30 cm from the objective lens. (*a*) How far apart are the lenses when the object is viewed with the lenses adjusted for minimum eyestrain? (*b*) What is the angular size of the magnified bug? (*c*) For classroom demonstration, a real image of the bug is formed on a screen 100 cm from the eyepiece. How far from its original position must the eyepiece be moved, and in which direction? (*d*) What is the linear size of the image of the bug on the screen?

27-C2 A laboratory (astronomical) telescope is used to view a scale that is 250 cm from the objective lens. The objective has focal length 20 cm, and the eyepiece has focal length 2 cm. Calculate the angular magnification when the telescope is adjusted for minimum eyestrain. (*Hint:* The object is not at infinity, and so the simple expression F/f is not quite accurate for this problem.)

27-C3 The lenses of an astronomical telescope are 122 cm apart when adjusted for viewing a distant object with minimum eyestrain. The angular magnification is 60. Compute the focal length of each lens.

27-C4 An opera glass (Galilean telescope) is used to view a singer's head that is 25 cm tall, 30 m from the objective lens. The focal length of the objective is +8.00 cm and that of the eyepiece is -2.00 cm. When the telescope is adjusted so that parallel rays enter the eye, compute (*a*) the size of the real image that would have been formed by the objective; (*b*) the virtual object distance for the negative lens; (*c*) the distance between the lenses; (*d*) the approximate overall angular magnification.

27-C5 An achromatic lens (Fig. 27-34) has a plano-convex crown glass lens of radius of curvature R_C in contact with a plano-concave flint glass lens of radius of curvature R_F. The indexes of refraction for red light (656 nm) and blue light (486 nm) are as follows: crown glass, $n_R = 1.5145$, $n_B = 1.5240$; flint glass, $n_R = 1.6221$, $n_B = 1.6391$. (*a*) Use the lens maker's formula and Prob. 25-C14 to show that the power $1/f$ of the

Figure 27-34

combination is $(n_C - 1)/R_C - (n_F - 1)/R_F$. (b) Show that the power is independent of color, for the two colors involved, if $\Delta n_C/R_C = \Delta n_F/R_F$, where $\Delta n_C = n_B - n_R$ for crown glass and $\Delta n_F = n_B - n_R$ for flint glass. (c) If the crown lens has $R_C = 15.00$ cm, calculate the radius of curvature R_F. (d) Calculate $1/f$ for the combination for each of the two colors, and verify that the lens is achromatic.

27-C6 A grating 2 cm long has 6000 slits per centimeter. When it is used in the 3rd order, could two lines of wavelengths 500.000 nm and 500.010 nm be separated?

27-C7 Could a person see as separate the two headlights of a compact car that is 8 km away if the lamps are 1.12 m apart? Assume yellow light for which $\lambda = 580$ nm (in air), and assume the pupil of the eye to be 4 mm in diameter. (*Hint:* Calculate the angular separation of the lamps, and compare with the angular radius of the Airy disk on the retina.)

27-C8 What must be the diameter of a radar dish antenna at an airport in order for it to distinguish between two planes 300 m apart at a distance of 15 km (a) if the wavelength is 10.2 cm? (b) if the wavelength is 3.0 cm?

References

1. G. Wald, "Eye and Camera," *Sci. American* **183**(2), 32 (Aug. 1950).
2. F. R. Hirsh, Jr., and E. M. Thorndike, "On the Pinhead Shadow Inversion Phenomenon," *Am. J. Phys.* **12**, 164 (1944).
3. E. G. Boring, *Sensation and Perception in the History of Experimental Psychology* (Appleton Century, 1942), p. 237. A description of the experiment of G. M. Stratton on the psychological adjustment of a person wearing inverting glasses.
4. T. Erismann and I. Kohler, *Upright Vision Through Inverting Spectacles* (film). The Pennsylvania State University, Audio Visual Aids Library, University Park, Pa.
5. C. H. Graham, ed., *Vision and Visual Perception* (Wiley, New York, 1965). See Chap. 1, "Light as a Stimulus for Vision," by L. A. Riggs; Chap. 2, "The Structure of the Visual System," by J. L. Brown; Chap. 11, "Visual Acuity," by L. A. Riggs.
6. S. F. Jacobs and A. B. Stewart, "Chromatic Aberration in the Eye," *Am. J. Phys.* **20**, 247 (1952).
7. H. W. Leibowitz and D. A. Owens, "Anomalous Myopia and the Intermediate Dark Focus of Accommodation," *Science* **189**, 646 (1975). The relaxed eye is not focused at infinity.
8. L. M. Herman, M. F. Peacock, M. P. Yunker, and C. J. Madsen, "Bottlenosed Dolphin: Double-Slit Pupil Yields Equivalent Aerial and Underwater Diurnal Acuity," *Science* **189**, 650 (1975). Surprisingly, although the dolphin's lens system is very powerful to allow underwater viewing, the dolphin is not myopic when viewing in air.
9. D. Davenport and J. M. Foley, "Fringe Benefits of Cataract Surgery," *Science* **204**, 454 (1979). When used in conjunction with a lensless eye, a simple converging lens can produce sharp retinal images with a wide range of magnification. Also, more light is transmitted than by the normal eye, especially in the blue, violet, and near-ultraviolet regions.
10. W. H. Price, "The Photographic Lens," *Sci. American* **235**(2) 72 (Aug. 1976).

11. E. N. daC. Andrade, "Robert Hooke," *Sci. American* **191**(6), 94 (Dec. 1954). Includes a description of Hooke's early microscopic investigations.
12. S. G. G. MacDonald and D. M. Burns, *Physics for the Life and Health Sciences* (Addison-Wesley, Reading, Mass., 1975), pp. 652–665. A discussion of numerical aperture and various types of optical microscopes.
13. J. R. Waaland, "Fraunhofer and the Great Dorpat Refractor," *Am. J. Phys.* **35,** 344 (1967). The first modern, achromatic, refracting telescope, of aperture 24 cm.
14. J. D. Kraus, "Radio Telescopes," *Sci. American* **192**(3), 36 (Mar. 1955).
15. K. I. Kellerman, "Intercontinental Radio Astronomy," *Sci. American* **226**(2), 72 (Feb. 1972). High resolving power is obtained by using interferometers consisting of widely separated dishes.
16. G. W. Swenson, Jr., and K. I. Kellerman, "An Intercontinental Array—A Next-Generation Radio Telescope," *Science* **188,** 1263 (1975).
17. D. J. Lovell, "Principles of Colorimetry," *Am. J. Phys.* **18,** 104 (1950).
18. P. Connes, "How Light Is Analyzed," *Sci. American* **219**(3), 72 (Sept. 1968).
19. F. Miller, Jr., *Resolving Power* (film).

28

Relativity

In the early years of the 20th century—some years after the wave theory of light had been successfully applied to the study of many aspects of radiation—an apparently contradictory corpuscular nature of light became evident. We shall study the corpuscular nature of both particles and light in Chaps. 29–31. First, however, we shall briefly study the theory of relativity, which pervades all of the physics of this century. We shall need some of the results of Einsteinian relativity in our later work.

The study of relative motion has a long history. We have already seen (pages 115–116) how Galileo and Newton made use of inertial frames of reference—frames in which Newton's laws of motion are valid. We also saw how inertial forces arise in a noninertial frame—one that is accelerated relative to the fixed stars or any other inertial frame. In the late 19th century there seemed to be no reason to challenge these ideas, but early in the 20th century Albert Einstein modified and extended the theory of relativity in a basic and dramatic fashion. Before we consider Einstein's theory of

relativity, we shall take a further look at the Galilean relativity that had been accepted by physicists for so long.

28-1 Galilean Relativity

In his *Dialogs Concerning Two New Sciences* (1638), Galileo describes a "thought experiment" in which he imagines that a sailor drops an object from the top of a mast of a moving ship. The ship is moving horizontally with constant velocity; where does the object strike the deck? Let us study Galileo's example using our terminology about frames of reference. The ship and the earth are both inertial frames (ignoring the small Coriolis acceleration due to the earth's rotation). The two frames have a *constant* relative velocity. In each frame of reference the laws of motion are perfectly normal. Our Problem 2-B22 (page 49) is just such a problem. A knife fell 19.6 m (which required 2 s), and during this time the ship (and knife) moved 16 m eastward (Fig. 28-1). Thus the knife hit the deck at the foot of the mast. In the

Joseph Hafele and Richard Keating with two atomic clocks prior to taking off on a round-the-world flight to test the time dilation predicted by the Einstein theory of relativity. The theory was verified, within experimental error.

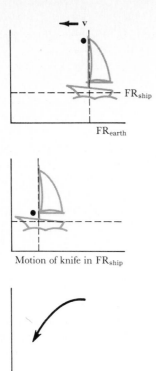

Motion of knife in FR$_{\text{ship}}$

Motion of knife in FR$_{\text{earth}}$

Figure 28-1

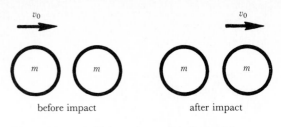

before impact   after impact

Figure 28-2 Collision of equal balls in laboratory frame FR$_{\text{lab}}$.

remain constant. These conclusions are true in the laboratory frame of reference FR$_{\text{lab}}$ (Fig. 28-2):

Momentum:

$$mv_0 + 0 \overset{?}{=} 0 + mv_0$$
$$mv_0 = mv_0$$

Kinetic energy:

$$\tfrac{1}{2}mv_0{}^2 + 0 \overset{?}{=} 0 + \tfrac{1}{2}mv_0{}^2$$
$$\tfrac{1}{2}mv_0{}^2 = \tfrac{1}{2}mv_0{}^2$$

Now let us look at the same event from a different frame of reference. The center of mass of the system is always midway between the two equal balls; thus the c.m. frame is moving to the right with velocity $\tfrac{1}{2}v_0$ relative to the laboratory frame. Figure 28-3 shows what the collision looks like in the new frame of reference (FR$_{\text{c.m.}}$). Note that the relative velocity of approach is v_0 in either frame, and the relative velocity of separation is also v_0 in either frame. Now we see if the

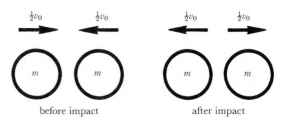

before impact   after impact

Figure 28-3 Collision of equal balls in center of mass frame FR$_{\text{c.m.}}$.

ship's frame of reference (denoted by FR$_{\text{ship}}$), the knife was released from rest and it fell straight down. In the earth's frame of reference, FR$_{\text{earth}}$, the knife was projected horizontally at 8 m/s by the ship's motion, and it followed the usual parabolic path downward and to the east. Although the appearance of the trajectory differed in the two frames, the same laws of motion—Newton's laws—were applicable in both inertial frames of reference.

Newton clearly stated that the *form* of any law of mechanics is not changed by the relative *uniform* velocity of an observer. Although a general proof is not difficult, we will be content with a simple illustration. In Sec. 6-9 we saw that during a *perfectly elastic* collision between a moving ball and an identical stationary ball, the balls simply exchange velocities; both the momentum and the kinetic energy of the system momentum and the KE of the system have changed during the collision.

Momentum:

$$m(\tfrac{1}{2}v_0) + m(-\tfrac{1}{2}v_0) \overset{?}{=} m(-\tfrac{1}{2}v_0) + m(\tfrac{1}{2}v_0)$$
$$0 = 0$$

Kinetic energy:

$$\tfrac{1}{2}m(\tfrac{1}{2}v_0)^2 + \tfrac{1}{2}m(-\tfrac{1}{2}v_0)^2 \overset{?}{=} \tfrac{1}{2}m(-\tfrac{1}{2}v_0)^2 + \tfrac{1}{2}m(\tfrac{1}{2}v_0)^2$$
$$\tfrac{1}{4}mv_0{}^2 = \tfrac{1}{4}mv_0{}^2$$

In the two frames of reference the numerical values of the total momentum are different (mv_0 in FR_{lab} and 0 in $FR_{c.m.}$). Likewise the numerical values of the total KE are different ($\tfrac{1}{2}mv_0{}^2$ in FR_{lab}, $\tfrac{1}{4}mv_0{}^2$ in $FR_{c.m.}$). Nevertheless, the *forms* of the laws are unaltered: during a perfectly elastic collision in any inertial frame, the total momentum is constant and the total KE is constant.

Galilean relativity, then, states that the forms of the laws of mechanics are the same in all inertial systems. This means that all inertial frames are equivalent. You would be unable to perform any experiment that would single out any one inertial frame as being at rest. For example, it is perfectly possible to play Ping-Pong on a jet plane, an inertial frame moving at 960 km/h (600 mi/h) relative to another inertial frame on the earth. You would not have to adjust your muscular reflexes to take account of the motion. The laws of mechanics would be the same as on the ground, and you would, in fact, be unaware of the plane's uniform motion. It would be impossible to tell which system was "actually" in motion: the plane moving forward at velocity v, with the earth stationary, or the ground moving backward at velocity v, with the plane stationary. However, the plane could easily become a noninertial frame of reference if it becomes accelerated relative to the earth or some other inertial frame. You would hardly be able to avoid knowing if the plane banked to the left, hit an air pocket, or accelerated along a runway. Similarly, fictitious or inertial forces arise in noninertial frames, such as the elevator considered in Example 5-8 on page 118.

Newton himself was greatly concerned about relativity, for he saw that the implications were tremendous. He felt it necessary to state what we have taken for granted—that space and time are "absolute." This means that even if two observers are in relative motion, they will always agree on the length of an object, or on the duration of a time interval. Nothing arose to challenge these very reasonable assumptions until late in the 19th century, when an attempt was made to fit the new science of electromagnetism—especially light waves—into this sytem.

28-2   Einsteinian Relativity

Historically, the modern relativity theory originated in an effort to dispose of certain difficulties in the Maxwell* equations. These equations describe the "field theory," which links light, radio waves, and x rays together as various forms of electromagnetic radiation. The measured speed of light and other electromagnetic waves in a vacuum is one of the most precisely known physical quantities:

$$c = 2.9979248 \times 10^8 \text{ m/s}$$

This constant, which we round off to three significant figures as

$$c = 3.00 \times 10^8 \text{ m/s} \quad \text{or} \quad 186,000 \text{ mi/s}$$

was obtained as part of the Maxwell theory of electricity and magnetism. However, it was recognized in the period 1890–1905 that fundamental inconsistencies existed between the Maxwell equations and the concepts of space and time used by Galileo, Newton, and their successors. False starts were made by a number of physicists, until finally Einstein succeeded in showing that Newton, not Maxwell, was at fault.

---

* James Clerk Maxwell (1831–1879), one of the great British physicists of the mid-19th century.

It was not a question of some one particular equation being wrong; rather, underlying concepts about space and time were found to be unnecessarily naive.

A century ago, as today, the wave nature of light was an accepted fact, but physicists then were thinking exclusively in mechanical terms. All *mechanical* waves need a medium in which to propagate. Thus, water waves exist on the surface of a medium, the water; sound waves are transmitted through gases, liquids, or solids, but not through a vacuum. So, too, it was felt necessary to postulate a medium—the "ether"—in which electromagnetic waves could propagate. This substance would have to be miraculous indeed, for it would have to be completely transparent to light, have negligible density, and yet be so very rigid that vibrations as rapid as 10^{15} per second could take place in it. Whatever the properties of the ether, it seemed reasonable to use Galilean relativity to measure the velocity (or "drift") of the earth through this medium. Perhaps by knowing the extent of our motion through the ether we could understand it a little better.

Such an experiment was devised in 1887 by Michelson and Morley,* who used a Michelson interferometer (Sec. 26-7). They naturally assumed the speed of light relative to the ether to be a constant, c, whose value would depend on the properties of the ether in much the same way that the speed of sound waves in air depends on the bulk modulus and density of air. Suppose that an observer on earth is moving toward the east through a stationary ether (Fig. 28-4). Relative to the observer, an "ether wind" is blowing past him, from east to west, and his measurements of the speed of light would be expected to depend on the ether wind's magnitude and direction. The experimenters planned to deduce the motion of the earth relative to the

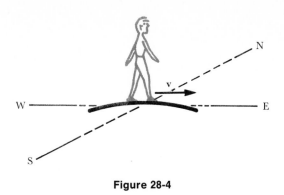

Figure 28-4

ether by comparing the speed of light in several directions. They set up an interferometer (Fig. 28-5) on a massive concrete slab that floated on a pool of mercury. In this way the orientation of the apparatus could be varied without introducing unwanted vibration.

In Sec. 2-9 (page 37), we showed how relative velocities can be combined by vector addition (with due regard to the order of subscripts). If the velocity of light relative to the ether is $\mathbf{v}_{LE}$ (of magnitude c), and the velocity of the ether relative to the observer is $\mathbf{v}_{EO}$ (of magnitude v), we can find $\mathbf{v}_{LO}$, the expected velocity of light relative to the observer:

$$\mathbf{v}_{LO} = \mathbf{v}_{LE} + \mathbf{v}_{EO}$$

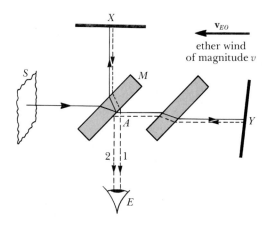

Figure 28-5 Michelson interferometer.

---

* Albert A. Michelson (1852–1931) and Edward W. Morley (1838–1923), American physicists who collaborated while at Case Institute of Technology and the neighboring Western Reserve University, respectively.

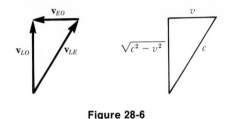

Figure 28-6

In Fig. 28-5, the vector $\mathbf{v}_{EO}$ is the velocity of the ether wind blowing past the observer. It is easy to see from Fig. 28-6 that for the "upstream" trip AY, the relative velocity is $c - v$; for the "downstream" trip YA, the relative velocity is $c + v$; and for the "broadside" trips AX and XA, the relative velocity is $\sqrt{c^2 - v^2}$ (Fig. 28-6, where it is assumed that $v < c$). We studied a similar situation in Prob. 2-B27, where we found that a swimmer going across stream and back took a shorter time for the round trip than did the same swimmer going the same distance upstream and back downstream.

Suppose, in Fig. 28-5, that $AY = AX$. If there were no ether wind, the time for light to travel from A to Y and back again would exactly equal the time for the round trip from A to X and back again. Rays 1 and 2 would be out of phase because of the extra 180° phase shift of one of the rays (which one?—see Fig. 26-27 on page 614), and a dark fringe would be seen. However, the earth revolves around the sun at about 30 km/s, and other sources of ether wind are also likely. Therefore, the times for the round trips would not be expected to be equal. The ratio of the times is easily calculated. (We denote by L the mirror separations AX and AY.)

$$AX + XA: \quad t_X = \frac{L}{\sqrt{c^2 - v^2}} + \frac{L}{\sqrt{c^2 - v^2}} = \frac{2L}{\sqrt{c^2 - v^2}}$$

$$AY + YA: \quad t_Y = \frac{L}{c - v} + \frac{L}{c + v} = \frac{2Lc}{c^2 - v^2}$$

The ratio is easily found to be $t_X/t_Y = \sqrt{1 - v^2/c^2}$, which is < 1. The round trip is indeed shorter for the broadside round trip, as was the case for the problem about the swimmer.

The Michelson-Morley experiment consisted of observing an interference fringe pattern, then rotating the apparatus 90° in a horizontal plane. The trip $AY + YA$, which was parallel to the ether wind, now takes place broadside to the wind; $AX + XA$ becomes parallel to the wind. The ratio t_X/t_Y is replaced by t_Y/t_X, which is > 1; this change in relative times of arrival should give rise to a shift of the interference fringes. An ether wind of 30 km/s would easily have been observed.

Experiments were made at different seasons of the year and at different times of day. The startling result was that no shift of fringes was ever observed. In other words, a straightforward experiment to measure the ether drift was a failure. One can surely find the speed of a river by measuring a motorboat's upstream and downstream velocities relative to the river bank. For light, it is as if no absolute space ("river bank," or ether) exists!

Albert Einstein (1879–1955) was led to a similar conclusion by quite different reasoning. Quite apart from the Michelson-Morley experiment—which Einstein later said had no direct effect on him, and of which he may not even have been aware—Einstein reasoned in 1905 that a wave without a wave velocity would be a logical inconsistency. Thus the velocity of light relative to an observer could never be zero, as would be predicted by the Doppler effect if the observer moved away from the source at a speed c. So he was led to postulate that the velocity of light is unaffected by the motion of the observer (and everybody agreed that it would be unaffected by the motion of the source). This, too, gives up the idea of absolute space and absolute time and makes the ether unnecessary.

We give this brief discussion about the origin of the relativity theory in order to indicate why c, the speed of *light*, enters into what seem to be essentially *mechanical* problems. If we remember that c is the speed of *any* electromagnetic wave, and if we realize that the Maxwell theory of electromagnetism had to be accepted on the basis of its internal consistency, we can see how the constant c (essentially an electromagnetic constant) might be expected to creep into the new equations of mechanics, which were designed specifically to avoid the conflict of mechanics with the existing electromagnetic theory of Maxwell. The constant c happened to be first discovered by workers in the field of electricity, long before electromagnetic waves were known to exist. Maxwell gave a *second* interpretation of this constant when he found it to be the speed of light and other electromagnetic waves; Einstein gave a *third* interpretation of this constant, as the maximum possible velocity of any body relative to any observer. Further interpretations may well be made by future investigators.

Evidently, a rigorous study of relativity theory must require thorough knowledge of electricity and magnetism; such study is beyond the scope of this book. However, it will be worthwhile to write down the two postulates on which Einstein based his theory of special relativity:

1. **The forms of the laws of physics are the same in all inertial frames of reference.**

2. **Measurements of the speed of light always give the same numerical value, regardless of the relative velocity of the observer and the source.**

The first postulate is, of course, Galilean relativity—but with an important change. Instead of referring to the laws of *mechanics*, Einstein broadened Galilean relativity to include *all* laws of physics, including the laws of electromagnetism.

The second postulate specifically gives up the idea of absolute space and absolute time but it is in agreement with the results of the Michelson-Morley experiment. The radical nature of the second postulate is apparent if we consider a "thought experiment" in which we imagine two searchlights that are uncovered for a brief instant by two experimenters A and B who are both at the same point in space (Fig. 28-7). A is stationary, and B is in motion to the right with a speed

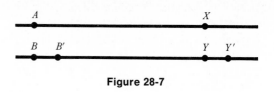

Figure 28-7

of 30,000 km/s, one-tenth the speed of light. After 1 s, a pulse of light has spread out from A's lamp a distance 300,000 km and has reached point X. However, after 1 s the light from B has spread out 300,000 km from B, and, since B has moved to B' during the second, the light has reached Y', which is 330,000 km from A. (Note that the second postulate requires that the velocity of the light away from B must be 300,000 km/s, regardless of B's velocity.) From A's point of view, the light from B's searchlight should only have reached to Y at this same instant, for otherwise its velocity relative to A would have been more than 300,000 km/s. How can the light be at both Y and Y' at the same time? Einstein's postulate seems to lead to contradiction; in the resolving of such contradictions Einstein was led to reconsider space-time relations in general. He found that A and B are measuring their time intervals of 1 s by clocks that do not agree with each other, and the light was not "simultaneously" at Y and Y', because of this difference in the rates of ticking of their clocks. Eventually, a set of physical laws was found that was consistent with the two postulates. It is because of the second postulate that we are justified in speaking of "the" speed of light, equal to 3.00×10^8 m/s (300,000 km/s or 186,000 mi/s) regardless of the motion of the source or the observer.

You will recall that our work in classical mechanics began with kinematics (the description of motion in terms of length and time) in Chap. 2; then we considered dynamics (mass, momentum, and energy) in Chaps. 3, 5, and 6. We shall organize our brief study of relativistic mechanics in much the same way. In the following sections we first discuss the kinematics of relativity (time dilation and length contraction);

then in our study of dynamics we introduce the ideas of mass increase and the equivalence of mass and energy. We shall make brief excursions into electromagnetism and the Doppler effect. Some further material about Einsteinian relativity is found in the For Further Study sections.

28-3 Time Dilation

One major consequence of the Einstein relativity is that the duration of a time interval between two events depends on the frame of reference in which the events are observed. Newton explicitly denied this possibility when he wrote: "Absolute, true, and mathematical time, of itself, and from its own nature, flows equably and without regard to anything external, and by another name is called duration." This view of the nature of time turned out to be wrong, but we cannot fault Newton, for the effect is observationally significant only for speeds comparable with that of light.

As a thought experiment, let us imagine two boys playing a game. One boy, A, is on the sidewalk (frame of reference FR), while the other boy, B, is on a truck (frame of reference FR') that is moving past A eastward at a uniform velocity v. The situation as seen by A is shown in Fig. 28-8a. The distance between the boys (at closest approach) is a, and they are equipped with identical clocks with which they can measure time intervals. The object of the game is for a short pulse of light from A's flashlight to be reflected from B's mirror back to A. We have two events—when A sends the pulse, and when A receives the reflected pulse. What is the duration of the time interval between these two events?

Suppose that A turns on his flashlight at just the right time to illuminate B when B is opposite A. The velocity of light relative to either A or B is c, independent of the motion of the source or the observer (Einstein's second postulate). In order to hit his target, the boy A must turn on his light just before B is opposite him (B at the dotted position in Fig. 28-8a); the light travels a distance $2a$ at velocity c, so the time required for the round trip, t_0, is given by

$$t_0 = \frac{\text{distance}}{\text{speed}} = \frac{2a}{c} \qquad (28\text{-}1)$$

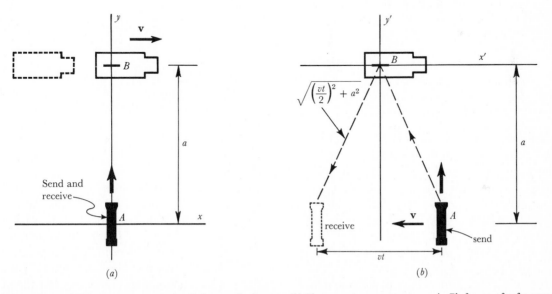

Figure 28-8 (a) Two events as seen in A's frame of reference. (b) The same two events as seen in B's frame of reference.

This is A's measurement of the interval between the two events. We call t_0 the *proper time* between the two events. This technical term, proper time, refers to the duration between two events that take place *at the same place* in some frame of reference (A's frame, in this case).

On the other hand, B describes the same experiment somewhat differently. As B sees it (Fig. 28-8b), *he* is stationary and A is moving backward (to the west) with a velocity v. As before, A points his flashlight northward, and, as before, he turns it on just before he is opposite B. The resultant velocity of the light relative to B is found by B to be equal to c, as it must be according to Einstein's second postulate. According to B, his friend A has moved a distance vt during the interval between the two events, where t is the duration of the interval *measured in B's frame*. B now uses the Pythagorean theorem to find the distance the light has traveled during this time; he obtains $2\sqrt{(vt/2)^2 + a^2}$. Since the speed of light in B's frame is c, B concludes that the time interval is given by

$$t = \frac{\text{distance}}{\text{speed}} = \frac{2\sqrt{(vt/2)^2 + a^2}}{c} \qquad (28\text{-}2)$$

Now a little elementary algebra is needed to solve Eq. 28-2 for t. We obtain

$$t = \frac{2a}{\sqrt{c^2 - v^2}} = \frac{2a}{c\sqrt{1 - v^2/c^2}}$$

$$= \frac{t_0}{\sqrt{1 - v^2/c^2}} \qquad (28\text{-}3)$$

We see that t is greater than t_0. A and B find different answers for the duration of the interval between the "sending" and the "receiving" events, even though their clocks are identical. There *is* a difference between their frames, however, because the two events do not take place at the same place in B's frame (see Fig. 28-8b). Thus, only t_0, measured at the same place in A's frame, is a proper time; t is an improper time. A considers that B's clock is "wrong," measuring an interval

that is "too long." In an exactly similar experiment, B could hold the flashlight and A the mirror. Now the interval measured in B's frame would be the proper time, and B would claim that A's clocks are giving too long an interval. Fortunately, we do not have to decide the argument, for it is impossible to tell which of the observers is "really" at rest (Einstein's first postulate). We are sure of one thing—there is no absolute time, and A and B are both correct, each according to his own standards.

In general terms, we conclude that if t_0 is the time interval between two events that occur at the same point in a frame of reference, then the time interval between the same two events has a longer duration t as measured by an observer in a frame of reference that is in uniform motion relative to the first frame.

In our example, the ratio of the two time intervals representing the same experiment is

$$\frac{t}{t_0} = \frac{1}{\sqrt{1 - v^2/c^2}} \qquad (28\text{-}4)$$

This same ratio, which is always greater than 1, holds true for any experiment viewed by two observers who are moving relative to each other. This is what is meant by *time dilation*, the stretching out of time.

Time dilation has been experimentally observed. For example, unstable particles called muons are created in the upper atmosphere by the action of cosmic rays (to be discussed in Sec. 33-11). The half-life of a muon is about 1.5×10^{-6} s; this means that on the average, half of any given group of muons will decay into an electron and a neutrino during this time interval. If N muons pass checkpoint P_1 at the top of a mountain, some of them will decay as they move downward and reach a lower checkpoint P_2. The experiment consists of finding the location of P_2 such that the number reaching that point is $\frac{1}{2}N$. One might expect that a very fast muon, traveling at nearly the speed of light, would move only about 450 m in its half-life of 1.5×10^{-6} s, so that P_2 would be about 450 m below P_1. However, experiment shows that P_2 is much lower than this; a typical experiment might give 1800 m for the position of this checkpoint.

The theory of relativity helps explain this result. (Refer to the summary in italics just before Eq. 28-4.) In the experiment, the first event occurs when the muon passes P_1, and the second event occurs when the same muon passes P_2 after one half-life has elapsed. What is the time interval between these two events? In the muon's own frame, the events occur at the same place (as the checkpoints rush past it), so the time interval of 1.5×10^{-6} s is a proper time, t_0. For the observers on earth, the two events occur at different places, separated by 1800 m, and the observers calculate the time interval to be about 1800 m/$(3 \times 10^8$ m/s), or 6×10^{-6} s, which is 4 times as long as t_0. This measured time interval t is an improper time, which is larger than t_0 according to Eq. 28-4. We could use Eq. 28-4 to calculate the relative velocity v for which $t/t_0 = 4$ (see Prob. 28-A6).

We see that a clock attached to the muon is running slow, and the muon reaching checkpoint P_2 is "younger" (hence more likely to be "alive") than it would have been if it had been at rest relative to the earth.

Time dilation is also involved in the paradox of the traveling twin. A space traveler moving at nearly the speed of light would not age as rapidly as her twin sister left behind on the earth. All bodily processes would slow down (as judged by the stay-at-home twin), although the traveling twin would be unaware of the difference until she greeted her sister after a seemingly short trip. The analysis is more complicated than for the muon, since the traveling twin changes frames of reference (from outward-bound to homeward-bound) during the turnaround, whereas the stay-at-home twin remains in the same frame throughout. A traveling-clock experiment is shown on page 661.

28-4 Length Contraction

Consider once more the muon described in the previous section. Let v be the magnitude of the velocity of the muon relative to the earth; v is also the magnitude of the velocity of the earth relative to the muon. This velocity is almost equal to c. How far did the muon travel (from checkpoint P_1 to checkpoint P_2) during its decay period? There are two answers to this question. In 1.5×10^{-6} s (the proper time t_0, measured in its own frame of reference), the muon traveled $t_0 v = 1.5 \times 10^{-6} v$ m. Experimenters on the earth saw the muon move for a much longer time $t = t_0/\sqrt{1 - v^2/c^2} = 6 \times 10^{-6}$ s. Earthbound observers thus found the distance to be $tv = 6 \times 10^{-6} v$ m. This is the *proper distance*, since observers on earth were at rest relative to both P_1 and P_2. In the muon's frame of reference, in which P_1 and P_2 were moving, the distance was only $t_0 v = 1.5 \times 10^{-6} v$ m. We see that there has been a *length contraction* along the direction of the muon's motion:

$$\frac{L}{L_0} = \frac{\text{measured length}}{\text{proper length}} = \frac{t_0 v}{tv} = \frac{t_0}{t}$$
$$= \sqrt{1 - v^2/c^2}$$

Thus,

$$L = L_0 \sqrt{1 - v^2/c^2} \qquad (28\text{-}5)$$

Note that v cancels out because it is the relative speed—that's the only thing the two observers can agree on! The length contraction ratio is numerically equal to the time dilation ratio.

Another way of looking at this is to consider it from the muon's viewpoint. The earth is rushing toward it at a speed v; because of the relative motion the 1800 m is contracted to only 450 m, and the muon can easily negotiate this foreshortened distance in the allotted time (1.5×10^{-6} s, by its own reckoning).

The length contraction given by Eq. 28-5 explains the negative result of the Michelson-Morley experiment. You will recall that in Fig. 28-5 the time for the broadside trip $AX + XA$

was predicted to be shorter than the time for the trip $AY + YA$, and the ratio of the times, t_X/t_Y, was predicted to be $\sqrt{1 - v^2/c^2}$. Now we see that the relativistic length contraction affects only the path AY that is parallel to the relative velocity v, so t_Y (only) is decreased by exactly the right amount. The two times t_X and t_Y are in fact equal, and no fringe shift is observed.

It is interesting to note that a length contraction, indeed of the identical amount derived here, was proposed about 1900 by Fitzgerald, to account for the negative result of the Michelson-Morley experiment, and independently by Lorentz, in an attempt to fix the theory of electromagnetism. However, their proposed contraction was not related to the fundamental concepts of space and time with which Einstein dealt about five years later.

28-5   Mass Increase

We shall use the law of conservation of momentum and the relativistic time dilation to derive the Einstein mass formula $m = m_0/\sqrt{1 - v^2/c^2}$. This equation concerns the *measurement of mass*, since m is the mass as measured by a moving observer or by a stationary observer who looks at a moving body,* and m_0, the *rest mass*, is the mass as measured by an observer who is at rest relative to the body. Since the denominator is < 1, this equation says that the mass of a moving body is always greater than its rest mass.

For our thought experiment let us perform a collision experiment with two identical, perfectly elastic, spherical balls. The two boys A and B arm themselves with balls instead of flashlights and throw the balls with equal speeds so as to cause them to collide just when they are halfway between the truck and the sidewalk. Even though the balls are identical, their masses may not seem to be the same as judged by the two

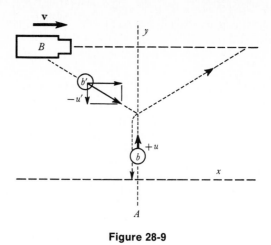

Figure 28-9

observers; each boy considers his own ball to have a mass m_0 (its rest mass) and his friend's ball to have a mass m.

As A describes the event (Fig. 28-9), the initial y component of the velocity of ball b' is $-u'$, and the y component of velocity of his own ball b is $+u$. After impact, the y components are reversed, $+u'$ for ball b' and $-u$ for ball b. The observer A now applies the law of conservation of momentum[†] to the elastic collision:

Total momentum in y direction before impact
= total momentum in y direction after impact

$$m(-u') + m_0(u) = m(u') + m_0(-u)$$
$$2m_0 u = 2mu' \qquad (28\text{-}6)$$

Now the boys agreed to throw the balls at the same speed u, but A believes that B's clocks are running too slowly, and so he believes that B failed to give the proper velocity to his ball. Hence he deduces that u' is less than u; in fact, A concludes that

$$u' = u\sqrt{1 - v^2/c^2}$$

where the square root (which is less than 1) is the reciprocal of the time dilation factor given in Eq. 28-4. Putting this value for u' into Eq. 28-6,

---

* It is the essence of relativity that only *relative* motion is important; the two measurements of m must give the same result, regardless of who is "stationary" (Einstein's first postulate).

[†] This is in accord with Einstein's first postulate, according to which the *forms* of the laws of physics do not depend on the relative motion of the two frames of reference. All momenta are judged by *one* observer, namely A.

we get

$$2m_0u = 2mu\sqrt{1 - v^2/c^2}$$

$$m = \frac{m_0}{\sqrt{1 - v^2/c^2}} \qquad (28\text{-}7)$$

This is the expression for the increase of mass with velocity. Here m_0 is the rest mass, as seen by the stationary observer A, and m is the mass of the ball that is moving relative to him.

It follows that mass is not a unique property of a body. Two observers, equally competent, will measure the inertial mass of a body differently if their velocities relative to the body are different. Note that Eq. 28-7 predicts that the mass becomes infinitely great as the reiative velocity approaches c, the speed of light. Therefore, according to Newton's second law, an ever-increasing force is required to accelerate the body, and a physical body can never reach the speed of light, since an infinite force is not possible.

Example 28-1

A hypothetical spaceship travels away from the earth at a velocity of 2×10^8 m/s. A projectile is shot forward from the spaceship at 2×10^8 m/s relative to the spaceship. What is the velocity of the projectile relative to the earth?

We do not have to know the equation for relative velocity in order to look at certain features of the solution. Of one thing we can be sure—the velocity of the projectile relative to the earth is certain to be less than 3×10^8 m/s. Offhand, we would have expected 4×10^8 m/s, but the simple vector treatment of relative velocity (Sec. 2-9) is not valid at high velocities.* The relativity result is 2.77×10^8 m/s (see Prob. 28-C10).

---

* Relative to the people on earth, the projectile is moving faster than it is moving relative to the people on the ship. As judged by the earthlings, the projectile therefore has more mass than as judged by the space travelers, and so the same force (reaction force of the fuel) produces less acceleration. The projectile is given a smaller final velocity, as viewed from earth, than would have been expected if the mass had not changed.

The increase in mass of a moving electron had been studied for several years before the theory of relativity appeared, but without clear-cut results. No satisfactory explanation was made until 1905, when the mass equation, Eq. 28-7, appeared as a sort of unexpected by-product of Einstein's fundamental reexamination of space and time and relative motion. And it was not until 1916 that precise experimental results confirmed the relativistic formula.

28-6   Mass and Energy

The most "practical" consequence of the theory of relativity is the possibility of interconversion between matter and energy. Einstein showed that whenever mass disappears and is converted into energy, and vice versa, the same amount of energy (9×10^{16} J) always buys the same amount of mass (1 kg), and vice versa. By now we are familiar with the conversion of matter into energy, for that is the source of the vast energy of the nuclear bombs. All the fragments, gases, ash, and so on, of a nuclear explosion would, if gathered together, have slightly less mass than the original uranium, plutonium, or hydrogen. The Einstein equation relating mass and energy is

$$\Delta E = (\Delta m)c^2 \qquad (28\text{-}8)$$
$$\text{J} = (\text{kg})(\text{m/s})^2$$

where c is the fundamental constant of nature that, among other things, equals the speed of light in a vacuum (3.00×10^8 m/s).[†]

Einstein reasoned thus: By giving a body kinetic energy, we increase its mass according to Eq. 28-7. He concluded that if *increase* of mass corresponds to *gain* in energy, the mass of a body even at rest must be equivalent to some energy. His conclusion (1905) was verified in 1919 when Rutherford observed the conversion of mass into

---

[†] If the history of physics had gone in a different way, or if the human race had evolved without sight, c might well have first been discovered and *defined* as the square root of the ratio of energy to mass equivalent.

energy during the first experiments in transmutation of elements.

There is no room for doubt that the theory of relativity correctly describes the interconversion of mass and energy. Nuclear physicists have verified the equation $\Delta E = (\Delta m)c^2$ in countless precise experiments. The law of conservation of energy must, therefore, be worded in such a way as to include the energy equivalent of mass. The law has been generalized in another way. An isolated system is one that does not have any interactions with objects outside the system, and the universe is, by definition, an isolated system. By extrapolation, and as a matter of scientific faith, the law of conservation of energy has been assumed to apply to the entire universe. Thus, finally, we state the law of conservation of energy in its most general form:

The total equivalent amount of energy in the universe remains constant.

This generalization, evolved by many scientists over a period of several centuries, has proved to be of tremendous importance for every branch of physical science. We shall use it frequently in our later study of nuclear physics.

The conversion of mass into energy is of utmost importance to our understanding of nature. Aside from the spectacular controlled release of energy in the nuclear reactor and in nuclear bombs, a less obvious but more important conversion of mass into energy takes place continually in the sun and other stars. As the end result of a chain reaction, 5×10^{11} kg of hydrogen is converted into nearly the same amount of helium in 1 s (see Sec. 33-3). The mass balance is not exact, however, and the sun's mass continually decreases, at the rate of 4×10^9 kg second. This is the mass that is converted into energy, keeping the sun hot. The sun has been radiating energy (and losing mass) at this rate for several billion years. It speaks for the tremendous mass of the sun (1.97×10^{30} kg) that, had it been losing 4×10^9 kg (4 million metric tons) every second for a billion years, over 99.99% of the sun's mass would still remain!

In everyday life, however, the effects of relativity and the increase of mass with velocity are too small to be noticed.

Example 28-2

A rocketship of rest mass exactly 4×10^3 kg is moving at 3000 m/s. (a) What is its increase in mass? (b) What is its KE?

(a) Its new mass m is given by Eq. 28-7:

$$m = \frac{m_0}{\sqrt{1 - v^2/c^2}}$$

$$= \frac{4000.0000000 \text{ kg}}{\sqrt{1 - \dfrac{(3 \times 10^3 \text{ m/s})^2}{(3 \times 10^8 \text{ m/s})^2}}}$$

$$= \frac{4000.0000000 \text{ kg}}{\sqrt{1 - \dfrac{9 \times 10^6 \text{ m}^2/\text{s}^2}{9 \times 10^{16} \text{ m}^2/\text{s}^2}}}$$

$$= \frac{4000.0000000 \text{ kg}}{\sqrt{1 - 10^{-10}}} = \frac{4000.0000000 \text{ kg}}{0.99999999995}$$

$$= 4000.0000002 \text{ kg}$$

$$\Delta m = 4000.0000002 \text{ kg} - 4000.0000000 \text{ kg}$$

$$= \boxed{2 \times 10^{-7} \text{ kg}} = 0.2 \text{ mg}$$

This increase is about equal to the mass of the ink used to write an address on a rocket-mail letter carried by the ship.

(b) The KE of the rocketship is

$$\text{KE} = \tfrac{1}{2}mv^2$$

$$= \tfrac{1}{2}(4 \times 10^3 \text{ kg})(3 \times 10^3 \text{ m/s})^2$$

$$= \boxed{18 \times 10^9 \text{ J}}$$

We may say that 18×10^9 J of KE is equivalent to 2×10^{-7} kg of mass. This checks with $\Delta E = (\Delta m)c^2$, as you can easily verify; each kilogram of mass increase corresponds to 9×10^{16} J of energy increase, as it should.

This example clearly shows that the mass changes accompanying ordinary energy changes are tiny indeed. It is for this reason that we can apply the law of conservation of energy to the solution of problems with a clear conscience, ignoring these small relativity effects. In other words, we usually need to consider only chemical, thermal, electrical, or mechanical energies. The law of conservation of energy, even in this limited form, is a profound generalization, and we saw in Sec. 6-6 how to use it in solving problems.

Let us look critically at the statement "mass can be converted into energy." We shall find that it is *rest mass* that can be so converted; the total (relativistic) mass of the universe remains constant. In this sense mass is conserved, and also energy is conserved. Unless we use words carefully and consistently, we shall run into some difficulty in interpreting the law of conservation of mass and the law of conservation of energy.

Surely Example 28-2 seems to be no example of the conversion of mass into energy, or vice versa, since both the ship's mass *and* its energy increased. This is like trying to keep books with all credits and no debits. In fact, the theory of relativity considers that *any* energy increase ΔE, not just a kinetic energy increase, is accompanied by an increase of mass Δm. The Einstein equation says that these changes are related by $(\Delta m)c^2 = \Delta E$, or $\Delta m = (\Delta E)/c^2$. Considering the whole system, ship plus fuel, we would find that the fuel had more mass before it was burned than afterward, because of its original PE.* In such problems mass and energy are each conserved, if by "mass" we mean total (relativistic) mass, including the mass equivalent of KE or PE.

Consider a 100-kg safe that is raised to the top of a building 100 m high. The *system*, earth + safe, gains PE, calculated from *mgh* as $(100 \text{ kg})(9.8 \text{ N/kg})(100 \text{ m}) = 98,000 \text{ J}$. Thus the change in mass of the system is

$$\Delta m = \frac{\Delta E}{c^2} = \frac{9.8 \times 10^4 \text{ J}}{(3.00 \times 10^8 \text{ m/s})^2}$$

$$= 1.1 \times 10^{-12} \text{ kg}$$

The safe and earth are at rest in an inertial frame, but their rest mass has increased by 1.1×10^{-12} kg. Here a change in energy has been accompanied by a change in rest mass.

You may ask, "Where did the extra 1.1×10^{-12} kg of mass come from in the first place?" Suppose the safe had been hoisted to the top floor by a pulley system activated by the muscular efforts of several workers. The mass of the workers must have *decreased* during the process, for they used up chemical PE, perhaps from a ham-and-eggs breakfast. In a sense, mass "flowed" from workers to safe as it was raised. No matter how far back you carry the questioning, you will find that the eventual source of the energy (and mass) was nuclear—more than likely from solar energy, which caused the plants to grow, which certain hogs and hens converted into ham-and-eggs-on-the-hoof, and so forth. But we know that the sun's energy comes to us as a result of nuclear changes. Thus it would seem that all or nearly all[†] of the energy changes on the earth ultimately come from mass changes in nuclear reactions, either in man-made nuclear reactors or in the gigantic natural reactor we know as the sun. What, then, is the difference, if any, between the changes taking place in the safe or in chemical reactions and the changes taking place in a nuclear reactor? The difference is one of degree only.

Let us compare the common chemical heating reaction mentioned on page 127 with some nuclear heating reactions to be discussed in Chap. 33:

Chemical fuel:

$$C + O_2 = CO_2 + \text{energy}$$

Nuclear fuel (fusion):

$$^1H + {}^3H = {}^4He + \text{energy}$$

Nuclear fuel (fission):

$$^{235}U = \text{fragment } A + \text{fragment } B + \text{energy}$$

---

* In an exact treatment of the rocketship problem, we would have to consider also the KE and the mass changes of the exhaust gases.

[†] The energy of the tides comes from the rotation of the earth; no one can yet say with certainty what was the original source of the earth's rotational KE, but this, too, may very likely have been a result of nuclear changes.

Before combustion, the C and the O_2 are at rest and have no KE, and they have PE because of their separation. Because of this PE, they also have some extra rest mass. Therefore the rest masses of the C and O_2 add up to more than the rest mass of the CO_2, and we say that some of the rest mass of the universe has changed to energy. Similarly, on a much grander scale, in nuclear fusion the rest mass of a ^{4}He nucleus is less than the total rest mass of the ^{1}H and ^{3}H nuclei, which have PE because of their separation. In nuclear fission the ^{235}U nucleus is unstable, which means that PE is stored in the nucleus ready to be "triggered off." This stored energy means that ^{235}U has more rest mass than its fragments after fission. In all these cases, it is *rest mass* that has been converted into energy; total (relativistic) mass (which includes the mass equivalent of all forms of energy) has been conserved. We use the term *mass energy* to represent m_0c^2, where m_0 is the rest mass of a body. The mass energy of a 100-g cup of water, for instance, is tremendous, for it equals the energy that could be obtained if *all* the rest mass of the water were converted to energy. However, the changes that occur in mass energy are only a tiny fraction of the total mass energy of a body, except in certain special cases (to be discussed in Chaps. 32 and 33, where we study nuclear physics in more detail).

We have written the relationship between mass changes and energy changes as

$$\Delta E = (\Delta m)c^2$$

where the increased energy of a body is accompanied by an increase in mass. We can also write

$$E = mc^2$$

where E is the total energy (including the mass energy) and m is the total (relativistic) mass given by $m = m_0/\sqrt{1 - v^2/c^2}$. This equation places energy in a prominent place, without distinguishing too closely between the various forms that energy can take. The kinetic energy due to motion is then total energy $-$ rest energy. Thus,

$$KE = mc^2 - m_0c^2 \qquad (28\text{-}9)$$

This is the relativistic formula for KE; it is derived in Sec. 28-9. We expect that at velocities that are small relative to c, Eq. 28-9 will reduce to the familiar $\frac{1}{2}m_0v^2$ (it does—see Prob. 28-C4). In fact, *all* relativity formulas must include the classical formulas of physics as special cases (Sec. 8-4). We see that the KE of a body is not a definite quantity, since it depends on m, which is not the same for all observers. This variability of m is new, but even in the classical relativity of Galileo and Newton the KE of a body depends on the observer. For example, consider a lunar lander propelled in a forward direction from a satellite that is already in orbit. The KE of the lander is very much less as viewed from the satellite than as viewed from the earth, and, of course, is zero as viewed from the lander itself.

28-7 Electromagnetic Force

To illustrate how relativity pervades all of physics and is not limited to mechanics, we give an example from electromagnetism. In Sec. 22-7 we stated as an experimental fact that like currents attract each other; this magnetic force was considered to be separate and distinct from the electrostatic (Coulomb) force between two charges. We now show how the attraction of like currents follows from applying the theory of relativity to Coulomb forces; no additional magnetic force need be postulated.

Consider two long wires in each of which there is a current toward the right (Fig. 28-10). The wires are electrically neutral, with equal numbers of + and − charges per unit length, equally

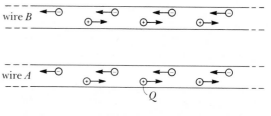

Figure 28-10

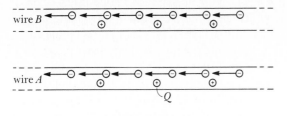

Figure 28-11

spaced. We choose to represent the currents by indicating in Fig. 28-10 that the + charges have a velocity to the right and the − charges have an equal velocity to the left.* To find the force on, say, the positive charge Q, we transform to a frame of reference in which Q is at rest and the negative charges are moving faster (Fig. 28-11). Because of the relativistic length contraction, when in motion relative to Q the negative charges are closer together than are the positive charges. Now we consider electrostatic forces, given by Coulomb's law, acting on the marked positive charge Q. (1) By symmetry, the force of repulsion on Q due to the + charges in A add up to 0, since there are equal numbers of + charges to the right of Q and to the left of Q. (2) Also by symmetry, the forces of attraction on Q due to − charges in A add up to 0. (3) The vector sum of all electrostatic forces due to the + charges in B is a force pushing charge $+Q$ away from wire B. (4) The vector sum of all electrostatic forces due to the − charges in B is a force pulling the charge $+Q$ toward wire B. Forces 1 and 2 are each zero; force 4 is greater than force 3 because there are (according to the observer at Q) more − charges per unit length in B than there are + charges per unit length. Since force 4 is greater than force 3, there is a net attraction on charge Q. This and other charges tend to move toward the boundary of the wire, and when they reach the boundary, the wire itself experiences the force of attraction. Therefore, like currents attract each other.

---

* The figure is simplified by showing only representative charges; actually, there are two interpenetrating clouds of + and − charges, moving in opposite directions.

A similar argument could be applied to oppositely directed currents. Reverse each velocity in wire B and again transform to a frame in which Q is at rest. Now the − charges in wire B are at rest relative to Q, and the + charges are moving faster toward the left and are closer together. Wire B has more + charges per unit length than it has − charges, and there is a net force of repulsion acting on charge Q. Unlike currents repel each other.

We see that magnetic forces arise quite naturally without any special postulates other than Coulomb's law, which applies to all charges whether or not they are in motion. We can say that magnetism is a relativistic aspect of electrostatics, connected with a transformation from one inertial frame to another. Relativity theory unified the previously separate fields of electrostatics and magnetism.

28-8 Relativistic Doppler Effect

According to Einstein's first postulate, only relative motion is observable. The theory of the Doppler effect developed in Sec. 10-8 seems to violate this postulate; by measuring the apparent frequency and the actual frequency, we could determine which of two bodies (source or observer) is "actually" in motion. Let us recall the appropriate equations for sound waves in air (see page 232).

sound waves
(source in motion)
$$f' = \frac{f}{1 - v/v_w} \qquad (10\text{-}5)$$

sound waves
(observer in motion)
$$f' = f(1 + v/v_w) \qquad (10\text{-}6)$$

In comparing Example 10-6 with Example 10-7 (page 231), suppose we know the relative velocity to be 34 m/s. If we measured the apparent frequency to be 272.7 vib/s, we could conclude that the source is moving relative to the air; if we measured the apparent frequency to be 270.0 vib/s, we could conclude that the observer is moving relative to the air. Thus a decision could be made as to which body is in motion. To resolve this paradox, we should note that for sound waves there is a medium (air) that can serve for a reference set of axes; we can find out which body is moving relative to the air. For light waves, there is no such medium. The Doppler effect for light waves is calculated in the relativity theory, and, as expected, only the relative velocity of source and observer enters into the equation. The result (for head-on motion) is

$$\text{light waves} \qquad f' = \frac{\sqrt{1 + v/c}}{\sqrt{1 - v/c}} f \qquad (28\text{-}10)$$

where v is the relative velocity of source and observer, and the symbol c (speed of light) is used instead of v_w. This interesting relativistic formula can be compared with the two formulas for sound waves; f' is the geometric mean of the values given by the two acoustic formulas.

An interesting sidelight on the relativistic Doppler effect is the prediction of a shift in frequency when a sideways motion takes place. In the acoustic effect there is no change in frequency at the moment when the moving source is just opposite the observer, since at this instant it is neither approaching nor receding. However, there *is* a small effect in the case of light, given by $f' = f\sqrt{1 - v^2/c^2}$. This "transverse Doppler effect" is related to the relativistic time dilation (Eq. 28-4). It has actually been observed in an experiment with moving ions. Since the factor v^2/c^2 is positive for either direction of motion (v^2 is always positive and less than c^2), the shift is always toward lower frequencies. There are still other experiments that show the validity of the relativistic concept of the Doppler effect.

In summary, we have seen that the time interval between two events is longer when measured by an observer who is moving relative to the frame in which the events take place. This time dilation is the basis for the relativistic mass increase, through the following argument: we must retain the form of the law of conservation of linear momentum, but because of time dilation the observed speeds of colliding bodies are al-

tered. Therefore a corresponding change in the observed mass of a moving body is predicted, to conserve the quantity mv. The mass increase, in turn, leads to the equivalence of mass and energy and thus to a profound generalization of the law of conservation of energy. We also found that the observed length of an object suffers a contraction along the direction of motion. All these relativistic refinements of Newtonian mechanics that we have considered in this chapter are significant only for large relative velocities, where v^2/c^2 is appreciable. Thus, Einsteinian relativity contains within itself the older Galilean-Newtonian relativity; it is entirely possible that Einsteinian relativity is a special case of some still more comprehensive theory yet to be discovered.

Before we leave the subject of relativity, a further point should be noted. Our study of Einsteinian relativity has been limited to the *special theory of relativity*, which applies to inertial frames of reference in *uniform* relative motion (no acceleration of one observer relative to another). By experiments performed inside a steadily moving train, you cannot tell whether the trees and ground outside your window are moving backward or whether you are moving forward. The laws of physics are the same within the train as on the ground. But try to remain standing upright in a bus that suddenly starts up, or try to eat soup in a plane that swerves in a sharp unbanked curve. The laws of physics appear quite out of the ordinary in such situations, and the observer can explain the strange happenings in either of two ways: (*a*) his frame of reference is accelerated, or (*b*) a force acts on him or on the soup. The *general theory of relativity* is an extension of the special theory to deal with accelerated motion, and in the general theory it is postulated that these two ways of looking at any experiment involving force are entirely equivalent. For instance, if a (weightless) astronaut in outer space is enclosed in a sealed box that is accelerated upward at 9.8 m/s^2, she experiences all the sensations of being on the earth subject to the downward force of gravity. No experiment she can perform inside the box can help her decide whether her frame of reference has been accelerated or whether gravity has suddenly been "turned on." The general theory of relativity tells us much about the nature of gravitation and other forces, but this is far beyond the scope of this text.

Summary In Galilean relativity the forms of the laws of mechanics are the same in all inertial frames. The Michelson-Morley experiment, which implicitly assumed Galilean relativity, failed to measure the expected velocity of the earth relative to the ether, leading to doubts about the universal validity of the Galilean relativity. Einstein broadened the theory of relativity to apply to all laws of physics, including electromagnetism. This implies that the speed of light in a vacuum is a constant c for any observer who is in uniform motion relative to the source of light.

Some consequences of relativity are time dilation, length contraction, mass increase, and the equivalence of energy and mass. When v/c is small compared to 1, Einsteinian relativity reduces to the earlier Galilean relativity.

The mass energy of a body is m_0c^2, where m_0 is the rest mass; the total (relativistic) energy is mc^2. When an isolated system undergoes change, the total (relativistic) mass of the system remains unaltered, and the total energy content of the system remains unaltered. These are the laws of conservation of mass and conservation of energy. No energy can be created or destroyed unless a corresponding amount of rest mass disappears or appears; this conversion of matter into energy is governed by the Einstein equation $\Delta E = (\Delta m)c^2$.

The theory of relativity applies to all of physics, not just mechanics. Electromagnetic forces are the result of viewing electrostatic (Coulomb) forces from a moving frame of reference.

Check List

Galilean relativity
Michelson-Morley experiment
Einstein's postulates
time dilation
proper time
length contraction
proper distance
mass increase
rest mass

energy equivalence of mass
$t = t_0/\sqrt{1 - v^2/c^2}$
$L = L_0\sqrt{1 - v^2/c^2}$
$m = m_0/\sqrt{1 - v^2/c^2}$
$\Delta E = (\Delta m)c^2$
$\mathrm{KE} = mc^2 - m_0c^2$
special theory of relativity
general theory of relativity

Questions

28-1 A passenger in a car is moving with constant velocity at 6 m/s toward the south. As he passes a child on the sidewalk, the child throws a ball straight up at 9.8 m/s. (a) How long is the ball in the air? (b) Relative to the car, where does the ball strike the ground? (c) Describe the path of the ball relative to the child. (d) Describe the path of the ball relative to the passenger in the car.

28-2 The projectile of Example 28-1 emits a beam of light that can be seen from both the spaceship and the earth. What is the velocity of this light as measured by a scientist on the spaceship? What is the velocity of the light as measured by a scientist on earth?

28-3 Since a formula for the speed of light in a vacuum can be derived from the laws of electromagnetism, does this mean that Einstein's second postulate is in fact just a special case of the first postulate, which concerns the invariant form of *all* laws of physics?

28-4 In the Michelson-Morley experiment (Fig. 28-5), why is the path AX not subject to the relativistic length contraction?

28-5 Do you think that the traveling twin (Sec. 28-3) upon her return to earth would "really" be younger than her sister?

28-6 An electron's speed cannot be greater than c; is there some upper limit to the momentum that an electron can have?

28-7 If mass is a form of energy, is it then a true statement that a wound-up watch has more mass than the same watch when it has run down?

28-8 In principle, is there *any* change of weight (however slight) when the flash bulb of Ques. 6-9 (page 139) is fired?

28-9 A nuclear particle in a large circular accelerator is traveling at $0.99c$; is it possible to double the kinetic energy of this particle by speeding it up?

28-10 In Example 7-8, was it legitimate to use the Newtonian form of the KE formula ($\frac{1}{2}mv^2$) for the KE of the proton, or should we have used the relativistic formula, taking account of the increase in mass due to its velocity? (*Hint:* How does the proton's velocity compare with the speed of light?)

28-11 Water flows over a dam, electric energy is generated, energy is transmitted over a line and through various step-up and step-down transformers, and eventually an elevator rises in an office building. Did any *mass* flow through the transformers?

28-12 Suppose a research worker invented a device capable of converting mass energy into electric energy, and vice versa. Would he be justified in calling his invention a seat of emf?

28-13 Kepler's laws of planetary motions assume that the planet has a constant mass. Explain why this condition is violated for the elliptical orbit of Mercury shown in Fig. 8-6 on page 186.

28-14 Is the equivalence of mass and energy a consequence of the special theory of relativity or of the general theory of relativity?

MULTIPLE CHOICE

28-15 The law of conservation of momentum is valid in all frames of reference if momentum is defined as   (*a*) m_0c;   (*b*) m_0v;   (*c*) $m_0v/\sqrt{1 - v^2/c^2}$.

28-16 The time interval Δt between event A and event B is the proper time if events A and B take place   (*a*) at the same time;   (*b*) at the same point;   (*c*) at two points separated by $v\Delta t$.

28-17 The special theory of relativity is applicable to the phenomena of   (*a*) mechanics;   (*b*) electromagnetic waves such as light;   (*c*) both of these.

28-18 A particle of rest mass m_0 moves at speed $0.60c$ relative to an observer. The observed mass is   (*a*) $1.25m_0$;   (*b*) $1.33m_0$;   (*c*) $1.67m_0$.

28-19 If an electron of rest mass m_0 moves at speed $0.80c$, its momentum is   (*a*) $0.80m_0c$;   (*b*) $1.25m_0c$;   (*c*) $1.33m_0c$.

28-20 The relativistic expression for KE, valid for all speeds, is   (*a*) $\frac{1}{2}m_0v^2/(1 - v^2/c^2)$;   (*b*) $(m - m_0)c^2$;   (*c*) $\frac{1}{2}mv^2$, if $m_0/\sqrt{1 - v^2/c^2}$ is used for m.

Problems **28-A1** (*a*) Using values from Appendix Table 7, verify the statement in Sec. 28-2 that the earth's orbital speed is about 30 km/s. (*b*) What is v/c for this orbital speed?

28-A2 A body is moving at 98% the speed of light. What is the ratio of its relativistic mass to its rest mass?

28-A3 An electron of rest mass m_0 is moving at a speed $0.5c$. What is its relativistic mass?

28-A4 How fast is a cosmic-ray particle moving when its mass is 3 times its rest mass?

28-A5 (*a*) How much energy would be obtained if the rest mass of 30 mg of coal were to be entirely converted into energy? (*b*) How high could this much energy lift an office building of mass 4 million metric tons?

28-A6 What is the value of v/c for the muon in Sec. 28-3?

28-B1 In Fig. 5-5 (page 114), the girl's mass is 35 kg. (*a*) What is her KE relative to the floor? (*b*) What is her KE relative to the suitcase? (*Note:* This illustrates that even at low speeds, where classical physics is applicable, the KE of a body depends on the frame of reference.)

28-B2 Suppose the collision between equal balls described in Sec. 28-1 is viewed from a frame that is moving toward the left at speed $\frac{3}{2}v_0$. Calculate (*a*) the total momentum, both before and after collision, and (*b*) the total KE, both before and after collision.

28-B3 In an x-ray tube an electron leaves a filament, moves with speed $0.99c$ down a tube 0.400 m long, and strikes a target. In the laboratory frame, how long, in nanoseconds (ns), is required for the trip? (*b*) From the electron's point of view, how long is the tube? (*c*) From the electron's point of view, how long is required for the target to come to meet the electron? (*d*) Why is the answer to part (*c*) so much less than that to part (*a*)?

28-B4 A space traveler takes off at 2×10^8 m/s, leaving his twin brother behind on the earth. One day later (measured on the ship), the traveling twin's heart has made a normal 100,000 beats. (*a*) What is the duration of the "experiment" as measured by an observer on the earth? (*b*) How many heartbeats has the stay-at-home twin's heart made?

28-B5 How fast would a rocket of mass 10^6 kg have to move in order for its mass to increase by 3 g?

28-B6 (*a*) What is the speed of a particle whose mass is twice its rest mass? (*b*) What is the speed of a particle whose KE is twice its rest energy?

28-B7 What is the KE of a cosmic-ray particle of rest mass 3×10^{-26} kg that is moving at 1.8×10^8 m/s?

28-B8 The Great Pyramid of Cheops has a mass of about 7×10^9 kg (see Prob. 1-A8), and its c.g. is about 37 m above the base. (*a*) What is the weight of the pyramid? (*b*) What is the PE of the stones that make up the pyramid? (*c*) How much mass flowed into the pyramid while the workers erected it in about 2890 B.C.? Where did this mass come from?

28-C1 Electrons move through a high-energy linear accelerator 1.5 km long at a speed so close to c that their masses are 10^4 times their rest masses. (*a*) How long is required for the trip, as measured in the laboratory frame of reference (use $v \approx c$)? (*b*) In the electron's frame of reference, what is the length of the tube through which they move? (*c*) What is the proper time for the trip?

28-C2 In 1969 James A. McDivitt and others spent 10 days in orbit around the earth, moving at about 7.8×10^3 m/s. (a) What was the time dilation factor? (*Hint:* Use $1/\sqrt{1-x} \approx 1 + \frac{1}{2}x$ for $x \ll 1$.) (b) Show that when he splashed down, Col. McDivitt was about 300 μs younger than he would have been if he had not been in orbit.

28-C3 An astronaut is stationary, 5×10^{12} m from earth, when a yellow light in the space-craft goes on, indicating that her life-support system will run out of oxygen in 10^4 s (2.78 h). A red light will come on when the last of the oxygen is used up. (a) Is the 10^4 s a proper or a nonproper time? (b) What velocity relative to earth must the astronaut give her spacecraft so that she can reach earth just as the red light goes on? (*Hint:* Consider that the earth is coming to meet the spacecraft; use the relativistic length contraction.) (c) The astronaut sends a message to earth when the yellow light goes on; according to the earthlings, how long after the message is received does the spacecraft arrive? (*Hint:* Imagine a rod 5×10^{12} m long rigidly attached to the earth, extending to the initial position of the spacecraft.)

28-C4 Combining Eqs. 28-9 and 28-7, show that $\mathrm{KE} = m_0 c^2 (1 - v^2/c^2)^{-1/2} - m_0 c^2$. Simplify this expression, using the binomial theorem, and show that $\mathrm{KE} = \frac{1}{2}m_0 v^2 + (\quad) + \cdots$. To prove that the correct expression for KE reduces to the classical formula as a special case, you will have to prove that the second and other terms in the infinite series become negligibly small compared with the first term, as v/c approaches zero.

28-C5 An electron at rest has mass 9.11×10^{-31} kg. (a) Calculate its mass when it is moving at a constant speed of 1.8×10^8 m/s. (b) Calculate the increase of mass (Δm) of the electron due to its motion. (c) Calculate from Δm the KE of the electron, using the Einstein formula $\Delta E = (\Delta m)c^2$. (d) Verify numerically that the formulas $\frac{1}{2}m_0 v^2$ and $\frac{1}{2}mv^2$ both give incorrect answers for the KE of this "relativistic" particle.

28-C6 A proton of rest mass 1.67×10^{-27} kg has a relativistic mass 4 times its rest mass. What is its KE?

28-C7 (a) Show that a PD of only 1.02×10^6 V (1.02 megavolts) would be sufficient to give an electron a speed equal to twice the speed of light if Newtonian mechanics remained valid at high speeds. (b) What speed would an electron actually acquire in falling through a PD of 1.02×10^6 V? (*Hint:* Find the mass of the electron from its total mc^2, and then find its speed from $m = m_0/\sqrt{1 - v^2/c^2}$.)

28-C8 Suppose that the projectile of Example 28-1 has a rest mass of 2000 kg. (a) What is its KE relative to the ship? (*Hint:* Use Eq. 28-9.) (b) What is its KE relative to the earth? (c) What is the value of the quantity $\frac{1}{2}mv^2$, where m is the relativistic mass and v is 2.77×10^8 m/s, the velocity relative to the earth? (d) Does this latter quantity have any physical significance?

For Further Study

28-9 Derivation of Equivalence of Mass and Energy

Einstein himself felt that the relationship between mass and energy was one of the most important results of the theory of relativity. Only a modest use of derivatives is needed to derive the relativistic formula for kinetic energy (Eq. 28-9), from which we can generalize the result that $E = mc^2$.

We begin by reaffirming some definitions that are already familiar from Newtonian mechanics.

Momentum is *defined* as $p = mv$, the product of the relativistic mass and the velocity. Thus a body's momentum approaches infinity as its speed approaches c. Just as in Newtonian mechanics, force is *defined* as $F = dp/dt$, the rate of change of momentum. When a force acts on a body, the momentum of the body changes. The change in kinetic energy is the work done by the force.

$$dE = F \, dx = \frac{dp}{dt} \, dx = \frac{dx}{dt} \, dp = v \, dp \quad (28\text{-}11)$$

Since $p = mv$, we have $dp = m \, dv + v \, dm$. Thus Eq. 28-11 becomes*

$$dE = mv \, dv + v^2 \, dm \quad (28\text{-}12)$$

To find $mv \, dv$, we differentiate the relativistic mass formula, using the chain rule (Appendix E).

$$m = m_0 \left(1 - \frac{v^2}{c^2}\right)^{-1/2}$$

$$\frac{dm}{dv} = -\tfrac{1}{2} m_0 \left(1 - \frac{v^2}{c^2}\right)^{-3/2} \left(-\frac{2v}{c^2}\right)$$

$$\frac{dm}{dv} = \left[m_0 \left(1 - \frac{v^2}{c^2}\right)^{-1/2}\right]\left(1 - \frac{v^2}{c^2}\right)^{-1}\left(\frac{v}{c^2}\right)$$

$$\frac{dm}{dv} = m\left(1 - \frac{v^2}{c^2}\right)^{-1}\left(\frac{v}{c^2}\right)$$

$$\frac{dm}{dv} = \frac{mv}{c^2}\left(\frac{c^2 - v^2}{c^2}\right)^{-1} = \frac{mv}{c^2 - v^2}$$

$$mv \, dv = (c^2 - v^2) \, dm \quad (28\text{-}13)$$

Now we substitute Eq. 28-13 into Eq. 28-12:

$$dE = (c^2 - v^2) \, dm + v^2 \, dm$$

or

$$dE = c^2 \, dm \quad (28\text{-}14)$$

This relates the increase of kinetic energy of a body to the increase of its mass. To obtain the

---

* In Newtonian mechanics m would be constant and $dm = 0$. Then Eq. 28-12 would be $dE = mv \, dv$, yielding, upon integration, kinetic energy $= E = m \int v \, dv = \tfrac{1}{2}mv^2 - \tfrac{1}{2}mv_0{}^2$. This derivation does not assume uniformly accelerated motion, as was the case in Prob. 6-B40.

kinetic energy, we integrate Eq. 28-14, starting from rest.

$$\int_{E=0}^{E=KE} dE = c^2 \int_{m=m_0}^{m=m} dm$$

or

$$KE = c^2(m - m_0)$$

This is Eq. 28-9.

28-10   The Lorentz Transformation

The mathematical framework of special relativity is a set of equations called the Lorentz transformation. In general, a *transformation* connects the values of the variables x, y, z, and t in one frame of reference FR to the corresponding variables x', y', z' and t' measured in another frame FR'.

In the classical mechanics of Galileo and Newton, relative motion is handled in a "common sense" fashion, as in Sec. 5-5 (page 114). If the motion is along the x axis and frame FR' is moving with velocity v relative to frame FR, then we see from Fig. 28-12 that the origin O' of frame FR' is displaced vt' from the origin O of frame FR; vt' increases steadily with time, and at any instant we have $x = x' + vt'$. The other coordinates, y and z, are not affected by motion along the x axis. Also, by the assumption of the uniform flow of "absolute" time, the readings of clocks in the two frames always agree

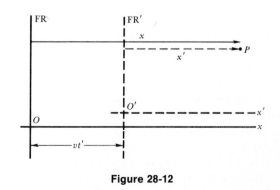

Figure 28-12

with each other. We summarize these results as follows:

Galilean transformation (valid for $v \ll c$)

$$x = x' + vt' \qquad (28\text{-}15a)$$

$$y = y' \qquad (28\text{-}15b)$$

$$z = z' \qquad (28\text{-}15c)$$

$$t = t' \qquad (28\text{-}15d)$$

We can show that the Galilean transformation is consistent with our earlier formulation for relative velocity. In Sec. 2-9 (page 37), a passenger P walked on the deck of a ship S, which in turn was moving through the ocean O. Let the ocean frame be FR and the ship frame be FR'; v is the velocity of the ship relative to the ocean and was called v_{SO} in the earlier formulation.

Taking the derivative of Eq. 28-15a with respect to time t' gives

$$\frac{dx}{dt'} = \frac{dx'}{dt'} + v$$

and since $t = t'$ in the Galilean transformation,

$$\frac{dx}{dt} = \frac{dx'}{dt'} + v$$

or

$$v_{PO} = v_{PS} + v_{SO} \qquad (2\text{-}15)$$

We now know that the Galilean transformation is an approximation—a very good one in "everyday" affairs, where v is very much less than c. The fact that the forms of the laws of mechanics (for example, the law of conservation of momentum) are the same in all inertial systems was illustrated in Sec. 28-1. Another way of stating this is to say that the laws of *mechanics* are invariant when a Galilean transformation is applied to the equations representing those laws. When it was found that the laws of *electromagnetism* are not invariant under a Galilean transformation, the search commenced for a better transformation. In 1904, H. A. Lorentz (1853–1928), a Dutch physicist who had already won the Nobel Prize for his work in electron theory, first wrote down the transformation that bears his name. His derivation was, however, based on the erroneous concept of an all-pervading ether. A year later, in 1905, Albert Einstein, unaware of Lorentz's equations, showed that the postulate of the invariance of the speed of light (page 666) led to the same result:

Lorentz transformation

$$x = \frac{x' + vt'}{\sqrt{1 - v^2/c^2}} \qquad (28\text{-}16a)$$

$$y = y' \qquad (28\text{-}16b)$$

$$z = z' \qquad (28\text{-}16c)$$

$$t = \frac{t' + vx'/c^2}{\sqrt{1 - v^2/c^2}} \qquad (28\text{-}16d)$$

Note that if $v \ll c$, the terms involving c^2 drop out and the Lorentz transformation reduces to the Galilean transformation, as it should. We shall not attempt to derive the equations of the Lorentz transformation (see, for example, Ref. 8 at the end of the chapter), but we shall use the equations to discuss several features of special relativity.

First, we note that Eq. 28-16d contains the time dilation previously studied in Eq. 28-4. The duration of an experiment can be written as $t_2 - t_1$ in frame FR and as $t_2' - t_1'$ in frame FR'. Let us assume that the experiment begins and ends at the same point x' in frame FR'.

$$t_2 - t_1 = \frac{t_2' + \dfrac{vx'}{c^2}}{\sqrt{1 - v^2/c^2}} - \frac{t_1' + \dfrac{vx'}{c^2}}{\sqrt{1 - v^2/c^2}}$$

or

$$t_2 - t_1 = \frac{t_2' - t_1'}{\sqrt{1 - v^2/c^2}} \qquad (28\text{-}17)$$

Now $t_2' - t_1'$ is the proper time t_0 for the duration of the experiment, because the start and finish of the experiment are measured at the same place x' in frame FR'. Thus we can rewrite Eq. 28-17 as

$$\Delta t = \frac{\Delta t_0}{\sqrt{1 - v^2/c^2}}$$

which is equivalent to Eq. 28-4 for the time dilation.

Equation 28-16d also shows that *simultaneity is a relative thing.* Suppose that a group of observers in frame FR observe that two events occur at the same time t. One event occurs at position x_1, and the other event occurs at position x_2. What does an observer in frame FR' conclude about these events? Using Eq. 28-16d, we calculate the times as follows:

$$t_1 = \frac{t_1' + \frac{vx_1'}{c^2}}{\sqrt{1 - v^2/c^2}}; \qquad t_2 = \frac{t_2' + \frac{vx_2'}{c^2}}{\sqrt{1 - v^2/c^2}} \qquad (28\text{-}18)$$

If $t_1 = t_2$ (simultaneity as judged in frame FR), then

$$\frac{t_1' + \frac{vx_1'}{c^2}}{\sqrt{1 - v^2/c^2}} = \frac{t_2' + \frac{vx_2'}{c^2}}{\sqrt{1 - v^2/c^2}}$$

or

$$t_1' = t_2' + v(x_2' - x_1')/c^2$$

Thus $t_1' \neq t_2'$ and the events are *not* simultaneous as viewed by an observer in frame FR'. It is this relativity of simultaneity that explains the paradox shown in Fig. 28-7, where light seemingly has to be in two different places at the same instant of time.

28-11 A Formula for Relative Velocity

What if two bodies approach each other, each moving with a speed $0.8c$ relative to their center of mass? "Common sense" would give $1.6c$ as the velocity of one body relative to the other, but we know this is impossible, for no relative velocity can exceed c, the speed of light. In the Einstein theory of relativity, the formula for relative velocity is

$$v_x = \frac{v_x' + v}{1 + \frac{v}{c^2} v_x'} \qquad (28\text{-}19)$$

In this equation* v_x is the velocity of a body as seen by an observer in one frame of reference FR; v_x' is its velocity as seen by an observer in another frame of reference FR'. The velocity of frame FR' relative to frame FR is v, and c is the speed of light. Let us see how this works out in an example where the velocities are large fractions of c.

<hr>

Example 28-3

Two bodies A and B approach each other, each moving at velocity $0.8c$ relative to their center of mass. What is the velocity of one body relative to the other?

Let A have velocity $+0.8c$ and B have velocity $-0.8c$, both relative to the center of mass (c.m.).

frame FR = body B;   frame FR' = c.m.

v_x = velocity of A relative to B = ?

v_x' = velocity of A relative to the c.m. = $+0.8c$

v = velocity of FR' relative to FR = $+0.8c$

c = velocity of light = 3×10^8 m/s

Equation 28-19 gives

$$v_x = \frac{0.8c + 0.8c}{1 + \frac{(0.8c)(0.8c)}{c^2}} = \frac{1.60c}{1.64} = \boxed{\frac{40}{41}c}$$

The velocity of A relative to B is just slightly less than c, as expected.

<hr>

If the relative velocity of the two frames is small compared with c, then the denominator in Eq. 28-19 becomes 1, and the equation becomes $v_x = v_x' + v$. This is the Galilean velocity addition equation discussed in Sec. 2-9 (page 37).

<hr>

* We assume that the relative velocity of the two frames is along the x axis, and we also assume that v_y' and v_z' are both zero, so that motion is along the x axis in each frame. A derivation of Eq. 28-19 based on the Lorentz transformation is outlined in Prob. 28-C12.

Example 28-4

A ship is moving southward at 15 m/s, and a deck-walker moves northward with a velocity of 1 m/s relative to the ship. What is the velocity of the walker relative to the water?

Here the water is one frame of reference (FR) and the ship is the other frame (FR′). Calling northward the positive direction, we have

frame FR = water; frame FR′ = ship

v_x = velocity of walker relative to FR = ?

v_x' = velocity of walker relative to FR′ = +1 m/s

v = velocity of FR′ relative to FR = −15 m/s

c = velocity of light = 3×10^8 m/s

Plugging these values into the formula, we get

$$v_x = \frac{(+1 \text{ m/s}) + (-15 \text{ m/s})}{1 + \dfrac{(-15 \text{ m/s})(+1 \text{ m/s})}{(3 \times 10^8 \text{ m/s})^2}}$$

$$= \boxed{-14.000000000000002 \text{ m/s}}$$

In carrying through the calculation, we see that the denominator is so close to 1 that it can be ignored; indeed we are not justified in carrying out the division to so many decimal places. Thus, on account of the smallness of v and v_x' in comparison with the velocity of light, the equation is essentially $v_x = v_x' + v$, the same as the *classical* result that would have been expected without relativity theory.

Problems

28-C9 A high-energy accelerator produces a pair of particles traveling in opposite directions, each with velocity $\frac{1}{2}c$ relative to the laboratory. What is the velocity of one particle relative to the other?

28-C10 Prove that the projectile's velocity relative to the earth in Example 28-1 is 2.77×10^8 m/s, as stated in the example.

28-C11 Experimenter B in Fig. 28-7, moving at velocity v relative to A, holds a flashlight that emits photons (corpuscles of light). The velocity of the photons relative to B is $+c$. Use the relative velocity formula (Eq. 28-19) to show that the velocity of the photons relative to A is also $+c$, in agreement with Einstein's second postulate.

28-C12 Derive the relative velocity formula (Eq. 28-19) from the Lorentz transformation. (*Hints*: Write Eq. 28-16a to give x_1 in terms of x_1' and t_1'; write the same equation to give x_2 in terms of x_2' and t_2'. Now form $\Delta x = x_2 - x_1$. Similarly, form $\Delta t = t_2 - t_1$ from the results given in Eq. 28-18. Finally, find $\Delta x/\Delta t$, which is v_x; the resulting equation can be simplified to give Eq. 28-19.)

References

1. R. S. Shankland, "Michelson: America's first Nobel Prize winner in science," *Phys. Teach.* **15,** 19 (1977).
2. R. S. Shankland, "The Michelson-Morley Experiment," *Sci. American* **211**(5), 107 (Nov. 1964).
3. I. B. Cohen, "An Interview with Einstein," *Sci. American* **193**(1), 68 (July 1955).
4. H. Dukas and B. Hoffmann, eds., *Albert Einstein: The Human Side* (Princeton University Press, Princeton, N. J., 1979). Letters and short writings by Einstein.
5. A. Einstein, *Relativity: The Special and General Theory* (Crown Pubs., New York, 1961). A nontechnical exposition in 32 short chapters of a few pages each, by the man who made it happen.
6. C. Kacser, *Introduction to the Special Theory of Relativity* (Prentice-Hall, Englewood Cliffs, N. J., 1967). A thorough treatment at the mathematical level of algebra and simple calculus. The Lorentz transformation is derived.
7. H. Bondi, *Relativity and Common Sense* (Anchor Books, Garden City, N. Y., 1964).

8. L. R. Lieber, *The Einstein Theory of Relativity* (Holt, Rinehart & Winston, New York, 1945). Our proof of the relativistic mass formula is based on material in this remarkable little book.

9. The "twin paradox" has been a subject of debate among relativists. (Identical twins are separated at a certain instant when one of them climbs aboard a spaceship that leaves Earth at a speed nearly equal to c. When the traveling twin returns, will he actually be physically and biologically younger than his twin who stayed at home?) See *Sci. American*, **195**(6), 58 (Dec. 1956), **196**(3), 63 (Mar. 1957), **197**(1), 68 (July 1957); R. M. Frye and V. M. Brigham, "Paradox of the Twins," *Am. J. Phys.* **25,** 553 (1957); G. Builder, "Resolution of the Clock Paradox," *Am. J. Phys.* **27,** 656 (1959); J. Bronowski, "The Clock Paradox," *Sci. American* **208**(2), 134 (Feb. 1963).

10. D. E. Hall, "Intuition, time dilation, and the twin paradox," *Phys. Teach.* **16,** 209 (1978). Includes a description of the Hafele-Keating experiment on time dilation.

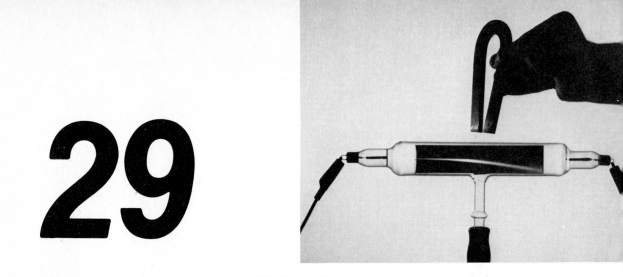

29

Electrons and Photons

You are familiar with the general idea of atomic structure—the massive positively charged nucleus surrounded by a cloud of negatively charged electrons that make up the outer atom. In the next five chapters we consider the finer details of the picture, study the ways in which knowledge of atomic structure has come to us, and see how that knowledge can be put to use. We shall find that the nucleus is complex and often unstable. We shall find that the emission and absorption of light and other e-m radiation are closely related to the electrical structure of the outer atom, but in a strange and "non-classical" way. To begin our study of the atom we consider the corpuscular nature of electric charge (the electron), and the corpuscular nature of light (the photon).

29-1 *The Charge of the Electron*

The facts of electrolysis (Sec. 18-6) strongly suggest that the charges of ions such as Cu^{2+} are multiples of some fundamental unit of charge, which we call the electron. Strictly speaking,

electrolysis experiments prove only that the *average* charge per univalent ion is 1.6×10^{-19} C. The individual charges might differ greatly among themselves and still average out to 1.6×10^{-19} C when many ions are considered. (Even in the most delicate weighing experiments, we are dealing with some 10^{16} ions.) We have already met with such a statistical averaging-out effect in our study of the kinetic theory of gases. At any given temperature, the molecules of a gas have many different speeds, ranging from very slow to very fast, and yet the average speed has a definite value.

In his oil-drop experiment, Robert A. Millikan (1868–1953) succeeded in measuring the effect of single electrons. Tiny drops of mineral oil sprayed from an atomizer are allowed to fall into the space between two charged plates; many of the drops are charged by friction during the atomization process. In Fig. 29-1 the observer has selected a negatively charged drop that is held stationary by applying a positive potential to the upper plate. The drop is subject to two forces: its weight W, due to the downward gravitational field, and an upward force $F = QE$,

Cathode rays (electrons) are deflected downward as they move toward the right through a region of magnetic field. Can you tell which horizontal direction the magnetic field supplied by the horseshoe magnet is in?

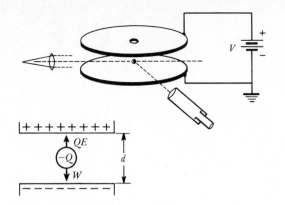

Figure 29-1 Millikan's oil-drop experiment to measure the charge of the electron.

due to the electric field E. The electric field's intensity, in newtons per coulomb, equals the potential gradient V/d, in volts per meter (Sec. 19-5). When the electric field is adjusted until the drop is motionless,

$$\text{net } F = 0$$

$$Q\frac{V}{d} - W = 0$$

$$Q = \frac{Wd}{V} \qquad (29\text{-}1)$$

Since d and V are easily measured, the problem of finding Q reduces to that of finding W. In the Millikan experiment the weight of each drop is measured indirectly by turning off the field and allowing the drop to fall by the action of gravity alone. The terminal velocity (Sec. 12-9) depends on the size of the drop and the viscosity of air. By judicious use of a switch to turn the electric field on or off, a single drop can be kept moving up and down in the field of view for hours at a time, and its radius can be computed.* Having determined the drop's size, we can calculate its weight from the density of the oil. Equation 29-1

---

* The drop is so small that diffraction effects prevent direct measurement of its radius. Only a point of light is seen in the microscope.

can then be used to compute the charge Q on any drop.

Millikan found the charge on each drop to be an integral multiple of 1.60×10^{-19} C. For instance, five successive experiments on five drops might give charges as follows (all in units of 10^{-19} C): -3.20; -8.00; $+4.80$; -1.60; $+3.20$. No drop having charge less than 1.60×10^{-19} C was ever observed. In addition, any given drop was occasionally seen to change its charge abruptly, always by a small multiple of 1.60×10^{-19} C. We interpret these facts as showing the existence of an elementary charge, which we call the electron. A drop gains or loses an integral number of electrons, either by friction or by some other means. If an ion bumps into a drop that is just balanced between the plates, one or more electrons may be transferred to or from the drop, which then takes off suddenly either upward or downward, depending on the sign of its charge and the direction of the electric field. The PD between the plates must then be adjusted to a new value to balance the drop with its new charge, and a new calculation can be made.

The oil-drop experiment shows the atomic† nature of charge. At least in this experiment and, by inference, in general, we conclude that electric charge is *quantized*, rather than subdivisible into infinitely small fragments. The experiments of electrolysis are consistent with the quantization of charge, but do not require it. The oil-drop experiment sharpens our concept of electric charge by showing it to come in small units or packages. This amount of charge is the electronic charge, of magnitude $e = 1.60 \times 10^{-19}$ C.

29-2 The Mass of the Electron

Nothing in the preceding section implies that electrons are themselves particles having mass. Evidence for small negatively charged particles comes from studies of the discharge of electricity through gases. A typical experiment is shown in

---

† We use the word in its general sense, meaning indivisible.

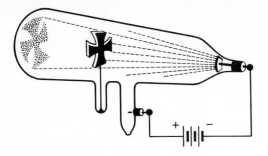

Figure 29-2 Cathode rays.

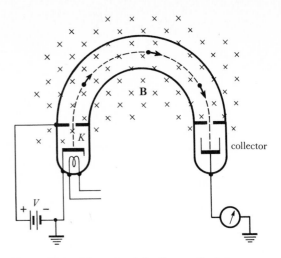

Figure 29-3 Magnetic deflection method for measurement of the electron's e/m ratio.

Fig. 29-2. The tube contains air at low pressure (about 10^{-4} torr). When the connections are as shown, charge flows through the tube, and the gas glows. Using the modern interpretation, we identify the carriers of charge as positive and negative ions of oxygen and nitrogen, as well as electrons. Positive ions move in one direction, and negative ions in the opposite direction. In addition, a stream of particles leaves the *negative* terminal (the cathode) and proceeds in straight lines, perpendicular to the surface of the cathode. The passage of the particles is made evident by a bluish glow in the gas, and their straight-line propagation is shown by the sharp shadow cast by an obstacle. The particles evidently have inertia, for they exert a measurable force on the obstacle, and they carry energy, since a thin metallic target can be made red-hot by the impact of the particles. The moving particles can be deflected by a magnetic field, and so they must carry electric charge (Sec. 22-2). The direction of the magnetic force on the moving particles shows them to be negative. We now know that these *cathode rays* are electrons that have been ejected from the cathode by the impact of the heavy ions. Exactly similar mechanical and heating effects are observed for a stream of electrons emitted by a heated metal. Cathode rays were studied extensively by the British physicist J. J. Thomson (1856–1940), who is commonly regarded as the discoverer of the electron.

To measure the mass of the electron, we must somehow apply Newton's second law. The essential features of the *deflection method* of measur-

ing the electron's e/m ratio are shown in Fig. 29-3. Electrons, emitted from an indirectly heated cathode K in an evacuated tube, are accelerated to some velocity **v** by an applied potential difference V. Some of the electrons pass through a small hole or slit in the accelerating electrode and enter a region where a magnetic induction **B** is applied at right angles to their motion. The magnetic force **F** serves as a centripetal force to cause the particles to move in a circle whose radius R is determined by the slits. To make a measurement, the magnetic induction **B** is adjusted until the electron current into the collector cup shows a sharp maximum.

The magnitude of the velocity v is found by the energy principle. Each electron (of charge e) falls through a PD equal to V, and hence loses an amount of PE equal to Ve.

Increase of KE = decrease of PE

$$\tfrac{1}{2}mv^2 = Ve$$

$$v^2 = \frac{2Ve}{m} \tag{29-2}$$

The centripetal acceleration is, therefore,

$$a = \frac{v^2}{R} = \frac{2Ve}{mR}$$

Now we apply Newton's second law to the circular motion:

$$\text{net } F = ma$$

$$Bev = ma$$

$$Be\sqrt{\frac{2Ve}{m}} = m\left(\frac{2Ve}{mR}\right)$$

$$\frac{e}{m} = \frac{2V}{B^2R^2} \qquad (29\text{-}3)$$

Using Eq. 29-3, we can determine the *ratio* of charge to mass for an electron by measuring V, B, and R. Experiments of this sort show that all electrons have the same e/m ratio for nonrelativistic speeds.* At higher speeds the ratio e/m decreases by just the right amount as demanded by relativity, since the mass increases according to the formula $m = m_0/\sqrt{1 - v^2/c^2}$. The experimental value of e/m_0 is 1.759×10^{11} C/kg. Combining this with the value of e gives for the rest mass of the electron

$$m_0 = \frac{e}{e/m_0} = \frac{1.602 \times 10^{-19} \text{ C}}{1.759 \times 10^{11} \text{ C/kg}}$$

$$= 9.11 \times 10^{-31} \text{ kg}$$

The combined result of the experiments described so far in this chapter show that (1) charge is quantized, and (2) free electrons exist, all of which are identical, each carrying one unit of charge always associated with the same amount of mechanical inertia.

29-3   The Photoelectric Effect

With the development of high-vacuum techniques in the late 19th century, it became possible to study the emission of electrons from a surface under the influence of light. In the early photocell circuit of Fig. 29-4, light shining on a

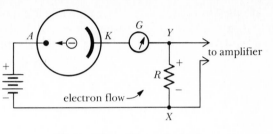

Figure 29-4   Photocell circuit.

cathode K may cause electrons to leave the surface and travel through the vacuum to the positive electrode A. The electrons then flow through the battery, the resistor R, and the galvanometer G back to K again. As might be expected, the current is proportional to the intensity (power per unit area) of the light, and so the photocell current can actuate a relay to open doors, actuate smoke detectors, and so on. The PD between X and Y is applied to the input terminals of an amplifier; this PD equals the IR drop in the resistor R. If R is large enough, even a weak current may cause the amplifier input to change by several volts. In a gas-filled photocell, the electrons create ions by collision as they move between the electrodes, and the total current is much greater than for a vacuum photocell.

Quantitative study of the photocell current reveals some surprising facts. For any surface, there is a *threshold frequency* for the light; if the frequency is lower than the threshold, no photoelectrons are emitted no matter how intense the light. Also, if the frequency is above the threshold, all the photoelectrons that are ejected from the top layers of a surface have the same KE, regardless of the intensity of the light.[†] One might have expected a bright light to eject electrons with a greater velocity than a less bright light, but experiment shows that a bright light merely ejects *more* electrons. To illustrate this

---

* If v is an appreciable fraction of c, the electron is "relativistic," and the calculation is somewhat more complicated since the classical formula for KE cannot be used in deriving Eq. 29-2.

[†] This is an idealization, true for thin metallic films. In actual experiments, an electron may lose some of its energy while traveling through the metal to reach the surface.

strange situation, let us imagine a photocell surface whose threshold frequency is 6×10^{14} Hz. This corresponds to green light of wavelength 5×10^{-7} m. If the surface is illuminated by strong *red* light, not a single photoelectron is ejected, since the frequency of red light is less than the threshold frequency. On the other hand, the faint light from a distant *blue* star can eject electrons, since the frequency of blue light is greater than the threshold frequency; the KE of these ejected electrons does not depend on the star's brightness. Classical physics was unable to cope with these experimental facts.

We discussed one of the early symptoms of the inadequacy of classical physics in Sec. 16-4. We saw that the idea of a quantum of energy explains the decreased specific heat capacity of hydrogen gas at very low temperatures, for the energy of the collisions is then insufficient to give a molecule even one quantum of rotational energy. In 1900 the German physicist Max Planck (1858–1947) was able to explain the shape of the blackbody radiation curves (Fig. 14-4) by making a quantum assumption. An incandescent solid body emits e-m radiation because it contains many oscillators (of atomic dimensions). According to Sec. 24-6, each oscillator serves as an antenna to radiate an e-m wave. Planck's new assumption was that any given tiny oscillator cannot vibrate with an arbitrary amplitude; its energy E is quantized according to the equation

$$E = nhf \qquad (29\text{-}4)$$

In this equation n is an integer, f is the frequency of the oscillator (determined by its mass and its force constant), and h is a universal constant that has since been known as *Planck's constant*. The value of h is 6.63×10^{-34} J·s.

Einstein boldly applied a quantum hypothesis to the radiation itself: he assumed (in 1905) that the energy of e-m radiation is absorbed one quantum at a time. Such a quantum of radiation energy is now called a *photon*. For light of any frequency f, the energy of a photon is given by

$$E = hf \qquad (29\text{-}5)$$

Thus, light comes in packages, and the size of the package depends on the color of the light.

Example 29-1

What is the energy of a photon of green light?
If we take 500 nm as the average wavelength of green light, then

$$f = \frac{c}{\lambda} = \frac{3 \times 10^8 \text{ m/s}}{5 \times 10^{-7} \text{ m}} = 6 \times 10^{14} \text{ s}^{-1}$$

The energy of a photon for this light is

$$E = hf$$
$$= (6.63 \times 10^{-34} \text{ J·s})(6 \times 10^{14} \text{ s}^{-1})$$
$$= \boxed{3.98 \times 10^{-19} \text{ J}}$$

Einstein's photoelectric equation states that the KE of the ejected electron equals the energy of the incoming photon, minus the work w, called the *work function*, needed to remove the electron from the surface. In symbols,

$$\tfrac{1}{2}mv^2 = hf - w \qquad (29\text{-}6)$$

We see from this equation that no electrons can be emitted unless hf is greater than w; hence we infer the existence of some threshold frequency. The equation also tells us that the KE of an ejected electron depends only on the frequency of the light and the work function of the surface. We interpret a bright red light as a stream of many relatively weak photons, and a faint violet light as a stream of a few, more energetic, photons.

At this point we introduce a new unit of energy called the *electron volt*, which we shall use repeatedly in our study of atomic physics. *An electron volt (eV) is the energy acquired by an electron in moving through a PD of one volt.* To calculate its value, we use Eq. 19-3.

$$W_{B \to A} = Q V_{AB}$$
$$1 \text{ eV} = (1.60 \times 10^{-19} \text{ C})(1 \text{ J/C})$$

or

$$1 \text{ eV} = 1.60 \times 10^{-19} \text{ J}$$

This is a handy unit for photoelectricity, since the work functions of most metals lie in the range of 1 to 10 eV. For the sake of convenience, we also calculate the energy of a photon in electron volts, as follows:

$$E = hf = h\frac{c}{\lambda}$$

$$= (6.63 \times 10^{-34} \text{ J} \cdot \text{s})\left(\frac{3 \times 10^8 \text{ m/s}}{\lambda}\right)$$

$$\times \left(\frac{1 \text{ nm}}{10^{-9} \text{ m}}\right)\left(\frac{1 \text{ eV}}{1.60 \times 10^{-19} \text{ J}}\right)$$

where λ is the wavelength *in nanometers*. Simplifying, we obtain*

$$E = \frac{1240 \text{ eV} \cdot \text{nm}}{\lambda} \tag{29-7}$$

If λ is in nanometers, E is in electron volts. The usefulness of the electron volt as an energy unit is seen in the following example.

Example 29-2

The work function for sodium metal is 2.46 eV. With what KE are photoelectrons ejected from a thin sodium surface illuminated by light of wavelength 400 nm?

First we calculate the photon energy in electron volts.

$$E = \frac{1240 \text{ eV} \cdot \text{nm}}{400 \text{ nm}} = 3.10 \text{ eV}$$

This is the energy of each photon of the light of wavelength 400 nm. Now, using Einstein's photoelectric equation (Eq. 29-6), we find

$$KE = hf - w$$

$$= 3.10 \text{ eV} - 2.46 \text{ eV} = \boxed{0.64 \text{ eV}}$$

In practice, this is an upper limit; some photoelectrons may have less than this energy if they

---

* If λ is in angstrom units (1 Å = 10^{-10} m), the numerator is replaced by 12,400 eV·Å.

---

lose energy by collisions as they make their way through the metal to the surface.

Example 29-3

What is the threshold wavelength for a sodium surface?

At the threshold, no energy is available for KE. Hence,

$$0 = hf - w$$

and the photon energy at the threshold is

$$E = hf = w = 2.46 \text{ eV}$$

$$\lambda = \frac{1240 \text{ eV} \cdot \text{nm}}{E} = \frac{1240 \text{ eV} \cdot \text{nm}}{2.46 \text{ eV}} = \boxed{504 \text{ nm}}$$

The threshold is in the green region of the spectrum (see Table 25-1).

Example 29-4

Through what PD would an electron have to fall in order to acquire as much energy as a dust particle has, if the dust particle has a mass of 0.0004 g and is falling at a speed of 1 cm/s?

The KE of the dust particle is first calculated in joules and then converted to electron volts.

$$KE = \tfrac{1}{2}mv^2 = \tfrac{1}{2}(4 \times 10^{-7} \text{ kg})(10^{-2} \text{ m/s})^2$$

$$= 2 \times 10^{-11} \text{ J}$$

$$= (2 \times 10^{-11} \text{ J})\left(\frac{1 \text{ eV}}{1.6 \times 10^{-19} \text{ J}}\right)$$

$$= 1.25 \times 10^8 \text{ eV} = 125 \times 10^6 \text{ eV}$$

$$= \boxed{125 \text{ MeV}}$$

The dust particle's KE is 125 million electron volts (125 MeV), and an electron would have to fall through a PD of 125 million volts to have as much energy.

The vacuum photocell has now largely been replaced by solid-state devices. The absorption of a photon's energy hf can in some n-type materials remove some bound electrons from

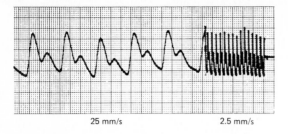

25 mm/s   2.5 mm/s

Figure 29-5 Photoelectric plethysmograph measures blood flow in the capillary bed that nourishes the skin of a patient's toe. The sensing probe contains a tiny light source and a photodiode that responds to light reflected from the arterial blood in the peripheral vascular bed over which it is placed. The amount of light that is registered on the recording chart is directly proportional to the amount of pulsating arterial blood. Pulse features can be studied at the faster chart speed (left). The instrument measures blood flow, not pressure, and is valuable as a routine check before surgery, as well as in assessing the likelihood of the onset of gangrene.

their parent atoms, creating an excess of free electrons. (In *p*-type material the liberated electrons cancel some of the positive holes already present.) In either case, the current through the material changes when light shines on it; we have a *photodiode*. Unlike vacuum-tube devices, a photodiode can be made sensitive to the low-energy photons of infrared radiation. Among the many uses are burglar alarms, the laser cane for the blind (Sec. 30-6), and the photoplethysmograph (Fig. 29-5).

29-4 The Photon as a Corpuscle

The first subatomic particle to be discovered was the electron. We have seen two rather different attributes of the electron: It is a quantum of charge (1.60×10^{-19} C); it is also a corpuscle of matter (9.11×10^{-31} kg), and hence, because of its inertial mass, an electron can have momentum and kinetic energy. Presumably, an electron also has gravitational mass (see the last paragraph of Sec. 8-1) and therefore has weight and can have gravitational potential energy. Let us

now consider the corpuscular nature of the photon—it is a quantum of *energy*. To be sure, some photons carry more energy than others (according to $E = hf$), but the photoelectric effect shows clearly that the energy is absorbed a quantum at a time. By itself, the photoelectric effect shows nothing one way or the other about the possibility that a photon is a corpuscle with localized *mechanical* properties such as momentum and mass. Such evidence came from x-ray photons.

We shall study the production of x rays in Sec. 31-7. At this time we need only know that x rays are electromagnetic radiation whose photons have high energy (typically 5000 to 100,000 eV) and correspondingly short wavelength (typically 0.1 to 2 Å). In his study of the scattering of x rays, the American physicist Arthur Compton in 1922 observed a small change in wavelength, now known as the *Compton effect*. In a typical case, the molybdenum $K\alpha$ line of wavelength 0.710 Å was changed to 0.734 Å when scattered by the electrons in a piece of carbon. Although this is a small difference (about 3%), Compton's analysis of his experiment was destined to play a major role in the development of the dual theory of light.

The experimental setup is shown in Fig. 29-6. X rays from a molybdenum target strike a car-

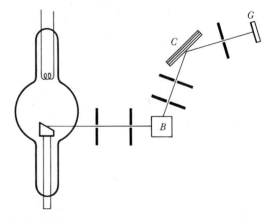

Figure 29-6 Compton effect: experimental arrangement for studying change of wavelength of x-ray photons scattered by carbon block *B*.

bon block B, and the scattered rays are analyzed by a crystal C, which serves as a grating. The x-ray beam is defined by a series of slits, and the scattered x rays are detected by a Geiger counter (Sec. 32-9), photographic plate, or ionization chamber G. A man-made grating would be too coarse to give a diffraction pattern, for to be effective the grating space must be comparable with the wavelength. Fortunately, the atoms in a crystal are about an angstrom apart, and so the layers of atoms—so-called crystal planes—can serve as a natural grating for x-ray diffraction. In England, before World War I, the Braggs, father and son,* had developed the crystal spectrometer used by Compton; as so often happens, improved instrumentation was a necessary prelude to a critical experiment.

Accurate data are not enough; physical insight and intuition are also needed to build a theory. Compton successfully applied the laws of mechanics to the elastic collision between an x-ray photon and a "free" electron in the carbon.[†] The photon is deflected by the electron, and the electron acquires a recoil velocity in some direction, as shown in Fig. 29-7. To apply the law of conservation of momentum to this collision, we must have a formula for the momentum of a photon. Such a formula comes from relativity theory. Recall that the mass of any particle is given by

$$m = \frac{m_0}{\sqrt{1 - v^2/c^2}} \qquad (28\text{-}7)$$

where m_0 is the rest mass. The rest mass m_0 of a photon is 0, since a photon cannot be at rest in any frame of reference (Einstein's second postulate). However, the total *relativistic* mass is not zero. This is consistent with Eq. 28-7, because then

$$m = \frac{0}{\sqrt{1 - c^2/c^2}} = \frac{0}{0}$$

---

* Sir William Henry Bragg and Sir William Lawrence Bragg.
[†] Since carbon has an atomic number of only 6, even an electron in an inner (K) shell is loosely bound, and the few hundred eV required to remove it is negligible compared with the energy of the x-ray photon.

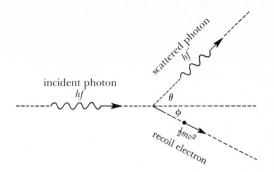

Figure 29-7 Conservation of energy during scattering of a photon by an electron.

which is an indeterminate expression. To calculate the mass of a photon we use $E = mc^2$; we see that a photon's mass is proportional to its energy: $m = E/c^2$. The momentum is, as always, the product of mass and velocity.

$$p = mc = \frac{E}{c^2} \cdot c = \frac{hf}{c} = \frac{h}{\lambda}$$

where we use the fact that the wavelength λ is c/f. Thus the magnitude of a photon's linear momentum is

$$p = \frac{h}{\lambda}$$

or

$$\text{Momentum} = \frac{h}{\text{wavelength}} \qquad (29\text{-}8)$$

Now we are ready to apply the conservation laws to the collision of a photon with an electron that is originally at rest.[‡] By the law of conservation of energy,

$$
\begin{array}{ccc}
\text{energy of} & \text{energy of} & \text{KE of} \\
\text{original} & = \text{scattered} & + \text{recoil} \\
\text{photon} & \text{photon} & \text{electron}
\end{array}
$$

$$hf = hf' + \tfrac{1}{2}mv^2 \qquad (29\text{-}9a)$$

[‡] For simplicity we use $\tfrac{1}{2}mv^2$ for the KE of the recoil electron, which is valid for $v \ll c$. Using the correct relativistic formula ($\text{KE} = mc^2 - m_0c^2$) leads to the same final results for the wavelength change in the Compton effect.

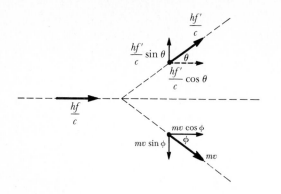

Figure 29-8 Conservation of momentum during scattering of a photon by an electron.

If the scattered photons are observed at $\theta = 90°$, Eq. 29-10 gives

$$\lambda' - \lambda = \frac{h}{m_0 c}(1 - 0)$$

or

$$\Delta\lambda = \frac{h}{m_0 c}$$

$$= \frac{6.63 \times 10^{-34} \text{ J} \cdot \text{s}}{(9.11 \times 10^{-31} \text{ kg})(3.00 \times 10^8 \text{ m/s})}$$

$$= 2.43 \times 10^{-12} \text{ m}$$

Since $1 \text{ Å} = 10^{-10}$ m, we see that at $90°$ the Compton shift is

$$\Delta\lambda = 0.0243 \text{ Å}$$

This equation has two unknowns, f' and v. The law of conservation of momentum gives two additional equations, one for the x components of momentum and one for the y components (Fig. 29-8).

$$\frac{hf}{c} = \frac{hf'}{c}\cos\theta + mv\cos\phi \qquad \text{(29-9b)}$$

$$0 = \frac{hf'}{c}\sin\theta - mv\sin\phi \qquad \text{(29-9c)}$$

Experimentally, the slits (Fig. 29-6) are adjusted to select a given angle θ for the scattered photon; we can then solve the three equations simultaneously to find the three unknowns, which are f', v, and the angle ϕ at which the electron recoils.

The result of this analysis is most easily written in terms of $\Delta\lambda$, which is $\lambda' - \lambda$, the difference in wavelength between the scattered photon and the original photon. According to Eq. 29-9a, the scattered photon always has less energy (hf') than the original photon (hf); therefore, since f decreases, λ increases. Compton showed that the change in wavelength is given by

$$\lambda' - \lambda = \frac{h}{m_0 c}(1 - \cos\theta) \qquad \text{(29-10)}$$

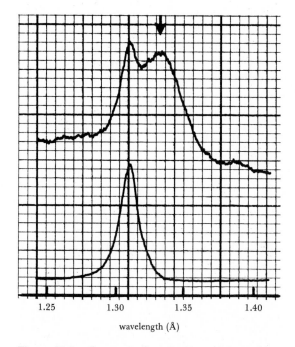

Figure 29-9 Compton effect. Incident photons (lower curve) with $\lambda = 1.314$ Å strike a paraffin block and are scattered through $90°$ to give the upper curve. The arrow indicates the wavelength of the scattered photons.

This is called the *Compton wavelength* of the electron—a "natural" unit of length in dealing with many processes on the atomic scale of things. Figure 29-9 shows the Compton shift for x rays scattered through 90° by a block of paraffin.

Example 29-5

X rays of wavelength 1.314 Å are scattered by a free electron. If the electron is knocked "head-on" in the forward direction, calculate (a) the wavelength of the scattered photon; (b) the energy of the scattered photon; (c) the KE of the recoil electron.

Since the electron recoiling in the forward direction has no y component of momentum, and the original photon had no y component of momentum, we conclude that the scattered photon must also have zero y component of momentum and hence must be scattered directly backward. We use Eq. 29-7 as an easy way of finding the energies of the photons involved (replacing the numerator by 12,400 eV·Å). The energy of the incoming photon is

$$E = \frac{12{,}400 \text{ eV·Å}}{1.314 \text{ Å}} = 9437 \text{ eV}$$

(a) Since the photon is scattered straight back, $\theta = 180°$ in Eq. 29-10.

$$\Delta\lambda = (0.0243 \text{ Å})(1 - \cos 180°) = 0.0486 \text{ Å}$$

$$\lambda' = 1.314 \text{ Å} + 0.049 \text{ Å} = \boxed{1.363 \text{ Å}}$$

(b) $\qquad E' = \dfrac{12{,}400 \text{ eV·Å}}{1.363 \text{ Å}} = \boxed{9098 \text{ eV}}$

(c) By conservation of energy, the KE of the recoil electron is

$$E_{\text{recoil}} = 9437 \text{ eV} - 9098 \text{ eV} = \boxed{339 \text{ eV}}$$

You may ask, why wasn't the Compton effect observed long ago, since visible light ($\lambda = 5000$ Å) is scattered, as by air molecules in the case of the blue sky (Fig. 26-35)? The maximum change of wavelength for scattering by a free electron is only 0.049 Å—a small amount, to be sure, but one that could be measured with a high-resolving-power grating spectrometer. However, for scattering by an air molecule, the m_0 in the denominator of Eq. 29-10 is that of a molecule—thousands of times more massive than a single electron—and $\Delta\lambda$ is correspondingly smaller than 0.049 Å. Similarly, scattering by intact, massive carbon atoms in the paraffin accounts for the unmodified line in Fig. 29-9. One might try to observe scattering of visible light by free electrons in an electron gas, but this is difficult because such a gas is electrically conducting and therefore opaque.

The next example shows that Compton scattering gives an overwhelmingly large change in wavelength and energy for the scattering of the high-energy photons known as gamma rays.

Example 29-6

A gamma ray emitted by the ^{60}Co nucleus, of energy 1.330 MeV (million electron volts), is scattered by an electron in a solid-state detector. What is the energy of a scattered photon that is at 90° from the direction of the incident photon?

$$\lambda = \frac{12{,}400 \text{ eV·Å}}{1.330 \times 10^6 \text{ eV}} = 0.0093 \text{ Å}$$

$$\lambda' = \lambda + (0.0243 \text{ Å})(1 - \cos 90°)$$
$$= 0.0093 \text{ Å} + 0.0243 \text{ Å} = 0.0336 \text{ Å}$$

$$E' = \frac{12{,}400 \text{ eV·Å}}{0.0336 \text{ Å}} = 0.369 \times 10^6 \text{ eV}$$

$$= \boxed{0.369 \text{ MeV}}$$

The scattered photon has less than $\frac{1}{3}$ the energy of the incident photon; the balance of the energy is taken up by the recoiling electron in the solid.

Compton not only measured the change in the wavelength of the scattered photon, verifying

Eq. 29-10 for various values of θ; he also measured the recoil electron's energy and direction, using a Geiger counter as a detector, and found them to be in precise agreement with the theory. Compton and his co-workers were also able to show that the scattered photon and the recoil electron entered their respective Geiger counters at the same instant (the time of flight was negligibly short), and so they must be the end results of a single collision process.

The Compton effect is important because it shows the photon to have corpuscular properties, in the same sense that a bullet or billiard ball has. This goes beyond the concept of a photon that was built up to explain the photoelectric effect. In addition to considering the photon to be a quantum of energy, now we must consider it also to have mechanical properties such as inertia and momentum. The model of light as a stream of bullets seems inescapable.

Let us stop for a moment and take stock of the properties of the electron and the photon, as they have been revealed by experiments described so far in this chapter. The electron has mass and a negative charge. The photon has no rest mass, but it has energy and momentum. These corpuscles can interact in collisions. It is perhaps useful to recall that the theory of relativity does not make a sharp distinction between mass and energy. Thus the corpuscular nature of mass and energy are no doubt intimately related. The quantum nature of charge is still not well understood, nor is it clear whether charge and energy (that is, mass) are fundamentally related to each other. These questions, like those about the basis of gravitation, are challenges to the best physicists of today.

29-5 The Duality of Light

The Compton effect (Fig. 29-6) illustrates the wave-particle duality of light in an extreme form. In the very apparatus by which the bullet-like character of x-ray photons is revealed, these same x rays show their wave character when reflected by the crystal grating. In Chap. 26 we took the phenomenon of interference to be an indisputable proof for a wave motion and were at great pains to build up a wave model for light (including x rays and all other e-m radiation). Now we see a corpuscular aspect of light. How can we reconcile these conflicting wave and particle models for e-m radiation?

In the first place, we note that the particle aspects of radiation come to the fore only when radiation interacts with matter. Photons are involved in the emission or absorption of light (blackbody radiation, spectrum lines, photoelectric effect), or in its scattering by an electron (Compton effect). On the other hand, the wave aspects are evident when light travels from one place to another without losing energy (Huygens' principle, interference, diffraction). The two aspects of light seem to be important at different times. One way out of the difficulty is to cry "modelitis." The function of a model is to make a new phenomenon intelligible in terms of familiar laws and events of our everyday, large-scale world. Note how we have used large-scale models in our description of e-m waves (comparing them with water waves) and also in our description of photons (comparing them with billiard balls). It may well be, however, that *no* intuitively satisfying model for the small-scale phenomenon of light can exist. There is no reason why the behavior of a single photon should be the same as that of *any* large-scale object with which we are familiar. Many physicists prefer to accept light for what it is and attempt no models. In fact, half a century of living with the strange duality of photons has bred a certain familiarity with light; advanced workers now use photons as "understandable" models for still more obscure phenomena!

Insofar as a model is possible, we can visualize the e-m waves of Maxwell as *probability waves*. There is a large probability of finding a photon at any point where the amplitude of the wave is large. If two probability waves interfere destructively to form a node, the probability of

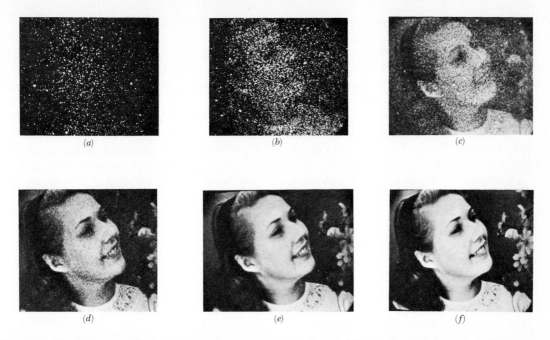

Figure 29-10 Photons of light needed to form a complete picture. The numbers given below tell how many photons were used to form each of these reproductions of the same image. (*a*) 3×10^3 photons. (*b*) 1.2×10^4 photons. (*c*) 9.3×10^4 photons. (*d*) 7.6×10^5 photons. (*e*) 3.6×10^6 photons. (*f*) 2.8×10^7 photons.

finding a photon at the node is zero. Suppose a pair of slits and a screen are placed near a source, as in Young's experiment (Fig. 26-3 on page 604). The mere presence of the slits and screen sets up a pattern of probability waves, which interfere constructively to form an antinode at A and a node at Y. The spacing of these nodes of probability depends on the spacing of the slits and on the wavelength. If now the source is turned on, photons are emitted, one at a time, and reach the screen. We cannot trace the path of any photon from S through either slit to A, but we know that it is highly probable that a photon will appear at A. Let us assume that the screen is a photographic plate with many silver bromide molecules; each photon that is absorbed gives up its energy hf to a single grain of the plate. As soon as the source is turned on, silver bromide

grains become activated,* more or less at random; after a while, the pattern becomes evident, and we realize that A is a probability antinode and Y a probability node. Experiments of this sort have been performed, using very weak sources of light. Individual photons were detected by a row of Geiger counters in place of the photographic plate. Figure 29-10 shows how an image formed by a lens is actually a probability distribution for the arrival of photons.

Probability waves are also involved in the photoelectric effect. Even before the Compton effect showed that photons have mass and are concentrated bundles of energy, it was known

---

* The photon's energy hf is stored in the grain as chemical PE and is used in the development process to help along the chemical reaction that makes metallic silver.

that very weak light can eject photoelectrons if the frequency is above the threshold frequency as determined by the work function. A serious difficulty arises when the process is considered from the wave point of view. For instance, let a very weak source shine on a metal surface of area 1 m². About 10^{20} atoms lie in a surface of this area. To eject a photoelectron might require about 10^{-18} J (~ 6 eV), and if the intensity of the light is 10^{-6} J/s·m², each of the 10^{20} atoms in the surface could absorb only 10^{-26} J/s. It would require 10^8 s (over 3 years!) for an atom to absorb enough energy to eject a photoelectron. Yet experiment shows that *some* photoelectrons are emitted within a fraction of a second. Once again, we are forced to assume that the energy of the incoming light is carried by small bulletlike photons, which, in the case of a very weak source, are powerful enough but few and far between. We visualize the e-m wave as a probability wave such that it is equally likely for a photon to be absorbed anywhere on the surface, and hence photoelectrons are emitted at random. This is the way in which a uniform beam of light is interpreted according to the probability wave model.

If a similar duality were true for water waves, we would have the following strange phenomenon: A Coke bottle dropped off a New York pier 3 m above water level sends out a wave front that spreads out in ever-widening circles until it reaches Lisbon, Bordeaux, and Dublin. At some unpredictable place, say Lisbon, a similar Coke bottle floating near a pier suddenly pops up 3 m out of the water, and simultaneously the very weak and undetectable water wave that was in the form of a large circular wave front disappears, all of its energy being given to the Lisbon bottle. If this sort of thing did happen, we would be tempted to talk about water waves as probability waves for the transmission of bundles of energy. The photoelectric effect for weak light is no stranger than this, and equally inexplicable by the classical wave model of e-m waves.

29-6   *The Duality of Matter*

According to the dual model of light, a probability wave is associated with every photon, and the wavelength of the probability wave can be measured by interference experiments with gratings and the like. In 1923 the French student Louis de Broglie proposed that probability waves are also associated with ordinary particles of matter, such as electrons and baseballs.* To pursue the analogy between photon waves and matter waves, let us first recall the relativistic result for a photon's momentum. Equation 29-8 can be written as

$$\lambda = \frac{h}{\text{momentum}}$$

This says that the wavelength of the photon's probability wave equals Planck's constant divided by the photon's momentum.

De Broglie assumed that the same equation holds for matter waves; that is, a probability wave is associated with *any* moving body that has momentum. For matter waves, the wavelength of the probability wave is given by

$$\lambda = \frac{h}{mv} \qquad (29\text{-}11)$$

As our next example shows, the wavelengths associated with ordinary bodies are much too small to be observed.

Example 29-7

Calculate the de Broglie wavelength associated with a 20-g Ping-Pong ball that is moving at 400 cm/s.

---

* De Broglie's term for the associated waves was "les ondes pilotes." However, a literal translation as "pilot waves" is excessively anthropomorphic, so we shall use the less vivid terminology of *probability waves*, or simply *de Broglie waves*.

$$\lambda = \frac{h}{mv} = \frac{6.63 \times 10^{-34} \text{ J} \cdot \text{s}}{(20 \times 10^{-3} \text{ kg})(4.00 \text{ m/s})}$$

$$= \boxed{8.29 \times 10^{-33} \text{ m}}$$

This fantastically short wavelength is unobservable. A Ping-Pong ball headed in the general direction of an open window proceeds in a straight line with no fuzzy "shadow." It either goes out the window or is reflected back, depending on how it is aimed. Relative sizes are important. The wave nature of visible light escaped notice for so long because λ is so much smaller than the size of an ordinary opening such as a window. Therefore, according to Eq. 26-2, the angle of diffraction (1st-order minimum given by $\sin \theta = \lambda/w$) is so small that the light travels in straight lines as far as one can tell in this case. The diffraction is there, to be sure, but it is too small to be observed; λ/w is too small. Only when λ approaches the size of the slit or obstacle does the diffraction of light become evident. The same argument applies with even greater force to the ball of Example 29-7. The probability wave associated with the ball has a wavelength that is only about 10^{-26} as large as a light wave. We are quite justified in using Newton's laws of motion for the motion of a Ping-Pong ball.

Passing now to particles of smaller mass, we see from Eq. 29-11 that the wavelengths associated with moving electrons may well be observable. In 1927, C. J. Davisson and L. H. Germer in the United States interpreted their experiments on the scattering of electrons by a metal surface as indicating the wave nature of the electron. Independently, in England, wave effects were sought for and found in 1927 by G. P. Thomson, son of J. J. Thomson (Sec. 29-2). The 1937 Nobel prize in physics was shared by Davisson and G. P. Thomson for their discovery of the *wave* nature of the electron; it is interesting to note that in 1907 J. J. Thomson had received the Nobel prize for his experimental proof of the *corpuscular* nature of the electron. The Thomsons, father and son, brilliantly demonstrated two opposite, but complementary, aspects of the nature of matter.

The wave nature of matter has also been shown experimentally for nuclear particles, such as the neutron.

Example 29-8

Calculate (*a*) the KE and (*b*) the wavelength of a neutron whose speed is 10^3 m/s.

(*a*) The neutron mass is 1.67×10^{-27} kg; hence,

$$\text{KE} = \tfrac{1}{2}mv^2$$

$$= \tfrac{1}{2}(1.67 \times 10^{-27} \text{ kg})(10^3 \text{ m/s})^2$$

$$= 8.35 \; 10^{-22} \text{ J}$$

$$= (8.35 \times 10^{-22} \text{ J})\left(\frac{1 \text{ eV}}{1.60 \times 10^{-19} \text{ J}}\right)$$

$$= \boxed{5.22 \times 10^{-3} \text{ eV}}$$

(*b*) The wavelength of the wave associated with the neutron is calculated by the de Broglie equation:

$$\lambda = \frac{h}{mv} = \frac{6.63 \times 10^{-34} \text{ J} \cdot \text{s}}{(1.67 \times 10^{-27} \text{ kg})(10^3 \text{ m/s})}$$

$$= 3.97 \times 10^{-10} \text{ m} = \boxed{3.97 \text{ Å}}$$

This wavelength is comparable with the distances between atoms in crystals, and hence a beam of such neutrons passing through a crystal can give rise to interference and diffraction just as a beam of x rays can. Figure 29-11*a* shows an x-ray diffraction pattern produced by a NaCl crystal, and Fig. 29-11*b* shows a neutron diffraction pattern of a NaCl crystal. The arrangement of spots is determined by the crystal structure of NaCl; the patterns show many similarities. Diffraction of x rays and neutrons is a powerful tool for crystal structure analysis.

R. E. Lapp

(a)

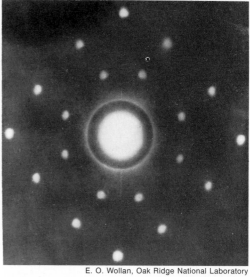

E. O. Wollan, Oak Ridge National Laboratory

(b)

Figure 29-11 Wave nature of photons and neutrons. (a) X-ray diffraction by a single crystal of NaCl. (b) Diffraction of neutrons by a single crystal of NaCl.

29-7 *The Electron Microscope*

The microscope discussed in Sec. 27-4 uses visible light of wavelength several thousand angstroms. We can call such an instrument a *photon microscope*, in recognition of the fact that a ray of light is a beam of photons. Now we consider the *electron microscope*, which uses a beam of electrons instead of a beam of light. The main advantage of the electron microscope is its potential for very high resolving power. This is based on the possibility of using electrons whose de Broglie wavelengths are less than 1 Å. Objects as small as 2.3 Å have been resolved, a feat forever beyond the capability of a microscope using visible light. We saw in Sec. 27-5 that high resolving power requires a small ratio of wavelength to diameter of lens. In the electron microscope the ratio λ/D can be made negligibly small, and the quality of the image is limited only by various aberrations and the stability of the power supplies.

The *transmission electron microscope* (TEM) is analogous to a conventional photon microscope. The whole specimen is illuminated with an electron beam. Electrons transmitted through the specimen are scattered by adjacent parts of the specimen and are brought to a focus at correspondingly adjacent points on a fluorescent screen or photographic film. The microscope is focused by adjusting the currents through the magnetic lens coils and by adjusting the accelerating voltage. To obtain high-speed electrons (with correspondingly small de Broglie wavelengths), the accelerating voltages are in the range 100 kV to 1 MV. The field of view can be relatively large (Fig. 29-12). But radiation damage to the specimen is bound to occur at the high energies necessary to give the greatest resolving power.

The *scanning electron microscope* (SEM) usually detects electrons that are back-scattered from a tiny portion of the specimen, or are knocked out (as secondary electrons) by the im-

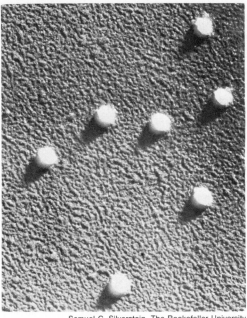

Figure 29-12 Electron microscope photograph of reovirus particles; total magnification about 60,000. Extremely high resolving power is possible because of the small wavelengths associated with electrons. Detail down to about 50 Å is clearly shown in this photograph.

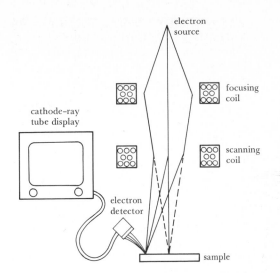

Figure 29-13 Scanning electron microscope (schematic).

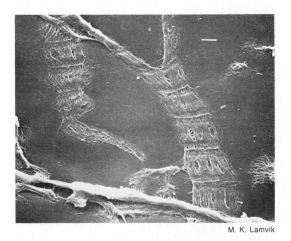

Figure 29-14 Isolated skeletal muscle myofibrils, disrupted by agitation in a relaxing solution, are imaged in this photo made using a scanning electron microscope. The sample was lightly coated with Pt-Pd alloy about 4 nm thick; scale bar = 1 μm.

pact of the primary electron beam on the specimen (Fig. 29-13). The electron beam is focused in the plane of the specimen to a small spot perhaps only 2 Å in diameter. By periodically changing the current in the scanning coil, the probe beam is caused to move across the length and breadth of the specimen. The narrow pencil of electrons causes secondary electrons to be emitted from the target region. These secondary electrons are detected by a device such as a solid-state electron detector, and electrical pulses are fed into a cathode-ray tube whose beam synchronously scans the face of the tube. Thus the picture built up on the TV tube is a magnified view of the specimen. More precisely, it is a representation of the relative electron-scattering powers of the different parts of the specimen. A three-dimen-

sional effect is obtained because more secondary electrons are produced when the primary beam strikes a sloping or contoured surface than if it strikes a similar surface that is perpendicular to the beam. Figure 29-14 illustrates the large amount of information that can be obtained using a scanning electron microscope.

It is also possible to use a scanning technique in the transmission microscope. An added advantage here is the production of x rays, localized at the point where the narrow beam strikes the sample. As we shall see in Chap. 31, high-energy electrons striking a target give rise to x rays whose wavelength is characteristic of the atomic number of the target atom. Thus by analyzing the x-ray wavelengths from a selected point in the specimen, the chemical composition of a tiny region can be studied.

The *scanning proton microscope* now under development offers some advantages over the electron scanning microscope. Bound hydrogen in organic material can be detected because protons are efficiently scattered by other protons (see the discussion of elastic collisions on page 134). Also, since protons are easily absorbed, they can give information about the material's internal structure. Figure 29-15 shows a proton microscope representation of two unstained human chromosomes.

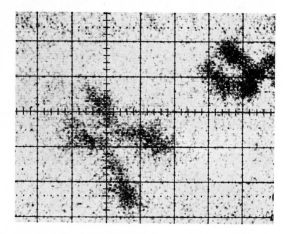

Figure 29-15 Two unstained human chromosomes appear in this photo using a scanning transmission proton microscope developed by R. Levi-Setti and collaborators. The dark areas indicate regions where incident beam protons, of energy 55 keV, have been stopped by the specimen. (Specimen courtesy of Dr. M. Golomb.)

Let us return to the physics of the electron microscope. To obtain maximum sharpness of the scanning beam, the deflecting forces given by $Q\mathbf{v} \times \mathbf{B}$ must be held constant. Therefore, the accelerating voltage (which determines velocity $\mathbf{v}$) and the magnet currents (which determine $\mathbf{B}$) must be stabilized to within 1 part in 10^6—a formidable task. Ultrahigh vacuums, of the order of 10^{-9} or 10^{-10} torr, are often used to prevent unwanted electron scattering and diffusion of the beams. Only if these difficult problems, and others, can be overcome will it be possible to realize the theoretically very high resolving power implicit in the smallness of the de Broglie wavelengths of the high-speed electrons.

29-8 The Uncertainty Principle

The dual nature of light implies an uncertainty principle that would have no meaning in a single theory, whether a wave theory or a particle theory. As applied to photons, the uncertainty principle can be described with the help of Fig. 29-16, in which we illustrate the diffraction of light waves by a single slit. According to Sec. 26-5, almost all the light energy goes into the central band BB' of the diffraction pattern. The narrower the slit, the wider the central broad diffraction band on the screen. In symbols, $\sin \theta = \lambda/\Delta x$, where Δx is the width of the slit.

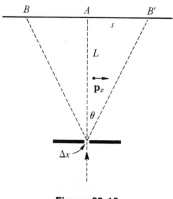

Figure 29-16

Let us look at this experiment from the photon point of view. The fact that the photons spread out means that while passing through the slit, they have some sideways momentum (p_x) as well as forward momentum.* Since photons reach all parts of the central diffraction band BB', they must have different values of p_x. In fact, the sideways momentum of a photon may lie anywhere between 0 (for a photon that reaches A) and Δp_x (for a photon that reaches B or B'). That is, there is an uncertainty of Δp_x in our knowledge of the x component of the momentum of the photons that pass through the slit. The position of the photons is also uncertain; they may pass through any portion of a slit whose width is Δx. If we try to locate the position of any given photon more precisely by narrowing the slit, we decrease the uncertainty in x, but, according to $\sin \theta = \lambda/\Delta x$, the beam spreads out, and the uncertainty in the sideways momentum becomes greater. Thus when Δx is small, Δp_x is large, and vice versa.

The relationship between Δp_x and Δx is a simple one, easily derived from Fig. 29-16. We treat the photon as a projectile, just as in Sec. 2-11. The x motion and the y motion are separate, both taking place during the same time. If t is the time of flight from slit to screen, we have $L = ct$, and $s = (\Delta v_x)t = (\Delta p_x/m)t$. We start from the experimentally verified formula for the sine of the angle of diffraction (see Eq. 26-2):

$$\sin \theta = \frac{\lambda}{\Delta x}$$

Hence, since $\sin \theta \approx \tan \theta$,

$$\frac{s}{L} \approx \frac{\lambda}{\Delta x}$$

Substituting for s and L, we get

$$\frac{(\Delta p_x/m)t}{ct} \approx \frac{\lambda}{\Delta x}$$

---

* We use the symbol p for linear momentum mv; p_x is the x component of the momentum.

whence

$$\Delta p_x \Delta x \approx mc\lambda$$

Since $\lambda = h/mc$ (see Sec. 29-4), we get

$$\Delta p_x \Delta x \approx mc\left(\frac{h}{mc}\right)$$

or, for this experiment,

$$\Delta p_x \Delta x \approx h \qquad (29\text{-}12)$$

The product $\Delta p_x \Delta x$ might be greater than h, but no experiment has ever been imagined that could give a value of $\Delta p_x \Delta x$ much less than Planck's constant.[†]

Werner Heisenberg (1901–1976) in about 1925 proposed that Eq. 29-12 represents a natural limitation to our knowledge. We have met with such natural limitations before: The maximum efficiency of a Carnot heat engine is $(T_H - T_C)/T_H$; the maximum speed of a particle is 3×10^8 m/s; the sharpness of the image of a distant star is limited by the ratio λ/D. These limitations are inherent in nature and are not due to heat losses, friction, lens imperfections, or the like. The *Heisenberg uncertainty principle* applies to matter as well as to photons; it is impossible to know *precisely* both the position and the momentum of a particle. As an armchair experiment, let us try to determine the exact position of an electron by looking at it through a microscope. To achieve high precision, we would try to avoid diffraction effects by using light of very short wavelength, perhaps x rays in an (imaginary) x-ray microscope. Photons of such light have high energy, according to $E = hf$, and would give the object a recoil velocity, thus rendering its position uncertain (Compton effect). If we use long-wavelength light to avoid recoil, then we lose resolving power. It turns out

---

[†] The precise formulation of the principle depends on how "uncertainty" is defined. If the uncertainties are defined by usual statistical procedures, the inequality reads $\Delta p_x \Delta x \geqq h/4\pi$. There is nothing special about the x direction; for example, it is also true that $\Delta p_y \Delta y \geqq h/4\pi$.

that here, too, the Heisenberg uncertainty principle gives the relationship between the uncertainty in position and the uncertainty in momentum for a particle of matter.

We have come a long way from the simple cause-and-effect mechanics of Newton's day. It was thought then that one could exactly predict the future position and momentum of any body if one knew its original position and momentum precisely enough. It should be sufficient to apply Newton's laws of motion, and the only reason that one could not predict the future course of the universe was thought to be that the mathematical solution of so many simultaneous equations (one for each particle in the universe) would be too difficult. Now we see that the future path of a particle cannot be exactly predicted. The unavoidable uncertainty in our knowledge of the initial position or momentum, or both, affects our knowledge of the future.

It has often been said that the law of causality has been abandoned; this is a true statement only when applied to a single particle's motion. The predictions of modern physics are probabilistic, as discussed at the end of Sec. 29-5. We cannot say where any given photon will go, but the *observable* photograph of a double-slit interference pattern is quite definite, being an average effect for many billions of photons. An exactly similar interpretation can be made when a stream of particles passes through a pair of slits; the de Broglie waves are waves of probability, and although the path of any single particle is subject to uncertainty, the average effect for many particles is exactly predictable (Fig. 29-17).

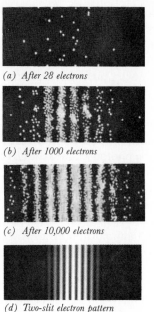

(a) After 28 electrons

(b) After 1000 electrons

(c) After 10,000 electrons

(d) Two-slit electron pattern

Figure 29-17 Double-slit interference pattern for electrons. (*a*), (*b*), (*c*) Computer simulation of growth of pattern. (*d*) Actual photograph of two-slit electron diffraction pattern. (*a, b, c* by Dr. Elisha Huggins; *d* by Dr. C. Jönsson.)

We have seen that the very act of measuring position introduces an unavoidable error in our knowledge of momentum. The uncertainty principle is a limitation on our *knowledge* of nature. Whether nature is "actually" determinate is an unanswerable question, since only measurable quantities (probabilities) can enter into a physical theory.

Summary All electrons are identical in both charge (Millikan) and rest mass (J. J. Thomson), regardless of their origin.

Electromagnetic radiation is emitted or absorbed a photon at a time, with the photon energy given by $E = hf$ (Planck). The work function is the energy required to remove an electron from a metal surface. The KE of a photoelectron equals the energy of the incoming photon minus the work function (Einstein). At the threshold, the photon energy just equals the work function, and the electrons are emitted with zero KE. Collision experiments show that x-ray photons are like concentrated bullets, having momentum as well as energy (Compton). The dual nature of light is shown by such experiments. It is not necessary to make a model for light, but if a model is desired, the interpretation of e-m waves as probability waves for the photons is as good as any.

Matter as well as light has both wave and particle properties. For each, the wavelength of the associated probability wave is given by $\lambda = h/$momentum (de Broglie, Davisson, G. P. Thomson). Matter waves have such small wavelengths that they can be ignored except when dealing with single elementary particles, such as electrons, protons, or neutrons. The de Broglie wavelengths of high-energy electrons and protons are small enough to make them useful in electron microscopes and proton microscopes, which have extremely high resolving power.

According to the uncertainty principle (Heisenberg), the product of the uncertainty in position and the uncertainty in linear momentum is never much less than Planck's constant, and may be much more. The principle applies both to photons and to particles of matter. The law of causality in physics is a statistical (probabilistic) one, and the motions of individual particles or photons cannot be predicted with certainty.

Check List

Planck's constant	$E = 1240 \text{ eV} \cdot \text{nm}/\lambda$	$\lambda = h/mv$
photon	$E = 12{,}400 \text{ eV} \cdot \text{Å}/\lambda$	$\Delta p_x \Delta x \gtrsim h$
$E = hf$	$\frac{1}{2}mv^2 = hf - w$	Heisenberg uncertainty
work function of a surface	Compton effect	principle
electron volt	probability waves	

Questions

29-1 Suppose there were two kinds of electrons: type A of charge 1.00×10^{-19} C, and type B of charge 2.20×10^{-19} C. Suppose also that the two kinds of electrons occurred at random in nature, equally plentifully in all kinds of atoms. Could this hypothesis about two kinds of electrons be tested by the oil-drop experiment? Could it be tested by electrolysis experiments?

29-2 In Fig. 29-2, in what direction (up or down) will the shadow move if a magnetic field directed toward the reader is turned on?

29-3 Do all electrons have the same mass? Do all electrons have the same rest mass?

29-4 Ordinary photographic film is least sensitive to red light, hence the use of red safelights in darkrooms. Explain, using the photon model of light.

29-5 It is harder to remove a free electron from copper than from sodium. Which metal has the greater work function? Which has the higher photoelectric threshold frequency?

29-6 If a proton is accelerated through a PD of 10^5 V, what is its KE, in electron volts?

29-7 If traveling at the same speed, which particle, a proton or an electron, has a probability wave of longer wavelength?

29-8 Has the highest theoretically possible resolving power of the electron microscope been achieved as yet?

29-9 Is the future of the physical world predictable? Discuss this question, both pro and con.

29-10 If, in the Millikan oil-drop experiment, the electric field that opposes the gravitational field is turned off, the drop is observed to move   (a) downward at constant speed;   (b) downward at steadily increasing speed;   (c) downward or upward, depending on the sign of the charge on the drop.

29-11 The dimensions of Planck's constant are   (a) $[MLT^{-1}]$;   (b) $[ML^2T^{-2}]$;   (c) $[ML^2T^{-1}]$.

29-12 If a metal surface is replaced by one whose work function is less, the threshold frequency for the photoelectric effect   (a) increases;   (b) decreases;   (c) remains the same.

29-13 An electron volt is a unit of   (a) electric field strength;   (b) work;   (c) potential difference.

29-14 When scattered by a free electron, an x-ray photon's wavelength   (a) increases;   (b) decreases;   (c) increases or decreases depending on the atomic number of the scattering atoms.

29-15 A certain proton and a certain electron have the same momentum. Compared with the electron, the proton has   (a) less kinetic energy;   (b) the same de Broglie wavelength;   (c) both of these.

Problems

29-A1 Calculate the energy of a photon of ultraviolet light whose wavelength is 248 nm. Express your answer in   (a) joules and   (b) electron volts.

29-A2 Calculate the wavelength of a photon whose energy is 2.45 eV. Express your answer in   (a) nanometers and   (b) angstroms.

29-A3 A particle has an energy of 10^{-12} J (1 pJ). Express this energy in million electron volts (MeV).

29-A4 A speck of dust of mass 40 μg is lifted 2 cm. Calculate the increase in gravitational PE in   (a) joules and   (b) million electron volts (MeV).

29-A5 Calculate the wavelength associated with a particle of mass 2×10^{-18} kg moving at a speed of 4×10^{-7} m/s. Express your answer in angstroms.

29-A6 Calculate the wavelength associated with an electron of mass 9.11×10^{-31} kg moving through a vacuum tube with speed 2.0×10^7 m/s.

29-B1 An oil drop for which the density is 900 kg/m^3 is balanced between two plates 5 mm apart. The PD between the plates is 400 V. If the drop has a charge of five electrons, calculate its radius. (*Hint:* First find the drop's mass, then its volume.)

29-B2 Calculate the mass of an oil drop carrying a charge of four electrons, which is just balanced between two plates that are 2 cm apart. The PD between the plates is 300 V.

29-B3 Using the experimental value of e/m_0 given in Sec. 29-2, calculate the radius of circular path in Fig. 29-3 for electrons that have been accelerated through a PD of 250 V. The magnetic field is 9×10^{-4} T. (Relativity effects can be ignored.)

29-B4 A particle of charge Q moves through a magnetic field B in a circle of radius R. Prove that the momentum of the particle is given by $p = BQR$.

29-B5 If the magnetic field in Fig. 29-3 is that of the earth ($B = 6 \times 10^{-5}$ T), what accelerating voltage V would give an electron beam a radius of curvature 0.20 m?

29-B6 What is the KE in pJ of a helium ion He^{2+} that is accelerated through a PD of 3 GV?

29-B7 Use Table 14-5 to find the approximate ratio of the energy of an x-ray photon to that of a photon emitted by an FM broadcasting station. Why is the photon nature of the radio radiation not easily observed?

29-B8 Revise the photoelectric equation (29-6) to be applicable to relativistic electrons with speeds not negligible compared to c.

29-B9 What is the maximum KE of the photoelectrons ejected from a piece of copper, for which the work function is 4.70 eV, by ultraviolet radiation of wavelength 220 nm?

29-B10 The work function of a surface is 3.00 eV. Calculate the maximum KE of electrons ejected from this surface by radiation of wavelength 300 nm.

29-B11 Light of wavelength 453 nm shines on a thin metal surface, and the ejected photoelectrons have KE equal to 0.70 eV. Calculate the work function of the surface.

29-B12 Calculate the work function of a surface from which red light of wavelength 633 nm from a helium-neon laser ejects photoelectrons of maximum KE 0.30 eV.

29-B13 What is the threshold wavelength for a gold surface whose work function is 4.82 eV?

29-B14 What is the work function of a surface whose threshold wavelength is 400 nm?

29-B15 Monochromatic visible light shines on the cathode of a photocell for which the work function is 1.39 eV. The negative "stopping potential" needed to prevent photoelectrons from reaching the anode is 1.22 V. What color is the light?

29-B16 Can red "laser light" ($\lambda = 633$ nm) eject photoelectrons from a clean metallic-sodium surface for which the work function is 2.36 eV?

29-B17 As the wavelength of light striking a surface is gradually decreased, emission of photoelectrons starts when λ reaches 560 nm. What is the maximum KE of the emitted photoelectrons when λ reaches 400 nm?

29-B18 (*a*) What is the momentum of an x-ray photon whose wavelength is 0.71 Å? (*b*) What is the (relativistic) mass of this photon?

29-B19 Calculate the momentum of a photon of blue light whose wavelength is 475 nm.

29-B20 X-ray photons of energy 9437 eV are scattered at 90° by a paraffin block. Calculate the wavelengths of (*a*) the original photons and (*b*) the scattered photons.

29-B21 An x-ray photon of wavelength 1.540 Å strikes a free electron, and the wavelength of the scattered photon is found to be 1.560 Å. What is the KE of the recoiling electron?

29-B22 (*a*) Calculate the wavelength of the de Broglie waves associated with a rifle bullet of mass 20 g moving at 500 m/s. (*b*) Does the bullet show any observable wave behavior?

29-B23 (*a*) Suppose that the position of the bullet in Prob. 29-B22 is known to within 1.5 cm (the diameter of the rifle barrel); what is the least uncertainty in the sideways component of its momentum? (*b*) If the gun is aimed at a target 200 m away (time of flight 0.4 s), by how much might it miss the target due to the Heisenberg uncertainty principle? (*c*) Why do most bullets miss the target by much more than the amount calculated in (*b*)?

29-B24 (*a*) Derive a formula for the de Broglie wavelength of a particle of charge Q moving in a circular path of radius R in a magnetic field B. (*b*) Make a dimensional check for your formula.

29-B25 Electrons in a certain electron microscope are accelerated through a potential difference of 40 kV. Calculate the de Broglie wavelength of these electrons. (*Hint:* The speed of the electrons is small enough to allow use of nonrelativistic formulas connecting energy and momentum without serious error.)

29-B26 What is the ratio of the de Broglie wavelength of an electron to that of a proton that has the same kinetic energy? Assume nonrelativistic speeds. (*Hint:* Solve first in symbols, then use data from the Appendix.)

29-B27 A proton scanning microscope uses protons accelerated through 55 kV (Fig. 29-15). Calculate the de Broglie wavelength of these protons, in femtometers (fm) (see Appendix Table 2).

29-B28 (*a*) Calculate the de Broglie wavelength of an average O_2 molecule at $0°C$ (see Example 16-1, page 356). (*b*) Compare your answer with the distance between the two atoms in the molecule, which is about 1.2 Å. Does the result of this problem indicate that the wave nature of matter is a significant factor in molecular structure?

29-C1 When the wavelength of light striking a certain surface is 200 nm less than the threshold wavelength, the maximum KE of the ejected photoelectrons is 2 eV. Calculate the threshold wavelength for this surface.

29-C2 An explorer on the moon turns on a 60-W lamp, and the light from the bulb spreads out equally in all directions. Assuming that 1% of the energy is radiated as visible light (of average wavelength 555 nm), about how many photons per second of visible light from the bulb would strike the mirror of a large telescope on the earth? (Diameter of the mirror = 4 m; moon's distance = 3.8×10^8 m; neglect absorption in the earth's atmosphere.)

29-C3 (*a*) What is the wavelength of a photon whose momentum is the same as that of a 50-eV electron? (*b*) In what region of the e-m spectrum (Table 14-5) is this photon?

29-C4 Example 29-2 surely illustrates conservation of energy, which is the basis of the photoelectric equation. Study this same example with respect to conservation of *momentum*. Calculate, in SI units, (*a*) the momentum of the incoming photon, and (*b*) the momentum of the ejected electron. (*c*) Explain how the electron's momentum can be greater than the photon's momentum, as in this example.

29-C5 Ideally, in Fig. 29-3 the electron's horizontal velocity v_x is 0 as it passes vertically through the first slit. The width Δx of the first slit introduces an uncertainty Δv_x, which in turn introduces an uncertainty $\Delta x'$ in the horizontal position of the beam at the second slit, which is 0.003 mm wide. (*a*) Using data from Prob. 29-B3, find the time of flight. (*b*) Calculate $\Delta x'$ if Δx is 0.002 mm. (*c*) From these results, decide whether the uncertainty principle is significant for this experiment.

For Further Study

29-9 Wave Mechanics— The Free Particle

A particle, such as an electron, that is in no way confined is a *free particle*. It moves with constant momentum p in a straight line (Newton's first law). We have seen that a de Broglie wave is associated with such a particle, with a wavelength given by $\lambda = h/p$. Let us now consider more closely the mathematical representation of such a de Broglie wave. For simplicity we confine our analysis to motion in one dimension, along the x axis.

In Sec. 10-1 we described a wave as a *propagation of a disturbance*. We studied mechanical waves where the "disturbances" are fairly obvious: for a string, positive or negative displacements of points on the string; for sound waves, positive or negative variations in air pressure. To represent *matter waves*, however, we use a non-mechanical disturbance, which we call ψ, the *wave function*. Just as we did for the water wave of Fig. 10-5, we can graph ψ as a function of x at a given time t, as in Fig. 29-18a, and we can graph ψ as a function of t at a given position x. Thus ψ is a function of two variables, written as $\psi(x, t)$.

We approach a physical interpretation of the wave function by noting that all we can observe is the *probability* that the particle will be at position x at time t. But probabilities are never negative. Therefore, we use the square of the absolute value of the wave function to represent the observable probability.

> **The probability of finding a particle at time t in a region of width Δx, centered on x, is proportional to $|\psi(x,t)|^2 \Delta x$.**

We denote the probability function $|\psi|^2$ as $P(x, t)$.*

In Fig. 29-18b, the probability of finding the particle in a small region centered at point A is zero at the instant shown. However, as the wave moves to the right or left, point A's probability goes through the whole range from 0 to 1. There is nothing particular about point A. Averaged over a long time, *every* point in space has the same periodic variation in probability as A.

We have arrived at a seeming paradox: A free particle, with a definite momentum, is equally likely to be found anywhere in space, from $-\infty$ to $+\infty$! Yet this is, after all, to be expected from the uncertainty principle. If the momentum is precisely known, $\Delta p = 0$; hence $\Delta x = \infty$, and the position must be completely indeterminate. Also, from $\lambda = h/p$, a precisely known momen-

---

* The probability is proportional to Δx as well as to $|\psi(x,t)|^2$. In the limit, as $\Delta x \to 0$, there is zero probability that a particle is exactly "at" a mathematical point, which has zero extent. This is nothing new; it is true in classical mechanics as well as wave mechanics. A similar concept regarding molecular speeds is implicit in Fig. 16-7 on page 368.

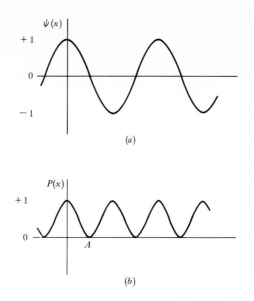

Figure 29-18 (*a*) Wave function for a free particle at a given time. (*b*) Probability function for a free particle at a given time.

tum p implies a precisely known wavelength λ. The only function of x that has a single well-defined λ is the sinusoidal function, and $\sin x$ is a function that extends from $-\infty$ to $+\infty$.

Our conclusion is that a free particle with a definite momentum cannot be localized, and its wave function extends over all space. A "particle," such as an actual electron, that is to a large extent localized must be represented mathematically not by a single ψ but by a group of ψ's, all nearly equal, which cancel the probability function to a resultant value of 0 everywhere except in the small region we call the electron.

29-10 Wave Mechanics— A Particle in a Box

A particle is placed in a box and bounces back and forth between two perfectly reflecting walls. This problem is solvable without the use of advanced mathematics, and from it we can gain an insight into the quantized energy levels of the system.

In contrast to the free particle discussed in the previous section, our particle is *bound* (within the box) and can travel in either direction. Because the walls are perfectly reflecting and energy is conserved, the magnitude of the momentum (and hence of λ) does not change during reflection. Thus we have exactly the condition for the formation of stationary waves (Sec. 11-1) by the interference or superposition of two waves of equal wavelength traveling in opposite directions through the same region of space. Since the particles cannot escape the box, $\psi(x)$ must be 0 outside the box. Hence $\psi(x)$ must be 0 at each wall, and each wall must be a node for the resulting wave function (Fig. 29-19), just as for the fixed ends of a vibrating string. We see an important analogy with the string whose modes of vibration are shown in Fig. 11-3. Because of the requirement of a node at each wall, *only certain wavelengths are possible* for the particle in the box. Therefore, the momentum p is quantized, and so is the kinetic energy.

Let the box have width L and consider the nth mode in which there are n "segments" of the resulting ψ function. Since nodes are separated by $\frac{1}{2}\lambda$,

$$n(\tfrac{1}{2}\lambda) = L$$

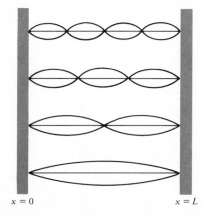

$$x = 0 \qquad\qquad x = L$$

Figure 29-19 Superposition of wave functions for particle in a box.

or

$$\lambda = \frac{2L}{n}$$

By the de Broglie formula,

$$mv = \frac{h}{\lambda} = \frac{nh}{2L}$$

whence

$$v = \frac{nh}{2Lm}$$

and

$$\tfrac{1}{2}mv^2 = \tfrac{1}{2}m\left(\frac{nh}{2Lm}\right)^2 = \frac{n^2h^2}{8mL^2}$$

or

$$E_n = \frac{n^2h^2}{8mL^2}$$

These are the "allowed" or "quantized" energies for a particle of mass m in a box of length L. Note that in this case the energies are proportional to n^2.

Example 29-9

Calculate the next-to-lowest value of the KE of an electron in a one-dimensional box of length 1 Å. Here the quantum number n is 2.

$$E_2 = \frac{n^2h^2}{8mL^2} = \frac{(2^2)(6.63 \times 10^{-34}\ \text{J}\cdot\text{s})^2}{8(9.11 \times 10^{-31}\ \text{kg})(10^{-10}\ \text{m})^2}$$

$$= \boxed{2.41 \times 10^{-17}\ \text{J}}$$

This energy is

$$(2.41 \times 10^{-17}\ \text{J})\left(\frac{1\ \text{eV}}{1.60 \times 10^{-19}\ \text{J}}\right) = \boxed{151\ \text{eV}}$$

The electron in the box of Example 29-9 can have only a discrete set of energy values—37.7 eV, 151 eV, 339 eV, 603 eV, ... For other bound systems the formulas are different; for example, in a hydrogen atom the energies E_n are proportional to $1/n^2$, as we show in the next chapter. But the main idea is the same: discrete energy levels arise because of the nodes and antinodes in the superposition of de Broglie waves in a bound system.

Problems   **29-C6** What is the only other allowed energy that is less than 1 keV, in addition to those listed above for the electron of Example 29-9?

29-C7 Calculate the size of a one-dimensional box in which an electron's lowest energy level is 7 eV.

References   **1.** W. F. Magie, *A Source Book in Physics* (McGraw-Hill, New York, 1935), pp. 578–579, discovery of the photoelectric effect by Hallwachs in 1888; pp. 583–597, discovery of the electron by J. J. Thomson.

2. G. P. Thomson, "J. J. Thomson and the Discovery of the Electron," *Physics Today* **9**(8), 19 (Aug. 1956). Reminiscences by his son.

3. D. J. Kelves, "Robert A. Millikan," *Sci. American* **240**(1), 142 (Jan. 1979).

4. R. A. Millikan, *The Electron: Its Isolation and Measurement and the Determination of Some of Its Properties* (University of Chicago Press, 1963). A revision of Millikan's original book, edited by J. W. M. DuMond.

5. K. K. Darrow, "The Quantum Theory," *Sci. American* **186**(3), 47 (Mar. 1952). The work of Planck, Bohr, and Compton.

6. G. Gamow, "The Principle of Uncertainty," *Sci. American* **198**(1), 51 (1958). See also a book review (pp. 111–116 of the same issue) by J. R. Newman of D. Bohm's *Causality and Chance in Modern Physics*—the argument of a physicist who does not accept the uncertainty principle.

7. K. K. Darrow, "Davisson and Germer," *Sci. American* **178**(5), 51 (May 1948). The experimental discovery of the wave nature of the electron, independently of the work of G. P. Thomson in England.

8. H. A. Medicus, "Fifty Years of Matter Waves," *Physics Today* **27**(2), 38 (Feb. 1974). A historical, nonmathematical account.

9. C. Jönsson, "Electron Diffraction at Multiple Slits," *Am. J. Phys.* **42**, 4 (1974). Gives experimental details of the production of the slits used in making the pattern shown in Fig. 29-17.

10. A. E. Walters, *The Photoelectric Effect* (film).

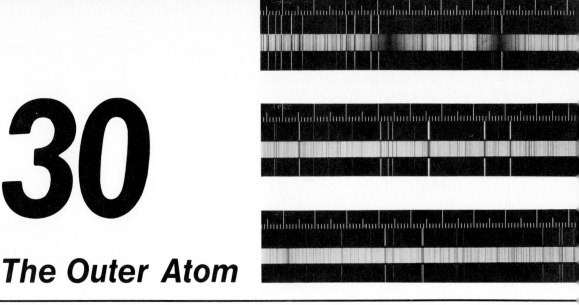

30

The Outer Atom

Almost all of our knowledge of the outer atom—the cloud of electrons surrounding the nucleus—has come to us through analysis of the light emitted or absorbed by an atom. From this analysis came the concept of energy levels and an orbital model that considered the atom to be a miniature solar system. This model is no longer capable of giving a full account of atomic behavior, but it is a valuable first stage that gives much useful insight. We start in the year 1911, at a time when it was known that electrons and photons somehow interact on the atomic scale of things, but it was not known just how the positive and negative charges are distributed in an atom. Through the Rutherford scattering experiment, the atomic nucleus was discovered.

30-1   *Distribution of Charge in the Atom*

One of the great experimental physicists of the past century was Ernest Rutherford (1871–1937), a New Zealander who worked in Canada at McGill University and in England at Man-

chester University, and later was head of the Cavendish Laboratory at Cambridge University. It was Rutherford and his co-workers who in 1908 identified the α particles shot out by radioactive atoms as helium ions. At Manchester, he directed the work of two of his students in a crucial experiment on the scattering of α particles by thin metallic foils.

It is hard for us to realize that only 75 years ago the accepted picture of an atom was quite different from the orbital model we know today. There was no doubt that the electrically neutral atom consisted of equal amounts of positive and negative charge, as shown by the existence of ions. It was also realized that electrons could be separated from an atom, as shown by the photoelectric effect and by J. J. Thomson's measurements of e/m_0 for the "corpuscle," as the electron was then called. The Bohr theory had not yet been worked out, and most physicists believed that the atom's structure must somehow allow the electrons to vibrate back and forth as they emitted e-m radiation. One model, due to Kelvin and Thomson, pictured a hydrogen atom

Absorption spectrum of the sun (dark lines) in the violet region, 3900 A to 4200 Å. The bright lines on either side of the sun's spectrum are from a laboratory iron arc, used for calibration purposes. Many lines coincide, showing the presence of iron vapor in the sun's atmosphere. The two wide dark lines in the upper portion are from calcium.

as a sort of spherical ball of jellylike positive charge in which were embedded one or more electrons. A single electron's normal position would be at the center of the ball, where the net force on it would be zero; if pulled aside, the electron would oscillate back and forth in SHM and give out an e-m wave. The main problem was to explain the existence of so many spectrum lines, each of a different frequency, for every element emits infinitely many lines having different frequencies and intensities.

Rutherford in 1911 explored the structure of the atom by shooting α particles (helium nuclei) through a thin metallic foil. According to the Thomson model, most of the particles would be scattered somewhat, due to the Coulomb force of repulsion between the positively charged α particles and the positive balls of atomic charge. An α particle could hardly miss coming close to some like (positive) charge as it passed through the metal, thus being deflected. According to Coulomb's law (Sec. 18-5), any repulsive force is inversely proportional to the square of the distance; in the Thomson model, the α particle is at no time very close to any concentrated bit of charge, and so it could experience no large repulsive force. Hence, according to this model, almost every α particle would suffer a moderate deflection, not exceeding a few degrees (Fig. 30-1a).

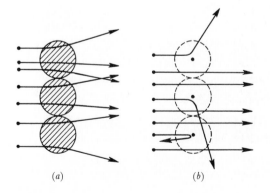

(a)   (b)

Figure 30-1 Scattering of α particles according to two models of atomic structure. (a) Thomson's model. (b) Rutherford's nuclear model.

The experiment showed an entirely different result. Most of the α particles went right on through the foil with no change of direction, but a few were deflected through large angles (Fig. 30-1b). Some even bounced straight back, a deflection of 180°. To explain the large forces that cause such large deflections, Rutherford had to assume that an α particle is itself small, and that it can, if aimed correctly, approach to within about 10^{-14} m of a concentrated "center of charge," which he called the *nucleus*. From the data of crystal structure, the atoms themselves were known to be about 10^{-10} m (1 Å) apart. Thus, Rutherford was led to the nuclear model of an atom: a small, heavy, positively charged central nucleus, with negative electrons separated from the nucleus by much empty space. To visualize these open spaces, let us imagine an atom blown up to the size of a football field. The nucleus would then be a small, heavy glob of matter the size of a lead-pencil eraser; the rapidly moving electrons would be still smaller, but (averaged over a period of time) there would be a reasonably uniform probability of finding an electron at any point within the football field. To complete the picture, imagine firing another lead-pencil eraser at the football field, and add a strong repulsive force that is effective only when the two erasers are within a meter or so of each other. Small wonder that so few α particles are scattered in the Rutherford experiment!

By mathematical analysis of his scattering experiments with gold, silver, and copper foils, Rutherford was able to determine the magnitude of the central charge in each case. Expressed in units of the electron charge e, the gold nucleus turned out to have a measured charge of $+77.4e$, silver $+46.3e$, and copper $+29.3e$. Within experimental error, these numbers are the atomic numbers (Z) of the elements; for instance, silver is the 47th element in the Periodic Table. Hence there must be Z electrons outside the nucleus, to make up the volume of the atom, and they must be in planetary orbits to keep from falling into the nucleus. The scattering of x-ray photons confirms this; such photons are scattered by the electron cloud, and Barkla (also in 1911) showed experimentally that the scattering of x rays is proportional to the atomic number Z.

The Rutherford experiment is an outstanding example of how important results could (in 1911) be obtained with simple apparatus. The scattered α particles struck a fluorescent screen, which was simply a layer of small zinc sulfide crystals, and the tiny flashes of light produced by impacts were observed with a magnifying glass. A speck of radioactive material, some slits, and the foil itself completed the apparatus by which Rutherford established the nuclear model for the atom. As so often happens, the experiment raised new problems. If the electrons move in orbits outside the nucleus, a serious question of stability arises: Why should not such an electron continuously emit radiation on account of its centripetal acceleration (Sec. 24-8)? Physicists at least were now asking the right questions; and among those physicists was a young Danish student in Rutherford's laboratory at Man-

chester named Niels Bohr (1885–1962). We shall see in the next sections how Bohr's model of the hydrogen atom followed in 1913, just two years after Rutherford's historic experiment.

30-2 Emission Spectra

The quantum hypothesis, which had proved so successful in explaining blackbody radiation (Planck, 1900) and the photoelectric effect (Einstein, 1905), was applied in 1913 by Bohr to achieve a breakthrough in still another perplexing problem for which classical physics was inadequate. Bohr succeeded in calculating the wavelengths of the lines in the spectrum emitted by hydrogen atoms.

In a solid, a liquid, or a dense gas, the atoms are close to each other, and they interact to give a *continuous spectrum* with all frequencies (and wavelengths) represented. However, the spectrum emitted by atoms of an incandescent gas is

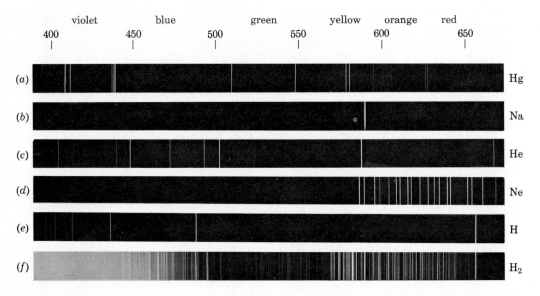

Figure 30-2 Typical emission spectra of gases; wavelengths in nanometers. The spectrum of the hydrogen molecule (*f*) shows two of the lines of atomic hydrogen (*e*), since some of the molecules in the discharge tube are dissociated into atoms; H_2 also shows a continuous spectrum in the blue-violet region.

Table 30-1 Emission Spectra

Source		Nature of Spectrum	Information Given by Spectrum Analysis
Hot solid, liquid, or dense gas		continuous	temperature of source
Hot gas (thermally or electrically excited)	atoms	discrete, well-separated lines	composition of source
	molecules	discrete, close groups of lines	

discrete; that is, there are sharp lines of definite wavelength, with no light emitted at the intermediate wavelengths. Some representative spectra are shown in Fig. 30-2. Presumably the atoms in a gas are free enough to be able to emit their own characteristic frequencies, much as a piano string vibrates with certain natural frequencies. Some energy must be supplied to start a string vibrating; likewise, an atom must absorb energy in order to emit radiation. There are many ways in which an atom can be given the necessary energy. For instance, the sodium spectrum is obtained if solid NaCl is placed in a flame; the substance breaks down into Na and Cl atoms, which absorb energy by thermal bombardment. The copper spectrum can be formed by passing an electric arc or a spark between metallic copper electrodes; the radiation comes from copper vapor between the electrodes. Gaseous substances such as helium or hydrogen can be excited electrically when enclosed in a glass tube at low pressure.

The spectrum of a *solid* or a *liquid* depends only on its temperature. Molten iron in a furnace and the ceramic bricks lining the furnace look about the same—"red-hot" in each case—and they emit the same blackbody spectrum, which is continuous (Sec. 14-7). On the other hand the frequencies emitted by any incandescent *gas* are characteristic of the substance, and temperature affects only the relative intensities of the lines. The line spectrum of a gas therefore can be used as a means of identification. The spectrum of

hydrogen molecules (H_2) is considerably more complex than that of hydrogen atoms (H); in addition to many closely spaced lines, there is a continuous emission in the blue, violet, and ultraviolet region (Fig. 30-2*f*). In general, the term *band spectrum* is used to describe the very closely spaced line spectrum emitted by molecules that contain more than one atom.

We summarize the main facts about emission spectra in Table 30-1.

30-3   The Bohr Theory for Hydrogen

In most spectra, the lines seem to be distributed almost at random, but hydrogen, the simplest atom, gives a spectrum that shows some regularity. For this reason, the hydrogen spectrum has long been the testing ground for new theories, and it was this spectrum that Bohr attacked. Pursuing the acoustical analogy, we might call the lowest frequency the fundamental, and the higher frequencies (shorter wavelengths) the overtones. However, the hydrogen frequencies tend toward a limit and cluster together near 8.2×10^{14} Hz; this cannot be explained on any classical model, according to which the frequencies of overtones increase without limit. Another difficulty with the classical model is that the electron in hydrogen could not be stationary, for it would then fall toward the positive nucleus. This leads to the expectation that the electron would have to move in some sort of orbit; but then its centripetal acceleration would cause e-m radiation (Sec. 24-8), and the electron would spiral into the nucleus as it lost energy. We know, however, that hydrogen atoms can exist for long times without radiating energy.

Bohr made a virtue of necessity by *postulating* that an electron can revolve around a nucleus without radiating. He assumed that an atom can exist in any one of a number of radiationless *energy states*, $E_1, E_2, E_3, \ldots$. Next, Bohr postulated that Einstein's photoelectric equation for the energy of a photon applies within the atom. When an atom changes from one energy state E_2 to another lower energy state E_1, the difference in energy is $E_2 - E_1$. Bohr's assumption was that all this energy is used in forming a single photon. Thus, the energy of the emitted photon is given by

$$hf = E_2 - E_1 \qquad (30\text{-}1)$$

An *energy-level diagram* such as Fig. 30-3 helps us visualize these postulates. Each allowed energy state is represented by a horizontal line; for the present, we choose our reference level of zero energy to be the normal (unexcited) state of the atom. This state of lowest energy is called the *ground state*. The hydrogen atom can exist in radiationless states having energies 0 eV, 10.20 eV, 12.09 eV, and so on, but an energy of 5.00 eV, for instance, is impossible. The level shown at the top of the diagram (dashed line) represents the *ionization energy*. The diagram shows that 13.60 eV is required to remove the electron completely from the atom, leaving a hydrogen ion (H^+), which is, of course, simply a bare proton.

The arrows represent *transitions*, or *jumps*, between energy states, and the length of any arrow is proportional to the frequency of the photon that is emitted during the transition, according to the equation $hf = E_2 - E_1$. Any given atom emits only one photon at a time, and the complete spectrum containing many lines is an average effect formed by many atoms emitting many photons. Imagine a series of photon counters replacing the photographic plate in a

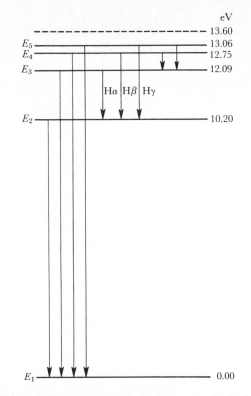

Figure 30-3 Energy levels for hydrogen. Excitation energies are shown in electron volts. The same levels, with a different choice of zero energy, are shown in the right-hand part of Fig. 30-6.

spectroscope, with the counters arranged so that each counter receives light of one of the characteristic wavelengths. The counters click at random, each click corresponding to an atomic process during which some one atom's electronic energy changes abruptly by some amount ΔE. If, say, the mercury green line is twice as bright as one of the yellow lines, this merely means that one transition is twice as probable as the other, and more atoms per second emit the green photons. On the average, the light of either color appears to be steady. In a similar fashion, heavy rain on a metal roof makes a steady sound, although the noise is actually due to a series of discrete raindrops falling at random.

The lines in the hydrogen spectrum occur in several groups known as "series" (Fig. 30-4),

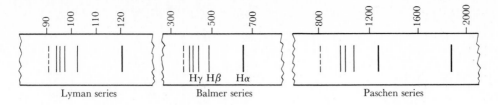

Figure 30-4 Spectral series in hydrogen; wavelengths in nanometers.

which are named after the scientists who first investigated them, before the Bohr theory. The energy-level diagram gives a ready explanation for these series. For instance, the *Balmer series* contains lines emitted during quantum jumps $E_3 \to E_2$, $E_4 \to E_2$, $E_5 \to E_2, \ldots$, always with the second energy level E_2 as the final state. For the lines of the *Lyman series*, it is the level E_1 that is the final state. The Lyman series lies entirely in the ultraviolet region, since the quantum jumps are large and the photons have high frequency and short wavelength. The *Paschen series*, for which E_3 is the final state, and other higher series lie entirely in the infrared region. It so happens that the entire *visible* spectrum of hydrogen contains only members of one series, the Balmer series. This is why the hydrogen spectrum appears so simple.

Example 30-1

From Fig. 30-3, estimate the wavelength of the second Balmer line, Hβ in Fig. 30-4.

Since $hf = E_4 - E_2$ for the second line in the series, the photon's energy is 12.75 eV − 10.20 eV = 2.55 eV.

$$\lambda = \frac{1240 \text{ eV} \cdot \text{nm}}{2.55 \text{ eV}} = \boxed{486 \text{ nm}}$$

This line is in the blue-green region of the spectrum; it can be identified in Fig. 30-2e.

So far, we have seen the implications of Bohr's really revolutionary first two postulates:

1. Radiationless energy states exist.

2. During a transition between states, the emitted or absorbed photon's energy is given by $hf = E_2 - E_1$.

These two postulates of Bohr are as good today as they were in 1913.

Bohr's third postulate gave a rule for finding the values of the "allowed" energy levels. Bohr was led to assume that the allowed energies correspond to certain allowed orbits for the electrons; those in orbits of larger radius have larger energies. The third Bohr postulate can be stated as follows:

3. The allowed orbits are those for which the angular momentum ($I\omega$; see Sec. 7-7) of the electron is an integral multiple of Planck's constant divided by 2π. That is,

$$\text{Angular momentum} = n\frac{h}{2\pi} \quad (30\text{-}2)$$

where n is an integer.

Using this postulate, Bohr was able to calculate the energies of the hydrogen atom's energy levels, and hence, using the differences between these energies, he could calculate the frequencies of the emitted photons. He obtained exact agreement, within experimental error, with the observed frequencies. We will now show that the theory also leads to exact agreement between the calculated ionization energy and the experimental value of 13.60 eV.

For generality, we will consider a *hydrogen-like atom* (Fig. 30-5) consisting of a single electron moving in a circular orbit around a nucleus of charge $+Ze$ (here Z is the atomic number). Such an atom might be ionized helium, $_2\text{He}^+$.

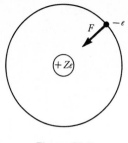

Figure 30-5

Helium is the second element in the periodic table and normally has two orbital electrons; the removal of one electron leaves an ion that, like hydrogen, has a single electron, but has a nuclear charge $+2e$ instead of $+1e$, which is the charge of a proton in the hydrogen atom. Similarly, other hydrogenlike ionized atoms, such as $_3\text{Li}^{2+}$ and $_4\text{Be}^{3+}$, contain only one orbital electron.

Bohr's assumption (Eq. 30-2) was that the angular momentum of the electron can only be an integral multiple of $h/2\pi$. In Chap. 7 we saw that angular momentum equals moment of inertia × angular velocity; here the moment of inertia is simply mr^2, and the angular velocity is v/r. Hence the angular momentum of the electron is $I\omega = (mr^2)(v/r) = mvr$. Bohr's postulate is, therefore,

$$mvr = \frac{nh}{2\pi} \qquad (30\text{-}3)$$

where n is an integer.

The centripetal acceleration of the electron is v^2/r, and the centripetal force is supplied by the Coulomb attraction between the nucleus and the electron (Eq. 18-1). Therefore, applying Newton's second law, $F = ma$, we find

$$\frac{k(Ze)(e)}{r^2} = \frac{mv^2}{r} \qquad (30\text{-}4)$$

We now have two equations with v and r as the two unknowns. Solving simultaneously* for v

---

* A good way to start is to solve Eq. 30-4 for mv^2r and then divide by Eq. 30-3 to get v.

and r, we obtain

$$v = \frac{2\pi kZe^2}{nh} \qquad (30\text{-}5)$$

$$r = \frac{n^2h^2}{4\pi^2 mkZe^2} \qquad (30\text{-}6)$$

We can immediately calculate the radii of the allowed orbits. From Eq. 30-6 we see that the radii are proportional to the square of the quantum number n. Using the numerical values of the various constants (and setting $Z = 1$ for hydrogen), we obtain for the radius of the smallest orbit

$$r_1 = \frac{(1)^2(6.63 \times 10^{-34}\,\text{J·s})^2}{4\pi^2(9.11 \times 10^{-31}\,\text{kg})(9.00 \times 10^9\,\text{N·m}^2/\text{C}^2)} \times (1)(1.60 \times 10^{-19}\,\text{C})^2$$

$$= 0.529 \times 10^{-10}\,\text{m} = 0.529\,\text{Å}$$

This is a remarkable achievement: The radius of the first Bohr orbit is calculated using only fundamental constants h, m, k, and e, and the value so obtained is in agreement with various indirect experimental estimates of the size of the H atom in its normal state.

To obtain the energy levels, we must calculate both KE and PE. The kinetic energy is $\frac{1}{2}mv^2$, and from Eq. 30-4 we find that

$$\text{KE} = \frac{kZe^2}{2r} \qquad (30\text{-}7)$$

The PE is electrical in nature. The system gains energy if the electron is moved from its orbit to infinity against an opposing (attractive) force kZe^2/r^2 of the positive nucleus. Similarly, the system loses PE if an electron falls down from infinity into some orbit.

Now we must decide on a reference level from which to measure the PE. We choose PE = 0 when the electron is at $r = \infty$ (removed from the atom). (Recall that only *changes* of PE are observable; we are free to select any convenient configuration of the system as the reference level.) The system has PE = 0 when the electron is at infinity; it loses PE when the electron falls toward the nucleus, so the PE when the electron is in a circular orbit of radius r is negative. The PE of the electron, when at a distance r from the nucleus, is found from the electric potential V given by Eq. 19-5 (page 419): $W_{r \to \infty} = VQ_1 = kQQ_1/r$. Substituting $-e$ for Q and $+Ze$ for Q_1 gives

$$\text{PE} = -\frac{kZe^2}{r} \qquad (30\text{-}8)$$

The PE is twice as large as the KE and is negative. The total energy is the sum of Eqs. 30-7 and 30-8:

$$E = \text{KE} + \text{PE}$$

$$= \frac{kZe^2}{2r} - \frac{kZe^2}{r}$$

or

$$E = -\frac{kZe^2}{2r}$$

Finally, we substitute the value of r for the nth orbit (Eq. 30-6) to obtain the Bohr energy formula:

$$E_n = -\frac{2\pi^2 mk^2 e^4 Z^2}{h^2 n^2} = -\text{constant}\left(\frac{Z^2}{n^2}\right)$$

To evaluate the constant, we use

$$m = 9.11 \times 10^{-31} \text{ kg}$$
$$k = 9.00 \times 10^9 \text{ N} \cdot \text{m}^2/\text{C}^2$$
$$e = 1.60 \times 10^{-19} \text{ C}$$
$$h = 6.63 \times 10^{-34} \text{ J} \cdot \text{s}$$

The result is

$$\frac{2\pi^2 mk^2 e^4}{h^2} = 2.17 \times 10^{-18} \text{ J}\left(\frac{1 \text{ eV}}{1.60 \times 10^{-19} \text{ J}}\right)$$

$$= 13.6 \text{ eV}$$

When values of m, k, e, and h to four significant figures precision are used, the calculation gives 13.60 eV. Thus, for any hydrogenlike atom,

$$E_n = -13.60\left(\frac{Z^2}{n^2}\right) \qquad (30\text{-}9)$$

For the ground state of hydrogen, $Z = 1$ and $n = 1$, yielding $E_1 = -13.60$ eV. This is exactly equal to the experimentally determined ionization energy for hydrogen as determined from the analysis of its spectrum.

From the point of view adopted in this section, the total energy of the atom is negative because we have chosen $E = 0$ for the ionized state, formerly called $E = +13.60$ eV. The two notations in Fig. 30-6 are equivalent, and the energy *differences* are identical. Thus the first Balmer series line (shown as a transition between E_3 and E_2) has a photon energy corresponding to 1.89 eV in either scheme. We can consider 13.60 eV to be the *binding energy* of the H atom, for this amount of work must be done to separate the atom into its constituent parts.

The energy of any transition is found from the difference in energy levels as given by Eq. 30-9. In the hydrogen spectrum, for which $Z = 1$, the

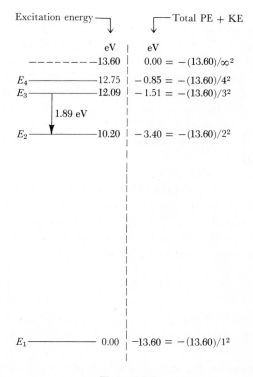

Figure 30-6

formula for the emitted photon's energy is, therefore,

$$hf = 13.60\left(\frac{1}{n^2} - \frac{1}{m^2}\right) \qquad (30\text{-}10)$$

where n and m are integers. The Lyman series has $n = 1$, with $m = 2$, 3, 4, ...; the Balmer series has $n = 2$, with $m = 3$, 4, 5, ...; and similarly for the other series.

A nice verification of the Bohr theory comes from a study of the spectrum of He^+, for which $Z = 2$. We can predict that some of the spectrum lines of He^+ should coincide with certain lines in the H spectrum. For example, we can use suitable values of n and m, found by trial, to calculate the following photon energies:

H spectrum:

$$hf = (13.60)(1)^2\left(\frac{1}{2^2} - \frac{1}{4^2}\right) = (13.60)\left(\frac{3}{16}\right)$$

He^+ spectrum:

$$hf = (13.60)(2)^2\left(\frac{1}{4^2} - \frac{1}{8^2}\right) = (13.60)\left(\frac{3}{16}\right)$$

The lines should therefore have identical energy (and wavelength), even though they are emitted by different substances. Actually there is a very slight difference (less than 0.1%), because the motion of the nucleus (which we have neglected) is less for the He^+ nucleus, which is more massive. Historically, helium* was discovered through its bright lines in the spectrum of the sun during a total eclipse in 1868; some workers thought it to be a form of hydrogen because of this near-coincidence of lines, which was explained later by the Bohr theory.

As so often happens, Bohr's third postulate (Eq. 30-2) turned out to be an approximation. Modern quantum mechanics uses a different, more general postulate, which (of course) contains Bohr's as a special case. This is the usual

---

* From the Greek *helios*, sun. The lines flashed into view for a few seconds just before totality.

course of events and should cause us no concern. Bohr's bold and imaginative theory of the hydrogen atom was a necessary first step toward a complete theory valid for all atoms. Considering what we now know about the duality of matter, we recognize that the model of an atom as a miniature solar system, with an electron in a definite orbit around the nucleus, must be incomplete. These ideas are developed further in the next chapter.

Energy-level diagrams for atoms with more than one electron are more complex than for hydrogen, and hence the spectra are also more complex. For instance, two of the mercury energy levels are very close together (E_8 and E_9 in Fig. 30-7). The transitions $E_8 \rightarrow E_5'$ and $E_9 \rightarrow E_5$ are therefore of almost equal energy. These transitions appear as a pair of yellow lines of wavelengths 577 nm and 579 nm in the mercury spectrum (Fig. 30-2a). The sodium spectrum (Fig.

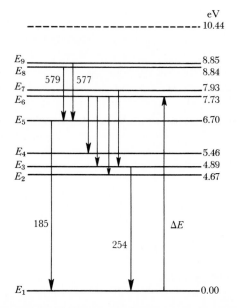

Figure 30-7 Energy levels for mercury. Excitation energies are given for each level. The ionization energy is 10.44 eV. Wavelengths of some of the transitions are shown in nanometers.

30-2b) has a strong pair of yellow-orange lines of wavelength 589.0 nm and 589.6 nm, and 13 other very weak lines in the visible region.

30-4 Absorption Spectra

An atom must, somehow, absorb energy if it is to be able to emit radiation at a later time. When light is absorbed by a gas atom, a photon disappears and the atom's internal energy increases. Since the atom's energy can change only by discrete amounts, such as ΔE in Fig. 30-7, only certain photons can be absorbed. In fact, it should be apparent that *an atom can absorb only those photons it is capable of emitting*, since the equation $\Delta E = hf$ applies equally well for a transition in either direction. It was discovered by Joseph von Fraunhofer in 1814 that the sun's spectrum consists of dark lines across a continuous background (page 712); these lines are caused by absorption by atoms in the sun's atmosphere, which contains Na, Fe, H, and over 60 other elements in gaseous form. The frequency of any absorption line is exactly equal to that of a line emitted by the same atom when excited. This reminds one of the phenomenon of acoustic resonance (Sec. 11-4), but the quantum explanation avoids using a mechanical model such as a resonating string or air column. The *absorption spectrum* is a powerful tool for chemical analysis, since the characteristic dark lines are caused by a gas that is not necessarily hot enough to emit light.

To understand the relation between absorption and emission, we must pay close attention to probabilities and lifetimes. At ordinary temperatures, an atom spends almost all its life in its lowest energy state (E_1), where its total energy is least. If by chance an incoming photon gives up its energy to the atom, or if energy is available during a collision, the atom can be *momentarily* raised to one of its allowed excited states. An atom usually stays in an excited state for only about 10^{-8} s, and then if left to itself emits a photon and spontaneously "jumps" to a lower state. After a very short time, the atom has been "shaken down" to its normal state, and one or more photons have been emitted in the process. For example, after mercury vapor has been illuminated with photons of energy $E_6 - E_1$ (Fig. 30-7), an atom that has been excited to state E_6 can radiate any one of several photons in returning (by stages) to its lowest level. The emission of light that is thus dependent on the absorption of other light is called *fluorescence*.

We see from the energy-level diagram that fluorescent radiation always has a longer wavelength than the radiation that caused the excitation, since the fluorescent photons are emitted during smaller energy jumps. Fluorescence in solids is more complex than in gases, and if the fluorescent photons are emitted after a measurable time interval, we speak of the phenomenon as *phosphorescence*. There is no sharp dividing line between fluorescence and phosphorescence. The fluorescent lamps used in homes and factories are coated with certain phosphors, which absorb ultraviolet photons (having large hf values) from the mercury arc in the gas within the tube and later on emit photons of visible light (low hf values).

Certain energy states have "long" lifetimes of 10^{-3} s or even, in some cases, 1 s or more. These energy levels are called *metastable states* (they are "almost stable"). In practice, an atom in a metastable state is more likely to return to a lower state by giving up its excitation energy in a collision with a gas atom, ion, or free electron than by radiating a photon. In mercury, energy level E_2 at 4.67 eV is a metastable state; note that no photon transition $E_2 \rightarrow E_1$ is drawn in Fig. 30-7. The mercury atom can, of course, absorb 4.67 eV in a collision process, or if it is already in state E_2, it can give up 4.67 eV of energy in a collision and thus return to the lowest state. A famous example of metastable states occurs in ionized oxygen atoms. High in the earth's atmosphere the density of air is so small that collisions are very infrequent, and the excited metastable state can endure long enough (1 s, on the average) for photons to be emitted. This accounts for spectrum lines first observed in the aurora borealis that cannot easily be observed in sources in the laboratory, where collisions are more frequent.

If enough energy is available either from collision or from an incoming photon, an electron can be removed from an atom, leaving behind a positive ion. The dotted lines at the top of the energy-level diagrams of Figs. 30-3 and 30-7 represent ionization energies. For instance, 12.75 eV of energy is insufficient to ionize hydrogen, but this amount of energy can be absorbed and retained for a short time. Using the planetary model, we say that the electron has been placed in a larger orbit. If exactly 13.60 eV is absorbed, the electron is just removed from the atom, and if 13.80 eV is absorbed, the electron is removed to infinity (which in this case means 10^{-8} m or so) and given an extra 0.20 eV of KE as it flies away. We speak of 13.60 V as the *ionization potential* of hydrogen; according to the diagram, it is just equal to the photon energy (in eV) of the most energetic lines of the Lyman series (right at the series limit). Experiment confirms the spectroscopic value of the ionization potential. Conduction of electricity through a gas depends on the presence of ions, and the current through a hydrogen-filled tube suddenly increases when the applied PD reaches 13.60 V.

30-5   Coherent Light—The Laser

Let us look a little more closely at the way in which a photon interacts with an atom. If a photon's energy hf is not equal to the energy difference ΔE between two allowed energy states of an atom, either the photon goes right on by the atom and there is no interaction, or it collides with the atom, in which case the collision is perfectly elastic, and no energy is transferred. If, however, the photon's energy hf exactly matches the energy difference ΔE, it will be able to cause a transition in *either* direction. Such a transition caused by a photon is called a *stimulated transition*. This is a form of resonance; the

incoming photon is "tuned" (in energy) to a pair of energy levels in the atom. In discussing absorption, we assumed that the incoming photon of the proper energy induces or stimulates an upward transition, for instance $E_1 \rightarrow E_6$ in Fig. 30-7. But the same photon would be equally effective in causing a downward transition $E_6 \rightarrow E_1$, because in this case also the photon's energy matches the energy-level difference. In a downward stimulated transition, after the encounter there are two identical photons (the incoming one and the emitted one) each of energy hf (= $E_6 - E_1$); the energy of the extra photon has come from the stored PE of the atom as it changes from state E_6 to the lower energy state E_1.

Suppose now a beam of photons of the proper energy passes through a container in which atoms are enclosed. Some photons will stimulate certain atoms to absorb energy ($E_1 \rightarrow E_6$); other photons will stimulate other atoms to emit radiation ($E_6 \rightarrow E_1$). Even though the probability of a stimulated emission exactly equals the probability of a stimulated absorption, there is a *net* absorption because normally the population of a higher energy level is much less than that of a lower energy level. In a *normal population* there are more atoms in state E_1 (ready to absorb) than there are in state E_6 (ready to emit); the relative numbers depend strongly on the temperature. If, somehow, an *inverted population* can be achieved so that there are more atoms in the upper state than in the lower state, then instead of absorption there can be a net emission of photons; this buildup of the number of photons in the container amounts to an amplification process.

Our understanding of this light amplification process has led to the development of a new source of extremely intense coherent light—the *laser*. To obtain laser action (the word is an acronym for *light amplification by stimulated emission of radiation*), three conditions must be met: (1) The atoms must have an upper energy state in which electrons linger longer than usual—a metastable state—so that there is time for the stimulating impact before the normal spontaneous emission takes place. (2) There must be a means of raising far more electrons to this excited metastable state than would be there normally as a result of ordinary temperature-limited collision processes, so that this upper state *alone* becomes as highly populated as it would be at an extremely high temperature, but without actual heating of the material. This is an inverted population. Finally, (3) the emitted photons must be reflected back and forth through the excitable atoms very many times so that stimulated emission can make up for the escape of photons and keep the process going.

The third condition is met in practice by enclosing the excitable atoms in a box or tube—a resonant cavity—with both ends consisting of perfectly parallel mirrors, but with one or both mirrors slightly transparent to "leak out" enough photons to be used as an external beam. The other two conditions for laser action are illustrated by the popular helium-neon (He-Ne) laser so useful in elementary physics laboratories. Energy levels are shown in Fig. 30-8. When an electrical discharge is maintained through a tube that contains about 85% neon and 15% helium at low pressure, some helium atoms reach a metastable level, E_1, by absorbing 20.61 eV of energy during collisions. It happens, by chance, that there is a level E_3 at 20.66 eV in the neon atom—close enough to E_1 that the

excitation energy of the helium atom can be transferred to a neon atom, raising it from the ground state to 20.66 eV (with the extra 0.05 eV coming from the thermal energy of the moving atoms). In this way the inverted population is achieved in the neon atoms, with more atoms having energy E_3 than E_2. The neon atom loses some of its energy to state E_2 by spontaneous emission; the red light is from those photons of energy $hf = E_3 - E_2$, which in turn stimulate other neon atoms in state E_3 to emit similar photons. As with all amplifiers, there is no net gain in *energy*; here the amplification is powered by the electrical discharge, which keeps forming helium in the E_1 state.

Laser radiation has several valuable features. Because of the reflecting box system, the beam has a very small angular divergence, limited only by diffraction effects related to the diameter of the mirror considered as an aperture (see Eq. 27-2 on page 645); here λ/D is extremely small because D is large. Other sources of radiation emit from exposed surfaces in all directions with considerable divergence. This parallelism of the laser beam allows greatly increased intensity; that is, more energy per second from a laser beam can be focused on a smaller spot. In addition, the phase of the stimulated emission is identical with that of the incident radiation, so that the entire emerging beam consists of coherent radiation (Sec. 26-8). The laser wave

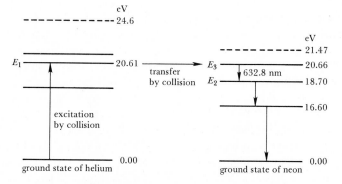

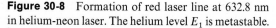

Figure 30-8 Formation of red laser line at 632.8 nm in helium-neon laser. The helium level E_1 is metastable.

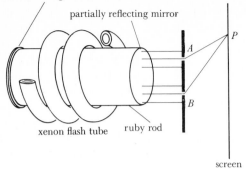

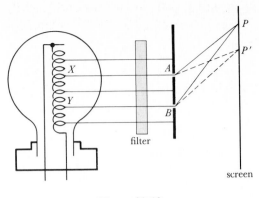

Figure 30-9

Figure 30-10

fronts are at all times surfaces of constant phase, in contrast to the wave fronts of ordinary, "incoherent" light.

A simple experiment demonstrates the coherence of a broad laser beam: An interference pattern can be obtained merely by putting two slits in front of a laser (Fig. 30-9). A maximum occurs at P if the path difference $BP - AP$ equals an integral number of wavelengths. Even though rays passing through A and B come from different parts of the laser, their phase relationship remains constant. In exactly similar fashion, the phases of two radio transmitters can be maintained constant (by electrical means) to give an interference pattern such as was discussed in connection with Fig. 26-10 on page 607. If, instead of a laser, we used a broad source of ordinary light (made monochromatic by use of a filter), no interference would be observed (Fig. 30-10). To be sure, a momentary pattern on the screen might endure for a nanosecond or so (10^{-9} s) if we could see the effect of just two separate atoms in the broad source, located at X and Y, which were at that moment in proper phase relative to each other. Due to random changes in phase caused by atomic collisions in the source, atoms at X and Y would surely have an entirely different phase relationship a microsecond later (10^{-6} s), so a new point P' would be a maximum, with a change in path difference compensating for a change in phase. The com-

bined effect of the superposed interference patterns of many atoms in the source emitting waves having random phases would average out to give a uniformly illuminated screen, and no pattern would be seen. Interference *can* be observed with ordinary light, but only if a single wave front is split into several portions that travel different path lengths and are recombined. With this point in mind, study again the various arrangements illustrated in Chap. 26 (Figs. 26-3, 26-4, 26-7, 26-8, 26-25, 26-29, 26-31).

Practical lasers (Fig. 30-11) fall into two classes. Systems with atoms in the gaseous phase, such as the helium-neon laser, may be operated so as to emit continuously. These have rather low total power output (although they still have beam intensities exceeding those of the best searchlights, per unit power input, owing to their much smaller angular divergence). Such lasers are stable and convenient to use, and they emit very sharp spectrum lines. Systems with atoms in the solid state, such as the ruby laser (Fig. 30-9), usually have to be operated in discrete flashes. In these lasers the electrons are often raised to metastable states by bathing them in an intense flash of (incoherent) white light from a xenon flash tube similar to a photographic "strobe" light. Solid-state lasers have been employed to produce phenomenal intensities (when focused down to small areas) capable of such feats as vaporizing diamonds, drilling

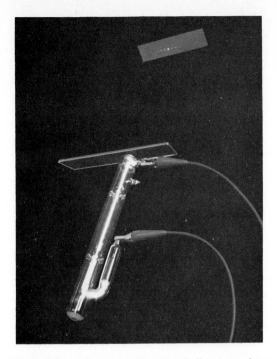

Figure 30-11 Coherent light from helium-neon laser tube passes through a grating ruled on glass. Several orders of interference are seen on each side of the central image.

fine holes in steel or tungsten, and repairing detached retinas. On a less spectacular scale, the large electric field intensities in laser beams have opened up entirely new fields of fundamental physics. For instance, it has been found that the dielectric constant of any given substance is not constant, but varies with the electric field strength.

Some other applications of laser radiation are in telecommunications and in photography. The phase coherence of laser light makes it exactly similar to e-m radiation generated by electrical circuits, except for a tremendously higher frequency. Many new channels of communication can be utilized for voice, pictures, or data transmission by modulating the electric field of a coherent laser beam. The coherence of laser light makes possible photographic images without lenses. These images, formed with the help of

holograms, are discussed in Sec. 26-10. For further applications of lasers and masers to a variety of problems, see the references at the end of this chapter.

30-6 The Laser Cane

From the many applications of physics to engineering, we have selected the laser cane as a typical device making use of many of the concepts we have touched on in this book. Do not consider this section to be just an illustration of a laser device. Rather, use it as a chance to review and integrate material already studied earlier in mechanics, acoustics, electricity, and optics. (See also Problems 30-C5–30-C10.)

A great deal of engineering and physics went into the design of the laser cane for the blind, developed by Bionic Instruments, Inc. (Fig. 30-12). Three small gallium arsenide solid-state lasers, each only 5 mm in diameter, emit beams of infrared radiation (905 nm) that are focused by thin plastic lenses to beam widths of only 25 mm at a distance of 4 m (Sec. 30-5). The return beams, diffusely scattered by obstacles, are received by photodiodes (Sec. 29-3) located down the cane 35 cm from the emitting lasers. To avoid interference from sunlight, lamps, or even other laser canes, each beam is amplitude modulated (Sec. 24-7), and each of the three receiving silicon photodiodes is synchronized to be sensitive only to the desired emitted beam. Modulation is in the form of pulses of

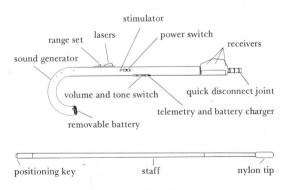

Figure 30-12 The laser cane.

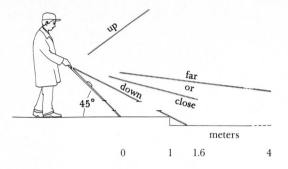

Figure 30-13 Protection zones of the laser cane.

duration 200 ns, repeated 40 times per second. The received pulses are amplified by transistor circuits (Sec. 23-10) in the top section of the cane, and information is given to the user by both audible and mechanical means. Three audiofrequency tones are generated: 2600 Hz for the UP reflection (Fig. 30-13), 1600 Hz for the FORWARD reflection, and 200 Hz for the DOWN reflection. The ratio of 2600 Hz to 1600 Hz was chosen *not* to be a simple ratio, in order to avoid possible interpretation of the combination as a single musical tone (Sec. 11-7). The 200-Hz tone is given a complex overtone structure (Sec. 11-5) to make an easily recognizable "rasping" sound compared with the "pure" tones for the other two channels. To build up the intensity of the audible 200-Hz tone, the loudspeaker cavity is shaped to give resonance at 200 Hz (Sec. 11-4). In addition, the returned signal from the 1600-Hz FORWARD beam is made to actuate a tactile stimulator that tickles the finger to warn of a forward obstacle.

The cane is powered by a 6-V rechargeable battery that can deliver 600 mW for 3 h on one charge. Since production is on a small scale (less than 100 canes so far), it has proved more economical to use discrete components except where mass-produced IC's (Sec. 33-10) have been available cheaply. The whole cane weighs about a pound (mass 450 g), and its mass is distributed to have the "feel" of a conventional "long cane" used by blind travelers. This requires attention both to the center of gravity (Sec. 4-6) and the moment of inertia (Sec. 7-6) of the cane. The lasers are sufficiently low powered that even a 30-minute exposure to the eyes of rhesus monkeys caused no observable effects.

It might be expected that distance ranging would employ some sort of time-of-flight measurement, as for sonar (Example 10-3) and radar, but it proved simpler and more precise to use optical triangulation. The shorter wavelength of infrared light (compared with that of ultrasonic acoustic waves) allows for a narrow beam, since λ/D can be made small (Sec. 27-5). The angle made by the reflected ray passing through the receiving lens is an indication of the distance to the object detected. Thus, for example, the narrow FORWARD beam strikes a wall at a distance d_1 from the cane (Fig. 30-14). Light is not reflected from the wall as from a mirror; rather it is diffusely scattered. Some of the scattered rays will return to the receiving photodiode R at an angle θ_1 and be focused by a lens system to strike the diode's sensitive surface. (Other rays in the diffuse bundle of rays leaving that point of the wall miss the diode and are not registered.) If, however, the wall is at a distance d_2 greater than the desired range, the only scattered rays that reach the diode are coming in at a slightly greater angle θ_2, and the focused image fails to reach the sensitive region of the diode. Thus, the cane ignores obstacles that are too far away to be of im-

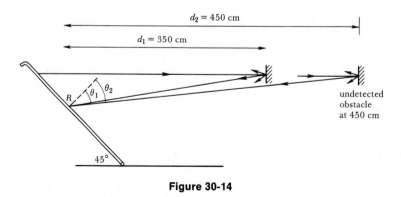

Figure 30-14

portance, presenting the user with a warning that is neither too soon (and therefore confusing) nor too late (and therefore dangerous). Similarly, the UP and DOWN beams are sensitive to obstacles within predetermined ranges. A twig the thickness of a lead pencil, or a clothesline, is usually just detectable 45 cm in front of the tip of the cane, 180 cm above the ground.

The training period for the use of the laser cane is about 2 hours per day for 4 weeks. The cost is about half that of a trained dog. The remaining problems are not technical; instead they cover such areas as identifying the travelers who can best use the cane, developing a training program for instructors as well as users, and establishing funding channels.

Summary   The nucleus of an atom is of the order 10^{-14} m in size, which is about 10^{-4} the size of the electron cloud. When α particles bombard a thin metallic foil, only a few are scattered, but they are scattered through large angles. This shows the concentration of charge in a central nucleus (Rutherford).

A hot solid, liquid, or dense gas emits a continuous spectrum that depends only on temperature. A hot gas emits a line spectrum that depends on the nature of the gas. The Bohr theory of the line spectrum of hydrogen is based on three postulates: (1) An atom can exist without radiating; (2) when the energy state of an atom changes by ΔE, the frequency of the photon emitted or absorbed is given by $\Delta E = hf$; (3) the allowed energies are those for which the electron's angular momentum is an integral multiple of $h/2\pi$.

An atom can remain in an excited state for less than a microsecond, except in rare cases when it is in a metastable state having a considerably longer lifetime. Hence, at ordinary temperatures, most of the atoms in a gas are in their lowest energy state (normal population). Absorption of a photon can raise an atom to a higher state, and one or more fluorescent photons can be emitted as the atom returns (in steps) to its normal state. An atom can absorb only those wavelengths it is capable of emitting. Stimulated emission of radiation leads to light amplification if photons interact with atoms that are in an inverted population condition. The laser uses stimulated emission to produce an intense beam of coherent light.

Check List   continuous spectrum   hydrogenlike atom   metastable state
                  discrete spectrum   Lyman series   ionization potential
                  Bohr's postulates   Balmer series   stimulated transition
                  energy state or energy   Paschen series   normal population
                    level   absorption spectrum   inverted population
                  ground state   fluorescence   laser
                  ionization energy   phosphorescence

Questions   **30-1**   What sort of spectrum, line or continuous, gives information about the chemical composition of a source? How can the temperature of a source be found from a study of its spectrum? Is this always possible for all sources?

                 30-2   Using Fig. 30-2, explain why a neon sign is red. What color do you expect a helium sign to have?

                 30-3   Why do so many dark lines in the sun's spectrum on page 712 coincide in position with adjacent bright lines? How do you account for the dark lines that do *not* match up?

30-4 A dancer's costume coated with fluorescent material is illuminated with ultraviolet radiation, so-called "black light," and visible light is emitted. Could visible fluorescence also be produced by infrared radiation shining on a suitable substance?

30-5 Try this experiment in a room that can be darkened, in which there is a TV set. Without turning on the set, turn out the room lights. Explain why the screen of the TV set glows visibly for a while. (This works best with a black-and-white set.)

30-6 What is the difference between a downward spontaneous transition and a downward stimulated transition?

30-7 By which, if any, of the following methods can a mercury atom reach the excited metastable state E_2 shown in Fig. 30-7? (a) absorption of a photon of energy $E_2 - E_1$; (b) spontaneous emission of a photon of energy $E_6 - E_2$; (c) absorption of energy in a collision process; (d) emission of a photon of energy $E_6 - E_2$ in a stimulated transition.

30-8 Explain the operation of a laser, and describe the characteristics of the light emitted by a laser.

30-9 The Balmer series lines appear in the absorption spectrum found in hot stars, but not in absorption spectra obtained in room-temperature experiments on the earth. Explain.

MULTIPLE CHOICE

30-10 The Rutherford scattering experiment showed the existence of (a) the nucleus; (b) energy levels; (c) isotopes.

30-11 A source that emits a line spectrum is (a) solid; (b) gaseous; (c) either of these.

30-12 The frequencies of the lines in a spectral series, such as the Balmer series in hydrogen, (a) extend to higher frequencies without limit; (b) have some lower limit; (c) have some upper limit.

30-13 Compared with the energy of the emitted spectrum line of longest wavelength, the ionization energy of an atom is (a) greater; (b) the same: (c) less.

30-14 Iron is chemically different from aluminum because (a) its atomic number is different. (b) the number of orbital electrons is different; (c) both of these.

30-15 Exactly half of a collection of 10^{10} atoms are in an excited state with energy $E_1 = 3.65$ eV, and the other half of the atoms are in an excited state with energy $E_2 = 3.95$ eV. This population (a) is normal; (b) is inverted; (c) could be either, depending on the temperature.

Problems　**30-A1** Using the photograph on page 712, determine to five significant figures the wavelengths of the three most intense iron spectrum lines in the region 400 nm to 410 nm.

30-A2 In state A the electronic energy of an atom is 5.97 eV, and in state B its electronic energy is 3.46 eV. (a) What is the energy of the photon emitted in the transition from A to B? (b) What is the wavelength of this photon?

30-A3 Calculate (a) the energy and (b) the wavelength of the photon emitted when a mercury atom's energy changes from E_7 to E_3 (Fig. 30-7).

30-A4 Calculate the maximum KE (in J) of an electron for which a collision with a hydrogen atom in its normal state is surely an elastic one.

30-A5 How much energy is needed to ionize a mercury atom that is already in the metastable state E_2 (Fig. 30-7)?

30-B1 Use the energy-level diagram of Fig. 30-3 to calculate the wavelength of the Lyman α line (the line in the Lyman series having the longest wavelength). Is this line visible?

30-B2 Using Fig. 30-3, calculate the wavelength of the limit of the Lyman series (Fig. 30-4) in hydrogen. In what region of the e-m spectrum is this limit?

30-B3 Using the energy-level diagram of Fig. 30-3, estimate the energy and wavelength of the line of longest wavelength in the Paschen series of hydrogen (all transitions in this series terminate on E_3). Check your answer by referring to Fig. 30-4. Is this line visible?

30-B4 The fourth line in the hydrogen Balmer series ($H\delta$) has a wavelength of 410.1 nm. (a) Identify the line in Fig. 30-4. (b) Identify the line in the hydrogen emission spectrum in Fig. 30-2e. (c) Identify the line in the sun's absorption spectrum on page 712. (d) One of the energy levels involved in the emission or absorption of this line is shown in Fig. 30-3; use the wavelength of the line to compute the value of the other energy level involved.

30-B5 Determine which two energy levels in the mercury atom (Fig. 30-7) are involved in the emission of the violet line whose wavelength is 408 nm. (*Hint:* First find the energy of the photons, in electron volts.) Identify the line in Fig. 30-2a.

30-B6 A mercury line has wavelength 546 nm. Identify the two levels in Fig. 30-7 involved in the production of this line. Identify the line in Fig. 30-2a.

30-B7 What is the energy of the least energetic photon that can be absorbed by a hydrogen atom at room temperature? (Use Fig. 30-3.) Identify the line in Fig. 30-4.

30-B8 What is the fractional decrease in mass of a mercury atom when a photon of the ultraviolet line of wavelength 185 nm is emitted (Fig. 30-7)?

30-B9 (a) Calculate the angular momentum of the earth in its orbital motion around the sun (use data from Appendix Table 7). (b) Suppose the orbital angular momentum of the earth were quantized according to the Bohr relation (Eq. 30-2). What would the quantum number n be? Could such a quantization be detected experimentally?

30-B10 Show that the ratio of the wavelengths (which is the same as the ratio of the frequencies) of the first two lines in the Balmer series is $27:20$.

30-B11 What is the electric potential in volts at a point in the first Bohr orbit of a hydrogen atom? Indicate whether the potential is $+$ or $-$ (relative to infinity).

30-B12 There is a gravitational attraction between the nucleus and the electron. Explain quantitatively why we can ignore the contribution of gravitational PE to the total energy of the hydrogen atom.

30-B13 What would be the quantum number for a circular orbit in a hydrogen atom for which the radius of the atom is large enough to be seen in a visible-light microscope that can resolve details down to 500 nm? (*Hint:* Use a proportion; how is r_n related to r_1?

30-B14 Solve Eqs. 30-3 and 30-4 to obtain the equations for v and r given in Eqs. 30-5 and 30-6.

30-C1 A large ruby laser with end mirror of diameter 8 cm (Fig. 30-9) sends out a not-quite-parallel beam of wavelength 694.3 nm. (a) Calculate the angular divergence of the beam, which arises from the diffraction pattern of the end mirror serving as a circular aperture (use Eq. 27-2 on page 645). (b) If the laser is pointed at the moon, 3.8×10^8 m away, how large is the diameter of the central part of the circular diffraction pattern on the moon's surface?

30-C2 Show that for an electron in the lowest-energy circular orbit in hydrogen, v/c is approximately $1/137$. How does this verify the approximate validity of using non-relativistic mechanics in the Bohr theory?

30-C3 What is the approximate wavelength of the H and He^+ lines that were discussed on page 720? Identify the line in Fig. 30-2e. Why is this line not found in Fig. 30-2c?

30-C4 (a) Derive a formula for the orbital frequency of revolution of the electron in the hydrogen atom when the quantum number is n. (b) Use your formula to calculate the orbital frequency, in rev/s, when $n = 2$ and also when $n = 3$. (c) Calculate the frequency of the radiation emitted when n changes from 3 to 2 (Balmer α line), and show that this frequency is of the same order of magnitude as the average of the orbital frequencies found in (b).

Note: The following problems on the laser cane can serve as a review of several topics in your physics course. Use data from Sec. 30-6.

30-C5 (a) Calculate the number of waves contained in a single infrared pulse of duration 200 ns. (b) How long, in meters, is this train of waves?

30-C6 The receiving lens system for the FORWARD beam is a pair of lenses, each of focal length $+16$ mm, separated by 10 mm. How far behind the second lens is the focal point for light coming from a distant object?

30-C7 The angles in Fig. 30-15 are found by trigonometry to be $\theta_1 = 40.96°$ and $\theta_2 = 41.85°$. The diameter of the receiving lens is 10.0 mm. Calculate the resolving power of the lens system, that is, the smallest detectable difference in angle between two incident beams, and verify that the beams in Fig. 30-15 are, in fact, well resolved.

30-C8 How far apart (less than 1 mm) are the two images formed at θ_1 and θ_2, respectively, in the plane of the receiving diode, if the effective focal length of the lens is 8.8 mm?

30-C9 How much energy, in kJ, can be stored in the rechargeable battery in the laser cane?

30-C10 The laser cane is 120 cm long and weighs 3.6 N. Its center of gravity is 80 cm from the lower tip. If the cane is supported by a vertical force supplied by the hand, and if the tip rests on a smooth horizontal pavement, calculate (a) the force exerted by the hand; (b) the force on the pavement.

References

1. W. F. Magie, *A Source Book in Physics* (McGraw-Hill, New York, 1935), pp. 354–356, Kirchhoff's analysis of the origin of Fraunhofer lines; pp. 360–365, Balmer's empirical formula for the wavelengths of the visible hydrogen lines.
2. E. N. daC. Andrade, "The Birth of the Nuclear Atom," *Sci. American* **195**(5), 93 (Nov. 1956).
3. T. H. Osgood and H. S. Hirst, "Rutherford and His Alpha Particles," *Am. J. Phys.* **32**, 681 (1964).
4. O. Struve, "The Fraunhofer Lines," *Sky and Telescope* **10**, 117 (1951); see also the back cover of this issue, which has a 56-cm-long photograph of the solar spectrum.
5. C. E. Behrens, "Atomic Theory from 1904–1913," *Am. J. Phys.* **11**, 60 (1943); "The Early Development of the Bohr Atom," *Am. J. Phys.* **11**, 135 (1943); "Further Development of Bohr's Early Atomic Theory, *Am. J. Phys.* **11**, 272 (1943).
6. W. Thumm, "Some Thoughts on Teaching about Ultraviolet Radiation," *Phys. Teach.* **13**, 135 (1975). Emphasis on the biological effects of ultraviolet radiation.

7. T. W. Hansch, A. L. Schawlow, and G. W. Series, "The Spectrum of Atomic Hydrogen," *Sci. American* **240**(3), 94 (Mar. 1979). All you ever wanted to know about hydrogen. See especially pages 94–99.

8. A. L. Schawlow, "Laser Light," *Sci. American* **219**(3), 120 (Sept. 1968).

9. D. R. Herriott, "Applications of Laser Light," *Sci. American* **219**(3), 141 (Sept. 1968).

10. R. Tinker, "The Safe Use of Lasers," *Phys. Teach.* **11,** 455 (1973).

11. H. Weichel, W. A. Danne, and L. S. Pedroth, "Laser Safety in the Laboratory," *Am. J. Phys.* **42,** 1006 (1974).

12. J. M. Coakley, "Probability of Laser Injury to the Eye—Further Considerations," *Phys. Teach.* **13,** 388 (1975).

13. National Film Board of Canada, *Thomson Model of the Atom; Rutherford Scattering* (films).

31

Atomic Structure

31-1 *Quantum Mechanics and the Limitations of the Bohr Model*

The Rutherford-Bohr model of the atom gave, as we have seen, a dramatically successful initial interpretation of the hydrogen spectrum and the general processes of emission and absorption of photons by atoms. However, in the dozen years following 1913, it became apparent that the details of the spectra of atoms other than hydrogen—even of such a simple one as $_2$He—could not be explained. The Bohr theory gave no explanation of the probabilities of transition between energy levels, which are related to the intensities of spectral lines. Other basic phenomena were outside the framework of the Bohr theory; these included the chemical bond and solid-state phenomena, superconductivity, electron spin and the related magnetic properties of matter, and radioactive decay. In addition, the Bohr theory was nonrelativistic. In 1923, de Broglie's proposal of the wave-particle duality of matter set the stage for a development in

physics comparable to the Newtonian synthesis of 1666 in its generality and power.

About 1925, Heisenberg, Schroedinger, Born, Jordan, Dirac, and others applied the idea of duality to develop a new science of mechanics called wave mechanics, or quantum mechanics. Newton's laws are a special case of quantum mechanics, valid for collections of many particles, such as a baseball. Bohr's third postulate is also a special case, valid for circular orbits in the hydrogen atom only. Quantum mechanics, which is the mechanics of matter waves, is highly mathematical. It is a successful theory, and it is fair to say that the physicist's knowledge of the outer atom is, in principle, complete. The only remaining difficulties seem to be mathematical.

To illustrate the contrast between the Bohr quantum theory and modern quantum mechanics, consider the description of the first allowed orbit of the electron in the hydrogen atom. According to Bohr, the electron moves in a circular orbit with a definite momentum and a definitely assigned position at all times. According to

The Zeeman effect is studied by placing a vertical spectrum source in the horizontal magnetic field supplied by an electromagnet. Only a portion of the spectroscope (black tube) is shown; the knob at the right adjusts the slit width.

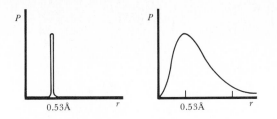

Figure 31-1 (a) Hydrogen atom in ground state according to Bohr model. (b) Probability density according to quantum mechanics.

quantum mechanics, since the momentum is known precisely, the position must be highly uncertain. There is *some* probability of finding the electron in any small region of space, even as far as a kilometer or so from the nucleus; but upon working out the probabilities, it turns out that the electron is *most likely* to be at a distance of 0.53 Å from the nucleus, which is exactly the radius of the definite orbit that Bohr calculated (Fig. 31-1a). The uncertainty principle leads, therefore, to the concept of a "smeared-out" atom. The electron is still supposed to be a point charge, but when joined with a proton to form a H atom, the probability of finding it in any given small volume is less than 100%. Thus, the electron in a H atom can never be assigned a precise position. According to our probabilistic model, it is the probability waves, not the electron, that are smeared out.

Although the *position* of the electron in a hydrogen atom can only be described by probabilities, it is nevertheless true that modern quantum mechanics does give exact, discrete *energy states* for the atom. The "allowed" energy states come from the wave nature of the electron, in somewhat the same way that certain "allowed" modes of vibration of a string lead to discrete frequencies known as the fundamental and the overtones. The string of Example 11-2 on page 243 just *cannot* vibrate at 205 Hz; the allowed frequencies are determined by the geometry of the situation, which leads to a discrete set of

stationary waves, with a node at each fixed end of the string.*

The mathematical analysis of the H atom by the methods of quantum mechanics leads to a *probability cloud* representation, where we show in a graph or drawing (Fig. 31-1b) the probability P that the electron will be found within a small region near any given place.[†] As an example, we consider the H atom in its ground state ($n = 1$) for which the energy is $-13.60/1^2 = -13.60$ eV. Quantum mechanics agrees exactly with the Bohr formula (Eq. 30-9) for this energy, but the picture is quite different.

31-2 Quantum Numbers

The detailed working out of the quantum-mechanical theory of the H atom is rather complicated because it is a three-dimensional problem, with the force on the electron variable both in magnitude and direction. It turns out that *three* quantum numbers are needed to describe the probability cloud for the electron's position; in addition, a fourth quantum number is necessary to describe the direction of the electron's "spin" (Sec. 31-4).

The *principal quantum number*, denoted by n, can have any integral value, 1 to ∞. The total energy of the atom depends on n in exactly the same way as in the Bohr model—that is, $E_n = -13.60Z^2/n^2$ for the energy (in eV) of a hydrogenlike atom of nuclear charge $+Ze$.

The *orbital quantum number*, denoted by l, is related to the angular momentum of the electron, and can have any value 0, 1, 2, . . . , $(n - 1)$. Thus, if $n = 5$, l can be 0, 1, 2, 3, or 4. Roughly speaking,

---

* See Sec. 29-10, where the method of stationary waves is used to derive the allowed energy levels for a simple one-dimensional mechanical system, the particle in a box.

[†] $P \Delta r$ is proportional to the probability of finding the electron between r and $r + \Delta r$; the probability that the electron is "at a point" is zero, because a point is a mathematical abstraction having zero volume.

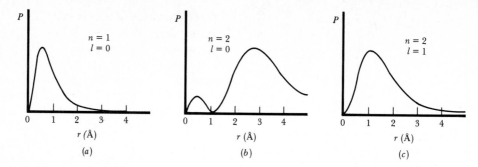

Figure 31-2 Quantum-mechanical representation of probability density, as a function of distance from nucleus, for hydrogen atom in several states. (a) 1s state with $n = 1$, $l = 0$. (b) 2s state with $n = 2$, $l = 0$. (c) 2p state with $n = 2$, $l = 1$.

insofar as the Bohr model is useful, l may be said to be related to the shape of the orbit. (In later versions of the Bohr model, elliptical orbits as well as circular ones were allowed.) In quantum mechanics, the magnitude of the angular momentum is given by

$$L = I\omega = \sqrt{l(l + 1)}(h/2\pi) \qquad (31\text{-}1)$$

so it is possible to have the electron in the H atom with no angular momentum ($l = 0$); this is not possible in the Bohr model, where an electron in a circular or elliptical orbit must have *some* angular momentum. When $l = 0$, the electron cloud is spherically symmetrical and has no inherently definable axis of rotation.

The *magnetic quantum number*, denoted by m_l, has integral values ranging from $-l$ to $+l$. Thus if $l = 3$, m_l can be $-3, -2, -1, 0, 1, 2,$ or 3. If $l \neq 0$, the electron's probability cloud is not spherically symmetrical, and we can visualize the electron as being in some sort of orbit, with angular momentum **L**.* The quantum number m_l is related to the orientation of **L** relative to any given axis. If we call this axis the z axis, the z component of **L** is given by $L_z = m_l(h/2\pi)$. If we know L_z, we cannot also know the x or y component of **L**, for then the remaining component

would become known,[†] and the plane of the orbit would be precisely determined. But this is not possible; according to Heisenberg's uncertainty principle, momentum and position cannot *both* be determined at the same time. If there is an external magnetic field, this determines the "up" or z axis. In the absence of any field, the z axis could be in any preassigned direction; we would not expect a measurement of the probability cloud to depend on the direction selected for the z axis.

In Fig. 31-2, graphs of the probability density P are given for the H atom in various states. Wherever $P = 0$ there is a node of probability, where the electron can never be found; this is not surprising, since the whole distribution is formed as a stationary wave system of de Broglie waves, and the presence of nodes is characteristic of stationary waves. Thus, if $n = 2$ and $l = 0$, the electron can never be found at a distance 1.06 Å from the nucleus.

31-3 The Zeeman Effect

When the electron in an H atom has $l \neq 0$, there is a resultant angular momentum, and we can consider the atom to be a current loop (Sec.

---

* Angular momentum is discussed in Sec. 7-7 and 7-8; the vector **L** is parallel to the axis of rotation or revolution.

[†] If we know the magnitude of a vector and two of its components, we can use the Pythagorean theorem to find the third component.

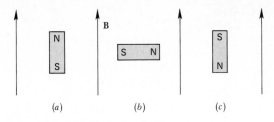

Figure 31-3 A small magnet has more PE when oriented as in (c) because work must be done to rotate it from its normal position (a).

22-5). The atom behaves like a small magnet, with its axis antiparallel* to **L**; its total energy depends on the orientation of the atomic magnet relative to the applied magnetic field (Fig. 31-3). Therefore, when the atom is placed in a magnetic field, its energy levels are split according to the value of m_l, and therefore the spectral lines are split. This effect, discovered by Pieter Zeeman[†] in 1896, was the first evidence for the presence of moving charges within atoms.

The Zeeman effect is small; in an external magnetic field of 1 T, supplied by a strong laboratory magnet, the difference between the $m_l = 1$ and $m_l = 0$ levels of the $(n = 2, l = 1)$ hydrogen state is only about 6×10^{-5} eV, tiny compared with the 1.89-eV energy difference between the $n = 3$ and $n = 2$ levels (Balmer red line, Hα).

The detailed study of the Zeeman effect has been important in the unraveling of many of the complexities of energy levels, and calls for spectroscopes of the highest quality with the greatest possible resolving power.

31-4 Electron Spin

The fourth quantum number is m_s, the *spin magnetic quantum number*. In contrast to the other quantum numbers, m_s can have only two values: $+\frac{1}{2}$ and $-\frac{1}{2}$. We are getting into a gray area where large-scale models are not possible. The word "spin" conjures up an image of the earth, spinning on its axis while at the same time having orbital motion around the sun. Here, too, there are just two possible directions of spin—clockwise and counterclockwise—which is perhaps a justification for using the term "spin" in describing this aspect of an electron's behavior. But the uncertainty principle tells us that the electron's "orbit" is not in a well-defined plane, so the geometrical interpretation of spin, though helpful, must be imprecise. *Electron spin is essentially nonclassical* and has no exact large-scale counterpart; it can be treated adequately only by the methods of quantum mechanics. Nevertheless, we can use the spinning electron model as an aid in interpreting some of the features of atomic spectra.

The effect of electron spin shows up in the fine structure of spectral lines—details not considered in the original Bohr theory. For example, the red Hα line at 656.28 nm is a transition between levels for which $n = 3$ and $n = 2$. Experiment shows that the line has a complex structure, being essentially a very close pair of lines (656.273 nm and 656.285 nm). From this splitting we deduce that one or both of the energy levels is split by a tiny amount. We explain the splitting as follows: From the electron's point of view, the charged nucleus (a proton) is moving in a circle around the electron, and hence a magnetic field exists at the electron's position. The energy of the electron therefore depends on which way the spin magnet is aligned with this internal magnetic field. The splitting is an internal Zeeman effect. In hydrogen, these spin doublets are too close to be resolved in Fig. 30-2e (page 714). The similar spin-orbit splitting in the sodium spectrum (Fig. 30-2b) is clearly seen in the photograph; these wavelengths are 588.995 nm and 589.592 nm.

We summarize the four quantum numbers of an electron in the hydrogen atom in Table 31-1. In a collection of atoms, say in a discharge tube, each H atom will be, at any instant, in a *state*

---

* The orbiting electron is negative, whereas the current loops of Sec. 22-5 are for conventional (positive) current.

[†] A Dutch physicist (1865–1943) (pronounced *Zay´mahn*).

Table 31-1   Quantum Numbers for an Electron in a Hydrogen Atom

Name	Symbol	Range of Values	Physical Quantity Determined
principal	n	1, 2, 3, . . .	energy of the atom
orbital	l	0, 1, 2, . . . , $(n-1)$	magnitude of angular momentum
magnetic	m_l	$-l, . . . , 0, . . . , +l$	orientation of orbital angular momentum
spin magnetic	m_s	$-\frac{1}{2}, +\frac{1}{2}$	orientation of electron spin

characterized by four quantum numbers selected from this table. For any given state, the energy depends only on n, unless there is an external or internal magnetic field, in which case, the energy also depends on m_l and m_s. The shape of the electron probability cloud—the chance of finding the electron at any position—is determined by the value of l.

31-5   The Pauli Exclusion Principle

We now leave "square one" and turn to the description of atoms that contain more than a single electron: $_2$He, $_3$Li, $_4$Be, . . . , up to $_{92}$U and beyond. These atoms have atomic numbers Z greater than 1.

Shortly after the discovery of electron spin about 1925, the Austrian physicist Wolfgang Pauli (1900–1958) introduced a powerful new postulate that sharply limits the possible ways in which the electron clouds of more than one electron can be put together to form atoms:

No two electrons in the same atom can have all their quantum numbers the same.

As an example of the Pauli exclusion principle, we consider in some detail the neutral lithium atom, $_3$Li. This atom, with its nuclear charge $+3e$, must have 3 electrons in the electron cloud. For each electron we assign a set of four quantum numbers from Table 31-1.

	n	l	m_l	m_s
1st electron	1	0	0	$+\frac{1}{2}$
2nd electron	1	0	0	$-\frac{1}{2}$
3rd electron	?	?	?	?

For $n = 1$ we *must* have $l = 0$ and hence $m_l = 0$; there are only two possible values for m_s, and we see that adding a third electron to form $_3$Li would violate the Pauli exclusion principle unless we change the value of n. Therefore, we conclude that this third electron has $n = 2$, or 3, or 4, . . . , anything but 1. Suppose we try the combination

Lithium, state A	1	0	0	$+\frac{1}{2}$
	1	0	0	$-\frac{1}{2}$
	3	0	0	$+\frac{1}{2}$

This would surely be possible, but so also is the combination

Lithium, state B	1	0	0	$+\frac{1}{2}$
	1	0	0	$-\frac{1}{2}$
	2	0	0	$-\frac{1}{2}$

We realize from our discussion of the Bohr model for hydrogen that the energy of an electron depends primarily on the principal quantum number n. Thus, when the lithium atom is in state A, with one electron having $n = 3$, the atom has more energy than when it is in state B, with the third electron having $n = 2$. State A is an excited state, and state B is, in fact, the ground state of $_3$Li.

The exclusion principle gives a logical explanation for *electron shells*. Only two electrons in an atom can have $n = 1$ (their spins are opposite). Only 8 electrons in an atom can have $n = 2$, according to the following table:

n	l	m_l	m_s
2	0	0	$\pm\frac{1}{2}$
2	1	-1	$\pm\frac{1}{2}$
2	1	0	$\pm\frac{1}{2}$
2	1	1	$\pm\frac{1}{2}$

Following this scheme, you can easily find that the $n = 3$ shell can contain a maximum of 18 electrons, the $n = 4$ shell a maximum of 32 electrons, and so forth. It is the dependence of energy primarily on n that allows us to group the electrons into shells in this way. A filled shell has the maximum number of electrons for any given value of n. As early as 1910, C. G. Barkla had deduced (from x-ray experiments) the existence of electron shells, and named them the K shell ($n = 1$), the L shell ($n = 2$), the M shell ($n = 3$), and so on. In lithium, the third electron, with $n = 2$, is in the L shell and is much more easily removed from the atom than either of the two K-shell electrons; it is a *valence electron*. This easily detached electron takes part in vigorous chemical reactions; lithium is an alkali metal.

In the following example we introduce the concept of *screening*—the partial cancellation of the nuclear charge by electrons in inner orbits.

Example 31-1

Make an approximate calculation of the binding energy of the $_3$Li atom (*a*) when the third (valence) electron has $n = 2$; (*b*) when it has $n = 3$.

The approximation consists of using Bohr's circular orbits and the equations of Sec. 30-3; the results should be of the right order of magnitude. From Eq. 30-6, the radius of the $n = 2$ orbit is 2^2 times that of the $n = 1$ orbit (0.53 Å), so the lithium atom in its ground state is about 2 Å in radius. The two electrons with $n = 1$ (K shell) by their negative charge cancel approximately 2/3 of the positive nuclear charge, so the valence electron, in

its larger orbit, "sees" an effective charge of about $+1e$, the same as seen by an electron in a $_1$H atom.

(*a*) From Eq. 30-9, for the $n = 2$ state, the energy would be $-13.60/2^2 = -3.40$ eV.

(*b*) For the $n = 3$ state, the energy would be $-13.60/3^2 = -1.51$ eV.

We see that as a first approximation we can draw the $_3$Li energy-level diagram (Fig. 31-4*a*) exactly like the upper part of the $_1$H diagram (Fig. 30-6). The calculations in Example 31-1 are, of course, not quantum mechanical, and spin is not taken into account. We predicted the binding energy to be about 3.4 eV, but experiment gives 5.4 eV for this energy (Fig. 31-4*b*). Evidently the valence electron is more tightly bound than we calculated. It is not hard to explain this discrepancy. The electron in the L shell ($n = 2$) is not definitely localized but is described by a probability cloud. A glance at Fig. 31-2 shows that in any state there is *some* probability that the electron will be found much closer to the nucleus than its average position. In effect, for some of

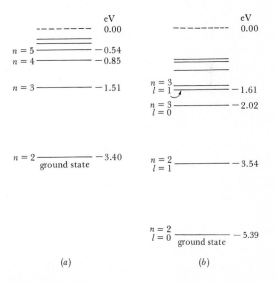

Figure 31-4 Energy levels for lithium. (*a*) Bohr model. (*b*) Experimental values, illustrating effect of screening.

the time, the L electron is inside the K shell, and the nuclear charge is no longer screened out by the two K electrons to have the value $+1e$ that we assumed. More work is needed to remove the L electron because of the larger Coulomb force exerted (some of the time) by the entire nuclear charge $+3e$. So we expect the ionization energy for $_3$Li to be greater than the simple Bohr model predicts, and experiment bears this out.

The exclusion principle can be formulated quantum mechanically as a statement about the symmetry of the wave functions, and it has a far greater generality than is apparent from our simple examples. The principle is valid for *any* particle of half-integral spin, including protons and neutrons as well as electrons. The difference between a metal and an insulator is a consequence of the exclusion principle as applied to a mole-sized chunk of material (containing perhaps 10^{23} electrons), where we speak of energy bands instead of energy shells (for an isolated atom). We will explore next a magnificent application of the exclusion principle—the arrangement of the atoms in the periodic table.

Table 31-2   Periodic Table of the Elements*

1 H 1.0080																	2 He 4.0026
3 Li 6.941	4 Be 9.0122											5 B 10.81	6 C 12.011	7 N 14.0067	8 O 15.9994	9 F 18.9984	10 Ne 20.179
11 Na 22.9898	12 Mg 24.305											13 Al 26.9815	14 Si 28.086	15 P 30.9738	16 S 32.06	17 Cl 35.453	18 Ar 39.948
19 K 39.102	20 Ca 40.08	21 Sc 44.956	22 Ti 47.90	23 V 50.941	24 Cr 51.996	25 Mn 54.9380	26 Fe 55.847	27 Co 58.9332	28 Ni 58.71	29 Cu 63.54	30 Zn 65.37	31 Ga 69.72	32 Ge 72.59	33 As 74.9216	34 Se 78.96	35 Br 79.909	36 Kr 83.80
37 Rb 85.467	38 Sr 87.62	39 Y 88.906	40 Zr 91.22	41 Nb 92.906	42 Mo 95.94	43 Tc (99)	44 Ru 101.07	45 Rh 102.906	46 Pd 106.4	47 Ag 107.870	48 Cd 112.40	49 In 114.82	50 Sn 118.69	51 Sb 121.75	52 Te 127.60	53 I 126.9045	54 Xe 131.30
55 Cs 132.906	56 Ba 137.34	57 La 138.906	72 Hf 178.49	73 Ta 180.948	74 W 183.85	75 Re 186.2	76 Os 190.2	77 Ir 192.2	78 Pt 195.09	79 Au 196.967	80 Hg 200.59	81 Tl 204.37	82 Pb 207.2	83 Bi 208.981	84 Po (210)	85 At (210)	86 Rn (222)
87 Fr (223)	88 Ra (226)	89 Ac (227)	104 Rf (261)	105 Ha (262)													

Lanthanide Series

58 Ce 140.12	59 Pr 140.908	60 Nd 144.24	61 Pm (147)	62 Sm 150.4	63 Eu 151.96	64 Gd 157.25	65 Tb 158.925	66 Dy 162.50	67 Ho 164.930	68 Er 167.26	69 Tm 168.934	70 Yb 173.04	71 Lu 174.97

Actinide Series

90 Th (232)	91 Pa (231)	92 U (238)	93 Np (237)	94 Pu (242)	95 Am (243)	96 Cm (248)	97 Bk (249)	98 Cf (249)	99 Es (254)	100 Fm (257)	101 Md (258)	102 No (259)	103 Lr (260)

* Atomic weights of stable elements are those adopted in 1969 by the International Union of Pure and Applied Chemistry. For those elements having no stable isotope, the mass number of the "most stable" well-investigated isotope is given in parentheses.

31-6 The Periodic Table of the Elements

You are no doubt familiar with the periodic table of the elements (Table 31-2). About a century ago, the Russian scientist Dmitri Mendeleev (1834–1907) arranged the then-known elements into groups according to their chemical properties, including valence. We now know that chemical properties depend on the ease with which electrons in outer shells can interact with the electrons of nearby atoms. The elements in the right-hand column are the *noble gases* $_2$He, $_{10}$Ne, ... All these gases are highly inert; only in 1962 did chemists succeed in making compounds containing $_{54}$Xe, a noble gas. Why are these gases so inert? (Or, putting it another way, why are their orbital electrons so tightly bound?) Other regularities in the periodic table are easily recognized. Elements just before the noble gases are *halogens* $_9$F, $_{17}$Cl, . . . , and those just after the noble gases are *alkalis* $_3$Li, $_{11}$Na, ...

In broad outline, the exclusion principle accounts for these regularities. Neon ($_{10}$Ne) is inert because there are 2 K electrons and 8 L electrons—full shells in each case. Sodium ($_{11}$Na) is an alkali because the valence electron must be an M electron with $n = 3$; this electron is easier to remove than the other 10 electrons for which $n = 1$ (in the K shell) or $n = 2$ (in the L shell). It is instructive to plot the ionization energies of the elements as a function of atomic number Z (Fig. 31-5). We see large peaks at the noble gases, and small values of ionization energy for the alkali metals.

There are some subtle features of the periodic table that deserve further discussion. To avoid lengthy tables, we use a convenient notation to represent the orbital quantum number l of an electron in an atom. The symbols are s, p, d, f, g, h, i, . . . , to represent $l = 0, 1, 2, 3, 4, 5, 6, \ldots$, respectively. (The origin of these particular symbols lies in the early history of spectroscopy.) Principal quantum numbers n are given as numbers; superscripts tell *how many* electrons in the electron cloud have a given type of orbit. For

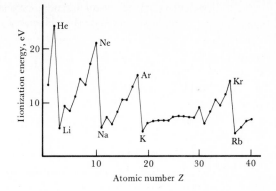

Figure 31-5 Ionization energy (binding energy) for the first 40 elements.

example, for $_3$Li (Sec. 31-5) we could use the notation $1s^2 2s^1$ to represent lithium in state B, in which there are two K electrons (for which $n = 1$ and $l = 0$) and one L electron (for which $n = 2$ and $l = 0$). Similarly, we use the notation $1s^2 3s^1$ for lithium in state A, in which there are two K electrons ($n = 1$ and $l = 0$) and one M electron ($n = 3$ and $l = 0$). The ground state of $_{11}$Na can be represented by the following table:

	n	l	m_l	m_s	
K shell	1	0	0	$\frac{1}{2}$	$1s^2$
	1	0	0	$-\frac{1}{2}$	
	2	0	0	$\frac{1}{2}$	$2s^2$
	2	0	0	$-\frac{1}{2}$	
L shell	2	1	-1	$\frac{1}{2}$	
	2	1	-1	$-\frac{1}{2}$	
	2	1	0	$\frac{1}{2}$	$2p^6$
	2	1	0	$-\frac{1}{2}$	
	2	1	1	$\frac{1}{2}$	
	2	1	1	$-\frac{1}{2}$	
M shell	3	0	0	$\frac{1}{2}$	$3s^1$

All this can be summarized in a single line called the *configuration* of the atom: $1s^2 2s^2 2p^6 3s^1$ (as shown in the right-hand column of the table).

Table 31-3 Electronic Configurations for the Ground States of the First 37 Elements

n	1	2		3			4				5			
shell	K	L		M			N				O			
l	0	0	1	0	1	2	0	1	2	3	0	1	2	
	s	s	p	s	p	d	s	p	d	f	s	p	d	
1 H	1													
2 He	2													noble gas
3 Li .	2	1												alkali
4 Be	2	2												
5 B	2	2	1											
6 C	2	2	2											
7 N	2	2	3											
8 O	2	2	4											
9 F	2	2	5											halogen
10 Ne	2	2	6											noble gas
11 Na	2	2	6	1										alkali
12 Mg	2	2	6	2										
13 Al	2	2	6	2	1									
14 Si	2	2	6	2	2									
15 P	2	2	6	2	3									
16 S	2	2	6	2	4									
17 Cl	2	2	6	2	5									halogen
18 Ar	2	2	6	2	6									noble gas
19 K	2	2	6	2	6		1							alkali
20 Ca	2	2	6	2	6		2							
21 Sc	2	2	6	2	6	1	2							
22 Ti	2	2	6	2	6	2	2							
23 V	2	2	6	2	6	3	2							
24 Cr	2	2	6	2	6	5	1							
25 Mn	2	2	6	2	6	5	2							
26 Fe	2	2	6	2	6	6	2							
27 Co	2	2	6	2	6	7	2							
28 Ni	2	2	6	2	6	8	2							
29 Cu	2	2	6	2	6	10	1							
30 Zn	2	2	6	2	6	10	2							
31 Ga	2	2	6	2	6	10	2	1						
32 Ge	2	2	6	2	6	10	2	2						
33 As	2	2	6	2	6	10	2	3						
34 Se	2	2	6	2	6	10	2	4						
35 Br	2	2	6	2	6	10	2	5						halogen
36 Kr	2	2	6	2	6	10	2	6						noble gas
37 Rb	2	2	6	2	6	10	2	6			1			alkali

This is the ground state* of $_{11}$Na, but it is not the only possible configuration; for example, $1s^2 2s^2 2p^6 3p^1$ and $1s^2 2s^2 2p^6 4s^1$ are two of the infinitely many excited states of the $_{11}$Na atom.

Now let us explain the curious gap at the top of the chart of the periodic table. Calcium ($_{20}$Ca) is chemically very similar to magnesium ($_{12}$Mg), so calcium "belongs" beneath magnesium. Similarly, on chemical evidence, $_{31}$Ga "belongs" beneath $_{13}$Al. But there are 10 extra elements—scandium through zinc—sandwiched in between $_{20}$Ca and $_{31}$Ga, whereas there are no such elements between $_{12}$Mg and $_{13}$Al. These extra elements are called *transition elements* and are chemically somewhat similar to each other.

Table 31-3 offers a ready explanation for the transition elements. All is straightforward until the $3p$ subshell is filled with its six electrons at $_{18}$Ar. The configuration of argon is $1s^2 2s^2 2p^6 3s^2 3p^6$, 18 electrons in all (add up the superscripts). In the next atom, $_{19}$K, it turns out that a $4s$ electron ($n = 4$, $l = 0$) is more tightly bound than a $3d$ electron ($n = 3$, $l = 2$) would be. This is related to the screening effect discussed in the last section; when $l = 0$, as for a $4s$ electron, the electron probability cloud extends closer to the nucleus, screening is less, and the electron feels a greater attractive force on the average, so it is more tightly bound than a $3d$ electron would be, with its larger angular momentum and more nearly "circular" orbit. Thus, for $_{19}$K we have

Ground state: $1s^2 2s^2 2p^6 3s^2 3p^6 4s^1$

Excited state: $1s^2 2s^2 2p^6 3s^2 3p^6 3d^1$

Therefore, $_{19}$K is an alkali, similar to $_{11}$Na and the other alkalis, all of which have one s electron outside a filled subshell.

---

* The ground state's outer electron is $3s^1$ rather than $3p^1$ because in the p state, for which $l = 1$, the electron is on the average farther from the nucleus and is more completely screened by the inner electron shells. See Fig. 31-4b, where the $3p$ state in $_3$Li is shown less tightly bound than the $3s$ state.

For elements 21 through 29 there is a competition. Other small effects, previously ignored, make the difference, tipping the balance one way or the other. Note that for $_{24}$Cr, five $3d$ electrons and one $4s$ electron are more tightly bound than four $3d$ and two $4s$ electrons would be; a similar inversion takes place at $_{29}$Cu. The "valence" of a transition element is not well defined, since the outer one, two, or three electrons are all about equally tightly bound. It is worthy of note that $_{26}$Fe, $_{27}$Co, and $_{28}$Ni are ferromagnetic metals, forming domains easily (Sec. 22-9).

Next we must explain the group of 14 elements that begin with $_{57}$La and are known as the *lanthanide series*. Also called "rare earths," they are not really scarce in the crust of the earth, but their chemical properties are so similar that they seem to fit into no space at all—between $_{57}$La and $_{72}$Hf. As with the transition elements, the explanation lies in the "delayed" filling of inner shells (Table 31-4). Although there is a little competition between a $4f$ and a $5d$ electron around $_{57}$La and again at $_{64}$Gd, the differences between the lanthanides are mainly in the N shell ($n = 4$) as the $4f$ electrons are added. The elements are so similar because the *outer* electrons that take part in chemical reactions are in similar $5s^2 5p^6 6s^2$ configurations.

Another group of chemically similar elements, the *actinides*, begins at actinium ($_{89}$Ac). This group contains 14 elements in which the $5f$ shell finally fills up.

We have seen how the Pauli exclusion principle furnishes the basic framework within which the periodic table and much of chemistry can be understood. Here is a good example of the interplay between physics and chemistry, for it is through the study of atomic spectra and the energy levels of atoms that the ground-state configurations are established.

We can speculate on a universe in which there is no exclusion principle. Elements would differ in mass, but not in chemical properties. There would be no chemistry, no biology, no "life as we know it."

Table 31-4 Electronic Configurations for the Ground States of Elements 52 through 74

Note: For each of these atoms, the K, L, and M shells, not listed in the table, are filled with a total of 28 electrons.

n shell		4 N				5 O					6 P			
l		0 s	1 p	2 d	3 f	0 s	1 p	2 d	3 f	4 g	0 s	1 p	2 d	
52	Te	2	6	10		2	4							
53	I	2	6	10		2	5							halogen
54	Xe	2	6	10		2	6							noble gas
55	Cs	2	6	10		2	6				1			alkali
56	Ba	2	6	10		2	6				2			
57	La	2	6	10		2	6	1			2			
58	Ce	2	6	10	2	2	6				2			
59	Pr	2	6	10	3	2	6				2			
60	Nd	2	6	10	4	2	6				2			
61	Pm	2	6	10	5	2	6				2			
62	Sm	2	6	10	6	2	6				2			
63	Eu	2	6	10	7	2	6				2			
64	Gd	2	6	10	7	2	6	1			2			lanthanides
65	Tb	2	6	10	9	2	6				2			
66	Dy	2	6	10	10	2	6				2			
67	Ho	2	6	10	11	2	6				2			
68	Er	2	6	10	12	2	6				2			
69	Tm	2	6	10	13	2	6				2			
70	Yb	2	6	10	14	2	6				2			
71	Lu	2	6	10	14	2	6	1			2			
72	Hf	2	6	10	14	2	6	2			2			
73	Ta	2	6	10	14	2	6	3			2			
74	W	2	6	10	14	2	6	4			2			

31-7 X Rays

With the background provided by the exclusion principle and the concept of filled electron shells, we are ready to consider the origin of x rays. The lines of the visible spectrum of an atom arise when the *outer* electrons gain and then lose a few electron volts of energy. Far greater photon energies can be obtained by first removing one of the inner electrons. This can be accomplished in an x-ray tube (Fig. 31-6), in which high-speed electrons strike a target. The energy required to remove an electron from an atom is the binding energy of that electron. To create an ion by knocking out an inner electron of an atom of the target material requires much work; for instance, the binding energy of a K electron in molybdenum is some 20,000 eV (hence the need for a high voltage to operate an x-ray tube with a molybdenum target). Once a K electron has been removed, an energy level is available, and an L, M, or N electron can fall down. During this rearrangement of the electrons, a photon is emitted whose energy is given by $\Delta E = hf$.

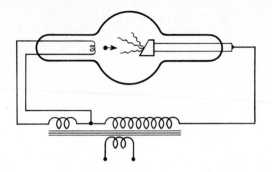

Figure 31-6 X-ray tube. Electrons, emitted from the heated filament, are accelerated during the half-cycles in which the target is positive relative to the filament.

In molybdenum, some of the many possible transitions have the energies shown in Table 31-5. The corresponding wavelengths, found from 12,400/E, are about 1 Å, which is about the size of the interatomic spacings in crystals (see Fig. 1-1 on page 14). Thus, a crystal can serve as a grating for the analysis of x-ray spectra. We have already noted that Compton used a crystal grating spectrometer in his analysis of the wavelengths of scattered x rays (Fig. 29-6).

The x-ray photons we have described are members of the *K series* of lines, resulting from a vacancy in the *K* shell. Different x-ray photons are obtained if an *L* electron is knocked out; this requires less work than to knock out a *K* electron. When an *M* or *N* electron falls to an empty place in the *L* shell, ΔE is less than for the jump to a *K* shell; hence these *L-series* photons have considerably less energy (and longer wave-

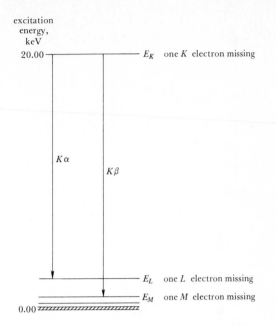

Figure 31-7 Energy-level diagram for x-ray transitions in molybdenum. Not all levels are shown. The shaded region (not to scale) represents "optical levels" corresponding to excitation of outer electrons without ionization.

length) than those of the *K* series listed in Table 31-5.

A simplified energy-level diagram for x-ray production is given in Fig. 31-7. Note that E_K is not the "energy of a *K* electron"—it is the energy of the entire atom when one *K* electron has been removed. Thus, E_K, E_L, E_M, ... are the binding energies of the *K*, *L*, *M*, ... electrons. The transition is between energy states *of the atom*. The "hole" in the *K* shell is replaced by one in the *L* shell, as indicated by the downward arrow labeled $K\alpha$.

X-ray spectra (Fig. 31-8) give valuable information about atomic structure, since the wavelengths of x rays, like the wavelengths of optical spectra, are characteristic of the atom. X-ray spectra depend on the inner parts of the electron cloud, whereas visible, ultraviolet, and infrared spectra depend on the outer parts of the electron cloud, where the binding energies are small. X

Table 31-5 Approximate Wavelengths of *K*-series Lines in the X-ray Spectrum of Molybdenum

Transition	Energy Change, eV	Wavelength, Å	Designation of Line
$L \rightarrow K$	17,500	0.71	$K\alpha$
$M \rightarrow K$	19,800	0.63	$K\beta$
$N \rightarrow K$	20,000	0.62	$K\gamma$

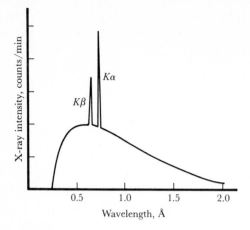

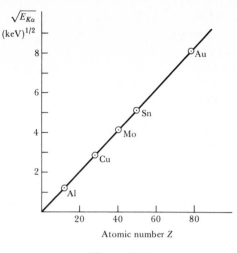

Figure 31-8 X-ray spectrum of molybdenum in a tube operated at 50 kV. The $K\alpha$ and $K\beta$ lines are shown superposed on the bremsstrahlung.

Figure 31-9

rays can be produced by fluorescence if a photon is absorbed by an inner electron. The absorbed photon's energy must be sufficient to take the inner electron all the way out, leaving the atom ionized with an inner vacancy; thus, x-ray photons are needed to give x-ray fluorescence.

We can use the Bohr model to see how the energies of x-ray lines depend on the atomic number Z of the target material. Consider the energy $E_K - E_L$ that is released when a hole in the K shell is replaced by a hole in the L shell. This is equivalent to the energy released when the electron in a hydrogenlike atom changes from an $n = 2$ state to an $n = 1$ state. But this "falling" electron feels a nuclear charge not of $+Ze$, but rather $+(Z - 1)e$, since one unit of the positive nuclear charge is screened by the other K electron that is not involved in the transition. Now we use Eq. 30-9, with the effective nuclear charge $(Z - 1)e$, and obtain

$$E_n = \frac{13.60(Z - 1)^2}{n^2}$$

for the ionization energy of the nth shell in eV. The energy of the $K\alpha$ line is, therefore,

$$E_{K\alpha} = 13.60(Z - 1)^2 \left(\frac{1}{1^2} - \frac{1}{2^2} \right)$$

or

$$E_{K\alpha} = 13.60(\tfrac{3}{4})(Z - 1)^2 \text{ eV} \tag{31-2}$$

Since $E_{K\alpha}$ is proportional to $(Z - 1)^2$, the graph of $\sqrt{E_{K\alpha}}$ versus Z is a straight line (Fig. 31-9).

Example 31-2

Estimate the energy and wavelength of the $K\alpha$ line of $_{42}$Mo.
Here,

$$Z - 1 = 42 - 1 = 41$$

$$E_{K\alpha} = (13.60)(\tfrac{3}{4})(41)^2 = \boxed{17{,}150 \text{ eV}}$$

$$\lambda_{K\alpha} = \frac{12{,}400 \text{ eV} \cdot \text{Å}}{17{,}150 \text{ eV}} = \boxed{0.72 \text{ Å}}$$

This agrees well with the experimental value given in Table 31-5.

The $K\beta$ line would correspond to a transition from $n = 3$ to $n = 1$, and the factor $\tfrac{3}{4}$ would be replaced by $(1/1^2 - 1/3^2)$, or $\tfrac{8}{9}$.

It is evident that Eq. 31-2 can be used to find the atomic number Z of an element by measuring the wavelength or energy of its x-ray lines. It was through such an analysis by H. G. J. Moseley in 1914 that the concept of atomic number was placed on a firm experimental basis. In the modern scanning x-ray microscope, the chemical composition of a tiny region of the specimen can be studied by measuring the energies of the x rays emitted when an electron beam strikes the region of interest. In another method of analysis, after an inner electron is ejected by electron impact, the excitation energy of the atom is given not to an emitted x-ray photon but to one of the outer electrons whose energy is measured. This electron, called an *Auger* * electron, has a KE that is characteristic of the element from which it came.

In an x-ray tube, an electron that strikes and enters the target undergoes many deflections as it passes near the positively charged nuclei. During each deflection the electron's path is curved, giving rise to a centripetal acceleration. We saw in Sec. 24-8 that acceleration of charge is associated with the production of e-m radiation. Thus, one or more photons are radiated during each encounter; this radiation is called *bremsstrahlung.* [†] The electron gradually loses its KE in this series of collision events, eventually becoming slow enough to be captured by an ion or to become a free electron in the metallic target material.

The bremsstrahlung (Fig. 31-8) is a base on which the sharp x-ray lines of the $K, L, \ldots,$ series are superposed. Bremsstrahlung is a continuous function of wavelength, independent of the target material, analogous to the continuous optical spectrum of a blackbody (Fig. 14-4), which is also independent of the source material. Note, however, that for any given PD across the x-ray tube, the bremsstrahlung has a short-wavelength limit, corresponding to a highest-energy photon. This limit represents the conversion of the electron's initial KE entirely into the energy of a single photon. X-ray diagnosis and therapy make use of the bremsstrahlung, not the characteristic spectral lines which are of much less total intensity.

___
* Pierre-Victor Auger (1899–), a French physicist. The name is pronounced *Oh-zhay'*.

___
[†] German, meaning "braking radiation."

Summary Quantum mechanics, or wave mechanics (Heisenberg, Schroedinger, Born, Jordan, Dirac), is a logical development based on the duality of matter. Newtonian mechanics is a special case of quantum mechanics. Electrons in atoms are represented by a probability density that is proportional to the probability that an electron will be found in any given small region of space.

Four quantum numbers (n, l, m_l, m_s) are used to specify the state of an electron. For hydrogen, the simplest atom, the energy depends only on n, unless the atom is in a magnetic field. Spectral lines are split by the presence of an external magnetic field (Zeeman); this is due to a splitting of energy levels for which l is not zero. Internal magnetic fields interacting with electron spins also cause a splitting of energy levels, as illustrated by the yellow pair of lines in the sodium spectrum.

The exclusion principle (Pauli), which applies to atoms with more than one electron, states that no two electrons in the same atom can have the same set of quantum numbers. As a result, orbital electrons can be grouped into shells; all electrons in any shell have the same value of n and roughly the same energy, except for differences due to screening by inner electrons. The maximum number of electrons in any shell is limited by the exclusion principle: 2 in the K shell, 8 in the L shell, and so forth.

Chemical properties of atoms are determined by the shell structure and also by the finer details of the energy levels of the states that are permitted by the exclusion principle.

X rays are produced when an inner electron is removed. This requires a high-speed impact, or absorption of an energetic (high-frequency) photon. X-ray photons ordinarily have energies of tens of thousands of electron volts and wavelengths of the order of magnitude of an angstrom. Since atoms in a crystal are about an angstrom apart, a crystal can be used as a grating for the study of an x-ray spectrum. The wavelengths of x-ray spectral lines are characteristic of the atomic number of the target material. A continuous spectrum, the bremsstrahlung, is also emitted when high-speed electrons strike a target.

Check List

probability cloud
principal quantum number
orbital quantum number
magnetic quantum number

spin magnetic quantum number
exclusion principle
K, L, M shells
configuration

K series
L series
Auger electron
bremsstrahlung

Questions

31-1 Are any of Bohr's postulates of 1913 retained in the quantum mechanics of 1925?

31-2 What is the symbol for the principal quantum number of an electron in an atom? What is the physical significance of this quantum number?

31-3 What is the symbol for the orbital quantum number of an electron in an atom? What is the physical significance of this quantum number?

31-4 What is the symbol for the magnetic quantum number of an electron in an atom? What is the physical significance of this quantum number?

31-5 What is the symbol for the spin magnetic quantum number of an electron in an atom? What is the physical significance of this quantum number?

31-6 When a hydrogen atom is in a certain energy state, the electron's probability cloud is spherically symmetric. What does this imply about the average result of a large number of measurements of the angular momentum of the electron?

31-7 Why does the yellow spectrum line of Na (Fig. 30-2b) have two very close components?

31-8 Why are the chemical properties of the lanthanides so similar?

31-9 Describe a sequence of events, on an atomic scale, that might result in the production of an x-ray photon.

31-10 Explain why an L electron in, say, a copper target cannot fall to the K shell, thus emitting the $K\alpha$ x-ray line, without the prior ejection of a K electron by collision or by other means.

31-11 Could a photon of wavelength 0.71 Å be absorbed by a molybdenum atom that is in its ground state? (See Table 31-5.)

31-12 Is any bremsstrahlung radiation emitted during the α-particle scattering events shown in Fig. 30-1b?

MULTIPLE CHOICE

31-13 Consider a group of H atoms, each in its ground state. Measurement of the energies of these atoms gives (a) -13.60 eV on the average, with some values greater and some values less; (b) -13.60 eV every time; (c) zero.

31-14 The Zeeman effect is a splitting of energy levels due to an externally applied   (a) gravitational field;   (b) electric field;   (c) magnetic field.

31-15 The configuration of a certain neutral atom is $1s^2 2s^2 2p^6 3s^1$. This atom is   (a) a halogen in an excited state;   (b) an alkali in the ground state;   (c) an alkali in an excited state.

31-16 The configuration of a certain neutral atom is $1s^2 2s^2 2p^5 3s^1$. This atom is   (a) an alkali in the ground state;   (b) an alkali in an excited state;   (c) a noble gas in an excited state.

31-17 The configuration $1s^2 2s^2 2p^3 3s^3$ is   (a) a possible configuration for a noble gas in an excited state;   (b) a possible configuration for an alkali in an excited state;   (c) forbidden by the exclusion principle.

31-18 The wavelength of the $_{50}$Sn $K\alpha$ x-ray line is about 0.5 Å. From this, the wavelength of the $_{25}$Mn $K\alpha$ line is expected to be about   (a) 0.25 Å;   (b) 1 Å;   (c) 2 Å.

Problems

31-A1 List the quantum numbers for each of the electrons in a beryllium atom $_4$Be that is in its ground state.

31-A2 Calculate the magnitude of the angular momentum, in SI units, of an electron in the H atom for $n = 3$, $l = 2$.

31-A3 At what value of Z does the graph of Fig. 31-9 cross the Z axis?

31-A4 Consider the states in an atom, for each of which $n = 5$. What is the maximum number of electrons, with different values of l, m_l, and m_s, that *could* be in this shell? Is this likely?

31-B1 Consider the valence electron in a $_{11}$Na atom when it has the least possible value of n and the maximum possible angular momentum (for that value of n). In such a state, the electron can be pictured (approximately) as being in a circular orbit.   (a) What is the value of l for this orbit?   (b) What is the approximate radius of this orbit? (*Hint:* Most of the nuclear charge is screened by the inner electrons.)

31-B2 Using the values of wavelength given in Sec. 31-4 for the two components of the yellow sodium doublet, calculate the energy for each photon, in eV, and thus find the splitting, in eV, of the upper energy level.

31-B3 For copper, a table similar to Table 31-5 would show the following approximate energies: $L \to K$, 8050 eV; $M \to K$, 8970 eV; $N \to K$, 8980 eV. Calculate the energies and wavelengths of   (a) the $K\alpha$ line;   (b) the $K\beta$ line;   (c) the $L\alpha$ line (which results from the transition of an electron from the M shell to a vacancy in the L shell).

31-B4 What is the short-wavelength limit, in Å, of the bremsstrahlung radiation used in cancer therapy, if the x-ray tube is operated at 150 kV?

31-B5 What is the wavelength, in Å, of the highest-energy x rays emitted when an electron beam is accelerated through a PD of 32 kV and strikes the face of a color TV tube?

31-B6 Show that the third Bohr postulate, Eq. 30-2, can be obtained by assuming that there are stationary de Broglie waves in the orbit, such that n wavelengths are contained in the circumference of the orbit. *Note:* This result, while interesting, is not as powerful as it seems. According to wave mechanics, the electron cannot be in a plane orbit, for then both its angular momentum and the corresponding position coordinate (orientation of the axis of revolution) would be known precisely, in violation of the uncertainty principle. The extension of the idea of this problem to three dimensions involves complicated mathematics and leads to three quantum numbers instead of one.

For Further Study

31-8 Molecular Rotation and Vibration

Our study of atomic structure has shown that an atom has a ground state (normal state) and also infinitely many excited states that are a few electron volts above the ground state. These are called *electronic states* because they represent the energy of the orbital electrons. Spectrum lines resulting from transitions between electronic states have wavelengths in the visible region or the near ultraviolet or near infrared region. For example, from Fig. 31-4 we find that the energy difference between the 2*p* and 2*s* states of Li is 5.39 eV − 3.54 eV = 1.85 eV (the exact value is 1.849 eV). We use Eq. 29-7 to calculate the wavelength of a photon emitted during a transition between these states:

$$\lambda = \frac{1240 \text{ eV} \cdot \text{nm}}{1.849 \text{ eV}} = 671 \text{ nm}$$

This is the strong red line that is the basis for the well-known flame test used in chemistry to detect the presence of a lithium compound. In general, because the energy levels are well separated, the spectrum of an *atom* consists of discrete, well-separated lines.

Next we consider a simple diatomic molecule such as O_2 or HCl. When the density is small, as for gases, the molecules are far apart and are not acted on appreciably by their neighbors. They are then free to rotate or vibrate with certain characteristic frequencies. An oxygen molecule O_2 consists of two atoms, each of atomic mass 16 (Fig. 31-10). Because of their relatively small mass, we can for present purposes ignore the 16 electrons in the electron cloud surrounding the two nuclei and think of the molecule as a pair of identical balls connected by a stiff spring. This model is suggested by the known facts that molecules tend to resist both being compressed and being stretched, and there is some equilibrium distance of the one oxygen atom from the other. As a result of molecular collisions in the gas, we expect that a typical molecule both rotates and vibrates. The spectrum analysis confirms this, and actually allows determination of the force constant for the hypothetical spring and the moment of inertia of the rotating system (thus allowing measurement of the separation of the atoms, since $I = \Sigma mr^2$). A very pretty interaction is evident, for experiment shows that those molecules that are rotating faster are also stretched a little bit, by a centrifugal force effect.

It is not difficult to use Eq. 31-1 to find the allowed energy levels for a rotating HCl molecule. This energy is KE, given by $\frac{1}{2}I\omega^2$ (Sec. 7-7). Recalling that angular momentum L is given by $I\omega$, where I is the moment of inertia and ω the angular speed of rotation, we write

$$\text{KE} = \tfrac{1}{2}I\omega^2 = \frac{1}{2I}(I\omega)^2 = \frac{1}{2I}L^2 \quad (31\text{-}3)$$

According to quantum mechanics, the magnitude of the angular momentum L is quantized to have values that depend on an integral quantum number (see Eq. 31-1). We now call this

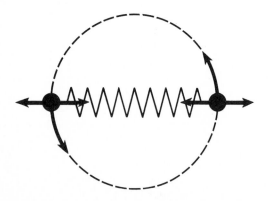

Figure 31-10 Simultaneous rotation and vibration of an oxygen molecule.

quantum number J (reserving l for the electronic orbital quantum number). Thus, as on page 734,

$$L = I\omega = \sqrt{J(J+1)}\left(\frac{h}{2\pi}\right)$$

$$L^2 = J(J+1)\frac{h^2}{4\pi^2}$$

Substituting into Eq. 31-3 gives

$$KE = \frac{J(J+1)h^2}{8\pi^2 I} \qquad (31\text{-}4)$$

We see that the rotational energy levels are spaced proportionally to $J(J+1)$ and the energies of rotation are proportional to 0, 2, 6, 12, 20, ... The second excited rotational state has 3 times as much rotational KE as the first rotational state.

Example 31-3

What is the KE of HCl when the molecule is in its first excited state of rotation, with $J = 1$? The distance between the atoms is 1.17 Å.

Because the Cl atom is so much more massive than the H atom, we can consider the Cl atom to be approximately stationary. The moment of inertia is, therefore, that of the H atom (of mass 1.67×10^{-27} kg) rotating in a circle of radius 1.17 Å.

$$I = mR^2 = (1.67 \times 10^{-27} \text{ kg})(1.17 \times 10^{-10} \text{ m})^2$$
$$= 2.29 \times 10^{-47} \text{ kg} \cdot \text{m}^2$$

$$KE = \frac{J(J+1)h^2}{8\pi^2 I} = \frac{1(2)(6.63 \times 10^{-34} \text{ J} \cdot \text{s})^2}{8\pi^2(2.29 \times 10^{-47} \text{ kg} \cdot \text{m}^2)}$$

$$= 4.86 \times 10^{-22} \text{ J}\left(\frac{1\text{eV}}{1.60 \times 10^{-19} \text{ J}}\right)$$

$$= \boxed{3.04 \times 10^{-3} \text{ eV}}$$

This example shows that the rotational energy of the HCl molecule is a few thousandths of an electron volt (meV), which is very small compared with the electronic energy of an atom.

Example 31-4

Calculate the wavelength of a photon emitted when the rotational quantum number of HCl changes from $J = 2$ to $J = 1$.

The rotational energies are proportional to $J(J+1)$, so here the ratio of energies is 6:2, or 3:1. The previous example has given $E_1 = 0.003$ eV, so $E_2 = 3 \times 0.003$ eV $= 0.009$ eV. Therefore,

$$\Delta E = E_2 - E_1$$
$$= 0.009 \text{ eV} - 0.003 \text{ eV} = 0.006 \text{ eV}$$

$$\lambda = \frac{1240 \text{ eV} \cdot \text{nm}}{0.006 \text{ eV}} = 2.1 \times 10^5 \text{ nm} = \boxed{0.21 \text{ mm}}$$

This is an emission of e-m radiation in the far infrared or short microwave region. The transition is represented by the short arrow in Fig. 31-11.

Molecular *vibration* is also quantized, but we cannot develop the theory here. It turns out that the energies of the vibrational states of a diatomic molecule such as HCl are given by

$$E = \frac{h}{2\pi}\sqrt{\frac{k}{m}}\,(v + \tfrac{1}{2}) \qquad (31\text{-}5)$$

where k is the force constant of the "spring" that connects the atoms, m is the effective mass of the atoms, and v is the vibrational quantum number 0, 1, 2, 3, 4, ... The energies of vibration are therefore proportional to the odd integers 1, 3, 5, 7, ...; for HCl the lowest vibrational state (with $v = 0$) has an energy of 0.18 eV. Even at absolute zero, then, there is this much vibrational energy—a seeming contradiction to the concept of absolute zero. But we must realize that this energy is not "available," since there is no lower vibrational level and no downward transition is possible. If the molecule had no vibrational energy, both the position and momentum of the atoms making up the molecule would be known exactly (both would be zero); this would violate the Heisenberg uncertainty principle.

In practice, rotational energy levels are deduced from the observed spectra, and the moment of inertia is then calculated from Eq. 31-4.

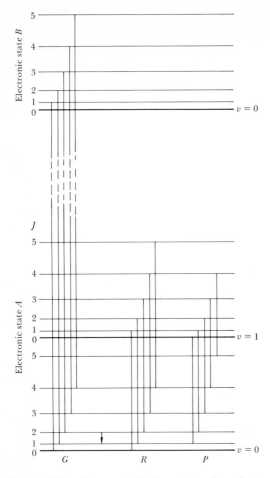

Figure 31-11 Energy levels in a diatomic molecule (not to scale). Vibrational states shown by heavy lines; rotational states shown by light lines.

Similarly, the force constant (strength of inter-atomic binding) can be calculated from the observed vibrational levels, using Eq. 31-5. The study of molecular spectra is a powerful tool for the study of the structure of molecules. The spectra are usually observed in absorption rather than emission (Fig. 31-12).

We summarize our study of molecular spectra by noting that there are three kinds of energy change, each of rather different order of magnitude. (1) Electronic changes, if they occur, are of the order of a few eV; (2) vibrational changes, if they occur, are smaller, of the order of a few

tenths of an eV; (3) rotational changes, if they occur, are still smaller, of the order of a few thousandths of an eV. The energy-level diagram (Fig. 31-11) shows that many lines can occur close together in the visible or infrared region. The major change is electronic, as the atom goes from electronic state A to electronic state B, or vice versa. The transitions shown as group G are all in the visible or ultraviolet region, because of the large difference between states A and B, but the spacing between the lines in the group is very small, caused by the close spacing of the rotational states (calculated by Eq. 31-4). Such a close group of lines, split due to molecular rotation and/or vibration, is called a *band spectrum*. Figure 31-12 is a recording of the absorption spectrum of HCl in the infrared region near $\lambda = 3.4$ μm; the recorded transitions are absorption lines in the groups labeled P and R in Fig. 31-11. Note that Fig. 31-11 is consistent with Fig. 31-12 in showing that the lines in the R branch have shorter wavelengths than the lines in the P branch.

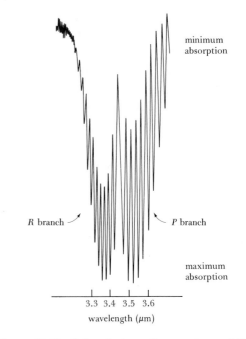

Figure 31-12 Infrared absorption spectrum of HCl molecule.

Problem

31-C1 (*a*) Use the method of Example 31-4 to find the HCl energies E_3, E_4, and E_5, for which $J = 3$, 4, and 5, respectively. (*b*) Note that in Fig. 31-11, the only transitions that are allowed are those for which $\Delta J = 0$ or ± 1. In Example 31-4, the energy of the transition $E_2 \rightarrow E_1$ was found to be 0.006 eV. Calculate the energies of the transitions $E_3 \rightarrow E_2$, $E_4 \rightarrow E_3$, and $E_5 \rightarrow E_4$. Indicate these transitions by drawing arrows on Fig. 31-11. (*c*) The lines found in part (*b*) are equally spaced in energy. Prove that this must always be true. (*Hint*: Use Eq. 31-4 for two values of J, which can be called J' and J, with $J' = J + 1$.)

References

1. "Fifty Years of Spin," *Physics Today* **29**(6), 40 (June 1976). A pair of articles by the discoverers of electron spin: S. A. Goudsmit, "It Might as Well Be Spin," and G. E. Uhlenbeck, "Personal Reminiscences."
2. E. C. Watson, "The Discovery of X Rays," *Am. J. Phys.* **13**, 281 (1945). A full translation of Roentgen's original papers; see also W. F. Magie, *A Source Book in Physics* (McGraw-Hill, New York, 1935), pp. 600–610.
3. W. Thumm, "Röntgen's Discovery of X Rays," *Phys. Teach.* **13**, 207 (1975). See also an exchange of letters, *Phys. Teach.* **13**, 324, 325 (1975).
4. G. E. M. Jauncey, "Birth and Infancy of X Rays," *Am. J. Phys.* **13**, 362 (1945).
5. J. F. Hyde, "Let the Elements Teach Periodic Law," *Phys. Teach.* **13**, 538 (1975). Includes a full-color foldout of an unusual periodic table.
6. B. Crawford, Jr., "Chemical Analysis by Infra Red," *Sci. American* **189**(4), 42 (Oct. 1953).
7. F. Miller, Jr., *Absorption Spectra* (film).

32

The Nucleus

The closing years of the 19th century were extraordinarily fruitful in several ways. In 1890, atoms were thought to be indivisible, and light was thought to be an e-m wave with no corpuscular properties. By the year 1900, J. J. Thomson had discovered the free electron, Roentgen had produced x rays, and Planck had postulated a quantum mechanism for the emission of light by a blackbody. This busy decade marked the beginning of a new era in physics. No longer was it possible to describe nature entirely with models based on Newton's laws and with appeal to everyday experience.

32-1   *A Brief History of Nuclear Physics*

In 1896 the French physicist Henri Becquerel was studying the fluorescence and phosphorescence of various minerals that had been exposed to the newly discovered x rays. When, in the course of his work, he came to the uranium compounds, he discovered, by chance, that even unexposed uranium compounds emit penetrating radiations that can blacken a photographic plate or ionize a gas such as air. These properties are shared by x rays, but Becquerel's "rays" were continuously emitted even in the absence of a source of energy. At first, Becquerel suspected that the radiations were a form of invisible fluorescence. The uranium might merely be releasing energy that it had stored up in some way from sunlight. To test this hypothesis, he examined a piece of uranium ore that had been kept in a lead-shielded, light-tight box for several months. The sample was able to produce continuous ionization of a gas and to affect a photographic plate wrapped in black paper, and yet there was no apparent source for this energy. The phenomenon was called radioactivity, and it posed a serious problem, for it apparently violated the law of conservation of energy.*

---

* The discovery of radioactivity did clear up one mystery. Although geological evidence showed the age of the earth to be of the order of a billion years, physicists had protested that the earth could not be this old; according to various estimates, it would have cooled off from a molten state to its present temperature in 20 to 200 million years, unless there were a source of heat inside the earth. The radioactive minerals supply such a source, and so the great debate of 1862–1900 was resolved in favor of the geologists.

High-energy particles traversing a portion of a cloud chamber located 59 m from a brass target that is inside a particle accelerator of 1 GeV energy. Tracks of widely different density are visible.

Many workers pushed forward the study of radioactivity in the next decade. Pierre and Marie Curie processed many tons of impure uranium ore and isolated a fraction of a gram of two new elements, polonium and radium, both of which are considerably more active (and less abundant) than uranium. Several dozen radioactive substances, all of high atomic weight, were discovered, and it was found that just three kinds of radiation are emitted. These radiations were labeled α, β, and γ rays, in order of increasing penetrating power. We can describe these "rays," using modern terminology.

Alpha rays are a stream of positively charged particles. The α particle is a doubly charged helium ion $_2^4\text{He}^{2+}$, that is, a bare helium nucleus. Alpha particles are easily stopped by a sheet of paper or by a few centimeters of air.

Beta rays are a stream of high-speed electrons. The β particles have the same e/m_0 ratio as do the free electrons that evaporate from a heated filament. Beta rays are much more penetrating than alpha rays; they can usually penetrate thin metal sheets, such as the wall of a Geiger counter, but cannot pass through a millimeter of lead or a few centimeters of flesh. We attribute the greater penetrating power of β particles to their high speeds; the slower α particles are more likely to lose energy by creating ions during their travel through matter, since they take a longer time to pass through an atom. Both α and β particles penetrate farther if they have greater KE to start with. Ultimately, an α particle picks up two electrons and becomes a neutral helium atom,* and a β particle attaches itself to the electron cloud of some atom or ion, or enters a metal to become a free electron.

Gamma rays are high-energy photons. The waves associated with γ-ray photons have such short wavelengths that diffraction and interference can be observed only with difficulty; hence the corpuscular aspects of γ rays dominate

the picture. For instance, the γ rays from radium are photons of energy about 2×10^5 eV, or 0.2 MeV. Like all photons, γ rays are not deflected by magnetic or electric fields.

By 1900 stable isotopes (of neon) had been discovered by J. J. Thomson, and their masses were measured with great precision by Aston about 1915. The nucleus itself was discovered by Rutherford in 1911, as described in Sec. 30-1. Reactions between colliding nuclei were achieved by Rutherford, Chadwick, and others in 1919. The mass-energy relation of Einstein's relativity theory was thus for the first time a matter of laboratory experience. The free neutron was looked for, and was finally found by Chadwick in 1932. The discovery of the fission of uranium in 1938 may be considered the culmination of the "classical period" of nuclear physics. More recent developments, to be discussed in Chap. 33, are concerned with the intimate details of nuclear structure and involve gigantic particle accelerators that produce projectiles with billions of electron volts of KE. Experimenters are also concerned with the elusive neutrinos, which have zero charge and very small or zero rest mass.

32-2 Nuclear Structure

Before studying unstable nuclei, radioactive decay, and nuclear transmutation, we will take a look at the properties of stable nuclei. Some of these properties were discussed in Secs. 1-5 and 1-7.

According to our present understanding, the nucleus contains heavy particles, protons and neutrons, which are collectively called *nucleons* (Sec. 1-7). Any given combination of protons and neutrons is called a *nuclide*; several nuclides all of the same atomic number but of different mass numbers are called *isotopes* of the element in question. The symbol Z is used for the *atomic number*, or the number of protons in the nucleus, and A represents the *mass number*, or total number of nucleons. Hence the number of neutrons

---

* The spectrum of helium is obtained from a sealed glass tube that has contained an α emitter for some time.

in any nucleus is given by $A - Z$. The mass number A also represents the atomic mass (to the nearest integer) of the nucleus. Atomic numbers Z are given in the Periodic Table (Appendix Table 5). The symbol for a nuclide has a subscript for Z and a superscript for A: $^{56}_{26}\mathrm{Fe}$ has $Z = 26$ and $A = 56$. Information about Z is, of course, contained in the chemical symbol; in advanced work the subscript for Z is usually omitted.

Not all combinations of protons and neutrons can stick together to form a stable nuclide. Roughly speaking, stable isotopes of light elements have about 1 neutron for every proton (example: $^{16}_{8}\mathrm{O}$ has 8 protons and 8 neutrons). In the heavier elements, the trend is toward somewhat more neutrons than protons (example: $^{200}_{80}\mathrm{Hg}$ has 80 protons and 120 neutrons). If a nuclide deviates too much from the norm, it is certain to be unstable and radioactive.

Sizes and shapes of nuclides are found by analyzing the way in which very high energy electrons are scattered by the nuclei in a target (Sec. 33-8). The radius of a nucleus (in meters) is given experimentally by

$$r \approx (1.2 \times 10^{-15})A^{1/3} \qquad (32\text{-}1)$$

The femtometer (fm) is 10^{-15} m, a convenient unit for nuclear sizes:

$$r \approx 1.2A^{1/3} \text{ fm}$$

The nuclear radius is proportional to the cube root of the mass number, that is, to the cube root of the number of nucleons that are gathered together. Since the volume of a sphere is $\frac{4}{3}\pi r^3$, this means that the volume V of a nucleus is very nearly proportional to the mass number A. As pointed out in Sec. 1-7, these experimental facts show that the density of nuclear matter is nearly constant. The nucleus may be thought of as an incompressible bundle of nucleons—protons and neutrons—that are in close contact with each other. Refined measurements of several kinds show that some nuclides are slightly ellipsoidal in shape, instead of being spherical.

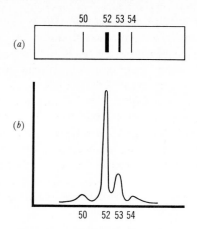

Figure 32-1   Chromium isotopes.

Nuclear Masses; The Unified Atomic Mass Unit (u).   Consider the isotopes of chromium, a rather typical element. Ordinary chromium is a mixture of four stable isotopes. The masses of the individual isotopes can be found by use of the *mass spectrometer*, which is similar in principle to the e/m apparatus pictured in Fig. 29-3 for electrons. Many types of mass spectrometer have been developed, in which combinations of magnetic and electric fields focus ions on a collector. Figure 32-1a is a record of the chromium mass spectrum; each Cr^+ ion sticks to a glass plate at a place determined by its e/m ratio. In this way, isotopes of different masses are brought to different places.* In another type of mass spectrometer, the Cr^+ ions strike a metallic collector, they are neutralized by free electrons, and the electron current is a measure of the abundance of the isotope (Fig. 32-1b). Such experiments show four stable isotopes of chromium, with relative abundances as follows: $^{50}_{24}\mathrm{Cr}$ (4.3%), $^{52}_{24}\mathrm{Cr}$ (83.8%), $^{53}_{24}\mathrm{Cr}$ (9.5%), and $^{54}_{24}\mathrm{Cr}$ (2.4%). The chemical atomic weight of chromium (51.996) is the weighted average of these values; it turns out to be so close to 52 because the

---
* See Fig. 33-1 on page 785 for a mass spectrogram of cadmium isotopes.

isotope of mass number 52 is by far the most abundant in nature. Careful use of a precise mass spectrometer reveals that the atomic masses of the isotopes are not exactly integers. Thus a neutral atom of ^{50}Cr has mass 49.94606, in terms of 12.00000 for the mass of $^{12}_6$C, a neutral carbon atom of mass number 12. We shall see in our later work that precise knowledge of nuclear masses is essential to our understanding of radioactive decay and nuclear transmutations. Masses of some nuclei from $_1$H to $_{12}$Mg are given in Appendix Table 9.

Conservation of mass and conservation of energy are intimately related, as we saw in Sec. 28-6. If we use relativistic mass in our calculations, mass and energy are each (separately) conserved. Usually, however, we deal with *rest mass*, and we find that rest mass (m_0) can be converted into energy according to the Einstein equation $\Delta E = (\Delta m_0)c^2$. Instead of using joules for energy and kilograms for mass, we introduce units that are much more convenient for nuclear physics. For energy, we use MeV (million electron volts). According to the definition in Sec. 29-3, 1 MeV $= 10^6 \times 1.60 \times 10^{-19}$ J $= 1.60 \times 10^{-13}$ J. For mass, we use the *unified atomic mass unit* (u), which is defined so that the mass of the neutral $^{12}_6$C atom (nucleus plus 6 electrons) is exactly 12.000000 u. The Einstein formula is used to obtain the conversion factor between MeV (an energy unit) and u (a mass unit). One unified atomic mass unit is $\frac{1}{12}$ the mass of a carbon-12 atom, which is found by use of Avogadro's number:

$$1 \text{ u} = \frac{1}{12}\left(\frac{12.000000 \text{ g}}{6.02 \times 10^{23}}\right) = 1.66 \times 10^{-24} \text{ g}$$

$$= 1.66 \times 10^{-27} \text{ kg}$$

Hence,

$$\Delta E = (\Delta m_0)c^2$$
$$= (1.66 \times 10^{-27} \text{ kg})(3.00 \times 10^8 \text{ m/sec})^2$$
$$= 1.49 \times 10^{-10} \text{ J}$$
$$= 9.31 \times 10^8 \text{ eV} = 931 \text{ MeV}$$

Thus,

the energy equivalent of 1 u is 931 MeV.

The electron's mass is 9.11×10^{-31} kg, which is 0.000549 u, equivalent to 0.511 MeV.

32-3   Binding Energy

Work must be done to "take apart" any assemblage of particles; the amount of energy that must be supplied is called the *binding energy* (BE) of the system. A large-scale example of binding energy is the work required to separate a spacecraft from the earth. In this case, the BE can be calculated relatively easily, since the gravitational force is a well-known function of distance, Newton's law of gravitation.

On the atomic scale of things, the binding energy of the hydrogen atom is found by considering the energy levels of its electron. According to Fig. 30-3 on page 716, if the electron is originally in its lowest energy state, 13.60 eV is needed to remove the electron from the electrical influence of the proton. Thus, the BE of the H atom is 13.60 eV. The binding energies of organic molecules are more difficult to calculate, but they can be measured in the laboratory and are from 2 eV to 20 eV.

On the nuclear scale of things, binding energies are much larger; they are not easily calculated because the nuclear forces are not as well understood. Fortunately, experimental values of nuclear masses can be used to find nuclear binding energies.

Example 32-1

What is the binding energy of the least strongly bound neutron in a $^{12}_6$C nucleus?

In effect, we are asked for the difference in energy between $^{12}_6$C and ($^{11}_6$C + $^1_0$n). We use the nuclear masses from Appendix Table 9.

Before taking apart

$^{12}_{6}C$ 12.000000 u

After taking apart

$^{11}_{6}C$ 11.011432 u

$^{1}_{0}n$ 1.008665

                             12.020097 u

Mass difference = 0.020097 u

$$= (0.020097 \text{ u})\left(\frac{931 \text{ MeV}}{1 \text{ u}}\right) = \boxed{18.7 \text{ MeV}}$$

Since the mass of the fragments is greater than the mass of the original $^{12}_{6}C$ nucleus, energy must be supplied in the amount of 18.7 MeV to remove the neutron from the nucleus.

_____ **Example 32-2**

Calculate the binding energy of an *average* nucleon in $^{12}_{6}C$.

Let us imagine breaking up the $^{12}_{6}C$ nucleus into its 12 constituent nucleons, that is, into 6 protons and 6 neutrons.

Before

$^{12}_{6}C$ 12.000000 u

After

$6\,^{1}_{1}H = 6(1.007825)$ 6.046950 u

$6\,^{1}_{0}n = 6(1.008665)$ 6.051990

                           12.098940 u

Mass difference = 0.098940 u = 92.1 MeV

BE per nucleon = 92.1/12

                   = $\boxed{7.7 \text{ MeV per nucleon}}$

To remove only one neutron from $^{12}_{6}C$ (Example 32-1) would require 18.7 MeV, more than twice the average amount per nucleon to break up $^{12}_{6}C$ into 12 separate nucleons.

Binding energies per nucleon (Fig. 32-2) show a rising trend for the lightest nuclides, and are remarkably constant at about 8 MeV per nucleon for most nuclides above about mass number $A = 20$. The pronounced peak at $A = 4$ (for the alpha particle $^{4}_{2}He$) shows that the combination of 2 protons and 2 neutrons is exceptionally stable (has large BE). This is correlated with the fact that α particles are ejected as a single stable entity during radioactive decay of heavy elements such as radium. The peaks at 4, 12, and 16

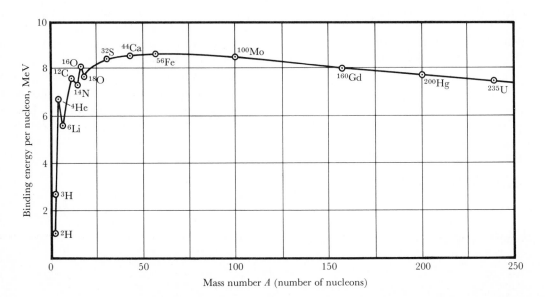

Figure 32-2   Nuclear binding energy curve.

in the *nuclear* binding energy curve are reminiscent of the peaks in the *atomic* BE curve (Fig. 31-5), where ionization energies of the atoms show peaks at the noble gases due to closed shells in the electron probability clouds. By analogy, we can say that two protons and two neutrons (an α particle) represent a closed shell in the nucleus. The nuclear binding energy curve has a broad maximum at about $A = 50$ to 60. Thus, $^{56}_{26}$Fe, a common constituent of metallic meteorites,* is more stable than either lighter or heavier nuclides.

These experimental facts about nuclear binding energies are data that must serve as a basis for any theory about nuclear structure. When we talk about binding energy, we are also talking about nuclear forces. Experiment shows that the $^{56}_{26}$Fe nucleus does not spontaneously fly apart, or *fission*, into fragments, so the Coulomb repulsion of the positively charged protons evidently is balanced by an attractive force between nucleons. This attractive *nuclear force*, of very short range, is exerted alike between protons and protons, between protons and neutrons, and between neutrons and neutrons. We shall return to these matters in Chap. 33 when we discuss interactions between the nucleons (protons and neutrons) that are the constituents of the "nuclear matter" discussed first on pages 13–14.

Example 32-2 shows that energy is needed to separate a light nuclide such as $^{12}_{6}$C into its component parts. The reverse is also true: energy is released when $^{12}_{6}$C is built up out of its parts. Such a process, called *fusion*, releases energy because the BE curve is rising—has a positive slope—for light elements. The opposite is true for heavy elements. Here the BE curve is decreasing—has a negative slope—and energy is released when a nuclide such as $^{235}_{92}$U splits into two or more fragments during fission. The BE curve is thus of utmost importance for the controlled release of nuclear energy by fission (Sec. 33-2) and by fusion (Sec. 33-3).

---

* Iron seems to be widely distributed in the solar system.

32-4 Radioactive Decay

The fact that some nuclides are stable, whereas others are radioactive and may spontaneously break down into fragments, can be understood through the conservation laws. It has been stated that "Anything that *can* happen *will* happen" . . . unless prevented by some conservation law. The important laws in radioactive decay are conservation of charge, conservation of mass-energy, conservation of linear momentum, and conservation of angular momentum (including orbital and spin momentum). During every decay process these conservation laws are satisfied, as will be apparent from what follows. Conversely, we may say that any given nuclide is stable if its decay would violate some conservation law.

One outstanding early observation about radioactivity was its independence of external conditions. The nature of the radiations from uranium, for instance, and the strength (activity) of a given amount of uranium were found to be almost completely† unaffected by temperature, pressure, and other physical changes; neither could any effect of chemical combination be found. The activity of a mole of UCl_3 exactly equals that of a mole of UO_2 or a mole of UF_6 gas.

The nuclear or planetary model of an atom developed by Rutherford in 1911 explains why radioactivity is so unaffected by external conditions. Chemical reactions take place when the outer parts of electron shells come close together. During collisions between atoms in a gas, the outer shells may become momentarily distorted. However, during such physical or chemical

---

† A very minor effect of chemical binding (valence) was discovered, with great difficulty, half a century after the discovery of radioactivity. This effect depends on the extremely small (but not zero) penetration of an outer electron's probability cloud to within the nuclear radius (some 10^{-5} Å). See Fig. 31-2, where a $2p$ cloud has a different probability density than a $1s$ cloud, indicating that the penetration of the nucleus depends slightly on the electronic state of an atom.

changes the nucleus is protected by the outer electron cloud; the nuclei never get close enough to each other to react, except at high speeds corresponding to temperatures of over 10^6 K. Since radioactivity is practically unaffected by chemical state, temperature, pressure, and other conditions, we conclude that it is a nuclear property. The α, β, and γ rays are, therefore, emitted from the nucleus.

Before studying specific modes of decay, we will discuss a general property of radioactivity—the rate at which a nuclide spontaneously disintegrates. What follows refers equally to all modes of decay, but for definiteness we might imagine an experiment in which a sample of radioactive chromium, $^{55}_{24}$Cr, emits β particles (electrons) that are detected by some detector such as a Geiger counter.

The *half-life* $T_{1/2}$ of a radioactive decay process equals the time required for half the nuclei in a sample of a nuclide to make the transition. For the decay of $^{55}_{24}$Cr, experiment shows the half-life to be 3.52 min. This means that if, say, 8 million $^{55}_{24}$Cr nuclei are present at a certain time, 3.52 min later only 4 million will still be $^{55}_{24}$Cr; after another 3.52 min, the number will be down to 2 million; and so on. The nuclei are much like birds sitting on a fence. Assuming that the birds take off at random, we can estimate quite closely how many will be left after, say, 3.52 min or 7.04 min, but we cannot say *which* bird will take off first or which will be the last to go. We see that radioactive decay is intimately related to probability, and only a statistical model is possible. Half-lives for β decay are 10^{-2} s or longer. Considering all types of decay, the measured half-lives of known nuclides vary from less than a microsecond (high probability of decay) to more than 10^{17} years (very low probability of decay).

Example 32-3

A microgram of the nuclide $^{25}_{11}$Na (half-life 60 s) is in a test tube at noon. The material decays with emission of a β particle. (*a*) How many ^{25}Na atoms are present at the start? (*b*) How many are pres-

ent at 12:10 P.M.? (*c*) Estimate how long it will be before all the atoms are transformed.

(*a*) The atomic weight is 25, and we use Avogadro's number to find how many atoms are present.

$$\frac{6.02 \times 10^{23}}{25 \text{ g}} = \frac{N}{10^{-6} \text{ g}}$$

$$N = \boxed{2.408 \times 10^{16} \text{ atoms}}$$

(*b*) At the end of one half-life, which in this case is 1 min, only half the original number of atoms is present, and so on. We can make a table of the number present at any time:

12:00	2408×10^{13}
12:01	1204×10^{13}
12:02	602×10^{13}
12:03	301×10^{13}
⋮	⋮

After 10 half-lives (10 min) the number is $1/2^{10}$ or 1/1024 of the original number. That is,

$$N' = \left(\frac{1}{1024}\right)(2408 \times 10^{13})$$

$$= \boxed{2.35 \times 10^{13} \text{ atoms}}$$

(*c*) To within a few percent, each 10 half-lives reduce the number by a factor of 10^3. Thus, approximately, the amount of ^{25}Na decays as follows:

12:00	2.4×10^{16}
12:10	2.4×10^{13}
12:20	2.4×10^{10}
12:30	2.4×10^7
12:40	2.4×10^4
12:50	$2.4 \times 10 = 24$ atoms
12:51	12
12:52	6
12:53	3
12:54	?

By 12:54, only one or two ^{25}Na atoms remain, and the chances are 50-50 that if *one* is left at 12:55, it will decay during the next 60 s. We are running into a statistical situation here, and exact predictions cannot be made. (It would be meaningless to say that $1\frac{1}{2}$ atoms remain at 12:54!) However, we can be reasonably sure that by 1:15 P.M. there are *no* remaining ^{25}Na atoms, unless by a wild million-to-one chance.

The decay of a radioactive nucleus is a matter of chance. If N nuclei are present at a time t, we cannot tell when a given nucleus will decay, but there is a definite probability that ΔN of these nuclei will decay during a small time interval Δt that extends from t to $t + \Delta t$. The number decaying is proportional to the number present (N) and to the time interval (Δt). The proportionality constant is the *disintegration constant* λ, defined by the equation

$$\Delta N = -\lambda N \, \Delta t \qquad (32\text{-}2)$$

The $-$ sign is used because N decreases when t increases.

In Sec. 32-10 it is shown how Eq. 32-2 leads to a formula for N as a function of t:

$$N = N_0 e^{-\lambda t} \qquad (32\text{-}3)$$

where e is 2.718, the base of natural logarithms; N_0 is the number of atoms present at the start ($t = 0$); and N is the number remaining after a time t has elapsed. Note how the equation shows these facts: If we substitute $t = 0$, we get

$$N = N_0 e^{-\lambda(0)}$$

and since anything to the 0th power is 1, it checks out that N is equal to N_0 at the start.

We see that radioactive decay is governed by an exponential equation (Fig. 32-3a). Also, it follows from Eq. 32-3 that

$$\ln N = \ln N_0 - \lambda t \qquad (32\text{-}4)$$

or

$$\log N = \log N_0 - (\lambda/2.30)t$$

[Note that $\ln N$ means $\log_e N$, and $\log N$ means $\log_{10} N$; the two are related by the equation $\log N = (\ln N)/2.30$.] Thus a graph of $\log N$ versus t is a straight line (Fig. 32-3b). A graph such as Fig. 32-3b is called a semilog graph because only one variable is plotted logarithmically. For ease of plotting, special semilog paper can be used (Fig. 32-3c).

The half-life $T_{1/2}$ is the time required for the number of nuclei to decrease from N_0 to $\frac{1}{2}N_0$. To obtain a relationship between the half-life and the disintegration constant, we first multiply Eq. 32-4 by -1 and then substitute the values $t = T_{1/2}$ and $N = \frac{1}{2}N_0$:

$$-\ln \tfrac{1}{2}N_0 + \ln N_0 = \lambda T_{1/2}$$

$$\ln \frac{N_0}{\frac{1}{2}N_0} = \lambda T_{1/2}$$

$$\ln 2 = \lambda T_{1/2}$$

$$T_{1/2} = \frac{\ln 2}{\lambda} = \frac{0.693}{\lambda} \qquad (32\text{-}5)$$

This equation makes sense; for a large λ (high probability of decay), the half-life is short.

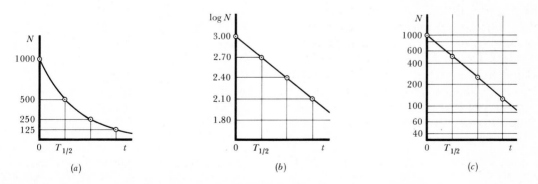

Figure 32-3 Radioactive decay. (a) Graph of N as a function of t. The value of N decreases by a factor of 2 in any time interval equal to a half-life $T_{1/2}$. (b) Graph of $\log N$ versus t for the same four observations as in part (a). Note that $\log 1000 = 3.00$, and so on. Since $\log 2 = 0.30$, the value of $\log N$ decreases by 0.30 in any time interval equal to a half-life $T_{1/2}$. (c) Same straight line as in part (b), plotted on semilog paper.

The activity A of a sample is the rate of decay measured, say, in disintegrations per second.* Thus, A is the absolute magnitude of $\Delta N/\Delta t$, which from Eq. 32-2 is

$$A = \lambda N \qquad (32\text{-}6)$$

The activity is the product of the disintegration constant and the number of nuclei present. Since A is proportional to N, the activity of a sample decreases in the same fashion as does N, and we can write

$$A = A_0 e^{-\lambda t} \qquad (32\text{-}7)$$

_____ **Example 32-4**

Calculate the activity on May 15 of a sample of ^{32}P that gave 2400 counts per minute in a Geiger counter on May 5. The half-life of ^{32}P is 14 days.
First we calculate the disintegration constant:

$$\lambda = \frac{0.693}{T_{1/2}} = \frac{0.693}{14 \text{ d}} = 0.0495 \text{ d}^{-1}$$

The elapsed time during decay has been 10 d.

$$A = A_0 e^{-\lambda t}$$
$$= 2400\, e^{-(0.0495 \text{ d}^{-1})(10 \text{ d})}$$
$$= 2400\, e^{-0.495} = 2400(0.610)$$
$$= \boxed{1463 \text{ counts/min}}$$

To find $e^{-0.495}$, an electronic calculator or a table of exponentials can be used.

_____ **Example 32-5**

How many disintegrations per second take place in a 2-μg sample of ^{226}Ra for which the half-life is 1622 years?
The number of nuclei in the sample is calculated using Avogadro's number; 1 mole of ^{226}Ra is 226 g.

_____

* The SI unit for radioactive disintegration rate is the becquerel, which is 1 disintegration per second (1 Bq = 1 s^{-1}). Activities can also be expressed in counts per second or counts per minute; for any given counter system there is a definite proportionality between the number of disintegrations and the number of recorded counts.

$$\frac{N}{2 \times 10^{-6} \text{ g}} = \frac{6.02 \times 10^{23} \text{ nuclei/mol}}{226 \text{ g/mol}}$$
$$N = 5.33 \times 10^{16} \text{ nuclei}$$

The disintegration constant is found from the half-life: $T_{1/2} = 1622 \text{ y} = 5.12 \times 10^{10}$ s. Hence

$$\lambda = \frac{0.693}{5.12 \times 10^{10} \text{ s}} = 1.35 \times 10^{-11} \text{ s}^{-1}$$

From Eq. 32-6 we find the activity:

$$A = \lambda N$$
$$= (1.35 \times 10^{-11} \text{ s}^{-1})(5.33 \times 10^{16} \text{ nuclei})$$
$$= \boxed{7.20 \times 10^5 \text{ nuclei/s}}$$

From this invisibly small sample, almost a million radium nuclei disintegrate each second.

32-5   Modes of Decay

Beta Decay.   We mentioned the four stable isotopes of chromium in Sec. 32-2. Turning now to the unstable isotopes of Cr, let us first consider $^{55}_{24}$Cr, which has a half-life of 3.52 min. This nuclide has more than the normal number of neutrons; there is a high probability that one of its neutrons will spontaneously change into a proton, increasing the atomic number of the nuclide by 1 unit. According to the law of conservation of charge, a negative charge is created to balance the newly created extra positive charge in the nucleus. This negative charge is emitted as a β particle, or negative electron, also called a _negatron_[†] ($_{-1}^{0}$e, or β^-). Since the nucleus now has an atomic number of 25, we know that it has become an isotope of manganese. The total number of nucleons has remained unaltered during the changeover of a neutron into a proton, so the mass number is still 55. Symbolically,

$$^{55}_{24}\text{Cr} \rightarrow\,^{55}_{25}\text{Mn} +\,_{-1}^{0}\text{e} \quad (3.52 \text{ min})$$

Note how this equation balances in two ways. The charge balances: $24 = 25 + (-1)$; and the

_____

† From now on, we speak of the "ordinary" electron, with which you are familiar, as a negatron, paving the way for introduction of the positron, or positive electron.

total number of nucleons balances: $55 = 55 + 0$. This sort of spontaneous change is called *radioactive decay*.

Positron Emission; Electron Capture. Another of the unstable isotopes of chromium is $^{49}_{24}Cr$, which disintegrates according to the equation

$$^{49}_{24}Cr \rightarrow ^{49}_{23}V + ^{0}_{+1}e \quad (41.9 \text{ min})$$

Here the emitted particle is a *positron*, which is a positive electron ($^{0}_{+1}e$, or β^+). Although positrons were first discovered as a component of cosmic rays, many unstable nuclides such as $^{49}_{24}Cr$ emit positrons spontaneously. All such nuclides have rather short half-lives and have to be made artificially.

Still another chromium isotope decays thus:

$$^{51}_{24}Cr + ^{0}_{-1}e \rightarrow ^{51}_{23}V \quad (27.8 \text{ days})$$

This is an example of electron capture, represented by the symbol ϵ, in which the ^{51}Cr nucleus "sucks in" one of the electrons from its external electron cloud.* The electron disappears, and at the same time (thus conserving charge) a proton changes into a neutron inside the nucleus, making an isotope of element number 23, vanadium.

In a sense, positron emission and electron capture are alternative processes, since in each case the daughter nucleus would be the same. Some nuclides decay in both ways at once; that is, each process has a certain probability of occuring. Positrons ($^{0}_{+1}e$) and negatrons ($^{0}_{-1}e$) are identical except for the sign of their charge. We consider them to be the same particle, the electron, in two different states of charge; more precisely, the positron is the "antiparticle" of the negatron.† To complete the picture, we list in Table 32-1 the eleven isotopes of chromium. Appendix Table 9 gives similar data for the light elements, $_1H$ to $_{12}Mg$.

Table 32-1   Isotopes of $_{24}Cr$

Nuclide	Abundance or Mode of Decay	Half-Life
$^{46}_{24}Cr$	β^+	1.1 s
$^{47}_{24}Cr$	β^+	0.4 s
$^{48}_{24}Cr$	ϵ	23 h
$^{49}_{24}Cr$	β^+	42 min
$^{50}_{24}Cr$	4.3%	
$^{51}_{24}Cr$	ϵ	27.8 d
$^{52}_{24}Cr$	83.8%	
$^{53}_{24}Cr$	9.5%	
$^{54}_{24}Cr$	2.4%	
$^{55}_{24}Cr$	β^-	3.5 min
$^{56}_{24}Cr$	β^-	5.9 min

Alpha Decay; Fission. The unstable isotopes of chromium illustrate the three major ways in which a light nucleus decays: negatron emission, positron emission, and electron capture. Most unstable nuclides decay in one or more of these three ways. Other modes of decay are possible, however, especially among the heavy elements. *Alpha-particle emission* was among the first to be observed by Becquerel and the Curies; a typical example is the decay of one of the radium isotopes into radon‡ (Rn):

$$^{226}_{88}Ra \rightarrow ^{222}_{86}Rn + ^{4}_{2}\alpha \quad (1622 \text{ years})$$

The α particle is the same as a helium nucleus $^{4}_{2}He$, and the equation balances both "above and below." Apparently the combination of two protons and two neutrons is so stable that such a group can stick together inside the nucleus and be emitted as a single superparticle, the α particle. Some heavy nuclei (rarely) break into two large fragments; this is called *spontaneous fission*. A typical example is $^{238}_{92}U$, whose half-life for α emission is 4.5×10^9 years, and whose half-life for spontaneous fission is 8.3×10^{15} years.

---

* Capture from the K shell is usually much more likely than capture from the L shell.

† The emission of either a negatron or a positron, or electron capture, is called beta decay.

---

‡ This element, of atomic number 86, is a noble gas (see the Periodic Table, Appendix Table 5). All the isotopes of radon are unstable; among them are $^{219}_{86}Rn$ (actinon), $^{220}_{86}Rn$ (thoron), and $^{222}_{86}Rn$ (radon). It is unfortunate that the word "radon" refers both to the element and to one specific isotope of the element.

Notes on Cobalt-60. We illustrate the use of Fig. 32-4 by discussing in detail the hexagons for ^{60}Co and its daughter ^{60}Ni. All of the following information is packed in two hexagons, each of total area 3.4 cm². ∎ A metastable state of ^{60}Co (left half of the hexagon) can decay in two ways, with a half-life of 10.5 minutes. 99.7% of the time it undergoes an isomeric transition (I.T.) to the ground state of ^{60}Co (horizontal arrow to the right half of the hexagon). The excitation energy of 0.0589 MeV is released 3% of the time by a gamma ray and 97% of the time by internal conversion (e^-). Very rarely (only 0.3% of the time) does the isomeric state of ^{60}Co decay to the excited state of ^{60}Ni, emitting a negatron (β^-) whose maximum energy is 1.54 MeV (downward arrow). ∎ The ground state of ^{60}Co decays with a half-life of 5.27 y by the emission of β^- particles of several energies. The decays always lead to excited states of ^{60}Ni, as shown by the arrow labeled 100% pointing downward to the left. ∎ One excited state of ^{60}Ni is immediately deexcited by emission of a 1.17-MeV γ ray; another excited state of ^{60}Ni lives for a short time (0.8 picosecond), then settles down to the ground state by emitting a 1.33-MeV γ ray. ∎ The ^{60}Co nuclide that is so useful in medicine and industry has a half-life of 5.27 y because the 10.5-min isomer fully decays during shipment and storage. It is usually stated that ^{60}Co emits two γ rays, of energies 1.17 MeV and 1.33 MeV, both governed by the 5.27-y half-life of the parent; but actually both of these rays represent the deexcitation of the intermediate ^{60}Ni isomer. ∎ The symbol σ shows the cross sections (probabilities) for capture of slow neutrons: the 10.5-min ^{60}Co isomer, while it exists, has a cross section of about 100 barns (100×10^{-28} m²; see Sec. 33-2); the more stable ground state has a cross section of about 6 barns; and the ground state of the stable ^{60}Ni nuclide has a capture cross section of 2.6 barns. ∎ The magnetic moment μ of the ground state of ^{60}Co is +3.80 units. ∎ The stable isotope ^{60}Ni comprises 26.23% of naturally occurring nickel. ∎ The mass of ^{60}Co is 59.93381 u and that of ^{60}Ni is 59.93079 u, both masses being measured when the nuclides are in their ground states.

Gamma-Ray Photons from Nuclei. We have seen in Chap. 30 that the energy of the outer electron cloud of an atom is quantized; energy-level diagrams such as Fig. 30-7 show the allowed energies of the entire system of electrons. When the energy of the electron cloud changes by ΔE, a photon is radiated whose frequency is given by $\Delta E = hf$. These changes take place in the electron cloud. In a similar fashion, the nucleus, too, can exist in various energy states, and a transition from one state to another results in emission of a photon whose energy is ΔE. The *nuclear energy levels* are separated by thousands or millions of electron volts, and so the photons that are emitted have high energy. These high-energy photons that result from nuclear deexcitation or "shakedown" are called gamma rays.* Most γ rays accompany α or β decay or electron capture. For instance, the $^{238}_{92}$U nucleus emits γ-ray photons of energy 0.045 MeV as well as several groups of α particles. (In some cases it happens that the energy of the excited nucleus is given directly to one of the $K, L, M, \ldots$ orbital electrons, which is then ejected from the atom with a definite KE. This deexcitation process is called *internal conversion.*) Among the chromium isotopes, the nuclide $^{48}_{24}$Cr emits two γ rays when it captures a K electron; $^{49}_{24}$Cr emits three γ rays during positron decay; $^{51}_{24}$Cr emits one γ ray during K capture; but $^{55}_{24}$Cr emits no γ ray during its negatron emission.

Sometimes a nucleus can remain in an excited state for a measurable time; such an excited nucleus (in a metastable state) is called an *isomer.* An isomer can decay to a stable state (with emission of a γ-ray photon), or if its lifetime is long enough, the isomer may decay in any of the standard ways without first going to its stable state. Nuclear γ rays offer one of the best means of mapping the structure of nuclear energy levels.

A portion of a chart of nuclides is shown in Fig. 32-4; both negatron (β^-) and positron (β^+) emission are represented, as well as electron capture (ϵ). Four of the nuclides, ^{58}Co, ^{60}Co, ^{60}Ni, and ^{62}Co, have isomeric states with measurably long half-lives. The relative abundances of the stable nuclides are shown, in percent. We have chosen this form of nuclear chart because so much information can be presented in a small space.

---

* There is no difference between a 0.100-MeV γ-ray photon and a 0.100-MeV x-ray photon; the distinction, if any, is usually based on origin.

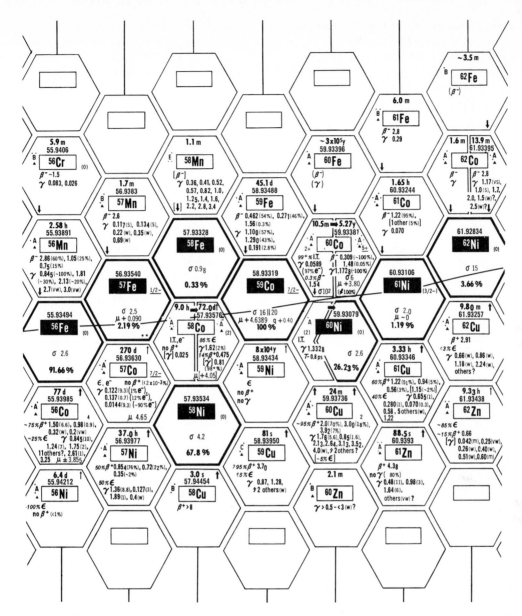

Figure 32-4 Stable and unstable nuclides with mass numbers 56 through 62. Masses, in u, are relative to $^{12}_{6}C = 12.00000$. For each stable nuclide, relative abundance is shown in percent. For each unstable nuclide, modes of decay are given, and half-lives are shown in years (y), days (d), hours (h), minutes (m), or seconds (s). Numbers following β or γ are energies in MeV of emitted β particles or γ photons. Other symbols: ϵ, electron capture; I.T., isomeric transition; σ, cross section in barns for slow neutron capture; e^-, internal-conversion electrons; μ and q refer to magnetic properties of nuclide; A, B, C, E, at left of nuclide symbol refer to method of experimental identification. [Figure adapted from W. H. Sullivan, *Trilinear Chart of Nuclides*, Oak Ridge National Laboratory (U.S. Government Printing Office, 1957).]

Let us summarize the various modes of disintegration of a nucleus. It may emit an α particle ($^4_2$He nucleus), a β^- particle (negatron, or negative electron), or a β^+ particle (positron). It may capture an orbital electron, usually from the K shell. It may undergo spontaneous fission. During any of these changes, a γ ray (photon) may be emitted from the nucleus. A γ ray is also emitted during an isomeric transition when a nucleus falls from an excited state to a stable state. About the only imaginable mode of decay that has not been observed is the direct emission of a proton or neutron from the nucleus, with a measurably long half-life, although these particles are emitted by some nuclides promptly after β decay to an excited state of the daughter nuclide.

32-6 Conservation Laws in Radioactivity

Our study of classical physics revolved around four conservation laws: the laws of conservation of linear momentum (Sec. 5-3), angular momentum (Sec. 7-8), charge (Sec. 18-4), and mass-energy (Sec. 28-6). All these laws are valid during nuclear processes; indeed, this universality is one reason we have stressed the conservation form for our general principles.

Conservation of charge is illustrated by disintegration equations such as $^{55}_{24}$Cr $\rightarrow$ $^{55}_{25}$Mn + $^{0}_{-1}$e. Here, the essential physical process is the creation of a negative electron while the neutron changes into a proton. It is easy to see *mathematically* that charge is conserved in the process, but you may feel uneasy that an electron's charge has been created out of nothing, so to speak. For this reason you may have the impression that a neutron is a "proton plus an electron," and that the electron is simply ejected from the neutron. This is incorrect, however, for there are several reasons for believing that an electron cannot exist inside a nucleus. If an electron is ejected with, say, 1 MeV of KE, its momentum is about 10^{-22} kg·m/s. For this much Δp, the Heisenberg uncertainty principle

gives $\Delta x \approx h/\Delta p \approx 10^{-12}$ m. This is 100 times larger than the size of a typical nucleus of diameter about 10^{-14} m (see Eq. 32-1), so the electron could not be confined in the nucleus. Another approach is to calculate the de Broglie wavelength of a 1-MeV electron, which also works out to be about 10^{-12} m.

We can consider the β-decay process to be the change of a nucleon from one state to another. When it is neutral, we call it a neutron, and when it is charged positively, we call it a proton. The situation is somewhat analogous to the emission of a photon of yellow light when a sodium atom's electron cloud changes from a certain energy state to a certain lower energy state. The photon is materialized during the transition. Similarly, the electron does not exist inside the nucleus, ready to be ejected like a bullet from a gun; rather, it is created when a nucleon changes its state during radioactive decay. Even this picture may suffer too much from modelitis. A nucleus may be just a sort of melting pot in which individual nucleons lose their identity.

Conservation of mass and energy is illustrated in radioactive decay. For every nuclear event, a mass-energy balance is possible. To illustrate this, let us once more study the decay $^{55}_{24}$Cr $\rightarrow$ $^{55}_{25}$Mn + $^{0}_{-1}$e, this time considering the rest masses in detail.

Before decay

$^{55}_{24}$Cr nucleus	54.9279 u
24 electrons	0.0132
	54.9411 u

After decay

$^{55}_{25}$Mn nucleus	54.9244 u
24 electrons	0.0132
1 negatron (emitted from nucleus)	0.0005
	54.9381 u

The 24 outer electrons of $^{55}_{24}$Cr still surround the $^{55}_{25}$Mn nucleus just after the decay process; thus, $^{55}_{25}$Mn is formed as a positive ion, having a deficiency of one electron. Eventually, it picks up an electron, or it takes part in some chemical reaction. Note that the mass total after decay includes 25 electrons, just enough for a neutral Mn atom. For this reason, masses of *neutral* atoms can be used in calculations of this sort. (A special treatment is needed for positron decay; see Prob. 32-C1.) Nuclear data charts such as Fig. 32-4 and tables such as Appendix Table 9 give neutral atomic masses, which include the mass of Z orbital electrons.

Example 32-6

Calculate the mass difference and energy available in the decay of ^{55}Cr to ^{55}Mn by negatron emission. The masses of neutral atoms are used:

Before decay

neutral ^{55}Cr atom   54.9411 u

After decay

neutral ^{55}Mn atom   54.9381

Mass difference   0.0030 u

Multiplying by 931 MeV/u gives for the available

energy $\boxed{2.8 \text{ MeV}}$

The mass balance is not exact. As we see, the total rest mass before decay is greater than the total rest mass after decay. The difference is 0.0030 u, which is equivalent to 2.8 MeV. This is the energy available for KE of the ejected particle (or particles) and for γ-ray photons. Experiment shows that the maximum KE of the emitted negatrons is 2.8 MeV, which agrees with the available energy calculated from the mass data. No γ rays are observed to accompany the decay of $^{55}_{24}$Cr, but many other nuclides emit γ rays during the decay process.

The decay of ^{55}Cr is typical of the conversion of rest mass into energy.* Detailed calculations

of this sort for hundreds of nuclear processes strengthen our confidence in the theory of relativity.

One major problem arises in the study of β-ray energies: particles are emitted with a continuous range of energies from zero up to the maximum given by the known mass difference. At first it was supposed that all the particles start out with the same KE (2.8 MeV in the example we are considering), but that some of them give up a portion of their KE to the nucleus on the way out. This would show up as thermal energy, which should be observable by a calorimeter experiment. Surprisingly, no increase of temperature was observed. Furthermore, the missing energy cannot be radiated in the form of γ rays, as shown by experiments in which the calorimeter cup was made of thick lead, which would absorb γ rays and produce thermal energy. For a while it was seriously proposed that the law of conservation of energy be given up for β decay.

The present view is that the missing energy is carried away by *neutrinos*, neutral particles of small (probably zero) rest mass.[†] Such particles do not affect a counter or a cloud chamber and are almost impossible to detect. Every β decay is, therefore, the simultaneous emission of an electron-neutrino pair. If 2.8 MeV is the energy released during the decay of a nucleus of $^{55}_{24}$Cr, and if the observed electron has 2.5 MeV, then the neutrino carries the other 0.3 MeV. Another $^{55}_{24}$Cr nucleus might eject a negatron of energy 2.1 MeV, along with a neutrino of energy 0.7 MeV. In this way the law of conservation of energy can be satisfied for each individual disintegration event. Only the maximum energy of any given β-particle group is listed in tables such as Fig. 32-4, or in diagrams such as Fig. 32-5.

Although neutrinos are elusive, they do carry energy and hence have momentum. Neutrinos are also emitted during electron capture, and the

---

* The *total* (relativistic) mass of the system has been conserved, for the fact that the emitted negatron has 2.8 MeV of KE means that this negatron's mass is greater than if it were at rest.

[†] If you are bothered by the idea of a particle of zero rest mass that has energy and momentum, recall that the photon is just such a particle, as shown by the Compton effect (Sec. 29-4).

recoil of the nucleus can be observed. Neutrinos can be captured (very rarely) by protons or neutrons. The reality of the neutrino is no longer in question. There are about 10^6 neutrinos in every cubic meter of so-called empty space throughout the universe.

The complete equation representing a β decay includes the neutrino:

$$^{49}_{24}\text{Cr} \rightarrow \,^{49}_{23}\text{V} + \,^{0}_{+1}\text{e} + \nu_e$$

or

$$^{55}_{24}\text{Cr} \rightarrow \,^{55}_{25}\text{Mn} + \,^{0}_{-1}\text{e} + \bar{\nu}_e$$

where ν_e and $\bar{\nu}_e$ (the Greek nu) represent the electron's neutrino and *antineutrino* (which differ only in the direction of their spins). However, the neutrinos are often omitted from the equations, their presence being tacitly understood.

Conservation of linear momentum in a nuclear collision is illustrated in Fig. 32-10 (page 773), which shows the elastic scattering of an α particle by a heavier nucleus of fluorine. The vector sum of the momenta before collision equals the vector sum after collision. Note how similar this event is to the collision of billiard balls; the resemblance is more than accidental, since linear momentum is conserved in each process. During radioactive decay, conservation of linear momentum is easily satisfied by the recoil of the nucleus that emits an α, β, or γ ray.

Conservation of angular momentum is valid for the resultant of the spin and orbital angular momentum of the nucleus. The intrinsic angular momentum of a particle such as a proton, electron, or neutron is considered to be a spin, analogous to the rotation of the earth on its axis. In the Bohr theory (Sec. 30-3), the *orbital* angular momentum is quantized, in multiples of $h/2\pi$. Similarly, *spin* angular momentum of a particle is quantized, but in multiples of $\frac{1}{2}(h/2\pi)$.* In addition to spin, there is the possibility that the nucleons within the nucleus may have some "orbital" angular momentum relative to the center of the nucleus. Total angular momentum is conserved in nuclear processes, as it is in large-scale physics. Even without the problem of energy balance, the neutrino, with its spin of $\frac{1}{2}$ unit, would have to be invented in order to conserve angular momentum. For further discussion of this point, see Sec. 32-11.

Still another conservation law is the law of *conservation of parity*; this deals with the symmetry properties of the particles or atoms and has no counterpart in large-scale physics, being essentially quantum mechanical in nature. Conservation of parity requires that if an experiment takes place, the mirror image of the same experiment can also take place. Thus, if a "left-handed" neutrino emitted during β^+ decay spins counterclockwise when viewed from behind (spin vector opposite to linear momentum vector), then we would expect "right-handed" neutrinos (spin vector parallel to linear momentum vector) also to be emitted during β^+ decay. Until recently, it was supposed that conservation of parity applied to all nuclear processes, as it is known to do with regard to the outer electrons of an atom. In 1956 it was shown, as a result of experiments suggested by the theoretical physicists T. D. Lee and C. N. Yang, that this law is not true during certain "weak interactions"; β decay is such an interaction. Experiment shows that neutrinos emitted during β^+ decay are *all* left-handed, and antineutrinos emitted during β^- decay are *all* right-handed. The nonconservation of parity for weak interactions was sensational news, and Lee and Yang received the 1957 Nobel prize in physics for their imaginative work.

We have been considering the validity of five conservation laws: those dealing with charge, mass-energy, linear momentum, angular momentum, and parity. Of these, the law of conservation of parity has been shown to be less generally valid than had been previously thought. Far from being a blow to progress, the breakdown of parity conservation opened the way to bold speculation about the validity of other

---

* Since an orbital electron has a spin quantum number $m_s = \pm \frac{1}{2}$, Eq. 31-1 implies that the magnitude of the spin angular momentum is $\sqrt{\frac{1}{2}(\frac{1}{2}+1)}(h/2\pi)$.

conservation laws. So far, the remaining four conservation laws appear to be valid in all nuclear processes, as they are in large-scale, everyday life. It is entirely possible, of course, that new experiments may some day reveal areas in which the law of conservation of charge, for instance, is not true. As always, a well-established law must be modified or abandoned if even one experimental fact is at variance with it. Physicists welcome the challenge of these new ideas.

Nuclear energy-level diagrams can be constructed that show the ground states and excited states of several adjacent elements that have the same mass number A (Fig. 32-5). The ground state of $^{23}_{10}$Ne, with half-life 37.6 s, can decay directly to the ground state of $^{23}_{11}$Na by emitting a β particle of energy 4.39 MeV (this happens 67% of the time). An alternative process is the emission of a β particle of energy 3.95 MeV (32% of the time) or 2.3 MeV (1% of the time). When the product nucleus is left in an excited state, it rapidly "shakes down" by emitting γ-ray photons. In this example, the half-life of the 0.44-MeV excited state of $^{23}_{10}$Na has been measured to be about 1.3 ps. Conservation of mass-energy is illustrated in several ways by the data shown in Fig. 32-5. The mass difference between the two ground states is given by $22.994473 - 22.989771 = 0.00470$ u (see Appendix Table 9), which corresponds to 4.38 MeV at the rate of 931 MeV per u. This is equal to the observed β-particle energy, within experimental error. Also,

the observed γ-ray energies add up correctly: 0.44 MeV + 1.64 MeV = 2.08 MeV, which is the same as the difference of the β-ray energies given by 4.39 MeV − 2.3 MeV = 2.1 MeV.

32-7 Stable and Unstable Nuclides in Nature

You may well wonder just why some isotopes of any given element are stable and others are unstable. This is indeed a puzzling question on which much work is being done. Using chromium again as an illustration, we find that isotopes of mass number 50, 52, 53, and 54 are stable, and 46, 47, 48, 49, 51, 55, and 56 are unstable. However, the terms "stable" and "unstable" are relative. When we say that ^{50}Cr is stable, we mean that the probability of a disintegration is zero (as is the case here, since the mass balance is unfavorable for β emission), or else is so small that no decay is observable. (See Prob. 32-B17.) A nuclide having a half-life of more than 10^{18} years would, for all practical purposes, be stable. At the other extreme, when we say that ^{57}Cr does not exist, we mean that if it does exist, its half-life is so short that it decays before it can be observed. From this point of view all nuclides are *possible*, although most of them are so unstable as to be unobservable.

In the beginning—at a time estimated to be some 10^{10} years ago, when the universe acquired its present form—probably all nuclides were formed in equal abundance and the highly unstable ones immediately decayed, becoming eventually long-lived "stable" or "nearly stable" nuclides. The sun and earth were formed about 5×10^9 years ago. Thus, any ^{55}Cr (half-life 3.52 min) that was in the solar system at that time has long since decayed. At the other extreme, ^{238}U has a half-life of 4.5×10^9 years; therefore, about half the ^{238}U originally present is still in existence. With such a long half-life, the probability of its decay is small, and so ^{238}U is not particularly "active." The extremely active

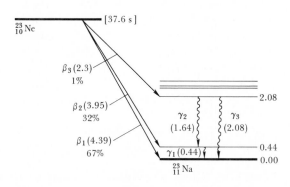

Figure 32-5

radium isotope $^{226}_{88}$Ra, discovered and isolated with so much effort by Madame Curie, has a half-life of 1622 years; any "primeval" radium would have long since decayed.

To understand why ^{226}Ra and some nuclides of even shorter half-life are still found in nature,

we consider the *radioactive series* that starts with the long-lived nuclide $^{238}_{92}$U (Fig. 32-6). This nuclide emits an α particle and becomes $^{234}_{90}$Th; half the original atoms undergo this decay in 4.5×10^9 years. We indicate this process by a diagonal line on the diagram. The

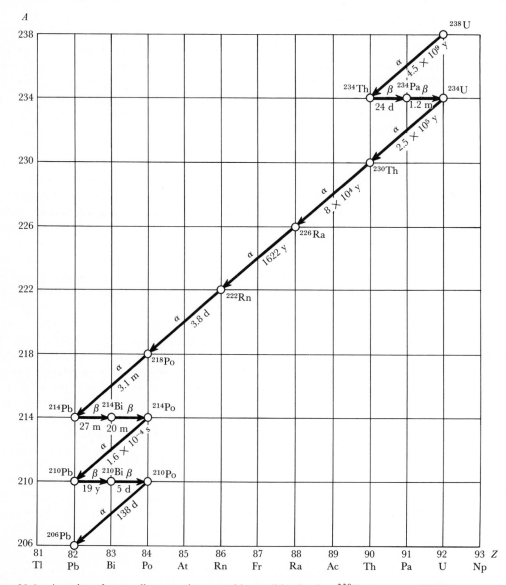

Figure 32-6 A series of naturally occurring unstable nuclides having ^{238}U as parent. Half-lives are given in years (y), days (d), hours (h), minutes (m), or seconds (s).

$^{234}_{90}$Th in turn decays by β^- emission to $^{234}_{91}$Pa, an isotope of protactinium. This negatron emission is represented by a short horizontal line on the diagram. The half-life for this decay is 24 days. Next, $^{234}_{91}$Pa decays into $^{234}_{92}$U by another negatron emission, this time with a half-life of 1.2 min. Now a long series of α emissions takes place, with $^{226}_{88}$Ra as one stage in the life history of the nucleus. The end result is the stable lead isotope $^{206}_{82}$Pb (some of the alternative β decays of small probability are omitted). The Curies found small amounts of radium in the uranium ore (pitchblende) that they purified. This radium, of relatively short half-life, was continually being replenished by radioactive decay of nuclei higher in the series. The whole series is dependent on one nuclide ($^{238}_{92}$U) having a half-life comparable with the age of the earth.

Other radioactive series start with $^{235}_{92}$U (half-life 7.1×10^8 years) and $^{232}_{90}$Th (half-life 1.4×10^{10} years). Many radioactive series exist whose longest-lived member's half-life is relatively short; such series can now be studied by making the necessary parent nuclide artificially. Before the nature of radioactivity was well understood, special names were given to certain nuclides occurring in the various natural series. For instance, in the uranium series of Fig. 32-6, $^{234}_{90}$Th was called uranium X1 (UX$_1$); $^{210}_{82}$Pb was called radium D (RaD); $^{210}_{83}$Bi was called radium E (RaE); and so on. These names are now of only historical interest.

Every element has at least one unstable isotope. The hydrogen isotope $^3_1$H, of mass number 3, is known as tritium; it is a negatron emitter with a half-life of 12.3 years. Even the neutron itself is unstable when it is not bound together in a nucleus with other particles; the free neutron is a negatron emitter with a half-life of 11 min. On the other hand, several elements of medium mass and all the heavy elements have *no* stable isotopes. These include technetium (whose isotopes of longest half-life are $^{98}_{43}$Tc and $^{99}_{43}$Tc, each of half-life about 10^5 years), promethium (whose stablest isotope is $^{145}_{61}$Pm, about 30 years), and all elements of atomic number 84 and higher.

32-8 Transmutations

The artificial production of gold from mercury was one of the dreams of the alchemists, but by 1895 it had generally been agreed that chemical elements are stable and unalterable. With the discovery of naturally occurring radioactivity, it was realized that the elements can and do change. The modern search for artificial transmutation of the elements began with the use of high-speed particles as projectiles. Rutherford in 1919 allowed α particles from $^{214}_{84}$Po (KE = 7.68 MeV) to pass through air, and he observed scintillations that were caused by protons. The reaction is with the nitrogen atoms in the air:

$$^{14}_7\text{N} + {}^4_2\text{He} = {}^{17}_8\text{O} + {}^1_1\text{H}$$

Here, as in radioactive decay, the equation balances for both charge and mass number.

The α particle (which has a charge of $+2e$) is strongly repelled as it approaches the $^{14}_7$N nucleus (which has a charge of $+7e$). This is one reason a charged projectile must have a high energy, for otherwise it will never get close enough to its target to react. The critical distance is about 10^{-14} to 10^{-15} m, the approximate radius of a nucleus.

A detailed calculation using exact masses shows that some energy is available for the KE of the proton and for the KE of the ^{17}O nucleus as it recoils. (In the calculation, all masses are for the neutral atoms of ^{14}N, ^{4}He, and so on; thus 9 outer electrons in all are included both before and after.)

Before collision

$^{14}_7$N	14.003074 u
$^4_2$He	4.002603
Original KE of particle (equivalent to 7.68 MeV)	0.00825
	18.01393 u

After collision

$^{17}_8$O	16.999133 u
$^1_1$H	1.007825
	18.006958 u

The difference is 0.00697 u, or 6.49 MeV, and this energy is shared between the proton and the ^{17}O nucleus. The cloud chamber photograph of Fig. 32-11 (page 773) shows the collision. Even if the mass balance is right for a reaction to take place, we still have to deliver the missile to the target, which requires that the projectile have high energy. In the nitrogen reaction described above, we see that the initial KE of the α particle is also needed for another reason: without some of the 0.00825 u represented by this KE, the total mass before collision would have been insufficient to supply the total mass after collision, as simple arithmetic shows. This is an example of a reaction for which there is a threshold energy.

As an illustration of nuclear transmutation, we have studied one reaction in detail. To save space, nuclear reactions are often written in abbreviated form; the bombardment of ^{14}N by α particles can be written as

$$^{14}\text{N}(\alpha, \text{p})^{17}\text{O}$$

The atomic numbers are omitted, since they are really superfluous. The necessary information about atomic number is given by writing the chemical symbol. The first symbol inside the parentheses represents the projectile, an α particle, and the second symbol represents the emitted particle, a proton. Many other types of reaction are possible. Bombardment with neutrons is very effective, since the neutron has no charge and can easily come close to the target nucleus. In fact, simple capture of slow neutrons works best of all, because the projectile spends more time passing by the target. In this connection a "slow" neutron means one whose KE is less than 1 eV. (Compare with the millions of electron volts usually required for a charged particle to cause a transmutation!)

A few examples will serve to illustrate the variety of nuclear reactions. Our symbols are as follows: α for alpha particle ($^4_2$He); p for proton ($^1_1$H); d for deuteron ($^2_1$H); n for neutron ($^1_0$n); γ for gamma-ray photon (hf).

1. ^{9}Be(α, n)^{12}C. This is the reaction by which neutrons were first produced and recognized as such by Chadwick, in England (1932). Convenient, portable neutron sources are made by mixing powdered beryllium with a long-lived α emitter such as plutonium.

2. ^{50}Cr(n, γ)^{51}Cr. This is a simple capture reaction and illustrates one way of producing the unstable nuclide ^{51}Cr, which we discussed in Sec. 32-5. Ordinary chromium metal, containing stable ^{50}Cr, is placed in the vicinity of a fission reactor, which produces large numbers of neutrons.

3. ^{2}H(d, n)^{3}He. This is the bombardment of deuterons (in the form of a target of heavy ice) by deuterons. This reaction is remarkable because of its large yield, even for bombarding energies as low as 25,000 eV. The product ^{3}He is stable.

4. ^{7}Li(p, α)^{4}He. This is an abbreviation for ^{7}Li + ^{1}H = 2 ^{4}He. Two α particles are produced, with a total energy release of 17 MeV.

5. ^{14}N(n, p)^{14}C. The ^{14}C, a negatron emitter with half-life of 5730 years, is useful as a tracer in studying chemical reactions and as an atomic clock in dating by the radiocarbon method (Sec. 33-1).

Over 1000 different nuclear reactions have been studied. Whether or not any proposed reaction is possible depends on the mass balance, so the importance of precise knowledge of nuclear masses is obvious. Conversely, masses of unstable nuclides can be determined by study of the energy balance for reactions that are known to take place.

Our notation for transmutations is really a shorthand for a complex process. In many cases a *compound nucleus* is formed, which has an unobservably short lifetime, estimated to be about 10^{-15} s. Thus reaction 5 above should be written in two steps:

$$^{14}_{7}\text{N} + ^1_0\text{n} = [^{15}_{7}\text{N}]$$
$$[^{15}_{7}\text{N}] \rightarrow ^{14}_{6}\text{C} + ^1_1\text{H}$$

The brackets around $[^{15}_{7}N]$ indicate that this is a compound nucleus formed in a highly excited state having extremely short half-life. We should realize that any description or model for a nuclear transmutation is subject to the limitations of all models, and should not be taken too literally.

For any reaction, the nature of the target, the product, or the particles can be found by applying the laws of conservation of charge and conservation of mass number. The Periodic Table (Appendix Table 6) must be used to identify the element when the atomic number is known, and vice versa.

Example 32-7

The phosphorus isotope $^{32}_{15}P$ has a half-life of 14.3 days and is useful in biological studies. It can be produced by a (n, p) reaction. What target must be used?

The equation must read

$$^{A}_{Z}X + ^{1}_{0}n = ^{32}_{15}P + ^{1}_{1}H$$

and it is evident that $Z = 16$ and $A = 32$. The target nuclide is, therefore, $^{32}_{16}S$.

Example 32-8

When ^{24}Mg is bombarded with deuterons, α particles are emitted. Identify the product nuclide.

We look up the atomic number of magnesium in the Periodic Table and write $^{24}_{12}Mg + ^{2}_{1}H = ^{A}_{Z}X + ^{4}_{2}He$. The equation balances for $Z = 11$ and $A = 22$. Hence the product is $^{22}_{11}Na$.

Example 32-9

The sodium isotope produced in the reaction of Example 32-8 is unstable. Make a guess as to the nature of its decay.

If $^{22}_{11}Na$ is unstable, it probably has too few neutrons, since the trend is toward slightly more than one neutron for every proton. Hence we expect a proton to turn into a neutron, and the expected decay is $^{22}_{11}Na \rightarrow ^{22}_{10}Ne + ^{0}_{+1}e$ (positron emission), or perhaps $^{22}_{11}Na + ^{0}_{-1}e \rightarrow ^{22}_{10}Ne$ (electron capture). Since the nuclide is a light one, compared with uranium or radium, we rule out the possibility of an α decay. Reference to Appendix Table 9 shows that ^{22}Na is indeed a positron emitter, with a half-life of 2.6 years.

32-9 *Experimental Techniques*

Over the years, physicists have developed many devices for detecting the particles and photons emitted by radioactive atoms. Most of these methods depend for their success on the ions produced by a moving charged particle. As a charged particle, such as an α particle or an electron, passes through air, its electric field distorts the air molecules, and some molecules become ionized. As a consequence, the air becomes a better conductor of electricity. In the early days, a radioactive sample was placed directly inside a charged gold-leaf electroscope, and the leaves collapsed at a rate proportional to the strength of the sample. Modern electronic amplifiers can be used to detect the ionization pulse due to a single, heavy, charged particle such as an α particle or a proton passing through an *ionization chamber*.

A solid-state version of the ionization chamber is the *surface-barrier detector*. In one form, the detector is simply a reverse-biased diode (Sec. 23-10) in which a very thin region of p-type silicon is backed by a thicker region of n-type silicon. A large electric field of perhaps 10^6 V/m is produced in the p-type material by, say, a PD of 100 V across a thickness of 10^{-4} m. When a charged particle or a photon is absorbed, charges are released in the p-type silicon that are collected by electrodes at the faces of the crystalline material. The detectors are fast acting (as little as a few nanoseconds to collect the charge), sensitive (as little as 2 to 4 eV to produce an electron-hole pair in the silicon), and small.

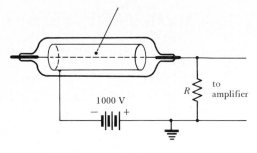

Figure 32-7   Geiger counter.

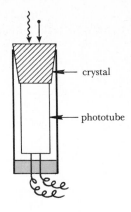

Figure 32-8   Scintillation counter.

The Geiger-Müller tube counter (Fig. 32-7) (invented in 1928, and now called simply a *Geiger counter*) makes use of an avalanche effect. The tube contains a gas such as argon at a pressure of, say, 0.1 atm, and the thin central wire is about 1000 V positive with respect to the metal wall. If the incoming particle creates just one ion pair, the negative ions are attracted to the + wire, creating more ions by collision. A current rapidly builds up, and the *IR* drop in the resistor *R* may amount to several volts, to be further increased by the amplifier. The Geiger counter responds only to charged particles. A photon is uncharged and therefore produces practically no ions in the gas. Photons may, however, eject photoelectrons from the metal of the wall, and these photoelectrons may initiate an avalanche. The efficiency of a Geiger counter may be as high as 99% for electrons that penetrate the outer wall, but less than 1% for x rays or γ rays.

We have mentioned earlier the scintillation detector of Rutherford. The modern development of this device is the *scintillation counter* of Fig. 32-8, which uses a single crystal such as NaI. The crystal, perhaps a few centimeters in depth, is so much more dense than air that there is a good chance that a γ-ray photon will be absorbed and will emit a number of fluorescent photons of visible light. The crystals are usually sensitized by controlled addition of an impurity such as thallium. The crystal is in a light-tight enclosure, and the faint splash of light is picked up by a sensitive phototube to produce an electric signal that is amplified. No avalanche process is involved, and the scintillation counter is fast-acting, rugged, and highly efficient both for charged particles and for γ rays.

The *cloud chamber* (Fig. 32-9) makes visible the path of an ionizing particle. When a charged particle passes through air, it leaves a trail of ions, and if the region is supersaturated with water vapor, the ions serve as centers on which visible drops of water may condense (Figs. 32-10 and 32-11). The effect is similar to the formation of vapor trails in the sky by a passing jet aircraft.

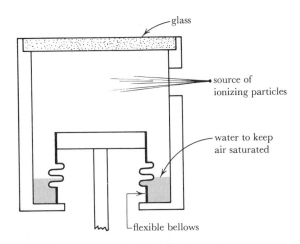

Figure 32-9   Cloud chamber. If the ionizing particles have insufficient energy to penetrate the thin window, the source would be placed inside the chamber.

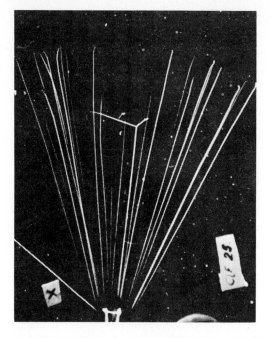

Figure 32-10 Scattering of α particle by a heavier nucleus of fluorine. Note also that all the unscattered α particles have about the same range in the gas of the cloud chamber. (I. K. Bøggild, from *An Atlas of Typical Expansion Chamber Photographs*, by W. Gentner, H. Maier-Leibnitz, and W. Bothe, Pergamon Press.)

Figure 32-11 Transmutation of nitrogen by α particle, according to the reaction $^{14}_{7}N + ^{4}_{2}He \rightarrow ^{17}_{8}O + ^{1}_{1}H$. The ejected proton travels downward and to the left, and the oxygen nucleus leaves a short track, upward and slightly to the right. The source of α particles is a mixture of two nuclides: ^{212}Bi (emitting α particles of shorter range) and ^{212}Po (emitting α particles of longer range). (P. M. S. Blackett and D. S. Lees, from *An Atlas of Typical Expansion Chamber Photographs*, by W. Gentner, H. Maier-Leibnitz, and W. Bothe, Pergamon Press.)

The *bubble chamber* uses a liquid such as propane or liquid hydrogen, and the passage of an ionizing particle is made evident by a trail of tiny bubbles (Fig. 33-11, page 806). The *wire chamber* is gas filled and has criss-crossing sets of fine wires between which the PD is almost enough to sustain a discharge. The ionization left by a particle serves as a path for many tiny discharges that trace out the particle's path. The electrical pulses from the hundreds of wires are analyzed by a computer to reconstruct the trajectory of the particle in three dimensions. Wire chambers are faster acting than bubble chambers, and unlike bubble chambers they can be triggered to be sensitive only when an event of interest has just occurred.

All three of these chambers give a great deal of information about the particle that makes a track. If the chamber is placed in a magnetic field, moving charged particles are bent in arcs of circles, and so their energies can be found. Particles can be identified by the density of drops or bubbles along the path. The slow, heavy α particles create many more ion pairs per centimeter of path, and give a denser trail, than the light electrons.

Another detection device is the *photographic emulsion;* upon development, grains of silver are deposited where ionizing particles have passed. The tracks are viewed with a microscope (Fig. 33-13, page 811).

Summary A stable or unstable nucleus contains A nucleons that are closely packed in such a way that all nuclei have about the same density. The nuclear radius is approximately proportional to $A^{1/3}$. The binding energy (BE) of a nucleus is the energy required to separate it into its component nucleons. The BE per nucleon is between 7 and 9 MeV for most nuclides.

Radioactive decay takes place when a neutron changes into a proton (negatron emission) or when a proton changes into a neutron (positron emission). A group of two neutrons and two protons is exceptionally stable and can be emitted as an α particle by a heavy nucleus. During electron capture a nucleus captures an orbital electron, usually from its K shell. Less frequent modes of decay are spontaneous fission and neutron emission. During any nuclear change, γ rays may be emitted; these are high-energy photons whose energy equals the difference between two energy levels in the nucleus. Nuclear disintegration is a statistical process described by the disintegration constant, which is the probability per unit time that a disintegration will take place. The half-life is the time required for half of any given number of nuclei to decay. Natural radioactive series begin with ^{238}U, ^{235}U, and ^{232}Th, whose half-lives range from 0.5 to 14 billion years. Since these lives are comparable with the age of the earth, some of the parent nuclides still remain and are continually producing a chain of daughter products having shorter half-lives.

The laws of conservation of charge, mass-energy, linear momentum, and angular momentum are obeyed during nuclear events. In nuclear disintegrations and in transmutations, the energy released as KE comes from the transformation of rest mass into energy according to $\Delta E = (\Delta m_0)c^2$. Conversely, if the rest mass of the system increases during a reaction, energy must be supplied. One unified atomic mass unit (u) is equivalent to 931 million electron volts (MeV). During β decay, some of the available energy may be given to a neutrino, which has no charge and no rest mass. Many types of transmutation reactions are possible if the mass balance is favorable. Capture of slow neutrons is effective because of the absence of Coulomb repulsion between projectile and target.

Check List

alpha ray	binding energy	radioactive series
beta ray	half-life	compound nucleus
gamma ray	disintegration constant	ionization chamber
nucleon	negatron	surface-barrier detector
nuclide	positron	Geiger counter
isotope	neutrino	scintillation counter
atomic number	antineutrino	cloud chamber
mass number	internal conversion	bubble chamber
MeV	isomer	wire chamber
u		

Questions

32-1 Explain why the early observations of radioactivity seemed to contradict one of the conservation laws. Is this objection still valid?

32-2 Compare and contrast α, β, and γ rays with respect to (a) penetrating power; (b) charge; (c) mass; (d) deflection by a magnetic field.

32-3 What is the approximate ratio of the rest mass of an α particle to that of a β particle?

32-4 Why are the chemical atomic weights of some elements not very close to integers?

32-5 The noble gas radon has half-life 3.8 days and has been used in cancer therapy for many years. Explain how a nuclide of such short half-life can occur naturally. (See Fig. 32-6.)

32-6 In the portion of the chart of nuclides shown in Fig. 32-4, there is no example of two stable nuclides in vertically adjacent hexagons. Show that this must be true throughout the entire chart of nuclides.

32-7 Describe three devices for detecting the passage of a charged particle.

32-8 Does a γ ray have (*a*) any mass? (*b*) any momentum? (*c*) any charge?

32-9 What is the precise meaning of the statement "$^{16}_{8}O$ is a stable isotope of oxygen"?

32-10 In Fig. 32-7, when an ionizing particle passes through the counter, does the potential of the wire in the counter (relative to ground) momentarily increase or decrease?

32-11 Does a neutrino have (*a*) mass? (*b*) rest mass? (*c*) linear momentum? (*d*) KE?

32-12 What is the evidence for believing that radioactivity is a nuclear process?

32-13 Explain why bombardment of a nucleus with neutrons is so much more effective than bombardment with protons.

MULTIPLE CHOICE

32-14 Isotopes of an element have the   (*a*) same number of electrons but a different number of protons; (*b*) the same number of protons but a different number of neutrons; (*c*) the same number of neutrons but a different number of protons.

32-15 All nucleons have (*a*) exactly the same rest mass; (*b*) almost the same rest mass; (*c*) rest masses that depend on the nuclide in which they are found.

32-16 The number of elementary charge units in a nuclide determines the atomic (*a*) weight; (*b*) mass; (*c*) number.

32-17 The dimensions of disintegration constant are (*a*) $[MT^{-1}]$; (*b*) $[T]$; (*c*) $[T^{-1}]$.

32-18 When $^{55}_{24}Cr$ disintegrates into $^{55}_{25}Mn$, it emits (*a*) a negatron; (*b*) a positron; (*c*) an α particle.

32-19 When $^{66}_{29}Cu$ decays into $^{66}_{30}Zn$ by negatron emission, the zinc atom (just after the decay) is formed as a (*a*) neutral atom; (*b*) positive ion; (*c*) negative ion.

Problems

Note: Refer when needed to the Periodic Table and the table of masses of light nuclides in the Appendix.

32-A1 Identify the elements that have the following isotopes: $^{118}_{50}X$; $^{3}_{2}X$; $^{50}_{22}X$; $^{197}_{79}X$.

32-A2 Identify the elements that have the following stable isotopes: $^{12}_{6}X$; $^{20}_{10}X$; $^{200}_{80}X$; $^{107}_{47}X$.

32-A3 How many neutrons are in each nuclide of Prob. 32-A2?

32-A4 Consider the nuclide $^{60}_{28}Ni$. (*a*) How many protons does it have? (*b*) How many neutrons? (*c*) How many nucleons?

32-A5 What is the approximate radius of a nucleus of $^{64}_{29}Cu$?

32-A6 What is the approximate atomic mass number A of a nuclide whose radius is 3.6 fm?

32-A7 During a certain transmutation, 0.015 u of rest mass has disappeared. How much nuclear energy has been released (in MeV)?

32-A8 How many u of rest mass must disappear in a nuclear reaction in which 25 MeV of energy is released?

32-A9 The half-life of ^{55}Fe is 3 years. What fraction of a sample of this nuclide remains after 12 years?

32-A10 A sample of radioactive material gives 6000 counts per minute at 1:00 P.M., and it gives 750 counts per minute at 4:00 P.M. What is its half-life?

32-A11 A reactor produces 256 μg of a certain unstable nuclide, and only 16 μg remains after 5 days. What is the half-life of the nuclide?

32-A12 A sample of ^{100}Pd (half-life 4 days) has mass 1.776 g at 3:00 P.M. on July 4; what mass of ^{100}Pd remains at 3:00 P.M. on July 12?

32-A13 Extend the table in Example 32-3 to 1:15 P.M., and verify the "million-to-one-chance" mentioned in the last line of the example.

32-A14 Identify the daughter nuclides formed, and write the equation, for each of these decays: (*a*) ^{59}Ni captures a *K* electron; (*b*) ^{40}K emits a negatron; (*c*) ^{25}Al emits a positron; (*d*) ^{210}Po emits an α particle; (*e*) ^{58}Co makes an isomeric transition and emits a γ ray.

32-A15 Write an equation for each of the following decays: (*a*) ^{20}Na emits a positron; (*b*) ^{32}P emits a negatron; (*c*) ^{226}Ra emits an α particle; (*d*) ^{55}Fe captures a *K* electron.

32-A16 The nuclide $^{57}_{27}$Co can decay by β^- emission, by β^+ emission, and by electron capture. What is the final nuclide in each case?

32-A17 ^{52}Mn decays by an isomeric transition to ^{52}Mn, with a half-life of 21 min, emitting a γ ray of energy 1.46 MeV. (*a*) Which isomer of ^{52}Mn has greater mass? (*b*) What is the mass difference between these two isomers?

32-A18 Does the nuclide ^{60}Mn, discovered in 1978, decay by β^- or by β^+ emission?

32-B1 Calculate the average BE per nucleon in a $^{14}_{7}$N nucleus.

32-B2 Calculate the BE of the least strongly bound proton in a $^{12}_{6}$C nucleus.

32-B3 The binding energy curve (Fig. 32-2) shows peaks of stability at $^{4}_{2}$He, $^{12}_{6}$C, and $^{16}_{8}$O, corresponding to 1, 3, and 4 α-particle groups. There is no peak at $^{8}_{4}$Be, as might have been expected. (*a*) Use mass data to show that $^{12}_{6}$C is stable against decay into three α particles. (*b*) Show that $^{8}_{4}$Be is unstable against decay into two α particles. The mass of $^{8}_{4}$Be is 8.005308 u.

32-B4 Compare the stability of ^{4}He and ^{7}Li as follows: (*a*) Using nuclear mass data, calculate the work required to remove one neutron from $^4_2$He to form $^3_2$He. (*b*) Again using nuclear mass data, calculate the work required to remove one neutron from $^7_3$Li to form $^6_3$Li. (*Note:* The fact that a neutron is almost three times as strongly bound to a helium nucleus as it is to a lithium nucleus is evidence for a "closed shell" in helium and explains the stability of the α particle.)

32-B5 Use Fig. 32-2 to find the approximate total binding energy of the $^{75}_{33}$As nucleus.

32-B6 4.00×10^{20} nuclei are in a sample of ^{60}Cu at 4:00 P.M.; how many ^{60}Cu nuclei are present at 4:30 P.M.? (See Fig. 32-4 for the half-life.)

32-B7 How long is required for a sample of ^{131}I (half-life 8 days) to lose $\frac{1}{4}$ of its activity?

32-B8 What fraction of the ^{235}U atoms in existence at the formation of the earth 4.5×10^9 y ago still survive? The half-life of ^{235}U is 7.1×10^8 y.

32-B9 A quantity of the iodine nuclide ^{123}I, of half-life 13 h, is manufactured in a cyclotron and shipped by air to a hospital. If the shipment requires 4 h, what fraction of the produced ^{123}I reaches the hospital?

32-B10 A certain nuclide has one chance in 4×10^5 of disintegrating during an interval of 1 s. What is the half-life, in days?

32-B11 There are 10^{11} atoms of ^{59}Ni in a sample. (a) What is the disintegration constant, in s^{-1}? (Use Fig. 32-4.) (b) About how many nuclei in this sample will disintegrate in 1 h?

32-B12 Answer the following questions for Fig. 32-4. (a) Which of the elements have only one stable isotope? (b) Which unstable nuclide has the longest half-life? (c) For which nuclide are positron emission and electron capture equally likely? (d) Are any nuclides shown that decay only by β^- decay? only by β^+ decay? only by electron capture? only by γ-ray emission? (e) Do any nuclides decay by β^--particle emission in more than one way?

32-B13 Compute the maximum energy of the negatrons emitted during the decay of a free neutron. The reaction is ${}^1_0n \rightarrow {}^1_1H + {}^{\ 0}_{-1}e + \bar{\nu}_e$.

32-B14 Tritium (hydrogen of mass number 3) decays by negatron emission. Identify the daughter nuclide, and compute the maximum energy of the emitted negatrons.

32-B15 Compute the maximum energy of the negatrons emitted during the decay of ^{14}C.

32-B16 Using data from Fig. 32-4, show that the energy of the most energetic β^- particle from ^{59}Fe is consistent with the mass difference between ^{59}Fe and ^{59}Co.

32-B17 Show that ^{50}Cr is stable in the sense that the probability of β^- decay is zero. How about α decay? Electron capture? Neutron emission? Masses of neutral atoms are as follows: ^{50}Cr, 49.94606; ^{50}Mn, 49.95422; ^{50}V, 49.94716; ^{49}Cr, 48.95127; ^{46}Ti, 45.95263.

32-B18 (a) Using mass data from Fig. 32-4, calculate the maximum energy available in the negatron decay of ^{56}Mn. (b) For this decay, find a β ray and a γ ray whose combined energy is within 0.01 MeV of the maximum energy found in part (a).

32-B19 What is the KE (in MeV) of an α particle whose speed is 1.60×10^7 m/s? (*Hint:* First convince yourself that the classical formula for KE is valid in this case.)

32-B20 The first isotope widely used in therapy was $^{226}_{88}$Ra, which decays by α-particle emission to radon, $^{222}_{86}$Rn (see Fig. 32-6). Masses of neutral atoms are as follows: $^{226}_{88}$Ra, 226.0254; $^{222}_{86}$Rn, 222.0175. What energy is released during the decay?

32-B21 According to mass spectrometer data, the masses of ^{209}Bi and ^{205}Tl are 208.9804 and 204.9745, respectively. Would these masses allow α decay? If so, with what energy would the α particle be ejected?

32-B22 Calculate the energy threshold for the ^{14}N$(\alpha, p)^{17}$O reaction that was studied at the start of Sec. 32-8.

32-B23 Show that the ^{2}H$(d, n)^3$He reaction (item 3 of Sec. 32-8) has no energy threshold and can take place for very low bombarding energies. What is the minimum total KE of the fragments in this reaction?

32-B24 Fill in the missing symbols: ^{12}C$(d, ?)^{13}$N; ^{32}S$(n, p)^?$?; $^?$Fe$(p, n)^{56}$?; $^?$?$(d, 2n)^{65}$Zn; ^{107}Ag$(n, ?)^{108}$Ag.

32-B25 Fill in the missing symbols: ^{11}B$(n, \alpha)^?$?; ^{27}Al$(\alpha, ?)^{30}$P; $^?$?$(n, \gamma)^3$H; ^{25}?$(p, \alpha)^?$Na; ^{113}Cd$(?, \gamma)^{114}$Cd.

32-B26 Slow neutrons create no ions and thus do not affect an ordinary Geiger counter. If a counter is filled with the gas boron trifluoride, the capture of neutrons by ^{10}B nuclei is made evident by a ^{10}B(n, α) reaction; the α particles are counted with high efficiency. (a) Identify the product nuclide in this reaction. (b) Calculate the energy of the particles.

32-B27 Calculate the minimum energy of an α particle that could cause the $^{14}N(\alpha, p)^{17}O$ reaction shown in Fig. 32-11 to take place. (For masses, see Sec. 32-8 or Appendix Table 9.)

32-B28 Neutrons were discovered by Chadwick using the $^9Be(\alpha, n)^{12}C$ reaction; the α particles (from a ^{210}Po source) had KE of 5.30 MeV. Calculate the total energy of the products of this reaction.

32-C1 (a) Use mass data from Fig. 32-4 to show that ^{57}Co can decay by electron capture but cannot decay by positron emission. (*Hint:* Make a table similar to that on page 764. *Caution:* Masses in Fig. 32-4 are for neutral atoms.) (b) Make a similar calculation to show that ^{56}Co can decay by positron emission as well as by electron capture.

32-C2 A possible nuclear energy-level diagram for ^{56}Fe has levels at 0 MeV (normal state of ^{56}Fe), 0.84 MeV, 2.65 MeV, and 2.96 MeV. The normal state of ^{56}Mn is 3.70 MeV above the normal state of ^{56}Fe. (a) Verify that the maximum energy available in the decay of ^{56}Mn is consistent with the mass difference given in Fig. 32-4. (b) Draw the energy levels to scale, and show numerically that all eight of the β^- and γ emissions listed in Fig. 32-4 for ^{56}Mn are accounted for by these levels.

32-C3 The nuclide ^{60}Co is commonly used in cancer therapy. (a) What total energy is available when the 5.27-y isomer decays, in two stages, to stable ^{60}Ni? (Use masses from Fig. 32-4.) (b) Show quantitatively how the total available energy is apportioned among the products of the decays.

32-C4 The 9.58-min ^{27}Mg decays with negatrons of maximum energies 1.59 MeV and 1.75 MeV. Gamma rays of energies 1.01 MeV and 0.84 MeV are also observed. The masses of the neutral atoms are as follows: ^{27}Mg, 26.98435; ^{27}Al, 26.98154. (a) What is the total energy of the decay? (b) Draw an energy-level diagram in which ^{27}Al has a ground state and two excited states.

32-C5 Using data from Fig. 32-4, construct a possible energy-level diagram in which ^{59}Co has a ground state and two excited states that can account for all six of the β^- and γ emissions from ^{59}Fe. (*Hint:* First use the mass difference to find the maximum available energy.)

32-C6 A nuclide of mass number A has an energy equivalent, in MeV, of about $931A$, and its mass can be expressed as $931A$ MeV/c^2. (a) What is the conversion factor by which masses in MeV/c^2 can be converted to kg? (b) Using Eq. 32-1, calculate the approximate density of nuclear matter in (MeV/c^2)/fm^3; in kg/m^3.

32-C7 The cloud chamber photograph (Fig. 32-11) shows the 7.68-MeV α particle to be traveling in a well-defined path. To see whether the Heisenberg uncertainty principle would predict a measurable uncertainty in position for such a particle, calculate the de Broglie wavelength of the waves associated with the motion of an α particle of energy 7.68 MeV. (*Hint:* The velocity is low enough so that you need not use any relativistic formulas.)

32-C8 Alcohol vapor is placed in a cloud chamber to provide carbon nuclei that can serve as targets for bombardment by α particles. What is the total KE of the fragments in a $^{12}_6C(\alpha, p)^{15}_7N$ reaction if the energy of the α particles is 5.66 MeV?

32-C9 A cobalt "bomb" containing 3 g of ^{60}Co is used for cancer therapy. At what rate, in watts, is heat produced in a thick-walled storage container that absorbs all radiation? (*Hint:* See the answer to Prob. 32-C3 for the energy released per disintegration.)

32-C10 How many disintegrations per second occur in a 48-g sample of pure carbon in the form of charcoal that contains radioactive carbon ^{14}C to the extent of 1 ^{14}C atom for every 10^{12} atoms of ^{12}C? The half-life of ^{14}C is 5730 years.

32-C11 This problem explores the possibility of confining slow neutrons in a metal bottle. It might seem that a neutron would easily pass through a metal wall since the atomic spacings (a few angstroms) are some 10^4 the size of a neutron. However, the neutron will see a solid wall if the atomic spacings are much smaller than the neutron's de Broglie wavelength. (*a*) Verify these ideas by calculating the de Broglie wavelength of a neutron moving at 6 m/s. (*b*) What would be the KE of the neutrons, in eV? (*c*) What would be the temperature of a collection of neutrons moving at this speed? (*Hint:* The KE per particle is $\frac{3}{2}kT$; see Sec. 16-2.)

For Further Study

32-10 Formula for Radioactive Decay

The disintegration constant λ for a radioactive decay was defined in Sec. 32-4 as the probability per unit time that a nucleus will decay. Thus,

$$\Delta N = -\lambda N \Delta t \qquad (32\text{-}2)$$

We now derive an equation for N as a function of t. First we rearrange Eq. 32-2 to obtain

$$\frac{\Delta N}{N} = -\lambda \Delta t$$

$$\lim_{\Delta N \to 0} \sum \frac{\Delta N}{N} = -\lambda \left(\lim_{\Delta t \to 0} \sum \Delta t \right)$$

$$\int_{N_0}^{N} \frac{dN}{N} = -\lambda \int_0^t dt$$

The initial number of nuclei is N_0, and the lower limits are chosen to conform to the fact that $N = N_0$ when $t = 0$.

We carry out the integrations using the table of integrals in Appendix E of the Mathematical Review (page 839). The result is

$$[\ln N]_{N_0}^{N} = -\lambda [t]_0^t$$
$$\ln N - \ln N_0 = -\lambda t \qquad (32\text{-}8)$$
$$\ln \frac{N}{N_0} = -\lambda t$$
$$\frac{N}{N_0} = e^{-\lambda t}$$
$$N = N_0 e^{-\lambda t}$$

where e is 2.718, the base of natural logarithms. This is Eq. 32-3.

Much the same analysis can be applied to exponential *growth*. Suppose the number of bacteria in a culture increases in proportion to the number already present. We need only replace $-\lambda$ by a growth constant $+k$ to obtain

$$N = N_0 e^{kt}$$

Such an expression has wide application in biology, ecology, population studies, and finance.

Growth and decay curves are most conveniently plotted on semilog paper to obtain straight lines for the graph of $\log N$ versus t (Fig. 32-3c).

32-11 Conservation of Angular Momentum in β Decay

Let us consider the decay of $^{55}_{24}\text{Cr}$ once again, this time paying attention to the angular momentum of the particles. The nucleus consists of 24 protons and 31 neutrons. In terms of $h/2\pi$, the intrinsic angular momentum (spin) of each proton is $\frac{1}{2}$, and the spin of each neutron is also $\frac{1}{2}$ (see Table 33-5). We make a model in which the spin axes of the particles inside the nucleus are parallel to each other but can point in either direction. Thus, the total spin of all 55 particles *could* be $\frac{1}{2}, \frac{3}{2}, \frac{5}{2}, \ldots, \frac{55}{2}$, always an odd multiple of $\frac{1}{2}$ since there is an odd number of particles. However, it has been found that in the ground state (the state of lowest energy), the spins of the nucleons in any nucleus are paired, that is, aligned as nearly as possible in opposite directions. In the ground state, the spin is either 0 (for an even number of nucleons) or $\frac{1}{2}$ (for an odd number of nucleons). Therefore, the resultant *spin* of $^{55}_{24}$Cr is $\frac{1}{2}$. In addition, the nucleons have a possible orbital angular momentum represented (just as for outer electrons) by a quantum number (denoted by J) with integral values 0, 1, 2, 3, ..., so that the total *orbital* angular momentum must be an integral multiple of $h/2\pi$. Therefore, considering both spin and orbital angular momentum, the total must be $\frac{1}{2}, \frac{3}{2}, \frac{5}{2}, \frac{7}{2}, \ldots$, some odd multiple of $\frac{1}{2}$. Of all these possibilities, it appears that the angular momentum of $^{55}_{24}$Cr is $\frac{3}{2}$. Similarly, the total angular momentum of the $^{55}_{25}$Mn nucleus must be an odd multiple of $\frac{1}{2}$; experiment shows it to be $\frac{5}{2}$.

Now we consider the process of β decay. It is known that the electron has $\frac{1}{2}$ unit of spin (this applies to both the negatron and the positron). We now write the equation for the β decay of $^{55}_{24}$Cr:

$$^{55}_{24}\text{Cr} \rightarrow {}^{55}_{25}\text{Mn} + {}^{0}_{-1}e + \bar{\nu}_e \qquad (32\text{-}9)$$

$$\tfrac{3}{2} \rightarrow \tfrac{5}{2} + \tfrac{1}{2} + ? \qquad (32\text{-}10)$$

We have already postulated neutrinos (ν_e) and antineutrinos ($\bar{\nu}_e$) to help conserve energy in β decay; now we see that these particles are equally important for the conservation of angular momentum. Equation 32-10 is a vector equation, and angular momentum can be conserved if the antineutrino's spin is $\frac{1}{2}$ and if the negatron and the antineutrino are each emitted with their spins opposite to that of the $^{55}_{25}$Mn nucleus. Thus, using $+$ and $-$ signs to indicate possible directions of the spins, we can write

$$\tfrac{3}{2} \rightarrow \tfrac{5}{2} + (-\tfrac{1}{2}) + (-\tfrac{1}{2})$$

While it is true that a spin of $\frac{3}{2}$ for the antineutrino would also be possible for the reaction of Eq. 32-9 ($\frac{3}{2} = \frac{5}{2} + \frac{1}{2} - \frac{3}{2}$), other evidence indicates the value $\frac{1}{2}$ for the spin of both the neutrino and the antineutrino. It is worthy of note that the neutrino must have *some* half-integral spin; under no circumstances could a spin of 0 or 1 satisfy the vector equation 32-10, which illustrates the law of conservation of angular momentum for a nuclear process.

Problems

32-C12 Use the law of conservation of angular momentum to show that a neutrino is emitted during electron capture.

32-C13 The total angular momentum of $^{49}_{22}$Ti in its ground state is known to be $\frac{7}{2}$ units. What is the orbital angular momentum of the nucleons in this nucleus?

References

1. W. F. Magie, *A Source Book in Physics* (McGraw-Hill, New York, 1935), pp. 610–613, Becquerel's experiments with uranium; pp. 613–616, discovery of polonium and radium by the Curies.
2. S. Devons, "Recollections of Rutherford and the Cavendish," *Physics Today* **24**(12), 39 (Dec. 1971).
3. G. E. M. Jauncey, "The Early Days of Radioactivity," *Am. J. Phys.* **14**, 226 (1946).
4. G. B. Collins, "Scintillation Counters," *Sci. American* **189**(5), 36 (Nov. 1953).
5. D. A. Glaser, "The Bubble Chamber," *Sci. American* **192**(2), 46 (Feb. 1955).
6. G. K. O'Neill, "The Spark Chamber," *Sci. American* **206**(2), 37 (Aug. 1962). Discusses Geiger counter, cloud chamber, and bubble chamber.
7. G. Charpak, "Multiwire and drift proportional chambers," *Physics Today* **31**(10), 23 (Oct. 1978).
8. P. Morrison, "The Overthrow of Parity," *Sci. American* **196**(4), 45 (Apr. 1957).
9. F. Miller, Jr., *Radioactive Decay; Scintillation Spectroscopy* (films).

33

Applied Nuclear Physics

33-1 Some Uses of Radioactivity

We have described the basic physical facts about the nucleus in Chap. 32; now we look at some of the ways in which this knowledge has been put to practical use. Some of the most spectacular engineering of the past few decades has been nuclear engineering, and more is yet to come. In this chapter we shall study devices capable of producing large-scale conversion of rest mass into energy. Our first topic in applied nuclear physics, however, deals with some applications of radioactivity to scientific investigation and therapy.

Every element has at least one unstable isotope; if its half-life is not too short or too long, such an isotope can be used in many ways. All high-energy radiations to some extent damage biological tissues. These effects depend in a complicated way on the nature of the tissue and on the kind and energy of the radiation. In general, the damage is related to the production of ions in the tissue. It was early found that rapidly growing cancerous cells are more susceptible than normal cells to γ rays. Thus, from

the beginning of the 20th century, impure radium and its daughter radon were used for cancer therapy (most of the effect was from the γ rays emitted by ^{214}Pb and ^{214}Bi, which are descendants of ^{226}Ra; see Fig. 32-6). The patient may recover if the dose is sufficient to kill all the cancerous cells while killing only a fraction of the intermingled normal cells. Since the advent of cyclotrons and nuclear reactors, strong samples of many radioactive nuclides, such as ^{60}Co (half-life 5.3 years), can be manufactured and used for therapy. In fact, ^{60}Co can replace radium for almost all purposes; its γ rays have energies of 1.2 and 1.3 MeV (see Fig. 32-4). The iodine isotope ^{131}I (8 days, 0.4 MeV) is used to treat cancer of the thyroid gland; it is selectively absorbed by the thyroid, and so the radiation is concentrated right where it is needed. These are just a few of the nuclides now being used for therapy.

Many unstable nuclides are used as *tracers*. One example is in the study of the wear of piston rings in an automobile engine. Before the piston ring is placed in the engine, it is irradiated with

Experimental Breeder Reactor II, shown prior to adding liquid coolant. Six vertical control rods and hundreds of fuel elements are visible here. The EBR-II was designed by Argonne National Laboratory for experimental use rather than as a demonstration plant.

neutrons, and some ^{59}Fe is formed by the $^{58}Fe(n, \gamma)^{59}Fe$ reaction. As the piston rings wear during the test, some of the iron atoms that rub off into the oil are ^{59}Fe, which can be detected when the oil is later tested with a Geiger counter. The method is far more sensitive than chemical analysis of the oil would be, and only a few minutes are required. In addition, the only radioactive iron in the oil must have come from the piston ring, and iron from other parts of the engine does not influence the analysis.

Many compounds can be "tagged" with an unstable atom and traced through a chemical reaction. For instance, the methane-producing bacterium *Methanobacterium omelianski* produces acetic acid and methane from ethyl alcohol and carbon dioxide according to the reaction

$$2C_2H_5OH + CO_2 \rightarrow 2CH_3COOH + CH_4$$

Into which of the product molecules has the carbon of the CO_2 gone? By making the CO_2 with a radioactive isotope of carbon, it has been shown that the "labeled" carbon goes into the methane (CH_4), none of it appearing in the acetic acid (CH_3COOH). The long-lived carbon isotope ^{14}C (half-life 5730 years) is of great importance for biological research, as is tritium, the hydrogen isotope ^3H (half-life 12 years).

In the following example, the number of "labeled" red blood cells decreases as the old cells die and are replaced by new (unlabeled) ones. Both biological and radioactive decays are involved in this clinical procedure, since for medical reasons the diagnosis must be completed in only a few days, which is less than the half-life of the injected radioisotope tracer element.

_____ **Example 33-1**

The red cells of a patient are labeled by injecting into the bloodstream a compound containing ^{51}Cr (radioactive half-life 27.8 d, as given in Table 32-1). The labeling takes place because some of the Fe in the hemoglobin is replaced by ^{51}Cr. After allowing about a day for labeling and mixing to take place in

the bloodstream, the following data are taken for the activity A in cpm (counts/min) of 5 cm^3 of blood at various times: $t = 0.0$, $A = 210$ cpm; 2.0 d, 188 cpm; 4.0 d, 167 cpm; 6.2 d, 148 cpm. Calculate the biological half-life of a red cell in this patient's bloodstream.

First we must correct each observed activity A to what it would have been at $t = 0.0$ d if there had been no radioactive decay. The disintegration constant of ^{51}Cr is

$$\lambda = \frac{0.693}{T_{1/2}} = \frac{0.693}{27.8 \text{ d}} = 0.0249 \text{ d}^{-1}$$

Hence, since $A = A_0 e^{-\lambda t}$, we have $A_0 = A e^{+\lambda t}$ For example, at $t = 4.0$ d,

$$A_0 = 167 \, e^{(0.0249)(4.0)} = 167 \, e^{0.0996}$$
$$= 167(1.105) = 184 \text{ cpm}$$

In this way we construct the following table:

t, days	A, cpm	$e^{\lambda t}$	A_0, cpm	$\log A_0$
0.0	210	1.000	210	2.32
2.0	188	1.051	198	2.30
4.0	167	1.105	184	2.27
6.2	148	1.167	173	2.24

If you make a graph of $\log A_0$ versus t, as in Fig. 32-3c, you will find that $\log A_0$ decreases by 0.30 (that is, log 2) in about 24 d. This result lies within the normal range of 20 to 40 days for the biological half-life of a red cell in the bloodstream.

Tracer experiments can also be performed using stable isotopes; for instance, a protein can be enriched in ^{15}N, which is normally present to an extent of only 0.4% (see Appendix Table 9). The biological utilization of the protein can be determined by analyzing nitrogen from various tissues, using a mass spectrometer. In general, work with radioactive tracers is much easier than with stable isotopes.

The unchanging disintegration rate of an unstable nuclide can serve as an atomic clock reaching far back into the past. An early example was the determination of the age of the earth (more precisely, of the time that has elapsed since the solidification of the crust). As shown in

Fig. 32-6, ^{238}U (half-life 4.5×10^9 years) ultimately forms the stable lead isotope ^{206}Pb. Thus if a uranium ore contains only a little ^{206}Pb, it must be relatively young. Careful analysis using several long-lived nuclides, including the potassium isotope ^{40}K (half-life 1.3×10^9 years), gives a consistent figure of about 5×10^9 years for the age of the earth's crust. The carbon isotope ^{14}C is of shorter half-life (5730 years) and is useful for dating archeological finds, such as a wooden coffin in a Pyramid or prehistoric charcoal or cloth. Because of its short half-life, no primeval ^{14}C remains, but it is constantly being created in the atmosphere by the action of cosmic rays in a $^{14}N(n, p)^{14}C$ reaction. Thus, a certain small fraction of CO_2 molecules in the air contain ^{14}C; most of them, of course, contain the ordinary stable isotope ^{12}C. Living plants take up this same small fraction of ^{14}C, and animals feed on the plants. Thus all *living* organisms contain about 1 atom of ^{14}C for every 10^{12} atoms of ^{12}C. When the organism dies, however, the ^{14}C decays to ^{14}N by negatron emission and is not replaced. The time since the specimen died can be determined by measuring the ratio of ^{14}C to ^{12}C.

One more example illustrates the usefulness of radioactive-dating methods. Certain meteorites fell in Kansas and Nebraska in 1948. One of the stony fragments was analyzed by the ^{40}K method and found to be about 4.2×10^9 years old, approximately the same as the age of the earth. By studying the amount of ^3He formed by cosmic-ray action during the stone's flight through space, it was determined that the meteorite had been exposed to cosmic rays for only 0.28×10^9 years. The conclusion was that the rock had been formed a few billion years ago and lay buried in a planet or asteroid, which broke into fragments a few hundred million years ago. Thus, nuclear science in this case supports the idea of a catastrophic collision in space, perhaps the breakup of a planet to form the asteroids that lie mostly between the orbits of Mars and Jupiter.

33-2 Fission

As we all know, nuclear fission releases energy, and the process can be made self-sustaining in a chain reaction. Fission was discovered in 1939 by the German chemists and physicists Hahn, Strassmann, Meitner, and Frisch; Hahn's own account (see Ref. 5 at the end of the chapter) makes interesting reading and shows that other workers of the period were on the verge of the discovery. The "liquid drop" model of the nucleus, due to Bohr, pictures the nucleus as a sort of drop of liquid able to become constricted and to oscillate. The addition of one more nucleon might cause a nucleus to oscillate too wildly, resulting in a breaking-up process called *fission*—the splitting of a nucleus into relatively large fragments. The typical and most familiar fission reaction takes place when slow neutrons strike ^{235}U:

$$^{235}_{92}U + ^1_0n \rightarrow \left[^{236}_{92}U \right] \rightarrow$$
$$A + B + \text{neutrons} + \text{energy}$$

The compound nucleus $\left[^{236}_{92}U \right]$ splits into two fragments A and B, with some neutrons also released during the fission.

To see how this works out, let us first make the simple assumption that the nucleus splits into two equal fragments and that no neutrons are emitted. Half of 92 is 46, and the element of atomic number 46 is palladium. Thus a conceivable reaction would be

$$^{235}_{92}U + ^1_0n \rightarrow ^{118}_{46}Pd + ^{118}_{46}Pd$$

Reference to a table of nuclides shows that the stable Pd isotopes range from $^{102}_{46}Pd$ to $^{110}_{46}Pd$; $^{118}_{46}Pd$ would have at least eight extra neutrons and in fact is so unstable that it has never been observed. A more likely possibility would be

$$^{235}_{92}U + ^1_0n \rightarrow ^{110}_{46}Pd + ^{110}_{46}Pd + 16\,^1_0n$$

This reaction does take place, but experiment shows that the two fragments are most likely to be of somewhat different mass.

Another fission reaction is

$$^{235}_{92}U + ^1_0n \rightarrow ^{90}_{38}Sr + ^{136}_{54}Xe + 10\,^1_0n$$

This is just one of hundreds of possible fission reactions; it is important for human survival because the strontium isotope $^{90}_{38}Sr$ is a negatron emitter with a half-life of 28 years and is a major constituent of fallout from nuclear explosions. To see whether the reaction is possible, we check the masses.

Before fission

1 neutron	1.0087 u
$^{235}_{92}U$	235.0439
	236.0526 u

After fission

$^{90}_{38}Sr$	89.9073 u
$^{136}_{54}Xe$	135.9072
10 neutrons	10.0867
	235.9012 u

Mass difference

$$0.1514\ u = 141\ MeV$$

Since the total rest mass of the system *decreases*, the reaction is possible. The mass calculation shows that 0.1514 u of rest mass disappears during this fission. At the rate of 931 MeV/u, this is equivalent to about 141 MeV of energy for each fissioning ^{235}U nucleus. Other fission reactions release more or less energy from ^{235}U; the average is about 200 MeV per atom of ^{235}U.

It is the release of neutrons that makes the chain reaction possible. The number of neutrons per fission is considerably less than ten, for in many cases the fission products convert their excess neutrons into protons by negatron emission. Nevertheless, more than one neutron is formed, on the average, and so the process builds up. If the multiplication factor were exactly 2, then one incoming neutron would release two neutrons, which in turn would strike two ^{235}U nuclei and release four neutrons, and so on until an explosion resulted. For controlled release of nuclear power, which is what we want from a nuclear reactor, after startup the multiplication factor is adjusted to be exactly equal to 1. Uranium-235 is most easily fissioned by a neutron that is moving slowly, for in this way the neutron has more time to interact with the nucleus. We saw in Chap. 16 that the average translational KE per molecule of any solid, liquid, or gas is given by $\frac{3}{2}kT$, which works out to be about 0.038 eV at room temperature. Neutrons of about this energy are called slow neutrons, or *thermal neutrons*.

To understand the operation of a nuclear reactor, we first consider what might happen when a neutron strikes a nucleus. Several possibilities exist. (1) The neutron may simply bounce off, transferring some of its KE to the target nucleus. This is called a scattering process. (2) The neutron may be captured, probably with emission of γ rays, in a (n, γ) reaction. (3) The neutron may induce some nuclear reaction such as (n, α) or (n, p). (4) Finally, the neutron may be like the straw that broke the camel's back and induce fission. The probabilities of these events are measured in terms of *cross sections*. The atoms of a solid are about 3×10^{-10} m apart, and so to each atom there belongs an area (for an oncoming neutron) of about $(3 \times 10^{-10}\ m)^2$, or about $10^{-19}\ m^2$. Similarly, the "geometrical cross sections" of the nuclei range from about $0.4 \times 10^{-28}\ m^2$ for $A = 20$ to about $1.8 \times 10^{-28}\ m^2$ for $A = 240$, where A is the mass number. To deal with such small areas we use the *barn*, defined as a unit of area equal to $10^{-28}\ m^2$. We see that a *nucleus* has a geometrical cross section of about 1 barn, and an *atom* has a geometrical cross section of about 10^9 barns. If only 1 out of 1000 nuclei scatters a slow neutron, we say that the cross section for scattering is $\frac{1}{1000}$ of the geometrical cross section, that is, about $\frac{1}{1000}$ barn. The cross section thus is the "effective area" for any given process, and the same nucleus may have a number of different cross sections for different processes.

Table 33-1 gives some cross sections that are important for reactor design. Nuclei differ widely in their capture cross sections; the very large capture cross section for $^{113}_{48}Cd$ means that in

Table 33-1   Nuclear Cross Sections

Nuclide	Cross Section for Capture of Slow Neutrons, barns	Cross Section for Fission by Slow Neutrons, barns	Half-Life for α Emission
$^{1}_{1}\text{H}$	0.33		
$^{2}_{1}\text{H}$	0.0006		
$^{10}_{5}\text{B}$	4000		
$^{12}_{6}\text{C}$	0.0037		
$^{113}_{48}\text{Cd}$	27000		
$^{232}_{90}\text{Th}$	7.5	0	1.4×10^{10} y
$^{233}_{92}\text{U}$	50	530	1.6×10^{5}
$^{235}_{92}\text{U}$	100	580	7.1×10^{8}
$^{238}_{92}\text{U}$	2.7	0	4.5×10^{9}
$^{239}_{94}\text{Pu}$	270	740	2.4×10^{4}

effect this nucleus "reaches out" and captures slow neutrons that do not really strike within its geometrical area. This property of one cadmium isotope (comprising 12% of naturally occurring cadmium) makes cadmium useful as a shield or absorber of slow neutrons (see Fig. 33-1). The boron nuclide $^{10}_{5}\text{B}$ is also used for the same pur-

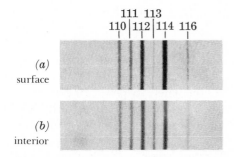

(a)
surface

(b)
interior

Figure 33-1   Mass spectra of cadmium, showing that the large cross section for absorption of neutrons (Table 33-1), is due to a $^{113}\text{Cd}(n, \gamma)^{114}\text{Cd}$ capture reaction. (a) Cadmium scraped from the surface of an irradiated sample of the metal; ^{113}Cd has been changed to ^{114}Cd by neutron capture. (b) Cadmium from interior of sample, to which neutrons have not penetrated; normal abundance of isotopes. (A. J. Dempster, from R. E. Lapp and H. L. Andrews, *Nuclear Radiation Physics*, Prentice-Hall, 1948.)

pose; in this case the neutrons are removed by a (n, α) reaction. The high fission cross section of ^{235}U indicates that this isotope is useful in nuclear reactors. Unfortunately, it occurs only to the extent of 1 part in 140; ordinary uranium is almost all ^{238}U, and ^{233}U has such a short half-life that it is not found in nature. Several nuclides, including ^{238}U, undergo fission when bombarded with *fast* neutrons, but such reactions cannot be controlled and are useful only for explosive purposes.

We are ready now to consider briefly some problems to be overcome in constructing a nuclear reactor. The fuel elements of commercial reactors in the United States are slugs of fissionable material, usually uranium oxide that has been enriched to contain about 3% ^{235}U. The fuel slugs are surrounded by a *moderator*, such as water or carbon (graphite), whose function is to slow down the fast neutrons released during fission. As discussed in Sec. 6-9, the most effective way to slow down a neutron to thermal velocity is to let it strike another particle of mass equal to that of the neutron. The proton would be ideal for the purpose if it did not have such a large capture cross section (Table 33-1). Using ordinary water as a moderator is inefficient, for many of the precious fission neutrons are lost by absorption in the reaction $^1H(n, \gamma)^2H$, forming deuterium. The best practical moderator is heavy water containing deuterium, for which the capture cross section is small. The Canadian deuterium-uranium reactors (CANDU) successfully use ordinary (unenriched) uranium and a heavy-water moderator. No isotopes of He, Li, Be, or B are suitable moderators. The abundant isotope of carbon, ^{12}C, has a satisfactorily low capture cross section, and some reactors use graphite moderators. Carbon of extreme purity is required; traces of boron will use up neutrons and poison the carbon as a moderator.

Before the reactor can operate, it must be built up to some critical size, for otherwise too many neutrons will escape before they have had a chance to cause a second fission. The critical size for a ^{235}U reactor using heavy water as a moderator is surprisingly small; one design is 21 cm in radius and uses only 242 g of ^{235}U. Rods of cadmium or boron are inserted in the reactor to control the multiplication factor and hence control the power level. To turn the reactor on, these rods are pulled out just far enough to allow the multiplication factor to become equal to 1; to shut down, the rods are dropped back in. Neutron-absorbing control rods are visible in the photo on page 781, which shows the interior of a reactor before filling with moderator.

As a research tool, nuclear reactors supply intense beams of neutrons. Thus, many useful unstable nuclides can be prepared simply by placing a flask of the target material in the vicinity of an operating reactor for a few days or months. The uranium slugs (which are sealed in cans) must be removed periodically and purified of the fission products; these products can be separated to give many radioactive nuclides of medium atomic weight.

Until 1940, the Periodic Table ended with uranium, element number 92. Thirteen new elements have since been produced and identified, the latest being hahnium, of atomic number 105. The production of two of these new elements, neptunium and plutonium, is of interest. In a reactor using natural uranium, the chain reaction is kept going by the fission of the relatively few ^{235}U nuclei present. The abundant isotope ^{238}U does not undergo slow neutron fission, but these nuclei do capture neutrons, forming unstable ^{239}U, which forms neptunium and plutonium as follows:

$$^{238}_{92}U + ^1_0n \rightarrow ^{239}_{92}U$$
$$^{239}_{92}U \rightarrow ^{239}_{93}Np + ^{0}_{-1}e \quad \text{(24 min)}$$
$$^{239}_{93}Np \rightarrow ^{239}_{94}Pu + ^{0}_{-1}e \quad \text{(2.33 days)}$$

The plutonium isotope $^{239}_{94}Pu$ is an α emitter of half-life 24,000 years; it is therefore stable enough to be separated chemically from uranium, neptunium, and all the fission products in the reactor. It happens that $^{239}_{94}Pu$ is fissionable by slow neutrons (see Table 33-1), and so it can serve the same purposes as ^{235}U. A uranium reactor could thus convert all the "inert" ^{238}U into "useful" ^{239}Pu. This process, called *breeding*, would effectively extend the world supply of fissionable uranium by a factor of over 100. A similar breeding process could convert the plentiful $^{232}_{90}Th$ into fissionable $^{233}_{92}U$ of half-life 160,000 years. However, the technical problems of large-scale breeder reactors have not yet been overcome.

The application of fission to nuclear weapons is well known. A firing mechanism brings together several small pieces of ^{235}U or ^{239}Pu metal, and an explosive release of energy takes place when the critical mass is thus formed. The production of nuclear power for peaceful uses is indirect. In one design, water pipes are buried in the reactor, and the thermal energy of the moderator is used to make steam that turns a generator. The technology of nuclear power is being rapidly developed, and electric power generated by nuclear reactions has become economically competitive with conventional sources of power, which are also being developed to higher efficiencies. During the next few decades we look forward to a breakthrough in this field. What is needed is a method of converting thermal energy directly into electric energy without the intervening low-efficiency processes. Even more desirable would be a method of converting nuclear energy directly into electric energy without becoming involved at all with heat engines and their inherently low (Carnot-limited) efficiencies.

33-3 *Fusion*

Release of nuclear energy also takes place when the lighter nuclei are joined together. This process of *fusion* is represented by the simple case of the capture of a neutron by a proton, the process that lowers the efficiency of ordinary water as a moderator in a fission reactor. The reaction is

$$_1^1H + _0^1n \to _1^2H$$

Using masses from the table of light nuclides in Appendix Table 9, it is easy to show that 0.00239 u disappears during this capture. Therefore, the energy release is 931×0.00239, or 2.23 MeV; this energy is available for KE of the deuteron and for the emission of γ rays. The conversion of rest mass into energy also takes place during other fusion reactions between light nuclei. Some reactions involving deuterium and tritium are

(a) $_1^2H + _1^2H \to _1^3H + _1^1H$ (4.0 MeV)

(b) $_1^2H + _1^2H \to _2^3He + _0^1n$ (3.3 MeV)

(c) $_1^2H + _1^3H \to _2^4He + _0^1n$ (17.6 MeV)

(d) $_1^3H + _1^3H \to _2^4He + 2_0^1n$ (11.3 MeV)

(e) $_1^2H + _2^3He \to _1^1H + _2^4He$ (18.4 MeV)

A desirable fusion reaction is one in which four protons combine to form helium plus two positrons:

$$_1^1H + _1^1H + _1^1H + _1^1H \to _2^4He + 2_{+1}^0e$$

$$(26.7 \text{ MeV})$$

The reaction is really between nuclei, not neutral atoms. That is,

$$4p \to \alpha + 2_{+1}^0e + \text{energy}$$

To obtain an equation involving the masses of the neutral atoms, we add 4 electrons to each side

of this equation:

$$(4p + 4e) \to (\alpha + 2e) + (2_{+1}^0e + 2_{-1}^0e)$$

$$+ \text{energy}$$

or

$$4_1^1H \to _2^4He + \text{energy} + \text{energy}$$

In writing the last equation, we have made use of the fact that the two positrons and the two negatrons will come together and be annihilated, transforming their masses entirely into energy. Thus the total rest mass converted into energy during this fusion can be computed as the difference between the mass of four neutral hydrogen atoms and one neutral helium atom.

Before fusion

     4_1^1H 4.03130 u

After fusion

     $_2^4He$ 4.00260

Mass difference $\overline{0.02870 \text{ u}} = 26.7$ MeV

Spectroscopic evidence shows that the sun consists almost entirely of hydrogen, with some helium and traces of heavier elements. The energy radiated by the sun and other stars must come from nuclear fusion reactions, but the simultaneous collision of four protons would be rare indeed, even inside the sun. However, several chain reactions are possible in which the $_2^4He$ is built up one step at a time. The *proton-proton chain* forms helium out of hydrogen as follows:

(f) $_1^1H + _1^1H \to _1^2H + _{+1}^0e$

(g) $_1^2H + _1^1H \to _2^3He$

(h) $_2^3He + _1^1H \to _2^4He + _{+1}^0e$

       or

(i) $_2^3He + _2^3He \to _2^4He + 2_1^1H$

Another possible chain starts with $_6^{12}C$ and builds up helium out of hydrogen in a series of steps. The $_6^{12}C$ is regenerated in the last step, and so it is not used up in the reaction. This *carbon cycle* is discussed in Prob. 33-B4.

The basic reactions *f*, *g*, and *h* have been studied in the laboratory, and the cross sections are known. We must bear in mind that these experiments use projectiles of energies of 1 MeV or so;

such energies are needed so that protons can approach within nuclear distances of other protons that are also positively charged. At best, in the laboratory, we use beams of low intensity, perhaps just a few billion projectiles per second, and make up in efficiency (that is, in probability or cross section) what we lack in numbers. Inside the sun, the situation is reversed. The energies are very low in comparison with those of protons produced by a cyclotron, and so the cross sections are extremely small. However, the sun is almost all hydrogen, and the collisions occur so often that many reactions take place per second, in spite of the small probability. To fix these ideas, let us compute the average energy of a proton inside the sun.

Example 33-2

Calculate the average energy of a proton at the center of the sun where the temperature is 15,000,000 K.

Using Eq. 16-6, we find

$$\text{KE per mole} = \tfrac{3}{2}RT$$

$$= \tfrac{3}{2}(8.31 \text{ J/mol·K})(1.5 \times 10^7 \text{ K})$$

$$= 18.7 \times 10^7 \text{ J/mol}$$

$$\text{KE per atom} = \frac{18.7 \times 10^7 \text{ J/mol}}{6.02 \times 10^{23} \text{ atoms/mol}}$$

$$\times \left(\frac{1 \text{ eV}}{1.60 \times 10^{-19} \text{ J}}\right)$$

$$= \boxed{1.94 \times 10^3 \text{ eV/atom}}$$

The average KE of a proton inside the sun is only 1940 eV, or 0.00194 MeV; the sun is a puny atom-smasher indeed. Nuclear reactions at such low energies are unobservable on earth. It is a triumph of nuclear theory that the predicted cross sections for such reactions give just the observed amount of radiation from the sun. To supply this energy, the sun loses 4 million tons of mass per second, through fusion reactions such as we have described.

33-4 Nuclear Power

In this section we compare and contrast fission and fusion power reactors in the light of both present technology and long-range outlook.

Fission Power. In Sec. 33-2 we described the physical principles of the nuclear reactor that uses the fission of ^{235}U (or, in principle, ^{239}Pu or ^{233}U). Control of the process demands use of slow neutrons that can be absorbed by cadmium or boron control rods. The power appears as thermal energy, and hence a series of auxiliary heat exchangers, steam boilers, and turbines are necessary to produce easily managed electrical or mechanical power. We list some of the uses of fission reactors: (1) production of ^{239}Pu and ^{233}U by breeding, starting from inert ^{238}U and ^{232}Th; a breeder reactor could actually produce more ^{239}Pu than the amount of ^{235}U used to keep the reactor going; (2) production of artificially radioactive nuclides for tracer studies and medical therapy; (3) production of neutron beams of high intensity for use in experimental work; (4) production of power for turning electric generators or for propelling ships and submarines; (5) production of vast quantities of thermal energy for desalinization of sea water.

In the immediate future, the only viable source of power to supplement that from fossil fuels is nuclear fission. In 1981, about 72 reactors had the capacity of producing 50 GW of electric power in the United States; the capacity of more than 100 reactors then having construction permits was an additional 100 GW. Fission reactors are an important part of the *short-term* solution to the energy shortage, but this shortage must also be approached through greater emphasis on conservation and on the development of solar power.

The prospects are not good for the *long-term* success of nuclear fission as a source of nuclear power. Those who argue against reliance on fission list a number of disadvantages: (1) Uranium ore is not inexhaustible; it is limited in quantity, and the price is rising sharply. (2) For efficiency in the reactor, natural uranium should be enriched in ^{235}U, but a typical gaseous-diffusion separation plant to do this would cost \$2 or \$3 billion to construct, would consume enough power to supply a city of half a million, and would utilize millions of gallons of cooling water each day. (3) The risk of a nuclear accident is small but not negligible. Even if such an accident does not harm the public, the decontamination and reconstruction of a damaged reactor is costly and time consuming. (4) The safe handling of fission products has not yet been achieved on a commercial scale and will prove to be extraordinarily difficult, if indeed it is possible at all. It is estimated that if all the power needs of the United States today were supplied by fission reactors, the nuclear waste produced each year would have an activity equal to the fallout from some 200,000 atomic bombs.* It is not yet known how to dispose of this nuclear waste, keeping it safe for perhaps 10^6 years into the future. (5) Most deadly, and at the same time most valuable, of the nuclear wastes is ^{239}Pu, of half-life 24,000 y. There is considerable uncertainty about the effects of plutonium, but it is one of the most lethal substances known, primarily because of its long-lived radioactivity. If the fission economy comes into existence, one can anticipate many shipments each year of 50- to 75-kg of loads of plutonium, transported by trucks or by rail. (6) Human nature being what it is, it is impossible to rule out theft or diversion of plutonium while in transit. The prospect of nuclear blackmail or terrorism must be considered since absolutely perfect security is next to impossible. (7) A final and very serious objection to nuclear fission power is that it brings into existence the uranium and plutonium isotopes used in nuclear weapons. This increases the risk of proliferating nuclear weapons in all

countries, making catastrophic nuclear warfare more likely.

Fission power from *breeder reactors* is ingenious and theoretically of great potential. The breeder reactor is the only long-term solution to the energy problem that uses fission. However, according to critics, the attempt to develop the breeder reactor as a significantly large energy source may pose insuperable problems. With its reliance on the recycling of plutonium, the liquid-metal fast breeder reactor[†] is inherently less desirable in every aspect except the economic one (by which it extends the fuel supply to include the plentiful ^{238}U as well as the rare ^{235}U). In any case, breeder reactors will not be in operation in the United States before 1993 or later, although demonstration breeder reactors went into operation in the mid-1970s in France, the United Kingdom, and the Soviet Union.

The consumption of electric power in the United States is expected to double by the year 2000, and it is doubtful if oil, coal, gas, and hydroelectric installations can supply this much power. It is also unlikely that users will voluntarily cut back their increasing demand for power. Therefore, it is reasoned by many that fission power, despite its hazards, is a necessary stopgap while fusion power is being developed.

Fusion Power. Controlled fusion reactors present a different set of problems that have not yet been solved. *Fission* chain reactions are possible at low temperatures only because of the release of neutrons during fission, and because of the fortunate circumstance that slow neutrons can approach close to the nucleus without suffering electrostatic repulsion. In a *fusion* reactor, charged particles will have to have high energies to be able to come close enough to react. This calls for high temperatures—in fact, the problem is nothing less than constructing an artificial star, at a temperature of millions of degrees. At such temperatures, all atoms break down into

---

* The standard of comparison is the bomb exploded at Hiroshima in 1945, which was equivalent in energy to 20 kilotons of TNT (8×10^{13} J).

[†] Fast neutrons are used; the coolant is liquid sodium metal.

plasma, which consists of a mixture of nuclei and free electrons. Fortunately, the density of such a plasma need not be great, and it is estimated that a powerful fusion reactor would use a plasma that is less dense than the air we breathe. The problem is how to contain such a hot plasma long enough that the particles can interact before touching any material object (such as a wall) where they would lose their thermal energy.

One broad class of containment schemes uses magnetic fields shaped to exert suitable forces on the moving plasma particles (through the basic equation $\mathbf{F} = Q\mathbf{v} \times \mathbf{B}$). It was shown by J. D. Lawson that for a self-sustaining reaction to be possible, the product $N\tau$ must be at least 10^{14} s/cm^3, where N is the plasma density in particles/cm^3 and τ is the containment time. In 1976, a value of $N\tau$ of about 1.3×10^{13} s/cm^3 was achieved at MIT in a device called Alcator; this device is used primarily for basic plasma research but gives a valuable clue to reactor design.

An entirely different containment scheme uses solid pellets of fuel rather than a plasma. The pellets are suddenly heated and vaporized when they absorb energy from a laser flash or a beam of electrons. The confinement time is related to the inertia (mass) of the vapor particles, which simply cannot disperse fast enough. This inertial confinement scheme, in effect, generates many controlled microexplosions per second. So far, scientists have not been able to reach the Lawson criterion of 10^{14} s/cm^3, although the goal seems to be within grasp, for both magnetic confinement and inertial confinement.

The fusion processes are called *thermonuclear reactions* because of their resemblance to the ordinary burning of coal or wood. There is no critical size, except possibly to allow for heat leakage, and no theoretical upper limit to the rate at which energy can be produced from mass. It is only necessary to produce a high temperature in a controlled thermonuclear reactor (CTR), and then (figuratively) to shovel in the fuel, as much or as little as may be desired.

The rate of release of nuclear energy depends on the temperature, and the fuel "ignites" if this rate equals or exceeds the rate of energy loss through radiation, conduction through the walls, and other processes. The ignition temperatures are calculated to be 350×10^6 K for the deuteron-deuteron reaction, and only 50×10^6 K for the deuteron-triton reaction. The nuclear fuel of the future will in all likelihood be deuterium; even though this isotope of hydrogen is rare, it is present to the extent of 1 part in 7000 in naturally occurring hydrogen (Appendix Table 9). As shown in Probs. 33-C5 and 33-C6, each liter of seawater contains enough deuterium to give the energy equivalent of about 200 liters of high-test gasoline.

A further advantage of fusion power over fission power is its freedom from long-lived radioactive waste as far as fuel is concerned. In the equations of Sec. 33-3, only tritium ($^3_1$H) is unstable. It emits no γ rays, and its negatrons are easily shielded; in any case, the half-life is 12 years, and so it is stable enough to be used up as a fuel element yielding stable $^4_2$He (Eqs. *c* and *d*), or it can be stored long enough to decay to a harmless radiation level. Containment of a large amount of radioactive material such as tritium *will* pose a serious problem, however, since structural failure may occur due to radiation damage caused by induced radioactivity in the container material. Similarly, maintenance of the reactor chamber will be difficult and hazardous.

Of the various nuclear reactions listed as (*a*) through (*e*) in Sec. 33-3, it appears that (*c*), the deuterium-tritium (D-T) reaction, will be the first to be exploited because of its lower ignition temperature and large energy release. However, there are two potential problems associated with this reaction. First, tritium is most easily formed in a ^{6}Li(n, ^{4}He)^{3}H reaction, and on an energy basis the supply of lithium is not much more than 20 times the reserves of coal. Second, note that in this thermonuclear reaction, as in some of the others, much of the energy is released as KE of neutrons, which are a copious by-product. These neutrons, when captured in (n, γ) reactions, will render the reactor walls highly radioactive, possibly even to the point of structural failure. Reaction (*e*), the $^2H(^3He, ^4He)^1H$ reaction, is especially interesting because it produces energetic charged particles (protons). This opens up the possibility of generating electric power directly in the reactor without the necessity of first creating thermal energy to run heat engines (with their inherently low efficiency) to turn the electric generators. However, it is not yet known how to do this.

Consideration is being given to a fusion-fission hybrid reactor, in which the fusion neutrons would cause fission in a surrounding blanket of uranium; this would, in turn, result in the production of fissile ^{239}Pu, as well as fission energy. If this policy is followed, fusion energy would not be "clean" but would suffer from all the problems associated with fission reactors. It is to be hoped that the designers of fusion power plants will use restraint and bypass the superficially attractive hybrid concept; but in that case, some other, less dangerous, way must be found to use the energy of the fusion neutrons.

To summarize, the advantages of fusion power are (1) inexhaustible and cheap fuel supply and (2) no waste disposal problem, if the container problem can be solved. Among the disadvantages are (1) extremely high ignition temperature; (2) the difficulty of confining the reaction in a vessel; and (3) the size and complexity of installations, unsuited for small-scale uses such as propulsion. Fusion power generation is sure to be extremely difficult. It is not yet certain that the formidable engineering problems (including many not mentioned here) can be overcome in the next two or three decades to give a safe, environmentally acceptable, economically attractive fusion power plant.

The Costs of Energy. Let us look briefly at some of the costs of producing useful energy from fossil or nuclear fuel. These costs can be broken down into several categories.

The *direct costs* of fuel, salaries, amortization of plant, and distribution to the consumer are well recognized. Of these, the present trend of economic development indicates that the cost of coal is about half the total cost of conventional power generation. For fission plants of the type now in operation, the cost of uranium is a small fraction of the total cost of power, but this cost will rise as reserves of uranium are depleted in the coming decades. Breeder reactors can alleviate this cost problem, as can the reactor of the CANDU (Canadian deuterium-uranium) type, which uses natural (unenriched) uranium with a more efficient heavy-water moderator.

The *energy cost* of producing energy has often been overlooked. For example, the energy cost of extracting oil or gas from shale may well exceed the energy output obtained. The enrichment of uranium ore to yield the rare isotope ^{235}U requires an enormous expenditure of energy; is it certain that the energy ultimately made available from this uranium will repay the separation energy? Also, one must consider the energy costs of mining, refining, and transporting the metals used in constructing a nuclear power plant, the energy cost of construction, the energy cost of safely disposing of spent fuel (if that is even possible), even the energy cost of destroying obsolete reactors. So far, the nuclear power program based on fission reactors does not appear to be a drain on the rest of the economy. The important thing is *net energy* delivered to the consumer (output minus input). This is by no means a trivial or obvious point. In 1975, a British study showed that the largest consumer of energy in the United Kingdom is the energy industry itself. The solar energy industry, if it comes about, will surely not be immune to this problem.

Finally, the *environmental costs* of producing useful power are manyfold; some are illustrated in Fig. 33-2, in which the environmental costs of producing 1 MWh (3.6×10^9 J) are shown for both a coal economy and a fission economy. In each case, the heat rejected to cooling water or air is an inescapable consequence of the second

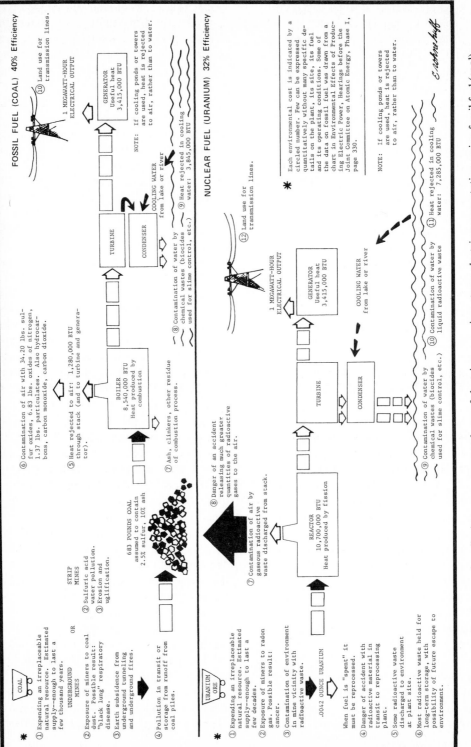

Figure 33-2 Environmental costs of producing one megawatt-hour of electric energy in a steam electric power station, using fossil fuel (coal) or nuclear fuel (^{235}U). Thermal energy is given in Btu, as is engineering practice. Note that 1 MWh is equivalent to 3,415,000 Btu. (Chart by D. E. Abrahamson, Scientists' Institute for Public Information.)

law of thermodynamics; this heat flow is a source of thermal pollution. It is not generally appreciated that a coal economy extracts a much larger death toll due to atmospheric pollution (chiefly from use of high-sulfur coal) than from underground mining accidents. We have accepted as many as 70 deaths per year for each urban coal-fired plant of 1-GW capacity, chiefly because the deaths are scattered and not readily identifiable. In contrast, a properly constructed nuclear reactor is almost free of pollution except in case of a major accident (for which the probability is a matter of sharp debate). Fission energy poses other problems that are broadly environmental: reactor safety; disposal of radioactive wastes; and safeguard of plutonium in transit, especially for breeder reactors. The costs of dealing with these problems are not yet known, if indeed solutions are possible. Some scientists and engineers are cautioning against starting construction of new nuclear fission plants unless the technical problems are overcome; others caution against the dangers of developing the breeder reactor. Many rightly propose an increased emphasis on research and development of alternative energy sources such as solar energy and fusion power.

The interaction between science and society is nowhere more complex than in the area of energy utilization. You should study the sources (see References at the end of the chapter) and as an informed citizen take part in the ongoing debate. One thing is certain—power consumption is growing in our technological society, and some form of nuclear power from clean fusion reactors will be an economic necessity within the foreseeable future.

33-5 Biological Effects of Radiation

In a purely physical sense, the activity of a radioactive source is measured in *curies*: 1 curie (Ci) is defined as exactly 3.7×10^{10} disintegrations per second (dis/s). This is approximately the activity of 1 gram of $^{226}_{88}$Ra. Radiation sources in the laboratory are often measured in microcuries (μCi). [The SI unit for activity is the becquerel (Bq), defined as 1 disintegration per second.]

Example 33-3

In a tracer experiment in physiology, a sample of muscle tissue has 0.022 μCi of the radioactive phosphorus isotope $^{32}_{15}$P, of half-life 14 days. What counting rate would be registered by a Geiger counter that receives 40% of the β particles emitted by the $^{32}_{15}$P in the sample? Assume that each β particle is counted with 95% efficiency.

Rate
$$= (0.022 \times 10^{-6} \text{ Ci})\left(\frac{3.7 \times 10^{10} \text{ dis/s}}{1 \text{ Ci}}\right)(.40)(.95)$$

$$= \boxed{310 \text{ counts/s}}$$

The *biological* effect of a source, or dose, of radiation is determined in part by the total energy absorbed; this energy, delivered by an incoming particle, causes the breakup of a protein or other complex molecule, or it causes genetic damage through altering a gene in a chromosome. Since it takes less than 100 eV to create an ion pair in a complex organic molecule, it is obvious that a single β particle or γ-ray photon of energy up to several MeV can create many ion pairs. Thus, radiation damage depends on the energy of the radiation and also on the ease of absorption of the radiation by the type of tissue (such as muscle or bone) considered. Because of its mass, an α particle, for example, of a given energy is more effective than a β particle of the same energy.

We see the need for a factor, the *relative biological effectiveness* (RBE), also called the *quality factor* (Q), so that different kinds of radiation can be compared. The RBE for x rays, γ rays, β rays, and high-energy protons is about 1, for thermal neutrons about 3, for fast neutrons about 10, and for α particles about 10 or 20. The *radiation absorbed dose* (rad) is the usual unit of measurement for absorbed dose of ionizing radiation; it corresponds to the absorption of 10^{-5} J (100 ergs) of energy per gram of tissue. [The SI unit for absorbed dose is the gray (Gy), defined as 1 J absorbed per kg of tissue.] The purely physical unit *roentgen* (R) is also used; it is approximately the same as the rad, for x rays or γ rays. A useful unit for dose equivalent is the *radiation equivalent, man* (rem), which is the product of the number of rads and the RBE. [The SI unit for dose equivalent is the sievert (Sv), which is the absorbed dose in Gy multiplied by the RBE.]

Biological radiation damage falls into two categories. *Somatic damage* (to parts of the body other than the reproductive cells) is in most (not all) cases tolerable in small doses. For example, the body recovers from the somatic damage of a minor sunburn or a cut finger. This means that there is often a *threshold* for somatic damage; presumably organisms have evolved so as to be only slightly affected somatically by the normal, ever-present background of radiation. This background is due to several causes. At sea level, cosmic rays subject the body to about 100 mrad (millirads) per year (but as much as 170 mrad at mile-high Denver, Colorado). In addition, there is a natural radiation in the air due to the decay products of the gas $^{222}_{86}$Rn (Sec. 32-7). Equally important is the body burden of β radiation from $^{40}_{19}$K (half-life 1.3×10^9 years). A 70-kg adult has about 0.13 kg of potassium, chiefly in the muscle, and the normal $^{40}_{19}$K content of the human body is about 0.1 μCi, which delivers about 20 mrad/year to the gonads and other soft tissues and about 10 mrad/year to bone. Other natural sources of radiation include drinking water and foods. We have lived with these and other somatic challenges for millions of years; we should be concerned at any increased radiation levels that might, say, double the body burden of radioactivity.

We have mentioned the use of radioactive nuclides in cancer therapy (Sec. 33-1). The relation of radiation to cancer is twofold, for it has become apparent that exposure to low levels of radiation actually causes certain cancers of the blood (leukemia), bone, thyroid, and lung.

Genetic damage seems to have no threshold; *any* dose to a gonadal chromosome, however small, produces *some* genetic damage. Most mutations caused by gene damage are harmful, although these effects, even lethal ones, may not become overt until many generations have gone by. Thus very minor exposure to low levels of radiation can, in the long run, cause large genetic effects in a population. A little mutation is a good thing; it is probable that the course of evolution depends in part on mutations caused by the cosmic-ray background. But there is considerable concern about genetic damage to the human race from excessive medical use of diagnostic x rays. Radiation thus far introduced into the atmosphere by the testing of nuclear weapons is of lesser magnitude, but if such testing leads to actual warfare, the somatic and genetic effects will be catastrophic. The nuclides $^{90}_{38}$Sr (a fission product, half-life 28 years) and $^{14}_{6}$C (a fusion by-product, half-life 5730 years) are especially troublesome in this regard. Genetically speaking, for the population as a whole it is the *total* dose to the gonads that counts; it probably makes little difference whether radiation exposure of an individual is at a low level for a long time, or at a high level for a short time (as in nuclear warfare). Contamination of atmosphere, water, or food that affects large segments of the world's population is harmful precisely because of the large number of persons involved. The calculated risk taken by workers in nuclear energy installations is, by contrast, less important, but no one is quite sure that these imponderables are being correctly assessed.

In spite of biological dangers, man-made radiation sources have so far conferred far more identifiable benefit than harm. Medical use of x rays has saved countless lives. Nuclear power plants hold out the promise of cheap power, with tolerable environmental contamination, and this power can be used for social goods such as massive desalinization of seawater. Industrial x-ray inspection of aircraft wings for hidden flaws reduces the hazard of air travel. There are many similar examples. Nevertheless, society must eliminate excessive or needless exposure to ionizing radiation. Groups of scientists and engineers in several countries are active in presenting these issues to the general public.

33-6 Particle Accelerators

In Sec. 33-1 we assumed that many unstable isotopes are available and that they can be produced in quantity when needed. Broadly speaking, nuclear reactions take place in particle accelerators (or "atom-smashers") and in the chain-reacting nuclear reactors discussed in Sec. 33-2. We shall discuss accelerators in the next two sections.

The simplest accelerator is the *Van de Graaff generator*, shown in Fig. 33-3. The pointed electrode A is connected to a rectifier circuit V, which maintains a PD of perhaps 50,000 V, and positive charge is sprayed onto a moving silk belt. The charge is removed at C and replaced by negative charge at D. In an endless process, electrons are moved from the sphere S to ground, and the sphere becomes charged to a high positive potential. To use the generator as an atom-smasher, a source of H^+ or He^+ or He^{2+} ions is mounted inside the sphere, and the positive ions are repelled toward a grounded target by the positively charged sphere. Protons of energy up

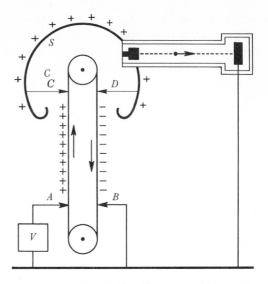

Figure 33-3 Van de Graaff generator (schematic).

to about 30 MeV can be produced if three Van de Graaff generators are used in tandem.

The *cyclotron*, developed in 1930 by E. O. Lawrence and M. S. Livingston, uses a magnetic field to bend the particles in orbits (Fig. 33-4). In this way a modest PD of, say, 50,000 V can be applied over and over again to the same particle, to build up a large final energy. The proton source injects H^+ ions at low speed, and the particles move inside two semicircular flat cavities, called "dees" because of their shape. The dees are enclosed in an evacuated chamber

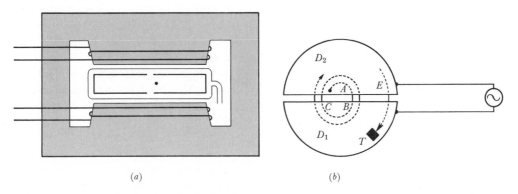

(a) (b)

Figure 33-4 Cyclotron (schematic). (a) Side view. (b) Top view of dees.

placed between the pole pieces of a powerful electromagnet, so that the particles cut across a vertical magnetic field. As we saw in Chap. 22, the magnetic force on a moving charge is perpendicular both to the field and to the motion; therefore each proton moves in a circular orbit, for which the centripetal force is supplied by the magnetic force. The dees are connected to a rapidly varying source of alternating potential, say, $\pm 50,000$ V at a frequency of 10^7 Hz. If a proton reaches A at just the right time, when D_1 is negative relative to D_2, it will be accelerated across the gap and receive an additional 50,000 eV of energy. Since it is now moving faster, the magnetic field bends it in a semicircle of larger radius. By the time the proton reaches C, the polarity of the dees has reversed, D_2 now is negative relative to D_1, and the proton gains another 50,000 eV. In this way the proton spirals outward in a series of semicircles, and eventually strikes a target T.

The cyclotron works only because of a lucky coincidence: the time for each semicircular journey is the same for all particles, whether they are moving slowly (near the center) or fast (near the outside). To prove this, we use Newton's second law and Eq. 22-4 for the magnetic force on a moving charge.

$$\text{net } F = ma$$

$$BQv = m\frac{v^2}{r}$$

whence

$$v = \frac{BQr}{m}$$

This equation is valid for particles of any speed if the relativistic mass m is used. The time for half a revolution is given by

$$T = \frac{\text{distance}}{\text{speed}} = \frac{\frac{1}{2}(2\pi r)}{BQr/m}$$

or

$$T = \frac{\pi m}{BQ} \qquad (33\text{-}1)$$

The radius has canceled out; therefore, the time for one semicircular trip depends only on B, m, and the charge Q. If these quantities are constant, then T is constant, and an oscillator of fixed frequency can accelerate *simultaneously* both the slow protons at A and the faster ones at E, which have already made many trips in their spiral path.

One might think that there is no limit to the energy that can be given to a particle by cyclotron action. To obtain large final energy, a strong magnetic field must be used, and the diameter of the pole pieces must be large so that each particle can make many trips as it spirals outward. However, aside from cost and other practical limitations on size, there is an unavoidable limitation due to the relativistic increase of mass of the high-speed particles. Equation 33-1 shows that the time for a semicircular trip increases if mass increases, and so as the particles gain energy—and mass—they fall behind and get out of step with the oscillating voltage applied to the dees. A 20-MeV proton moves at about $\frac{1}{5}$ the speed of light, and its mass is 2% greater than its rest mass. This is enough to spoil the regularity of the cyclotron action. The difficulty can be overcome in the *synchrocyclotron* by letting protons move through the machine in short bursts, increasing the time between alternations of the dees as each group of particles spirals out. In this way, Eq. 33-1 remains satisfied throughout the spiral path of any group of particles. Another solution is to increase the magnetic induction B while the group of particles is spiraling outward, allowing T to remain fixed. By such methods, proton energies up to 700 MeV have been obtained.

Still higher energies are equivalent to 10^9 eV or more.* To reach the giga-electron-volt (GeV)

---

* The term "billion" is ambiguous—it means 10^9 in the United States and often means 10^{12} in Great Britain—so the metric prefix giga- (G) is in general scientific use to represent 10^9. However, British usage is now changing toward "billion" to mean 10^9, as in the United States. The first "g" in the prefix *giga* is soft, as in "giant."

Table 33-2 Dependence of Speed of Various Particles on Their Kinetic Energy*

Kinetic Energy	Electron Speed	Proton Speed	Deuteron Speed
0.1 MeV	.553	.0146	.0103
1 MeV	.941	.0462	.0326
10 MeV	.9988	.145	.103
100 MeV	.999987	.429	.316
1 GeV	.99999987	.876	.760
10 GeV	.9999999987	.9963	.988
100 GeV	.999999999987	.999957	.99983

*All speeds are expressed as fractions of the speed of light.

Figure 33-5 Early synchrotron shown while under construction at Argonne National Laboratory. The gap in the magnet is for the vacuum chamber containing the proton beam. Protons gain energy each time they pass through the radiofrequency cavity at upper left. Before injection into the main ring, protons are accelerated by a linac, whose exit portal is seen at top center. The man on the ladder is using an optical alignment device.

range, physicists took advantage of the facts of relativity, which previously were a limitation. No particle can have a speed greater than the speed of light. A 2-MeV electron is already moving at a speed of $0.98c$, and further increase of energy goes mainly into increasing its mass. Thus, *all* very high energy particles move at the same speed, about 3×10^8 m/s, regardless of rest mass or energy (Table 33-2). The *synchrotron* takes advantage of this simplicity. If the particles are somehow brought up to nearly the speed of light, then as they gain energy, their speed remains constant and the time for each revolution remains constant. The magnetic induction must be gradually increased as the particles pick up energy and mass (see Eq. 33-1). The great advantage of the synchrotron principle is that the radius of the orbit is practically constant, and so the magnetic induction need be applied only at the circumference, instead of throughout the entire area of the orbit, as in the cyclotron. The beam of particles is confined to a doughnut-shaped chamber that lies between the poles of a series of strong electromagnets (Fig. 33-5; see also Fig. 7-11 on page 158). In each circular trip the beam goes through a number of radiofrequency cavities (operating at, say, 50 MHz) in which energy is imparted to the particles during

each pass. Before entering the main ring the beam is brought up to the injection energy by a series of auxiliary accelerators of various types, including a linear accelerator (to be discussed later in this section).

As of 1981, 9 electron synchrotrons yielding electrons or positrons of energy greater than 1 GeV had been constructed in 7 countries, and 13 proton accelerators in the 1- to 500-GeV range had been constructed in 10 countries. At Batavia, Illinois, the main ring of the Fermi National Accelerator Laboratory (Fermilab) has 1014 conventional magnets to keep the 500-GeV proton beam in a path of radius 1.000 km (Fig. 33-6). High intensity (a beam of many particles per second) is as important as high energy per particle; the Fermilab accelerator gives 3.3×10^{12} protons per second.

For a given radius, the time for one revolution is essentially constant (since $v \approx c$). Then Eq. 33-1 shows that the mass and energy of a particle

Figure 33-6 Aerial view of the Fermi National Accelerator Laboratory, Batavia, Illinois. The largest circle is the main ring, of diameter 2.00 km. The smaller circle, with a cooling-water pond in the center, is the booster accelerator. Experimental areas lie at a tangent to the left.

are proportional to the magnetic induction B. In the early 1980s superconducting magnets were added to the Fermilab system in the space left for that purpose below the main-ring magnets shown in Fig. 7-11. A large liquid-helium plant installed on the premises supplies coolant to bring more than 10^6 m of NbTi cable to a temperature of 4.6 K. Each conductor, only 1.3 mm × 7.6 mm, carries a current of 4000 A without ohmic losses. By thus doubling the magnetic induction to more than 4 teslas, the energy is doubled, to reach 1000 GeV or 1 TeV (10^{12} eV). The new arrangement, called the Tevatron, also gives a significant savings in the cost of the power—several megawatts—needed

to maintain the large magnetic field in the main ring.

The synchrotron at CERN (European Organization for Nuclear Research) at Geneva operates at 500 GeV. For reasons of stability, the main ring, 1.1 km in radius, is in a tunnel through bedrock, 25 to 60 m below ground level.

A natural limitation still remaining in all accelerators, even if the relativistic increase in mass is taken care of, is the loss of energy by radiation. A proton or other charged particle moving in a circular orbit in an accelerator has a centripetal acceleration, and so it radiates energy in the same way as discussed in Sec. 30-3 for a pre-Bohr atom. In a circular accelerator, however, no radiationless states exist to save the day, as was the case for an electron's orbit around the nucleus. The maximum energy of a circular accelerator is reached, therefore, when the rate of e-m radiation, called *synchrotron radiation*, equals the power input to the beam of particles. These radiation losses are inevitable as long as the particles have centripetal acceleration.

Usually considered an unfortunate and wasteful by-product, synchrotron radiation has become valuable as a source of high-intensity e-m radiation in the x-ray and ultraviolet region. The radiation is usually extracted from an electron synchrotron or storage ring built for other purposes. More than a dozen existing accelerators have been modified to produce synchrotron radiation; such use of the machine for two purposes simultaneously is picturesquely called the "symbiotic mode" of operation. "Dedicated" machines designed for the purpose of generating photons are highly desirable and are now coming into operation as this new and growing field of high-energy physics is being exploited.

Synchrotron radiation has a continuous spectrum, like bremsstrahlung, with a range of wavelengths depending on the electron's energy E and the radius r of the electron orbit. For example, photons from the small storage ring TANTALUS at the University of Wisconsin ($E = 0.24$ GeV, $r = 0.65$ m) have maximum intensity at 300 Å; those from the larger ring SPEAR at Stanford University ($E = 4.0$ GeV, $r = 13$ m) have peak intensity at 3 Å. In many experiments the continuous spectrum is desirable; if a single wavelength is needed, a monochromator can be used to provide a narrow "slice" from the continuous spectrum. One of the chief advantages of synchrotron radiation is the availability of an intense beam of well-collimated photons from a small source. Exposure times for x-ray micrography of a biological specimen can be a few minutes instead of many hours. Other applications of synchrotron radiation are found in chemistry, microbiology, solid-state physics, surface physics, and protein crystal structure, to

name just a few. A scanning x-ray microscope has been constructed, and x-ray holography may be attainable.

In the *linear accelerator* (linac), particles move in a straight line between many properly spaced electrodes, instead of returning over and over to the same gap; no magnetic field is needed. There is no centripetal acceleration, and radiation losses occur only while the particles are crossing the narrow gaps where they gain energy. The Stanford linear accelerator (SLAC) is 3.2 km long; it became operational in 1966 and produces an intense beam of electrons of energy above 30 GeV. The Heavy Ion Linear Accelerator (Hilac) at Berkeley, California, can accelerate ions of all masses, 1 through 238, to an energy that is variable between 2.5 and 8.5 MeV per nucleon. A search is being made for $^{298}_{114}X$, a superheavy nuclide predicted to have exceptional stability,* by bombarding uranium with heavy-ion projectiles such as partially stripped uranium ions $^{238}_{92}U^{39+}$.

We have given only the broad outlines of the various particle accelerators. Their successful operation calls for precise and imaginative design. Not the least problem is that of focusing the particles into a narrow, stable beam of high intensity. Much current research is devoted to such questions of stability and intensity.

33-7 Colliding Beams

There is a difficulty with all accelerators that is clearly apparent even in a classical nonrelativistic analysis. In the accelerators so far described we are dealing with an inelastic collision, say between a proton that is moving at high speed and a target proton that is at rest. We saw in Chap. 6 that in such a collision, momentum is conserved, but some kinetic energy inevitably disappears. In a nuclear experiment this "lost"

---

* Nuclear theory suggests that such a nuclide would be an α emitter with a half-life of perhaps 1000 years.

kinetic energy is available for the rest mass of new particles, and it would be desirable to use *all* of the KE of the incoming particle in this way. Unfortunately, to conserve the linear momentum of the system, the combined mass (or the masses of the fragments) must recoil in the forward direction. Especially at high energies, this recoiling target or the recoiling products of the collision carry away a large fraction of the incoming KE. For example, calculation shows that if a 31.5-GeV proton collides with and sticks to a stationary target proton, the combined mass moves forward with 25.5 GeV of KE; only 6.0 GeV is available for transformation into the mass of new particles.

The total KE of a collision can be made "useful" by a clever stratagem, if two equal particles moving with equal speeds in opposite directions collide and stick together. In the laboratory frame of reference the momentum of the system is zero both before and after the collision; there is no KE after the collision and 100% of the KE of the colliding particles is available.

Two powerful colliding-beam machines that utilize light particles are already in operation. They are the Positron-Electron Project (PEP) at SLAC in California, and PETRA in West Germany. Since the two particle beams are oppositely charged and travel in opposite directions, the same magnetic field can be used to supply centripetal force. These machines have total collision energies of about $18+18=36$ GeV; because of radiation losses, this is about the upper limit for present-day electron machines of 1 kilometer radius.

Proton machines can reach much higher energies before radiation losses become a limiting factor. The Intersecting Storage Ring (ISR) is located at CERN at the border between Switzerland and France near Geneva. A 31.5-GeV proton synchrotron is used to inject beams of protons into two almost-circular rings about 300 m in diameter that intersect at 8 points (Figs. 33-7 and 33-8). The beams can be stored and used for about a day before they must be replenished from the synchrotron. The

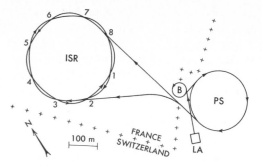

Figure 33-7 Beam paths for Intersecting Storage Rings (ISR) at CERN. Shown are the linear accelerator (LA), which operates at 50 MeV, the booster (B) at 800 MeV, and the proton synchrotron (PS) at 28 GeV.

two colliding proton beams, each of energy 31.5 GeV, release 63 GeV of energy, equivalent to a 2200-GeV beam striking a stationary target. Since both beams are positively charged, two sets of magnets are required (Fig. 33-8).

The next step in the development of colliding beams is to use heavy particles of opposite charge. The idea was tested at CERN, where proton-antiproton collisions were achieved in 1981 in the ISR, with a collision energy of

Figure 33-8 Intersection point 2 at the CERN Intersecting Storage Rings. Here the tunnel is widened and the floor lowered to accommodate experimental equipment later installed at the intersection. At the right, beneath the number 2 on the wall, is the end of a beam-transfer tunnel that brings protons from the synchrotron to the ISR.

26 + 26 = 52 GeV. Two accelerator laboratories are now modifying their largest proton synchrotrons to give proton-antiproton collisions of very high effective energy. The collision energy at CERN's Super Proton Synchrotron (SPS) will be 270 + 270 = 540 GeV; to obtain this amount of useful energy would require a fixed-target proton synchrotron operating at 155,000 GeV. By the mid-1980s, at Fermilab, an antiproton storage ring added to the Tevatron will provide proton-antiproton collisions with effective energy of 1000 + 1000 = 2000 GeV. The engineering problems associated with such machines are great, but the reward, in effective energy, is tremendous.

33-8 High-Energy Physics

Contemporary physics seems to be preoccupied with bigger and bigger machines that give higher and higher energies to particles. Let us look at the motivation for such a trend and see what some of the results have been. A particle accelerator is a tool with three fairly distinct uses: (1) to probe the structure of nuclear matter; (2) to create new particles; and (3) to study the interactions between particles. We shall consider the first two uses in this section. The third is the subject of Sec. 33-9.

The Structure of Nuclear Matter.
High-energy particles are shot at nuclei whose structure is to be determined. In order to "see" clearly, we must use a fine enough probe. More specifically, to obtain high resolving power we need to use a wavelength that is much smaller than the object we are investigating, for only in this way can the effects of diffraction be made negligible. From this point of view, we can interpret Rutherford's scattering experiment in a new fashion. He used α particles of energy about 9 MeV and was able to deduce something about the nucleus. The de Broglie wavelength of any particle is given by $\lambda = h/mv$, where mv is the momentum of the particle; it is not hard to show

that 9-MeV α particles have a velocity of about 2.1×10^7 m/s and a wavelength of about 5×10^{-15} m ($= 5$ fermis).* We now know that the radius of a gold nucleus is about 7 fermis, and so we see that Rutherford's probing beam that penetrated a gold foil was just fine enough to reveal some detail. In fact, Rutherford was able to deduce the existence of the nucleus, but not much more.

As higher energies have become available, the de Broglie wavelengths of particles used have become shorter, and the probing action has become more delicate. A probe of high-energy electrons is preferable to a probe using high-energy protons, because electrons respond only to electromagnetic forces, which are better understood than nuclear forces. The newest accelerators take the projectiles far into the relativistic range, where a particle's mass is much larger than its rest mass. In a sense, relativistic mechanics is here simpler than Newtonian mechanics, since the velocity $\approx c$ and cannot increase. Under these circumstances, the de Broglie wavelength is

$$\lambda = \frac{h}{mv} \approx \frac{h}{mc} = \frac{hc}{mc^2} = \frac{hc}{E}$$

and so the wavelength of any fast particle is inversely proportional to its energy. If the electrons produced in a linear accelerator have energy 1.3 GeV (2.1×10^{-10} J), their wavelength is given by

$$\lambda = \frac{hc}{E}$$

$$= \frac{(6.63 \times 10^{-34} \text{ J·s})(3.00 \times 10^8 \text{ m/s})}{2.1 \times 10^{-10} \text{ J}}$$

$$= 0.95 \times 10^{-15} \text{ m} = 0.95 \text{ fm}$$

Since the wavelength is $\frac{1}{7}$ the radius of a gold nucleus, the resolving power is sufficient to show

---

* The metric prefix for 10^{-15} is femto-; thus, the abbreviation fm (femtometer) also stands for the fermi, in honor of Enrico Fermi (1901–1958).

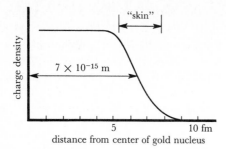

Figure 33-9 Distribution of electric charge in a gold nucleus; graph based on high-energy electron scattering experiments of Hofstadter. The electron energy was about 1.3 GeV, and the de Broglie wavelength about 1 fm.

some detail in the structure, as shown in Fig. 33-9, which reveals a uniform charge density in the interior and a "skin" at the outer edge. The electric structure of the proton and the neutron have been studied, using electron beams of energy up to 21 GeV. Experiment shows that the proton itself does not have a sharp boundary; its effective radius is about 0.8 fm. Experiment also shows that protons and neutrons have a complex internal structure consisting of scattering centers much smaller than the size of the nucleon. These pointlike entities are thought to be the *quarks* that are postulated in the theory of the structure of subatomic particles. It is all very reminiscent of Rutherford's original study of atomic structure.

Production of Subatomic Particles. This is a second use of very high energies. The rest mass of an electron is 0.00055 u, which is equivalent to 0.51 MeV, so we might expect that a 0.51-MeV photon can be converted into an electron according to $\Delta E = (\Delta m_0)c^2$. This would be an impossible process, however, since charge would not be conserved. A possible process is *pair production*, in which a negatron and positron are formed from a γ ray of energy 1.02 MeV or more. (Any excess photon energy is shared as KE of the two electrons.) Figure 33-10 illustrates this process; the incoming photons leave no tracks,

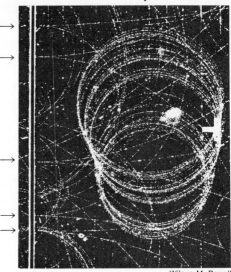

Wilson M. Powell

Figure 33-10 Conversion of energy into mass during the production of negatron-positron pairs. The cloud chamber is in a strong magnetic field, perpendicular to the plane of the photograph; negatrons curve clockwise, and positrons curve counterclockwise. Gamma rays enter the chamber from the left. At least five pair-production events can be seen originating in the absorbing plate, which runs vertically through the chamber. Another pair is created in the gas at top center of the photo about 6 mm below the upper arrow; the negatron member of the pair spirals about 36 times in the magnetic field before coming to rest.

since they are uncharged and produce no ions. When the negatron and positron are materialized, they are both traveling in a forward direction, thus carrying most of the initial momentum of the photon, which has disappeared.* The cloud chamber is in a magnetic field, which gives opposite curvatures to the two tracks, showing that they are oppositely charged. In this one photograph we see illustrated three of the con-

---

* Pair production must take place near some particle, such as a nucleus, that takes up a small part of the photon's momentum. The track of this nucleus is not seen in Fig. 33-10.

servation laws: the conservation of mass-energy, linear momentum, and charge. Angular momentum (spin) is also conserved, but this fact cannot be determined from the photograph.

The rest mass of a proton is about 1.008 u, equivalent to 938 MeV (almost 1 GeV). A proton can be materialized if sufficient energy is available. Once again, conservation of charge requires that two particles be produced. The proton-antiproton pair was first produced in 1955 by the 6-GeV synchrotron at Berkeley, California, in a high-energy collision between two protons. A total of 6 GeV is needed: 2 GeV to create the mass of the proton-antiproton pair, and 4 GeV for the kinetic energies of the various particles after collision. The positron and antiproton are *antimatter*—exactly similar to their normal counterparts except for the sign of their charge and the direction of spin. These particles are stable, in the sense that they do not undergo radioactive decay; but their lifetimes are very short, at least under laboratory conditions on the earth. A positron that comes near an electron will be annihilated, and the mass equivalent of the two particles becomes *annihilation radiation*, which usually consists of two γ rays, each of energy 0.51 MeV. Similarly, an antiproton cannot long endure in an environment that contains matter made up of ordinary protons and neutrons. Atoms made of antimatter actually exist; for instance, a positron can circulate around an antiproton to form antihydrogen. Nuclear antimatter has been created; thus, an antideuteron is a combination of an antiproton and an antineutron. However, there is no evidence for antigravity; antimatter behaves normally in the earth's gravitational field. Many other particles have been materialized artificially with the energies now available. No one yet knows what new particles may come into existence when experiments are performed with colliding proton beams having effective energies in the TeV (10^{12} eV) range.

33-9 Interactions Between Particles

More than 200 subatomic particles having masses up to several times that of the proton have been discovered. Almost all of them are unstable, with lifetimes as short as 10^{-23} s. The key to understanding and classifying this array of particles is the way in which they interact. In Sec. 19-3 we introduced a model in which the electric field is described by the exchange of field particles, which for the e-m interaction are photons. Physicists now recognize four different types of field, each with its own field particle. These interactions have radically different strengths. In Table 33-3 we summarize the basic interactions and their field particles. It is noteworthy that the field particles have integral spins: 1 or 2, in units of $h/2\pi$.

First let us consider the more familiar interactions, the long-range electromagnetic (E) and the gravitational (G) interactions. The first field particles to be discovered were photons, which are easy to produce. All you have to do is to accelerate a charge, and some photons are

Table 33-3 Interactions Between Particles

Symbol	Interaction	Relative Strength	Range of Force	Particles Acted On	FIELD PARTICLES		
					Name	Mass	Spin
S	strong	1	short (~ 1 fm)	quarks	gluons	?	1
E	electromagnetic	10^{-2}	long ($1/r^2$)	all charged particles	photons	0	1
W	weak	10^{-5}	short (~ 1 fm)	leptons, quarks	$W^\pm$, Z^0	> 80 GeV	1
G	gravitational	10^{-39}	long ($1/r^2$)	all particles	graviton	0	2

Table 33-4   Leptons

Particle	Symbol	Spin	Rest Mass, MeV	Charge	Mean Life, s	Typical Decay Products
electron	e^- (e^+)	$\frac{1}{2}$	0.511	1	stable	
e-neutrino	ν_e ($\bar{\nu}_e$)	$\frac{1}{2}$	0	0	stable	
muon	μ^- (μ^+)	$\frac{1}{2}$	106	1	10^{-6}	$e^- + \bar{\nu}_e + \nu_\mu$
μ-neutrino	ν_μ ($\bar{\nu}_\mu$)	$\frac{1}{2}$	0	0	stable	
tau	τ^- (τ^+)	$\frac{1}{2}$	1783	1	10^{-12}	$\mu^- + \bar{\nu}_\mu + \nu_\tau$; $e^- + \bar{\nu}_e + \nu_\tau$
τ-neutrino	ν_τ ($\bar{\nu}_\tau$)	$\frac{1}{2}$	0	0	stable	

"shaken loose" from the swarm that is continually being exchanged between that charge and some other charge. This is how photons associated with radio waves come into existence as electrons move back and forth in a transmitting antenna; and it is how x-ray photons come into existence when a beam of high-speed electrons strikes a stationary target in an x-ray tube. The model can be extended to cover the emission of photons during quantum jumps in an atom; one of the photons being exchanged between the nucleus and the orbital electrons is materialized when an electron changes its orbit.

Gravitational forces are inherently extremely weak compared with e-m forces (see Prob. 18-B1), and the gravitational field particles (gravitons) are beyond the reach of experimental techniques. Both the e-m and the gravitational interactions are "long range," decreasing gradually, inversely as the square of the distance between the interacting particles. It can be shown that the long range of the E and G interactions is due to the fact that their field particles have no rest mass.

The two nuclear interactions S (strong) and W (weak) are of short range, decreasing abruptly to practically zero when the distance between the interacting particles exceeds a few fermis. This distance is approximately the size of a nucleus; thus, nuclei attract or repel each other only when they are essentially in contact.

33-10   Subatomic Particles

We group the subatomic particles into two broad categories according to the interactions they experience. Table 33-4 gives 6 *leptons*,* which do not experience the strong interaction; they interact via W and G (and E, if charged). Table 33-5 gives 18 *hadrons*,[†] which experience the S interaction as well as W and G (and E, if charged). Each lepton and most of the hadrons have antiparticles, indicated by a bar over the symbol; the antiparticles have opposite charge and opposite direction of spin. (The positron, e^+, is the antiparticle of the negatron, and the positive muon is the antiparticle of the negative muon. A few neutral hadrons are their own antiparticles.) The relatively few hadrons listed are stable against decay via S, and they decay, if at all, via the E or the W interaction. In addition to the 18 hadrons shown in the table, about 200 are known that are highly unstable and decay spontaneously via the strong interaction, with lifetimes of 10^{-20} s or less. By definition, hadrons experience the S interaction, whereas leptons do not. Another distinction between the particles is that hadrons have size and structure, as shown by the scattering experi-

---

* From the Greek *leptos*, small, tiny; the smallest Greek coin is a lepton, equivalent to about 0.02 cent.
† From the Greek *hadros*, thick, stout, strong.

Table 33-5   Hadrons

Group	Members	Symbol	Spin	Rest Mass, MeV	Quark Content	Half-Life, s	Typical Decay Products
Mesons	pion	π^0	0	135	*	10^{-16}	$\gamma + \gamma$
		$\pi^+ \ (\pi^-)$	0	140	$u\bar{d}$	10^{-8}	$\mu^+ + \nu_\mu$
	kaon	$K^+ \ (K^-)$	0	494	$u\bar{s}$	10^{-8}	$\mu^+ + \nu_\mu; \ \pi^+ + \pi^0$
		K_S^0	0	498	*	10^{-10}	$\pi^+ + \pi^-; \ \pi^0 + \pi^0$
		K_L^0	0	498	*	10^{-8}	$\pi^\mp + e^\pm + {}^{(\bar{\nu})}{}_e; \ \pi^\mp + \mu^\pm + {}^{(\bar{\nu})}{}_\mu;$ $\pi^+ + \pi^- + \pi^0; \ \pi^0 + \pi^0 + \pi^0$
	eta	η	0	549	*	$< 10^{-16}$	$\gamma + \gamma; \ \pi^0 + \pi^0 + \pi^0; \ \pi^+ + \pi^- + \pi^0$
	charmed mesons	$D^0 \ (\bar{D}^0)$	0	1863	$c\bar{u}$	10^{-13}	$K^- + \pi^+ + \pi^0$
		$D^+ \ (D^-)$	0	1868	$c\bar{d}$	10^{-13}	$K^- + \pi^+ + \pi^+$
Baryons	nucleons:						
	proton	p $(\bar{p})$	$\frac{1}{2}$	938.3	uud	stable	
	neutron	n $(\bar{n})$	$\frac{1}{2}$	939.6	udd	10^3	$p + e^- + \bar{\nu}_e$
	hyperons:						
	lambda	$\Lambda \ (\bar{\Lambda})$	$\frac{1}{2}$	1116	uds	10^{-10}	$p + \pi^-; \ n + \pi^0$
	sigma	$\Sigma^+ \ (\bar{\Sigma}^+)$	$\frac{1}{2}$	1189	uus	10^{-10}	$n + \pi^+; \ p + \pi^0$
		$\Sigma^0 \ (\bar{\Sigma}^0)$	$\frac{1}{2}$	1192	uds	10^{-19}	$\Lambda + \pi^0$
		$\Sigma^- \ (\bar{\Sigma}^-)$	$\frac{1}{2}$	1197	dds	10^{-10}	$n + \pi^-$
	xi	$\Xi^0 \ (\bar{\Xi}^0)$	$\frac{1}{2}$	1315	uss	10^{-10}	$\Lambda + \pi^0$
		$\Xi^- \ (\bar{\Xi}^-)$	$\frac{1}{2}$	1321	dss	10^{-10}	$\Lambda + \pi^-$
	omega	$\Omega^- \ (\bar{\Omega}^-)$	$\frac{3}{2}$	1672	sss	10^{-10}	$\Xi^0 + \pi^-; \ \Lambda + K^-; \ \Xi^- + \pi^0$
	charmed lambda	$\Lambda_c^+ \ (\overline{\Lambda_c}^+)$	?	2270	udc	10^{-12}	$\Lambda + \pi^+ + \pi^+ + \pi^-$

Notes:
1. Particles (resonances) that decay by the strong interaction, with half-lives less than 10^{-20} s, are not listed.
2. Antiparticles are shown in parentheses; π^0, K_S^0, K_L^0, and η are their own antiparticles.
3. All spins are in multiples of $h/2\pi$. All charges are numerically equal to the electron charge, or zero. All masses are in MeV.
4. Several of the mesons, indicated by * in the table, must be represented as combinations or superpositions of several different quark-antiquark pairs. For example, π^0 is a mixture of $u\bar{u}$ and $d\bar{d}$.
5. Decay products are shown for the particles; decays for the antiparticles are analogous. The most prevalent modes of decay are shown.

ments described in Sec. 33-8. Leptons, on the other hand, appear to be truly pointlike and have no size or structure.

Leptons come in three families, each with its own type of neutrino. For example, the ordinary electron e^- has an associated e-neutrino ν_e; their antiparticles are the positron e^+ and the antineutrino $\bar{\nu}_e$. The muon μ^- and the tau particle τ^- are simply heavy electrons, and unstable ones at that. Associated with the muon is the μ-neutrino, and with the τ particle is the τ-neutrino (not yet observed, as of 1981).

Neutrinos that appear in β decay are e-neutrinos, as in the decay of ^{55}Cr (Sec. 32-6), or in the decay of the free neutron:

$$n \rightarrow p + e^- + \bar{\nu}_e$$

This illustrates the law of conservation of leptons: there are no leptons on the left, and the lepton plus the antilepton on the right also add up to no leptons. The μ and the τ are unstable, but the lightest lepton, the electron, is stable because there is no lighter lepton except the neutrino for it to decay to, and $e^- \rightarrow \nu_e$ would not conserve electric charge. The μ-neutrino

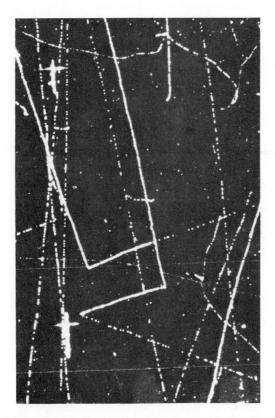

Figure 33-11 Bubble chamber photograph showing two $\pi \rightarrow \mu \rightarrow e$ decay processes (heavy tracks with right-angle bends). The electron tracks are characterized by a small density of bubbles.

appears in the decay of a *pion*:

$$\pi^+ \rightarrow \mu^+ + \nu_\mu$$

and two neutrinos are involved in the decay of a muon (Fig. 33-11):

$$\mu^+ \rightarrow e^+ + \nu_e + \bar{\nu}_\mu$$

Free neutrinos react (very rarely) with matter, and can be observed to initiate reactions such as

$$\bar{\nu}_e + p \rightarrow n + e^+$$

An electron, which is a charged lepton, has only the G, W, and E interactions; it is unaffected by the strong interaction, and it plows right through a nucleus except for effects caused by the electric forces—this is why Fig. 33-9 gives the *charge* distribution in a nucleus. In effect, the electron ignores all nuclear forces because the weak interaction that it and the nucleus experience is much weaker than the e-m interaction that dominates the picture.

Turning now to the hadrons, we see that Table 33-5 includes 8 *mesons* and 10 *baryons*. Baryons that are heavier than the nucleons are called *hyperons*.* By definition, all hadrons experience the strong interaction, yet some hadrons are much more stable than others. To understand this, we introduce the idea that the strength of an interaction determines the stability of a particle.

A strongly interacting particle, left to itself, tends to break up (disintegrate) as rapidly as it can. We can estimate this time to be about 10^{-23} s, which is the time for a high-speed particle to travel across the diameter of a nucleus. During this time, the strong interaction will cause a particle to decay *unless something prevents this from happening*. We therefore expect the half-life of any strongly interacting particle

* Meson is from the Greek *mesos*, in the middle, moderate; as in "mezzo-soprano" and many biological terms such as "mesoderm." Baryon is from the Greek *barys*, heavy; as in "barium," "baritone." Hyperon is from the Greek *hyper*, over, above, beyond; as in "hypersensitivity," "hypertrophy."

to be about 10^{-23} s unless it is in something like a metastable state and is prevented from breaking up. Such a particle could then live long enough for a less likely process to take place, governed by the e-m interaction, which is 10^{-2} as strong as the strong interaction. Such a particle would break up in about 10^{-20} s unless once again it is in a metastable state relative to the e-m interaction. The weak interaction eventually takes over if the first two decays are "forbidden"; typical lifetimes for particles that decay by the weak interaction are about 10^{-10} s. Even then, if the available energy (that is, the mass difference) is small, the lifetime of a weakly interacting particle can be much longer than 10^{-10} s; this is how the measurable half-lives of radioactive nuclei range from a fraction of a second to as long as 10^{18} years. We conclude that the π^+, for example, is immune to decay by the S and E interactions, and therefore it decays via the W interaction, with the "long" lifetime of 10^{-8} s. We shall see later how the quark theory of subatomic particles leads to this result.

The field particles for weak interactions, the W particles and the Z^0 particle, have not yet been found; the energy needed for their creation is more than can be supplied even by a 500-GeV accelerator. It is hoped that the effective energy of 2200 GeV available from the CERN colliding beams (Sec. 33-7) will be sufficient. Finally, the stable nuclides, such as $^{12}_6\mathrm{C}$, $^1_1\mathrm{H}$ (the proton itself), and the electron, are stable against decay by any of the first three interactions. It is a matter of speculation whether the very weak gravitational interaction can give rise to spontaneous decay; at any rate, none has yet been detected.

We list in Table 33-5, somewhat arbitrarily, only particles that are stable against decay via the strong interaction. Most of the particles in Table 33-5 endure for at least 10^{-10} s. This is a "long" time—a nucleus moving at almost the speed of light could, in this time, travel about 3 cm, which is 10^{12} times its own diameter. But the table by no means exhausts the list of known particles, if we are willing to accept particles of

very short lifetime. As a typical example, pions are fired at protons (in the liquid hydrogen of a bubble chamber). A pion and a proton may stick together as a unit, becoming (momentarily) a particle that breaks up in about 10^{-23} s, because of the strong interaction; in this case no metastable state exists to save the day. We have the rest masses of the pion and the proton plus some definite amount of KE that is transformed into mass to make up the mass of the new entity called a *resonance* (an unstable state of strongly interacting matter). This resonance is observed experimentally by measuring the energies and momenta of the particles after breakup; there is a strong preference for the total mass-plus-energy of the fragments to add up to a constant, which is the mass of the super-unstable particle. This mass is characteristic of the resonance, independent of the exact way the resonance is produced.

Some of the resonances produced in pion-proton collisions are particles with double charge; other resonances are singly charged or neutral. One group of resonances, the nucleon resonances (denoted by N), includes at least 12 well-established pairs of particles (one positive, the other neutral) similar to the familiar proton-neutron pair of nucleons. To these excited states $N_1{}^+$, $N_1{}^0$, $N_2{}^+$, $N_2{}^0$, ..., which have masses 1470 to 2220 MeV, we can add the nucleon's ground state, which we can denote as $N_0{}^+$ (the ordinary proton) and $N_0{}^0$ (the ordinary neutron), each of mass about 939 MeV. We are reminded of the analysis of the hydrogen optical spectrum in terms of energy levels (Fig. 30-3). The analogy would be complete if we were to give different names to the H atom in its various energy states—for instance, if we called it H′ in the ground state E_1 and H″ in the excited state E_2 (which has a lifetime of about 10^{-8} s relative to the emission of a photon of ultraviolet light). There *is* a mass difference between H′ and H″ because 10.20 eV has been added, but the mass difference (given by $\Delta E/c^2$) is so small as to be unmeasurable. In the case of the subatomic particles, the excited states and the resonances have considerable extra mass, and their lifetimes are $\approx 10^{-23}$ s. The decay processes differ: the H atom decays from its excited state by emitting photons, but resonances decay by emitting charged or neutral particles rather than photons. For example, two of the nucleon resonances decay as follows:

$$N_1{}^+ \to N_0{}^+ + \pi^0 + KE \qquad (10^{-23}\ \text{s})$$

and

$$N_4{}^- \to N_0{}^0 + \pi^- + KE \qquad (10^{-23}\ \text{s})$$

The rapid strong-interaction decay of $N_0{}^0$ (the neutron) is forbidden, so it decays by the weak interaction to a proton and a lepton-antilepton pair:

$$N_0 \to N_0{}^+ + e^- + \bar{\nu}_e \qquad (640\ \text{s})$$

The other low state of the nucleon, $N_0{}^+$ (the proton) cannot decay at all, because there is no baryon state of lower energy (lower mass) for it to "fall" to.

Physicists believe that hadrons are composed of a very few basic entities known whimsically as *quarks*. We can do no more than mention a few features of this theory (see References at the end of the chapter for further information about quarks). The 6 quarks are categorized by "flavors," and come in three families, as do the leptons. Their charges are not integral multiples of the electronic charge e, but are $+\frac{2}{3}e$ or $-\frac{1}{3}e$, with opposite charges for the antiquarks. The rather fanciful flavor names of the quarks, and their electric charges (in multiples of e), are as follows:

u (up)	$+\frac{2}{3}$	d (down)	$-\frac{1}{3}$
c (charmed)	$+\frac{2}{3}$	s (strange)	$-\frac{1}{3}$
t (top or truth)	$+\frac{2}{3}$	b (bottom or beauty)	$-\frac{1}{3}$

Each quark has a spin of $\frac{1}{2}(h/2\pi)$. As of 1981, the t quark has not yet been observed.

Now we use the quark model to describe the hadrons in Table 33-5. *Mesons* are combinations of one quark and one antiquark; for example, the positive pion π^+ is $u\bar{d}$. Since $\bar{d}$ has charge $+\frac{1}{3}$ (opposite of d), the total charge of π^+ is

$\frac{2}{3} + \frac{1}{3} = 1$. You can verify the charges of the other mesons in the table. Many other mesons exist, such as the resonance ρ^+, which is also $u\bar{d}$, but this can be regarded as an excited state of π^+, which decays to π^+ by the strong interaction in about 10^{-23} s. Evidently the two quarks making up the π^+ combine their spins to give a net spin of 0.

Baryons are combinations of three quarks. For example, a proton (p) is uud and has charge $\frac{2}{3} + \frac{2}{3} + (-\frac{1}{3}) = +1$; a neutron (n) is udd and has charge $\frac{2}{3} + (-\frac{1}{3}) + (-\frac{1}{3}) = 0$. The beta decay $n \to p + e^- + \bar{\nu}_e$ represents udd $\to$ uud. One d quark changes to a u quark; this proceeds via the W interaction and as a side-product produces the leptons e^- and $\bar{\nu}_e$, which do not interact strongly but are needed to conserve charge. Evidently the strong interaction is not able to change the flavor of a quark; this requires the W or the E interaction.

Any antiparticle has a quark content that is similar to that of the corresponding particle, with each quark replaced by its antiquark. Thus, the antineutron is $\bar{u}\bar{d}\bar{d}$, and the π^- (the antiparticle of π^+) is $\bar{u}d$.

It is a triumph of particle-physics theory that every one of the hundreds of observed hadrons fits into this scheme, and every allowed combination of quarks has been observed (at least for the u, d, and s flavors). For example, a baryon such as $uu\bar{d}$ or a meson such as ud are impossibilities and are not observed. The basic experimental observations are those of high-energy scattering experiments, and this explains the eagerness with which physicists await the operation of machines that utilize colliding proton beams. The highest effective collision energy thus far reached is 63 GeV, at CERN. The Tevatron, giving an effective collision energy of 2000 GeV, will come along in the mid-1980s.

Hadrons, which feel the strong interaction, are made up of quarks. The quarks themselves feel the strong interaction by means of field particles called *gluons*. No free quark or free gluon has been observed, and it is believed that an infinite energy would be required to remove a quark from a hadron within which it is confined. The strong force acts on the "color charge" property of a quark, just as the e-m force acts on the "electric charge" property of an electron. (The term "color charge" is fanciful and has nothing to do with color as perceived by the eye.)

The theory is complex. There are 6 flavors of quark (u, d, c, s, t, and b), each with three possible colors (red, blue, and green). The forces between the color charges are *mediated* (transmitted) by 8 kinds of gluons, 6 of which themselves bear color. (Recall that the e-m force between *electric* charges is mediated by photons that are free of electric charge.) Color charge and electric charge are entirely independent of each other. Along these lines ordinary mass, which feels the gravitational interaction, can be called "mass charge." Just as for electric charge, the force between like color charges is repulsive, and the force between unlike color charges is attractive. However, there are no unlike mass charges, and the force between like mass charges is always attractive.*

The theory of subatomic particles that we have outlined is a mind-stretching intellectual enterprise having much aesthetic value. The theory allows us to see deeply into the nature of things. But in everyday life, we could get along perfectly well with only one family of leptons, e and e-neutrino, and with only one family of quarks, u and d. The atoms and molecules we are built out of are made up of nuclei containing protons and neutrons (uud and udd), with electrons in the outer clouds that give the atoms their chemical properties.

Consider a complete stable nucleus, such as a helium nucleus $^4_2$He (an α particle). The Coulomb repulsion between like electric charges (the protons) tends to cause the nucleus to fly apart. We know, therefore, that there must be a net *nuclear* force of attraction between the nucleons. This

force is the strong interaction between the quarks that make up the nucleons, and this force is mediated by gluons. Since a single gluon cannot leave the hadron within which the quarks are confined, we must consider that the effective field particles being exchanged between a proton and a proton, or a proton and a neutron, are mesons—pairs of gluon-swapping quarks—such as a π^+ (which is u$\bar{\text{d}}$). We picture a nucleon as surrounded by a cloud of pions or other mesons that venture out a little way and are then reabsorbed within about 10^{-23} s. Traveling at nearly the speed of light, a meson could move only about 3×10^{-15} m (3 fm) from its parent nucleon during this short interval of time. This explains the short range of the nuclear force, which is effective only if the nucleons are essentially in contact. The "boys playing catch" analogy (page 414) is applicable here: the repulsive force between the two boys would become zero if their separation were greater than the distance to which they could throw the ball.

Free pions and other mesons can be materialized if enough energy is concentrated near a nucleus; for example, the pion's rest mass is equivalent to 135 MeV, and if a fast projectile such as a proton comes within a few fermis of a nucleus, some of the energy of the projectile may be transformed into a pion's rest mass. The pions are shaken loose from the nucleons, just as photons are shaken loose when electrons are accelerated or undergo collision. The existence of pions was predicted by the Japanese physicist H. Yukawa several years before they were first found in the cosmic-ray debris at sea level. When sufficient energy became available from accelerators, pions and other mesons were created artificially in copious quantities.

What of the future? On the theoretical side, there have been some advances and hints of things to come. A long-standing goal of physicists has been to unify the seemingly different interactions, in the same way that Maxwell long ago unified the electric and magnetic interactions into one inclusive e-m interaction. Two of the four interactions in Table 33-3 have,

---

* The fact that like electric charges repel each other, whereas like mass charges attract each other, is correlated with the spins of the field particles: an odd integer, 1, for photons, and an even integer, 2, for gravitons.

indeed, been unified. The E and W interactions have been theoretically merged into a single electroweak (EW) interaction; the γ, W^+, W^-, and Z^0 are combinations of the four field particles used in the EW interaction. This breakthrough won the 1979 Nobel prize in physics for S. Glashow, A. Salam, and S. Weinberg. Theoretical physicists are now working on a "grand unification" scheme that would also include the S interaction. Some dare to envision a supergravity that would merge G into S and EW, so there would be only one basic force in the universe. On the experimental side, there is a search for the possible decay of the proton, which may have a lifetime of some 10^{31} years, according to some grand unification schemes—and this would require a new type of very weak interaction linked with the possible existence of a small, but nonzero, mass for the neutrinos. It is hoped that the $W^\pm$ and Z^0 will be materialized as free particles when sufficient energy is available in collisions. It is possible that at higher energies new particles will appear that would require a new lepton family or a new quark family.

33-11 Cosmic Rays

The powerful accelerators we have described are expensive and complicated, but fortunately we are living in a perpetual high-energy laboratory, which operates day and night free of charge. We are referring to the cosmic rays, first discovered about 1900 when it was found that ions are continually being produced in the atmosphere, making the air a slight conductor of electricity. Since then, cosmic rays have been studied at high and low altitudes, with electroscopes, Geiger counters, cloud and bubble chambers, and photographic emulsions. Before the development of large accelerators, the only source of high-energy collisions was in the upper atmosphere, and many of the particles listed in Tables 33-4 and 33-5 were first discovered in cosmic rays.

The *primary cosmic-ray particles* that strike the earth from outer space are mostly extremely high energy protons, with a sprinkling of deuterons and other positively charged light nuclei. The track of a primary particle is shown in Fig. 33-12. The photographic plate was exposed during a high-altitude balloon flight; it recorded the stopping of an iron nucleus in the emulsion. The whole track is about 1.5 mm long. As the iron nucleus, originally a bare nucleus with $26+$ charges, traveled through the emulsion, it picked up electrons and finally came to rest at the right as a neutral atom. The track is thicker at the start because the highly charged nucleus caused more ionization, evidenced by the many side tracks of secondary electrons. Very few cosmic-ray primaries reach sea level; instead, we observe a complex mixture of *secondary particles*, including negatrons and positrons, photons, neutrons, protons, and a variety of other particles. Among these are the charged pion (π^- or

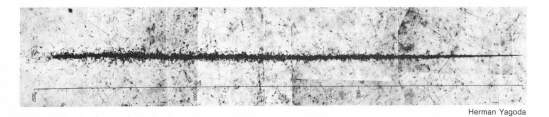

Herman Yagoda

Figure 33-12 Primary cosmic-ray particle. An iron nucleus with 26 positive charges enters a photographic emulsion from the left, and the track tapers to a point as the ion captures electrons and finally comes to rest as a neutral iron atom. Total path length, 1.5 mm.

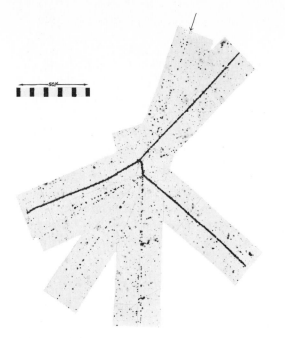

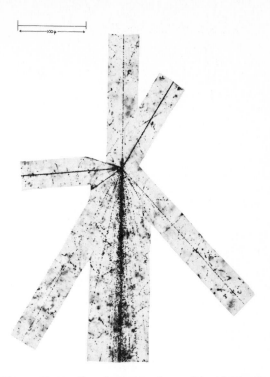

Figure 33-13 Cosmic-ray star in photographic emulsion. A primary cosmic-ray proton (arrow at top) initiates a star, consisting of about five light particles and two heavier particles. One of these is a pion, which, after a short upward journey, strikes another nucleus in the emulsion and causes a second disintegration.

Figure 33-14 Cosmic-ray star in emulsion, initiated by an α particle of about 3000-GeV energy. In addition to several heavy particles, a jet of about 150 secondary charged mesons is observed.

π^+) and the muon (μ^- or μ^+). Charged pions, with half-life about 2×10^{-8} s, decay into muons, and the muons, with half-life about 2×10^{-6} s, in turn decay into electrons.

Cosmic-ray stars originate when a primary particle strikes a nucleus head-on (Figs. 33-13 and 33-14). Such collisions are, naturally, more frequent at high altitudes where there are more primary particles. Many primary cosmic-ray particles have energies of more than 1000 GeV. At sea level, we see only the secondary effects. Often *cosmic-ray showers* of many thousands of electrons are observed, apparently originating from a single point above the observer. Some extensive showers originate from single primary particles of energy as high as 10^{20} eV. The source of this tremendous energy is unkown; mere conversion of rest mass to energy seems

unlikely, since the complete annihilation of a heavy nucleus such as ^{238}U would give only 238×931 MeV, about 200,000 MeV or 200 GeV. It is possible that a gigantic synchrotron action takes place in the weak magnetic fields that surround some stars, or in the stronger magnetic fields associated with sunspots. There is some suspicion that the gravitational interaction, weak as it is, may be the clue to understanding cosmological processes that operate on an extremely large scale. The origin of the cosmic-ray primaries is one of the most challenging problems of contemporary physics.

Instruments carried by balloons, artificial satellites, and lunar probes have measured the intensities of cosmic rays and other radiations at high altitudes. In 1958, intense radiation consisting of high-speed charged particles in the

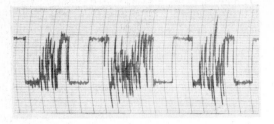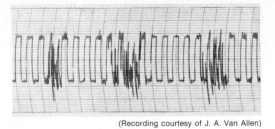

(Recording courtesy of J. A. Van Allen)

(*a*) $08^h20^m20^s$ Greenwich mean time. Satellite at 35° S. lat., 68° W. long., altitude 1500 km.

(*b*) $08^h26^m35^s$ Greenwich mean time. Satellite at 22° S. lat., 53° W. long., altitude 1165 km.

Figure 33-15 Radio signals from satellite 1958 Epsilon (Explorer IV). The signals were received on Aug. 1, 1958, at Santiago, Chile; they show the increased intensity (*b*) as the satellite entered the equatorial radiation belt. *Interpretation*: Time axis is horizontal; 1 division = 0.5 s. Vertical axis, frequency of an audio oscillator on board the satellite. A Geiger counter controls the oscillator's frequency, which changes abruptly from 675 Hz to 785 Hz, or vice versa, whenever 1024 counts have been accumulated. These audible tones modulate a carrier wave of frequency 108.03 MHz. The carrier wave is radiated by a dipole antenna consisting of the two halves of the satellite itself, which are electrically insulated from each other. As it moves along its orbit, the satellite tumbles end over end with a period of about 7 s. When it is pointing head-on or away from the observer, reception is poor (see Sec. 24-6), and "hash" appears every 3.5 s.

MeV range of energies was discovered at very high altitudes. The earth is surrounded by two layers of trapped particles—electrons, protons, or both—that are kept from reaching the surface by the earth's magnetic field. There are two doughnut-shaped regions of radiation, called the *Van Allen belts*, with maximum intensities at about 5000 km and 16,000 km above the earth's magnetic equator. The radiation extends to about 65,000 km above the equator but is much less intense over the magnetic poles. Figure 33-15 shows how radiation data were obtained from an early, specially instrumented satellite. Two sources for these trapped particles have been suggested: (1) the decay of energetic cosmic-ray neutrons (n → p + e⁻ + $\bar{v}_e$) at low altitude, and (2) bombardment by solar plasma (protons,

electrons, and some helium nuclei) shot out from the sun. It is probable that the protons in the inner and outer zones have different physical origins. The inner zone has protons with energies up to more than 500 MeV (formed by neutron decay); the stronger and larger outer zone has protons up to 1 MeV injected by the solar wind. Electrons up to 1 MeV energy are in both zones, and for electrons the line of demarcation between the zones is not sharply defined. Space probes have established that the planet Venus does not have a trapped radiation belt; this is consistent with the observed absence of a magnetic field for that planet. On the other hand, Jupiter has a strong magnetic field and is surrounded by a complex system of intense radiation belts.

Summary Radioactive nuclides find use in therapy, as tracers, in industry, and in dating experiments.

Slow neutrons of thermal energy (about 0.04 eV) cause several heavy nuclides to undergo fission. Other heavy nuclides are fissionable by fast neutrons only. Chain reactions are possible because of the production of excess neutrons during fission;

in a reactor, a moderator reduces the speeds of these excess neutrons to thermal values.

The cross section of a nucleus for a given process is the effective area of the nucleus for that process. A large cross section means a high probability for the process to happen. Only three nuclides have high cross sections for fission by slow neutrons and also a usefully long half-life. These nuclides are ^{235}U, which can be separated from natural uranium, and ^{239}Pu and ^{233}U, which can be manufactured in breeder reactors.

Fusion of light nuclides releases energy, since rest mass disappears. In the sun, four protons are built up into an α particle; the proton-proton chain and the carbon cycle are possible mechanisms for the process. Fusion reactors on earth will require high temperatures, but they will have the advantages of plentiful fuel and freedom from radioactive waste products.

Somatic and genetic radiation damage to tissue depends on the nature and energy of the absorbed radiation, as well as on the amount of dosage.

Particle accelerators of several types are used to achieve high energies. The Van de Graaff is an electrostatic generator. The cyclotron and synchrotron use magnetic fields to bend the particles into orbits. No magnetic field is used in the linear accelerator. Limitations on maximum energy are size of machine and cost; fundamental theory limits energies in some machines because of the relativistic increase of mass at high speeds, and in all machines radiation losses from accelerated particles constitute a limit. Synchrotron radiation (x-ray and ultraviolet photons from particles moving in circular orbits) has many uses. Use of colliding beams makes the full KE of the particles available for causing reactions or creating new particles.

Particles of very high energy are used as probes to explore the structure of matter. High resolving power is obtained if the de Broglie wavelength of the probability waves associated with the projectiles is much smaller than the object being studied. Particles can be created, according to $\Delta E = (\Delta m_0)c^2$, if enough energy is available. Easily remembered energy equivalents of certain rest masses are as follows: electron, about 0.5 MeV; proton, about 1 GeV. In pair production or annihilation, mass-energy, linear momentum, angular momentum, and charge are all conserved.

The four types of interaction and the field particles that are exchanged are, in order of decreasing strength, the strong interaction (gluons basically, mesons observationally), the e-m interaction (photons); the weak interaction (W and Z^0 particles); and the gravitational interaction (gravitons). The field particles for the W and G interactions have not been observed. Almost 40 particles are stable against the strong interaction; their lifetimes are at least 10^{-20} s, and most are 10^{-10} s or longer. Resonances are highly unstable states of strongly interacting matter; they can be considered to be particles of lifetime about 10^{-23} s.

Quarks are pointlike entities that make up the strongly interacting mesons and baryons. There are six flavors of quark, each with three colors. Quark charges are $+\frac{2}{3}e$ and $-\frac{1}{3}e$, but free quarks and their field particles (gluons) cannot be directly observed.

Primary cosmic-ray particles are light nuclei of atomic number from 1 up to about 26. The primaries create many secondary particles including mesons and baryons, as they pass through the earth's atmosphere. The Van Allen belts consist of high-energy charged particles emitted from the sun, trapped in the upper atmosphere of the earth by the earth's magnetic field.

Check List

fission	rad	baryon
thermal neutron	rem	pion
cross section for capture	Van de Graaff generator	hyperon
barn	cyclotron	resonance
moderator	synchrotron	quark
breeding	linear accelerator	gluon
fusion	colliding beams	color charge
proton-proton chain	pair production	cosmic-ray primaries
carbon cycle	annihilation	cosmic-ray stars
plasma	lepton	cosmic-ray showers
curie	hadron	Van Allen belts
RBE	meson	

Questions

33-1 The overall result of photosynthesis is that a living plant takes in CO_2 and H_2O and releases oxygen to the air. Describe an experiment by which you could determine if the released oxygen comes from the carbon dioxide or from the water. What isotope or isotopes of oxygen would be suitable for this experiment?

33-2 What are some of the peaceful uses of fission reactors? What are some of the disadvantages?

33-3 Why is ordinary water not the best moderator for use in a fission reactor?

33-4 The nuclide ^{227}Th is an α emitter of half-life 18 days and has a cross section of 1500 barns for fission by slow neutrons. Why is this nuclide not a useful fuel for a nuclear reactor?

33-5 What are some advantages and disadvantages of fusion power as compared with fission power?

33-6 A dentist who x-rays your teeth steps behind a lead shield before making the exposure. Why is the dentist apparently not equally concerned with the radiation risk to the patient?

33-7 What is meant by a threshold for radiation damage?

33-8 Explain why the facts of relativity limit the maximum energy that a proton can acquire in a cyclotron. Tell how this limitation is overcome in the synchrotron.

33-9 Why are high-energy synchrotrons constructed with a large radius?

33-10 Two proton beams, each of 1 GeV energy, collide head-on. Explain why this collision is more useful than if a beam of 2 GeV energy strikes a stationary target.

33-11 How can the beam tubes near the point of intersection in Fig. 33-8 be straight rather than curved?

33-12 About how much energy is needed to create a neutron-antineutron pair, exclusive of the kinetic energies of the particles after their formation?

33-13 Verify that each of the decay processes listed in Table 33-5 is consistent with the law of conservation of charge.

33-14 Which of the particles of Table 33-5 decay by the e-m interaction? (*Hint:* Consider the half-lives and the decay products.)

33-15 Tell whether the following quark combinations are mesons, baryons, or impossibilities: (*a*) uū; (*b*) duc; (*c*) uūs; (*d*) du; (*e*) ddd.

33-16 (*a*) What is the quark content of an antiproton? (*b*) Explain how the π^0 is its own antiparticle. (*Hint:* See Note 4 to Table 33-5.)

33-17 How many neutrinos, of which kinds, are associated with the two $\pi \rightarrow \mu \rightarrow$ e events in Fig. 33-11? Why do the neutrinos not leave visible tracks in the photograph?

33-18 What is the difference between a cosmic-ray star and a cosmic-ray shower?

MULTIPLE CHOICE

33-19 The ^{14}C in living organisms was originally formed in a nuclear reaction that took place in (*a*) the sun; (*b*) the earth's crust; (*c*) the earth's atmosphere.

33-20 The release of energy by the sun is the result of (*a*) fission; (*b*) fusion; (*c*) both of these.

33-21 The control rods in a reactor depend for their action on (*a*) absorption of γ rays; (*b*) fission of heavy nuclei; (*c*) absorption of neutrons.

33-22 A nuclear fusion reactor releases 10 MeV per reacting particle. If a mole of particles reacts, the energy released is about (*a*) 1 kJ; (*b*) 1 GJ; (*c*) 1 TJ.

33-23 A particle in a synchrotron has a speed of $0.99c$. To increase the energy of this particle, you must (*a*) increase the magnetic induction; (*b*) increase the mass of the particle; (*c*) both of these.

33-24 Relative to the gravitational interaction, the strength of the weak interaction is (*a*) greater; (*b*) about the same; (*c*) less.

Problems

(Use nuclear masses, where needed, from Appendix Table 9.)

33-A1 Show that the reaction ^{238}U(n, $\gamma)^{239}$U is possible. (Masses: ^{238}U = 238.0508; ^{239}U = 239.0543.)

33-A2 Calculate the energy released in the fission process $^{235}_{92}$U + $^1_0$n → $^{139}_{53}$I + $^{95}_{39}$Y + 2 $^1_0$n. (Masses: $^{235}_{92}$U = 235.0439; $^{139}_{53}$I = 138.9350; $^{95}_{39}$Y = 94.9134; $^1_0$n = 1.008665.)

33-A3 For one of the reactions *a*, *b*, *c*, or *d* of Sec. 33-3, verify the stated value of the energy release.

33-A4 Calculate the time for one semicircular revolution of a proton in a magnetic field of 1.8 T, assuming that the particle is moving slowly enough so that its rest mass (Appendix Table 1) can still be used.

33-A5 Calculate the time for one complete revolution of a very high energy particle in the synchrotron shown in Fig. 33-6 (radius of path, 1.00 km).

33-A6 In a large accelerator, to about what energy must protons be accelerated in a booster so that thereafter, as they gain energy, their speed differs from *c* by no more than 0.4%?

33-A7 A proton makes 1.5×10^7 revolutions per second in a cyclotron; what magnetic induction is required? (Assume the particle's speed is much less than *c*, so the non-relativistic mass of the proton in Appendix Table 1 can still be used.)

33-A8 A proton in a cyclotron gains 0.05 MeV each revolution. (*a*) How many revolutions must it make to reach a final energy of 0.2 GeV? (*b*) About how far will it travel during this time if the average path radius is 1 m?

33-A9 Calculate the wavelength, in fm, of the waves associated with an electron of energy 18 GeV that is moving in the Stanford Linear Accelerator (SLAC) almost at the speed of light. (*Hint:* The particle is highly relativistic; see Sec. 33-8.)

33-A10 Give the electric charge and the quark content of the following antiparticles: (*a*) antiproton; (*b*) anti-π^+; (*c*) anti-Σ^-.

33-B1 (*a*) If 201 MeV of energy is released, on the average, during fission of a ^{235}U nucleus, how many fissions per second take place in a reactor operating at a power level of 50 MW? (*b*) If the reactor operates for 1 year, how many ^{235}U nuclei undergo fission during this time? (*c*) What is the mass of ^{235}U fuel consumed during 1 year? (*d*) In how much natural uranium was this ^{235}U originally contained? (The abundance of ^{235}U is 0.72%.)

33-B2 Assuming an average release of 201 MeV per fission, show that only about 0.1% of the mass of the uranium fuel is actually transformed into energy in a reactor.

33-B3 Fission of ^{239}Pu releases 180 MeV per nucleus. (*a*) Express this energy yield in J/kg. (*Note:* For comparison, the heat of combustion of a good grade of coal is about 30×10^6 J/kg.) (*b*) How much ^{239}Pu must be used up, at 20% efficiency, to generate 10 million kWh of electric energy?

33-B4 The carbon cycle builds up ^{4}He out of four ^{1}H nuclei, in steps as follows: ^{12}C captures a proton, forming A; A decays by positron emission to B; B captures another proton to form C; C captures another proton to form D; D decays by positron emission, forming E; E captures a proton in a (p, α) reaction. Write out these reactions in detail; identify the nuclides A, B, C, D, and E; and show that the ^{12}C is regenerated at the end of the cycle, and thus is not used up.

33-B5 Calculate the average energy, in MeV, of a particle at the center of a star where the temperature is 40,000,000 K.

33-B6 Suppose that a fusion reactor is constructed that uses reaction *c* of Sec. 33-3. How much tritium per day must be used as fuel if the power output of the installation is to be 500 MW?

33-B7 The fuel value of garbage is about 11×10^6 J/kg (about $\frac{1}{3}$ that of a good grade of coal). The city of Paris burns 1.6×10^9 kg of garbage each year, 1.2×10^9 kg to supply electric power and 0.4×10^9 kg for heating purposes. (*a*) What is the annual monetary value, at 3¢/kWh, of the electric energy so generated, if the power station has 30% efficiency? (*b*) How much uranium would be used in a year to supply this thermal energy in a reactor? (The fuel value of natural uranium is 5.9×10^{11} J/kg.)

33-B8 Show that for thermal neutrons an absorbed dose of 1 rad is the same as 0.03 sievert.

33-B9 How many β decays take place per second due to ^{40}K in the 70-kg adult mentioned in Sec. 33-5? (*Hint:* You can use the value of 0.1 μCi given there, or you can find the disintegration constant from the half-life and then apply Eq. 32-2 on page 759. The isotope ^{40}K is rare, with abundance of only 0.012% of all potassium atoms.)

33-B10 The whole-body dose of γ radiation that is lethal to half the people receiving it (called LD50) is about 500 rem. (*a*) Express this dose in rad and in Gy. (*b*) Calculate the total energy of this dose, in J, for an 80-kg human, and compare with the KE the same human acquires by falling from a height of 1 m.

33-B11 How much time is required for a proton in the main synchrotron ring of Fig. 33-6 (radius 1.00 km) to increase in energy from an injection value of 8 GeV to the final value of 500 GeV, if it gains 2.5 MeV during each revolution?

33-B12 A proton in a cyclotron moves in a spiral path whose maximum radius (at point E in Fig. 33-4) is 0.50 m. If the proton at this point has a speed of 1.0×10^7 m/s, calculate (a) its energy, in MeV; (b) the vertical magnetic induction B needed to supply the necessary centripetal force on the moving proton.

33-B13 When the accelerator of Fig. 33-6 operates with a beam current of 2×10^{13} protons/s, at 500 GeV, what is the average current, in μA? What thermal energy, in MW, must be dissipated in the target?

33-B14 For a proton whose total energy is 17.00 GeV, calculate (a) the rest-mass energy; (b) the kinetic energy.

33-B15 (a) What is the rest-mass energy of a deuteron, in GeV? (b) What is the total energy of a deuteron whose KE is 16.30 GeV?

33-B16 (a) Use the average radius given in Fig. 33-9 to obtain the approximate volume of a gold nucleus. (b) Check the value for the density of nuclear matter ($\sim 2 \times 10^{14}$ g/cm^3) given on page 16. (c) Estimate the radius of the heaviest known nucleus, $^{262}_{105}$Ha. (*Hint:* If the density is constant, the volume is proportional to the mass number A.) (d) Identify a stable nuclide for which the radius is about 0.4 that of a gold nucleus.

33-B17 How much energy, in MeV, is released in reaction i of Sec. 33-3?

33-B18 Is the reaction $p + p \rightarrow d + \beta^+ + \nu_e$ a possible one, if the proton projectile has negligible KE? If so, what energy is released? If not, how much energy must be supplied?

33-B19 (a) What minimum energy must a neutrino ν_μ have in order for the reaction $\nu_\mu + n \rightarrow p + \mu^-$ to take place? (b) Repeat, for the reaction $\bar{\nu}_e + p \rightarrow n + e^+$.

33-B20 Using Fig. 33-15 and the data given there, compute the radiation intensity, in counts/s, at each of the two points of observation.

33-C1 An organic archeological specimen contains 84 g of carbon, with a ^{14}C activity of 720 dis/min. How long ago did the organism die?

33-C2 Any recently fallen meteorite has an activity of ^{26}Al ($T_{1/2} = 7.3 \times 10^5$ y), due to previous cosmic-ray exposure in space, of about 50 disintegrations per minute (dpm) per kg. A meteorite that was found in Antarctica had an activity of only 11.4 dpm per kg. What do these data indicate for the age of the ice sheet on which the meteorite was found?

33-C3 What average magnetic induction must the bending magnets in the main ring of the accelerator of Fig. 33-6 supply to keep a 1-TeV proton beam in an orbit of radius 1.00 km?

33-C4 Singly charged particles in a large accelerator move with orbital radius R. The bending magnets supply an induction B, and the particle energy in eV is V. If the particles are highly relativistic ($v \approx c$), show that $V = RBc$.

33-C5 (a) Calculate the energy, in joules, contained in 20 liters of ordinary water if the deuterium in the water is converted according to reaction a of Sec. 33-3. (*Hint:* Only 0.015% of the H atoms in the water are ^{2}H.) (b) Compare your answer with the energy that can be derived from 20 liters of gasoline which yields about 4×10^7 J/liter. (Give a ratio.)

33-C6 In the previous problem, you considered reaction a of Sec. 33-3. Can any more nuclear energy be extracted from the end-products of this reaction? If so, how, and how much?

33-C7 Calculate the energy available from 1 metric ton of Chattanooga shale that contains 50 parts per million of uranium by weight. Assume that a perfectly efficient breeder reactor is available, so all the U (or the Pu derived from it) is fissile, yielding 200 MeV/fission. Compare your answer with the energy content of coal, which is about 2×10^{10} J/ton.

33-C8 Using the cross section from Table 33-1, calculate the average distance L a slow neutron travels in carbon before being captured. (*Hint:* For an approximate calculation it will be sufficient to neglect the motion of the target atoms. Imagine a column of carbon 1 m^2 in cross-section and L m long. If all the carbon nuclei within this column were moved to the front face of the column, the total area would be obscured, each carbon nucleus obscuring an area equal to its cross section for capture. In this way, calculate the total number of atoms, and hence the mass, volume, and length of the column of carbon. The specific gravity of carbon is 2.25, and its atomic weight is 12.00.)

33-C9 Verify that each of the decay processes listed in Table 33-5 is consistent with the law of conservation of angular momentum (spin), using the model described in Sec. 32-11.

33-C10 Calculate the electric potential, relative to infinity, at a point that is 12 femtometers (12 fm) from a copper nucleus. (*Hint:* See Eq. 30-8.)

References

1. M. D. Kamen, "Tracers," *Sci. American* **180**(1), 31 (Jan. 1949).
2. O. Hahn, "Radioactive Methods for Geologic and Biologic Age Determinations," *Sci. Monthly* **82,** 258 (1956).
3. C. Emiliani, "Ancient Temperatures," *Sci. American* **198**(2), 54 (Feb. 1958). The ratio $^{16}O:^{18}O$ depends on the temperature that existed when a fossil carbonate was laid down.
4. K. G. McNeill, "Photonuclear Reactions in Medicine," *Physics Today* **27**(4), 75 (Apr. 1974). [A printing error was corrected in **27**(5), 71 (May 1974).] Activation of biological samples using photons instead of the more common neutrons.
5. O. Hahn, "The Discovery of Fission," *Sci. American* **198**(2), 76 (Feb. 1958).
6. E. B. Sparberg, "A Study of the Discovery of Fission," *Am. J. Phys.* **32,** 2 (1964).
7. H. G. Graetzer, "Discovery of Nuclear Fission," *Am. J. Phys.* **32,** 9 (1964).
8. J. D. Macdougall, "Fission-Track Dating," *Sci. American* **235**(6), 114 (Dec. 1976).
9. G. A. Cowan, "A Natural Fission Reactor," *Sci. American* **235**(1), 36 (July 1976). A vein of uranium ore in Africa went critical 2×10^9 years ago; the fission fragments are still there.
10. H. A. Bethe, "The Necessity of Fission Power," *Sci. American* **234**(1), 21 (Jan. 1976); see also letters to the editor, Apr. 1976, pp. 8–21.
11. H. C. McIntyre, "Natural-Uranium Heavy-Water Reactors," *Sci. American* **233**(4), 17 (Oct. 1975). Describes the CANDU type of reactor.
12. G. T. Seaborg and J. L. Bloom, "Fast Breeder Reactors," *Sci. American* **223**(5), 13 (Nov. 1970).
13. R. F. Post and F. L. Ribe, "Fusion Reactors as Future Energy Sources," *Science* **186,** 397 (1974).
14. K. Z. Morgan, "Comments on Radiation Hazards and Risks," *Phys. Teach.* **9,** 415 (1971); "Never Do Harm," *Environment* **13**(1), 28 (Jan./Feb. 1971).
15. R. Wilson and W. J. Jones, *Energy and the Environment* (Academic Press, New York, 1974). Reviewed in *Am. J. Phys.* **43,** 284 (1975). Includes a balanced study of various nuclear power options as of the mid-1970s.
16. B. L. Cohen, "The Disposal of Radioactive Waste from Fission Reactors," *Sci. American* **236**(6), 21 (June 1977).

17. B. L. Cohen, "Health risks of nuclear power," *Phys. Teach.* **16,** 526 (1978). A generally optimistic view of the problem.
18. M. C. Day, "Nuclear Energy: A Second Round of Questions," *Bull. Atomic Scientists* **31**(10), 52 (Dec. 1975).
19. A. M. Weinberg, "The future of nuclear energy," *Phys. Today* **34**(3), 48 (Mar. 1981).
20. M. K. Hubbert, "The Energy Resources of the Earth," *Sci. American* **225**(3), 60 (Sept. 1971).
21. P. W. Quigg, A two-part series on alternative energy sources in "World Environment Newsletter," appearing in *Saturday Review/World:* "New Energy Sources," Feb. 9, 1974, p. 47, and "Alternative Energy Sources, Feb. 23, 1974, p. 29.
22. T. D. Miner, "Energy Alternatives," *Phys. Teach.* **13,** 292 (1975). A tabular summary of an article by two industrial engineers.
23. P. H. Rose and A. B. Wittkower, "Tandem Van de Graaff Accelerators," *Sci. American* **223**(2), 24 (Aug. 1970).
24. E. L. Ginzton and W. Kirk, "The Two-Mile Accelerator," *Sci. American* **205**(5), 49 (Nov. 1961).
25. G. K. O'Neill, "Particle Storage Rings," *Sci. American* **215**(5), 107 (Nov. 1966).
26. R. R. Wilson, "The Batavia Accelerator," *Sci. American* **230**(2), 72 (Feb. 1974). Describes the Fermilab accelerator.
27. R. R. Wilson, "The Tevatron," *Phys. Today* **30**(10), 23 (Oct. 1977).
28. C. F. Tsang, "Superheavy Elements," *Phys. Teach.* **13,** 279 (1975). The island of stability.
29. C. E. Swartz, "A Temporary Organization of the Subatomic Particles," *Phys. Teach.* **13,** 542 (1975). Includes a large four-color chart.
30. M. L. Perl and W. T. Kirk, "Heavy Leptons," *Sci. American* **238**(3), 50 (Mar. 1978). Discovery of the tau particle.
31. M. Jacob and P. Landshoff, "The Inner Structure of the Proton," *Sci. American* **242**(3), 66 (Mar. 1980). See especially the first six pages.
32. S. L. Glashow, "Quarks with Color and Charm," *Sci. American* **233**(4), 38 (Oct. 1975).
33. K. A. Johnson, "The Bag Model of Quark Confinement," *Sci. American* **241**(1), 112 (July 1979).
34. "Quark Glue," *Sci. American* **241**(5), 84 (Nov. 1979). Indirect evidence for the gluon.
35. "QCD," *Sci. American* **240**(5), 90 (May 1979). A good, brief, nontechnical account of quantum chromodynamics.
36. "Quarkprints," *Sci. American* **243**(5), 82 (Nov. 1980). A proposed search for free quarks in a jet of falling drops of liquid mercury.
37. W. H. K. Panofsky, "Needs versus means in high-energy physics," *Phys. Today* **33**(6), 24 (June 1980). A census of particle accelerators and the outlook for the future.
38. G. Burbidge, "The Origin of the Cosmic Rays," *Sci. American* **215**(2), 32 (Aug. 1966).
39. F. Reines and J. P. F. Sellschop, "Neutrinos from the Atmosphere and Beyond," *Sci. American* **214**(2), 40 (Feb. 1966). Includes an illustration of an exchange of boomerangs to give an attractive force.
40. G. Gamow, "Half an Hour of Creation," *Physics Today* **3**(8), 16 (Aug. 1950).
41. J. A. Van Allen, "Radiation Belts Around the Earth," *Sci. American* **200**(3), 39 (Mar. 1959).

Mathematical Review

The study of physics will be of very little value to a student who is unsure of arithmetic and algebra, for by its nature physics deals with measurable quantities expressed in numbers. To study physics while encumbered with an inability to use numbers fluently is like trying to read a foreign language without an easy familiarity with vocabulary and syntax.* Mathematics is the language of science, and algebra is its elementary syntax.

For the main part of our study, you are expected to bring with you nothing beyond elementary high-school algebra. A certain small amount of trigonometry may prove helpful, but it is not necessary. The trigonometry that is used in this book is developed in the text itself as needed. Starting from algebra, we also define the basic concepts of calculus in Chap. 2, so that we can, throughout our study of physics, take advantage of the terminology of derivative (limit of a ratio) and integral (limit of a sum). The more detailed formulas of calculus are not used in the main part of the book; such formal calculus is developed and used in some of the optional For Further Study sections. Some additional material on elementary calculus is included in this Appendix for use by students reading these optional sections.

Sections A, B, and C in this Appendix have been prepared for those who wish to review their previous work in arithmetic, algebra, and geometry. Not everyone will need this review; you are encouraged to sample the portions dealing with topics about which you feel unsure. Many problems have been included so that you can test your mastery of each technique. *Solve as many of these problems as you need to, until you are sure that you could do one more of them with little hesitation or anguish.* Then you will be able to devote your valuable time and energy to thinking about *physics*, instead of continually wasting your efforts on what should be merely routine.

---

* The difficulty is not limited to foreign languages; many persons who are "educated" have failed to grasp ideas of politics, economics, or philosophy simply because they fail to understand and write their own language.

Learn the language and grammar so that you can understand and appreciate the concepts and ideas with which we are concerned.

It is suggested that you study Secs. A-1 through A-3, and B-1, within the first week of the course, filling in your areas of uncertainty and strengthening your mathematical muscles by repetition and practice.

Appendix A / Arithmetic

A-1 Elementary Operations

We assume that you know how to add, subtract, multiply, and divide. Some of these operations, especially long division, are tedious, and so we strongly recommend that you use a calculator for multiplication and division. There is a method resembling division for extracting the square root of a number, but you are strongly urged to *forget* this method, if you ever knew it. Instead, use a calculator, or *guess* the answer to 1 or 2 figures, which is often sufficiently accurate for the purpose at hand.

Multiplication is indicated in several ways: 31×42.6, or $31 \cdot 42.6$, or $(31)(42.6)$. We prefer the latter form, where the parentheses are used without any cross or dot between the numbers. The first parentheses are often omitted; thus $(31)(42.6)(3.1416)$ may be written as $31(42.6)(3.1416)$. The multiplication of algebraic quantities is indicated simply by putting the symbols next to each other, unless a pair of parentheses or brackets is needed to indicate one term that has several parts:

$$2.7x^2 \qquad 2x^2y^3 \qquad 31(41 + x)$$

Division is also indicated in several ways: $24 \div 6$, or $\frac{24}{6}$, or $24/6$. In order to facilitate canceling out common factors, we prefer the *horizontal line* between numerator and denominator. However, to save space, some quotients are written with the slant line, especially in the case of units, such as kg/m^3 or m/s^2.

A fraction may be simplified by canceling out equal factors from numerator and denominator.

Thus,

$$\frac{24}{36} = \frac{6(4)}{9(4)} = \frac{6}{9}$$

(dividing numerator and denominator by 4); also $\frac{24}{36} = \frac{2}{3}$ (dividing through by 12). Often we can apply the rule for division: *invert the divisor and multiply*. Thus

$$42 \div \tfrac{6}{11} = 42 \times \tfrac{11}{6} = 77$$

Rounding off. In this text we almost always "round off" the answers to three significant figures (see Sec. A-4).

$$\frac{(10)(81)}{14} = \frac{405}{7} = 57.857142857 \ldots = 57.9$$

Here we call $57.85714 \ldots$ equal to 57.9, since it is closer to 57.9 than to 57.8. What to do about rounding off 42.450000000 to three figures is not a world-shaking problem; either way is acceptable, but usually the higher value (42.5) is preferred. To *two* figures, 42.450000000 becomes 42, since it is surely closer to 42 than to 43. The number π can be rounded off in various ways, all equally "correct":

3.141592654 ...
3.14159265
3.1415927
3.141593
3.14159
3.1416
3.142
3.14
3.1
3.

(Answers are on page 842.)

Multiply:

1 3(42.1)
2 30(0.421)
3 $6(1 + \frac{12}{3})$

4 $(4 + 0.4)[111 + 0.1(110)]$
5 $3.1(20)^2$
6 0.0015(200)

Divide:

7 70/35
8 700/3.5
9 0.70/(3.0 + 0.5)

10 (32)(0.15)/8
11 720/0.072

Calculate:

12 $\frac{22}{7}$

13 $\frac{(14)(5)(4)}{(3)(7)(2)}$

14 $\pi(21)^2$

15 $\frac{(57.8573)(\frac{16}{3})}{2}$

16 $\frac{1}{(1 + \frac{1}{2} + \frac{1}{4})}$

17 $2\sqrt{9}$

18 $(\frac{1}{3})(\frac{6}{5})(\frac{1}{21})(84)$

19 $.02\sqrt{900}$

20 30% of 600

21 2.5% of 4.4

22 0.5% of the difference between 43.0 and 43.6

A-2 Powers and Roots—Exponents

When we write $y = x^n$, we call n the *exponent* and say that we have raised x to the nth power. Thus $2^4 = 2 \times 2 \times 2 \times 2 = 16$; we use 2 as a factor 4 times to raise it to the 4th power. The exponent 1 means only one factor, so $n^1 = n$, where n is any number. Another special case is the exponent 0; this is *defined* so that $n^0 = 1$, where n is any nonzero number. We illustrate, by examples, some rules of exponents.

Multiplication. The rule for multiplying numbers that are expressed as powers is summarized by the formula

$$(x^a)(x^b) = x^{a+b}$$

For example,

$$(2^4)(2^3) = 2^7$$

or

$$(16)(8) = 128$$
$$128 = 128$$

Note that exponents are *added*, but only if they have the same base (2 in the above example). There is no way of simplifying an expression such as $(2^5)(3^2)$, where the exponents have different bases that do not have a common factor. It is also true that

$$x^n y^n = (xy)^n$$

which allows some simplification if the same exponent is involved. Thus,

$$(2^3)(3^3) = (6)^3$$

or

$$(8)(27) = 216$$
$$216 = 216$$

The expression $(x^a)^b$ means (x^a) raised to the bth power. That is, $(x^a)(x^a)(x^a)(x^a) \ldots$ (b times), which equals $x^{a+a+a+a+\cdots}$, or x^{ba}. Therefore we can write

$$(x^a)^b = x^{ab}$$

For example,

$$(5^2)^3 = 5^6$$
$$(25)^3 = 15{,}625$$
$$15{,}625 = 15{,}625$$

Division. The rule for dividing numbers that are expressed as powers is

$$\frac{x^a}{x^b} = x^{a-b}$$

The rule is valid only if the exponents have the same base. Thus,

$$\frac{2^5}{2^3} = 2^2$$

$$\frac{32}{8} = 4$$

$$4 = 4$$

Also,

$$\frac{y^n}{x^n} = \left(\frac{y}{x}\right)^n$$

which is illustrated by

$$\frac{6^3}{2^3} = \left(\frac{6}{2}\right)^3$$

$$\frac{216}{8} = 3^3$$

$$27 = 27$$

Negative exponents are used to indicate division. By definition,

$$x^{-n} = \frac{1}{x^n}$$

Thus, $2^{-3} = \dfrac{1}{2^3} = \dfrac{1}{8}$. Let us write one of the previous examples, using negative exponents and multiplication:

$$\frac{2^5}{2^3} = (2^5)(2^{-3}) = 2^{5+(-3)} = 2^2 = 4$$

A more complicated example is this one involving powers of 3:

$$\frac{(3^3)(3^2)}{3^4} = (3^3)(3^2)(3^{-4}) = 3^{3+2-4} = 3^1 = 3$$

As a check, straightforward division gives the same result:

$$\frac{(27)(9)}{81} = \frac{243}{81} = 3$$

Fractional exponents are used to indicate roots. By definition,

$$x^{1/n} = \sqrt[n]{x}$$

Thus,

$$16^{1/2} = \sqrt{16} = 4$$

By combination of our previous rules for exponents, any fractional exponent can be interpreted.

$$16^{3/2} = (16^{1/2})^3 = (\sqrt{16})^3 = 4^3 = 64$$
$$27^{2/3} = (27^{1/3})^2 = (\sqrt[3]{27})^2 = 3^2 = 9$$

$$100^{-3/2} = \frac{1}{100^{3/2}} = \frac{1}{(100^{1/2})^3} = \frac{1}{(\sqrt{100})^3}$$

$$= \frac{1}{10^3} = \frac{1}{1000} = 0.001$$

PRACTICE PROBLEMS

Simplify the following:

23 $(x^5)(x^2)(x^3)$

24 $(x^{20})(y^4)(x^3)$

25 $(a^3)(a^{-1/2})$

26 $\dfrac{2(2^6)}{2^3}$

27 10^{-2}

28 $(3t)^2$

29 $[(1+5)^2 - 32]^2$

30 $(16^{-3/2})(2^7)$

31 $p^{1.4}\sqrt{p}$

32 $46^1 - 23^0$

33 $\sqrt{\dfrac{x^n}{x^{n-2}}}$

34 $3(10^{10}) \div 2(10^7)$

35 $(x^{1/n})^n \div x$

36 $(x^n)^{1/n} - x$

A-3 Powers of Ten—Scientific Notation

Although all that has been said about exponents applies, whatever the base may be, our decimal system makes *powers of ten* of major importance. We have

10^0	= 1	10^0	= 1
10^1	= 10	10^{-1}	= 0.1
10^2	= 100	10^{-2}	= 0.01
10^3	= 1000	10^{-3}	= 0.001
10^4	= 10,000	10^{-4}	= 0.0001
10^5	= 100,000	10^{-5}	= 0.00001
10^6	= 1 million	10^{-6}	= one millionth
$\vdots$		$\vdots$	
10^9	= 1 billion	10^{-9}	= one billionth
10^{12}	= 1 trillion	10^{-12}	= one trillionth

Any number can be expressed as some power of 10 multiplied by another number between 1 and 10. This is often called *scientific notation*. Thus $4163.8 = 4.1638 \times 1000 = 4.1638 \times 10^3$. (Moving the decimal point 3 places to the left decreases the number by a factor of 10^3, so we multiply by 10^3 to keep the value the same.) If you have trouble in placing the decimal point, perhaps the following scheme will help. Suppose we have the number 0.0000367, which we wish to express in scientific notation.

$$0.0000367 = 3.67 \times 10^?$$

One way to proceed is to place the point of your pencil between the 3 and the 6 (thus pointing out the number 3.67, which is between 1 and 10). Think of this as 3.67×10^0 (say it to yourself). Now move the pencil 5 places to the left, one zero at a time, stopping when your pencil reaches the actual decimal point. Your number is therefore 3.67×10^{-5}. The same method can be used for expressing large numbers. The number of seconds in a year is about $(365\frac{1}{4})(24)(60)(60)$; the exact figure is 31,556,925.9747. Start with the pencil point between the 3 and the 1, so that the number is $3.15569259747 \times 10^?$. Moving over to the right and counting, you reach the actual decimal point after saying "seven." Hence there are $3.15569259747 \times 10^7$ seconds in a year.

Multiplication and division. To multiply or divide numbers expressed in scientific notation, combine the powers of 10 by addition or subtraction of the exponents, and use ordinary multiplication or division for the "ordinary" numbers preceding the powers of 10.

$$(2.5 \times 10^8)(3 \times 10^6) = (2.5 \times 3) \times 10^{8+6}$$
$$= 7.5 \times 10^{14}$$

$$\frac{42 \times 10^{-16}}{7 \times 10^{10}} = \frac{42}{7} \times 10^{-16-10} = 6 \times 10^{-26}$$

$$\frac{(2 \times 10^6)(8 \times 10^{11})(3 \times 10^{-4})}{(4 \times 10^{-5})(10^3)}$$

$$= \frac{(2)(8)(3)}{(4)(1)} \times \frac{10^{13}}{10^{-2}} = 12 \times 10^{15}$$

Sometimes it pays to revise a numerator or denominator to make the division come out easier.

$$\frac{1.8 \times 10^{16}}{9 \times 10^{-4}} = \frac{1.8 \times 10^{16}}{0.9 \times 10^{-3}} = 2 \times 10^{19}$$

or

$$\frac{1.8 \times 10^{16}}{9 \times 10^{-4}} = \frac{18 \times 10^{15}}{9 \times 10^{-4}} = 2 \times 10^{19}$$

Considering the great differences among physical magnitudes, it is not surprising that powers of 10 are widely used. Objects range in size from a nuclear particle (about 10^{-14} m) up to whole galaxies (about 10^{21} m), with man more or less in the middle at about 2×10^0 m. Aside from convenience, scientific notation is often more informative than ordinary numbers. When we say that a light-year is 9,460,000,000,000,000 m, we immediately see that this length unit is a tremendous number of meters, but just how tremendous we don't immediately comprehend. Writing it as 9.46×10^{15} m allows us to make immediate comparison with other large numbers.

Example

Compare the light-year with the distance from earth to the sun (about 150,000,000 kilometers, or 93,000,000 miles).

We want the ratio of these two numbers. Expressing them both in scientific notation (remembering that 1 km equals 10^3 m), we have

$$\frac{9.46 \times 10^{15} \text{ m}}{1.50 \times 10^{11} \text{ m}} = 6.3 \times 10^4 = 63 \times 10^3$$

That is, a light-year is 63 thousand times as far as the distance from the earth to the sun. When it is expressed in this way, we can begin to "feel" the relationship, which remains obscure unless we use the scientific notation.

PRACTICE PROBLEMS

Express in scientific notation:

37 142.63	41 86,400
38 1,500,000	42 0.0000000000000144
39 0.00336	43 0.12500
40 4,663,310.56	44 0.00125

Express in ordinary notation:

45 1.63×10^7	47 1.01×10^6
46 4.781354×10^3	48 9.81×10^1

49 3.11×10^{-4} **51** 8.001×10^{-4}
50 6.55×10^{13} **52** 9.3×10^{-6}

Evaluate the following:

53 $(1.6 \times 10^2)(3 \times 10^6)(2 \times 10^{11})$

54 $\dfrac{(1.6 \times 10^2)(3 \times 10^6)}{2 \times 10^{11}}$

55 $\dfrac{(3200)(200)}{(1.6 \times 10^3)(1000)}$

56 $\dfrac{(0.0022)(600)}{3 \times 10^{-10}}$

57 $\dfrac{16 \times 10^{-2}}{16 \times 10^2}$

58 $(6.02 \times 10^{23})(1.60 \times 10^{-19})$

59 $\dfrac{1.08 \times 10^{10}}{(2 \times 10^6)(3 \times 10^{-2})(9 \times 10^{20})}$

60 $\dfrac{(5.00 \times 10^{-5})(2400)}{(0.00008)(3.00 \times 10^{10})}$

Square roots and cube roots require careful consideration in scientific notation. To take a square root we must divide the exponent by 2; but since a fractional power of 10 is not useful to us, we must first see to it that the exponent is an even number.

$$\sqrt{9 \times 10^{16}} = \sqrt{9} \times \sqrt{10^{16}} = 3 \times 10^8$$
$$\sqrt{4.9 \times 10^5} = \sqrt{49 \times 10^4}$$
$$= \sqrt{49} \times \sqrt{10^4} = 7 \times 10^2$$
$$\sqrt{0.00000025} = \sqrt{25 \times 10^{-8}} = 5 \times 10^{-4}$$

Similarly, a cube root is taken only if the exponent of 10 is divisible by 3:

$$\sqrt[3]{6.4 \times 10^8} = \sqrt[3]{640 \times 10^6}$$
$$= 8.62 \times 10^2$$
$$\sqrt[3]{1.25 \times 10^{-13}} = \sqrt[3]{125 \times 10^{-15}}$$
$$= 5 \times 10^{-5}$$

PRACTICE PROBLEMS

Find the square roots of the following:

61 $810,000$ **64** 2.5×10^7
62 $81,000$ **65** 0.004
63 1.69×10^{-12} **66** 0.0004

Find the cube roots of the following:

67 8000 **70** 10^8
68 0.008 **71** 6.4×10^{22}
69 1.25×10^{-19} **72** 0.000000027

A-4 Significant Figures

A number that represents the result of a physical measurement does more than merely indicate the *value* of the measured quantity. Some indication of the *precision* of the measurement is given by the form in which the number is written. For example, if a car's speed is reported by observer A to be 65.3 km/h, he might consider this to mean that the speed is somewhere between 65.25 km/h and 65.35 km/h; for if it were 65.36 km/h, for instance, he would report it as 65.4. Another observer B might assign a greater uncertainty to the last figure. He might consider that 65.3 indicates a number lying between 65.25 km/h and 65.35 km/h; for if it range of a few tenths of a kilometer per hour on either side of the stated value of 65.3 km/h.

A *significant figure* is one that has *some* significance but does not necessarily denote a certainty. In the example just mentioned, the 3 of 65.3 is "significant," but the exact degree of significance depends on the experimental techniques, habits, and temperaments of observers A and B. Thus we cannot make any simple statement about significant figures. About all we can do is to lay down the following rule:

The last figure written down should be the first uncertain figure.

When we read that the velocity of light is 2.99793×10^8 m/s, we know that the 3 is uncertain, but just how uncertain we cannot tell. On the other hand, the last 9 is certain. It would be wrong to write 2.99793×10^8, *knowing* that the last 9 is uncertain, for then the 3 would be entirely meaningless. This would be like trying to balance a checkbook to the nearest penny when you are uncertain about the dollars or dimes.

In the beginning physics laboratories, we often strive for "three-figure accuracy." We measure the density of iron as 7.86 g/cm³, and by writing it in this way we indicate that the 6 is uncertain by some (unspecified) amount. If the apparatus is not capable of three-figure accuracy, then the answer must be rounded off as necessary, to 7.9 g/cm³ or even to 8 g/cm³. Your laboratory manual (if you are using one) will cover this point more fully; at present we wish only to emphasize that the exact degree of significance of the final significant figure must remain somewhat vague in our elementary treatment.

Example

The dilemma of significant figures is illustrated by a hypothetical experiment in which two observers measure the time required for a stream of water to fill a bucket. Observer *A* gets 98 s; observer *B* gets 102 s. Each measures to the nearest second or so, and so the last figure in each measurement is "significant." However, *A*'s result has two figures, and *B*'s has three. A simple criterion of "two significant figures" or "three significant figures" would be misleading here. A trained physicist might report her result thus: 98.1 s $\pm$ 2.2 s, meaning that 98.1 is her best estimate of the value, and that there is a 68% chance that the correct value is within 2.2 s (one standard deviation) of 98.1 s. This violates our rule that the last figure written down is the first uncertain figure; but by stating the standard deviation to be 2.2 s, the observer makes clear just *how certain* her value is, without any mention of significant figures at all.

Significant zeros. A special problem is posed by zeros, for zeros are used to indicate the location of the decimal point as well as to indicate a value. In the following example, only the zeros that are underlined are significant; the other zeros are used to fill out the space to the decimal point and are not significant figures.

Example

4.2$\underline{0}$1

4.2$\underline{0}$1$\underline{00}$ (Here the final two zeros need not have been written; hence the fact that they *are* written means that they are significant. The uncertainty is in the 5th decimal place.)

0.00420$\underline{1}$0

42.$\underline{0}$1

A number such as 86,400 g is ambiguous, for we cannot tell whether the two zeros are just to locate the decimal point (in which case the error might be 100 or 200 g), or whether they are really significant (in which case the true value is within several grams of being *exactly* 86,400 g). There are three ways in which this ambiguity can be avoided: (1) Use scientific notation, and write 8.64×10^4 g *or* 8.6400×10^4 g to distinguish between the two possibilities. (2) Use a different unit for the mass, which will place the decimal point elsewhere. Write 86.4 kg *or* 86.400 kg to distinguish the two cases. (3) Use a standard deviation or limit of error; thus, 86,400 $\pm$ 8 g is a definite statement, provided we know the precise meaning of the author's $\pm$ sign. In this book, we shall prefer methods (1) and (2) as useful compromises that are sufficiently precise for most purposes.

Arithmetic operations and significant figures. Consider the circumference of a circle whose diameter is 12.3 cm. In order to determine this quantity, we need to multiply 12.3 by π, which is 3.1416.... If the 3 of 12.3 is uncertain, then all numbers following the 3 are unknown and can be denoted by x. Carrying out the multiplication,

$$
\begin{array}{r}
12.3x \\
3.1416 \\
\hline
738x \\
123x \\
492x \\
123x \\
369x \\
\hline
38.64168x
\end{array}
$$

we see that the product should be rounded off to 38.6, since the 4, 1, 6, and 8 are really *unknown* because an unknown "x" has been added to each of these digits. Even the 6 of 38.6 is uncertain, because it depends on a 9 arising from the uncertain 3 of 12.3. Writing the answer as 38.6 cm expresses this correctly; the last figure written down is the first uncertain one.

Similar results are obtained for division.

Addition and subtraction behave somewhat differently. For example, an orange weighing 3.1 oz is added to a basket containing 12.71 oz of lemons and 58.2 oz of grapefruit. The total weight is 74.0 oz of citrus fruit:

$$
\begin{array}{r}
12.71xx \\
58.2xxx \\
3.1xxx \\
\hline
74.0xxx
\end{array}
$$

We can generalize as follows: In multiplication or division, the result has the same number of significant figures as the least accurate number entering into the multiplication or division. In addition or subtraction, the last significant figure of the result is in the same column relative to the decimal point as is the last significant figure of the least accurate number entering into the sum or difference. These are approximate, rough-and-ready rules; an exact treatment would involve standard deviations, and "significant figures" would not even be mentioned.

PRACTICE PROBLEMS

In each case, express the answer to the proper number of significant figures.

73 What fraction of a minute is 45.0 s?

74 What fraction of a minute is 45.00 s?

75 How many days are contained in 2.1 years?

76 What is the area of a square, each edge of which is 3.041 m in length?

77 If a car travels for 3.00 h at 61.34 km/h, how far does it go?

78 If a car travels for 3.0 h at 61.3415 km/h, how far does it go?

79 What is the height (in feet) of a boy who measures up at 66.00 in.? (*Note:* The conversion factor, 12 in. equals 1 ft, is as exact as you please, since this is a matter of definition.)

80 Do not work out, but state how many significant figures would be proper for the answers to each of the following:

$$
\frac{(144)(2.06)}{11.8943}; \quad \frac{(3624.8)(28.1)}{(44)(100.23)};
$$

$$
(3.652 \times 10^8)(42.8 \times 10^{-6})
$$

81 A tank contains 42.8 liters of water; how much will be left if 3.72 liters are removed?

82 How many seconds are there in a day?

83 Add: $4.22 \times 10^5 + 3.11 \times 10^7 + 6.003 \times 10^6$.

84 The mass of an electron is 9.11×10^{-28} g. What is the total mass of 6.02×10^{23} electrons?

85 One student measures a length as 84.62 cm, and another measures the same length as 84.70 cm. By how much do these measurements differ? What is the percent difference (the difference divided by the average value)?

86 In a commendable effort to give exposure to metric units, the center-field wall at Riverfront Stadium in Cincinnati, Ohio, is marked "404" [feet] and also "123.13" [meters]. Comment on these markings.

Significant figures in textbooks. Ideally, problems in a textbook should be formulated with strict attention to standard deviations and significant figures. However, we have adopted a rule that is easy to apply, even though not theoretically sound. In examples and problems in this text, all numbers are assumed to be *precise enough* to yield a final answer having three significant figures, and final significant zeros are often omitted from both the statements of problems and their answers. Purists may object, but we justify our course of action on the following grounds:

1. The rules about significant figures are themselves not rigid. Exceptions can easily be found, and some have already been illustrated. (Does 98 have the same number of significant figures as 102?) We see no reason to abandon one wrong treatment in favor of another wrong treatment. The three-figure rule we use, though known to be wrong, is clearly stated and is definite enough to apply consistently.

2. Examples and problems—hundreds of them—are scattered through the text to help you understand the principles of physics. Their purpose is *not* primarily to give a model for recording experimental data in the laboratory. Your laboratory manual will, rightly, be concerned with this aspect of quantitative science, but in the text we have adopted a working rule for the sake of giving a more immediately understood illustration of laws and principles.

3. Where it serves a purpose, as in tables of experimentally observed properties of matter, physical data are given with the number of significant figures indicating (roughly) the precision of measurement.

4. There is real value in simplifying the routine arithmetic associated with problem solving. This is why "simple" numbers are often used. On the other hand, students wish to check their work against something more definite than an answer that has been rounded off to one or two significant figures. We give answers to three significant figures and *assume* or *imply* that the given data warrant such precision.

As examples of our treatment of significant figures, consider two examples in the text. In Example 3-7 we find it necessary to compute (6)(9.8), and we give the answer as 58.8. Here we assume the acceleration due to gravity is 9.80 m/s^2, even though we write it and speak of it as simply 9.8 m/s^2. In solving Problem 13-B13, we find that 4.3 liters of turpentine overflows from a 200-liter drum. In this case, we must state the volume of the drum to be 200.0 liters, giving the necessary significant figures so that the subtraction $200.0 - 4.3$ has meaning.

Appendix B / Algebra

B-1 The Rules of Algebra

The methods of algebra are used in physics for much the same reason that arithmetic is used. In each case we are dealing with quantities, but if a quantity is unknown in magnitude, we represent it by a letter or other symbol. For some reason, the symbol x has come to be the commonest one for an unknown, but you should realize that *any* symbol will do. Generally the symbol should be chosen to give some hint as to what it represents: F for force, m for mass, p for pressure, t for time, and so on. Greek letters are often used for angles, especially θ (theta) and ϕ (phi). The symbols x, y, and z usually refer to distances or coordinates.

The "ground rules" of algebra can be summarized by a number of axioms, many of which are no doubt familiar to you. We illustrate these rules by solving some typical equations with their aid.

Rule 1: An equation remains true if equal quantities are added to or subtracted from each side.

$$x + 3 = 4$$

Hence,

$$x = 1 \quad \text{(subtracting 3 from each side)}$$

Another way of looking at this rule is to say that any separate term can be moved from one side of the equal sign to the other, if the sign of that term is reversed.

$$x + 3 = 4$$
$$x = 4 - 3 \quad \text{(moving the 3 to the other side, and changing its sign)}$$
$$x = 1$$

Rule 2: An equation remains true if both sides are multiplied or divided by the same quantity.

$$4F = 17.0$$
$$F = 4.25 \quad \text{(dividing each side by 4)}$$

Rule 3: An equation remains true if both sides are raised to the same power.

$$v^2 = 441$$
$$v = \sqrt{441} \quad \text{(raising each side to the } \tfrac{1}{2} \text{ power—}$$
$$\text{that is, taking square roots)}$$
$$v = 21$$

Rule 4: In a proportion, such as $\dfrac{a}{b} = \dfrac{c}{d}$**, we can cross-multiply to obtain the new equation** $ad = bc$**.**

$$\frac{4}{30} = \frac{22}{p}$$
$$4p = 30(22) = 660$$
$$p = 165$$

To solve most equations, several of the basic procedures are used, one after the other. Thus, the distance s traveled by a bullet while slowing down is to be found from the equation

$$(300)^2 = (400)^2 + 2(-5000)s$$
$$90{,}000 = 160{,}000 - 10{,}000s$$
$$-70{,}000 = -10{,}000s \quad \text{(Rule 1)}$$
$$s = \frac{-70{,}000}{-10{,}000} = 7 \quad \text{(Rule 2)}$$

Here is an equation that is to be solved for the elapsed time t, where it is known on physical grounds that t is greater than 2:

$$4 = \tfrac{1}{2}(32)(t-2)^2$$
$$8 = 32(t-2)^2 \quad \text{(Rule 2)}$$
$$\text{(multiply each side by 2)}$$
$$0.25 = (t-2)^2 \quad \text{(Rule 2)}$$
$$\text{(divide each side by 32)}$$
$$0.50 = t - 2 \quad \text{(Rule 3)}$$
$$\text{(take the positive square root of each side)}$$
$$t = 2.50 \quad \text{(Rule 1)}$$
$$\text{(add 2 to each side)}$$

Solve for the unknown:

87 $3x + 42 = 0$

88 $1 - 0.4p = 2$

89 $(T + \tfrac{1}{2})4 = 5$

90 $\dfrac{50}{y} = \dfrac{y}{2}$

91 $\dfrac{1 + 2m}{2m - 1} = \dfrac{14}{10}$

92 $13^2 = 25 + 2(a)(45)$

93 $\tfrac{1}{2}(L^3 - 4) - 30 = 0$

94 $\sqrt{\dfrac{h}{100}} = 0.8$

Solve for x in terms of the other variables and the constants:

95 $3x + y - 14 = 2 + 3y$

96 $mg - 2x = F + 200$

97 $y = \tfrac{1}{2}x - y^2$

98 $\dfrac{x-3}{x-1} = \dfrac{2y}{y-3}$

B-2 Simultaneous Equations

If there are two unknowns, then two equations are required to solve a problem. In general, if there are n unknowns, n independent equations are needed for solution. To solve a pair of simultaneous equations, a routine method is to isolate one unknown by solving one equation for this unknown in terms of the other one. Then we substitute this expression into the second equation. For example, we wish to find x and y from the pair of equations

$$\begin{cases} 4x + y = 7 \\ x - 2y = 4 \end{cases}$$

We first work on the top equation to find y in terms of x:

$$4x + y = 7$$
$$y = 7 - 4x \quad \text{(Rule 1)}$$

We then substitute this expression for y into the second equation:

$$x - 2y = 4$$
$$x - 2(7 - 4x) = 4$$
$$x - 14 + 8x = 4$$
$$9x = 18 \quad \text{(Rule 1)}$$
$$x = 2 \quad \text{(Rule 2)}$$

Now that we know x, we can use either equation to find y:

$$4x + y = 7$$
$$4(2) + y = 7$$
$$8 + y = 7$$
$$y = -1 \quad \text{(Rule 1)}$$

As a final check, we substitute these values of x and y into each equation and verify that the original equations are satisfied:

$$4(2) + (-1) \overset{?}{=} 7; \qquad 7 = 7 \quad \text{(O.K.)}$$

and

$$2 - 2(-1) \overset{?}{=} 4; \qquad 4 = 4 \quad \text{(O.K.)}$$

This is not the only way to solve a pair of simultaneous equations; refer to any algebra text for further details of several other methods, including one involving determinants. However, you will not go wrong by following the steps above.

B-3   Quadratic Equations

Almost all the equations so far discussed are called "linear," since the unknowns are raised only to the first power (such equations have straight *lines* as their graphs). From time to time we will have occasion to solve a more complex equation known as a *quadratic equation* because the unknown appears raised to the second power as well as to the first power. As an example, the time t required for a racing sled to move 225 m might be given by the following quadratic equation:

$$225 = 20t + 5t^2$$

A routine method of solution is to use the *quadratic formula*: If

$$ax^2 + bx + c = 0$$

then

$$x = \frac{-b \pm \sqrt{b^2 - 4ac}}{2a}$$

The $\pm$ (plus-or-minus) sign means that there are *two* solutions. Using the $+$ sign gives

$$x = \frac{-b + \sqrt{b^2 - 4ac}}{2a}$$

and using the $-$ sign gives

$$x = \frac{-b - \sqrt{b^2 - 4ac}}{2a}$$

Usually only one of the two solutions fits the physical problem, although both solutions are, of course, algebraically correct. Naturally, b^2 must be greater than $4ac$ so that $\sqrt{b^2 - 4ac}$ gives a real number.* If b^2 is less than $4ac$, there is no real solution.

Applying the quadratic formula to our example, we first rearrange the terms in the equation to place the term in t^2 first, the term in t second, and the constant term last.

$$5t^2 + 20t - 225 = 0$$

$$t = \frac{-20 \pm \sqrt{(20)^2 - 4(5)(-225)}}{2(5)}$$

$$t = \frac{-20 \pm \sqrt{400 + 4500}}{10}$$

$$t = \frac{-20 \pm \sqrt{4900}}{10} = \frac{-20 \pm 70}{10}$$

The two answers are

$$t_1 = \frac{-20 + 70}{10} = \frac{50}{10} = 5 \text{ s}$$

and

$$t_2 = \frac{-20 - 70}{10} = \frac{-90}{10} = -9 \text{ s}$$

If this were a problem in pure algebra, we would stop here, with two equally correct answers, 5 s and -9 s. However, from the physics of the situation we know that the desired time is posi-

---

* If $b^2 = 4ac$, the square root is 0, and the two solutions coincide.

tive, so we reject* the solution $t = -9$ s, keeping only the solution $t = 5$ s.

These brief remarks and illustrations give a sufficient guide for a review of algebra as it is used in this text. For further information, consult any high-school algebra text. Our review has been of techniques only, with no attempt to set up "word problems." The business of transferring the data of physical problems into mathematical form is a necessary first step, and a most important step; algebra cannot be used successfully if the equations are wrong to begin with. However, discussion of how to set up the equations belongs in the physics text itself, rather than in this Appendix.

PRACTICE PROBLEMS

99 Solve for u and v: $3u + v = 7$; $2u - v = 3$.
100 Solve for p and q: $q = p + 2$; $p + q = 8$.
101 Solve for A and B: $A - 4B = -6$; $2B + A = 18$.
102 Solve for x and y: $10x + 2y = 11$; $20(x - y) = -8$.

Find the real solutions, if any, of the following quadratic equations:

103 $2t^2 - 3t + 1 = 0$
104 $v^2 + 5v + 6 = 0$
105 $x^2 - 10 = 3x$
106 $y^2 + 1 = -y$
107 $3 + z^2 + 2z = 2$
108 $4\theta^4 - 5\theta^2 + 1 = 0$ (*Hint:* Let θ^2 be the variable; call it x if you wish. Solve for θ^2 by the quadratic formula, and then find θ.)

B-4   Logarithms

The logarithm of a number is the power to which a base must be raised to obtain the number. Ordinary or "common" logarithms are to base 10:

$$\text{If } y = 10^x, \text{ then } x = \log y$$

For example, $\log 10^5 = 5.000$; $\log 2 = 0.3010$. The base of natural logarithms is e, a number

---

whose value is $2.718218 \ldots$:

$$\text{If } y = e^x, \text{ then } x = \ln y$$

For example, $\ln 10 = 2.3026$; $\ln 1000 = 6.9078$. Logarithms to base e are written as "ln" to distinguish them from logarithms to base 10, which are written as "log." For any number N, the two logarithms are related as follows:

$$\ln N = 2.3026 \log N$$

There are several definitions of e, given in any calculus book. One interesting and useful property is related to the slope of a graph. If $y = e^x$, then the slope, given by dy/dx, is also e^x. This is not true for any other base; the slope of $y = 10^x$ is *not* 10^x.

What follows applies to logarithms to any base.

Multiplication.

If $N = AB$, then $\log N = \log A + \log B$

Division.

If $N = A/B$, then $\log N = \log A - \log B$

Exponentiation.

If $N = A^n$, then $\log N = n \log A$

Some special cases of this last rule are: $\log 1 = 0$; $\log (10^x) = x$; $\ln (e^x) = x$.

Logarithms are no longer used for routine numerical calculations (although many pocket calculators perform exponentiation by internally calculating and using logarithms to base 2). In our study of physics, logarithms are used to interpret graphs of experimental data. For example, the activity (in counts/min) of a radioactive sample might be decreasing according to an exponential equation:

$$N = N_0 e^{-\lambda t}$$

where N_0 is the initial activity and t is the elapsed time. Taking logarithms of both sides of this equation gives

$$\ln N = \ln N_0 + \ln (e^{-\lambda t})$$
$$\ln N = \ln N_0 - \lambda t$$
$$2.30 \ln N = 2.30 \ln N_0 - 2.30 \lambda t$$
$$\log N = \log N_0 - 2.30 \lambda t$$

---

* A physical interpretation of a negative time is possible; see Prob. 2-C7 of the text.

Thus a graph of either $\ln N$ or $\log N$ is a straight line. From the intercept the value of N_0 can be found, and from the slope the value of λ can be found.

PRACTICE PROBLEMS

109 From the value of $\log 2$ given above, calculate $\ln 2$.

110 Starting with the value of $\log 100$, calculate the value of $\ln 100$ and verify that $\ln 100 = \ln 1000 - \ln 10$.

111 Calculate the value of $\ln [(2e^5)^3]$.

112 If $p = 4e^{5x}$ and $q = 2e^{-x}$, calculate $\ln (p/q)$ if $x = 3$.

113 If you have a calculator with exponentiation capability, find the value of y if $\log y = 2.4350$.

Appendix C | Plane Geometry and Trigonometry

C-1 Plane Geometry

As a brief review of plane geometry, we give without proof several useful theorems.

1. The sum of the angles of any triangle is 180°.

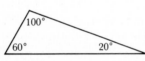

2. Two triangles are similar if two of their angles are equal.

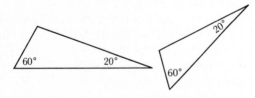

3. The corresponding sides of similar triangles are proportional.

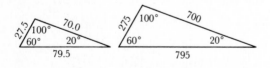

4. The two acute angles in a right triangle add up to 90°. (Such angles are called complementary angles.)

5. Two angles are equal if any of the following is true:

(a) Their sides are parallel and in the same sense.

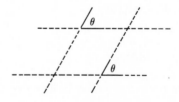

(b) Their sides are mutually perpendicular.

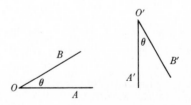

(*c*) They have the same complement.

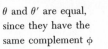

θ and θ' are equal, since they have the same complement ϕ

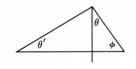

(*d*) They are vertical angles.

6. Some special theorems apply to right triangles. Most important is the Pythagorean theorem: The sum of the squares of the two legs equals the square of the hypotenuse. Some right triangles of importance are the 30-60-90 triangle, in which the short side is half the hypotenuse, and the 45-45-90 triangle, whose sides are in the ratio $1:1:\sqrt{2}$. The 3-4-5 triangle has acute angles that are approximately $37°$ and $53°$.

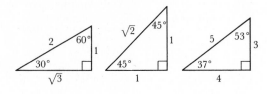

C-2   Trigonometry

The elementary functions of sine, cosine, and tangent are defined (for acute angles in a right triangle) in Sec. 2-10.

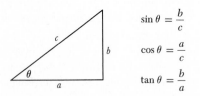

$$\sin \theta = \frac{b}{c}$$

$$\cos \theta = \frac{a}{c}$$

$$\tan \theta = \frac{b}{a}$$

The general definitions for angles of any size are made in terms of the abscissa, ordinate, and radius:

$$\sin \theta = \frac{\text{ordinate}}{\text{radius}}$$

$$\cos \theta = \frac{\text{abscissa}}{\text{radius}}$$

$$\tan \theta = \frac{\text{ordinate}}{\text{abscissa}}$$

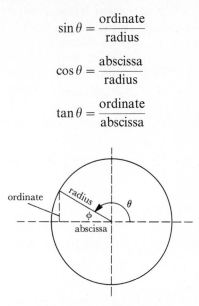

The radius is always $+$, but the ordinate and abscissa can be either $+$ or $-$. In the illustration, $\sin \theta$ is $+$ and $\cos \theta$ is $-$. No elaborate formulas are needed to remember sign rules, because each case can be calculated when required. If $\theta = 150°$, as in the figure, the acute angle ϕ is $180° - 150° = 30°$. Thus $\cos 150°$ is numerically equal to $\cos 30°$, except for a minus sign, because the abscissa is negative. Hence, $\cos 150° = -\cos 30° = -0.866$ (table on page 847).

For completeness, we include two formulas that are sometimes useful, although they are not required for any of the physics problems in this book. Their application is usually to triangles that do not have a right angle.

Law of cosines. In the shaded triangle,

$$c^2 = a^2 + b^2 - 2ab \cos \phi$$

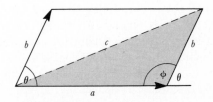

Here ϕ is an angle interior to the triangle, at the vertex where sides a and b join. If $\phi = 90°$, the formula reduces to the Pythagorean theorem, because $\cos 90° = 0$. An alternative statement of the law of cosines, well suited to problems involving vector addition, makes use of the angle θ, which is the angle between the directions of two vectors. Since $\cos \theta = -\cos \phi$ (they are supplementary angles), the law of cosines becomes

$$c^2 = a^2 + b^2 + 2ab \cos \theta$$

Law of sines. The relationship between the sides of a triangle and the sines of the opposite angles is expressed as a proportion:

$$\frac{a}{\sin A} = \frac{b}{\sin B} = \frac{c}{\sin C}$$

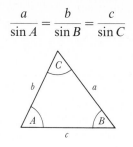

Appendix D | Functions and Graphs

D-1 Functions

The concept of "function" perhaps epitomizes the spirit of scientific investigation, for a function is nothing more or less than a *relationship* or a *correlation* between observables. In physics we seek to correlate cause and effect. Often we try to relate our observations of complex phenomena to our observations of factors that are under our own control. For example, it is found that more engine power is required to move a car at high speed than at low speed. Power is a function of speed, and a test engineer who makes a series of observations of the required power at a number of different speeds can describe the function by means of a table of values or by a graph. A function can also be theoretical, in contrast to the experimental function just described. For instance, a relationship exists between the reading of a photographer's exposure meter and the necessary diameter of lens opening for a given type of film. Here the nature of the function is known from other theories (themselves based eventually on experiment); the function serves a useful purpose and allows the photographer to take advantage of previously acquired knowledge.

The information that describes a function may be contained in a table of values:

Exposure-Meter Reading, current in amperes	Proper Lens Opening, diameter in millimeters
0.5×10^{-6}	20
1	14
2	10.0
4	7.1
6	5.8
8	5.0
10	4.5
20	3.2

Although the photocell currents are all in the range of 0.5 to 20×10^{-6} A, the 10^{-6} is written only after the first entry in the table. It is understood that the other entries are in the same units. To write the first entry as 5×10^{-7} A, and the last two entries as 1.0×10^{-5} and 2.0×10^{-5} A, would also be correct. However, in tables of values it is customary to write all values in any one column with the same power of ten, even though some of the values are not in scientific notation. This is to help present the whole trend more clearly.

In an experiment, the *independent variable* is the one that is under our control. Thus, the engineer testing the effect of air resistance set the speed of the car at some one of a number of predetermined values and observed the horsepower necessary to maintain this speed. In this case, the speed was the independent variable. On the other hand, if the engineer had chosen to adjust the motor to one of a number of predetermined horsepowers and observed the resulting speed, then the power would have been the independent variable and the speed would have been the *dependent variable*. It is a matter of emphasis, based on experimental procedure. If a function is expressed as a table listing corresponding values, perhaps obtained by a theoretical calculation, then the independent variable is the one that is most likely to be known. Our exposure-meter table represents a function of this sort, and so would a table that might be used to convert inches into centimeters, or one that would give a parcel's postage as a function of its weight. In the tables of functions below, the independent variable is in the first column, and the dependent variable is in the second column.

Temperature, °C	Density of Water, g/cm^3
0	1.000
20	0.998
40	0.992
60	0.983
80	0.972
100	0.958

Time of Day	Air Temperature, °C
8:00	3
10:00	4
12:00	8
13:00	11
15:00	12
18:00	10

Year	Consumer Price Index (Urban)
1930	50
1935	41
1940	42
1945	54
1950	72
1955	80
1960	89
1965	95
1970	116
1975	161
1980	248

Elapsed Time, min	Depth of Rainwater in a Gauge, cm
0	0.00
10	0.51
20	0.69
30	0.79
40	0.79
50	0.79

Applied Force, N	Stretch of a Spring, m
0	0.00
6	0.17
12	0.33
18	0.50
24	0.61
30	0.82

D-2 Graphs

Graphs are often used to display the behavior of functions. A good graph must be plotted carefully, but it is even more important to choose scales judiciously and to label axes clearly. Note the features of the accompanying graph (Fig. D-1), which represents the rate at which the lung muscles do mechanical work during

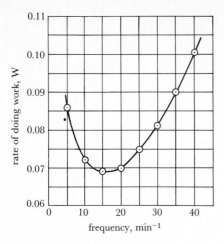

Figure D-1 Rate of doing work during forced breathing at various respiratory frequencies. Dead space, 200 cm³; alveolar ventilation constant, 6 liters/min. [After Otis, Fenn, and Rahn, *Journal of Applied Physiology*, **2**, 592 (1950).]

breathing. The subject breathes a fixed amount of air per minute (called the ventilation rate); he can do this by taking a few deep breaths per minute or many shallow breaths per minute.

1. A table of values is written down near the graph.
2. The independent variable (rate of breathing) is plotted on the horizontal axis; the dependent variable (rate of doing work) is plotted on the vertical axis.

Frequency, breaths per min	Rate of Doing Work, watts
5	0.086
10	0.072
15	0.069
20	0.070
25	0.075
30	0.081
35	0.090
40	0.101

3. Axes are well labeled, including units of measure.
4. On the frequency axis, 2 squares = 10 breaths per minute; on the power axis, 2 squares = 0.01 W. (Avoid multiples such as 3 squares = 0.01 W or 1 square = 7 breaths per minute. These ratios make it hard to plot points and also hard to read values off the graph.)
5. Consistent with item 4, the graph is plotted to fill up as much of a full sheet of graph paper as possible.
6. Plotted points are located by dots surrounded by small circles. A smooth curve is drawn through the points.
7. There is adequate description of the circumstances of the experiment, or the assumed circumstances for which the graph is applicable. This information is contained in a short legend below the graph. References to source of data are included.
8. Since these data involve only rates lying between about 0.06 and 0.11 W, the zero of the vertical axis is "suppressed." It would make the graph harder to read if we included a full range of 0 to 0.11 W, using only the upper part of the range.

The purpose of a graph is to give us information about a relationship. For instance, we see from this graph that the least work is required when about 15 breaths are taken per minute; this is the optimum frequency of breathing for a swimmer or runner who wishes to ventilate at the rate of 6 liters/min, if his dead space (fixed volume of throat and mouth cavity) is 200 cm³. It is important to note that the graph does not give unlimited information. According to the legend beneath the graph, the subject is ventilating at a certain rate. The graph (or table) gives no information at all for a ventilation rate of 8 liters/min or 10 liters/min, or for a subject whose dead space is not 200 cm³. Additional experimentation or calculation would be required to answer questions pertaining to these altered conditions.

Further discussion of graphs will be found in Sec. 2-2 of the text, where slopes are discussed, and in your laboratory manual.

PRACTICE PROBLEMS

Refer to the tables in Sec. D-1.

114 Plot the graph of the density of water from the table. Is it wise to suppress the zero of either axis? What is the density (to three significant figures) at 30°C?

115 Plot the air temperature from the table. At what time of day was the temperature highest? When was the temperature changing most rapidly?

116 Plot the graph of the cost of living. Estimate from the graph what the price index was in 1948; in 1973. What do you expect it to be in 1987? What can you say about 2000?

117 Plot the graph of the depth of water in the rain gauge. How long did the rain last?

118 Plot the stretch of the spring. Should any measurement be repeated?

Often a functional relationship can be expressed in mathematical form. Thus, the information contained in the table of values for the exposure meter can be written as $D = \sqrt{200/I}$, where I is the reading of the exposure meter in microamperes and D is the diameter of the lens opening in millimeters. None of the given values is very different from that which would be calculated from this equation. (Try it.)

A functional relationship is made more explicit by writing the independent variable in parentheses, such as $f(x)$, which is read "f-of-x." The equation

$$f(x) = (x + 1)^2$$

tells us to "add 1 to the independent variable and square the result." The symbols used for the variables are unimportant, and $F(z) = (z + 1)^2$ conveys the same message. If $x = 3$, then $f(x) = (3 + 1)^2 = 16$; this is indicated by writing $f(3) = 16$. As another example, using the same

function, we find that $f(y^3) = (y^3 + 1)^2 = y^6 + 2y^3 + 1$.

Expressing a function in mathematical form is convenient, but it is important to realize that mathematical expressions are not *necessary* for the existence of a function. Indeed, in many cases a table of values will serve as the only definition of a function. *Any* rule for finding the value of one variable when the value of another one is known may serve to define a function. Consider the mathematical function represented by the following table of values:

x	1	2	3	4	5	6	7
y	1	6	5	11	21	20	?

At first glance, no relationship is apparent. What is the value of y when $x = 7$? Inspection of the table will probably be of little help, yet a perfectly definite procedure exists by which each y is found.* Hence, y is a function of x.

PRACTICE PROBLEMS

119 Which of the functions described in the tables in Sec. D-1 do you think might be represented by relatively uncomplicated mathematical expressions?

120 From Fig. D-1, estimate to two significant figures the rate of doing work if the subject breathes once every 5 s.

121 If $f(y) = y^2 - 1$, find $f(4)$; $f(-4)$; $f(2x)$; $f(\sqrt{t})$; $f(\sqrt{t}) - f(-\sqrt{t})$.

122 If $s(t) = 1 + 3t$, find $s(-1)$; $[s(t)]^2$; $s(t^2)$.

123 Calculate the next two values of y, corresponding to $x = 7$ and $x = 8$, for the function described in the footnote below.

---

* In this case our rule is as follows: Form an integer out of the numbers 1, 2, 3, 4, ... x; divide by x; add up the digits of the result (ignoring fractions). Thus, for $x = 4$, $1234/4 = 308\frac{1}{2}$; $3 + 0 + 8 = 11$. This function is defined only for integral values of $x > 0$.

No formal calculus is required in the main body of this book. However, some formulas of calculus will be of use in certain of the For Further Study sections.

The derivative of a function $f(x)$ is called $f'(x)$ or df/dx, and is defined in sec. 2-4 as the limit of a ratio:

$$f'(x) = \frac{df}{dx} = \lim_{\Delta x \to 0} \left(\frac{\Delta f}{\Delta x} \right)$$

At any point, the value of the derivative is the slope of the tangent to the graph of $f(x)$ (see the discussion in Sec. 2-4). Derivatives are found by taking a limit, and the results for some simple functions are given in the following table, where a represents any constant:

$f(x)$	$f'(x)$
a	0
x^n	nx^{n-1}
$\sin ax$	$a \cos ax$
$\cos ax$	$-a \sin ax$
$\tan ax$	$a \sec^2 ax$
e^{ax}	ae^{ax}
$\ln x$	x^{-1}
$\log x$	$0.434\, x^{-1}$

Some of the rules of formal calculus follow:

(a) The derivative of a sum of terms is the sum of the derivatives; the derivative of $a[f(x)]$ is a times the derivative of $f(x)$.

(b) The derivative of a product of two functions $u(x)$ and $v(x)$ is given by

$$\frac{d}{dx}(uv) = u\frac{dv}{dx} + v\frac{du}{dx}$$

For example, if $y = 3x^2 \sin x$, let $u = 3x^2$ and $v = \sin x$. Then

$$\frac{du}{dx} = 6x \quad \text{and} \quad \frac{dv}{dx} = \cos x$$

Then, by the rule,

$$\frac{dy}{dx} = u\frac{dv}{dx} + v\frac{du}{dx} = 3x^2 \cos x + 6x \sin x$$

To differentiate a quotient of two functions, first express the quotient as a product, using a negative exponent; then use rule (b).

(c) *The chain rule:* If $z = F(y)$, and $y = f(x)$, we consider z to be a function of a function. The chain rule states that

$$\frac{dz}{dx} = \left(\frac{dz}{dy} \right)\left(\frac{dy}{dx} \right)$$

The method can be extended to any number of intermediate functions. For example, if $z = (x + \sin x)^3$, first think of $(x + \sin x)$ as an entity (call it y if you wish). Differentiate the "cube" function first (getting $dz/dy = 3y^2$); then multiply by the derivative (dy/dx) of the expression inside the parentheses.

$$\frac{dz}{dx} = 3(x + \sin x)^2(1 + \cos x)$$

The derivative of a derivative is the *second derivative*. The derivative of $f'(x)$ is written as $f''(x)$, or as d^2f/dx^2.

Example

Calculate $f'(2)$ and $f''(2)$ if $f(x) = 4x^5 - x$.

First we find that $f'(x) = 20x^4 - 1$, and $f''(x) = 80x^3$. Substituting $x = 2$ gives $f'(2) = 319$, and $f''(2) = 640$.

PRACTICE PROBLEMS

Find the derivatives of the following functions:

124 $7x^4 - 3x + 4$
125 $\sin 3x + e^{5x}$
126 $1/x^3$
127 $4x^2 e^{-2x}$
128 $(9 - x^2)^{1/2}$
129 $\sin^2 x + \cos^2 x$

Calculate the following:

130 $y''(t)$, if $y(t) = 10 \sin 2t$
131 $y'(2)$, if $y(u) = 5e^u + 5u^5$
132 dv/dt at $t = 1$, if $v(t) = 41t^2 + 41t - 41$

The definite integral is defined in Sec. 2-5 as the limit of a sum:

$$\lim_{\Delta x \to 0} \sum y \, \Delta x = \int_{x_1}^{x_2} y \, dx$$

The integral is the area under the portion of the graph of $y(x)$ that lies between $x = x_1$ and $x = x_2$. The *fundamental theorem of calculus* uses the antiderivative function $F(x)$, which is defined in such a way that $f(x)$ is the derivative of $F(x)$. The theorem states that the definite integral between the limits x_1 and x_2 is the difference between the values of the antiderivative at the two points:

$$\int_{x_1}^{x_2} f(x) \, dx = F(x_2) - F(x_1)$$

which is also written as

$$\int_{x_1}^{x_2} f(x) \, dx = \left[F(x) \right]_{x_1}^{x_2}$$

The brief table of antiderivatives given below can be used to evaluate definite integrals, as illustrated in the examples. The antiderivative is also called the indefinite integral of $f(x)$.

$f(x)$	$F(x)$
A	Ax
Ax^n	$\dfrac{Ax^{n+1}}{n+1}$ $(n \neq -1)$
Ax^{-1}	$A \ln x$
$\sin ax$	$-\dfrac{1}{a} \cos ax$
$\cos ax$	$\dfrac{1}{a} \sin ax$
$\tan ax$	$-\dfrac{1}{a} \ln \cos ax$
$\ln x$	$x \ln x - x$
e^{ax}	$\dfrac{1}{a} e^{ax}$

Examples

(a) $\displaystyle\int_1^3 (x^2 + 2) \, dx$

$$= \left[\frac{x^3}{3} + 2x \right]_1^3 = \left[\frac{(3)^3}{3} + 2(3) \right] - \left[\frac{(1)^3}{3} + 2(1) \right]$$

$$= 12\tfrac{2}{3}$$

(b) $\displaystyle\int_{0°}^{60°} \sin 2\theta \, d\theta = \left[-\tfrac{1}{2} \cos 2\theta \right]_{0°}^{60°}$

$$= \left[-\tfrac{1}{2} \cos 120° \right] - \left[-\tfrac{1}{2} \cos 0° \right]$$

$$= (-\tfrac{1}{2})(-\tfrac{1}{2}) - (-\tfrac{1}{2})(1) = \tfrac{3}{4}$$

In Example (b), the integral (area under the curve) is 0.75, as given by the formula, if θ is in radians; but the same area is 43.0 units if θ is in degrees. It is immaterial whether radians or degrees are used when substituting into the indefinite integral ($\cos 120°$ is $-\tfrac{1}{2}$ whether the angle is expressed as $120°$ or as $2\pi/3$ radians). The integral can be used to find the average height of the curve over the interval 0 to 1.05 rad:

$$\bar{y} = \frac{\text{area}}{\text{base}} = \frac{0.75}{1.05} = 0.717$$

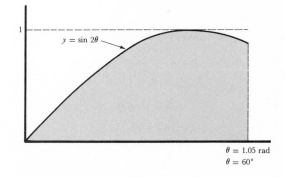

PRACTICE PROBLEMS

133 Is it possible for two different functions to have the same derivative? Is it possible for two different functions to have the same value of the definite integral evaluated (for each function) between the same limits a and b?

Compute the values of the following definite integrals:

134 $\int_0^3 (1 - x + x^2)\, dx$ **136** $\int_0^\infty e^{-bt}\, dt$

135 $\int_{0°}^{30°} \sin 3\phi\, d\phi$ **137** $\int_0^1 e^{-3x}\, dx$

138 $\int_0^{\pi/2} \sin x\, dx$

139 $\int_2^5 \dfrac{dz}{z^3}$

140 $\int_0^V 10 V^{3/2}\, dV$

141 $\int_1^5 (x^4 + x)\, dx$

142 $\int_0^B \tfrac{1}{2}\sqrt{x}\, dx$

143 $\int_a^b \ln x\, dx$

Appendix F | Approximate Calculations

If mathematics is to be our servant rather than our master, we must be able to make a quick estimate when the tedious work of an exact solution is not needed. Often a rough idea of the result of a calculation can give important insight into a physical problem. Quick approximate calculation also helps guard against gross errors such as a misplaced decimal point.

What we have in mind is a calculation to "one-figure accuracy" or even worse. We round everything off to one figure, combine the digits first, and then place the decimal point by combining whatever powers of ten are left over.

Example

How many seconds are there in a year?

(365.24)(24)(60(60)

$$= (4 \times 10^2)(2 \times 10)(6 \times 10)(6 \times 10)$$
$$= (4)(2)(6)(6) \times 10^5$$
$$= (8)(36) \times 10^5$$
$$= (8)(40) \times 10^5$$
$$\doteq 300 \times 10^5$$
$$= 3 \times 10^7$$

The exact answer, to four significant figures, is 3.156×10^7 s in a year. Don't be alarmed at setting (8)(40) equal to 300. To be sure, 320 is an exact value for (8)(40), but we are interested in a quick and easy estimate, not an exact value.

Example

How many years are there in one 50-min class period?

$$\frac{50 \text{ min}}{(365.24)(24)(60) \text{ min}} = \frac{\cancel{50}}{(4 \times 10^2)(20)(\cancel{60})}$$

$$= \frac{1}{80 \times 10^2}$$

$$= \frac{100 \times 10^{-2}}{80 \times 10^2}$$

$$= 1.2 \times 10^{-4}$$

A more exact value is 0.951×10^{-4} year. Several points are to be noted in the last example: (*a*) We canceled 50 and 60 just as if they were equal. This is not right, but not far wrong either. Any shortcut or approximation of this sort is permissible in a rough calculation. (*b*) To divide by 80 with a minimum of pain, we changed the 1 in the numerator to 100×10^{-2} (really the same thing). Then 100/80 is easily seen to be just a bit more than 1—call it 1.2. (*c*) Our answer, 1.2×10^{-4}, is not very close to the correct value of 0.951×10^{-4}; however, if it were a question of deciding between 0.951×10^{-4} and 0.951×10^{-3} we could easily see from our approximate answer that the former is correct. Any approximate answer between, say, 0.7×10^{-4} and 2×10^{-4} would allow us to fix the decimal point if an exact calculation gave an answer of the form $0.951 \times 10^{-?}$.

Example

What is the volume of a droplet of water in a fog if the radius of the droplet is 2.73×10^{-3} cm?

$$V = \tfrac{4}{3}\pi r^3 = \tfrac{4}{3}\pi(2.73 \times 10^{-3})^3$$
$$= \tfrac{4}{3}\pi(3 \times 10^{-3})^3 = 4(27 \times 10^{-9})$$
$$= 100 \times 10^{-9} = 1 \times 10^{-7} \text{ cm}^3$$

A more precise value for the volume is 0.85×10^{-7} cm^3.

PRACTICE PROBLEMS

Estimate the approximate value of the following:

144 $\dfrac{(1660)(2\pi)}{41.7}$

145 $(468)(5280)(2.0813 \times 10^6)$

146 $\dfrac{(0.0332)(1.01 \times 10^4)^2}{\pi(18)^2}$

147 $\dfrac{(1.6 \times 10^{-19})(6.02 \times 10^{23})}{\sqrt{84.5}}$

148 $\dfrac{(1.15 \times 10^{50})^{1/5}(3\pi \times 10^{-10})}{220}$

149 $\sqrt{x^2 - 1}$, where $x = 20$

Answers to Practice Problems in Mathematical Review

1 126.3

2 12.63

3 30

4 536.8

5 1240

6 0.3

7 2

8 200

9 0.20

10 0.6

11 10,000

12 3.14

13 6.67

14 1390

15 154

16 0.571

17 6

18 1.6

19 0.6

20 180

21 0.11

22 0.003

23 x^{10}

24 $x^{23}y^4$

25 $a^{5/2}$

26 2^4

27 0.01

28 $9t^2$

29 16

30 2

31 $p^{1.9}$

32 45

33 x

34 1.5×10^3

35 1

36 0

37 1.4263×10^2

38 1.5×10^6

39 3.36×10^{-3}

40 4.66331056×10^6

41 8.64×10^4

42 1.44×10^{-14}

43 1.25×10^{-1}

44 1.25×10^{-3}

45 16, 300,000

46 4781.354

47 1,010,000

48 98.1

49 0.000311

50 65,500,000,000,000

51 0.0008001

52 0.0000093

53 9.6×10^{19}

54 2.4×10^{-3}

55 4×10^{-1}

56 4.4×10^9

57 10^{-4}

58 9.63×10^4

59 2×10^{-16}

60 5×10^{-8}

61 900

62 285

63 1.3×10^{-6}

64 5×10^3

65 0.063

66 0.02

67 20

68 0.2

69 5×10^{-7}

70 464

71 4×10^7

72 0.003

73 0.750 min

74 0.7500 min

75 7.7×10^2 days

76 9.248 m^2

77 184 km

78 1.8×10^2 km

79 5.500 ft

80 3; 2; 3

81 39.1 liters

82 8.6400×10^4 s
or 86.400 ks

83 3.75×10^7

84 5.48×10^{-4} g

85 0.08 cm; 0.1%

87 -14

88 -2.5

89 0.75

90 ± 10

91 3

92 1.6

93 4

94 64

95 $(16 + 2y)/3$

96 $(mg - F - 200)/2$

97 $2y^2 + 2y$

98 $(9 - y)/(3 + y)$

99 $u = 2, v = 1$

100 $p = 3, q = 5$

101 $B = 4, A = 10$

102 $x = 0.85, y = 1.25$

103 $1, \frac{1}{2}$

104 $-2, -3$

105 $5, -2$

106 no real solution

107 -1

108 $+1, -1, +\frac{1}{2}, -\frac{1}{2}$

109 0.693

110 4.605

111 17.079

112 18.693

113 272.27

114 0.995 g/cm^3

115 15:00; 10:00

116 63; 135; 350?; unknown

117 30 min

118 Check data at 24 N

120 0.070 W

121 $15; 15; 4x^2 - 1; t - 1; 0$

122 $-2; 1 + 6t + 9t^2; 1 + 3t^2$

123 29; 24

124 $28x^3 - 3$

125 $3 \cos 3x + 5e^{5x}$

126 $-3x^{-4}$

127 $8x(1 - x)e^{-2x}$

128 $-x(9 - x^2)^{-1/2}$

129 0

130 $-40 \sin 2t$

131 436.95

132 123

133 yes; yes

134 7.5

135 $\frac{1}{3}$

136 $1/b$

137 -0.573

138 1

139 0.105

140 $4V^{5/2}$

141 636.8

142 $\frac{1}{3}B^{3/2}$

143 $a - b + \ln(b^b/a^a)$

144 250

145 5×10^{12}

146 300

147 1.1×10^4

148 0.05

149 20

List of Reference Tables

Table 1 Important Constants

velocity of light, c	3.00×10^8 m/s
gravitational constant, G	6.67×10^{-11} N·m²/kg²
rest mass of electron, m_e	9.11×10^{-31} kg
rest mass of proton, m_p	1.67×10^{-27} kg
rest mass of neutron, m_n	1.67×10^{-27} kg
unified atomic mass unit, u	1.66×10^{-27} kg
energy equivalent of 1 u	931.5 MeV
Avogadro's number, N_A	6.02×10^{23} mol^{-1}
Joule equivalent, J	4.186 J/cal
absolute zero, 0 K	-273.15°C
volume of 1 mole of ideal gas at STP, V_0	22.4 liters
Boltzmann constant, k	1.38×10^{-23} J/molecule·K
gas constant per mole, R	8.31 J/mol·K
charge of electron, e	1.60×10^{-19} C
Faraday constant, Q_F	9.65×10^4 C/g.e.w.
Planck's constant, h	6.63×10^{-34} J·s
electrostatic constant, k	8.99×10^9 N·m²/C²
electromagnetic constant, k'	10^{-7} N/A²

Table 2 Important Metric Prefixes*

Prefix	Abbreviation	Meaning	Typical Examples
tera	T	$\times 10^{12}$	1 TeV (particle accelerator energy) = 10^{12} electron volts
giga	G	$\times 10^9$	1 gigahertz (radar frequency) = 10^9 Hz = 10^9 cycles/s
mega	M	$\times 10^6$	1 megaton (equivalent TNT strength of nuclear weapon) = 10^6 tons
kilo	k	$\times 10^3$	1 kilogram = 1000 g
deci	d	$\times 10^{-1}$	1 decibel = 0.1 bel
centi	c	$\times 10^{-2}$	1 centimeter = 0.01 m
milli	m	$\times 10^{-3}$	1 milliampere = 0.001 A
micro	μ	$\times 10^{-6}$	1 microvolt = 10^{-6} V
nano	n	$\times 10^{-9}$	1 nanosecond = 10^{-9} s
pico	p	$\times 10^{-12}$	1 picofarad = 10^{-12} F
femto	f	$\times 10^{-15}$	1 femtometer (approximate size of a proton) = 10^{-15} m

* Other metric prefixes not in common use are peta (P) for 10^{15}, exa (E) for 10^{18}, and atto (a) for 10^{-18}. The micron, a unit of length used in older literature, is $1 \mu m = 10^{-6}$ m = 0.001 mm.

Table 3 Greek Alphabet

A	α	alpha	H	η	eta	N	ν	nu	T	τ	tau
B	β	beta	Θ	θ	theta	Ξ	ξ	xi	Υ	υ	upsilon
Γ	γ	gamma	I	ι	iota	O	o	omicron	Φ	ϕ	phi
Δ	δ	delta	K	κ	kappa	Π	π	pi	X	χ	chi
E	ε	epsilon	Λ	λ	lambda	P	ρ	rho	Ψ	ψ	psi
Z	ζ	zeta	M	μ	mu	Σ	σ	sigma	Ω	ω	omega

Table 4 Equivalents and Conversion Factors*

Length
1 ft = 30.48 cm = 0.3048 m
1 mi = 5280 ft = 1.609 km
1 in. = 2.540 cm
1 angstrom (Å) = 10^{-10} m = 10^{-8} cm
1 fermi = 1 femtometer (fm) = 10^{-15} m = 10^{-13} cm
1 micron (μ) = 1 μm = 10^{-6} m = 10^{-4} cm = 10^{4} Å
1 light-year = 9.461×10^{12} km = 5.88×10^{12} mi

Area
1 ft^2 = 929.0 cm^2 = 0.09290 m^2
1 in.^{2} = 6.452 cm^2 = 645.2 mm^2
1 barn (b) = 10^{-28} m^2 = 10^{-24} cm^2

Volume
1 liter = 1000 ml = 1000 cm^3 = 0.001 m^3 = 1.0576 qt = 61.03 in.^{3} = 0.0353 ft^3
1 ft^3 = 7.481 gal = 28.32 liters = 0.02832 m^3
1 m^3 = 1000 liters = 10^6 cm^3 = 1.308 yd^3 = 35.31 ft^3

Velocity
60 mi/h = 88 ft/s = 26.82 m/s = 96.5 km/h
1 mi/h = 0.4470 m/s = 1.609 km/h = 1.47 ft/s = 0.85 knot

Mass
1 metric ton = 1 tonne = 1000 kg = 10^6 g

Force
1 newton (N) = 10^5 dynes = 0.2248 lb
1 lb = 4.448 N
1 ton = 2000 lb = 8.896×10^3 N
An object of mass 1 kg weighs[†] 9.807 N = 2.205 lb
An object of weight[†] 1 lb has mass 453.6 g = 0.4536 kg

Pressure
1 bar = 10^6 dyn/cm^2 = 10^5 N/m^2 = 10^5 Pa = 14.50 lb/in.^{2}
1 lb/in.^{2} = 6.89×10^4 dyn/cm^2 = 6.89×10^3 N/m^2 = 6.89 kPa
1 atm = 760 torr = 760 mm Hg = 76.0 cm Hg = 14.70 lb/in.^{2} = 2116 lb/ft^2
    = 1.013×10^5 N/m^2 = 101.3 kPa = 1.013 bar = 1013 mbar = 1.013×10^6 dyn/cm^2

Work and energy
1 joule (J) = 10^7 ergs = 0.239 cal = 0.7376 ft·lb
1 cal = 4.186 J = 3.088 ft·lb
1 Btu = 252 cal = 778 ft·lb = 1054 J = 2.93×10^{-4} kWh
1 kilowatt·hour (kWh) = 3.60×10^6 J = 860 kcal
1 electron volt (eV) = 1.60×10^{-19} J

Power
1 horsepower (hp) = 0.746 kilowatt (kW) = 550 ft·lb/s = 3.30×10^4 ft·lb/min
1 watt (W) = 1 J/s = 0.738 ft·lb/s

Specific heat capacity
1 cal/g·°C = 1 Btu/lb·°F = 4.186 J/g·°C = 4186 J/kg·°C

Latent heat and heat of combustion
1 cal/g = 4.186 J/g = 4186 J/kg = 1.80 Btu/lb

Gas constant
R = 8.314 J/mol·K = 1.99 cal/mol·K = 0.0821 (atm·liter/mol·K)

---

* See also Table 2 for metric prefixes.
[†] All weights are at a point on the surface of the earth where g has the standard value 9.80665 m/s^2.

Table 5 Sines, Cosines, and Tangents

Angle	Sine	Cosine	Tangent	Angle	Sine	Cosine	Tangent
0°	0.000	1.000	0.000				
1°	.017	1.000	.017	46°	0.719	0.695	1.036
2°	.035	0.999	.035	47°	.731	.682	1.072
3°	.052	.999	.052	48°	.743	.669	1.111
4°	.070	.998	.070	49°	.755	.656	1.150
5°	.087	.996	.087	50°	.766	.643	1.192
6°	.105	.995	.105	51°	.777	.629	1.235
7°	.122	.993	.123	52°	.788	.616	1.280
8°	.139	.990	.141	53°	.799	.602	1.327
9°	.156	.988	.158	54°	.809	.588	1.376
10°	.174	.985	.176	55°	.819	.574	1.428
11°	.191	.982	.194	56°	.829	.559	1.483
12°	.208	.978	.213	57°	.839	.545	1.540
13°	.225	.974	.231	58°	.848	.530	1.600
14°	.242	.970	.249	59°	.857	.515	1.664
15°	.259	.966	.268	60°	.866	.500	1.732
16°	.276	.961	.287	61°	.875	.485	1.804
17°	.292	.956	.306	62°	.883	.469	1.881
18°	.309	.951	.325	63°	.891	.454	1.963
19°	.326	.946	.344	64°	.899	.438	2.050
20°	.342	.940	.364	65°	.906	.423	2.145
21°	.358	.934	.384	66°	.914	.407	2.246
22°	.375	.927	.404	67°	.921	.391	2.356
23°	.391	.921	.424	68°	.927	.375	2.475
24°	.407	.914	.445	69°	.934	.358	2.605
25°	.423	.906	.466	70°	.940	.342	2.747
26°	.438	.899	.488	71°	.946	.326	2.904
27°	.454	.891	.510	72°	.951	.309	3.078
28°	.469	.883	.532	73°	.956	.292	3.271
29°	.485	.875	.554	74°	.961	.276	3.487
30°	.500	.866	.577	75°	.966	.259	3.732
31°	.515	.857	.601	76°	.970	.242	4.011
32°	.530	.848	.625	77°	.974	.225	4.331
33°	.545	.839	.649	78°	.978	.208	4.705
34°	.559	.829	.675	79°	.982	.191	5.145
35°	.574	.819	.700	80°	.985	.174	5.671
36°	.588	.809	.727	81°	.988	.156	6.314
37°	.602	.799	.754	82°	.990	.139	7.115
38°	.616	.788	.781	83°	.993	.122	8.144
39°	.629	.777	.810	84°	.995	.105	9.514
40°	.643	.766	.839	85°	.996	.087	11.43
41°	.656	.755	.869	86°	.998	.070	14.30
42°	.669	.743	.900	87°	.999	.052	19.08
43°	.682	.731	.933	88°	.999	.035	28.64
44°	.695	.719	.966	89°	1.000	.017	57.29
45°	.707	.707	1.000	90°	1.000	.000	

Table 6   Periodic Table of the Elements*

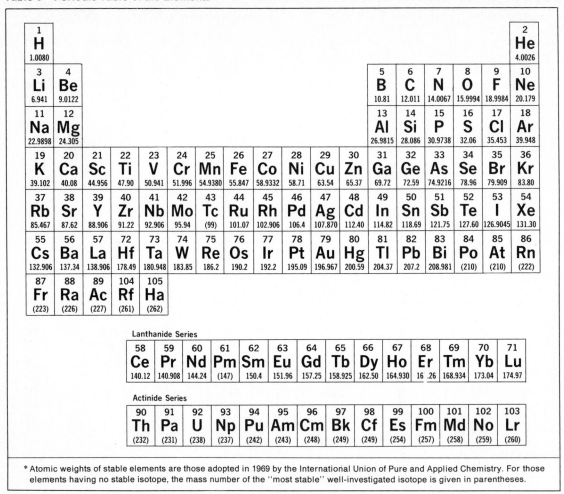

1 **H** 1.0080																	2 **He** 4.0026
3 **Li** 6.941	4 **Be** 9.0122											5 **B** 10.81	6 **C** 12.011	7 **N** 14.0067	8 **O** 15.9994	9 **F** 18.9984	10 **Ne** 20.179
11 **Na** 22.9898	12 **Mg** 24.305											13 **Al** 26.9815	14 **Si** 28.086	15 **P** 30.9738	16 **S** 32.06	17 **Cl** 35.453	18 **Ar** 39.948
19 **K** 39.102	20 **Ca** 40.08	21 **Sc** 44.956	22 **Ti** 47.90	23 **V** 50.941	24 **Cr** 51.996	25 **Mn** 54.9380	26 **Fe** 55.847	27 **Co** 58.9332	28 **Ni** 58.71	29 **Cu** 63.54	30 **Zn** 65.37	31 **Ga** 69.72	32 **Ge** 72.59	33 **As** 74.9216	34 **Se** 78.96	35 **Br** 79.909	36 **Kr** 83.80
37 **Rb** 85.467	38 **Sr** 87.62	39 **Y** 88.906	40 **Zr** 91.22	41 **Nb** 92.906	42 **Mo** 95.94	43 **Tc** (99)	44 **Ru** 101.07	45 **Rh** 102.906	46 **Pd** 106.4	47 **Ag** 107.870	48 **Cd** 112.40	49 **In** 114.82	50 **Sn** 118.69	51 **Sb** 121.75	52 **Te** 127.60	53 **I** 126.9045	54 **Xe** 131.30
55 **Cs** 132.906	56 **Ba** 137.34	57 **La** 138.906	72 **Hf** 178.49	73 **Ta** 180.948	74 **W** 183.85	75 **Re** 186.2	76 **Os** 190.2	77 **Ir** 192.2	78 **Pt** 195.09	79 **Au** 196.967	80 **Hg** 200.59	81 **Tl** 204.37	82 **Pb** 207.2	83 **Bi** 208.981	84 **Po** (210)	85 **At** (210)	86 **Rn** (222)
87 **Fr** (223)	88 **Ra** (226)	89 **Ac** (227)	104 **Rf** (261)	105 **Ha** (262)													

Lanthanide Series

58 **Ce** 140.12	59 **Pr** 140.908	60 **Nd** 144.24	61 **Pm** (147)	62 **Sm** 150.4	63 **Eu** 151.96	64 **Gd** 157.25	65 **Tb** 158.925	66 **Dy** 162.50	67 **Ho** 164.930	68 **Er** 16 .26	69 **Tm** 168.934	70 **Yb** 173.04	71 **Lu** 174.97

Actinide Series

90 **Th** (232)	91 **Pa** (231)	92 **U** (238)	93 **Np** (237)	94 **Pu** (242)	95 **Am** (243)	96 **Cm** (248)	97 **Bk** (249)	98 **Cf** (249)	99 **Es** (254)	100 **Fm** (257)	101 **Md** (258)	102 **No** (259)	103 **Lr** (260)

* Atomic weights of stable elements are those adopted in 1969 by the International Union of Pure and Applied Chemistry. For those elements having no stable isotope, the mass number of the "most stable" well-investigated isotope is given in parentheses.

Table 7   Astronomical Data

	Sun	Earth	Moon
Mass	1.99×10^{30} kg	5.98×10^{24} kg	7.36×10^{22} kg
Radius*	6.96×10^8 m	6.38×10^6 m	1.74×10^6 m
Density*	1460 kg/m^3	5500 kg/m^3	3340 kg/m^3
Period of rotation	2.2×10^6 s*	8.62×10^4 s	2.36×10^6 s*
Radius of orbit*		1.49×10^{11} m	3.84×10^8 m
Period of revolution		3.16×10^7 s	2.36×10^6 s*

* Average value.

Table 8 The Elements (Alphabetical According to Symbol)

Symbol	Name	Atomic Number	Symbol	Name	Atomic Number	Symbol	Name	Atomic Number
Ac	Actinium	18	Ge	Germanium	32	Po	Polonium	84
Ag	Silver	47	H	Hydrogen	1	Pr	Praseodymium	59
Al	Aluminum	13	Ha	Hahnium	105	Pt	Platinum	78
Am	Americium	95	He	Helium	2	Pu	Plutonium	94
Ar	Argon*	18	Hf	Hafnium	72	Ra	Radium	88
As	Arsenic	33	Hg	Mercury	80	Rb	Rubidium	37
At	Astatine	85	Ho	Holmium	67	Re	Rhenium	75
Au	Gold	79	I	Iodine	53	Rf	Rutherfordium	104
B	Boron	5	In	Indium	49	Rh	Rhodium	45
Ba	Barium	56	Ir	Iridium	77	Rn	Radon[†]	86
Be	Beryllium	4	K	Potassium	19	Ru	Ruthenium	44
Bi	Bismuth	83	Kr	Krypton	36	S	Sulfur	16
Bk	Berkelium	97	La	Lanthanum	57	Sb	Antimony	51
Br	Bromine	35	Li	Lithium	3	Sc	Scandium	21
C	Carbon	6	Lr	Lawrencium	103	Se	Selenium	34
Ca	Calcium	20	Lu	Lutetium	71	Si	Silicon	14
Cd	Cadmium	48	Mg	Magnesium	12	Sm	Samarium	62
Ce	Cerium	58	Mn	Manganese	25	Sn	Tin	50
Cf	Californium	98	Mo	Molybdenum	42	Sr	Strontium	38
Cl	Chlorine	17	Mv	Mendelevium	101	Ta	Tantalum	73
Cm	Curium	96	N	Nitrogen	7	Tb	Terbium	65
Co	Cobalt	27	Na	Sodium	11	Tc	Technetium	43
Cr	Chromium	24	Nb	Niobium[†]	41	Te	Tellurium	52
Cs	Cesium	55	Nd	Neodymium	60	Th	Thorium	90
Cu	Copper	29	Ne	Neon	10	Ti	Titanium	22
Dy	Dysprosium	66	Ni	Nickel	28	Tl	Thallium	81
Es	Einsteinium	99	No	Nobelium	102	Tm	Thulium	69
Er	Erbium	68	Np	Neptunium	93	U	Uranium	92
Eu	Europium	63	O	Oxygen	8	V	Vanadium	23
F	Fluorine	9	Os	Osmium	76	W	Tungsten[†]	74
Fe	Iron	26	P	Phosphorus	15	Xe	Xenon	54
Fm	Fermium	100	Pa	Protactinium	91	Y	Yttrium	39
Fr	Francium	87	Pb	Lead	82	Yb	Ytterbium	70
Ga	Gallium	31	Pd	Palladium	46	Zn	Zinc	30
Gd	Gadolinium	64	Pm	Promethium	61	Zr	Zirconium	40

* In older literature, symbol A is sometimes used instead of Ar.
† Niobium is also known as columbium; radon is also known as emanation (Em); tungsten is also known as wolfram.

Table 9 Properties of Light Nuclides

Symbol	Mass, u*	Abundance or Mode of Decay†	Half-Life†	Symbol	Mass, u*	Abundance or Mode of Decay†	Half-Life†
e	0.000549			$^{14}_{8}O$	14.008597	β^+	71 s
p	1.007277			$^{15}_{8}O$	15.003070	β^+	2.05 m
$^{1}_{0}n$	1.008665	β^-	10.6 m	$^{16}_{8}O$	15.994915	99.76%	
				$^{17}_{8}O$	16.999133	0.04%	
$^{1}_{1}H$	1.007825	99.985%		$^{18}_{8}O$	17.999160	0.20%	
$^{2}_{1}H$	2.014102	0.015%		$^{19}_{8}O$	19.003578	β^-	29.1 s
$^{3}_{1}H$	3.016050	β^-	12.3 y	$^{20}_{8}O$	20.004079	β^-	14 s
$^{3}_{2}He$	3.016030	0.0001%		$^{17}_{9}F$	17.002096	β^+	67 s
$^{4}_{2}He$	4.002603	~100%		$^{18}_{9}F$	18.000937	β^+,ϵ	1.83 h
$^{6}_{2}He$	6.018893	β^-	0.80 s	$^{19}_{9}F$	18.998405	100%	
$^{8}_{2}He$	8.0341	β^-	0.12 s	$^{20}_{9}F$	19.999987	β^-	11.6 s
				$^{21}_{9}F$	20.99995	β^-	4.4 s
$^{6}_{3}Li$	6.015125	7.4%		$^{22}_{9}F$	22.004	β^-	4.0 s
$^{7}_{3}Li$	7.016004	92.6%					
$^{8}_{3}Li$	8.022487	β^-	0.84 s	$^{17}_{10}Ne$	17.0364	β^+	0.1 s
$^{9}_{3}Li$	9.02680	β^-	0.18 s	$^{18}_{10}Ne$	18.005711	β^+	1.5 s
				$^{19}_{10}Ne$	19.001881	β^+	17.4 s
$^{7}_{4}Be$	7.016929	ϵ	53.6 d	$^{20}_{10}Ne$	19.992440	90.92%	
$^{9}_{4}Be$	9.012186	100%		$^{21}_{10}Ne$	20.993849	0.26%	
$^{10}_{4}Be$	10.013534	β^-	2.5 My	$^{22}_{10}Ne$	21.991385	8.82%	
$^{11}_{4}Be$	11.02167	β^-	13.6 s	$^{23}_{10}Ne$	22.994473	β^-	37.6 s
				$^{24}_{10}Ne$	23.99361	β^-	3.38 m
$^{8}_{5}B$	8.024609	β^+	0.77 s				
$^{10}_{5}B$	10.012939	19.6%		$^{20}_{11}Na$	20.0089	β^+	0.4 s
$^{11}_{5}B$	11.009305	80.4%		$^{21}_{11}Na$	20.997655	β^+	23 s
				$^{22}_{11}Na$	21.994437	β^+,ϵ	2.62 y
$^{9}_{6}C$	9.0312	β^+	0.13 s	$^{23}_{11}Na$	22.989771	100%	
$^{10}_{6}C$	10.01681	β^+	19.5 s	$^{24}_{11}Na$	23.990962	β^-	15.0 h
$^{11}_{6}C$	11.011432	β^+	20.3 m	$^{25}_{11}Na$	24.98996	β^-	60 s
$^{12}_{6}C$	12.000000	98.89%		$^{26}_{11}Na$	25.9917	β^-	1.04 s
$^{13}_{6}C$	13.003354	1.11%					
$^{14}_{6}C$	14.003242	β^-	5730 y	$^{20}_{12}Mg$	20.017	β^+	0.6 s
$^{15}_{6}C$	15.010600	β^-	2.5 s	$^{21}_{12}Mg$	21.0117	β^+	0.12 s
$^{16}_{6}C$	16.01470	β^-	0.74 s	$^{23}_{12}Mg$	22.994125	β^+	12 s
				$^{24}_{12}Mg$	23.985042	78.7%	
$^{13}_{7}N$	13.005738	β^+	10.0 m	$^{25}_{12}Mg$	24.985839	10.1%	
$^{14}_{7}N$	14.003074	99.63%		$^{26}_{12}Mg$	25.982593	11.2%	
$^{15}_{7}N$	15.000108	0.37%		$^{27}_{12}Mg$	26.984345	β^-	9.5 m
$^{16}_{7}N$	16.006103	β^-	7.14 s	$^{28}_{12}Mg$	27.98388	β^-	21.2 h
$^{17}_{7}N$	17.00845	β^-	4.16 s				
$^{18}_{7}N$	18.0141	β^-	0.63 s				

* Except for the first three particles, the masses given are those of the neutral atoms.
† Abbreviations: β^-, negatron decay; β^+, positron decay; ϵ, electron capture; y, year; My, megayear; d, day; h, hour; m, minute; s, second. No nuclides are listed that have half-lives of less than 0.1 s.

Answers
to
Odd-Numbered
Problems

CHAPTER 1

1-A1 3.156×10^9

1-A3 23.3 metric tons

1-A5 7.24 kg

1-A7 904 cm^3

1-A9 200 kg

1-B1 1.07

1-B3 27 km

CHAPTER 2

2-A1 (*a*) 87.3 km/h; (*b*) 54.2 mi/h

2-A3 (*a*) 2.08×10^{-4} km/h; 1.29×10^{-4} mi/h; 1.90×10^{-4} ft/s; (*b*) -2.00 m/day·min; -3.86×10^{-7} m/s^2

2-A5 (*a*) 20 m/s; (*b*) 0

2-A7 0.02 m/s^2, westward

2-A9 (*a*) 10.0 m/s; 9.62 m/s; (*b*) 3.8 m

2-A15 (*a*) 20 blocks; (*b*) 1.4 blocks SW from the starting point

2-A17 (*a*) 198 km/h; (*b*) -198 km/h

2-A19 (*a*) 4070 m/s; (*b*) 3130 m/s; (*c*) so they can be aimed eastward, out over the Atlantic Ocean

2-B1 (*a*) 486 km/h; (*b*) 50 km/h

2-B3 2.08 s

2-B5 135 m

2-B7 7.00 m/s

2-B9 (*a*) 11.03 m; (*b*) 1.50 s; (*c*) 3.00 s; (*d*) 14.7 m/s

2-B11 (*a*) 17.6 m/s, downward; (*b*) 19.6 m

2-B13 (*a*) 4 s; (*b*) 29.4 m/s; (*c*) 39.2 m

2-B15 0.163 s

2-B17 1.03 m

2-B19 31 s

2-B21 (*a*) 46 m/s; (*b*) -4.4 m/s^2

2-B23 181 km/h, northward

2-B25 7.37 m/s, 16° N of W

2-B27 (*a*) 30° upstream; (*b*) 1.73 m/s; (*c*) 115 s; (*d*) 36 s longer

2-B29 0.404 s; 0.364 m

2-B31 78 cm

2-B33 (*a*) 2.35 m; (*b*) $v_x = 3.00$ m/s; $v_y = 7.67$ m/s

2-B35 (*a*) 2.86 s; (*b*) 14 m/s; (*c*) 27° from the vertical

2-B37 15.0 m

2-B39 17.9 m/s, 33° above the horizontal

2-B41 (*a*) 10 s; (*b*) 245 m downrange; (*c*) 103 m above ground level, 98 m downrange

2-B43 (*a*) 9.81 m/s; (*b*) 7.14 m

2-C1 yes, in 6 s

2-C3 (*a*) 6 s; (*b*) 12 m/s; (*c*) 156 m; (*d*) 120 m

2-C5 (*a*) 1 s or 3 s; (*b*) 9.8 m/s

2-C7 1.20 s

2-C9 13.1 s

2-C11 8.69 m/s

2-C13 (*a*) horizontal, 23.8 m/s; vertical, 11.6 m/s; (*b*) 2.36 s; (*c*) 56.1 m

2-C15 $H = (v_0{}^2/2g) \sin^2 \theta$

2-C17 $\tan \phi = \frac{1}{2}\tan \theta$

2-C19 (*a*) m; m/s; m/s^2; (*b*) $+12$ m/s; (*c*) 7 m; (*d*) -10 m/s^2

CHAPTER 3

3-A1 2000 dyn, downward
3-A3 10.2 kN
3-A5 8 g
3-A7 0.033 N
3-A9 12 N

3-B1 4 m/s^2, downward
3-B3 (a) 1.2 m/s^2; (b) 2.4 m
3-B5 (a) 5.25 km; (b) 1320 kN
3-B7 24 kN
3-B9 3.24 kN
3-B11 (a) 11.5 kN; (b) 11.5 kN
3-B13 (a) 19.8 kN; (b) 1.62 kN
3-B15 (a) 60.9 m/s^2, upward; (b) 5.30 kN, upward
3-B17 370 N
3-B19 249 N
3-B21 5.2 m/s^2
3-B23 (a) 2500 N; (b) 1300 N
3-B25 (a) 147 N; (b) 5 m/s
3-B27 (a) 2.39 s; (b) 33.6 N; (c) 33.6 N

3-C1 (a) 20 m/s; (b) yes, with 0.33 s to spare
3-C3 4.40 m
3-C5 (a) 1.5 m/s^2; (b) 0.730 s; (c) 83 N; (d) 98 N
3-C7 (a) 4 m/s^2; (b) 4.8 N
3-C9 (a) 0.625 m, downward; (b) 72 N
3-C11 9.95 s

CHAPTER 4

4-A1 (a) 56.6 N; (b) 69.3 N
4-A3 100 units, 37° above the horizontal
4-A5 88.3 N
4-A7 1000 N
4-A9 0.10
4-A11 clockwise
4-A13 800 N·m; clockwise; vector into the plane of the paper

4-B1 3.38 N, 20° N of E
4-B5 453 N, 24.5° N of W
4-B7 (a) 179 N; (b) 146 N
4-B9 84.9 N
4-B11 289 N
4-B13 111 N
4-B15 12.8 N

4-B17 46.7 cm
4-B19 1.25 m
4-B21 60 N; no; 0.42 m/s^2
4-B23 2.12 m
4-B25 54.3 N
4-B27 39.7 N; $H = 39.7$ N toward the right; $V = 29.4$ N upward
4-B29 (a) 37.7 N; (b) 28.3 N; (c) 9.4 N toward the left
4-B31 (a) 1.21 kN; (b) 1.45 kN, 33° above the horizontal
4-B33 345 N; 555 N
4-B35 0.23
4-B37 (a) 540 N; (b) $H = 468$ N; $V = 170$ N
4-B39 1.08 m/s

4-C5 65.6 N·m
4-C7 (a) 80 N; (b) 125 N; (c) 75 N; (d) 127.5 N, downward and forward, 79° below the horizontal
4-C9 (b) $C = 2743$ N; $T = 1981$ N
4-C11 1.33 m
4-C13 3.50 s
4-C15 (a) $+5$ m; (b) $+4$ m; (c) 1 m toward the left; (d) -0.125 m/s^2; (e) -0.125 m/s^2
4-C17 $-\frac{1}{6}R$

CHAPTER 5

5-A1 240 kg·m/s
5-A3 2.1×10^4 kg·m/s^2; 21 kN; 21 kN
5-A5 0.0375 m/s
5-A7 0.4 m/s

5-B1 250 N, by the water
5-B3 33.6 N
5-B5 (a) -300 kg·m/s^2; (b) -300 N; (c) $+300$ N
5-B7 0.556 m/s, in the direction of the fullback's motion
5-B9 7220 m/s
5-B11 4 cm higher
5-B13 (a) 110 N; (b) 429 N
5-B15 320 N, westward

5-C1 (a) 4 kg; (b) 76 kg; 9.12 m/s
5-C3 621 N, downward and to the right, 15° from the vertical
5-C5 (a) backward; (b) 2.23 m/s^2

CHAPTER 6

6-A1 (a) car, 1500 kg·m/s; bullet, 25 kg·m/s;
(b) car, 1125 J; bullet, 12,500 J
6-A3 (a) 24 J; (b) 90 J
6-A5 500 J
6-A7 75 m
6-A9 (a) 12.4 N; (b) 18.6 J
6-A11 4×10^5 J
6-A13 6400 J
6-A15 6.5 m
6-A17 0.03 g; 100 cm/s

6-B1 22.4 J
6-B3 (a) 30 m/s; (b) 45.9 m; (c) 150 N
6-B5 960 J
6-B7 24.5 m/s
6-B9 3 m/s
6-B11 7.92 m
6-B13 0.72 m
6-B15 322 J
6-B17 1.18 m/s
6-B19 3.70 m/s
6-B21 70%
6-B23 727 kJ
6-B25 706 W
6-B27 187 kJ
6-B29 (a) 172 kW; (b) 2.07×10^{10} J; (c) 3.81 kN
6-B31 (a) 180 kJ; (b) 300 kJ; (c) 18 s
6-B33 27.2 kW
6-B35 29.4 J
6-B37 (a) 5 m/s, westward; (b) 6 J lost
6-B39 heavy ball stops; light ball 4 m/s, eastward

6-C3 (a) 1.48 m/s; (b) 80 cm
6-C5 6 kg
6-C7 (a) 7.5; (b) 6.0; (c) 65.4 N
6-C9 (a) 25; (b) 9; (c) 36%
6-C11 about $\frac{2}{3}$
6-C13 25.5%
6-C15 0.167
6-C17 (a) both balls move in same direction, heavy
ball at 0.2 m/s, light ball at 3.4 m/s;
(b) 4.32 J lost

CHAPTER 7

7-A1 (a) 3.14 rad; (b) 0.52 rad; (c) 1.57 rad;
(d) 0.87 rad; (e) 1.88 rad
7-A3 5 m/s
7-A5 128 m
7-A7 0.0167 rad
7-A9 (a) 7.29×10^{-5} rad/s; (b) 0.0339 m/s^2
7-A11 (a) 1.745×10^{-3} rad/s;
(b) 9.14×10^{-5} cm/s^2
7-A13 $P = \tau\omega$
7-A15 377 kg·m^2/s

7-B1 0.772 rev/s^2; 4.85 rad/s^2
7-B3 (a) 0.024 rev/s^2; (b) 7.5 rev
7-B5 79.9 dyn
7-B7 $\sqrt{\mu_s R g}$; no
7-B9 7200 N
7-B11 (a) 1199 N, inward; (b) will not slip
7-B13 (a) 5584 N; (b) 5.58 N
7-B15 (a) 0.738 rad/s; (b) 1960 N
7-B17 20°
7-B19 15.1 m/s
7-B21 0.83 kg·m^2
7-B23 2.95 s
7-B25 (a) 40 kg·m^2/s; (b) 29.4 rad/s; (c) 160 J;
588 J
7-B27 (a) 711 J; (b) 125 rad/s
7-B29 4.73×10^4 rev/min

7-C1 12.6 m
7-C3 (a) 1200 kg·m^2; (b) 0.4 rad
7-C7 8.85 m/s
7-C9 (a) $\sqrt{10gH/7}$; (b) $\sqrt{2gH}$
7-C11 (a) 0.0143 m; (b) 7.38×10^{-6} kg·m^2;
(c) 0.108 N
7-C13 upward; no
7-C15 5.42×10^{22} N·m

CHAPTER 8

8-A1 1.01×10^{-7} N
8-A3 22 N
8-A5 365 days

8-B1 1.4×10^{-13} N
8-B3 0.074 dyn
8-B5 (a) 7.91 km/s; (b) 84.5 min
8-B7 (b) 4.1 cm farther
8-B9 3.9×10^{-3} N/kg, toward the sun
8-B11 (a) 2.72×10^{-3} m/s^2; (b) 2.72×10^{-3} m/s^2
8-B13 4 a.u.
8-B15 (a) $T = 2\sqrt{r^3/Gm_E}$; (b) 101 min

CHAPTER 9

9-A1 4.9×10^7 Pa
9-A3 0.001
9-A5 18.7 N/m
9-A7 (*a*) 0.8 vib/s; (*b*) 40 vib
9-A9 25 s
9-A11 (*a*) 8 cm; (*b*) 8 cm
9-A13 3.14 s
9-A15 11.0 s

9-B1 2 mm
9-B3 1.22×10^7 N/m^2
9-B5 0.0025
9-B7 0.5 m
9-B9 8.00 rad/s
9-B11 yes
9-B13 0.16×10^{10} Pa
9-B15 0.186 m^3; 3.7 cm^3
9-B17 253 N
9-B19 (*a*) 0.942 m/s; (*b*) 888 m/s^2
9-B21 (*a*) 0.314 s; (*b*) 0.500 m/s
9-B23 1.99 s
9-B25 3.1 J
9-B27 0.316 J
9-B29 (*a*) 60 J; (*b*) 49.0 m/s
9-B31 0.8 s
9-B33 (*a*) at the top of the motion; (*b*) 2.23 cm
9-B35 3.67 s

9-C1 1.24 cm
9-C3 $k = \pi r^2 E/L$
9-C5 4.39 mm^2
9-C7 (*a*) 25.5 N; (*b*) 19.6 N; (*c*) 13.7 N
9-C9 (*a*) 1.32 s; (*b*) 5.42 m/s; (*c*) 3.68 m
9-C11 28.1 s
9-C13 (*a*) 6 J; (*b*) yes; (*c*) yes
9-C15 1.64 s
9-C17 1.56 s
9-C19 (*a*) 0.628 s; (*b*) 1.76 m; (*c*) 20 m/s;
      (*d*) -12.9 m/s; (*e*) -176 m/s^2; (*f*) 0.207 s

CHAPTER 10

10-A1 2:1
10-A3 3.35 km/s
10-A5 (*a*) 1410 m/s; (*b*) 4640 ft/s
10-A7 5.27×10^{10} Pa
10-A9 (*a*) 15 vib/min; (*b*) to the right; (*c*) 6 m/s

10-A11 250 Hz
10-A13 1.5 m
10-A15 8.5 mm
10-A17 360 beats/min

10-B1 18 s
10-B3 upward
10-B5 0.002 s
10-B7 25.5 kHz
10-B9 (*a*) 405 m; (*b*) 1.11 s
10-B11 6 beats/s
10-B13 (*a*) 3.27 m; (*b*) 2000 beats/s
10-B15 (*a*) 537 Hz; (*b*) 463 Hz
10-B17 (*a*) 200 m/s; (*b*) 120 m/s

10-C1 (*a*) 680 m; (*b*) 21.7 m/s toward the observer;
      train is between cliff and observer
10-C3 (*a*) 12.5 cm; (*b*) 79 m/s
10-C5 (*a*) 111 km/h; (*b*) yes
10-C7 279 m/s

CHAPTER 11

11-A1 91.3 m/s
11-A3 0.15 m
11-A5 300 Hz; 600 Hz; 900 Hz
11-A7 200 Hz
11-A9 600 Hz
11-A11 220 Hz; 440 Hz; 660 Hz
11-A13 430 Hz; 1290 Hz; 2150 Hz
11-A15 about 100 Hz

11-B3 128 N
11-B5 503 N
11-B7 (*a*) 0.15 g; (*b*) 3 N
11-B9 2.00 m
11-B11 7 beats/s
11-B13 (*a*) open pipe; (*b*) 0.795 m
11-B15 3.2 beats/s
11-B17 (*b*) 1:3:5:7:..., (*c*) 4950 m/s
11-B19 (*b*) 3288 m/s; (*c*) 9.12×10^{10} Pa
11-B21 no
11-B23 5000:1
11-B25 (*a*) 4×10^{-9} W; (*b*) 250 s

11-C1 17 cm; 51 cm; 85 cm; ...
11-C3 148 N
11-C5 126 Hz

CHAPTER 12

12-A1 2.04×10^6 N/m^2; 20.2 atm
12-A3 81.0 kN; 81.0 kN
12-A5 1.10×10^5 N/m^2; 441 lb/in.^{2}; 1.47 lb/in.^{2}; 76 mm Hg; 7.6×10^{-8} torr
12-A7 (a) 2.79 torr; (b) 3720 dyn/cm^2; (c) 3.7 mbar
12-A9 15.1 kPa
12-A11 (a) 3.8 kPa; (b) 95.3 kPa
12-A13 1.25×10^4 dyn

12-B1 1.77 cm
12-B3 723 kN
12-B5 110.4 kPa
12-B7 178 torr
12-B9 yes; 1870 kPa required
12-B11 (a) 12.7 cm^3; (b) 61.7 cm^3
12-B13 (a) 2.5 cm^3; (b) 2.0
12-B15 0.24 g/cm^3
12-B17 (a) 0.088 m^3; (b) 0.014 m^3
12-B19 8.79 m^3
12-B21 (a) 2.70; (b) 0.92
12-B23 34.1 dyn/cm
12-B25 55 dyn, downward
12-B27 (a) 905 dyn; (b) 288 dyn/cm
12-B29 0.0408 cm
12-B31 (a) $h = 2\sqrt{\gamma/dg}$; (c) 0.555 cm
12-B33 (a) 7.00 m/s; (b) 2.33; (c) 1.53
12-B35 24 cm
12-B37 (a) 13.8 kPa; (b) 276 kN

12-C1 (a) 3.6%; (b) 1.07
12-C3 (a) 878 cm^3; (b) 68 N

CHAPTER 13

13-A1 36.0°C; 40.0°C
13-A3 20°C; 10°C; −200°C; 37°C
13-A5 68°F; 932°F; −40°F; −459.4°F
13-A7 0.28°C to 0.50°C
13-A9 8.3 mm
13-A11 1.25 cm
13-A13 brass
13-A15 31.2 liters

13-B1 160°C
13-B3 yes

13-B5 166°C
13-B7 61.7°C
13-B9 41 kN
13-B11 0.012 cm
13-B13 195.7 liters
13-B15 408×10^{-6} (°C)$^{-1}$
13-B17 20°C

13-C1 (a) 2.0003 s; (b) 14 s
13-C3 (a) 145.1×980 dyn; (b) 147.7×980 dyn

CHAPTER 14

14-A1 6000 cal
14-A3 215 cal
14-A5 0.30 cal/g·°C
14-A7 312 kcal
14-A9 24 kcal
14-A11 1500 cal
14-A13 59 cal/g
14-A15 4.35 m
14-A17 1000 Å; 0.1 μm; 100 nm
14-A19 3×10^4 Å

14-B1 0.47 cal/g·°C
14-B3 (a) 4:21 P.M.; (b) 4:48 P.M.
14-B5 60 g of water at 40°C
14-B7 −196°C
14-B9 18 g
14-B11 540 cal/g
14-B13 1003 kg
14-B15 1.89 MJ; 452 kcal
14-B17 (a) 875 J; (b) 164°C
14-B19 4670 times
14-B21 3.77 min
14-B23 37 g
14-B25 1.84°C
14-B27 (a) 1.58×10^5 J; (b) 175 s
14-B29 (a) 90 W; (b) 270 W
14-B31 1.69°C
14-B33 56.4 h

14-C1 (a) 5.5 cal/g
14-C3 131 cal
14-C5 (a) 14.6 metric tons; (b) 10.0 m^3
14-C7 (a) 2600 MW; (b) 38.4%; (c) 2.8°C
14-C9 203 kW
14-C11 1.2×10^{24} J

CHAPTER 15

15-A1 18 cm
15-A3 400 tanks
15-A5 10 mol
15-A7 1.2×10^6 N/m^2
15-A9 40 liters
15-A11 4.67×10^5 N/m^2
15-A13 92%
15-A15 90°C

15-B1 1.67×10^{-27} kg
15-B3 1.7×10^{21} atoms
15-B5 (a) 1.34×10^{27} molecules;
(b) 67,000 molecules
15-B7 200 molecules
15-B9 28.7 cm
15-B11 7.5 cm
15-B13 (a) 400 kPa; (b) 33.6 m
15-B15 (a) 1 atm; 3 atm; 6 atm; (b) 10 atm;
(c) 12.41 kg
15-B17 1.5 g
15-B19 (a) 9.06 mol; (b) 399 g;
(c) 5.45×10^{24} molecules
15-B21 10.2 atm
15-B23 1.20 g/liter
15-B25 6.44 g/liter
15-B27 (a) 0.560 m^3; (b) 5.33 N
15-B29 574.61°
15-B31 55.1%
15-B33 2.21×10^7 Pa; 218 atm; 3210 1b/in.^{2};
0.318 g/cm^3
15-B35 (a) 44.7 m^3; (b) 152 g; (c) 18.3%; (d) no

15-C1 92.5 torr
15-C3 69.8 mm^3
15-C5 2.96 kg
15-C7 7.4×10^4 g/m^3

CHAPTER 16

16-A1 the can
16-A3 800 m/s

16-B1 1200 N
16-B3 310 m/s
16-B5 (a) 5.62×10^{-21} J; (b) 1.56×10^{-24} J;
(c) no
16-B7 (a) 492 m/s; (b) 32
16-B9 $\frac{1}{3}v$
16-B11 140°C

16-B13 29.9 kJ
16-B15 2.07×10^{-23} J/molecule·K
16-B17 8.82 cal/mol·K; 6.77 cal/mol·K
16-B19 (a) 269 cm^3; (b) 39°C
16-B21 618 cal

16-C5 1.0043
16-C7 (a) 2.0; (b) 2.4; (c) 2.83
16-C9 270 m/s
16-C11 −49°C
16-C13 379 K

CHAPTER 17

17-A1 40%
17-A3 30%
17-A5 2200 J

17-B1 27.0 kW
17-B3 (a) 300°C; (b) 34.9%; (c) 27.9%
17-B5 809 K
17-B7 (a) 30%; (b) 3.6×10^8 J; (c) 25%
17-B9 (a) 6.78%; (b) 1100 MW
17-B11 (a) 7.95; (b) 2.62; (c) 9.15 kW
17-B13 (a) 3.46 W; (b) 1.31; 12.0

17-C1 4.2×10^{-4} °C
17-C5 (a) 17,600 cal; (b) 45.0 cal/K
17-C7 J/K; cal/K; kWh/K

CHAPTER 18

18-A1 12 A
18-A3 6.25×10^9 electrons
18-A5 1.6 pA
18-A7 64 mN
18-A9 134 m
18-A11 6.02×10^{24} electrons

18-B1 (a) 8.2×10^{-8} N; (b) 3.6×10^{-47} N
18-B3 5.08 m
18-B5 1.50×10^{-9} N
18-B7 6.68×10^{-7} N
18-B9 13.5 mN, toward the right
18-B11 (a) 2.0 N; (b) 0
18-B13 0.249 N, away from the triangle, perpendicular to the opposite base
18-B15 839 g
18-B17 Cu$^+$
18-B19 (a) 473 A; (b) 47.4 liters

18-C1   2.42 μC
18-C3   1 nC; 5 nC
18-C5   73.4 mN, 15° from the y axis
18-C7   0.00153 cm
18-C9   7.38 A
18-C11   148 min^{-1}

CHAPTER 19

19-A1 2×10^6 N/C, downward
19-A3   2000 N/C, downward
19-A5   0.8 V
19-A7   300 V/m
19-A9   12.5 μF
19-A11   0.018 C

19-B1   (*a*) 5.4×10^4 N/C;   (*b*) 4.0×10^4 N/C;
       (*c*) 9.4×10^4 N/C
19-B3 2.7×10^4 N/C, toward the empty corner
19-B5   (*a*) 4.8×10^{-16} N;   (*b*) 5.27×10^{14} m/s^2
19-B7   40 kV/m
19-B9   (*a*) 60 MV;   (*b*) Y is higher than X
19-B11 8.79×10^6 m/s
19-B13   106 pF
19-B15   150 kV/m
19-B17   2.63 mg
19-B19   11.01 μC
19-B21   (*a*) 30 V; 30 V;   (*b*) 180 μC; 90 μC

19-C1   (*a*) 9.58×10^{11} m/s^2;   (*b*) 4.79 mm
19-C3   (*a*) 5×10^{-10} s;   (*b*) 3.2×10^{-23} N·s;
       (*c*) 3.51×10^7 m/s
19-C5   1.88 cm^2
19-C7   1.59 μC
19-C9   2.25 J
19-C11   (*a*) $x = -5$ m;   (*b*) $x = +1$ m or -1.25 m
19-C13   1800 μF
19-C15   (*a*) 1.013 cal/g·°C;   (*b*) 0.995 to 1.031
       cal/g·°C
19-C17   (*a*) 15 V;   (*b*) 45 μC; 75 μC;   (*c*) 540 μJ

CHAPTER 20

20-A1   8 kV
20-A3   50 mA
20-A5   1.8 W
20-A7   yes
20-A9   0.133 kWh

20-B1   (*a*) 6 V;   (*b*) 30 W
20-B3   (*a*) \$2.92;   (*b*) 20 kg/y

20-B5   93%.
20-B7   11.4 kJ
20-B9   22.9 kg/min
20-B11   (*a*) 41.9 Ω
20-B13   405 J; 1.12×10^{-4} kWh
20-B15   459 μV
20-B17   aluminum
20-B19   192°C
20-B21   (*a*) 31.2 Ω;   (*b*) 31.2 Ω

20-C1   21.4 m
20-C3   (*a*) 1.22 Ω;   (*b*) 0.06%
20-C5   (*a*) 480 W;   (*b*) 83.3%;   (*c*) 3 Ω
20-C9   (*a*) 51.8 MJ;   (*b*) 0.026 MJ/kg;   (*c*) 31.0°C
20-C11   (*a*) 9.76×10^5 MJ;   (*b*) 21.7 MW

CHAPTER 21

21-A1   9 V
21-A3 -12 V
21-A5   35 Ω
21-A7   (*a*) 18 Ω;   (*b*) 2 Ω;   (*c*) 9 Ω;   (*d*) 4 Ω
21-A9   11.24 V
21-A11   50.6 Ω
21-A13   6 Ω

21-B1   10 Ω
21-B5   6 A
21-B7   (*a*) 6 A through 5 Ω; 4 A through 6 Ω; 2 A
       through 12 Ω;   (*b*) 432 W
21-B9   (*a*) 12 A;   (*b*) 96 W;   (*c*) 114 V; 30 V
21-B11   (*a*) 7.56 V;   (*b*) 2.16 V
21-B13   (*a*) 16 V;   (*b*) 0 V;   (*c*) 2 V;   (*d*) 14 V
21-B15   1.7 V
21-B17   9800 Ω in series
21-B19   0.0200 Ω in parallel
21-B21   (*a*) 600 Ω;   (*b*) 10 V
21-B23   (*a*) 16.5 V;   (*b*) 16.0 V;   (*c*) A to B;
       (*d*) 22 Ω
21-B25   4.71 cm to the right

21-C1   144 Ω
21-C3   2304 W
21-C5   (*a*) 4.5 A;   (*b*) 162 W
21-C7   15 Ω
21-C9   (*a*) 0.0372 S;   (*b*) $G_p = G_1 + G_2 + \cdots$
21-C11   (*a*) $R_x = R_2 - \frac{1}{4}R_1$;   (*b*) 2.25 Ω; not quite
       adequate
21-C13 4.08×10^{-6} m^2
21-C17   (*a*) 15 A;   (*b*) 2.07 Ω
21-C19   (*a*) 4 A;   (*b*) 8 V

CHAPTER 22

22-A1   0.25 T
22-A3 3.6×10^{-13} N
22-A5   240 N
22-A7   1.07 A
22-A9   38.2 A
22-A11   180 mV
22-A13 -30 V

22-B1   (a) 1.2×10^{-10} N;   (b) 7.33×10^{15}
22-B3   (a) 6.4×10^{-13} N;   (b) 1.92×10^{14} m/s^2;
   (c) 2.09 m
22-B5   (a) 2.5×10^4 m/s;   (b) toward the south;
   (c) yes
22-B7   (a) upward;   (b) 6.98×10^{-5} T;   (c) a bit
   stronger
22-B9 2.67×10^{-5} N, toward the west
22-B11   (a) repulsion;   (b) 8 mN
22-B13   12.1 V
22-B15   4.62 V
22-B17   0.302 A
22-B19   (a) 1.33 m/s;   (b) no

22-C1   7.54 C
22-C3   18 kA
22-C5   (a) 0.0779 V; A is positive;   (b) 39.0 A;
   (c) 6.08 N;   (d) 3.04 W;   (e) 3.04 W
22-C7   (a) 0.40 V;   (b) 10 s;   (c) 1.6 V

CHAPTER 23

23-A1 $16°$
23-A3   36 V
23-A5   75 V
23-A7   0.03 A

23-B1   4.2 A
23-B3   (a) 5.5 A;   (b) 0.5 A;   (c) 240 Ω
23-B5   (a) clockwise;   (b) counterclockwise
23-B7   (a) 1200 kW;   (b) 18 kW;   (c) 1182 kW;
   (d) 98.5%
23-B9   (a) 23.5 kW;   (b) 97.7%;   (c) 12.28 kV
23-B11   800 V
23-B13   (a) 66.3 Hz;   (b) 14.9 kHz
23-B15   318 Hz
23-B17   (a) -4.20 V; -3.15 V;   (b) 70

23-C1   (a) 51%;   (b) 2.16 kWh;   (c) motorcycle,
   0.135¢/km; car, 4¢/km;   (d) 0.26¢/km
23-C3   33.3 A

23-C5   0.00512 m^2
23-C7   (a) 23.6 sin 377t;   (b) 18.9 V;   (c) 13.4 V
23-C9   205 Ω at 67°; current lags voltage
23-C11   bulb, 2880 Ω; circuit, 2883 Ω
23-C13   253 pF
23-C15   (a) 0.8 A;   (b) 1.76 μF;   (c) 1.71 kV
23-C17   (a) 33.6°;   (b) 250 Ω;   (c) 0.440 H

CHAPTER 24

24-A1   (a) 2 cm;   (b) 1.5 cm
24-A3   12 GHz
24-A5 4×10^{14} Hz
24-A7   3.34 m
24-A9   (a) 6.328×10^{-7} m;   (b) 5.6×10^{-11} m;
   (c) 1.5×10^{-5} m;   (d) 5.93×10^{-7} m

24-B1   360 J
24-B3   600 kHz
24-B5   142 Hz
24-B7   (a) 10.977 m;   (b) 21.2 pF
24-B9   2:1
24-B11   (a) 54.4 nH;   (b) 39.9 pF

24-C1   (a) $\sqrt{L/C}$;   (b) 913 Ω
24-C3   (a) 2.53 pF;   (b) 1.69 cm;   (c) 6.28 Ω

CHAPTER 25

25-A1 1.54×10^8 m/s
25-A3   1.60
25-A5   345 nm
25-A7 2.25×10^8 m/s
25-A9   (a) 4 diopters;   (b) positive

25-B1   45°
25-B3   1.25
25-B5   43.3°
25-B7   37°
25-B9   (a) 400 cm, diverging;   (b) 500 cm,
   converging
25-B11   (a) 3 cm to left of lens;   (b) virtual;
   (c) erect;   (d) 1.5 mm
25-B13   120 cm below the object, virtual, erect, 16 mm
   wide
25-B15   (a) -75;   (b) 7.89 cm
25-B17   0.46 mm
25-B19   50 cm
25-B21 -60 cm
25-B23   (a) 6 cm behind the surface;   (b) $+0.4$

25-B25 (a) $+8$ cm; (b) 10 cm in front of the mirror; (c) real; inverted; 1.25 cm

25-B27 (a) 48 cm behind the mirror; (b) concave; (c) 16 cm

25-C1 (a) $53°$; (b) glass

25-C3 (a) $n = \sqrt{1 + 16t^2/D^2}$; (b) 1.75 cm; (c) violet

25-C5 144 cm

25-C7 (a) converging; (b) 60 cm; (c) 45 cm and 75 cm from the lens

25-C9 (a) 105 cm; (b) 8 mm

25-C11 (a) 21 cm; (b) 84 cm

25-C13 (a) $(n-1)/R$; (b) $2(n-1)/R$; (c) $2n/R$

25-C15 (a) 50 cm to the left of the lens; (b) real; erect; $m = +\frac{1}{3}$

25-C17 3.72 diopters

CHAPTER 26

26-A1 450 nm

26-A3 3.0 μm

26-A5 1.18 μm

26-B1 1.25 m

26-B3 19.3 kHz; 57.9 kHz; 96.5 kHz

26-B5 27 images

26-B7 (a) 31.90 cm; (b) 4.36 cm

26-B11 1.74 m

26-B13 4.04 m

26-B15 (a) 1.6 mm; (b) 51 fringes

26-B17 constructively

26-B19 2.8 μm

26-B21 190 fringes

26-B23 $37°$

26-C1 $n\lambda = a(\sin\theta_1 - \sin\theta_2)$

26-C3 radius inversely proportional to $\sqrt{n}$

26-C5 1.59

26-C7 (a) 413.7 nm; 409.7 nm; (b) $12.8°$

26-C9 yes

26-C11 (a) 1.116; (b) extraordinary ray

CHAPTER 27

27-A1 the player

27-A3 $f/2.4$; yes

27-A5 about 200

27-A7 160

27-A9 30; 80

27-B1 $\frac{1}{50}$ s

27-B3 1.82 cm

27-B5 top, -0.8 diopters; bottom, $+2.5$ diopters

27-B7 $+75$ cm

27-B9 8

27-B11 21

27-B13 4.13

27-B15 (a) 11.2 cm; (b) 290

27-B17 (a) 50; (b) 20 m

27-B19 (a) 0.571 cm; (b) 87.5

27-B21 (a) 6.4; (b) 92.5 cm

27-B23 19 cm

27-B25 $15.6°$

27-C1 (a) 17.0 cm; (b) 0.171 rad; (c) 0.0199 cm, outward; (d) 16.9 cm

27-C3 120 cm; 2 cm

27-C5 (c) 26.8421 cm; (d) 1.1124 diopters

27-C7 yes, just barely

CHAPTER 28

28-A1 (b) 10^{-4}

28-A3 $1.15m_0$

28-A5 (a) 2.7 TJ; (b) 68.9 m

28-B1 (a) 25.2 J; (b) 11.2 J

28-B3 (a) 1.35 ns; (b) 0.0564 m; (c) 0.190 ns

28-B5 2.32×10^4 m/s

28-B7 6.75×10^{-10} J

28-C1 (a) 5 μs; (b) 0.15 m; (c) 0.5 ns

28-C3 (a) proper time; (b) 2.57×10^8 m/s; (c) 2770 s

28-C5 (a) 11.39×10^{-31} kg; (b) 2.28×10^{-31} kg; (c) 2.05×10^{-14} J; (d) $\frac{1}{2}m_0v^2$ gives 1.48×10^{-14} J; $\frac{1}{2}mv^2$ gives 1.85×10^{-14} J

28-C7 (b) 2.83×10^8 m/s

28-C9 $\frac{4}{5}c$

CHAPTER 29

29-A1 8.0×10^{-19} J; 5.0 eV

29-A3 6.25 MeV

29-A5 8.29 Å

29-B1 1.20 μm

29-B3 5.92 cm

29-B5 12.7 V

29-B7 3×10^{10}

29-B9 0.94 eV

29-B11 2.04 eV

29-B13 257 nm

29-B15 blue

29-B17 0.89 eV

29-B19 1.40×10^{-27} kg·m/s

29-B21 103 eV

29-B23 (*a*) 4.4×10^{-32} kg·m/s; (*b*) 9×10^{-31} m

29-B25 6.14×10^{-12} m

29-B27 122 fm

29-C1 466 nm

29-C3 (*a*) 1.74 Å; (*b*) x-ray region

29-C5 (*a*) 1.98×10^{-8} s; (*b*) 7.2 μm; (*c*) yes

29-C7 2.32 Å

CHAPTER 30

30-A1 404.59 nm; 406.36 nm; 407.18 nm

30-A3 (*a*) 3.04 eV; (*b*) 408 nm

30-A5 5.77 eV

30-B1 122 nm; no

30-B3 0.66 eV; 1880 nm; no

30-B5 $E_7 \to E_3$

30-B7 10.24 eV; Lyman α

30-B9 (*a*) 2.64×10^{40} kg·m²·rad/s; (*b*) 2.5×10^{74}

30-B11 $+27.2$ V

30-B13 97

30-C1 (*a*) $\pm 1.06 \times 10^{-5}$ rad; (*b*) 8.0 km

30-C3 486 nm

30-C5 (*a*) 6.6×10^7 waves; (*b*) 60 m

30-C7 1.1×10^{-4} rad (0.006°)

30-C9 6.48 kJ

CHAPTER 31

31-A1 n, l, m_l, m_s values are 1, 0, 0, $\frac{1}{2}$; 1, 0, 0, $-\frac{1}{2}$; 2, 0, 0, $\frac{1}{2}$; 2, 0, 0, $-\frac{1}{2}$

31-A3 at $Z = 1$

31-B1 (*a*) $l = 2$; (*b*) 4.8 Å

31-B3 (*a*) 1.54 Å; (*b*) 1.38 Å; (*c*) 13.5 Å

31-B5 0.39 Å

CHAPTER 32

32-A1 Sn; He; Ti; Au

32-A3 6; 10; 120; 60

32-A5 4.8 fm

32-A7 14.0 MeV

32-A9 $\frac{1}{16}$

32-A11 30 hours

32-A17 (*a*) the parent; (*b*) 0.00157 u

32-B1 7.47 MeV

32-B5 650 MeV

32-B7 3.32 days

32-B9 0.81

32-B11 (*a*) 2.75×10^{-13} s^{-1}; (*b*) about 100 atoms

32-B13 0.78 MeV

32-B15 0.16 MeV

32-B19 5.32 MeV

32-B21 yes; 3.1 MeV

32-B23 3.27 MeV

32-B27 1.19 MeV

32-C3 (*a*) 2.81 MeV

32-C5 decay energy, 1.57 MeV; excited states of ^{59}Co at 1.10 MeV and 1.29 MeV

32-C7 5.2 fm

32-C9 56 W

32-C11 (*a*) 660 Å; (*b*) 2×10^{-7} eV; (*c*) 0.0015 K

32-C13 3

CHAPTER 33

33-A5 20.9 μs

33-A7 0.98 T

33-A9 0.07 fm

33-B1 (*a*) 1.55×10^{18} fissions/s; (*b*) 4.91×10^{25} fissions/y; (*c*) 19.1 kg; (*d*) 2660 kg

33-B3 (*a*) 7.25×10^{13} J/kg; (*b*) 2.48 kg

33-B5 0.00518 MeV

33-B7 (*a*) \$33,000,000; (*b*) 22.4 metric tons

33-B9 about 4000 dis/s

33-B11 4.12 s

33-B13 (*a*) 3.2 μA; (*b*) 1.6 MW

33-B15 (*a*) 1.88 GeV; (*b*) 18.18 GeV

33-B17 12.9 MeV

33-B19 (*a*) 104 MeV; (*b*) 1.8 MeV

33-C1 2460 y

33-C3 3.33 T

33-C5 (*a*) 6.4×10^{10} J

33-C7 4×10^{12} J/ton

Index

Italic numbers refer to illustrations.
n refers to footnotes.

Angular size, 647
Angular velocity, 152
  vector representation of, 166
Angular vibration, 198
Anisotropic crystal, 572, 620,
  630–632
Annihilation radiation, 803
Antennas, 560–562
  loop, 562
Antimatter, nuclear, 803
Antineutrinos, 766, 780
Antinodes, 241, 244, 608, 697
Antiproton, 801, 803
Apparent coefficient of volume
  expansion, 303
Apparent weight, 117
Applied nuclear physics, 781–814
Applied optics, 634–654
Approximate calculations, 842–843
Arago, Dominique, 613
Archimedes, 276
Archimedes' principle, 268, 275–278
Arithmetic, review of, 823–830
Armature, 516
Artificial gravity, 119
Astigmatism:
  in the eye, 638
  off-axis, 635, 638
Aston, Francis W., 753
Astronomical telescope, 647–649
Atmosphere (unit of pressure), 268
Atom, 13–16
  distribution of charge in, 713–714
  early models of, 712–713
  energy levels of, 716–722
  hydrogenlike, 717–718
  impurity, 444
  mass of, 14, 334–336
  nuclear model of, 713–714
  outer, 712–746
  planetary model of, 13–16, 81n
  Rutherford model of, 713–714
  structure of, 13–16, 732–746,
    748–750
  Thomson model of, 712–713
Atom smashers, *see* Particle
  accelerators
Atomic clock, 8
Atomic mass unit, unified, 754–755
Atomic number, 14, 753
Atomic weight, 15
Atwood's machine, 64–65
Audibility graph, 253, *253*
Auger, Pierre-Victor, 745n
Auger electron, 745
Available work, 377–379
Average velocity, 26–27
Avogadro's hypothesis, 357
Avogadro's number, 334–336

B

B-field, strength of, *see* Magnetic
  induction
Back electromotive force, 517–519
Back torque, 518–519
Balance:
  current, 496
  mass, 205
Ballistics, 44
Ballistocardiograph, 112
Balmer series, 717, 719, 720
Band spectrum, 715, 750
Banking of curves, 159–161
Bar (unit of pressure), 268
Barkla, C. G., 714, 737
Barn (unit of area), 784
Barometers, 273–274
Barrow, Isaac, 55
Baryons, 805 (*table*), 808
Base (transistor), 532
Batteries, 277, 452, 461–463
  in parallel, 462–463
  in series, 461–463
  symbol for, 454
  See also Cells (electric)
Beats (musical), 229–231, 239
Beccaria, Giovanni, 399n
Beck, W. S., 219n, 637
Becquerel (unit of radiation), 760n
Becquerel, Henri, 752, 761
Bel (unit of sound intensity), 250
Bell, Alexander Graham, 250
Bernoulli, Daniel, 283n, 396n, 406
Bernoulli's equation, 269n, 282–285,
  294–295
Bessel, F. W., 207n
Beta decay, 758–760, 760–761
  conservation of angular momentum
    in, 779–780
  conservation of energy in, 765–766
  conservation of parity in, 767
Beta particles, 758–761
Beta rays, 753
Bias, forward and reverse, 530
Bias voltage, 533
Bifocal glasses, 639–640
Binding energy:
  atomic, 719, 755
  nuclear, 755–757
Binoculars, 649–650
Biological effects of radiation,
  793–795
Bipolar junction transistor (BJT),
  531–532
Black, Joseph, 310, 311
Blackbody, *310*, *322*, 324
Blackbody radiation, 690, 715
Bohr, Niels, 714
Bohr energy formula, 719
Bohr theory for hydrogen, 714,
  715–721

limitations of, 732–733
Boiling point, 345
Boltzmann, Ludwig, 368
Boltzmann's constant, 357
Bonds, covalent, 11
Bore, 302
Born, Max, 732
Bound particles, 709–710
Bourdon gauge, 274
Boyle, Robert, 334
Boyle's law, 333–334
  effect of mass in, 336–337
  effect of temperature in, 337
Bragg, Sir William Henry, 693
Bragg, Sir William Lawrence, 693
Brahe, Tycho, 178
Breeder reactors, 786, 788–789
Bremsstrahlung, 745
Brewster's angle, 620
Bridgman, Percy W., 1, 271
British engineering system, 9
British thermal unit (Btu), 313
Broglie, Louis de, 698, 732
Brown, Robert, 135
Brownian motion, 135, *135*
  Einstein and, 135n
Btu (British thermal unit), 313
Bubble chamber, 773
Bulk modulus, 196–198, 197 (*table*)
Buoyant force, 268, 275

C

Calculations, approximate, 842–843
Calculus:
  applied to accelerated motion,
    52–53
  mathematical review of, 840–842
Caloric theory, 310–311
Calorie, 313
Camera, 634–636
Candela (unit of luminous intensity),
  7
CANDU, 786, 791
Capacitance, 420–422
Capacitors, 420–426
  in electric circuits, 424–426
  energy storage in, 432–434
  impedance of, 526–527
  in parallel, 420, 421, 426
  reactance of, 543–545
  in series, 426
  symbol for, 454
  working voltage of, 425
Capillary action, 281
Carbon cycle, 787, 816
Carbon-14, 782, 783
Carnot, N. L. Sadi, 375
Carnot cycle, 375
Carnot efficiency, 376
Carnot engine, 375–376

Carrier wave, 563–564
Cassegrain focus, 648
Cathode, 535
Cathode-ray tube, 537–538, 688
Cathode rays, *686*, 688
Cavendish, Henry, 177, 396, 407
Cavendish's experiments:
    on electric force law, 396*n*, 407
    on gravitation, 177–178, *178*, 179
Cells (electric):
    dry, 457, 462
    electromotive force of, 457*n*
    internal resistance of, 457
    in parallel, 462–463
    polarization of, 457
    in series, 461–463
    standard, 470–471
    symbol for, 454
    terminal voltage of, 457–459
    Weston, 471
Celsius, Anders, 298*n*
Celsius temperature scale, 298
Center of gravity, 75, 104, 105–106
    computation of, 89–90
    distinguished from center of mass, 104
    solution of problems involving, 91–93
    of a uniform body, 75
Center of mass, 104–106, 112*n*
Centigrade temperature scale, *see* Celsius temperature scale
Centime, 8
Centimeter, 6
Centimeter-gram-second (cgs) system of units, 9, 61–62
Central force, 183
Centrifugal force, 158–159
Centripetal acceleration, 154–156
Centripetal force, 156–158
CERN, 798, 800–801, 808
Cgs (centimeter-gram-second) system of units, 9, 61–62
Chadwick, Sir James, 753, 770
Chain reaction, 783–784
Change of phase, 344–346
Charge, *see* Electric charge
Chemical forces, 11
Chemical potential energy, 127–128
Chromatic aberration, 635–636
Chromium, isotopes of, 761 (*table*)
Circuits, *see* Electric circuits
Circular motion, uniform, 153–156
    simple harmonic motion and, 200–206
Clausius, Rudolf Julius Emanuel, 373*n*
Clock, atomic, 8
Cloud chamber, 770, 772, 773, *802*
Cobalt-60, 762, *763*

Coefficient:
    of expansion, 299–305, 300 (*table*)
    of kinetic friction, 79, 80 (*table*)
    of performance (COP), 380
    of restitution, 147
    of static friction, 79, 80 (*table*)
    of surface tension, 279, 281 (*table*)
Coherent light, 619, 722–725
Cohesive forces, short-range, 278–280, 301–302, 333
Coils:
    energy storage in, 569
    impedance of, 524–526
    reactance of, 543–545
    self-inductance of, 525
Collector (transistor), 531
Colliding beams (in particle accelerators), 799–801
Collimator lens, 650
Collisions:
    conservation of momentum in, 111, 113
    elastic, 134–136
    energy changes in, 134–137
    inelastic, 111, 119, 134
    partially elastic, 134, 147–149
Color charge (of quarks), 809
Colors:
    relation to speed in a medium, 574
    of thin films, 615–616
    wavelengths of, 571 (*table*)
Combustion, heat of, 320–321, 320 (*table*)
Commutator, 516–517
Compass needle, 490, 491
Component method, in equilibrium problems, 76, 77
Compound microscope, 644
Compound nucleus, 770–771
Compression waves (pulses), 220–221
    superposition principle and, 229
    velocity of, 221–223, 238
    See also Sound waves
Compton, Arthur H., 692
Compton effect, 692–696
Compton wavelength, 695
Conant, James B., 1
Concave mirror, 590–591
Condensation:
    in a pressure wave, 228
    of a vapor, 344
Conduction, thermal, transfer of heat by, 321–322, 321 (*table*)
Conductivity:
    electric, 393
    thermal, 321–322, 321 (*table*)
Conductors:
    electrical, 392
    liquid, 399–401
    thermal, 321–322, 321 (*table*)

Configuration, electronic, 739, 740 (*table*), 741, 742 (*table*)
Conjecture, 3
Conservation of angular momentum, 113–114, 163–166
    in nuclear processes, 766, 779–780
Conservation of charge, 392–396
    in nuclear processes, 764
Conservation of energy, 108, 125–138, 312–314, 671–674, 693–694
    law of, *see* Law of conservation of energy
    in nuclear processes, 764–766
    in relativity, 671–674
Conservation of linear momentum, 110–113, 694–695
    in collisions, 111, 113, 135–137, 148
    in nuclear processes, 766
    in relativity, 670–671
    significance of, 137
Conservation of mass:
    in nuclear processes, 764–766
    in relativity, 671–674
Conservation of momentum, 108–120
    See also Conservation of angular momentum; Conservation of linear momentum
Conservation of parity, 766–767
Constant-pressure gas thermometer, 305
Constant velocity, 24
Constant-volume gas thermometer, 305
Constructive interference, 606–608
Contact, angle of, 281, 281 (*table*)
Contact potential difference, 528
Continuous spectrum, 714
Convection, heat transfer by, 321–322
Conventional current, 439–440
Converging lens, 579, 584
Converging mirror, 590–591
Converging wave front, 583
Convex mirror, 590
Coolness storage, 449
COP, *see* Coefficient of performance
Coriolis, Gaspard, 116
Coriolis forces, 116
Cornea, 637, 641
Corpuscular theory of light, 602–604
Cosines, 38, 39, 835–836
Cosmic-ray showers, 811
Cosmic-ray stars, 811
Cosmic rays, 810–812
Coulomb (unit of charge), 397–398
Coulomb, Charles-Augustin, 396, 407
Coulomb forces, 11, 390
Coulomb's law, 11, 396–398
    scientific priority and, 405–409
Covalent bonds, 11

of a rigid body, 84–89
  stable, 90, 91, 133
  static, 74
  unstable, 90, 91, 133
Equipartition of energy, law of, 358
Equipotential surfaces, 418–420
Erg, 125, 126
Eta particle, 805 (*table*)
Ether, 664
Evaporation, 342–343
Exclusion principle, Pauli, 278,
  736–738
Expansion:
  coefficients of, 300 (*table*)
  thermal, 299–305
Exponents, review of, 824–825
Extraordinary ray, 631
Eye, human, 636–641
Eyepiece, 644
Eyestrain, minimum, and adjustment
  of instruments, 643–644

F

f-number of a lens, 636
Fahrenheit temperature scale, 298
Faller, J. E., 408
Farad (unit of capacitance), 421
Faraday (unit of charge), 400–401
Faraday, Michael, 395*n*, 408, 410,
  412, 501, 523
Faraday-Maxwell theory of electric
  field, 413–414, 433
Faraday's law for induced emf,
  501–502
Farsightedness, 638, 639
Fermi (unit of length), 801
Fermi, Enrico, 801*n*
Fermilab, 797, *798*
Fictitious forces, 116
Field:
  electric, *see* Electric field
  gravitational, *see* Gravitational
    field
  magnetic, *see* Magnetic field
  scalar, 412
  vector, 410
  velocity, 412
Field effect transistor (FET), 533–535
Field particles, 414, 803–804
Field theory, of electromagnetic
  radiation, 663
Fifth (musical interval), 249
Films, thin, colors of, 615–616
Filter, electric, 531
First-law efficiency, 374
Fission, 757, 761–762, 783–787
  in nuclear reactors, 785–786,
    788–789
  spontaneous, 761
Fitzgerald, G. F., 670

Flow:
  plastic, 286–287
  streamline, 282, 283
  turbulent, 282
Fluids, 267–287
  defined, 267–268
  in motion, 282–285
  at rest, 267–281
  See also Gases; Liquids
Fluorescence, 721
Fluorescent screen (in a cathode-ray
  tube), 537
Flux, magnetic, 501–503
FM broadcast band, 323 (*table*),
  563–565
Focal length:
  of a lens, 579
  of a mirror, 590
Focal plane, 579
Focal point, 579
Focus:
  Cassegrain, 648
  Newtonian, 648
  prime, 648
  principal, 579
Foot (unit of length), 9
Foot-pound-second (fps) system of
  units, 9
Force, 10–11, 56
  action-reaction pairs, 58, 59
  adhesive, 280
  buoyant, 268, 275
  as a cause of acceleration, 56
  central, 183
  centrifugal, 158–159
  centripetal, 156–158
  chemical, 11
  cohesive, 278–280, 301–302, 333
  Coriolis, 116
  Coulomb, 11
  defined, 56
  elastic, 11
  electric, *see* Electric forces
  electromagnetic, 485*n*, 674–675
  electromotive, *see* Electromotive
    forces
  electrostatic, 11
  equilibrium of, 74–94
  fictitious, 116
  frictional, *see* Friction
  gravitational, 10–11, 117, 803–804
  inertial, 116, 118–119
  lines of, 412–413
  magnetic, *see* Magnetic forces
  nature of, 56
  net, 56, 109
  normal, 81, 82
  nuclear, 11, 757
  reaction, 128
  restoring, 199
  units of, 59–63

variable, work done against,
    126–127
Force constant, 199
Force diagram, 76, *77, 78, 79, 84, 87,
  88*
Force-displacement curve, *126, 127,
  127*
Forward-biased diode, 530
Foucault, Jean, 603–604
Fourier, Joseph, 252
Fps (foot-pound-second) system of
  units, 9
Frame of reference, 114–115, *114*
  accelerated, 117–119
  inertial, 115–116
  noninertial, 116
  rotating, 158–159
Franklin, Benjamin, 391, 406
Fraunhofer, Joseph von, 721
Free-body method, 64
Free electrons, 392
Free particle, 708–709
Free pole, 496*n*
Frequency, 227
  angular, 544
  of circular motion, 155
  in Doppler effect, 231–233
  fundamental, 243
  natural, 242, 554
  pitch and, of musical sounds,
    248–249
  resonant, 548–549
  threshold, 689
  unit of, 228
Frequency modulation (FM), 323
  (*table*), 563–565
Friction, 79–84
  centripetal force supplied by, 156
  coefficients of, 79, *80*
  electrification by, 392
  fluid (viscosity), 282, 285–287
  kinetic, 79–80
  "laws" of, 79–80
  model for, 81
  rolling, 80
  static, 79–80
  temperature rise due to, 316–317
Frisch, O. R., 783
Fuel cell, 451
Functions, mathematical review of,
  836–837
Fundamental frequency, 243
Fusion:
  heat of, 318–320, 319 (*table*)
  of nuclei, 757, 787–788, 789–791

G

Gabor, Dennis, 630
Galilean relativity, 115, 661–663
Galilean telescope, 649–650

Galilean transformation, 115, 682
Galileo Galilei:
  and classical relativity, 661–663
  and dynamics, 55, 180
  and kinematics, 22
  and the telescope, 649–650
  and the thermoscope, 297
Galvanometer, 466–467, 513–515
Gamma rays, 219, 695, 753
  in electromagnetic spectrum, 323
    (*table*)
  emission from nuclei, 762–764
Gas law:
  adiabatic, 362–363
  general, 340–341, 356
Gas thermometers, 305
Gases:
  adiabatic gas law, 362–363
  compression waves in, 220–221
  cooling of, during expansion, 342
  elasticity of, 197 (*table*)
  emission spectra of, 714–715, *714*
  general gas law, 340–341, 356
  ideal, 333
  kinetic theory of, 354–363
  noble, 739, 740 (*table*), 742 (*table*)
  partial pressure of, 341
  pressure coefficients for, 337–338,
    338 (*table*)
  specific heat capacity of, 315
    (*table*), 359–362, 360 (*table*)
  speed of sound in, 368–369
  thermal behavior of, 333–346
  thermal expansion of, 300 (*table*),
    304–305
  viscosity of, 286*n*
Gate (in transistors), 533
Gauge pressure, 274
Gauss (unit of magnetic induction),
    506
Gauss, Karl Friedrich, 407*n*
Gaussmeter, 492
Geiger counter, 772
General gas law, 340–341
General theory of relativity, 676–677
Generator of sound, in musical
    instruments, 255, 256
Generators, electric, 515–517
  of electromagnetic waves, 323
    (*table*), 560–562
  electromotive force of, 543
  Van de Graaff, 795
Genetic damage, 794
Geometrical optics, 570–593,
    599–600
Geometry, review of, 834–835
Germer, L. H., 699
Gilbert, William, 499, 500
Glashow, S., 810
Gluons, 415*n*, 803 (*table*), 808–809
Gram, 8

Graphical method, in equilibrium
    problems, 76
Graphs:
  audibility, 253, *253*
  mathematical review of, 837–839
  vibration, 225, *225*, 226, *226*, 227
  wave-form, 225, *225*, 226, *226*, 227
Grating, 608–610
Grating constant, 609
Grating space, 609
Grating spectroscope, 651–653
Gravitation:
  Newton's law of universal, 177
  and planetary motions, 177–188
  See also Gravity
Gravitational field, 181–182, 220*n*
Gravitational field strength, 181
Gravitational forces, 10–11, 117,
    803–804
Gravitational mass, 180
Gravitational potential, 415–416
Gravitational potential energy, 127,
    129
  storage of, 415–416
Gravitons, 415, 803 (*table*), 804
Gravity:
  acceleration due to, 33–34
  artificial, 119
  center of, *see* Center of gravity
  specific, 13
Gray (unit of absorbed radiation
    dose), 794
Grid (of vacuum-tube triode), 536
Grimaldi, Francesco, 612
Ground state, 716
Grounded point, symbol for, 422, 454
Grounding circuits, 422
Gyroscopes, 173–175

H

Hadrons, 804, 805 (*table*), 808–810
Hafele, Joseph, *661*
Hahn, Otto, 783
Hahnium, 786
Hale reflector, 648
Half-life, 758–760
Hall effect gaussmeter, 492
Halley, Edmund, 55, 182
Halogens, 739, 740 (*table*), 742 (*table*)
Hariot, Thomas, 576
Harmonic motion:
  non-simple, 208–209
  simple, *see* Simple harmonic
    motion
Harmonics, 243
Head-to-tail method of vector
    addition, 35, *35*
Heat, 310–325
  of combustion, 320–321, 320 (*table*)
  of dissociation, 318

distinguished from temperature,
    354–355
  as energy, 311–314
  of fusion, 318–320, 319 (*table*)
  latent, 318*n*, 448
  low-temperature, 378
  mechanical equivalent of, 314
  quantity of, 313
  sensible, 448
  solar, 379
  specific, *see* Specific heat capacity
  theory of, 310–314
  transfer of: by conduction,
    321–322, 321 (*table*); by
    convection, 321–322; by
    radiation, 322–325, 323 (*table*)
  of vaporization, 318–320
  waste, 378
Heat capacity, 314
Heat death, 373
Heat engines, 373–377
Heat pumps, 379–383
Heavy hydrogen (deuterium), 15,
    782, 786, 787, 790
Heavy water, 136, 786
Heisenberg, Werner, 703, 732
Heisenberg uncertainty principle,
    703–704
Helium, 717–718, 720
Helium-neon laser, 724–725
Helmholtz, Hermann von, 314
Henry (unit of inductance), 525
Henry, Joseph, 501, 523
Hero, 110
Hertz (unit of frequency), 228
Hertz, Heinrich Rudolf, 228*n*
Heterodyning, 230
High-energy physics, 801–803
Hilac, 799
Hill, H. A., 408
Holes, positive, 530
Hologram, 628
Holography, 628–630
Hooke, Robert, 182, 193, 405
Hooke's law, 193
Horsepower, 133
Human ear, 253–257
Human eye, 636–641
Humidity, relative, 343–344
Huygens, Christian, 405, 571
Huygens' principle, 571–572
Hydrodynamics, 282–285
Hydrogen:
  Bohr theory for, 714, 715–721
  heavy, 15, 782, 786, 787, 790
  isotopes of, 15
  spectrum of, 714–721
Hydrogenlike atom, 717–718
Hydrometer, 277
Hypermetropia, 638, 639

Negative lens, 584
Negatron-positron pairs, 802–803
Negatrons, 760
Neptunium, 786
Net force, 56, 109
Neutral equilibrium, 90–91, *91*
Neutrinos, 668, 753, 765–766, 780
Neutron, 753–754
    defined, 14
    emission of, in fission, 783–786
    thermal (slow), 784
Neutron capture, 784–786
Newton (unit of force), 59–60
Newton, Sir Isaac:
    and classical relativity, 662, 663
    and dynamics, 55–56
    and gravitation, 177–178
    and optics, 602–604, 650
Newtonian focus, 648
Newtonian synthesis, 182–186
Newton's constant, 179
Newton's corpuscular theory of light,
    602–604
Newton's first law, 56–57
Newton's law of universal
    gravitation, 177, 182–186
Newton's rings, 615
Newton's second law, 57–58
    for rotational motion, 161
    status of, in quantum mechanics,
        187, 732
    in terms of momentum, 109–110
Newton's third law, 58–59
Noble gases, 739, 740 (*table*), 742
    (*table*)
Nodes, 241, 244, 608, 696–697
Non-inertial forces, 116
Non-inertial frames of reference, 116
Non-simple circuits, 461, 481–482
Non-simple harmonic motion,
    208–209
Normal force, 81, 82
Normal population (energy levels),
    722
n-p-n transistors, 531–532
Nuclear antimatter, 803
Nuclear binding energy curve,
    755–757
Nuclear cross sections, 784–786, 785
    (*table*)
Nuclear energy levels, 762–764
Nuclear fission, 674, 757, 783–787,
    788–789
Nuclear forces, 11, 757
Nuclear fusion, 674, 757
Nuclear masses, 754–755
Nuclear matter, structure of, 801–802
Nuclear physics, 752–774
    applied, 781–814
Nuclear power, 788–793
Nuclear reactions, 769–771

Nuclear reactors:
    breeder, 786
    fission, 785–786
    fusion, 789–791
Nuclear transmutation, 769–771
Nuclear weapons, 786
Nucleon resonances, 807
Nucleons, 14, 753
Nucleus, 13–14, 713, 752–774,
    779–780
    compound, 770–771
    structure of, 753–755
Nuclides, 753–754
    stable and unstable, 767–769

O

Object-image relationships, 580–589
Objective lens, 644, 645–646, 649
Octave (musical interval), 249
Ocular, 644
Oersted, Hans Christian, 501
Off-axis astigmatism, 635, 638
Ohm (unit of resistance), 439
Ohm, Georg Simon, 439
Ohm's law, 439
Oil-drop experiment, Millikan, 392,
    686–687
Omega particle, 805 (*table*)
One-fluid theory of electricity, 391
Onnes, Kamerlingh, 443
Optic axis, 631–632
Optical activity, 621–622
Optical density, 575
Optical flat, 614
Optics:
    applied, 634–654
    geometrical, 570–593, 599–600
    wave, 602–623, 628–632
Orbital quantum number, 733, 736
    (*table*)
Order of interference, 604
Order of magnitude, 10
Ordinary ray, 631
Oscillations, electric, 554–556
Outer ear, 254
Overtone structure, and quality of
    sound waves, 251–253
Overtones, 243

P

Pair production, 802–803
Parallax, 628&*n*
Parallel resonance, 557–558
Parallelogram method for vector
    addition, 37, *37*
Parity, conservation of, 766–767
Partial pressure, of a gas, 341
Partially elastic impact, 134, 147–149
Particle accelerators, 487, 795–799

applications of, 799
Particles:
    bound, 709–710
    equilibrium of, 75–79
    free, 708–709
    primary cosmic-ray, 810
    reference (in simple harmonic
        motion), 200
    secondary cosmic-ray, 810–811
    speed of, related to kinetic energy
        of, 797 (*table*)
    subatomic, 802–810, 804 (*table*),
        805 (*table*)
Pascal (unit of pressure), 194, 268
Pascal, Blaise, 270
Pascal's principle, 270–272
Paschen series, 717
Pauli, Wolfgang, 736–738
Pauli exclusion principle, 278,
    736–738
Peak-shaving, 448
Pendulum:
    physical, 216–217
    simple, 206–207
PEP, 800
Performance, coefficient of, 380
Period:
    of circular motion, 155
    of simple harmonic motion, 200
    of a wave, 227
Periodic motion, 155
Periodic table of the elements, 738
    (*table*), 739–741
Periodic waves, 220, 227–228, *229*
Permanent magnets, 496–499, 500,
    515
Perturbations, 186
PETRA, 800
Phase (states of matter), 317
    changes of, 317–320
Phase angle, 546
Phase contrast microscope, 644
Phase difference, interference and,
    606–608
    microscopes and, 644
Phase of vibration, 200, 227
Phasors, 545–548
Phosphorescence, 721
Photocell circuit, 689
Photodiode, 692
Photoelastic stress analysis, *621*
Photoelectric effect, 689–692,
    697–698
Photographic emulsion, 773
Photon-field theory, 414–415
Photon microscopes, 700
Photons:
    as corpuscles, 692–696
    as electromagnetic field particles,
        414, 803, 803 (*table*)
    energy of, 690